D0026156

Nondestructive Characterization of Materials VIII

Nondestructive Characterization of Materials VIII

Edited by

Robert E. Green, Jr.
The Johns Hopkins University
Baltimore, Maryland

PLENUM PRESS • NEW YORK AND LONDON

Library of Congress Cataloging-in-Publication Data

Nondestructive characterization of materials VIII / edited by Robert E. Green, Jr.
p. cm.
Proceedings of the 8th International Symposium on Nondestructive Characterization of Materials, held 6/16-20/97, in Boulder, Colo.
Includes bibliographical references and index.
ISBN 0-306-45900-0
1. Non-destructive testing--Congresses. 2. Materials--Testing--Congresses. I. Green, Robert E. (Robert Edward), 1932- II. International Symposium on Nondestructive Characterizationof Materials (8th : 1997 : Boulder, Colo.)
TA417.2.N65679 1998
620.1'127--dc21
98-28222
CIP

Proceedings of the Eighth International Symposium on Nondestructive Characterization of Materials, held June 15 – 20, 1997, in Boulder, Colorado

ISBN 0-306-45900-0

A Division of Plenum Publishing Corporation
233 Spring Street, New York, N.Y. 10013

http://www.plenum.com

10 9 8 7 6 5 4 3 2 1

Printed in the United States of America

The papers published in this proceedings represent the latest developments in nondestructive characterization of materials and were presented at the **Eighth International Symposium on Nondestructive Characterization of Materials** held June 15-20, 1997 in Boulder, Colorado, USA.

SYMPOSIUM CO-CHAIRMEN:

Robert E. Green, Jr.
Center for Nondestructive Evaluation
The Johns Hopkins University
Baltimore, MD, USA

Harry I. McHenry
National Institute of Standards and Technology
Boulder, CO, USA

Clayton O. Ruud
Pennsylvania State University
University Park, PA, USA

ORGANIZING COMMITTEE

SPONSORS:

American Society for Nondestructive Testing
Federal Aviation Administration
Federal Highway Administration
NASA Langley Research Center
National Institute of Standards and Technology
National Science Foundation
US Air Force, Wright Laboratories
US Navy, Naval Surface Warfare Center

HISTORY OF PREVIOUS SYMPOSIA:

LOCATION: Hershey, Pennsylvania, USA
TITLE: Nondestructive Methods for Material Property Determination
EDITORS: C.O. Ruud and R.E. Green, Jr.
PUBLISHER: Plenum Press, New York, USA (1984)

LOCATION: Montreal, Canada
TITLE: Nondestructive Characterization of Materials II
EDITORS: J.F. Bussiere, J-P Monchalin, C.O. Ruud and R.E. Green, Jr.
PUBLISHER: Plenum Press, New York, USA (1986)

LOCATION: Saarbrucken, Germany
TITLE: Nondestructive Characterization of Materials
EDITORS: P. Holler, V. Hauk. G. Dobmann, C. Ruud and R.Green
PUBLISHER: Springer-Verlag, Berlin, Germany (1989)

LOCATION: Annapolis, Maryland, USA
TITLE: Nondestructive Characterization of Materials IV
EDITORS: Clayton O. Ruud, Jean F. Bussiere, and Robert E. Green, Jr.
PUBLISHER: Plenum Press, New York, USA (1991)

LOCATION: Karuizawa, Japan
TITLE: Nondestructive Characterization of Materials V
EDITORS: Teruo Kishi, Tetsuya Saito, Clayton Ruud and Robert Green, Jr.
PUBLISHER: Iketani Science and Technology Foundation, Tokyo, Japan (1992)

LOCATION: Oahu, Hawaii, USA
TITLE: Nondestructive Characterization of Materials VI
EDITORS: Robert E. Green, Jr., Krzysztof J. Kozaczek, and Clayton O. Ruud
PUBLISHER: Plenum Press, New York, USA (1994)

LOCATION: Prague, Czech Republic
TITLE: Nondestructive Characterization of Materials VII
EDITORS: Anthony L. Bartos, Robert E. Green, Jr., and Clayton O. Ruud
PUBLISHER: Transtec Publications, Zurich-Uetikon, Switzerland (1996)

ACKNOWLEDGEMENTS:

The conference organizers would like to thank the following individuals for their assistance in making the symposium a success: the authors for their excellent presentations and manuscripts; the session chairpersons for keeping the sessions on time and stimulating lively discussions; Boro Djordjevic, Moshe Rosen, Jim Spicer, Jim Wagner, and John Winter, from the Center for Nondestructive Evaluation at Johns Hopkins University for their support.

The symposium and these proceedings would have not been possible without the enthusiasm and extremely hard work of Debby Manley, Secretary at the Johns Hopkins University Center for Nondestructive Evaluation, who not only prepared material for all of the notices and publications about the symposium, worked industriously at the registration desk in Boulder, and played a major role in preparing the proceedings for publication.

The ***Ninth International Symposium on Nondestructive Characterization of Materials*** will be held in late *June of 1999 in Sydney, Australia* and the ***Tenth International Symposium on Nondestructive Characterization of Materials*** is planned to be held in *Karuizawa, Japan in summer of the year 2000.*

Robert E. Green, Jr.
Center for Nondestructive Evaluation
The Johns Hopkins University

CONTENTS

LASER ULTRASONICS I

Laser-Ultrasonics with a Single Laser for Generation and Detection 1
J.-P. Monchalin and D. Drolet

Pulse Parameter Influence on the Laser Generation of Acoustic Waves 7
T.W. Murray, K.C. Baldwin and J.W. Wagner

Optical Detection of Ultrasound Using Two-Wave Mixing in Semiconductor Photorefractive Crystals and Comparison With the Fabry-Perot 13
A. Blouin, P. Delaye, D. Drolet, L.-A. de Montmorillon, J.C. Launay, G. Roosen and J.-P. Monchalin

Optimizing the Photo Induced-EMF Response for High-Speed Compensation and Broadband Laser-Based Ultrasonic Remote Sensing 21
G.J. Dunning, D.M. Pepper, M.P. Chiao, P.V. Mitchell and T.R. O'Meara

High Frequency Laser-Ultrasonics 27
A. Moreau and M. Lord

Measurements of Sound Velocities at High Temperatures by Using Laser Generated Ultrasound 33
H. Nakano, Y. Matsuda and S. Nagai

A Novel Laser Ultrasound Source and its Implementation in the Drinks Canning Industry 39
S. Dixon, C. Edwards and S.B. Palmer

Laser Ultrasound Source for NDE Applications: Calibration in a Liquid 47
S. Egerev, L. Lyamshev and Y. Simanovskii

LASER ULTRASONICS II

Sub-Micron Materials Characterization Using Near-Field Optics 53
D.W. Blodgett and J.B. Spicer

Laser Ultrasonic Testing of Highly Attenuative and Weakly Generating Composites ... 59
K.R. Yawn, T.E. Drake, Jr. and M.A. Osterkamp

Laser Ultrasound Imaging of Lamb Waves in Thin Plates 67
P.W. Lorraine

Detection of Corrosion in Aging Aircraft Structures by Laser-Ultrasonics 73
M. Choquet, D. Lévesque, C. Néron, B. Reid, M. Viens, J.-P. Monchalin, J.P. Komorowski and R.W. Gould

Photorefractive Laser Ultrasound Spectroscopy for Materials Characterization 79
K.L. Telschow, V.A. Deason, K.L. Ricks and R.S. Schley

Application of Laser-Ultrasonic Method to Evaluate Elastic Stiffness Degradation in Ceramic Matrix Composites 85
Y.M. Liu, T.E. Mitchell and H.N.G. Wadley

Measurement of the Stiffness Tensor of Orthotropic Materials From Line Source Point Receiver Laser Ultrasonic Method 91
B. Audoin, C. Bescond and M. Qian

Sagnac Interferometer for Ultrasound Detection on Rough Surfaces 97
P. Fomitchov, S. Krishnaswamy and J.D. Achenbach

Laser-Ultrasound Tissue Characterization Methods for Potential use in Laser Angioplasty Procedures 105
K.A. Roome, R.F. Caller, P.A. Payne and R.J. Dewhurst

Inspection of Rocket Engine Components Using Laser-Based Ultrasound 111
A.D.W. McKie and R.C. Addison, Jr.

Modeling Heterogeneities and Elastic Anisotropy in Single Crystal Zinc and Carbon Fiber Epoxy Composites 117
D.H. Hurley, J.B. Spicer, J.W. Wagner and T.W. Murray

NONLINEAR EFFECTS

Feasibility Study of a Nonlinear Ultrasonic Technique to Evaluate Adhesive Bonds .. 125
T.P. Berndt and R.E. Green, Jr.

Nonlinear Vibro-Acoustic Nondestructive Testing Technique 133
A.M. Sutin and D.M. Donskoy

Noncontact Ultrasonic Spectroscopy for Detecting Creep Damage in 2.25Cr-1Mo Steel 139
T. Ohtani, H. Ogi, T. Morishita and M. Hirao

Ultrasonic Damping and Velocity During Recovery and Recrystallization of Aluminum 145
W. Johnson

Investigation of Material Properties by Acoustic Methods on the Base of Inelastic Mechanical Processes 151
O.V. Abramov and O.M. Gradov

Study of Microplasticity of Solids by Laser Interferometry Method 157
V.V. Shpeizman, N.N. Peschanskaya and P.N. Yakushev

Interatomic Interaction and Alloying Criterion for Ferritic Alloys 163
I.S. Golovin

Internal Friction Nondestructive Evaluation of Plastic Properties of Crystals 169
A.B. Lebedev

PROCESS CONTROL I

Physical Property Determination for Process Monitoring and Control 175
G. Dobmann

NDE for Strength Determination of Diffusion Bonds 183
O. Buck, C.G. Ojard, D.J. Barnard and D.K. Rehbein

Monitoring of Texture Development in Copper-Alloy Sheet 189
G.A. Alers, P. Purtscher, J.F. Breedis and F.N. Mandigo

Magnetostrictive EMAT Efficiency as a Nondestructive Evaluation Tool 197
B. Igarashi, G.A. Alers and P.T. Purtscher

Ultrasonic Detection of Unstable Plastic Flow in Metal Cutting 205
M.A. Davies, S.E. Fick, C.J. Evans and G.V. Blessing

Ultrasonic Determination of Case Depth and Surface Hardness in Axles 211
R.C. Addison Jr., A. Safaeinili and A.D.W. McKie

High Energy X-Ray Diffraction Technique for Monitoring Solidification
of Single Crystal Castings 217
D.W. Fitting, W.P. Dube and T.A. Siewert

PROCESS CONTROL II

Characterizing Martensitic Steel With Measurements of Ultrasonic Velocity 223
K.W. Hollman, P.T. Purtscher and C.M. Fortunko

Ultrasonic Characterization of Microstructural States in Aluminum Welds
Depending on the Welding Parameters 229
N.M. Mourik, E. Schneider and K. Salama

Development of On-Line Lamb Wave Testing Technique Using Real-Time
Digital Signal Processing 235
M. Nakamura, H. Yokoyama, M. Yamano and R. Murayama

Ultrasonic and Acousto-Ultrasonic Inspection and Characterization of Titanium
Alloy Structures 239
A.L. Bartos, J.O. Strycek, R.J. Gewalt, H. Loertscher and T.C. Chang

Ultrasonic On-Line Monitoring and Mapping of Low-Temperature Diffusion
Bonding of 6061-T6 Aluminum 245
G. Kohn, Y. Greenberg, O. Tevet, O. Yeheskel, Y. Feuerlicht, U. Admon and D. Itzhak

Application of Ultrasonic Image to the Evaluation of Temperature Distribution
in Metal Powder Compacts During Spark Plasma Activated Sintering 251
T. Abe, H. Hashimoto, Y.-H. Park, T.-Y. Um and Z.-M. Sun

Reaction-Kinetic Model for the Calculation of the Chemical Composition
of Weld Metals in Shielding Gas Arc Welding Processes 257
N. Meyendorf

The Characterization of Rolled Sheet Alloy Using Infrared Microscopy 263
M.L. Watkins

Determination of Mechanical Properties of Steel Sheet by Electromagnetic Techniques ... 269
D. Stegemann, W. Reimche, K.L. Feiste and B. Heutling

Resonant Frequency Testing ... 277
R.H. Gassner

Rolled Steel Surface Inspection Using Microwave Methods ... 285
R. Zoughi, C. Huber, S.I. Ganchev, R. Mirshahi, E. Ranu and T. Johnson

MICROWAVES

Two-Port Network Analyzer Dielectric Constant Measurement of Granular or Liquid Materials for the Study of Cement Based Materials ... 291
K. Bois, A. Benally and R. Zoughi

Microwave Dielectric Properties of a Slab Determined by a Multi-Frequency Free-Space Technique ... 297
J.M. Liu and J.H. Wasilik

Preliminary Evaluation of Microwave Techniques for Inspection of Thick Layered Composite Deck Joints ... 305
L.M. Brown, J.J. DeLoach, R. Zoughi and E. Ranu

Analysis of Loading vs. Microwave Fatigue Crack Detection Sensitivity Using Open-Ended Waveguides ... 311
N. Qaddoumi, R. Mirshahi, E. Ranu, V. Otashevich, C. Huber, P. Stepanek, R. Zoughi and J.D. McColskey

Nondestructive Characterization of Materials: Applications for Humanitarian Demining ... 317
G.W. Carriveau

LASER ULTRASONICS III

Monitoring of Attenuation During Phase Transformations in Steel Using Laser-Ultrasonics ... 323
M. Dubois, A. Moreau, M. Militzer and J.F. Bussiére

Rapid Microstructure Assessment in Rolled Steel Products Using Laser-Ultrasonics ... 329
M. Dubois and J.F. Bussiére

Compensated Laser-Based Ultrasonic Receiver for Industrial Applications 335
G.J. Dunning, D.M. Pepper, M.P. Chiao, P.V. Mitchell and T.R. O'Meara

Industrial Demonstrations of Laser-Ultrasonics Based Process Control, Utilizing the Textron Laserwave™ Analyzer 341
P. Kotidis and D. Klimek

PROCESS CONTROL III

Advances in Ultrasonic Inspection Methods for PIM Process Monitoring and Control 347
J.L. Rose, R.M. German, D.D. Hongerholt

Anisotropic Electric Conductivities in Graphite Fiber Reinforced Composites 353
J.M. Liu and S.N. Vernon

Acoustic Characterization of Morphologically Textured Short-Fiber Composites: Estimation of Physical and Mechanical Properties 359
M.L. Dunn and H. Ledbetter

Laser Scattering Detection of Machining-Induced Damage in Si_3N_4 Components 365
J.G. Sun, M.H. Haselkorn and W.A. Ellingson

Nondestructive Characterization of the Nucleation and Early Vertical Bridgman Crystal Growth of $Cd_{1-x}Zn_xTe$ 371
B.W. Choi and H.N.G. Wadley

Multi-Channel Pyrometry and Thermography - A Non-Contact Method for the Determination of the Spectral Emissivity of High-Temperature Materials 377
T. Vetterlein, G. Walle and N. Meyendorf

A Comparison of Dielectric and Ultrasonic Cure Monitoring of Advanced Composites 383
D.D. Shepard, K.R. Smith and D.C. Maurer

Noninvasive Measurement of Acoustic Properties of Fluids Using an Ultrasonic Interferometry Technique 393
W. Han, D.N. Sinha, K.N. Springer and D.C. Lizon

Characterization of Fiber-Waviness in Composite Specimens Using Deep Line-Focus Acoustic Microscopy 401
W. Sachse, K.Y. Kim, D. Xiang and N.N. Hsu

X-RAY METHODS

New X-Ray Refractography for Nondestructive Evaluation of Advanced Materials 409
M.P. Hentschel, D. Ekenhorst, K.-W. Harbich, A. Lange and J. Schors

CCD-Cameras for X-Ray Investigations 417
F. Fandrich, R. Köhler and F. Jenichen

Development of X-Ray Diffraction Methods to Examine Single Crystal Turbine Blades 423
K.G. Lipetzky, R.E. Green, Jr. and P.J. Zombo

White Beam Transmission Topography of Nickel-Based Alloy Single Crystal Turbine Blades Using Synchrotron Radiation 431
J.M. Winter, Jr., R.E. Green, Jr. and G. Strabel

Non-destructive Evaluation of Plastic Strain in Deformed Layer Using X-Ray Diffraction 437
M. Katoh, K. Nishio and T. Yamaguchi

Nondestructive Testing of Ceramic Automotive Valves 443
U. Netzelmann, H. Reiter, Y. Shi, J. Wang and M. Maisl

Advancement in Nondestructive Investigation of Liquid YAG at Very High Temperature by Synchrotron Radiation 449
C. Landron, X. Launay, J.-P. Coutures, M. Gailhanou and M. Gramond

Multi-Energy Radioscopy - An X-Ray Technique for Materials Characterization 455
N. Meyendorf, G. Walle, H. Reiter, M. Maisl, H. Bruns, A. Hilbig, F. Heindörfer, R. Pohle and S. Ehlers

In-Situ, Positive Materials Identification at Room and Elevated Temperatures with Modern, Portable, Optical Emission and X-Ray Fluorescence Analyzers .. 461
S. Piorek, J. Ojanpera, E. Piorek and J.R. Pasmore

Problem of Characterization of Vacuum Vessel Shell for Spherical Tokamak Globus-M at Different Stages of its Fabrication 469
G.P. Gardymov, V.K. Gusev, N.Ya. Dvorkin, V.M. Komarov, E.G. Kuzmin, V.V. Mikov, V.B. Minaev, V.I. Nikolaev, A.N. Novokhatsky, K.A. Podushnikova, I.E. Sakharov, N.V. Sakhorov, S.V. Shatalin and V.V. Shpeizman

CIVIL STRUCTURES I

Measurement of Material Properties for Early Detection of Fatigue Damage in Highway Bridge Steels 475
R.A. Livingston, W. Johnson, D. McColskey, C. Fortunko and G. Alers

Ultrasonic Measurement of Stress in Bridges 481
A.V. Clark, P.A. Fuchs, M.G. Lozev, D. Gallagher and C.S. Hehman

Development of Scanning High-T_C SQUID-Based Instrument for Nondestructive Evaluation in a Magnetically Noisy Environment 487
N. Tralshawala, J.R. Claycomb, H.-M. Cho and J.H. Miller, Jr.

A Detection Method for Fine Plastic Deformations of Steel Using a High Performance Magnetic Gradiometer 493
H. Yamakawa, N. Ishikawa, K. Chinone, S. Nakayama, A. Odawara and N. Kasai

Distributed Real-Time System for Acoustic Emission Waveform Capture 499
J.D. Gentry

Acoustic Emission Method for Partial Discharges Measurements of Electric Power Components 507
J. Sikula, B. Koktavy, P. Vasina, Z. Weber, M. Korenska, L. Pazdera, T. Lokajícek and F. Matejka

The Recent Results of Computer Processing of Digitized Signals Gained During Flaw Detection Type Checking of Steel Cables 511
M. Lesnak and O. Lesnák

Detection of Crack Position by AE and EME Effects in Solids 517
T. Lokajícek, J. Sikula and P. Vasina

Analysis of Acoustic Emission From Rock Samples Loaded to Long-Term Strength Limit 523
V. Rudajev, J. Vilhelm and T. Lokajícek

Neutron Diffraction and Ultrasonic Sounding: A Tool for Solid Body Investigation 529
T. Lokajícek, Z. Pros, K. Klíma, A.N. Nikitin and T.I. Ivankina

CIVIL STRUCTURES II

Development of an Epithermal Neutron Detector for Non-destructive Measurement of Concrete Hydration 535
R.A. Livingston and H. Saleh

Ultrasonic Assessment of Damage in Concrete Under Cyclic Compression 541
Z. Radakovic, K. Willam and L.J. Bond

Assessment of Acoustic Travel Time Tomography (ATTT) at Barker Dam 549
W.F. Kepler, L.J. Bond and D.M. Frangopol

A Nonlinear Acoustic Technique for Crack and Corrosion Detection in Reinforced Concrete 555
D.M. Donskoy, K. Ferroni, A. Sutin, and K. Sheppard

Evaluation of Wood Products Based on Elastic Waves 561
M.L. Peterson

Numerical Modeling of Elastic Wave Propagation in Random Particulate Composites 567
F. Schubert and B. Koehler

Gas-Coupled Ultrasonics for High-Pressure Pipeline Inspection 575
C.M. Fortunko, R.E. Schramm and J.L. Jackson

Elastic Modulus and Damping of Concrete Elements 581
R. Kohoutek

ACOUSTIC EMISSION

Nondestructive Evaluation of Connection Stiffness 587
R. Kohoutek

Renewing Original Shape of Acoustic Emission Signals Changed While Propagation in a Material 595
O.V. Abramov and O.M. Gradov

Nondestructive Evaluation of the Residual Life of Steel-Belted Radial Truck Tires ... 601
H.L.M. dos Reis and K.A. Warmann

Application of the Pseudo Wigner-Ville Distribution to the Measurement of the Dispersion of Lamb Modes in Graphite/Epoxy Plates 609
W.H. Prosser, M.D. Seale and B.T. Smith

Stochastic Microfracture Process Analysis of SiC Particle Dispersed Glass Composites by Acoustic Emission 615
M. Enoki, H. Fujita and T. Kishi

Predictive Measurement Using AE: Fracture and Lifetime of Al-Alloys 621
X. Lu, W. Sachse and I. Grabec

AE Characterization of Microfracture Process in Ceramic Materials 629
S. Wakayama and B.-N. Kim

Characterization of Fiber Fracture Via Quantitative Acoustic Emission 635
D.J. Sypeck and H.N.G. Wadley

Acoustic Emission Characteristics During Ring Burst Test of FW-FRP Multi-ply Composite 641
A. Horide, S. Wakayama and M. Kawahara

The Study of PVDF Acoustic Emission Sensor 647
Z. Wang, L. Li, M. Wu and H. Hou

STRESS MEASUREMENT

Microstructure and Texture Influences on Ultrasonic Quantities for Welding Stress Analysis 653
U. Arenz and E. Schneider

Residual Stress Measurements in Front of a Crack Tip With High Spatial Resolution by Using Barkhausen Microscopy 659
I. Altpeter, G. Dobmann, N. Meyendorf, H. Blumenauer, D. Horn and M. Krempe

Nondestructive Stress Measurement in Steel Using Magnetostriction 665
T. Yamasaki and M. Hirao

Electromagnetic Acoustic Spectroscopy in the Bolt Head for Evaluating the Axial Stress 671
H. Ogi and M. Hirao

Detection of Residual Stresses and Nodular Growth in Thin Ferromagnetic Layers with Barkhausen and Acoustic Microscopy 677
I. Altpeter, G. Dobmann, S. Faßbender, J. Hoffmann, J. Johnson, N. Meyendorf and W. Nichtl-Pecher

Fatigue Characterization of AISI 321 Austenitic Steel by Means of HTC-SQUID 683
M. Lang, H.-J. Bassler, and J. Johnson

Nondestructive Characterization of Materials at the NIST Research Reactor 689
H.J. Prask

MATERIALS CHARACTERIZATION I

An Ultrasonic Comb Transducer for Guided Wave Mode Selection in Materials Characterization 695
J.L. Rose, S.P. Pelts, J.N. Barshinger and M.J. Quarry

Characterization of the Microstructure of Laser-Hardened Carbon Steels by Means of Positron Lifetime Measurements and Micromagnetics 701
B. Somieski, N. Meyendorf, R. Kern and R. Krause-Rehberg

Determination of Mass Density and Elastic Constants in the Surface Layer with Leaky Surface Waves 707
K. Kawashima, I. Fujii and N. Takenouchi

Determination of Porosity in Fiber Reinforced Plastics by Computer Aided Ultrasonic Signal Analysis (CAMPUS) 713
A. Kück, W.J. Bisle and G. Tober

Application of Neural Networks in Auto-Recognition of Eddy Current Testing Signal On Line 719
S. Keren, Z. Ping and H. Zhaohui

Initial Results of Applying Single Transducer Thickness-Independent Ultrasonic Imaging to Tubular Structures 725
D.J. Roth and D.V. Carney

Flaw Detection in Steel Wires by Electromagnetic Acoustic Transducers 733
T. Yamasaki, S. Tamai and M. Hirao

Nondestructive Characterization of the Thermal Diffusivity of Poor Conductors by a Low Cost, Flash Technique 739
J.M. Liu, R.A. Brizzolara and D.N. Rose

MATERIALS CHARACTERIZATION II

Rapid, Contactless Measurement of Thermal Diffusivity 747
Z. Ouyang, L. Wang, F. Zhang, L.D. Favro and R.L. Thomas

Effects of Texture on Plastic Anisotropy in Sheet Metals 751
C.-S. Man

Applying Phase Sensitive Modulated Thermography to Ground Sections of a Human Tooth 757
C. John, D. Wu, A. Salerno, G. Busse and C. Löst

Recovering Echo Signals for *In-Vitro* Characterization of Hard Dental Tissues 763
C. John

Inversion of Eddy Current Data for Recovery of the Electromagnetic Properties of Materials in Layered Flat and Tubular Products 769
C. González and R. Martín

MECHANICAL PROPERTIES

Ultrasonic Leaky Wave Measurements for Materials Evaluation 775
D. Xiang, N.N. Hsu and G.V. Blessing

Modeling Rayleigh Wave Dispersion Due to Distributions of One and Two Dimensional Micro-Cracks: A Review 781
C. Pecorari

Sizing of 3-D Surface Breaking Flaws From the Distribution of Leakage Field 787
D. Minkov and T. Shoji

Characterizing the Surface Crack Size by Magnetic Flux Leakage Testing 793
L. Li and J. Zhang

Time Domain Waveforms of a Line-Focus Transducer Probing Anisotropic Solids 799
N.N. Hsu, D. Xiang and G.V. Blessing

Rapid Infrared Characterization of Thermal Diffusivity in Continuous Fiber Ceramic Composite Components 805
J. Stuckey, J.G. Sun and W.A. Ellingson

Raman Spectroscopy for In-Situ Characterisation of Steam Generator Deposits 811
P.A. Rochefort, D.A. Guzonas and C.W. Turner

Characterization of Near Surface Mechanical Properties of Ion-Exchanged Glasses Using Surface Brillouin Spectroscopy 817
M. Puentes, J. Bradshaw, A. Briggs, O. Kolosov, K. Bowen and N. Loxley

Nonlinear Ultrasonic Properties of As-Quenched Steels 825
D.C. Hurley, P.T. Purtscher, K.W. Hollman and C.M. Fortunko

INDEX 831

Nondestructive Characterization of Materials VIII

LASER-ULTRASONICS WITH A SINGLE LASER FOR GENERATION AND DETECTION

Jean-Pierre Monchalin and Denis Drolet
Industrial Materials Institute, National Research Council of Canada
75 de Mortagne Blvd, Boucherville, Québec, J4B 6Y4, Canada

INTRODUCTION

The generation and detection of ultrasound with lasers (laser-ultrasonics) presents several important advantages compared to the conventional approach based on piezoelectric generation and detection.[1-4] Among those is the generation and detection without contact and at a distance (10 cm to meters), which allows easy probing of parts at elevated temperature. Also, since generation and detection occur at the surface of the material, parts of complex geometry can be readily tested without any need for transducer orientation. This unique feature is important for numerous industrial applications since parts made of metals, polymers or composites of various kinds have often curved surfaces and complex shapes.

Laser-ultrasonics is usually applied by using two lasers, a high power short pulse laser for ultrasound generation and a long pulse receiving laser for detection.[2-4] We are reporting a simpler approach in which a single long pulse laser is used for both generation and detection.

PRINCIPLE

The principle consists first in splitting the laser beam into two parts. One part is intensity modulated and is used for ultrasound generation. The other part is unmodulated and is used for detection. The laser should have the characteristics necessary for providing sensitive detection, such as being single frequency and of sufficient power. High power is also necessary for strong generation. It should also be very stable both in frequency and in intensity so as to avoid any additional noise on top of the fundamental photon noise at detection. Its pulse duration should be also sufficiently long to capture all the ultrasonic echoes of interest, which means for most practical cases a length exceeding 10 μs. In the work reported below we have used a single mode Nd-YAG laser operating at 1.06 μm with 4 kW peak power and 60 μs pulse duration.

A second aspect of the approach is the type of modulation used and the associated signal processing. The laser pulse being long (for example in the case of a plate, equal to

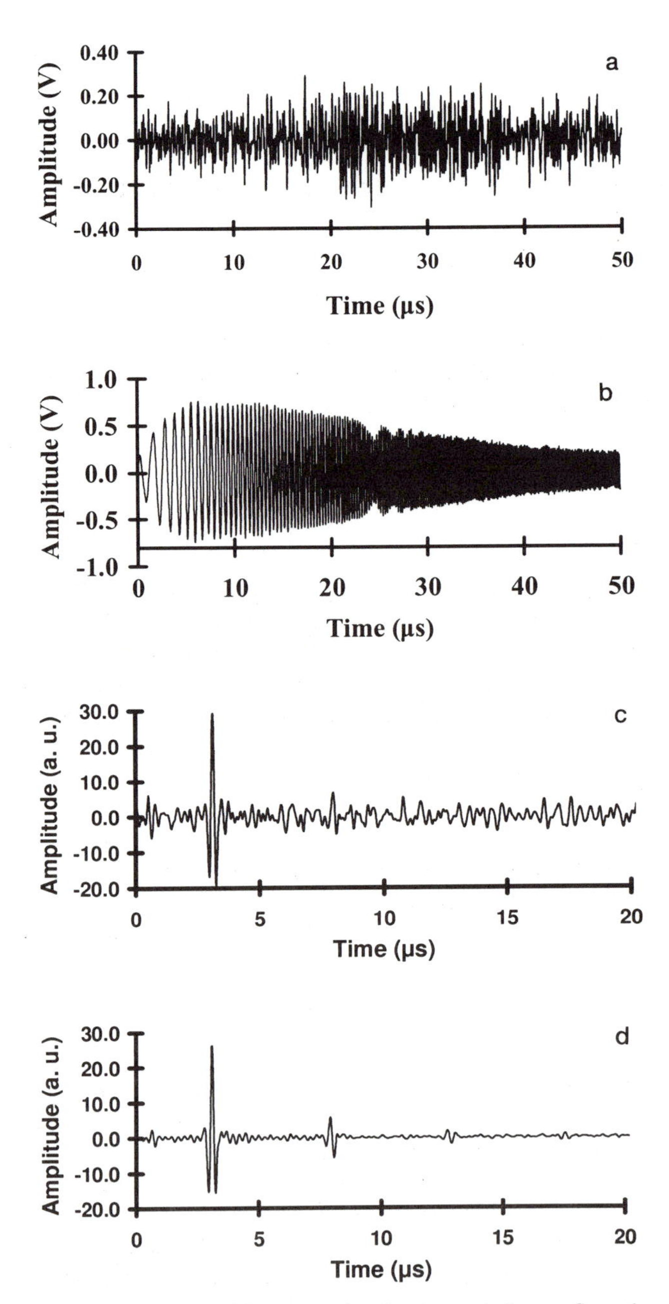

Figure 1. Results for a 7.5 mm [illegible] graphite-epoxy plate in a transmission configuration; a) single shot raw signal, b) modulated intensity (chirp from 0.5 to 8 MHz), c) compressed single shot signal, d) compressed averaged signal (100 times).

several times the through-thickness propagation time) the various ultrasonic echoes overlap and no clear information is obtained from the raw detected signal. What is needed is to perform pulse compression on this raw signal, as it has been done for radar.[5] A practical way to compress the signal is to perform its digital cross-correlation with the signal used to modulate the generation beam or the modulated laser intensity. In the work reported here the laser is modulated by a frequency chirp, but other modulation codes are possible.

The technique is readily applicable when generation and detection are performed on opposite sides of a specimen or when the detection location is offset from the one of generation. When these two locations overlap, as in a pulse-echo experiment, the optical detector located after the phase demodulator receives a strong parasitic signal coming from the light scattered from the modulated generation beam. This signal gives a very strong initial peak after correlation. This parasitic contribution can be minimized by polarizing differently the generation and detection beams and by using a phase demodulator essentially insensitive to intensity modulation (differential Fabry-Perot[6] or two-wave mixing photorefractive demodulator[7-8]).

EXPERIMENTAL RESULTS

We are reporting experiments performed on graphite epoxy samples. The samples were covered by a peel-ply layer, as often used in composite fabrication to protect surfaces. The layer used here has proper absorbing properties at 1.06 μm for sufficiently strong ultrasound generation. The generation beam was modulated by an electrooptic modulator

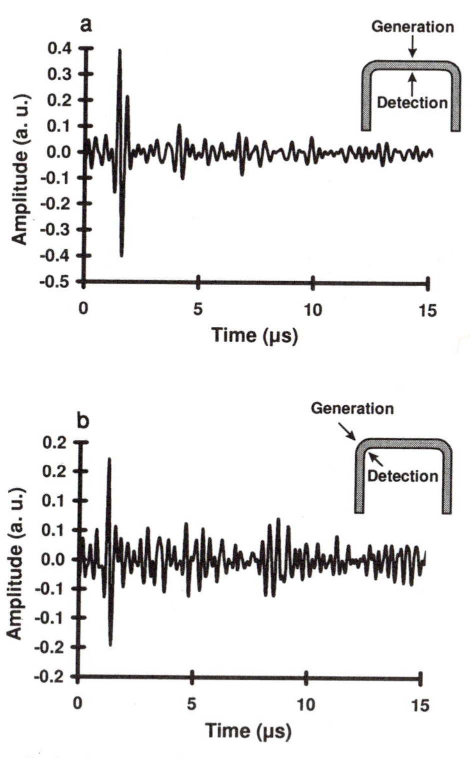

Figure 2. Single shot results obtained on a U-shape graphite epoxy specimen; a) on a flat surface, b) through a corner (spot sizes at generation and detection = 2 mm and frequency chirp from 0.1 to 10 MHz).

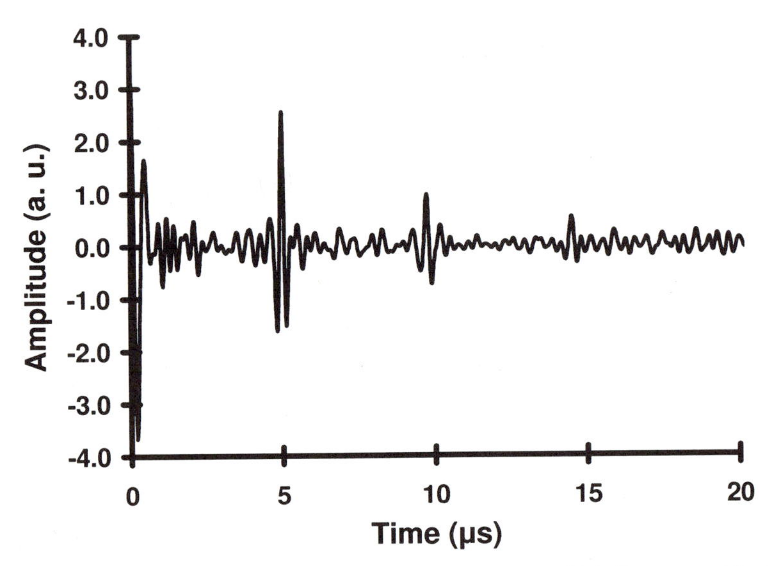

Figure 3. Result obtained in pulse-echo on graphite-epoxy plate 7.5 mm thick (frequency chirp from 0.5 to 8 MHz, 100 times average).

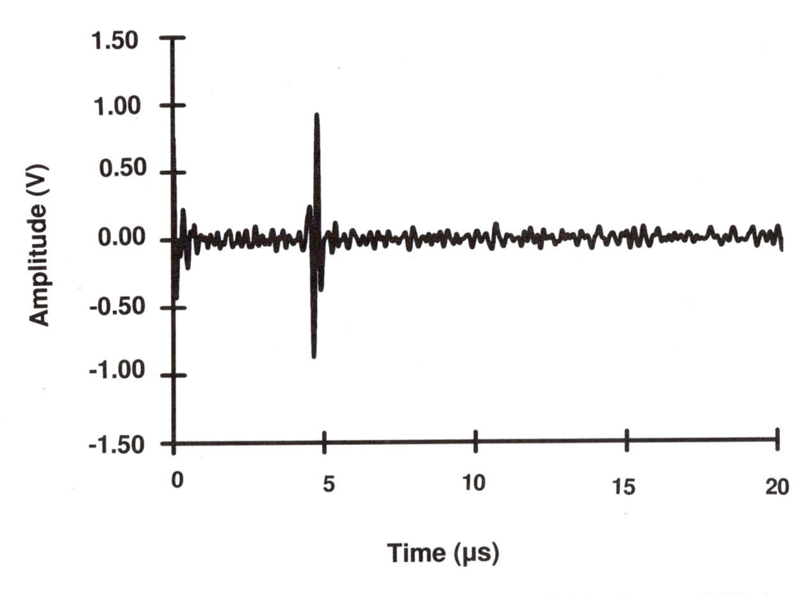

Figure 4. Detection of a surface wave generated on an aluminum block (signal averaged 100 times, frequency chirp from 0.5 to 8 MHz).

driven by a frequency chirp signal. The scattered beam from the unmodulated detection beam was collected by a lens system and then sent onto an InP:Fe photorefractive demodulator.[8] In a first series of experiment, a plate specimen of thickness 7.5 mm was used, the generation beam being sent onto the back surface of the specimen, opposite to generation (transmission configuration). Spot sizes of 2 mm in diameter were used both in generation and detection and the frequency chirp was from 0.5 to 8 MHz. The results are shown in Figure 1a)-d). Figure 1a) shows the single shot raw signal, in which no echo is visible. Figure 1b) shows the modulated generation beam intensity, which was cross-correlated with the raw signal to give the final compressed single shot signal shown in Figure 1c). After compression, echoes corresponding to reverberations through the plate should be observed. In Figure 1c), because of limited signal-to-noise, only one is clearly identified. When signal averaging is used, the other echoes are seen, as shown in Figure 1d).

The technique was also shown to work well when a strongly curved laminate part is tested. Figure 2a)-b) shows the results obtained on a U-shape specimen of thickness 4 mm. As can be seen, echoes are observed not only on the flat sections of the U, but also through the corner regions (inner radius ~3 mm). These results demonstrate that the technique could be used to inspect in transmission composite parts of complex geometry. To make a convenient system, the generation and detection beams will be coupled through optical fibers, which is known to be readily feasible in this case since the beams have low peak power and are in the near infrared (coupling through a fiber length of 40 meters has been demonstrated). Such an inspection system could be made by retrofitting the existing mechanical scanners widely used in the aeronautic industry with focusing and collecting optics linked to optical fibers for beam delivery and transmission, replacing the piezoelectric transducers and the water jets.

Single side generation and detection is also possible when a differential demodulator is used. In the example whose results are shown in Figure 3 a two-wave mixing photorefractive demodulator using an InP:Fe crystal was used.[8] The specimen is the same graphite-epoxy plate as the one used above for taking the data shown in Figure 1. To minimize further the parasitic collection of the modulated generation beam, the generation and detection spots of 2 mm size were offset by 2 mm.

Finally we present an example of the generation and detection of a surface wave on a thick aluminum sample. The generation beam was projected onto the surface to a small line 1 mm wide and 15 mm long using a cylindrical lens. The detection beam was focused to a round spot of 0.4 mm in diameter 15 mm away from the line. Figure 4 shows the result obtained after signal compression, which demonstrates that Rayleigh surface waves can be generated and detected by this single laser technique.

CONCLUSION

We have introduced a novel laser ultrasonic technique that uses only one laser. This technique has the advantage by comparison to the usual two-laser technique to provide a reduction of the complexity and cost of the hardware needed. Another advantage is the possibility of using fiber coupling for generation and detection, so the laser, which could be bulky, can be located far away from the tested part. A third advantage is the ease of choosing the range of ultrasonic frequencies used to probe the material by a simple electronic setting. We have presented several examples of application of the technique, particularly to the inspection of polymer matrix composite materials. Future work will include evaluation of the sensitivity, particularly in comparison with the traditional two-laser approach and the investigation of other modulation codes, such as a Golay code,[9] which may improve the signal-to-noise ratio by diminishing some self noise of the cross-correlation.

REFERENCES

1. C. B. Scruby and L. E. Drain, *Laser-Ultrasonics: Techniques and Applications*, Adam Hilger, Bristol, U.K (1990).
2. J.-P. Monchalin, Progress towards the application of laser-ultrasonics in industry, in: *Review of Quantitative Nondestructive Evaluation*, vol. 12, D. O. Thompson and D. E. Chimenti, ed., Plenum Press, New York, pp. 495-506 (1993).
3. J.-P. Monchalin, C. Néron, J. F. Bussière, P. Bouchard, C. Padioleau, R. Héon, M. Choquet, J.-D. Aussel, C. Carnois, P. Roy, G. Durou, J. A. Nilson, Laser-ultrasonics: from the laboratory to the shop floor, *Physics in Canada*, 51:122 (1995).
4. J.-P Monchalin, C. Néron, P. Bouchard, M. Choquet, R. Héon and C. Padioleau, Inspection of composite materials by laser-ultrasonics, *Canadian Aeronautics and Space Journal*, 43: 34 (1997).
5. M. I. Skolnik, *Introduction to Radar Systems*, McGraw-Hill, New-York (1980).
6. J.-P. Monchalin and R. Héon, Laser optical ultrasound detection using two interferometer systems, *US patent* # 5,080,491 (1992).
7. A. Blouin and J.-P. Monchalin, Detection of ultrasonic motion of a scattering surface by two-wave mixing in a photorefractive GaAs crystal, *Appl. Phys. Lett.*, 65 : 932 (1994).
8. P. Delaye, A. Blouin, D. Drolet, L. A. de Montmorillon, G. Roosen and J.-P. Monchalin, Detection of ultrasonic motion of a scattering surface by photorefractive InP:Fe under an applied dc field , *J. Opt. Soc. Am.*, 14 : 1723 (1997).
9. M. J. E. Golay, Complementary series, *IRE Trans. Inform. Theory*, IT-2 : 82 (1961).

PULSE PARAMETER INFLUENCE ON THE LASER GENERATION OF ACOUSTIC WAVES

Todd W. Murray, Kevin C. Baldwin, and James W. Wagner

Department of Materials Science and Engineering
Johns Hopkins University
Baltimore, MD 21218

INTRODUCTION

The principles and uses of optical techniques for the generation and detection of ultrasound have been discussed extensively in the literature. Several excellent reviews of the progress in the field of laser ultrasonics are available.[1,2,3] Laser ultrasonic techniques have several advantages over other inspection methods, making them an attractive option for select applications. Unfortunately, poor sensitivity and high cost of laser ultrasonic systems are currently limiting industrial application. In general, implementation is limited to situations where laser inspection methods present the only available solution, or the few cases that prove cost effective. The goal of the current work is to increase the sensitivity of optical generation and detection systems. The work focuses on the laser/materials interaction that occurs during the generation of ultrasonic waves, and considers a number of methods by which laser ultrasonic systems can achieve greater sensitivity through control of the laser generation source.

The acoustic signal generated by a laser source depends on the thermal, optical, and elastic properties of the specimen, and on the characteristics of the laser source. For a given materials system, the laser pulse parameters can be chosen such that the signal-to-noise ratio (SNR) of the detection system is maximized. There are four laser pulse parameters to consider: temporal profile, spatial profile, energy, and wavelength. Of these, the spatial profile of the generating laser pulse and the energy used for generation will be considered.

LASER GENERATION OF ULTRASOUND IN THE ABLATIVE REGIME

Laser generation of acoustic waves in the thermoelastic regime has been well characterized.[4,5,6,7] In the low irradiance regime, before vaporization takes place, the amplitude of laser generated acoustic waves shows a linear increase with pulse energy.[8] The laser generation process is somewhat less well characterized in the ablative regime. Experimentally, it is well known that there is a large enhancement in the amplitude of the longitudinal wave generated in the ablative regime. Qualitative models have been presented which indicate that the shape of the ablative waveform can be understood as being a superposition of a thermoelastic (point expansion) source and a force acting normal to the surface.[8,9] The physical origin of the normal force is the momentum transfer from the evaporating species. Although the shape of the displacements predicted in the literature agrees well with experiment, indicating that the assumed forcing functions are correct, the amplitude and shape of the normal forcing function is deduced qualitatively without solution of the laser vaporization problem.

A more quantitative analysis of the generation of acoustic waves in the ablative regime was recently presented.[10] Many of the important issues concerning laser vaporization were addressed. Still, assumptions made considering, for example, the distribution of forces exerted within the laser spot, indicate the need for further study in this area. Unfortunately, there are few comparisons of experimental and theoretical displacements and the agreement between experiment and theory presented is only moderate.

An understanding of laser generation of acoustic waves in the ablative regime is important for several reasons. Generating in the thermoelastic regime, it is often important to use as much laser energy as possible while remaining below the vaporization threshold. Thus it is desirable to be able to predict the laser irradiance at which a detectable amount of vaporization takes place. Next, there are many ultrasonic inspection applications in which a small amount of surface damage can be tolerated. In these cases, it is important to have a good understanding of the vaporization process to take advantage of the large acoustic wave enhancement possible in the ablative regime. This may allow for the generation of high amplitude acoustic signals, while staying below some specified damage tolerance level. If the relationship between acoustic signals produced in the ablative regime and surface damage incurred is well characterized, then the acoustic signature can be used simultaneously for process monitoring / inspection and surface damage monitoring. It also allows for the possibility of using acoustic signals for direct monitoring of the laser/material interaction region to track vaporization, surface temperature, etc.

A schematic view of the laser ablation process is given in Figure 1. The surface is first heated to the melting point, at which time the melt front begins to propagate into the material. Continued heating brings the material to the vaporization point. Adjacent to the surface is a thin layer (on the order of a few molecular mean free paths) known as a Knudsen layer, where the vapor is not in translational equilibrium. Across this layer, there are discontinuities in temperature, pressure, and density. Beyond the Knudsen layer is a region of expanding vapor. In this analysis it is assumed that vaporization takes place in vacuum. In this case, the gas dynamic processes outside of the Knudsen layer need not be considered and the vapor expands freely into vacuum. Continued heating well beyond the vaporization point will lead to vapor breakdown and absorption of the incoming laser light through photo-ionization and inverse bremsstrahlung processes. For simplicity, vapor breakdown is not considered in this model. Thus the applicability of the model holds up to the point when laser light absorption by the vapor becomes significant.

The pressure exerted on the surface due to vaporization in vacuum was calculated using an implicit finite difference technique. The positions of the melt and vaporization fronts were tracked. The temperature dependences of the thermal conductivity and heat capacity were accounted for. The displacment caused by thermoelastic expansion was calculated with a model following that of Spicer.[5] The final waveform was taken as the superposition of surface displacements caused by thermoelastic expansion and vaporization processes. Experiments were performed in a vacuum chamber in order to verify the working

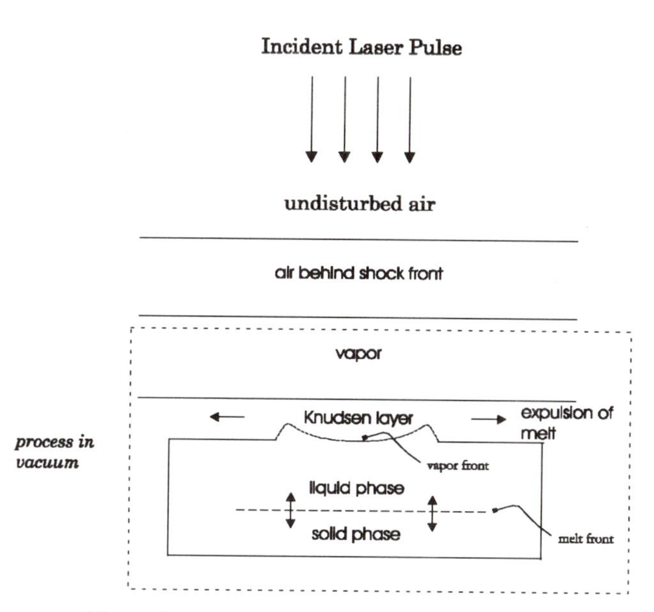

Figure 1. Schematic view of laser ablation processes.

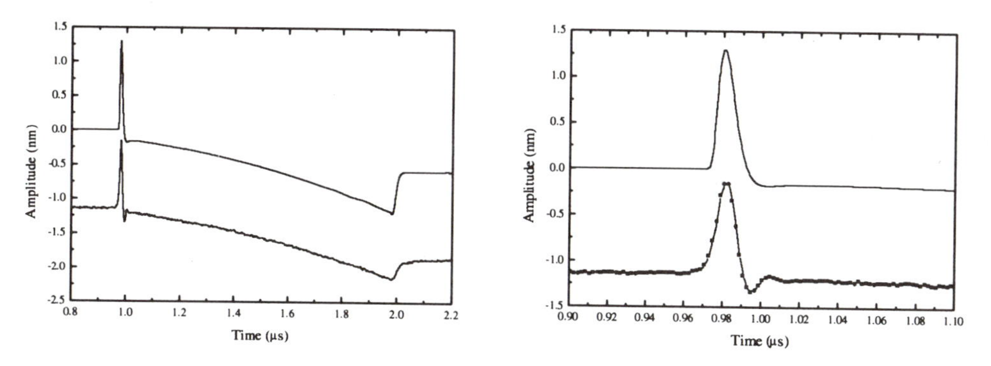

Figure 2. Theoretical and experimental waveforms at 300 MW/cm^2. The figure at the right shows a comparison of the first longitudinal arrival.

of the model. The surface displacement in aluminum was detected using a Michelson interferometer. Experimental results agree reasonably well with theoretical predictions for irradiance levels below about 350 MW/cm^2. Above this irradiance, the model predicted significantly larger displacements than were observed experimentally. It is believed that this is the point at which processes in the vapor become important and absorption in the vapor reaches significant levels. A comparison of theory and experiment at 300MW/cm^2 is given in Figure 2. The shape of the waveforms as well as the temporal extent of the first longitudinal arrivals show good agreement.

SPATIAL MODULATION OF LASER SOURCES

As an alternative to single pulse excitation, phased array systems have been proposed and demonstrated which can enhance surface and bulk wave amplitudes to acceptably high levels without requiring any individual element of the array source to exceed the material ablation threshold.[11,12,13,14] Alternatively, periodic spatial arrays[15,16,17,18,19] and temporal modulation of a single source[20,21] have been demonstrated to produce tone burst acoustic signals to which a receiving detector can be tuned, thus providing enhanced immunity to broadband noise. Among these three methods for modifying the laser source for enhanced detectability, the spatial array methods for generation of narrowband, tone burst ultrasound are the simplest and least expensive to implement. The disadvantage of these narrowband methods, however, is that the temporal extent of the wave packet generated by such sources can lead to uncertainty in time-of-flight measurements important for many practical laser inspection and measurement applications.

A new laser ultrasonic generation technique has been investigated which uses a spatial mask in a manner similar to that which may be employed for generation of narrowband, tone burst ultrasound. Instead of periodic spatial modulation at the surface of a test object, however, the mask employed for this technique is linearly frequency modulated as a function of position and correspondingly generates ultrasonic wave packets which are linearly frequency modulated (FM) as a function of time (chirped). Although the chirped mask launches an ultrasonic wave packet which is extended in time, not unlike an ultrasonic tone burst, the nature of the chirped signal is such that once detected, it can be processed by a matched filter algorithm which generates a single narrow spike corresponding to the time of arrival of the wave packet.

The experimental setup for the generation of FM surface waves is given in Figure 3. The FM surface waves were detected with a Michelson interferometer. The signals were then transferred to a computer where the signal processing was performed. A surface wave generated on an aluminum specimen with this setup is shown in Figure 4a. The waveform was then filtered with the result given in Figure 4b. An apparent 15-fold increase in the SNR is observed in the filtered signal.

CONCLUSIONS

Laser generation of ultrasound in the ablative regime has been modeled. Agreement was seen between experiment and predictions for low level laser ablation. The current model

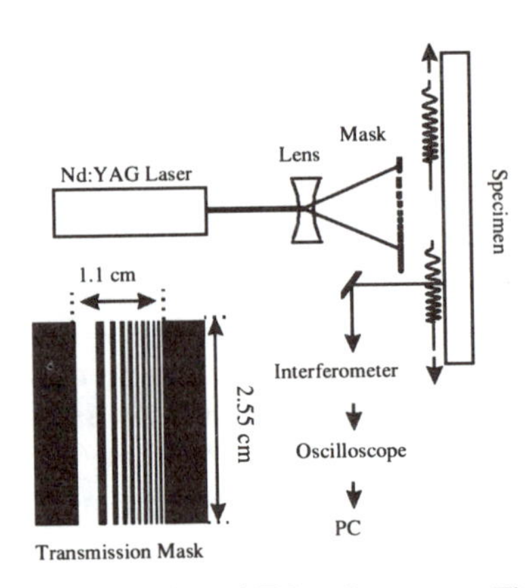

Figure 3. Experimental setup for the generation of FM surface waves. The transmission mask is shown in the lower right hand corner.

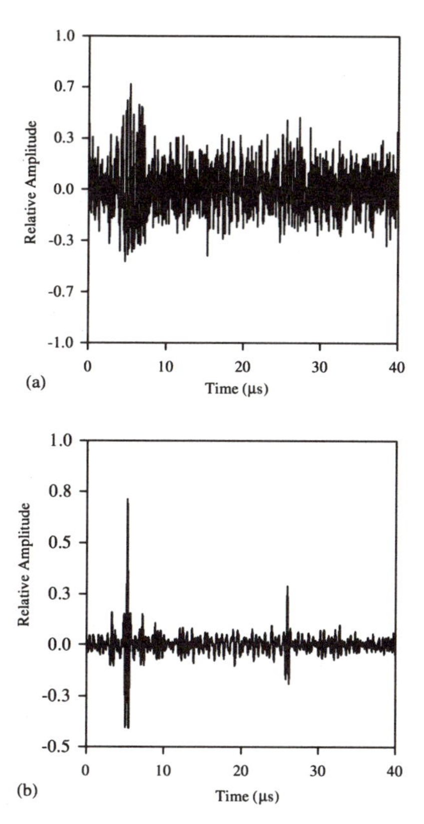

Figure 4. a) Single shot FM surface waveform and b) the waveform after matched filtering showing a 15 fold increase in signal to noise ratio and compression of the wave packet.

can be expanded to incorporate processes occurring in the vapor as well as the effects of a backing gas. A new spatial modulation technique for the generation of surface waves has also been presented. This allows for surface damage to be avoided by illuminating the specimen over a broad area. Temporal resolution is maintained through the compression of the FM signal in the matched filtering process.

REFERENCES

1. C.B. Scruby and L.E. Drain, *Laser Ultrasonics, Techniques and Applications*, Adam Hilger, N.Y. (1990)
2. S.J. Davies, C. Edwards, G.S. Taylor and S.B. Palmer, "Laser-generated ultrasound: its properties, mechanisms, and multifarious applications," *J. Phys. D.* **26** 329-348 (1993)
3. D.A. Hutchins, "Mechanisms of pulsed photoacoustic generation," *Can. J. Phys.* **64** 1247-1264 (1986)
4. L.R.F. Rose, "Point-Source Representation for Laser Generated Ultrasound, " *J. Acoust. Soc. Am* **75** (3), 723-732 (1984)
5. J.B. Spicer, "Laser Ultrasonics in Finite Structures: Comprehensive Modeling with Supporting Experiment," Ph.D. Dissertation, The Johns Hopkins University (1991)
6. F.A. McDonald, "Practical quantative theory of photoacoustic pulse generation," *Appl. Phys. Lett.*, **54** (16) 1504-1506 (1989)
7. J.C. Cheng, S.Y. Zhang, and L. Wu, "Excitations of thermoelastic waves in plates by a pulsed laser," *Applied Physics A*, **61**, 311-319 (1985)
8. R.J. Dewhurst, D.A. Hutchins, S.B. Palmer, and C.B. Scruby, "Quantitative measurements of laser generated acoustic waves," *J. Appl. Phys.*, **53** (6) 4064-4071 (1982)
9. J.D. Aussel, A. Le Brun, and J.C. Baboux, "Generating acoustic waves by Laser: theoretical and experimental study of the emission source," *Ultrasonics* **27** 165-177 (1988)

10. A. Hoffman and W. Arnold, "Calculation and measurement of the ultrasonic signals generated by ablating material by a Q-switched laser pulse," *Applied Surface Science*, **96-98** pp.71-75 (1996)
11. M-H. Noroy, D. Royer, and M. Fink, "The laser-generated phased array: Analysis and experiments," *J. Acoust. Soc. Am.* **94**(4) 1934-1943 (1993)
12. T.W. Murray, J.B. Deaton Jr., and J.W. Wagner, "Experimental evaluation of enhanced generation of ultrasonic waves using an array of laser sources", *Ultrasonics* **34**, 69-77 (1996)
13. J.S. Steckenrider, T.W. Murray, J.W. Wagner, and J.B. Deaton Jr., "Sensitivity enhancement in laser ultrasonics using a versatile laser array system", *J. Acoust. Soc. Am.* **97**(1), 273-279 (1995)
14. Y. Yang, N. DeRidder, C. Ume, and J. Jarzynski, "Noncontact optical fibre phased array generation of ultrasound for non-destructive evaluation of materials and processes," *Ultrasonics* **31**, 387-394 (1993)
15. J. Huang, S. Krishnaswamy, and J.D. Achenbach, "Laser generation of narrow-band surface waves," *J. Acoust. Soc. Am.* **92**, 2527-2531 (1992)
16. S. Nagai and H. Nakano, "Laser generation of antisymmetric lamb waves in thin plates," *Ultrasonics* **29**, 230-234 (1991)
17. A. Harata, H. Nishimura, T. Sawada, " Laser-induced surface acoustic waves and photothermal surface gratings generated by crossing two pulsed laser beams," *Appl. Phys. Lett.* **57** (2), 132-134 (1990)
18. H. Nishino, Y. Tsukahara, Y. Nagata, T. Koda, and K. Yamanaka, " Excitation of high frequency surface acoustic waves by phase velocity scanning of a laser interference frings," *Appl. Phys. Lett.* **62** (17), 2036-2038 (1993)
19. A.D.W. McKie, J.W. Wagner, J.B. Spicer, and C.M. Penny, "Laser generation of narrowband and directed ultrasound," *Ultrasonics* **27**, 323-330 (1989)
20. J.W. Wagner, J.B. Deaton Jr., and J.B. Spicer, "Genaration of ultrasound by repetitively Q-switching a pulsed Nd:YAG laser," *Applied Optics* **27**(22) 4696-4700 (1988)
21. J.B. Deaton Jr., A.D.W. McKie, J.B. Spicer, and J.W. Wagner, "Generation of narrowband ultrasound with a long cavity mode-locked Nd:YAG laser," *Appl. Phys. Lett.* **56**, 2390-2392 (1990)

OPTICAL DETECTION OF ULTRASOUND USING TWO-WAVE MIXING IN SEMICONDUCTOR PHOTOREFRACTIVE CRYSTALS AND COMPARISON WITH THE FABRY-PEROT

Alain Blouin*[(1)], Philippe Delaye[(2)], Denis Drolet[(1)], Louis-Anne de Montmorillon[(2)], Jean Claude Launay[(3)], Gérald Roosen[(2)], Jean-Pierre Monchalin[(1)].

(1) Institut des Matériaux Industriels, Conseil National de Recherches du Canada, 75 de Mortagne, Boucherville, Quebec, J4B 6Y4, Canada.

(2) Institut d'Optique Théorique et Appliquée, Unité de Recherche Associée 14 au Centre National de la Recherche Scientifique, Bat. 503, Centre Scientifique d'Orsay, B.P. 147, 91403 Orsay Cedex, France.

(3) Action Aquitaine de recherche en apesanteur, B.P. 11, 33165 Saint Médard en Jalles Cedex, France.

INTRODUCTION

Optical detection of ultrasound presents several advantages over conventional ultrasonic techniques; performed without contact and at a distance it can be used to probe parts at elevated temperature, in particular on a production line, and parts of complex shape. However, optical techniques are typically less sensitive than conventional piezoelectric-based techniques [1]. Hence, efforts have been done to improve the sensitivity of optical techniques while maintaining their useful features such as a large detection bandwidth.

The principle of the optical detection of ultrasound consists in measuring the phase modulation induced by the small ultrasonic surface displacement upon a probe laser beam impinging on the surface. An interferometer, which receives the scattered light from this probe beam and is followed by an optical detector, is used to transform the phase-modulated light scattered from the surface into a modulated electrical signal. Since usual material surfaces are not polished to optical quality and are rough, such an interferometer should operate with the speckle beam collected from these rough surfaces. For interferometers with small throughput or étendue, the solution is to collect only one speckle, which requires to focus the probe beam onto the surface and results into a light collection strongly varying from one location to the other[2]. Several years ago, a device based on a confocal Fabry Perot (CFP) was introduced for efficient detection of ultrasound from rough surfaces[2]. More recently, an interferometer based on real-time holography or two-wave mixing in photorefractive crystals was proposed and successfully demonstrated [3,4]. The purpose of this paper is to report the latest developments regarding this photorefractive detection scheme and to compare its characteristics (sensitivity, bandwidth, étendue) to those of the Fabry-Perot. Regarding sensitivities, it is useful to compare them to the ultimate sensitivity which can be obtained from an interferometric method.

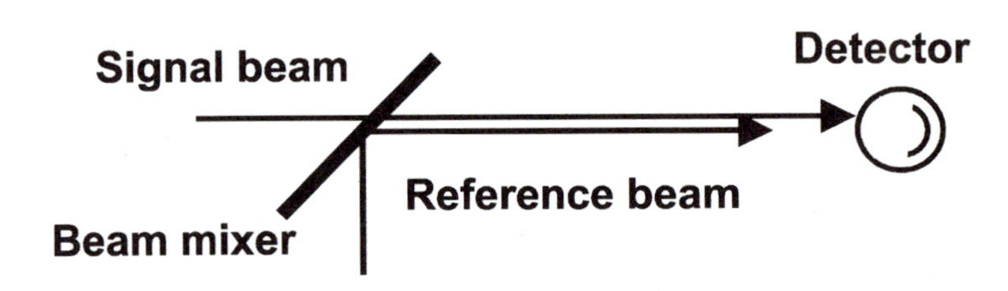

Figure 1: Basic principle of homodyne phase demodulation. A phase-modulated signal beam and a reference beam or local oscillator are mixed to produce a beating measured by a photodetector.

This ultimate sensitivity is obtained with the basic scheme for homodyne detection shown in Figure 1. It consists in mixing the phase-modulated signal beam with a reference beam or local oscillator onto a photodetector. The ultimate sensitivity for a bandwidth of 1 Hz and an incident power of the signal beam of 1 W is given by:

$$\delta_{min} = \frac{\lambda}{4\pi}\sqrt{\frac{h\nu}{2\eta}}$$

where h is the Planck constant, λ and ν are the wavelength and the optical frequency respectively and η is the quantum efficiency of the photodetector used. This ultimate sensitivity is obtained: 1) for $I_R >> I_S$ where I_R and I_S are the reference and signal beam intensities received by the detector respectively, 2) for $\pi/2$ out of phase or in quadrature signal and reference waves and, 3) for perfectly matched signal and reference wavefronts, 4) for a very high transmissivity of the beam mixer, ideally close to 100% (alternatively, a 50% transmission and reflection beam mixer could be used with two detectors mounted in a differential configuration) 5) photon noise limited detection. For example, for 1.06 μm wavelength and for a quantum efficiency of η= 0.3, the ultimate sensitivity is $5\text{x } 10^{-8}\, nm\sqrt{\frac{W}{Hz}}$. It should be noted that this sensitivity is the ultimate which can be obtained with the classical light fields produced by all existing lasers, but that it can be improved in principle by using squeezed states or other non classical states of light [5]. When working with a speckled signal beam, the sensitivity is reduced well below the ultimate limit unless the interferometer also acts as a signal-to-reference wavefront adapter or only one speckle is collected.

The sensitivity depends on the light power injected in the interferometer, and hence rely on the light gathering efficiency or étendue of the interferometer. In a typical application, the light collected from the surface is brought to the interferometer through a large-core multimode optical fiber. Hence, the étendue of the interferometer should at least be equal to the étendue of such an optical fiber which is about 0.4 mm^2 sr (numerical aperture of 0.39 and core diameter of 1mm) to process coherently all the light collected by the fiber.

CONFOCAL FABRY PEROT INTERFEROMETER

The confocal Fabry-Perot is now a well established device for detection of ultrasound. Its large étendue follows from the confocal nature of the cavity: a ray entering the cavity travels along the same path after multiples reflections upon the mirrors. For example, a meter long cavity with mirrors with 85% reflectivity provides an étendue exceeding that of the collecting optical fiber mentioned above (0.4 mm^2 sr). The interferometer can be used under different versions providing different bandwidths and sensitivities. Examples of sensitivity spectra for these various configurations (referenced to the ultimate sensitivity) are shown in Figure 2. A first configuration is obtained with mirrors of identical reflectivities, the detector being located either on the transmission side or the reflection side. The use in reflection provides high sensitivity at high ultrasonic frequencies (higher than the cavity bandwidth) except at a few rejection bands. When high sensitivity is only required in this range of frequencies, the back mirror could be made totally reflecting. This leads to improved sensitivity, as can be seen in Figure 2. This can be explained by the fact that there are four output beams exiting a confocal cavity, two in transmission side and another two in reflection side, giving four independent output ports (the incoming beam has speckle and the wavefronts are not adapted between these ports). There is multiple interference on each port and each port gives an output ultrasonic signal. The signal observed in transmission or reflection is the sum of the signals collected over the two transmission or reflection ports. In

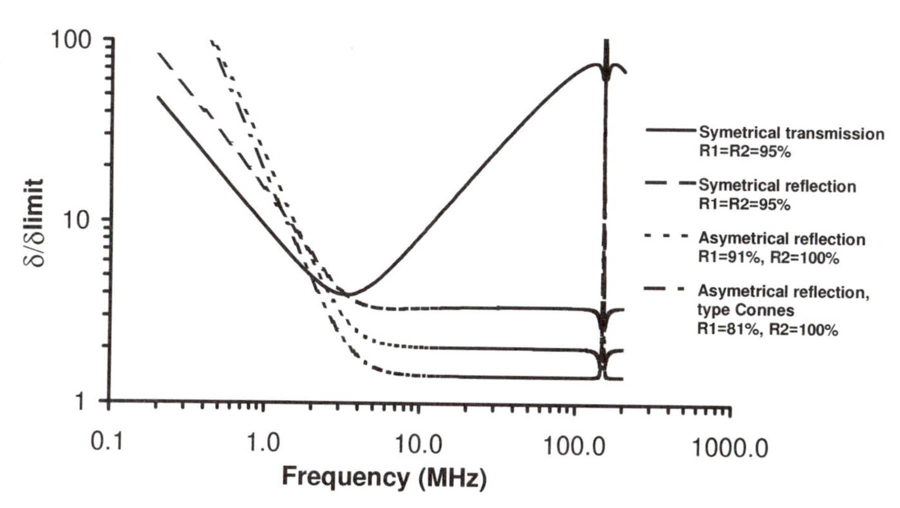

Figure 2: Theoretical sensitivity versus ultrasonic frequency for several confocal Fabry-Perot configurations operating either in the transmission or in reflection (cavity length = 50 cm).

the case of the configuration using a totally reflecting back mirror, there are only two output ports on the reflection side, which has the consequence to increase the intensity of the interfering terms (the total intensity is shared between two ports instead of four) and finally contributes to a better sensitivity.

A further improvement can be obtained with a configuration with only a single output port on the reflection side. This can be realized with a totally reflecting back mirror and a front mirror which is totally reflecting over half of its surface, as in the original confocal Fabry-Perot design [6]. This configuration which will be designated "Fabry-Perot type Connes" realized an ideal homodyne demodulator, but since half the incoming light is discarded its detection limit (at high frequencies) is √2 the ultimate limit. The combination of two systems will provide a limit equal to the ultimate limit with the penalty of higher complexity. It should be also noted that the transmission scheme could operate with unpolarized light whereas the use in reflection requires polarizing optics for optimum operation. Therefore often in practice, especially if a large core multimode fiber is used to transmit light to the Fabry-Perot, the transmission configuration gains a sensitivity factor of about √2 with respect to the others. If the range of frequencies of interest is between 1 and 10 MHz, which is often the case in nondestructive testing, this configuration will be the one usually selected with a proper choice of mirror reflectivity and cavity length to give adequate étendue and frequency response. The main weakness of the Fabry-Perot demodulators, this is obvious from Figure 2, is its lack of sensitivity at low ultrasonic frequencies (below 2 MHz), which is circumvented by the devices based on two-wave mixing in photorefractive materials.

PHOTOREFRACTIVE TWO-WAVE MIXING INTERFEROMETER

In the two-wave mixing approach, wavefront adaptation is performed actively, by opposition to the confocal Fabry-Perot in which adaptation is performed by passive or linear optical components; the technique used is also known as real-time holography. This active wavefront adaptation eliminates the need of an external stabilization device against thermal drift or ambient vibrations, required for the confocal Fabry-Perot. The basic setup of the two-wave mixing interferometer is sketched in figure 3. A signal beam which acquires phase shift and speckle after reflection on a surface in motion, is mixed in a photorefractive crystal with a pump plane wave to produce a speckle adapted reference wave that propagates in the same direction as the transmitted signal wave and interferes with it. The quadrature is assured by passive optical components after the crystal.

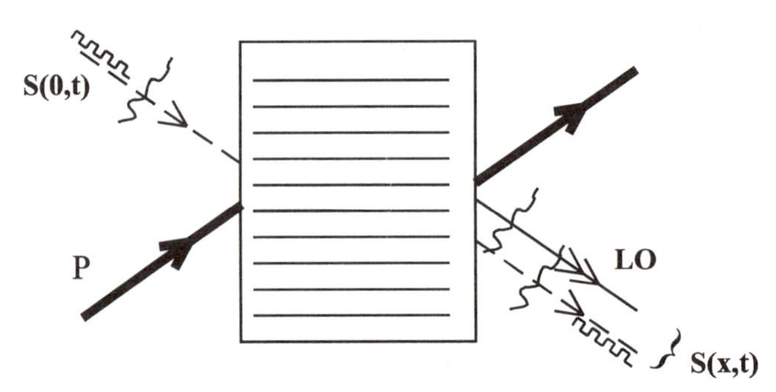

Figure 3: Basic setup of the two-wave mixing interferometer; Signal beam (S), Pump beam (P), Reference beam or local oscillator (LO).

The étendue, fixed by the size of the crystal and the angle between the signal and pump beams, is easily made larger than the étendue of the fiber mentioned above [7]. In figure 4, the sensitivities of the TWM interferometer operated with different semiconductor photorefractive crystals are compared to the confocal Fabry-Perot used in transmission (R= 85 % , 1 m length). It is shown that the sensitivity of the two-wave mixing device is comparable to the maximum sensitivity of the CFP. The best sensitivity is obtained with a CdTe:V crystal due to its higher electro-optic constant [8]. This best sensitivity is about 3 times less than the ultimate sensitivity, and can be improved by about a factor of 2 with a better tailored CdTe:V crystal. The signal field at the output of the crystal is given by equation 1 where α is the optical absorption coefficient, γ is the photorefractive gain, x the crystal length and ϕ the phase modulation [9]. Contrary to the Fabry-Perot, the photorefractive device cannot operate without optical losses since it is necessary to absorb photons to write the grating. Hence the relevant figure of merit for the sensitivity of the photorefractive interferometer is the ratio of the photorefractive gain to the optical absorption [3]. This ratio should be optimized while maintaining the étendue and the response time (or bandwidth) to the required values [7].

$$S(x,t)=e^{-\frac{\alpha x}{2}}S_{input}\left[\left(e^{\gamma x}-1\right)+e^{i\phi(t)}\right] \tag{1}$$

The photorefractive device is more sensitive than the CFP for ultrasonic frequency below 1 MHz. To confirm this increased sensitivity at low frequencies, laser-ultrasonic data was taken on a ½" thick graphite epoxy sample. The laser-ultrasonic system was composed of a CO_2 laser for generation of ultrasound and a long pulse Nd:YAG laser for detection. A symmetrical CFP (length: 1m, R= 85 %) used in transmission mode and a photorefractive interferometer using an InP:Fe crystal under an applied field [7] were used to compare the performance of both techniques. B-scans shown in figure 5 clearly indicated that more echoes are seen with the TWM than with the CFP. Since distant echoes are composed essentially of low frequencies, because material attenuation increases with frequency, this demonstrates the better sensitivity of the two-wave mixing interferometer at low frequencies than the Fabry-Perot. The photorefractive interferometer also has the advantages of a flat frequency response up to at least a few GHz without the periodic high frequency notches of the Fabry-Perot [10].

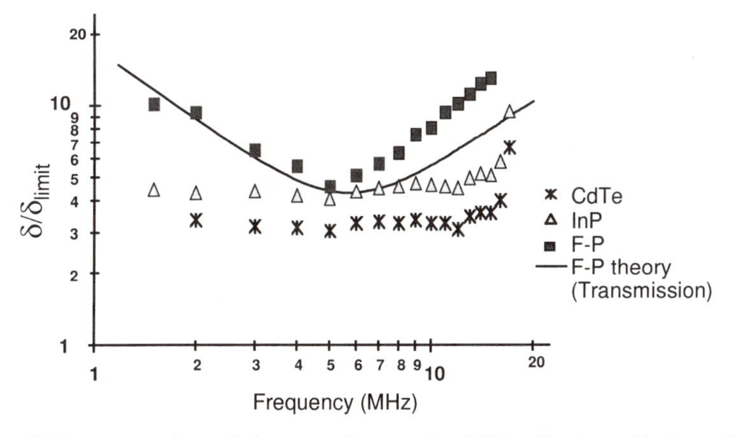

Figure 4: Sensitivities versus ultrasonic frequency for a confocal Fabry-Perot used in transmission mode and photorefractive two-wave mixing devices operated with various crystals.

Photorefractive

Fabry-Perot

Figure 5: Laser-Ultrasonics B-scans of ½" thick graphite epoxy sample obtained with a symmetrical confocal Fabry-Perot (L= 1m, R=85 %) used in transmission and a photorefractive interferometer operated with an InP:Fe crystal.

Finally, a differential or balanced detection scheme is easily inserted in the photorefractive interferometer to eliminate the amplitude fluctuations of the detection laser. Such a scheme is possible with the Fabry-Perot but with a greater complexity.

CONCLUSION

The sensitivities of two types of interferometers developed for optical detection of ultrasound on rough surfaces have been compared and are close to the ultimate sensitivity expected from an interferometric technique (operating with plane waves). The two interferometers work with beam with speckle and have similar large étendue. As an active device, the TWM does not need the external stabilization system required for the CFP. For low ultrasonic frequencies, the TWM presents a better sensitivity than the CFP and therefore should be chosen when this range of frequencies is important.

REFERENCES

1. E.S. Boltz, C.M. Fortunko, M.A. Hamstad, M.C. Renken, Absolute sensitivity of air, light and direct-coupled wideband acoustic emission transducers, *Review of Progress in Quantitative Nondestructive Evaluation*, 14: 967, (1995), Edited by D.O. Thompson and D.E. Chimenti, Plenum Press, New-York.

2. J.-P. Monchalin, Optical detection of ultrasound, *IEEE Trans. Ultrason. Ferroelectrics and Freq.Cont.* UFFC-33:485,1986.

3. R.K. Ing, J.P. Monchalin, Broadband optical detection of ultrasound by two-wave mixing in a photorefractive crystal, *Appl. Phys. Lett.* 59: 3233, (1991).

4. A. Blouin, J.P. Monchalin, Detection of ultrasonic motion of a scattering surface by two-wave mixing in a photorefractive GaAs crystal, *Appl. Phys. Lett.* 65:932, (1994).

5. C.M. Caves, Quantum-mechanical noise in an interferometer, *Phys. Rev D*, 23:1693, 1981.

6. P. Connes, L'étalon de Fabry-Pérot sphérique, *J. Physique et le Radium*, 19:262, (1958).

7. P. Delaye, A. Blouin, D. Drolet, L.-A. de Montmorillon, G. Roosen, J.P. Monchalin, Detection of an ultrasonic motion of a scattering surface by photorefarctive InP:Fe under an applied DC field, *Journal of the Optical Society of America*, 14:1723, 1997.

8. L.A. de Montmorillon, I. Biaggio, Ph. Delaye, J.C. Launay, G. Roosen, Eye-safe large field-of-view homodyne detection using a photorefractive CdTe:V crystal, *Opt. Comm.* 129:293 (1996).

9. P. Delaye, L.A. de Montmorillon, G. Roosen, Transmission of time modulated optical signals through an absorbing photorefractive crystal, *Opt. Comm.*, 118:156, (1995).

10. A. Moreau, M. Lord, High frequency Laser-Ultrasonics, *This proceeding.*

OPTIMIZING THE PHOTO INDUCED-EMF RESPONSE FOR HIGH-SPEED COMPENSATION AND BROADBAND LASER-BASED ULTRASONIC REMOTE SENSING

G. J. Dunning, D. M. Pepper, M. P. Chiao, P.V. Mitchell* and T.R. O'Meara

Hughes Research Laboratories
3011 Malibu Canyon Road
Malibu, California, USA

*Melles Griot, Inc.
1770 Kettering Street
Irvine, CA, USA

Keywords: Laser based ultrasound, Photo induced-emf, Nondestructive.

ABSTRACT

We present results from a parametric study on the performance of photo-emf sensors made from GaAs and GaAs:Cr utilizing different electrode processing and geometries for the detector elements. In order to improve the device performance and ease of manufacturing we produced and tested sensors constructed in a compact hybrid design.

INTRODUCTION

There is a need in manufacturing environments to nondestructively evaluate components and control processes in real time. Laser-based ultrasound (LBU) [1] has proved to be an attractive diagnostic which can circumvent the constraints imposed by the manufacturing environment on traditional ultrasonic techniques. In LBU systems lasers replace both the ultrasound launching and receiving transducers of conventional ultrasonic techniques. Therefore the nondestructive inspection can be performed while the sample is at high temperature, under vacuum, or in motion.

In this paper we present a study to optimize the performance of a simple and economical Compensated Laser Ultrasonic Evaluation (CLUE™) system [2] sensor based on the photo-induced-electromotive (PI-EMF) effect [3,4]. We characterized the bandwidth, linearity, and sensitivity of the PI-EMF sensor as a function of the detector materials and investigated several electronic and optical architectures. Sensors were fabricated using GaAs and GaAs:Cr crystals, utilizing both surface and indiffused electrodes. The output signal was characterized as a function of the electrode spacing and aspect ratio, number of grating periods across the detector element, and the input signal amplitude and frequency.

Nondestructive Characterization of Material VIII
Edited by Robert E. Green Jr., Plenum Press, New York, 1998

As a means to further improve device performance we packaged the sensor elements and amplifiers into a compact hybrid subassembly with many of the components integrated onto a single substrate. A major advantage of this design is that the compact geometry enables a reduction of stray impedances and increases the operating gain-bandwidth product. By making gain-bandwidth trade-offs the signal-to-noise ratio can be improved and therefore the sensitivity of the module for a given probe laser fluence. These hybrid sensors have demonstrated bandwidths greater than 50 MHz and subnanometer displacement sensitivity.

CLUE™ 1000 SENSOR DESIGN AND CONSTRUCTION

The philosophy in the design and optimization of the CLUE™ 1000 photo induced-emf sensor was to make a testbed to characterize the sensor elements and electronics independently. We designed a baseline module consisting of a compact, easily replaced detector element, a three-stage amplifier, and a shielded power supply with filter networks. The detector element is located on an interchangeable detector mini-circuit board. The current generated by the detector element is coupled into a transimpedance amplifier, buffer amplifier and output power amplifier sufficient to produce an output signal compatible with the diagnostic equipment. The entire housing (3x5.5x1.9 inches) is made from aluminum and is designed to shield the internal electronics from external electromagnetic interference. An additional BNC connector is provided for applying an external bias level to the detector or can also be configured for calibration purposes.

The sensor active area was either square or rectangular depending upon the aspect ratio of the electrode length to electrode spacing (1:1 or 2:1) with the electrode spacings varying between 0.25 mm to 5.00 mm. Each detector die is attached to the interchangeable circuit board with multiple wire bonds. The wirebonds are connected to gold coated copper pads which were custom printed on the mini-circuit boards to accommodate the various electrode spacings. There is physical hole in the circuit board directly behind the active area of the detector. This hole allows laser energy below the bandgap of the semiconductor to pass through the detector element and out from the assembly. The overall design of the interchangeable detector board enabled us to keep the amplification circuit the same for all electrode spacings or material studies and therefore maintain the other sensor module parameters for a given set of experiments.

The typical values for the conventional photo-emf detector current are in the sub-100 nA range and with frequencies form 10 kHz to 100 MHz or more. In the CLUE™ 1000 series sensors we used a chain of three discrete commercially available amplifiers. A low-noise transimpedance amplifier was followed by a buffer amplifier and then a power amp. The electronics were tested by simulating a photo-emf current transient. An RC network amplifier which consisted of a 1 kΩ load resistor in series with a 0.1 μf capacitor was placed on the input to the transimpendance amplifier. From these simulations it was determined that the bandwidth of the amplifier circuit was approximately 40 MHz, with an output approaching 80 mV/mA. This performance was considered adequate, since the circuitry consists of discrete off-the-shelf electronic components. The fabrication of the photo-emf detector element required a multi-step process, involving a series of depositions. Three different masks were designed for the processing: one for producing the electrodes, one for an (optional) anti-reflection coating to minimize Fresnel losses, and one for producing an aperture to define the detector active area. By incorporating the various electrode geometries, we were able to evaluate the performance of the detector as a function of the laser fringe spacing and the number of fringes across a specific electrode separation.

PHOTO INDUCED-EMF SENSOR CHARACTERIZATION

A characterization facility was constructed in order to quantify the sensitivity, linearity and frequency response of the detectors. A plane wave reference beam was interfered on the detector with a probe beam which had a calibrated phase shift. The various test were conducted by introducing a variable phase difference using an electro-optic modulator. A portion of the phase modulated and reference beams were used to form a Mach Zehnder interferometer in order to calibrate the phase retardation introduced by the E-O modulator. A maximum phase retardation of 54 mrad, and a maximum frequency range of 30 Mhz (limited by the frequency synthesizer) were used in the characterization measurements. A schematic of the experimental apparatus is shown in Figure 1. For the experiments that immediately follow a cw argon-ion laser operating at 514 nm was used to generate the probe and reference beams, with an intensity of about 60 mW incident on the crystal.

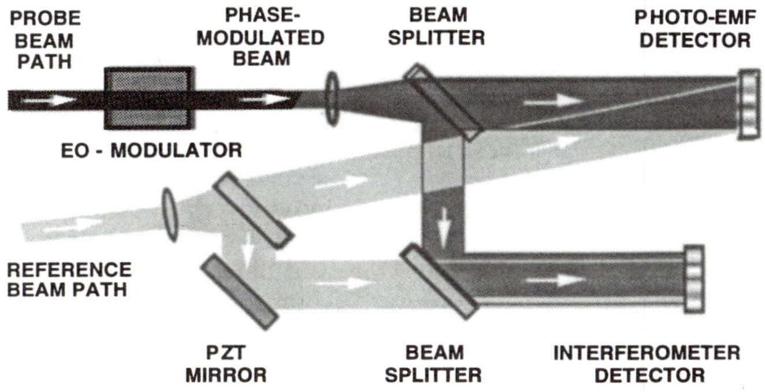

PROBE BEAM PATH
PHASE-MODULATED BEAM
BEAM SPLITTER
PHOTO-EMF DETECTOR
EO - MODULATOR
REFERENCE BEAM PATH
PZT MIRROR
BEAM SPLITTER
INTERFEROMETER DETECTOR

Figure 1. Schematic of PI-EMF detector characterization testbed.

A parameterized graph of the normalized signals versus fringe spacing are shown in Figure 2. The elements were made from GaAs and GaAs:Cr with the electrodes annealed at low and high temperatures to promote indiffusing. The optimum grating period is determined by the diffusion length of the carriers and is seen to be approximately 49 degrees. It is also clear that the GaAs:Cr annealed elements have the best performance of the materials tested.

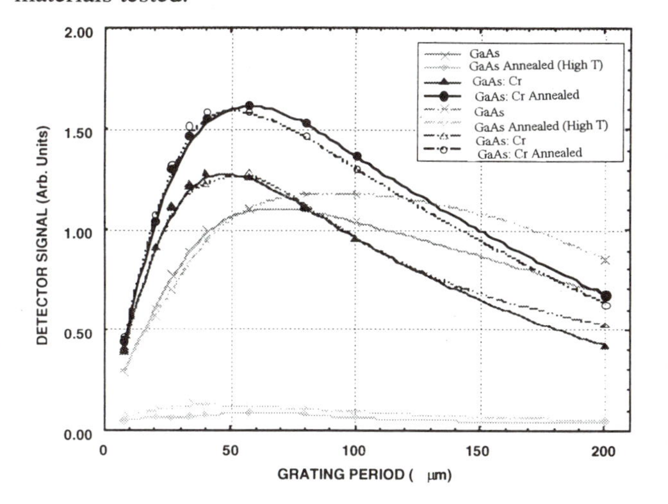

Figure 2. Detector output signal versus grating period.

A topic of interest for the optimization of the photo induced-emf detector performance is how the electrode geometry effects the device performance. A system advantage could be derived by decreasing the response time of the detector by increasing the incident flux. In operation it would be advantageous for the electrode spacing to become infinitesimally small so that for a given energy incident on the detector the highest flux possible would be produced.

However, not only is it impossible to make an infinitely small detector but it is also difficult to focus the aberrated probe beam to a diffraction limited spot. Furthermore, it is essential for robust operation to have several grating periods present on the active area between the electrodes. Therefore a compromise must be reached balancing the grating criteria, focusing constraints and element manufacturability. These issues were addressed by manufacturing several different electrode config-urations with varying electrode separation, and electrode lengths. We measured the current produced for different separation between the electrodes. This was done for aspect ratios of 2:1 and 1:1 for the electrode length to the separation between the electrodes. The experimental results of the normalized signal versus the electrode spacing are graphed in Figure 3. The circles are for the aspect ratios of 1:2 and the diamonds are for the aspect ratios of 1:1.

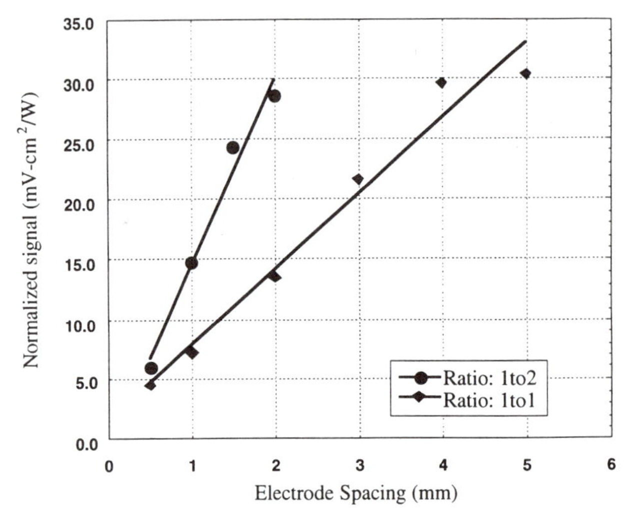

Figure 3. Sensor output signal versus electrode spacing.

This graph clearly shows that the amount of photo induced-emf current depends upon the length of the electrodes and not on the number of grating periods between the electrodes. This can be seen by observing the value of the current for a particular electrode spacing on the 1:1 graph and then finding the current obtained for the same electrode spacing on the 2:1 graph. The current is seen to double. In this case the same number of grating periods are present between the electrodes and the results demonstrate that the current is independent of the number of grating periods. By examining the 1:1 aspect ratio curve, the current also doubles when the value of the electrode spacing doubles because for this case the electrode length also doubles. In conclusion the photo-induced emf current produced using uniform illumination is not a function of the total number of grating periods between the electrodes. However, it is a function of the electrode length for a given electrode separation.

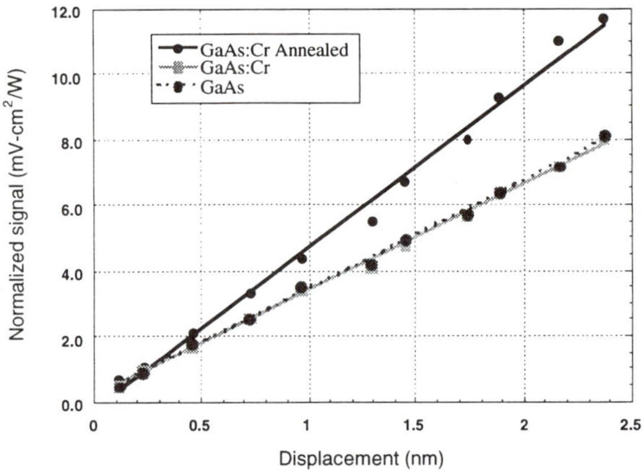

Figure 4. Graph of detector output signal versus displacement.

The linearity and sensitivity of the device were measured by simulating different amplitudes of the ultrasonic surface displacement with the amplitude of the phase shift impressed on the probe beam. The sensor output signal was measured as a function the E-O modulator drive voltage. For these measurements the drive frequency of the E-O modulator was fixed at 1 MHz. In these experiments, the grating period was 57 μm, and the interference modulation depth was close to 1.0. A graph of the detector output signal (normalized to the beam average intensity at the detector) versus the surface displacement is plotted in Figure 4.

The output signal is again parameterized by the detector material and electrode anneal condition. The data shows the linearity of the detector response in the displacement range of interest for most commercial applications. A typical trans-ducer characteristic indi-cates a signal-to-noise ratio of unity at a 1.0 MHz with 10 kHz bandwidth for phase shifts on the order of 0.046 mrad. Such a phase shift corresponds to a surface displacement of 0.039 Å, which is on the order of typical induced displacements generated in a laser-based ultrasonic inspection system. Given the active area of our sensor (2 mm x 2 mm), and light intensity of 66.5 mW if the detector were limited by noise from the sensor the normalized sensitivity would be about 1.0×10^{-4} Å $(W/Hz)^{1/2}$. However, most of the noise is from the amplifiers. This value is not optimum because the testbed was designed for testing the detector elements, and is therefore not the fundamental limit, especially considering the equivalent current input noise of the amplifiers.

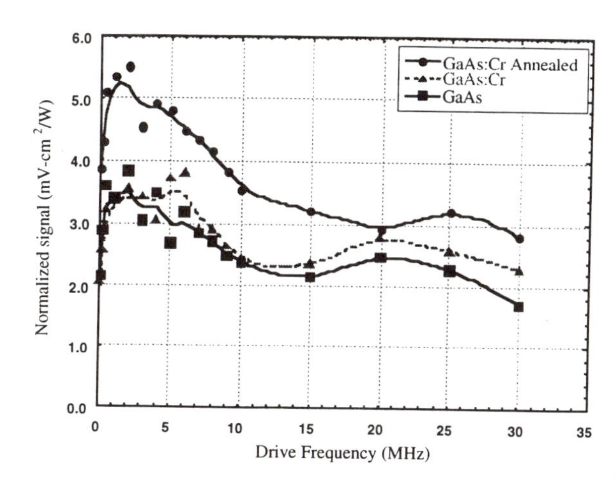

Figure 5. Graph of sensor output signal versus frequency.

The frequency response of the detector and receiver electronics were measured with a fixed drive voltage on the E-O modulator while the drive frequency was scanned from 100 kHz to 30 MHz. The PI-EMF detector output signal (normalized to the beam intensity at the detector) versus frequency is plotted in Figure 5. The low frequency roll-off is due to the ability of the gratings formed by the space charge field to track the moving intensity pattern. The high frequency roll-off is related to the bandwidth of the electronics. The bandwidth of the sensor is approximately 20 MHz using the full-width at half-maximum criterion. The low frequency limit was set by the last stage amplifier at 100 kHz while the 30 MHz upper limit represents the upper bound of the frequency synthesizer and not the ultimate limit of the detector material response. Since the material upper limit is determined by the dominant carrier lifetime. Under these experimental conditions the detector response yielded a fractional detection bandwidth in excess of 99%.

CLUE™ 2000 HYBRID DEVICE CHARACTERIZATION

The sensor elements and amplifiers were packaged into a compact hybrid subassembly with many of the components integrated onto a single substrate in an effort to further improving device performance. A major advantage of this design is that the compact geometry enables a reduction of stray impedances. This reduction facilitates extending the gain-bandwidth product leading to an optimization of the signal-to-noise ratio. Therefore it follows that the sensitivity of the sensor could be increased for a given probe laser fluence. Additional benefits of a hybrid device include isolating the sensor from the factory environment, reduction in overall size, power consumption and manufacturing cost. These hybrid sensors have demonstrated bandwidths greater than 50 MHz, subnanometer displacement sensitivity and are designated as the CLUE™-2000 series.

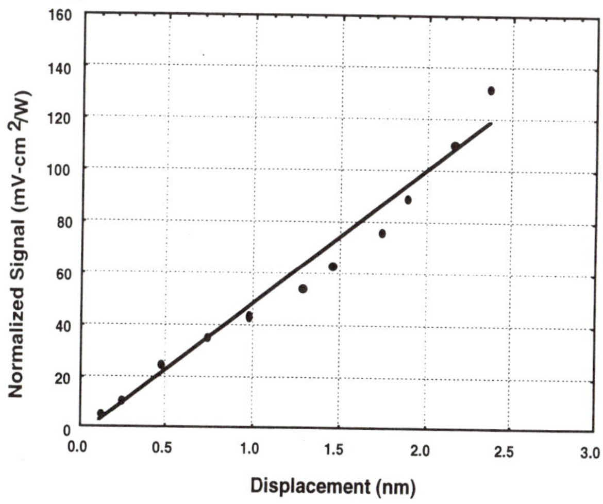

Figure 6. Graph of hybrid detector signal versus displacement.

In order to characterize the preliminary performance of the hybrid devices we conducted linearity and frequency response measure-ments using the charac-terization facility described earlier in this paper. In these experiments the grating period was 57 μm with a modulation depth of 1.0. The drive frequency of the E-O modulator was 1.0 Mhz for the linearity measure-ments. A graph of the hybrid sensor output signal (normalized to the beam average intensity at the detector) is shown in Figure 6. The data is linear in the region of interest for many commercial applications.

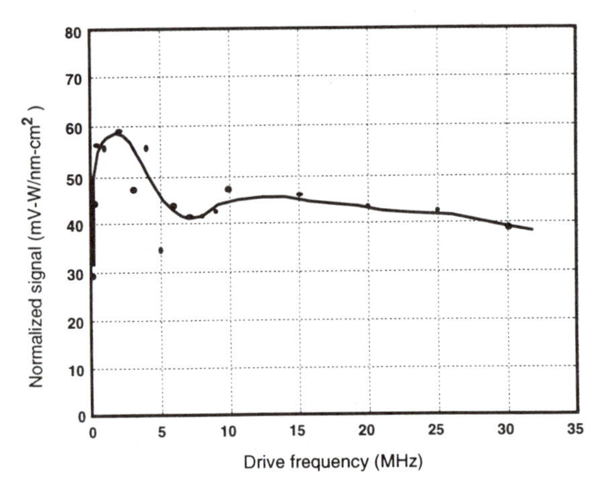

Figure 7. Hybrid detector output signal versus frequency.

We then measured the frequency response of the detector. The drive voltage on the E-O modulator was fixed at 10 V(P-P) while the drive frequency was scanned from 100 kHz to 30 MHz. The hybrid sensor output signal (normalized to the beam intensity at the detector) versus frequency is plotted in Figure 7. As can be seen the bandwidth is greater than 30 Mhz using the full width at half maximum criterion.

One of the design options for optimum system performance is the trade-off between the system bandwidth and gain. The data in Figures 6 and 7 were made with a 1 KΩ feedback resistor in the transimpedance amplifier and an overall multistage gain of 500 KΩ.

In many applications it is not necessary to have a bandwidth greater than 10 MHz. Exercising this option should improve the system signal to noise ratio for a fixed incident laser intensity. We are currently assembling additional hybrids with different values of feedback resistors, 1, 2.5 , 5 and 10 KΩ, for the transimpedance amplifiers. These units will also have the ensuing amplifiers stages tailored to match the bandwidth of the first stage. In the near future we will characterize these devices and investigate the trade-offs between the bandwidth and gain with the goal of improving the signal to noise ratio.

CONCLUSIONS

In conclusion we have characterized laser-based ultrasonic receivers employing the photoinduced-emf effect. Experiments revealed that detector elements made from GaAs doped with chromium together with indiffused electrodes gave the best performance. These type of detector elements were then incorporated into compact, rugged and cost effective hybrid devices which exhibited greater operating bandwidths than the discrete amplifier configurations. The demonstrated frequency response and sensitivity are well suited for many manufacturing applications in the areas of quality assurance and process control.

ACKNOWLEDGMENTS

The authors wish to thank D. Bohmeyer, R. Harold, S. Bourgholtzer, R. Doty, R. Lohr, and D. Saboe for their excellent technical assistance and are grateful to D. Sipma, G. Valley and A. Hunter for technical discussions. This work was supported in part by DARPA and the Hughes Research Laboratories.

REFERENCES

1. CB Scruby and L.E. Drain, *Laser Ultrasonics: Techniques and Applications* (Adam Hilgar Press, Bristol, 1990).
2. P.V. Mitchell, G.J. Dunning, T.R. O'Meara, M.B. Klein and D.M. Pepper, Rev. of Prog. in Quant. Nondest. Eval. **15A,** D. Thompson & D. Chimenti, Eds. (Plenum Press, N.Y., 1995); D. M. Pepper, G.J. Dunning, P.V. Mitchell, S.W. McCahon, M.B. Klein, and T.R. O'Meara, SPIE Proc. **2703**, 91 (1996).
3. M.P. Petrov, S.I. Stepanov, and G.S. Trofimov, Sov. Tech. Phys. Lett. **12,** 379 (1986); I.A. Sokolov and S.I. Stepanov, J. Opt. Soc. Am. **B10,** 1483 (1993).
4. S.I. Stepanov, Appl. Opt. **33,** 915 (1994).

HIGH FREQUENCY LASER-ULTRASONICS

A. Moreau and M. Lord

Industrial Materials Institute
National Research Council of Canada
75 de Mortagne Blvd.
Boucherville, Québec, J4B 6Y4

ABSTRACT

The usual frequency bandwidth of laser-ultrasonics is in most cases of order 1 to 30 MHz. Although this is sufficient for most applications, applications involving attenuation measurements in weakly attenuating materials, or velocity measurement in thin samples or coatings may require a much higher detection bandwidth. By shortening the generation light pulse and improving the interferometer electronics, laser-ultrasonic bandwidths of 500 MHz were achieved. Much higher bandwidths (up to 100 GHz) have been achieved by others using delayed femtosecond laser pulses, but what distinguishes our technique is the capability to acquire a full A-scan with a single generation pulse. The capabilities of high frequency laser-ultrasonics will be illustrated with measurements in aluminum foil and sheet, fused quartz, and galvanized steel.

INTRODUCTION

The usual frequency bandwidth of laser-ultrasonics is in most cases of order 1 to 30 MHz. The most important factors that determine the ultrasonic bandwidth are the optical penetration depth of the material and the generation laser pulse duration. In metals, the penetration depth is very small and it has been shown that very short pulses of light may be used to generate high frequency ultrasound. In particular, frequencies as high as 50 to 100 GHz were obtained using laser pulses of 100 to 200 fs duration and delayed probing techniques.[1,2] In this paper, a 2.5 ns fwhm excimer laser was used to generate ultrasound in the frequency bandwidth of 10 to 500 MHz. The experimental setup will be discussed first with a special emphasis on the characterization of the Fabry-Pérot interferometer used to detect the high frequency ultrasound. Then, selected examples will be shown to illustrate the characteristic behavior of the system, and to indicate potential applications of high-frequency laser-ultrasonics to weakly attenuating or thin materials

EXPERIMENTAL SETUP

All ultrasonic measurements presented in this paper were made by generating on one side of a sample, and detecting on the other side. The excimer (KrF) generation laser produced UV (248 nm) light pulses of roughly 100 mJ which were weakly focused to a square or rectangular area of 4 to 10 mm to a side. This produced an irradiance of roughly

Nondestructive Characterization of Material VIII
Edited by Robert E. Green Jr., Plenum Press, New York, 1998

40 to 250 MW/cm^2, which corresponds to the beginning of the ablation regime in metals. The detection light source was a 3 kW (peak power) long-pulse (50μs fwhm), Nd:YAG (1.06 μm) laser focused to a circular area of 4 mm diameter. An attenuator keeps the received light peak power below 300 mW to avoid saturating the photodetector.

To detect the ultrasound, a 50 cm confocal Fabry-Pérot interferometer was used in the reflection mode. Its detection electronics was carefully redesigned. A fast photodiode was coupled to a high-pass filter followed by a DC-500 MHz amplifier. The first low-pass RC time constant of the detection circuit (300 MHz) was set by the internal capacitance of the photodiode and the 50 Ohm load impedance. However, given that simple RC filters have a rolloff of only 6 dB per octave, the detection interferometer worked well up to the cutoff frequency of the amplifier (more than 500 MHz).

The confocal Fabry-Pérot interferometer has the great practical advantage of being able to process multiple speckles without loss of signal to noise ratio.[3] Therefore, it is especially well suited for detection on rough surfaces. However, the amplitude and phase response depend on the frequency of the ultrasound. The response predicted in the literature[4] was computed with the parameters of our interferometer (mirror reflectivities of 91 an 99.5%, 50 cm cavity length). This was compared to that measured using a laser beam modulated in phase and amplitude using a electrooptic cell.[5] The results are shown in Figure 1. One can see that the interferometer has a nearly flat amplitude and phase response with rejection bands beginning at zero frequency and every 150 MHz thereafter. These rejection bands are approximately 10 MHz wide, except near zero frequency, where the rejection bandwidth is approximately 5 MHz wide.

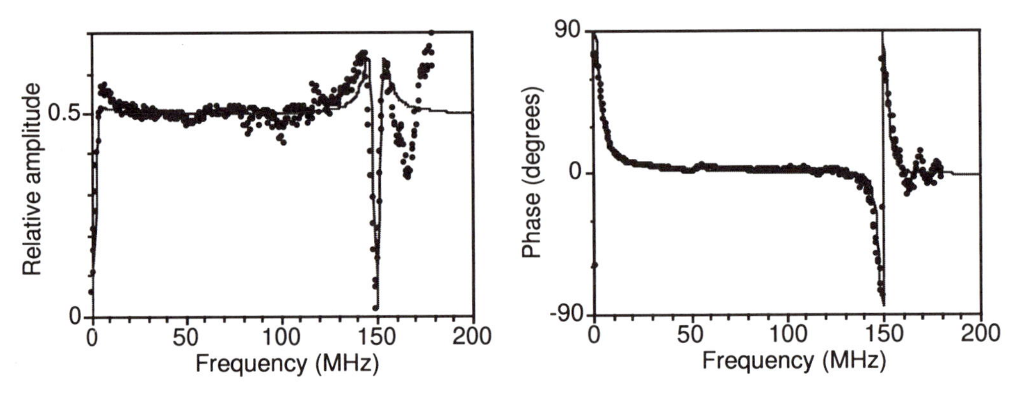

Figure 1. Amplitude and phase response of the Fabry-Pérot interferometer: theory (solid line) and experiment (dots). The experimental signal to noise ratio was very poor above 160 MHz.

The effect of these rejection bands on the ultrasonic signal is observed when the bandwidth exceeds 150 MHz. Figure 2 compares the ultrasonic signal measured using the Fabry-Pérot interferometer to that obtained using a photorefractive interferometer.[6] The rejection bands of the Fabry-Pérot interferometer cause oscillations to follow the laser-generated longitudinal acoustic pulse (negative displacements correspond to outwards motion of the surface). These oscillations are not observed with the photorefractive interferometer because its response is completely linear below the cutoff frequency of the photodetectors and above the response frequency of the photorefractive crystal (100 kHz).

Effects of the response of the Fabry-Pérot interferometer can also be observed on the ultrasonic signal obtained in a 6061-T6 aluminum plate of 800 μm thickness (Figure 3). Ripples similar to those observed in Figure 2 are observed in the time domain. However, in the frequency domain, the first four rejection bands are clearly observed. Although the theory predicts zero amplitude response at the bottom of the rejection bands, the finite resolution of the Fourier transform techniques utilized to analyze the data limits the sharpness of the measurement. The zero-frequency rejection band is also widened by a 100 MHz high-pass filter. This filter reduces the low-frequency amplitude because otherwise, the dynamic range of the digitizer would be exceeded. The theory also predicts 174°, 288°, and 174° phase shifts at the 150, 300, and 450 MHz rejection bands. This corresponds to delay shifts of 3.2, 2.7, and 1.1 ns. This is experimentally observed in Figure 3 as three sudden discontinuities in the arrival time of the first echo as a function of frequency. Again, the finite resolution of the Fourier transform reduces the amplitude of the discontinuities below the predicted values. However, the relative amplitudes of the three discontinuities, 2.3, 1.7, and 0.7 ns, agree with the prediction. No explanation is offered for another apparent discontinuity in the amplitude and delay spectra near 350 MHz.

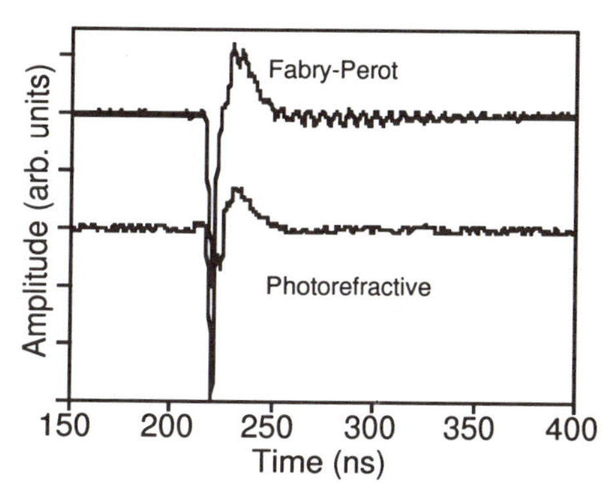

Figure 2. High frequency ultrasonic signal in a fused quartz sample measured with a Fabry-Pérot interferometer and a photorefractive interferometer. The 150 MHz oscillations are due to the filtering effect of the Fabry-Pérot interferometer. The generation laser area measured 10x12 mm and the generation surface was lightly oiled. This A-scan is a single trace sampled at 5 GSamples/second using an 8-bit digitizer.

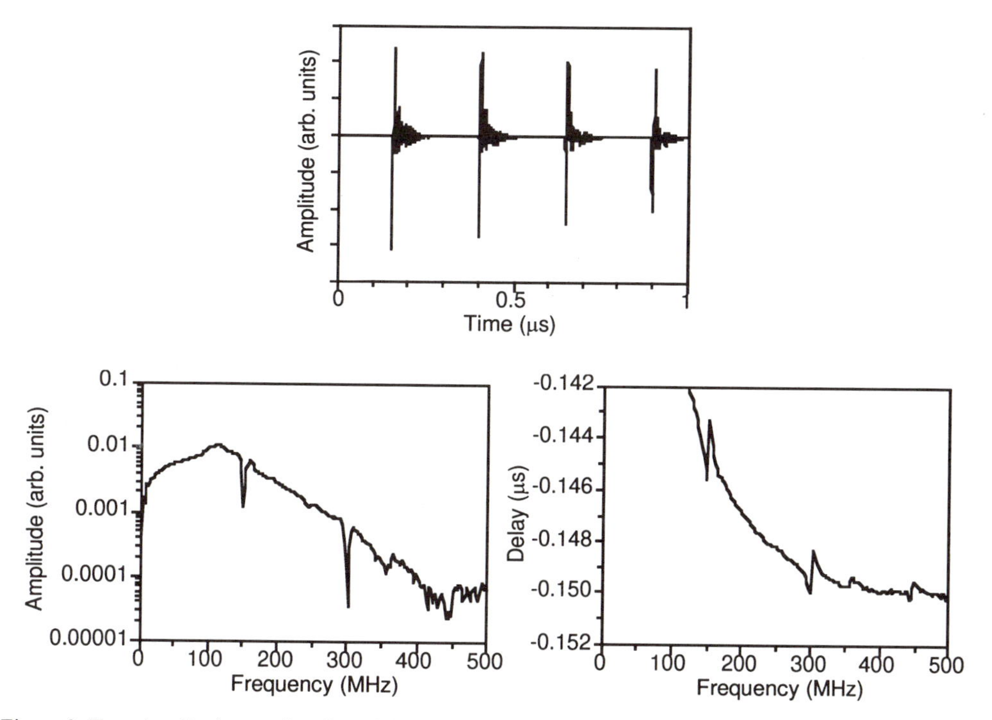

Figure 3. Top: Amplitude as a function of time (A-scan) showing the first four acoustic echoes in an 800 μm thick aluminum plate (grade 6061-T6). Left: Amplitude spectrum and Right: Delay spectrum of the first echo. Signal to noise ratio reaches unity near 500 or 600 MHz. The two spectra were calculated from the A-scan using a 350 ns rectangular window and Fourier transform techniques. The generation laser area measured 4x8 mm. Average of 400 traces sampled at 5 GSamples/second using an 8-bit digitizer.

APPLICATIONS

Ultrasonic Attenuation in an 800 μm Thick Aluminum Plate

The fine grain size and low single crystal anisotropy of aluminum cause low ultrasonic attenuation in this material. However, ultrasonic attenuation caused by grain scattering, absorption, or other effect, generally increases with frequency. Therefore, high frequency laser-ultrasonics may be used to measure ultrasonic attenuation in aluminum and other weakly attenuating materials. Figure 4 shows the attenuation spectrum obtained using Fourier transform techniques on pairs of echoes in the same 800 μm aluminum plate as shown in Figure 3. It is best to utilize the first two echoes to measure attenuation at high frequencies because the highest frequency components are below the noise level of subsequent echoes. At low frequencies, however, the attenuation is so low that measurement accuracy is much improved if the first echo is compared to a more distant echo. The effect of diffraction must also be considered. The large dimensions of our source (4x8 mm) insures that the measurement is made in the near acoustic field. The Fresnel parameter $s = \lambda z/a^2$, where λ is the acoustic wavelength ($\lambda = 640$ μm at 10 MHz from ref. 7), z is the distance from the source ($z = (2n - 1) * 800$ μm, where n is the echo number), and a is the radius of the source (3 mm) is approximately equal to 0.057 for the first echo at 10 MHz. One may assume that no diffraction correction is required when $s < 0.3$ and detection occurs over an area comparable to the generation area.[7] Here, this inequality is verified for frequencies higher than 32 MHz for the ninth or earlier echoes.

It is interesting to note the absence of anomaly in the attenuation spectrum near 150 and 300 MHz. In theory, no signal is detected at these frequencies and the measured attenuation should be random. However, the finite resolution of the Fourier transform methods used to obtain the attenuation spectrum smoothes out this anomaly. Also, as mentioned earlier, the amplitude response of the Fabry-Pérot interferometer is *nearly* flat between the rejection bands, but it is not *exactly* flat. However, because these distortions apply in the same way to both echoes, they have no effect on the attenuation spectrum.

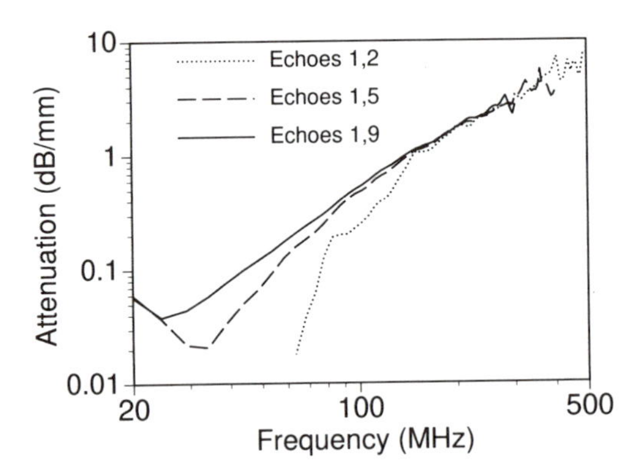

Figure 4. Attenuation spectra in an 800 μm thick aluminum plate. Blackmann-Harris windows of 200 ns and Fourier transform techniques were used to obtain these spectra from Figure 3.

Thickness and Attenuation Measurement in a 17 μm Thick Aluminum Foil

Thin materials present a challenge for ultrasonic measurement through the thickness because, at low frequencies, the echoes cannot be resolved in the time domain. In the frequency domain, the simple through-transmission configuration utilized here excites the natural resonances in the foil thickness. Because the foil is very thin, these resonances occur at high frequencies.

Figure 5 shows the signal obtained in an aluminum foil of the type sold in supermarkets. The amplitude spectrum shows a strong resonance at 184 MHz in the thickness of the foil (the second harmonic could also be observed at 365 MHz). Using an ultrasonic velocity of 6410 m/s for an isotropic aggregate,[8] this resonance can be used to calculate the foil thickness. The value obtained, 17.4 μm, agrees with that measured using a hand-held micrometer (17 ± 1 μm). Because a single frequency is present, and because the experiment is in the very near acoustic field ($s << 0.1$), the ultrasonic attenuation may be measured from the decay of the signal envelope in the time domain. This procedure yields a value of approximately 100 dB/μs, or 16 dB/mm. This is substantially more than that observed in the 800 μm plate.

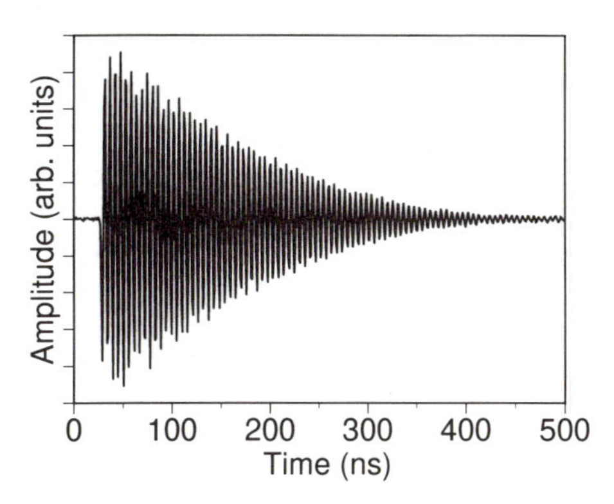

Figure 5. Ultrasonic signal in a 17 μm thick aluminum foil. The generation laser area measured 6x7 mm. Average of 100 traces sampled at 1 GSamples/second using an 8-bit digitizer.

Coating Thickness Measurement for Galvanized Steel

Another area of potential application of high frequency laser-ultrasonics is to obtain information about coatings. Again, multiple echoes in the coating thickness occur at high frequencies. Figure 6 compares the ultrasonic signal obtained in a hot-dipped galvanized, high-strength low-alloy (HSLA) steel sheet of 1.38 mm thickness, to that obtained in the same sheet after the two coatings were removed by dissolving them in dilute HCl acid.

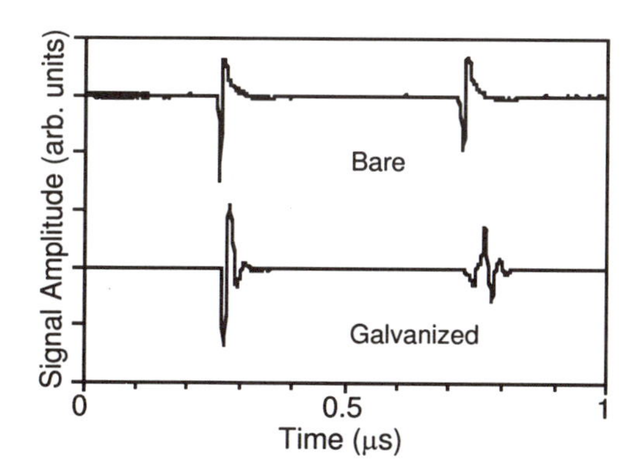

Figure 6. Laser-ultrasonic signal in a bare and hot-dipped galvanized sample of HSLA steel. The generation laser area measured 5x4 mm. Average of 25 traces sampled at 1 GSamples/second using an 8-bit digitizer.

Inspection of the first echo in the galvanized samples reveals a second negative pulse of 10 to 40 % of the amplitude of the first pulse. This second pulse occurs 23 ns after the arrival of the first pulse. Because the acoustic impedance of zinc is less than that of iron, the generated acoustic pulse interferes constructively with echoes that travel an even number of times through the coating thickness. Therefore, this second negative pulse would have traveled four times the coating thickness. In between these two negative echoes, there should be a positive echo corresponding to the first round trip through the zinc thickness. This may be observed by comparing the signal obtained on the coated sheet to that obtained on the bare sheet: the positive echo manifests itself in the higher "positive overshoot" for the coated sheet. Using the ultrasonic velocity of the isotropic aggregate (approximately 4100 m/s),[7] one then calculates that the zinc coating had a thickness of 24 μm. On the other hand, the coating is known to have a preferential [0001] texture. Assuming an ideal [0001] texture, the ultrasonic velocity would be 2983 m/s and the estimated coating thickness would be 17.5 μm. The actual thickness was measured by weighting a sample before and after removing the coatings, and by dividing half the missing mass by the mass density of zinc. A value of 16.4 μm was obtained. This is somewhat lower than that measured using ultrasonics, but other factors such as chemical and phase composition of the coating may affect the measurement.

In Figure 6, it may also be noticed that the arrival time of the various echoes is earlier in the bare sample because the two zinc coatings were removed. Assuming that only the coatings (no steel) were removed, then the arrival time of the first echo in the bare sample should precede that of the galvanized sample by the travel time through twice the coating thickness, i.e. by 11.5 ns. The measured value was 11 ns.

CONCLUSION

The extension of the capabilities of laser-ultrasonics to frequencies as high as 500 MHz was demonstrated. Examples were shown using weakly attenuating or thin samples. In particular, ultrasonic attenuation was measured in an 800 μm aluminum plate to frequencies as high as 500 MHz; the thickness and ultrasonic attenuation at 184 MHz of a 17 μm thick aluminum foil was obtained from the first two acoustic through-thickness resonances; and the thickness of zinc coatings was measured on a hot-dipped galvanized HSLA steel sheet. The non-contact nature of laser-ultrasonics is a great advantage in doing such high-frequency measurements because no coupling material needs to be characterized and because no fancy alignment of transducers is required. Moreover, the measurements are very quick: all A-scans presented for metallic samples may be obtained without sample preparation and within seconds after the sample is mounted in front of the lasers. Finally, the authors believe that with shorter laser pulses and faster electronics, much higher frequencies could be obtained.

ACKNOWLEDGMENT

This work was funded in part by the American Iron and Steel Institute and the United States Department of Energy.

REFERENCES

[1] C. Thomsen, H. T. Grahn, H. J. Maris, and J. Tauc, Surface generation and detection of phonons by picosecond light pulses, *Phys. Rev. B*, 34:4129-4138 (1986).

[2] Curtis Jon Fiedler. *The Interferometric Detection of Ultrafast Pulses of Laser Generated Ultrasound*, Report WL-TR-96-4057, Materials Directorate, Wright Laboratory, Air Force Material Command, Wright-Patterson Air Force Base, OH 45433-7734, Final report, 1 April 1996.

[3] J.-P. Monchalin, Optical detection of ultrasound, *IEEE Trans. Sonics, Ultrasonics, Freq. Control.* UFFC-33:485-499 (1986).

[4] J.-P. Monchalin, R. Héon, P. Bouchard, C. Padioleau, Broadband optical detection of ultrasound by optical sideband stripping with a confocal Fabry-Perot, *Appl. Phys. Lett.* 55:1612-1614 (1989).

[5] P. Basséras, A. Moreau, and J.-P Monchalin, *unpublished* (1994).

[6] A. Blouin and J.-P. Monchalin, Detection of ultrasonic motion of a scattering surface by two-wave mixing in a photorefractive GaAs crystal, *Appl. Phys. Lett.*, 65:932-934 (1994).

[7] J.-D. Aussel and J.-P. Monchalin, Measurement of ultrasound attenuation by laser ultrasonics, *J. Appl. Phys.*, 65:2918-2922 (1989).

[8] G. Simmons and H. Wang. *Single Crystal Elastic Constants and Calculated Aggregater Properties: A Handbood.* Second Edition, M.I.T. Press, Cambridge MA.

MEASUREMENTS OF SOUND VELOCITIES AT HIGH TEMPERATURES BY USING LASER GENERATED ULTRASOUND

Hidetoshi NAKANO, Youichi MATSUDA, Satoshi NAGAI

National Research Laboratory of Metrology
1-1-4 Umezono, Tsukuba-shi, Ibaraki, Japan

INTRODUCTION

Laser ultrasonics has been accepted as a unique tool for material testing. The technique can probe samples in a noncontact manner, which offers considerable advantages in high temperature experiments such as sound velocity measurements for advanced materials. A serious drawback of the technique is the poor sensitivity of the detection, especially for the samples with optically rough surfaces. Recently, phase conjugate methods have been proposed to overcome the problem [1,2]. In the method, a distorted optical beam from the rough surface is changed to a well collimated beam, thereby a conventional interferometer can be used retaining the sensitivity.

In this study, a Michelson type interferometer combined with the phase conjugator is presented. The interferometer can be applicable for many types of materials, and achieve the faithful and broad-band detection, which is essential to precise measurements. To investigate the effects of experimental conditions on the velocity measurement, a single Si crystal is examined. In addition, the experimental results are compared with those obtained by a classic pulse echo method to confirm the reliability of the noncontact method. At elevated temperatures, sound velocities for ceramics are also compared. Finally, the uncertainty on the sound velocity measurements is estimated.

OPTICS LAYOUT

A Q-switched YAG laser with a second harmonic generator was used for generation of the ultrasound source. The laser provides a light pulse of 10ns in duration at a repetition rate of 10Hz. The interferometer is shown in Figure 1. A diode pumped laser with 200 mw output was used. The intensity of the prove beam on the sample was adjusted by a half-wave plate. Scattered light from the sample was collected with an aspherical lens and focused into a multimode fiber, through which the light was transmitted into a $BaTiO_3$(A) crystal. A bird wing configuration, in which signal beam k1 and pump beam k2 were

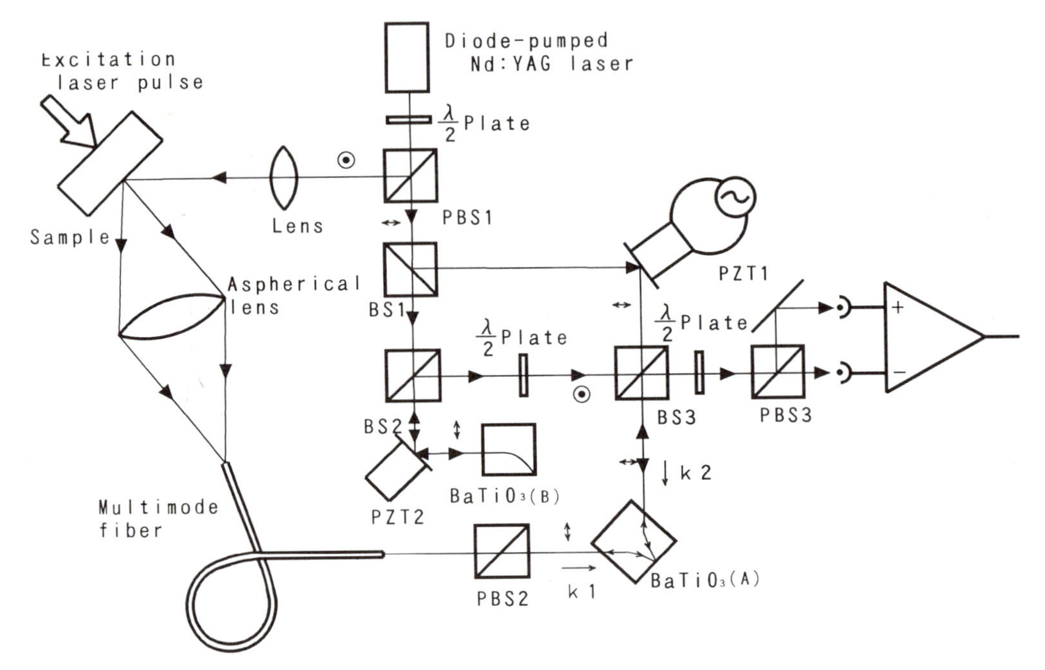

Figure.1 Schematic diagram for the detection of ultrasonic signals using a phase conjugator: PBS1-PBS3, polarized beam splitters; BS1-BS3, beam splitters; PZT1-PZT2, piezoelectric supported mirrors.

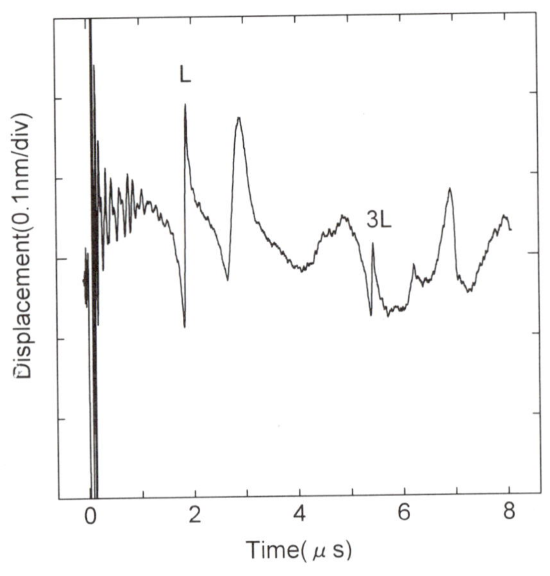

Figure.2 Displacement waveform for a Si sample generated by 532 nm source at epicenter. The spot diameters are 4.5 mm and 50 μ m, for source and probe beams, respectively.

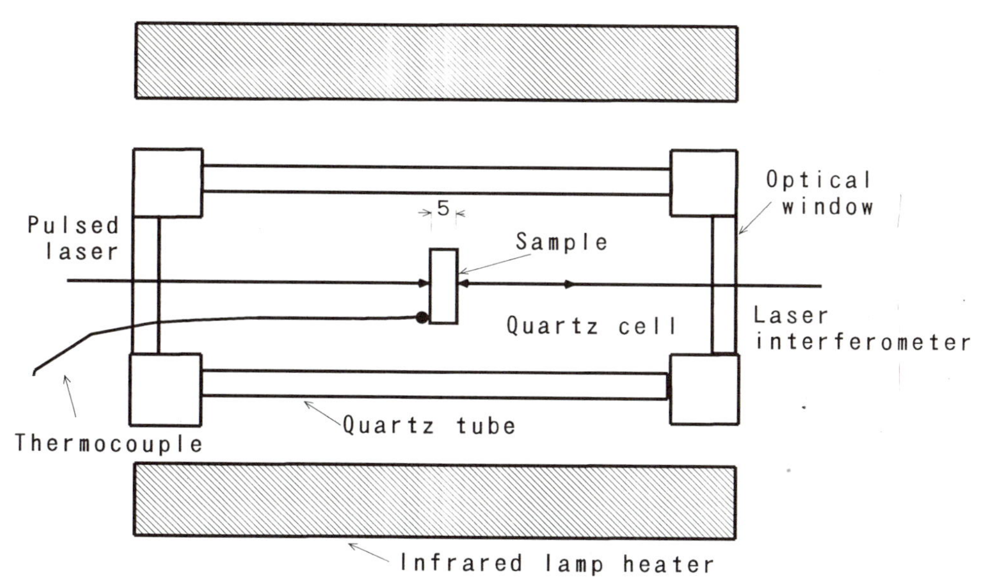

Figure.3 Experimental setup for high temperature measurements.

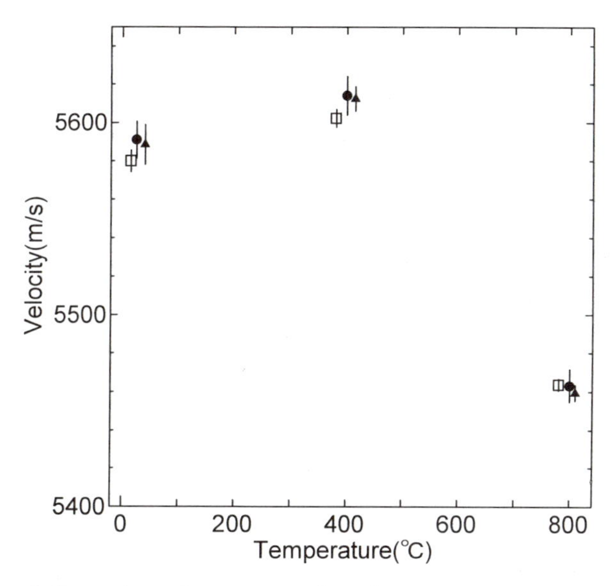

Figure.4 Comparison of temperature dependence of longitudinal velocity measured different methods in a machinable ceramic: ●▲, laser ultrasonic; □, pulse echo overlap using a stepped sample.

introduced into the opposite sides of the crystal, was adopted. Pump beam k2 was modulated by PZT1 at the frequency above the response time of the crystal, because beam k1 and k2 should be mutually incoherent to suppress undesired reflection beams from backscatter gratings in the crystal [3]. Beam k1 was converted into the phase conjugate replica of beam k2, and vice versa. In other words, beam k1 was changed to the well collimated signal beam and counterpropagated to beam k2. A Michelson interferometer was used to detect the phase change of the signal beam. The reference beam in the interferometer was passed through BS1 and BS2, and then reflected by another $BaTiO_3$(B) crystal to match the beam diameters between the signal and the reference beams. The optical path of the reference beam was adjusted by PZT2 to keep the maximum sensitivity. The reference beam was reflected by BS2 and its polarization was changed by a half-wave plate. The signal beam and reference beam were orthogonally polarized, thereby the balance detection arrangement was employed to minimize the common mode noise. The half-wave plate in front of PBS3 was rotated to make the outputs of both photo diodes balanced.

RESULTS

To analyze the influence of the experimental conditions on the velocity measurement, 4 parameters were selected as controllable factors which were taken 2 levels each. The factors were 1)the excitation wavelength of 1064 nm and 532 nm, 2)the epicenter and off-epicenter of 2 mm, 3)the diameter of the source beam of 7 mm and 4.5 mm, and 4)the diameter of the prove beam of 2 mm and 50 μ m. There were 3 runs of repetition, total 48 measurements were carried out. The sample was a single crystal Si 15.021 mm thick, which was measured by a optical linear gauge. The ultrasonic waves traveled along [100] direction. At each measurements, the sample was removed from a stage and placed on it again, thus the sample setup error including optical alignments could be estimated. The source energy was adjusted to be about 14mJ, which was low enough to avoid surface damage. Figure 2 shows a typical waveform excited by 532 nm source. The direct longitudinal and the second-reflection echoes were detected. The waveforms were recorded by a digitizing oscilloscope at the sampling rate of 2 ns. The round trip time from the first to the second echo was measured by overlapping the rising edges of the echoes on the oscilloscope.

For all of 48 measurements, the analysis of variance was calculated. By using F-testing, only the wavelength was the significant factor. Next these data were split into two groups according to the wavelength. Table 1 shows the analysis of variance table. The error values mean the repetition errors, which are associated with the optical alignment and the overlapping process. The error for 1064 nm source was much greater than 532 nm source. Since Si is transparent for 1064 nm, the echo pulse widths are longer. These result suggest that the pulse width has the large influence on the measurements, and 532 nm source is suitable for the velocity measurements. In contrast, the source and probe beam diameter are not significant for 532 nm source. It seems the velocity measurements are fairly robust against the beam diameters. Thus these two factors were treated as the error terms. Finally, the round trip-time for 532 nm source at epicenter was estimated to be 3559 ns, and the standard deviation was 2.8 ns.

Next the same sample was measure by a pulse echo overlap method. A X-cut quart transducer of 10MHz and phenyl salicylate were used as a couplant. At each measurements, the transducer was removed from the sample, then attached to it again with the new bond layer. One of the error sources of the method is the phase shift in the bond layer. The shift in the bond were calculated from the difference of the round trip times which were

Table.1 Analysis of variance

Source of variance	Degrees of freedom	Wave length 1064		Wave length 532	
		Size of factor effect	Variance	Size offactor effect	Variance
epicenter or not	1	10.7	10.7	66.7	66.7*
source beam diameter	1	10.7	10.7	18.7	18.7
prove beam diameter	1	384	384*	0	0
Error	20	1177.6	58.88	151.6	7.58
Total	23	1583		237	

* the variance ratio is greater than 5 %.

measured at 9MH and the resonance of 10MHz [4]. In this study, the phase shift was estimated to be very small and negligible. The average round trip time of 6 measurements was 3562ns, the standard deviation was 2.2ns. Two measurements by the laser ultrasonics and the pulse echo overlap were in good agreement.

Figure 3 shows the experimental set-up of the laser generated ultrasound at high temperatures. The furnace has a heat chamber, in which a fused quartz tube is inserted. Two optical windows are equipped at the both sides of the tube. The temperature of the sample was measured by a thermocouple, which was attached to the sample. The quartz tube was evacuated to pressures in the range of 0.1 Pa to prevent oxidation of the samples. A machinable ceramic Macor was chosen as the sample at high temperature experiments. Two samples of 5 mm in thickness were cut from the same rot, and measured by the laser ultrasound, where the proved faces were gold coated to improve the reflectivity. In the conventional pulse echo method, a stepped specimen was made to avoid a bonding problem at high temperatures. Also taking advantage of the machinablity, screw threads were cut at the walls to suppress undesired reflection. Figure 4 shows the experimental results. The solid circle and triangle show the data by the laser generated ultrasound, and open square presents those by the pulse echo method. The experiments were repeated 4 times for each samples and standard deviation were calculated as shown in the figure. The differences between two method were less than 0.2%. That was almost equal to the standard deviation of the laser method.

CONCLUSION

Sound velocities in a Si crystal were measured by the laser generated ultrasound. The statistical analysis showed that the optical beam diameters in the usual ranges were not significant on the velocity measurements. The same sample was also measured by the pulse echo overlap method, and the velocities by two methods agreed within their standard deviations. The uncertainty of the velocity measurements was estimated as follows. The standard deviation of the sample thickness measurements was 0.4μm. The uncertainty of the optical gauge was assumed to be 1μm. The measurements were carried out at temperature of 23$\pm$1℃, so that the effect of thermal expansion and the temperature dependence of velocity could be negligible. The standard deviation of the round trip measurements was 2.8 ns as mentioned before. The time base error of the digitizing oscilloscope was estimated to be 0.5 ns. Adding up these factors, the expanded uncertainty with the coverage value k of 2 was 13.3 m/s without the correction of the diffraction effect. The uncertainty was 0.16 % of the sound velocity of 8441 m/s in [100] direction. At high temperatures, ceramic samples were measured by both methods, and the data were in good agreement within the measurement resolution.

REFERENCES

1. H.Nakano, Y.Matsuda, S.Shin and S.Nagai, Optical detection of ultrasound on rough surfaces by a phase-congugate method, Ultrasonics 33:261(1995).
2. P.Delaye, A.Blouin, D.Drolet, and J.-P. Monchalin, Heterodyne detection of ultrasound from rough surfaces using a double phase conjugate mirror, Appl. Phys. Lett. 67:3251(1995).
3. S.-C.D.L.Cruz, S.MacCormack, J.Feinberg, Q.B.He, H.-K.Liu, P.Yeh, Effect of beam coherence on mutually pumped phase conjugators, J.Opt.Soc.Am.B 12:1363(1995)
4. H.J.McSkimin, Pulse superposition method for measureing ultrasonic wave velocities in solids, J.Acoust.Soc.Am 33:12(1961).

A NOVEL LASER ULTRASOUND SOURCE AND ITS IMPLEMENTATION IN THE DRINKS CANNING INDUSTRY

Steven Dixon, Christoper Edwards and Stuart B. Palmer

Department of Physics,
University of Warwick,
Coventry CV4 7AL,
United Kingdom.

INTRODUCTION

The majority of conventional liquid level monitors in the drinks canning industry operate using a gamma-ray absorption technique. Gamma-rays with energies utilised in these machines are strongly absorbed by water/liquids. In a gamma ray measurement a collimated beam is momentarily exposed to the can that is to be measured, with some of the beam below and some above the nominal liquid level. The gamma-rays are detected using a scintillation tube after passing through the can and its contents. This is shown in the schematic diagram of fig.1. Thus if more of the gamma-ray beam is absorbed by a higher liquid level, the counts drop, and vice-versa. The type of machine is in the main a pass/fail monitor - that is it passes a can if the liquid level is above a predetermined height, and rejects it if it is not. There are several problems associated with the use of gamma ray based systems, most notably environmental considerations, the fact they are potentially hazardous and the strict legislation required.

Ultrasonic measurements can be used in much the same way as the gamma-ray system to determine if a liquid is up to or below a certain level. This can be done by generating the ultrasound at a fixed height, just below the nominal liquid level and measuring the ultrasound that is transmitted through the liquid. This is shown in the schematic diagram of fig.2.

As with the case of gamma-ray machines the measurement is susceptible to small errors in the amplitude measurement in the area between pass/fail. Thus such machine would be set with a safety margin to allow for slight variation in detected ultrasonic amplitude. Such ultrasonic measurements will be prone to variations in the signal amplitude as there is no sudden cross-over between signal and no-signal.

ACCURATE NON-CONTACT ULTRASONIC MEASUREMENT OF LIQUID LEVEL

The most reliable ultrasonic measurement that can be made is generally obtained from a measurement in the time domain, rather than a measurement in the amplitude domain. It is

Nondestructive Characterization of Material VIII
Edited by Robert E. Green Jr., Plenum Press, New York, 1998

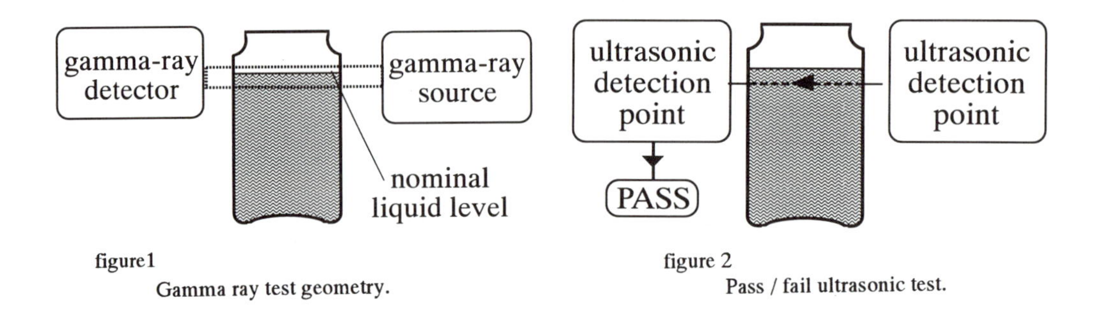

figure1
Gamma ray test geometry.

figure 2
Pass / fail ultrasonic test.

generally more accurate to measure the time between two signals than to measure their amplitude or the amplitude of one signal alone. To employ a 'time' measurement for liquid level a different technique must be employed.

The most obvious method of ultrasonically monitoring liquid level in a container is to position the ultrasonic probe on the bottom of the container as shown in fig.3. The time taken for ultrasound to travel from the top of the liquid to the bottom of the liquid can be measured and if the velocity in the liquid is known, the liquid level can be calculated. The problem with this technique is either obtaining access to the centre of the can dome or the uneven can top (if cans are run upside-down). The only way to couple such probes is to use water jets. If an accurate measurement is to be performed then the ultrasound needs to be generated and detected in the centre of the bottom of the can. As the ultrasound takes a finite time to travel through the liquid this measurement takes a significant time in terms of the velocity of the can on the conveyor. Thus this type of measurement may be implemented on stationary cans, but is too problematic to perform dynamically. There are also difficulties associated with using water jet coupled techniques on rapidly moving targets.

The ideal way to ultrasonically monitor a liquid level is to use non-contact techniques. Containers will pass a test point at rates of up to 20 per second, making any contact method difficult to implement. The technique described here uses a new concept in ultrasonic liquid level monitoring.

Consider the two paths (fig.4) that ultrasound waves may take in travelling through a can from one side to the other. One path travels directly through the can, perpendicular to the can wall. This is the shortest route that ultrasound can take through the can and thus it is the first longitudinal signal that can be observed (see fig.5). The other path is reflected from the liquid-air interface and arrives later than the direct signal as it has travelled a further distance. This is the second ultrasonic signal shown in fig.5. Using the information of the time difference between these two ultrasonic signals, the liquid height above the generation/detection position (h_L) can be calculated if the ultrasonic velocity in the liquid, V_L is known.

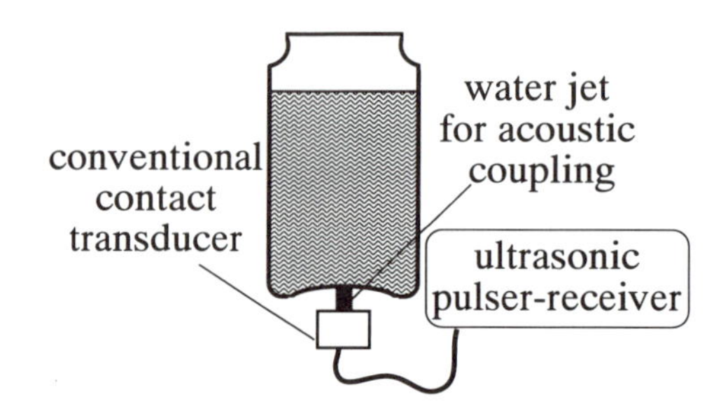

figure 3
Conventional ultrasonic fill level test using water coupled probes.

This can be expressed by the following equation.

$$h_L = \left(\frac{\delta t \, . V_L}{2}\right) \sqrt{\left(1 + \frac{2D}{\delta t \, . V_L}\right)} \qquad \text{equation 1}$$

Where D is the can width and δt is the temporal separation of the two ultrasonic signals in fig.5, and V_L is the velocity in the liquid.

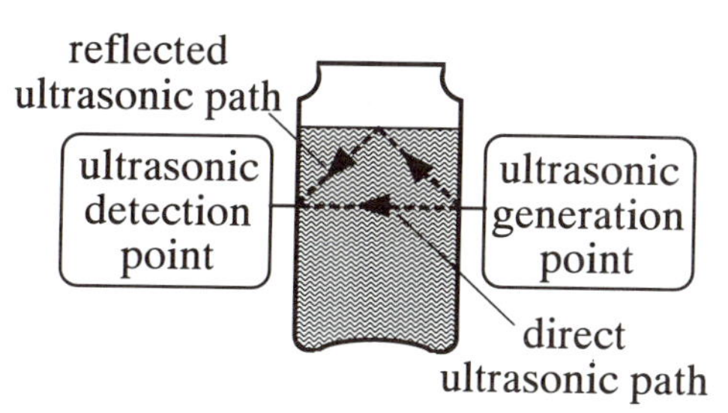

figure 4
Ultrasonic paths in level measurement.

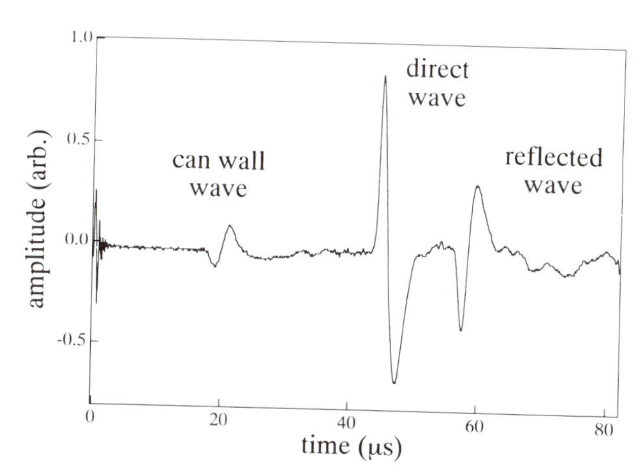

figure 5
Ultrasonic waveform corresponding to geometry of figure4.

If the can width is known to a reasonable accuracy, the ultrasonic velocity can be calculated from the ultrasonic reverberations in the can. This means that the technique will measure the actual liquid level, independent of can contents or temperature, which is valuable where products have been pasteurised as there may be a significant difference in content temperature (and thus ultrasonic velocity) when the can is tested. Thus once the system has been initially calibrated, no further re-calibrations are required for change of contents.

Dynamically, this ultrasonic system is less prone to error due to liquid 'slop' within the can, as most of the ultrasonic waves are reflected from the centre of the can. However, the disadvantage is that it is prone to error when there is severe vibration of the liquid surface, and

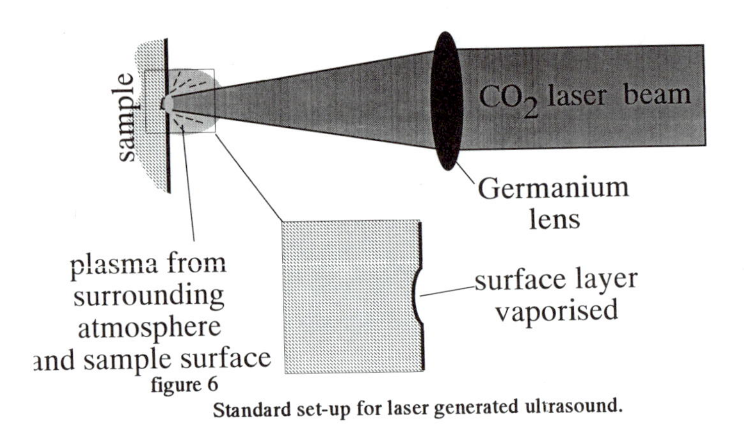

figure 6
Standard set-up for laser generated ultrasound.

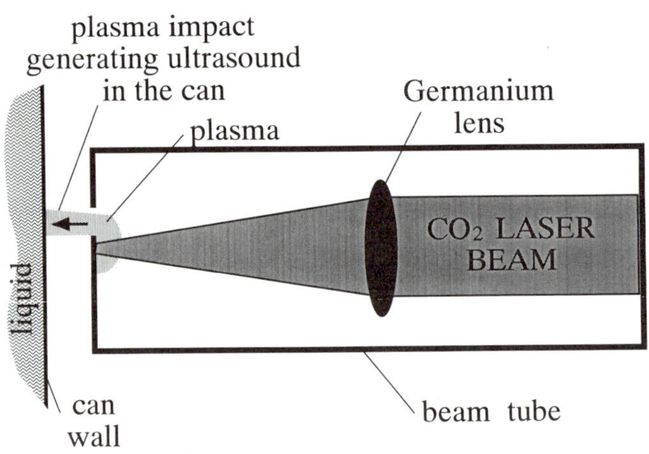

figure 7
Set-up for plasma generated ultrasound.

when a carbonated liquid has been strongly agitated. The former effect causes the ultrasonic reflection from the liquid surface to be scattered which can result is a poor or complicated signal. The latter effect causes absorption of the ultrasonic waves in the liquid, which can be so strong in some cases that no ultrasonic signals are observed. This however only occurs in certain liquids under extreme and abnormal agitation.

EXPERIMENTAL SET-UP

Pulsed lasers can be used to generate longitudinal (compressional) ultrasonic waves[1,2,3] in a container by focusing a laser onto the container surface forming a plasma. The plasma can be formed by vaporising the surface of the sample (such as a painted layer - see fig.6). One problem with this type of ultrasonic generation of longitudinal waves is that the surface of the sample may absorb significant amounts of the incident energy and be in some way damaged. This is the case for instance when a TEA CO_2 laser[4,5] is focused onto a painted metal surface, where the paint would be damaged or removed from the surface.

A technique has been developed that uses the impulse from a plasma as it impacts a sample to generate ultrasound without damaging the sample surface[6]. The method exploits the fact that the laser generated plasma is a hot expanding gas. If the plasma is generated on a 'dummy' target

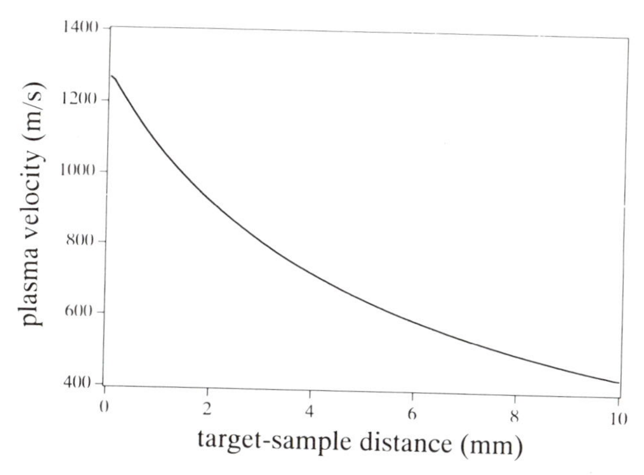

figure 8
Velocity of plasma as it travels from the target.

in front of or close to the experimental sample it can impact onto the experimental sample surface without exposing that surface to the laser beam. In one possible configuration as shown in fig.7, the laser beam is focused onto the 'dummy' target in front of the sample, and the beam is thus totally blocked from the container. The plasma is however free to expand away from the point of impact on the 'dummy' target and may impact on the container at supersonic speeds (see fig.8). Thus only the plasma is incident on the surface where ultrasound is to be generated. This technique will predominantly generate broadband longitudinal ultrasonic waves.

Both the direct laser impact and the plasma source generate a divergent ultrasonic wavefront in the can. The geometry of the set-up 'selects' the two possible ultrasonic paths shown in fig.4.

When metallic containers are used, the ultrasound that has passed through the container may be detected using an electromagnetic acoustic transducer (EMAT)[7,8]. The broadband EMATs used in this system detect 'out-of-plane motion' of the container wall, due to the arrival of longitudinal waves at the wall. As the ultrasonic generation mechanism produces broadband ultrasound and the detector also has a broadband frequency response the ultrasonic signals are sharp features in the time domain. Measuring the time difference between temporally sharp features in the ultrasonic waveform means that the measurement will have good resolution.

RESULTS

The most basic measurement that can be made on the waveform is the time separation between the direct acoustic signal and that reflected from the liquid surface. In the limits of the geometry used in this system the relationship between the time separation and the liquid level is linear. The liquid height algorithm (equation 1) will of course work for any liquid at any temperature as shown in fig.9 In these measurements water, sugar solution and a bitter beer sample were measured by successively adding a fixed small volume of liquid to the can. The measured points closely correlate to the predicted liquid level, and in-fact some of the small divergence is due to the non-uniform can cross-section.

Temperature of the can contents can vary on production lines and as stated earlier will affect the ultrasonic velocity. The relative arrival time (and hence velocity) variations of ultrasonic signals travelling through a sample of beer are shown in fig.10. The exact form of this temperature dependant velocity will vary from product to product. Thus it is clear that some self

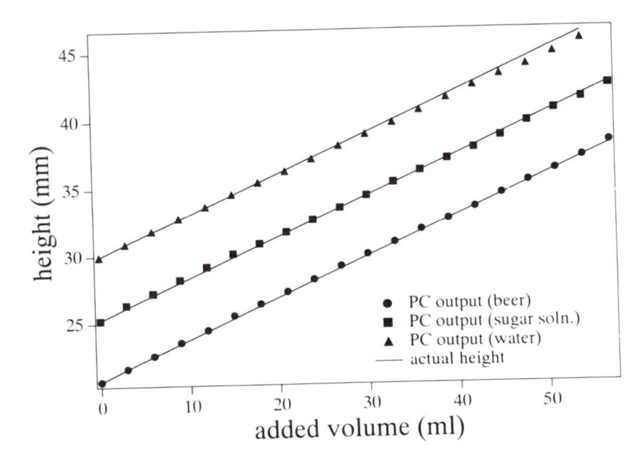

figure 9
Height measured by computer compared to predicted height.

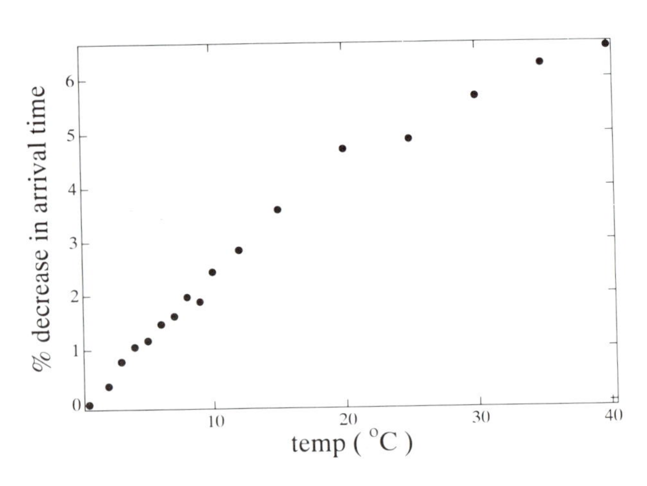

figure 10
Variation in ultrasonic velocity in beer with temperature.

calibration would be required by an ultrasonic measurement system as monitoring temperature of can contents and setting a pre-defined 'product temperature-velocity dependence' would be highly impractical.

Typical vaules for the accuracy of the method are +/- 0.1mm liquid level height statically (corresponding to +/- 0.3ml volume) and +/- 0.25mm liquid level height dyanmically (corresponding to +/- 0.8ml volume) at can speeds of 1 m/s.

FURTHER WORK

It is expected that the system is shortly to go into commercial production and will initially be used solely for a liquid level measurement. There are other possible areas where this type of technology may be applied such as beverage can insert detection, food and non-food product monitoring. Work is currently under-way to find a suitable alternative transducer for examining non-metallic containers.

ACKNOWLEDGEMENTS

The work presented was performed under an EPSRC funded 'Teaching Company Scheme' between the University of Warwick and M&A Packaging, Malvern, UK. The work described in this paper is subject to various patent applications.

REFERENCES

1. R.M. White, Generation of elastic waves by ransient surface heating, J. Appl. Phys., 34:3559 (1963)
2. G. Birnbaum and G.S. White, Laser Techniques in NDE, in:Research Techniques in Nondestructive Testing, R.S. Sharpe, ed., Academic Press, New York, 7:259, (1984)
3. D.A. Hutchins, Ultrasonic generation by pulsed lasers, in:Physical Acoustics VII, W.P. Mason and R.N. Thurston, ed., Academic Press, New York., 18:21, (1988)
4. J.F. Ready, Impulse produced by the interaction of CO2 TEA laser pulses, Appl. Phys. Lett., 25:558, (1974)
5. G.S. Taylor, C. Edwards and S.B. Palmer, The CO2 laser - a new ultrasound source, J. Nodestr. Test. Eval., 5:135, (1990)
6. S. Dixon, C. Edwards and S.B. Palmer, Generation of ultrasound by an expanding plasma, J. Phys. D: Appl. Phys., 29:3039, (1996)
7. P.K. Larsen and K. Saermark, Helicon excitation of acoustic waves in aluminium, Phys. Lett., A24:374, (1967)
8. H.M. Frost, Electromagnetic-ultrasonic transducer : principles, practice and applications, in:Physical Acoustics XIV, W.P. Mason and R.N. Thurston, ed., Academic Press, New York, 18:179, (1988)

LASER ULTRASOUND SOURCE FOR NDE APPLICATIONS: CALIBRATION IN A LIQUID

Sergey Egerev[1], Leonid Lyamshev[1] and Yaroslav Simanovskii[2]

[1]N.N.Andreev Acoustics Institute, 4 Shvernika st., Moscow 117036 Russia
[2]General Physics Institute, 38 Vavilova st., Moscow 117942 Russia

INTRODUCTION

The laser ultrasound source emerges when radiation is absorbed by a body. At a small energy deposit the linear thermoacoustical effect dominates in the process of sound generation. At higher energy deposits the conversion from optical to acoustic energy happens via various channels. We present calibration studies of laser ultrasound source for the media with non-zero optical penetration depths. It is known that models of laser generation of ultrasound taking into account optical penetration depth proved to be useful tools for NDE applications. There is a necessity to implement so-called "buried ultrasound sources". Also, there is a growing interest in laser ultrasound testing of non-metals [1]. Liquid solutions can serve as testing media to predict the influence of the optical penetration depth upon the fine structure and the magnitude statistics as well of the acoustic signal. Calibrating measurements of laser-induced stress waves help to reveal some new mechanisms of optoacoustic (OA) conversion.

MULTIPLE MECHANISMS CONCEPT

Various mechanisms of OA conversion contribute to the acoustic field. In the present paper we describe the fate of the deposited energy in terms of co-existence of contributions to the outgoing pressure transient from laser-induced thermal expansion as well as from laser-induced phase transition in the sample. The role of simultanious contributions to overall acoustic waveforms and signal statistics is especially important if one deals with inhomogeneous liquid media of various kinds.

Hence, one of the fields of application of multiple conversion mechanisms concept of a laser ultrasound source is OA spectroscopy of inhomogeneous liquids [2]. The

Nondestructive Characterization of Material VIII
Edited by Robert E. Green Jr., Plenum Press, New York, 1998

presence of light absorbing microparticles (dust or colloidal particles, for example) in a transparent liquid sample changes essentially the pattern of sound generation for bulk conversion geometry due to a "delayed" heat deposit from particles to liquid as well as to a cavitation. The development of below-threshold fluence sound generation theory when no laser spark formation is observed in diluted suspensions should take into account the co-existence of thermoacoustic and non-linear photoacoustic conversion mechanisms. The statistical properties of photoacoustic signal are of special interest.

The interaction of laser beam with weakly absorbing inhomogeneous liquid leads to strong heating of immersed particles. Their temperature can exceed boiling temperature of host liquid.

The contribution due to bulk absorption of the irradiated host liquid and its subsequent thermal expansion is the "N-shaped" acoustic pulse having magnitude:

$$p_m \approx (\alpha\beta c^2 E)/(\pi a^{3/2} c_p r^{1/2}). \tag{1}$$

where α is the absorption coefficient of the probe, c, β and c_p are velocity of sound in liquid, isobaric volume expansion and specific heat coefficients, a is laser beam cross-section, r is the distance of the observation point from the source axis, and E is pulse energy.

The additional response of the liquid due to the thermal relaxation of microparticle system at low energy density levels within the monodisperse approximation is described by the following magnitude:

$$p_p \approx (\beta N k_a E/c_p R) \cdot (\pi c \chi_p^3/8r)^{1/2}, \tag{2}$$

where χ_p is thermal diffusivity of particle, N is the number concentration of inhomogeneities.The cavitation mechanism contribution in the vicinity of a certain energy threshold ϵ_0 is marked off by extreme magnitude instability, distinct spectrum changes and a non-linear energy dependence deposit of the mean magnitude value.

THRESHOLDS

It is possible to roughly estimate the corresponding energy threshold for a single absorbing particle as follows:

$$\epsilon_0 = (4R/k_a) \cdot (T_b - T_l) \cdot (\rho c_p)_{\text{particle}},$$

where T_l is the initial temperature of the probe, T_b is the boiling point of the probe, ρ is density and c_p is the specific heat of the particle. R is particle radius and k_a is the dimensionless absorption efficiency of particle absorption which depends on optical constants of particle matter as well as on laser wavelength λ to R ratio. For dielectric particles having refraction index $n = n' + in''$ with $n \geq 1$ and $n'' < 1$ it is possible to put down for k_p:

$$k_a = [1 - \exp(-8\pi n'' R/\lambda)] \cdot \exp[-0.2((n'^2 - n''^2)^{1/2} - 1)].$$

For small particles ($8\pi R n''/\lambda \ll 1$) $k_a \sim R$ and ϵ_0 is independent of their radii. Let us estimate threshold ϵ_0 for typical atmosphere aerosol microparticles which are most often found in water, assuming $(T_b - T_l) = 80°$. For strongly absorbing carbon particles it is valid $k_a \approx 1$. The corresponding values ϵ_0 comprise $2 \cdot 10^{-3} \mathrm{J/cm}^2$ (for $R = 0.1\mu$) or $2 \cdot 10^{-2} \mathrm{J/cm}^2$ (for $R = 1\mu$). Weakly absorbing dust microparticles (quartz ones, for example) having $n'' = 0.06 - 0.025$, $R = 0.1\mu$ and $k_a = 0.9$ are associated with $\epsilon_0 = (2-5) \cdot 10^{-3} \mathrm{J/cm}^2$. In the case of high quality polystirol latex $\epsilon_0 \approx 10^{-1} \mathrm{J/cm}^2$.

The threshold ϵ_0 thus defined can't be regarded as something of an energy value responsible for mechanisms switching. On the contrary, it defines the lower limit of the energy band where one can observe the coexistence of various mechanisms.

THE ROLE OF MICROSCOPIC AIR BUBBLES

Dealing with real liquids it is necessary to take into account the presence of numerous cavitation nuclei. The major mechanism for initiation of cavitation in water by a laser pulse involves interaction of the stress transients with nanoscopic air bubbles (of about 10 nm in diameter). The major force that resists cavitation is the force of surface tension of microscopic air bubbles in aqueous solution that serves as nucleation centers for cavitation. The simplified condition for initiation and growth of cavitation bubbles under a steady tensile stress can be put down as

$$p \geq 2\sigma/R_b,$$

where R_b is the minimal size of a nucleation bubble which can expand up to a critical size in the tensile wave and sustain after stress dissipation, $\sigma = 7.26 \cdot 10^{-6} J/cm^2$is the surface tension coefficient for water at 25°C. It can be shown that for typical conditions a negative stress might expand the 100-120 nm bubbles to create macroscopic cavitation bubbles even if the fluence is not enough for superheating the liquid. A tensile acoustic wave loses its energy upon propagation due to production of bubbles. Therefore a laser-induced cavitation takes place in a "limited "zone of the irradiated liquid sample.

By taking into account the simultaneous radiation of sound by numerous bubbles satisfying the threshold condition we can explain non-linear growth of the mean pressure magnitude $\langle p \rangle$ with E. Within the approximation of monodisperse suspension we obtain for the far field magnitude in the direction perpendicular to the laser beam axis [3]:

$$\langle p \rangle \sim (1/r) \cdot N k_a R^2 a_0^3 \epsilon_0 \cdot ((E/\pi a_0^2 \epsilon_0) - 1)^{3/2}, \quad (3)$$

where a_0 is focal point radius. The contribution Eq.(3) is due to the radiation of a short cylinder region occupied by expanding cavities. It is "enclosed" into the bodiless extended cylinder responsible for thermoacoustic contributions of both kinds. If there are particles of different size immersed in the host liquid, a stronger than of power 3/2 dependence energy is expected. Here we are to take into account concrete form of $n(R) = dN/dR$ and energy threshold condition for ϵ_0 which now depends on the particle size. It is known from literature that a uniform law of carbon particle size distribution provides the dependence $\langle p \rangle \sim E^4$. At high-above threshold energy deposits source region tends to an extended cylinder configuration, individual cavities contributions become incoherent and the expression Eq.(3) is no more valid. One faces a new portion of linear dependence, $\langle p \rangle \sim E$.

STATISTICS

In order to separate cavitation induced contribution we use the following statistical approach. One can represent the results of a series of acoustic response magnitude measurements in the form of a histogram. By analyzing the form of the histogram and its approximation by means of some known probability distribution law it is possible to estimate the sample content. It is known that in the case of homogeneous liquid sample thermal expansion contribution is of a good repeatability. The resulting histogram can be approximated within a good accuracy by Gauss probability distribution law; mean magnitude value $\langle p_m \rangle$ coincides with Eq.(1), also standard deviation σ is defined by the level of reception system noise, $\sigma \approx \sigma_N$.

As follows from Eq.(2) that the ratio of standard deviation σ_p of the low-energy particles contribution to the mean magnitude can be put down as $\frac{\sigma_p}{\langle p_p \rangle} \approx (V_0 N)^{-1/2}$, where V_0 is the volume of the source. It is possible to detect the component Eq.(2) if N is not too small, hence the inequality $\frac{\sigma_p}{\langle p_p \rangle} \ll 1$ is valid. One faces good repeatability of the total response at low energy deposits as a result.

The cavitation mechanism considered is characterized by a significant contribution of individual particles satisfying the threshold condition. In the case of monodisperse suspension it is easy to detect the phenomenon by studying the resulting histograms which suffer broadening thus giving $\frac{\sigma_p}{<p>} \sim 1$. In the case of polydisperse suspension it is naturally to expect more complicated, non-Gaussian types of response-magnitude probability distribution.

THE EXPERIMENT ON MULTI-MECHANISM LASER ULTRASOUND SOURCE

The experimental studies were conducted with the help of pulsed XeCl excimer laser beam ($\lambda = 308$ nm and $\tau_l = 12$ nS) which passed through diaphragm, attenuator and quartz lens and then was focused into the photoacoustic measurement cell filled with the analyzed suspension probe. The cell of stainless steel having quartz windows was analogous to that described in [2]. Piezoelectric transducer having responsivity $7\,\mu$V/Pa and frequency band up to 200 kHz was attached to the cell, the distance between transducer and source axis comprised $r = 2$ cm. The cell was filled with a filtered distilled water probe containing particles with radii less than $0.3\,\mu$, other probes represented unfiltered water; these probes represented polydisperse suspensions. Diluted Dow Chemical latex suspension ($R = 2.2\,\mu$) added to distilled water with initial particle content served to increase the contribution of monodisperse particles ensemble. The laser energy measurements were provided by means of calorimeter and parameter $\epsilon = E/\pi a_0^2$ was calculated. The original software was used for signal series (1000 shots) recording and histogram analyzing. The mean series magnitude and standard deviation were calculated at this very stage.

Figure 1 represents the results of two series of measurements for distilled water. The first series obtained at low energy deposit where thermal expansion mechanism dominates is described by a symmetric histogram for pressure magnitudes. It, in turn, can be approximated by Gauss distribution law. The two scenarios of histograms transformation (as energy deposit grows up) are possible depending on the particle size distribution. For polydisperse solution the energy growth leads to the appearance of a "long-tail" histogram, such as curve 2. The tail is due to rare events of particles induced cavitation. The most probable magnitude value still defined by thermal expansion mechanism (light arrow) is no longer coinciding with mean magnitude influenced by cavitation mechanism (dark arrow). This is just the case of coexistence of axisymmetric acoustic sources driven by different mechanisms. Over-threshold particles of different size give rise to random acoustic radiation of many-factor type. Such random processes can be described by Zipf's distribution law popular in social sciences. The probability density can be put down as follows:

$$n_z(p) \approx 1/p^{1+\xi}, \xi > 0. \tag{4}$$

It is necessary to plot histograms in "log-log" coordinates to become certain that right edges of histograms can be approximated by straight line of Eq.(4), its slope being equal to $\arctan(1+\xi)$. It is evident from figure 2 that the energy growth leads to slope steepening. Finally, at well-above threshold conditions $\xi \to \infty$ and histogram returns

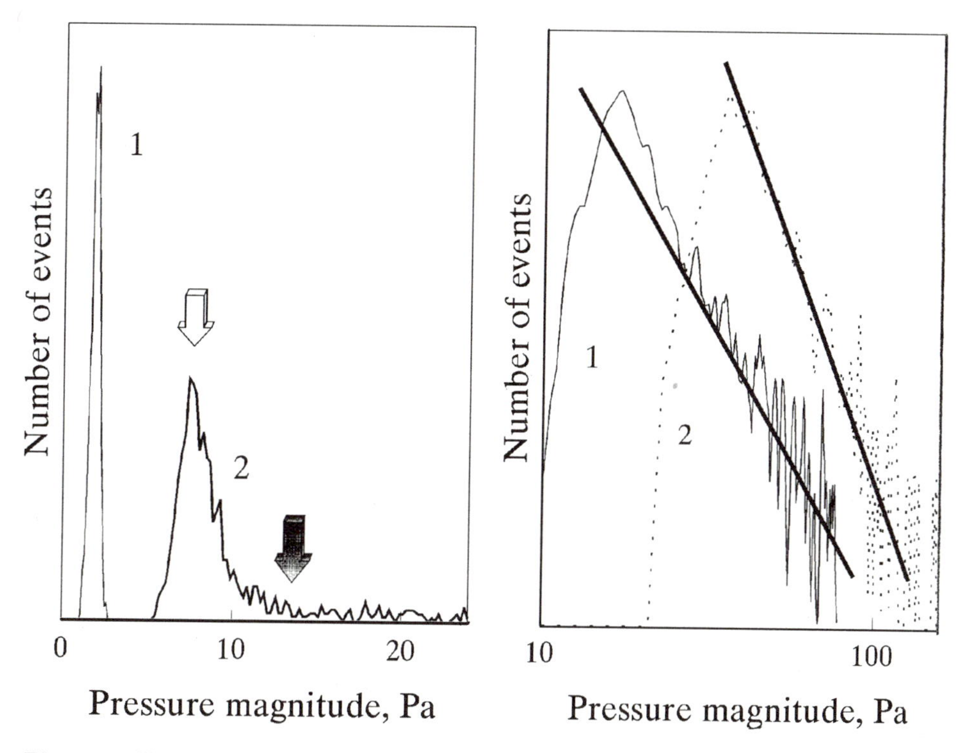

Figure 1: Histograms of pressure magnitude at 0.08 J/cm^2 (1) and 0.2 J/cm^2 (2)

Figure 2: Histograms in "log-log" coordinates at 0.36 J/cm^2 (1) and 0.45 J/cm^2(2)

to the Gauss form (not shown).

The second scenario valid for a large portion of identical particles is presented in figure 3. The addition of diluted latex suspension influences the left edge of histogram thus reducing the thermal expansion radiation (only small spike remains). Notice that Zipf's law approximation lines are of equal slopes for both samples, since energy deposits are equal.

And, finally, figure 4 represents the energy dependence of mean pressure magnitude which is the additional characteristic of the cavitation source. Linear and non-linear portions of the dependence alternate as predicted. Further increase of energy deposit leads to especially high pressure response.

CONCLUSION

These results form the basis against which changes in acoustic response of irradiated microinhomogeneous probe can be understood. Actually, axisymmetric configuration of the source and random character of the response are to be taken into account when testing the NDE applications of laser ultrasound source in the samples having a significant optical penetration depth.

ACKNOWLEDGMENT

The work was supported, in part, by the grants 96-05-64538 and 96-02-16160 of the Russian Foundation of Basic Research.

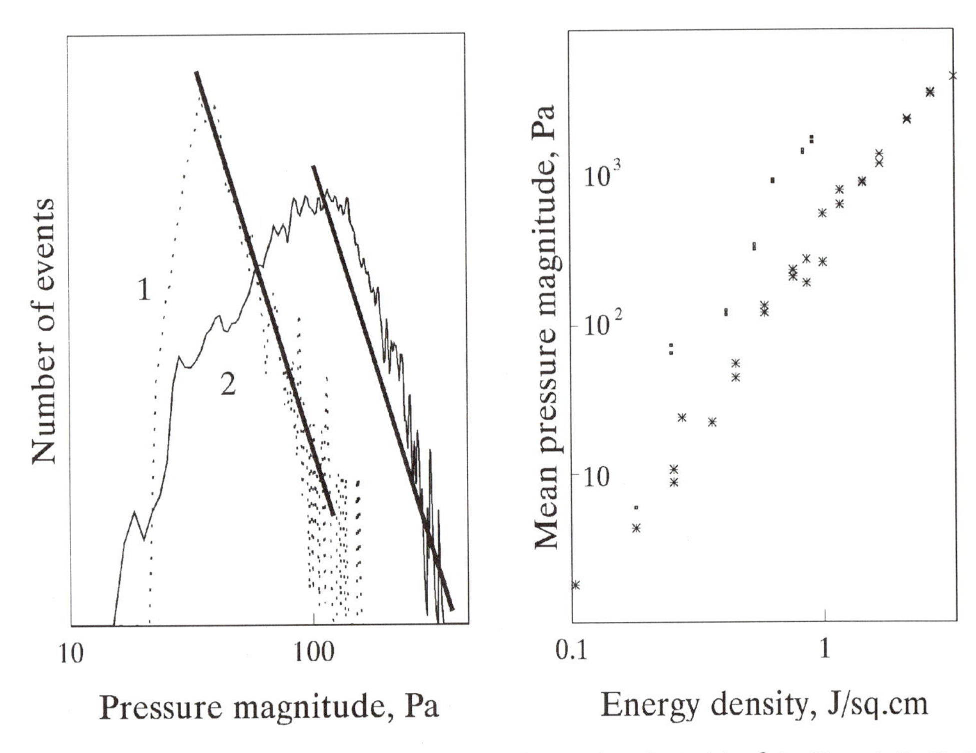

Figure 3: Histograms in "log-log" coordinates observed at 0.45 J/cm^2 in filtered distilled water (1) and in the latex suspension (2)

Figure 4: Plot of mean pressure magnitude versus the laser energy density: distilled water (asterisks) and latex suspension (dots)

References

[1] M.Dubois, F.Enguehard, L.Bertrand, M.Choquet, J.-P.Monchalin. Numerical and experimental studies of the generation of ultrasound by laser, *J.de Physique IV, Colloque C7,* 4:C7-689 (1994)

[2] S.Egerev, L.Lyamshev, O.Puchenkov. Time-resolved optoacoustic diagnostics. *Sov.Phys.Usp.* 33:739 (1990).

[3] S.Egerev, K.Naugol'nykh, A.E.Pashin, Ya.O.Simanovskii, Sound radiation from pencil shape optoacoustic arrays, in: *Nonlinear Acoustics*, The Institute for Advanced Physics Studies, La Jolla, 76 (1994).

SUB-MICRON MATERIALS CHARACTERIZATION USING NEAR-FIELD OPTICS

David W. Blodgett and James B. Spicer

Department of Materials Science and Engineering
The Johns Hopkins University
Baltimore, MD 21218

INTRODUCTION

There is currently a need for the ability to perform high-resolution sub-surface materials characterization and inspection. In the microelectronics industry, with the decrease of line widths and the increase of component densities, sub-surface voids become increasingly detrimental. Any voids along an integrated circuit (IC) line can lead to improper electrical connections between components and cause failure of the device. In the thin film industry, the detection of impurities is also important. Any impurities can detract from the film's desired optical, electrical, or mechanical properties. To study these effects, we have combined the sub-surface inspection capabilities of ultrasonics with the high resolution surface characterization capabilities of the near-field scanning optical microscope (NSOM).

When an ultrasonic wave encounters any defect, void, or phase variation in a material with different mechanical properties than the bulk material, it is diffracted. Measurement of these diffracted waves in the ultrasonic near-field of the defect allows for the detection of sub-micron sub-surface defects. Both piezoelectric transducers and optical interferometers are commonly used to detect ultrasonic displacements, with the optical interferometer providing increased spatial resolution. However, the optical interferometer, even when focused on the sample, is diffraction limited in its spatial resolution. If the spatial extent of the focused spot is larger than variations in the arrivals of the scattered ultrasonic wave, the sub-surface defect will remain undetected. Figure 1 illustrates this effect and shows the spatial resolution of the focused spot exceeding any variations in the ultrasonic arrivals due to the sub-surface defect. The minimum required optical resolution for resolving a sub-surface defect must be less than the dimensions of the defect itself (i.e. for a micron sized defect, sub-micron spatial resolution is required). Typical resolutions for focused interferometric measurements are on the order of tens of microns, far too large to resolve a one micron defect.

Near-field optics provides a means of exceeding the diffraction limit for optical resolution. The diffraction limit for conventional optical systems is determined by the inspection wavelength, λ, and approaches $\lambda/2$ for an ideal system. At visible

Nondestructive Characterization of Material VIII
Edited by Robert E. Green Jr., Plenum Press, New York, 1998

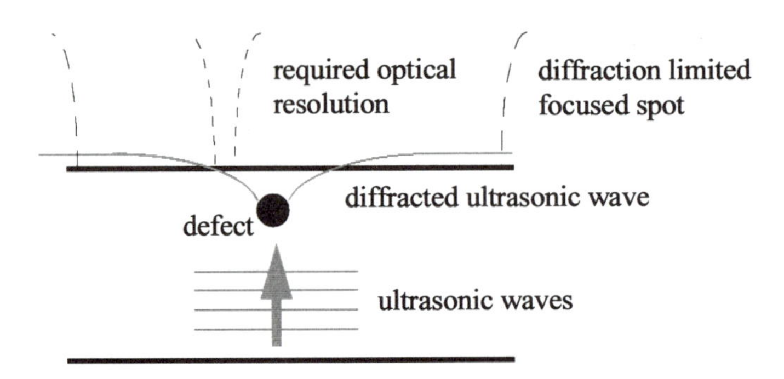

Figure 1. Effects of spatial resolution on detection of sub-surface defects.

wavelengths, this corresponds to approximately 200 nm. NSOMs exceed this diffraction limit by scanning a sub-wavelength aperture in close proximity to the sample's surface. This aperture defines a "near-field" source or receiver for the near-field microscope [1]. The resolution of these systems is not determined by the inspection wavelength, but only by the diameter of the sub-wavelength aperture and the aperture-sample separation. In NSOMs, a tapered optical fiber coated with aluminum is used to define the sub-wavelength aperture. Typical apertures range from 10 to 100 nm in radius and are scanned 5 to 100 nm off the surface. Betzig [2] and others [1] have reported resolutions approaching 10 nm, far exceeding the diffraction limited resolution of conventional microscopes. Of course, accompanied with these higher resolutions is a decrease in signal power.

Recently, a new form of NSOM has been developed. In this NSOM, the light scattered from the end of a standard atomic force microscope (AFM) tip defines the light source such that the light is no longer transmitted through a sub-wavelength aperture. This method was first proposed and implemented by Zenhausern *et al.* [3,4]. Another configuration, based upon the same principles, was implemented by Bachelot *et al.* [5,6]. The two configurations differ in the collection method for the scattered light, but both attain resolutions in excess of $\lambda/600$. In fact, Zenhausern *et al.* report resolutions of 3 nm and predict near atomic resolutions.

These techniques have several names including: apertureless NSOM (ANSOM), scanning tunneling optical microscopy (STOM), and scanning interferometric apertureless microscopy (SIAM). Each of these offers potential advantages over the standard NSOM. Because they are based upon an AFM, they provide very precise topographical information of the sample. The NSOM is regulated using a shear or lateral force microscope, the "tip" size being the sum of both the diameter of the optical fiber tip and the thickness of the aluminum coating. In addition, an NSOM tip behaves as a waveguide operating below cut-off, which results in high signal loss. The apertureless technique, which detects the scattered field from the AFM tip, does not suffer from this condition and appears to offer much larger signals for equivalent optical resolutions.

NEAR-FIELD SCANNING OPTICAL MICROSCOPE

Due to the potential benefits in signal power of the ANSOM, a microscope similar in design to that implemented by Bachelot *et al.* [5] has been constructed. As with other ANSOM systems, the design is based upon a tapping mode AFM [7]. In a tapping mode AFM, the probe tip is vibrated (dithered) at or near its resonant frequency with a piezoelectric tube or bimorph with an amplitude between 50 and 100 nm. As the tip

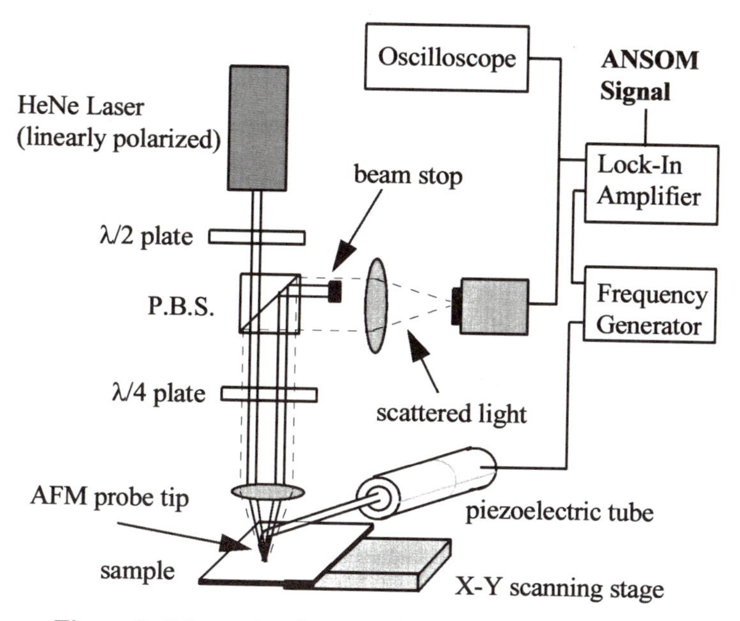

Figure 2. Schematic of apertureless near-field microscope.

approaches the sample it interacts with surface forces which lower its resonant frequency. Measurement of this frequency shift or the corresponding decrease in dither amplitude at the original dither frequency provides a means of stabilizing the tip a fixed distance above the sample.

A schematic of the ANSOM is shown in figure 2. In this design, the λ/2 plate and polarizing beam splitter (P.B.S.) control the amount of power reaching the sample to avoid saturation of the Si avalanche photodiode (APD). The light is passed through a λ/4 plate, focused on the tip, collected, and finally focused again onto the Si APD. Off-axis focusing on the tip removes light reflected from the back of the tip that could mask the near-field signal. In addition, a beam stop used after the P.B.S. ensures that only the scattered light, which contains the near-field signal, is collected and maximizes the amount of scattered light passed through the P.B.S. to the Si APD.

The output of the Si APD is monitored using a lock-in amplifier, referenced at the driving frequency of the AFM tip. Since the tip is vibrated perpendicularly to the sample, the ANSOM sugnal is effectively amplitude modulated, allowing it to be separated easily from any static background components. The near-field signal is highly dependent on tip-sample separation, with intensity scaling as $1/r^6$ [4]. Scanning the sample below the tip generates both a high resolution topographical and optical image. Optical resolution for this system is estimated at less than 100 nm.

DETECTION OF ULTRASOUND

Because of the high frequency nature of ultrasound, lock-in detection cannot be used. For ultrasonic arrival detection, the output from the Si APD is passed to a digitizing oscilloscope. A 5 MHz contact transducer generated the ultrasound on the back of the sample as shown in figure 3. The frequency of the ultrasonic pulses was synchronized with the AFM probe dither frequency so that pulses arrived at the sample surface for a constant tip-sample separation. The high frequency nature of the ultrasonic arrival effectively causes the dither to appear quasi-static with the ultrasound modulating

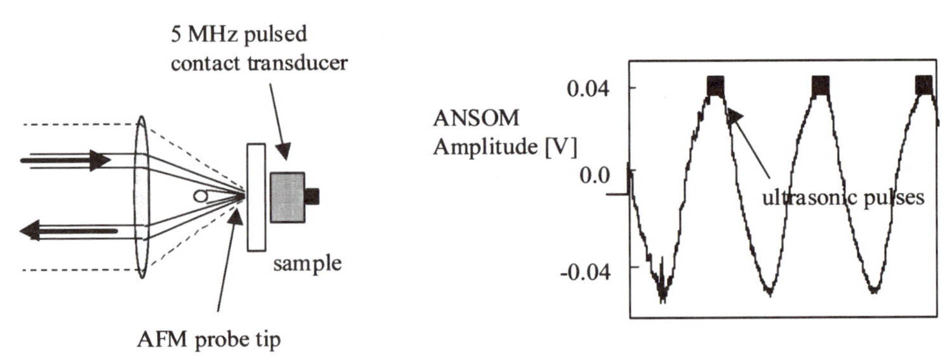

Figure 3. Ultrasonic generation configuration and referencing of ultrasonic arrivals with modulation of near-field signal.

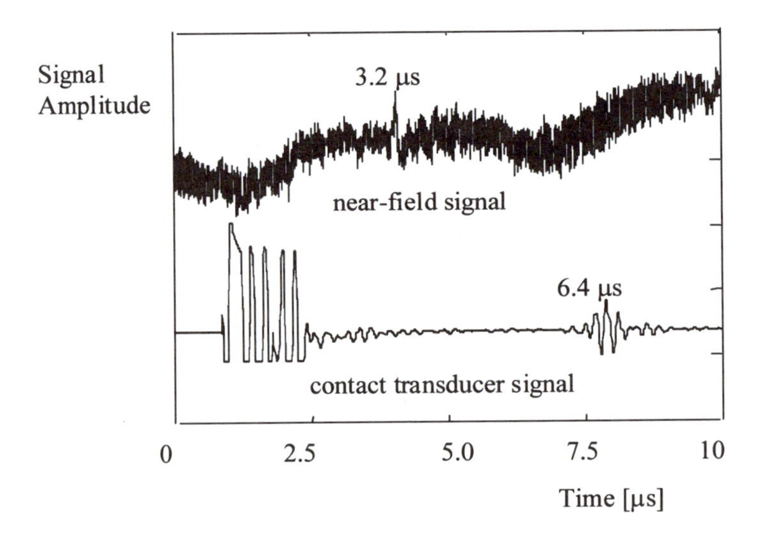

Figure 4. Ultrasonic arrival on near-field signal.

the tip-sample separation. The near-field microscope is highly sensitive to variations in tip-sample separation. It is this sensitivity that leads to the amplitude modulation of the near-field signal due to the dither amplitude of the tip. The expected ultrasonic effect on the near-field signal can be measured by averaging the signal, allowing for sub-nanometer surface displacements to be detected. This experiment was run using an aluminum sample. Figure 4 shows both the contact transducer signal and the accompanying ultrasonic arrival on the near-field signal, after 1000 averages. Note that the peak in the ANSOM signal occurs at half the time of that for the transducer signal. This is expected since the ultrasound travels half the distance in reaching the ANSOM probe.

Regardless, this result is not sufficient to show that the ANSOM detection of ultrasound has occurred for regions immediately under the probe (i.e. the near-field region). To verify the near-field detection of ultrasound, a sample was scanned with sub-micron dark spots on its surface. A loss of near-field signal over these spots would similarly remove any ultrasound signal in the near-field. Unfortunately, when scanned, the amplitude of the ultrasonic arrival never decreased, as would be expected. This indicated that the ultrasonic signal was being carried by a far-field component of the scattered field. Varying the size and location of the optical beam stop failed to remove this far-field component.

CONCLUSIONS

An apertureless near-field microscope has been successfully designed and built. The high resolution optical scanning capabilities of this microscope are being combined with the sub-surface inspection capabilities of ultrasonics to allow for sub-surface sub-micron materials characterization and inspection. One detection scheme has already been implemented with results indicating that improvements to the ANSOM must be implemented for the near-field detection of ultrasound.

REFERENCES

1. Paesler and Moyer, Near-Field Optics: Theory, Instrumentation, and Applications, John Wiley & Sons (1996).

2. Betzig, Lewis, Harootunian, Isaacson, and Kratschmer, "Near-field scanning optical microscopy (NSOM): development and biophysical applications", Biophys. J. 49, pp267-279 (1986).

3. Zenhausern, O'Boyle, and Wickramasinghe, "Apertureless near-field optical microscope", Appl. Phys. Lett. 65(13) pp1623-1625 (1994).

4. Zenhausern, Martin, and Wickramasinghe, "Scanning interferometric Apertureless microscopy: optical imaging at 10 Angstrom resolution", Science, Vol. 269 pp1083-1085 (1995).

5. Bachelot, Gleyzes, and Boccara "Near-field optical microscope based on local perturbation of a diffraction spot", Optic Lett., Vol. 20, No. 18 pp1924-1926 (1995).

6. Lahrech, Bachelot, Gleyzes, and Boccara, "Infrared-reflection-mode near-field microscopy using an apertureless probe with a resolution of λ/600", Optic Lett., Vol. 21, No. 17 pp1315-1317 (1996).

7. Sarid, Scanning Force Microscopy, With Applications to Electric, Magnetic, and Atomic Forces, Oxford University Press (1994).

LASER ULTRASONIC TESTING OF HIGHLY ATTENUATIVE AND WEAKLY GENERATING COMPOSITES*

Kenneth R. Yawn and Thomas E. Drake Jr.

Laser Ultrasonics Laboratory
Lockheed Martin Tactical Aircraft Systems
P.O. Box 748, MZ 5984
Fort Worth, TX 76101

Mark A. Osterkamp

Computer Sciences Corporation
6100 Western Place
Fort Worth, TX 76107

INTRODUCTION

Current fighter aircraft contain on the order of 10% composite structure, and future fighter aircraft may contain upwards of 50% composite structure. These structures are often large, complex in shape, and flight critical. Stringent engineering requirements mandate 100% inspection prior to installation of flight critical parts as well as periodic inspections in service. Conventional automated ultrasonic systems have been used by the aerospace industry for inspecting flat or mildly contoured parts in the production environment. These systems are slow, require significant setup time for highly contoured parts, and are generally inappropriate for in-service inspections where access is limited to a single side. Lockheed Martin Tactical Aircraft Systems (LMTAS) has gained extensive practical experience over the past decade applying laser ultrasonic testing (Laser UT) methods to advanced aerospace composite materials (Drake, 1994; Chang et al., 1993). Laser UT greatly reduces the time and cost of inspecting complex composite materials by reducing or eliminating part fixtures and long setup times. It is capable of scanning complex contoured composite structures more than 10-times faster than conventional systems. This paper will address some of the problem areas associated with the laser ultrasonic testing of advanced composite materials.

* This work was partially supported by LMTAS IR&D and the Air Force Materials Laboratory, Contract# F33615-92-C-5981.

Nondestructive Characterization of Material VIII
Edited by Robert E. Green Jr., Plenum Press, New York, 1998

THE LASER ULTRASONIC TESTING SYSTEM

LMTAS has designed and built a highly flexible scanning laser ultrasonic system for inspecting large complex composite structures. The LMTAS Laser UT system has successfully evaluated many types of composite structures from advanced aircraft programs, including: complex contoured skins, bulkheads, stiffeners, pivot shafts, and radar absorbing coatings. An outline of the system operation is shown in Fig. 1.

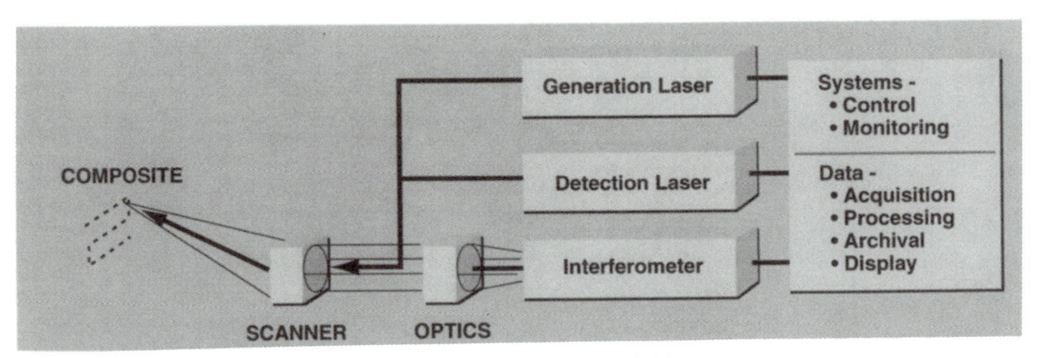

Figure 1. Laser UT System Block Diagram

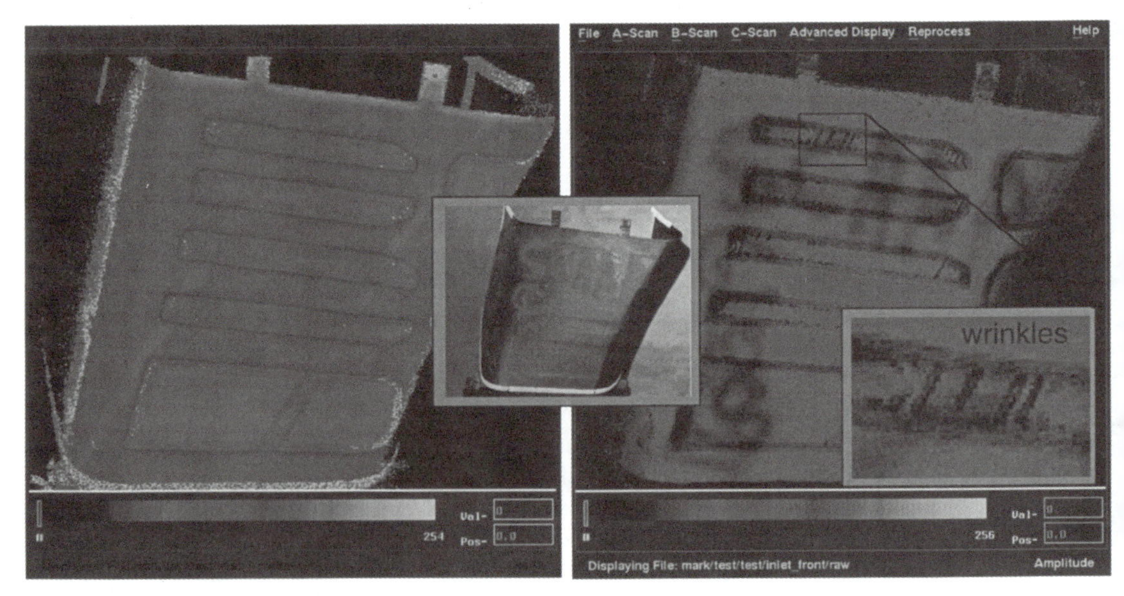

Figure 2. Laser Ultrasonic C-scan of Complex Inlet Duct Skin

The basic system consists of pulsed generation and detection laser beams which run coaxially in the system out to the target point. These beams are indexed across the target at precise intervals using an optical scanner. The generation pulse rapidly heats a local area of the surface producing an ultrasonic wave through the process of thermoelastic expansion.

This wave propagates into the material perpendicular to the surface, independent of the incident beam angle, and it interacts with the bulk material in the same manner as waves produced with conventional piezoelectric transducers. The detection laser light is modulated by vibrations at the material surface and is scattered back to the collection optics. The ultrasonic vibrations carried by the detection light are demodulated by the interferometer and converted into electrical signals by photodetectors (Monchalin, 1985). These ultrasonic signals are generally equivalent or superior to those of conventional automated systems for moderately thick parts (< 0.5 inch) and are evaluated using standard ultrasonic methods. Therefore, no additional ultrasonic training is required for factory inspectors beyond that used for current conventional systems. Fig. 2 shows a typical laser ultrasonic inspection of a large complex inlet skin from an advanced fighter aircraft.

APPLICATION OF LASER ULTRASONICS

The Detection Process

The signal-to-noise ratio of the detection process places fundamental limitations on the ability of the system to detect defects in a material. Eqn. 1 illustrates the essential elements of the signal-to-noise ratio in the detection process.

$$SNR = 4\pi R(f_u)U(P_{Gen})\sqrt{\frac{\eta I_{Det}}{2hc\lambda BW}} \tag{1}$$

Where, $R(f_u)$ is the response of the interferometer to signal f_u, assuming that it holds for an extended source, i.e. large e′tendue, U is the magnitude of the displacement of the vibration at the surface and is a function of the generation laser power density P_{Gen}, η is the detector quantum efficiency, I_{Det} is the detector laser power intensity, λ is the detection laser wavelength, h is Planck's constant, c is the speed of light in vacuum, and BW is the signal bandwidth. Although somewhat simplified, Eqn. 1 shows the key parameters that affect the SNR of the laser ultrasonic process. Broadband detection of pulsed ultrasound requires a bandwidth on the order of 10^7 Hz for composite materials. From Eqn. 1, there are obviously several routes that can be taken to improve the SNR. For example, one can 1) increase the amount of energy in the generation laser, 2) shorten the wavelength of the detection light, 3) increase the response of the interferometer to signal, or 4) increase the detector quantum efficiency and the amount of detection laser light. Each of these methods has its limitations and we will briefly outline some of them here. (1) Increasing the energy density of the generation spot increases the SNR linearly until damage occurs to the material. On typical composite materials this limit is not very large and is on the order of 1 J/cm^2 (Drake, 1985; Yawn, 1995). The LMTAS laser ultrasonic system typically operates near this point. (2) Reducing the detection laser wavelength will increase the SNR at the risk of signal distortion and reduced dynamic range if the ratio of U/λ becomes large and in some cases signal clipping occurs. (3) The signal response of the spherical Fabry-Perot interferometer is shown in Fig. 4. The transmission, reflection, and combined response curves are shown relative to an ideal interferometer. Obviously, for any interferometer a response of 1 is the maximum and our configuration of the Fabry-Perot has a response greater than 0.5 between 1.25 and 10 MHz with a peak of 0.7 near 2 MHz. Improved interferometer approaches can only provide, at best, a factor of 2 increase in SNR. (4) Increasing the amount of detection laser light can be done by either increasing the power of the detection laser or increasing the collection efficiency of the system. This choice provides the greatest potential for significant SNR improvements.

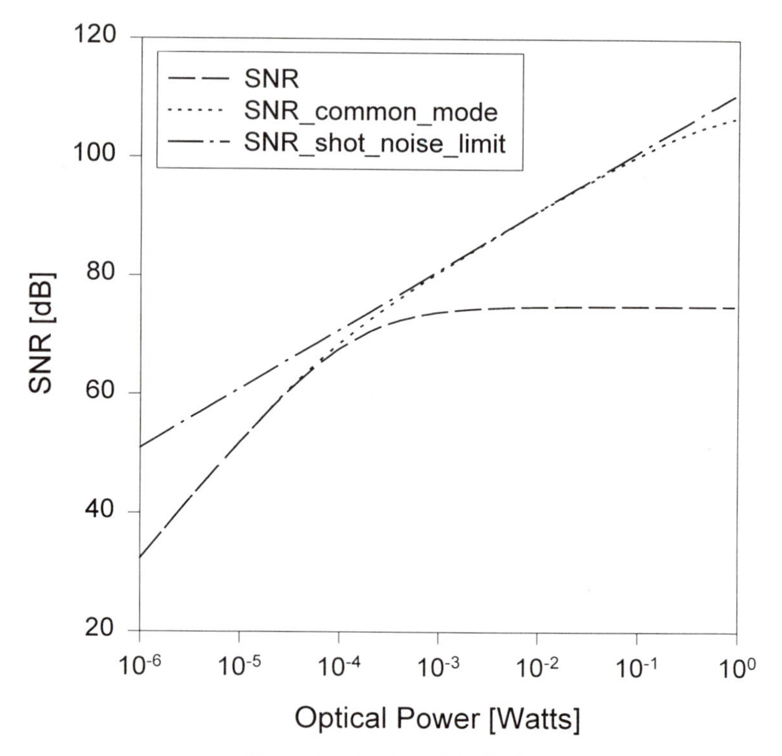

Figure 3. Signal-to-Noise Ratio

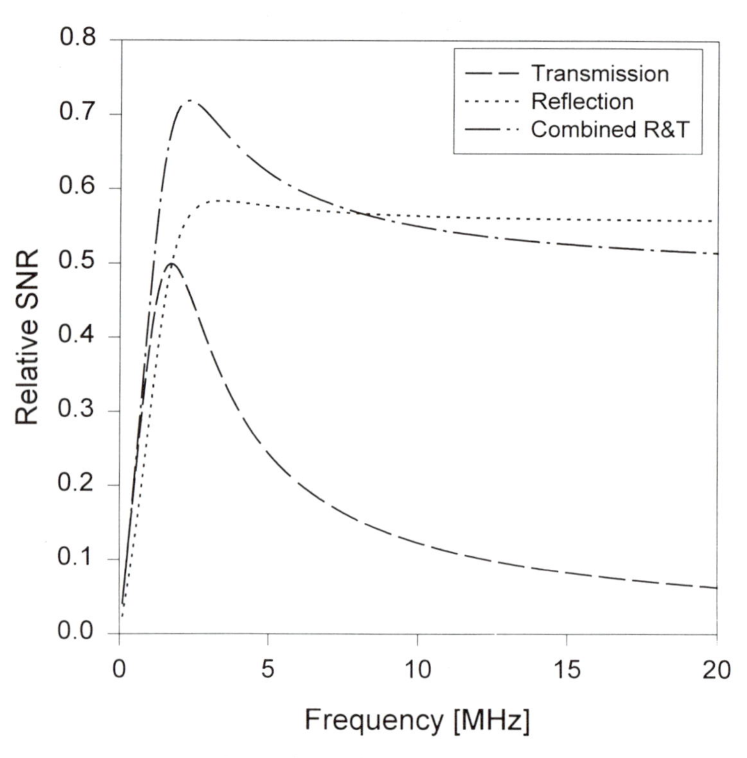

Figure 4. Interferometer Response

Fig. 3 shows the noise limits and the detection SNR before and after laser noise elimination and Fig. 4 shows the interferometer signal response curve for both the reflection and transmission ports.

The measured signal is a convolution of the true signal and the signal response of the system, amplitude noise convoluted with the noise response of the system, and other sources of uncorrelated noise. This measured signal is shown in Eqn. 2.

$$s(t) = u(t) * r(t) + a(t) * r'(t) + UncorrelatedNoise \quad (2)$$

where $u(t)$ is the ultrasonic signal, $a(t)$ is amplitude noise, and $r(t)$ and $r'(t)$ are the response functions of the interferometer to signal and noise respectively. The LMTAS Laser UT system removes common-mode noise, a(t), by differential processing of the interferometer signals. This preserves the signal and allows shot-noise limited detection in the presence of large amplitude laser noise.

Effects of Surface Conditions on Detection Efficiency

The detection process is very sensitive to the surface condition of the material under test for several reasons. Firstly, the surface condition directly affects the amount of light scattered back to the collection system. Smooth resin surfaces, for example, are highly specular and scatter much less light back to the collection system except when the laser beams are directly on axis. This reduced amount of scattered light when off axis lowers the detection signal-to-noise ratio, and the highly specular return on axis must be attenuated in order to prevent damage to the optical detectors.

Secondly, the resins used in advanced composites are typically transparent to the detection laser wavelength. In materials composed of multiple layers of woven fabric the surface resin layer can have varying thickness. In application, we have observed a phenomena where the detection laser spot partially penetrates the surface resin layer and reflects off the underlying fibers. The ultrasonic vibration is modulating light that is returning from areas on the material at different depths and thus have different phases. This phase mixing causes broadening and interference effects on the ultrasonic signals as a function of angle.

Effects of Surface Conditions on Generation Efficiency for Partially Cured Materials

Since the primary mechanism for the generation of ultrasound in a composite material is thermoelastic expansion (White 1963; Rudd et al. 1983), the efficiency of generation is highly dependent on the surface condition and absorption properties of the bulk material very near the surface. Aerospace composite structures are typically manufactured with several possible types of surface conditions depending on the application. Examples include the surface application of removable glass or organic peel plies for paint and bonding applications as well as bare composite surfaces. Each of these conditions can have varying effects on the generation of ultrasound in a composite.

As an example, comparisons of signal strengths for several surface conditions tested are shown in Table 1 (Yawn, 1995). Listings are for a B-staged (50% cured) graphite/epoxy laminate with the following different surface conditions: bare graphite/epoxy composite surface, 128 weave glass peel ply, 180 weave glass peel ply, and polyester peel ply. The ultrasonic wave was induced by a 0.389 J/cm^2, 100 nsec, CO_2 laser pulse. The data shows approximately a five-fold increase in signal strength for the polyester peel ply over the glass. The existence of the large glass fibers near the surface appears to prevent the penetration and thus absorption of the 10.6μm radiation into a depth sufficient to efficiently produce ultrasound of the desired frequencies (1-5 MHz).

TABLE 1. Relative Signal Strength by Surface Condition

Surface Condition	Ultrasonic Signal Strength (arb.)
Bare Graphite/Epoxy	0.073
Polyester Peel Ply	1.000
128 Weave Glass Peel Ply	0.166
180 Weave Glass Peel Ply	0.273

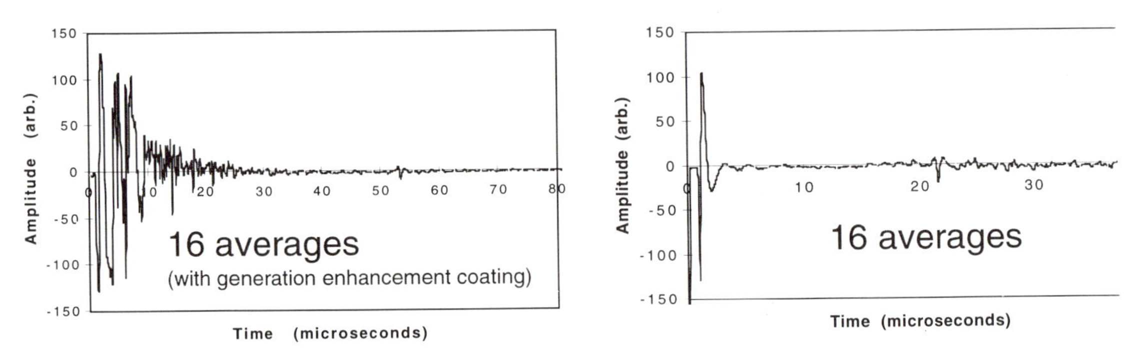

Figure 5. Waveforms from 1.5 inch and 2.3 inch thick graphite/epoxy

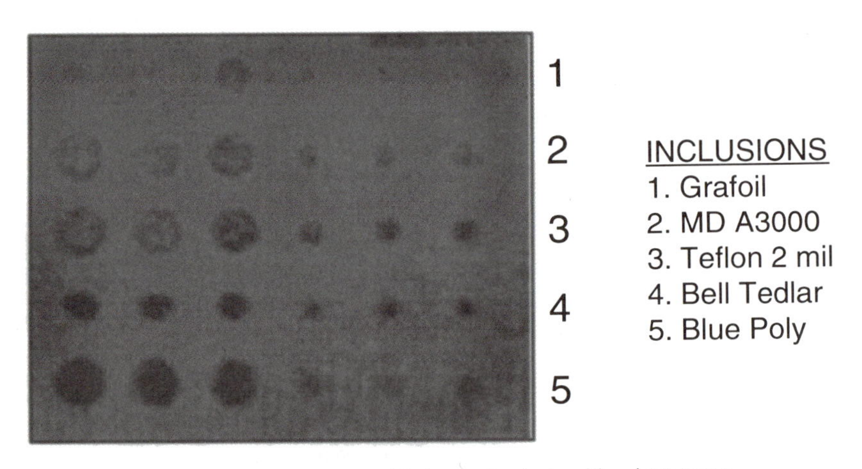

Figure 6. RAM coated graphite/epoxy laminate with paint topcoat

Effects of Bulk Material

The attenuation of ultrasonic signals in bulk material is an effect which often limits a systems material inspection capability. In particular, very thick contoured composite structures, which are becoming more commonplace in advanced applications, can be very difficult even for mature ultrasonic techniques. This is more difficult for laser ultrasonic techniques where material damage thresholds limit the amount of laser energy that can strike the surface and therefore establish an upper limit on the amount of ultrasonic wave energy that can be transferred into the material. Fig. 5 shows two waveforms taken from 1.5 inch and 2.3 inch thick composite laminates using the laser ultrasonic system.

The waveforms shown are an average of 16 waveforms and the 2.3 inch region was coated with a generation enhancement material. Here we see that attenuation in very thick composite structure can cause the waveforms to be severely degraded causing reduced defect detection sensitivity. This problem is compounded if the structure is also highly contoured or has non-parallel surfaces. For materials up to 1/2 inch in thickness a time-dependent attenuation correction factor of 2 dB/μsec is typically applied to the signals. Substantially thicker materials tend to propagate only low frequency components (0.5-1.0 MHz) where a correction factor of 1 dB/μsec is sufficient.

Another example is the case where the material itself has highly attenuative intrinsic characteristics such as radar absorbing materials (RAM) or B-staged laminates (50% cure). The RAM coatings are typically made of a polyurethane or other rubber matrix that highly attenuate sound, especially the higher frequencies (> 1 MHz). Fig. 6 shows a laser ultrasonic c-scan of a RAM coated graphite/epoxy laminate in which defects were implanted into the laminate as a test. This test is also very difficult with conventional ultrasonic techniques.

REFERENCES

Chang, F.H., et al., 1993, *Review of Progress in QNDE* **12**, pp. 611-616.
Drake, T.E., 1985, unpublished General Dynamics IR&D report, ERR-FW-2444.
Drake, T.E., 1994, *SAMPE* **39**, pp. 725-739.
Monchalin, J.P., 1985, *Can. J. Lett.* **47**, pp. 14-16.
Rudd, M.J., Doughty, J.A., 1983, *Laser Generation of Ultrasound*, NADC-81067-60
White, R.M., 1963, *Journal of Applied Physics* **34**, pp.3559-3567.
Yawn, K.R., 1995, unpublished IR&D data.

LASER ULTRASOUND IMAGING OF LAMB WAVES IN THIN PLATES

Peter W. Lorraine

General Electric Corporate Research and Development Center,
P.O. Box 8,
Schenectady, NY 12301

INTRODUCTION

Laser ultrasound offers many advantages over conventional piezoelectric ultrasound including the potential for rapid wide-area scanning, non-contacting (no couplant) generation and sensing, and large bandwidth[1,2]. Ultrasonic surface waves may be easily generated by a laser and can travel extended distances when the part is not immersed and loss to a surrounding water bath is eliminated. In addition, the geometric attenuation is significantly less as the sound energy spreads out in a circular annulus rather than in a spherical shell giving rise to an amplitude decay proportional to $r^{-1/2}$. We have shown that synthetic focusing of laser ultrasound data[3] permits us to use this information to create images of near-surface defects outside the scan area. A single scan line can be used to image the complete surface of part with high speed, resolution, and sensitivity (Figure 1).

The benefits of focused ultrasonic imaging are better spatial localization of flaw signals and increased sensitivity to small defects. A physically focused transducer sums wavefronts arriving across the face of the piezoelectric material. A coherent sum is produced for signals arriving in phase from a localized region (the focus), and an incoherent sum is produced for all other signals. The physical focus can be realized either with a shaped lens with sound

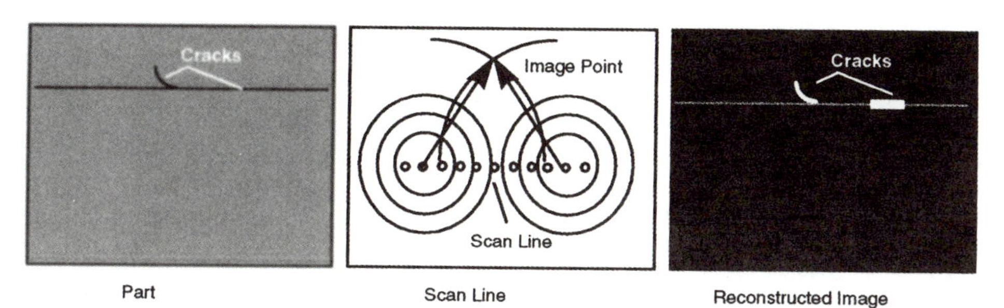

Figure 1. Rapid imaging of surface defects. A wide area image can be reconstructed from a limited scan area.

Nondestructive Characterization of Material VIII
Edited by Robert E. Green Jr., Plenum Press, New York, 1998

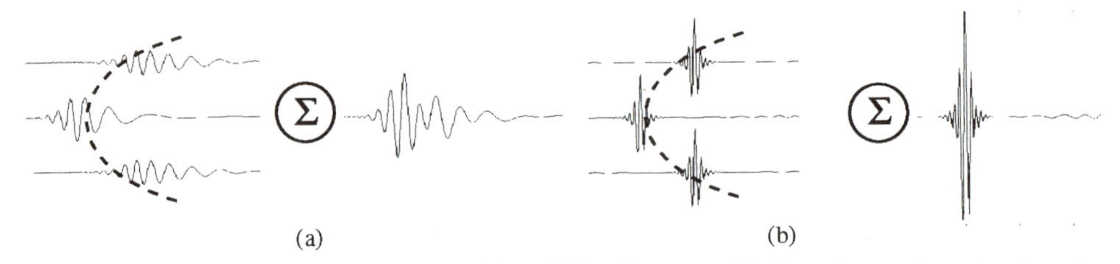

Figure 2. Synthetic aperture imaging difficulties with Lamb waves. (a) dispersion produces incoherent summation of signals. (b) a coherent image may be formed following dispersion compensation;

velocity different than the propagation medium or with a shaped transducer element. If the single physical transducer is replaced with multiple small elements, a generalized or synthetic transducer can be formed by creating arbitrary delays. The conventional synthetic aperture focusing technique (SAFT) experiment[3,4] typically involves a single transmitter with a diverging beam and a single receiver which are scanned to cover the desired aperture. In order to create a correctly focused image, it is necessary to know the transducer location accurately and the appropriate speed of sound. Surface wave focusing forms an image by summing the detected waveforms $U(x_j, y_j, t)$ along lines of either constant propagation delay or constant distance across the reception aperture as shown if Eq. 1

$$I(x_i, y_i) = \sum_j U(x_j, y_j, \Delta t_{ij}) = \sum_j A(x_j, y_j, r_{ij}), \tag{1}$$

$$\Delta t_{ij} = 2\left[(x_i - x_j)^2 + (y_i - y_j)^2\right]^{1/2} / v_{material}$$

$$r_{ij} = 2\left[(x_i - x_j)^2 + (y_i - y_j)^2\right]^{1/2}$$

where Δt_{ij}, r_{ij}, $v_{material}$ are the round trip propagation delay, round trip distance, and velocity of the material, respectively.

A complication arises in thin plates where sound propagates as dispersive Lamb waves[5,6]. Although sensitive to interesting defects such as corrosion or cracking, their dispersive nature has made direct imaging challenging although approaches with limited resolution have been demonstrated[7]. In this paper we describe two approaches which compensate for dispersion to produce rapid, high resolution images (Figure 2).

DISPERSION COMPENSATION USING BACK PROPAGATION

Dispersion results in a waveform that evolves with time or distance from the source. This dispersion causes a spreading of the pulse energy and a resulting reduction in peak amplitude giving rise to an apparent attenuation dependence that goes as r^{-1}. The first dispersion compensation approach involves mathematically handling the frequency dependence on phase velocity to back propagate the observed waveform $u(t)$ and its Fourier transform $U(\omega)$ to obtain the distance varying amplitude function $A(r)$:

$$A(r) = \int G(\omega, r) d\omega, \tag{2}$$

$$G(\omega, r) = \int e^{ik(\omega)r} U(\omega) d\omega$$

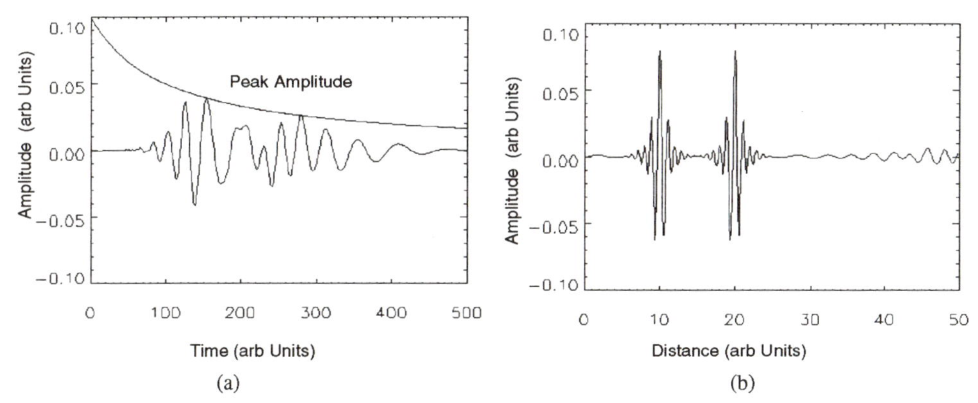

(a) (b)

Figure 3. Dispersion and dispersion compensation. (a) Two pulses located 10 and 20 distance units from the receiver, (b) the same waveform following dispersion compensation.

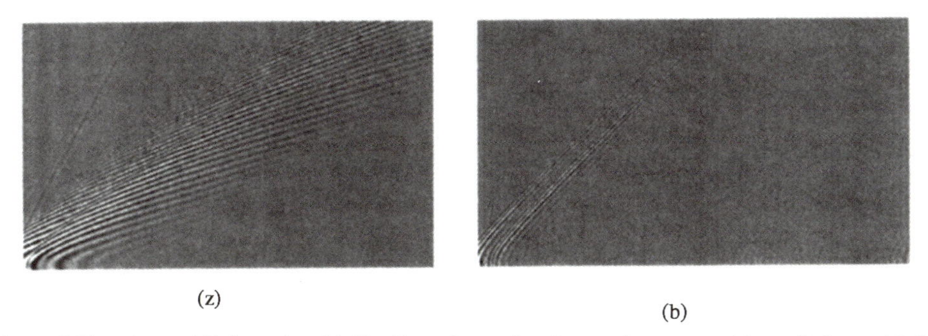

(z) (b)

Figure 4. Experimental B-Scans in a highly dispersive regime from a ring source. (a) raw B-Scan with S_0 and strongly dispersed A_0, (b) dispersion compensated B-Scan

The propagation wavevector is $k(\omega) = k_i + k_j \; i, j = S_0, A_0, \ldots S_n, A_n$ and is chosen to include the desired modes for the incident and scattered wave. A model calculation for a dispersion relation approximating the low frequency limit of the lowest order antisymmetric mode A_0 is shown in Figure 3. Figure 3(a) shows the waveform produced by two sources located at 10 and 20 units from the observation point. Figure 3(b) shows the results of the backpropagation dispersion compensation algorithm, which has separated the two points and eliminated the dispersion peak attenuation. Figure 4 shows raw and dispersion compensated results for experimental B-Scans acquired on a 0.075 mm thick Al foil in a strongly dispersive regime using a ring source[7,8]. The ring source produces two waves - the direct wave and a second due to the initial inward going component after passing through the center. In Figure 4(b) the dispersion compensation algorithm is focused on the A_0-A_0 mode pair and clearly shows the two waves.

EMPIRICAL FILTERING

An alternative approach to obtaining *A(r)* is to measure the time-dependent sound wave surface displacement over a range of positions $r_0, r_1, \ldots, r_m$ use that to produce an empirical filter matched to the specific combination of modes produced by the source in that specific material. Matrix multiplication of a raw waveform by the matched filter produces a

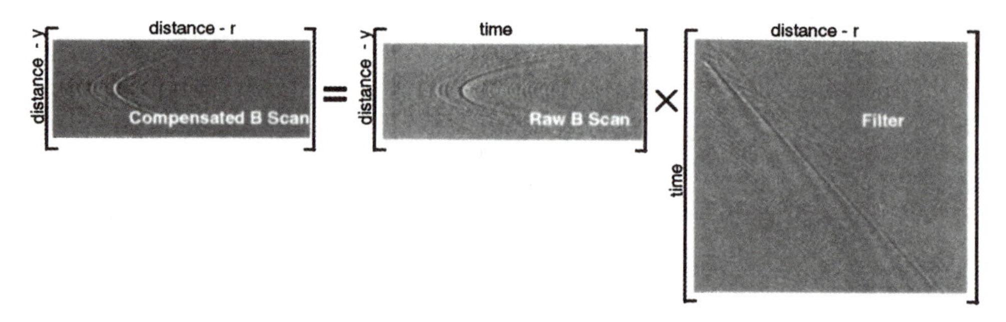

Figure 5. Graphical depiction of empirical matched filter approach. Raw B-Scan and filter were acquired on a 0.5 mm thick Al sample with a ring source in a regime above cutoff with multiple modes and dispersion.

waveform vector with dispersion removed To image a surface, waveforms are acquired for a range of positions $y_0, y_1, \ldots, y_n$, corresponding to the scan positions shown in Figure 1. In this case, the resulting matrix of data $\bar{u}(y_i, t_j)$, (a B-Scan) is multiplied by the filter:

$$\begin{bmatrix} A(y_0,r_0) & \cdots & A(y_0,r_m) \\ \vdots & \ddots & \vdots \\ A(y_n,r_0) & \cdots & A(y_n,r_m) \end{bmatrix} = \begin{bmatrix} u(y_0,t_0) & \cdots & u(y_0,t_m) \\ \vdots & \ddots & \vdots \\ u(y_n,t_0) & \cdots & u(y_n,t_m) \end{bmatrix} \begin{bmatrix} f(t_0,r_0) & \cdots & f(t_0,r_m) \\ \vdots & \ddots & \vdots \\ f(t_n,r_0) & \cdots & f(t_n,r_m) \end{bmatrix}. \tag{3}$$

This process is illustrated in Figure 5 from an experiment using a thermoelastic ring source and a 0.5 mm thick aluminum plate. The raw B-Scan shows both multiple modes and strong dispersion. Following the processing, the compensated B-Scan shows a single signal without dispersion suitable for computational focusing. With this approach, the image is tuned to a specific combination of modes present in the filter.

IMAGING EXPERIMENTS

The approaches described in the previous two sections have been successfully applied to experimental situations in several distinct Lamb wave regimes. In the low frequency limit on an isotropic free plate, only the fundamental symmetric S_0 and antisymmetric A_0 modes are supported. In a thin, 0.075 mm thick Al sheet, three 0.75 mm holes were formed. Waveform data was acquired along a single scan line 1.5 cm from the nearest defect. Figure 6 shows the reconstructed images of these defects for an effective area scan rate of 10 cm^2 sec^{-1}. Focusing of the raw data provides only a marginal improvement over the raw B-Scan. The two compensation schemes both provide significant gains in SNR and resolution.

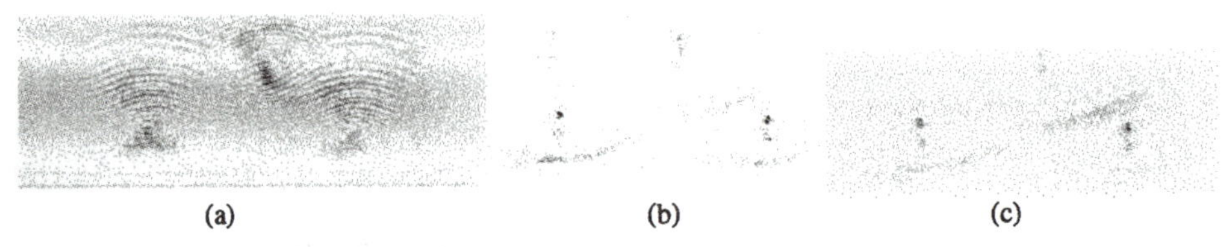

Figure 6. Reconstructed images of three 0.75 mm diameter holes on 0.075 mm thick Al sheet in strongly dispersive regime with a thermoelastic source. (a) focused raw data, (b) focused dispersion compensated data, (c) focused empirically filtered data.

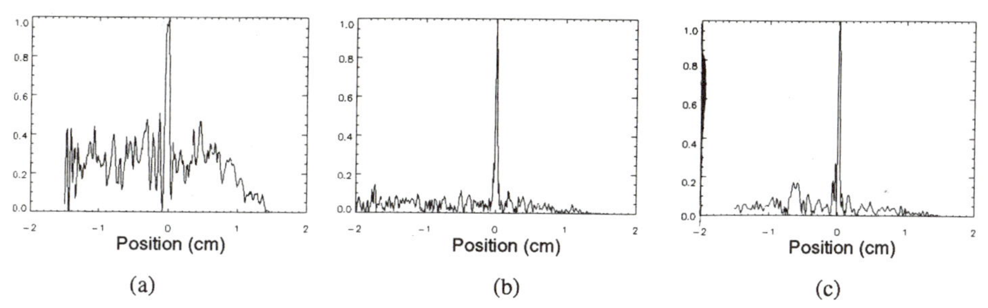

Figure 7. Reconstruction image amplitude through center of flaw. (a) focused raw data, (b) focused dispersion compensated data, (c) focused empirically filtered data.

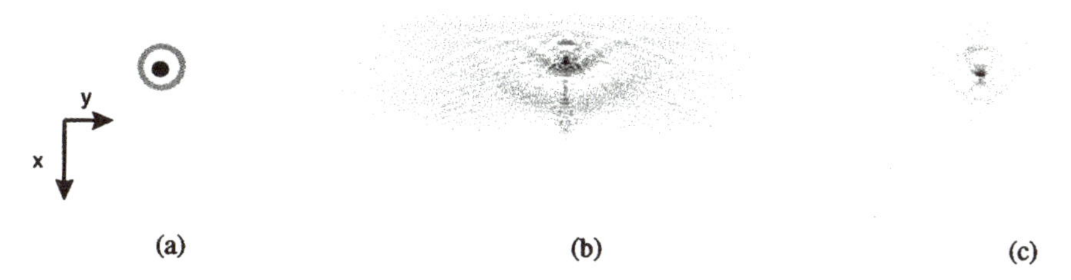

Figure 8.. SAFT reconstructed images of a thermoelastic ring source in multi-modal regime on 0.5 mm Al sheet. (a) ideal image, (b) uncompensated data shows poor focusing due to dispersion, (c) empirically filtered data shows dramatic improvement in resolution and sensitivity.

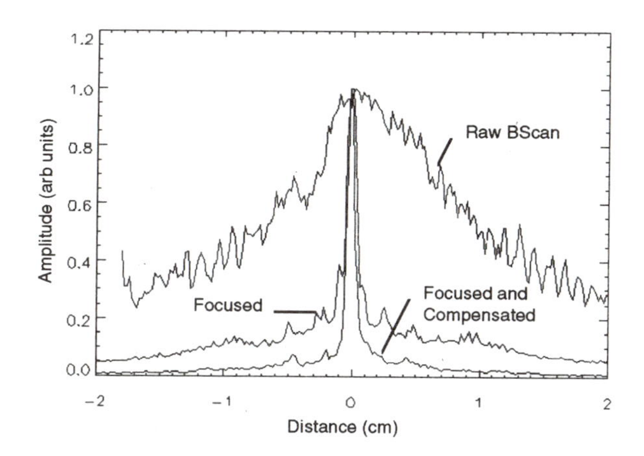

Figure 9. Reconstruction image amplitude through center of target for 0.5 mm thick Al sheet.

Figure 7 shows the reconstruction image amplitudes through the centers of the defects showing an SNR of approximately 10:1 for both compensated approaches versus the SNR of 2:1 for focused raw data. Range lobe information, not reproduced here but visible in Figure 6, shows a strong advantage for the compensation approaches. The back propagation dispersion compensation algorithm is focused on the strongly dispersive A_0 A_0 mode pair using an empirically derived dispersion relation.

Above certain critical frequencies, additional modes with unique dispersion curves appear. These modes generally overlap and interfere, giving rise to confusing signals. The strength of the matched filter technique is that it is not necessary to understand the modal properties in depth - the measured filter provides a match for the modal distribution for the given source in the given material. As a preliminary validation we have imaged a ring laser source by scanning a receiver along a line remote from the source on a 0.5 mm thick Al sheet. The results of this approach are shown in Figure 8 where the improvements in resolution and contrast are dramatic. Figure 9 shows reconstruction image amplitudes through the target center for the raw data, focused raw data, and empirically filtered compensated data. The matched filter becomes less effective when mode conversion occurs during scattering as the interfering waves decrease the correlation of the filter to the observed waveform, however refinements to the technique have been demonstrated.

SUMMARY

High resolution images can be made over wide areas using laser ultrasonic techniques and computational focusing following dispersion compensation. Two distinct approaches for dispersion compensation have been demonstrated in both the highly dispersive regime below cutoff and in the high frequency regime where multiple modes exist. Dispersion compensation through back propagation provides an ability to image specific mode pairs for the incident and scattered sound while empirical filtering handles multiple modes simultaneously.

ACKNOWLEDGEMENTS

The author is grateful for the contributions of Robert Filkins, Grace Li, and Harry Ringermacher and support by GE IR&D funds. Aspects related to titanium matrix composites were supported by TMCTECC cooperative agreement F33615-94-2-4439.

REFERENCES

1. J. W. Wagner, “Optical Detection of Ultrasound”, in *Physical Acoustics*, Vol XIX, 201 Academic Press, New York, (1990)
2. J-P Monchalin, and R. Heon, “Laser ultrasonic generation and optical detection with a confocal Fabry-Perot interferometer”, *Mat Eval*, **44** 1231:1237 (1986)
3. P. W. Lorraine, R. A. Hewes, and D. Drolet, “High resolution laser ultrasound detection of metal defects”, in *Review of Progress in Quantitative NDE*, edited by D.O. Thompson and D.E.Chimenti, Vol. 16, 555:562, Plenum, New York, (1997)
4. P. J. Howard and R. Y. Chiao, "Ultrasonic Maximum Aperture SAFT Imaging", in *Review of Progress in QNDE*, edited by D. O. Thompson and D. E. Chimenti, Vol 14 901:908, Plenum, New York, (1995).
5. A. Viktorov, *Rayleigh and Lamb Waves,* Plenum Press, New York (1967)
6. D. A. Hutchins, K. Lundgren, and S. B. Palmer, “A laser study of transient Lamb waves in thin materials”, *J. Acoust. Soc. Am.*, **85** (4), 1441:1448 (1989)
7. Y. Nagata, J. Huang, J. D. Achenbach, and S. Krishnaswamy, “Lamb wave tomography using laser based ultrasonics”, in *Review of Progress in Quantitative NDE*, edited by D.O. Thompson and D.E.Chimenti, Vol. 14, 561:568, Plenum, New York, (1995)
8. H. I. Ringermacher, F. A. Reed, and J. R. Strife, “Laser ultrasonics for coating thickness evaluation at 1200 C”, in *Review of Progress in Quantitative NDE*, edited by D.O. Thompson and D.E.Chimenti, Vol. 12, 549:552, Plenum, New York, (1993)

DETECTION OF CORROSION IN AGING AIRCRAFT STRUCTURES BY LASER-ULTRASONICS

M. Choquet[1], D. Lévesque[1], C. Néron[1], B. Reid[1], M. Viens[1], J.-P. Monchalin[1], J. P. Komorowski[2] and R. W. Gould[2]

[1] Industrial Materials Institute (IMI), National Research Council Canada, Boucherville, Québec.
[2] Institute for Aerospace Research (IAR), National Research Council Canada, Ottawa, Ontario.

ABSTRACT

Corrosion has been recognized to be a serious problem for the aging aircraft fleet currently in operation. In this work, laser-ultrasonics is shown to be a promising technique for the rapid and quantitative evaluation of corrosion in lap joints. By analyzing the broadband laser-ultrasonic signal obtained on a lap joint, we show that bonded areas can be discriminated from disbonded areas, which could be affected by corrosion. In the disbonded areas, corrosion thinning is evaluated by monitoring, in the frequency domain, the position of a resonant peak. Using a specimen simulating material loss, we show that an accuracy of 1% in the thickness determination is feasible.

INTRODUCTION

Corrosion of aircraft structures has been recognized as a serious problem for the aging aircraft fleet currently in operation. For economic reasons, numerous aircraft are expected to fly well beyond their initially designed life time. For example, the US Air Force C/KC 135's and the commercial DC-9's will fly well into the 21st century. However, the limited original corrosion protection measures have not remained effective over such a long operational life. In response, the FAA has recently established compulsory programs for corrosion monitoring and repair. A concern over aging aircraft issues has generated an interest in rapid and accurate corrosion detection methods.

Corrosion is a serious problem as it may directly affect the airworthiness of an aircraft. In lap joint structures, corrosion causes thinning and pillowing of the lap joint skins which in turn lead to very high stresses and may cause degradation of structural integrity. Recent studies at IAR indicate that corrosion at thickness loss levels lower than is currently accepted may require aggressive remedial action[1]. Since the by-product of corrosion causes bulging of the lap joint skins, visual inspection is applicable and is currently the main method for the detection of corrosion. Unfortunately, visual inspection results are highly dependent on the operator and can only detect a

Nondestructive Characterization of Material VIII
Edited by Robert E. Green Jr., Plenum Press, New York, 1998

high level of corrosion. Also, visual inspection does not produce documentation that can be easily used for comparison at a later date. Hand-held eddy current probes are also used to evaluate specific areas suspected to have corrosion problems. However, inspection of large aircraft such as the C/KC-135 with hand-held probes is very time consuming. Recently, improved and automated eddy current methods have shown promise along with an enhanced visual inspection technique known as D Sight™ [2]. The D Sight™ Aircraft Inspection Systems (DAIS) is currently the fastest and one of the most sensitive methods of inspecting lap joints for corrosion. However it can not quantify corrosion thinning in individual layers. A technique based on thermal waves has been developed [3] and is currently implemented for tearstrap disbond detection. Thermal wave capability to measure thickness of external skin layer has also been demonstrated. There is still, however, a need for a technique which provides both direct measurement of the residual thickness and rapid inspection.

Direct evaluation of residual metal skin thickness can be obtained with ultrasonics which, when implemented with lasers for the detection and the generation of the ultrasound (laser-ultrasonics), can provide a rapid, large area inspection technique. Laser-ultrasonics has already demonstrated the ability for rapid, automated, large area ultrasonic inspection on composite materials structures. [4,5] It has also been shown to be applicable in the aircraft maintenance environment of a hangar [6]. This paper presents the on-going work at the National Research Council of Canada on the use of laser-ultrasonics for the detection of low level corrosion in aircraft structure, primarily in a lap joint assembly.

LASER ULTRASONIC DETECTION OF CORROSION IN A LAP JOINT

Typical lap joint in an aircraft structure is formed by splicing together two aluminum skins and a stiffener, and fastening these with rivets. Some lap joints have adhesively bonded layers, while some have only a layer of sealant for corrosion protection. Sealant layers and some adhesive layers do not necessarily, according to the manufacturers, function as a structural bond. Lap joints which have been opened for low level corrosion removal and had the original adhesive removed, are typically rebuilt with a faying surface sealant between the skins. Different configurations of lap joints with tapers, doublers and triplers are found in the aerospace industry. For example, finger doublers are sometimes included between the skins and the stringer to provide for the distribution of in-flight stresses. Also, there is often only a sealant bead on the outboard, bottom edge of the outer skin. There may be a sealant bead on the inboard, upper edge of the inner doubler.

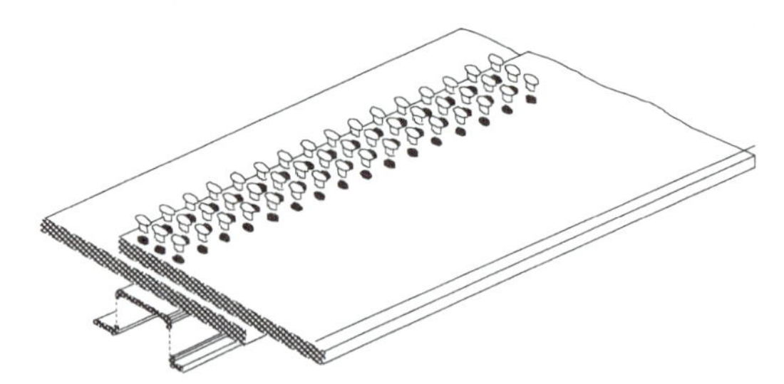

Figure 1: Typical lap joint configuration

The present study has been limited to a simple lap joint assembly (shown in Figure 1). Before corrosion occurs between the skins of this lap joint, the bond between the two skins must degrade. As corrosion progresses, the skins may become totally disbonded, leading to the creation of an air gap. An ultrasonic pulse will be totally reflected by this air gap. The outer skin can then be considered as *acoustically isolated* at the location of corrosion (or disbond). The measurement of the outer skin thickness, and hence of material loss, can be made by monitoring the ultrasonic resonance frequencies of the isolated outer skin. The resonance frequencies are directly related to the skin thickness and provide a precision measurement of skin thickness.

Laser-ultrasonic detection of material loss in a painted aluminum plate

The first step before using laser-ultrasonics for the detection of corrosion in an actual lap joint is to demonstrate its ability for precise measurement of thickness in a simple aluminum plate (representative of a typical lap joint skin). For this demonstration, a 12 inch x 10 inch, 62 mils thick 2024-T3 aluminum panel is used. Three rows of flat bottom holes of different depths and diameters are milled in the test panel. The diameters are 0.5 inch, 0.25 inch and 0.375 inch for the first, second and third rows, respectively. The flat bottom holes simulate material loss ranging from 15% (9.5 mils thickness reduction) to less than 1% (0.6 mils thickness reduction). The location, diameter and simulated material loss in mils are indicated in Figure 2a. The panel is painted with a white, water based, peelable latex paint to allow for efficient laser generation. It should be noted that the presence of paint on the aluminum skin does have some impact on the resonance ultrasonic frequencies. The overall resonant frequencies of the paint/skin structure are slightly shifted with respect to the resonant frequencies of a bare plate. This could yield a measurement result for the thickness, which is different from its actual value. However, using a model of the ultrasonic propagation in a layered media (see below), we have been able to ascertain that the high ultrasonic frequencies (higher than 3 MHz) are not, in this case, appreciably affected by the presence of paint.

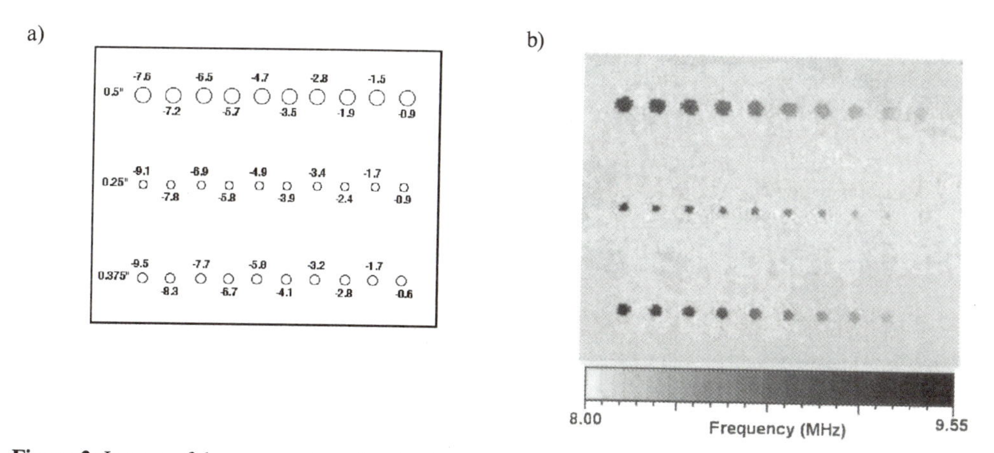

Figure 2: Images of the test panel: a) sketch of the test panel indicating thickness reduction and diameters, b) ultrasonic frequency C-scan image of the test panel with the corresponding gray-scale code.

For the inspection, ultrasound is generated with a short pulse CO_2 laser (pulse width of 150 nsec, energy per pulse of ~300 mJ) focused onto the painted surface to a spot size of ~5 mm. Given the energy density, generation occurs in a nondestructive thermoelastic regime within the paint layer. The laser generated ultrasonic pulse has a bandwidth of ~15 MHz. Ultrasound is detected with a long pulse Nd:YAG laser (pulse width of 55 μsec, 2.5 kW peak power) coupled to a stabilized confocal Fabry-Perot. The signal at the output of the Fabry-Perot is then digitized by a PC digitizing card and gives, for each acquisition, a data array of 1024 points, with a 40 nsec sampling step. The frequency spectrum of the waveform is obtained by numerical FFT of the sampled signal. Given the sampling step of the digitizer, a frequency resolution of 0.025 MHz is achieved. A typical recorded signal and its frequency spectrum are shown in Figure 3a-b. Notice that the ultrasonic resonance peaks are clearly observable in Figure 3b. By measuring the values of the resonance frequency between 8.0 and 9.55 MHz, which corresponds to the fourth harmonic, we can determine the plate's thickness at each inspection point. The resonance frequency value for each acquisition point on the test panel is then coded with a gray-scale value to produce the ultrasonic image shown in Figure 2b.

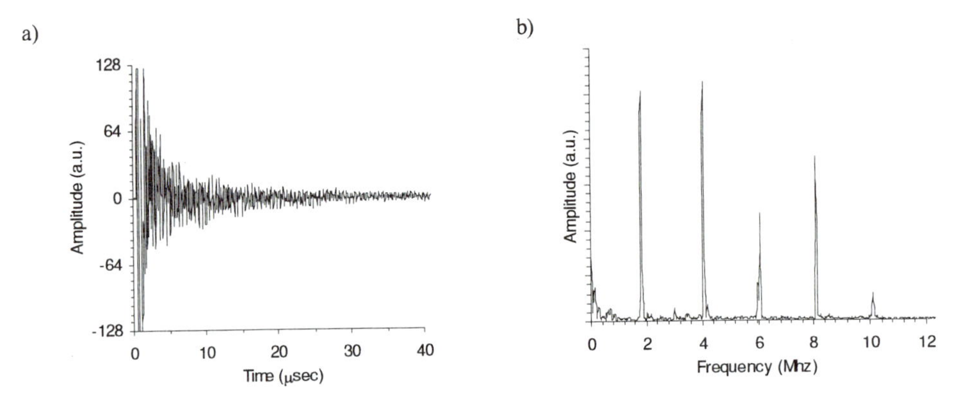

Figure 3: Laser-ultrasonic signal from the test specimen simulating skin thinning; a) typical waveform, b) corresponding frequency spectrum.

Knowing the acoustic velocity of aluminum (6320 m/sec), we can calculate the thickness of the plate at a given position from the value of the peak resonance frequency at this position. We can therefore measure the thickness of the plate for all the flat bottom holes simulating corrosion loss. Tables 1, 2 and 3 indicate the plate thickness measured by laser-ultrasonics for each row of flat bottom holes and the corresponding actual thickness measured with a micrometer. Notice that the maximum relative error is approximately 1%.

Table 1: Comparison of laser-ultrasonic thickness measurements to actual values for the flat bottom holes of 0.25 inch in diameter (in mils).

Actual	52.9	54.2	55.1	56.2	57.1	58.1	58.6	59.6	60.3	61.1
Laser-ultrasonic	52.8	54.3	55.2	56.1	56.7	57.6	58.3	59.2	60.3	61.2
Error %	0.17%	0.18%	0.16%	0.16%	0.68%	0.81%	0.48%	0.69%	0.08%	0.21%

Table 2: Comparison of laser-ultrasonic thickness measurements to actual values for the flat bottom holes of 0.375 inch in diameter (in mils).

Actual	52.5	53.7	54.3	55.3	56.2	57.9	58.8	59.2	60.3	61.4
Laser-ultrasonic	52.7	53.9	54.3	55.2	56.3	57.4	58.6	59.4	60.1	61.1
Error %	0.34%	0.43%	0.07%	0.20%	0.20%	0.81%	0.36%	0.35%	0.27%	0.46%

Table 3: Comparison of laser-ultrasonic thickness measurements to actual values for the flat bottom holes of 0.5 inch in diameter (in mils).

Actual	54.4	54.8	55.5	56.3	57.3	58.5	59.2	60.1	60.5	61.1
Laser-ultrasonic	54.3	55.3	56.1	56.8	57.6	58.4	59.4	60.1	60.1	60.7
Error %	0.11%	0.97%	1.10%	0.82%	0.49%	0.12%	0.35%	0.03%	0.69%	0.74%

Laser ultrasonic detection of corrosion in the outer skin of a lap joint

The work described above has shown that, if the outer skin of the lap joint is acoustically isolated from the rest of the structure, a very precise measurement of material loss can be made. Since corrosion occurring in a bonded lap joint may start with a disbond of the outer skin from the rest of the assembly, the first step should be to discriminate bonded areas from disbonded areas. The Industrial Materials Institute has recently developed an ultrasonic method to determine the quality of bonds within a multi-layered structure[7]. The method is based on modeling the propagation of ultrasonic waves in the bonded structure and on introducing a very thin interface layer of proper stiffness to describe adhesion strength. Using this model, we have calculated the theoretical ultrasonic frequency response of this lap joint assembly. The results for three levels of adhesion strength are shown in Figure 4. The frequency spectrum of laser-ultrasonic signal obtained from a corroded lap joint are also shown for comparison in the same figure. We observe excellent correspondence between the location of the resonance peaks given by the model and those observed in the test data.

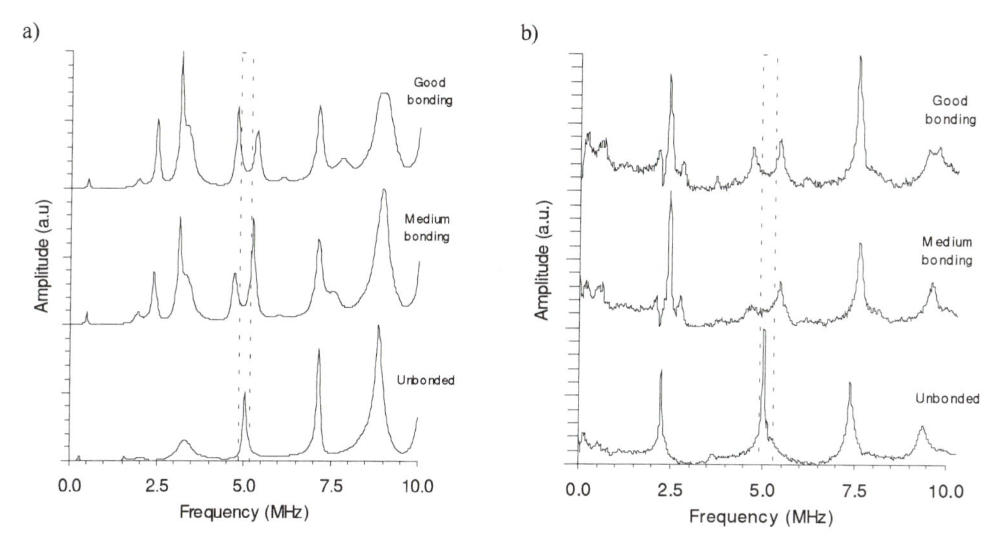

Figure 4: Laser-ultrasonic frequency spectrum as a function of bond quality: a) numerical model results, b) laser-ultrasonic data.

As can be seen in Figure 4, the model provides us with a method to discriminate bonded areas from disbonded ones. For disbonded areas, there is a sharp resonance peak near 5 MHz, whereas for bonded areas, there is a valley at about the same frequency. We also observe that a peak near 7 MHz is always present, independent of the bond quality. This peak's position can be shown to relate to the thickness of the outer skin of the lap joint.

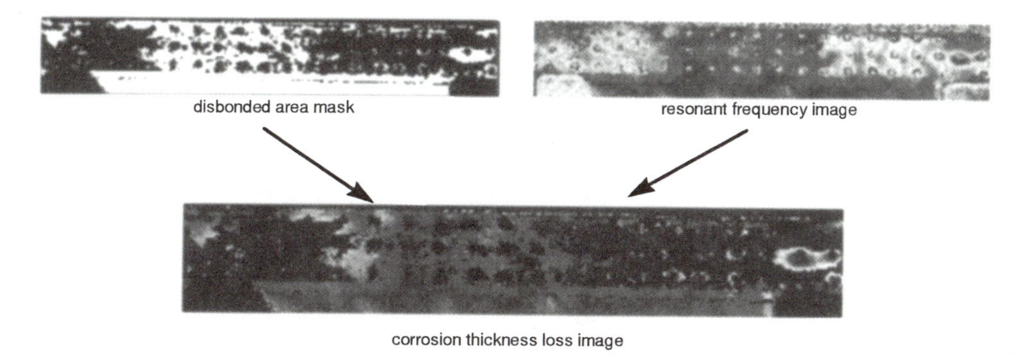

Figure 5: Procedure for corrosion identification and residual thickness mapping in a lap joint.

The procedure to identify corrosion and image material loss in the lap joint under study is therefore composed of three steps: First, we make a mask (black and white image) using a narrow bandwidth centered at 5 MHz. For every acquisition point, we verify if the maximum value in the narrow bandwidth is above or below a given threshold determined by the noise level in the signal. Points for which the value is above the threshold are marked with a white pixel in the image mask, while points for which the value is below the threshold are marked with a black pixel. The result is a mask which blocks out the regions of the lap joint with intact bonds, i.e., with no corrosion. Second, we make an ultrasonic frequency image of the lap joint by color coding the resonance frequency value near 7 MHz. Third, both images, the mask and the frequency image, are superimposed resulting in a thickness image of the corroded area of the lap joint. Figure 5 shows the mask, the initial ultrasonic image and the final corrosion image. These results, obtained by laser-ultrasonics, are in good agreement with measurements done by an eddy-current system and a D Sight™ image analysis. At this time, however, we have not been able to correlate these results with a direct measurement of the skin thickness. The lap joint is presently being disassembled for skin thickness measurement using a precision X-ray method. The results of the comparison will be reported at a later date.

CONCLUSION

We have demonstrated that laser-ultrasonics is a promising approach for the detection of corrosion in aircraft structures. We have shown that a frequency analysis of the laser-ultrasonic signal allows the discrimination of bonded areas from disbonded areas in a lap joint. We have also shown that, for an isolated aluminum skin, the thickness can be measured with an accuracy better than 1% by monitoring the variation of one or several ultrasonic resonance peaks in the laser-ultrasonic signal. Work is presently under way to validate these results obtained on one lap joint sample with a destructive analysis. Future work will involve the analysis of more specimens, the study of the effect of variations of the paint thickness and the study of the effect of the roughness of the corroded faying surface.

REFERENCES

1. J.P Komorowski, N.C. Bellinger and R.W. Gould, The Role of Corrosion Pillowing in NDI and in the Structural Integrity of Fuselage Joints, *Fatigue in New and Aging Aircraft* – Proc. Of the 19th Symposium of the International Committee on Aeronautical Fatigue Edinburgh, 16-20, June 1997 (to be published).
2. J.P Komorowski, N.C. Bellinger, R.W. Gould, A. Marincak, R. Reynolds, Quantification of Corrosion in Aircraft Structures with Double Pass Retroreflection, *Canadian Aeronautics and Space Journal*, Vol. 42, No.2, June 1996, pp76-82.
3. L.D. Favro, Xiaoyan Han, T. Ahmed, P.K. Kuo and R. L. Thomas, Measuring corrosion thinning by thermal wave imaging, in *Nondestructive Evaluation of Aging Aircraft, airports, and Aerospace Hardware* , SPIE vol. 2945, pp. 374-379, SPIE the International Society for Optical Engineering, Bellingham, Washington,1996.
4. J.-P. Monchalin, C. Néron, P. Bouchard, M. Choquet, R. Héon and C. Padioleau, Inspection of composite materials by laser-ultrasonics, *Canadian Aeronautics and Space Journal*, vol. 43, no. 1, pp. 34-38, 1997.
5. C. J. Fiedler, T. Ducharme and J. Kwan, The laser ultrasonic inspection system (LUIS) at the Sacramento Air Logistic Center, in: *Review of Progress in Quantitative Nondestructive Evaluation*, vol. 16, pp. 515-522, Plenum Press, New-York, 1997.
6. M. Choquet, R. Héon, C. Padioleau, P. Bouchard, C. Néron and J.-P. Monchalin, Laser-ultrasonic inspection of the composite structure of an aircraft in a maintenance hangar, in: *Review of Progress in Quantitative Nondestructive Evaluation*, vol. 14, pp. 545-552, Plenum Press, New-York, 1995.
7. D. Lévesque and L. Piché, A robust transfer matrix formulation for the ultrasonic response of multi-layered absorbing media, *J. Acoust. Soc. Am.*, vol. 92, pp. 452-467, 1992.

ACKNOWLEDGEMENTS

This work was partly funded by the Canadian Department of National Defense. We acknowledge Mr. Douglas Froom, at that time at McClellan US Air Force Base, Sacramento, California for providing the flat bottom holes test specimen.

PHOTOREFRACTIVE LASER ULTRASOUND SPECTROSCOPY FOR MATERIALS CHARACTERIZATION

K.L. Telschow, V.A. Deason, K.L. Ricks and R. S. Schley

Idaho National Engineering and Environmental Laboratory
Lockheed Martin Idaho Technologies Company
Idaho Falls, ID 83415-2209

INTRODUCTION

Ultrasonic elastic wave motion is often used to measure or characterize material properties. Through the years, many optical techniques have been developed for applications requiring noncontacting ultrasonic measurement. Most of these methods have similar sensitivities and are based on time domain processing using interferometry[1]. Wide bandwidth is typically employed to obtain real-time surface motion under transient conditions. However, some applications, such as structural analysis, are well served by measurements in the frequency domain that record the randomly or continuously excited vibrational resonant spectrum. A significant signal-to-noise ratio improvement is achieved by the reduced bandwidth of the measurement at the expense of measurement speed compared to the time domain methods. Complications often arise due to diffuse surfaces producing speckle that introduces an arbitrary phase component onto the optical wavefront to be recorded. Methods that correct for this effect are actively being investigated today.

Adaptive interferometry, which utilizes the photorefractive effect in optically nonlinear materials, offers a potentially powerful method to automatically correct for environmental and speckle effects.[2,3] Photorefractivity employs optical excitation and transport of charge carriers to produce a hologram of the interference pattern developed inside the material. A spatially and temporally modulated charge carrier distribution results that is a direct measure of the phase information impressed onto the optical object beam by the vibrating surface. This hologram stores phase information from all the surface points on the vibrating specimen simultaneously. The hologram developed then can be detected through diffraction of a readout beam off this photoinduced volume grating. Several optical frequency domain measurement methods of vibration have been proposed using photorefractive two-wave and four-wave mixing in select materials.[4,5,6] These have provided a response that is a nonlinear function of the specimen vibration displacement amplitude.

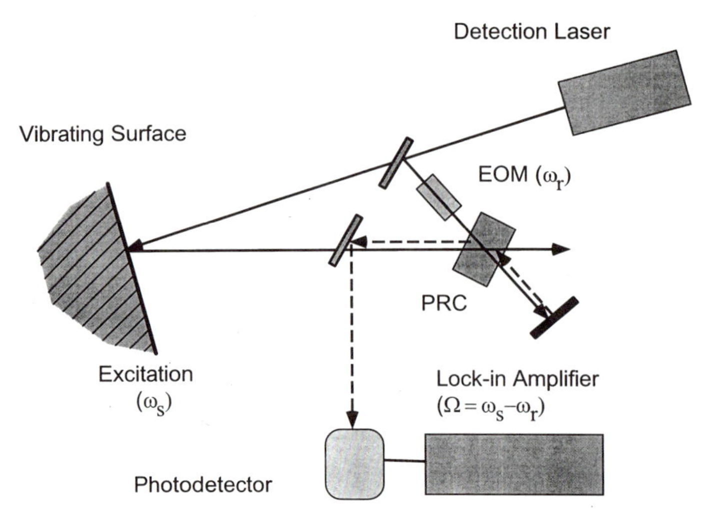

Figure 1. Photorefractive detection of resonant vibration, BS (beamsplitter), EOM (electro-optic modulator), (PRC) photorefractive crystal.

Recently, a method was developed for vibration detection[7,8] that employs the photorefractive effect in an optical lock-in synchronous detection manner.[9] This method phase modulates the object and reference beams such that a moving photorefractive grating at a fixed beat frequency is established within the material regardless of the specimen vibration frequency. The intensity of a readout beam scattering off the photoinduced grating is a direct measure of the vibrational amplitude and phase. It can be used for spectral analysis with a response proportional to the Bessel function of order one, providing a linear output for small amplitudes. The method accommodates rough surfaces and exhibits a flat frequency response above the photorefractive cutoff frequency. A minimum detectable displacement of 2×10^{-3} nm has been recorded; however, the limitation was from optical phase noise, not shot noise, suggesting that further improvement is possible. The corresponding demonstrated detectability was 4.5×10^{-6} nm $(W/Hz)^{1/2}$, in terms of the light power received at the photodiode.

In this paper, the ability of the optical lock-in process to provide absolute calibration of measured displacement amplitude is described and application of the method to full-field imaging of vibrating square plates is presented. Typically, the nonphotorefractive interferometric methods do not allow imaging of more than one surface point at a time. However, since the photorefractive process records a volume hologram of all vibrating surface points simultaneously, imaging can be achieved. Both four-wave and two-wave mixing configurations have been employed for reading out the vibration-induced phase grating image. Results are presented for vibration modes of a free square stainless steel plate with diffusely reflecting surfaces that is driven piezoelectrically at one corner.

PHOTOREFRACTIVE OPTICAL LOCK-IN METHOD

Figure 1 shows the experimental setup for optical detection of a specimen undergoing continuous vibration. A diode-pumped Nd:YAG laser source (532 nm, 200 mW) is split into object and reference beams. The excited vibrational modes of the specimen determine the frequency-dependent displacement amplitude of the surface, which is transferred into phase modulation of the object beam. The reference beam is frequency modulated by an electro-optic modulator at a fixed modulation depth. The modulated

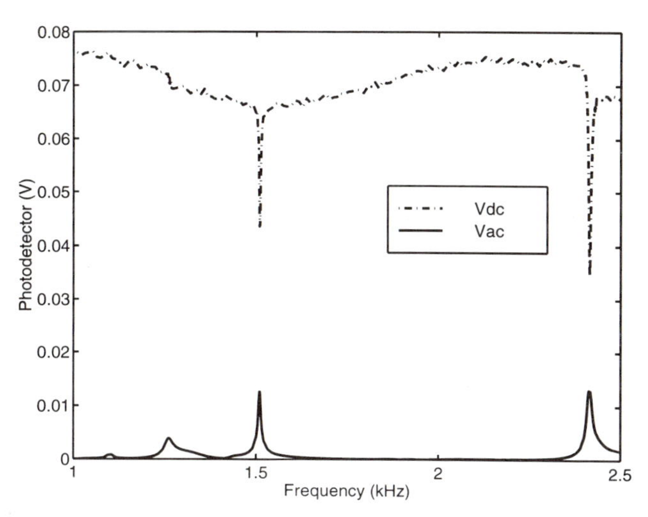

Figure 2. Photodetector DC and AC voltages from the vibration spectrum of two modes of a square stainless steel plate.

beams are then combined and interfered inside a Bismuth Silicon Oxide (BSO) photorefractive crystal. A four-wave mixing configuration is used for readout of the photorefractive index grating for the single-point measurements. The reference beam is reflected back into the crystal along a counter-propagating path that matches the Bragg angle of the photorefractive grating in the medium. The resulting scattered wave is then sampled at the plate beamsplitter and deflected toward the photodetector. Subsequently, the photodetector signal is processed with conventional electrical lock-in methods to provide a measurement bandwidth of 1 Hz. A recorded spectrum showing three spearate modes of a 1.5 inch square, 0.015 inch thick, stainless steel 304 plate is shown in figure 2; the plate is driven piezoelectrically at one corner.

A model of the lock-in process has been developed using one-dimensional plane wave coupled mode analysis. Both optical beams reaching the photorefractive crystal experience phase modulation due to the path lengths. In addition, the signal beam is modulated by the vibrating surface according to $\Phi_s = \Phi_{s0}\sin(\omega_s t + \varphi_s)$, $\Phi_{s0} = \frac{4\pi\xi_0}{\lambda}$. The reference beam is phase modulated by the EOM according to $\Phi_r = \Phi_{r0}\sin(\omega_r t + \varphi_r)$. The interference distribution and subsequent charge migration within the crystal generates a corresponding space charge electric field distribution, E_{sc}. The dynamic behavior of this field is controlled by the charge carrier mobility and trapping that produces, in the diffusive operation regime, a single relaxation time response given by

$$\frac{\partial E_{sc}}{\partial t} + \frac{E_{sc}}{\tau} = \frac{iE_q}{\tau}\frac{\Delta I(\vec{r},t)}{I_0} \tag{1}$$

where τ is the material response time, $\frac{\Delta I(\vec{r},t)}{I_0}$ is the interference fringe contrast, and E_q is the maximum achievable space-charge field, controlled by the concentration of available charge trapping sites and the fringe spacing. In the above configuration, the photorefractive crystal acts as a mixing and low-pass filtering element, providing the benefits of lock-in

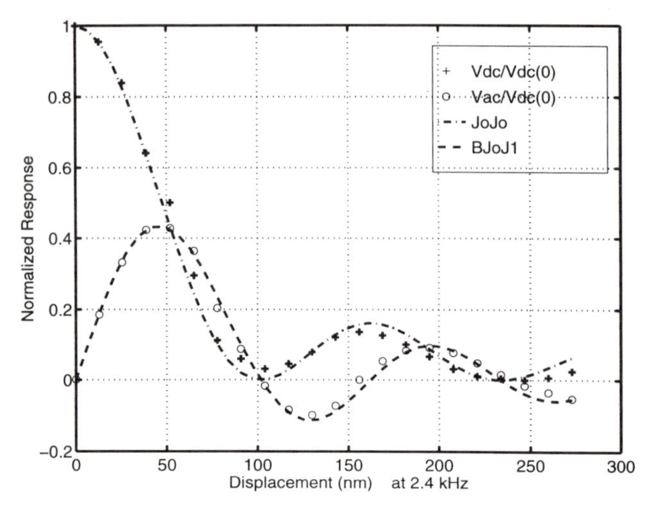

Figure 3. Measured DC and AC photodetector voltages for a square plate resonance as a function of the vibration amplitude.

detection. Therefore, the space charge field responds to slowly varying phase modulations occurring within the material response time, allowing only the terms around the difference frequency $\Omega = \omega_s - \omega_r$ to be important. The space-charge field modulates the local refractive index through the linear electro-optic effect. This effect creates a diffraction grating within the crystal that contains the low frequency phase information desired. Several methods can be used to read out the space charge field and diffraction grating, including four-wave mixing, two-wave mixing with polarization selection, and electrical measurement through conduction of photoexcited carriers.

In the four-wave mixing arrangement, the reference beam that passes through the crystal is reflected back and diffracts into the signal beam direction, see figure 1. The magnitude of the index of refraction grating produced is proportional to the space charge field and the orientation-dependent electro-optic coefficient in BSO. The diffracted beam intensity is a direct measure of the grating established and produces DC and AC output photodetector voltages given by

$$\frac{V_{DC}(\Phi_{s0})}{V_{DC0}} = J_0^2(\Phi_{s0}), \qquad V_{DC0} = V_{DC}(\Phi_{s0}=0) = C\,J_0^2(\Phi_{r0}) \tag{2}$$

$$\frac{V_{AC}(\Phi_{s0})}{V_{DC0}} = J_0(\Phi_{s0})\,J_1(\Phi_{s0})\left[\frac{J_1(\Phi_{r0})}{J_0(\Phi_{r0})}\frac{4}{\sqrt{1+\Omega^2\tau^2}}\right]\cos\!\left(\Omega t+\Psi+(\varphi_s-\varphi_r)\right) \tag{3}$$

where $\tan(\psi) = \Omega\tau$. Since the reference modulation and the photorefractive time constant are known, measurement of the DC and AC voltage magnitudes provides a direct calibration of the vibration amplitude in terms of the known optical wavelength through

$$\frac{2J_1(\Phi_{s0})}{J_0(\Phi_{s0})} = \left|\frac{V_{AC}}{V_{DC}}\right| B(\Phi_{r0},\Omega\tau) \approx \Phi_{s0} = \frac{4\pi\xi_0(\omega)}{\lambda}, \quad \text{for} \quad \Phi_{s0} << 1 \tag{4}$$

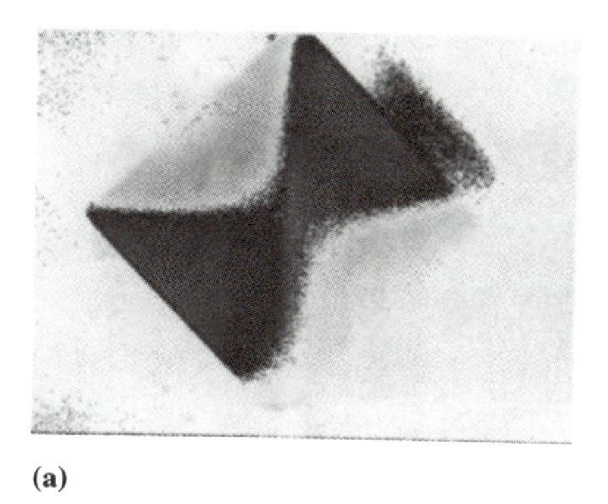

(a)

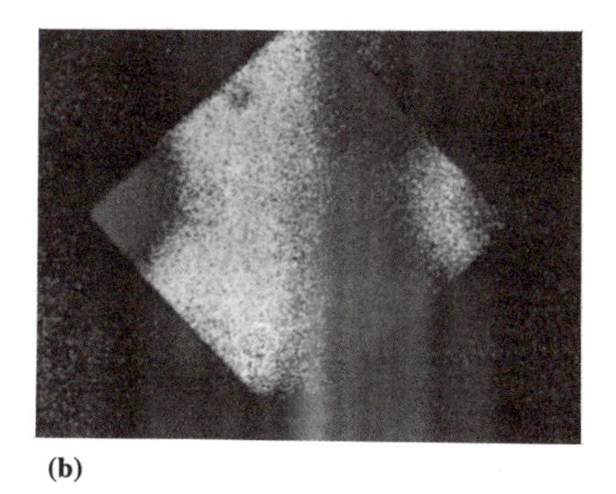

(b)

Figure 4. Vibrational mode images of a free square plate at (a) 1.5 kHz, (b) 2.4 kHz

where $B(\Phi_{r0}, \Omega\tau)$ is the controllable parameter given by the bracketed term in eq. 3. The measured vibration amplitude dependence support this result as shown in figure 3, which shows results from a single mode with $\Phi_{r0} \approx 1.1$ and $\Omega\tau \approx 1$. In this figure, the vibration amplitude was determined by the best fit of the data to eq. 3 at small drive amplitudes. The overall consistency of the results supports the model and the calibration procedure.

VIBRATION IMAGING

A two-wave approach, based on polarization rotation through anisotropic self-diffraction,[10,11] was used for imaging as it offers improved light throughput from diffusely reflecting surfaces compared to the four-wave method. Figure 4 shows images of two modes of a free square stainless steel plate excited by a contact piezoelectric transducer at one corner. The nodal lines are clearly defined, and the relative phases of the vibration displacements are indicated by the light and dark areas. The entire modal pattern can be made to switch from light to dark by varying the offset frequency, $\Omega/2\pi$ (in this case, equal to 2 Hz) between the object and reference excitations. This provides a powerful tool for visual mode searching and suggests processing methods that can be employed to enhance the detectability of specific modes. The minimum detectable displacement in the imaging mode (~1 nm) is much larger than for the point detection method as no post electronic lock-in processing was performed for the data of figure 4.

CONCLUSIONS

A photorefractive optical lock-in vibration spectral measurement method has been described that provides a simple method for quantitative determination of the vibration amplitude. The method uses optical synchronous or lock-in detection which can also include conventional electrical lock-in detection for narrow bandwidth high sensitivity measurements. A minimum detectable displacement of 2×10^{-3} nm has been demonstrated using the single-point vibration detection method with a 1 Hz bandwidth. Readout methods employing four-wave or two-wave mixing produce an output intensity directly proportional to the amplitude of the vibration being measured, for amplitudes small compared with the optical wavelength, and provide the capability for mechanical phase measurement if synchronous excitation is used. Vibration imaging was shown from a diffusely scattering surface showing detectable displacements in the 1-45 nanometer range. The method is capable of flat frequency response over a wide range above the cutoff of the photorefractive effect and is applicable to rough surfaces.

ACKNOWLEDGMENTS

This work was supported through the INEEL Laboratory Directed Research & Development program under DOE Idaho Operations Office Contract DE-AC07-94ID13223.

REFERENCES

1. J. W. Wagner, "Optical Detection of Ultrasound," *Physical Acoustics*, Vol.XIX, Eds. Thurston, R.N., and Pierce, A.D., (Academic Press, New York, 1990) Chp. 5.
2. P. Yeh, *Introduction to Photorefractive Nonlinear Optics*, (John Wiley, New York, 1993).
3. S. I. Stepanov, *International Trends inOptics*, (Academic Press, New York, 1991) Chp. 9
4. J. P. Huignard and A. Marrakchi, "Two-wave mixing and energy transfer in $Bi_{12}SiO_{20}$ crystals: application to image amplification and vibration analysis," Opt. Lett., **6**, (12), 622 (1981).
5. H. R. Hofmeister and A. Yariv, "Vibration detection using dynamic photorefractive gratings in KTN/KLTN crystals," Appl. Phys. Lett., **61** (20), 2395 (1992).
6. H. Rohleder, P. M. Petersen and A. Marrakchi, "Quantitative measurement of the vibrational amplitude and phase in photorefractive time-average interferometry: A comparison with electronic speckle pattern interferometry," J. Appl. Phys., **76** (1), 81 (1994).
7. T.C. Hale and K. Telschow, "Optical lock-in vibration detection using photorefractive frequency domain processing," Appl. Phys. Lett. **69**, 2632 (1996).
8. T.C. Hale and K.L. Telschow, "Vibration modal analysis using all-optical photorefractive processing," Proc. SPIE Vol. 2849 , Photorefractive Fiber and Crystal Devices: Materials, Optical Properties, and Applications II, Francis T. Yu; Shizhuo Yin; Eds., p. 300 (1996).
9. J. Khoury, V. Ryan, C. Woods and M. Cronin-Golomb, "Photorefractive optical lock-in detector," Opt. Lett., **16**, 1442 (1991).
10. R.C. Troth and J.C. Dainty, "Holographic interferometry using anisotropic self-diffraction in $Bi_{12}SiO_{20}$," Opt. Lett., **16** (1), 53 (1991).
11. T.C. Hale, K.L. Telschow and V.A. Deason, "Photorefractive optical lock-in vibration spectral measurement," submitted for publication to Applied Optics.

APPLICATION OF LASER-ULTRASONIC METHOD TO EVALUATE ELASTIC STIFFNESS DEGRADATION IN CERAMIC MATRIX COMPOSITES

Y. M. Liu,[1] T. E. Mitchell[1], and H. N. G. Wadley[2]

[1] Center for Materials Science, Mail Stop K765
Los Alamos National Laboratory, Los Alamos, NM 87545, U. S. A.
[2] Department of Materials Science and Engineering
School of Engineering and Applied Science
University of Virginia, Charlottesville, VA 22903, U. S. A.

INTRODUCTION

Damage in composites can be defined as the more or less gradual development of microdefects such as matrix or fiber cracks that eventually leads to final fracture of a component.[1] Associated with damage is a reduction of elastic stiffness.[2,3,4] Knowledge of the overall stiffness degradation in composites is of great importance for predicting its serviceability, especially if the component is under a multiaxial stress state.

Ultrasonic measurements provide an efficient way of determining the complete elastic stiffness tensor of a composite material.[5,6,7] For example, Baste et al. have immersed samples in a water bath to conduct *in situ* conventional ultrasonic measurements during mechanical testing and found that the ultrasound velocity was a good indicator of composite damage.[6] In this paper, we describe a more convenient laser-ultrasonic (LU) approach to measure and evaluate the anisotropic stiffness degradation during uniaxial tensile loading. *In situ* measured ultrasonic velocities along different propagation directions were used to obtain the elastic stiffness tensor components by fitting the experimental data to solutions of the Christoffel equation using a nonlinear optimization procedure. The results will show that this NDE technique provides an efficient means to assess the damage status of composite components in engineering applications.

FUNDAMENTALS OF LASER-ULTRASONIC APPROACH

The ultrasonic evaluation of anisotropic damage is based on a reduction in the elastic stiffness tensor components as a result of the collective effects of microcracks. The elastic wave propagation velocities in anisotropic media are given by solutions to the well-known Christoffel equation[8] :

Nondestructive Characterization of Material VIII
Edited by Robert E. Green Jr., Plenum Press, New York, 1998

$$\det\left|C_{ijkl}\, n_j n_l - \rho V^2 \delta_{ik}\right| = \Omega(V,\boldsymbol{n}) = 0,$$

where C_{ijkl} are the elastic stiffness constants of the material, $\boldsymbol{n}$ is a unit vector in the wave propagation direction, ρ is the density of the medium, V is the phase velocity of elastic wave and δ_{ik} is the Kroneker delta. Using wave velocities measured along suitable propagation directions, solutions to the above equation can be inverted to determine the elastic constants, C_{ijkl}, through a nonlinear curve fitting process.

The use of a pulsed laser to generate a broad band acoustic pulse is well known. Point source LU techniques enable accurate measurements of wave propagation speeds between precisely defined locations on a sample. In the LU approach, one strictly measures a group velocity. However, for the unidirectional and 0°/90° calcium aluminosilicate reinforced by Nicalon SiC fibers (CAS/SiC) composites studied here, elastic anisotropy is small [9] and the difference between phase velocity and group velocity is negligible. Therefore, measured wave-speeds can be used directly in the inversion scheme.

It should be noted that solutions to the Christoffel equation are applicable to homogenous materials only. Composites are, by definition, heterogeneous. However, when the ultrasonic wave length in the material is long compared with the length scale of the inhomogeneity, the medium to a good approximation responds as an equivalent homogeneous medium.[10] This appears to be the case here.

EXPERIMENTAL

Unidirectional and 0°/90° CAS/SiC composites were provided by Corning, Inc. The Nicalon SiC fiber volume fraction was 0.35 ~ 0.40, the average fiber diameter was 15 μm, the Young's modulus of the fiber was 200 GPa and that of the matrix was 97 GPa. Tensile specimens were cut to dimensions of ~ 150 mm × 10 mm × 3 mm for the unidirectional composites, and ~ 150 mm × 10 mm × 2.7 mm for the 0°/90° composites. Sample ends were bonded with fiber-glass tabs of low modulus. The side (edge) surface of specimens was polished and acetate replicas were taken at various stages of loading to reveal damage evolution under load.

The coordinate systems used for this fiber architecture are shown in Fig. 1. For the unidirectional composite, *3*- (the fiber direction) is the loading direction, *1*- is the thickness direction. For the 0°/90° cross-ply, *1*- is the loading direction, *3*- is the laminate thickness direction, and the continuous SiC fibers are aligned along the *1*- and *2*- directions. Plies with fibers in the *1*- direction are defined as the 0° plies.

Details of tensile testing and LU measurement setup have been given previously.[11] The specimen was loaded to a pre-set stress level, held at that level for LU measurements, and the stress then reduced to 10 MPa to measure a mechanical modulus. The above procedure was repeated at successively higher stress levels. Laser pulses generated by a Q-switch 1.064 μm wavelength Nd:YAG laser were delivered onto the loaded sample surface by an optical fiber. The laser source position was controlled with a computer and could be moved parallel and perpendicular to the loading direction. Ultrasonic wave arrivals were detected by two broadband piezoelectric transducers in contact with the sample.

DAMAGE EVOLUTION AND ANISOTROPIC STIFFNESS EVALUATION

Fig. 2 shows typical loading/unloading stress-strain curves for unidirectional and 0°/90° CAS/SiC composites. Matrix cracking initiated in the weak 90° plies at a much lower

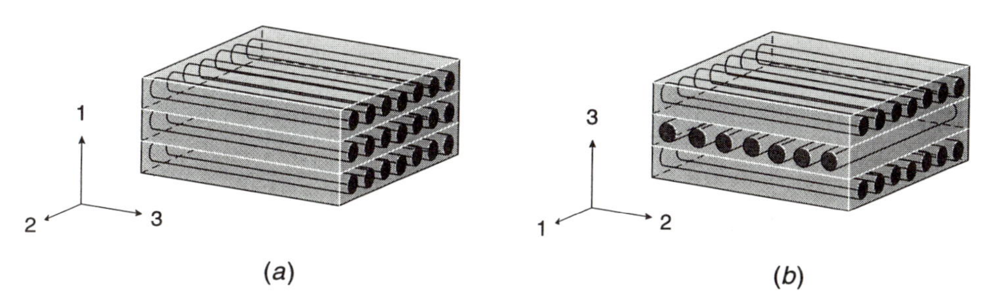

Figure 1. Coordinate system for (a) unidirectional (*3-* is the loading direction) and (b) 0°/90° laminated composites (*1-* is the loading direction).

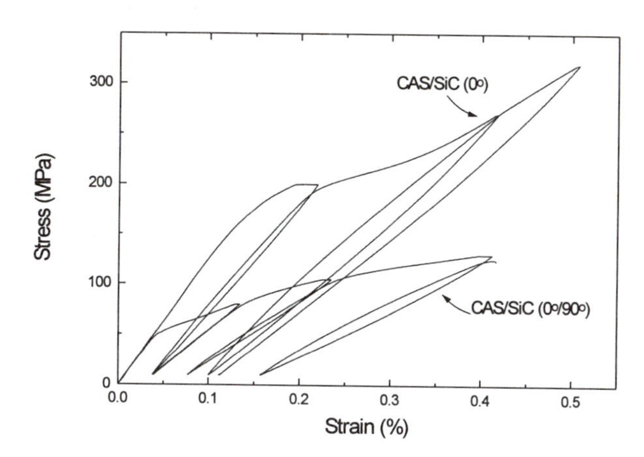

Figure 2. Loading/unloading stress-strain curve for unidirectional and 0°/90° CAS/SiC.

applied stress (~50 MPa) compared to that in unidirectional material (~140 MPa). This is mainly due to residual stresses, and the details are presented elsewhere.[12] Fig. 3 shows waveforms propagating over the same flight distance in the *1-3* plane of unidirectional CAS/SiC at different applied loads. With an increase of applied stress, the flight time increased. In the example shown in Fig. 3, the flight times of the longitudinal wave at 0 MPa, 200 MPa, 270 MPa and 320 MPa were 830, 835, 950 and 970 ns respectively. The largest time increase occurred between 200 and 270 MPa. As shown by the stress-strain curve (Fig. 2), this was the region where significant inelastic strain (damage) had accumulated. These observations correlate well with surface replicas: matrix crack density is substantially higher at 270 MPa than at 200MPa.[11] Wave velocity experienced a very small decrease when stress was increased to 320 MPa from 270 MPa, because matrix cracks were nearly saturated at this stage.[11] These results show that the ultrasound velocity is sensitive to the damage state.

Unidirectional composites can be considered to have a transversely isotropic symmetry. The symmetry of the as-received 0°/90° composite with even numbers of 0° and 90° plies can be regarded as tetragonal, but is reduced to orthotropic after damage. Using the ultrasonic velocities measured within the principal *1-3* and *1-2* planes for the unidirectional composite, or the *1-3* and *2-3* planes for the 0°/90° cross-ply, elastic constants at different damage states can be determined using a nonlinear curve fitting method.[11] Because of the approximations involved, the off-diagonal elastic constants had

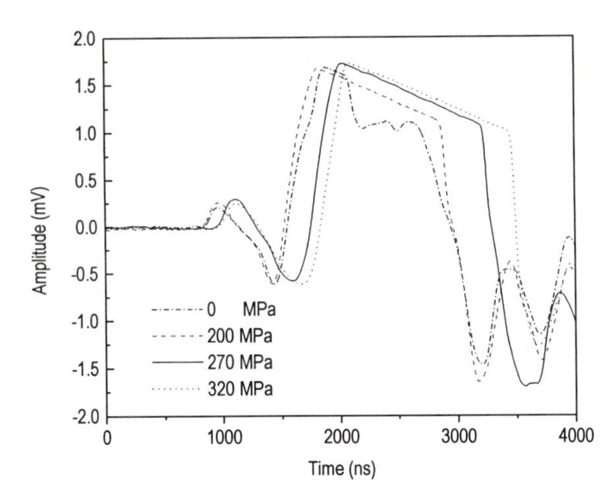

Figure 3. Waveforms over the same distance in the *1-3* plane of unidirectional CAS/SiC at four stress levels.

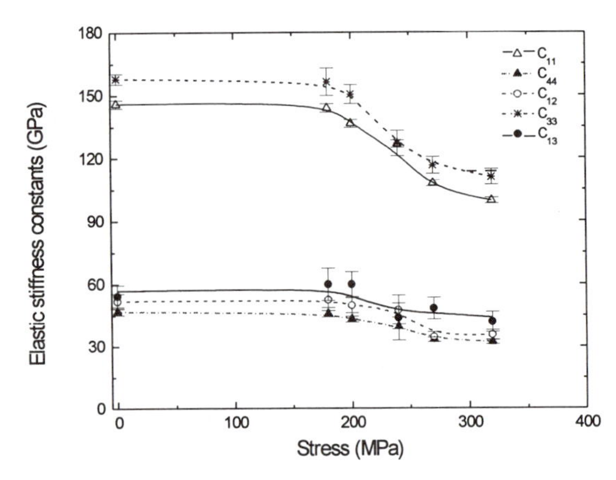

Figure 4. Stiffness constants determined from LU velocities as a function of applied stress for the unidirectional CAS/SiC.

relatively larger scatter compared to the diagonal elastic constants. Thus, C_{11}, C_{22} and C_{33} were chosen for 0°/90° laminate to represent the degree of damage in the three principal directions. Elastic stiffness constants are plotted against stress in Fig. 4 (unidirectional) and Fig. 5 (0°/90°). For both composites, substantial axial stiffness degradation is observed, due to multiple matrix cracking running nearly perpendicular to the loading direction. There is a pronounced transverse softening accompanying the stiffness reduction in the axial (loading) direction,. This arises because in the CAS/SiC matrix cracks are bridged by the fibers and deflected along the interface to induce interfacial debonding, rather than fiber breakage. For the 0°/90° composite, debonding occurs both in the 0° and 90° plies. Because the elastic constants in the direction parallel to the crack plane are insensitive to the crack density,[13] softening in the transverse plane is mostly affected by debonding in the 0° plies. Since the 0°/90° laminates have only half the fibers aligned in the loading direction compared to the unidirectional laminates, the stiffness reduction in the transverse direction is less than in the unidirectional CAS/SiC, but bigger in the loading direction due to longer matrix cracks in the 90° plies.[12]

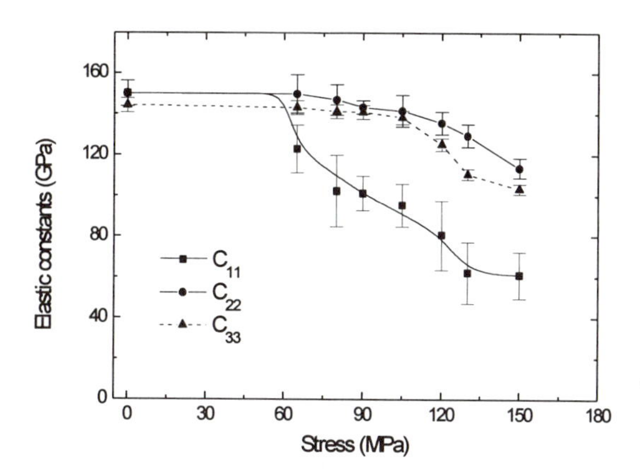

Figure 5. Stiffness constants C_{11}, C_{22}, C_{33} determined from LU velocities as a function of applied stress for the 0°/90° CAS/SiC.

CONCLUSIONS

Applying a LU approach, anisotropic elastic stiffness degradation has been evaluated for unidirectional and 0°/90° CAS/SiC composites. Elastic stiffness reduction in the axial (loading) direction is due to multiple matrix cracking and interfacial debonding/sliding induced by matrix cracking. "Softening" in the transverse direction is dominated by interfacial debonding. For the 0°/90° CAS/SiC cross-ply, decreases of the elastic stiffness constants C_{22} and C_{33} in the transverse plane are found to be similar, while C_{11} (along the loading direction) exhibits the largest decrease. Transverse elastic stiffness reduction for the unidirectional materials is greater than for the 0°/90° cross-ply. This is because only half of the fibers are aligned in the loading direction for the 0°/90° cross-ply and there is consequently less fiber/matrix interfacial damage. Results indicate the close correlation among damage states, elastic stiffness constants and ultrasound velocities. Most importantly, the *in situ* LU technique shows great potential for the routine inspection of structural components, predicting the failure location and serviceability, and therefore helping prevent catastrophic failure during service.

ACKNOWLEDGMENTS

This work is co-supported by the Defense Advanced Research Projects Agency through the University Research Initiative at University of California at Santa Barbara under ONR Contract No. N00014-92-J-1808 (S. Fishman, Program Manager) and the US Department of Energy, Office of Basic Energy Sciences. We gratefully acknowledge Dr. D. C. Larsen and K. Chyung (Corning Inc.) for supplying the material for this study.

REFERENCES

1. P. Ladeveze, A. Gasser, and O. Allix, Damage Mechanisms Modeling for Ceramic Composites, *J. Engineering Mater. Technology* 116: 331-336 (1994).
2. D. S. Beyerle, S. M. Spearing, F. W. Zok, and A. G. Evans, Damage and Failure in Unidirectional Ceramic-Matrix Composites, *J. Am. Ceram. Soc.* 75 [10]: 2719-2725 (1992).

3. M. Drissi-Habti, Damage Development and Moduli Reduction in a Unidirectional SiC-MAS. L. Composite Tested Under Uniaxial Tensile Loading, *Scripta Metallurgica et Materials* 33: 967-973 (1995).
4. P. Ladeveze, Inelastic Strains and Damage, *Damage Mechanics of Composite Materials*, edited by R. Talreja, Elsevier Science B. V. (1994).
5. B. Castagnede, K. Y. Kim, W. Sachse and M. O. Thompson, Determination of the Elastic Constants of Anisotropic Materials using Laser-generated Ultrasonic Signals, *J. Appl. Phys* **70** [1]: 150-157 (1991).
6. S. Baste, R. E. Guerjouma and B. Audoin, Effect of Microcracking on the Macroscopic Behavior of Ceramic Matrix Composites: Ultrasonic Evaluation of Anisotropic Damage, *Mechanics of Materials* 14: 15-31 (1992).
7. J. H. Gieske, and R.E. Allred, Elastic Constants of B-A1 composites by Ultrasonic Velocity Measurements, *Exp. Mech.* 14: 158-165 (1974).
8. B. A. Auld, *Acoustic Fields and Waves in Solids*, Krieger Publishing Company, Malabar, Florida (1990).
9. Y. M. Liu, Y. He, F. Chu, T. E. Mitchell, and H. N. G. Wadley, Elastic Properties of Laminated CAS/SiC Composites Determined by Resonant Ultrasound Spectroscopy, *J. Am. Ceram. Soc.*, **80** [1]: 142-148 (1997).
10. R. M. Christensen, *Mechanics of Composite Materials*, Wiley, New York (1979).
11. Y. M. Liu, T. E. Mitchell, and H. N. G. Wadley, Anisotropic Damage Evolution in Unidirectional Fiber Reinforced Ceramics, *Acta. Mater.*, in press (1997).
12. Y. M Liu, T. E. Mitchell, and H. N. G. Wadley, Anisotropic Damage in Laminated 0°/90° Glass-Ceramic Composites Subjected to Uniaxial Tension, to be submitted (1997).
13. Y. Huang, K. X. Hu, and A. Chandra, The Effective Elastic Moduli of Microcracked Composite Materials, *Int. J. Solids Structure* 30: 1907-18 (1993).

MEASUREMENT OF THE STIFFNESS TENSOR OF ORTHOTROPIC MATERIALS FROM LINE SOURCE POINT RECEIVER LASER ULTRASONIC METHOD

Bertrand Audoin[(1)], Christophe Bescond[(1)] and Menglu Qian[(2)]

[(1)] Laboratoire de Mécanique Physique, URA CNRS n° 867, Université de Bordeaux I, 351 cours de la Libération, 33405 Talence, France.
[(2)] Institute of Acoustics, Tongji University, Shanghai 200092, P.R. China.

INTRODUCTION

The propagation of an ultrasonic wave through a sample is widely used for the measurement of the stiffness tensor of orthorhombic materials. The immersion technic[1-3] allows the identification of all the nine coefficients of the above mentioned tensor, from phase velocity measurements of the plane waves propagating through the sample. LASER generation of ultrasound[4], associated with an optical interferometric detection, is a non contact methodology offering an alternative solution for the characterization of materials.

The LASER induced ultrasonic source is a broadband repetitive source which is readily amenable to operation in a scanning mode. When it is operated in the ablation regime, at a high power level, it generates ultrasound by vaporising a small amount of surface material. It can be represented as a force normal to the surface of the specimen. The radiated field of such a source resembles a monopole radiating strongly in all directions from the source[4,5].

The source generates transient divergent waves that propagate through the sample at group velocity[6,7]. The phase and group velocities of acoustic waves in elastically anisotropic solids are not in general equal, even in the absence of dispersion and attenuation. Actually, the direction of group velocity or energy flux at any point of the slowness surface[8] is parallel to the surface normal at that point[6]. Moreover, the projection of the group velocity on the phase direction equals the associated phase velocity. One of the important consequences of this fact is the phenomenon of phonon focusing, whereby the energy flux radiated is much more highly concentrated in some directions than in others[9]. In many materials, focusing is capable of building up a superposition of waves with high energy, with fruitful consequences in the area of material characterization.

Previous studies were limited to the measurements of four stiffness coefficients from the inversion of group velocities measured in a principal plane of symmetry of an orthorhombic material[10,11]. In this paper, the method to measure all the nine coefficients, by including data measured in non principal plane of symmetry, is briefly described. Details can be found in reference 10. Measurements of the stiffness tensor of a silicon cristal are presented.

FORMULATION OF THE INVERSE PROBLEM

Let consider a line source as obtained with a LASER beam focused with a cylindrical lens at the sample surface. The surface of the sample is assumed to include two axes of symmetry denoted by 2 and 3, i.e. the interface is a plane of symmetry. Axis 1 is thus normal to the surface. The line lies in any direction in plane (2, 3) and scanning is performed by translating the line in a direction normal to it. Let axis 2' be the direction of the displacement and let ϕ be the angle between axes 2 and 2', As shown in figure 1, the relative position of the source and receiver defines the observation angle ϑ_g in plane (1, 2').

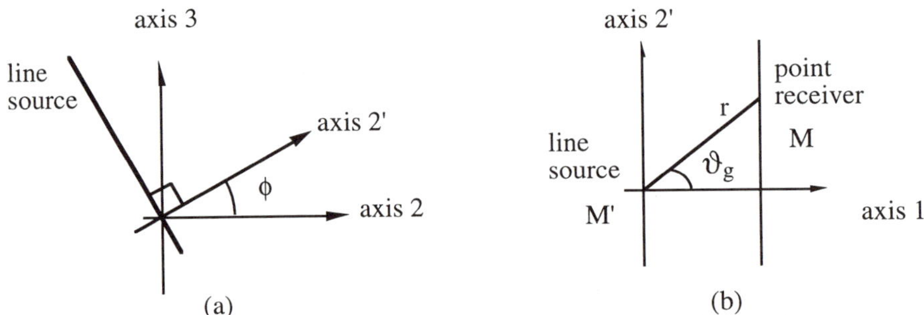

Figure 1. Geometry: displacement of the line source (a), and observation angle (b).

The purpose is to identify the stiffness coefficients from a set of group velocities measured in various directions of observation. This inverse problem shall be processed numerically, since there is no equation available that relates the elastic constants and the group velocities[10]. The inversion process is based on considerations yielding from the direct problem of the calculation of the wave forms generated by a line impact at the boundary of an orthotropic half space.

The displacement field $U(u_1, u_2, u_3)$ verifies a system (Σ) of partial differential equations[6,12] issued from i/ the wave equation, ii/ the boundary and initial conditions for a line source, iii/ the Sommerfield condition and iv/ the causality principle. It can be calculated, according the Cagniard De Hoop method[13], on performing a simple Fourier transform on the spatial variable $x_{2'}$ and a Laplace transform over time t. This allows a partial linearization of the above mentioned system (Σ) of differential equations. The contribution of both longitudinal or shear mode, to the Laplace transform of the normal component of the displacement field is, for any location M of the half space, except on the free surface:

$$u_1(M,t) = \frac{1}{2i\pi}\int_{-i\infty}^{+i\infty} F(S_p^2)\exp[-sr\left(S_p^1\cos\vartheta_g + S_p^2\sin\vartheta_g\right)]\,dS_p^2. \tag{1}$$

where s is the Laplace parameter, ϑ_g and r are the polar coordinates of the point receiver in plane (1, 2'), and S_p^2 and S_p^1 are the phase slowness components along axes 2' and 1, respectively. $F(S_p^2)$ denotes the transmission of the source which is obtained from the linearized boundary equations included in (Σ), for any value of S_p^2. On using, as De Hoop suggested, the change of variable:

$$t = r\left(S_p^1\cos(\vartheta_g) + S_p^2\sin(\vartheta_g)\right). \tag{2}$$

the argument of the exponential appears as the argument of the Laplace transform, and the contribution of the mode to the normal displacement is directly identified from (1).

Introducing the De Hoop change of variable (2) in Christoffel equation yields a polynomial expression, in terms of S_p^2 such that:

$$\mathcal{F}(t, S_p^2, \mathbf{C}, \vartheta_g) = 0. \qquad (3)$$

The complex solutions S_p^2 of equation (3) define the Cagniard-de Hoop contour of integration.

A second analytic form of the same variables is established in[10] on noting that the time of flight t_0 such that $V_g = r / t_0$ corresponds, on the t parametric contour, to the point at which the contour leaves the imaginary axis $\mathrm{Im}(S_p^2) = 0$. It is shown that this yields the following relation between the unknown $\mathbf{C}$, the experimental group data t_0 and ϑ_g, and the parameter S_p^2:

$$\left.\frac{\partial \mathcal{F}}{\partial S_p^2}\right|_{t_0} = \mathcal{G}(t_0, S_p^2, \mathbf{C}, \vartheta_g) = 0. \qquad (4)$$

The system of Eq. (3-4) is numerically inverted to optimally recover the elastic constants.

The process minimizes the L_2 norm of a set of functions $\left(\mathcal{F}(t_{0n}^{exp}, S_{p,n}^2, \mathbf{C}, \vartheta_n^{exp})\right)_N$ where index n refers to an experimental data. N experimental time t_{0n}^{exp} and experimental angles ϑ_n^{exp} are consirered. Since the parameters $S_{p,n}^2$ vary with the variable $\mathbf{C}$, the inversion algorithm has to be double iterative. For each increment of the variable $\mathbf{C}$, the optimal values of each slowness component $S_{p,n}^2$ is calculated. The correct value of each $S_{p,n}^2$ minimizes the quadratic sum:

$$\mathcal{F}(t_{0n}^{exp}, S_{p,n}^2, \mathbf{C}, \vartheta_n^{exp})^2 + \mathcal{G}(t_{0n}^{exp}, S_{p,n}^2, \mathbf{C}, \vartheta_n^{exp})^2. \qquad (5)$$

RESULTS

Experimental set up

A Nd:Yag LASER is used for the ultrasonic wave generation. Green light emission at 532 nm is chosen, and the pulse duration is of about 10 ns, with a medium power output typically ranging from 60 to 100 mJ per pulse. The collimated optical beam is focused by means of a cylindrical lens. The transverse size of the spot is approximately 0.1 mm giving the source a dimension smaller than the acoustic wavelength. In those conditions, ablation occurs at the sample surface, and marks it slightly. The LASER interferometric probe for the detection is a Mach-Zehnder, heterodyne type[14]. The bandwidth extends from 200 kHz to 45 MHz, and the sensitivity is 10 mV/Å. A periscope device permits to move the line impact location in order to change the observation angle[15].

Results in Principal Planes of Symmetry

For the plane problem of an impacting line source along a principal axis, two waves are generated, which are quasi longitudinal and quasi shear polarized. Consequently,

Christoffel equation is not a cubic but a quadratic equation in square slowness[8]. The convexity of the $\mathcal{F}$ and G is improved by factorizing both functions. In that way, the polarization associated with each measured data is taken into account[10].

Group velocities have been measured[15] for line source orientations $\phi = 0$ and $\phi = 90°$. These data are denoted by crosses in figure 2. The solid lines represent the optimum fit calculated from these data. Four stiffness coefficients are identified in each principal plane. They are reported in tables 1 and 2. A 90 % confidence interval is associated with each of the identified stiffness coefficient. Its calculation[16] results from the analysis of the residual squares of the functions derived from Eq. (3). It does not provide an exact calculation since some systematic errors cannot be taken into account. Nevertheless, it characterises the quality of the identification of the stiffness coefficients from the group velocity data.

The identified stiffness coefficients are in accordance with the intrinsic values issued from the literature. However, it is noticeable that the off-diagonal elastic constants, C_{12} and C_{13}, differ from the intrinsic module. Since the anisotropy of Silicon is not important, compared with other anisotropic materials, few cusp data are measured and they are spread over a small sector[15]. These cusp data are highly representative of anisotropy, and consequently, they have an important effect on the identification of the off-diagonal elastic constants. Since the sensitivity is few, small systematic errors may yield to the difference between the measured and standard values of off diagonal coefficients. For instance, a bad positioning of the epicenter has important repercussions for the determination of these constants. The confidence intervals which characterize the quality of the minimization of the function, can not quantify these systematic errors.

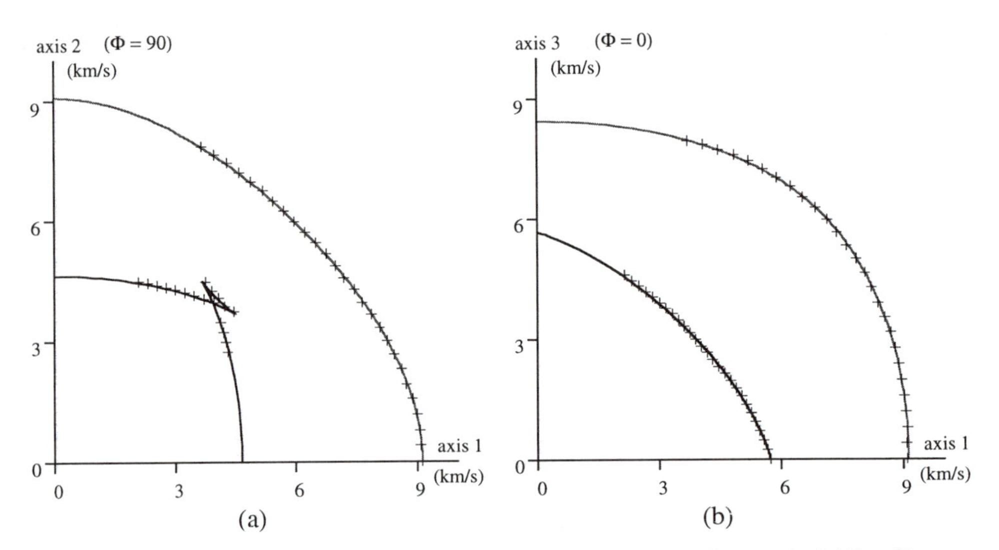

Figure 2. group velocity curves in silicon: (a) line source along axis 3 ($\Phi = 0$) and (b) along axis 2 ($\Phi = 90$). Solid lines represent the optimal fit calculated from data marked by crosses.

Table 1. Stiffness coefficients and 90 % confidence interval (GPa) measured with line source along axis 3 ($\Phi = 0$).

	C11	C22	C12	C66
Standard	194.4	194.4	35.2	50.9
Measured	194.2 ± 0.6	192.1 ± 2.3	39.2 ± 0.9	50.4 ± 0.4

Table 2. Stiffness coefficients and 90 % confidence interval (GPa) measured with line source along axis 2 ($\Phi = 90°$).

	C_{11}	C_{33}	C_{13}	C_{55}
Standard	194.4	165.7	63.9	79.6
Measured	194.3 ± 0.4	166.1 ± 1.4	67.4 ± 0.7	76.9 ± 0.3

Results in Non Principal Planes of Symmetry

The two remaining unknown elastic coefficients C_{23} and C_{44} are identified from group velocity data measured with a line source laying in a non principal direction. Because of line source generation, the propagation still operates in two dimensions, i.e., phase and group velocity vectors both belong to plane (1, 2'). The projection of the phase velocity on the interface is along axis 2' and the angular dependence of group velocities can be discussed from that of phase velocities in plane (1, 2'). Three waves are possibly generated in such a non principal plane which polarizations are quasi longitudinal and quasi transverse. Thus, the whole set of coefficients interfere in the propagation equation. The system of equations to minimize, equations (3-5), is constructed on considering the known values of the seven coefficients identified previously. Measurements have been performed on using a line source direction of 22.5°, 45° and 67.5°. The group velocity data for the three waves, and corresponding optimum fits, are plotted in figure 3. The corresponding coefficients are reported in table 3. The small discrepancies between the measured and reference values yield from the errors on the measurements performed in principal planes. Moreover, the anisotropy of velocities appears small and the cusps are spread over small sectors. Consequently, as discussed in the previous section, the sensitivity of the coefficients identification is weak, mostly concerning off diagonal coefficient C_{23}. Measurements quality could be increased if they were performed with material of higher anisotropy, such as copper crystal, or composite material.

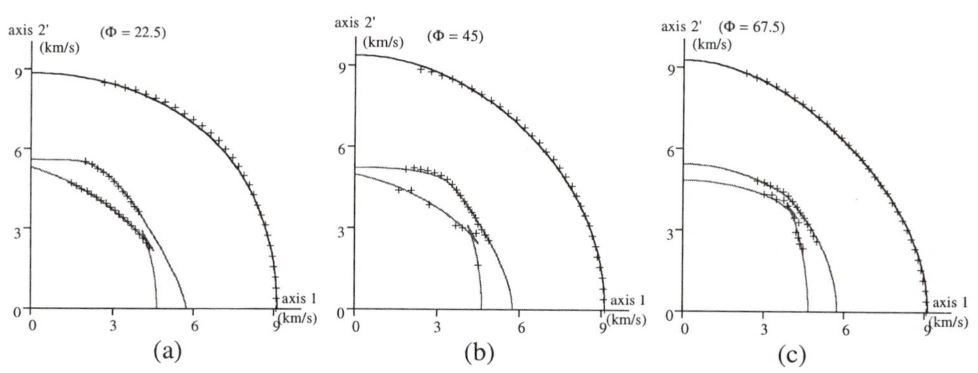

Figure 3. group velocity curves in silicon: (a) line source along axis $\Phi = 22.5$, (b) along axis $\Phi = 45$ and (c) along axis $\Phi = 67.5$. Solid lines represent the optimal fit calculated from data marked by crosses.

Table 3. Measured stiffness coefficients and 90 % confidence interval (GPa). Standard values are: $C_{23} = 63.9$ GPa and $C_{44} = 79.6$ GPa.

line source direction	C_{23}	C_{44}
$\Phi = 22.5$	63.6 ± 3.7	80.0 ± 2.0
$\Phi = 45$	60.5 ± 2.7	84.2 ± 2.42
$\Phi = 67.5$	54.1 ± 2.1	83.9 ± 1.3

CONCLUSIONS

An inversion process to measure the stiffness coefficients from group velocities data has been presented. The method has first been implemented for the measurement of four stiffness coefficients from the analysis of velocities collected in a principal plane of symmetry. It has been successfully used in previous works to analyse the changes of four stiffness coefficients of a composite material at elevated temperature up to 300°C[17]. The degradation of the material properties, induced by temperature, could be identified. However, complete measurement of the stiffness tensor is required.

In this paper, the method was extended to identify the lacking coefficients from data measured in a non principal plane of symmetry. The whole stiffness tensor of a Silicon crystal could be measured from a single sample.

Various applications are ahead in the field of material behaviour investigation, since the stiffness tensor changes are required to identify the anisotropic damage induced by temperature in anisotropic materials intended to be used at elevated temperatures.

REFERENCES

1. M. F. Markham, Measurement of the elastic constants of fibre composites by ultrasonics, *Composites 1*, pp. 145-149 (1970).
2. J. Roux, B. Hosten, B. Castagnede, and M. Deschamps, Caractérisation mécanique des solides par spectro-interférométrie ultrasonore, *Revue Phys. Appl.*, **20**, pp. 351-358, (1985).
3. S. Baste and B. Hosten, Evaluation de la matrice d'élasticité des composites orthotropes par propagation ultrasonore en dehors des plans principaux de symétrie, *Revue Phys. Appl.*, **25**, pp. 161-168, (1990).
4. C. B. Scruby and L. E. Drain, *Laser Ultrasonics Techniques and Applications* , Bristol, Philadelphia and New York, (1990).
5. D. A. Hutchins, *Ultrasonic Generation by Pulsed Laser* ,New York, (1988).
6. K. Aki and P.G. Richards, *Quantitative Seismology* , vol. 1, Freeman, San Francisco, (1980).
7. A. G. Every, W. Sachse, K. Y. Kim, and M. O. Thompson, Phonon focusing in silicon at ultrasonic frequencies, *Phys. Rev. Lett.* **65,** pp. 1446-1449 (1990).
8. B. A. Auld, *Acoutic Fields and Waves in Solids,* Second Edition Vol.I, Krierger, (1990).
9. H. J. Maris, Enhancement of heat pulse in crystal due to elastic anisotropy, *J. Acoust. Soc. Am.* **50,** pp. 812 - 818 (1971).
10. M. Deschamps and C. Bescond, Numerical method to recover the elastic constants from ultrasound group velocities, *Ultrasonics.* **33,** pp. 205-211 (1995).
11. A.G. Every and W. Sachse, Determination of the elastic constants of an anisotropic solids from acoustic-wave group-velocity measurements, *Phys. Rev. B*, **42**, pp. 8196-8205, (1990)
12. A. Mourad and M. Deschamps, Lamb's problem for an anisotropic half space studied by the Cagniard de Hoop method, *J. Acoust. Soc. Am.*, **97**, pp. 3194-3197, (1994).
13. A. T. D. Hoop, A modification of Cagniard's method for solving seismic pulse problem, *Appl.sci.res.*, **B-8**, pp. 349-356, (1960).
14. D. Royer and E. Dieulesaint, Optical detection of sub-Angstrom transient mechanical displacement, *IEEE Ultrason. Symp. Proc.*, pp. 527-530 (1986).
15. B. Audoin, C. Bescond, and M. Deschamps, Recovering of stiffness coefficients of anisotropic materials from point-like generation and detection of acoustic waves, *Journal of Applied Physics* **80** (7), pp. 3760-3771 (1996).
16. B. Audoin, S. Baste, and B. Castagnéde, Estimation de l'intervalle de confiance des constantes d'élasticité identifiées à partir des vitesses de propagation ultrasonores, *C.R. Acad. Sci. Paris 312*, **II**, pp. 679-686, (1991).
17. B. Audoin and C. Bescond, Measurement by LASER generated ultrasound of four stiffness coefficients of an anisotropic material at elevated temperatures. *Journ. of Nondest. Eval.*, **16** (2), (1997).

SAGNAC INTERFEROMETER FOR ULTRASOUND DETECTION ON ROUGH SURFACES

Pavel Fomitchov, Sridhar Krishnaswamy, and Jan D. Achenbach

Center for Quality Engineering and Failure Prevention
Northwestern University
Evanston, IL 60208

INTRODUCTION

Laser-based ultrasonic (LBU) systems have been developed by a number of researchers[1-8]. In this paper, we present a compact fiberized Sagnac-type laser ultrasonic detection system that can be used on rough surfaces. The device uses a low cost, long coherence He-Ne laser that has better intensity noise characteristics than typically used laser diodes. A frequency shifting technique has been introduced to eliminate problems arising from parasitic interferences that may occur with long coherence sources. A random speckle modulation scheme has been incorporated to overcome the "speckle problem" that arises when the test object is rough. Examples of applications are also reported in this paper.

A Sagnac interferometer has many advantages, which make it very attractive for ultrasound detection: (i) it is truly path-matched and does not require active stabilization and it can thus be used in a noisy environment, (ii) it can be completely fiberized and (iii) it can be designed as a small portable device suitable for field applications, (iv) additional signal demodulation is not required.

The compact fiberized frequency-shifted Sagnac interferometer shown in Figure 1a is described in detail in reference[9]. The laser source used was a 6 mW HeNe laser with center wavelength 633 nm. A pigtailed acousto-optic modulator with a working frequency of 40 MHz was used. All the fibers and the directional couplers were polarization-maintaining. The loop length difference was kept at 20 m which is optimal for 5 MHz ultrasonic signals (the corresponding time delay between reflections of the sampling beams from the object surface is equal to 100 ns). As discussed in reference[6], a GRIN lens is glued to the end of the probing fiber to focus the light beam onto the specimen surface and also to collect the reflected light back into the fiber probe. The fiber probe shown in Figure 1b, which can be hand-held, incorporates the mechanical part of the speckle hunting system.

RANDOM SPECKLE MODULATION TECHNIQUE FOR ULTRASOUND DETECTION ON ROUGH SURFACES

Since the light scattered from a rough surface is speckled, it is possible for the optical fiber probe (fiber core diameter about 5 μm) to be in a dark speckle region (Figure 2a) thereby not coupling any light back for detection. There are several approaches to solve this speckle problem including: (i) using a speckle insensitive Fabry-Perot interferometer with a multimode fiber light delivery system (fiber core diameter about 50 um) which improves light collection; (ii) using interferometers with a phase-conjugation mirror for speckle removal; or (iii) using a random speckle modulation technique for bright speckle hunting by

Nondestructive Characterization of Material VIII
Edited by Robert E. Green Jr., Plenum Press, New York, 1998

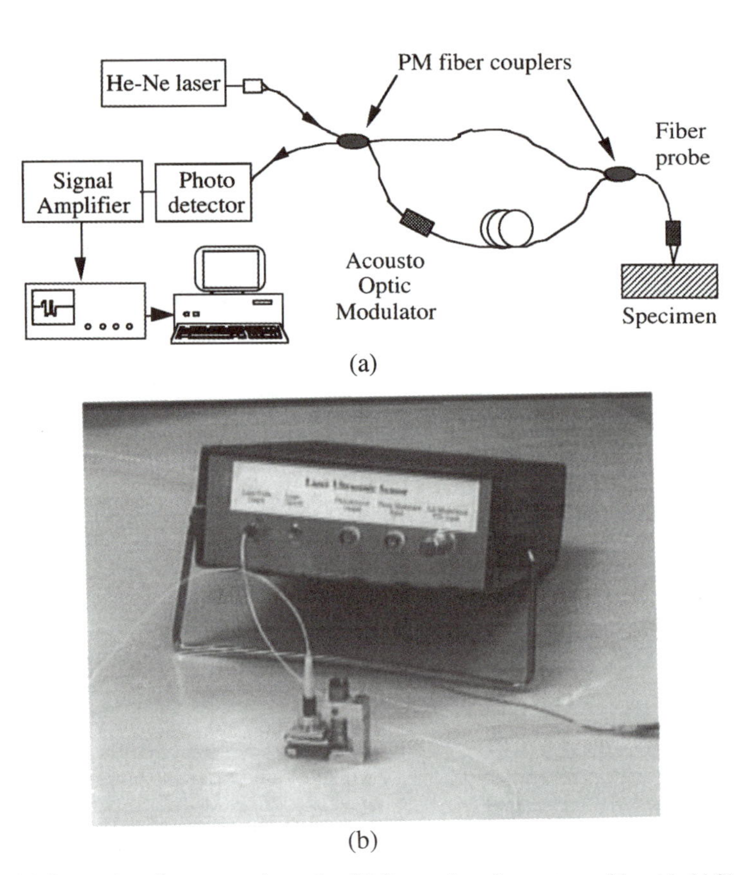

Figure 1. (a) Sagnac interferometer schematic , (b) Sagnac interferometer and hand-held fiber probe.

fast motion of a speckle field around a detector. However, the Fabry-Perot approach requires active stabilization of the cavity length, and systems based on a phase conjugation mirror require some expensive optical components. Therefore we have incorporated a variation of the random speckle modulation technique that has previously been used for Michelson-type interferometers[10]. This scheme is particularly suited for our common-path Sagnac interferometer since it is insensitive to the random phase variation between different bright speckles. In contrast, the performance of the technique on actively stabilized homodyne and heterodyne interferometers depends critically on the electronic demodulation.

The technique works as follows. The fiber probe that illuminates the surface and collects the backscattered light is continuously moved by an electromechanical device around any desired point of measurement along a straight scan line (see Fig 6b) . The amplitude of this motion (probe jitter) ranges between 0.3-0.5 mm at a frequency of 20 -100 Hz. We have found experimentally that at least one bright speckle can be found along a scan line if the surface roughness is within 1-30 μm range. Since the DC component of the photodetector signal is proportional to the intensity of the light backcoupled into the fiber, it is used to detect the times when the probe is moving across a bright speckle. A typical example of the DC signal as the probe head moves into and out of a bright speckle region is shown in Fig. 3a. When the amplitude of the DC component of the photodetector signal is above a threshold a bright speckle region is indicated, and an ultrasonic source (piezoelectric transducer in our case) is fired using the trigger circuitry shown schematically in Fig. 3b.

The typical time that the probe remains within a bright speckle (> 100 μs) is sufficient to send the ultrasonic wave and to detect it at the point of interest. The fact that the Sagnac interferometer is insensitive to the random phase fluctuations between different bright speckles allows for averaging of ultrasound signals obtained from different bright speckles

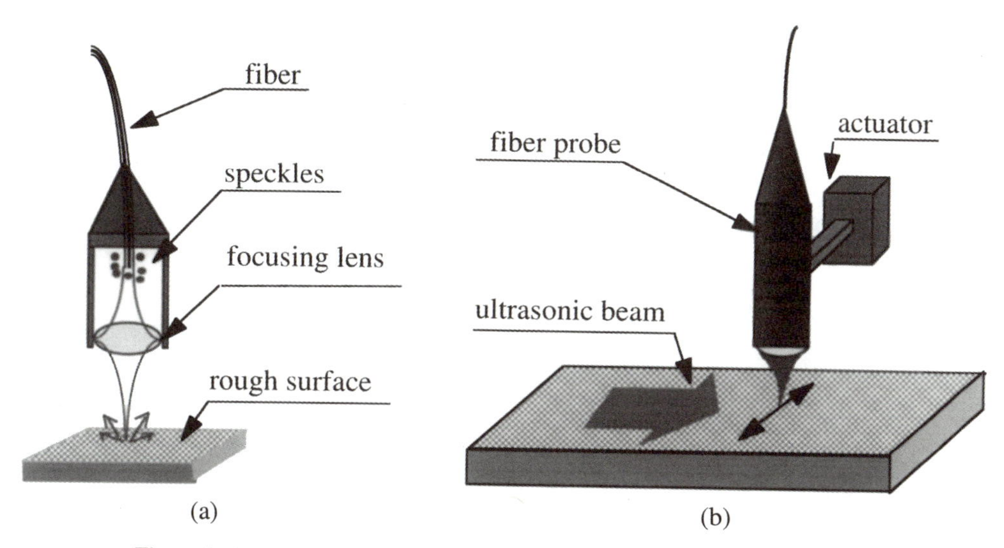

Figure 2. (a) Basic concept of diffuse reflection, (b) speckle hunting system design.

that appear during the probe jitter. Even if only one bright speckle is found along a scan line, a number of ultrasonic measurements can be repetitively acquired. The spatial resolution of the technique is reduced due to the probe jitter, but it is still higher than that of conventional piezoelectric transducers.

RESULTS AND DISCUSSION

Detection of Laser Generated Ultrasonic Waves

A Q-switched Nd:YAG pulsed laser was used for generating ultrasonic waves in a thick aluminum specimen (25x25x150 mm size) with a partially polished surface. The laser pulses were of 10 ns duration with up to 200 mJ energy per pulse at a 30 Hz repetition rate. The laser beam was delivered to a generation point through a high power optical fiber (200 μm core diameter) and then focused on to a specimen surface by a concave cylindrical lens to form a single line-focused illumination. In another experiment the laser beam was split by a binary diffraction grating into 10 beams of equal intensity for line-array illumination. Each beam was then coupled into an optical fiber (350 μm core diameter) and focused to a line on the specimen surface. The spacing between the lines of the array on the specimen, as well as the width of the lines can be adjusted by changing the distances between the specimen and the focusing lenses[4]. All results reported here were obtained for the nondestructive thermoelastic mode of generation.

The signal detected by the Sagnac interferometer for single line generation is shown in Figure 4a. The duration of the ultrasonic pulse is about 0.3 μs. This generation mode provides a high temporal resolution for ultrasonic testing. The ultrasonic signal generated in this mode is broadband, and the spectrum of the signal depends on the width of the line source [4]. Note that even though the Sagnac interferometer's response is peaked at 5 MHz, it is capable of detecting the broad band signal generated by a single laser line source. The signal detected by the Sagnac interferometer for the line-array generation mode is shown in Figure 4b. A narrowband acoustic wave is generated in this case. Therefore, the best signal-to-noise ratio of detection can be obtained in this mode if the center frequency of the generated signal is matched to the maximum of the frequency response of the Sagnac sensor. All signals were obtained in a digital oscilloscope over a 10 MHz detection bandwidth and with 10 averages.

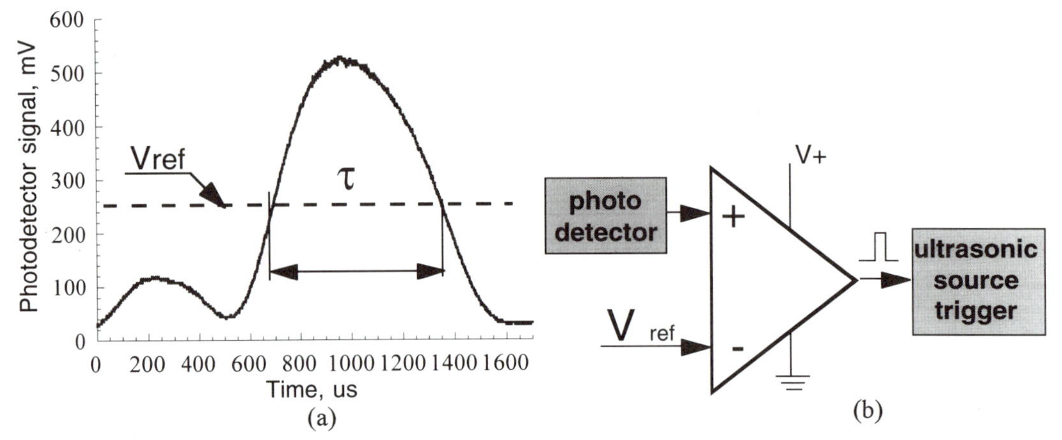

Figure 3. Random speckle modulation scheme: (a) DC signal indicating region of bright speckle; (b) schematic of trigger circuit

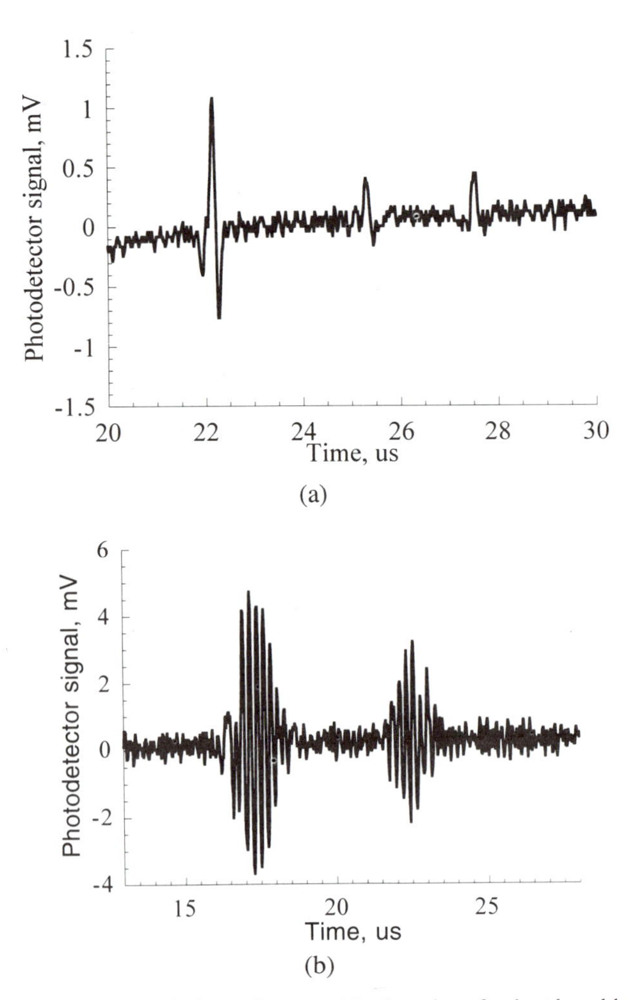

Figure 4. Detection of laser generated ultrasonic waves (a) - detection of a short broad-band ultrasonic wave, (b) detection of a narrow-band ultrasonic wave and it's echo.

2D Ultrasonic Imaging

The Sagnac interferometer has been applied to the detection of simulated cracks. The specimen was made of an aluminum alloy plate, 0.4 mm thick. An EDM notch of length 1.0 and width 0.1 mm was made to simulate a crack. A PZT transducer was used to generate 5 MHz tonebursts. The transducer was aligned with the center axis of the notch as shown in Fig. 5a. Because the thickness of the plate is comparable to the acoustic wavelength, Lamb waves were generated. A 2D acoustical scan over an area of 5x5 mm was made with the

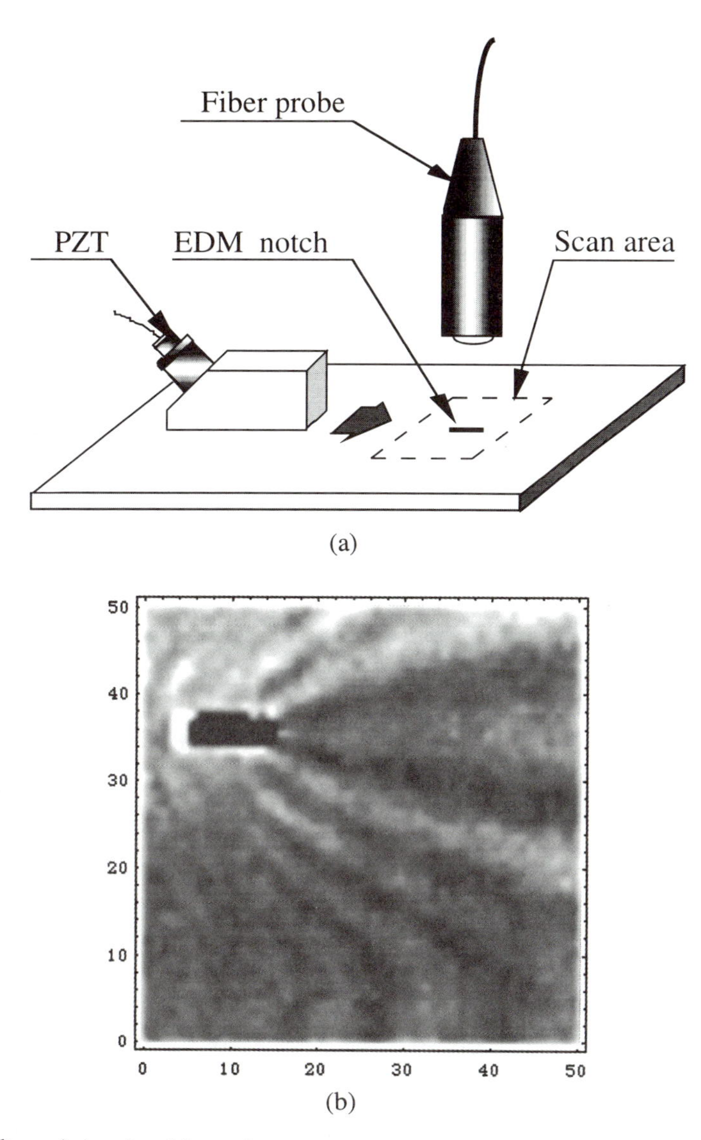

Figure 5. 2D ultrasonic imaging: (a) transducer - receiver arrangement, (b) ultrasonic image of EDM notch .

Sagnac interferometer. This scan is shown in Fig. 5b. The scattering of the ultrasound as it impinges on the narrow edge of the EDM notch is clearly seen. The shadow region indicates the presence of the EDM notch. This result illustrates the point detection capability (and the corresponding high spatial resolution) of the interferometric sensor.

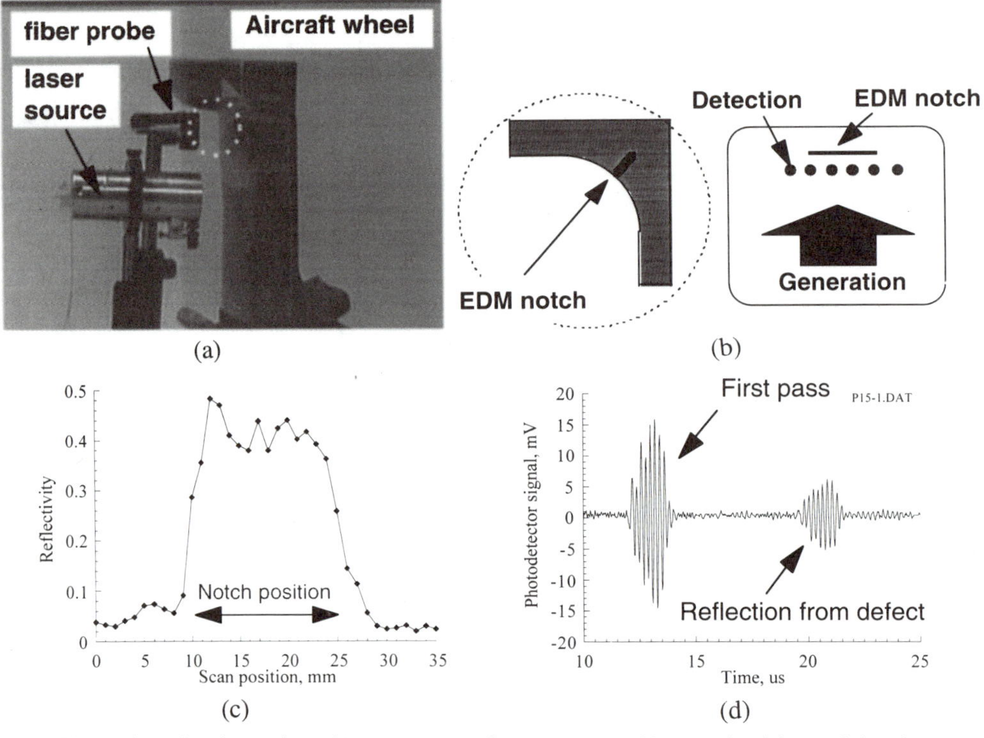

Figure 6. (a) Specimen view, (b) arrangement of measurements, (c) example of detected signal, (d) reflectivity plot.

Detection of Cracks in Aircraft Structures

The Sagnac interferometer was also used to detect an artificial defect on a rough curved surface. The specimen was part of aluminum aircraft wheel with an EDM notch on a curved surface (Figure 6a). The pitch -catch technique was applied to detect this defect. The amplitudes of the transmitted and reflected signals were measured for each point along a scan line parallel to the notch (see Figure 6b). An example of the detected waveform with a strong reflected signal is shown in Figure 6c. The length of the defect can be obtained by measuring the ratio between the amplitudes of the transmitted and the reflected signals (Figure 6d).

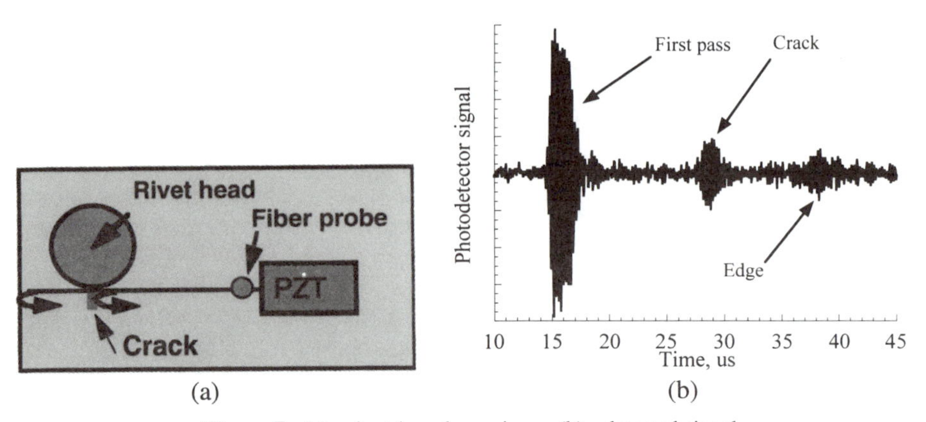

Figure 7. (a) - rivet / crack specimen, (b) - detected signal.

In a final experiment, the Sagnac interferometer was applied to the detection of fatigue cracks emanating from rivet holes. The specimen was made of two aluminum alloy plates, 1.8 mm thick, that were riveted together. A fatigue crack was made to emanate from the countersunk rivet hole in the top plate. The same PZT transducer described above was used to generate 5 MHz tonebursts. The orientations of the transducer and the Sagnac detector with respect to the rivet are shown in Figure 7a. The detected signal shows the transmitted signal and also the reflections of ultrasonic wave from the crack and the edge of the specimen (see Figure 7b).

ACKNOWLEDGEMENTS

This material is based upon work performed at Northwestern University for the FAA Center for Aviation Systems Reliability operated by Iowa State University and supported by the Federal Aviation Administration under Grant No. 93-G-018.

REFERENCES

1. C. B. Scruby and L. E. Drain, *Laser Ultrasonics* , Adam Hilger, New York (1990).
2. J. B. Spicer and J. W.Wagner, *NDE. of Materials,* edited by P. Holler, et al, Springer-Verlag, New York,, p. 691, (1989).
3. J. Huang and J. D. Achenbach, *J. Acoust. Soc. Am.*, 90(3), p. 1269, (1991).
4. J. Huang, S. Krishnaswamy and J. D. Achenbach, *J. Acoust. Soc. Am.*, vol. 92(5), p. 2527 (1992).
5. J. Huang, S. Krishnaswamy and J. D. Achenbach, in *Review of Progress in QNDE*, vol. 13A, eds. D.O.Thompson and D.E.Chimenti, Plenum Press, New York, p. 485, (1993).
6. J. Huang, S. Krishnaswamy and J. D. Achenbach, in *Review of Progress in QNDE*, vol. 12A, eds. D.O.Thompson and D.E.Chimenti, Plenum Press, New York, p. 503, (1992).
7. J. E. Bowers, *Applied Physics Letters*, vol. 41(3), 231 (1982).
8. P. Fomitchov, S. Krishnaswamy and J.D. Achenbach, in *Review of Progress in QNDE*, vol. 15A, eds. D.O.Thompson and D.E.Chimenti, Plenum Press, New York, p. 645, (1995).
9. P. Fomitchov, J.S. Steckenrider, S. Krishnaswamy and J.D. Achenbach, in *Review of Progress in QNDE*, vol. 16B, eds. D.O.Thompson and D.E.Chimenti, Plenum Press, New York, p. 645, (1996).
10. P. B. Nagy, G. Blaho and L. Adler, in *Review of Progress in QNDE*, vol. 12A, eds. D.O.Thompson and D.E.Chimenti, Plenum Press, New York, p. 527, (1993).

LASER-ULTRASOUND TISSUE CHARACTERISATION METHODS FOR POTENTIAL USE IN LASER ANGIOPLASTY PROCEDURES

K.A. Roome, R.F. Caller, P.A. Payne and R.J. Dewhurst

DIAS, UMIST,
P.O. Box 88
Manchester M60 1QD
UK

INTRODUCTION

Angioplasty has become a common procedure for the treatment of arteriosclerosis, and a popular form of angioplasty is balloon angioplasty. The procedure involves the insertion of a balloon into the artery at the location of the blockage. The balloon is then inflated so that the blockage is compressed. This method has disadvantages in that the occlusion is likely to reoccur, necessitating further treatment. It is also difficult to insert the balloon in the case of total or near total occlusions. More recently lasers have been used to ablate arterial blockages. However successful treatment is dependent on adequate guidance in order to ensure both precise alignment and correct identification of diseased areas.

This paper reports on the development of a combined probe for potential use in laser angioplasty procedures with a laser ultrasound guidance capability. Its design is capable of tissue characterisation and is also compatible with that required for a therapeutic laser beam to be delivered to the same site.

CONCEPT OF THE COMBINED LASER ULTRASOUND PROBE (CLUP)

The probe consisted of a 2.5mm diameter probe head and a combined fibre optic and electrical form (Figure 1). Laser energy was delivered to the irradiation site via a 600μm core diameter optical fibre running down the centre of the probe. Ultrasound was detected by a 28μm thick PVDF transducer located on the probe head, with signals forming the basis of materials characterisation. Probe design details have been published elsewhere [1].

TISSUE CHARACTERISATION WITH A FORWARDS LOOKING PROBE

In order to interpret ultrasonic signals obtained from tissue, it was necessary to understand probe behaviour. Time evolution of signals from materials in a transparent fluid medium have already been studied by several authors [2-3]. For studies in layered material,

Nondestructive Characterization of Material VIII
Edited by Robert E. Green Jr., Plenum Press, New York, 1998

the optical absorption coefficient, layer thickness and transducer response are crucial to the understanding of time-resolved measurements of ultrasonic signals. We have therefore developed a one-dimensional model, using optical neutral density filters as phantoms for real biotissue [4] since, for the purposes of experimental verification, such filters possess well-defined optical and thermoelastic properties which facilitated careful modelling.

With short pulse irradiation of an optical absorption layer within a transparent liquid medium, solutions were developed for both the forward and backward travelling waves, expressed in terms of time-resolved pressure changes. These were then combined with a model of piezoelectric transducer performance to provide predictions of time-resolved voltage signals. Predictions were compared with experimental waves generated using the CLUP. Both predicted and experimental waveforms are shown in Figure 2. The water medium, coupling the sample to the transducer, was assumed to be non-absorbing to ultrasonic signals and laser light.

Samples of fixed healthy and diseased post mortem human aorta were irradiated to investigate the use of the CLUP. The experimental arrangement is shown in Figure 3. Thermoelastic ultrasound was generated using a frequency doubled Nd:YAG laser of wavelength 532nm and pulse duration 10ns. From healthy tissue, the ultrasonic signal (Figure 4a) was typified by a bipolar pulse arising from near uniform optical absorption within the tissue. The resulting photoacoustic frequency spectrum is shown in Figure 4b .

Diseased tissue samples contained several forms of arteriosclerosis, with predominant forms consisting of varying thicknesses of fat. Experiments using the CLUP were carried out on a variety of diseased tissue samples. A signal from a sample with a surface fat layer approximately 0.5mm thick (Figure 4c) showed a narrowing of the positive peak of the bipolar signal from the CLUP. The corresponding frequency spectra (Figure 4d) showed anomalous features when compared with healthy tissue. Experiments were also conducted on a section of diseased tissue showing an ulcer-like feature. A typical signal obtained (Figure 4e) showed more structure than those from healthy or fatty tissue samples. Multiple peaks visible in this signal suggested that the sample consisted of a multi layered structure with layers having varying optical absorption. When the waveform was transformed to the frequency domain (Figure 4f) the spectra showed a complex structure that had some similarities to the structure seen from fatty tissue. There was an increase in the signal obtained between 10MHz and 15MHz with a sudden drop in signal at about 16MHz. These results suggest that the CLUP could be used to determine tissue state.

A CONCEPT FOR A SIDEWAYS VIEWING PROBE

We have also suggested that in the case of a finite laser spot size, a significant amount of ultrasonic energy is emitted in directions other than normal to the surface. An adaptation of the probe is being considered where this energy can be used for guidance purposes.

Figure 5 illustrates how a sideways probe would operate. The diagram on the left shows a forwards looking probe inserted inside an artery. Pulsed laser light is delivered to the tissue via an optical fibre located in the centre of the probe. If the probe is far enough away from the tissue surface so that it is irradiated with a wide laser beam, then returning ultrasonic wavefronts are assumed to be plane-like and can be viewed by a transducer on the head of the probe.

The diagram on the right shows a possible sideways-looking probe. The probe is placed closer to the tissue surface than in the previous case so that it is irradiated with a smaller laser spot size. As the incident laser spot size is reduced, significant ultrasound can be expected to propagate sideways. If this sideways-travelling ultrasound is reflected from the artery wall and viewed by an additional transducer on the side of the probe, it will hopefully provide information on the orientation of the probe in the artery.

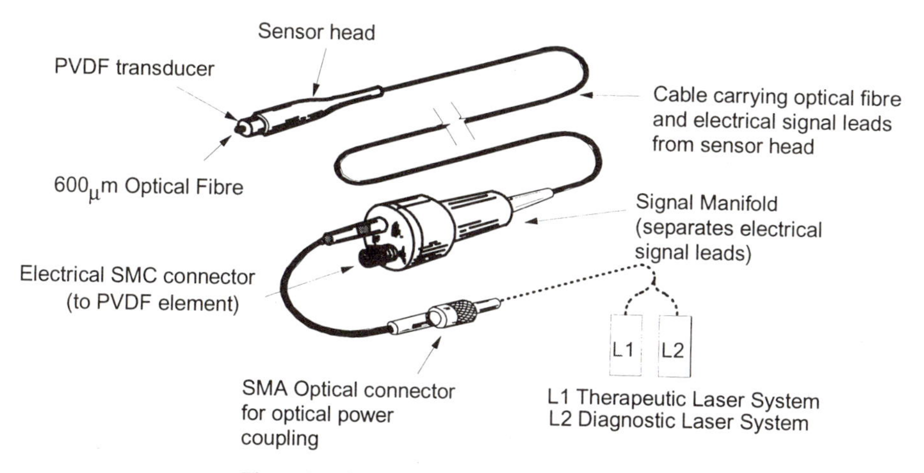

Figure 1. The combined laser ultrasound probe.

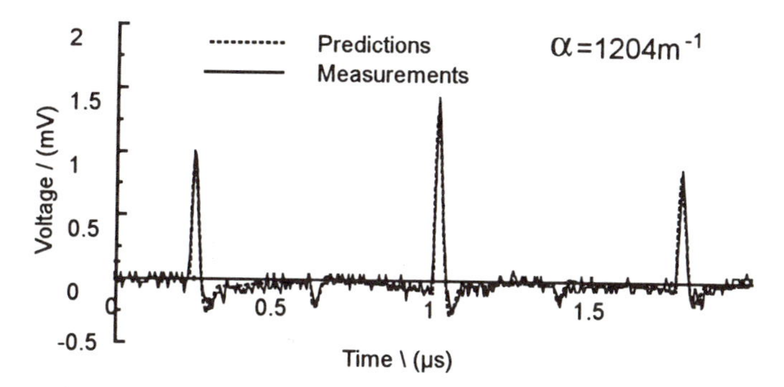

Figure 2. Output of the PVDF transducer in response to backward travelling pressure waveforms. Measured waveforms are indicated by the solid lines, whilst predicted waveforms based on the model are indicated by dashed lines.

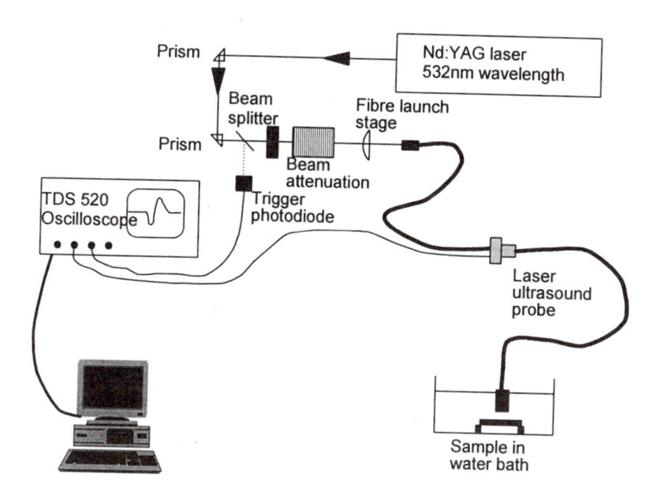

Figure 3. The experimental arrangement used for examining tissue samples using the CLUP.

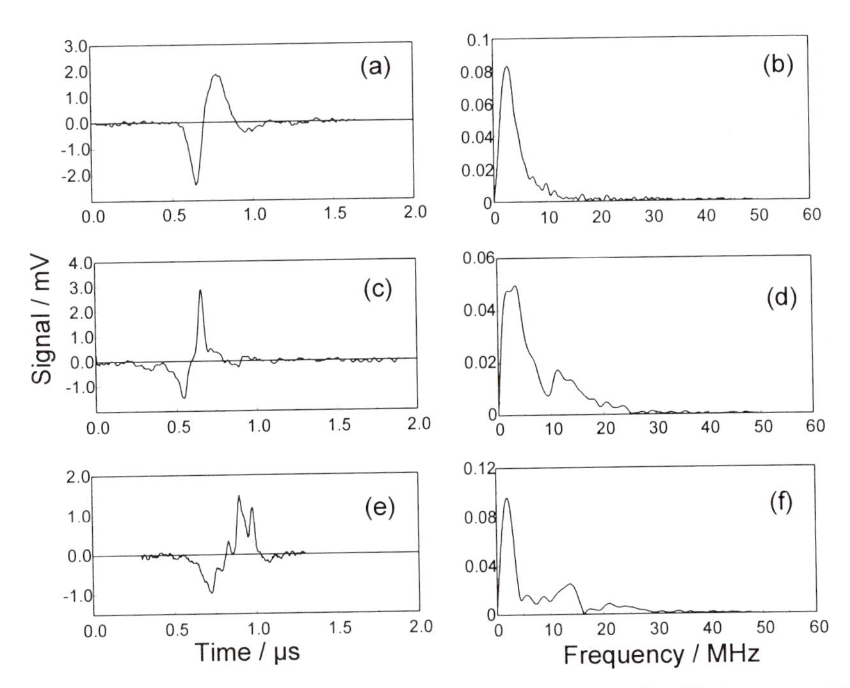

Figure 4. Ultrasonic signals from a range of tissue samples: (a) from a sample of healthy human arterial tissue, with corresponding FFT (b); (c) from a sample of diseased human arterial tisssue exhibiting fatty deposits, with a corresponding FFT (d); and (e) from an ulcer like section of diseased human arterial tissue, with its corresponding FFT (f).

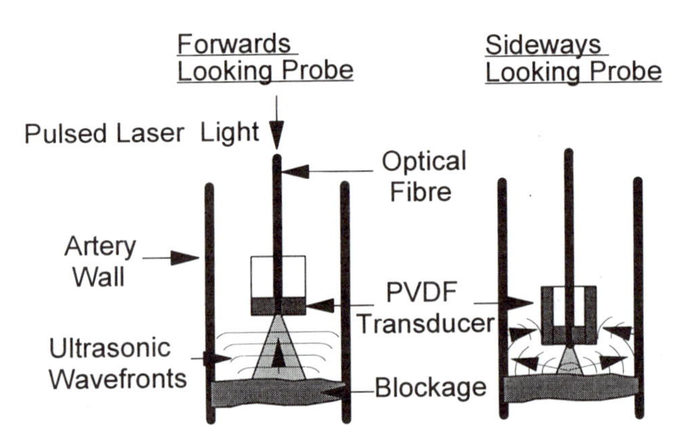

Figure 5. Illustrating the concept of the sideways viewing probe. The left-hand probe shows how ultrasonic wavefronts are generated using the forwards looking probe. The right-hand probe illustrates how sideways propagating waves would be used for probe guidance.

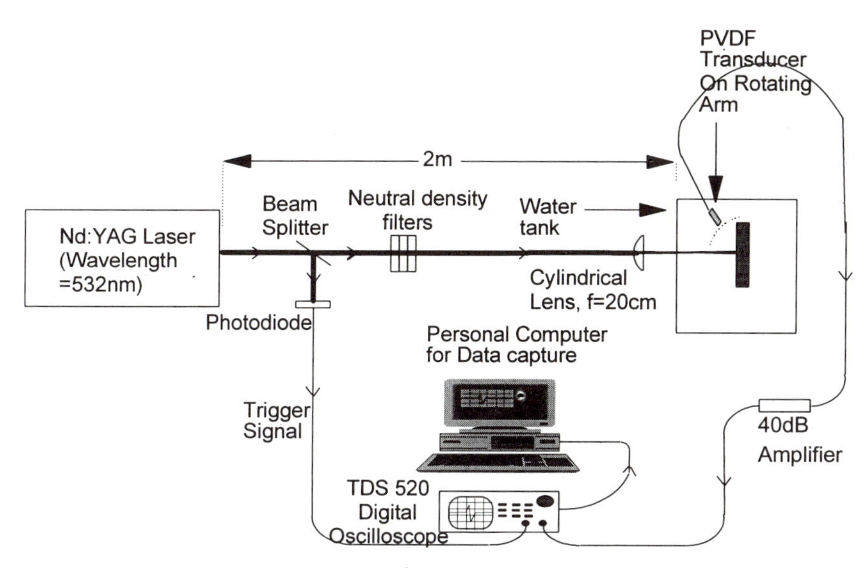

Figure 6. Schematic diagram of apparatus used for directivity pattern measurements.

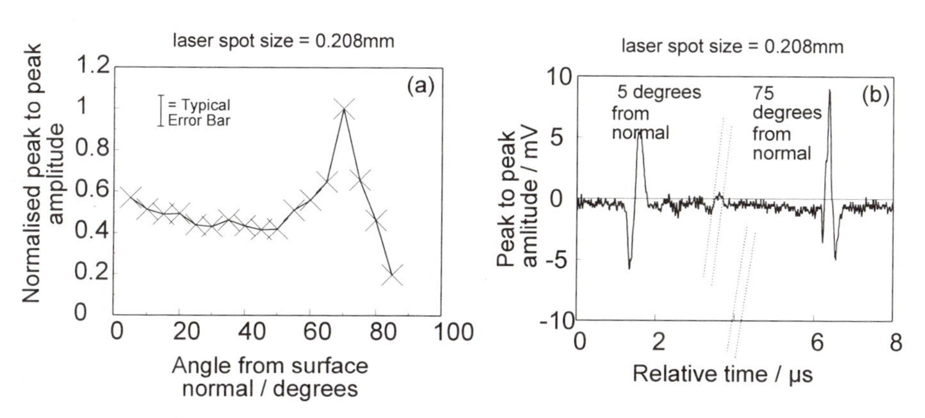

Figure 7. (a) A backward going directivity pattern generated from fixed healthy human aorta. (b) Corresponding laser-ultrasound signals generated from angles of 5° and 75° .

To test the concept it was necessary to investigate the magnitude of ultrasonic energy emitted at different angles, generating directivity patterns. A backwards-going pattern was produced for the case of a sample of human aorta, as described below.

MEASUREMENT OF DIRECTIVITY PATTERNS

Figure 6 shows a schematic diagram of the apparatus used for directivity pattern measurements. Ultrasound was generated by a frequency doubled Nd:YAG laser of wavelength 532nm and pulse duration 10ns. The sample was placed inside a tank full of water, on a specially designed table and a cylindrical lens of focal length 20cm was used to produce a thermoelastic photoacoustic line source on the sample surface.

A directivity pattern was measured from a piece of fixed healthy human aorta. This pattern is shown in Figure 7(a). A lobe is visible, centred around 70° from the sample normal. Example waveforms are shown in Figure 7(b). The wave recorded close to the sample normal is bipolar whereas the wave at 75° is tripolar. The fact that it is so well defined indicates that it may be useful for sideways imaging purposes.

CONCLUSIONS

A combined laser ultrasound probe incorporating a PVDF transducer of diameter 2.5mm has been constructed for materials characterisation. Its behaviour has been modelled using phantoms based on neutral density optical filters with well defined properties. From this basis, it has subsequently been used to examine diseased arterial tissue. Compared with healthy tissue, photoacoustic signals are developed in diseased tissue which show additional structure in both the time and frequency domain. Whilst this forward viewing probe may therefore examine materials directly in front of the probe head, it does not provide any information from sidewalls. However, recently, further experimental work has shown that from optical absorption in the forward direction, photoacoustic signals are emitted sideways with a peak amplitude at about 70° from the surface normal. It is suggested that this may serve as the basis of a sideways viewing imaging system.

ACKNOWLEDGEMENTS

The authors would like to thank both the Engineering and Physical Science Research Council and The National Heart Research Fund, grant number 004/93, for studentship support of this work.

REFERENCES

1. R.J. Dewhurst, Q.X. Chen, G.A. Davies, A.Kuhn, K.F. Pang, P.A. Payne and Q. Shan Acoustical Imaging 21, 523-532 (1995).
2. A.A. Oraevsky, S.L. Jacques and F.K. Tittel SPIE Vol 1882, 86-101 (1993).
3. A.A. Oraevsky, S.L. Jaques, R.O. Esenaliev and F.K. Tittel SPIE Vol 2323, 37-46 (1995).
4. Q. Shan, A. Kuhn, P.A. Payne and R.J. Dewhurst, Ultrasonics 34, 629-639 (1996).
5. P.C. Beard and T.N. Mills Physics In Medicine And Biology 42, 199-217 (1997).

INSPECTION OF ROCKET ENGINE COMPONENTS USING LASER-BASED ULTRASOUND

A.D.W. McKie and R.C. Addison, Jr.

Rockwell Science Center
1049 Camino Dos Rios
Thousand Oaks, CA 91360

INTRODUCTION

High-performance rocket engines, which require active cooling of the thrust surfaces, are being developed for the next generation of reusable launch vehicles. These engines require an effective and rapid method for inspecting the critical braze bond between the liner assembly and the stainless steel outer jacket of the thrust chamber, to determine its integrity and prevent catastrophic failure during hot fire testing. Currently, a series of conventional ultrasonic techniques are used to inspect the thrust cell chamber. However, the complexly contoured surfaces prevent 100% inspection. Laser-based ultrasound (LBU) techniques have previously been shown to be an effective method for rapid inspection of contoured polymer-matrix composite structures[1-3]. The transition to inspection of metallic components requires increased spatial resolution to resolve defects on the order of 1 mm diameter, in addition to modification for limited access inspection from inside the thrust cell chamber. This paper reports progress on inspection of a heat exchanger NDE standard which demonstrates the resolution capability of the LBU system. In addition, C-scan results for LBU inspection of a three-quarter section aluminum rocket nozzle proof part and a quarter-section throat bond demonstrator are presented.

EXPERIMENTAL APPROACH

The ultrasonic generation efficiency of the thermoelastic process has long been known to depend on the surface boundary conditions of the material being inspected[4,5]. Whilst polymer-matrix composite materials lend themselves to inspection using thermoelastic laser generation and detection, metallic structures do not. To suitably modify the surface boundary conditions to obtain forward propagating longitudinal waves, the surfaces of the metallic samples inspected in this study were coated with a white polyurethane paint. Although elimination of the need for painting these structures is desirable, the stringent inspection demands outweigh the inconvenience of the additional coating procedure. Thus painting the thrust cell liner is an acceptable step to achieve 100% inspection of a flight-ready system.

Nondestructive Characterization of Material VIII
Edited by Robert E. Green Jr., Plenum Press, New York, 1998

For many practical NDE applications, limited access to a component dictates that inspection be performed from one side, requiring the generation and detection laser beams to be incident on the same side of the part. Inspection of a rocket engine thrust cell also adds the requirement for limited access inspection. The concept for the inspection of the thrust cell liner is shown in Figure 1, where the generation and detection laser beams from the LBU system are scanned over the inner surface of the thrust cell by rotating a mirror that is positioned axially within the thrust chamber. The mirror is rotated 360° and translated linearly along the length of the thrust cell, thereby performing a 100% inspection.

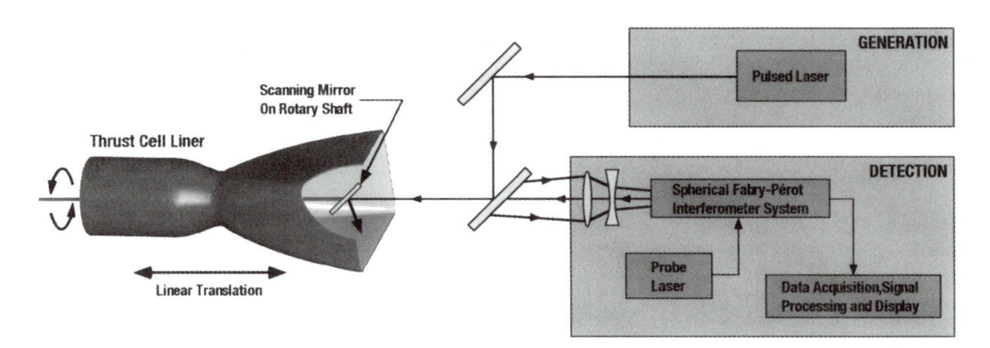

Figure 1. Concept for the laser-based ultrasound inspection of a rocket engine thrust cell liner.

The essential features of the LBU system have been described in detail previously[1,2]. Briefly, a CO_2 laser operating at 10.6 μm is used to generate longitudinal ultrasonic waves in the painted metallic components. A long-pulse Nd:YAG probe laser is aligned colinearly with the CO_2 generation laser and is phase modulated by the arrival of ultrasonic waves at the surface of the part. A 1 m long spherical Fabry-Pérot interferometer is used to demodulate the ultrasonic signal from the probe laser beam and provides an intensity modulation which is detected by a photodiode. For the initial proof-of-concept demonstration, rapid scanning of the stationary rocket engine thrust cell liner is achieved by angular deflection of the generation and probe laser by means of a computer controlled pair of galvanometer scanning mirrors. To obtain sufficiently high spatial resolution, the generation and probe laser beams were focussed to spot sizes of ~1 mm diameter.

EXPERIMENTAL RESULTS

Figure 2a shows the experimental configuration for a reflection mode ultrasonic C-scan inspection of a heat exchanger NDE standard using the LBU system. The heat exchanger NDE standard was fabricated to have cooling channel spacings and defect diameters representative of the inspection requirements of a complete thrust chamber. Six defects were located along the center line of the specimen, at the braze bond interface, and ranged in diameter from ~3.2 mm to ~0.8 mm. Figure 2b shows an amplitude mode C-scan image resulting from ultrasonic inspection of the heat exchanger NDE standard using the LBU system. The ultrasonic gate was set to include all ultrasonic signals occurring from the near surface to the backwall of the part so that both defects and channels were visualized in the C-scan image. All six defects were detected with good resolution. The seven closely spaced cooling channels are also clearly visible.

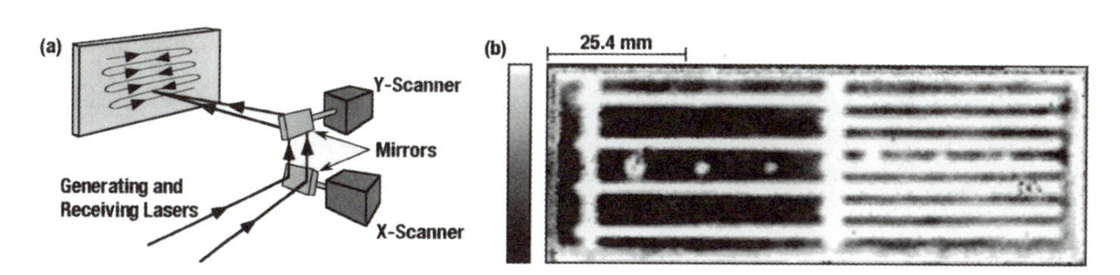

Figure 2. (a) Experimental configuration for LBU inspection of a heat exchanger NDE standard, and (b) resulting amplitude mode C-scan image.

Figure 3a shows the experimental configuration for a reflection mode ultrasonic inspection of a three-quarter section aluminum rocket nozzle proof part using the LBU system. This part simulates an actual thrust cell liner and allowed the capability of the LBU system, to inspect complexly contoured surfaces, to be evaluated. A quarter-section was cut away so that inspection from the inside was not required for the proof-of-concept demonstration. Figure 3b shows an amplitude mode ultrasonic C-scan image resulting from a full-area scan of the region accessible through the cut-away section. The area inspected transitioned from a flat region, around a corner having a variable radius of curvature (ranging from ~23 mm to ~37 mm) and back to a flat region, as shown in Figure 3a. The C-scan results (Figure 3b) demonstrate that the LBU system was able to accurately resolve the individual cooling channels. The change in contour of the part had no adverse effect on the inspectability when using the LBU technique.

Several regions of increased ultrasonic amplitude were observed in the C-scan image. The thrust cell uses a pass-and-a-half cooling design, and an abrupt increase in ultrasonic amplitude was detected as the cooling channels became more widely spaced. A portion of this region is highlighted in Figure 3d. Figure 4 shows ultrasonic waveforms acquired from opposite sides of the transition region (Figure 3d). When the channels are widely spaced, a single longitudinal pulse echo (Figure 4a) is detected from the back wall of the thrust cell liner, followed by multiple reverberations. However, when the channels are closely spaced, the initial longitudinal pulse (Figure 4b) has a double pulse characteristic. Thus, as the channels become closely spaced, spreading of the ultrasonic beam from the laser generation location results in reflections from the side-wall of the channel, thereby reducing the ultrasonic amplitude in the first detected pulse. A similar effect is evidenced near the top of the channels (Figure 3c), although the increase in ultrasonic amplitude is predominantly caused by a thicker layer of paint in the corner region. It is expected that a more uniform paint layer could be applied to the thrust cell which would eliminate this latter ultrasonic amplitude variation.

During the C-scan inspection, the generation and detection laser beams experienced incident angles ranging from 0 to ~54°. Although the galvanometric method of scanning is very different from what would be used for a real inspection (Figure 1), the angles of incidence are very similar, and so provided a suitable evaluation of the LBU system capability for inspecting the contoured geometry of a thrust cell liner.

Figure 5a shows a photograph of the quarter-section throat bond demonstrator, which has contour variations in several directions. This part was manufactured with four known simulated defects located at the braze bondline and again allowed an evaluation of the contour handling and resolution capabilities of the LBU system. Defects A and C had diameters of ~1.2 mm, with B and D slightly larger with diameters of ~2.0 mm. Figure 5b shows the resulting reflection mode ultrasonic C-scan image obtained when the LBU system

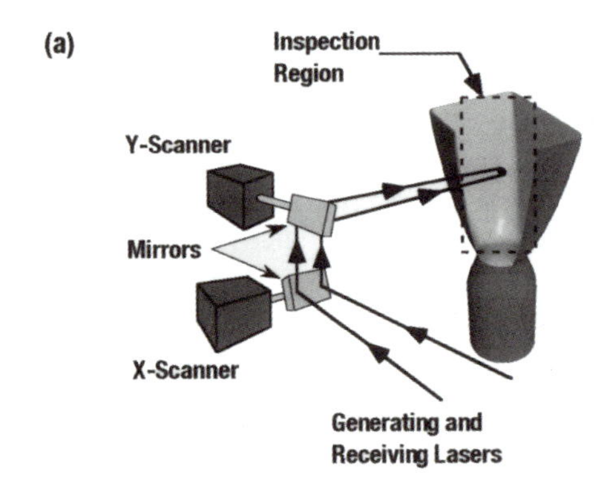

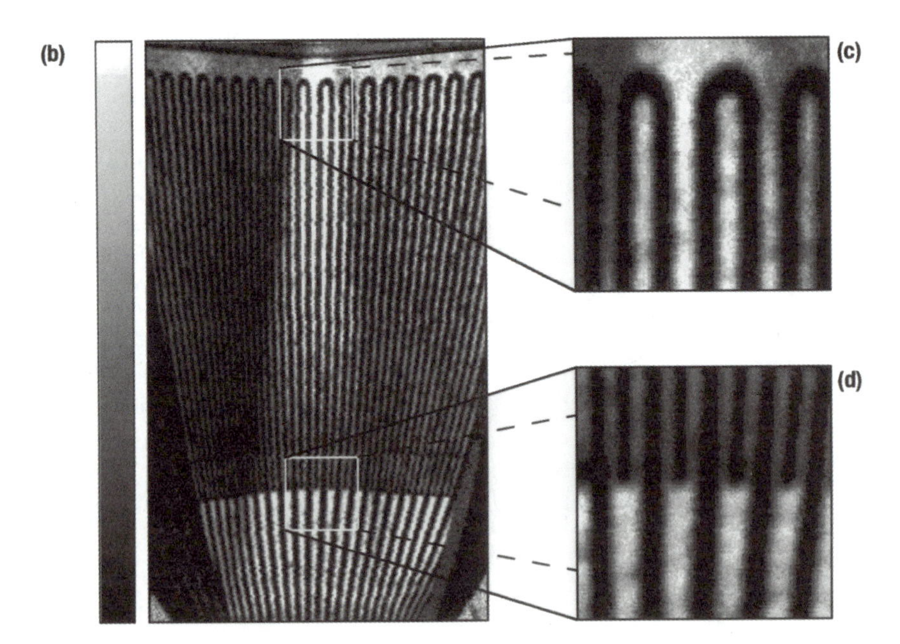

Figure 3. (a) Experimental configuration for LBU inspection of a three-quarter section aluminum rocket nozzle proof part, (b) resulting amplitude mode C-scan image of the region accessible through the cut-away section, (c) zoomed view of the top of the channels, and (d) zoomed view of the pass-and-a-half transition region.

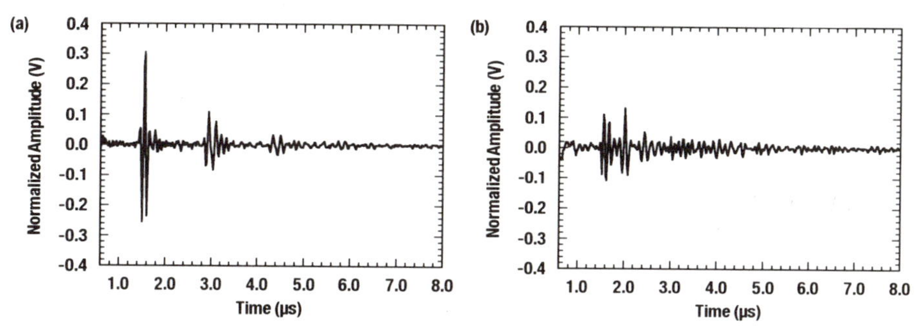

Figure 4. Ultrasonic waveforms acquired from opposite sides of the pass-and-a-half cooling channel transition region. (a) Single pass region , and (b) double pass region.

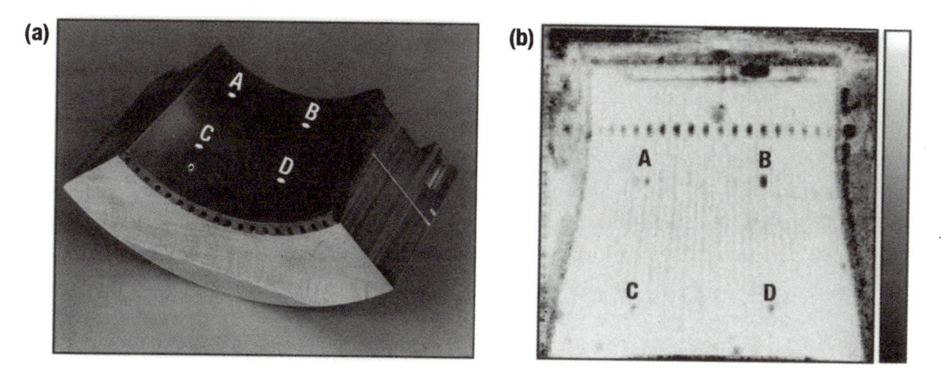

Figure 5. (a) Photograph of the quarter-section throat bond demonstrator and (b) the corresponding LBU C-scan inspection which clearly shows detection of the reference defects with good resolution.

was used to inspect the quarter-section throat bond demonstrator. The four defects are clearly visible with good resolution. A manifold, located one third of the way down the quarter-section throat bond demonstrator was also clearly imaged.

CONCLUSIONS

Reflection mode ultrasonic C-scan inspections were performed on a heat exchanger NDE standard to determine the laser-based ultrasound (LBU) system resolution. These inspections demonstrated that the LBU system was able to clearly resolve all of the defects which had diameters ranging from ~3.2 mm to ~0.8 mm.

LBU inspection of a three-quarter section aluminum rocket nozzle proof part and a quarter-section throat bond demonstrator were successfully completed, with the cooling channels and reference defects clearly resolved. These inspections mark the first time these structures have been inspected ultrasonically, since the tight radii and complexly varying contours of these parts preclude the use of conventional ultrasonic techniques. These preliminary experiments have demonstrated the applicability of the laser-based ultrasound system as an inspection methodology for complicated geometry metallic rocket engine components. The current method of scanning the generation and detection laser beams covered a range of incident angles similar to those expected for a complete flight-ready hardware inspection from the inside of the thrust chamber. To fully demonstrate the technology, a rotary scanning system will be implemented that can traverse the axis of a "work horse" thrust chamber, thereby realizing the limited access inspection capability.

ACKNOWLEDGMENTS

This work has been supported by Boeing Independent Research and Development funds.

REFERENCES

1. A.D.W. McKie and R.C. Addison, Jr., Inspection of components having complex geometries using laser-based ultrasound, *in: Review of Progress in Quantitative Nondestructive Evaluation,* 11:577, D.O. Thompson and D.E. Chimenti, eds., Plenum Press, New York (1992).

2. A.D.W. McKie and R.C. Addison, Jr., Practical considerations for the rapid inspection of composite materials using laser-based ultrasound, *Ultrasonics* 32:333 (1994).

3. M. Choquet, R. Héon, C. Padioleau, P. Bouchard, C. Néron and J.-P. Monchalin, Laser-ultrasonic inspection of the composite structure of an aircraft in a maintenance hangar, *in: Review of Progress in Quantitative Nondestructive Evaluation,* 14:545, D.O. Thompson and D.E. Chimenti, eds., Plenum Press, New York (1995).

4. C.B. Scruby, R.J. Dewhurst, D.A. Hutchins and S.B. Palmer, Laser Generation of Ultrasound in Metals *in: Research Techniques in Nondestructive Testing,* 5:281, R.S. Sharpe, ed., Academic Press, New York (1982).

5. C.B. Scruby and L.E. Drain. *Laser Ultrasonics – Techniques and Applications*, Adam Hilger, New York (1990).

MODELING HETEROGENEITIES AND ELASTIC ANISOTROPY IN SINGLE CRYSTAL ZINC AND CARBON FIBER EPOXY COMPOSITES

David H. Hurley, James B. Spicer, James W. Wagner, and Todd W. Murray

Johns Hopkins University
Dept of Materials Science and Engineering
102 Maryland Hall
Baltimore, MD 21218

INTRODUCTION

Laser generated ultrasound has been used to determine material properties and to characterize material defects [1-3]. To a large extent, the success of laser ultrasonics has been the researcher's ability to correctly predict the temporal evolution of the displacement waveform resulting from pulsed laser irradiation. Theories that assume isotropic elastic properties work well for crystalline materials that have randomly oriented grains with grain sizes that are small compared to the wavelength of the interrogating ultrasonic wave [4-5]. For single crystal samples or carbon epoxy composites, the elastic anisotropic nature must be taken into account. A number of researchers have shown that the behavior of single crystal materials in the presence of an ultrasonic disturbance differ markedly from their isotropic counterparts [6-13]. Mourad *et al.* [6] used the Cagniard-de Hoop method [14] to numerically obtain the solutions to Lamb's [15] problem in an anisotropic half-space. In their paper, Mourad *et al.* assumed that the laser source could be modeled as a shear stress dipole applied at the bounding surface. In addition, Weaver *et al.* [7] have studied the elastodynamic response of a thick transversely isotropic plate to a normal point source applied at the bounding surface. Of particular interest is the work by Payton [13], who has treated a general class of problems for crystals that exhibit transverse isotropy.

In this paper a set of boundary conditions which are equivalent to a thermoelastic point-source in a strongly absorbing material are developed. Analytical expressions for the surface wave and for waves traveling along the symmetry axis are given for materials that exhibit transverse isotropy. These analytical expressions are compared with experimental waveforms generated in a sample of single crystal zinc and in a sample of unidirectional carbon fiber epoxy composite. In addition, inhomogeniety due to variations in optical properties is considered by modeling a sub-surface source. The epicentral displacement resulting from a buried source is obtained numerically.

Nondestructive Characterization of Material VIII
Edited by Robert E. Green Jr., Plenum Press, New York, 1998

THEORY:

The first work to give a quantitative scientific basis to pulsed laser ultrasonics was that of Scruby *et al.* [1]. In this work, the thermoelastic source for strongly absorbing materials, such as metals, was reported to be equivalent to a shear stress dipole applied at the bounding surface. Later, Rose [4] gave a systematic derivation for a point-source representation for laser generated ultrasound. In this presentation, Rose showed that by neglecting the effects of heat conduction, the laser source can be approximated by a surface center of expansion (SCOE). In addition, Rose demonstrated that the SCOE is equivalent to the shear stress dipole source proposed by Scruby *et al.* [1].

In this paper, a method of images is used to aid in the development of an equivalent set of boundary conditions for an anisotropic half-space, Nowacki [16]. To understand the method of images consider an infinite, transversely isotropic elastic medium. The axis of symmetry is parallel to the x_3 axis. The medium is subjected to a line center of expansion and a line center of contraction as shown in Fig 1. The plane of separation, formed by the x_2 and x_3 coordinate axes, is midway between the two sources and is perpendicular to the line joining the two sources. This type of arrangement will result in a nonzero shear stress state, and a zero normal stress state at the plane of separation. If at the separation plane, a shear stress of opposite sign to the shear stress resulting from the source sink combination is applied, a stress free state will be obtained at the separation plane. The separation plane can now be identified with the bounding surface of an elastic half space. Upon bringing the source and sink to the boundary, the equivalent stress boundary conditions become

$$(\sigma_{23})\big|_{x_3=0} = \tilde{F}\,\delta'(x_2)H(\tau),\quad (\sigma_{33})\big|_{x_3=0} = 0 \tag{1}$$

$$\tilde{F} = (F_3 + F_2\alpha - F_3\kappa).$$

where $\delta'(x_3)$ is the spatial derivative of the Dirac delta function and $H(\tau)$ represent a unit step function. F_2, F_3, α, and κ are functions of the elastic stiffnesses and the thermal constants and are represented as follows:

$$\alpha = \frac{c_{33}}{c_{44}}, \beta = \frac{c_{11}}{c_{44}}, \tau = t\left(\frac{c_{44}}{\rho}\right)^{1/2}, \kappa = (1+\alpha\beta-\gamma)^{1/2},$$
$$\gamma = 1+\alpha\beta-\left(\frac{c_{13}}{c_{44}}+1\right)^2, F_2 = \frac{B_{22}T_o}{C_{44}}, F_3 = \frac{B_{33}T_o}{C_{44}} \tag{2}$$

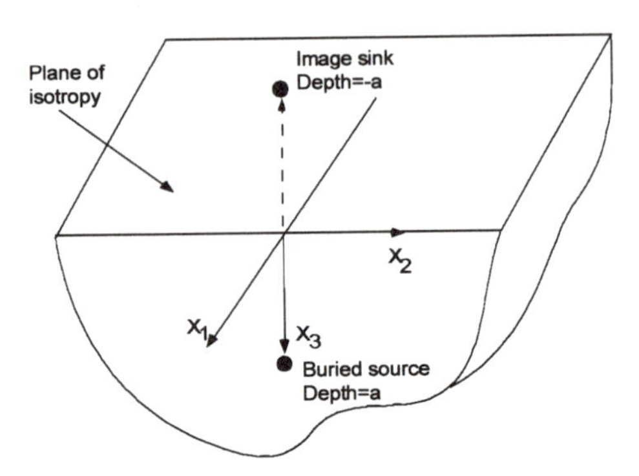

Fig. 1. Problem geometry with source and sink locations.

where B_{ij} are the components of the thermal expansion tensor, T_o is the ambient temperature and C_{ij} are the components of the stiffness tensor.

HALF SPACE SUBJECTED TO SHEAR STRESS DIPOLE

A line source representation for laser generated ultrasound is presented using the equivalent boundary conditions listed in Eq. 1. The solution technique proceeds by applying Fourier and Laplace transforms to eliminate dependence on the spatial and temporal variables, respectively, and then inverting the transformed equations using the Cagniard-Pekeris [14-17] inversion technique. Borrowing notation used by Payton [13], the behavior along the symmetry axis of crystals that exhibit elastic transverse isotropy can be divided into three categories according to the nature of their anisotropy:

$$\begin{aligned}
&\text{(i)} \quad (\alpha+\beta) < \gamma < (1+\alpha\beta), \\
&\text{(ii)} \quad (\beta+1) < \gamma < (\alpha+\beta) \ \text{and} \ (\gamma^2 - 4\alpha\beta) < 0, \\
&\text{(iii)} \quad \gamma < (\beta+1) \ \text{and} \ (\gamma^2 - 4\alpha\beta) < 0 \ \text{also} \ \beta > \alpha
\end{aligned} \tag{3}$$

For crystals belonging to class (i), the roots of the slowness equation are purely imaginary, while crystals belonging to classes (ii) and (iii) have complex roots and the wave-front curves for these crystals have cuspidal triangles. For class (iii) crystals, the triangluar portion of the wave-front is centered on the symmetry axis.

SOLUTIONS ALONG THE BOUNDING SURFACE

For solutions along the bounding surface, the Fourier inversion path is along the real ω axis, and the Cagniard path is along the imaginary ω axis. Presently, only the inversion of $\bar{u}_3$ is being considered. Since $\bar{u}_3$ is an even function of $\sqrt{\phi(\omega)}$, the solution technique is the same regardless of the crystal class. The analytical expression for the solution can be written as:

$$u_3(x_2,0,\tau) = \tilde{F}\left\{\frac{\bar{\psi}_r}{|x_2|}\delta(T_2-\omega_r) + \frac{1}{|x_2|\pi}g(T_2)\left[H(T_2-1/\sqrt{\beta}) - H(T_2-1)\right]\right\}, \tag{4}$$

where

$$g(T_2) = \frac{\tilde{F}T_2^{\,2}\alpha\zeta_1\zeta_3\left[(1-\kappa)(1-2T_2^{\,2}) - (\alpha-\gamma T_2^{\,2})\right]}{\left[-2(1-\kappa)T_2^{\,2} + \gamma T_2^{\,2} - \alpha\right]^2(1-T_2^{\,2}) + \alpha(\beta T_2^{\,2}-1)}, \quad T_2 = \frac{\tau}{|x_2|}. \tag{5}$$

SOLUTIONS ALONG EPICENTRAL AXIS

Another class of solutions that can easily be inverted using the Cagniard-Pekeris technique is that for displacements along the epicentral axis. The Cagniard path depends on the category of crystal being investigated. For materials belonging to class (i), the Cagniard path is along the real ω axis. For class (ii) and (iii) materials, the Cagniard path is off the real ω axis. The analytical expression for the epicentral displacement for class one crystals is given by,

$$u_3(0,x_3,t)=\frac{\tilde{F}}{\pi}\mathrm{Re}\left\{\left(\bar{A}_3\frac{\partial\varpi_1}{\partial\tau}\right)H(T_3-1)+\left(\bar{A}_3\frac{\partial\varpi_1}{\partial\tau}\right)H(T_3-1/\sqrt{\alpha})\right\},\quad T_3=\frac{\tau}{x_3},\tag{6}$$

where $\bar{A}_{3/4}$ are obtained by satisfying the boundary. For class (ii) and class (iii) materials, the Cagniard path no longer lies on the real ϖ axis. Applying the Cagniard technique to class (ii) and class (iii) materials gives

$$u(0,x_3,\tau)=\tilde{F}J(T_3)=\frac{\tilde{F}}{\pi}\left\{\begin{array}{l}\bar{A}_4\left(\frac{\partial\varpi_{1\to\bar{2}}}{\partial\tau}\right)\{H(T_3-T_e)-H(T_3-1)\}+\\ \bar{A}_3\left(\frac{\partial\varpi_{\bar{2}\to\bar{3}}}{\partial\tau}\right)\{H(T_3-1/\sqrt{\alpha})-H(T_3-T_e)\}+\\ \bar{A}_4\left(\frac{\partial\varpi_{\bar{3}\to\bar{4}}}{\partial\tau}\right)\{H(T_3-T_+)\}\end{array}\right\}\tag{7}$$

SOLUTIONS ALONG EPICENTRAL AXIS DUE TO A BURIED LINE SOURCE

Inhomogeniety due to variations in optical properties is considered by modeling a sub-surface source. Due to length requirements, only a brief summery of the solution technique for a buried line source in a transversely isotropic half-space is given. Since the source is buried, the expression for the Cagniard path must be obtained numerically. If Beryl is used as an example, the numerical solution can be represented by the following expression:

$$u(0,x_3,\tau)=DL+DS+RL+RS+ML+MS,\tag{8}$$

where the expressions DL, DS, RL, RS, ML, MS represent the direct longitudinal, the direct shear, the reflected longitudinal, the reflected shear, the mode converted longitudinal and the mode converted shear wave respectively. It should be noted that for a buried line source in a transverly isotropic material, there are six wave arrivals while, for isotropic materials, there are only three wave arrivals. The presence of six waves is due to the thermoelastic production of shear waves. It is commonly known that a thermoelastic source in an unbounded isotropic material does not produce shear waves. The production of shear waves in an anisotropic material is a result elastic and thermal anisotropy.

EXPERIMENT

The experimental setup used to generate surface waves was shown in Fig. 2. The generation of the ultrasonic disturbance was accomplished by irradiating the sample with a pulsed Nd:YAG laser operating at 1.064 μm, with a Gaussian transverse spatial profile. The pulse duration was 10 ns with a typical pulse energy 20 mJ . The ultrasonic disturbance was detected on the same side as the generation beam with a Michelson-type interferometer operating at 632.8 nm. The detection bandwidth of the interferometer was estimated to be in excess of 50 MHz.. The experimental setup for detection of epicentral waves was similar to the one shown in Fig. 2, except that the generation and detection points were on opposite sides of the sample.

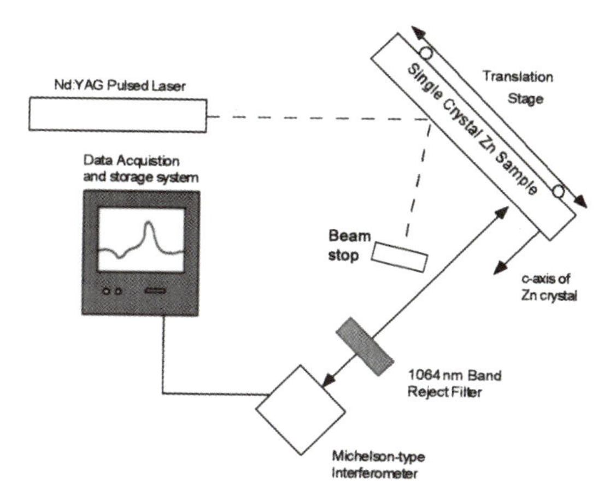

Fig. 2. Experimental setup. A Nd:YAG laser is used to generated the ultrasound and a Michelson interferometer is used to detect the ultrasound.

RESULTS AND DISCUSSION

In Figs. 3 and 4, a comparison between theorry and experiment for same side detection is presented for zinc and graphte epoxy respectively. In Figs 3 and 4, the theoretical results are convolved with a Gaussian function (FWHM 400 ns) in order to mimic the finite generation pulse duration and the transit time of the ultrasonic disturbance across the detection spot. In Fig. 3, the first disturbance turns on and off at times corresponding to the arrival of the longitudinal wave and shear wave respectively. The largest disturbance corresponds to the Rayleigh pole and is in the form of a traveling delta function. In Fig. 4, the experimental and theoretical results for a sample of unidirectional graphite epoxy are presented. The fiber direction is perpendicular to the sample surface and hence the bounding surface is a plane of isotropy. The overall character is similar to that of the zinc sample but the pulse appears to be temporally broadened. The pulse broadening or frequency dependent attenuation in Fig. 4 is most likely due to scattering effects or viscoelastic effects.

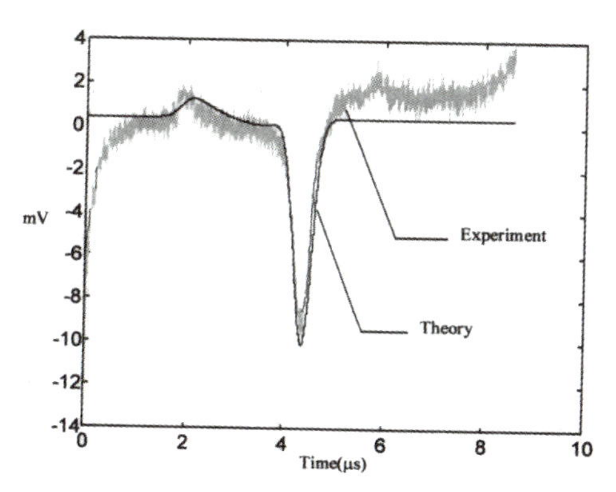

Fig. 3. Comparison between experiment and theory for surface waves generated with a line source in zinc.

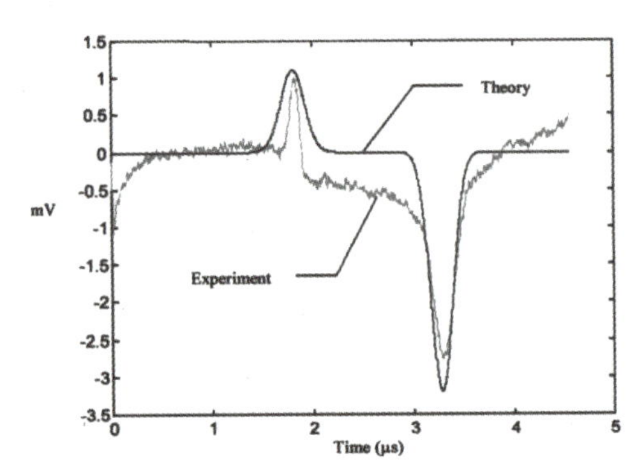

Fig. 4. Comparison between experiment and theory for surface waves generated with a line source in graphite epoxy.

In Fig. 5, an epicentral wave-form is shown for a sample of unidirectional graphite epoxy. The fiber direction is parallel to the sample's surface and also parallel to the line source. In order to account for the finite generation pulse duration, the theoretical result was convolved with a Gaussian pulse with a FWHM of 20 ns. The sample of graphite epxoy for this type of configuration should appear elastically isotropic. The close agreement between theory and experiment, reaffirms the claim of transverse isotropy.

A comparison between the theoretical and experimental epicentral displacements resulting from a laser line source in zinc is shown in Fig. 6. The crystallographic c axis was perpendicular to the bounding plane. Again, the theoretical result was convolved with a Gaussian pulse with a FWHM of 20 ns. Zinc is a class three crystal and as a result, the displacement character differs markedly form its isotropic counterpart. In a fashion similar to that of the surface wave case, the first disturbance turns on and off at times corresponding to the arrival of the longitudinal wave, t_l, and shear wave, t_s, respectively. After t_s, the Greens tensor is identically zero until the arrival of the majority of the acoustic energy at t_+. The disturbance that arrives a t_+ results from a reciprocal square root singularity displauing behavior not found in isotropic materials or class (i) materials.

Figure 7 compares theoretical results for a surface line source and a buried line source in the mineral Beryl. For this case, the plane of transverse isotropy is the bounding plane and the detection location is located along the symmetry axis. The ratio of the sample thickness to the source depth, X_a, is 5000. For Beryl, the cuspidal.

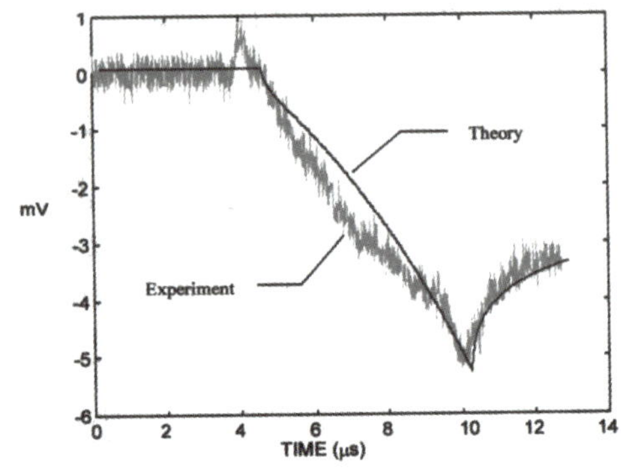

Fig. 5. Epicentral waveform in graphite epoxy. The fibers are parallel to the surface and the line source. This type of arrangement leads to isotropic behavior.

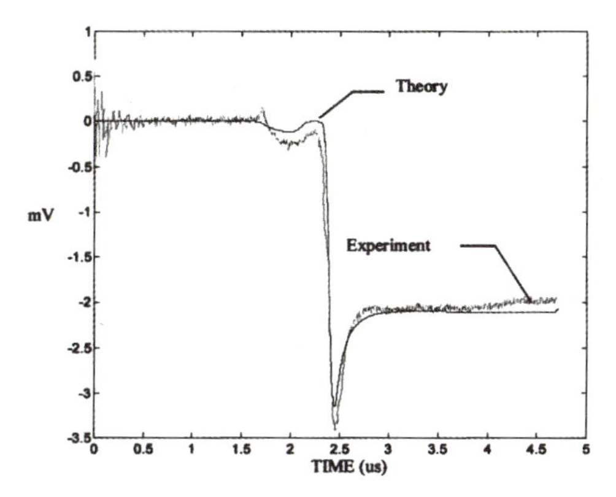

Fig. 6. Experiment and theory for epicentral waves generated with a line source in zinc.

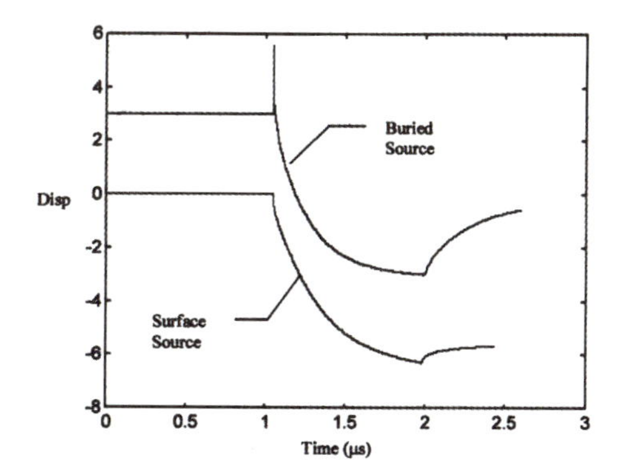

Fig. 7. Comparison between theoretical waveforms generated with a surface source and a buried source.

triangular portion of the wave-front does not intersect the symmetry axis. As a result, the wave form exhibits characteristics similar to its isotropic counterpart. The waveform for the buried source shows a precursor spike that is a result of bulk waves reflecting off the stress free boundary. Since X_a is large, it is impossible to discern 6 distinct wave arrivals

CONCLUSION

Analytical results for the displacement resulting from a laser line source was obtained for materials that exhibit transverse isotropy. The theoretical expressions for the surface and epicentral cases were compared to the experimental results for a sample of single crystal zinc. For both cases, the theory and the experiment showed close agreement. In addition, experimental results for unidirectional graphite epoxy were obtained. The carbon epoxy sample exhibited homogeneous behavior when the wave vector is perpendicular to the fiber direction.

REFERENCES

1. Scruby, C. B., Dewhurst, R. J., Hutchins, D. A., Palmer, S.B., Quantitative Studies of Thermally Generated Elastic Waves in Laser-Irradiated Metals, *Journal of Applied Physics* (1980) **51** 6210-6216.
2. Scruby, C. B., Smith, R. L., Moss, B. C., NDT Int (1986) **19**, 307.
3. Telschow, K., *Review of Progress in Quantitative Nondestructive Evaluation*, (Plenum, New York, 1988) Vol. 7b, p.1211.
4. Rose, R. L. F., Point Source Representation for Laser Generated Ultrasound, *Journal of the Acoustical Society of Americal* (1984) **73** 723.
5. Telschow, K., Conant, R., Optical and Thermal Parameter Effects on Laser Generated Ultrasound, *Journal of the Acoustical Society of America* (1990) **88**, 1494-1502.
6. Mourad, A., Deschamps, M., Castangnède, B., Acoustic Waves Generated by a Transient Line
   Source in an Anisotropic Half-Space, *Acustica* (1996) **82** 839-851.
7. Weaver, R. L., Sachse, W., Kwang, Y.K., Transient Elastic Waves in a Transversely Isotropic Plate, *J Applied Mech* (1996) **63** 337-346.
8. Royer, D., Dieulesaint, E., Rayleigh Wave Velocity and Displacement in Orthorhombic, Tetragonal, Hexagonal, and Cubic Crystals, *Journal of the Acoustical Society of America* (1984) **76** 1438-1444.
9. Kraut, E. A. *Rev. Geophys* (1963) **1** 401.
10. Burridge, R., Lamb's Problem for an Anisotropic Half-Space, *Quarterly Journal of Mechanics and Applied Mathematics*, **24**, 81-98, (1970).
11. Willis, J. R., and Bedding, R. J., Arrivals Associated with a Class of Self-Similar Problems in Elastodynamics, *Mathematical Proceedings of the Cambridge Philosophical Society*, **77**, 591-607, (1975).
12. Musgrave, M. J. P. *Crystal Acoustics*, Holden-Day, Inc., San Francisco (1970).
13. R.G. Payton, *Elastic Wave Propagation in Transversely Isotropic Media*, Martinus Nijhoff Publisher, The Hague (1983).
14. Cagniard, L., *Reflection and Refraction of Prog. Seismic Waves* New York, McGraw-Hill (1962).
15. Lamb, H. *Phil. Trans. Roy. Soc.* (1904) A203 **1.**
16. Nowacki, W. *Dynamic Problems of Thermoelasticity*, Polish Scientific Publishers, Warszawa (1975).
17. Pekeris, C.L. *Proc. Natn.Acad Sci.* (1955) **41** 629

FEASIBILITY STUDY OF A NONLINEAR ULTRASONIC TECHNIQUE TO EVALUATE ADHESIVE BONDS

Tobias P. Berndt and Robert E. Green, Jr.

Center for Nondestructive Evaluation
The Johns Hopkins University
Baltimore, Maryland 21218

INTRODUCTION

To date, the nondestructive determination of the integrity and true strength of an adhesively bonded structure still represents a significant challenge for the NDE community. While much progress has been made in recent years, theoretically as well as experimentally,[1-4] which has led to increased interest in nonlinear ultrasonic techniques, it is often times difficult to experimentally realize some of these potential methods.

In this paper, we report on the first experimental results of an on-going research effort aimed at the nondestructive evaluation of adhesive bonds. We have used a setup where the sample is immersed in water, and high-power mode-converted shear wave bursts are through-transmitted. In a first study, we have investigated the feasibility of using water coupling to measure nonlinear ultrasonic parameters of the sample. Subsequently, we have inspected two sets of adhesive lap joints by means of oblique incidence finite amplitude ultrasound, where samples in the one set contain an embedded polyester peel ply which weakens their effective strength.

EXPERIMENTAL

Our experimental setup bears some resemblance to a typical water immersion ultrasonic C-scan system. However, up to three transducers as well as a test specimen can be mounted simultaneously, each to its own high precision rotation and/or translation stage, and thereby independently positioned inside a transparent tank filled with deionized water. Depending on the particular nature of the experiment, either the sample, or one of the transducers may be mounted to a computer-controlled high-resolution XYZ translation stage. The entire system is placed onto a vibration isolated optical table.

In our setup, ultrasonic signals can be transmitted, received, and processed by either a conventional C-scan system based on broadband pulses (SONIX, Inc.), or by a tunable narrowband ultrasound generation and detection system (RITEC, Inc.). The latter is also entirely computer-controlled and comprises two independently controllable, yet fully coherent, high-power tone burst generators, a broadband amplifier, and a tunable tracking superheterodyne receiver with quadrature phase sensitive detectors, and gated integrators. Figure 1 shows a block diagram of the essential part of the electronics, which may be briefly explained as follows. A master clock synchronizes two synthesizers, labeled f_1 and f_2, and the timing of two gated amplifiers on the generator side, as well as a third synthesizer, labeled f_3, and the timing of the integrator gates in the receiver. Two transmitting transducers, T_1 and T_2, can thus be independently, yet coherent, driven at frequencies f_1 and f_2, respectively. The receiver can then be programmed to process only signals at the frequency f_3, which could, for example, correspond to harmonics of f_1 or f_2, or to combination frequencies such as f_1+f_2 and f_1-f_2. Therefore, our system is optimized for nonlinear measurements. Specially written software permits us to acquire and process various types of ultrasonic data, such as frequency- and power-dependent data, as well as images of nonlinear parameters. In addition, two different digital storage oscilloscopes permit a wide variety of waveform and spectrum analyses of the RF signal, before it enters the heterodyne stage.

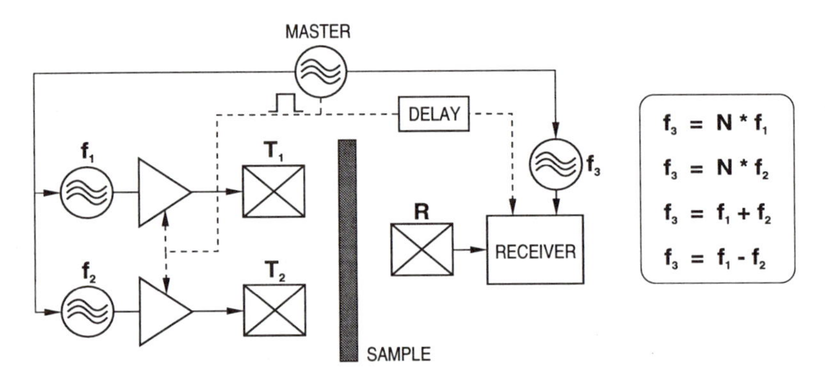

Figure 1. Essential part of the electronics which is optimized for nonlinear measurements.

We have used two sets of lap shear samples which were made at Boeing Commercial Airplanes. These samples are 4" wide, 1/16" thick aluminum sheets bonded with 1" overlap. Three of these were prepared with the normal Boeing structural bonding process using an adhesive designated BMS 5-101. The remaining three samples were made with one layer of a polyester peel ply bonded into them. According to Boeing, the reason for this peel ply arose from its use to protect the surfaces of certain graphite stiffeners that are cobonded to empennage skins. Furthermore, if the peel ply is inadvertently left on such a stiffener, standard pulse echo inspection techniques do not seem to be able to reveal the presence of these plies, which lead to a reduction in the tensile loads required to separate a stiffener from the skin.

Mechanical tests performed at Boeing on two sets of samples, with the same configuration as the ones provided to us, showed that the presence of a peel ply reduces the achievable peak load by up to 50% compared to normal bond samples. However, a previous inspection of these samples with water jet through-transmission ultrasound at 1, 3, and 5 MHz revealed little difference in the attenuation data between both sets of samples.

Additional ultrasonic tests performed at Johns Hopkins on these samples, using normal incidence 5-MHz through-transmission and 15-MHz pulse-echo setups, were found to be similarly inconclusive. Recently, tests performed by Don Price at CSIRO, Australia, suggest the possibility to detect the presence of the peel ply in Al-Al bonds by means of guided waves.[5]

Unfortunately, the use of water as a coupling medium may considerably impair the measurement of nonlinear elastic parameters of an immersed sample, because water itself can easily behave nonlinearly. Since the acoustic power required to cause finite amplitude displacements in a test specimen is generally high, this can lead to the generation of an extraneous number of harmonics in the couplant. Despite this potential problem, we chose to start off using a water immersion setup, because of various other advantages.

Much research has been done on the subject of nonlinear wave propagation in water, and on the nonlinear acoustic fields generated by various types of sources.[6-8] Nonetheless, to find out if we are indeed able to cause sufficiently high fundamental-frequency displacements in the vicinity of an adhesive bond, and subsequently, to calibrate our system, we set up an experiment illustrated in Figure 2. In this setup, the RF drive voltage into the transmitting transducer was monitored via the signal sampler, SS. Tone bursts at 5.37 MHz, 29 cycles long, were generated with a narrowband 5 MHz transducer (1" focus and 0.37" diameter), and focused at the back surface of a single aluminum plate of a lap joint. A calibrated 0.2 mm needle hydrophone (Precision Acoustics, Ltd.) was used to determine the acoustic pressure near this surface. The displacement, $\boldsymbol{\delta}$, was subsequently approximated using the relation

$$\delta = \frac{p}{\rho\, v\, 2\pi f}, \tag{1}$$

where p is the pressure, ρ and v are the density and the sound velocity in water, respectively, and f is the frequency. Equation 1 applies more correctly to small-amplitude longitudinal sound waves.

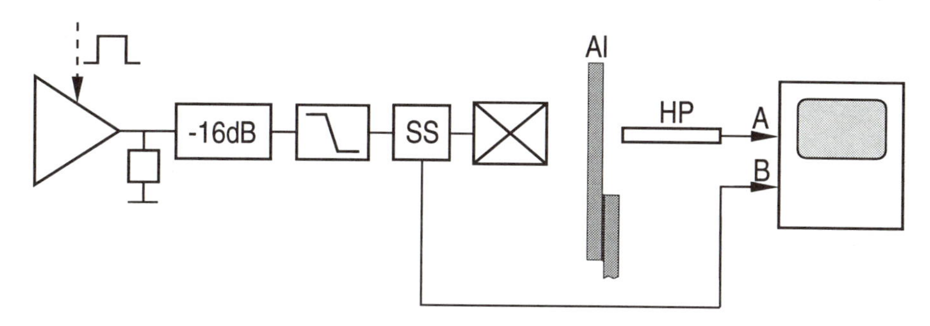

Figure 2. Setup used to calibrate the out-of-plane surface displacement as a function of ultrasonic input power by means of a calibrated needle hydrophone placed behind a single plate of a lap joint sample.

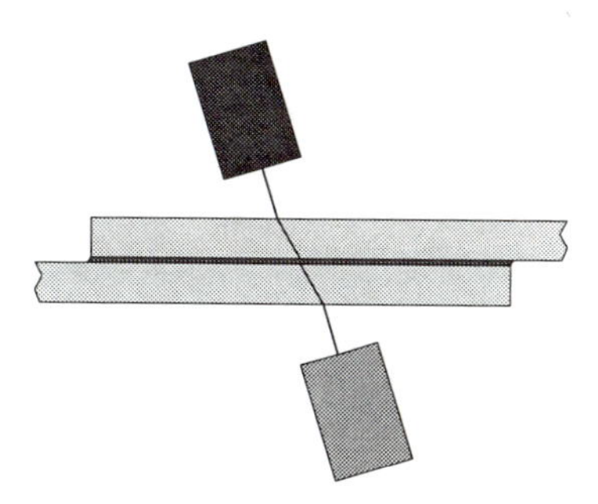

Figure 3. Transducer arrangement for the mode-converted shear wave through-transmission setup.

To study the feasibility of using nonlinear ultrasound to inspect the Boeing samples, we set up a through-transmission experiment, as shown in Figure 3. In this setup, both transducers were positioned at an angle of about 17.5° with respect to the normal of either surface. Otherwise, the setup for the transmitter was the same as in our calibration procedure. Tone bursts were then propagated through each Boeing sample into a 10 MHz receiving transducer (same focus and diameter as generating transducer), at various power levels. At each level, the tunable superheterodyne receiver was first set to 5.37 MHz, and then to 10.74 MHz.

RESULTS

The ultrasonic displacement amplitude at the back surface of a single plate of a bond sample, determined in our calibration procedure, is shown as a function of the RF drive voltage in Figure 4. Several conclusions may be drawn from this data. First, the deviation from linearity is almost negligible over a fairly wide range of drive voltages, which is emphasized by the straight line drawn along the linear portion of the data. This suggests that at the corresponding acoustic power levels system-related nonlinearities, possibly arising in the transmitter, in the water, at the water-sample interface, and in the aluminum plate itself, are small. Second, even though we have only determined the out-of-plane displacement at the aluminum surface into water, the effective displacement in an adjacent adhesive would be comparable, due to their similar acoustic impedances. At most, the displacement amplitude in the adhesive may drop to about half of the ones shown in Figure 4, as this would correspond to the displacement in the bulk of aluminum, which has a significantly higher impedance. This means that the achievable displacements in the adhesive of our lap joints are expected to be on the order of several nanometers. To verify our data, we set up a path-stabilized Michelson interferometer outside of the immersion tank, such that we could measure the out-of-plane displacement on the aluminum surface also by a more established technique. We found good agreement between the final data from both methods.

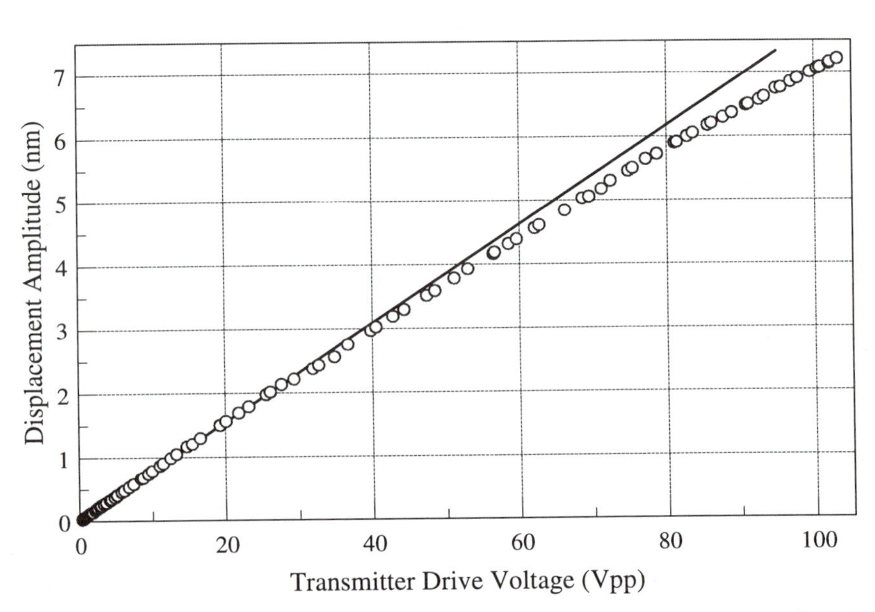

Figure 4. Calibration of the surface displacement into water, by measuring the pressure amplitude behind a single plate of a lap joint as a function of the RF drive voltage applied to the transmitting transducer.

At an incidence angle of 17.5° only a mode-converted shear vertical wave is expected to propagate through the aluminum layer, but an accompanying longitudinal mode may evolve.[9] The effective surface displacements at the aluminum-adhesive interface could be calculated by considering the in-plane and the out-of-plane components of the impinging shear wave. Thus, a first-order approximation suggests that the out-of-plane displacement is about as large as in the longitudinal wave case, whereas the in-plane displacement should be even larger, due to an expected increase in the power transmission coefficient, when using this angle of incidence. A detailed mathematical treatment of our configuration is rather complex, and will therefore be part of future work.

Nevertheless, we were able to measure the received power at the fundamental as well as the second harmonic through all of our six samples, over a fairly large range of acoustic input levels. This data is shown in Figures 5a and 5b, as the relative power in terms of the logarithm of the transmitter drive voltage. This type of representation permits us to verify that we have indeed measured a nonlinear acoustic behavior, since there appears to be a constant relationship between the data of the fundamental and the second harmonic over a range of input levels. Fitting each set of data over this range, we find that the slopes differ by a factor of 2.007 ± 0.0003 for the samples having a normal bond, indicating a strong quadratic nonlinearity. Interestingly, performing the same analysis on the data for the peel ply samples, we find a factor of 1.920 ± 0.0001. Another feature that may be noticed by a visual inspection of the plots, is that in the case of the samples comprising the additional polyester peel ply the second harmonic approaches the fundamental at a lower input power. This fact alone would suggest a higher nonlinearity in these samples. Since polymers are known to be easily nonlinear elastic when exposed to some stress, this result would be plausible. However, whether the increase in nonlinearity is due to the mere presence of the polyester or perhaps indeed due to weak interfaces, namely between the adhesive and the peel ply, will have to be addressed in future work as well.

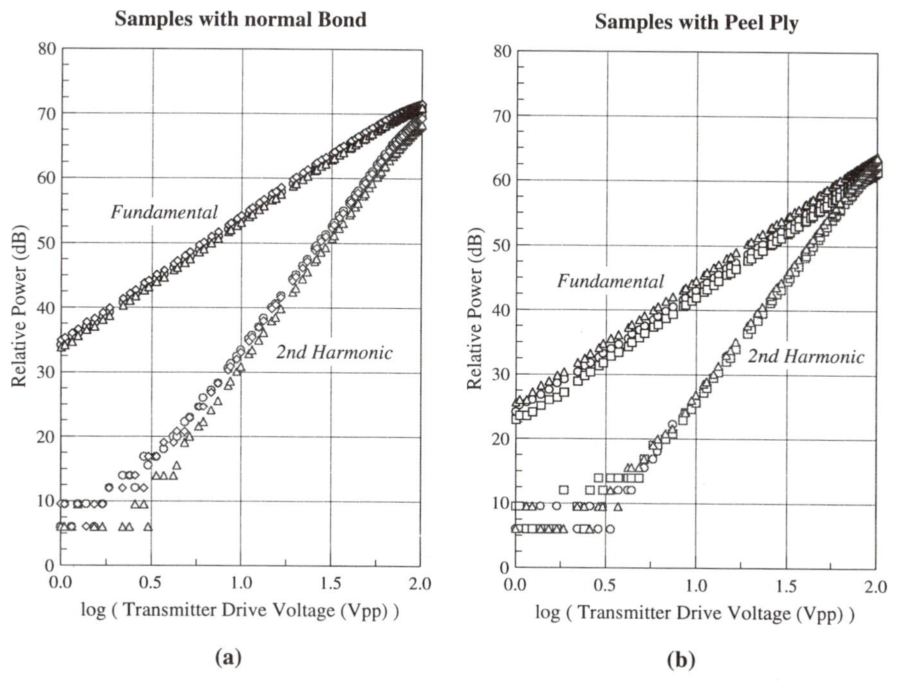

Figure 5. Received acoustic power versus input power for three lap joints with normal bond (a) and three with an embedded polyester peel ply (b), using the mode-converted shear wave through-transmission setup.

To better visualize the difference between all six bond samples, the amplitudes of the received second harmonic, A2, is plotted versus the square of the amplitudes at the fundamental, A1, for the various power levels, see Figure 6. This time, the slopes differ much more distinctively between the two sets of data. For each data point, the coefficient ($A2 / A1^2$), which is proportional to the parameter of nonlinearity, **β**,[10] may be calculated. A plot of these coefficients versus the transmitter drive voltage is shown in Figure 7. This representation indicates that already at a voltage of about 5 Vpp samples having an embedded peel ply could be readily distinguished from normal samples.

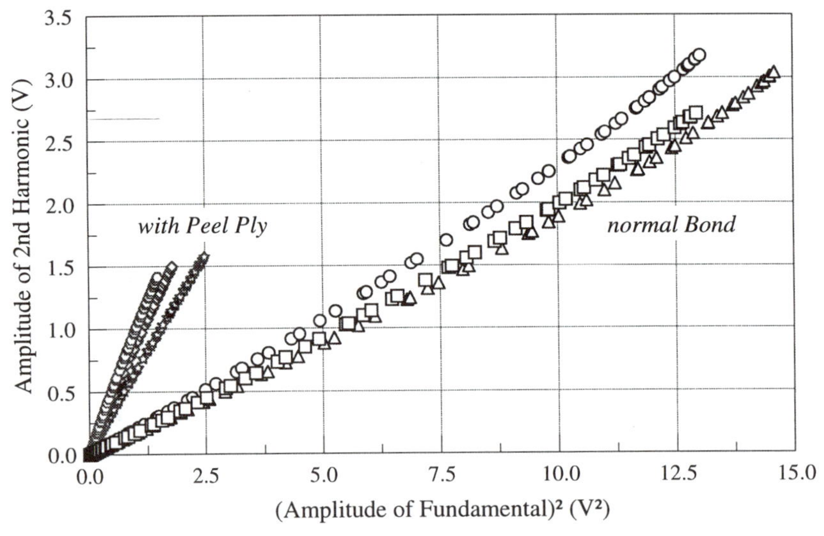

Figure 6. Comparison of all six lap joint samples, using only the received ultrasonic data, obtained with the mode-converted shear wave through-transmission setup.

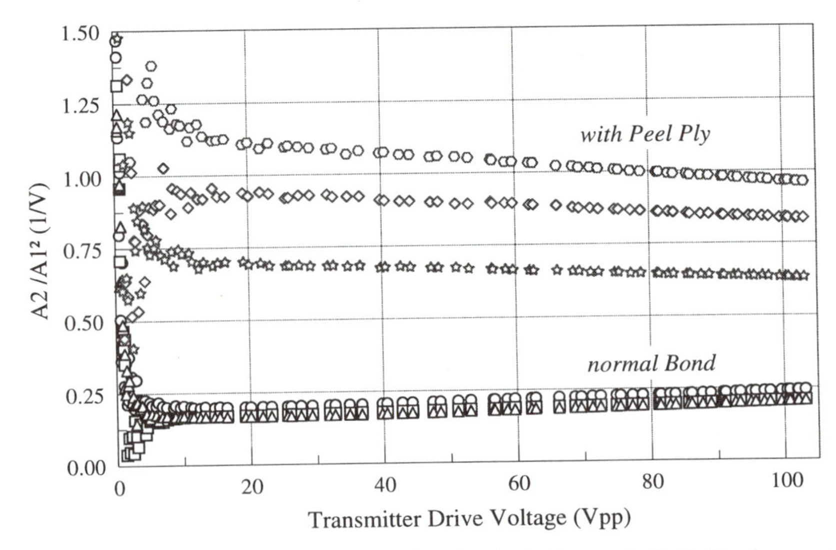

Figure 7. Comparison of all six lap joint samples, showing the 2nd harmonic divided by the square of the fundamental versus input power, obtained with the mode-converted shear wave through-transmission setup.

CONCLUSION

In this paper, we report on first results of an on-going research effort in which we are investigating various techniques to nondestructively evaluate adhesive bonds. First, we have calibrated our ultrasound generation system and thereby verified that within the nearly linearly range we should be able to produce displacement amplitudes on the order of several nanometers in the adhesive of a lap joint, which is immersed in water. We have then used a water immersion mode-converted shear wave through-transmission setup, and shown that samples containing an additional polyester peel ply can be clearly differentiated from normal bond samples by comparing their nonlinear acoustic behavior.

ACKNOWLEDGEMENTS

This research project is in part funded by NASA LaRC, Nondestructive Evaluation Sciences Branch, as part of a larger program, managed by Dr. Mark Roberts. Additional support is provided by the Center for Nondestructive Evaluation at The Johns Hopkins University. The authors are particularly grateful to Wayne Woodmansee at Boeing Commercial Airplanes for providing us with the samples used for this paper, as well as everyone at RITEC, Inc. for their outstanding and continuous technical support.

REFERENCES

1. J. M. Richardson, Harmonic generation at an unbonded interface - I. planar interface between semi-infinite elastic media, *Int. J. Engng. Sci.* **17**:73 (1979).
2. J. D. Achenbach, O. K. Parikh, and D. A. Sotiropoulos, Nonlinear effects in the reflection from adhesive bonds, in: *Review of Progress in Quantitative Nondestructive Evaluation* **8B**: 1401, D. O. Thompson and D. E. Chimenti, eds., Plenum Press, New York (1989).
3. P. B. Nagy, P. McGowan, and L. Adler, Acoustic nonlinearities in adhesive joints, in: *Review of Progress in Quantitative Nondestructive Evaluation* **9B**: 1685, D. O. Thompson and D. E. Chimenti, eds., Plenum Press, New York (1990).
4. D. J. Barnard, G. E. Dace, D. K. Rehbein, and O. Buck, Acoustic harmonic generation at diffusion bonds, *J. Nondestr. Eval.* **16**:77 (1997).
5. Wayne Woodmansee, Private communication
6. A. C. Baker, K. Anastasiadis, and V. F. Humphrey, The nonlinear pressure field of a plane circular piston: Theory and Experiment, *J. Acoust. Soc. Am.* **84**:1483 (1988).
7. T. S. Hart and M. F. Hamilton, Nonlinear effects in focused sound beams, *J. Acoust. Soc. Am.* **84**:1488 (1988).
8. S. Nachef, D. Cathignol, J. N. Tjøtta, A. M. Berg, and S. Tjøtta, Investigation of a high intensity sound beam from a plane transducer. Experimental and theoretical results, *J. Acoust. Soc. Am.* **98**:2303 (1995).
9. R. E. Green, Jr., Ultrasonic investigation of mechanical properties, in: *Treatise on Materials Science and Technology* **Vol. 3**, H. Herman, ed., Academic Press, New York and London (1973).
10. M. A. Breazeale and J. Philip, Determination of third-order elastic constants from ultrasonic harmonic generation measurements, in: *Physical Acoustics* **XVII**, W. P. Mason and R. N. Thurston, eds., Academic Press, New York (1984).

NONLINEAR VIBRO-ACOUSTIC NONDESTRUCTIVE TESTING TECHNIQUE

Alexander M. Sutin and Dimitri M. Donskoy

Davidson Laboratory
Stevens Institute of Technology
Castle Point on Hudson
Hoboken, NJ 07030, USA

INTRODUCTION

Conventional active acoustic methods of NDT are based on the principles of linear acoustics[1,2]. These include effects of reflection, scattering, transmission, absorption of probe acoustic energy. The presence of a defect leads to phase and/or amplitude variation of received signals while the frequencies of the received signals are the same as of emitted probe signals.

The principal difference between the nonlinear technique and linear acoustic NDT technique is that the nonlinear technique correlates the presence and characteristics of a defect (or material) with acoustical signals whose frequencies differ from the frequencies of the emitted probe signals. These signals with different frequencies are an outcome of a nonlinear transformation of the probe acoustic energy by a defect.

Nonlinear acoustic technique has been recently introduced as a new tool for nondestructive inspection and evaluation of defective and fractured materials[3-5]. The studies have shown strong correlation between nonlinear interaction of acoustic signals and various defects of a structure. The basic concept of this technique is straightforward: materials containing defects (cracks, fractures, unbondings, etc.) have a much larger nonlinear response than materials with no such defects. The nonlinear response can be understood analytically as a deviation of material stress-strain relation from the linear Hook's law. This is explained by asymmetry of stress-strain dependence with respect to compression and tensile stresses of a defected area. A perfect structure with integrity, behaves as a quasi-linear system. However fractured, cracked, or unbonded structure creates much higher deviation from the linear Hook's law leading to various nonlinear acoustic effects. It is important to note that the these nonlinear effects are observed at extremely small strains (less than 10^{-5}) and do not alter material structure.

The nonlinear NDT methods have a number of advantages as compared with the linear acoustic techniques. Among them are high sensitivity and applicability to highly non-

Nondestructive Characterization of Material VIII
Edited by Robert E. Green Jr., Plenum Press, New York, 1998

homogeneous structures, such as composites, engine components, etc. The present paper is aimed to explain the concept and methods of the nonlinear NDT technique, to present same results of its applications, and to demonstrate its advantages.

PHYSICAL MECHANISM OF THE NONLINEARITY DUE TO DEFECTS

In this paper we consider contact-type defects such as cracks, unbonds, delaminations, etc. The physical nature of the defect-related nonlinearity can be most easily understood when we consider two simple models. The first is illustrated in Fig.1, where a defect is modeled as a contact between two flat solid surfaces. If dynamic (acoustic) stress is applied to this defect, there is no contact at all (full opening) during the elongation phase of the stress, and full contact (closure) during the compression phase. The elastic deformation of medium containing such a defect will be different for elongation and compression leading to a piecewise-linear (nonlinear) stress-strain relationship[6], as shown in Fig.1b.

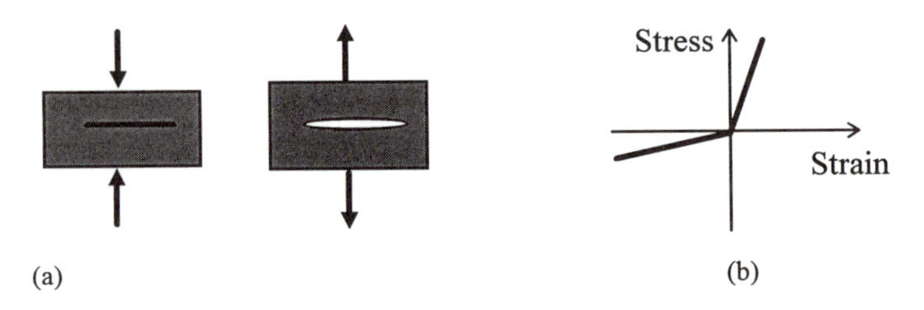

Figure 1. Contact/(no contact) defect (a) and its stress-strain dependence (b).

A more realistic model[4] of a defect is a contact between two rough elastic surfaces (Fig.2). The applied stress will vary the contact area within the defect, leading to its nonlinear elastic deformation.

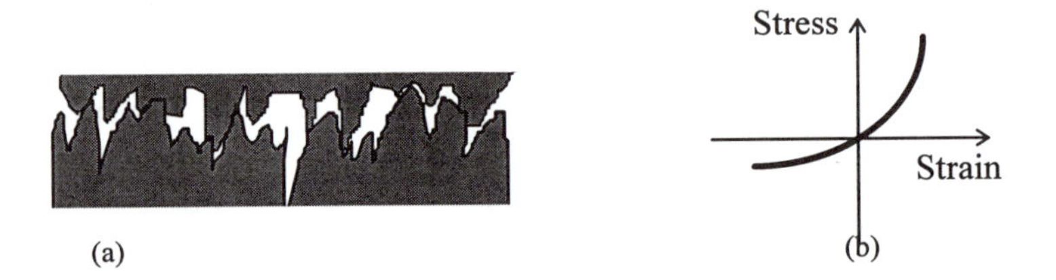

Figure 2. A defect with rough surfaces (a) and its stress-strain dependence (b).

The stress-strain dependence of a medium containing such contact-type defects will also be nonlinear and can be written in the form of the Taylor's expansion with respect to strain. For simplicity we write this relationship for one-dimensional longitudinal deformations:

$$\sigma = E(\varepsilon + \beta\varepsilon^2 + \gamma\varepsilon^3 + \dots)\,, \qquad (1)$$

where σ is the stress, ε is the strain, E is the modulus of elasticity, $\beta, \gamma, \dots$ are the nonlinear parameters, which characterize the nonlinearity of the medium. For the small strains used in acoustic NDT, the cubic and higher terms in this expansion commonly can be neglected and the equation (1) retains only linear and quadratic terms of ε. This is the case of so called quadratic nonlinearity. Typical values of the nonlinear parameter β as a rule do not exceed 10 for homogeneous media without any defects, so the contribution of the nonlinear quadratic term into the relationship (1) is very small (for small strains) and the media exhibit quasi-linear behavior. The defects may increase the parameter β up to two to three orders of magnitude. Even though the value of the nonlinear term may still be small compared with the linear term ($\beta\varepsilon << 1$), its contribution and, consequently, acoustical manifestations are much more visible.

There are various nonlinear acoustical manifestations of the contact-type nonlinearity. Thus, when a sinusoidal (with the frequency f_0) probe acoustic wave meets a defect, the wave changes the contact area within the defect, increasing the area in the compression phase and decreasing it in the rarefaction phase. Because of this, the phase and the amplitude of the transmitted wave varies accordingly, leading to the nonlinear distortion of this probe signal: generation of harmonics with the frequencies $2f_0$, $3f_0$, etc.

Another nonlinear effect due to the contact-type defect is the modulation of a probe ultrasonic wave by lower frequency vibration. In this case, again, vibration varies the contact area (or alters defect opening) modulating the phase and the amplitude of the higher frequency probe wave.

NONLINEAR ACOUSTIC METHODS OF DETECTION OF DEFECTS

Method of Second Harmonic

The simplest way to utilize the nonlinear elastic response of a medium with a defect for NDT purposes is to measure the level of the second harmonic of the probe sinusoidal signal. The level of this nonlinear signal indicates presence and size of the defect.[6,7]

The following two examples illustrate the second harmonic method. The first example deals with the assessment of the bonding quality of the thermo-protective coating (tiles) of the Soviet Space Shuttle "Buran"[6]. The coating consists of relatively rigid tiles (mass) attached to the ship's metal skin through an elastic bonding layer (spring). In the presence of bonding defect this spring has a piecewise-linear (nonlinear) stiffness characteristic similar to that as shown in Fig.1b. Then a harmonic vibration applied to the skin, such a spring leads to the nonlinear oscillation of the tile. Fig.3 shows the spectra of the oscillation of the tile without (a) and with (b) the bonding defect in the elastic layer between protective tiles and the hull of the ship. These spectra demonstrate significant increase of the amplitude of the second harmonic in the tile oscillation. It is interesting to note that the unique feature of the piecewise-linear system is a linear relation between the second and the first harmonic amplitudes, which was confirmed in this study.

Another example of the application of the second harmonic method is the detection of cracks in large (over 3 meter long, 0.65 x 0.45 m^2 cross-section) graphite electrodes used in production of aluminum[8]. The electrodes were longitudinally excited at their lowest natural modes (half or one wavelength with the frequencies near 200 Hz and 400 Hz, respectively). Transmitted vibrations were received with a laser vibrometer from the opposite from excitation end of the electrode. Similar spectra (as shown on Fig.3) were obtained for the samples with and without cracks. A sharp increase in the level of the second harmonic was observed in the electrodes with cracks. The tests also showed that the second harmonic amplitude A_2 is proportional to the square of the level A_1 of the first (fundamental) harmonic indicating that there is the quadratic type of nonlinearity due contact-type defect as shown in Fig.2(b). A parameter proportional to the ratio $A_2/(A_1)^2$ was used as the criterion. In the presence of cracks, this parameter exceeded the corresponding parameter for a non-defective electrode by a factor of 3 to 8.

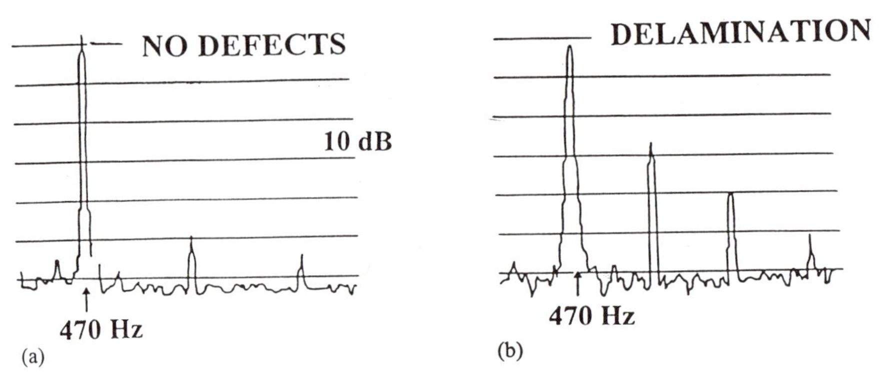

Figure 3. Spectra of nonlinear signal without (a) and with (b) a defect.

The method of the second harmonic proved the concept of nonlinear acoustic NDT: contact-type defects substantially increase the nonlinear response of the structure. The method is simple and responsive to many types of defects. However, this method has a significant drawback. The generating electronic and electro-mechanical equipment, such as signal generators, power amplifiers, and transducers, have their own nonlinearity and generate the second harmonic at certain level. This creates a background nonlinear signal, thereby, limiting the sensitivity of the method. In order to avoid this problem we have developed more advanced modulation methods.

Modulation Methods

The modulation methods utilize the effect of the modulation of the higher frequency ultrasonic probe wave by the lower frequency vibration. In the frequency domain the result of this modulation manifests itself as sidebands. Thus, if the probe and modulating waves are sinusoidal signals with the respective frequencies f_p and f_m, the resulting sidebands have the frequencies $f_p + f_m$ and $f_p - f_m$. The modulation methods do not suffer from the above mentioned drawback of the second harmonic method, since the sources of the probe and modulating signals are independent, so there is no background nonlinear signal due to the nonlinear interaction in the generating equipment.

We have developed two modifications of the modulation methods. One utilizes sinusoidal modulating vibration (vibro-modulation method) and the other uses vibration produced by an impact (impact-modulation method). Both of these methods proved to be very reliable and sensitive to crack detection even for the structures where the conventional acoustic methods do not work at all. Recent examples of the applications of these method include detection of

- unbonding of titanium plates used for airspace applications
- cracks in Boeing 767 steel fuse pins
- cracks in combustion engine cylinder heads
- cracks in a weld in a steel pipe at a nuclear power station
- adhesion flaws in bonded composite structures
- cracks and corrosion in reinforced concrete
- cracks in rocks

As an example of the application of the modulation methods, some results of the study performed for the Automotive Composite Consortium (ACC) to assess adhesion strength in a bounded composite structure are presented below.

The test was performed with two identical samples (defective and control) which were fabricated and supplied by ACC. Each specimen was fabricated from one corrugated and one flat piece of composite. Both are made out of a urethane resin matrix combined with short glass fiber strands. The defective sample had visible bonding defects. The control sample had no such defects.

Vibro-modulation (VM) and impact-modulation (IM) methods were applied to test the samples. The samples were firmly attached (clamped) to a shaker. We studied two cases: the transmitting and receiving ultrasonic sensors were positioned on the same side and on opposite sides of the sample. The frequency f_p of the sinusoidal ultrasonic probe signal was 82 kHz and the frequency of vibration f_m was 267 Hz. It was observed that the presence of the defect causes the modulation of the probe signal which can be seen on the spectra (Fig.4) as the sidebands with the frequencies $f_p \pm f_m$. It appeared that both same-sided and opposite-sided placement of the sensors can be used to inspect the defective sample. The difference in modulation level between the control and defective samples was very large (over 30 dB) on certain frequencies. An IM test using an impact hammer instead of the shaker yielded a similar result.

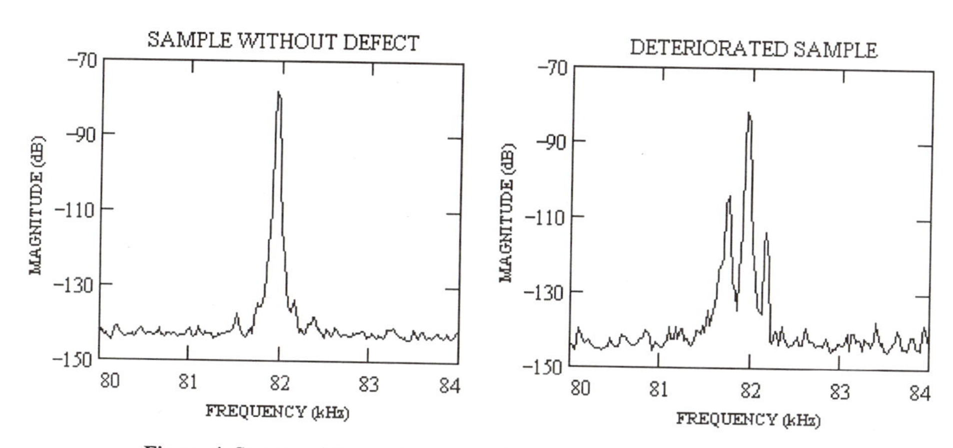

Figure 4. Spectra of the nonlinear modulated signal without and with a defect.

Similar results were observed for the other applications, mentioned above. Thus, the VM method was applied to evaluate quality of welded joints in steel (1 meter in diameter, 5 cm thick) pipe used in a nuclear power station[9] and to detect corrosion and cracking in a concrete structure.[10] The frequencies of the probe signal/vibration were 20kHz/30Hz and 28kHz/250Hz for the pipe and concrete structure, respectively. The Impulse-Modulated method was applied to detect defects in a combustion engine cylinder head, Boeing 767 steel fuse pins, and bonded titanium plates. These applications employed frequencies in the range 60 to 90 kHz.

Conducted investigations demonstrated capability of the modulation methods to detect flaws in highly non-homogeneous structures where conventional acoustic methods are not applicable.

CONCLUSIONS

The study demonstrated strong correlation between presence of contact-type defects such as cracks, debondings, delaminations and measured sideband spectral components. In some experiments the level of the sideband components (in presence of a defect) exceeded the reference signal (without defect) by over 30 dB. Such a high ratio couldn't be achieved using conventional linear methods.

The nonlinear technique separates nonlinear signals (caused by the defects) from the probe signals scattered and reflected from boundaries, layers, and other linear (from an acoustical point of view) non-homogeneities. Such a separation is possible due to the difference between the frequencies of the input probe signal and output nonlinear signal. This feature is especially advantageous for flaw detection in highly non-homogeneous materials and structures such as composite materials, concrete structures, airframes and engine components where conventional techniques often do not work at all.

Ongoing research is directed toward quantitative characterization of defects and their localization.

REFERENCES

1. J. Krautkramer and H. Krautkramer. *Ultrasonic Testing of Materials*, Springer-Verlag, Berlin - Heidelberg - New York (1977).
2. R.E. Green. *Ultrasonic Investigation of Mechanical Properties,* Academic Press, New York & London (1973).
3. O. Buck. Material characterization and flaw detection by acoustic NDE JOM, 44:17-23 (1992).
4. A.M. Sutin and V.E. Nazarov. Nonlinear acoustic methods of crack diagnostics. *Radiophysics & Quantum Electronics*, 38(3-4):109-120 (1995).
5. A.M. Sutin. Nonlinear acoustic non-destructive testing of cracks, in: *Nonlinear Acoustics in Perspective, 14th-Intern. Symp. on Nonlinear Acoustics*, R.J. Wei, ed., Najing University Press, China, 328-333 (1996).
6. V.A. Antonets, D.M. Donskoy, and A.M. Sutin. Nonlinear vibro-diagnostics of flaws in multilayered structures. *Mechanics of Composite Materials*, 15:934-937 (1986).
7. O. Buck, W.L. Morris, and J.N. Richardson. Acoustic harmonic generation at unbonded interfaces and fatigue cracks. *Appl.Phys.Letters*, 33(5):371-373 (1978).
8. A.M. Sutin, C. Delclos, and M. Lenclud. Investigations of the second harmonic generation due to cracks in large carbon electrodes, in: *Proc. 2-nd Symp. Acoustical and vibratory surveillance methods and diagnostic techniques*, Senlis (France), 725-735 (1995).
9. A.S. Korotkov, M.M. Slavinskii, and A.M. Sutin. Vibro-Acoustic methods for diagnostics of metal strength properties. *Advances in Nonlinear Acoustics*, H.Nobak, ed., World Scientific, Singapore-New Jersey-London, 370-375 (1993).
10. D.M. Donskoy, K. Ferroni, A.M. Sutin, and D.M., Sheppard, Nonlinear acoustic technique for crack and corrosion detection in reinforced concrete, *Eight International Symposium on Nondestructive Characterization of Materials*, Boulder, CO, (1997).

NONCONTACT ULTRASONIC SPECTROSCOPY FOR DETECTING CREEP DAMAGE IN 2.25CR-1MO STEEL

Toshihiro Ohtani,[1] Hirotsugu Ogi,[2] Tomohiro Morishita,[3] and Masahiko Hirao[2]

[1] Advanced Material Lab., Ebara Research Co., LTD.
Fujisawa, Kanagawa 251, Japan
[2] Faculty of Engineering Science, Osaka University, Toyonaka, Osaka 560, Japan
[3] Dept. of Mech. Eng., Akashi College of Technology, Akashi, Hyogo, Japan

INTRODUCTION

Many of fossil power plants, which were constructed during 1960's and 70's and exceed more than 10,000 working hours, have presently been operating while they have been receiving progressive damage like creep as the time proceeds.[1, 2] Furthermore, by shifting the base load of power from fossil power plants to nuclear power plants, they required the severe operating condition like daily or weekly start-up and shutdown in order to correspond to rapid change of the demand for power. As the consequence of the above trends, the degradation of materials in these plants was accelerated. Therefore, to safely operate plants, a technique for predicting the life and evaluate the damage of materials with nondestructive techniques is essential to assess the current state of these plants. Moreover, it is necessary for the technique to be simple and quick to apply to many objects.

In this study, we present an ultrasonic technique which is based on EMAR (Electromagnetic Acoustic Resonance) [3], for detecting the creep damage of the fossil plant components. This is a combination of the resonant technique and a noncontacting electromagnetic acoustic transducer (EMAT). Incorporation of EMATs in a resonant measurement contributes to improving the weak coupling efficiency of EMATs to a large extent. The attenuation measurement using an EMAT is inherently free from energy losses associated with intimate contact, resulting in the pure attenuation in a sample after correcting for the diffraction losses.

Nondestructive Characterization of Material VIII
Edited by Robert E. Green Jr., Plenum Press, New York, 1998

This paper describes the ultrasonic technique for evaluating creep damage in 2.25Cr-1Mo steel exposed to the temperature of 650℃ at various stresses. We obtain the ultrasonic velocity from the resonant frequency and the attenuation coefficient from the ringdown curve at the resonance. Two types of EMATs are used: the bulk-wave EMAT for plate samples and the axial-shear-wave EMAT [4] for cylindrical samples. The attenuation showed much larger sensitivity to the damage accumulation than the resonant frequency. Approaching the rapture, it becomes ten time as large as the initial value. The frequency dependence of the attenuation also showed a remarkable change with the damage. The EMAR technique has been proven to possess a large potential for the practical nondestructive/noncontact damage monitoring; the EMAR is capable of sensing the attenuation evolution and indicating the damage advance to predict the creep life of the high-temperature metals.

CREEP TEST

Two kinds of specimens, plate and cylindrical ones, of commercial 2.25Cr-1Mo steel (ASST. A182F22) were used in the creep test. The gauge section of plate samples is 3 mm thick, 35 mm wide and 45 mm long. For the cylindrical samples, it is 14 mm diameter and 60 mm long. They were manufactured by hot rolling and the rolling direction was parallel to the longitudinal direction of the specimen. Heat treatment was carried out as follows; after holding 920 ℃, 1 hr, cooling to 650 ℃ by 40 degree/hr and then furnace cooling. The chemical composition is shown in Table 1. The mechanical properties are as follows; 0.2% proof stresses are 237 and 266MPa, ultimate tensile strength 512 and 490 MPa and elongation 33 and 34.8%, for the plate and cylindrical ones, respectively.

Creep test with a vertical single lever type of creep testing machine and electrical furnace was carried out under 650 ℃ in air. The applied stresses is 70, 60 and 50MPa to the plate type of specimens, and 85 and 65 MPa to the cylindrical ones. As a reference, other specimens was attached to creep samples to observe only the heat effect.

EMAT

The bulk-wave EMAT [5] applied to the plate specimens is composed of a pair of permanent magnets, which have the opposite magnetization direction normal to the sample surface, and a spiral elongated coil (Fig. 1). The EMAT transmits and receives the polarized shear wave

Table 1. Chemical composition of 2.25Cr-1Mo steel (mass%)

	C	Si	Mn	P	S	Ni	Cr	Mo	Cu
Plate	0.12	0.21	0.52	0.01	0.01	0.1	2.13	0.88	0.04
Cylinder	0.13	0.25	0.56	0.02	0.02	0.08	2.2	0.9	0.13

propagating in the thickness direction of the sample mainly with a magnetostrictive effect. The effective area of the EMAT is 7 x 6 mm^2.

Figure 2 shows the EMAT transmitting and receiving axially polarized SH wave. It consists of a solenoidal coil which applies the static magnetic field in the axial direction of the sample and the meander line coil surrounding the surface of the sample which induces the periodic dynamic field normal to the axial direction. The principle of the EMAT was described in Reference 4. Axial SH wave is a kind of surface waves, which propagates along the circumference of cylindrical rod or pipe with the axial polarization. Each resonant mode of the axial SH wave is described by the normal function of radius showing the peak amplitude at specific radius. The spatial resolution in the radial direction is determined by to the spacing of the meander line coil. The spacing used in this study is 0.9 mm and the resolution is 0.3 mm from the surface.

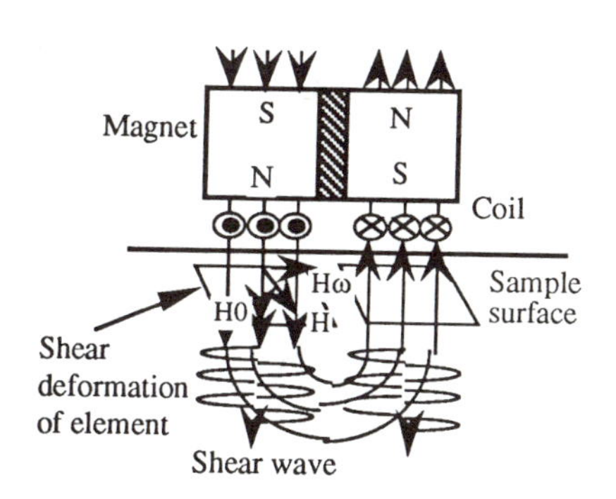

Figure 1. Structure and mechanism of bulk wave EMAT

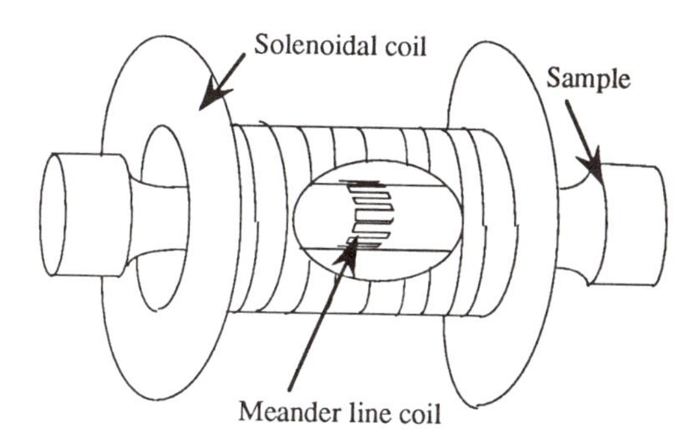

Figure 2. Generation and detection of the axially polarlized shear wave by the meander line coil and a solenoidal coil

THE MEASUREMENT OF ATTENUATION WITH EMAR

The ultrasound is transmitted to the sample by exciting EMAT by high power rf bursts (~1000V, ~50μs). In the plate sample, it propagates back and forth within the thickness and then ultrasound is received by EMAT. In the cylindrical one, propagating along the circumferential surface, ultrasound is received at straight parts of the meander line coil whenever it passes through below the coil. With the superheterodyne spectrometer, a resonant spectrum is obtained by sweeping the frequency of the rf burst. At a resonant frequency, the received signal is intensified after being coherently overlapped and an extremely high amplitude is obtained. [3]

In the plate sample, resonant modes are shown at the same interval of spectrum and the n-th resonant frequency f_n is given by $f_n = C n/(2d)$, where, C is the wave speed and d is the plate thickness. In the cylindrical sample, resonant modes appeared at unequal intervals and are determined by the following equation; $nJ_n(KR)-KRJ_{n+1}=0$, where J_n is the n-th Bessel function of the first kind and R is radius of sample. K is the wavenumber expressed by the m-th resonant frequency $f_m^{(n)}$ and wavespeed C as $K=2\pi f_m^{(n)}/C$. In the meander line coil used in this study,

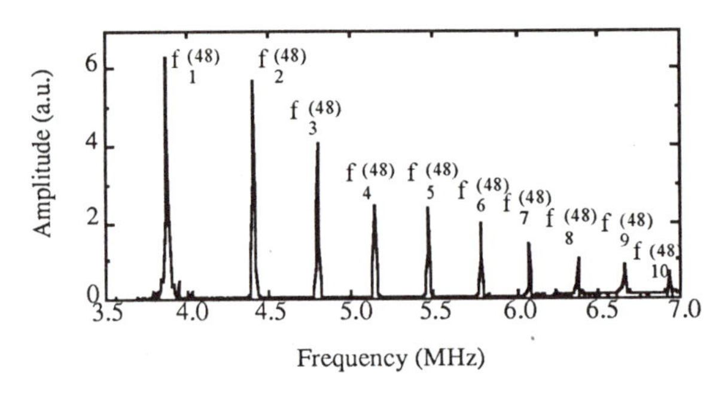

Figure 3. Measured resonant spectrum for 2.25Cr-1Mo steel cylinder of ϕ14 mm with n=48. Static field is 0.16KOe.

integer n is equal to 48. Figure 3 shows a resonant spectrum for 2.25Cr-1Mo steel of ϕ14 mm. The oscillation of the first mode is concentrated near the surface and as the mode becomes higher overtone, the peak amplitude moves inward. Consequently, it is capable of evaluating the property change in the radial direction using different modes. To measure the attenuation, the a resonant spectrum is first obtained by sweeping the frequency of the rf burst like in Figure 3. Second, the resonant frequency is determined from the center axial of the Lorentzian function fit to the spectrum around the peak. Third, we measured the ringdown curve by driving the EMAT at the resonant frequency. Figure 4 shows the measured ringdown curve at the first mode ($f_1^{(48)}$) shown in Figure 3. Finally, the attenuation coefficient is obtained by fitting an exponential decay to the ringdown curve and extracting the time constant. [6]

RESULTS AND DISCUSSION

We measured the resonant frequencies for the range 2 to 10 MHz and their attenuations during the creep test of 2.25Cr-1Mo steel plates with the bulk-wave EMAT. Figure 5 displays the frequency dependence of the attenuation under creep test (650℃ and stress: 70MPa). This measurement was carried out with a single sample, for which the creep test was interrupted for measurement until rupture. The time to rupture was 234 hr. The polarization of the shear wave was parallel to the load. As a reference data, the change of the attenuation due only to the heat

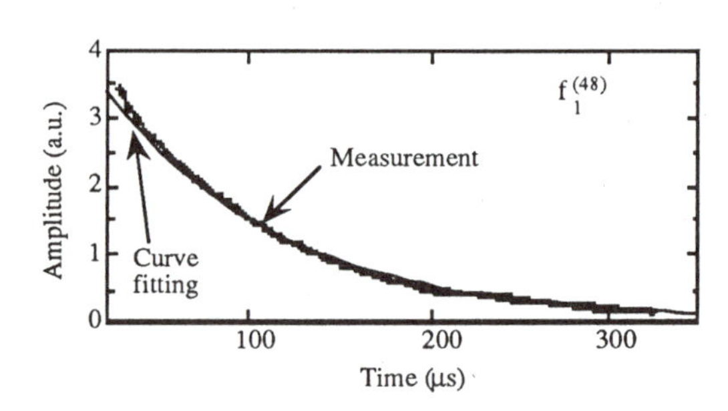

Figure 4. Measured ringdown curve of the first mode $f_1^{(48)}$ in Figure 3.

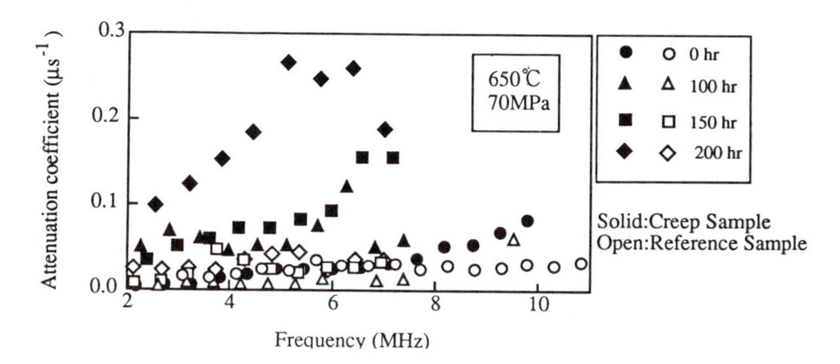

Figure 5. Frequency dependence of the shear-wave attenuation polarized in the stress direction (plate sample).

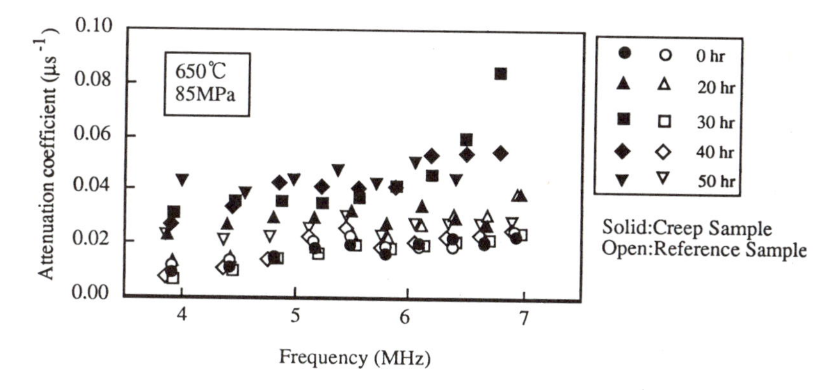

Figure 6. Change of attenuation coefficient of the axial shear resonance of cylindrical samples.

effect is also shown. We observe that the attenuation α always increase with the resonant frequency in an accelerating manner in course of the creep advance. The attenuation α just before the rupture is ten times larger than that before the creep test. The change of α by the only heat effect is much less than that by the creep damage. In the case of the normal polarization normal to the load, the similar behavior was observed.

We also shows the change of the attenuation α during the creep test of a cylindrical sample with the axial-shear EMAT in Figure 6. The resonant frequencies were measured to 10-th mode. Like the plate sample, creep test (650℃, stress: 85MPa and rupture time: 63hr) was carried out with a single sample. The attenuation α displays the frequency dependence and α at each mode increases equality as the creep advance. It is clear that the creep damage occurs uniformly in the cross section of the sample. As before, the effect of heat is very small.

In two kinds of samples, though the surfaces after being creep test in air were covered with the scale (oxide), it was possible to measure with slightly polishing by an emery paper.

CONCLUSION

Creep damage in 2.25Cr-1Mo steel plates and rods under 650℃ in air was evaluated by ultrasonic attenuation with the method of the EMAR. The attenuation extremely increased as the creep damage progresses, amounting to ten time larger than initial values. Prior to the measurement, the surface oxide was removed with slightly polishing. The EMAR technique has been proven to possess a large potential for the practical nondestructive/noncontact damage monitoring. It is capable of sensing the attenuation evolution and indicating the damage advance to predict the creep life of the metals. The relation between the metallurgical aspect and the attenuation evolution is currently under investigation.

REFERENCES

1. D.S.Drinon, P.K. Liaw, R.D.Rishel, M.K.Devine and D.C.Jiles, ASME PVP-Vol.239/MPC-Vol.33 (1992) p5-14
2. H.Yoneyama, M.Nakashiro K.Murakami, S.Shibata and A.Ohtomo, IHI Engineering Review Vol.22 No.1 (1989) p1-6
3. M. Hirao, H. Ogi and H. Fukuoka, Rev.Sci. Instrum., 64, 3198 (1993)
4. H. Ogi, M. Hirao and K. Minoura, J. Applied Phys., 81, 3677 (1997)
5. H. Ogi, M. Hirao, and T. Honda, J. Acoust.Soc.Am., 98, 458 (1995)
6. H. Ogi, M. Hirao and T. Honda, Rev.Prog. in QNDE, D.O.Thompson and D.E.Chimenti, eds, 14 (1996) p1601-1608

ULTRASONIC DAMPING AND VELOCITY DURING RECOVERY AND RECRYSTALLIZATION OF ALUMINUM†

Ward Johnson

Materials Reliability Division
National Institute of Standards and Technology
325 Broadway, Boulder, CO 80303

INTRODUCTION

Vibrational damping at 0.4-2.0 MHz in fully annealed pure polycrystalline aluminum has been found to increase dramatically and monotonically with temperature between 92 °C and 465 °C.[1] This temperature dependence and a frequency dependence that changes systematically with temperature are consistent with a physical damping mechanism involving the anelastic response of pinned dislocations to ultrasonic stress.[1] The apparently dominant contribution of dislocations to the damping at elevated temperatures leads to the suggestion that ultrasonic measurements may be used to monitor the microstructural evolution of commercial alloys during industrial annealing processes that induce partial or complete recovery or recrystallization. (The definitions of "recovery" and "recrystallization" are discussed by Cahn.[2]) Measurements presented in this report show that two large irreversible drops in the damping of shear modes and corresponding increases in the velocity occur in cold-worked aluminum during ramped anneals and that the dependences of these changes on temperature, time, and deformation are consistent with recovery and recrystallization.

MEASUREMENT TECHNIQUE

Noncontacting electromagnetic-acoustic transduction is used to generate and detect resonant vibrations of spherical samples in a vacuum furnace, and an optical pyrometer is used to measure the temperature. As described elsewhere,[3,4] a sample is surrounded by a solenoid coil between two permanent magnets which produce a static field transverse to the axis of the coil. The physical coupling mechanism is the same as that employed by conventional electromagnetic-acoustic transducers, the coil acting as both a transmitter and a receiver. A gated amplifier generates 1-5 ms driving tone bursts from a continuous sine wave, and a phase-sensitive receiver enables calculation of the log decrement δ and resonant frequency f from the resonant ringdown following the tone burst.[3]

The transducer couples to a variety of resonant modes of a sphere,[4] but only $T_{1,1}$ torsional modes[5] are considered here. In isotropic material, the frequencies of these modes depend only on the plane-wave shear velocity and the diameter.[5] However, since the samples in this study are not perfectly isotropic or spherical, modes that would be degenerate in an isotropic sphere are slightly split in frequency. The three $T_{1,1}$ modes for each sample have resonant frequencies that differ by 0.3-0.6%.

SAMPLE PREPARATION

Ultrasonic samples were prepared from 99.999% pure polycrystalline aluminum that was received from the manufacturer as cold-worked pieces approximately 12.5 mm in diameter and 14.5 mm long. The material contained 1.0 atomic parts per million (ppm) Si, 0.6 ppm Cu, and 0.6 ppm Ca as unintentional impurities. The concentrations of other elements were below the detection limit of the analysis: Fe (< 0.7 ppm), Mn (< 0.2 ppm), Cr (< 0.4 ppm), Zn (< 0.7 ppm), Ti (< 0.3 ppm), B (< 0.2 ppm), P (< 1.0 ppm), Ag (< 0.1 ppm), Ga (< 1.0 ppm), Ni (< 0.6 ppm), Mo (< 0.5 ppm), Na (< 0.2 ppm), K (< 0.2 ppm), and Li (< 0.1 ppm). The material was annealed in air at 400 ± 5 °C for 10 minutes to recrystallize the material and then slowly cooled. It was then cold-worked under uniaxial compression in a mechanical testing machine. The compression strokes were performed at constant speed with an initial strain rate of 1.0 ± 0.1/s. Spherical samples were fabricated from this deformed material by machining an approximate sphere on a lathe and then rotating by hand in a brass cone filled with a series of alumina polishing compounds (starting with 25 μm grit and ending with 3 μm grit). The fractional deformation ϵ and finished mean diameter $\bar{d}$ of each sample is given in Table 1. The uncertainty in ϵ is 0.005. As a consequence of slight microstructural anisotropy, the sphericity of AL-5 and Al-7 changed during the recovery and recrystallization that occurred in the high-temperature experiments. Δ_i in Table 1 is the maximum deviation of the diameter from the mean of each sample before the experiments, measured with an optical comparator, and Δ_f is the corresponding deviation after the experiments.

Table 1. Fractional deformation ϵ, mean diameter $\bar{d}$, and maximum deviations of the diameter from the mean before heating, Δ_i, and after heating, Δ_f.

Sample	ϵ	$\bar{d}$ (mm)	Δ_i (mm)	Δ_f (mm)
AL-5	0.095	6.29	0.02	0.09
AL-7	0.105	6.12	0.02	0.05
AL-8	0.211	6.317	0.002	0.002

RESULTS

Ramped anneals

Samples AL-7 and AL-8 were subjected to a series of continuously ramped anneals during which δ and f of one of the $T_{1,1}$ modes were continuously measured. The heating rate for all anneals was 1.5 ± 0.4 °C/min. The power of the driving tone burst was kept low enough that the resonant ringdown was exponential. That is, the vibrational amplitudes were below the range where δ depends significantly on amplitude.[1]

As shown in Fig. 1(a), δ during the first heating increased up to 210 °C, dropped irreversibly between 210 °C and 255 °C, and then increased again between 255 °C and 280 °C. After this heating, δ decreased monotonically as the sample was cooled to room temperature. During the second heating, δ followed the previous cooling curve and continued increasing with further heating up to 320 °C. Then, a second irreversible drop occurred between 320 °C and 380 °C. During the subsequent cooling, δ decreased results from elastic anharmonicity.[6]

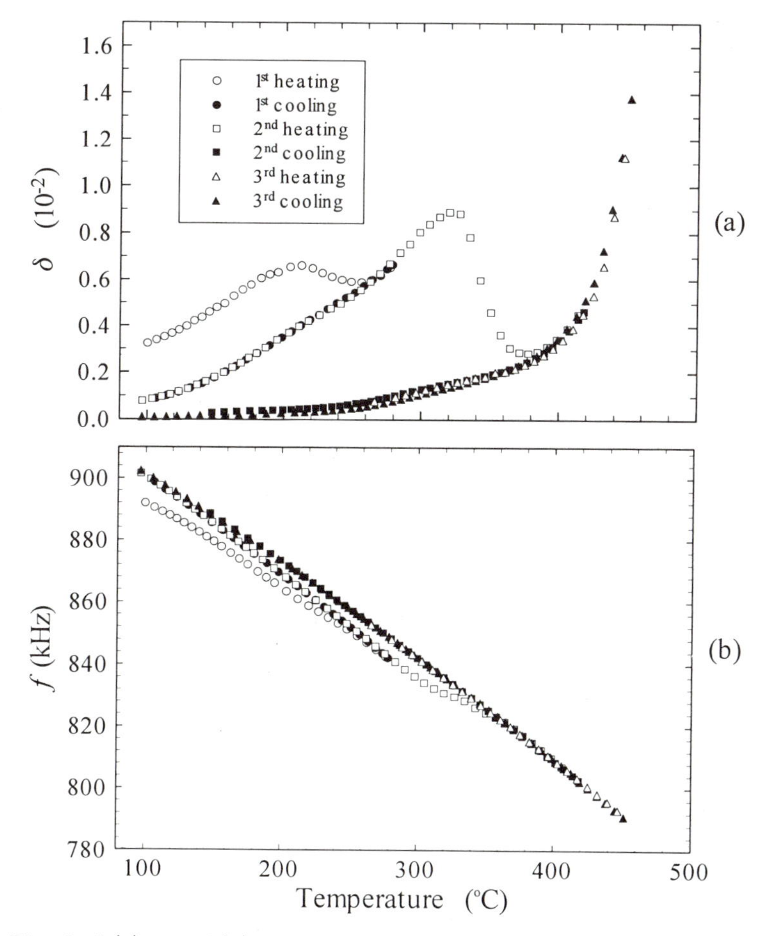

Fig. 1. δ (a) and f (b) of a $T_{1,1}$ mode of sample AL-7 vs temperature.

monotonically. The damping during the third thermal cycle followed the curve for the second cooling and showed no irreversible changes, increasing monotonically up to the highest measured temperature (465 °C).

Measurements of f during this annealing sequence are shown in Fig. 1(b). Relative increases in f during the first and second heatings coincide with the drops in δ. These are superimposed on the normal approximately linear decrease with temperature that

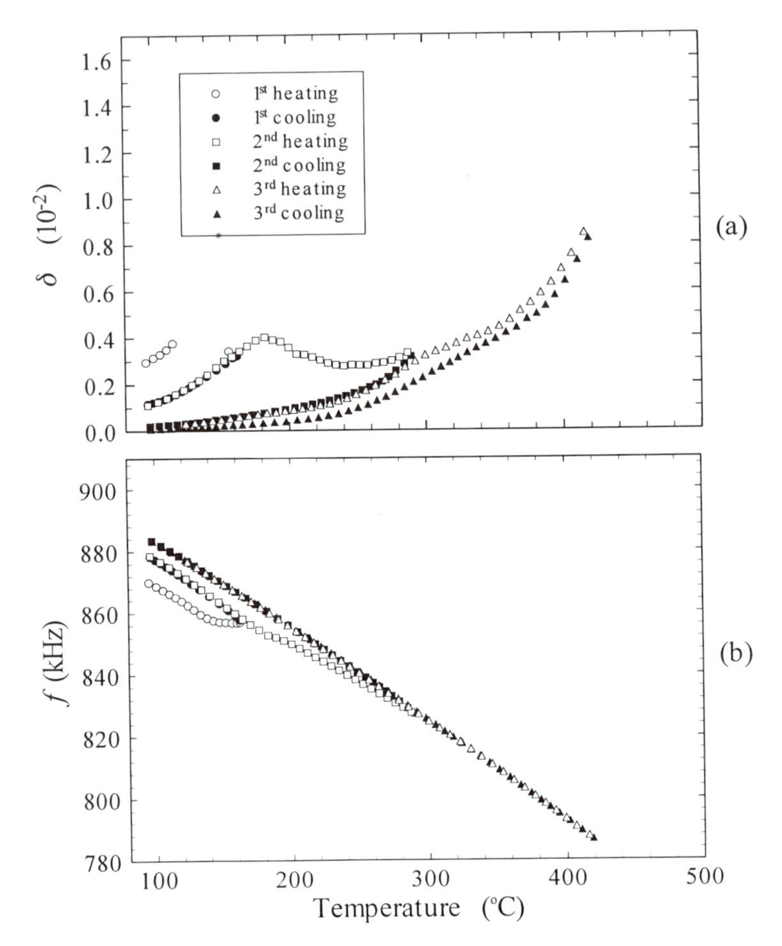

Fig. 2. δ (a) and f (b) of a $T_{1,1}$ mode of sample AL-8 vs temperature.

Similar measurements of δ and f during ramped anneals of sample AL-8 are shown in Fig. 2. As with AL-7, large irreversible changes in δ and f occur in two stages. However, the temperature ranges of these changes are lower than those for AL-7. (Reliable measurements of δ between 115 °C and 156 °C during the first heating were not obtained, but the corresponding measurements of f are shown.) During the third thermal cycle, f was essentially reproducible, but δ decreased slightly.

The irreversible changes in δ and f during the first thermal cycle are believed to result from recovery, and those during the second thermal cycle are believed to result from recrystallization. The temperature ranges of the changes during the second cycle are consistent with the ranges for recrystallization determined by Anderson and Mehl[7] for 99.97% aluminum and Bay and Hansen[8] for 99.4% aluminum, considering that recrystallization is more rapid for higher-purity or more heavily deformed material.[2] Also, the fact that the irreversible changes occurred at lower temperatures in AL-8 than AL-7 is consistent with the known dependence of recovery and recrystallization on the degree of deformation. Since the anelastic response of dislocations results in damping and a decrease in elastic constants,[9] the signs of the changes in δ and f are consistent with the decrease in dislocation densities expected during recovery and recrystallization.

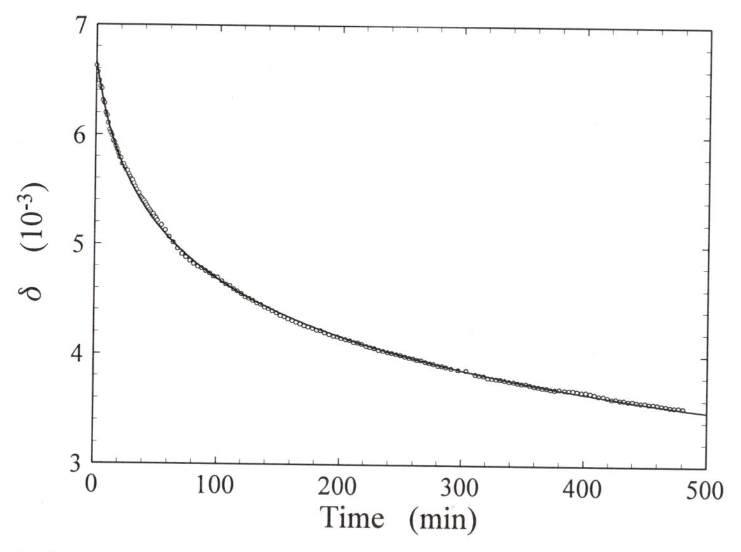

Fig. 3. δ of a $T_{1,1}$ mode of sample AL-5 vs time at 186.3 °C (first heating).

Isothermal recovery anneal

The association of the first irreversible change with recovery is supported by measurements of δ during an isothermal anneal of sample AL-5. The cold-worked sample was heated to 186.3 °C in 75 minutes and held at this temperature (±1.2 °C) for 480 minutes. As shown in Fig. 3, δ decreased monotonically with time by approximately the same total amount as that induced by the ramped anneal of sample AL-7, which had a similar amount of cold work (Fig. 1(a)).

A simple exponential does not provide an accurate fit to the data in Fig. 3, because it cannot reproduce the relatively rapid drop at early times. The standard deviation of an exponential least-squares fit (not shown) is 7.5×10^{-5}. The general shape of the curve is typical of that found for mechanical properties such as hardness and yield stress during recovery.[10] Property changes proceed relatively rapidly in the initial stage of recovery as dislocations are annihilated and then proceed more slowly as more stable structures are formed from the surviving dislocations. Several studies[10] support the idea that recovery kinetics in polycrystalline metals are dominated by thermally activated glide or cross-slip and that the internal stress σ_i is related to the time t according to

$$\sigma_i = \sigma_0 - A \ln\left(1 + \frac{t}{t_0}\right), \tag{1}$$

where σ_0 and A are functions of temperature.[11] For example, Barioz *et al.*[12] obtained results on Al-1%Mg that are consistent with this equation. The dislocation density is found to be approximately proportional to σ_i^2 in a wide range of materials.[10] Therefore, if the high-temperature damping were caused by a dislocation interaction with a magnitude proportional to the dislocation density, one would expect the time dependence of δ during recovery annealing to be given by

$$\delta = \left[\delta_0 - B \ln\left(1 + \frac{t}{t_0}\right)\right]^2, \tag{2}$$

where δ_0 and B are temperature dependent. The solid line in Fig. 3 is a least-squares fit of this form with δ_0=8.18×10^{-2}, B=6.20×10^{-3}, and t_0=13.13 min. The fit closely follows the data, having a standard deviation of 1.5×10^{-5}.

CONCLUSION

Irreversible changes in δ and f are observed in two stages during heating of pure polycrystalline aluminum that has been cold-worked 10% and 21%. The dependences on temperature, time, and degree of cold work indicate that these changes result from recovery and recrystallization. These results support the conclusion that high-temperature damping in aluminum is caused by an anelastic interaction with dislocations, and they lead to the suggestion that ultrasonic measurements have potential for monitoring recovery and recrystallization during production of commercial alloys. The very large magnitude of the irreversible changes in δ may make such measurements feasible in an industrial environment.

REFERENCES

1. W. Johnson, Ultrasonic damping in pure aluminum at elevated temperatures, *J. Appl. Phys.* (submitted for publication).
2. R. W. Cahn, Recovery and Recrystallization, in: *Physical Metallurgy*, R. W. Cahn and P. Haasen, ed. North-Holland, New York (1996).
3. W. Johnson, Ultrasonic resonance of metallic spheres at elevated temperatures, *J. Phys. IV* (supplement to *J. Phys. III (France)*) 6:C8-849 (1996).
4. W. Johnson, S. J. Norton, F. Bendec, and R. Pless, Ultrasonic spectroscopy of metallic spheres using electromagnetic-acoustic transduction, *J. Acoust. Soc. Am.* 91:2637 (1992).
5. Y. Satô and T. Usami, Basic study of the oscillation of a homogeneous elastic sphere, *Geophys. Mag.* 31:15 (1962).
6. J. A. Garber and A. V. Granato, Theory of the temperature dependence of second-order elastic constants in cubic materials, *Phys. Rev. B* 11:3990 (1975).
7. W. A. Anderson and R. F. Mehl, *Trans. AIME* 161:140 (1945).
8. B. Bay and N. Hansen, Recrystallization in commercially pure aluminum, *Metall. Trans. A* 15A:287 (1984).
9. A. V. Granato and K. Lücke, The vibrating string model of dislocation damping, in: *Physical Acoustics, Vol. IVA*, W. P. Mason, ed., Academic, New York (1966).
10. F. J. Humphreys and M. Hatherly, *Recrystallization and Related Annealing Phenomena*, Elsevier Science, Tarrytown, NY (1995).
11. A. H. Cottrell and V. Aytekin, The flow of zinc under constant stress, *J. Inst. Metals* 77:389 (1950).
12. C. Barioz, Y. Brechet, J. M. Legresy, M. C. Cheynet, J. Courbon, P. Guyot, and G. M. Ratnaud, in: *Proc. 3rd Int. Conf. on Aluminium*, Trondheim, p. 347 (1992).

INVESTIGATION OF MATERIAL PROPERTIES BY ACOUSTIC METHODS ON THE BASE OF INELASTIC MECHANICAL PROCESSES

O.V.Abramov, O.M.Gradov,

Laboratory of Ultrasound Technology,
N.S.Kurnakov Institute of General and Inorganic Chemistry of Russian Academy of Sciences, Leninsky prosp.,31, Moscow, 117907, Russia.,

Tel.: (095)-955-48-38, Fax: (095)-954-12-79.

INTRODUCTION

Nonelastic mechanical behaviour of a material is determined by the kinetics of load-activated processes of variation of the initial structure, more particularly, by movements of atoms between locally stable equilibrium positions along a self-agreed potential relief. As a rule, the change of atomic structure in that case reduces to the formation and subsequent synchronous transfer of a plurality of atoms of given types with retention of the short-range order. Such a stable atomic structure, a defect, can be regarded as a single quasi-particle. In terms of defective structure, the state of material is uniquely determined by the distribution of defects of various types and the mechanism of macro behaviour of the material under load is determined by the dynamics of formation and transfer of defects.

A general set of equations which describe the generation and interaction of elastic waves during the occurrence of collective inelastic processes of the evolution of defect structure is necessary to define the material properties under the certain load. The set has to consist of kinetic equations for density distributions of defects of various types on loading and of the equation for the elastic motion of a material describing the generation and propagation of elastic waves in a material the amplitude of which together with the level of a static stress applied corresponds to the magnitude of total force applied to the material. The elastic waves are generated in a material while the evolution of many defects of different types which act in this volume as sources or receivers of ultrasound impulses.

The above set describing the material behaviour in the general case of the inelastic mechanical process makes it possible to study non-linear acoustic effects which are related to the processes of the defect structure development in materials. In this approach, the general model of the material structure evolution on ultrasound loading used in applied problems opens up the way to the constructive analysis of the problem of improving the material properties with using the effect of the high intensity ultrasound as well as the questions of the degradation of material properties and a material failure as a

Nondestructive Characterization of Material VIII
Edited by Robert E. Green Jr., Plenum Press, New York, 1998

result of ultrasound loading of a special type or vibrations. The generation of acoustic emission signals accompanied the defect evolution gives the possibility to detect the changes and to determine the corresponding material properties.

BASIC RELATIONSHIPS

Let us separate a region in the material in the initial state and analyse the time dynamics of the formation of various defects under the action of stress. The general case of a system of interacting defects will be considered. In the frames of the present approach, the description of the behaviour of material under load is based on the use of a system of kinetic equations for the densities of defects (quasi-particles) of various types [1]. Averaging by the characteristics of collisions of defects (considering only pair collisions), we can write the following equation:

$$\dot{\rho}_i = k_i + k_{ij}\,\rho_j + k_{ijl}\,\rho_j\,\rho_l + \mathrm{div}\mathbf{J}_i \qquad (1)$$

where $\mathbf{J}_i$ is the flow of defects of the given type, $\mathbf{J}_i = \mathbf{v}\rho_i + D_j \ \mathrm{grad}\ \rho_i$; $\mathbf{v}$ is the drift velocity of a defect; D_j is the coefficient of diffusion; k_{ijl} is the intensity of interaction of defects of the given types j, l with the formation of a defect of type i. In cases when the formation of defects occurs by collision of defects of different types, $k_{ijl} = \langle v\rangle S_{ijl}$, where $\langle v\rangle$ and S_{ijl}, are the average values of the relative velocity of motion of defects and the cross section of collisions and the point upstairs denotes the derivative on the time. In cases when the direct generation of a defect takes place, k_i has the sense of the intensity of defect formation in the initial medium. According to thermodynamic models,

$$k_i = k_{i0}\exp[-(V - a\sigma / kT)] \qquad (2)$$

where k_{i0} and a are the parameters of the material corresponding to the formation of a defect of the given type; V is the energy of defect formation; k is Boltzmann's constant; T is the temperature; and σ is the stress in the given point.

Since the movement of a defect is associated with the transfer of a plurality of atoms between successive stable states, the expression for the velocity of moving defects $v(\sigma)$ has essentially different form depending on the type of defect and on stress. In a particular case $a\sigma < V_i$ (where V_i is the magnitude of energy barrier between successive stable states), the fluctuating nature of defect movement is inessential and the equation for defect velocity should have the standard form:

$$\dot{\mathbf{v}} + (\mathbf{v}\,\mathrm{grad})\mathbf{v} + \gamma\ \mathbf{v} = \mathbf{F} \qquad (3)$$

where γ and $\mathbf{F}$ are respectively the resistance coefficient and the force acting on the defect, both normalised for the "mass" of defect.

The stress distribution in the material is given by considering the boundary conditions and elastic motion in the material [2]. In the simplest case with $\sigma \sim u$ where $\mathbf{u}$ is the displacement vector

$$\ddot{\mathbf{u}} - c_l^2 \Delta \mathbf{u} - (c_l^2 - c_t^2)\mathrm{grad\,div}\,\mathbf{u} = \mathbf{P}(\rho, t) \qquad (4)$$

Here $c_{l,t}$ are sound velocities and **P** is the force acting on the material, in particular, associated with the presence of a defect.

The system of equations (1) - (4) makes it possible to describe movements in the material in the general case of inelastic mechanical processes, analyse the excitation of ultrasound in the material in the course of mechanical processes, in particular, at a constant stress and the evolution of a defective structure under the action of powerful ultrasound. It should be noted that, according to system (1) - (4), the passage of ultrasound through a substance determines the activation of oscillations and waves both in the initial medium and at the plurality of densities of quasi-particles. It is then possible to separate two limiting cases of the behaviour of a system of quasi-particles.

The intensity of the processes of "inelastic" collisions of defects and of the processes of formation of defects under load is negligibly small.

On the contrary, the intensity of interaction of defects is high. The effect of ultrasound determines the formation of defective structures in the fields of variable stresses. The behaviour of a system of defects is described by a set of kinetic equations of the type of (1) for the densities of the defects whose formation is permissible in the material of the given type.

In the frames of the given limiting case, let us consider the model of a process in which the structure of metallic material is changed under the action of powerful ultrasound.

An ultrasonic wave $u = u_0 \exp\{i(\omega t\text{-}kx)\}$ is introduced into a specimen from the surface and causes periodic movement of mobile dislocations. For determining the qualitative nature of the structures that are formed, we consider the simplest case of an one-dimensional model. In the process, mobile dislocations can be locked on collision with an obstacle; collision of two dislocations of opposite sign leads to the formation of a dipole, i.e. a stable dislocation structure, whereas a common fixed dislocation simply departs from the locking point if the direction of stresses changed to the opposite one. The corresponding set of equations for the densities of mobile, ρ , and fixed dislocations, ρ_1 , and dipoles, ρ_D , has the form:

$$\dot{\rho} = S(\rho) - \alpha_1\rho\rho_1 - \alpha_2\rho\rho_D - \alpha_3\rho^2 + \beta\rho_1 + D\Delta\rho$$

$$\dot{\rho}_1 = \alpha_2\rho\rho_D - \alpha_1\rho\rho_1 - \beta\rho \qquad (5)$$

$$\dot{\rho}_D = \alpha_1\rho\rho_1 + \alpha_3\rho^2 - \gamma\rho_D$$

where $S(\rho)$ is the velocity of work of dislocation sources; $\alpha_{3,2,1}$ are the velocities of collision of mobile defects with one another, with dipoles, and with fixed defects; β is the velocity of breakage of a fixed defect from an obstacle; and γ is the velocity of disappearance of dipoles due to rupture and annihilation. The densities of dislocations of opposite signs are assumed to be the same so that the dislocation flow is determined solely by the diffusion term. The reaction rate constants in (5) are considered to be averaged for the period of passing wave $t_0 = 2\pi/\omega$. Assuming for the dislocation velocity $v \sim \sigma$ (as for instance in [3]), we have

$$\alpha_i \cong S < \sigma >; \qquad < \sigma >= \sigma / \sqrt{2}$$

In accordance with [3]

$$\gamma = \exp[-(V_D - a < \sigma >) / kT]$$

$$\beta = \frac{1}{t_0} \int_0^{t_0} dt \ \exp\left\{ - \frac{V_1 - a_1 < \sigma >}{kT} \right\} \approx \frac{1}{2}$$

Thus, with the given approximation when the velocity of transient processes is neglected, the coefficients in (5) depend on the amplitude characteristics of ultrasonic wave.

If a dislocation emerges onto the surface of specimen or onto the boundary of crystallite situated at planes $x = 0$ and $x = x_0$ it is absorbed. This corresponds to the boundary conditions $\rho(x=0) = \rho(x=x_0) = 0$.

Noting that $\beta >> \gamma$, we can make an adiabatic approximation in (5) in the vicinity of equilibrium state

$$\rho_1 = \frac{\alpha_2}{\beta^2} \rho \rho_D (\beta - \alpha_1 \rho) \tag{6}$$

Substituting from (6) into (5), we obtain, after redesignation of variables, the following set of equations

$$\begin{aligned} \dot{\rho} &= S(\rho) - b_1 \rho^2 \rho_D + b_2 \rho^3 \rho - \alpha_3 \rho^2 + \beta \rho_1 + D \Delta \rho \\ \dot{\rho}_D &= b_3 \rho^2 \rho_D - b_3 \rho^3 \rho_D + \alpha_3 \rho^2 - \gamma \rho_D \end{aligned} \tag{7}$$

Consider the case of formation of new dislocations on movement of the existing ones, in particular, by the mechanism of cross slip. The velocity of a single act of cross slip is determined according to (3). We consider in that case that the dislocation structures which are formed in the material create an internal stress

$$\sigma = \sigma_{ex} - \sigma_1 \rho_D$$

Substituting the expression for σ into (3) and expanding by the lowest value of ρ_D we obtain

$$S(\rho) = \rho(S_0 - S_1 \rho_D); \qquad S_0, S_1 = \text{const}$$

From analysis of the behaviour of the linearized system for Fourier images ρ_i

$$\rho_i = \int dx\ \rho_i(x)\exp\{i\,qx\}$$

we find that the system is in a stable state if $q > q_c = \sqrt{S_0/D}$. It should be noted that, in accordance with the boundary conditions, the formation of a periodic structure with the discrete spectrum $q = nk_0$ ($n = 0, 1, 2, \ldots, k_0 = \pi/2x_0$) is permissible (for determining the values of q, we consider the approximation $\rho_D \sim \rho$).

In the vicinity of q_c, it is permissible to use the adiabatic approximation by the variable ρ_D

$$\rho_D = \frac{\alpha_3}{\gamma^2}\rho^2(\gamma - b_2\rho^3 + b_3\rho^2) \tag{8}$$

Substituting from (8) into (7) and redesigning the variables, we obtain with an accuracy to ρ^5

$$\dot{\rho} = S(\rho) - c_1\rho^4 + c_2\rho^5 - \alpha_3\rho^2 + \beta\rho_1 + D\Delta\rho \tag{9}$$

The solution of (9) is sought in the form

$$\rho = A(t)\exp\{iqx\} + \text{c.c.} \tag{10}$$

Substituting from (10) into (9) and separating the terms corresponding to the periodic distribution $\rho \sim \exp\{iqx\}$, we obtain

$$\dot{A} = (S_0 - Dq^2)A - S_1|A|^2 A + c_2|A|^4 A \tag{11}$$

A similar expression can be derived for A*.

Depending on the values of parameters in equation (11), its solution may correspond to the following cases:

- (I) absence of a periodic dislocation structure;
- (II) formation of a periodic stationary dislocation structure of dipole nature ($4S_0c_2 < S_1^2$):

$$\rho_D = \sum_{n=1}^{n_0} A_n \sin(nk_0x); \qquad n_0 = \left[\frac{S}{k_0D}\right];$$

$$A_n = \sqrt{\frac{S_1 - \sqrt{S_1^2 - 4c_2(S_0 - Dk_0n)}}{2(S_0 - Dk_0n)}} \tag{12}$$

- (III) development of instability; in that case, dislocation density increases unlimitedly ($S_1^2 < 4c_2(S_0 - Dk_0n)$)

$$A_n(t) = A_{n_0}\exp\{(S_0 - Dk_0n)t\} \tag{13}$$

Thus, depending on the amplitude characteristics of ultrasound, its effect in a material can result in either the formation of stable periodic dislocation structures, which gives strengthening of the material, or in an unlimited (within the frames of the approximations used in the model) rise of dislocation density, which in the general cage corresponds to degradation of the mechanical properties of materials.

On another side, the internal stress formed in the material has, evidently, the same space structure as (12) and produces the elastic waves in accordance with (4) where this stress defines the right side of the equation. So, it means that the evolution of such a defect structure in the material may be easily detected by the measure of the acoustic emission signal with the correspondent wave-length and frequency.

CONCLUSIONS

The specific case of the wave propagation over a multitude of quasi particles which are represented by the defects of various types which are formed and move on loading produced by the external wave has been studied in details. The relevant system of equations includes also the elastic phenomena so that it becomes possible to describe the response caused by the defect structure evolution. The solution of this system determines the behaviour of the defect structure, the redistribution of defects and the formation of their agglomerates during high intensity ultrasound loading. These processes have been analysed and their characteristics examined as functions of material parameters and external effects. The possibility to detect the effects discussed with help of acoustic emission signals is pointed out.

REFERENCES

1. A.K.Head. Dynamics of dislocation groups. *Phil.Mag.*, vol. 26, № 1, p.43 - 53 (1972).
2. L.D. Landau, E.M.Lifshits. *Theory of Elasticity, Nauka*, Moscow (1987) (Russia).
3. *Physical Metallurgy*. Ed. by R.W.Cahn. North-Holland, Amsterdam (1983).

STUDY OF MICROPLASTICITY OF SOLIDS BY LASER INTERFEROMETRY METHOD

Vitaly V. Shpeizman,[1] Nina N. Peschanskaya,[2] Pavel N. Yakushev[2]

[1]Department of Physics of Plasticity
[2]Department of Materials Dynamics
A.F.Ioffe Physico-Technical Institute of the Russian Academy of Sciences
S. Petersburg 194021, Russia.

INTRODUCTION

The laser interferometry method has a considerable opportunity for nondestructive characterization of materials. The advantages of this method are: inertialessness, high stability to external action (magnetic or electric fields, radiation), absence of mechanical contacts with measured subject, remote measuring, i.e. all components of device except one reflector may be placed at the large distance from measured subject; it may combine large range of registered displacement (from one tenth of micron up to 1 meter) with a constant division of deformation scale. The method is precise and absolute, calibration is not needed. And the main advantage is its high sensitivity and very small deformation base of measurements.

Starting about 20 years ago we continue to find the new directions of the application of laser interferometry method. At present we have the following directions:

1. Brittle - ductile transition
2. Initial stage of plastic deformation
3. Relaxation transitions in polymers
4. Brittle fracture of ceramics
5. Superconducting transition and small deformations in high-Tc superconductors
6. Internal stresses
7. Thermal expansion
8. Instability of deformation

Three last of them were added recently. More than a half of mentioned applications are connected with microplasticity.

Nondestructive Characterization of Material VIII
Edited by Robert E. Green Jr., Plenum Press, New York, 1998

EXPERIMENTAL PROCEDURE

We constructed a few installations using the laser interferometry method. One of them is shown in Figure 1. Its operation is following. The sensor of deformation and deformation rate is the mirror 2 rigidly connected with a sample 14. The type of loading may be various - tension, compression (as in the Figure 1), bending - the displacement of one part of a sample in relation to its immobile part is measured. The principle of the optic device operation is similar to those of Maikelson's interferometer. The laser beam is divided in two parts by the semitransparent mirror 6. The first part reflects from the sensor mirror 2 and then reaches the semitransparent mirror 5 where it meets the second part of the beam which by this time has passed the delay line including mirrors 3 and 4. The first part of the beam has a frequency changed in accordance with Doppler's effect if a strain rate of the sample is not a zero. The second part of the beam has an initial frequency of laser 1. The interference of two high-frequency oscillations with slightly distinguished frequencies leads to formation of low-frequency beatings which are transformed by means of photoreceiver 9 in electric signal. Its frequency is proportional to the deformation rate and the number of oscillations is proportional to the deformation value. The coefficient of proportionality is equal to a half of the laser wave length (for a helium-neon laser it is 0.3μm, approximately). So, a half-wave period of a sine-shaped interferogram corresponds to 0.15μm of deformation. The calculation procedure is reduced to the counting of half-waves in a definite time interval. If one needs a higher precision it may be possible to use a computer treatment of interferograms that is in the mathematical description of their deflection from the sine shape. Mirror 7 is served for subtracting the constant component of the beam.

SPECTRA OF THE SMALL INELASTIC DEFORMATION RATES

An original technique of measurements called "the temperature spectrum of the small inelastic deformations rates" was developed. Measurements carried out as follows: temperature was increased by steps in some interval and a sample was loaded by short impulses (3-30 sec) of small stresses (10-50% of the yield or fracture stresses) on each step. Thus, series of short-term creep curves were obtained. Then if any way of the strain rate calculation is chosen one can draw its temperature dependence. The reversibility of microplastic deformation is the physical base of this technique, so in each new loading we deal with undistorted structure and the strain rate-temperature dependence presents the properties of studied structure itself. It appeared that these dependences for various materials are nonmonotonic if measured in a wide temperature interval. It was the reason for us to call them as "spectra of small inelastic deformation rate".

The spectra of various steels was studied[1]. It was shown that the position of spectrum peaks didn't depend on the value of stresses, on the time of their action, on type of stress state, on the presence of stress concentrators. On the contrary, the spectra and their maxima differed if samples with various structure were compared (see Figure 2). These properties of spectrum gave us the opportunity to use it for determination of the tendency of steels and other materials to brittle fracture. It is obvious that the local deformation and brittle fracture are competitive processes. If the local deformation develops actively the brittle fracture doesn't occur. Thus, the temperature decrease below some peak of the spectrum may result in ductile-brittle transition. In Table 1 the critical brittle temperatures obtained from the spectra and in shock tests and by fracture mechanics methods are compared.

It follows from the comparison of traditional and new characteristics that they are closed. The method of spectrum is preferable because it is the single method, it substitutes the different ones and gives the single characteristics. Besides it is a physical method, giving an exact point while the former methods are conditional and their criteria are often arbitrary. At last the method of

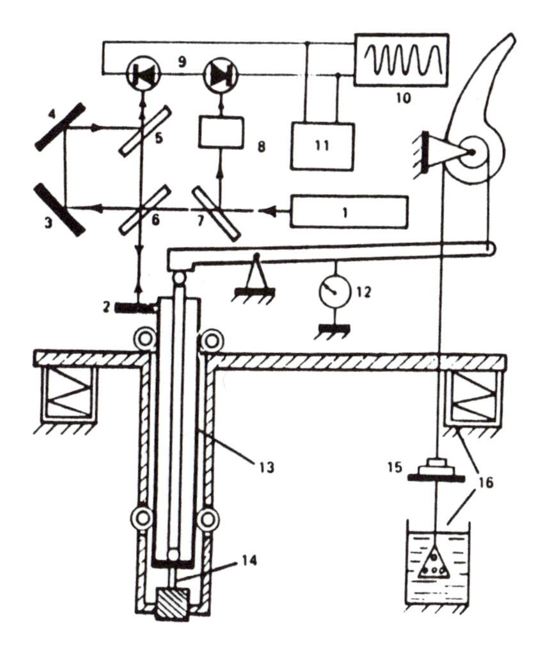

Figure 1. Scheme of the installation: (1) laser, (2-7) mirrors. (8) polarizer, (9) photodiods, (10) recorder, (11) block of data processing. (12) displacement indicator, (13) plunger, (14) sample, (15) weight, and (16) dampers.

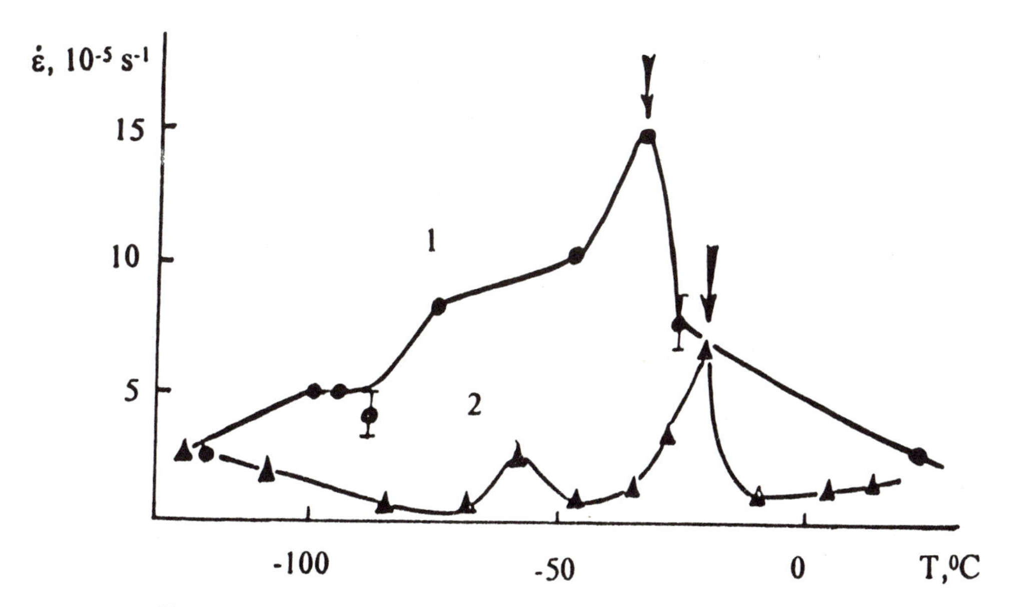

Figure 2. Spectra of the small inelastic deformation rate for the cast steel with 0.2%C,1 - annealed at 900°C, compression stress 200MPa, 2 - quenched from 900°C into water 280MPa. The arrows correspond to the critical temperature of brittleness in shock tests.

Table 1. Comparison of the critical temperatures of brittleness

N	Steel	Thermal treatment	Mechanical properties				T_{cr}^{ε},K	T_{50},K	T_{cr}^{K1c} K
			σ_b,MPa	σ_y,MPa	δ,%	ψ,%			
1	0.1%C 1%Mn 0.1%Ti	normalized	410	263	34.0	70.7	237	233	--
2	0.1%C 1%Mn 0.1%Ti 1%Cu	normalized	439	289	36.0	72.7	241	233	--
3	0.12%C 1%Mn 0.1%Ti 0.2%V	normalized	563	391	34.2	68.9	253	251	--
4	0.2%C 1%Mn 0.1%Ti	double normalized	571	436	21.9	42.0	258	261	--
5	0.4%C 1.5%Cr 0.2%Mo	quenched (850^{0}C) tempered (650^{0}C)	970	860	13.0	71.0	218	--	200
6	0.4%C 1.5%Cr 0.2%Mo 0.15%V	quenched (877^{0}C) tempered (550^{0}C)	920	770	14.0	61.0	300	--	310

Note: 1-4 - cast steels, 5,6 - deformed steels

spectrum requires only one sample of small size whereas the former ones require a lot of special samples and a large amount of tests.

Another example of the inelastic strain rate spectrum application is the study of microplasticity of extremely brittle solids[2]. Due to the high resistivity of the method it is possible to observe insignificant deformation (less than 1μm) of ceramics (Figure 3). As it was observed for steels the deformation rate of porous ceramics decreased rapidly. However, deformation didn't stop, but developed in an extremely nonuniform manner. Seldom oscillations with different period (average deformation rate - 10^{-4} -10^{-5} s^{-1}) appear suddenly in the interferograms at random times. These bursts of deformation correspond to hundredths and thousandths per cent, after which deformation doesn't develop. The nature of deformation in this case can be considered to be a reflection of crack The most interesting is the last stage of endurance before the final fracture of porous ceramics. When the ceramics are held under a constant compressive load, some bursts of deformation correspond to visible cracks near the ends of the sample. The laser interferometry method makes it possible to study the details of delay fracture process: the crack initiation time (incubation period) as well as nonuniformity of the rate, anomalies in the development, and the duration of deformation associated with cracks[3].

The spectra of small inelastic deformation for superconductive ceramics are studied as well. It was shown[4] that they have a peak near the temperature ofthe superconductive transition. The fact that the superconductive transition influences on the strain rate was directly established in experiments with breaking of superconducting state by electric current or magnetic field. Figure 4 shoes that strain rate decreases strongly (down almost to zero) at the moment when the current larger than the critical one is switched on. The subsequent deformation in normal state occurs at a

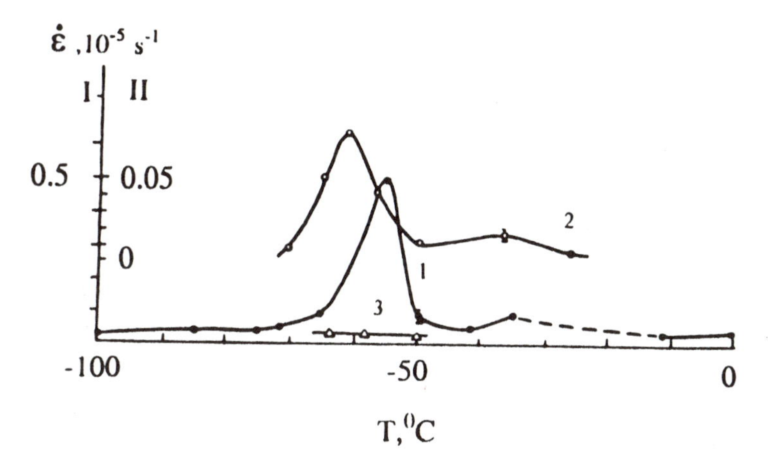

Figure 3. Spectra of creep rate of glass ceramics (1, scale), silicon nitride (2, scale II) and base line of installation (3). Stress σ = 26 MPa.

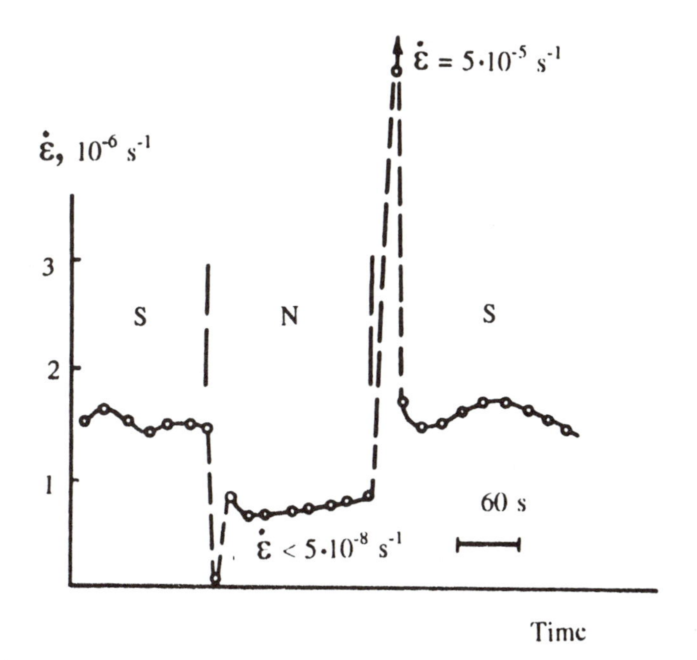

Figure 4. Changes of deformation rate of Y-Ba-Cu-o ceramics under switching on and off of a 30 A/cm2 current. S is the superconducting state, N is the normal state. Stress σ = 12 MPa, T = 77 K

lower rate than before the current flow when the sample was in superconducting state. The opposite effect is observed when the current is switched off. In this case the strain rate increases abruptly and the subsequent deformation goes on with the same rate as before switching of the current on. From the framework of the physics of deformation the strain rate increase in superconducting state may be connected with easier breaking of dislocations with impurities or other centers of pinning due to diminishing of the electron component of viscous drag of dislocations or due to changing the frequency of dislocation attack of obstacles if one takes into consideration the dislocation motion as thermal activated process.

CONCLUSION

Thus, the laser-interferometric method developed provides new possibilities for studying microplasticity due to its high precision and very small strain base of measurements. It is demonstrated that spectra of small inelastic strain rates give the information on various structural transformations in solids. The physical nature of some of them is known (e.g. the superconductive transition), other peaks of spectra correspond to structural (relax) transitions determining the process of deformation. It is practically important that the new technique is useful in the prediction of anomalous changes in mechanical behavior of solids, including the brittle-ductile transition. On the basis of this technique the new characteristics of the tendency to brittle fracture may be proposed.

REFERENCES

1. V.V.Shpeizman, N.N.Peschanskaya, A.K.Andreev, G.E.Kodzaspirov, Yu.P.Solntsev, P.N.Yakushev, Otsenka sklonnosti stalei k hrupkomu razrusheniyu, *Probl. prochn.* 7:115 (1987), in Russian .
2. N.N.Peschanskaya, N.A.Zlatin, V.V.Shpeizman, Peculiarities of deformation of porous brittle materials, *Zh.Tekh.Fiz* 57:1438 (1987).
3. N.A.Zlatin, N.N.Peschanskaya, P.N.Yakushev, Microplasticity in ceramics, *Zh.Tekh.Fiz.* 57:1419 (1987).
4. B.I.Smirnov, T.S.Orlova, V.V.Shpeizman, Defect structure and physico-mechanical properties of ceramic high temperature superconductors, *J.Mech.Behav.Mater.* 5:325 (1994).

INTERATOMIC INTERACTION and ALLOYING CRITERION for FERRITIC ALLOYS

Igor S. Golovin

Moscow Aviation Technology University - MATI, Moscow 103767 Russia

INTRODUCTION

The loss maximum in b.c.c. metals due to stress-induced reorientation of interstitial atoms (IA) known as Snoek relaxation is described by the Debye equation:

$$Q^{-1} = \Delta \cdot \frac{\omega\tau}{1+(\omega\tau)^2}. \qquad (1)$$

The loss maximum is centered at $\omega \cdot \tau = 1$ with height $Q^{-1} = \Delta/2$, where Δ is the relaxation strength, $\omega=2\pi f$; f is the measuring frequency. The Snoek relaxation strength in solid solution of b.c.c. metals is proportional to the concentration C_0 of the solute IA, to the difference of main values of the dipole tensor with tetragonal symmetry $(\lambda_1 - \lambda_2)$ and to the orientation parameter Γ [1]:

$$\Delta = 2Q_m^{-1} = \frac{4}{3}\frac{C_0 V}{kTG^{-1}(\Gamma)}(\lambda_1 - \lambda_2)^2 \Gamma, \qquad (2)$$

where V is the molecular volume, G is the shear modulus.

An Arrhenius equation gives the temperature dependence for the inverse relaxation time for IA in metals. The reorientation of IA under the applied stress is the elementary step of diffusion "under the stress" and leads to the corresponding internal friction (IF) peak [2]:

$$Q^{-1}(T) = Q^{-1}\frac{T_m}{T}\cosh^{-1}\left\{\frac{H}{r_2(\beta)k}\left(\frac{1}{T}-\frac{1}{T_m}\right)+\ln\frac{f(T)}{f_m}\right\}, \qquad (3)$$

where H is the activation enthalpy, k is Boltzman's constant, T is temperature, T_m is temperature of maximum, $r_2(\beta)$ is the parameter of peak broadening. Equation (3) includes additional correction from the temperature dependence of the relaxation strength, the peak broadening and the temperature variation of the elastic modulus which is important for high temperatures. These corrections become important if temperature of peak is high.

The carbon Snoek relaxation parameters in Fe are [3]: $H \approx 0{,}87$ eV; τ^{-1}, $10^{14} = 5.3 \pm 2\ s^{-1}$; $T_m = 314$ K; $|\lambda_1 - \lambda_2| = 0.83$. The effect of substitutional solutes on the Snoek relaxation of carbon (the decrease in peak height) was first reported for Fe-0,5% V, Cr, Mn, Ni and Mo by C.Wert[4] in 1952. Later on this effect was confirmed in many papers and the additional high-temperature Snoek-type peak appearance was established[5,6] for iron-based alloys with Cr and

Nondestructive Characterization of Material VIII
Edited by Robert E. Green Jr., Plenum Press, New York, 1998

Mo content more than 1%. The study of nitrogen influence on temperature dependent IF (TDIF) spectra in iron and N and O in V, Nb and Ta is recently carried out by Koiwa et.al.[7,8] The carbon Snoek relaxation for ternary iron based alloys (Fe-C-Me) is not yet well analyzed. The reason for that is rather complicated processes in substitutional solid solution of iron based alloys which takes place due to low temperature treatment.

EXPERIMENTAL PROCEDURE

IF measurements were carried out with different low frequency techniques:

1) free decay inverted torsion pendulum (f≈1 Hz, magnetic field $2\cdot10^4$ A/m),
2) forced, computer controlled, inverted torsion pendulum (0.01-9 Hz).

Activation enthalpy were determined with the use of temperature-frequency shift and/or Marx-Wert equation. In order to study interatomic interaction of carbon (C) and substitutional atoms (s) in b.c.c. solid solution for ferritic steels the TDIF spectra of the following binary alloys (with minimum N,%) were studied: Fe-Cr (Cr[5,6,9] up to 6% and from 16 to 35% Cr[10,11]), Fe-Mo[5,6,12] (up to 5%), Fe-Al (up to 24%[11], to 30%[13,14]), Cr-Fe[16,17] (up to 50%). The details of the alloys chemical composition, technique used and calculation methods are given in the references cited.

RESULTS

Two Snoek-type peaks are observed in the low alloyed (0,5÷5 at.%) by carbide-forming elements ternary systems (Fe-Cr-C, Fe-Mo-C (Fig. 1)): the first (≈39°C, 1 Hz) peak with the activation enthalpy close to the activation enthalpy of carbon in pure Fe ($H_0 \approx 0.8$ eV), the second one at higher temperature with the activation enthalpy close to the activation enthalpy of the carbon diffusion in b.c.c. lattice of the second component. The only one broadened Snoek peak is observed in the high alloy steels with the concentration of Cr or Mo more than 5 at.%. One broadened Snoek peak is observed in b.c.c. Fe-Al alloys at temperatures higher than that is in iron. Corresponding activation energies are generalized in Fig. 2.

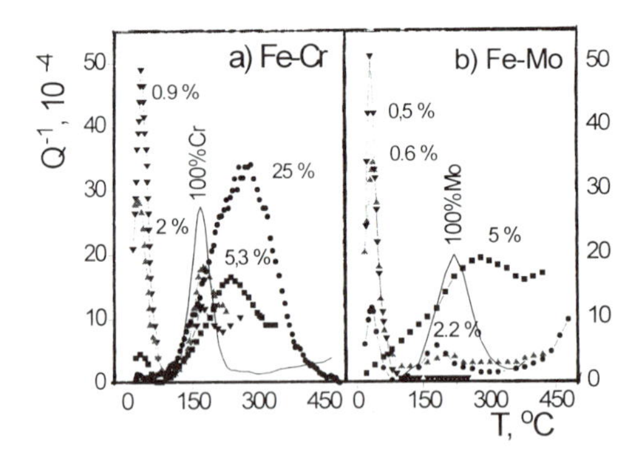

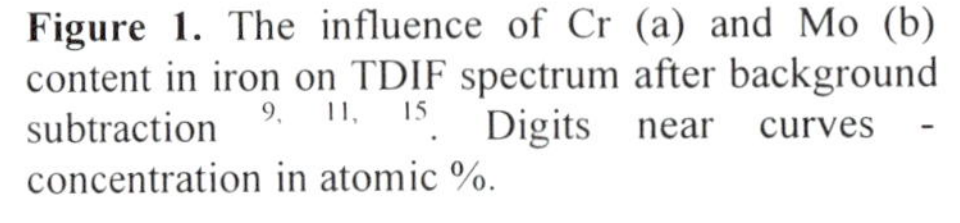
Figure 1. The influence of Cr (a) and Mo (b) content in iron on TDIF spectrum after background subtraction [9, 11, 15]. Digits near curves - concentration in atomic %.

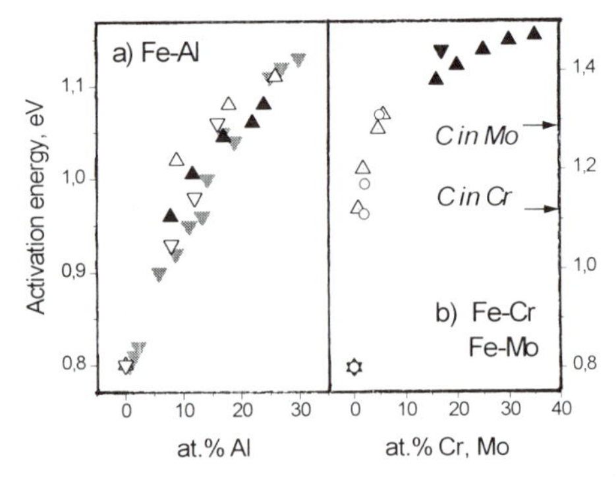

Figure 2 The influence of (a) Al (up solid triangles, down solid[13], down open [14], up open[20]); (b) Cr (down triangle[11], up triangles open [4,5,8], solid[9,11,15]) and Mo (circles) on the activation energy of Snoek peak in alloy iron.

Low temperature ageing (T_a) of Fe-(15...35%)Cr (T_a=400÷550°C) and Fe-(15...25%)Al (T_a= 200÷400°C) and cold-work deformation (α−Fe÷35%Cr) leads to the significant change of TDIF spectrum. The ageing of quenched Fe-Cr samples leads to the Snoek peak splitting (Fig.3), the ageing of quenched Fe-Al samples leads to the decrease in activation energy and narrowing of Snoek peak (Fig.4) and cold-work plastic deformation of low-carbon α–Fe[26] and Fe-Cr[27] leads to the dislocation-enhanced Snoek effect. The afore mentioned peculiarities should be explained from the viewpoint of interatomic interaction in b.c.c. iron-based alloys.

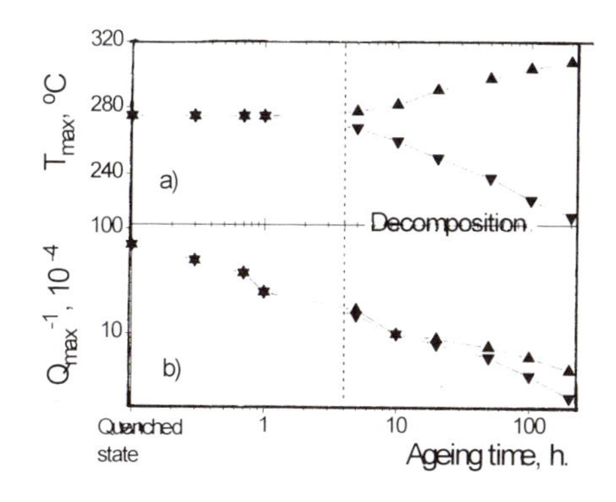

Figure 3. Splitting of temperature and height of Snoek peak for Fe-25 at.%Cr due to 475°C ageing (decomposition).

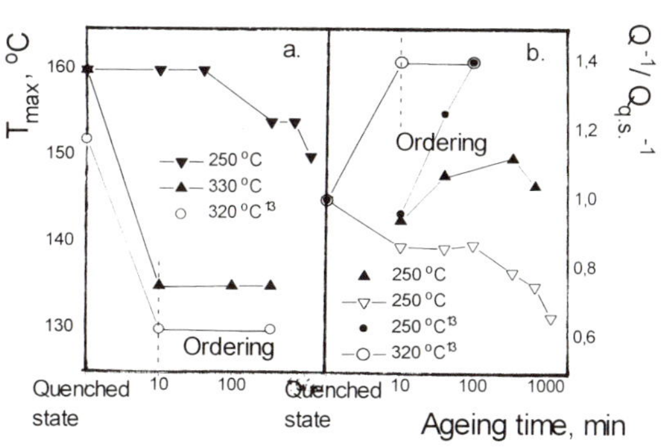

Figure 4. Temperature (a) and relative height (b) of Snoek peak for Fe-20 at.% Al due to 250 and 330°C ageing (ordering).

DISCUSSION

Equation (2) describes Snoek peak at $T = T_m$ for one interstitial species in pure b.c.c. metal. The presence of several species in metal leads to a superposition of a few partial IF maxima. In binary alloys the situation is much more complicated due to the formation of a set of non-equivalent energy position for interstitials. Two types of physical models for relaxation in binary alloy have been developed to explain the afore mentioned results.

First model is based on short-range interatomic interactions (i.e. regular solid solution approach) and takes into account a number of s-atoms in the first coordinate shell around C-atom as: $\{n{*}\text{Fe-}\ i\ \text{-}(n\text{-}6){*}\text{Me}\}$ with n from 0 to 6 and with the linear change of the activation enthalpy with concentration of s-atoms[10,24] : $H_n = H_o + n\,\Delta H$, where ΔH is the constant with certain scattering. The part of octahedral interstitial positions with different s-atoms content (P_i) is calculated for disordered distribution of s-atoms according to binomial distribution:

$$P_i = 6!\,N^i(1-N)^{(6-i)} / i!(6-i)!\ , \tag{4}$$

where ‘i’ is a number of s-atoms in lattice points around octahedral interstice in the first shell; N is Me concentration in alloy in the parts of the unit. The degree of filling of the different interstices by C-atoms from the condition of minimum free solid solution energy is:

$$\Theta = \sum_{i=0}^{6} \frac{6!\,N^i(1-N)^{6-i}\cdot\Theta_0}{i!(6-i)!(\exp(i\Delta E/RT)+(1+\exp(i\Delta E/RT)\Theta_0} \tag{5}$$

where Θ_o is fullness degree in pure Fe; $\Theta_0 = \Sigma P_i\Theta_i$; Θ_i is filling degree of interstices surrounded by i s-atoms. Resulting maximum is formed by summarizing partial peaks of IF (seven kinds of octagonal interstitial positions in high chromium ferrite) (Fig. 5):

$$Q^{-1} = \sum_{i=0}^{6} Q^{-1}{}_i \cosh^{-1}(\frac{H_i}{k}(1/T - 1/T_i)). \tag{6}$$

The obvious disadvantage of this model is that it does not separate elastic and chemical interatomic interactions. Also it takes into account the interatomic interaction in the first coordination shell which works well for the regular solid solution only [10]. That is why the model comes across the problem and can not adequately explain the situation in high alloy steels or gives unreasonably high ΔH values.

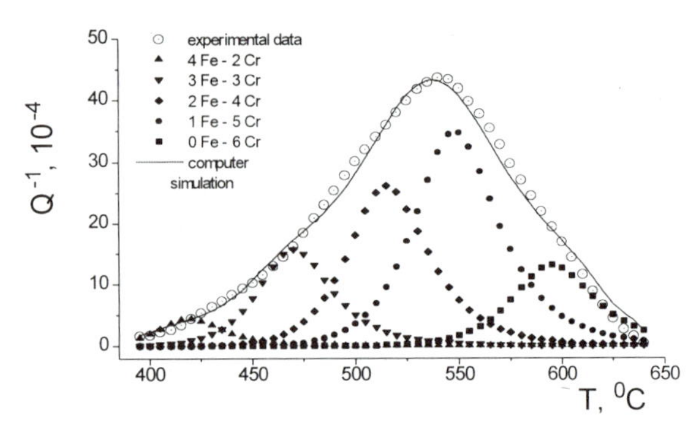

Figure 5. Computer simulations of Snoek relaxation Fe-25%Cr in the frames of short-range model (eq. 6).

The second model based on the long-range atomic interactions is elaborated to explain the influence of composition on the Snoek peak. The atoms distribution in crystal is simulated by a Monte Carlo method [16, 22, 25], using C-C and C-s atom interaction energies with the determination of the energy gap ΔE_P for each C-atoms in the potentials of other C and s-atoms (Fig.6). A model of long-range strain-induced interaction supplemented by the "chemical" interaction in a few coordination shells around an immobile s-atom is used for C-s interaction. The relaxation time is given as: $\tau_p = \tau_o \exp (H_p / kT)$ for p-th C atom, where: $H_p = H_D - \Delta E_p$. The IF at a given temperature (T) is calculated by summing up all the interstitials contributions given by equation (1),where δ is the relaxation strength per C-atom:

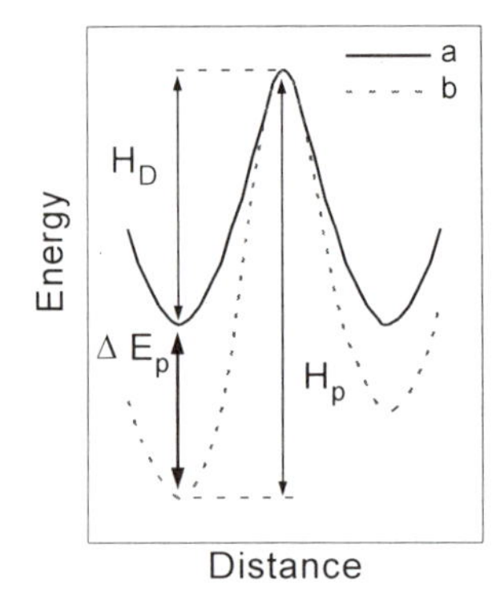

Figure 6. Scheme of a diffusion barriers: a - without solute interaction; b - with solute interaction; ΔE_p is the change of activation energy for IA in solution due to interaction with other atoms, H_p is the barrier for p-atom, H_D is the barrier value for binary alloy.

$$Q^{-1} = (\delta / T) \cdot \sum_{p=1}^{N} [(\omega \cdot \tau_p) / (1 + (\omega \cdot \tau_p)^2]. \qquad (7)$$

The simulation performed [16] shows that C-atoms in high alloy ferritic steels (Cr≥5%) are mainly situated in the first and second shells around Cr-atom. The amount of C-atoms in the 2-nd shells is nearly two times higher than that in the first one in spite of higher interaction energies in the 1-st shells due to the higher amount of interstices in the 2-nd shell and due to the 2-nd shells around the nearest Cr overlap. That means that the C-atoms move between the 1-st and the 2-nd shells around Cr atoms.

The attraction in the first two shells doesn't completely determine C-Cr interaction. The weak interaction in the distant shells considerably influences the temperature of the Snoek peak (Fig. 7). With an increase in maximal distance the average change of energy of C-atom due to its interaction with Cr attraction increases and passes through maximum in the vicinity of the 3-rd

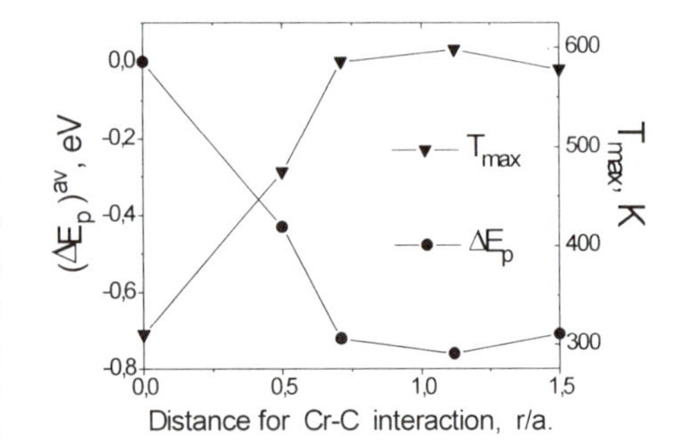

Figure 7. The influence of Cr-C interaction distance on $(\Delta E_P)^{av}$ value and peak temperature [16]

coordination shell. The T-shift takes place due to the change in ΔE_p. This point is important as it provides a new viewpoint at high alloy state of b.c.c. solid solution: any probated atom (in our case C in α–Fe) being put in the lattice undergoes the interaction of a few substitutional atoms simulteneously. Similar results are recently discussed for oxygen interatomic interaction in b.c.c. Nb-Ti alloys[22].

The splitting of Snoek peak in Fe-Cr system due to ferrite decomposition and the decreasing of activation energy for Snoek peak in Fe-Al due to ordering are the result of substitutional atoms redistribution in lattice. The splitting of Snoek peak into low- and high-temperature peaks is due to different average activation energy for carbon diffusion in zones enriched in Cr or in Fe. The resulting IF peak is the combination of these two Snoek peaks. Each of these peaks should be calculated according to the equation (6) with similar set of i, H_i, T_i and ΔE values but with different N values, i.e. for different Cr content. The corresponding calculations (short-range model) based on data presented in Fig. 4 allow to estimate the decomposition kinetics (N_1 and N_2 as a function of ageing time, i.e. the decomposition in Cr) of Fe-Cr ferrite[10]. The ordering in Fe-Al system (DO_3 - type) due to ageing leads to the significant decrease in elastic distortions in solid solution (it leads to the corresponding decrease in activation energy for carbon diffusion - Fig. 3a) and to the limitation of interstitial positions available for carbon atoms (it leads to the peak narrowing and corresponding temperary increase in peak height - Fig.3b). The kinetics of ordering strongly depends on ageing temperature of quenched ferrite with chaotic distribution of Al atoms: it takes less than 10 min for 330°C and at least 1 hour for 250°C. The detailed study of interatomic interaction and corresponding IF effect in Fe-Al will be reported soon[25].

The last point of this study is to make a distinction between "pure" Snoek effect and dislocation-enhanced Snoek effect in plastically deformed alloys. Contrary to Zeger model of dislocation-enhanced effect used[26] we estimate the influence of elastic distortions around immoveable dislocation to the Snoek effect. It is suggested that due to dislocation-impurities interaction (with energy E_d varying from zero far away from dislocation to 0,6 eV in the edge dislocation core) the concentration of C-atoms and corresponding distortion factor $|\lambda_1 - \lambda_2|$ are changed exponentially:

$$Q^{-1} = p(C_o \exp(-E_d / kT)) \cdot \{\lambda_1 - \lambda_2 \cdot C_o \exp(-E_d / kT)\}^2 \cdot ch^{-1}\left\{\frac{H}{k}\left(\frac{1}{T} - \frac{1}{T_m}\right)\right\} \quad (8)$$

Corresponding estimation shows (Fig. 8) that the experimentally observed IF peak height increase may be explained from the veiwpoint of classical Snoek peak model modified in respect to the dislocation influence and the dislocation kinks movement is of no use to discribe dislocation-enhanced Snoek effect for wide range of plastic deformation. Similar approach is of some use for austenitic steels[27] in respect to 260° peak (f = 1 Hz), nature of which is rather close to the Snoek peak in ferritic steels.

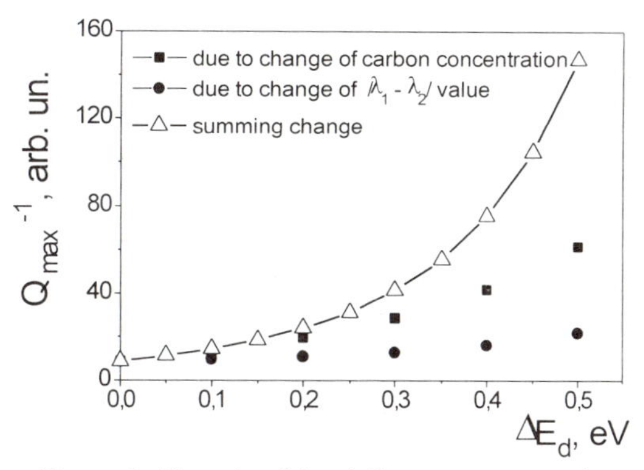

Figure 8. The role of local C-atoms concentration (squares) and λ (circles) factors and its sum (up triangles) to the dislocation-enhanced Snoek effect.

So the low frequency mechanical spectroscopy based on the study of relaxation effects on TDIF curves for ferritic steels and b.c.c. alloys gives the unique and rather simple opportunity for the study of atomic distribution in quenched, annealed, ordered and disordered states. Moreover it gives the chance to have a new look at the principles of metals alloying and alloying criterion.

SUMMARY

Different physical models for the Snoek-type relaxation in ternary systems (Fe-C-Me) are analyzed from the viewpoint of a distance of interatomic interaction taken into account:

For non-saturated from the viewpoint of overlapping of interatomic interaction in b.c.c. alloys the physically sufficient and optimal for the computer simulation is the short-range model, which takes into account the interatomic interaction and the average amount of substitutional atoms in the first coordination shell, only.

For high alloyed b.c.c. systems (i.e. with the overlapped interatomic interaction) the carbon atom undergoes an interaction of a few substitutional atoms simultaneously. That leads to the appearance of one broadened Snoek peak. Activation energy of such a peak is summed from the "elastic" and "chemical" interatomic interactions.

Experimental results for alloys with b.c.c. solid solution structure and its computer simulations allow to introduce the new criterion for the high alloy state of monophase steels: the high alloyed state corresponds to the situation when substitutional atoms can not be considered any longer as the isolated atoms. From the viewpoint of mechanical spectroscopy this situation corresponds to the appearance of one broadened IF Snoek-type peak instead of two peaks existed for the steels with lower substitutional atom concentration.

Decomposition, ordering and plastic deformation change the parameters of interatomic interaction in b.c.c. solid solution and the corresponding relaxation effect. Nevertheless the use of adequate structure parameters allows to explain IF effects and, vice versa, the relaxation IF spectrum allows to quantitatively estimate atoms redistribution in ferrite.

ACKNOWLEDGMENT

Author is grateful to Prof. M.S.Blanter for valuable discussions and previously published joint results used in this paper and to Mle T.Pozdova for the IF study of Fe-Al system. Special thanks are to the Organizing Committee of this Symposium for the financial support.

REFERENCES

1. A.S. Nowick, B.S.Berry. Anelastic Relaxation in Crystalline Solids.Academic Press, NY and London (1972).
2. M. Weller. J. de Physique IV, C.7, suppl. J. de Physiqie III, V.5, p. 199, (1995).
3. M. Weller. J. de Physique IV, C.8, suppl. J. de Physiqie III, V.6, p. 63, (1996).
4. C.Wert. Trans. metall. Soc. AIME: J. of Metals. 194, p. 602, (1952).
5. M.A.Krishtal, U.V.Piguzov, S.A.Golovin. Internal Friction in Metals&Alloys. Moscow, Metallurgy, (1964).
6. S.A.Golovin. In: Heat Treatment and Physical Metallurgy, p. 67, (1978).
7. H.Numakura, M.Koiwa, J. de Physique IV, C.8, suppl. J. de Physiqie III, V.6, p. 97, (1996).
8. H.Numakura, M.Miura, M.Koiwa et.al. ISIJ International, V.36, N.3, p.290, (1996).
9. M.A.Krishtal, V.I.Baranova. Fis. Met. and Metalloved, v.18, N3, p. 464, (1964).
10. I.S.Golovin, V.I.Sarrak, S.O.Suvorova, Met.Trans. 23A, 2567, (1992).
11. B.Dubois, F.Hernandez, M.Bouhafs. Proc. 6-th ICSMA. Australia, V.1, p.167, (1982).
12. I.S.Golovin. To be published in: Fis. Met. and Metalloved, N 6 (1997).
13. K.Tanaka. J. of the Phys. Soc. of. Jap., v. 30, N.2, p. 404, (1971).
14. N.P.Kulish, V.M.Mandrika, P.V.Petrenko. Fis. Met. and Metalloved, v.51, N6, p. 1229, (1981).
15. M.S.Blanter, U.V.Piguzov, et. al. The Method of Internal Friction in Metallurgical Research, Moscow: Metallurgia, 246 p., (1991).
16. I.S.Golovin, M.S.Blanter, R.Schaller. Phys. Stat. Sol.(a), V. 160, p. 49 (1997).
17. S.V.Zemsky, M.N.Spassky. Fis. Met. and Metalloved., v.21, p. 129, (1966).
18. Z.C.Szkopiak, J.T.Smith. J.Phys.D.: Appl. Phys., Vol. 8, p.1273 (1975).
19. N.P.Kushnareva, S.E.Snejko. J. of Alloys & Compounds , 211/212, p. 75 (1994).
20. N.Ya.Rochmanov. Private communications, unpublished data.
21. O.Florencio, W.J.F.Botta, C.R.Grandini e.a. J. of Alloys & Compounds , 211/212, p.41 (`1994).
22. I.S.Golovin, M.S.Blanter, A.V.Vasiliev. J. de Physique, C.8, V.6, p.107-111 (1996).
23. O.N.Carlson, H.Indrawirawan, C.V.Owen, O.Buck. Met.Trans.A., V. 18A, p.1415 (1987).
24. P.Gondi, R.Montanari. J.Alloys. and Comp., 211/212, p. 33, (1994).
25. I.Golovin, T.Pozdova, K.Tanaka, M.Blanter M. Proc. IIAPS'97 Conference, Russia, 22-24 / 09 / 1997.
26. J.Rubianes, L.B.Magalas, G.Fantozi, J.San Juan. J. de Physique, 1987, v.48, C.8, p. 185 (1987).
27. G.V.Serzhantova, I.S.Golovin, S.A.Golovin. To be published in J.of Metals Science and Heat Treatment.

INTERNAL FRICTION NONDESTRUCTIVE EVALUATION OF PLASTIC PROPERTIES OF CRYSTALS

Alexander B. Lebedev

Solid State Physics Division, Ioffe Physico-Technical Institute,
Russian Academy of Sciences
26 Polytekhnicheskaya, St. Petersburg 194021, Russia

INTRODUCTION

The problem of ultrasonic nondestructive evaluation of plastic properties is the problem of linkage between acoustic and plastic properties of crystals. Obviously, both are functions of structure and external conditions, and if one knows these functions, then it is possible to obtain a "plastic vs. acoustic law", which could be used for nondestructive evaluation.

Several approaches can be applied to obtain information on the macroscopic yield stress (one of the most important characteristics of plasticity) from acoustic measurements. For example, one can evaluate the yield stress dependence on the grain size from the data on ultrasonic attenuation due to scattering within the Rayleigh region[1]. In this case both the attenuation and the yield stress are functions of the grain size d (the former is proportional to d^3 while the latter obeys the Hall-Petch relation), thus, it is possible to calibrate the yield stress-attenuation curve. However, if external conditions (e.g. temperature) are changed, new calibration is necessary. Moreover, this method is not applicable to single crystals.

Another way to obtain information on macroyield stress behavior is to use acoustic data on dislocation damping. Amplitude-dependent internal friction (ADIF) is well known[2-8] to reflect a hysteresis of microplastic deformation (or dislocation hysteresis). If the amplitude of acoustic vibration is not very high, the dislocation hysteresis is completely reversible and the influence of vibrations is nondestructive. It is possible to define microyield stress from ADIF data, to evaluate it and to compare it with the conventional macroyield stress σ_c. Various definitions of microyield stress have been discussed elsewhere[9,10]. The comparison has shown that only the stress σ_ε at a constant level of dislocation strain ε_d leads to a proportionality (similarity law) between $\sigma_c(T)$ and $\sigma_\varepsilon(T)$ [9–13] at low temperatures (approximately at $T < 0.3\ T_m$, where T_m is the melting temperature).

As the strain ε_d plays an important role, let us consider first the effect of the dislocation hysteresis type on the evaluation of ε_d from the amplitude-dependent internal friction and the amplitude-dependent modulus defect (ADMD). Then, some example of the similarity law will be presented.

EVALUATION OF DISLOCATION STRAIN

Baker[14] in 1962 suggested a simple algorithm to evaluate the reversible microplastic (or dislocation) strain. He used an approximation for the dislocation modulus defect (see, e.g., Chapter 12 in monograph by Nowick and Berry[8])

$$\Delta M/M \approx \varepsilon_d/\varepsilon_0\,, \tag{1}$$

where $\varepsilon_0 = \sigma_0/M$ is the elastic strain amplitude (σ_0 is the stress amplitude). Then

$$\varepsilon_d \approx (\Delta M/M)\ \varepsilon_0 = \delta_h \varepsilon_0 / r\,. \tag{2}$$

Eqn.(2) takes into account a proportionality between the amplitude-dependent decrement δ_h and modulus defect $\Delta M/M$ frequently observed in experiments (see Read[3], Nowick[4] and, for a review, Lebedev[13,15]):

$$\delta_h = r\ (\Delta M/M), \tag{3}$$

where r is a coefficient of the order of unity.

Let us estimate the degree of approximation (1) for three types of dislocation hysteresis (Figure 1): (a) Granato-Lücke loop; (b) Davidenkov loop; and (c) NRF (no restoring force) loop. The first one reflects the well-known breakaway model[5], while, the second and the third can be described within the framework of a frictional model[6,7] (more details one can find in recent papers by Lebedev[13,15,16]). The dashed lines on Figure 1 show the difference ΔM_{sec} between the modulus without the hysteresis M and the secant modulus M_{sec} (the term "secant" is by Lazan[6]). Following Lazan[6], it is natural to define the secant modulus defect $(\Delta M/M)_{sec}$ exactly equal to $\varepsilon_{dm}/\varepsilon_0$, where $\varepsilon_{dm} = \varepsilon_d(\sigma_0)$ is the amplitude value of dislocation strain (ε_d in Eqn.(2) is, in fact, ε_{dm}). Then, $r_{sec} = \delta_h / (\Delta M/M)_{sec}$.

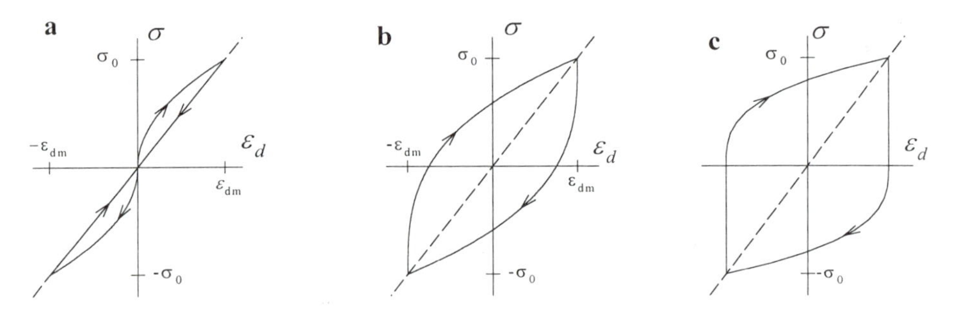

Figure 1. Schematic representation of various types of dislocation hysteresis (elastic part being subtracted): (a) Granato-Lücke loop; (b) Davidenkov loop; and (c) NRF (no restoring force) loop. The slope of the dashed lines corresponds to $\Delta M_{sec} = M_0 - M_{sec}$, where M_0 is the modulus of a crystal without dislocation hysteresis.

Nishino and Asano[17] (based on the $\delta_h(\varepsilon_d)$ formula derived by Asano[7] for the generalized Davidenkov loop) obtained:

$$r_{sec-Dav} = 4 - \frac{4}{\sigma_0 \varepsilon_d(2\sigma_0)} \int_0^{2\sigma_0} \varepsilon_d(\sigma) d\sigma\,. \tag{4}$$

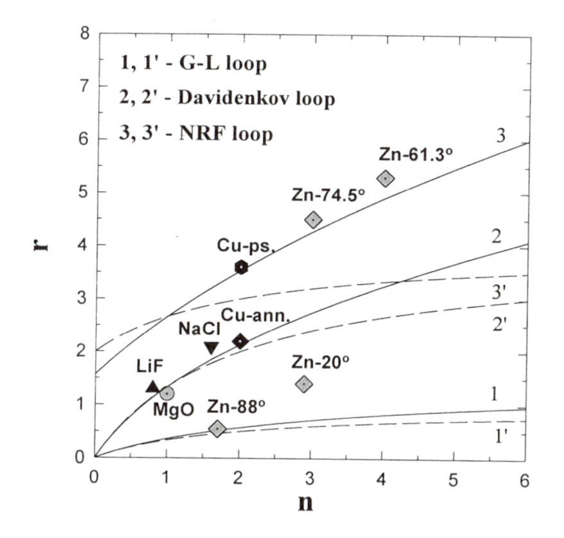

Figure 2. Comparison of calculated values of r (curves 1-3, Eqns. 9-11) and r_{sec} (curves 1'-3', Eqns.6-8) as a function of n with the experimental data for zinc[3] (orientation is indicated), copper[4], ionic crysals[20], and magnesium oxide[21].

Equations similar to (4) can be derived for both the G-L loop and NRF loop. At r_{sec}=const. these integral equations can be easily solved by simple differentiation and the solution is a power function $\varepsilon_d \propto \sigma^m$, which yields a power law for the decrement and modulus defect:

$$\delta_h = r_{sec}\,(\Delta M/M)_{sec} \;\propto \sigma_0^{\,n}\,, \quad (5)$$

with $n=m-1$ and r_{sec} is a function of n only:

$$r_{sec\text{-}GL} = n/(n+2), \quad (6)$$

$$r_{sec\text{-}Dav} = 4n/(n+2), \quad (7)$$

$$r_{sec\text{-}NRF} = 4(n+1)/(n+2). \quad (8)$$

It should be noted that Eqn.(7) have been obtained previously by Kustov[18], Lebedev and Kustov[19], and Nishino and Asano[17]. In addition, it is pointed out here, that for any symmetrical hysteresis loop, if nonlinear parts of the loops $\varepsilon_d(\sigma)$ are continuos functions and r_{sec} is a constant (independent of the amplitude), then there is no alternative for $\delta_h(\sigma_0)$, but Eqn.(5), i.e.,the condition $r=const$ is necessary and sufficient for the power law (5).

In order to derive $\Delta M/M$ and r correctly, Fourier analysis should be applied[4]. This was done recently by Lebedev[15,16] and for the power law $\delta_h = r(\Delta M/M) \propto \sigma_0^{\,n}$ the following results were obtained:

$$r_{G-L} = 2n\Big/\left[n+2+\frac{2(n+2)}{\pi}B\left(\frac{n+3}{2},\frac{1}{2}\right)\right], \quad (9)$$

$$r_{Dav} = 2^{n+1}n\pi\,/\,[(n+2)\int_0^{\pi}(\cos\Theta+1)^{n+1}\cos\Theta d\Theta], \quad (10)$$

$$r_{NRF} = \left[(n+1)^2\sqrt{\pi}\,\Gamma\left(\frac{n+1}{2}\right)\right]\Big/\left[(n+2)\Gamma\left(\frac{n+2}{2}\right)\right], \quad (11)$$

where $B(x,y)=\Gamma(x)\Gamma(y)/\Gamma(x+y)$ is the beta function and $\Gamma(x)$ is the gamma function.

Figure 2 shows the calculated curves, expressions (6)-(11), together with some experimental data (more data is given elsewhere[16]). It is clear, that the evaluation of ε_d with help of Eqn.(2) gives some error, which is especially noticeable in the case of the NRF loop (curves 3 and 3' on Figure 2). However, within the practically interesting region ($n < 6$), the relative error is of the order of unity and, which is more important, it is the same for the same exponent n.

Application of the procedure by Asano[7] to evaluation of the dislocation strain has advantages and disadvantages compare to the Baker[14] procedure. Asano postulated the Davidenkov-type loop; in this case one can evaluate not only the amplitude value of dislocation strain, but also the whole $\varepsilon_d(\sigma)$ response[7]. On the other hand, the loop shape is normally unknown, so the absolute error may be more considerable. However, in the case of the power law (Eqn.5), the error is again a function of n only.

THE SIMILARITY LAW

Figure 3 shows schematic representation of applying the procedure by Baker[14] for evaluation of temperature dependence of the microyield stress $\sigma_\varepsilon(T)$. This method was first used by Chelnokov and Kuzmin[22] for the grain size dependence of σ_ε in impure aluminum polycrystals (99.7%) at room temperature, which exhibited a pronounced amplitude hysteresis, i.e., the curves $\delta_h(\sigma_0)$ measured on increasing and decreasing the amplitude did not coincide with each other. In what follows, all $\sigma_\varepsilon(T)$ curves are evaluated from the ADIF data, which obey two conditions: (I) reversibility, i.e., there is no amplitude hysteresis; (II) there is a separation of variables, $\delta_h = f_1(T)f_2(\sigma_0)$, in the amplitude-temperature spectrum of the decrement. The separation, first pointed out by Nowick[4], has frequently been observed in many experiments[9-13]. Taking into account Eqn.(5), it can be written

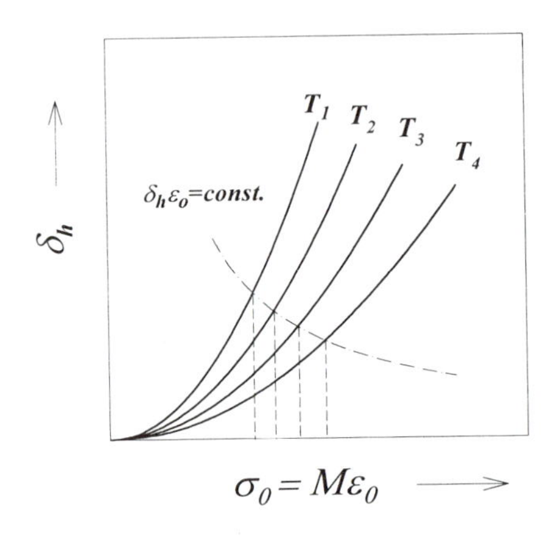

Figure 3. Schematic evaluation of the micro-yield stress temperature dependence $\sigma_\varepsilon(T)$ from a set of $\delta_h(\sigma_0)$ curves measured at various temperatures (T_1 - T_4). The broken line is a hyperbola $\delta_h\sigma_0$= *const.* The points of interception of the hyperbola with the $\delta_h(\sigma_0)$ curves, indicated by the dashed lines, give $\sigma_\varepsilon(T)$.

$$\delta_h = f_1(T)\ \sigma_0^{\,n}. \tag{12}$$

Then, $\sigma_\varepsilon(T)$ curves, measured within the range of validity of Eqn.(12) at various levels of ε_d, differ only by a scaling factor, i.e., they can be coincided in relative units.

Figures 4-6 give some examples of a comparison between the microyield and macroyield stresses as a function of temperature. Figure 4 shows the comparison for MgO single crystal. The macroyield ad microyield stresses here are the shear stresses $\tau=\Omega\sigma$, where Ω is the Schmid factor (in this particular case, Ω = 0.5). It is clear from the graph that, in spite of the one order of magnitude difference, the temperature trends are the same for both $\tau_c(T)$ and $\tau_\varepsilon(T)$. That gives us possibility to use this similarity law for nondestructive prediction of plastic properties.

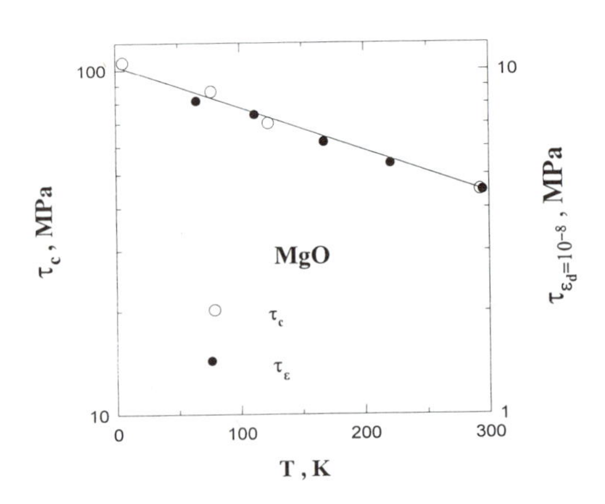

Figure 4. Effect of temperature on the macroyield τ_c and microyield τ_ε shear stresses in MgO single crystal. Data on τ_c (except for T=4.2K after Nikiforov[24]) are by Berezhkova et al.[23] Data on τ_ε are from the ADIF data measured by Kardashev et al.[21] at 100kHz (see also Lebedev and Kustov[9]).

Figure 5 demonstrates the validity of the similarity law for polycrystalline Al:Zn solid solution at low temperatures (data for $\sigma_{0.2}$ and σ_ε are by Podkuyko and Pustovalov[25] and by Ivanov et al.[26] correspondingly).

Figure 6 shows σ_c and σ_ε in Al-Si-Fe alloy[27]. The similarity law $\sigma_c(T) \propto \sigma_\varepsilon(T)$ holds well at low temperatures (T<150K). At T>150K, when an aging process starts in Al-based alloys, the microyield stress, which is detected with no dislocation multiplication, is more sensitive to the aging (due to the locking of dislocations), then the macroyield stress, when the multiplication takes place.

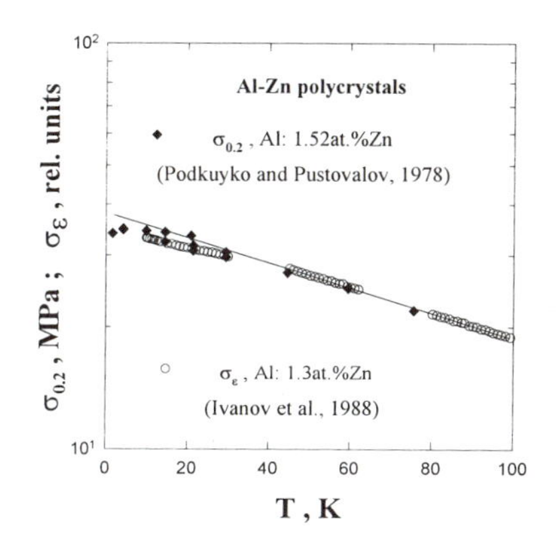

Figure 5. Macroyield $\sigma_{0.2}$ (after Podkuyko and Pustovalov[25]) and microyield σ_ε (obtained from ADIF data by Ivanov et al.[26]) stresses in polycrystals of Al:Zn solid solution of similar composition as a function of temperature.

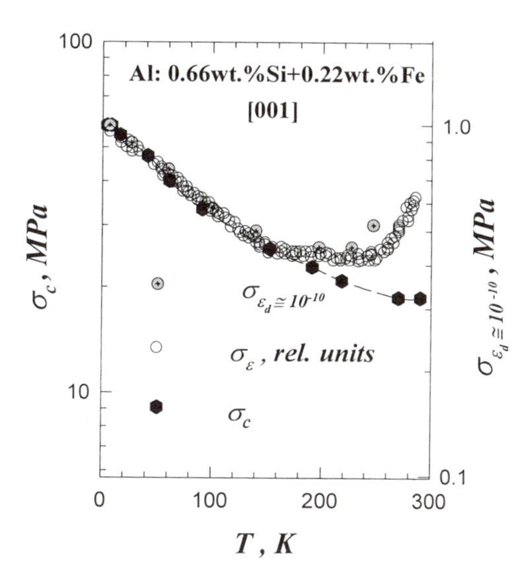

Figure 6. Comparison between $\sigma_c(T)$ and $\sigma_\varepsilon(T)$ in Al-Si-Fe alloy[27].

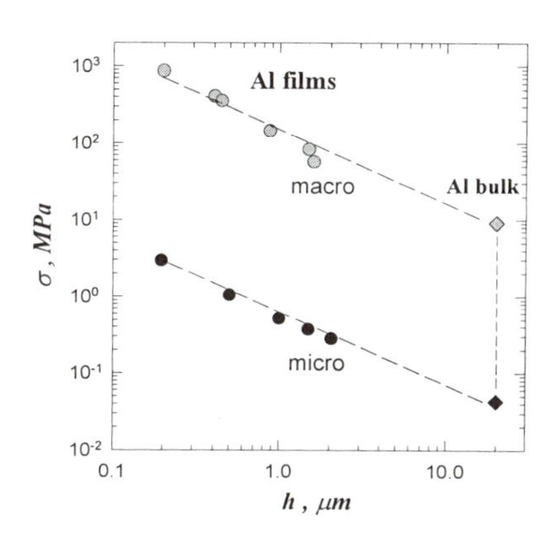

Figure 7. σ_ε (back circles) and σ_c (gray circles) stresses in 99.999% Al thin films on a substrate as a function on the film thickness[31]. Black and gray diamonds correspond to bulk samples (the former is from the ADIF data by Nishino and Asano[31] and the latter was measured by Didenko and Pustovalov[32] in polycrystalline 99.9997%Al).

The similarity law $\sigma_c(T) \propto \sigma_\varepsilon(T)$ has been observed in many crystalline materials: ionic crystals[9] (NaCl, NaF, LiF), metals with b.c.c. (Mo[9,11], W[9]), f.c.c. (Al of different purity[9,13,27] , and silicon bronze[12,13]) and h.c.p. (Mg[9,13]) lattice. The deviations from the similarity law can be used to analyze mechanisms of high and low temperature anomalies of the flow stress[13]. Physical reasons for the similarity law are discussed elsewhere[9,10,12,13].

It should be noted, however, that there is no similarity between the microyield and macroyield stresses as a function of impurity concentration[28,29] and grain size[13,30]. On the other hand, Nishino and Asano[31] for the aluminum films on a substrate observed the similarity $\sigma_\varepsilon \propto \sigma_c$ as a function of the film thickness, which might be also used for nondestructive testing (see Figure 7).

Acknowledgments

The work was partly done during my stay as a visiting researcher at the University of Illinois at Urbana-Champaign (UIUC) and was partly supported by NSF Grant DMR 93-19773. It is my pleasure to thank Dr. A.V. Nikiforov for measuring τ_c at T=4.2K in MgO and also Prof. S. Asano, Prof. A.V. Granato, Prof. B.K. Kardashev, Dr. S.B. Kustov, Prof. S.P. Nikanorov, and Dr. Y. Nishino for valuable discussions.

REFERENCES

1. B.E. Droney and R. Klinman, Ultrasonic techniques for determining the mechanical properties of steels, in: *Applications of Physics in Steel Industry, Bethlehem, 5-7 Oct.1981, AIP Conf. Proc. no.84* (1982) p.210.
2. N.N. Davidenkov, Dissipation of energy during vibrations, *Zh. Tech. Fiz.* 8: 483 (1938) (in Russian).
3. T.A. Read, Internal friction of single crystals of copper and zinc, *Trans. AIME* 143: 30 (1941).
4. A.S. Nowick, Variation of amplitude-dependent internal friction in single crystals of copper with frequency and temperature, *Phys. Rev.* 80: 249 (1950).
5. A.V. Granato and K. Lücke, Theory of mechanical damping due to dislocations, *J. Appl. Phys.* 27: 583 (1956).
6. B.J. Lazan *Damping of Materials and Members in Structural Mechanics*, Pergamon Press, Oxford (1968).
7. S. Asano, Theory of nonlinear damping due to dislocation hysteresis, *J. Phys. Soc. Japan* 29: 952 (1970).
8. A.S. Nowick and B.S. Berry *Anelastic Relaxation in Crystalline Solids,* Academic Press, New York (1972).
9. A.B. Lebedev and S.B. Kustov, Effect of temperature on amplitude-dependent internal friction and yield stress in crystals, *Phys. Stat. Sol. (a)* 116: 645 (1989).
10. A.B. Lebedev, The similarity law between the temperature dependences of yield stress and microyield stress evaluated from internal friction, *J. Alloys. Comp.* 211/212: 177 (1994).
11. A.B. Lebedev, B.K. Kardashev, U. Hofmann, H.-J. Kaufmann, and D.Shulze, Dislocation internal friction and temperature dependence of yield stress in high-purity molybdenum single crystals, *Cryst. Res. Technol.* 24: 1143 (1989).
12. A.B. Lebedev and S. Pilecki, Similarity law between the temperature dependences of microyield and macroyield stresses in polycrystalline silicon bronze, *Scripta Metall. Mater.* 32: 173 (1995).
13. A.B. Lebedev, Application of internal friction to analysis of plastic behaviour of crystals, *J. de Physique IV.* 6: C8-255 (1996).
14. G.S. Baker, Dislocation mobility and damping in LiF, *J. Appl. Phys.* 33:1730 (1962).
15. A.B. Lebedev, Amplitude-dependent decrement to modulus defect ratio in breakaway models of dislocation hysteresis, *Philos. Mag. A* 74: 137 (1996).
16. A.B. Lebedev, Ratio of decrement to modulus change for amplitude-dependent internal friction, *J. de Physique IV.* 6: C8-325 (1996).
17. Y. Nishino and S. Asano, Further discussion on the dislocation strain evaluated from amplitude-dependent internal friction, *Phys. Stat. Sol.(a)* 138: K9 (1993).
18. S.B. Kustov, PhD Thesis, Ioffe Phys. Tech. Institute Academy of Sciences of the USSR, Leningrad (1989) (in Russian).
19. A.B. Lebedev and S.B. Kustov, Comments on the dislocation strain evaluated from amplitude-dependent internal friction, *Phys. Stat. Sol.(a)* 136:.K85 (1993).
20. S.B. Kustov, S.N. Golyandin, and B.K. Kardashev, Inelastic deformation and amplitude-dependent internal friction in LiF and NaCl crystals at low loading frequencies, *Sov. Phys. Sol.State* 30:1248 (1988).
21. B.K Kardashev, S.B. Kustov, A.B. Lebedev, G.V. Berezhkova, P.P. Perstnev, F. Appel, and U. Messerschmidt, Acoustic and electron microscopy study of the dislocation structure in MgO crystals, *Phys.Stat.Sol.(a)* 91: 79 (1985).
22. V.A. Chelnokov and N.L. Kuzmin, Some distintive features of amplitude-dependent internal friction, *Sov. Phys. Sol. State* 24: 1796 (1982).
23. G.V. Berezhkova, P.P. Perstnev, N.R.Turaeva, and N.A. Batrakov, Mechanical behaviour of magnesium oxide single crystals at temperatures from -196 to 2400^{0}C, *Crys. Res. Technol.* 18: 1199 (1983).
24. A.V. Nikiforov, unpublished data.
25. V.P. Podkuyko and V.V. Pustovalov, Peculiar mechanical properties of aluminium-magnesium, aluminium-copper and aluminium-zinc alloys at low temperatures, *Cryogenics* 18: 589 (1978).
26. V.I. Ivanov, A.B. Lebedev, and N.L. Kuzmin, Microplastic deformation of Al-Zn solid solutions under ultrasonic excitation, in: *Plastic deformation of Materials under the Influence of External Energy Effects,* L.E. Popov, ed., Siberian Metallurgy Inst., Novokuznetsk (1988) p.268 (in Russian).
27. A.B. Lebedev and V.I. Ivanov, The temperature dependence of internal friction and yield stress in Al alloys, *Mater. Sci. Forum* 119-121: 245 (1993).
28. Y. Nishino, Y. Okada, and S. Asano Microplasticity and dislocation mobility in copper-nickel single crystals evaluated from strain-amplitude-dependent internal friction, *Phys. Stat. Sol.(a)* 129: 409 (1992).
29. M. Morita and S. Asano, Internal friction caused by microplasticity in aluminum-magnesium alloys, *J.Japan Inst. Metals* 57: 1006 (1993).
30. H. Goto, Y. Nishino, and S. Asano Effect of grain size on amplitude-dependent internal friction in polycrystalline copper, *J.Japan Inst. Metals* 55: 848 (1991).
31. Y. Nishino and S. Asano Dislocation damping and microplasticity of aluminum thin film on a substrate, in: *Strength of Materials (ICSMA 10),* H. Oikawa et al.,eds., Japan Inst. Metals, Sendai (1994) p.857.
32. D.A. Didenko and V.V. Pustovalov, Singularities in the temperature dependence of flow stress down to 1.4K in aluminum single crystals of various purity and orientations, *J. Low Temp. Phys.* 11: 65 (1973).

PHYSICAL PROPERTY DETERMINATION FOR PROCESS MONITORING AND CONTROL

Gerd Dobmann

Fraunhofer-Institut für Zerstörungsfreie Prüfverfahren, IZFP
D-66123 Saarbrücken

INTRODUCTION

Cold rolled and recrystallised steel strips in general are applied where the fitness for use asks for a high degree in deep drawability and surface quality features like low roughness and waviness. The good characteristic deep drawability of the steel grades developed for this purpose is achieved by low carbon contents (< 0.05%) and is described today by the properties of the standard tensile test: yield strength, tensile strength, percentage elongation before reduction and elongation at rupture. Futhermore, properties for anisotropy induced by crystallographic texture (Bleck, Bode and Hahn, 1990), i. e. the vertical (r_m) and planar (Δr) anisotropy parameters are of interest. A good deep drawability - for instance for complex shaped car body parts - asks for a high value of the vertical anisotropy r_m describing the resistance of the material against a reduction in thickness during the deep drawing process. The planar anisotropy value Δr is a measure for the direction-dependent yield resistence of the material in the strip plane. The value should be as small as possible in order to avoid earing during deep drawing (Schmidt and Schaffrath, 1993). The mechanical properties of the cold rolled steel strips are determined by the recrystallisation process after cold rolling. After laser welding to a strip of some kilometer length in a hot dip galvanising line the coils of cold rolled steel run through the recrystallisation furnace followed by the zinc-coating-line.

One of the demands of total quality management (TQM) is that information to the state of quality has to be available at each moment during the production process. Therefore a continuous and automated materials property determination for the hot dip galvanised steel strips is needed. Performing spot tests by destructive sampling of tensile specimens at the beginning and at the end of the coil does not fulfill the TQM-demand. Furthermore the sample taken at random cannot be a representative measure of the properties along the 4 km length of the steel strip.

Together with Thyssen Steel in a European BRITE-EURAM project (Borsutzki. Thoma, Bleck and Theiner, 1993) during a 14-day-time period a nondestructive testing technique was applied in a continuous annealing and hot dip galvanizing line to demonstrate the ability for nondestructive material properties determination of yield strength ($Rp_{0.2}$ - 0.2% yield stress σ_y) and anisotropy parameters (r_m, Δr). The technique is based on a micromagnetic approach using the incremental permeability (Dobmann et al, 1989) concerning the yield prediction and a time of flight measurement of shear horizontal ultrasonic plate waves for characterising the anisotropy. In the following a pilot plant equipment was designed, assembled and tested on-line during the period of one year. The results presented here are a summary of the comprehensive final report for the European Community.

YIELD STRENGTH AND DEEP DRAWABILITY PREDICTION AT STEEL STRIPS

Physical Basics and Transducer Systems

Reliability tests with the above mentioned two nondestructive (nd) testing techniques were mainly performed at microalloyed IF (interstitial free)-steel grades (thickness range 0.5 - 2 mm). The inspection system consists of two modules. The first is the ultrasonic module for transmitting and receiving shear horizontal plate waves and measuring their time of flight (TOF-module). The second module is for local dynamic magnetisation (hysteresis) of the strip and measurement of the incremental permeability's profile curve. Both modules are controlled by a master PC which communicates by a serial interface with the process control system of the plant.

Figure 1. shows in a top view the arrangement of the transducers for the nd-procedures. The shear horizontal (SH) plate waves are transmitted as the first symmetric (SS_0) mode under 0° and 45° compared with the rolling direction. As for the SS_0 mode the displacements are homogeneously distributed in thickness direction and in the strip plane, the mode is not sensitive to thickness changes and therefore excellent for on-line monitoring. Under practical conditions the SH-waves can be reliably transmitted and received only by using electromagnetic transducers (EMAT). In the specific case the principle of transforming electrical energy into ultrasonic waves is by producing perodic magnetostrictive forces in the inspection plane, superimposing a horizontal magnetic field (sinusoidal alternating, 300 Hz) to the lengthwise direction of an eddy current meander-like coil (meander distance 4 mm, wavelength λ = 8 mm, frequency 400 kHz). The waves are transmitted by one transmitter-EMAT and received by two receiving-EMAT placed at a center-to-center distance (transmitter-1st receiver) d_{TR} = 160 mm and (1st receiver-2nd receiver) d_{RR} = 80 mm. The time of flight is measured by using the pulse overlap technique (Herzer and Schneider, 1989). Due to the theoretical approach of Sayers and Proudfoot (1986) we know (equation (1) and (2)) that by measuring the speed (v_{SS0}) of the SH-wave in the two angular directions two coefficients (C411 and C413) of the orientation distribution function can be derived which are normally calculated after evaluating X-ray pole-figures:

$$C411 = \frac{210}{2A}\sqrt{\frac{3}{7}}\left(\rho\left(v_{SS0}^2(0°) + v_{SS0}^2(45°)\right) - 2\mu\right) \quad (1)$$

ρ is the density, μ is the shear module, θ is an angle $\neq$ 0° to the rolling direction

$$C413 = \frac{210\mu}{32A\sqrt{35}}\sqrt{\frac{3}{7}}\left(1-2\cos^2\theta\right)^{-1}\left(1-\sqrt{1-16\frac{\left(1-2\cos^2\theta\right)^2}{\cos^2\theta\sin^2\theta}\frac{v_{SS0}(\theta)-v_{SS0}(0°)}{v_{ss0}(0°)}}\right) \quad (2)$$

A = C11 - C12 - 2C44 is the anisotropy factor of the iron single crystal. Whereas C411 is correlated with r_m, C413 is with Δr.

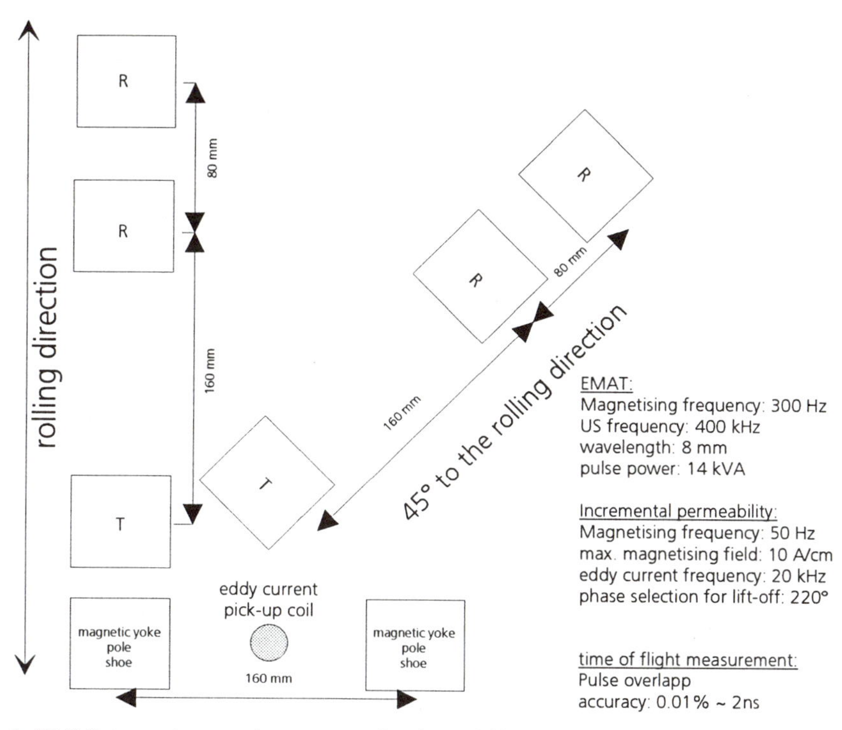

Figure 1. EMAT transmitter-receiver systems for time of flight measurements under 0° and 45° to the rolling direction; electromagnetic yoke with eddy current pick-up coil for incremental permeability determination.

The incremental permeability sensor is an electromagnetic yoke with 160 mm pole shoe center-to-center distance. The eddy current coil is placed in a symmetric position to the pole shoes. Because of the low coercivity of the steel grades of interest a maximum magnetising field strength of 10 A/cm is applied; the magnetising frequency is 50 Hz. The eddy current probe superimposes an incremental magnetisation at a frequency of 20 kHz on the dynamic magnetisation of the hysteresis. For suppression of lift-off changes of the probe a phase rotation of 220° is selected. The imaginary part of the impedence is then directly proportional to the incremental permeability ($\mu_\Delta(H)$) and is measured as a function of the dynamic magnetic field H in the hysteresis as a so-called profile curve (Dobmann, G., et al, 1988).

Figure 2. is a documentation of such a profile curve and the relevant parameters - derived from curve - are indicated: The peak maximum, peak minimum, peak separation and full width at half maximum (FWHM) amplitude. According to the 3MA-approach these quantities are a function of diverse information concerning the microstructure state of the steel grades. Therefore they are excellently selected to be used in a multiparameter multiregression model for property determination of the mentioned materials, i. e. the yield strength. The transducer system is built in a carriage device, which can be put in a roll table position. The transducer table is pressed against the running steel strip (speed 120 - 150 m/min) by two controllable contact guide rollers madeof austenitic stainless steel. This is a measure to quiet strip vibrations and to guarantee a controlled lift-off, which is 2±1 mm. The data acquisition for the two techniques is performed independently by hardware modules. Because of the 50 Hz magnetising frequency in the incremental permeability module during one hysteresis cycle a strip length of 4 cm has passed the transducer. Three hysteresis cycles are time-averaged to reduce noise and therefore one averaged incremental permeability profile curve is available for data evaluation and multiparameter analysis at every 12 cm strip length. In the ultrasonic module the two transmitter-receiver channels are multiplexed. In each channel the repetition rate is 300 Hz and 60 A-scans are time-averaged. Therefore a pair of (r_m, Δr)-values is available every 80 cm.

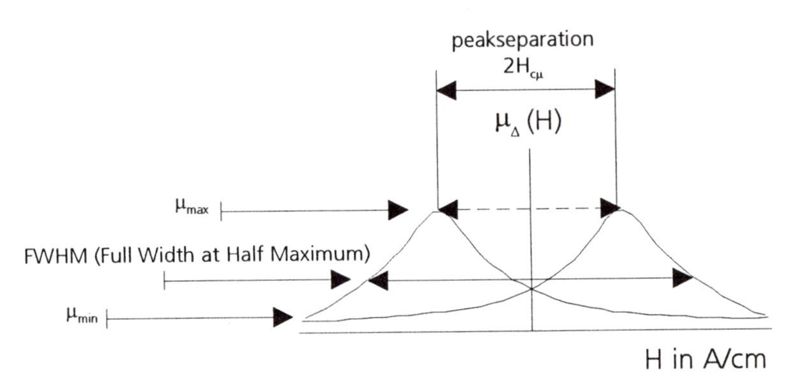

Figure 2. Incremental permeability and derivable parameters for a multiparameter multiregression model for material properties determination.

3MA-Approach - On-Line Monitoring and Calibration

The 3MA (Micromagnetic Multiparameter Microstructure and stress Analysis)-approach is based on the fact that all micromagnetic nd-parameters which can be derived from dynamic magnetisation of ferromagnetic materials in a hysteresis depend on microstructure parameters as well as on load-induced and residual stresses (Dobmann, G., et al, 1989). Micromagnetic techniques available for such a nd-testing approach are the magnetic and acoustic Barkhausen noise, the incremental permeability, the spectral analysis of the magnetic field, the dynamic magnetostriction as well as the eddy current inspection. For on-line monotoring in an electromagnetically disturbed environment - in this case a steel plant - the incremental permeability compared with the other techniques has the advantage to have a high signal-to-noise ratio because the information is an eddy current coil impedance for one single frequency as function of a superimposed dynamic magnetisation.

The microstructure parameters of interest here for the IF-steels are those which contribute mainly to the yield strength of these steels: grain size according to a Hall-Petch relation, carbon and nitrogene equivalent of the elements Nb or Ti which precipitate after hot rolling as carbides and nitrides, minimizing the solid solution effect of carbon and nitrogen as interstitial atoms and dislocation density which after recrystallisation is influenced mainly by the skin path rolling effect. Therefore we conclude:

$$Rp_{0.2} = Rp_{0.2}\,(P_1, P_2, P_3, \ldots),\ \{P_l\},\ l = 1,\ldots, L \text{ are microstructure parameters.} \qquad (3)$$

The micromagnetic techniques are of special interest for yield strength prediction because Bloch walls in ferromagnetic materials interact under magnetic load with the microstructure parameters such as dislocations under mechanical load. Therefore we assume for a set of micromagnetic parameter $\{M_k\}$, $k = 1, \ldots, K$:

$$M_k = M_k (P_1, P_2, P_3, \ldots). \qquad (4)$$

Combining (3) with (4) by eliminating the microstructure parameters it is assumed that there exists a unique relationship:

$$M_k = M_k (Rp_{0.2}) \text{ and vice versa } Rp_{0.2} = Rp_{0.2} (M_k), \; k = 1, \ldots, K. \qquad (5)$$

In multiregression algorithms, a mathematical model for the functional behaviour to be approximated - here the yield strength as a function of the micromagnetic Parameters M_k according to (5) - is found by developing $Rp_{0.2}$ with respect to a set of basis functions $\{\varphi_i\}$, $i = 1, \ldots, \infty$ in a complete infinite dimensional linear functional space (Mittra R., 1973), i.e.:

$$Rp_{0.2} = \sum_i^{\infty} \alpha_i \varphi_i (M_k), \; i = 1, \ldots, \infty, \; k = 1, \ldots, K. \qquad (6)$$

In (6) the set of coefficients $\{\alpha_i\}$, $i = 1, \ldots, \infty$ uniquely defines the function $Rp_{0.2}$. Reducing the set of coefficients to a finite number I means to project the function (6) in the I-dimensional subspace; in this case we have only an approximation $Rp_{0.2}^*$ of Rp0.2 and the so-called residual function res := $(Rp_{0.2}^* - Rp_{0.2})$ does not equal zero.

As $Rp_{0.2}^*$ is only an approximation of $Rp_{0.2}$, any norm of $\|res\|$ - will always be ≥ 0 and the strategy to find the α_i can only be to minimise a norm, where the norm is induced by the square root of a scalar product ($\| \cdot \| = \langle \cdot | \cdot \rangle^{1/2}$). By defining a scalar product, the vector space will be a Hilbert-space:

$$\underset{\alpha_i}{\mathrm{Min}} \left\{ \|res\|^2 \right\} = \underset{\alpha_i}{\mathrm{Min}} \langle res | res \rangle = \underset{\alpha_i}{\mathrm{Min}} \left\langle Rp_{0.2}^* - Rp_{0.2} \middle| Rp_{0.2}^* - Rp_{0.2} \right\rangle. \qquad (7)$$

As the $Rp_{0.2}$ function exists only in a set $\{Rp_{0.2\,s}\}$, $s = 1, \ldots, S$ of destructively determined values of a finite number S; these are the calibration values for the optimisation procedure (7). The number S has to be chosen much larger than I - the number of unknown - because the measuring data is always corrupted by random noise and by disturbing fluctuations caused by temperature changes, changes of tensile prestress from the coiler or lift-off of the transducer system by vibrations of the strip. However, solving (7) means solving a system of linear equations for the unknown α_i.

First calibration procedures were performed on 42 samples taken from coils of different IF-steel grades in the laboratory. Parallel the yield strength values were determined destructively. The incremental permeabilty was measured at different places at the front- and backside of the samples. In order to find a robust approximation lift-off variations of the tranducer system and temperature-variations in a climatic chamber were performed during the calibration procedure; 403 observations were put in the fit-algorithm to determine the α_i. This set of parameters was stored in the memory of the equipment for on-line monitoring in the hot dip galvanising line of the Thyssen Steel company. After a first period of experience it was obvious that laboratory calibration data are only a rough estimate for real steel qualities leaving a production line. Therefore an extensive calibration work was initiated where on-line measured nd-data was selected in a database and then off-line correlated with the results from the destructive tests from samples taken at the same position where the nd-measurement was performed. New sets of $\{\alpha_i\}$ were found, i = 1, ..., I, where in a first step the calibrations were made for each individual IF-steel grade. Figure 3. shows the continuous monitoring result along a strip length longer than 2250 m of the IF-steel M3A12 which is Ti-microalloyed. In addition to the nd-determined $Rp_{0.2}$-values 7 destructively determined values from samples taken at both ends of the strip are indicated. The strip was rejected because at the head end the values were outside the acceptance intervall. The continuously measured data shows a scatter of 2σ = 3% of the nominal value of Rp0.3 which is 160 MPa. This scatter is due to strip vibrations. However, the destructively determined values are within this scatterband of the nd-determined values.

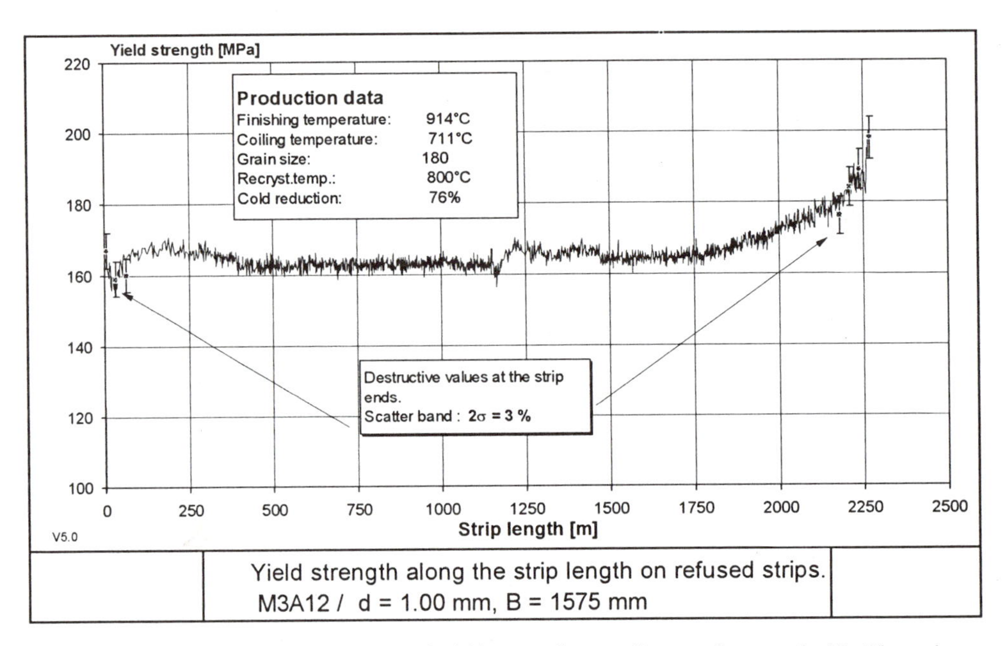

Figure 3. Continuous on-line monitoring of yield strength according a nd-approach (7). The strip was rejected because the yield value was outside the acceptance interval [130 MPa, 180 MPa].

At this point it should be mentioned that the residual standard deviation in the yield strength model according to Figure 3. is 1σ = 5.5 MPa which is a factor of 2 compared with the accuracy of the mechanical tensile test and therefore an enhancement is needed.

Figure 4. documents the on-line monitoring results concerning the nd-determined r_m and Δr anisotropy parameters according to (1) and (2) for the same strip as in Figure 3. For the steel manufacturer the r_m value is of special interest, correlating strongly - for these steel qualities -with the r-value in 90° to the rolling direction. For acceptance r_m has to be larger than 1.25. At the head-end - similiar to the indications in the yield strength - r_m is $\leq$ the acceptance value, therefore observing these quantity is also a criterion for strip rejection. The destructively determined values fit the nd-determined values very well within the 1σ = 0.05 scatterband caused by strip vibrations. The 1σ residual standard deviation of the model is 0.06 r_m units and comparable to the accuracy obtained in the mechanical tests. Therefore here no further enhancement is needed. It should be mentioned that the rejected strip was produced from a slab which was the first of a converter charge (run-in slab). Because of the higher impurity content the quality often tends to be lower.

FURTHER NUMERICAL STUDIES - CONCLUSION

In order to enhance the calibration approach for the yield strength model extensive further numerical analysis of the database was performed. It was found that in addition to the nd-determined parameters of the incremental permeability also external parameters are of interest. These are the thickness of the strip, the finishing and coiling temperature of the hot-rolled strip - influencing the precipitation process on the roller table - and the carbon and nitrogen equivalent for Nb and/or Ti. Taking into account these additional parameters - delivered by the plant system computer - a residual standard deviation of $1\sigma = 5$ MPa is achieveable. The most essential result for the production line is that this enhanced calibration procedure can be obtained independent of the individual IF-steel grade. By using a neural network approach (NeuralWorks, 1991) on the same data base instead of a multiregression model the residual standard deviation can be reduced to $1\sigma = 4$ MPa. A further enhancement can only be achieved if new and divers information is additionally selected. This can be performed by eddy current impedance measurements and by measuring and analysing the magnetic tangential field strength to find its higher harmonics (Dobmann, G. and Pitsch, H., 1989). A distortion factor can be derived from the higher harmonics amplitudes and a coercivity value is determined by analysing the time signal of the higher harmonics. Taking this additional information into account the residual standard deviation is reduced to 3 MPa.

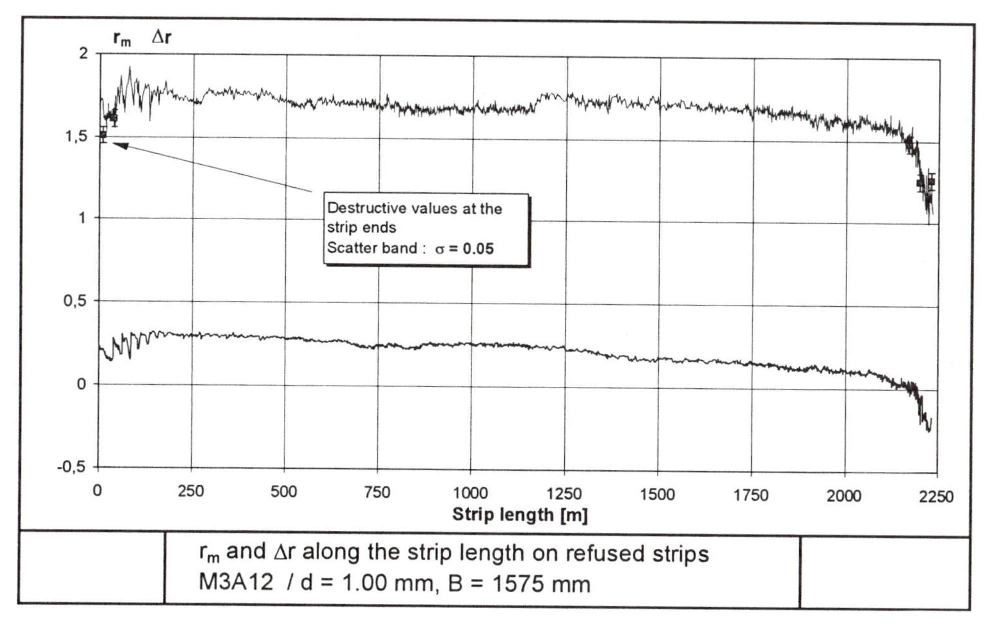

Figure 4. r_m and Δr at the same strip of Figure 3.

ACKNOWLEDGEMENTS

The project work was sponsored by the European Community for Coal and Steel under the contract number 7215-EB/102. All on-line measurements were performed by M. Borsutzki of Thyssen Steel as a part of his Ph.D. thesis. The extensive numerical studies to enhance the approximation model are from R. Kern. W.-A. Theiner was the responsible project manager.

REFERENCES

Bleck, W., Bode, R., and Hahn, F.J., 1990, Herstellung und Eigenschaften von IF-Stahl, *Thyssen Technische Berichte 1*: 69-85.

Borsutzki, M., Thoma, Ch., Bleck, W., and Theiner, W-A., 1993, On-line Bestimmung von Werkstoffeigenschaften an kaltgewalztem Feinblech, *Stahl Eisen 113, Nr. 10*:93-98.

Dobmann, G., et al, 1988, Quantitative eddy current variants for micromagnetic multiparameter microstructure and stress analysis, in: *Review of Progress in Quantitative NDE, Vol 7B*, D.O. Thompson and D. Chimenti, eds., Plenum Press, New York.

Dobmann, G., et al, 1989, Progress in the micromagnetic multiparameter microstructure and stress analysis (3MA), in: *Nondestructive Characterization of Materials III*, P. Höller, V. Hauk, G. Dobmann, R. Green, and C. Ruud, eds., Springer-Verlag, Berlin.

Dobmann, G and Pitsch, H., 1989, Magnetic tangential field strength inspection a further ndt-tool for 3MA, in: *Nondestructive Characterization of Materials III*, P. Höller, V. Hauk, G. Dobmann, R. Green, and C. Ruud, eds., Springer-Verlag, Berlin.

Herzer, R., and Schneider., E., 1989, Instrument for the automated ultrasonic time of flight measurement - a tool for materials characterisation, in: *Nondestructive Characterization of Materials III*, P. Höller, V. Hauk, G. Dobmann, R. Green, and C. Ruud, eds., Springer-Verlag, Berlin.

Mittra, R., 1973, *Computer Techniques for Electromagnetics*, Pergamon Press, Oxford.

NeuralWorks, 1991, Professional II/Plus, NeuralWare, Inc., Pittbourgh, PA 15276.

Sayers, C.M., and Proudfoot, G.G., 1986, *J Mech. Solids 34*: 579.

Schmidt, W., and Schaffrath, W, 1993, Kennzeichnung der Umformbarkeit von Flachprodukten durch r- und n-Wert, Bleche, *Bänder, Profile 10*.

NDE FOR STRENGTH DETERMINATION OF DIFFUSION BONDS

O. Buck, C. G. Ojard, D. J. Barnard, and D. K. Rehbein

Ames Laboratory and Materials Science
and Engineering Department
Iowa State University
Ames, IA 50011-3020

INTRODUCTION

Strength determination of diffusion bonds by nondestructive evaluation is a necessity for the qualification of such bonds in critical components. We have applied ultrasonic techniques for a number of years and come to the following conclusions. There is no problem in determining the bond strength (here defined as the ultimate tensile strength) by single frequency reflection coefficient measurements if the bond strength divided by the ideal strength (the ultimate tensile strength of the matrix material) is about 0.8 or less. Unfortunately, the reflection coefficient is not very sensitive to strength changes near the ideal strength[1] (bond strength/ideal strength between about 0.8 and 1.0). Near the ideal strength, the reflected energy can be more sensitive than the reflection coefficient (depending on the selection of certain acoustic parameters) since the reflected energy favors the high frequency end of the available acoustic spectrum.[2] Near the ideal strength, the reflected energy is about proportional to the strength degradation due to imperfect bonding. Acoustic harmonic generation is even more sensitive than the reflected energy.[3] However, the correlation between harmonic generation and the bond strength is nonlinear due to the transition of 3D-voids to "kissing bonds" which yield maximum acoustic harmonic generation about 10% below the ideal strength (bond strength/ideal strength about 0.9). Thus, a combination of reflected energy and harmonic generation measurements appears to be sufficient to quantify the strength of diffusion bonds of good quality. We are planning to use this information during in-situ measurements of a bonding apparatus to determine the progress of the diffusion bonding process.

BACKGROUND

Diffusion bonding is a thermally activated process. Therefore, appropriate process parameters can be chosen (temperature, time, pressure) such that a basically sound bond between two metallic alloys may come about. This is also true for reaction bonds, in which usually a ceramic material is bonded to a metal or another ceramic. Another parameter to be chosen is the geometry of the bonded piece. We have chosen cylindrical samples[1] as shown in Figure 1 with the bond line being symmetrical to the two end faces. This has the advantage that the bond plane can be looked at by acoustic reflection or through transmission experiments and that tensile specimens can be fabricated easily.

Reflected acoustic energy data taken from such a bonded specimen are shown in Figure 2. These data were taken using a 10MHz transducer focused in the bond plane, and scan-

Nondestructive Characterization of Material VIII
Edited by Robert E. Green Jr., Plenum Press, New York, 1998

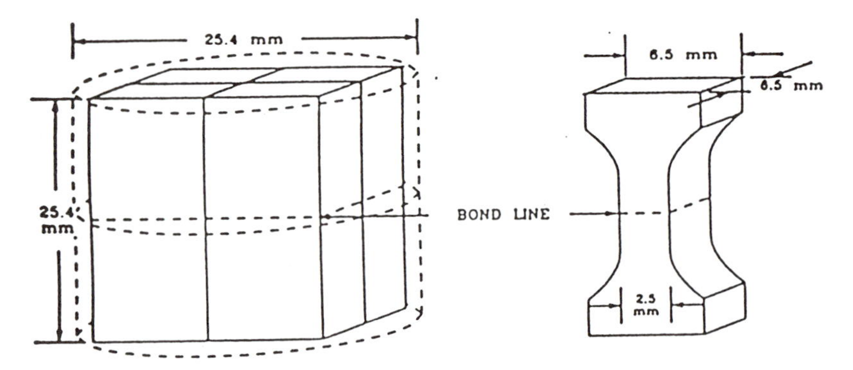

Figure 1. Diffusion bonded sample geometry and one of the four tensile specimens taken from the bonded sample (from Reference 1).

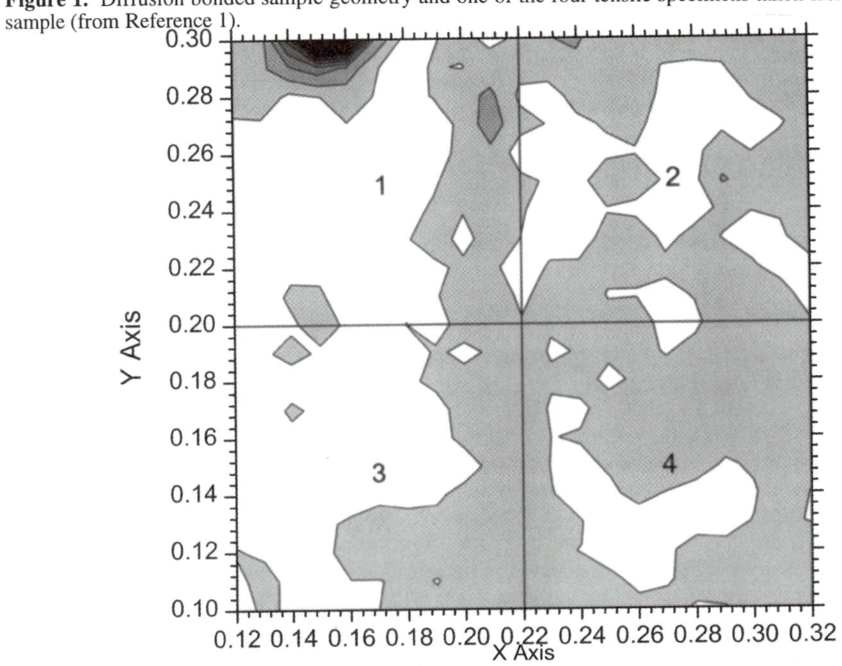

Figure 2. Reflected acoustic energy contour map of a Cu-Cu bond. White areas are well bonded, darker areas are signs of disbonding. The numbers indicate the specific tensile sample. In this case, sample 1 indicates some disbonded areas and therefore an indication of an average low bond strength. The area shown is 1.27cm x 1.27cm (from Reference 3).

ning at 0.65mm intervals in a water bath. Based on data, like that shown in Figure 2, we arrived at a strength versus total reflected energy curve,[3] as shown in Figure 3. Samples 1-1 and 2-3 show large errors in the total reflected energy since there are areas with obvious disbonds as shown, e.g., at the top of area 1 of Figure 2. The same is true for specimen 2-3 (not shown). As shown in Figure 4, the best bonds yield stress-strain curves that are practically identical to those of the matrix material.[1]

Figure 5 shows the progression of the changing microstructure in a Cu-Ni diffusion couple.[4] Ni is completely soluble in Cu (and vice versa), yet the compositions developing can be easily seen by metallographic means. After the bonding at 650°C for 4 hours and a pressure of 18.8MPa, this bond is nearly perfect and the specimen failed in the Cu matrix rather than along the original bond line since solid solution produces higher strength. Basically we understand the relatively fast diffusion (with respect to bulk diffusion) of the Ni into Cu due to grain boundary diffusion as well as enhanced bulk diffusion in Cu due to vacancy formation during creep. Furthermore, we observe the formation of a very uniform layer of Cu-Ni in the original Ni part of the sample. We also can see signs of "kissing" bonds.

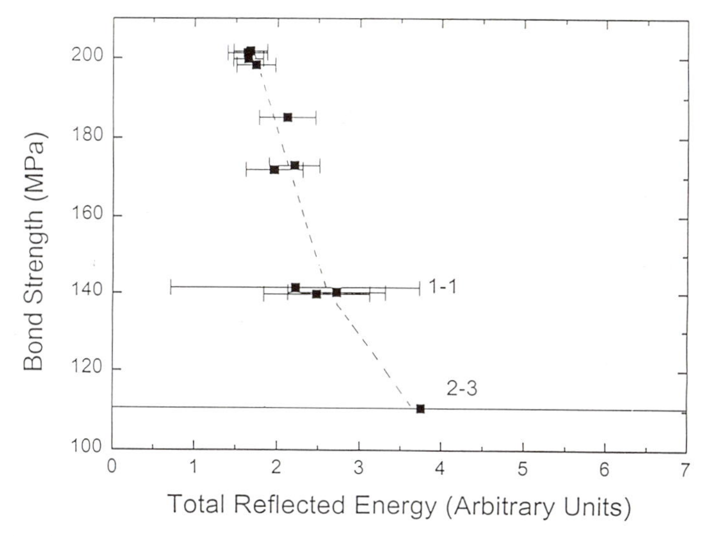

Figure 3. The bond strength of various diffusion bonded samples of a Cu-Cu bond. The large error bars of specimen 1-1 and 2-3 come about due to imperfectly bonded areas such as shown in Figure 2. The ideal strength is about 210 to 215 MPa. The values shown are the average over the cross-sectional areas of each tensile specimen (from Reference 3).

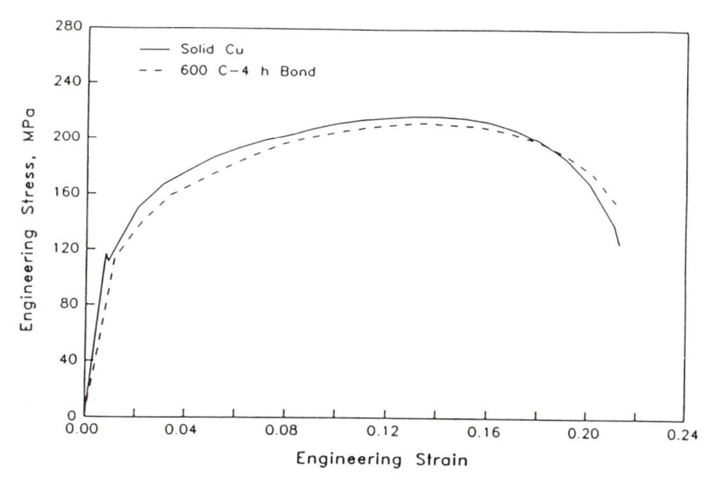

Figure 4. The engineering stress-strain curve of the matrix material (Cu) compared with that of a diffusion specimen of Cu versus Cu, bonded at 600°C for 4 hours (from Reference 1).

These are defined as areas of mechanical contact yet no obvious "bonding" due to inter-diffusion of Ni and Cu, as seen in Figure 5. This microstructural development may be explained by Figure 6. Enhanced surface diffusion participates strongly in filling the voids at the bond line. It has been found experimentally[5] that Cu diffuses on Ni four orders of magnitude faster than Cu on Cu, Ni on Ni, or Ni on Cu. Auger analysis[4] of fractured specimens showed Cu deposition on the Ni side, confirming the results in Reference 5 qualitatively.

Thus, the model that we use to describe our results can be envisioned as shown in Figure 7. Relatively large voids at the beginning of the bonding cycle become smaller. These voids are perfect ultrasonic scatterers as has been shown earlier.[1,6] Eventually, they

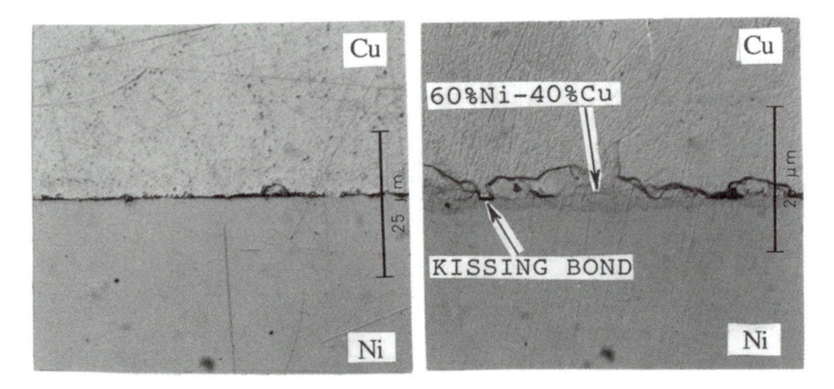

Figure 5. Cu-Ni diffusion bonds as a function of temperature. Common parameters: pressure = 18.8MPa, bonding time = 4h. Left: T = 500°C. Right: T = 650°C. At the original bond line the chemical composition is about 60 at% Ni - 40 at% Cu. After treatment at 650°C "kissing bonds" have been detected. Note that Ni diffuses into the Cu much faster than Cu into the Ni. Also note enhanced grain boundary diffusion of Ni in Cu and the relatively uniform layer of Cu-Ni as the Cu diffuses into the Ni (from Reference 4).

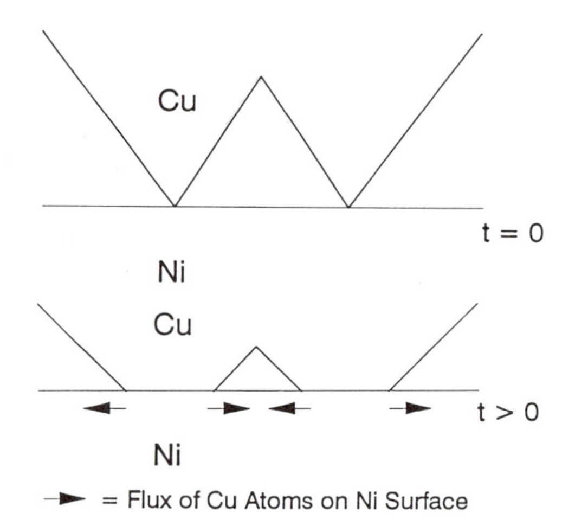

Figure 6. Top: Start of the diffusion bonding process. Asperity contacts form the first mechanical contacts. Bottom: As pressure is applied, Cu diffuses rapidly on the Ni surface. In addition, there is Cu diffusion into the Ni and Ni diffusion into the Cu as well as some creep in the Cu (from Reference 4).

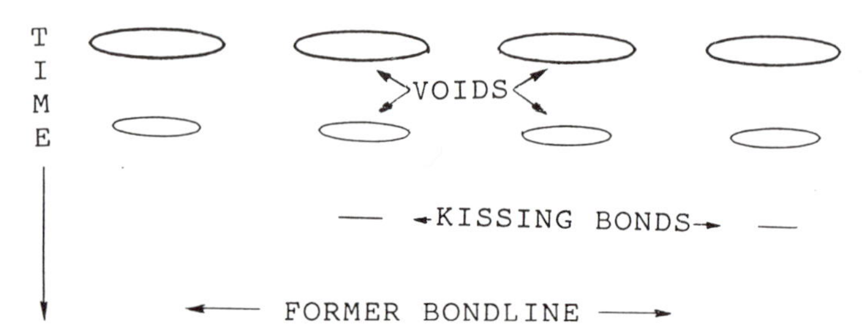

Figure 7. Sequence of the diffusion bonding process. After forming voids, "kissing" bonds are produced just before completion of a perfect bond.

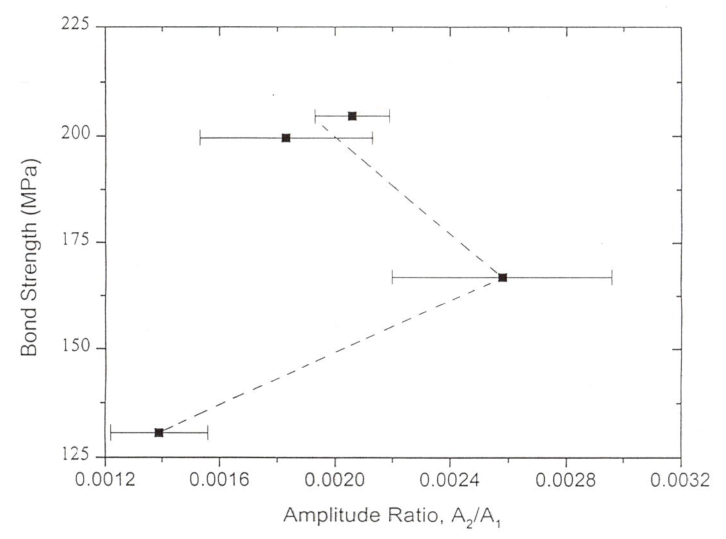

Figure 8. The strength of a Cu-Cu diffusion bond in relation to the measured, relative harmonic amplitude A_2/A_1 where A_1 is the fundamental amplitude. Harmonic generation is at a maximum at about 170 MPa bond strength, as expected if "kissing" bonds are present (from Reference 3) .

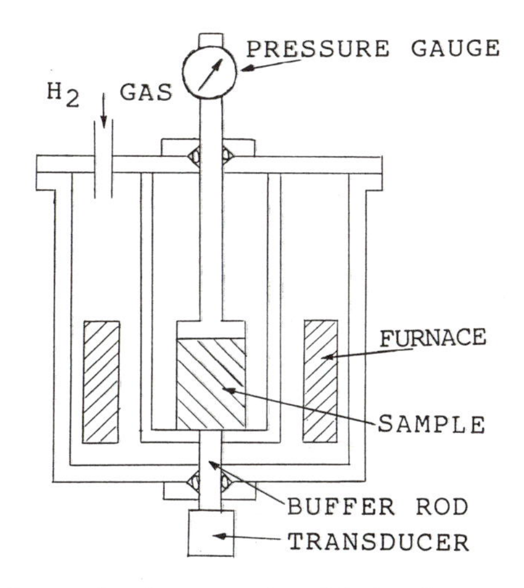

Figure 9. System envisioned to monitor the progression of the bonding state in-situ. The sample is exposed to H_2 gas to prevent an oxide layer from forming. The transducer is to be coupled with the sample via an alumina buffer rod. Two types of transducers will be used: one will be a focused transducer to monitor the reflected energy; the other will be to monitor the acoustic harmonic generation.

will collapse to create kissing bonds which are basically transparent to ultrasound. Their disappearance completes the bonding cycle. Note that the processing parameters for these cycles do not only depend on the chemical composition of the material but also on grain size, as can be seen from the enhanced diffusion of Ni into Cu by grain boundaries.

To determine if "kissing" bonds indeed are formed, we studied the effects of incomplete bonding by acoustic harmonic generation. Harmonic generation, whether from bulk materials or interfaces, manifests itself as a distortion of a sinusoidal wave. In bulk materials, the distortion can be the result of the nonparabolic potential of the atomic lattice or of anelasticity due to dislocation motion.[7] In the Richardson model,[8] distortion of a sinusoidal wave, impinging upon an unbonded interface, is caused by the inability of the interface to support the tensile phase of the longitudinal acoustic wave. This is easily seen in Richardson's model of a planar interface separating two semi-infinite elastic materials, with the surfaces in intimate contact but having no traction forces across the interface. As a longitudinally polarized elastic wave passes across this interface from one half space into the other, the surfaces will stay together and move as one during the compression phase of the wave and separate during the tensile phase. This nonlinear action, the opening and closing of the interface, will distort the wave and therefore generate harmonics. Figure 8 indeed indicates that voids do not produce acoustic harmonics; yet strong harmonic generation is observed at about 170MPa which we interpret as being due to the presence of "kissing" bonds. As the "kissing" bonds disappear, harmonic generation at the original interface disappears.

Presently, we are looking at possibilities to make use of the above observations during the bonding process. The system envisioned is sketched out in Figure 9.

CONCLUSIONS

It has been shown that the acoustic reflected energy in connection with harmonic generation can be used to quantify the state of diffusion bonds in terms of their tensile strength. Detailed metallographic observations helped to interpret the acoustic observations. We are presently designing bonding equipment which should allow the in-situ inspection of the bonding state of metallic samples.

ACKNOWLEDGMENT

This work was performed at Ames Laboratory under Contract Number W-7405-Eng-82 with the U. S. Department of Energy.

REFERENCES

1. D.D. Palmer, D.K. Rehbein, J.F. Smith, and O. Buck, Nondestructive characterization of the mechanical strength of diffusion bonds I -Experimental results, *J. Nondestr. Evaluation* 7:153 (1988).
2. O. Buck and G.C. Ojard, Strength, acoustic evaluation and metallurgy of diffusion bonds, in *Nondestructive Evaluation and Materials Properties II*, P.K. Liaw et al., eds, TMS, Warrendale, PA (1994).
3. D.J. Barnard, G.E. Dace, D.K. Rehbein, and O. Buck, Acoustic harmonic generation at diffusion bonds, *J. Nondestr. Evaluation* (in press).
4. G.C. Ojard, Strength, ultrasonic and metallurgical evaluation of diffusion bonds of dissimilar materials, Ph.D. Dissertation, Iowa State University, Ames, IA (1991).
5. B.Ya. Pines, I.P. Grebennik, and I.V. Gektina, *Sov. Phys. Cryst.* 12:556 (1968).
6. D.D. Palmer, D.K. Rehbein, J.F. Smith, and O. Buck, Nondestructive characterization of the mechanical strength of diffusion bonds II -Application of a quasi-static spring model, *J. Nondestr. Evaluation* 7:167 (1988).
7. A. Hikata and C. Elbaum, Generation of ultrasonic second and third harmonics due to dislocations I, *Phys. Rev.* 144:49 (1966).
8. J.M. Richardson, Harmonic generation at an unbonded interface I -Planar interface between semi-infinite elastic media, *Intl. J. Engineering Science* 17:73 (1979).

MONITORING OF TEXTURE DEVELOPMENT IN COPPER-ALLOY SHEET

George A. Alers[1], Pat Purtscher[1], John F. Breedis[2], Frank N. Mandigo[2]

[1] National Institute of Standards and Technology, Boulder, CO 80303
[2] Olin Corporation, New Haven CT 06511

ABSTRACT

Recent advances in the theory of ultrasonic-wave propagation in polycrystalline media have developed mathematical relationships between the velocity of Lamb waves propagating in the plane of rolled sheet metal and the lowest-order coefficients in the grain orientation distribution function. This paper describes the use of Lamb wave velocity measurements in thin sheets of three commercial copper-zinc alloys to determine three of these coefficients and documents how they change as the sheet is rolled and annealed. The results indicate that by measuring only three Lamb wave velocities at particular times during the processing of the sheet, it should be possible to monitor and control the development of a desired texture. By using electromagnetic transducers, the Lamb wave velocity measurements could be performed under on-line conditions.

INTRODUCTION

During the rolling of sheet metal, a strong preferred orientation (texture) in the grain structure is developed. This causes the properties of the final product to exhibit anisotropic behavior with a different response along the three natural directions in the sheet, namely, along the rolling direction, the transverse or width direction and the thickness or sheet-normal direction. Thus, rolled sheet can be expected to exhibit an ultrasonic response like an orthotropic single crystal and require nine elastic constants to fully describe its behavior. Since it is easy to excite and detect ultrasonic Lamb waves that propagate in the plane of the sheet, it should be possible to make accurate measurements of the Lamb wave velocity as a function of the angle between the propagation direction and the rolling direction under the harsh conditions that exist in a rolling mill operation. Because of recent advances in the theory of ultrasonic wave propagation in polycrystalline materials[1] and in its application to the particular case of Lamb waves in thin sheet metal[2,3], the angular dependence of the velocity of the zero-order symmetric Lamb wave can be expressed in terms of the two elastic moduli of a random distribution of grain orientations and three orientation distribution coefficients (ODCs). By using EMATs (electromagnetic acoustic transducers)[4], these three coefficients can be measured under rolling-mill conditions[5], and a prediction of a very useful engineering property, the formability, can be made at many stages of the rolling process.

This paper describes how the theoretical and experimental techniques employed by the formability instrument can be applied to the monitoring of texture development in the manufacture of copper, brass and bronze sheet-metal products. Prediction of a particular property such as formability will not be discussed explicitly here; instead, a technique for achieving a particular texture through an efficient choice of rolling reductions and annealing steps will be discussed. Definition of the texture needed to achieve a particularly desirable set of properties in the final product must be left as a proprietary goal for each manufacturer.

THEORETICAL FOUNDATIONS

If the individual grains in a polycrystalline aggregate possess cubic crystal symmetry (as is the case for copper and its alloys with zinc), only five individual quantities need be known to make a first-order approximation to the nine elastic moduli of an anisotropic rolled sheet. Two of these five quantities are the two elastic moduli of a random mixture of grains of the particular alloy being processed and are thus independent of the rolling reductions and annealing steps involved in the manufacture of a particular product. The other three are the lowest-order orientation distribution coefficients of the general orientation distribution function which describes the probability of finding a particular grain orientation among all the possible grain orientations in a polycrystal[6]. For an exact description of the elastic properties, more than the three lowest order terms would be needed. However, since only three ultrasonic measurements were needed to make a useful correlation with formability, it is anticipated that only three coefficients will be sufficient for monitoring texture development during the rolling of sheet metal products. With these three coefficients, the velocity of a Lamb wave propagating in the plane of the sheet can be described by a function with the form[7]

$$V(\theta) = A + B\cos 2\theta + C\cos 4\theta \tag{1}$$

where θ is the angle between the propagation direction and the rolling direction. The three coefficients A, B and C are directly related to the three lowest-order ODCs[6]. Our measurements of the phase velocity of a zero-order symmetric Lamb wave propagating in the plane of a thin sheet of brass show that this three-coefficient model is adequate for describing the observed anisotropy in the phase velocity of Lamb waves to within the $\pm 0.15\%$ experimental scatter in the data.

In the general theory for describing the elastic properties of a textured polycrystal, the three lowest-order ODCs are often written as W_{400}, W_{420} and W_{440}. For a random mixture of grain orientations, these three coefficients would have values equal to 0. For a nonrandom distribution, the mathematical relationship between the A, B and C coefficients and the three lowest-order ODCs are already available in the literature[7]. Inserting these relationships into Eqn. (1) and solving for the W coefficients yields

$$\begin{aligned} W_{400} &= 0.15672\{\rho V(0)^2 + \rho V(90)^2 + 2\rho V(45)^2 - 4L[1-(L/P)^2]\}/\{[3+8(P/L)+8(P/L)^2]c\}, \\ W_{420} &= 0.04956\{\rho V(90)^2 - \rho V(0)^2\}/\{[1+2P/L]c\}, \\ W_{440} &= 0.03746\{\rho V(0)^2 + \rho V(90)^2 - 2\rho V(45)^2\}/c, \end{aligned} \tag{2}$$

where V(0), V(45) and V(90) are the measured phase velocities of the zero-order symmetric Lamb wave mode (in the long-wavelength limit) propagating at 0, 45 and 90°, respectively, to the rolling direction. The quantity ρ is the mass density of the material and $c = (C^0_{11} - C^0_{12} - 2C^0_{44})$ where C^0_{11}, C^0_{12} and C^0_{44} are the three elastic constants of the individual cubic grains. The elastic constants L and P are the values of the diagonal elements of the elastic modulus tensor for a random distribution of grain orientations. They can be estimated from the single-crystal elastic constants of a grain by using one of several mathematical averaging techniques. For the measurements presented in this paper, we use an averaging procedure described by Ledbetter[8] and the single-crystal elastic constants of Rayne[9] for Cu-Zn alloys and of Moment[10] for the Cu-Sn alloys.

For monitoring the development of textures during rolling and annealing, we need only maintain internally consistent averaging procedures and not seek exact agreement with ODCs deduced from more rigorous techniques such as X-ray or neutron diffraction. Furthermore, it is more important that our measurement technique operate in the environment of a rolling mill than yield exact values for W_{400}, W_{420} and W_{440}. To meet the the environmental requirements of an operating rolling mill, EMATs[4] are suggested. These transducers are particularly well suited to the excitation and detection of individual Lamb-wave modes in thin sheet metal.

EXPERIMENTAL TECHNIQUE

EMATs for exciting and detecting the zero-order symmetric Lamb waves in thin metal sheets were assembled from 25 x 25 mm meander coils with a 3.05 mm spacing between adjacent wires. These were placed below the 25 mm square face of an NdFe permanent magnet 12 mm high. Such a meander coil will efficiently launch and detect the zero-order symmetric Lamb wave mode in copper and brass if it is tuned to operate at a frequency near 0.67 MHz and the sample sheets have thicknesses less than about 2 mm. For the measurements described here, the EMATs were driven by a high-power gated amplifier. The signals from the receiver EMAT were preamplified and displayed on a digital oscilloscope equipped with a delayed sweep circuit capable of measuring the arrival time of a signal to an accuracy of ± 17 ns.

Separate transmitter and receiver EMATs were mounted on a goniometer which allowed the line joining the pair to be rotated about a point by an angle measurable to within $\pm 1°$. The entire goniometer was mounted onto a frame equipped with a micrometer screw that would allow the pair of EMATs to be translated along a straight-line path over a distance determined by the pitch and length of the screw. For the apparatus used in the experiments described here, the micrometer screw had a pitch of 0.635 mm/turn. As a result, the distance moved by the translation device could be measured to within $\pm$ 0.01mm. For the Lamb-wave velocity to be measured to within $\pm$ 0.1% under the condition that the transit time had an uncertainty of ± 17 ns, the travel distance for the sound waves had to be at least 8 cm.

For measuring the phase velocity of a Lamb wave on a particular sample of sheet metal, the goniometer was placed on the sample so that the line joining the two EMATs was parallel to the rolling direction and perpendicular to an edge of the sheet that had been cut perpendicular to the rolling direction. The distance between the EMAT transmitter and the sheet edge was set to be approximately 50 mm. The motion axis of the translation device was also aligned parallel to the rolling direction and, therefore, parallel to the line joining the transducers. In this configuration, the oscilloscope screen showed two signals. The one arriving first was produced by the ultrasonic wave propagating directly from transmitter to receiver. Since simple EMAT transmitters produce waves that radiate from the transmitter in opposite directions, the second signal observed arrived by a path which included a reflection from the edge of the sample. With the micrometer screw used to translate the EMAT pair away from the cut edge, the arrival time of a zero crossing in the middle of the second signal could be observed to increase. A graph of that arrival time as a function of the change in separation distance from the sample edge displays a straight line whose slope is twice the phase velocity of the acoustic wave because the translation device moves both the transmitter and the receiver relative to the edge. Since the translation is parallel to the rolling direction, this velocity value is V(0) in Eqn. (2). To measure V(45) and V(90), the arrival time of a zero crossing in the first signal was measured as a function of the rotation angle of the goniometer. If the *change* in this arrival time as a function of angle relative to the rolling direction is defined as $\Delta t(\theta)$, where θ is the rotation angle, then it is easily shown that

$$V(\theta) = V(0)[\ 1 - V(0)\Delta t(\theta)/\ d]\ , \tag{3}$$

where d is the fixed separation between transmitter and receiver within the goniometer. For the measurements reported here, d=69.5 mm. Because of the reflection at the cut edge of the sample, the length of the acoustic wave travel path is twice the translation distance of the goniometer, which had to be only 40 mm to insure anuncertainty of ±0.1 % in the determination of the velocity.

SAMPLE PREPARATION

The objective of the experiments reported here was to establish an empirical relationship between the texture and the processing variables of rolling reduction and annealing temperature. Rolling reduction was measured as the total reduction in thickness accumulated during a series of rolling steps at room temperature expressed as a percentage of the initial thickness. The annealing temperature was taken as the temperature at which the material was held for one hour during an annealing step. The quantitative measure of texture is the set of three ODCs, W_{400}, W_{420} and W_{440}, as deduced from the Lamb-wave phase velocities V(0), V(45), V(90) with the aid of Eqns. (2).

The starting alloys were commercially cast and processed to a thickness of 0.686 mm and the annealing was performed by following standard industrial proceedures. Subsequent cold rolling produced samples reduced in thickness by 0%, 10%, 20%, 30%, 40%, 50% and 60%. The samples of the sheet whose thickness had been reduced by 50% were annealed for one hour at the temperatures listed below:

Alloy C110 - 150, 200, 250, 300, 350 and 450°C
Alloy C260 - 250, 300, 350, 400, 450 and 550°C
Alloy C510 - 300, 350, 400, 450, 500 and 600°C.

Chemical compositions were as follows (in units of mass%):

Copper: Alloy C110 -- Cu+0.021% O.
Brass: Alloy C260 -- Cu+31.7%Zn+0.029%Fe+0.043%Sn+0.009%Pb+0.06%Ni
Bronze: Alloy C510 -- Cu+4.47%Sn+0.065%P+0.15%Zn.

RESULTS

Measurements of the zero-order symmetric Lamb-wave phase velocity in three directions on 39 samples of sheet metal were performed by the translating goniometer device described above. The 39 samples were produced by seven different rolling reductions and six annealing conditions on three different alloy compositions. The measured wave velocities were converted into the three W coefficients through the use of Eqns. (2). Table I lists the three ODCs deduced in this way for the case of the commercial purity copper.

To be useful, these data must be presented in a way that conveys the progression of texture development as a function of processing variables. To accomplish this, we treat the three ODCs as coordinate axes of a three-dimensional space. In this representation, texture development by increasing the amount of reduction appears a single curve. The effect of annealing defines another curve. An intersection of these curves represents a branch point at which a different texture can be developed either by additional rolling or by initiating an annealing procedure. Although modern computer programs can make these 3-D presentations, it is difficult to extract quantitative information from them. To make a quantitative interpretation more practical, a two-dimensional presentation of the data has

TABLE I. Values of the three ODCs deduced from the angular dependence of the velocity of the zero-order symmetric Lamb wave in thin sheets of commercial-purity copper processed by the different rolling reductions and annealing temperatures indicated.

Copper - C110

% Thick. Red.	$100W_{400}$	$100W_{420}$	$100W_{440}$	1 h Anneal Temp	$100W_{400}$	$100W_{420}$	$100W_{440}$
0	0.247	0.102	0.326	20°C	0.839	0.101	-0.437
10	0.534	0.117	-0.069	150	0.828	0.107	-0.405
20	0.596	0.110	-0.166	200	0.394	0.0	-0.057
30	0.677	0.106	-0.258	250	0.373	0.0	0.026
40	0.683	0.102	-0.357	300	0.394	0.015	0.037
50	0.839	0.101	-0.437	350	0.384	0.0	0.045
60	0.633	0.101	-0.461	450	0.444	0.0	0.067

been developed. Here, the concept of using the three Ws as coordinate axes is retained but a transformation to spherical coordinates R, θ and ϕ has been chosen. This transformation of coordinates is accomplished by the equations:

$$\begin{aligned} R &= \{(W_{400})^2 + (W_{420})^2 + (W_{440})^2\}^{1/2} \\ \phi &= \arctan (W_{420}/W_{400}) \\ \theta &= \arccos (W_{440}/R). \end{aligned} \tag{4}$$

In this spherical-coordinate representation, a two-dimensional graph of θ versus ϕ with R as a parameter adequately displays the two intersecting curves that describe texture development by rolling reductions and by annealing treatments. Figure 1 shows all the ODC data obtained from the 39 sheet samples covering the three alloys presented in this spherical-coordinate representation. The three alloys, copper C110, brass C260 and bronze C510, are displayed separately; (1) for clarity and (2) because they would be treated separately in an actual rolling mill operation. Each data point on the graphs represents a texture and the line joining the points represents the path followed to produce that texture. For rolling reductions, the line is solid; for anneals, the line is dashed. The R coordinate associated with each texture point is presented as a number equal to 100R connected with its point by a light line. A two-digit number following the 100R value is the percentage reduction in thickness that produced the texture represented by the point. A three digit number is the temperature in degrees Celsius at which the sample was annealed for one hour to produce the texture represented by the point. For example, the notation (0.4,30) means that the associated point has a 100R value of 0.4 and the sheet thickness has been reduced by 30%. The intersection of the solid line with the dashed line in each graph represents the beginning of the series of annealing steps which followed the 50% rolling reduction.

DISCUSSION

Figure 1 shows how the three W coefficients that were deduced from ultrasonic Lamb wave measurements can be presented in a form suitable for monitoring the development of texture during the processing of sheet metal products. The changes in coordinate values that are displayed reflect unmistakable changes in the shape of the Lamb wave velocity versus angle curves and these, in turn, are clear evidence of changes in the elastic anisotropy that arise from changes in preferred orientations among the grains. No reference to a crystallographic description of the preferred orientation is needed. The figure shows that there are significant changes in the texture taking place with rolling reductions of only 10

COPPER - C110

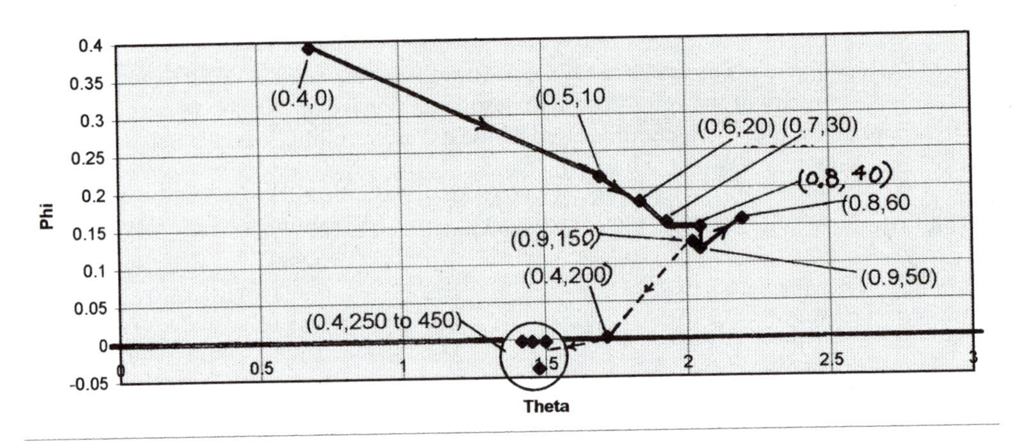

BRASS - C260

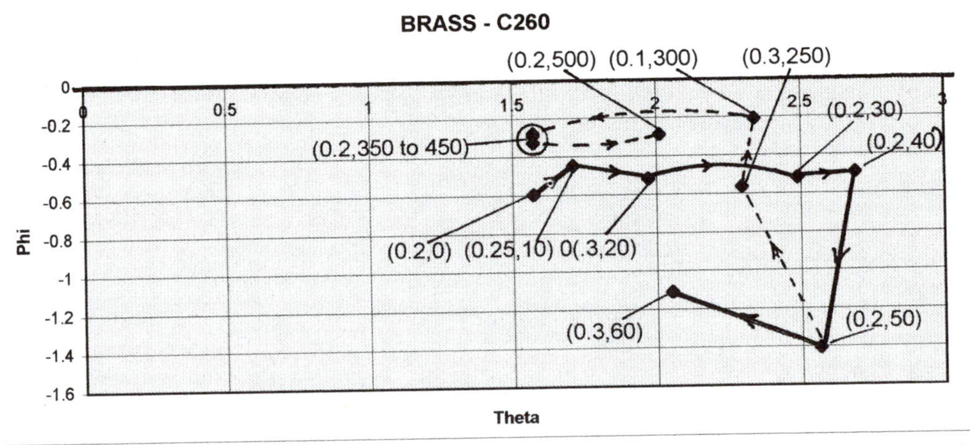

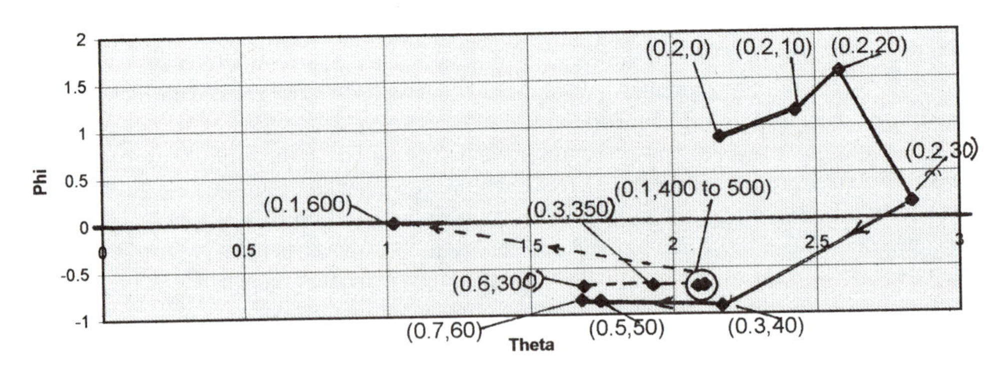

Figure 1. Graphical representation of texture development for three copper alloys as a function of rolling reduction (solid line) and annealing treatment (dashed line). The coordinates θ and ϕ are defined by Eqn. 4 relative to the three orthogonal coordinates, W_{400}, W_{420} and W_{440}. The first number in the parentheses associated with each point is 100 times the R coordinate for that point. The second number is either the percent thickness reduction (two digits) or the annealing temperature (three digits). For example, (0.4,30) indicates that 100R=0.4 and that there has been a 30% reduction in thickness.

and 20 percent from the initial annealed state. The clustering of data points in one region of (R,θ,ϕ) space indicates that the processing steps associated with those data points are not changing the texture. (They may be changing other properties, but not the texture.) Focusing attention on the numerical values of the coordinates shows that the textures developed in copper and brass are very distinctive. That is, copper is characterized by positive values of ϕ while brass displays negative values for this coordinate. Bronze shows a transition from copper-like to brass-like textures as the amount of deformation is increased. Rolling followed by annealing appears to return brass to near its initial texture while copper and bronze develop quite different textures following the annealing process.

ACKNOWLEDGMENTS

The authors would like to thank Ray Schramm of the NIST staff for his help in developing the techniques and in collecting the data from which the values of W_{400}, W_{420} and W_{440} were determined. Dr. Hassel Ledbetter of NIST and Prof. Martin Dunn of the University of Colorado deserve special thanks for their guidance in the theoretical descriptions of preferred orientation in polycrystalline materials.

REFERENCES

1. C.M. Sayers, Ultrasonic velocities in anisotropic polycrystalline aggregates, *J. Phys.- Appl. Phys.* 15:2157 (1982).
2. M. Hirao, K. Aoki and H. Fukuoka, Texture of polycrystalline metals characterized by ultrasonic velocity measurements, *JASA*, 81:1434 (1987).
3. R.B. Thompson, et al, A comparison of ultrasonic and x-ray determinations of texture in thin copper and aluminum plates, *Met. Trans.*, 20A:2431 (1989).
4. G.A. Alers and L.R. Burns, EMAT designs for special applications, *Materials Evaluation*, 45:1184 (1987).
5. E. Papadakis et al, Development of an automatic ultrasonic texture instrument and its transition from laboratory to market, *Materials Evaluation*, 51:77 (1993).
6. H.J. Bunge, Texture Analysis in Materials Science, Butterworth's, London (1982).
7. Y. Li and R.B. Thompson, Effects of dispersion on the inference of metal texture from S_o Lamb wave mode measurements. Part I, *J. Acoust. Soc. Am.*, 91:1298 (1992).
8. H. Ledbetter, Dynamic Elastic Modulus Measurements In Materials, Alan Wolfenden, ed., ASTM, STP 1045, Philadelphia (1990).
9. J.A. Rayne, Elastic constants of α-brasses: Variation with solute concentration from 4.2 to 300°C, *Phys. Rev.* 115:63 (1959).
10. R.L. Moment, Elastic stiffnesses of copper-tin and copper-aluminum alloy single crystals, *J. Appl. Phys.* 43:4419 (1972).

MAGNETOSTRICTIVE EMAT EFFICIENCY AS A NONDESTRUCTIVE EVALUATION TOOL*

B. Igarashi, G.A. Alers, and P.T. Purtscher

National Institute of Standards and Technology
Materials Reliability Division
Boulder, Colorado 80303

INTRODUCTION

Electromagnetic acoustic transducers (EMATs) are ideal for on-line inspection of hot metal sheets, because they provide a noncontacting means for generating and detecting ultrasound in metals. EMATs are especially attractive for probing steels and magnetic alloys, because their transduction efficiency is larger than in nonferromagnetic materials. The source of the extra efficiency is magnetostriction, the length change of a ferromagnetic material that accompanies magnetization.

This work demonstrates two novel applications of EMAT-generated ultrasound: (1) to determine key features of the magnetostriction of a steel sample and (2) to monitor strengthening precipitation states in a commercial high-strength low-alloy (HSLA) steel. This new technique of measuring magnetostriction offers the convenience of being noncontacting, in contrast to the standard strain gage method. The second application follows from the sensitivity of magnetostriction to the form and concentration of precipitates. Magnetostriction has the advantage over some other magnetic measurements in being sensitive to a broader range of potential effects of precipitates; it reflects effects of precipitates not only upon domain wall motion, but also upon rotation of domain magnetization.

The transduction mechanism behind EMAT waves in ferromagnetic materials is more complicated than in nonferromagnetic metals. In nonferromagnetic metals, EMATs generate and detect ultrasound through Lorentz forces, which are proportional to an applied static field. However, in ferromagnetic materials, the transduction efficiency is a nonlinear function of the applied static field, because there is a second mechanism for generating waves—magnetostriction. A significant aspect of magnetostrictive transduction

---

Nondestructive Characterization of Material VIII
Edited by Robert E. Green Jr., Plenum Press, New York, 1998

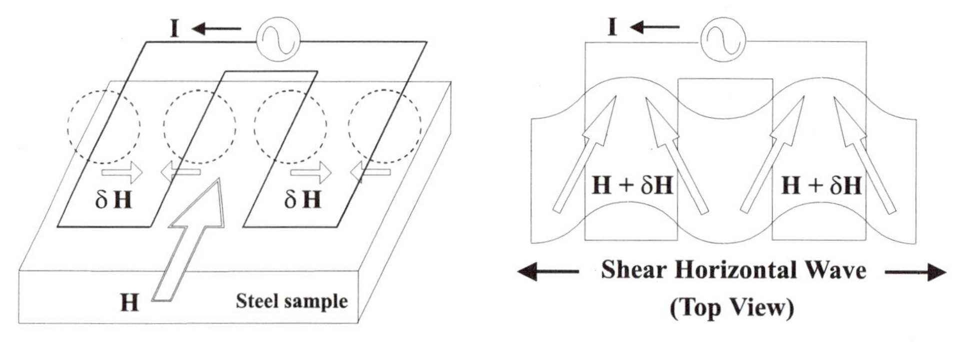

Figure 1. Setup for magnetostrictively generating SH ultrasonic waves with an EMAT. The meander coil carries ac current I, and H is applied by an external magnet. The resultant elastic displacements are parallel to the sample surface.

is that, at low static fields, it is much more efficient than transduction by Lorentz forces. In fact, for the range of fields used in this work, transduction by Lorentz forces is negligible, and measurements of the EMAT wave amplitude may be taken to reflect the magnetostrictive contribution only.

R.B. Thompson[1-3] has shown that shear horizontal (SH) waves may be generated magnetostrictively with the setup shown in Figure 1. A static magnetic field H establishes an axis of tension (or compression) due to magnetostriction. The ac current in the meander coil creates an oscillating field δH that adds perpendicularly to the static field and, in effect, rotates the axis of tension (or compression) in the surface plane, thus, creating a shear strain. The wavelength of the resultant SH wave is approximately twice the spacing between current elements of the meander coil that are parallel to H.

Thompson has shown that the generation efficiency of the SH wave is proportional to

$$|\lambda|\ \delta / H \tag{1}$$

where λ is the magnetostriction coefficient (the relative length change due to H), $\delta = \sqrt{2/(\mu\sigma\omega)}$ is the skin depth, μ is a "transverse" incremental permeability,[†] σ is the conductivity, and ω is the angular frequency. This expression applies best at high static fields and may not be valid at low fields. Since the detection efficiency of SH waves by EMATs also is proportional to expression (1), the amplitude of the signal from an EMAT receiving ultrasound that itself is generated by an EMAT may be approximated by the square of expression (1).

EXPERIMENTAL SETUP

Measurements were performed on ASTM A710 grade A, class 3 steel,[‡] a low-carbon, HSLA steel that was age-hardened by copper precipitates. The specimens measured in this work underwent two hardening heat treatments. The material received from the David

† The term "transverse" is appropriate here, because the oscillating field δH is perpendicular to the static field, H.

‡ In percent by mass, the chemical composition was:[4] [C]=0.04, [Mn]=0.58, [P]=0.001, [S]=0.004, [Si]=0.30, [Cr]=0.68, [Ni]=0.87, [Mo]=0.19, [Cu]=1.20, [Nb]=0.046.

Taylor Naval Ship Research and Development Center (Annapolis, Maryland) already had been heat-treated to increase its hardness.[§] At NIST-Gaithersburg, several 15 x 8 x 3 cm coupons subsequently were cut out and heat-treated again. The particular samples examined in this work were austenitized at 899 °C for 90 minutes, water-quenched, aged at either 482 °C or 593 °C for 90 minutes, and air-cooled.

Hicho et al.[4] determined the following properties of these samples:
(1) The austenitizing temperatures used in this study produced fine-grained ferrite with an average grain size of ASTM number 9. The same average grain size was obtained, whether the sample was austenitized for 30 or 90 minutes, or whether it was aged at 482 °C for 30 minutes or aged at 649 °C for 90 minutes.
(2) The inclusion content was very low, and microstructure analysis showed the inclusions to be spherical due to calcium additions.
(3) Small-angle neutron scattering (SANS) analysis indicated that the sample aged at 482 °C was left in the underaged condition; that is, the volume fraction of Cu precipitates was not maximized, and fine-scale Cu-rich precipitates were created. These precipitates were too small to be detected optically. (Typically the diameters are less than 5 nm.)[5]
(4) SANS analysis indicated that the sample aged at 593 °C was overaged; that is, the maximum volume fraction of precipitates was achieved and coarsening of Cu precipitates occurred.
(5) Impact tests conducted on standard (ASTM E-23) Charpy specimens indicated that the overaged sample had more than twice the toughness of the underaged sample at −18 °C (255 vs. 120 J absorbed).

For measurements in the present work, 130 x 20 x 8 mm bars were cut out. All measurements were performed on a single surface of each sample that was wet-ground

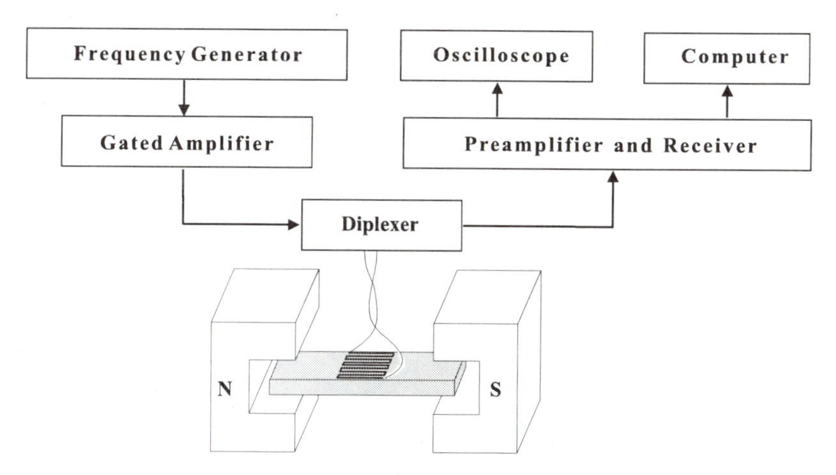

Figure 2. Experimental setup for measuring the amplitude of SH ultrasonic waves as a function of applied magnetic field. (Hall probe is not shown.)

[§] It was austenitized at 899°C for 30 minutes, water-quenched, aged at 593°C, and air-cooled.

using a sequence of sandpapers ranging in roughness between 60 and 800 grit. About 30 to 40 μm of material was removed by chemically polishing the surface with a solution of oxalic acid in hydrogen peroxide.

After two austenitizing heat treatments, the steel was expected to have little texture, and velocities measured as part of the present work bore out this expectation; for example, in the underaged sample, there was little anisotropy of velocities of orthogonally polarized shear waves propagating through the thickness of the sample. The velocity of waves polarized along the length of the sample was 3.223 ± 0.005 mm/μs and, along the width, 3.220 ± 0.005 mm/μs. The velocities were determined by mounting on one side of the sample an EMAT that issued ultrasonic pulses through Lorentz force transduction (for example, see Figure 7(c) of Reference 3) and then measuring the time between echoes with another EMAT mounted on the opposite side of the sample.

During other magnetic and ultrasonic measurements, samples were mounted between C-shaped pole pieces of an electromagnet (see Figure 2). Samples were positioned carefully in the electromagnet in such a way as to minimize pulling on the ends of the sample due to magnetic attraction between the sample and the pole pieces.

A single meander-coil EMAT of 1.5 mm wavelength was used both to generate and detect shear horizontal (SH_0) waves. A gated amplifier issued pulses consisting of 175 cycles of sinusoids at approximately 2 MHz. To boost the amplitude of received signals, the excitation frequency and EMAT position on the sample surface were selected to establish standing waves across the width of samples. As the magnetic field was varied, the frequency was adjusted constantly to compensate for velocity changes and maintain the resonance condition. (Velocity changes on the order of 0.1% originated from the ΔE Effect,[6] which is softening of the shear modulus caused by domain wall motion and rotation of domain magnetization.) Also, the output of the gated amplifier was adjusted constantly to compensate for changes in the coil inductance and to maintain a constant EMAT peak current of 5 A.

After the pulse, the EMAT was used to monitor acoustic ring-down. A diplexer served as a "switch box" to route microvolt signals from the EMAT to a preamplifier and superheterodyne receiver. The maximum amplitude of the standing wave built up by the pulse was recorded and may be taken as proportional to the amplitude of a traveling wave issued by the EMAT.** The wave amplitude was recorded as the magnetic field from the electromagnet was ramped down from about 80 kA/m (1000 Oe). The magnetic field was measured with a Hall probe resting upon the polished sample surface.

The magnetization, magnetostriction, and incremental permeability were measured separately. The B-H curve of a sample was obtained by measuring the voltage induced across a solenoid wound around a sample. A strain gage was glued onto a sample to give a contact measurement of the magnetostriction.

The "transverse" incremental permeability was measured through an eddy current technique. Rose et al.[8] have shown that the incremental permeability μ is proportional to the unique frequency f_0 that leaves the real inductance of a coil unaffected when the coil is brought close to a ferromagnetic material. In the present work, a flat, spiral, oval

---

** With this experimental setup, the proportionality constant does not depend strongly upon damping, since the duration of the pulse is short relative to the reciprocal of the ring-down rate. With a driving wave of amplitude A_0 starting at time $t = 0$, the resonant vibrations increase as

$$A(t) = A_0\left[1 - \exp(-n\Delta)\right] / \Delta \cong A_0 n$$

where Δ is the log decrement of damping ($\sim 10^{-3}$) and n is the number of cycles per pulse (n=175).[7]

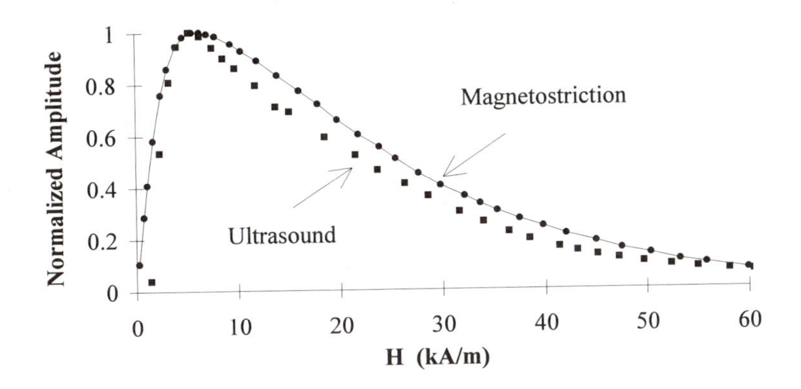

Figure 3. Amplitudes of the magnetostriction (–●–) and SH wave (■) vs. applied magnetic field in the underaged A710 sample. Curves are normalized with respect to their maximum values.

coil was used, and f_0 was measured as a magnetic field was applied. (The frequency f_0 varied between ~20 kHz (near saturation) and ~200 kHz (at low fields).) The coil was oriented in such a way as to measure the incremental permeability mostly along a direction perpendicular to the applied field, which is parallel to δH in Figure 1. With this configuration, μ/μ_0 was expected to be B/H near saturation, and μ was derived as a function of the applied field by scaling f_0 so that μ/μ_0 equaled B/H above ~40 kA/m (500 Oe) (with values of B taken from the B-H curve).

RESULTS AND DISCUSSION

Measurement of the amplitude of the SH wave provides a noncontacting means for predicting key features about the magnetostriction of a sample. This is demonstrated with the underaged sample in Figure 3, which shows normalized plots of the wave amplitude and magnetostriction. The wave amplitude reaches a maximum at nearly the same point as the magnetostriction. Furthermore, the wave amplitude successfully predicts an unusual aspect of the magnetostriction: it remains positive at high fields, unlike the case of most low-carbon steels, where magnetostriction becomes negative in the neighborhood of 16–24 kA/m (200-300 Oe). (For example, see plots of the magnetostriction in Reference 6.) From the similar behaviors of the wave amplitude and magnetostriction, we may infer that the amplitude of the SH wave always is proportional to the magnetostriction (or a positive power of the magnetostriction), even though expression (1) for the wave amplitude may not be valid at low fields.

Measurements of the wave amplitude also successfully predict differences between the magnetostrictions of the underaged and overaged samples. Figure 4(c) shows that, in the overaged sample, the magnetostriction is larger at all fields and that the peak shifts from 5.6 ± 0.8 kA/m (70 ± 10 Oe) to 3.2 ± 0.8 A/cm (40 ± 10 Oe). These changes are mirrored in the wave amplitudes (Figure 4(d)), although not as closely above ~40 kA/m (500 Oe).

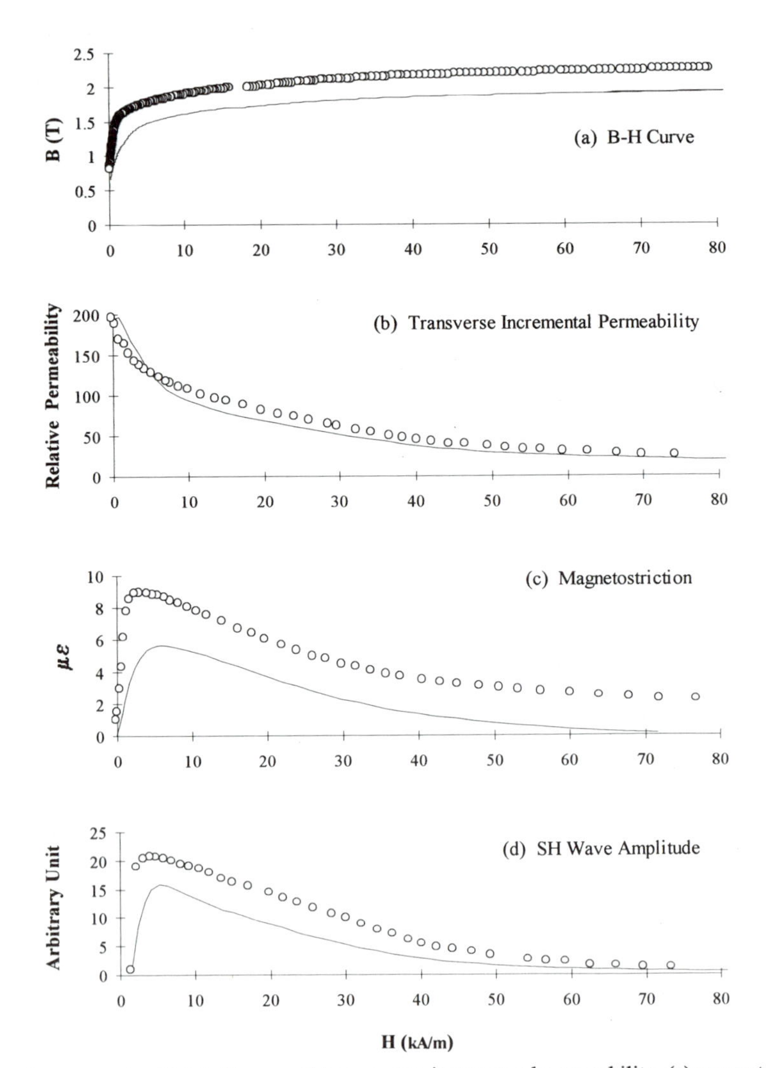

Figure 4. Measurements of (a) B-H curve, (b) transverse incremental permeability, (c) magnetostriction, and (d) SH wave amplitude in the underaged (———) and overaged (o) A710 samples.

Figure 4(b) shows that the transverse incremental permeability is not much different for the two samples; hence, the skin depths for the two samples should be similar, and differences in the wave amplitude can be attributed mostly to differences in the magnetostriction.

Significant differences also appear in the B-H curves (Figure 4(a)). Similar to magnetostriction, the magnetic induction at saturation is much larger in the overaged sample. It is likely that these increases in the saturation magnetic induction and magnetostriction are due to Cu precipitates, and ongoing experiments are attempting to confirm this hypothesis.

CONCLUSIONS

This study demonstrates that measurement of the amplitude of EMAT-generated SH waves provides a noncontacting method for predicting key features of the magnetostrictive curve. Also, this measurement may be used to monitor microstructural changes in steel. In the case of A710 steel, coarsening of Cu precipitates was detected. This result points to the possibility of using EMATs as an online NDE tool to indicate when the toughness of this steel is optimized. Future endeavors will attempt to determine whether this technique can be used to sense the formation of the small strengthening Cu precipitates during the early stages of aging.

ACKNOWLEDGMENTS

The authors thank R.J. Fields (NIST-Gaithersburg) for supplying us with the A710 samples used in this work. We also thank C. McCowan, W.L. Johnson, S.R. Schaps, and J.D. McColskey of NIST-Boulder for offering technical assistance and advice. We are grateful to G.E. Hicho (NIST-Gaithersburg) for offering helpful advice about the manuscript. B.I. was supported by a National Research Council Postdoctoral Research Associateship.

REFERENCES

1. R.B. Thompson, New configurations for the electromagnetic generation SH waves in ferromagnetic materials, in: *1978 Ultrasonics Symposium Proceedings*, Cat. No. 78CH1344-1SU, J. deKlerk and B.R. McAvoy, eds., IEEE, New York (1978).
2. R.B. Thompson, Generation of horizontally polarized shear waves in ferromagnetic materials using magnetostrictively coupled meander-coil electromagnetic transducers, *Appl. Phys. Lett.* 34:175 (1979).
3. R.B. Thompson, Physical principles of measurements with EMAT transducers, in: *Ultrasonic Measurement Methods,* Vol. 19 of *Physical Acoustics,* R.N. Thurston and A.D. Pierce, eds., Academic Press, Boston (1990).
4. G.E. Hicho, S. Sinhal, L.C. Smith, R.J. Fields, Effect of thermal processing variations on the mechanical properties and microstructure of a precipitation hardening HSLA steel, paper presented at the 1983 International Conference on Technology and Applications of High Strength Low Alloy (HSLA) Steels (In Conjunction with 1983 Metals Congress), Philadelphia, 1983, Cat. No. 8306-051, American Society for Metals, Metals Park, OH (1983).
5. S.W. Thompson and G. Krauss, Copper precipitation during continuous cooling and isothermal aging of A710-type steels, *Metall. Mater. Trans. A*, 27A:1573 (1996).
6. R.M. Bozorth, *Ferromagnetism*, Van Nostrand, New York (1955).
7. Private communication with W.L. Johnson (NIST-Boulder).
8. J.H. Rose, C.-C. Tai, and J.C. Moulder, Scaling relation for the inductance of a coil next to a ferromagnetic half-space, *J. Appl. Phys.*, in press.

ULTRASONIC DETECTION OF UNSTABLE PLASTIC FLOW IN METAL CUTTING

M. A. Davies, S. E. Fick, C. J. Evans, and G.V. Blessing

Manufacturing Engineering Laboratory
National Institute of Standards and Technology
Gaithersburg Maryland, USA

INTRODUCTION

Recently developed metal-cutting tools, made from such materials as ceramics and polycrystalline cubic boron nitride (PCBN), allow turning and milling operations to be applied to heat-treated steels in the hardened (50-65 Rockwell C) state. Understanding the linkage between controllable machining parameters and the associated mechanical phenomena is difficult because of the extreme complexity of the environment of the tool and workpiece. For most reasonable machining conditions, the flow of material is unstable, leading to oscillations in tool-workpiece stresses, local chip velocities, and temperatures. In precision hard turning, with chip dimensions on the order of tens of μm, the oscillation frequencies may approach a few hundred kHz. The goal of this work is to investigate ultrasonic measurement techniques suitable for characterization of material flow during machining, as an aid to understanding the effect of material flow on important process variables and product quality.

Chip segmentation due to adiabatic shear localization[1], a material flow instability due to the local heating and subsequent softening of material during deformation, has been observed in many materials[2-4]. Adiabatic shear band formation is accompanied by a rapid decrease in the local shear stresses required to cause further deformation[5-7]. The formation of discontinuous chips by adiabatic shear localization during machining has been found to cause measurable kHz-frequency oscillations in the cutting forces. However, to our knowledge, experimental measurement of shear band formation forces at much higher frequencies has not been previously reported.

EXPERIMENTAL PROCEDURE

Hard turning experiments were conducted using a two-axis diamond turning machine with air-bearing spindle to cut 52100 bearing steel through-hardened to approximately 62 Rockwell C. To establish a simplified orthogonal cutting geometry, a special workpiece and tools were used. The workpiece was a ring 94 mm in diameter (at the beginning of the

Nondestructive Characterization of Material VIII
Edited by Robert E. Green Jr., Plenum Press, New York, 1998

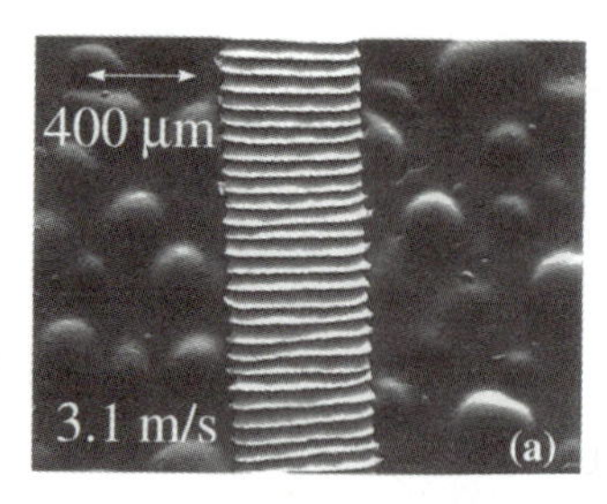

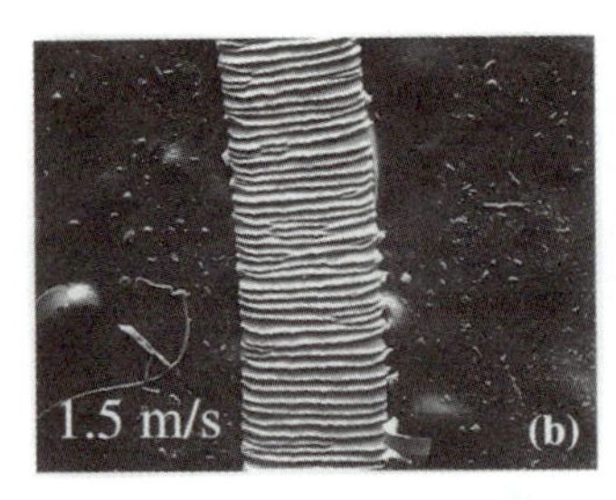

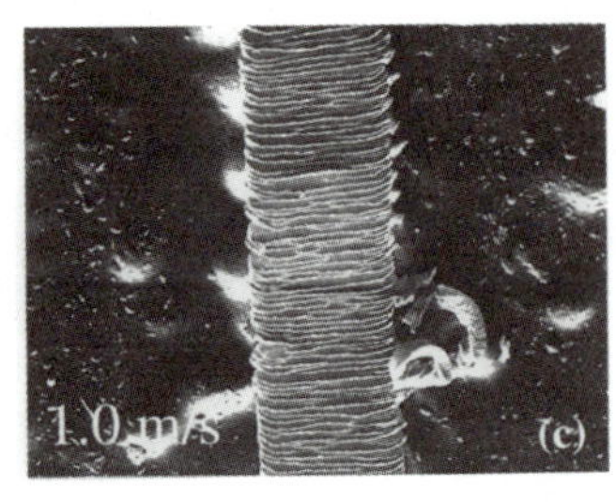

Figure 1. SEM photographs of chips obtained from orthogonal cutting with uncut chip thickness 31 μm, rake angle -27 degrees, and cutting speeds: (a) 3.1 m/s, (b) 1.5 m/s, and (c) 1.0 m/s

cut) and 0.5 mm wide. At 15° intervals around the ring, radial slots 125 μm deep were ground into the side of the ring to allow measurement of the changes in length of the chips formed during cutting. Lapped custom-ground flat-nosed PCBN tools were used with rake angle -27°. For the results reported in this paper, the depth of cut was 31 μm, total tool motion during a cut was less than 2 mm, and variations in geometry due to tool motion were negligible. Cutting velocity **V** was varied from 0.35 m/s to 4.3 m/s. High-frequency variations in tool loading were sensed by a mylar-backed polymer ($PVDF_2$) transducer assembly clamped between the insert and tool holder.

Chips were collected from each of the cutting experiments. Each chip was relatively flat in cross section; the surface which had originally been the free surface of the workpiece was always much rougher than the chip surface which had flowed over the rake face of the tool. Segment spacing, $\mathbf{d_s}$, was estimated by microscopic examination of the rough surface of each chip. Since each chip exhibited notches readily attributable to the radial slots in the workpiece, the ratio of its internotch spacing to the circumferential slot spacing at the appropriate workpiece radius was easily calculated to determine $\mathbf{r_c}$, the chip compression factor. Results of earlier experiments done with both slotted and unslotted workpieces had shown that the presence of slots has little effect on observed chip morphology.

Dimensional analysis invites the definition of a simple parameter, $\mathbf{f_s}$, the segmentation frequency, or number of segments formed per unit time, as the result of dividing the cutting velocity **V** by the average segment length *during formation*, with the latter determined by dividing $\mathbf{d_s}$ by $\mathbf{r_c}$, so that:

$$\mathbf{f_s = V(r_c/d_s)}$$

RESULTS

The chips from these experiments showed a clear trend - from closely spaced but disordered segmentation for lower velocities, to wider and more uniform segments for higher velocities. Shown in Figure 1 are scanning electron photomicrographs of three chips obtained with **V** ranging from 3.1 m/s to 1.0 m/s. Clearly both the magnitude and the uniformity of $\mathbf{d_s}$ decreases with decreasing **V**.

The sensor output voltage was large enough to obviate preamplification and was measured directly using a digital oscilloscope with sampling rates ranging from 2 to 20 megasamples per second. Based on independent knowledge of the frequency response of the sensor, the sampling rates exceeded the highest anticipated signal frequency by at least a factor of 10.

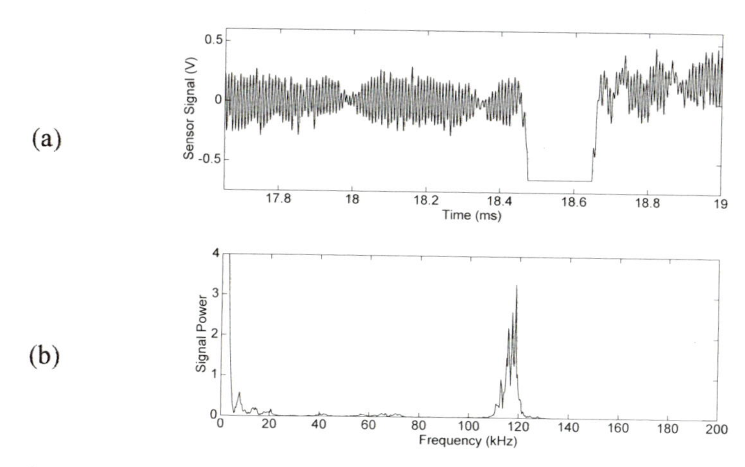

Figure 2. (a) Time trace and (b) power spectrum of sensor signal for **V** = 4.3 m/s.

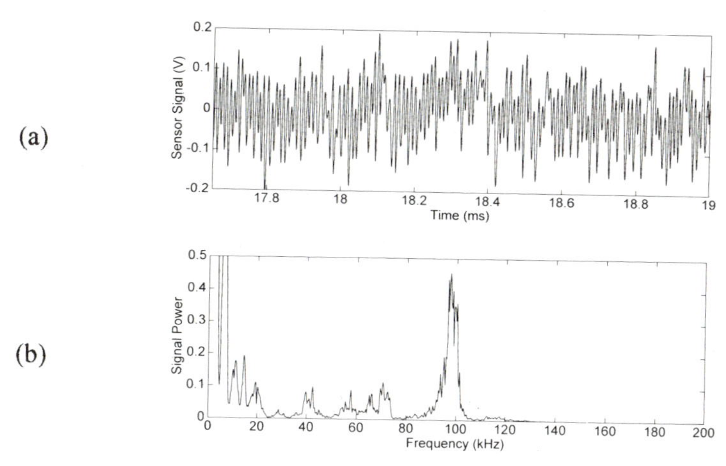

Figure 3. (a) Time trace and (b) power spectrum of sensor signal for **V** = 3.1 m/s.

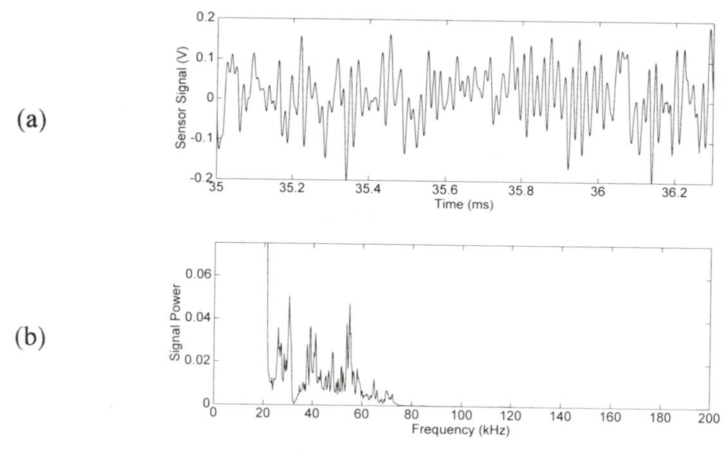

Figure 4. (a) Time trace and (b) power spectrum of sensor signal **V** = 1.5 m/s.

Output voltage waveforms and corresponding power spectra appear in Figures 2,3,and 4. In Figure 2, a strong component is evident at approximately 120 kHz. This frequency corresponds closely to the 119 kHz value of $\mathbf{f_s}$ calculated using the velocity (4.3 m/s) for this experiment and the characteristics ($\mathbf{r_c}$ =1.6, and $\mathbf{d_s}$ =58 μm) of the chips it produced. The (negatively) saturated part of the signal in Figure 2(a) is the result of the cutting tool traversing one of the radial notches in the workpiece. The periodicity of these notches is the cause of the dominant low frequency peaks in each of the power spectra. These low-frequency peaks were not seen in the data for other experiments using workpieces having no notches. From Figure 3 it is clear that a weaker signal can have an equally dominant spectral peak. Its frequency, about 100 kHz, corresponds reasonably well to the 104 kHz calculated using the velocity (3.1 m/s) for this experiment and the characteristics ($\mathbf{r_c}$ =1.6, and $\mathbf{d_s}$ =47 μm) of the chips it produced. Results for a lower cutting velocity, 1.5 m/s, are shown in Figure 4. The segmentation frequency for corresponding chip parameters, $\mathbf{r_c}$ =1.0, and $\mathbf{d_s}$ =30 μm, is 50 kHz. Examination of the voltage waveform shows numerous sequences of oscillations near 50 kHz, and the power spectrum is significant in the 40 kHz to 60 kHz range. The spectral broadening is consistent with the increased disorder observed in the chips cut at lower velocities.

CONCLUSIONS

We have demonstrated the use of a simple inexpensive $PVDF_2$ sensor to monitor the segmentation of chips in hard turning by detecting oscillations in tool stress due to thermoplastic instability in material flow during cutting. Several applications of this type of monitoring are apparent. First, because it represents underlying characteristics in real time, sensor data can be used to verify or refute models of material flow dynamics. Second, because results of other experiments indicate that chip segmentation is highly sensitive to cutting edge geometry, it is likely that variations due to wear will lead to measurable changes in the sensor signals. This evokes the possibility of schemes to detect tool wear by examining the spectra of sensor signals that result from segmentation. And finally, data from an improved appropriately calibrated sensor could be used to extend characterization of the dynamic tool environment beyond the envelope of conventional dynamometers. This would enable commensurate improvements of the accuracy and reliability of conventional process control methods.

ACKNOWLEDGEMENTS

The authors thank T. Burns, E. Whittenton, and K. Harper for their contributions.
M. Davies is very grateful to the U. S. National Research Council for the postdoctoral fellowship that supported his work.

REFERENCES

1. R. F. Recht, Catastrophic thermoplastic shear, *Journal of Applied Mechanics*. 31:189 (1964).
2. R. Komanduri and R. H. Brown, On the mechanics of chip segmentation in machining, *Journal of Engineering for Industry*. 103:33 (1981)
3. R. Komanduri and R. H. Brown, On shear instability in machining a nickel iron alloy,

4. R. Komanduri, T. Schroeder, J. Hazra, and B. F. VonTurkovich, On the catastrophic shear instability in high-speed machining of an AISI 4340 steel, *Journal of Engineering for Industry*. 104:121 (1982)
5. T. J. Burns, Approximate linear stability analysis of a model of adiabatic shear band formation, *Quarterly of Applied Mathematics*. 43:65 (1985)
6. L. S. Costin, E. E. Crisman, R. H. Hawley, and J. Duffy, On localization of plastic flow in mild steel tubes under dynamic torsional loading, *Institute of Physics Conference, No. 47, Oxford U. K*: 90 (1979)
7. K. A. Hartley, J. Duffy and R. H. Hawley, Measurement of the temperature profile during shear band formation in steels deforming at high strain rates, *Journal of the Mechanics and Physics of Solids*. 35:283 (1987)

ULTRASONIC DETERMINATION OF CASE DEPTH AND SURFACE HARDNESS IN AXLES

R.C. Addison, Jr.[1], A. Safaeinili[2], and A.D.W. McKie[1]

[1]Rockwell Science Center,
1049 Camino Dos Rios,
Thousand Oaks, CA 91360

[2]Radar Science Section
Jet Propulsion Laboratory
California Institute of Technology
4800 Oak Grove Dr
Pasadena, CA 91109

INTRODUCTION

Generally, NDI methods for determining case depth are based on inferring case depth indirectly through measuring electromagnetic or elastic properties of the part using eddy current or ultrasonic probes.[1-4] Eddy current systems are commonly used for case depth measurements and are known to be reliable for many applications[4]. However, they lack sensitivity if the case depth is deep (e.g. greater than 5 mm in steel parts) because of the compromise required between a sufficiently high frequency to obtain adequate coupling and a sufficiently low frequency to obtain adequate penetration of the part. Further, custom probes are required for inspection of components with different geometries. When the steel is hardened, the eddy current techniques are dependent on changes in the magnetic permeability and the electrical conductivity of the steel, whereas the ultrasonic techniques are dependent on changes in the elastic modulus of the steel. It is not generally known whether the changes in the electromagnetic properties or the elastic properties correlate better with the hardness. Further, since the same hardness can be achieved through different processing paths, it is not clear if the correlation for either set of properties is unique or is dependent on the specific processing path.

In this paper, the results of an investigation into the correlation of the elastic modulus with the hardness of the steel are presented along with a technique for using a laser-based ultrasound system to acquire the data. Over the past three decades, many studies have demonstrated that a relationship between the elastic moduli of steel and its hardness exists. Also, ultrasonic case depth measurement techniques were proposed based on dispersion of Rayleigh wave (SAW) velocity[5]. More recent works show how case depth can be obtained from the group velocity dispersion[1,2]. The success of any of these techniques is critically dependent on being able to measure the SAW velocity dispersion with sufficient accuracy to resolve changes in surface wave velocity of less than 1%. We have addressed this issue by first, measuring the phase velocity rather than the group velocity because it is easier to define at a single frequency in practice; and secondly, by developing a robust and precise technique

Nondestructive Characterization of Material VIII
Edited by Robert E. Green Jr., Plenum Press, New York, 1998

to measure the SAW phase velocity by accurately indexing the location of the generating laser beam.

The three requirements for developing a SAW dispersion-based case depth inspection method are: 1) existence of a strong correlation between hardness and the SAW phase velocity, 2) accurate measurement of surface velocity with accuracies of ~0.1%, and 3) development of a robust and fast inversion algorithm for case depth parameter extraction. An increase in SAW velocity in tempered martensite that had a good correlation with the hardness was observed in a recent measurement by Safaeinili et al.[6] Inversion of SAW dispersion data has been addressed by Richardson[7] who used a general inversion approach in which no *a priori* assumption about the case hardening profile was made. To achieve speed and robustness in inversion, we have specialized Richardson's solution to a model-based inversion algorithm with sufficient flexibility to match a variety of profiles that are likely to be observed in practice.

SAW PHASE VELOCITY MEASUREMENT

Figure 1 illustrates the experimental configuration for measurement of the SAW phase velocity on steel axles (1a) and gear teeth (1b) using a laser-based ultrasound (LBU) system. A Q-switched Nd:YAG laser is directed at the axle surface and is focused to a line, resulting in propagation of surface waves normal to the illuminated line. LBU relies on the transient thermal expansion at a material surface that results from the sudden heating caused by illumination with a high power, short duration laser pulse[8]. Relief of the surface induced thermal stresses results in the propagation of elastic waves that emanate from the heated region. The incident generation laser energy was ≈5 mJ, which was sufficiently low to avoid material damage. The generated SAW propagates along the surface of the part where it is detected with a piezoelectric transducer (for axles) or a continuous wave (CW) argon-ion probe laser beam (for gear teeth). For the gear teeth, the phase modulated probe laser beam is collected and directed through a 0.5 m spherical Fabry-Pérot interferometer (SFPI)[9,10] which detects the ultrasonic signal. Optical detection with a SFPI provides the most suitable means for noncontact detection from the curved surfaces typically encountered in industrial applications, such as the gear teeth. An LBU system has significant advantages over conventional contact piezoelectric techniques since it 1) eliminates the inconsistency problems encountered when using couplants, 2) provides spatially and temporally broadband generation and detection, hence removing the ambiguities associated with temporal and spatial transducer system response, 3) eliminates unnecessary propagation through contact wedges that can corrupt data, 4) provides a precise location for the generation point of the SAW, 5) provides access to small areas (e.g. gear teeth) that are not accessible with relatively large and bulky conventional piezoelectric contact transducers and 6) has the ability to inspect parts with curved surfaces and complex geometry since the laser beam automatically conforms to the surface.

As shown in Figs. 1a and 1b, the generation laser beam is scanned with respect to a detection probe, resulting in acquisition of a series of spatially and temporally wideband waveforms $S(x_i,t)$. The series of waveforms are acquired at a number of equally spaced increments along the axle shaft. From the series of single shot ultrasonic waveforms, the phase velocity of the SAW can be estimated.

If no dispersion is present, the envelope of the wideband signal remains constant, with the measured ultrasonic propagation time delay remaining the same for equal steps in position. However, when the medium is dispersive, a single velocity cannot be defined for the envelope of the wideband signal, since each frequency component of the signal travels with a different velocity. To separate these components, the time domain signals $S(x_i,t)$are Fourier transformed to yield;[6]

$$\hat{S}(x_i,\omega_j) = A(\omega_j)\exp[ip(x_i,\omega_j)] = \sum_{k=1}^{n} S(x_i,t_k)\exp[i\omega_j t_k]\Delta t \qquad (1)$$

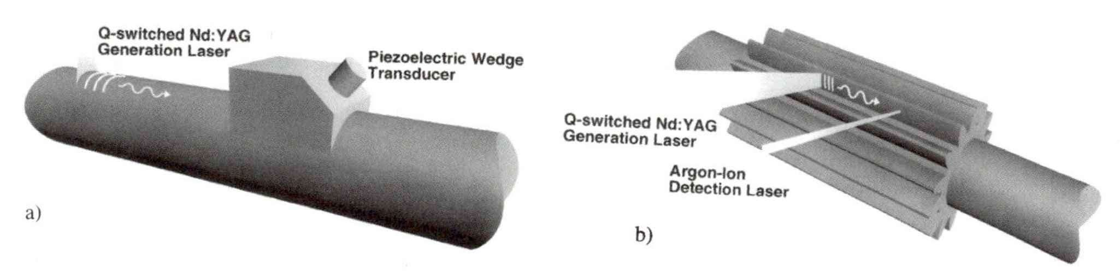

Figure 1. Experimental configuration for laser-based ultrasonic measurement of surface hardness in a a) steel axle shaft and b) gear tooth. The generation laser is linearly scanned while the detection probe remains stationary.

where, for a fixed frequency, the phase function $p(x_i,\omega_j)=\omega_j x_i / c(\omega_j)$ and $c(\omega_j)$ is the surface wave velocity for a given frequency ω_j. If time signals are acquired at sufficiently small steps (e.g. $\Delta x < \lambda_{min}/4$), the phase velocity can be measured directly from the slope of the unwrapped phase function[6].

Generally, this technique provides an accurate estimate of average phase velocity if the velocity variation is random and uncorrelated. This is due to redundant measurements and a least square estimation of the slowness. Alternatively, very slow (i.e. on the order of the scan length) variations in the SAW velocity in the direction of the scan may cause significant error in the phase velocity estimate. The existence of this condition is easily detected since the phase function $p(x_i,\omega_j)$ would have two or more different slopes corresponding to regions with different velocities. Furthermore, this method provides absolute phase velocity estimates for each frequency, with an accuracy limited by the spatial resolution of the stepper motor controller, without requiring the knowledge of source/receiver separation distance.

The least square estimate of the SAW phase velocity may be obtained in batch mode using a conventional line fitting algorithm. If there are a large number of acquired waveforms ($n > 10$), the error in the least square estimate of the SAW phase velocity, c, is given by[6]

$$\sigma_n^2 \approx \frac{3c^2\sigma_x^2}{n^3\Delta x^2} = \frac{3c^2\sigma_x^2}{nx_{max}^2} \tag{2}$$

The accuracy of this method is directly proportional to the scan length x_{max} and n, the number of waveforms acquired in the scan, and is inversely proportional to the stepper motor variance σ_x. For example, for a scan with n = 30 waveforms acquired over a 1 cm scan length with a stepper motor variance of less than 10 μm, the accuracy for the SAW phase velocity estimate is $\approx 0.1\%$.

CORRELATION BETWEEN HARDNESS AND SAW VELOCITY

An important step in the ultrasonic determination of the case depth is obtaining a correlation between the SAW velocity and hardness. Surface hardening is achieved typically through a carburizing process or induction heat treatment[11]. In both processes, a number of parameters such as the percentage of carbon, duration of treatment and temperature can be adjusted to control the surface hardness, its profile and the depth of the hardened case. In previously reported work[12], a correlation was obtained between the SAW velocity and the hardness for 1541H1 steel by subjecting a set of 10 nominally identical axles, having an initial surface hardness of 58 Rc, to different degrees of tempering to obtain a range of surface hardnesses ranging from 20 Rc to 58 Rc. Since this method of obtaining a range of hardness values is different than the way that the gradient in hardness within the case is achieved during the induction hardening process, and since it is unclear whether the correlation is dependent on the process used to achieve the hardness, a different method for measuring the correlation was devised. The hardness profile of an axle shaft that had been induction hardened was measured along a radius using a microhardness indenter. A section of the axle about 12" long was then cut along a diameter (Fig. 2) using an EDM technique.

The SAW phase velocity was measured along a series of lines parallel to the axis of the axle and at successively greater depths from the surface corresponding to hardness values ranging from 60 Rc to 20 Rc. The correlation obtained is shown in Fig. 3. The equation for the best least squares fit is given on the graph and has a correlation coefficient of 0.98. The slope of this line was noted to be ~9% less than the one previously obtained with the set of 10 axles using ultrasonic methods.

ESTIMATION OF THE CASE DEPTH

The SAW velocity dispersion depends on the profile of the hardened region (case). This profile can be found by fitting a model to the experimental data. An accurate and computationally efficient model, based on a general perturbation analysis done by Auld[13] and then specialized by Thompson, is found in ref 7. For an isotropic medium containing a layer with varying shear modulus, the change in the SAW velocity can be written as[7]

$$\Delta v = -\frac{(vk_1k_3^l)^2\mu}{P}\int_0^\infty (\exp[-k_3^l x_3] - \exp[-k_3^t x_3])^2 f(x_3)dx_3 \tag{3}$$

where v is the SAW velocity, P is the total unperturbed propagating energy per unit width, k_1 and $k_3^{l,t}$ are the projections of the wavenumbers in the directions parallel and normal to the propagation direction respectively, x_3 is the depth parameter (direction normal to the surface), and $f()$ is the profile function for the variation in shear modulus μ. The expression for P can be found in ref. 7.

The parameters in the profile function $f()$ can be found by an optimization algorithm that involves multiple calculations of Δv. The speed of obtaining a solution is related directly to the speed of calculation for Δv. Accurate results could be obtained by approximating the profile as a step discontinuity. The hardness was assumed to be the surface value until the case depth was reached, whereupon it was reduced to the core value. This can be modeled as a single homogeneous layer on a half-space. For this model, there are three unknown parameters: 1) SAW velocity parameter in the layer material, 2) SAW velocity parameter in the substrate, 3) the layer thickness parameter (or case depth). For measurements of the axle, the SAW is generated by a spatially and temporally broadband laser pulse and is detected by a piezoelectric transducer which is coupled to the axle using a wedge. The generation laser beam is scanned over a 57 mm line along the axis of the axle in 0.57 mm steps. The experimental dispersion curve is obtained using the processing scheme outlined above. The SAW velocity of the hardened layer is known from the high frequency region of the dispersion curve where most of the SAW energy is confined in the hardened layer.

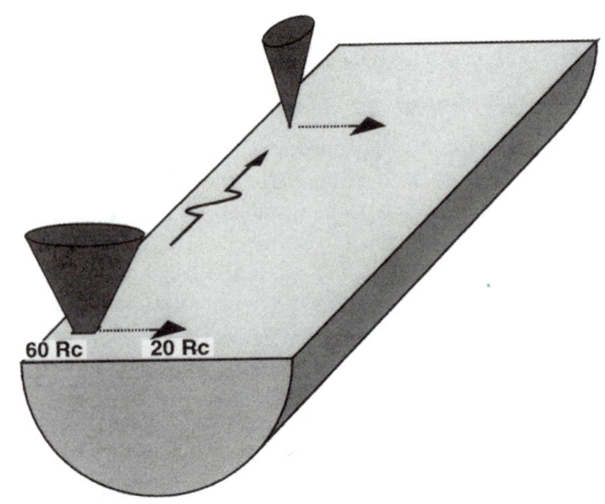

Figure 2. Method for measuring SAW phase velocity as a function of the hardness in an induction hardened axle.

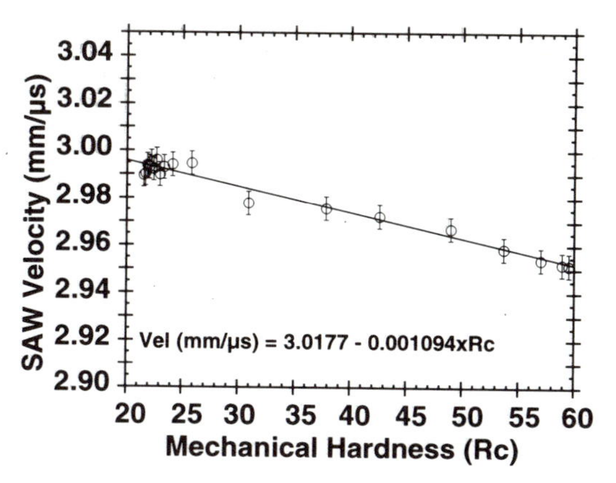

Figure 3. A comparison of surface hardness, measured with a microhardness indenter, and SAW phase velocity for segments of the steel truck axles. The results indicate a strong linear correlation.

Consequently, only the SAW velocity of the core and the case depth parameter remain to be found. The results for one axle (Fig. 4) estimated a case depth of 10.7 mm, which is within 4% of the average indenter measurement of 11.1 mm.

In interpreting the case depth parameters obtained from the dispersion curve, one has to realize that the accuracy is dependent on the existence of a strong correlation between hardness and SAW velocity. Consequently, for each application where metallurgy is different, this correlation needs to be verified. Furthermore, additional calibration may be needed to compensate for the inaccuracies of the theoretical model.

SUMMARY

This work presents a method for measuring the depth of the hardened layer (case) on steel parts based on velocity dispersion of SAW. A linear correlation between surface hardness and SAW phase velocity for induction hardened steel was established and experimental and data processing techniques for precise measurement of SAW phase velocity were developed. The spatial and temporal wideband nature of the LBU system allows SAW velocity measurements to be performed with accuracies of $\approx 0.1\%$. This method was successfully used to measure the effective case depth of hardened axles. Future work should be directed at understanding the changes in the correlation between the SAW phase velocity and the hardness for different processing methods such as carburizing, induction hardening, and tempering back a quenched part.

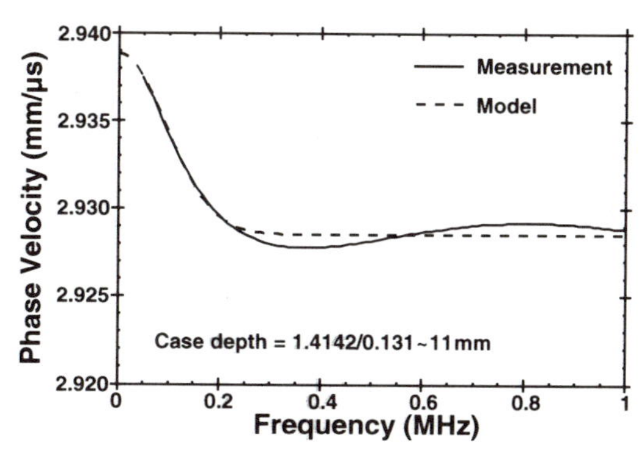

Figure 4. Experimental and theoretical SAW phase velocity dispersion curves for a 48 mm diameter axle with a case depth of 11 mm.

ACKNOWLEDGMENTS

This work was supported by the Rockwell Commercial Research and Development funds through Rockwell Automotive Heavy Vehicle Systems Division.

REFERENCES

1. A. Abbate, S.C. Schroeder, B.E. Knight, F. Yee and J. Frankel, *Review of Progress in Quantitative Nondestructive Evaluation*, Vol. 15, Eds. D.O. Thompson and D.E. Chimenti, Plenum Press, New York, 1996, p.585.
2. G. Gordon and B.R. Tittmann, *Review of Progress in Quantitative Nondestructive Evaluation*, Vol. 15, Eds. D.O. Thompson and D.E. Chimenti, Plenum Press, New York, (1996) p. 1597.
3. M. Rosen *Materials Analysis by Ultrasonics*, Ed. A. Vary, Noyes Data Corp. , 79, (1987).
4. C.H. Stephen and H.L. Chesney, *Materials Evaluation*, Vol. 42, (1984) p. 1612.
5. B.R. Tittmann and R.B. Thompson, *in Proc. of the 9th Symp. on NDT Tech*, SWRI, San Antonio, TX, p. 20, (1973).
6. A. Safaeinili, A.D.W. McKie, and R. C. Addison, Jr., *Material Research Bulletin*, Vol. 21, p. 53, (1996).
7. J.M. Richardson, *Journal of Appl. Physics*, Vol. 48(2), pp. 498-512, (1977).
8. C.B. Scruby and L.E. Drain, *Laser Ultrasonics - Techniques and Applications* , Adam Hilger, New York (1990).
9. J.M. Vaughan, *The Fabry-Pérot Interferometer: History, Theory, Practice and Applications* , Adam Hilger, New York (1989).
10. J.-P. Monchalin and R. Heon, *Materials Evaluation* Vol. 44, 1231 (1986).
11. T. Lyman ed., *ASM Metals Handbook* Vol. 2, 8th Edition, (1964) p. 93.
12. A. Safaeinili, A.D.W. McKie, and R. C. Addison, Jr., *Review of Progress in Quantitative Nondestructive Evaluation*, Vol. 16, Eds. D.O. Thompson and D.E. Chimenti, Plenum Press, New York, (1997) p. 1625.
13. B. A. Auld, *Acoustic Fields and Waves in Solids, II*, Krieger Publishing Company, Malabar, Florida 1990, Ch. 12.

HIGH ENERGY X-RAY DIFFRACTION TECHNIQUE FOR MONITORING SOLIDIFICATION OF SINGLE CRYSTAL CASTINGS

Dale W. Fitting, William P. Dube', and Thomas A. Siewert

Materials Reliability Division
Materials Science and Engineering Laboratory
National Institute of Standards and Technology
Boulder, CO 80303

INTRODUCTION

X-ray diffraction has been used successfully to study metal solidification and temperature-dependent phase changes[1-3]. However, this research used very thin specimens (a few mm at most), furnaces with low attenuation x-ray windows (beryllium, graphite, or polyimide), and low x-ray energies (< 50 keV). Since the penetration depth of low-energy x-rays is shallow, traditional x-ray diffraction is thus unable to probe the interior of thicker structures. Others have used higher energies to study thicker samples. Work, including that by Green[4] and by Kopinek, et al[5] , extended x-ray diffraction investigations to energies exceeding 150 keV.

In this paper, we show that by using higher x-ray energies (up to 320 keV) and a transmission configuration, solidification of a metal casting within a mold can be studied. The liquid-solid boundary in the casting can be precisely located from the spatial distribution of x-rays transmitted through the casting and diffracted from the liquid or solid. Diffraction spots arise when the probing x-ray beam encounters an crystalline solid. X-ray diffraction from the liquid metal is characterized by a diffuse ring. The high-energy x-rays used have sufficient energy to penetrate the furnace walls, mold, and specimen, while retaining a moderate cross-section for coherent scattering (which can produce diffraction).

PHYSICS OF TRANSMISSION DIFFRACTION

Probing the interior of a casting requires a transmission configuration. X-rays energies of over 100 keV are needed to penetrate the refractory oxide mold (5-10 mm wall thickness) and casting specimen (1 -20 mm thick). Below, we explore the physics of using high energy x-rays to produce transmission x-ray diffraction.

Nondestructive Characterization of Material VIII
Edited by Robert E. Green Jr., Plenum Press, New York, 1998

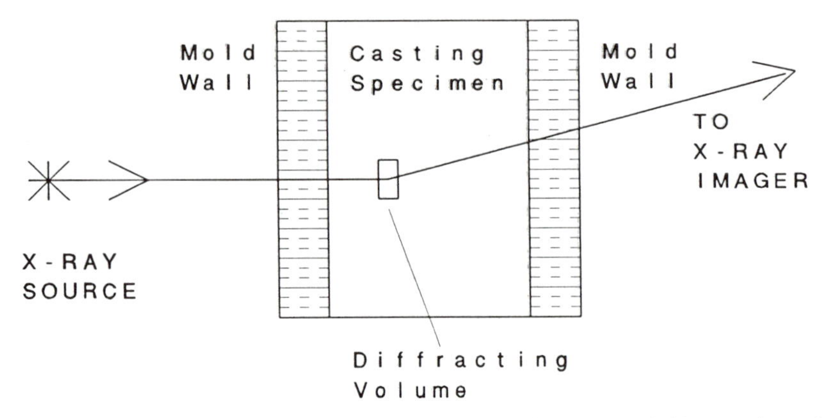

Figure 1. Geometry of transmission diffraction from a crystalline specimen surrounded by a casting mold.

Consider the situation shown in Figure 1, of an x-ray beam incident on a crystalline specimen contained within a casting mold. The primary x-ray beam is attenuated as it passes through the mold wall and a portion of the specimen to a location where coherent (elastic) scattering occurs. The scattered x-ray is attenuated along the exit path through the remaining specimen and exit mold wall. Losses along entrance and exit paths are minimized by raising the energy substantially above that used in conventional x-ray diffraction systems (5-20 keV) to 150-300 keV.

The amplitude of elastic scattering from a crystalline solid is given by the product of the structure factor (F) of the crystal and the Thomson scattering amplitude from an isolated electron[6]. F is a function of the types of atoms in the crystal (through the atomic scattering factor, f), the configuration of the unit cell of the crystal , and the particular lattice planes which are involved in the scattering. For a face-centered cubic (fcc) crystal, such as nickel, the structure factor is equal to $4f$ if the Miller indices (hkl) are unmixed (all odd or all even) and is equal to zero if the indices are mixed. The intensity of the scattering is the square of the amplitude. The structure factor is thus squared, to give $16f^2$ for an fcc material with lattice planes defined by unmixed Miller indices.

The dependence of the atomic scattering factor for nickel is plotted in Figure 2, as a function of the scattering angle for several x-ray energies. The empirical method proposed by Waasmaier and Kirfel[7] was used to calculate f. In the forward direction the scattering from all electrons in the atom is in phase and so f is equal to the number of electrons which scatter the

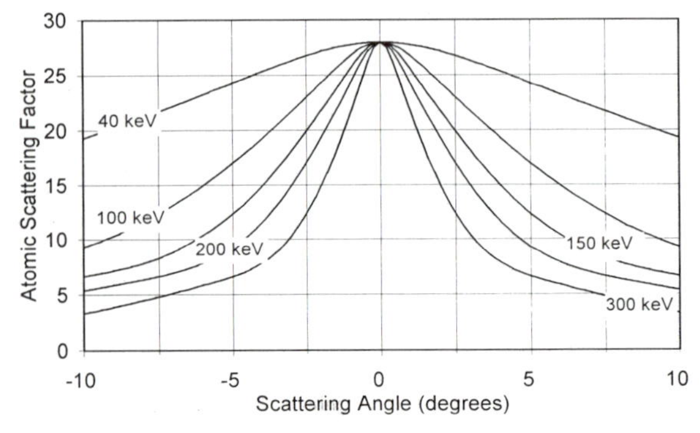

Figure 2. Atomic scattering factor calculated for nickel, plotted as a function of scattering angle for several x-ray energies.

x-rays, that is, the atomic number of the atom. At low x-ray energies, the atomic scattering factor is quite large, even for large scattering angles. However, with high x-ray energies, *f* is significant only for small scattering angles, that is for a transmission geometry. For example, the scattering factor for nickel is 16.5 for 150 keV x-rays scattered at 3.5-degrees from the direction of the incident x-ray beam (typical of our experimental geometry). The structure factor squared for this example is 16 x 16.5^2 , or 4356.

The large structure factor, the enormous number of atoms along the path of the primary beam (all with the same crystalline structure), in addition to the substantial intensity of high-energy x-rays which penetrate through a mold and specimen, account for the high efficiency of transmission diffraction. Figure 3 plots the transmission energy spectra recorded for a mold encased nickel alloy single crystal casting. The peak in the spectrum of the diffraction spot at 185 keV, confirms that the high energies required to penetrate the mold and casting can indeed produce intense transmission x-ray diffraction.

SENSING SOLIDIFICATION DURING CASTING

A directional solidification furnace, used for prototype single crystal turbine blade castings, was fitted with an x-ray source and real time x-ray imager as shown in Figure 4. The x-ray tube was a 320 kV tungsten target tubehead normally used for radiographic imaging. The emission from its 1.2 mm focal spot was collimated with lead apertures to produce a 1 mm diameter beam. The primary x-ray beam entered the vacuum furnace through a 10 mm thick borosilicate glass port. After transmission through the 9.5 mm thick alumina hot zone coil support, the beam entered the mold and specimen. Diffraction from the specimen emerged from the casting, and passed through the alumina coil support and furnace exit port. A real time x-ray imager (fiber optic glass scintillator coupled through an image intensifier to a CCD camera) was used to observe the diffraction pattern. Two-axis linear motion stages on the x-ray tube and imager were used to scan the probing x-ray beam over the casting. The distance from the x-ray focal spot to the casting specimen was 835 mm. The imager was 505 mm from the specimen.

In a series of casting experiments, the x-ray beam was first directed into the lower section of the casting, where the water-cooled ram of the DS furnace had caused the alloy to solidify. Generally, only one or two diffraction spots were observed in the XRD image; although others

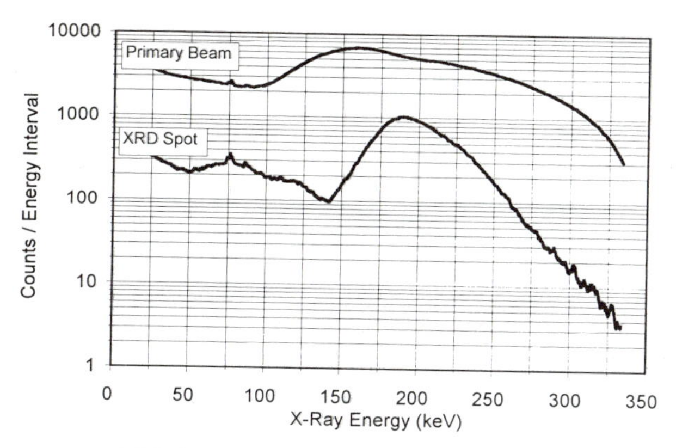

Figure 3. Energy spectrum recorded during a transmission diffraction experiment on a 6 mm thick nickel alloy (N5) single crystal casting in a mold (5 mm wall thickness). X-ray parameters were 320 kV, 0.75 mA. A 25 mm thick lead collimator, with a 2 mm clear aperture was placed on the 37 mm diameter x 13 mm thick intrinsic Ge detector to separately sample the two spectra by centering the aperture on the primary beam or the diffraction spot.

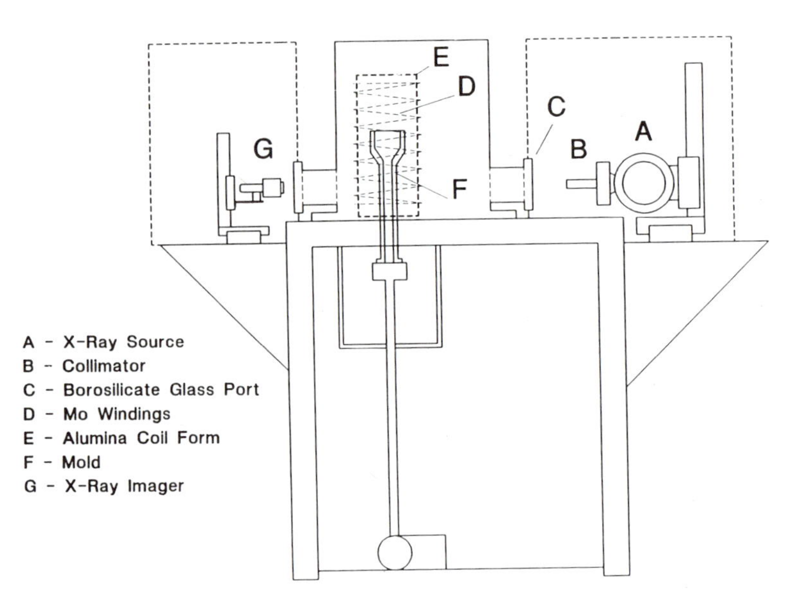

Figure 4. Resistively heated directional solidification furnace, fitted with a collimated x-ray source and real time x-ray imager for transmission x-ray diffraction studies during casting of single crystal nickel superalloys.

could be brought into the small field-of-view (50 mm diameter) of the x-ray imager by rotating the casting. We optimized (by rotating the specimen) the intensity of a particular diffraction spot and used the motion stages on the x-ray source and x-ray imager to scan the sensed area vertically.

Figure 5 plots the mean XRD spot gray level (spot brightness) of an intense transmission diffraction spot as the x-ray source and imager were scanned vertically with respect to the casting. A high spot intensity was observed when the x-ray beam probed a solid region of the casting. The diffraction spot intensity diminished as the x-ray beam entered the region of dendritic solidification (x-ray path traversing a region which is part liquid and part solid). When the x-ray beam was directed through a region of the casting containing only molten alloy, the diffraction spot vanished and the intensity in the region of interest dropped to a low level. Repeated vertical scans, both up and down, showed the same behavior of diffraction spot

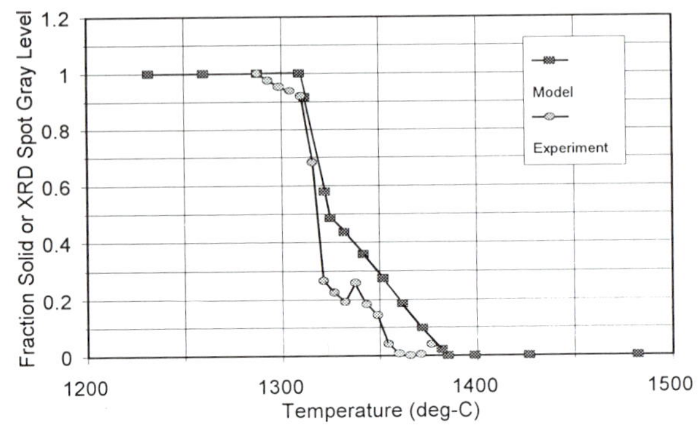

Figure 5. Transmission diffraction spot gray level plotted along with model (Lever) predictions of the fraction solid versus temperature for N5 alloy. XRD spot gray level (intensity) is plotted versus position as the XRD sensor is scannedvertically from a solid region of N5 upward into areas containing less solid and more liquid.

intensity increase (more solid) and decrease (more liquid). Other diffraction spots were rotated into view and additional vertical scans performed. The diffraction spot intensity varied in a similar manner.

A model (Lever) prediction of the fraction solid versus temperature for N5 alloy is also plottted in Figure 5 along with the measured transmission diffraction spot gray level. Since the temperature profile in the casting was not measurable in this experiment , we have adjusted the vertical and horizontal scale of the XRD data to coincide with the solidification model predictions. It is encouraging that the shape of the modeled and experimental curves are quite similar. The experimental data also appears to show formation of the predicted second phase (break in the curve).

SUMMARY

A method has been developed for sensing the physical state (liquid or solid) of a single crystal casting during withdrawal in a directional solidification furnace. A collimated high energy (320 kV) x-ray beam is directed, through the mold, into the metal alloy casting. X-rays diffracted from the crystalline solid produce an ordered array of high intensity transmission diffraction spots. Diffraction from the molten alloy produces a transmission image characterized by a diffuse ring of scattering. The vastly different diffraction patterns provide a means for locating areas of liquid and of solid during the casting process. We have also found that the intensity of a single diffraction spot varies with the fraction of the specimen volume probed by the x-ray beam which is solid. Plots of diffraction spot intensity versus vertical position in the casting are in good agreement with the modeled fraction solid versus temperature curve.

REFERENCES

1. S.E. Doyle, A.R. Gerson, K.J. Roberts, and J.N. Sherwood, Probing the structure of solids in a liquid environment: a recent in-situ crystallization experiment using high energy wavelength scanning, *J. Crys. Growth*, 112:302-307 (1991)

2. K.G. Abdulvakhidov and M.F. Kupriyanov, High-temperature unit for x-ray diffraction studies of single crystals, *Instr. & Exptl. Tech.*, 35(5): 942-943 (1992)

3. H.L. Bhat, S.M. Clark, A. Elkorashy, and K.J. Roberts, A furnace for in-situ synchrotron Laue diffraction and its application to studies of solid-state phase transformations, *J. Appl. Cryst.*, 23:545-549 (1990)

4. R.E. Green, Jr., Applications of flash x-ray diffraction systems to materials testing, *Proc Flash Rad. Symp., ASNT Fall Conf.*, 151-164 (1977)

CHARACTERIZING MARTENSITIC STEEL WITH MEASUREMENTS OF ULTRASONIC VELOCITY

K. W. Hollman, P. T. Purtscher, and C. M. Fortunko

National Institute of Standards and Technology
Materials Reliability Division
325 Broadway, Boulder, CO 80303

INTRODUCTION

The objective of this study was to examine the relationship between the longitudinal and transverse velocities of ultrasound and the microstructure of martensitic steel, which changes due to the tempering process. The hardness of a material is determined by its microstructure.

One of the most widely used methods for determining hardness involves creating an indentation in the surface of a specimen and measuring its depth. The indentation method has two major drawbacks. First, it is destructive, and, therefore, it is not applicable when the surface finish of a material is important. Second, it cannot determine hardness greater than the depth of the indentation. Ultrasonics, on the other hand, is nondestructive and, therefore, preserves surface finish. Also, ultrasound may be useful for determining hardness in the bulk of the material, not just at the surface.

Our approach was to accurately measure the longitudinal and transverse wave velocities in 9 specimens of O1 tool steel (produced as hardness calibration references) with hardness ranging from 24.2 to 55.7 HRC (Rockwell hardness C scale). Differences in the hardness were created by altering the temperature in the final step of the tempering process. Ultrasonic velocities in the specimens were compared with hardness as measured by the indentation method. The steel specimens were approximately 7.6 mm thick by 51 mm square.

We observed that as hardness increases from 24.2 to 55.7 HRC in martensitic steel, the longitudinal velocity of ultrasound decreases from 5942.6 to 5907.9 +/- 2.4 m/s. As the hardness increases from 30.5 to 51.6 HRC, the transverse wave velocity decreases from 3258.0 to 3218.4 +/- 4.4 m/s. A decrease in velocity with increasing hardness has been observed in other studies of steel.[1-3]

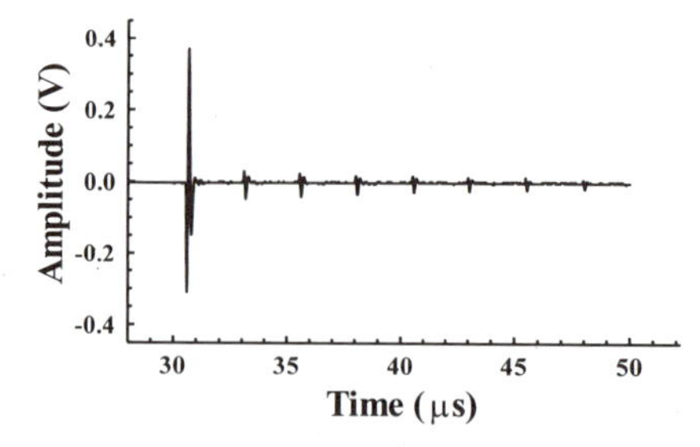

Figure 1. Typical ultrasonic signal from a specimen, showing the front reflection and multiple internal reflections.

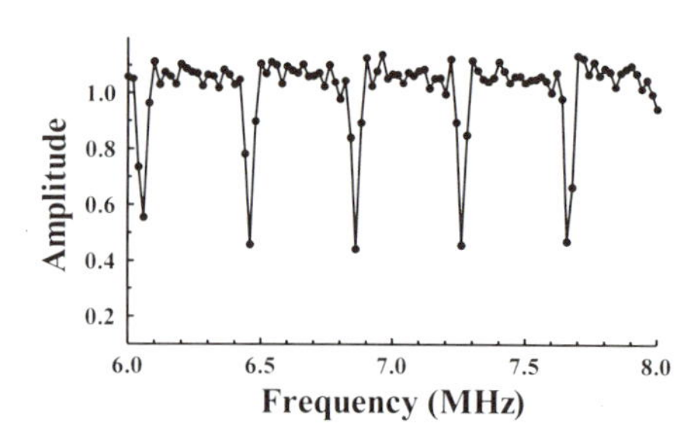

Figure 2. One portion of the amplitude spectrum obtained by Fourier transforming the typical signal in Figure 1.

EXPERIMENTS

To determine longitudinal velocity in a hardness specimen, a 10.2 mm diameter, PVDF, ultrasonic transducer insonified the specimen at a separation of 165 mm in an immersion tank. The transducer was aligned with the specimen so that the ultrasonic beam was incident perpendicular to the surface of the specimen. The broadband ultrasonic signal (1 to 14 MHz) produced by the transducer was reflected from the front surface of the specimen and multiply reflected within the specimen. Figure 1 shows the high-quality, reflected signals captured by the transducer and a broadband receiver. The received signals were displayed on an 8-bit digitizing oscilloscope with a sampling rate of 100 megasamples per second and were transferred to a computer via GPIB for analysis.

Longitudinal velocity was determined from the captured signals using the pi-point technique.[4] First, the amplitude spectrum was computed using a Fourier transform of the multiple reflections (for example, the entire time trace shown in Figure 1). The amplitude spectrum was normalized by dividing by the amplitude spectrum of the front surface reflection (the first pulse in Figure 1). The result is a generally flat spectrum near a normalized amplitude of 1 with periodically spaced, sharp dips as shown by a section of the spectrum in Figure 2. Within the bandwidth of the measurement system (1 to 14 MHz), we measured the location of the centroid of each dip to an accuracy of 0.0025 MHz. If the integer resonance number is plotted versus the frequency location of the dip, the transit time is equal to the slope of this relationship. We used the standard linear regression technique[5] to calculate propagation time from our data. Longitudinal velocity is twice the thickness of the specimen divided by this timing measurement.

The residuals to the fitted line are shown in Figure 3. Here, the residuals are the difference between the frequency location of the dip as predicted by the fitted line and the

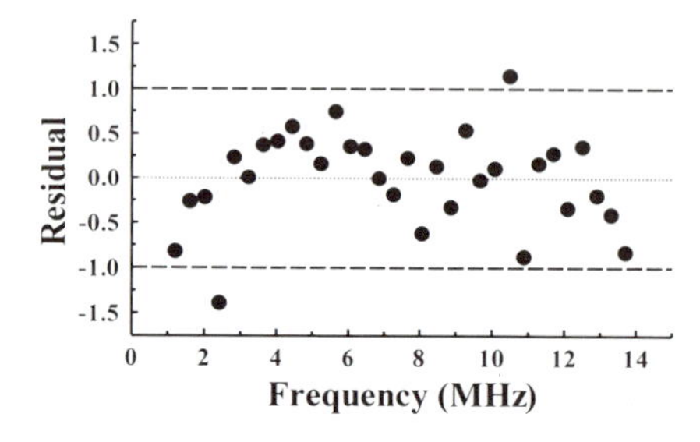

Figure 3. Residuals of the linear regression fit as a function of frequency.

actual location. This difference is then normalized by dividing it by the estimated error of the frequency location of the dips. Figure 3 indicates two important aspects of the data as analyzed by the pi-point method. First, the residuals are a random function of frequency. In other words, the residual is just as likely to be positive as negative. This aspect confirms the assumption that the relationship between resonance number and frequency location is linear. If the error in the velocity is due to the error in the frequency location, then the normalization of the residuals should cause them to lie between -1 and 1, which is what we observe in Figure 3. Consequently, we believe that the accuracy of 0.04% is correct for the longitudinal velocities.

To determine the transverse velocities in an immersion tank, we used the same transducer as before, but the specimen was tilted beyond the critical angle for longitudinal waves. Velocity was determined by the self-reference method[6] which involves transmitting through the specimen, reflecting from an acoustic mirror aligned perpendicular to the transducer, and transmitting back through the specimen. A time shift Δt can be measured by superimposing the leading edges of a pulse transmitted through the specimen at normal incidence and a pulse transmitted through the specimen at non-normal incidence. The velocity at a non-normal angle can be calculated from the equation developed by Chu and Rokhlin,[6]

$$V(\theta)=\left(\frac{1}{V_n^2}+\frac{t_0-(t_0+\Delta t)\cdot\cos(\theta)}{h\cdot V_0}+\frac{\Delta t\cdot(2\cdot t_0+\Delta t)}{4\cdot h^2}\right)^{-1/2}, \tag{1}$$

where V_0 is the velocity in water, V_n is the velocity in the specimen at normal incidence, h is the thickness of the specimen, and θ is the angle of the specimen. In this equation t_0 is

$$t_0=2\cdot h\cdot\left(\frac{1}{V_0}-\frac{1}{V_n}\right). \tag{2}$$

As an input to these equations, the velocity at normal incidence is simply the longitudinal velocity that was measured previously. The accuracy of the transverse velocity was determined as 0.13% from a propagation-of-error calculation.

RESULTS

For each of the 9 specimens, 12 independent measurements of longitudinal velocity were made. From the 12 measurements a mean and standard deviation were calculated. Figure 4 shows the mean longitudinal velocity as a function of hardness. The vertical error bars represent standard deviations (63% confidence interval) that were approximately equal

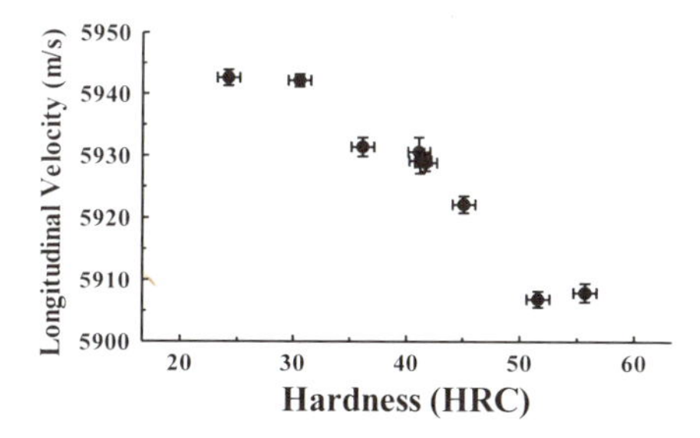

Figure 4. Measured longitudinal velocity as a function of hardness.

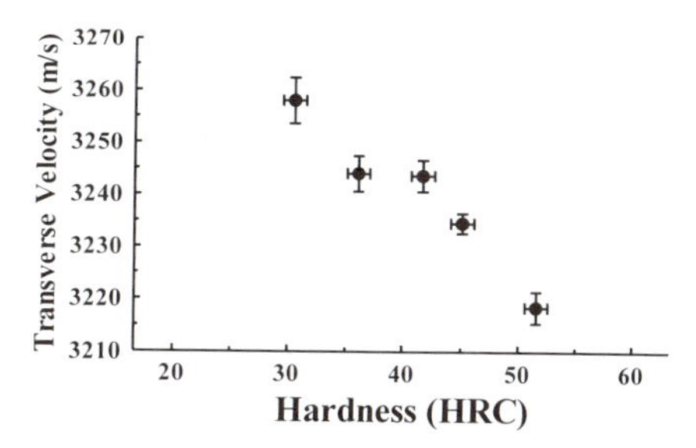

Figure 5. Measured transverse velocity as a function of hardness.

to the estimated error described earlier. Horizontal error bars represent a standard deviation in the measurement of hardness using the indentation technique. To obtain a measure of reproducibility, three of the specimens were tempered to a hardness of approximately 41 HRC. As the figure indicates, velocity measurements of these 3 specimens easily fall within the error bars indicating good reproducibility.

Transverse velocities were measured for 5 of the specimens at 8 refracted angles in the range of 40° to 50° and -40° to -50°. Beyond the critical angle for longitudinal waves, a phase shift occurs because of mode conversion.[7] To avoid correcting the velocities we limited our data to the listed refracted angles where the phase is approximately 0. The mean and standard deviation were calculated from the 8 measurements at these angles. Figure 5 shows the mean transverse velocity as a function of hardness. Vertical and horizontal error bars have the same meaning as described above.

For longitudinal velocity the change over the hardness range is 0.59%, which is significant compared to a measurement accuracy of 0.04%. Because of the more complicated measurement technique, the accuracy of the transverse velocity is 0.13%, but there is a greater change (1.23%) over the hardness range, so the measurement of transverse velocity is also significant compared to the accuracy.

CONCLUSIONS

We observed that as hardness increases from 24.2 to 55.7 HRC in martensitic steel the longitudinal velocity of ultrasound decreases from 5942.6 to 5907.9 m/s. As the hardness increases from 30.5 to 51.6 HRC the transverse wave velocity decreases from 3258.0 to 3218.4 m/s. Both of these relationships are significant when compared to the measurement accuracy.

Four main characteristics of the microstructure of martensitic steel can change due to the final temperature used in the tempering process: lattice distortion, dislocations, precipitates, and grain boundaries. More work needs to be done to separate the effects of these four microstructural changes on longitudinal and transverse velocities.

Heat treatment by quenching of tool steel produces a nonequilibrium phase, martensite, which is body-centered tetragonal (bct), because carbon atoms in interstitial sites distort the lattice, producing the tetragonal structure. Tempering this martensite reduces the lattice distortion by allowing the carbon to precipitate out of solution. This process should reduce the influence of lattice strains on our measurements. The biggest change in the lattice distortion occurs rapidly and at low tempering temperatures (20 to 150° C). At higher tempering temperatures, the structure is nearly bcc (body-centered cubic) rather than bct, and the influence of lattice strains is lost.[8]

Higher tempering temperatures reduce the dislocation density and coarsen carbide precipitates in the specimens. Dislocations interact with ultrasonic waves and deformation-induced dislocations have been shown to lower the velocity in NaCl crystals.[9] Tempering the martensite at high temperatures should reduce the dislocation density and thus reduce the velocity of sound.

Changes in the grain size or texture associated with the final tempering temperature can also change the hardness. As the final temperature increases the size of the grains may increase and the percentage of high-angle boundaries (defined as >20°) will increase. Because each grain defines a localized crystal lattice, and ultrasonic velocity is sensitive to the distribution of these lattice regions, the size and boundary angles of the grains should influence velocity.

It is beyond the scope of this paper to determine the relative contributions of each microstructural change to the change in velocity.

REFERENCES

1. G. A. Gordon and B. R. Tittmann, "Surface acoustic wave determination of subsurface structure," in *Review of Progress in QNDE*, edited by D. O. Thompson and D. E. Chimenti (Plenum Press, Brunswick, ME, 1993), Vol. 13, pp. 1595-1602.
2. G. S. Kino, D. M. Barnett, N. Grayeli, G. Herrmann *et al.*, "Acoustic measurements of stress fields and microstructure," Journal of Nondestructive Evaluation **1** (1), 67-77 (1980).
3. E. P. Papadakis, "Ultrasonic attenuation and velocity in SAE 52100 steel quenched from various temperatures," Metallurgical Transactions **1**, 1053-1057 (1970).
4. T. Pialucha, C. C. H. Guyott, and P. Cawley, "Amplitude spectrum method for the measurement of phase velocity," Ultrasonics **27** (5), 270-279 (1989).
5. W. H. Press, S. A. Teukolsky, W. T. Vetterling, and B. P. Flannery, *Numerical Recipes in C*, 2nd ed. (Cambridge University Press, New York, NY, 1994).
6. Y. C. Chu and S. I. Rokhlin, "Comparative analysis of through-transmission ultrasonic bulk wave methods for phase velocity measurements in anisotropic materials," J. Acoust. Soc. Am. **95** (6), 3204-3212 (1994).
7. A. I. Lavrentyev and S. I. Rokhlin, "Phase correction for ultrasonic bulk wave measurements of elastic constants in anisotropic materials," in *Review of Progress in QNDE*, edited by D. O. Thompson and D. E. Chimenti (Plenum Press, Brunswick, ME, 1996), Vol. 16, pp. 1367-1374.
8. G. Krauss, *Steels: Heat Treatment and Processing Principles* (ASM, Materials Park, OH, 1990).
9. A. V. Granato, J. DeKlerk, and Truell, Phys. Rev. **108**, 895 (1975).

ULTRASONIC CHARACTERIZATION OF MICROSTUCTURAL STATES IN ALUMINUM WELDS DEPENDING ON THE WELDING PARAMETERS

Nephi M. Mourik,[1] Eckhardt Schneider,[2] Kamel Salama[3]

[1]University of Houston, presently at IZFP participating in a Cooperative Program
[2]Fraunhofer Institut Zerstörungsfreie Prüfverfahren, IZFP, Saarbrücken, Germany
[3]University of Houston, Houston, Texas

INTRODUCTION

Aluminum alloys are increasingly used for light-weight construction in the transportation industry. Among the common joining techniques for such materials (riveting, adhesion bonding, clinching) welding is increasing in importance. In contrast to steel welding, there are restricted possibilities when applying post-welding treatments to optimize the material properties of the welded part. Welding parameters, such as the welding energy, determine changes in micro-structure, stress and texture states. Hence, the behavior of welded components under static or dynamic loading is strongly influenced by welding parameters. This investigation will show how easy-to-apply ultrasonic techniques can be used to support the optimization of the welding process.

WELDING PARAMETERS

The inspected aluminum sheets are made of the hardenable ISO grade designated alloy AlCu6Mn (equivalent to AL 2219). The weld feed material AlMg4.5Mn is non-hardenable and is considered easily weldable. The tungsten inert gas (TIG) welding process includes rigidly mounting and preheating the sheets to 150 °C in order to reduce thermal deflection. Table 1 details the welding parameters of flawlessly welded samples approximately 200 x 300 x 4 mm^3 in size.

Table 1. Welding parameters and compositions of inspected sheets.

	Sheet 1	Sheet 2	Sheet 3
Base Material	AlCu6Mn	AlCu6Mn	AlCu6Mn
Welding Process	TIG DC 100% He	TIG DC VP 100% He	TIG DC VP 100%Ar
Voltage [V]	15	15.5	10
Current [A]	115	112	275
Velocity [mm/min]	300	250	300
Welding Energy [kJ/m]	388	417	550
Weld Seam Width [mm]	9	9	10
Weld Feed Material	AlMg4.5Mn	AlMg4.5Mn	AlMg4.5Mn

Important welding parameters differences include the choice of helium or argon gas and the welding energy resulting from voltage, current and welding velocity. The stable arc voltage depends on the inert gas, which primarily prevents the oxydization of the melt. Helium gas increases the heat concentration and weld penetration.

EXPERIMENTAL RESULTS

The welding energy required for aluminum, which has a melting temperature of 660 °C, is roughly the same as that used in steel for the following reasons: aluminum has an increased heat conduction and the aluminumoxide layer has a melting point above 2000 °C. The welding parameters determine the energy transfered to the heat affected zone, where the temper of the hardenable aluminum alloy is changed. The resulting microstructure, stress and texture influence the elastic properties. Hence, the ultrasonic velocities, or times-of-flight (TOF), and their changes with location, sample temperature, elastic load and the direction of polarization are considered the easiest applicable nondestructive tool to investigate the influence of welding parameters on the mentioned structural properties.

TOF measurements of longitudinal and linear polarized shear waves are performed using the IZFP prototype system[1]. Since the ultrasound wavelength is an order of magnitude larger than grain sizes, scattering effects on TOF measurements are negligible. The longitudinal wave is generated piezoelectrically at 8 MHz and the shear wave electromagnetically at 1.7 MHz center frequency. A schematic of the weld (W), heat affected zone (HAZ) and base material (BM) as well as a TOF measurement trace are shown in Figure 1.

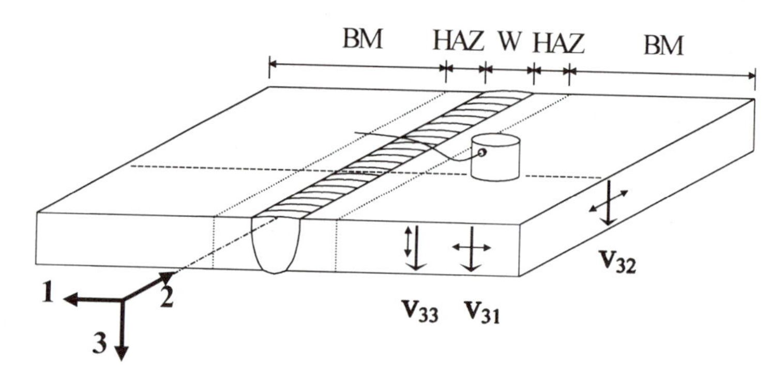

Figure 1. Welded sheet with schematic ultrasonic propagation and polarization direction.

By assuming a density of 2.71 g/cm^3 in the weld and BM, the longitudinal and shear velocities can be used to calculate the elastic moduli. In spite of the accumulated error, Figure 2 shows the softening tendency of the elastic modulus in the vicinity of the weld, which is generally seen in practical experience. Figure 3 shows similarities in the Vicker indentation hardness, even though the resistance to plastic deformation is compared to the elastic constants.

For this investigation, ideal surface conditions are prepared by machining off the weld reinforcement. In a real application, Figure 2 would have a data gap from -5 mm to +5 mm, which is the width of the weld reinforcement. However, one can still draw significant conclusions about the welding process by evaluating changes in the measuring quantities taken in the vicinity of the weld reinforcement.

In previous work the temperature dependence of sound velocity was found to be more sensitive to microstructural variations than were the absolute velocities and moduli[2]. The relative change in TOF using the shear wave is measured as a function of temperature at a specific position in the weld, HAZ and BM as shown in Figure 4. The slope is the temperature elastic constant (TEC), and is used to characterize the various material states due to changes in position and welding parameters as shown in Figure 5. There seems to be some tendency of change due to microstructure, but there is no significant effect due to welding parameters.

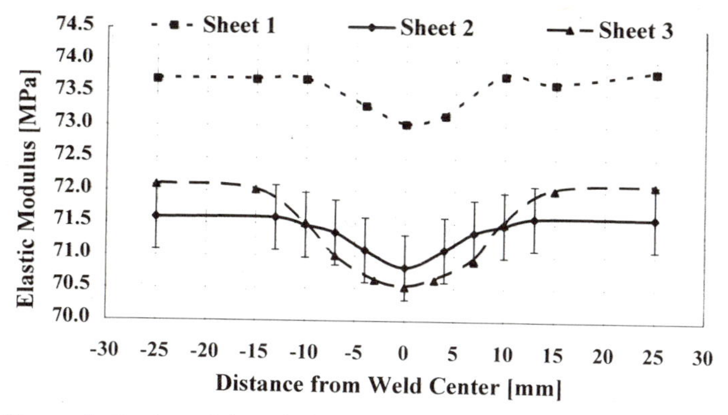

Figure 2. Elastic modulus calculated from longitudinal and shear velocities.

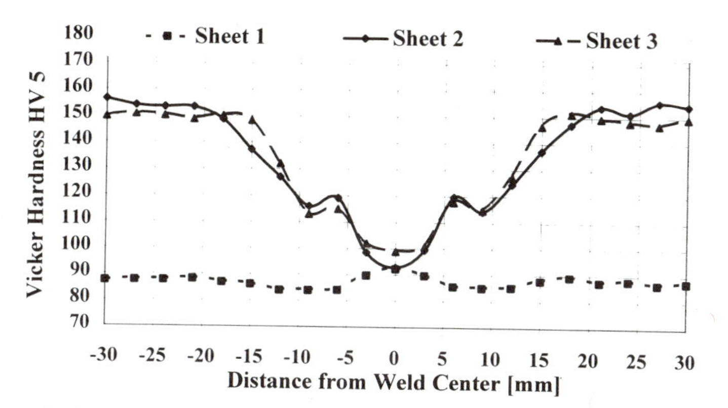

Figure 3. Indentation hardness across weld for sheets with different welding parameters.

It has been found[3], that the relative change in velocity as a function of strain does not differ significantly in the weld, HAZ or BM. This same result was found for different welding parameters, and is advantageous to the evaluation of residual stresses using ultrasonic techniques.

The relative change of the shear wave velocity can principally be used to determine the elastic anisotropy. The change of TOF is measured by rotating the polarization direction with respect to welding direction. The direction of the extrem values are the principal directions of the elastic anisotropy. Figure 6 shows a significant change in principal direction and magnitude at different weld seam positions. Though the welding process may not have changed at position A, the effect may be a change in welding temperature and direction of heat flow due to the near edge. Whether or not this

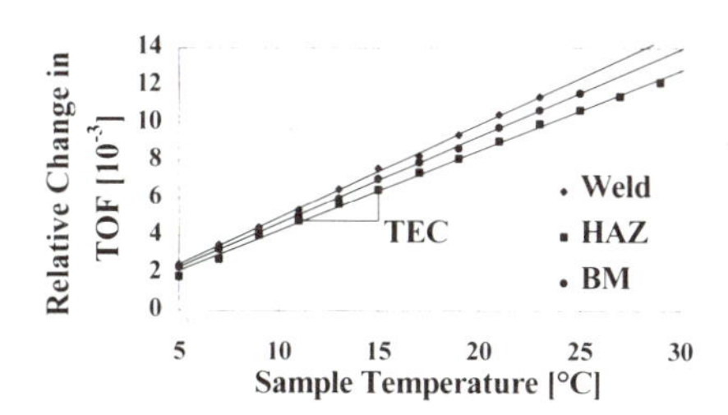

Figure 4. The relative change of TOF versus temperature is the TEC.

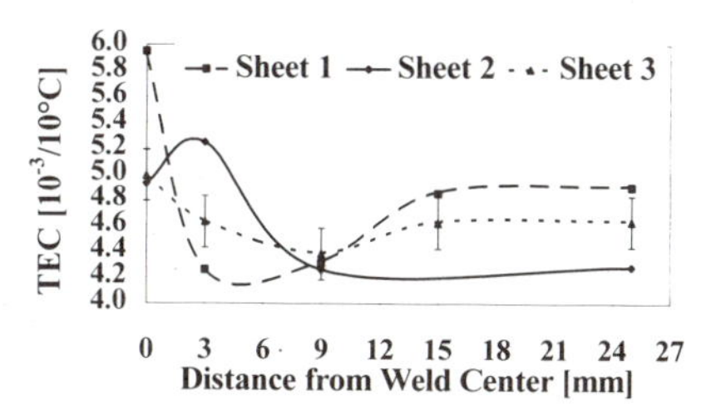

Figure 5. The TEC versus microstructural position for sheets with different welding parameters.

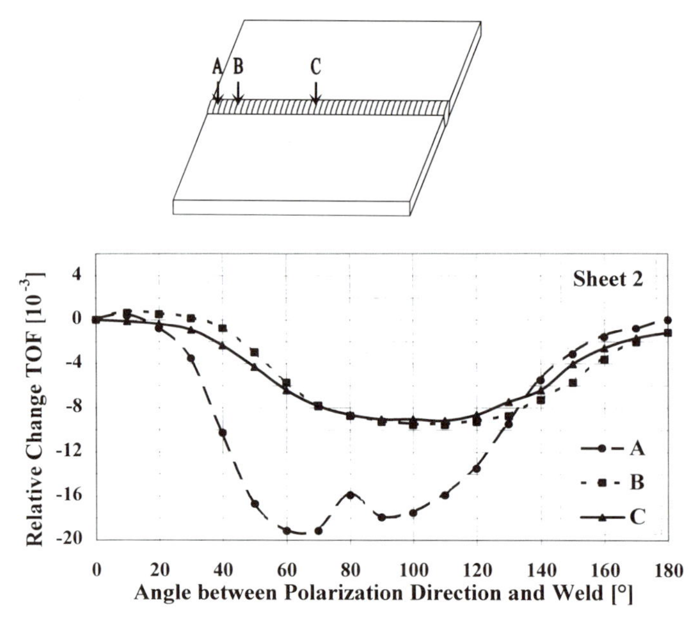

Figure 6. The elastic anisotropy measured in the weld center at three positions along the weld seam.

effect is due to texture, stress and/or microstructure, it can be concluded that such gradients negatively influence the quality and behavior of the weld.

In a similar way, the elastic anisotropy is determined at a specific position in the weld and at different distances from the weld center line. Measurements are made at 3 mm and 15 mm from the center line in the HAZ and at about 30 mm in the BM. Figure 7 shows that the principal directions in the weld and BM have shifted approximately 90°. This correlates with the cold rolling direction and the heat flow direction during welding. The shift in magnitude and principal direction in the HAZ can be a measure of the internal gradients.

The TOF profiles accross the weld seam are measured using a shear wave polarized parallel to the weld seam and perpendicular to it. Both profiles are shown in Figure 8 for each welded sheet. Contrary to manufacturing information, the pre-welding texture differs for each of the sheets. This is seen by the relative change of the two TOF profiles measured in the base material. This difference is about $-6\cdot10^{-3}$ in sheet 1 and $+2.5\cdot10^{-3}$ and $+3.5\cdot10^{-3}$ in sheets 2 and 3 respectively. The two TOF profiles differ only in sheet 1, where the direction of the pre-welding texture is rotated 90° relative to the weld and is higher in magnitude.

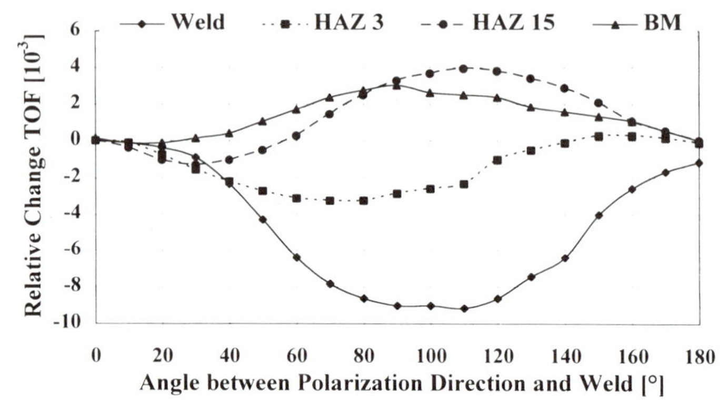

Figure 7. The change of direction and magnitude of the elastic anisotropy differ in the Weld, HAZ and BM. The result for sheet 2 is shown here.

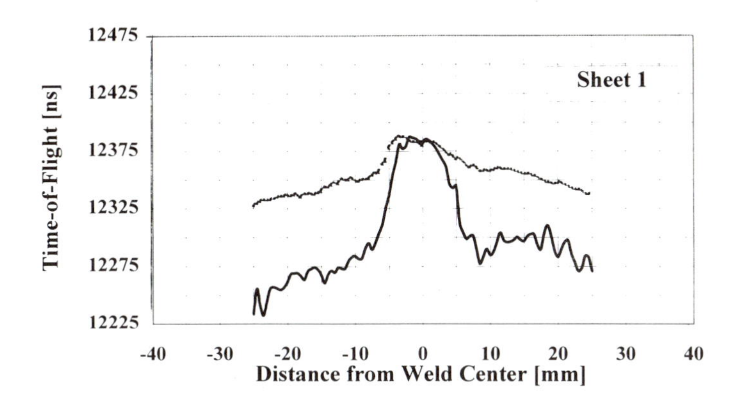

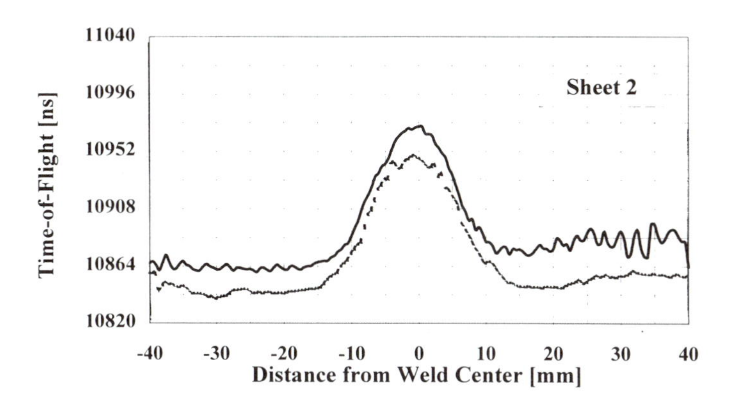

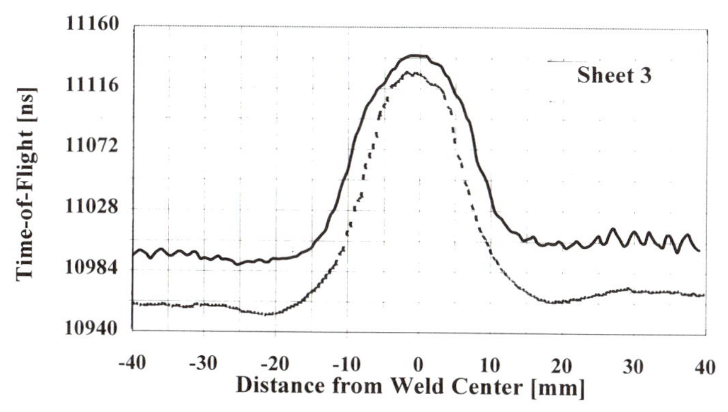

Figure 8. Shear wave TOF profiles measured with the linear polarization direction parallel (broken line) and perpendicular to weld (solid line).

The left part of Figure 9 shows the half-width of the TOF profile (based on the measuring results shown in Figure 8) measured with the shear wave polarized parallel to the weld seam. The relative difference between the TOF data measured in the BM and weld seam (Figure 8) are calculated and distinguished in the right part of Figure 9 as a function of welding energy. In real applications, the TOF profile width or height can be measured at a specific position in the vicinity of the weld to establish a similar correlation.

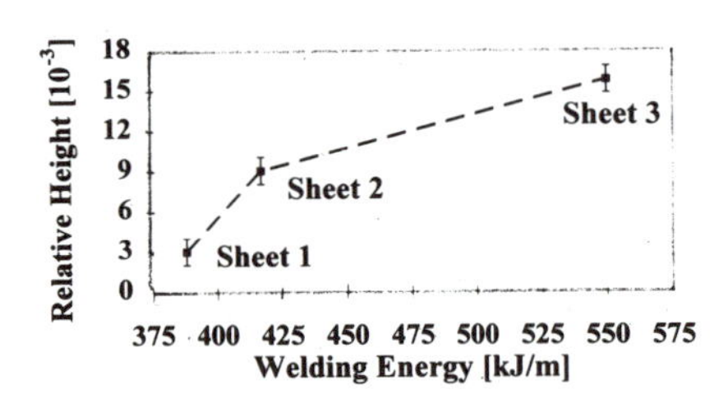

Figure 9. The half-width and height determined from the TOF profiles (Figure 8) of the shear wave polarized parallel to the weld.

CONCLUSION

Welding parameters influence the microstructure, texture and stress state in and around the weld; therefore they also influence the elastic properties. Although ideal measuring conditions were used on these aluminum welds, the significant changes in the ultrasonic time-of-flight encourage further investigations to build a statistical data base for different aluminum alloys. As it was shown, the robust and easy-to-apply ultrasonic technique holds high promise to support the optimization of the aluminum welding process.

ACKNOWLEDGEMENT

The authors acknowledge the cooperation of Dr. Kampmann, Alusuisse Technology & Management AG and of Mr. Dürr, Daimler-Benz Aerospace, Dornier GmbH.

REFERENCES

1. R. Herzer and E. Schneider, Instrument for the automated ultrasonic time-of-flight measurement -A tool for materials characterization-, *Nondestructive Characterization of Materials*, P. Höller, V. Hauck, G. Dobmann, C. Ruud, R. Green (eds.), Springer Verlag Berlin Heidelberg (1988) 673-680:
2. U. Arenz and E. Schneider, Microstructure and texture influences on ultrasonic quantities for welding stress analysis, *Nondestructive Characterization of Materials VIII*, R.G. Green, Jr. (ed.) (1997).
3. J.H. Cantrell and K. Salama, Acoustoelastic characterization of materials, *International Materials Review*, 36:125-145 (1991).

DEVELOPMENT OF ON-LINE LAMB WAVE TESTING TECHNIQUE USING REAL TIME DIGITAL SIGNAL PROCESSING

Masahiro Nakamura, Hirokazu Yokoyama, Masaki Yamano, Riichi Murayama

Control System Technology Department
SUMITOMO Metal Industries, Ltd.
1-8 Fuso-cho Amagasaki, Hyogo, Japan 660

Abstract

Lamb wave testing technique has been used for on-line defect inspection of steel strips. However, conventional Lamb wave testing technique has the following two large problems. One problem is the existence of the wide dead zones which appear in near and far field. Another one is the misdetection due to parasitic echo or electrical noise. We have developed the method which solves these problems by using the signal processing technique. We designed and constructed the new system which collected the received signals and change it into the two-dimensional signal. It distinguish defect signals from the other signals by the two dimensional signal characteristics and the if-then rule logic. This method does not require the inspection gate. As a result, the dead zone and the misdetection can be reduced. This paper describes the measurement principle, the algorithm of the signal processing and the distinction logic. The results showed that the dead zone was reduced to 100mm from the strip edge.

Introduction

Lamb wave testing technique has been used for the detection of an internal defect of steel strip in the production line.(Fig1) This method, Lamb wave propagates in the width direction of steel strips, and an echo from the defect is detected, using the wheel type probe as shown in Fig2.

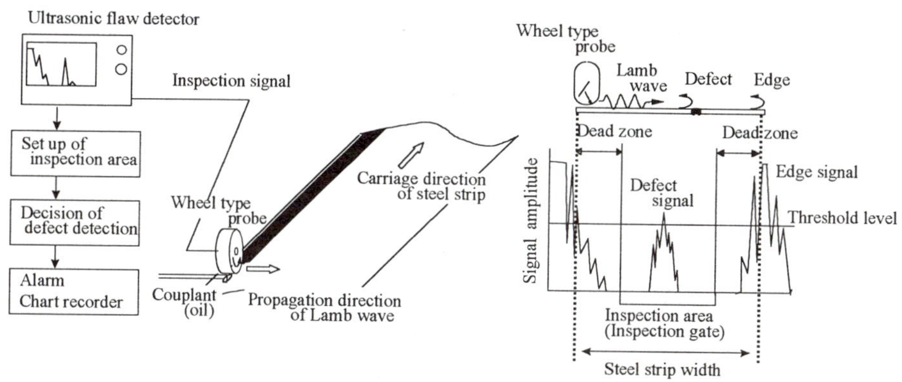

Fig.1 On-line Lamb wave inspection system in production line.

Fig.2 Principle of Lamb wave testing by using wheel type probe.

Nondestructive Characterization of Material VIII
Edited by Robert E. Green Jr., Plenum Press, New York, 1998

The defect is distinguished by signal amplitude and the position. But, the transmitting signal and edge signal appear at the same time, we must set the inspection gate so that it reject these noise signals. The position and the width of the inspection gate is set by the information of strip width and thickness from the process control computer. However, these signal positions fluctuate due to the probe coupling condition and the meandering of steel strip in the production line. Therefore, the inspection gate tends to set narrower than the most suitable width, and the dead zone in both strip edges increases.

Furthermore, the defect judgment is determined by only the signal echo level. So, in the conventional system misdetect the parasitic echo or electrical noise. We gathered the Lamb wave testing data in the picking line of hot strip rolling factory, and analyzed this data to solve these problems. We found that the cause of these problems was due to the use of the inspection gate. If the system could detect the defects without the inspection gate, these problems should be solved.

We developed a new method , named gate-less inspection(GLI), which detects a defect without the inspection gate. GLI makes two dimensional digital data of inspection signals as the result of the gate-less inspection, and distinguishes the defect signals from others using the characteristics of the two-dimensional data.(Fig.3)

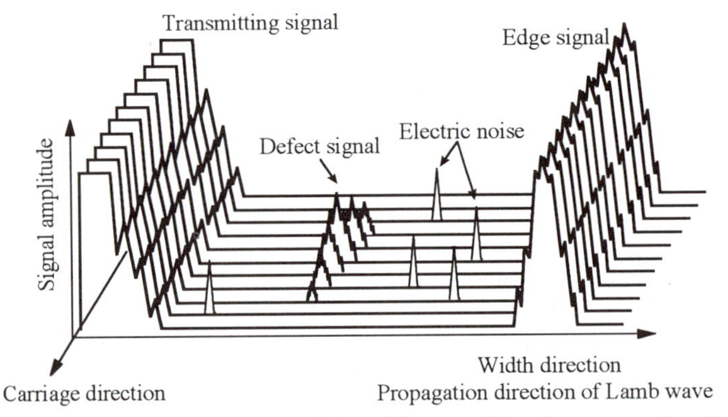

Fig3 The two dimensional signal image of Lamb wave testing data.

System configurations

The configuration of the new system is shown in Fig4. This system is attached to the conventional Lamb wave testing system. The signals are obtained in through the interface circuit and two signal processing board. This system is controlled by AT compatible personal computer. The signal processing boards include a 8-bit analog-digital converter. The signal data is sent to the digital frame memory (1Mbyte) at synchronously in the repetition signal of the ultrasonic transmitter. The data size is limited by the signal processing speed for real-time processing, the frame memory size and the inspection area. We realized a 500k samples per second in the production line at the test frequency of 2.25MHz. We designed the peak-hold circuit as the anti areas filter to prevent the arising and misdetection of the echo peak which is very important for the judgment of defect size.

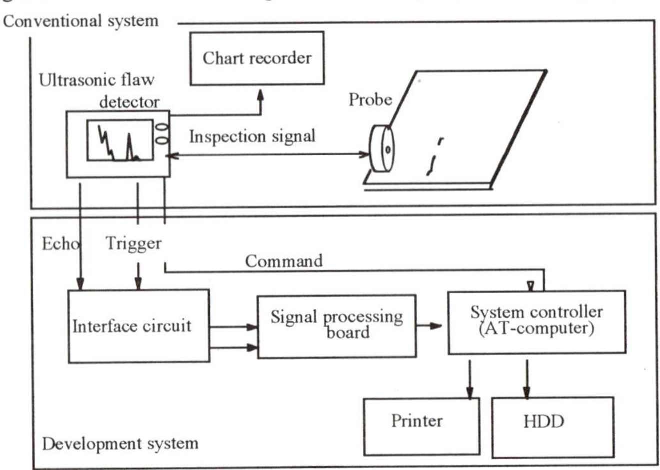

Fig.4 System configuration of development system.

It takes approximately two second to gather the 1Mbyte two dimensional data and to process data by high speed digital processing IC. During the one signal processing board is processing data, the other processing board works to store the data. This process made it possible to inspect the steel strip continuously. (Fig.5)

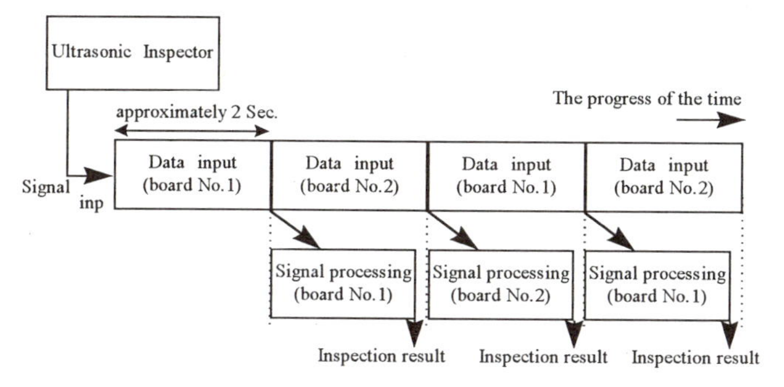

Fig.5 Time chart of data input and processing.

Result and discussions

We collected the Lamb wave inspection signal at pickling line which located after the hot strip mill using the developed apparatus. The maximum speed of this line is 150mpm and the steel strip 1.4mm-7.0 mm in thickness, 560mm-1,600 mm in width and over 1km in length is produced. As a result of analyzing collected data, we were found out that each signal has a two dimensional signal characteristic values as shown in table 1.

Table.1 Characteristics value of two dimensional Lamb wave inspection signals

The kind of the signal	Signal intensity E	Signal length L	Signal width W	Signal Position X
Defect signal	50—100%	20—250mm	16—60 μ sec	Un-known
Transmitting signal	Over 100%	Over 1m	60—120 μ sec	Known (approx.)
Edge signal	60—100%	Over 1m	60—120 μ sec	Known (approx.)
Electrical noise signal	Under 80%	Under 5mm	Under 8 μ sec	Random
Residual echo in the probe	30—60%	15mm—1m	Under 16 μ sec	Known (approx.)

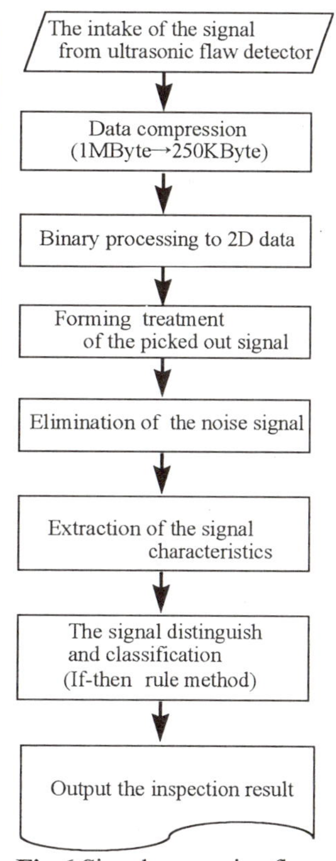

Fig.6 Signal processing flow

So, we devised ,an algorithm of signal processing in the figure 6 by using these result. In this developed logic, the two dimensional data is compressed to realize real-time signal processing. After that, the binary processing was used for finding the defect signal, transmitting signal, edge signal and other large intensity signal areas.

The signal which only a few distances leave in the length direction and which is on and off is connected by the forming processing, and the small area elimination processing remove the noise signals. And we use the If-then rule distinction to find a defect signal from other signals using the characteristic values of two dimensional.

It takes about a 1.1~ 1.5 seconds for the series of processing. Processing time is shorter than the time for data collection, so the long strip can be inspected continuously belong by using two signal processing boards, alternately.

It takes about a 1.1~ 1.5 seconds for the a series of processing. Processing time is shorter than the time for data collection, so the long strip can be inspected continuously belong by using two signal processing boards, alternately.

To evaluate a dead-zone in both steel strip edges, we used a test coil which contained artificial defects. Fig.7 show the distribution of the artificial defects which are through drill holes of 1mm diameter. The white and black circles show detected defects and undetected defect, respectively. It was possible to detect a defect without using the inspection gate. We confirmed that the dead-zone could be reduced at the wheel type probe side by 175mm (Incident point of ultrasonic from 75mm), the edge side is 100mm from edge.

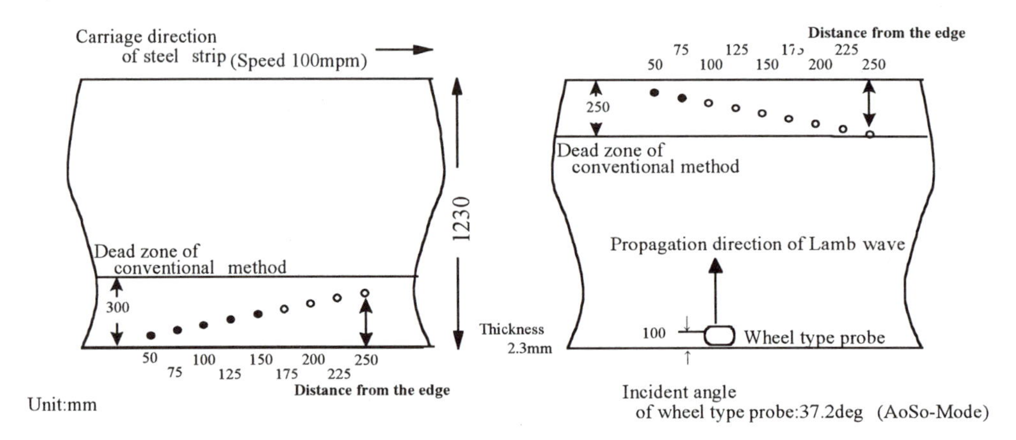

Fig. 7 Inspection result of the defect which are 1mm holes drilled through the sheet.
○:Detected with the developed system, ●:Undetected with the developed system

Conclusion

A new method (GLI) which doesn't require the inspection gate is presented in this paper. It could reduce the dead zone of Lamb wave testing and the misjudgment by the parasitic echoes. We described the configuration of this system, the algorithm of signal processing and distinction procedure in GLI method. We show some typical two dimensional data which was made by the developed system at the production line.

In the GLI method, the defect signal was distinguished from the transmitted echoes, edge echoes and noise signals using two dimensional signal characteristics, such as the continuity and observed position of these signals in stripes. Also, the digital signal processing technique was used for real-time estimation of the signal characteristics.

We show this new method effectively reduce the dead zone in the line Lamb wave testing by using a test coil which included drilled holes as artificial defects. The dead zone can be reduced to less than 100mm at the opposite side of probe and 175mm at the probe side. And we confirmed that, the misjudgment isn't caused by the parasitic echoes and the electrical noise signals by this new method.

References:
1)Ultrasonic Material Testing, Revised Edition, Edited by Japan Society for the Promotion of Science, 1974
2)Ultrasonic Testing using for Lamb Wave, Morio Onoue, Journal of Non-destructive Inspection. P135-144 10(3),1960

ULTRASONIC AND ACOUSTO-ULTRASONIC INSPECTION AND CHARACTERIZATION OF TITANIUM ALLOY STRUCTURES

A. L. Bartos[1], J. O. Strycek[1], R. J. Gewalt[1], H. Loertscher[1] and T. C. Chang[2]

[1]Quality Material Inspection, Inc.
919 Sunset Drive
Costa Mesa, CA 92627 (USA)

[2]McDonnell-Douglas Aerospace
1510 Hughes Way
Long Beach, CA 90810-1870 (USA)

INTRODUCTION

Research and development in Titanium (Ti) fabrication technology, such as Super Plastic Forming/Diffusion Bonding (SPF/DB), honeycomb (H/C) or composite Titanium-Graphite (TiGr) sandwich panels, requires parallel development of advanced nondestructive evaluation (NDE) techniques to establish high confidence levels in the critical airframe structures being produced. This particular effort is a subset of other NDE work directed towards developing a technique for inspecting and characterizing large area Ti aerospace wing structures. Ultrasonic NDE C-scans using pulse-echo or through-transmission test configurations, in general, are not able to quickly sort large sections of material into acceptable and flawed categories. Conventional ultrasonic configurations using plate-waves are more applicable to large area inspection, while Acousto-Ultrasonics[1] (AU) is a more unconventional NDE technique which has the potential for *rapidly* correlating extracted waveform attributes to flaw category and mechanical performance, for the area of material being insonified.

Beside identifying critical flaw categories endemic to the manufacturing process using the more conventional ultrasonic scanning approaches, the main purpose of this proof of concept NDE development effort was to assess the worth of using AU on Ti sandwich structures. AU test configuration development consisted of a series of trials using piezo-electric transducer (PZT) frequencies ranging from 1 to 10 MHz, and PZT separations from 1" to 10.5", applied to Ti flat plate two-sheet panels, as well as more complex four-sheet sandwich structures.

The results indicate that even when coupled to a straightforward pattern recognition classifier, AU can be used to rapidly inspect large Titanium alloy structures. Ultimately, when AU is coupled to more sophisticated signal processing and pattern recognition approaches, the final system has the potential to: 1) monitor various phases of Ti structure fabrication, in order to minimize scrap, 2) assess final product quality, possibly in terms of ultimate strength estimates, and 3) monitor fatigue in-service, in terms of residual strength prediction. Test times for conventional scanning NDE approaches are measured in minutes or hours; however, AU test times can be measured in seconds; therefore, repeated testing throughout the life cycle of the part then becomes both possible and practical to perform.

DISCUSSION

During the initial investigation, presented in this paper, various AU data acquisition configurations were tried in conjunction with a very preliminary signal analysis approach. AU signals were acquired on the same side from both simple and more complex geometrical structures with panels of acceptable and flawed diffusion bond integrity, to determine whether the AU NDE approach showed any promise of providing reliable signal separation.

The emphasis in conventional ultrasonic scanning NDE and Radiography is on test sensitivity and thoroughness; and therefore, performance is measured in terms of local flaw resolution at the cost of long inspection times. AU is being considered for this NDE application because mass production environments demand that test automation, test duration, and ultimately, overall material test throughput rates become *the* important considerations. With AU, lower frequencies enable larger area coverage, where the collective effects of globally occurring subcritical flaws can be readily detected. AU disadvantages are that localized flaw detection is not always possible and that PZT placement and propagation geometry must be consistent and repeatable. Also, more sophisticated signal analysis is generally required; however, these are not unacceptable or difficult constraints to satisfy in a mass production setting. All aspects of this effort were therefore, kept focused on how the AU concept will eventually be implemented in this manufacturing and in-service test setting.

Test Specimen Description, Preparation and Verification

The test material was Ti-6-4 (60% Aluminum-40% Vanadium), an α - β alloy, typically used for SPF/DB structures. Simple geometry two-sheet Ti panels were fabricated for the initial NDE investigations, because proving material classification is possible for simple flat plates was a pre-requisite for investigating more complex geometries. The two Ti sheets, making up the two-sheet panel, were first roll seam welded around the periphery (18" square) and then evacuated by diffusion pumping. They were then placed in a SPF/DB press die with blocks in the die cavity to limit the maximum thickness to approximately 0.125 inches. A Ti 6-4 (picture frame) sheet and a Ti 6-4 diaphragm sheet were used with the SPF/DB press die to provide the necessary bonding pressure, while the test sheets being fabricated rested at the bottom of the die. After fabrication a 1" by 18" rectangular strip was removed from the center portion of each panel for mechanical testing and visual bond quality assessment. This resulted in four flat, two-sheet Ti panels, with 6" x 18" (approx.) dimensions that were made available for NDE experimentation. The four panels had bond percentage estimates of 14%, 21%, 73%, and 83%, and were the initial AU test matrix specimens. For the AU NDE effort, two-sheet panels identified with 73 and 83 bond percentages represented the acceptable material categories, while panels with bond percentages identified to be 14 and 21 were assumed to represent flawed material.

The same SPF/DB press die was also used to fabricate several four-sheet panels using conventional approaches. Figure 1 illustrates the cross-section of both the two-sheet and four-sheet panels. Sheets 2 and 3 of the four sheet panel were welded prior to undergoing pressure bonding to face sheets 1 and 4, respectively, and evacuation by diffusion pumping, thus forming air-filled cells separated by webs made of sheets 2 and 3 joined by the welds. As with the two sheet panels, differences in fabrication quality were controlled by varying time and temperature process parameters. Imperfect bonding was achieved by shortening time and lowering temperature. Figure 1 illustrates the location of triangular rat-holes, located between sheet 1 and sheet 2 folded over (the same symmetry holds for sheets 4 and 3). In addition to having imperfect diffusion bonds between sheets 1-2 forming the upper face sheet of the panel, between sheets 3-4 forming the lower face sheet of the panel, and between sheets 2-3 folded over to form the web, flawed panels possessed rat-hole sizes that were considerably larger than normal.

Specimen verification was performed using immersion pulse-echo ultrasonics, as well as air-coupled ultrasound. Water immersion C-scans at 25 MHz, with the gate set to the bondline, produced marked differences between acceptable (73% and 83%) and flawed (14% and 21%) panel categories. Discernible differences were also visible between panels identified with 14% and 21% bonding, as well as between panels initially identified with 73% and 83% bonding; however, the C-scans showed the "73%" panel to have a better overall diffusion bondline than the "83%" panel. This was later confirmed by the AU test results.

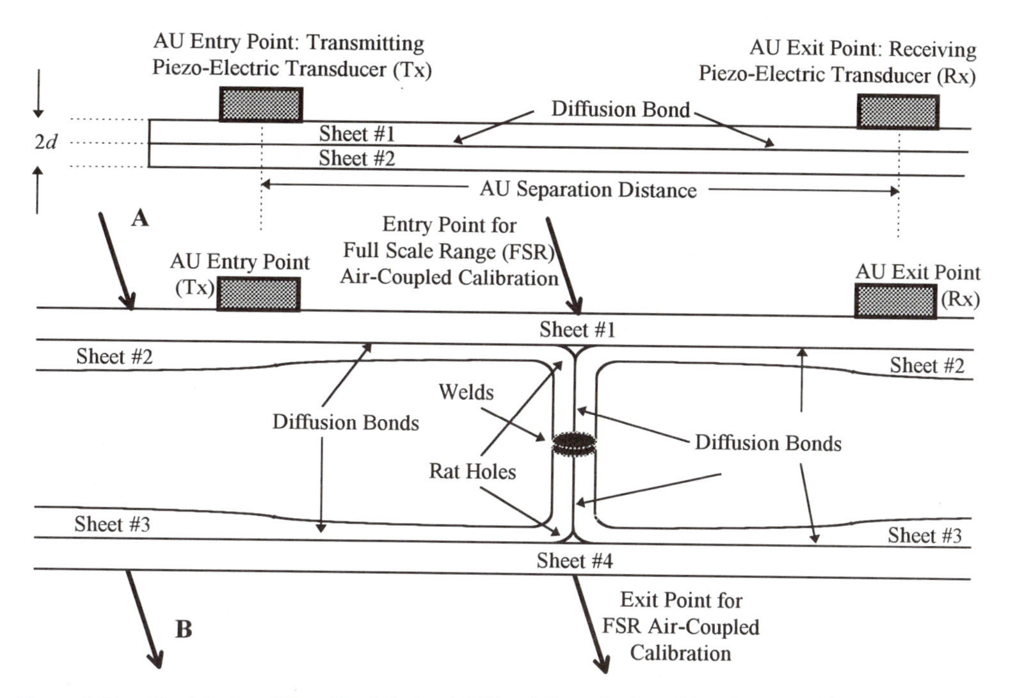

Figure 1. Two-Sheet (top) and Four-Sheet (bottom) Ti Panel Cross-Sections. Two-sheet panel C-scans showed discernible diffusion bond quality, while the four-sheet C-scans could not. Air-coupled ultrasound was used to launch plate-waves in order to detect diffusion bond differences between acceptable and flawed four-sheet material.

Four 9" by 9" sections of four-sheet panels were provided for NDE exploration, with an acceptable and flawed panel provided for two differing cell (web-spacing) geometries. Panel #1104 (acceptable) and #1105 (flawed) were also scanned with the gate set to monitor reflections at the four-sheet 1-2 bondline, illustrated in Figure 1. Differences in the vicinity of the rat-holes showed up clearly in the C-scan; however, bondline quality at the cell centers could not be determined, even though metallurgical specimens indicated disbonds in the #1105 panel. Microstructural results at 200X magnification indicated obvious differences in bondline quality at the cell centers, that were undetectable with immersion pulse-echo ultrasound. Confirmation of diffusion bond quality was finally attained by using air-coupled ultrasound to launch plate waves in a through-transmission configuration. The air-coupled system, operating at 400 kHz, was calibrated at full scale range (FSR) with equivalent entry and exit incidence angles focused at the rat-holes, as shown in Figure 1. The incidence angles of the transmit and receive air-coupled transducers, which must be non-zero to allow shear wave components to penetrate the material, were adjusted to maximize signal amplitude[2]. Once calibrated, both transducers were scanned over the panel. When the entry and exit points are at any other location than the vertical web (i.e. at A-B in Figure 1), the received amplitude is decreased due to attenuation from: entry point "A," propagation through the web, and exit point "B." Plate wave amplitudes received within the cells were significantly lower for panel #1105 than for panel #1104, because of bondline discontinuities parallel to the plate wave propagation direction from point A to point B.

AU Signal Acquisition

Simulated stress waves are generated by placing ultrasonic contact transducers on the same side of either type of panel, as shown in Figure 1. AU does not require two-sided panel access. The transmitting PZT issues a highly damped broad band impulse, which is scattered omnidirectionally throughout the medium, and the collective wave components are received by several remotely located PZTs. Each waveform, corresponding to a given propagation path, is individually analyzed and a go/no go decision is immediately available for subsections of or the entire material region covered by the transducer harness. This approach is being adapted from an AU NDE system[3] currently operational in a manufacturing environment.

Adequate experimentation to determine which waveform attributes act as discriminators for flawed material is a prerequisite to designing part-specific transducer harnesses needed for mass-production NDE. Tables 1 and 2 list the distances and frequencies evaluated for the two and four-sheet panels to determine what is optimum. The same signal analysis approach, based upon adherence to a prescribed spectral shape, was applied to all acquisition situations. AU laboratory instrumentation included an ultrasonic pulser-receiver, operated in through-transmission mode, ported to a digitizer, where signal classification was executed off-line on a PC.

Table 1. Two-Sheet Panel Probability of Correct Classification (%)

83% Bonded Panel vs. 14 % Bonded Panel				
PZT Separation Distance	PZT Frequency 1 MHz	2.25 MHz	5 MHz	10 MHz
2 3/8"			88	
6"	97.9		91.8	
10 1/2"	91.6	96	96.5	91.8
73% Bonded Panel vs. 21 % Bonded Panel				
PZT Separation Distance	PZT Frequency 1 MHz	2.25 MHz	5 MHz	10 MHz
2 3/8"			100	
6"	98		100	100
10 1/2"	100	100	100	91.6

Internal bond geometry of the two-sheet panels is uniform across the part; therefore, acquisitions (25 typically) could be uniformly distributed across the panel. Note that the "73%" panel outperforms the "83%" panel, confirming the prior C-scan results. Great care had to be exercised for acquisitions on the four-sheet panels because the plate wave modes generated were a function of transducer proximity and orientation with respect to the web structure. Signals could be acquired *perpendicular* to and *across* the web, as illustrated by the transmit and receiving PZTs in Figure 1, or both could be *within* a cell, or the PZT orientation could be *parallel* to the webs, as listed in Table 2. Since the thickness of sheets 2 and 3 change with distance from the web, *perpendicular* PZT orientations *across* the webs had the most repeatable propagation geometry, yielding the highest classification statistics in Table 2.

AU Signal Analysis And Classification

Two-sheet panel pulse-echo A-scans could be visually classified based on activity at the 1-2 bondline time gate. In contrast, there is nothing that would immediately indicate material state for time domain AU signals, such as those acquired near the center of the acceptable and flawed two-sheet panels, illustrated in Figure 2.a. However, by converting the entire signal set to the frequency domain, computing the power spectra, and averaging for each category, a subtle downward frequency shift is evident for flawed material, as shown in 2.b, c and d. All 25 power spectra from both material categories are plotted in 2.c and d. A relatively simple signal discrimination approach called *Cross-correlating Power Spectra* was selected to evaluate potential AU classification performance, for varying PZT separation, frequency, and, in the case of the four-sheet panels, PZT orientation and location. The time domain signals are first normalized by removing the means and dividing by the standard deviations, the spectra are computed (shown in 2.c and d), and the dot product (or zero-lag correlation) is taken with the two averaged spectral templates plotted below in 2.b. The template with the highest correlation identifies the material category. By running the entire waveform set through this classifier, the overall probability of correct classification is determined.

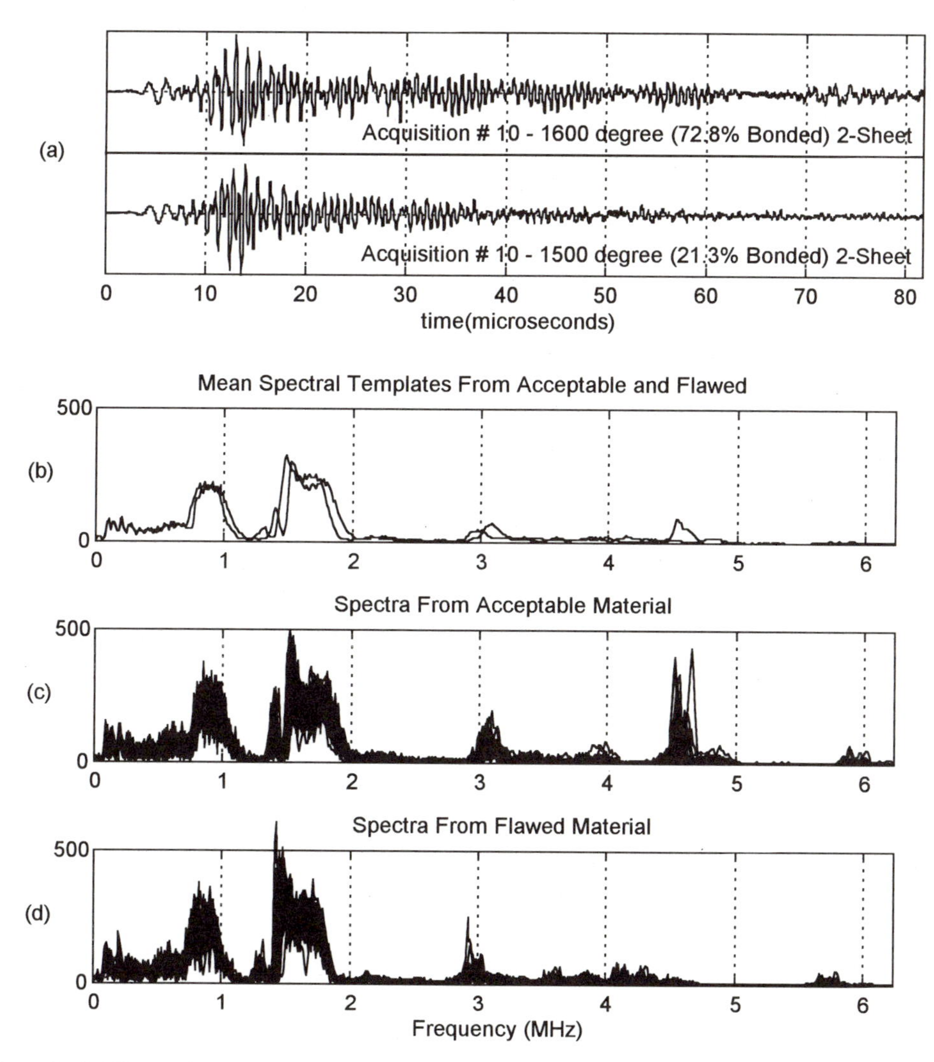

Figure 2. AU Time and Frequency Domain Signals for 5 MHz and a 6" separation. Part (a) Amplitude vs. time from acceptable/flawed two-sheet panels. Part (b) Averaged power spectra. Part (c) illustrates all spectra for the 1600 degree (73% bonded) panel, and Part (d), all spectra for the 1500 degree (21% bonded) panel.

The rationale for selecting this classifier was that poorly bonded panels have more air pockets or subcritical voids situated in the center of the propagation path, which attenuate the higher frequency components. For serious disbonds, this will guide energy into two separate channels, each supporting plate wave propagation compatible with a thickness *d*, rather than 2*d*, as illustrated in Figure 1. The other reason for selecting a weak classifier was to avoid "performance saturation," as exhibited by the 100% classifications for the 73%-21% panels in Table 1. More information on sensitivity to frequency and PZT separation is available from the 83%-14% matrix of the same table. The results in Table 2 confirm that consistent propagation geometry is necessary for AU to be successfully applied, because all signal sets having orientations that were *perpendicular* and *across* the webs resulted in high classification rates. PZT separations for the four-sheet panels were limited by the cell sizes, however, the overall results from both tables point to 5 MHz, as the optimum frequency.

Table 2. Four-Sheet SPF/DB Panel Classification - Probability of Correct Classification (%)

Freq. (MHz)	Panel Type	Distance (inches)	Classification (%)	Description of AU PZT Orientation and Positions.
5	1104 1105	2 3/8"	94.6 to 48.6%	Parallel* orientation and within the vertical webs. Performance varied with buffer length.
10	same	2 3/8"	75% (up to 100% for subsets)	Parallel and within vertical webs. 100% accuracy achieved for identical geometries (waveform types).
10	same	2 3/8"	92.1%	Perpendicular* orientation and across webs.
5	same	2 3/8"	95.3%	Perpendicular and across webs.
1	same	2 3/8"	91.9%	Perpendicular and across webs.
5	1103 1110	2 3/8"	90.6%	Perpendicular orientation and across the webs joining upper and lower face-sheets.
10	same	1 5/16"	85.7%	Perpendicular and across webs.
10	same	1 5/16"	77.8%	Perpendicular and within webs.

* "Parallel" and "Perpendicular" pertain to PZT orientations with respect to the vertical web, shown in Figure 1.

CONCLUSIONS

Approximately 700 waveforms from the two-sheet and 480 waveforms from the four-sheet panels were acquired during this initial investigation, where 650 and 270, respectively, were classified with 90% or greater accuracy. This proved the viability of the AU approach for designing a high throughput, Ti wing-skin inspection system. The next phase will apply AU to more complex geometrical shapes and material systems. Different insonification methods, such as: dry-contact PZTs, squirters, air-coupled and robust Laser-Based-Ultrasound[4] will be tried. Advanced discrimination using joint-Time-Frequency techniques will be developed and tested.

Acknowledgments

The authors gratefully acknowledge support from McDonnell-Douglas Aerospace, Mitchell Laboratories and the National Aeronautics and Space Administration.

REFERENCES

1. Vary, A., "The Acousto-Ultrasonic Approach," in *Acousto-Ultrasonics, Theory and Applications*, J. C. Duke, Jr., ed., Plenum Press, 1988.
2. Grandia, W. A., and C. M. Fortunko, "NDE Applications of Air-Coupled Ultrasonic Transducers," *Proceedings of the IEEE Ultrasonics, Ferroelectrics, and Frequency Control, International Ultrasonics Symposium*, Seattle, Washington, 1995.
3. Gill, Jr., T. J. and A. L. Bartos, "An Acousto-Ultrasonic Platform for the Quality Assessment of Thick Radial Ply Composite Structures," *Nondestructive Characterization of Materials VI*, R. E. Green, Jr., et al., eds., Plenum Press, 1994.
4. Pepper, D.M. , et al., "Double-Pumped Conjugators and Photo-Induced EMF Sensors: Two Novel, High-Bandwidth, Auto-Compensating, Laser-Based Ultrasound Detectors," *Nondestructive Characterization of Materials VII*, A. L. Bartos, et al., eds., Transtec Ltd., 1996

Ultrasonic On-Line Monitoring and Mapping of Low-Temperature Diffusion Bonding of 6061-T6 Aluminum

G. Kohn[1], Y. Greenberg[2], O. Tevet[2], O. Yeheskel[2], Y. Feuerlicht[2], U. Admon[2] and D. Itzhak[3]

[1]Rotem Industries Ltd., Beer-Sheva, Israel.
[2]Nuclear Research Centre-Negev, Beer-Sheva, Israel.
[3]Ben-Gurion University, Beer-Sheva, Israel.

INTRODUCTION

Diffusion bonding is a process in which both elevated temperature and pressure are applied to the interface between two samples until a permanent bond is established[1,2]. Diffusion bonding techniques are frequently used in applications where the high heat input of welding has to be avoided. Since the time required for the formation of a good bond depends on many parameters other than pressure and temperature, such as surface roughness, surface contamination, or the composition of the interlayer[3], it is difficult to know the exact time needed to complete the bonding process[4]. As a result, many trial and error experiments are needed to achieve good bonding parameters. Monitoring of the adhesion process is important in those applications where delicate components are being bonded and minimal heat input and applied pressure are sought.

The present paper describes a diffusion bonding process which was monitored on-line and in real-time using ultrasonic sound waves.

Ultrasonic sound waves, in the pulse-echo mode, were used during the bonding stage to monitor the progression of the adhesion process. A digital oscilloscope and a fast PC were used to detect and analyze the ultrasonic signals reflected from the specimen's walls and from the forming interface. Analysis of the amplitude of the reflected waves showed that bonding started at temperatures as low as 130 C. Off-line, post bonding C-scan was used for mapping bonding defects and misalignment of the system.

EXPERIMENTAL

Coating Procedure

Pairs of 20 mm diameter cylindrical specimens were used in this work. The length of the specimens was either 35 and 10 mm or 35 and 35 mm. The first pair was chosen so that the ultrasonic reflections do not overlap on the oscilloscope screen, while the

Nondestructive Characterization of Material VIII
Edited by Robert E. Green Jr., Plenum Press, New York, 1998

other pairs were used to machine standard tensile specimens after bonding. Prior to coating the surfaces of the bonded specimens were thoroughly cleaned and dried. A thin layer (~10 μm) of silver was coated on the aluminum specimens. The specimens were coated using either a vacuum-deposition sputtering system, or a commercial electrolytic coating process.

The coated specimens were kept in a protected atmosphere chamber during storage to avoid contamination of their surface.

Bonding System

The diffusion bonding system, see Fig. 1, consisted of the following components: A 20 ton press which was used to apply the pressure, a heater, an ultrasonic transducer and a sound velocity measurement system. The transducer was used in the pulse-echo mode to transmit and receive ultrasonic waves to and from the bonding specimens. A set of three K type thermocouples and a digital pressure transducer were used to feed the bonding parameters into a fast PC.

Bonding Process

6061-T6 Aluminum cylinders with parallel and polished faces were coated using either a proprietary multi-layer vacuum deposition, or a commercial electrolytic coating process. Pairs of coated cylinders were then bonded to one another by subjecting them to moderate axial pressure and temperature.

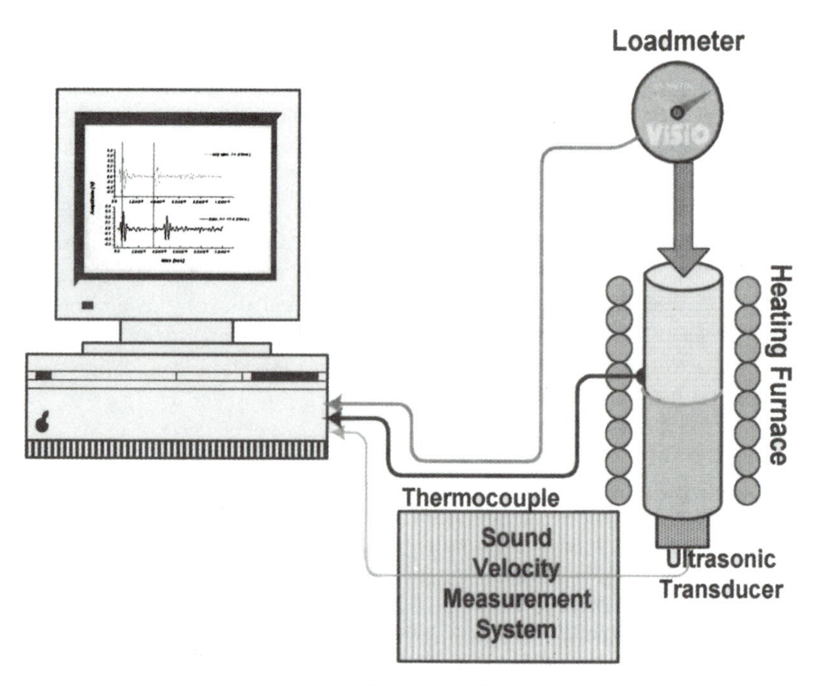

Figure 1. A schematic drawing of the diffusion bonding system.

The specimens were mounted in the bonding fixture and a pre-load of 0.3-1.5 MPa was applied. The specimens were heated at a constant rate up to the bonding pre-set temperature and were held at that temperature till the end of the bonding process. The furnace was then turned off letting the specimens to cool down to room temperature. As a result of heating the pressure on the specimens kept rising due to their thermal expansion. The maximum load on the specimens was kept below the room temperature micro-yield stress to prevent distortion.
The temperature, load readings and ultrasonic sound reflections were recorded at pre-set time-intervals and fed into the PC.

Ultrasound Monitoring System

A wave guide was used to couple between the acoustic transducer and the bonded specimens. The signals from the transducer were fed into a digital scope and to a PC for further analysis.

Figure 2 shows snapshots taken from two specimens at two different stages - one before bonding and the second after completion of the bonding process. It can be seen that the amplitude of the sound wave reflected from the end of the first specimen (bonding interface) is large before bonding, and small when the bond is complete. Upon completion of the bonding process the amplitude of the reflected wave from the far end (right side, Figure 2), is almost equal to the amplitude of the sound wave reflected from the interface between the wave guide and the first specimen. It is thus possible to monitor the progress of the bonding process by observing the gradual change in the ratio between the amplitudes of these reflections.

Ultrasound Mapping Process

Upon completion of the bonding process, a C-Scan comprising of 49 ultrasonic measurements was performed through one face of the bonded specimens, in order to evaluate the quality of the bond. The 49 measurements were carried out in a matrix like form (7 x 7). Bond quality is determined by the ratio between the amplitudes of the sound wave reflected from the far end of the second specimen and the sound wave reflected from the bonding interface. The higher the ratio between the reflections, the better is the bond at that particular spot. A perfect bond provides only a minimal reflection from the bonding interface.

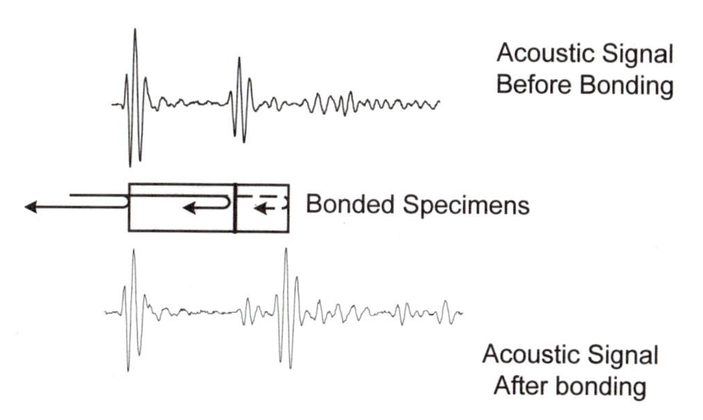

Figure 2: Snapshots of the acoustic signal, before bonding (above), and after bonding (below).

The ultrasonic mapping measurements were performed using a 3mm diameter, 20MHz transducer. The sound wave reflections were analyzed using a program which calculated the ratio between the reflections. The ratios, for each spot, were displayed on the screen and distinguished by different colors. The result was a map containing squares of different colors signifying the bond quality at each specific spot.

RESULTS

Two coated aluminum specimens were pre-loaded with a pressure of 0.3 MPa and subsequently heated to 230 C. Heating rate was 60 C per hour and soak time was 8 hours, the furnace was then turned off and the specimens were cooled down to room temperature.

The pressure on the specimens increased up to 22 MPa as a result of the expansion of the specimens upon heating. The pressure later dropped to the initial pressure when the specimens contracted upon cooling. A sequence of the ultrasonic signals is shown in Figure 3. The progress of the bonding process is estimated by comparing the amplitude of the wave reflected from the interface between the two specimens to that reflected from the far end of the second specimen. (Right-hand side in Fig. 2). At the initial stages ***a*** and ***b*** of the bonding process, the amplitude of the first reflection is much larger than that of the second one (fig. 3 ***a*** and ***b***). When the temperature reached 115 C, stage ***c***, the two amplitudes were almost equal, while at even higher temperatures, stages ***d-f***, the amplitude of the wave reflected from the far end of the bonded specimen, became much larger than that reflected from the bonded interface. The acoustic signals were recorded at 15 min. intervals and subsequently analyzed by calculating the ratio between the amplitudes of the first and second echoes of the ultrasonic waves. From the plot shown in Fig 4 it is evident that the bonding process has progressed rapidly during the

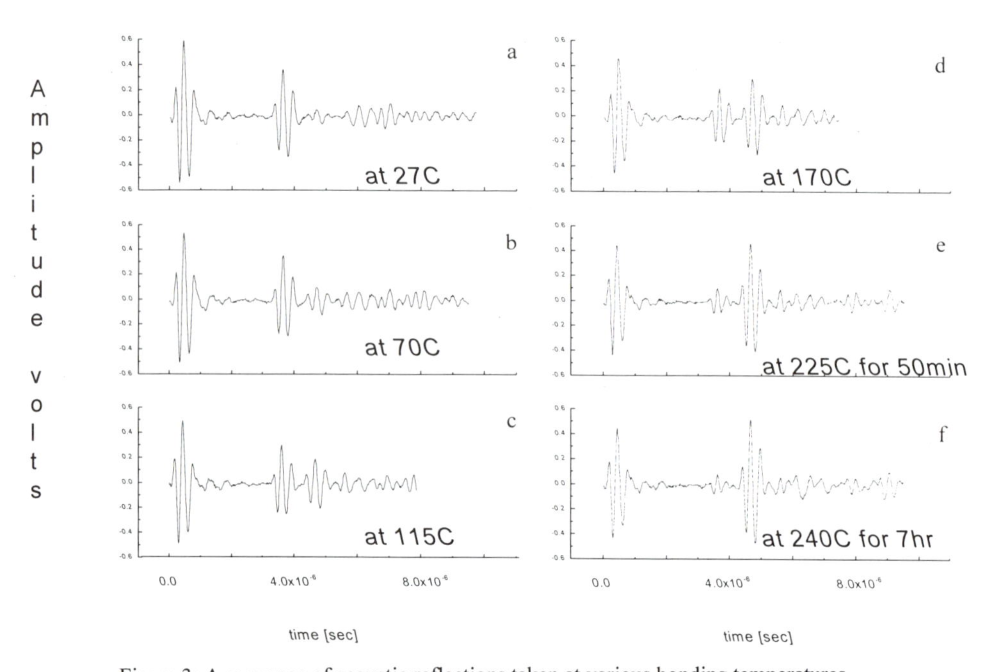

Figure 3: A sequence of acoustic reflections taken at various bonding temperatures.

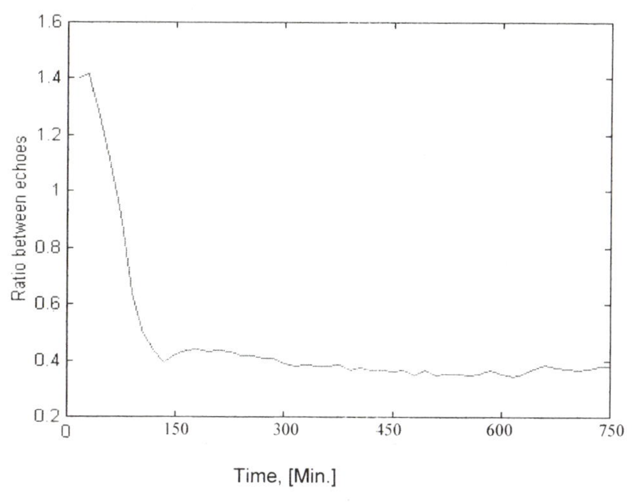

Figure 4: The ratio between the amplitudes of the reflected waves as a function of time (time interval between successive recordings -15 min.).

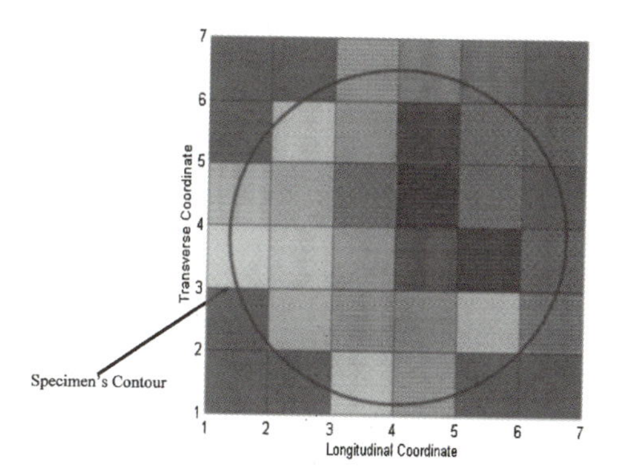

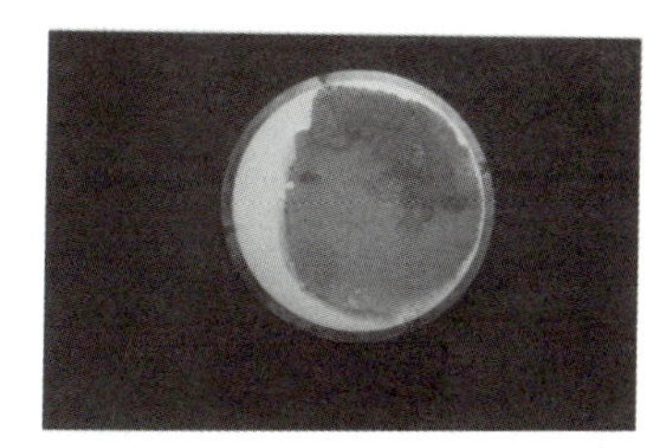

Figure 5a: A map displaying the bonding quality of diffusion bonded aluminum specimens. The darker colors on the upper right side correspond with weaker sound reflections from the bonded interface, indicating a good bond.

Figure 5b: A macrograph of the bonded area after separation. The darker area on the upper right side indicates better bonding between the specimens.

initial ten readings (150 min.) and then leveled off with only a slight decrease in the relative amplitudes up to the 50th reading (750 min.).

To verify this finding, another set of specimens was subjected to a similar bonding process but with a maximum temperature of 140 C.

After bonding the specimens were scanned with an ultrasonic transducer. The result of the scanning is shown in Figure 5a with various gray levels. The specimen's contour is marked as a circle on the map. The difference between the shades of gray indicates different bond quality for the various test areas. The darker the square, the better the bond quality. Figure 5b is a macrograph of the bonded area after separation. It is evident from this picture that there was no bond on the left-hand side of the sample (bright parts of Figures 5a and 5b), while good bonding was achieved on the right-hand side of the specimens (darker parts of Figures 5a and 5b). Bond deficiency was attributed to poor alignment of the specimens which was later corrected with a marked improvement of bond symmetry.

CONCLUSIONS

On-line ultrasound monitoring of diffusion bonding proved to be a viable instrument in the detection of the beginning, the progress and the end of the process. Bond defects could be detected and misalignment of the system could be adjusted using C-scan ultrasound mapping. Future work plans include optimization of both coating and bonding techniques and the development of a high resolution, automatic mapping system for defect analysis.

ACKNOWLEDGMENT

The assistance of H. Ettedgui and M. Ganor in the development of the computer software is highly appreciated.
This work was partially funded by the Inter-University Funding Committee.

REFERENCES

1. W.H. King and W. A. Owaczarski, *Weld. J. Res. Suppl.* **46,** 289 (1967).
2. A. Hill and E. R. Wallach, Acta. Metall. Vol. 37, No. 9, pp. 2425-2437, (1989).
3. M. S. Yeh and T. H. Chuang, Scripta Metallurgica *et* Materialia, Vol. 33, No. 8, pp. 1277-1281, (1995).
4. J. l. Knowles and T. H. Hazlett, *Weld. J. Res. Suppl.* July 1970, pp. 301-s-310-s.

APPLICATION OF ULTRASONIC IMAGE TO THE EVALUATION OF TEMPERATURE DISTRIBUTION IN METAL POWDER COMPACTS DURING SPARK PLASMA ACTIVATED SINTERING

Toshihiko Abe, Hitoshi Hashimoto, Yong-Ho Park, Tae-Young Um and Zheng-Ming Sun

Tohoku National Industrial Research Institute
Nigatake 4-2-1, Sendai 983, JAPAN

INTRODUCTION

Spark Plasma Activated Sintering, SPAS, is a newly developed sintering method for both metal and ceramic powders. An advantage of the SPAS is that a powder can be consolidated in a very short time within several hundreds seconds, because the powder is pressurized in a mold and is heated by direct electric resistance heating of the powder, using pulsative electric current . Another advantage of the SPAS is that the sintering temperature is lower than conventional sintering methods such as hot press and HIP. For instance, a titanium-aluminide powder has to be sintered at 1473K by HIP, while well sintered at 1223K by the SPAS. However, so called " sintering temperature" of the SPAS is usually measured at the mold in place of the powder itself, because direct measurement of the powder's temperature is much difficult from a technical point of view. It is considered that the powder is heated more rapidly than the mold and therefore the heat flows from the powder to the mold during sintering. Consequently, a temperature distribution is generated in the powder and therefore "true sintering temperature" cannot be clear.

In the past ten years, we have applied an ultrasonic imaging technology for material characterization and concluded this technology is very convenient to observe internal structure of metals [1], polymers, polymer composites [2], metal-ceramic composites [3] and many other materials. In particular, we have found the ultrasonic image is highly sensitive to grain size, porosity and microstructure of these materials. Therefore, we applied the ultrasonic imaging technique to evaluate the true sintering temperature and its distribution in a metal powder during sintering by the SPAS because the grain size, porosity and microstructure of the sintered metal compact are strongly affected by the sintering temperature.

BACKGROUND OF RESEARCH

Before applying the ultrasonic imaging technique to the temperature evaluation, we had been suffered from a queer result on the room temperature bending strength of titanium aluminide specimens consolidated by the SPAS. The titanium aluminide powder used for sintering had been synthesized by mechanical alloying process, using an Ar-filled ball mill. We consolidated two groups of specimens, group A and B, by means of two different SPAS machines supplied by two companies A and B, using the same powder and under the same

Nondestructive Characterization of Material VIII
Edited by Robert E. Green Jr., Plenum Press, New York, 1998

sintering condition. However, the three-point bending strength of the group B specimens was about 100MPa higher than that of the group A. According to an optical microscopic observation, crystal grain size was larger in the group B specimens than in the group A . These values are tabulated in Table 1. The strength in Table 1 is an average value of 17 specimens.

Table 1. Three-point bending strength and crystal grain size of specimens of group A and B consolidated by SPAS machines A and B.

Specimen	Sintered temperature	Three-point bending strength	Crystal grain size
Group A	1223K	533MPa	0.3μ m
Group B	1223K	632MPa	0.6μ m
Group A	1323K	797MPa	2μ m
Group B	1323K	893MPa	4μ m

To know a grain size-temperature relationship, a small chip of the consolidated titanium-alminide was encapsuled in an evacuated silica glass tube and heated in a muffle furnace for 300s. As the specimen was heated to higher temperature, the grain grew larger. The grain size of 4 μm appeared after heated above 1560K. Mother particle's boundary disappeared when the crystal grain of TiAl grew over 2μm, after heated above 1400K. The mother particle means the powder particle before sintering. From this result, we can easily suppose; 1) real sintering temperature was higher than the measured value, 1223K, 2) the group B specimens were heated to higher temperature than the group A , and 3)the resultant disappearance of the mother particle boundaries occurred in the group B specimens more intensively than the group A. Thus temperature discrepancy between the different SPAS equipment was recognized,

Subsequently, the temperature distribution in one specimen was investigated. Figures 1 a, b and c show optical micrographs taken at different positions on a radial cross section of titanium alminide specimen of 30mm in diameter, which was sintered at 1323K by the type B equipment. The crystal size is smaller than 1μm at an edge, Figure 1a, increased to about 2μm at an internal area, Figure 1b, and more than 5μm at a central area, Figure 1c. According to the result of the above mentioned reheating experiment, the grain size seen in Figure 1-c should be observed after heated to about 1600K. This result implies the real temperature was nearly 280K higher than the indicated value by thermocouple, or the temperature discrepancy ΔK=280K. N,ow, we need to know the real temperature and its distribution during sintering. We, therefore, tried to apply the ultrasonic imaging technique to detect the rea temperature and its distribution via crystal grain size and microstructural change.

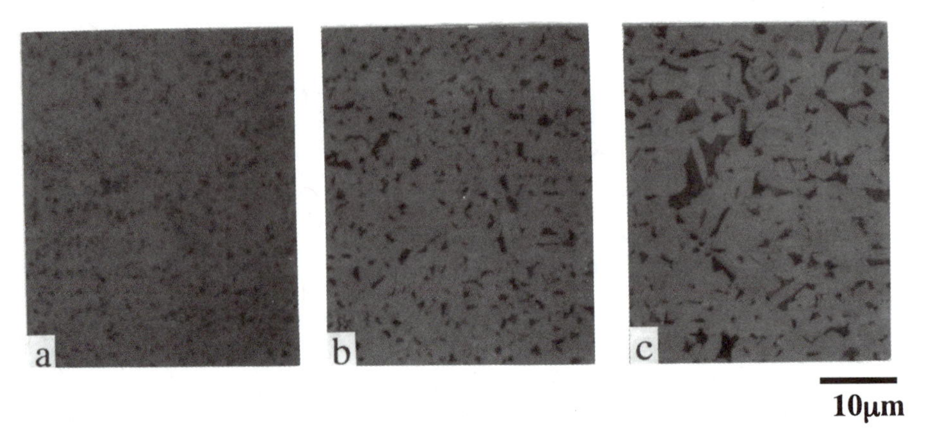

Figure 1. Optical micrographs showing grain size of titanium aluminide sintered by SPAS at 1323K; a: at an edge, b: 4mm apart from an edge, c: central area of the radial cross section of the specimen of 30mm in diameter.

SPECIMENS AND EQUIPMENTS

Two types of sintered compacts, iron and iron-copper, were prepared by the SPAS whose principle is shown in Figure 2. A resource powder is pressed by oil pressure through graphite punches and is directly heated in vacuum by pulsative electric current in a graphite die in which sintering temperature is measured by thermocouple. During the heating process, the pulsative current causes electric discharge between neighbor particles which is considered to generate a micro-welding. Through this process, a sintered compact with near full density is obtained rapidly as compared with either a conventional hot press or HIP. A typical heating rate is 50K/s and a hold time at the sintering temperature is 300s. After sintering was finished, top and bottom sides of the specimen were slightly ground and polished to observe its internal microstructure by the ultrasonic imaging technique . Figure 3 is a schematic diagram of the ultrasonic imaging equipment. A specimen was put in the water which transmits ultrasonic pulses. A point focusing pulse wave type probe with 25MHz was mechanically scanned above the specimen and back reflected echo intensity was converted to brightness of the image. Both spatial and depth resolution of the image by 25MHz is about 0.1mm. Incidental ultrasonic pulses are strongly scattered at defects, heterogeneous structure and large grain boundaries where the back echo image becomes dark and we can easily detect an area where structural change occurred during the sintering process.

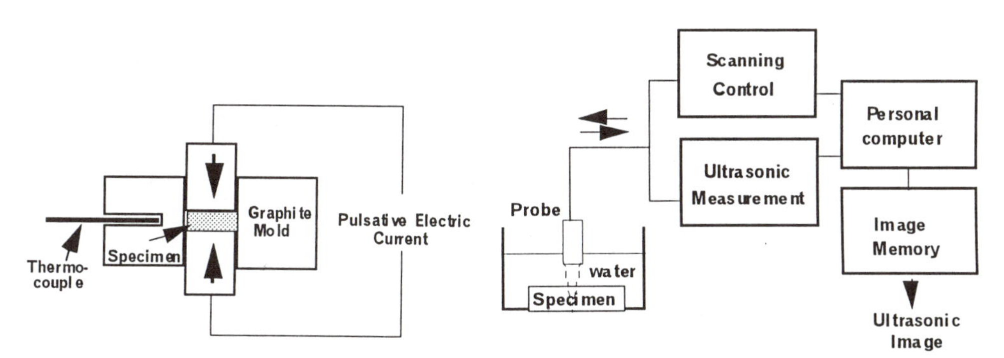

Figure 2. Outline of SPAS method.

Figure 3. Schematic diagram of ultrasonic imaging equipment.

RESULTS AND DISCUSSION

At first, an iron powder of over 37μm in particle size was sintered and observed by the ultrasonic imaging technique. Disc-shaped sintered compacts of 20mm in diameter and 5mm in thickness with near full density were made by sintering at 1053K, 1073K, 1093K and 1123K for 300s. Figures 4a, b, c and d show the back echo image of each specimen. No special structure is recognized in the specimen sintered at 1053K(Figure 4a). A dark area appears in the specimen sintered at 1073K(Figure 4b), which is 20K higher than Figure 4a. In the specimen sintered at 1093K, the dark area expands and becomes darker(Figure 4c). Whole image becomes dark in the specimen sintered at 1123K(Figure 4d). In order to know a mechanism of the dark image formation, the microstructure was observed at points A and B on a longitudinal cross section cut along a line in Figure 4b. Figure 5A is an optical micrograph showing the microstructure at the point A where the ultrasonic image is bright, and 5B is that at the point B where the image is dark. Small dark particles seen in Figure 5A are iron oxide particles which covered the iron powder particle's surface before sintering and therefore show

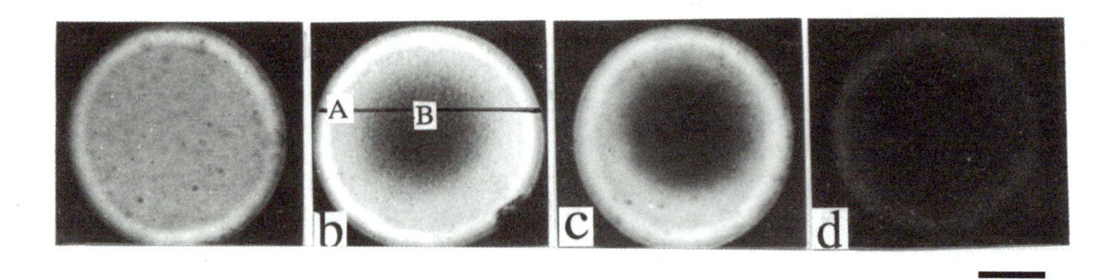

Figure 4. Ultrasonic back echo images of iron compact sintered at; a:1053K, b:1073K, c:1093K and d:1123K for 300s.(25MHz)

the mother particles' boundary. Iron particles which were deformed during pressurized sintering contain many small crystal grains less than 10 μm in diameter. On the other hand, the crystal grains grew more than 30 μm at the central area of the specimen, as seen in Figure 5B. These large crystal grains strongly scattered incident ultrasonic pulses and made the back echo image dark. Although this specimen was sintered at 1073K, it should be noted that pure iron transforms from ferrite(α) to austenite(γ), and therefore rapid grain growth occurs, over 1185K. The temperature of the dark area is, therefore, considered to have been higher than the α-γ transformation temperature, 1185K, and temperature discrepancy, ΔK, between the measured and the real value exceeded 110K in the dark area.

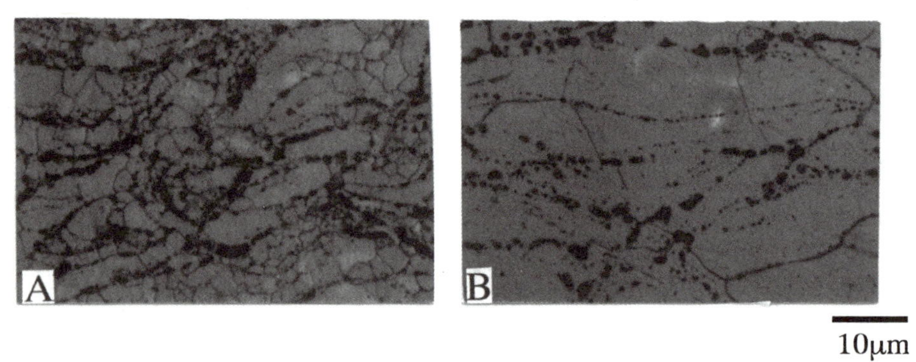

Figure 5. Optical micrographs showing the microstructures at poiunt A and B on a longitudinal cross section cut along a line in Figure 4b.

Next, we considered to apply this method for higher temperature with well known phase separation phenomena, that is, solid iron-copper mixture separates into mixed phase of solid iron and molten copper above 1369K. A well mixed 50 mol.% iron-copper powder was sintered by the SPAS for 300s and then observed by the ultrasonic image. Figure 6a, b, c and d are ultrasonic back echo images of the specimens sintered at 1163K, 1173K, 1193K and 1213K, respectively. The image of each specimen becomes brighter as the sintering temperature rose, in particular, a bright circle appears at a center of the specimens when sintered above 1173K. Ring patterns are recognized at a rim of the bright circle in Figure 6c. These patterns seems to reflect an internal structure change along the longitudinal direction as well as the radial direction of the specimen. According to an optical microscopic observation on the longitudinally cut and polished cross section, the central bright areas in Figures 6a-d were mostly iron single phase which shows that the specimen was heated over 1369K and molten copper was squeezed out through solid iron particles. However, the dark peripheral area was an iron-copper duplex phase which had not been heated to 1369K. Ultrasonic pulses are strongly scattered and attenuated at the iron-copper interface whose acoustic impedance is larger than

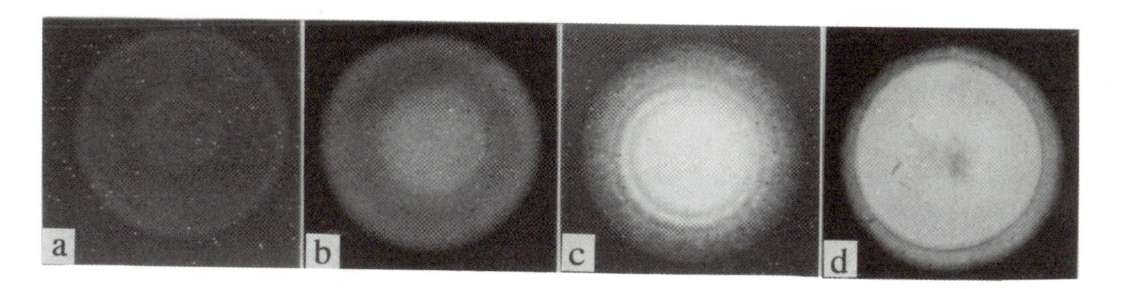

Figure 6. Ultrasonic images of iron-copper mixture compacts sintered at a:1163K, b:1173K, c:1183K and d:1213K for 300s (25MHz).

iron single phase. Therefore, the bright circle images seen in Figures 6a-d correspond to the area which was heated above 1369K during sintering. As the specimen in Figure 6b was sintered at 1173K, the temperature discrepancy, ΔK, reached about 200K at the central bright area.

Specimen shown in Figure 6c showed an ultrasonic ring pattern image. On the longitudinal cross section of the specimen, a parabolically curved boundaries between an iron single phase and an iron-copper mixture phase was observed at peripheral area of the disc-shaped specimen, as shown in Figure 7. From this structure, we can understand that the temperature near the specimen's surfaces which contacted with the upper and lower graphite punches was higher than inside of the specimen, because a contact resistance heating between the graphite punches and the specimen was supposed to be added to usual electric resistance heating. Furthermore, the parabolic boundaries shown in Figure 7 is unsymmetrical along the longitudinal direction. This shows that the longitudinal center of the specimen was not agree with the longitudinally central position of the graphite die where the highest sintering temperature along the longitudinal direction should be attained. From Figures 4 to 7, we can detect a relationship between the temperature distribution and the internal structure along the radial and longitudinal direction from the ultrasonic image.

Finally, we applied this technique to confirm the reproducibility of the sintering temperature by the SPAS method. For this purpose, the iron-copper powder was sintered several times under the same condition, at 1223K for 300s and was observed by the ultrasonic image. The result was contrary to our expectation. Figure 8a and b are two extreme examples of the ultrasonic images. The image in Figure 8a is dark, which shows that the real sintering temperature was lower than 1369K over whole area, or temperature discrepancy ΔK was less

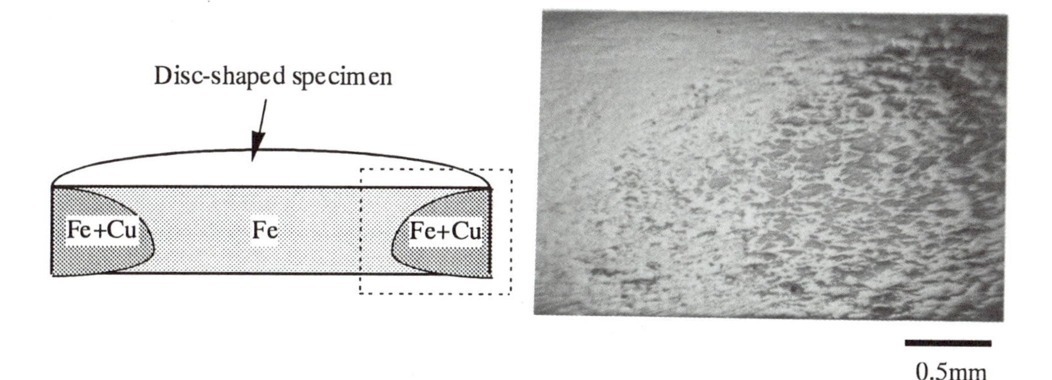

Figure 7. Schematic illustration of the microstructure on a longitudinal cross section of the specimen shown in Figure 6c and an optical micrograph of the area surrounded by dotted lines.

than 150K when sintered at 1223K. In Figure 8b, however, ΔK was larger than 150K at the central area where sharp and fine ring pattern is recognized.

The previously-mentioned values of ΔK in this research were ; larger than 110K at 1073K for iron ; about 200K at 1173K for iron-copper, 280K at 1323K for titanium-aluminide. The ΔK value of 150K at 1223K obtained in Figure 8a is obviously lower than the previous result. This poor reproducibility seems to have been caused by several factors such as the contact condition with the graphite punches, degree of oxidation, electric and thermal conductivities, and thickness of the powder, heating rate and so on. The poor reproducibility of the highest sintering temperature is revealed first time by the ultrasonic images in this research. This problem has a possibility to affect some properties of the products made by the SPAS, and its quality can be improved by controlling the sintering condition so that to make a homogeneous structure with good reproducibility by observing the ultrasonic image.

CONCLUSION

The ultrasonic imaging technique was applied to observe the temperature distribution in the metal powder compacts sintered by the Spark Plasma Activated Sintering, SPAS, method. The two criteria phenomena, α-γ phase transformation of iron at 1185K and the two phase separation of iron-copper at 1369K were used to estimate the real temperature and its distribution. In the iron powder compacts, the grain growth, which means the temperature rose above 1185K, was observed in the central area as the dark ultrasonic image. In the case of iron-copper powder compacts, the iron single phase, which means the temperature rose above 1369K, was observed as the ultrasonic bright image. The discrepancy, ΔK, between the temperature measured by thermocouple in the die and that indicated from the ultrasonic image was ; about 110K at 1073K and 200K at 1173K, but about 150K at 1223K in some specimen.

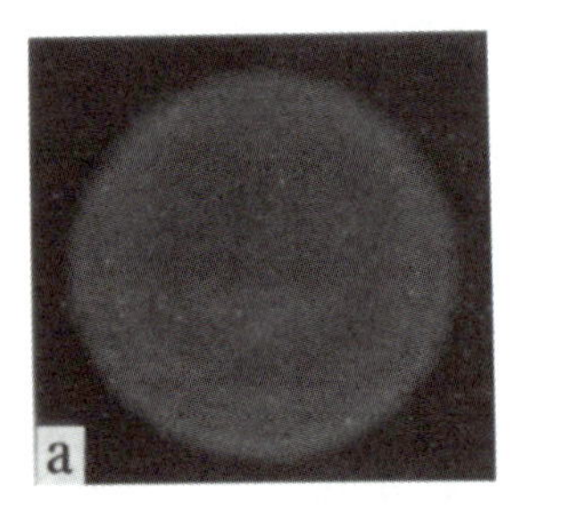

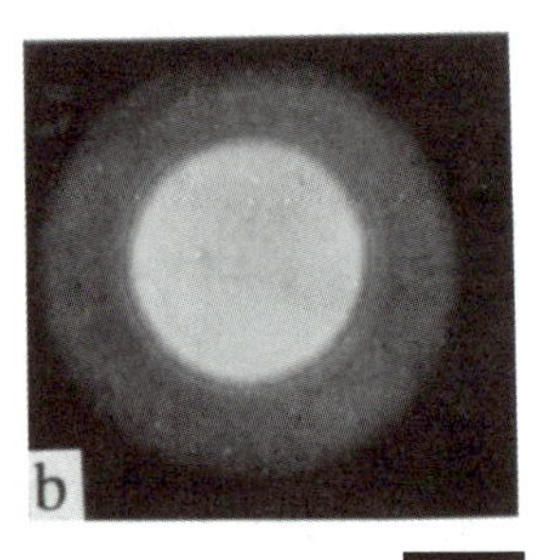

5mm

Figure 8. Two extremes of ultrasonic back echo image showing poor reproducibility of sintering temperature by SPAS. (Iron-copper powder compacts sintered at 1223K)

REFERENCES

1. T. Abe, M, Kawahara, K. Ikawa, Defect detection of cast iron by means of ultrasonic image analysis, Physical Metallurgy of Cast Irons IV. 549 (1989).
2. T.Abe, I. Hanada, T. Kuriyama and I. Narisawa, Application of ultrasonic imaging to warp analysis through flow pattern observation of injected FTRP products, Nondestructive Characterization of Materials VI. 231(1994).
3. T.Abe and S. Sumi, Observation of internal defect in functionally gradient PSZ-Ni by ultrasonic imaging, Characterization of Materials VII. 605(1996).

REACTION-KINETIC MODEL FOR THE CALCULATION OF THE CHEMICAL COMPOSITION OF WELD METALS IN SHIELDING GAS ARC WELDING PROCESSES

N. Meyendorf

Fraunhofer Institut für zerstörungsfreie Prüfverfahren (IZFP)
University, Bldg. 37
66123 Saarbrücken, Germany

INTRODUCTION

In gas metal arc welding processes the arc burns between a melting electrode wire and a weld pool (Figure 1). Shielding gases which are mixtures of inert gases (Ar), oxygen, and carbon dioxide (CO_2) are used to suppress the influence of air and especially nitrogen. Depending on the electric welding parameters (voltage and current) and the composition of the shielding gas two different types of metal transfer processes from the electrode into the weld pool can be distinguished.

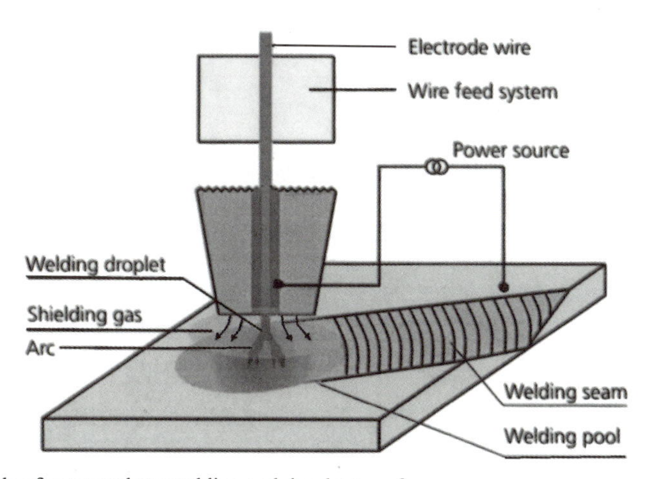

Figure 1. Principle of gas metal arc welding and droplet transfer

For a long arc, the metal droplets fall through the arc. There are no contacts between electrode tip and weld pool. For a short arc, there are a lot of contacts or electrical short circuits between electrode and weld pool. Metal transfer occurs during the short circuits. Due to physical, metallurgical and technological properties of the arc, metal-active gas processes are preferred (mixtures of argon, oxygen and/or carbon dioxide). For these active gases, long arcs need very high currents and voltages. Thus, most of the welding processes are performed in the short arc regime.

Nondestructive Characterization of Material VIII
Edited by Robert E. Green Jr., Plenum Press, New York, 1998

The active gases used are the reason for very intense chemical reactions during the droplet transfer. At the electrode tip the metal is overheated up to the boiling point (Figure 2). This leads to an evaporation of the metal and a reduction of the concentration of desoxidation elements such as silicon and manganese. The resulting chemical composition of the weld metal is important for the microstructure of the weld and the mechanical properties of the welding seam.

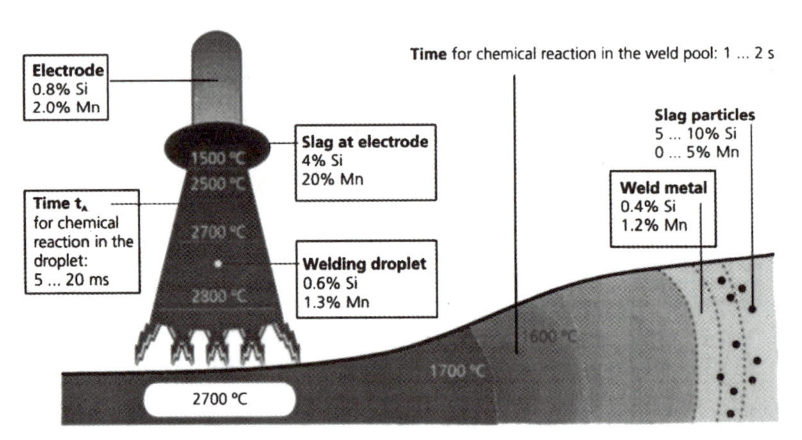

Figure 2. Temperatures and times for metallurgical reactions & concentration of Mn and Si for GMA-welding

In welding technology it is well established that metallurgical reactions can modify the droplet transfer by influencing the properties of the liquid metal, the metal evaporation and the metal circulation in the metal droplet at the electrode tip and in the weld pool. On the other hand, the droplet transfer time is a restrictive factor for metallurgical reactions. Using electric pulses, power sources with different inductivity or wire feed systems which generate mechanical pulses droplet transfer time can be influenced. Thus, it is possible to influence and to control metallurgy by the welding equipment.

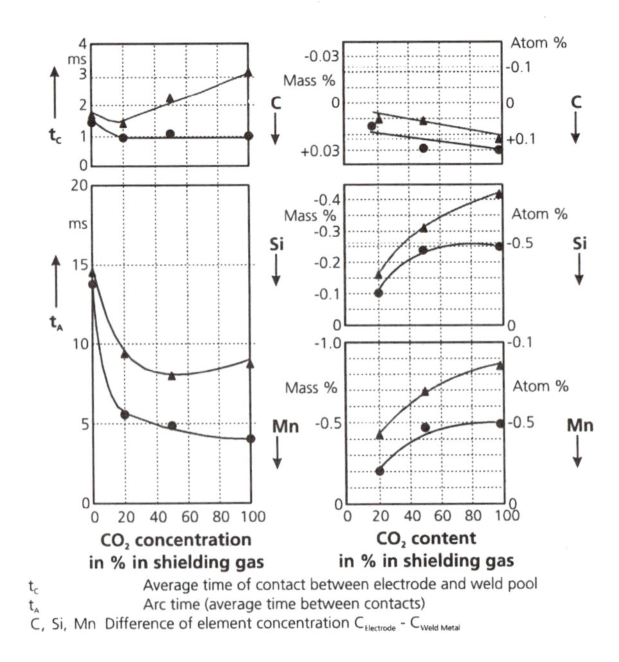

Figure 3. Effect of power source on droplet transfer and welding metallurgy
▲ frequency-converter; ● equalizer

Fig.3 compares welding processes with exactly the same electrode and electrical parameters but with different power sources. The shielding gas composition was varied from pure argon to pure carbon dioxide. The triangles are results for a transistor power source which uses the frequency-converter principle. The full circles are data for a conventional equalizer current source. It is obvious that droplet formation and droplet transfer time is drastically influenced by the parameters of the power source. Thus, we have very strong influences on the metallurgical reactions. If we compare the manganese loss, this is the same effect of 0.5 atom % for the equalizer and the gas with 80 % argon compared to the transistor frequency converter for pure carbon dioxide shielding gas.

MODEL FOR CHEMICAL REACTIONS

For a chemical reaction of the type

$$n_1\,A + n_2\,B \Leftrightarrow n_3\,C + n_4\,D$$

after a sufficient reaction time an equilibrium between the reaction partners A and B and C and D is reached. This equilibrium is described by the law of mass action:

$$(a_C^{n_3} \cdot a_D^{n_4}) / (a_A^{n_1} \cdot a_B^{n_2}) = \exp(-\frac{\Delta G}{RT})$$

a_i is the activity of the element i , ΔG is Gibb's free reaction enthalpy, R the gas constant and T the absolute temperature. Due to the very short reaction times during welding processes and especially during droplet transfer the equilibrium is only reached for fast gas-gas-reactions. Therefore, reaction kinetics must be taken into account which describes the velocity of the reactions. The reaction velocity is defined as the speed of the reduction of concentration c of partners A and B and the enhancement of concentration of current partners C and D. In a closed reaction system an ordinary reaction velocity can be defined:

$$v_R = -\frac{1}{n_1} \cdot \frac{dc_A}{dt} = -\frac{1}{n_2} \cdot \frac{dc_B}{dt} = -\frac{1}{n_3} \cdot \frac{dc_C}{dt} = -\frac{1}{n_4} \cdot \frac{dc_D}{dt}$$

It is assumed that the reaction velocity can be expressed by a product law

$$v_R = -K(t) \cdot c_A^{n_1} \cdot c_B^{n_2}$$

The sum of the exponents ($n_1 + n_2$) is called kinetic order and the coefficient $K(T) = K_0 \exp(-E/RT)$ can be expressed by an Arrhenius law . E is the activation energy for the process.

For complex gas-metal reactions during welding processes its necessary to consider the different steps of the process and to identify the slowest step. For the solution of oxygen in the liquid metal which is a strong exothermic reaction can be assumed that chemisorption is the slowest and thus the kinetic controlling reaction step. Therefore a kinetic order of one should be observed. Different authors have established a linear increase of an oxygen contend in the weld metal with the oxygen partial pressure.[1,2,3] The solution of hydrogen and nitrogen in steel are strong endothermic reactions and thus the transfer of the atoms through the surface should be the kinetic controlling reaction. Therefore the dissolved quantity of gas atoms in the liquid metal should be proportional to the squareroot of the partial pressure (following Sievert´s law). This was confirmed by Savoga et. al. and Kobayashi et. al. [4,5].

EXPERIMENTS

Welding experiments were carried out by varying the following parameters:

- shielding gas composition (Ar, CO_2, O_2 mixtures)
- electric parameters (mean) current I, (mean) voltage U
- length of free electrode (I^2RT-effect)
- electrode diameter and chemical composition of electrode material
- welding speed
- type of power source
- wire feed system (continuous, pulses with various frequencies)

For each welding process seven layer build-up welds were produced to get pure weld metal and to eliminate the influence of base metal. Chemical analyses were carried out from the electrode material, from the weld material and also from droplets at the electrode tip after switching off the current and from splashes which are usually emitted during droplet transition. The most important factors for a kinetic model are the reaction times. These are the times for droplet formation and droplet transfer and the mean time for the existence of liquid material in the weld pool. Droplet transfer times (t_A) were determined by high speed films, by oscillographic methods and by counting techniques. The mean life-time of the liquid metal in the weld pool was calculated from the welding velocity and the length of the weld pool which was determined by a thermographic system.

Weld pool temperatures and filler metal temperatures were measured by thermographic and pyrometric methods. The temperature at the electrode tip was found to be near to the boiling point of liquid steel, the temperature at the center of the droplet was 50 °C to 100 °C lower than the maximum temperature. The mean temperature of the weld pool ranges from 1600 °C to 1700 °C.

The chemical analyses of droplets and splashes lead to the result that nearly 100 % of the manganese loss is due to the droplet reactions but only 50 % of the silicon loss. Fig. 4 summarizes the results of all experiments. Each welding process represents one data point in the diagrams. It was found that shielding gas composition and the time for droplet formation at the electrode tip (t_A) are the most important parameters determining the chemical reactions.

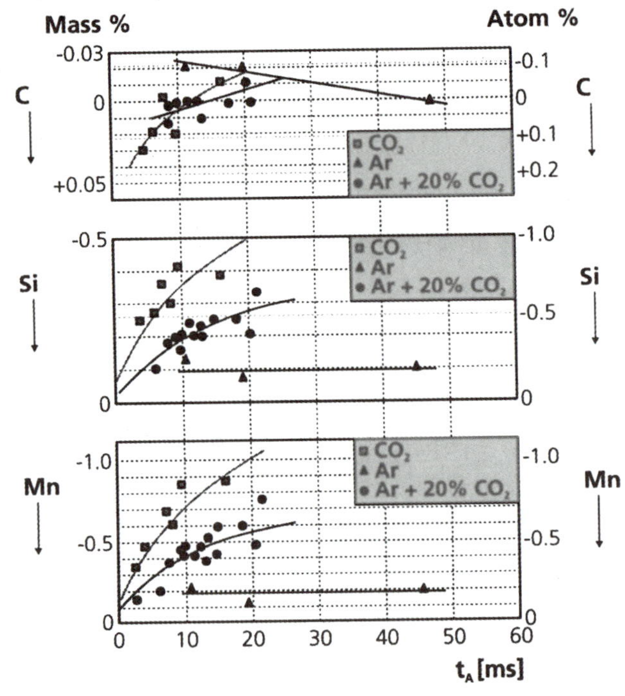

Figure 4. Difference of concentration of elements due to metallurgical reactions

DISCUSSION OF THE RESULTS

By analyzing experimental data and discussing additional results of the literature[6] the author concludes that the following processes are dominant for the reaction of oxygen with the weld metal in the droplet state:

1. CO_2 decomposes into CO and O_2. After physisorption and chemisorption oxygen enters the liquid metal.
2. For carbon dioxide at the metal surface a modified „Boudouard" reaction is preferred. With one carbon atom from the melt two CO-molecules are formed.
3. Metal evaporation from the surface is stimulated by the formation of fugitive oxides so that the loss of alloying elements is stimulated by the oxygen content in the material and at the surface.

Based on the described model above, the element loss in the weld metal is mainly due to the oxygen reactions at the metal surface. Thus for the reaction velocity the following formula can be assumed:

$$\frac{dc}{dt} = K_0 \cdot P_{O2}{}^{X} \exp\left(-\frac{E}{RT}\right)$$

(P_{O2}: oxygen partial pressure, x: kinetic order, E: activation energy, R: gas constant, T: absolute temperature)

It has to be considered that during the formation of droplets the surface-to-volume ratio varies. Calculations showed that a rough approximation is:

$$\frac{\text{surface}}{\text{volume}} \cong t_A^{-\frac{1}{2}}$$

Furthermore it was considered that during the formation of the metal droplet at the electrode tip new metal is molten continuously and mixed with the liquid droplet. The resulting equation for the manganese loss ΔMn was

$$\Delta Mn = -K_{Mn} \cdot t^{\frac{1}{2}} P_{O2}{}^{X} \exp\left(-\frac{E}{RT}\right)$$

The droplet temperature T was found to be about 2700 °C, the partial pressure P_{O2} at the metal surface was calculated by the dissociation reaction for the surface temperature of the droplet. In the above formula the kinetic coefficient K_{Mn}, the kinetic order x and the activation energy E have to be determined experimentally. Fig. 5 shows the logarithmic plot of the quantity $-(\Delta Mn)/t_A^{0,5}$ versus the partial pressure of oxygen at the surface. The slope for the lines in the plot leads to a kinetic order around 3/4. This is lower than the kinetic order of one which was assumed for the oxygen reaction. This can be explained by the reduction of oxygen partial pressure due to an increasing fraction of metal vapor for active gases compared to inert gases.

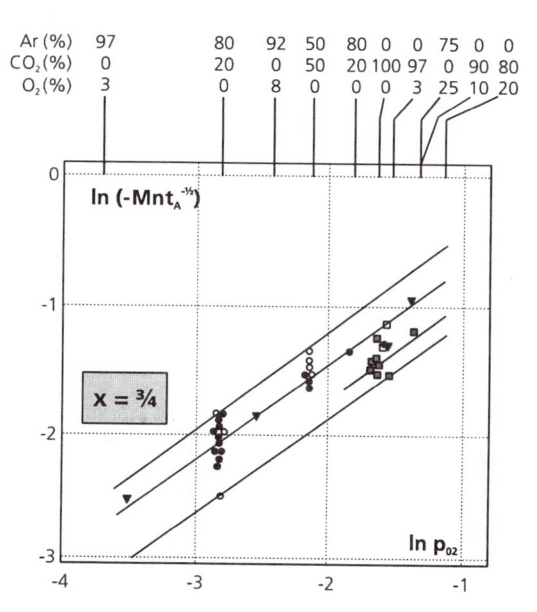

Figure 5. Plot to determine kinetic order

By fixing the kinetic order to 3/4 in fig. 6 the Arrhenius plots for two electrodes with different manganese content can be seen. In both cases the activation energy (E) for the manganese loss seems to be the same, but the reaction coefficient K_{Mn} differs. This is due to a different concentration of the other elements which can also react with the oxygen. It was found that the variation of the coefficient K_{Mn} was not so large, so that an average value can be assumed. From these results a rough formula for the manganese loss can be established.

$$\Delta Mn \approx -\sqrt{t_A} Po_2^{3/4} \quad (t_A \text{ in ms})$$

By similar analyses we also have got results for the silicon loss and the carbon reaction. This formula summarizes the results of all experiments carried out.

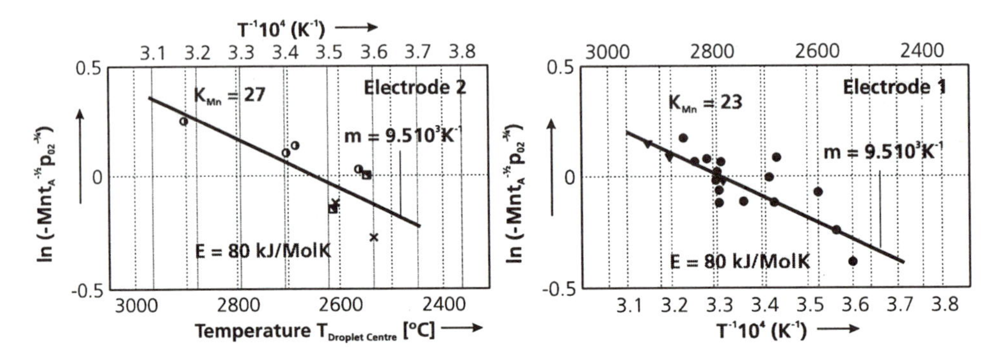

Figure 6. Arrhenius-Plots

SUMMARY AND OUTLOOK

1. Metallurgical reactions during arc welding are mainly influenced by the reaction partners (shielding gas, weld metal) and the reaction time (droplet transfer time).
2. The dominant reactions already appear during the droplet state at the electrode tip and can be described by reaction kinetic models.
3. Metallurgical reactions can be influenced by the welding equipment (power source, wire feed system) by varying the droplet transfer time.
4. Metallurgical reactions can be controlled by the droplet transfer rate. A simple way to measure this nondestructively is by analyzing the time dependence of welding current and welding voltage or fluctuation in the arc intensity.
5. The measurement of droplet transfer rate should be an additional parameter for quality assurance of arc welding processes. This comprises sensors, process observation and nondestructive testing.

REFERENCES

1. L.J. Allum, MIG-welding - time for a review, *Metal Constr.* 15:347 (1983).
2. O. Grong and N. Christensen, Factors controlling MIG-welding metallurgy, *Scandinavian J. of Metallurgy* 12:115 (1983).
3. T. Kuwana and Y. Sato, Abruption of oxygen by iron weld metal in Ar-O_2-mixed shielding gas atmosphere, *J. of Japan. Welding Soc.* 52:292 (1983).
4. W.F. Savaga, E.F. Nippes and E.T. Nasa, Hydrogen-induced cracks in HY-130 welds, *Welding J.* 61:233 (1982).
5. T. Kobayashi, T. Kojima and Ochiaik, Dynamic analysis of chemical reactions in the weld arc, *DVS-Berichte* 43:1 (1976).
6. N. Meyendorf. *Physikalisch-chemische Vorgänge beim Werkstoffübergang von Schutzgasschweißprozessen.* Dissertation B, Magdeburg (1986).

THE CHARACTERIZATION OF ROLLED SHEET ALLOY USING INFRARED MICROSCOPY

Michael L. Watkins

Research and Development
Philip Morris U.S.A.
Richmond, VA 23261-6583

The College of William and Mary
Department of Applied Science
Williamsburg VA, 23187-8795

INTRODUCTION

Iron aluminides are a group of intermetallic alloys containing primarily iron and aluminum and have a resistance to high temperature oxidation.[1] The focus of this work is nondestructive evaluation of thin (203 μm) rolled iron aluminide alloy using thermal microscopy techniques for quality control during manufacturing.

The initial step is the formation of a green sheet composite which is approximately 700 μm thick using either cold-rolling or tape casting of mixtures of the powdered starting material with a binding agent and solvent. The green sheet then undergoes a series of process steps (binder elimination, densification, sintering, annealing) to form the final FeAl sheet product. This final sheet product will be referred to simply as "sheet" to distinguish it from green sheet and intermediate products. Intermediate products are between 300 μm and 200 μm thick with densities as low as 60% of the theoretical density (about 6.0 g/cc for the fully dense material). Mass distribution and thickness are critical attributes throughout the sheet forming process. The relatively high viscosity of the green sheet composite and the characteristically low malleability of the iron aluminide particles inhibit modification of the mass distribution once the green sheet is formed. The impact of non-uniform mass distribution can vary from the formation of final sheet material that contains flaws to an intermediate sheet product which fails during processing. Specimens were fabricated using iron aluminide powder as the primary starting material.

EXPERIMENTAL APPROACH

Because the material of interest is thin, flaws with relatively small absolute size can constitute a significant fraction of the thickness of the material. This can compromise the integrity of the sheet and structures formed from it. Thus, it is desirable to evaluate the

Nondestructive Characterization of Material VIII
Edited by Robert E. Green Jr., Plenum Press, New York, 1998

material with high spatial resolution. A Thermal Analysis Microscope (TAM)[2] was used to investigate several alloy specimens. Characterization is accomplished by imaging the thermal response of materials subjected to a thermal stimulus. The advantage of the TAM is that it provides a unique combination of high magnification, data collection capabilities, and temperature sensitivity. Specimen heating may be induced by applying a current to electrically conductive samples (i.e., joule heating), external heating using a laser, flash lamp, or other means. The nature of the thermal excitation (i.e., duration and spatial distribution over the sample) in combination with the material properties (i.e., thermal conductivity and specific heat), sample homogeneity, and surface features, often results in a unique radiance field or thermal fingerprint. In some cases, the features of interest can be investigated using a steady state stimulus. In others cases, collecting data during the thermal transient is the most effective approach.

The TAM has three channels. The infrared channel consists of a diffraction-limited optical train, filter wheel, and an Amber AE4128 focal plan array camera. This camera nominally operates in the 3 to 5 μm band (InSb detector). Features in the infrared can be spatially resolved down to approximately 12.5 μm. The effective field of view is about 1.6 mm x 1.6 mm. Full-field, time-resolved radiometric data can be collected at frame rates up to 217 frames per second. The six position filter wheel permits wavelength selection over the spectral sensitivity of the camera. An optical channel provides a view of the sample surface via a black and white CCD camera. This allows thermal features to be correlated with physical landmarks. The optical design provides a long working distance (200 mm) which reduces the impact of optics heating from specimen radiation and allows small regions of irregularly shaped specimens to be easily observed. The third channel allows for localized heating of the sample surface using an argon-ion laser. The minimum laser spot size is approximately 50 μm and the power can be adjusted up to 5W. The microscope also features an external trigger output which can be used for the synchronization of data acquisition with thermal and/or mechanical specimen excitation.

RESULTS FOR IRON ALLUMINIDE SPECIMENS

In order to investigate the TAM's sensitivity, several sheet surface features were observed and investigated. These were pits, cracks, contaminants, inhomogeneity, and oxidation. All of these resulted in an observed variation in sheet radiance upon thermal stimulation of the specimen. Characteristic edge cracking results from the high stresses applied at the end of the rollers. Cracks in the bulk of the sheet were also observed but at a much lower incidence. These cracks provide excellent targets for infrared microscopy since they behave as good radiators, relative to the surrounding sheet material. Another feature was sheet inhomogeneity, which can result from inadequate mixing of the starting powders or contamination.

Figure 1 shows surface SEM micrographs of a flawed region of iron aluminide sheet. Full density was not achieved in this case. Cross-section microscopy revealed small voids throughout the volume. These can be seen as surface pits (10 μm or less) over the entire surface. In addition, separation may have occurred between the iron and aluminum species. Several large regions (greater than 50 μm) displayed as light colored centers with broad dark borders can be readily seen. The figure also shows a dark gray ring approximately 700 μm in diameter. X-ray analysis found the lighter regions to be aluminum and oxygen rich (probably aluminum oxides). The darker regions are associated with surface cracks and depressions.

Figure 2 is an infrared image of the same region obtained with the TAM. The image was obtained while heating the sample with a hot air blower. The main features seen in the SEM are readily picked up in the infrared image. This is probably due to a combination of two effects. First, the emissivity of aluminum oxides (in this temperature range) is typically higher than 0.7 while the emissivity of high iron containing metals is less than about 0.5.[3] Secondly, surface roughness also influences emissivity. The peaks and valleys of the oxide regions as well as the dimensions of the pits are on the order of the infrared wavelength. These regions may have emissivities greater than the surrounding areas and thus higher radiance.

Previous studies have demonstrated the ability to detect cracks using thermography.[4,5,6] In the present work, crack detection was investigated using two techniques. The first relies on the high in-plane thermal impedance of the crack. A thermal gradient was generated by applying a laser pulse near one side of the crack. The resulting image is shown in Figure 3. The crack is highlighted as a discontinuity in the

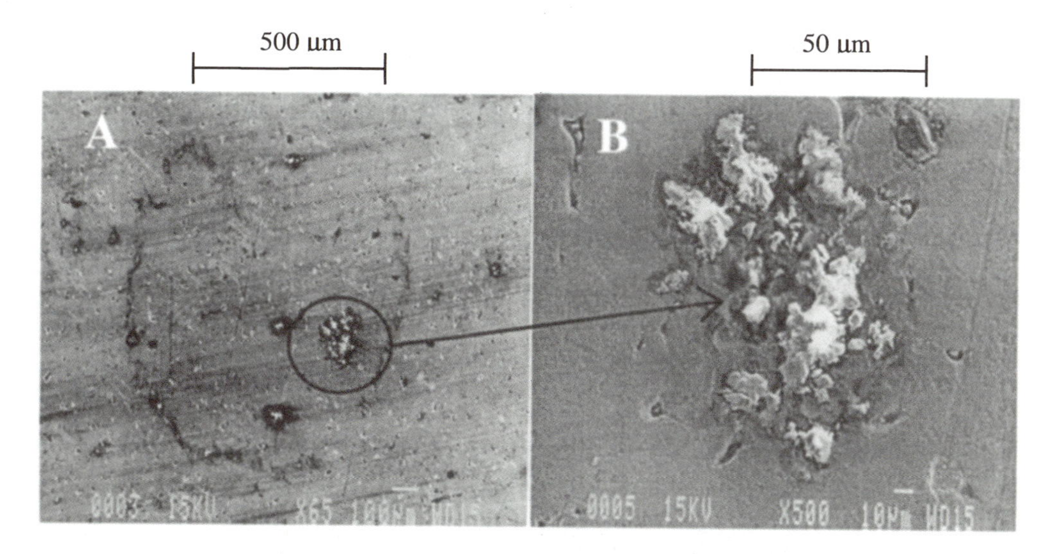

Figure 1. SEM of defect on the surface of an iron aluminide sheet. (A) shows surface pits, a ring, and aluminum rich regions. (B) is a magnified view of the central flaw, which consists primarily of aluminum oxide.

thermal gradient normal to the length of the crack. Optimization of the heating mode and data analysis techniques significantly enhances the signal to noise ratio for crack detection.[4]

The second method uses the high depth to width of the crack geometry to exploit it as a radiating cavity. In figure 4, the specimen was heated using a hot air gun. The crack was easily detected since the effective emissivity of the cavity resulted in a higher radiance or apparent temperature relative to the surrounding material. Figure 5, an SEM of the crack tip, shows the tip to be less than 2 µm in width. While this is about six times smaller than the resolvable pixel size (12.5 µm), it is clear that the radiance from the crack is capable of generating signals significantly higher than the background signal.

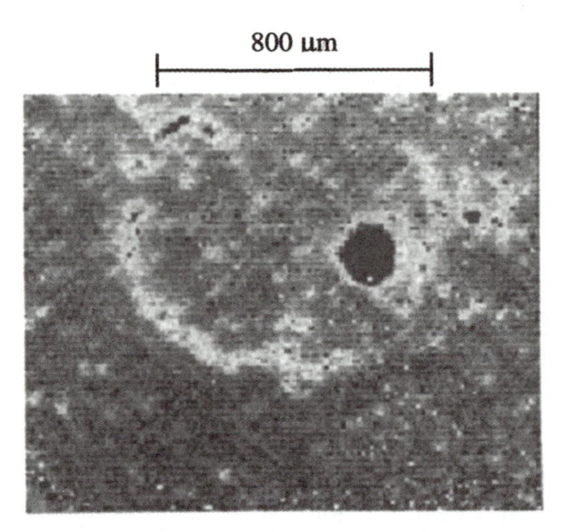

Figure 2. Infrared radiance micrograph generated by heating the defect region illustrated in figure 1 (A). Both aluminum oxide regions and pits/cracks exhibit higher radiance than the surrounding material.

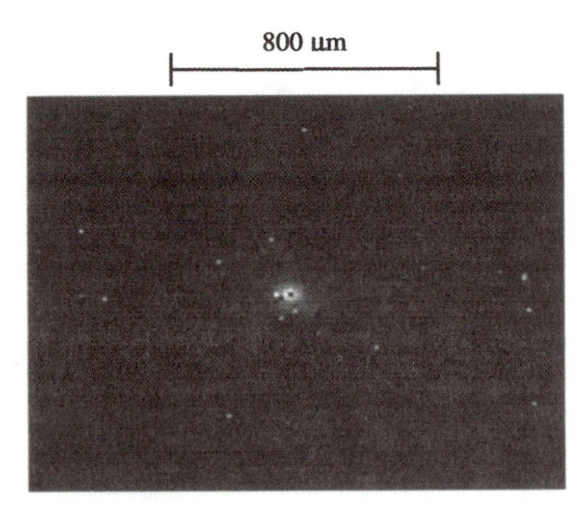

Figure 3. Infrared radiance micrograph of crack heated with laser spot. The in-plane discontinuity of conduction across the crack distorts the normal symmetry (in the absence of a crack) of the radiance field.

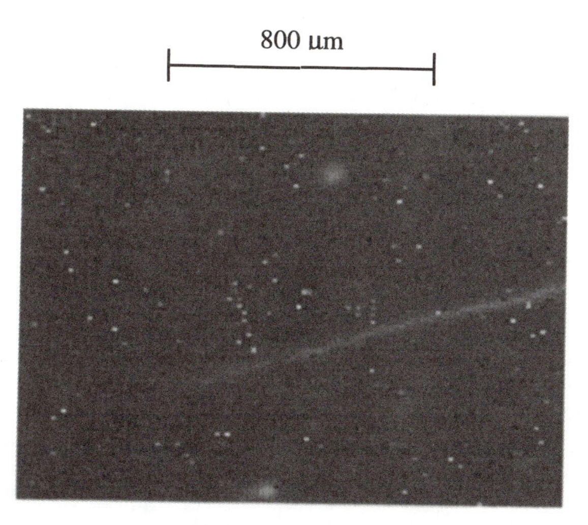

Figure 4. Infrared radiance micrograph of the region containing the crack in figure 3. The entire field of view was heated and allowed to equilibrate. The crack serves as a good radiating cavity with an effective emissivity much higher than the surrounding material.

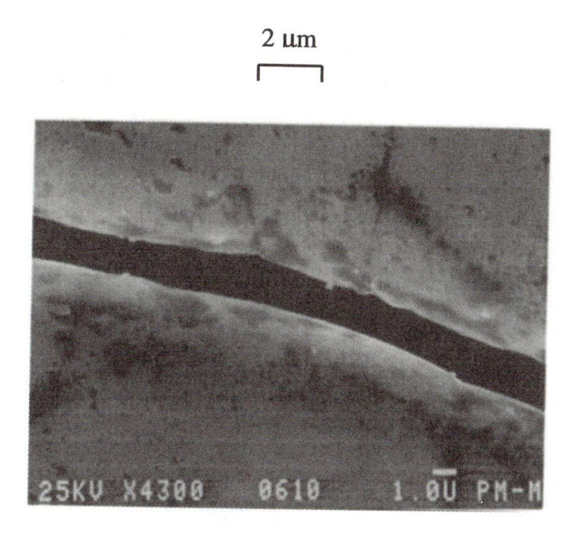

Figure 5. SEM of crack tip detected in Figure 4. The crack is less than 2 μm in width.

CONCLUSIONS

The detection of defects important to the quality of thin metallic sheet has been demonstrated using thermal microscopy. This technique can detect flaws that are small fractions of the iron aluminide sheet thickness. A feature depicted within two pixels (25 μm) is approximately equivalent to 15% of the material thickness. Features of this size are resolvable, and crack widths that are an order of magnitude smaller are detectable. Thermal microscopy provides several advantages in the evaluation of materials and structures. First, it is non-contact and requires no special specimen preparation or vacuum environment (as does SEM). In addition, thermal excitations can be used at energies low enough to preserve the state of the sample (i. e., nondestructive). Full field radiance data can be collected rapidly, correlated with physical features, and have a high potential use in quantitative evaluations.

While the features presented in this paper are surface flaws, it is also of interest to detect subsurface defects in thin metallic samples. Time-resolved infrared radiometry methods have been developed for thicker specimens.[7] It is anticipated that the through-plane thermal transient times will be very short for these samples (on the order of 10 ms). This, in combination with the nature of the flaws, will dictate the heating regime, data sampling rates, and data reduction methods necessary to acquire the desired detection capability.

ACKNOWLEDGMENTS

The author would like to thank Tom Hill and John Campbell for their assistance, Don Miser for providing SEM micrographs and elemental analysis, and the staff at Quest Integrated, Inc. for their enthusiastic partnership in the development of the instrument. I would also like to thank Professor Mark Hinders of the College of William and Mary for his helpful comments in preparing this manuscript.

REFERENCES

1. J.H. Schneibel and M.A. Crimp. *Processing, Properties and Applications of Iron Aluminides,* The Minerals, Metals & Materials Society, Warrendael PA (1994).
2. G. White et. al., Calibration issues affecting the operation of infrared microscopes over large temperature ranges, The International Society for Optical Engineering Vol. 2766, Thermosense XVIII, Bellingham, Washington (1996).
3. W.M. Rohsenow et. al. *Handbook of Heat Transfer Fundamentals,* McGraw Hill Book Company, New York (1985).
4. K.S. Stanley and P. O. Moore, P.McIntire. *Nondestructive Testing Handbook, Vol. 9,* American Society for Nondestructive Testing, Inc. Columbus, OH (1995).
5. K. E. Cramer et.al., Thermographic imaging of cracks in thin metal sheets, The International Society for Optical Engineering Vol. 1682 Thermosense XIV, Bellingham, Washington (1992).
6. G. White et. al., Crack detection using thermal microscopy, Materials Evaluation Vol. 53 (1995), American Society for Nondestructive Testing, Inc. Columbus, OH (1995).
7. W.M. Spicer and R. Osiander, Time-Dependent Temperature Distributions for Nondestructive Probing of Material Properties, Johns Hopkins APL Technical Digest, Vol. 16, Number 3, The Johns Hopkins University Applied Physics Laboratory, Laurel, Maryland (1995).

DETERMINATION OF MECHANICAL PROPERTIES OF STEEL SHEET BY ELECTROMAGNETIC TECHNIQUES

D. Stegemann, W. Reimche, K.L. Feiste, B. Heutling
Institute of Nuclear Engineering and Nondestructive Testing
University of Hannover, Germany
Tel.: ++49-511-7629321, Fax: ++49-511-7629353

ABSTRACT

In the domain of sheet metal production and manufacturing the harmonic analysis of eddy current signals is an appropriate nondestructive test method for process integrated determination of mechanical and technological material properties. This indirect magneto-inductive test method is based on the relation between mechanical and magnetic properties of ferromagnetic materials. It presents the possibility to create quantitative correlations between measured harmonics which depend on the material's magnetic properties and mechanical or technological material properties. This paper presents the measuring principle as well as the results of nondestructive determination of mechanical properties (tensile strength, yield stress) of sheet steels.
The main features of the developed test equipment for harmonic analysis are a high measuring velocity and a high measuring accuracy which is comparable to destructive test methods.

INTRODUCTION

Increasing quality requirements and low cost production of technical products demand additional quality assurance activities during the production of semifinished goods. Therefore, the importance of process-integrated non-destructive defect inspection which additionally offers the possibility of non-destructive material characterization is going to increase in the near future.
Important technological properties of cold rolled sheet steel result from the chemical composition, the crystalline structure, and the crystallographic texture of the material and are then strongly influenced by the subsequent forming methods and thermal treatment. The proven connections of technological properties and measured physical quantities are therefore the starting point for the development of nondestructive measuring systems (fig. 1).

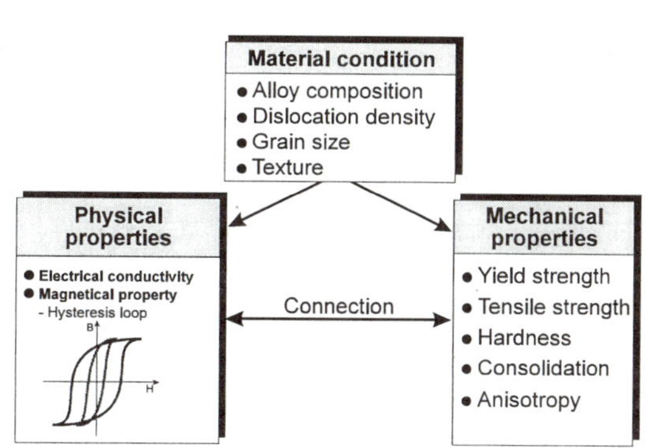

Figure 1: Indirect determination of mechanical-technological material properties

PRINCIPLE OF MEASUREMENT

A suitable nondestructive testing principle for the characterization of sheet steel is the harmonic analysis of eddy current signals [1, 2]. This indirect system determines measurable electromagnetic variables of sheet steel which are closely connected to the technological characteristics of the ferromagnetic material. The latter are then computed based on known connections between the electromagnetic and mechanical-technological properties of the material (fig. 1, 2, 3).

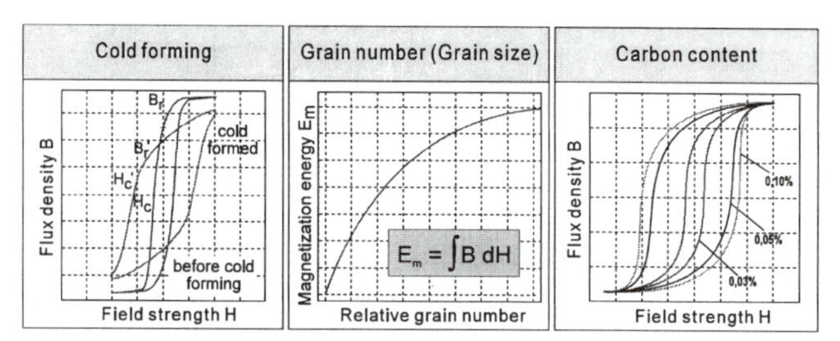

Figure 2: Factors which influence magnetic properties and hysteresis loop

In ferromagnetic materials the magentic domains are forced in the field direction, i.e. the axis of the magnetic moments are brought into line with the exciting magnetic field /1/. The physical relationship between the Weiss' spheres, the blochwalls and the exciting magnetic field can be described by the hysteresis loop:

$$B = \mu_0 \mu_r H$$

with:
- B = fluxdensity
- H = fieldstrength
- μ_0 = magnetic Field constant ($1{,}257 \cdot 10^{-6}$ Vs/Am)
- μ_r = relative permeability

The shape of a hysteresis loop is caused by different effects e.g. reversible and irreversible blochwall shiftings as well as rotations of molecular magnets out of their preferred crystallographic directions. In principle, the blochwalls only need low energies to shift through an ideal crystal lattice but in real crystal lattices of technical materials they are disturbed by dislocations, e.g. grain boundaries, displacements, inclusions, dispersions or internal elastic stresses.

MECHANICAL-TECHNOLOGICAL MATERIAL PROPERTIES

In the production of cold rolled steel sheets there are high quality standards for the cold-workability and the deep-drawing quality. A good cold-workability is generally described by means of mechanical properties, e.g. tensile strength, yield strength, and ductile yield, obtained from tensile tests. A good deep-drawing quality requires high values for the vertical anisotropy, which describes the resistance against unwanted reduction of thickness during the process of deep-drawing, and low values of the planar anisotropy characterizing the directional flow behaviour within the plane of the sheet. The strain-hardening which is proportional to the strain is described by the consolidation index.

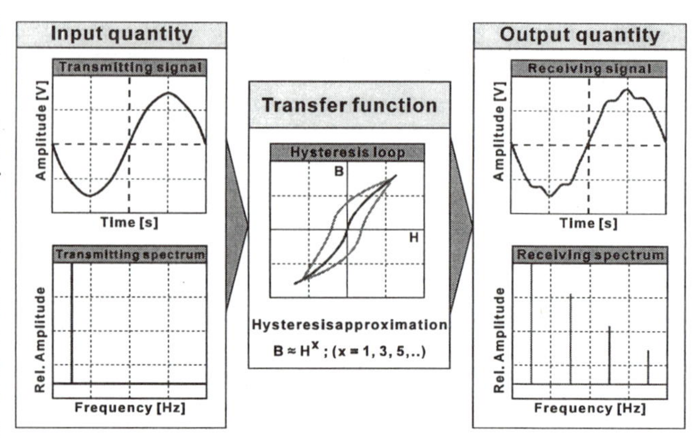

Figure 3: Origin of harmonics

The mechanical-technological as well as the ferromagnetic material properties are determined by the alloy composition, the crystallographic structure, and the structure of the material. Lattice imperfections and dislocation impediments have certain influences on the material properties such as the tensile strength resp. yield strength as well as the procedures of magnetic reversal. Thus they are important factors for the specific appearance of the hysteresis loop and therefore the development of harmonics during the eddy current analysis (fig. 1, 2, 3).

HARMONIC ANALYSIS OF EDDY CURRENT SIGNALS

The harmonic analysis of eddy current signals is based on the fact that a primary electromagnetic field caused by a sending coil will be influenced by the opposing secundary field which results from the distribution of eddy currents as well as processes of magnetic reversal within the test specimen (fig. 3). Changes of the electromagnetic properties show up in the signal's amplitude and phase /3, 4, 5/.
Furthermore, for ferromagnetic materials the measuring signal depends on the form of the hysteresis loop which is subject to the measuring frequency and the magnetic field intensity. Implied by nonlinearities of the hysteresis loop higher harmonics of the sending signal and displacements of the phase are caused in the measuring signal; these are characteristic for the material.

MEASURING SYSTEM

Figure 4 shows a measuring system as it is employed for the online determination of sheet steel properties within the cycle of work resp. the continuous operation in front of a press working line. Basically, it consists of a computer, an amplifier module, and a measuring coil system.
The eddy current plug-in card fitted into the measuring system generates a sinusoidal signal which is sent to the sending coil. The energizing field causes eddy currents within the test specimen and its magnetic field (the secondary field) is superimposed on the primary field. Both fields induce a measuring voltage within the receiving coil which then is amplified, filtered, digitized and analyzed.

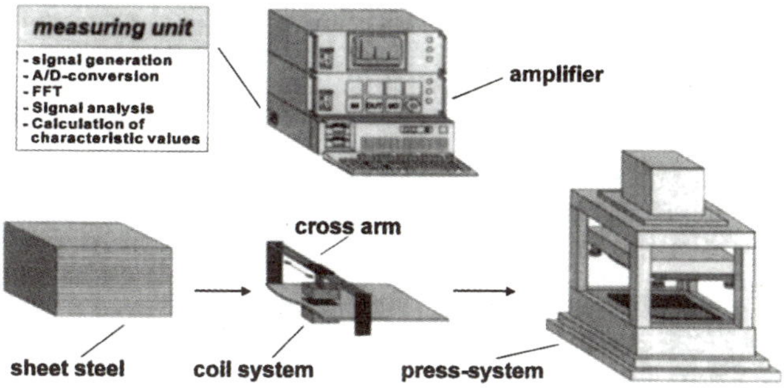

Figure 4: Development of an inline measuring system

MULTIDIMENSIONAL LINEAR REGRESSION ANALYSIS

By means of a variation of the measuring parameters, e.g. frequency and field intensity, it is possible to gain numerous measured quantities.
If there are several measured variables which correlate linearly with a material property, the latter can be described integrally as a function of several locally measured quantities by means of multi-dimensional linear regression analysis (fig.5). If the measured quantities are linearly independent from each other the result is improved.

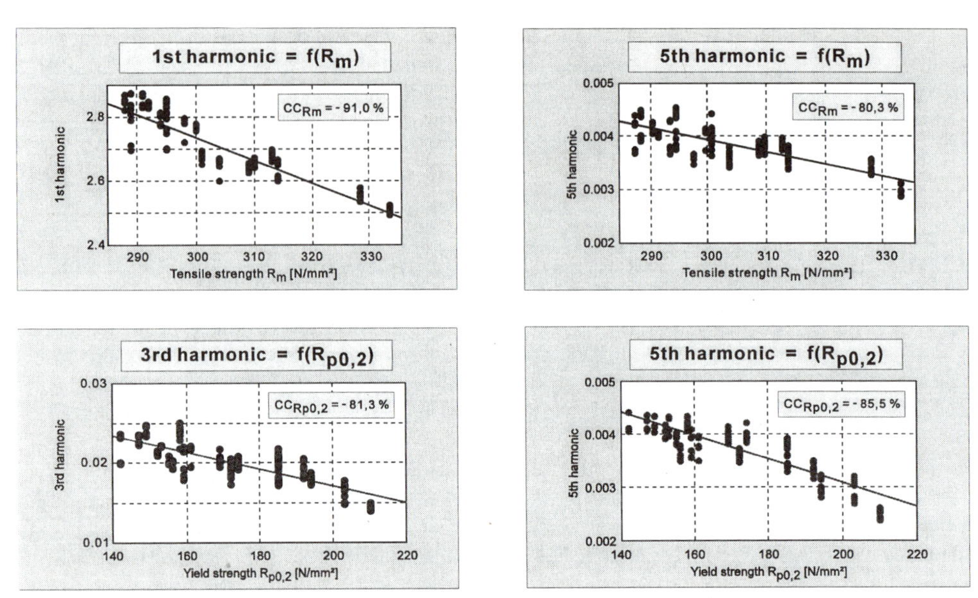

Figure 5: Measuring values of harmonic analysis as function of material properties

RESULTS

Mechanical properties and forming parameters are important quality characteristics when considering cold forming and deep drawing quality of sheet steel. To determine these sheet steel properties fast as well as nondestructively, extensive investigations were performed in co-operation with companies of the automobile and steel industries. The results of these investigations are given in the following section.

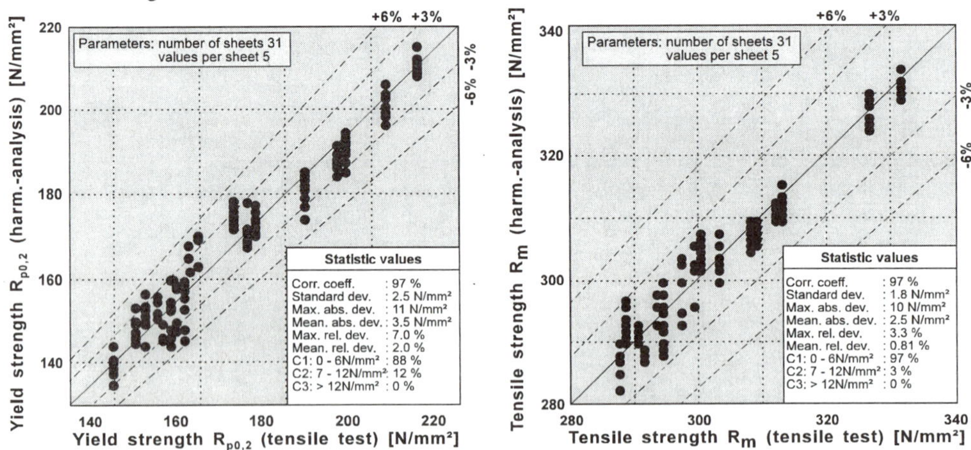

Figure 6: Result of a multidimensional regression of St14-sheets

The measuring system is based on multi-parameter eddy current measurement and additionally uses harmonic analysis algorithms and multidimensional linear regression analysis. Furthermore, it is specifically adapted to determine material properties of sheet steel.

Reducing the sheet steel qualities to the standard quality St14 (i.e. a common deep-drawing sheet steel used in the automobile industry) the measuring system has a high accuracy of measurement regarding the tensile strength R_m compared to the tensile test (fig. 6). Using five dimensions in the regression analysis the correlation coefficient amounts to 97% and nearly all measured values can be found in a narrow range of tolerance of ± 3 % resp. ± 6 N/mm^2. Similarly good results are gained for the yield strength (fig. 6) and the consolidation index (fig. 7).

Combining the sheet steel qualities of St12 up to St15 causes an increasing of the dispersion of non-destructively determined material properties when measured with a distance of 5 mm of the sensor to the sheet surface); this is compensated by expanding the number of measured values (fig. 8). Even when using the sheet steel quality St 14 provided by different producers and varying in sheet thickness, surface quality, coating, dress, etc. the dispersion of the nondestructively determined material properties is only slightly increased. Measuring 70 sheets a correlation coefficient of 91% is reached regarding the tensile strength and still 86% of the determined material properties lie

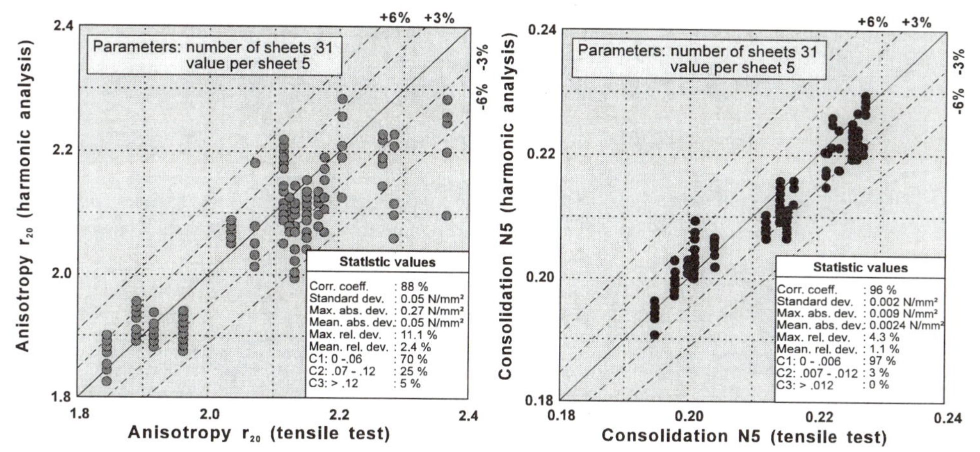

Figure 7: Result of a multidimensional regression of St14-sheets

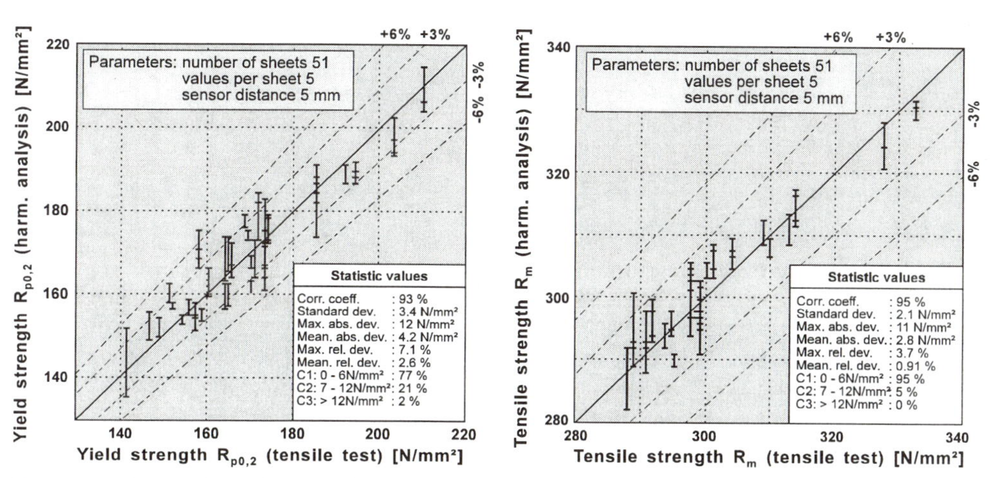

Figure 8: Result of a multidimensional regression for St12..St15-sheets (distance of sensor to sheet surface: 5 mm)

within a range of ± 6 N/mm² (fig. 9). This demonstrates additionally that the means of production and thermal treatment as well as the sheet steel parameters mentioned above are of minor influence on the nondestructive determination of sheet steel properties by means of harmonic analysis. Regarding the high-tensile micro-alloyed ZStE sheet steel only a few measured values are enough to reach correlation coefficients of more than 98 % because of the large extent of the material properties. The determined values lie in a small range of tolerance of less than ± 3 N/mm² compared to the tensile test (fig. 10).

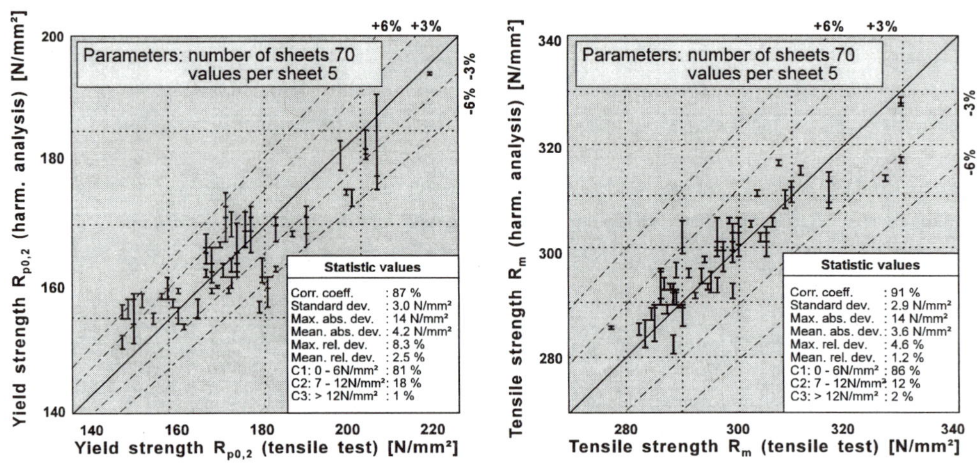

Figure 9: Result of a multidimensional regression of St14-sheets made by different producers

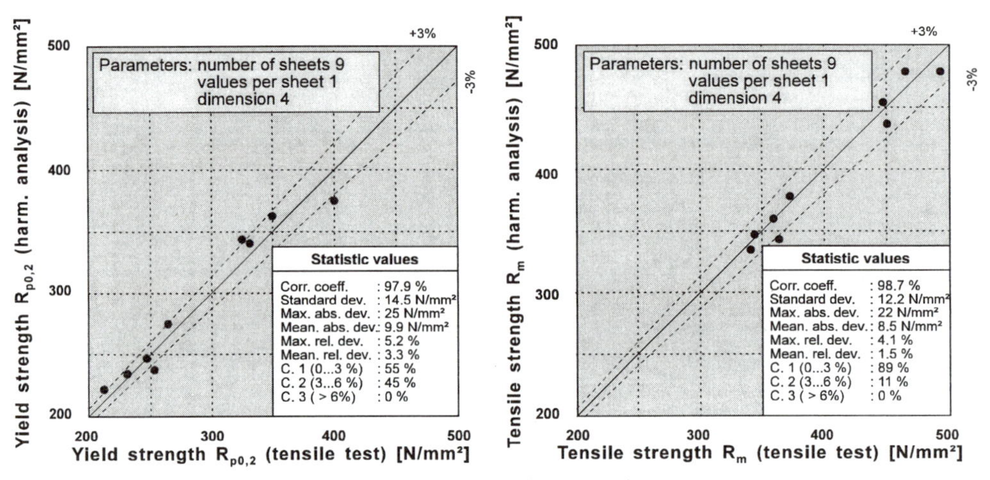

Figure 10: Result of a multidimensional regression for ZStE-sheets

Further investigations to determine material properties of mechanical-technological properties were conducted with high-tensile thick plates (TStE 690 and TStE 960). Despite of the (for eddy current systems) extreme plate thickness of 3 to 15 mm it was possible to attain high correlation coefficients of 99 %. According to the two groups of steel the measured values are concentrated in two groups with a variance of ± 3 % compared to the tensile test (fig. 11).

CONCLUSION

Extensive nondestructive investigations with different sheet steel qualities showed that the harmonic analysis of eddy current signals in combination with multi dimensional linear regression analysis is a suitable means to determine non-destructively mechanical-technological material properties of sheet steel and thus characterize material.

When restricting the range of sheets to one or similar sheet qualities high accuracies of measurement are attained compared to the tensile test. The variance of the measured values lies within a range of tolerance similar to that of destructive tests.

Considering the fact that despite the same sheet specification different producers use varying means of production (e.g. thermal treatment) these variations as well as distinct differences of one of the sheet's parameters such as sheet thickness, surface quality, coating, dress, etc., are of minor

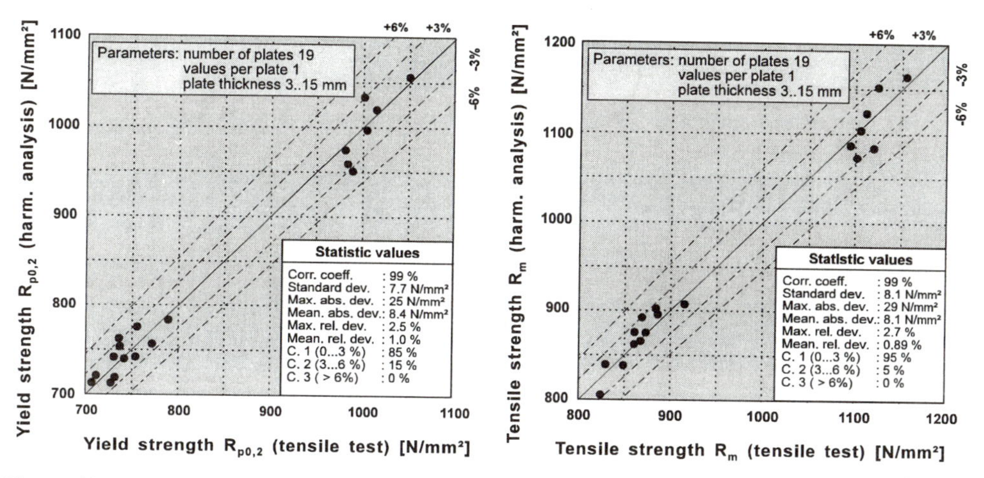

Figure 11: *Result of multidimensional regression for TStE-plates*

importance with this measuring system concerning the nondestructive determination of sheet steel properties.

Aside from the fact that the material properties are determined nondestructively the high velocity of measuring within a range of less than two seconds has to be emphasized. With these specifications the measuring system fulfills the main requirements raised for an online measuring system within a production sequence. Last but not least the costs for testing can be considerably reduced by this technique compared to destructive tests.

REFERENCES

/1/ Peterson, E.
Harmonic production on ferromagnetic material at low frequencies and low flux densities; Bell System Technical Journal 7 (1928), p762

/2/ Gruska, K.
Use of harmonic analysis for inspection of ferromagnetic materials; Soviet Journal NDT 19 (1983) p399

/3/ Feiste, K.L. / Reimche, W. / Stegemann, D.
Zerstörungsfreie Bestimmung mechanisch-technologischer Feinblecheigenschaften mittels Harmonischen-Analyse (Nondestructive determination of mechanical technological sheet steel properties by means of harmonic analysis); DGZfP-Tagungsband, 22.-24.5.1995, Aachen

/4/ Feiste, K.L. / Reimche, W. / Stegemann, D.
Zerstörungsfreie elektromagnetische Bestimmung mechanisch-technologischer Feinblech-eigenschaften (Nondestructive magneto-inductive determination of mechanical-technological sheet steel properties); Werkstoffprüfung 1995, 5. u. 6.12. 1995, Bad Nauheim

/5/ Feiste, K.L. / Reimche, W. / Stegemann, D.
NDT-determination of mechanical and technological sheet steel properties using harmonic analysis; XIV CONAEND abende RIO-95 - XIV Congresso Nacional de Ensaios Não Destrutivos 1995, p249

RESONANT FREQUENCY TESTING

R. H. (Bob) Gassner

CNS Co
2324 Camino Escondido
Fullerton, CA 92833

INTRODUCTION

Elastic constants are NOT CONSTANT!! They are NOT fundamental, immutable properties of a material. On the contrary, they are sensitive to variations in microstructure (e.g., due to heat treatment), crystallographic orientation and composition (even within prescribed limits). Therefore, since they can be readily measured, by way of resonant frequency, they provide a valuable tool for evaluating such variations and the properties which correlate with them.

TESTING AND CALCULATING

Using the impulse excitation technique, it is easy to rapidly determine the fundamental resonant frequency accurately and precisely. The square of that frequency is directly proportional to the modulus related to the mode of vibration, E - flexure and longitudinal modes, G - torsional mode. All one needs to perform the test is a non-dampening support for the sample, a light tap and a Grindo-Sonic instrument; the instrument detects the frequency of the very low amplitude vibrations remaining after the noise/harmonics have dissipated. That frequency, plus the sample weight and dimensions, are then entered into the appropriate equation to calculate the modulus. Because the dynamic moduli and correlating properties are derived from micromicrons of elastic strain, they probably reflect the structure/composition of the sample more truly than moduli determined by static testing methods.

The test normally takes less than three seconds and, when it is repeated many times on the same sample, there is practically no scatter in the results. This is a VERY IMPORTANT characteristic, i.e., the absence of scatter in successive readings. It enhances the researcher's confidence in the validity of very small frequency differences found among apparently identical samples, e.g., samples with visually undetectable differences in microstructure such as 17-4PH stainless (or 4340 alloy) steel aged (or tempered) at 1000°F

Nondestructive Characterization of Material VIII
Edited by Robert E. Green Jr., Plenum Press, New York, 1998

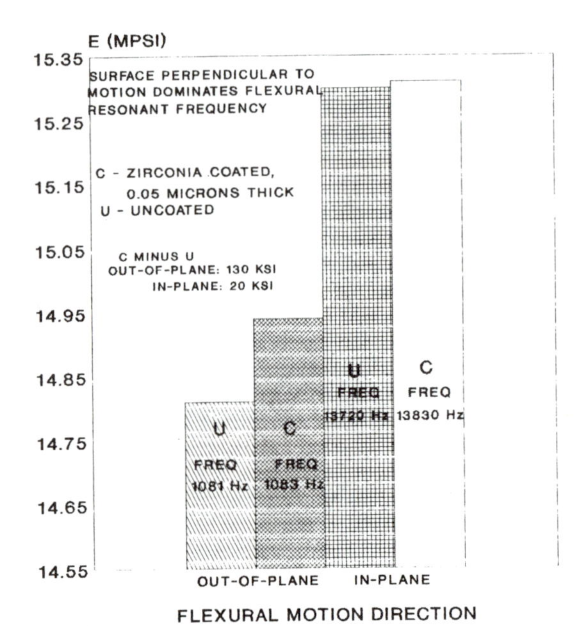

Figure 1. Thin (50nm) CVD Zirconia Coating on Titanium Strip (0.035" X 0.5" X 3")

and 1050°F. In Figure 1, for example, detection of the 50 nanometer thick coating was judged to be valid because the scatter of both the in-plane and out-of-plane frequencies was zero. When anomalies in results occur, they are invariably traced to errors in measurement of dimensions or weight, not resonant frequency.

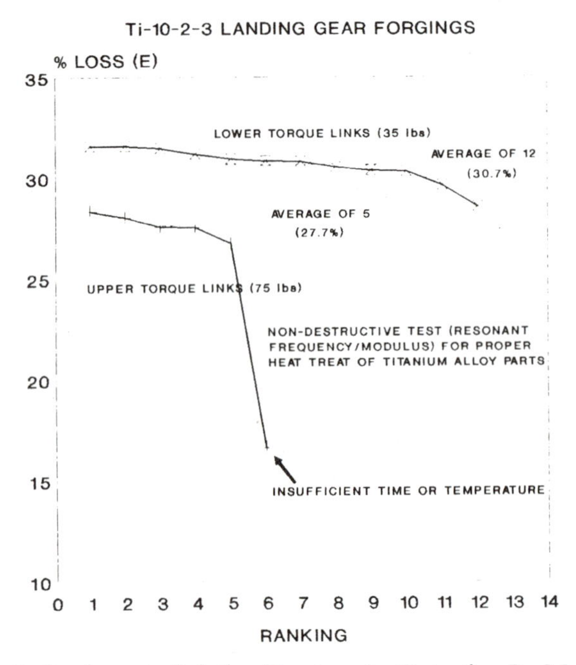

Figure 2. Inadequate Solution Treatment - Detection By Modulus Reduction

Materials and size are not constraints; this testing technique can be used on any sample whose resonant frequency is between 40 and 80,000 Hertz and is detectable by a supersensitive piezoelectric contact probe or microphone. (The probe is so phenomenally sensitive that it can detect the flexural resonant frequency of a 1" x 2" x 5" aluminum block when it is struck by a hollow, 3/4" long piece of plastic cocktail straw dropped from a height of 8 inches.) Materials have included metals, plastics, ceramics, glass, wood, concrete and composites. Sample sizes successfully tested have included CVD diamond specimens, 0.004" x 0.2" x 0.6", Nitinol wire, 0.020" diameter, 12 foot long aluminum alloy tubes, 4" OD x 0.080" W, and 500 lb titanium alloy forgings.

Computer programs permit rapid calculation of moduli for rectangular bars, cylindrical bars and disks. On bars, values can be obtained for E, G (torsion) and, by iteration, Poisson's Ratio based on flexure, transverse flexure, longitudinal and torsion modes of vibration. On some cylindrical bars, it is not possible to detect the torsional frequency because a glancing tap cannot introduce sufficient energy. This can be overcome, at a slight sacrifice in accuracy, by welding or glueing a small tit to the periphery of the bar to permit an increase in the torsion energy introduced. On disks, E, G and Poisson's Ratio are readily obtained based on the flexural resonant resonant frequency (tapping at the center) and torsional resonant frequency (tapping in the axial direction at the edge).

When samples are irregularly shaped and no modulus formula exists, it is frequently valuable to evaluate the relative moduli of like samples by comparing the squares of their resonant frequencies. In Figure 2, the relative modulus loss of each Y-shaped forging was calculated from the squares of its resonant frequencies before and after heat treatment. Alternatively, when actual modulus is desired for a group of irregularly shaped like samples, and the benefit outweighs the cost, a "shape factor" may be developed by sacrificing one sample to fabricate a regularly shaped specimen.

APPLICATIONS

A valuable application of a mechanical principle is illustrated by Figure 1, namely, the resonant frequency of an anisotropic sample in the flexural mode is dominated by the characteristics of the surface. In that case, it enabled the detection of a ultra-thin coating on the surface of a titanium specimen. In another case, a composite 0.048" x 1.2" x 5.4" specimen of copper, containing 25% longitudinal, high modulus graphite fibers in the surface layers, exhibited a modulus of 61.68 mpsi in flexure and 25.06 mpsi in transverse flexure. These experiences indicate a potential application for evaluation of plated, sprayed, and vacuum deposited coatings as well as diffused surface layers of carbon/nitrogen on steel etc. This principle is also illustrated in Figure 7 where the modulus based on the flexural vibration mode is lower than the modulus based on the longitudinal vibration mode for all three quenchants.

Figure 2 illustrates both mechanical and metallurgical principles. First, that the moduli of irregularly shaped production parts can be compared by the squares of their resonant frequencies and second, that the modulus changes can be used to monitor heat treatment of titanium alloys. Since use of heat treated titanium alloys has been inhibited by the lack of a nondestructive test for strength/heat treatment (comparable to hardness testing of steel), an example of a correlation between strength and modulus is shown in Figure 3.

Modulus of 30 grinding wheels from a single lot was found to vary from 4.9 to 6.1 mpsi (25%). This scatter proved to be the cause (an unexpectedly hard wheel causing high local surface temperatures) of hitherto unexplained occurrences of grinding burns/cracks.

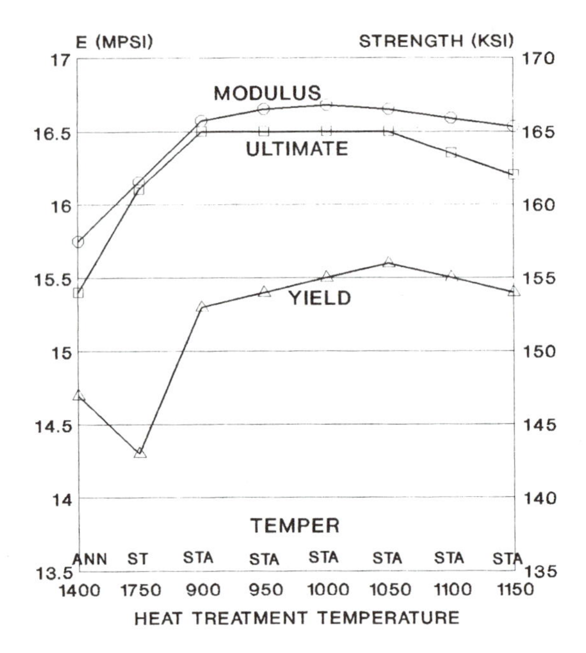

Figure 3. Ti-6Al-4V - Strength & Modulus Versus Heat Treatment

Wide variations have also been found in a similar product, honing stones. One manufacturer found that matching the resonant frequencies of the stones on each head allowed honing of 1500-2000 cylinders from each set of eight stones, instead of the previous capability of 400 cylinders, while reducing stone consumption from 133 sets per week to 148 sets per month; concomitantly, the job of cementing stones to heads was reduced from forty hours per week to four hours per month.

Because the maximum speed of a driveshaft is directly proportional to its modulus, resonant frequency testing was used to maximize the modulus of aluminum metal matrix composite driveshaft tubing (6061-T6/15%Al_2O_3). The tests showed that the modulus of the tubing, made by extrusion plus a 35% reduction by drawing, could be increased by changing the drawing operation from two passes to five passes; the loss in modulus due to drawing was decreased from 6% to 4%.

A remote sensing technique, and appropriate apparatus, has been developed to permit resonant frequency testing at temperatures up to 2000°F. It has been used to determine moduli at elevated temperature and to track aging curves of titanium and aluminum alloys. In one test on a superplastic aluminum alloy, Supral 100, an unexpecred retrogression was found in the temperature/modulus curve slightly below the superplastic forming temperature. With rising temperature, the steadily decreasing modulus abruptly changed direction at 829°F, increased by 10% to 835°F, and then resumed its downward slide.

To evaluate a correlation between modulus anisotropy and formability, moduli in the longitudinal (L), transverse (T) and diagonal (D) directions of seven 400 series stainless steel sheet materials were determined. For six materials, the difference found between the highest and lowest moduli was normal, varying from 1.5 to 3 mpsi. The seventh material, the one with suspect formability, exhibited a lower diagonal modulus than those in the other

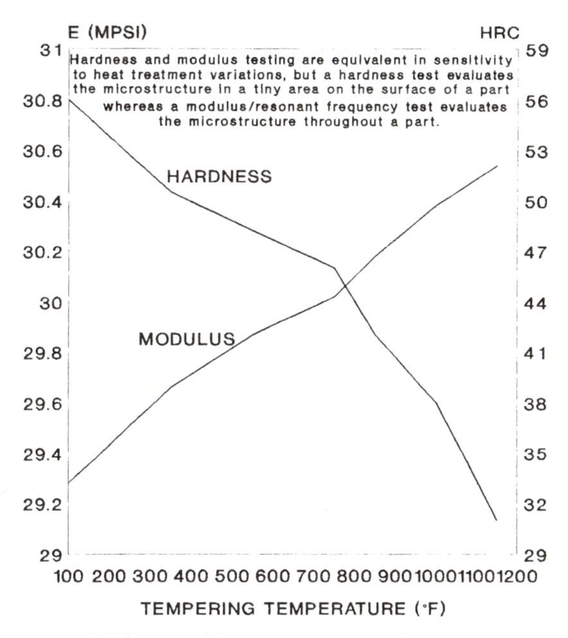

Figure 4. E Modulus & Hardness vs. Tempering 1.5 in DIA 4340 - Oil Quenched

directions and the difference between the diagonal modulus and the transverse modulus was found to be 5 mpsi.

A complex aircraft part was subjected to a spectrum of loads, simulating service conditions, introduced at five different locations and the resultant strains were read on eleven strain gages positioned at the critical locations. These readings were converted to stresses using the published modulus for the material. After completion of the test, specimens were excised from the critical locations of the part and their moduli were determined by resonant frequency testing. The actual stresses, based on the calculated dynamic moduli differed from those determined originally by 5%-12%.

A series of applications to steel take advantage of the metallurgical principle that the modulus of ferrite is higher than that of martensite. Figure 4 shows that the slopes of modulus and hardness vs. tempering temperature, up to 1150°F, of an alloy steel, are reciprocals, i.e., the two techniques are equivalent in reflecting the strength of properly heat treated material. However, as a replacement for, or a supplement to, hardness testing,, resonant frequency testing of parts offers a significant benefit, i.e., it provides a reflection of the microstructure of the entire part. In contrast, a hardness test reflects the microstructure at a single point on the surface. For example, a resonant frequency test will detect an attempt to mask a poor quench by a low tempering temperature whereas it would not be detected by a hardness test. Similarly, a local hard spot on a steel part (carburized due to presence of a grease spot during heat treatment) might show an acceptable hardness in spite of an overtempered, low hardness general microstructure which would be detected by a resonant frequency test.

Detection of a composition deficiency, resulting in excess ferrite in stainless steel castings, is illustrated in Figure 5. The resonant frequency of the part from the lot

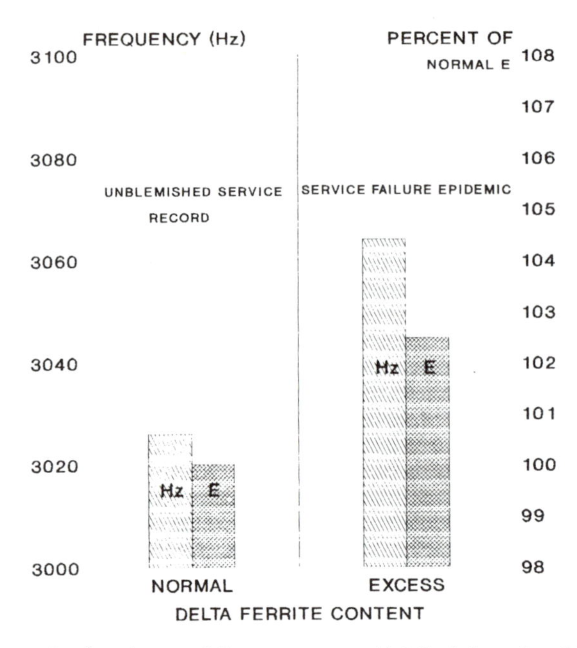

Figure 5. Service vs. Microstructure 416 Stainless Steel Pliers

experiencing service failures, due to the presence of excess, low-strength ferrite, was 36 Hz higher than that of the part from the good lot. Metallographic analysis revealed that the fracture path was predominantly through the ferrite grains.

The value of hardness testing, as currently practiced to verify proper heat treatment, is especially doubtful for parts made from precipitation hardening stainless steels. Although it is common practice to hardness test such parts, Figure 6 shows that the test adds little to part reliability, i.e., it shows that the acceptable range of values for the various aging treatments overlap each other, and even overlap the solution treated, unaged range (right edge of graph). Figure 6 also shows that resonant frequency/modulus tests offer the prospect of a much more reliable verification of proper heat treatment

One of the most important steps in the heat treatment of alloy steel is the quench. However, it is not practical to forecast the most suitable quenchant for each part configuration; there's a multiplicity of oils and polymer/water solutions available, offering a wide variety of cooling rates through the critical range. In lieu of this, specifications require quarterly testing of consistency of quench effectiveness; the quenchant's effectiveness must not change significantly due to deterioration, contamination etc.

Two methods of evaluation are specified, cooling curve tests.and mechanical properties tests. The former involves tracing the temperature of a heated probe after immersion in a sample of the test oil. The latter requires testing of a mechanical property (strength, hardness, modulus) after quenching a sample; it has the advantage of permitting evaluation of deterioration of both quenchant quality and production agitation on the type of material being processed. The most suitable property for the test is resonant frequency/modulus because of the absence of scatter of results - not a normal characteristic of results of tensile or hardness tests.

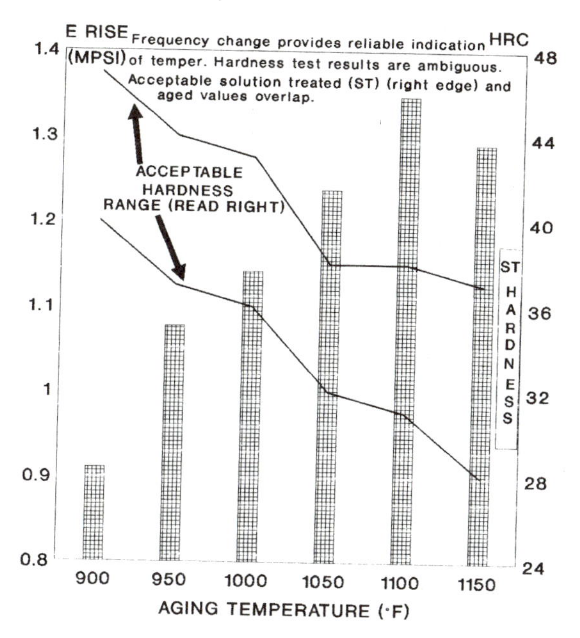

Figure 6. 17-4PH - Distinguishing Tempers by Modulus Rise vs. Hardness

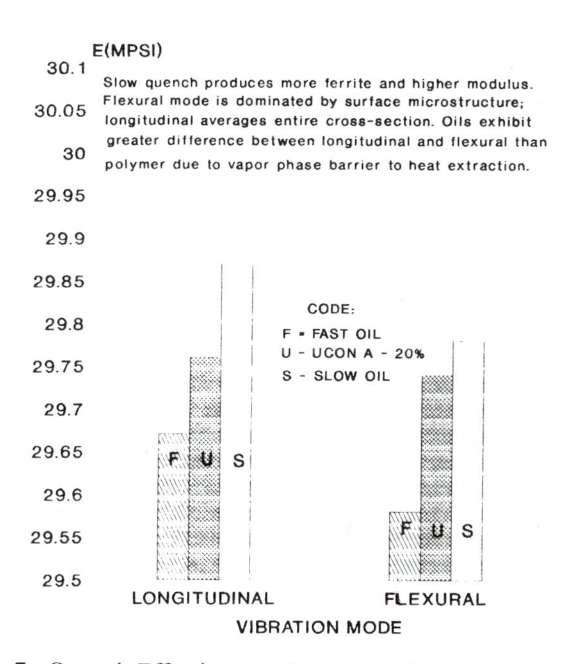

Figure 7. Quench Effectiveness Test 1.4" Diameter 4140 Steel Bar

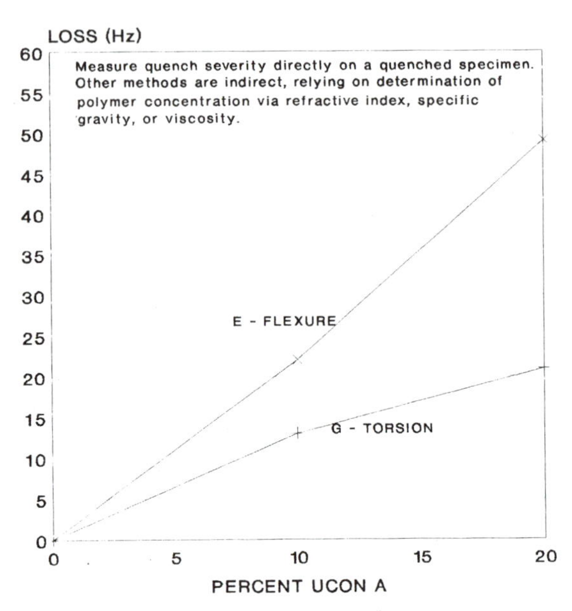

Figure 8. Percent Polymer in Quench Water vs. Loss in Resonant Frequency

Figure 7 shows that the slow oil produced a higher modulus (more ferrite) than the fast oil and the polymer/water solution produced an intermediate modulus. Furthermore, for each quenchant, it shows that the modulus based on the longitudinal mode of vibration is higher than that based on the flexural mode of vibration. It also suggests that the polymer/water produced a more uniform microstructure throughout the cross-section.

A similar challenge exists with polymer/water quenchants for aluminum. The concentration of polymer in a sample of the solution is normally checked periodically by tests of viscosity, specific gravity or refractive index. It is presumed that these properties, unaffected by contaminants, correlate with the quench effectiveness, but they are indirect. Far more assurance of production parts' cooling at the proper rate can be provided by resonant frequency testing of a heat treated sample. See Figure 8.

SUMMARY

The data and experiences presented indicate that the availability of an quick, user-friendly technique for accurately and precisely determining resonant frequency is not only attractive to those needing modulus but also to those wishing to determine other properties which can be correlated with modulus. Much additional development work is being planned to explore the wide range of possible applications in exotic research, engineering development and quality control.

ROLLED STEEL SURFACE INSPECTION USING MICROWAVE METHODS

Reza Zoughi[1], Christian Huber[1], Stoyan I. Ganchev[1], Radin Mirshahi[1], Emarit Ranu[1] and Thomas Johnson[2]

[1]Applied Microwave Nondestructive Testing Laboratory (*amntl*)
Electrical Engineering Department
Colorado State University
Ft. Collins, CO 80523

[2]US Steel Corporation
USS Group
Gary Works
One North Buchanan
Gary, IN 46402

INTRODUCTION

Rolled steel is used in a variety of products such as soup, paint and spray cans as well as automobile body panels (doors, hoods), ship structures, etc. For most of these applications it is important that the rolled steel be devoid of any defects, in particular surface defects. For example surface defects can leave an automobile door with unacceptable spot discoloration and container cans with holes. These defects can be separated into two broad categories of steel induced defects such as holes, laminations and slivers and roll mark defects such as pock marks, pick ups, bruises and wire marks. In general, roll mark defects are much more subtle that the steel induced type, and consequently harder to detect using any available detection method. Since microwave signals do not penetrate inside conductors such as steel and are only sensitive to surface features and perturbations, they are a good candidate for steel surface inspection. In particular, near-field microwave methods allow for optimization of the standoff distance for increased detection sensitivity. In recent years, microwave near-field inspection techniques have been used to study surface properties of metals and graphite composite materials for defects such as surface cracks and impact damage[1-5].

The purpose of this investigation is to demonstrate the capability and limitations of microwave approaches for steel surface inspection as well as understanding the role of measurement parameters.

DEFECT CATEGORIES

There are various defects that may exist in tin mill as well as hot strip mill rolled steel products. These defects are different in their physical characteristics (i.e. shape and size) as well as in the manner by which they are produced during the manufacturing of rolled steel products. The following are the primary defect categories for both the tin and the hot strip mill specimens (given a relatively large number of available specimens):

Tin Mill Defects
1. hole,
2. pock mark, pick up, wire mark, bruise, tailend mark, and bump mark (all roll marks),
3. scratch,
4. lamination, sliver, scale and scab (all steel induced),
5. gouge,
6. pincher and anode short (to be ignored for the purpose of current investigation).

Hot Strip Mill Defects
1. scratch,
2. gouge,
3. bruise, fire mark, pressure crack, skid mark, spall, scrap mark and pick up (all roll marks),
4. scale and sliver (all steel induced).

In this paper only the results for holes and surface laminations/slivers are presented.

MEASUREMENT APPARATUS

Figure 1 shows the typical measurement setup used to conduct the majority of the measurements. A horn antenna radiates the microwave energy/signal onto the steel surface at a given frequency and polarization. Subsequently, the reflected signal is picked up by the antenna. Since there is a relative motion between the sensor and the defect, a measured signal which is proportional to the reflected signal level and its phase is recorded as a function of the relative location of the strip. Figure 1 also shows all of the important measurement parameters such as the standoff distance, the incidence angle, the rotation angle, etc. It was shown that open-ended rectangular waveguides were also capable of detecting a large number of defects at reasonable standoff distances (i.e. several centimeters using less that ten milliwatts of incident power) without the need for a horn antenna.

SAMPLE RESULTS

Since it is not possible to report all measurement results in the limited number of pages here, only several representative results will be provided, as well as a summary section describing the attributes of some of the measurement parameters.

Holes

As a typical result, Figure 2 shows the detected signal as a function of the scanning distance for two adjacent holes (in the roll direction similar to those shown in Figure 1) with diameters of about 6 mm and 3 mm at a frequency of 24 GHz, a standoff distance of 30 mm, an incidence angle of 45 degrees, at vertical polarization and with a horn antenna. In this measurement (and other unless otherwise mentioned) a mixer diode was placed inside the transmitting waveguide. The reflected signal and the transmitted signal were then mixed in the waveguide. The output of the mixer diode, which is a simple dc voltage was then used as an indicator of the presence of a defect.

The solid line, in Figure 2, shows the results when the holes are in the center of the antenna footprint and the dashed lines show the results while the defect is moved out of the center of the footprint by 10 mm and 30 mm, respectively. The results indicate that the holes are clearly detected, in addition to the fact that if the antenna is not exactly lined up with the defect it may still be detected it as long as the defect is within its footprint. This fact renders the number of sensors needed to cover the entire width of a strip relatively small. Clearly, the results show an interference pattern caused by the presence of these holes (and their highly scattering edges).

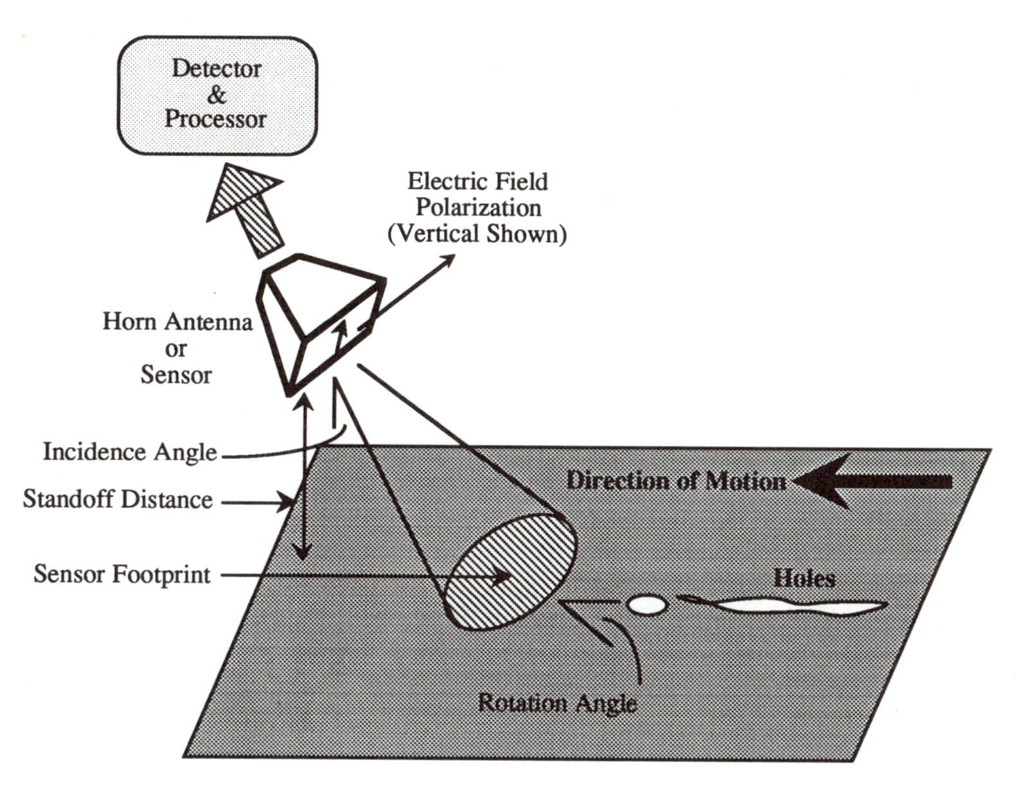

Figure 1: General schematic of the measurement setup.

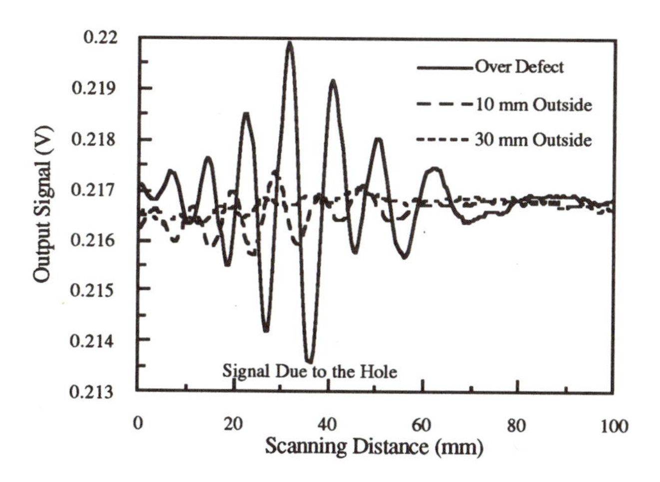

Figure 2: Detected signals due to two small adjacent (holes in the roll direction), at 24 GHz, vertical pol., standoff distance of 30 mm, incidence angle of 45° and with a horn antenna.

Surface Lamination & Sliver

Slivers and surface laminations are similar defects and are combined in one group for measurement purposes. These defects are more difficult to be detected than holes. Furthermore, in some cases there is no delamination and only a surface "discoloration" exists. Relatively speaking, the laminated defects and those where the lamination is peeled off are much easier to detect than the subtle ones. Another measurement challenge has to do with the continuous nature of these defects (unlike holes). This may require a comparison among the detected signals by adjacent sensors. However, since our measurements rely on the movement of the strip, the beginning of a sliver or surface lamination will also be a strong indication of its presence. Furthermore, the information about the start and end points of a sliver may be used to determine it approximate length and location on the strip both along its length and across its width.

Since these defects are generally elongated along the roll direction, they interact with the incident microwave signal differently than holes do. Our goal was to arrive at a set of common parameters (as much as possible) so that all or most of the defects may be detected. In our course of investigations we scanned slivers across the width of the strip and in all cases they were very nicely detected. Unfortunately however, this is not practical since the sensor will (most likely) need to be scanning along the strip length. For this reason it was decided to devise a slightly different type of measurement in which the sensor is looking along the strip length at a certain standoff distance and incidence angle but with a rotation angle as well, as shown in Figure 1. In this way the sensor is always illuminating both along the length and the width of the strip. The results of this approach have shown to be quite promising. Furthermore, this approach may be successfully applied to elongated holes with small widths.

Figure 3 shows the measured signals due to a surface lamination with a width of 1.5 mm at a frequency of 24 GHz, a standoff distance of 18 mm, an incidence angle of 30 degrees, at vertical polarization and with a horn antenna when the defect is in the center of the antenna footprint and when it is displaced by 40 mm (defect outside of the footprint), respectively. The defect is indeed detected but the detection is subtle. Consequently, a rotation angle of 45 degrees was introduced and the measurements repeated, as shown in Figure 4. As explained earlier, the defect is detected with much better measurement sensitivity compared with the previous case.

Standoff Distance Considerations

As the standoff distance changes so does the relative phase of the incident signal at the steel surface and the phase of the detected signal (also the properties of a signal which is composed of the combination of these two signals). Also as the standoff distance increases the magnitude of the reflected signal decreases. For large standoff distances this reflected signal is detectable only when the incident power is sufficiently large, rendering the reflected signal above the system noise level.

Figure 5 (standoff distance curve) shows the variations in the amplitude of the measured signal as a function of standoff distance (or sensor distance from the steel sheet surface) with no defect present at a frequency of 24 GHz, an incidence angle of 30 degrees, at vertical polarization and with a horn antenna. It is important to point out that this behavior - rising, falling and flat parts in the minimum and maximum - has significant implications on the actual detection sensitivity. Therefore, a flat sheet of steel or one with undulations dictate where on the standoff distance curve to operate. For example let us assume we have a very flat sheet with a defect on it. In this case it would be recommended to operate on the fastest rising or falling regions of the standoff distance curve since the presence of the defect would be maximally detected. However, one may also operate at the maximum or the minimum of the standoff distance curve and still see the defect quite nicely. For a sheet with surface warp/undulation it is best to operate at a minimum location on the standoff distance curve where some degree of standoff distance changes may be tolerated while the defect is still detectable.

SUMMARY

The smallest detectable hole at frequencies around 24 GHz is one with a diameter of approximately 1.6 mm. Smaller diameter holes may be detected at higher frequencies. These small holes are detected at relatively large standoff distances (up to 3-4 inches), despite the low microwave power used in our measurement setups. Larger standoff distances are possible using higher microwave power levels. These levels will still be relatively low and in the range of several tens of milliwatts. The optimum incidence angle is found to be close to

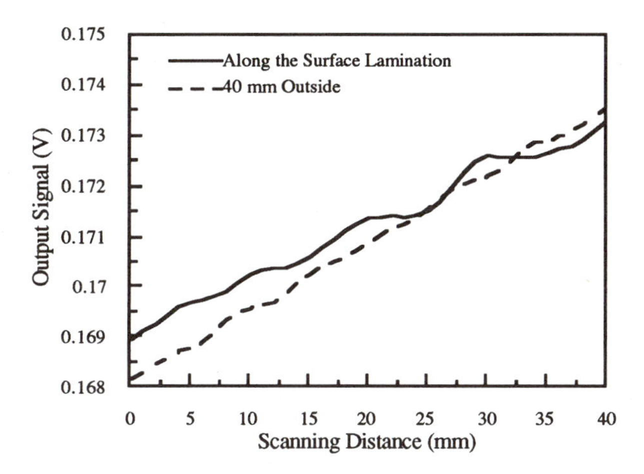

Figure 3: Detected signals for a 1.5 mm-wide surface lamination at 24 GHz, standoff distance of 18 mm, an incidence angle of 30°, at vertical pol. with a horn.

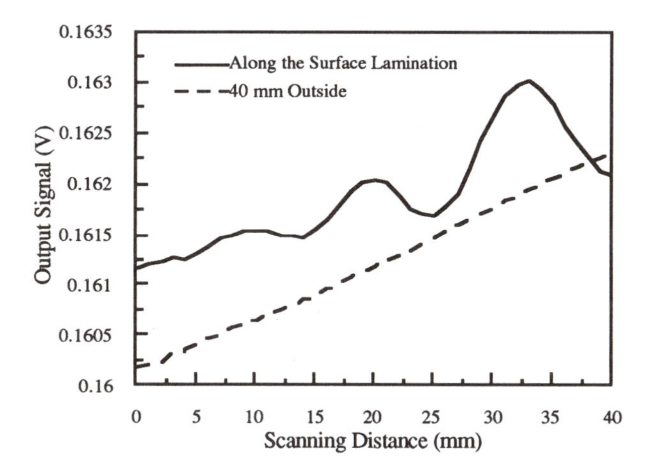

Figure 4: Detected signals for the surface lamination shown in Figure 3 but with a rotation angle of 45 degrees.

30 degrees, and the optimum polarization is vertical. A rotation angle of 45 degrees may be incorporated into the measurement setup for higher detection sensitivity of small but elongated holes.

The optimal set of measurement parameters that were found for detecting holes is also near-optimal for detecting steel defects such as slivers and surface laminations. At a frequency of 24 GHz a large percentage of these defects are detected, and for the more subtle defects higher frequencies may be necessary. The beginning and the end location of such defects may be determined as well. Defects such a slight slivers, scales, etc. are indeed difficult to detect. Some of these defects are nothing but discoloration which leave no electromagnetic basis for their detection at microwave and millimeter wave frequencies. Some of these defects depending on how they are metallurgically generated may still be detected at high enough frequencies and special detection mechanisms.

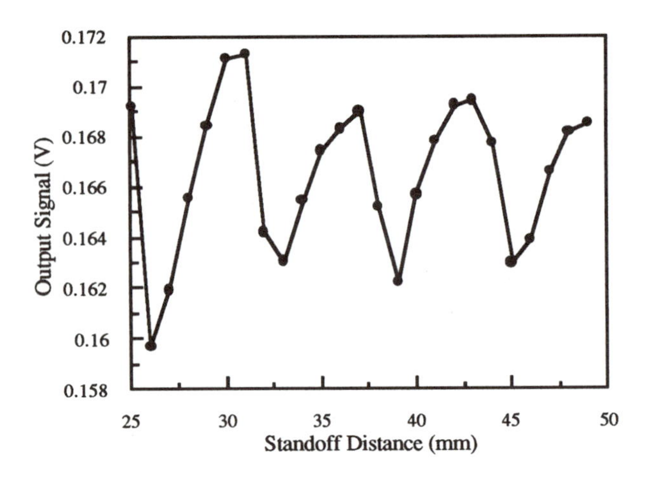

Figure 5: Influence of standoff distance on the measured signal (standoff distance curve).

It was also shown, during the course of these investigations, that it is possible to distinguish between a surface warp/undulation and a defect. Furthermore, defects in areas with surface undulation may be detected without the need to concern oneself with the effect of some surface warp. This is due to the fact that a hole or surface lamination has a distinct detected signal which is different than that of a surface warp/undulation. Ultimately, the use of smart signal processing schemes can aid in a more robust detection and classification of these defects.

Acknowledgment: This investigation was funded by US Steel Corporation, Gary Works, IN. The technical responsibility for the material presented in this paper rest with the investigators at Colorado State University.

REFERENCES

1. Huber, C., H. Abiri, S. Ganchev and R. Zoughi, "Analysis of the Crack Characteristic Signal Using a Generalized Scattering Matrix Representation," *IEEE Trans. on Microwave Theory and Tech.*, vol. 45, no. 4, pp. 477-484, April, 1997.
2. C. Huber, S.I. Ganchev, R. Mirshahi, J. Easter and R. Zoughi, "Remote Detection of Surface Cracks/Slots Using Open-Ended Rectangular Waveguide Sensors: An Experimental Investigation" To appear in the *NDT&E*, 1997.
3. Yeh, C., E. Ranu and R. Zoughi, "A Novel Microwave Method for Surface Crack Detection Using Higher Order Waveguide Modes," *Mat. Eval.*, vol. 52, no. 6, pp. 676-681, June, 1994.
4. Zoughi, R. and C. Lebowitz and S. Lukes, "Preliminary Evaluation of the Potential and Limitations of Microwave NDI Methods for Inspecting Graphite Composites," *Proceedings of the Seventh International Symposium on Nondestructive Characterization of Materials*, vol. 210-213, part 2, pp. 611-618, Prague, Czech Republic, June 19-23, 1995.
5. Bahr, A. J., "Microwave Nondestructive Testing Methods," Volume I, Gordon and Breach Science Publishers, 1982.

TWO-PORT NETWORK ANALYZER DIELECTRIC CONSTANT MEASUREMENT OF GRANULAR OR LIQUID MATERIALS FOR THE STUDY OF CEMENT BASED MATERIALS

Karl Bois, Aaron Benally, and Reza Zoughi

Applied Microwave Nondestructive Testing Laboratory (*amntl*)
Electrical Engineering Department
Colorado State University
Ft. Collins, CO, 80523

INTRODUCTION

The construction industry has a keen interest in using a nondestructive, real-time, reliable and inexpensive technique for the in-place evaluation of the compressive strength of concrete structures. Compressive strength of concrete is usually determined by drilling a core and testing it in a laboratory. This method is relatively expensive, and it may take a few days for the results to be known. In addition, this method is destructive. Consequently, several nondestructive techniques have been developed for this purpose. These include: pulse velocity method, surface hardness, penetration, pullout, breakoff and maturity techniques[1].

Microwave nondestructive testing methods are well suited for inspecting dielectric materials such as cement paste and concrete[2]. When applied to chemically reactive materials such as those in which curing occurs, microwave methods have shown the potential for cure state monitoring as well[3,4]. Porosity level estimation and constituent characterization in homogeneous materials is also possible with these methods[3,5].

Recently, a microwave nondestructive inspection method was used to measure the reflection and dielectric properties of cement paste specimens with various water/cement (w/c) ratios at several frequencies[6]. This method utilized an open-ended rectangular waveguide in-contact with the specimens. Subsequently, the microwave reflection properties of these specimens, at 5 and 9 GHz, were correlated to their w/c ratios and their compressive strengths[6]. The results of these preliminary investigations showed great promise for using a true nondestructive testing method for the in-place compressive strength evaluation of concrete based structural members.

To obtain the dielectric properties of cement based materials, a mixing model approach was proposed[7] in which the dielectric properties and the volume fraction of each constituent of a concrete specimen (e.g. cement, sand, coarse aggregate, water, fly ash and various catalyzers) are used to predict its effective dielectric properties. Since the reflection coefficient of a specimen is directly dependent on its dielectric properties and this parameter is subsequently used to estimate the compressive strength or the w/c of the specimen, the accurate knowledge of the respective dielectric properties of the concrete constituents is of utmost importance in the prediction of the effective dielectric properties of the concrete mixture.

For this purpose, numerous microwave dielectric property measurement techniques have been proposed. Of these, we include those involving the uses of coaxial lines[8,9] and rectangular waveguide[10,11,12,13]. All of these techniques posses their high points when measuring the dielectric properties of solid samples but require some form of approximation for the measurement of granular (sand, cement powder and coarse aggregate) and liquids (water based liquids and chemical additives). In most attempts at measuring the dielectric properties of liquids, the measurement setup is fixed in a vertical position and a dielectric plug, possessing properties closely related to air, whose effect is then not taken into account in the theoretical formulation of the problem, is used[4,13]. In addition to this unknown, a meniscus formed at the top of the sample deforms the measurement plane and induces an additional source of error. For the case of granular samples, the setup would be similar. Again, since it is tedious to ensure the compacting of the material such that the measurement plane is normal to the direction of propagation of the incident wave, errors in the determination of the dielectric properties will arise.

In this paper, a two-port network analyzer dielectric constant measurement of granular or liquid materials for the study of cement based materials is proposed. Through this study, we hope to increase the accuracy of the existing mixing models by better determining the frequency varying dielectric constant of the individual constituents that form a concrete specimen.

THEORETICAL APPROACH

The geometry of the problem is shown in Figure 1. In this setup, two dielectric plugs possessing identical lateral dimensions (equal to those of the waveguide), thicknesses and relative complex dielectric properties (ε_p^*) are used. In this fashion, the geometry is symmetrical and possesses characteristics that greatly simplify the formulation. The sample is then pressed between the two plugs such that no airgap, meniscus or irregularities in the measurement plane are present. The dielectric properties should be such that they present minimal attenuation of the incident wave to the samples under test (SUT).

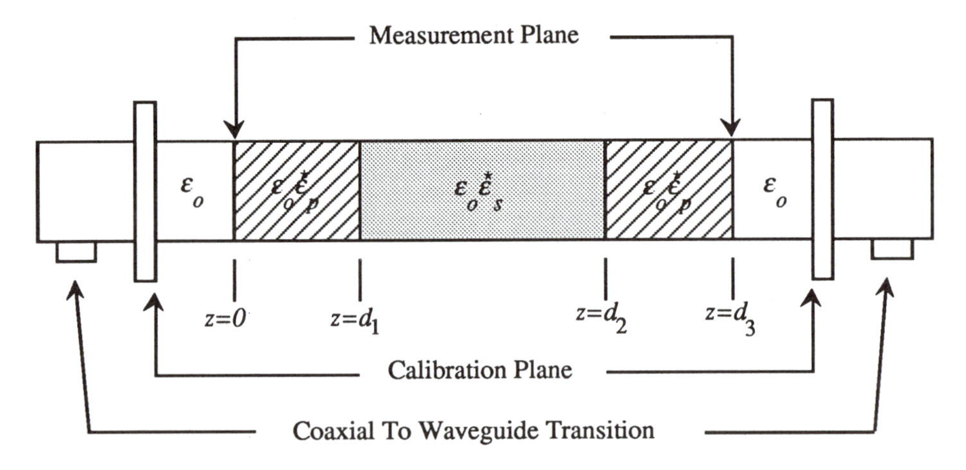

Figure 1: Measurement setup.

In a fashion similar to that established by Nicolson-Ross[10] and Weir[11] (known as the NWR procedure), the scattering parameters (S-parameters) of the SUT must be derived with respect to the dielectric properties of the plugs and the SUT. Through appropriate boundary matching, the following relationships are obtained:

$$S_{21}+S_{11}=\frac{\left(T_1^2\Gamma_2+T_1^2T_2+T_2\Gamma_1\Gamma_2+\Gamma_1\right)}{\left(T_1^2T_2\Gamma_1+T_1^2\Gamma_1\Gamma_2+T_2\Gamma_2+1\right)}=V_1 \tag{1}$$

$$S_{21}-S_{11}=\frac{\left(T_1^2\Gamma_2-T_1^2T_2-T_2\Gamma_1\Gamma_2+\Gamma_1\right)}{\left(T_1^2T_2\Gamma_1-T_1^2\Gamma_1\Gamma_2+T_2\Gamma_2+1\right)}=V_2 \tag{2}$$

$$\Gamma_1 = \frac{Z_p - Z_o}{Z_p - Z_o} \tag{3}$$

$$\Gamma_2 = \frac{Z_s - Z_p}{Z_s - Z_p} \tag{4}$$

$$T_1 = \exp(-\gamma_p d_p) \tag{5}$$

$$T_2 = \exp(-\gamma_s d_s) \tag{6}$$

$$\gamma_p = j2\pi\sqrt{\frac{\varepsilon_p^* \mu_p^*}{\lambda_o^2} - \frac{1}{\lambda_c^2}} \tag{7}$$

$$\gamma_s = j2\pi\sqrt{\frac{\varepsilon_s^* \mu_s^*}{\lambda_o^2} - \frac{1}{\lambda_c^2}} \tag{8}$$

$$\gamma_o = j2\pi\sqrt{\frac{1}{\lambda_o^2} - \frac{1}{\lambda_c^2}} \tag{9}$$

$$Z_p = \frac{j\omega\mu_o\mu_p^*}{\gamma_p} \tag{10}$$

$$Z_s = \frac{j\omega\mu_o\mu_s^*}{\gamma_s} \tag{11}$$

$$Z_o = \frac{j\omega\mu_o}{\gamma_o} \tag{12}$$

where S_{11} and S_{21} are the reflection and transmission scattering parameters and can readily be obtained using (1) and (2); Γ_1, Γ_2, T_1 and T_2 are the first reflection and transmission coefficients in the plugs and in the SUT respectively; (γ_o, Z_o), (γ_p, Z_p) and (γ_s, Z_s) represent respectively the propagation constants and the wave impedances of the air, plug and SUT filled cells; λ_o and λ_c correspond to the free-space and the waveguide cutoff wavelength; $d_p = d_1$ and $d_s = d_2 - d_1$ are the lengths of the plugs and SUT; and $(\varepsilon_p^*, \mu_p^*)$ and $(\varepsilon_s^*, \mu_s^*)$ are the complex relative to free-space electromagnetic properties of the plug and the SUT respectively. The plugs and the SUT are assumed non-ferromagnetic relative to free-space here.

SOLVING OF THE ELECTRICAL PARAMETERS

Proceeding with the measurement of the S-parameters of the setup described in Figure 1, the dielectric properties of the SUT are calculated in a two-step approach: 1) the permittivity is calculated via the NWR[10,11] procedure shown in (13)-(17):

S_{11}, S_{21} : measured scattering parameters.

$\Downarrow$

$$X=\frac{1}{2}\frac{\left(1-\Gamma_1^2S_{21}^2+\Gamma_1^2S_{11}^2-2\Gamma_1S_{11}\right)T_1^4+\Gamma_1^2-2\Gamma_1S_{11}+S_{11}^2-S_{21}^2}{T_1^2\left(\Gamma_1-\Gamma_1S_{21}^2+\Gamma_1S_{11}^2-\Gamma_1^2S_{11}-S_{11}\right)} \tag{13}$$

$$\Gamma_2=-X\pm\sqrt{X^2-1} \tag{14}$$

$$T_2=\frac{T_1^2\Gamma_2+\Gamma_1-V_1\left(1+T_1^2\Gamma_1\Gamma_2\right)}{V_1\left(\Gamma_2+T_1^2\Gamma_1\right)-T_1^2-\Gamma_1\Gamma_2} \tag{15}$$

$$\frac{1}{\Lambda^2}=\left[\frac{-j}{2\pi d_s}\ln\left(\frac{1}{T_2}\right)\right]^2 \tag{16}$$

$$\varepsilon_s^*=\lambda_o^2\left(\frac{1}{\Lambda^2}+\frac{1}{\lambda_c^2}\right) \tag{17}$$

2) the value of ε_s^* obtained from step 1 is used as an initial value for a mean square error solving of the following equation[12],

$$S_{21}^2-S_{11}^2=V_1V_2 \tag{18}$$

where V_1 and V_2 are functions of ε_s^* and S_{11}, S_{21} are the measured scattering parameters. The combination of both of these techniques will hereafter be referred to as the NWR/Baker-Jarvis procedure.

The reason for embedding these methods is twofold. Although the NWR derivations provide explicit equations for the permittivity, the solution is not straightforward since a phase ambiguity exists between the measured and calculated group delay. Also this procedure does not work well at frequencies where the combined effect of the plugs and the sample's dielectric properties and lengths result in a quarter-wave transformer (i.e. $S_{11}\rightarrow 0$). To circumvent this problem, short plugs and samples must be used but this has the disadvantage of decreasing the sensitivity of the measurement[12]. Equation (18) is not as sensitive to this behavior and provides a robust fine tuning of the NWR solution in the particular cases where the precision of the latter method decreases.

EXPERIMENTAL RESULTS

Two samples, air and dry sand, were measured at X-band (8.2 - 12.4 GHz) to evaluate the validity of the adaptation of the NWR and Baker-Jarvis procedures to the proposed geometry of the problem. In both cases, two Plexiglas plugs with thickness d_p=1 cm, whose dielectric properties were previously measured through the above two-step method setting T_1=1 and Γ_1=0 (i.e. air plug with zero thickness), were used to contain and compact the samples. The samples possessed identical thicknesses of d_s=8.86 cm. Figure 2 and 3 present the extracted frequency dependent relative (to free-space) permittivity of the samples through the NWR procedure and the further refined values obtained through the Baker-Jarvis procedure. As seen the combination of the two procedures offer a stable broad band dielectric measurement procedure. It also can be appreciated from the measurement of air that the effect of the plugs is adequately calibrated out of the measurement.

CONCLUSION

An exact method for measuringthe dielectric properties of granular materials (applicable to liquids as well) was presented. It involves a simple boundary matching procedure in which the S-parameters are related to the dielectric properties and lengths of the plugs and the

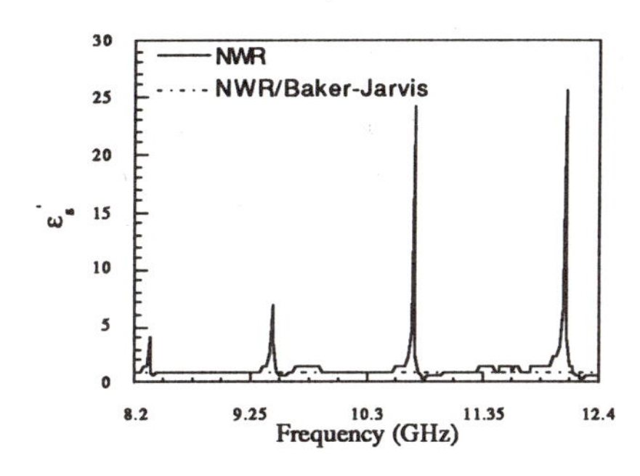

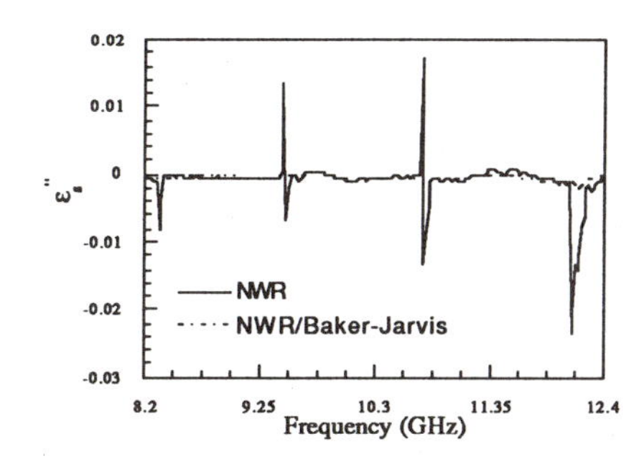

Figure 2: Measured X-band real (ε'_s) and imaginary (ε''_s) parts of the relative dielectric constant of air.

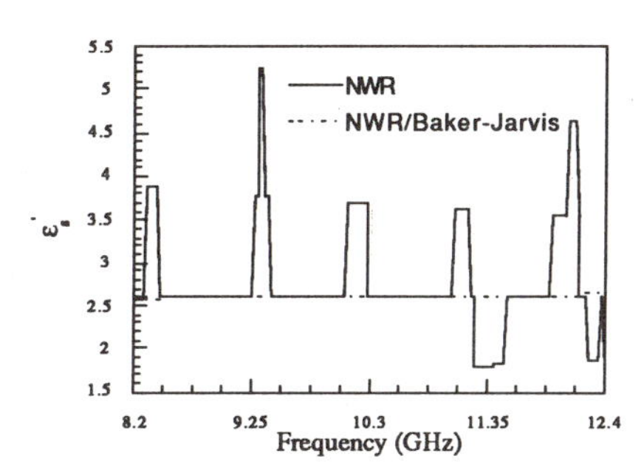

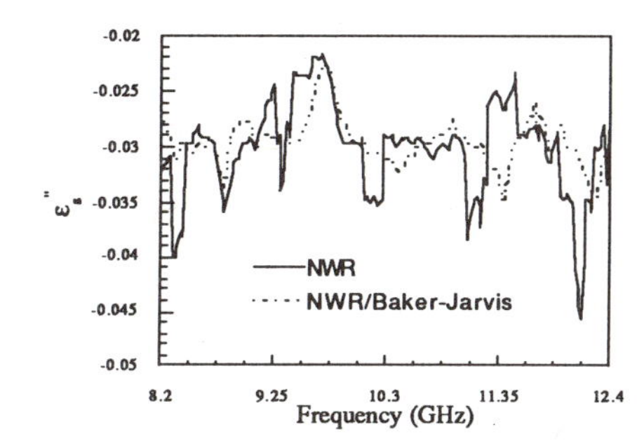

Figure 3: Measured X-band real (ε'_s) and imaginary (ε''_s) parts of the relative dielectric constant of dry sand.

sample. As opposed to the case where the effect of a single low permittivity and loss-loss dielectric plug is assumed to be calibrated out of the setup, our derivation analytically and effectively removes the effect of the two dielectric plugs on the dielectric property measurement. Also, the choice of the plugs is not restricted to low-permittivity materials. This would be useful in applications where the SUT could be chemically reactive to the plug or be at high temperatures for plugs such as Plexiglas or Teflon. The setup also readily ensures an exactlt determined plane of measurement.

ACKNOWLEDGMENTS

This study was funded by the joint National Science Foundation (NSF) (contract CMS-9523264) and the Electrical Power Research Institute (EPRI) (contract WO 8031-09) Program on Sensor Systems and Other Dispersed Civil Infrastructure Systems.

REFERENCES

1. M. Malhotra, and N.J. Carino, Editors, *Handbook on Nondestructive Testing of Concrete*, CRC Press, 1991, p.343.
2. R. Zoughi, "Microwave and Millimiter Wave Nondestructive Testing: A succinct Introduction," *Materials Evaluation*, vol. 53, no. 4, pp.461-462, April 1995.
3. S. Ganchev, N. Qaddoumi, D. Brandenberg, S. Bakhtiari, R. Zoughi and J. Bhattacharyya, "Microwave Diagnosis of Rubber Compounds," *IEEE Transaction on Microwave Theory and Techniques*, vol. 42, no. 1, pp. 18-24, January 1994.
4. Qaddoumi, N., S. Ganchev and R. Zoughi, "Microwave Diagnosis of Low Density Glass Fibers with Resin Binder," *Research in Nondestructive Evaluation*, vol. 8, no. 1, pp. 177-188, 1996.

5. S. Gray, S. Ganchev, N. Qaddoumi, G. Beauregard, D. Radford and R. Zoughi, "Porosity Level Estimation in Polymer Composites Using Microwaves," Materials Evaluation, vol. 53, no. 3, pp.404-408, March 1995.
6. R. Zoughi, S. Gray and P. S. Nowak, "Microwave Nondestructive Estimation of Cement Paste Compressive Strength," *ACI Materials Journal*, vol. 92, no. 1, pp. 64-70, January-February, 1995.
7. K. Bois, R. Mirshahi and R. Zoughi, "Dielectric Mixing Model for Cement Based Materials," *Proceedings of the Review of Progress in Quantitative NDE, vol. 16, Plenum Press*, NY, 1997.
8. M. A. Stuchly and S. S. Stuchly, "Coaxial Line Reflection Method for Measuring Dielectric Properties of Biological Substances at Radio and Microwave Frequencies-A Review," *IEEE Trans Instrum. Meas.t*, vol. 29, pp. 176-183, Sept. 1980.
9. J. Baker-Jarvis, C. Jones, B. Janezic, R. G. Geyer, J. H. Grosvenor, Jr., and C. M. Weil, "Dielectric and Magnetic Measurement: A Survey of Nondestructive, Quasi-Nondestructive and Process-Control Techniques," *Res. Nondestructive Eval.*, no. 7, pp. 117-136, 1995.
10. A. M. Nicholson and G. Ross, "Measurement of Intrinsis Properties of Materials by Time Domain Techniques," *IEEE Trans. Instrum. Meas.*, vol. 19, pp. 377-382, Nov. 1970.
11. W. B. Weir, "Automatic Measurement of Complex Dielectric Constant and Permeability at Microwave Frequencies," *Proc.* IEEE, vol. 62, pp. 33-36, Nov. 1970.
12. J. Baker-Jarvis, E. J. Vanzura, and W. A. Kissck, "Improved Technique for Determining Complex Permittivity with Transmission/Reflection Method," *IEEE Trans. Microwave Theory and Technique*, vo. 38, pp. 1096-1103, August 1990.
13. P. I. Somlo, "A Convenient Self-Checking Method for the Automated Microwave Measurement of μ and ε," *IEEE Trans. Intrum. Meas.*, vol. 42, no. 2, pp. 213-216, April 1993.

MICROWAVE DIELECTRIC PROPERTIES OF A SLAB DETERMINED BY A MULTI-FREQUENCY FREE-SPACE TECHNIQUE

John M. Liu,[1] and John H. Wasilik[2]
[1]Carderock Div. Naval Surface Warfare Center
West Bethesda, Md. 20817-5700
[2]1307 Sarah Drive
Silver Spring, Md.

INTRODUCTION

Many techniques exist for the determination of the real and imaginary parts of the relative permittivity of dielectric solids[1-5]. Among these are free-space techniques based on quasi-optics[6]. A nominally flat piece of material can be evaluated with this approach without cutting and shaping into precise dimensions, as are required by some approaches based on a waveguide or a cavity. Some techniques are based on coupling the E & M field from an open-ended waveguide or coaxial line to the material. One such device available commercially has a 3.5 mm active central element, and the calibration is based on the response of water. Our experience in using this probe indicates its limitation in accuracy for loss tangent less than 0.05.

A need arises in our laboratory for the measurement of the dielectric constant and loss tangent in low loss composites with the value of the dielectric constant between 4 and 10, and loss tangent less than 0.05. We report here an approach using a single sided, free-space technique which combines frequency domain and time domain measurement concepts for relatively thick materials. It is an extension of work reported previously by one of the authors for the detection of defects in dielectrics[7].

FREQUENCY DOMAIN CONSIDERATIONS

In Figure 1 are shown a plane wave propagating towards a dielectric slab. Successive reflections at the front and rear surfaces give rise to the interference phenomena in the frequency domain, as the input plane wave is swept through a band of frequencies. At normal incidence, the following expression holds, based on elementary physics, for the frequencies at which the amplitude of the multiply reflected waves is a minimum.

$$kd - \pi = n\,\pi \qquad n = 1, 3, 5.....$$

where $k = 2\pi\,(e')^{1/2} / \lambda$ and $\lambda = V_o / f$

Nondestructive Characterization of Material VIII
Edited by Robert E. Green Jr., Plenum Press, New York, 1998

d is the slab thickness, e′ the dielectric constant which is the real part of the relative permittivity, λ the wavelength, V_o the speed of light in free space, and f the frequency. From this, Δf, the difference in frequency between adjacent amplitude minima, is given by,

$$\Delta f = V_o / 2\, d(e')^{1/2} \qquad [1]$$

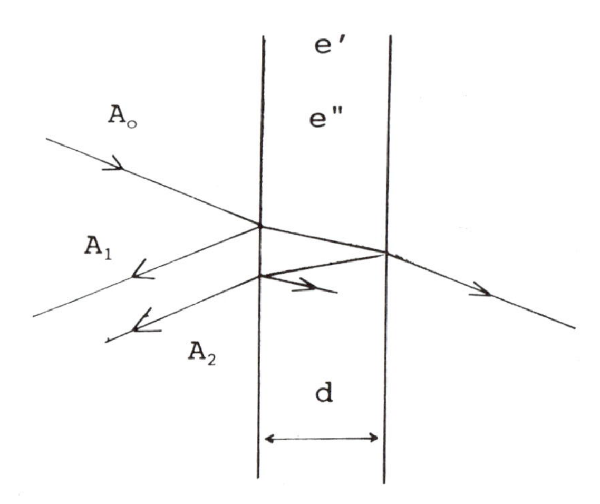

Figure 1. Wave reflection and interference in a dielectric slab.

Eq. [1] indicates that the first minimum occurs when the electrical thickness of the material equals half a wavelength. For a fixed thickness, as the frequency is increased, successive minima occur at those frequencies when this thickness equals multiples of half wavelength. The π phase shift at the air/dielectric interface has been included in our consideration. In general, e′ should be replaced by the complex, relative permittivity, even though the imaginary part is small compare to the real part for the dielectric materials considered in this work.

TIME DOMAIN CONSIDERATIONS

Suppose a short electromagnetic pulse impinges on the dielectric slab as shown in Fig. 1. The pulses partially reflected from the front and rear surfaces will be separated in time when they reach an external receiver. For a single-sided experiment, the time between these two pulses will be,

$$\Delta t = 2\, d\, (e')^{1/2} / V_o$$

Thus, in principle the dielectric constant can be determined by measuring these arrival times. However, in our experiments, instead of short pulses, the time domain information was reconstructed from the multi-frequency reflection coefficient via Fourier transformation. For the thickness and bandwidth available, the peak-to-peak separation in time was not large enough to make this determination of e′ accurate, compared to the frequency domain approach indicated above.

For the determination of the loss tangent we proceed as follows. Let a plane wave propagating along +x axis be represented by,

$$A_o = \exp^{j(kx-\omega t)}$$

The propagation constant $k = \alpha + j\beta$. Following Stratton[8], the attenuation coefficient β in units of Neper/cm and the phase constant α are related to the material properties by the following relationships,

$$(\alpha^2 - \beta^2) = \mu e \omega^2$$

$$\alpha \beta = \mu \sigma \omega / 2$$

where μ is the permeability, e the real part of the permittivity, and σ the conductivity. With the following definitions, appropriate for non-magnetic materials,

$$\mu = \mu_o$$

$$e = e_o e'$$

$$e'' = \sigma / \omega e_o$$

$$\tan(\delta) = e''/e'$$

$$\text{permittivity} = e_o(e' + je'')$$

and assuming $\alpha >> \beta$, the following relationships hold,

$$\alpha = 2\pi f \, (e')^{1/2}/V_o$$

$$\beta = \alpha \tan(\delta)/2 \qquad [2]$$

The following relationship holds for the ratio of the amplitudes of the front surface reflection, A_1, to the first rear surface reflection, A_2

$$A_2 \,/\, A_1 = (1 - r^2) \, e^{-2\beta d}$$

Thus, $$\beta = \{ -2.3*20\log(A_2/A_1) + 2.3*20\log(1-r^2)\}/40d \qquad [3]$$

where r is the Fresnel reflection coefficient for the air/dielectric interface. It is shown by Stratton[9] that, for normal incidence,

$$r^2 = \{[1-(e')^{1/2}]^2 + X^2\} \,/\, \{[1+(e')^{1/2}]^2 + X^2\} \qquad [4]$$

where $X = (e')^{1/2} \tan(\delta)/2$

EXPERIMENTAL PROCEDURE

The experimental set-up is shown in Fig. 2. Continuous microwave launched from the antenna was focused at the specimen. Waves back reflected were captured by the same antenna. The heart of the measurement system is a vector network analyzer which allows the amplitude and phase of the microwave to be analyzed simultaneously. The frequency was stepped from 22 through 40 GHz. After digitization, the system computer converted the frequency domain data via Fourier Transformation to the time domain, generating a pulse-echo type representation. In this representation, the signal associated with the front and back surfaces of the slab are separated in time, depending on the wave speed inside the dielectric. The attenuation coefficient determined from this pulse-echo train is an average over the frequency band used.

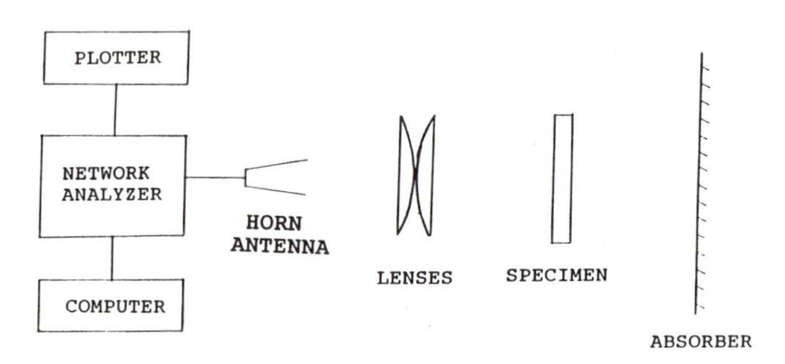

Figure 2. Schematic of free-space measurement set-up.

One aspect of our technique is to measure the separation in frequency of adjacent minima of the amplitude of the reflection coefficient, as given by Eq. (1), and the thickness d of the slab. One problem in converting these frequencies to dielectric constant is the fact that in general the permittivity in Eq. [1] is complex. However, we calculated the reflection coefficient numerically by well-known procedures, taking into account the material loss tangent, and compared these results to those obtained when the loss tangent was taken to be zero. These results show that the frequency separations at the amplitude minima given by Eq. [1] are virtually independent of material loss when the loss tangent is less than 0.05. Thus, the imaginary part in the permittivity in Eq. [1] can be ignored, and the real part is extracted directly from the observed Δf's for successive amplitude minima. For the material thickness studied, a large number of minima was observed when the frequency was stepped from 22 through 40 GHz, allowing accurate determination of an averaged Δf and e'.

We determine the attenuation coefficient (averaged over the frequency band 22-40 GHz) in the following manner. First, the difference in db between the reflection in the time domain from the front surface and the first reflection from the rear surface is measured. Then, Eq. [3] and [4] are applied using the value of e' determined in the frequency domain from the amplitude minima and assuming $\tan(d)=0$. This "0th order" attenuation coefficient, β, is then used to calculate a $\tan(\delta)$ using Eq. [2]. With this $\tan(\delta)$, new values of r and β (the 1st order β) are obtained using Eq. [3] and [4]. This new β is in turn used to obtain a new $\tan(\delta)$ and then new values of r and β (2nd order β). For the materials of low loss tangent studied, the value of β converges after 1 or 2 iterations.

RESULT AND DISCUSSIONS

Some typical plots of the reflection coefficient versus frequency are shown in Fig. 3 a and 3b for composites. The

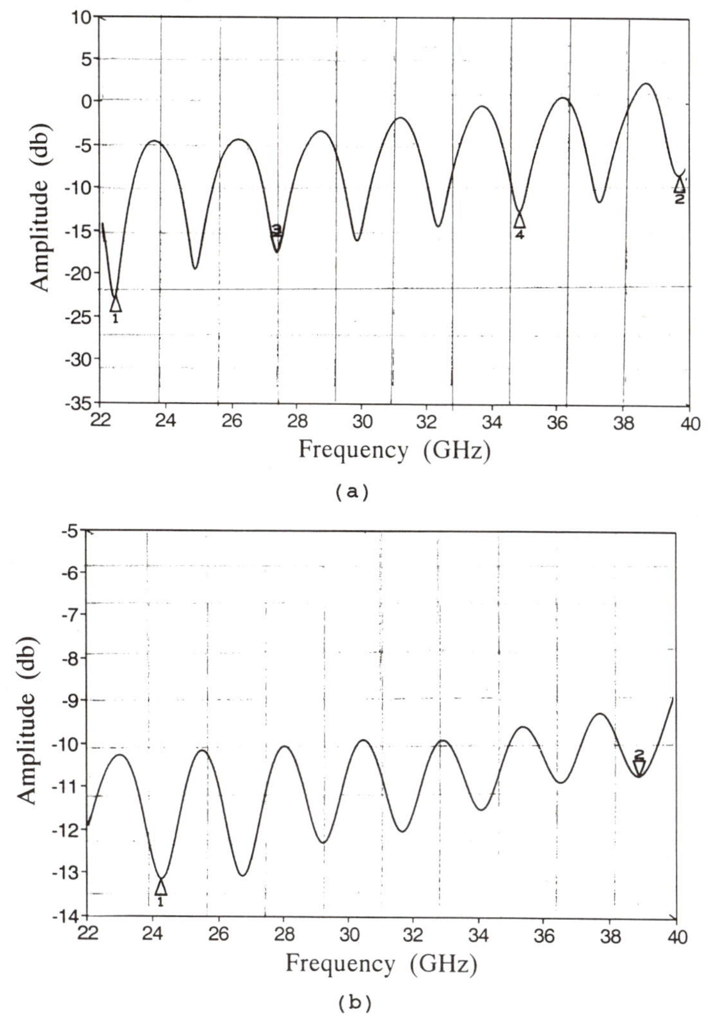

Figure 3 Frequency dependence of the amplitude of normally incident, perpendicularly polarized, back-reflected signal from composite slabs: (a) glass/epoxy composite, 2.858 cm thick, and (b) glass/melamine composite, 2.578 cm thick.

corresponding time domain data are shown in Fig. 4a and 4b. From data such as these, the Δf's, e', β, and tan(δ) were determined following the procedure stated above. Table I shows a summary of the averaged values of e' and tan(δ) together with RMS errors for several measurements for each specimen.

Table I. Summary of Results (22-40 GHz).

Material	e'	tan(δ)
Plexiglass	2.6 +/-0.1	0.007 +/-0.001
glass/epoxy	4.6 +/-0.2	0.015 +/-0.003
glass/melamine	5.5 +/-0.2	0.042 +/-0.003
glass/melamine	5.8 +/-0.2	0.043 +/-0.002

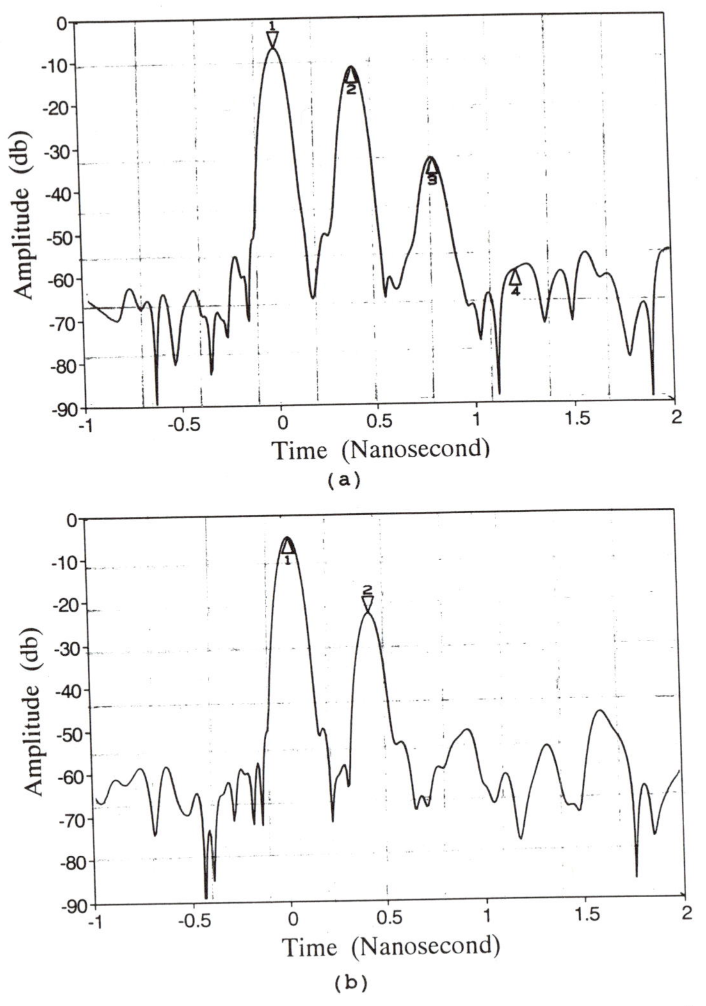

Figure 4. (a) and (b) Time domain, pulse-echo representation of the amplitude information shown in Fig. 3a and 3b.

The amplitude of the reflection coefficient in the frequency domain contains information on the loss tangent, in the form of an amplitude modulation as frequency increases. However, we found this amplitude modulation to be significantly affected by alignment problems experimentally, even though the frequencies at amplitude minima were not. Thus, we have not attempted to obtain dielectric loss information from the amplitude information in the frequency domain.

CONCLUSIONS

A free-space approach for the determination of the real and imaginary parts of the relative permittivity of dielectrics and composites is outlined. Successful applications of this technique to dielectrics with loss tangent in the range of 0.01 to 0.05 are shown.

REFERENCES

1. H. M. Altschuler, in Handbook of Microwave Measurements, Vol. 2, Polytechnic Institute of Brooklyn, p. 495 (1963).

2. M. A. Stuchly and S. W. Stuchly, IEEE Trans. Instr. & Meas. IM-29 (3), 176 (1980).

3. M. N. Afsar, et al, Trans. IEEE ,74 (1), 183 (1986).

4. R. Zoughi and S. Ganchev, Microwave Nondestructive Evaluation, Nondestructive Information Analysis Center, Texas Research Institute Austin, Inc. Austin, Tx (1995).

5. J. Baker-Jarvis, et al, Res. Nondrstr. Eval. 7, 117 (1995).

6. D. K. Ghodgaonkar, et al, IEEE Trans. Instr. & Meas. IM-38 (3), 789 (1989).

7. J. M. Liu, in Rev. Prog. QNDE, Vol 15, eds. D.O. Thompson and D.E. Chimenti (Plenum, NY 1996), p. 705.

8. J. A. Stratton, Electromagnetic Theory, (McGraw-Hill, NY 1941), p. 276.

9. J. A. Stratton, loc. cit. p. 506.

Preliminary Evaluation of Microwave Techniques for Inspection of Thick Layered Composite Deck Joints

L.M. Brown[1], J.J. DeLoach[1], R. Zoughi[2], and E. Ranu[2]

[1]Naval Surface Warfare Center, Carderock Division
Survivability, Structures and Materials Directorate
West Bethesda, MD 20817

[2]Colorado State University
Department of Electrical Engineering
Fort Collins, CO 80523

INTRODUCTION

The Carderock Division, Naval Surface Warfare Center (NSWCCD) is conducting a program to develop, install, and evaluate an Advanced Enclosed Mast/Sensor (AEM/S) System. The overall program objective is to develop an enclosed, composite mast for Navy surface ship combatants.

The AEM/S structure is comprised of composite skins with foam or balsa cores. On the upper half of the mast, the walls are constructed of foam cores sandwiched between composite skins of vinyl ester resin and E-glass fibers. The core is comprised of three sections of foam separated by signature control material bonded to the foam using film adhesive. The various sections of the structure are connected at complex joints joined by a combination of adhesive bonds, secondary bonds, and bolts.

To ensure desired performance of the AEM/S, NSWCCD and Ingalls Shipbuilding Division (ISD) are conducting extensive fabrication exercises and structural testing of large-scale mockups. After fabrication of mockups for Deck Joint test samples NSWCCD personnel observed voids and disbonds along the base of the triangular foam section. Figure 1 presents a schematic of the joint and shows the location of a representative void.

The joint geometry, core materials, and defect orientation render conventional techniques unable to detect planar flaws lying below the core material. Radiography is inadequate because the defect is planar and its thickness is oriented normal to the incident beam. A significant obstacle in applying ultrasonics is the attenuative nature of the laminate (skin) and core materials. Experimental trials on foam and honeycomb cored panels indicate that, at frequencies as low as 500 kHz, any reflected signal from the flaw will be attenuated completely before reaching the transducer [1].

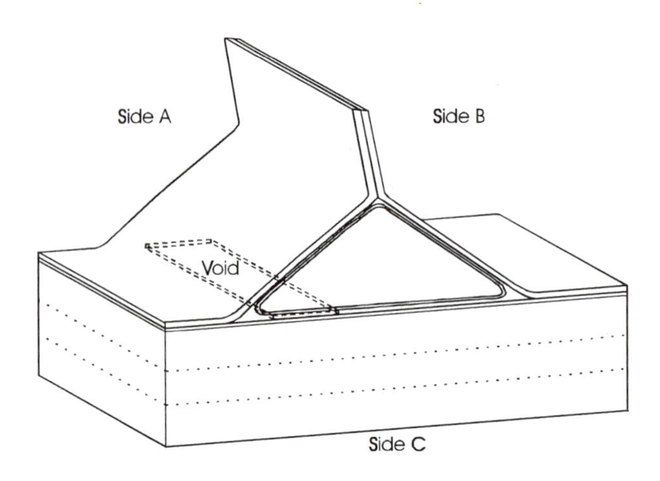

Figure 1. Schematic of Deck Joint showing the location of the defect.

Near-field microwaves are well suited for inspection of composite and laminate materials, because microwaves can easily penetrate low-loss dielectric materials. In a near-field microwave inspection, the signature control material in the Deck Joint establishes an infinite plane boundary condition. The reflected energy at the boundary sets up a standing wave pattern in the material and waveguide. A diode detector in the waveguide can measure the amplitude and/or phase of the standing wave. Material changes, either due to geometry changes or defects, will modify the phase and amplitude of the standing wave which can be observed at the diode detector. The main concern in a near-field microwave inspection of the Deck Joint is whether sufficient energy can be reflected back to the waveguide for detection of defects at critical locations in the material.

APPROACH

In order to evaluate near-field microwaves for inspection of the Deck Joint, a laboratory specimen was taken to the Applied Microwave Nondestructive Testing Laboratory at Colorado State University. The specimen, with a width of 250 mm, represents a full-scale section of the Deck Joint as manufactured at ISD. A full-width void, approximately 2.6 mm thick in height, was located along the base of the triangular foam section as illustrated in Figure 1.

In the experimental set-up, the open-ended waveguide is mounted at a fixed position above the specimen, and the face of the waveguide is oriented parallel to the base of the Deck Joint. The specimen is placed on a x-y scanning table and measurements are recorded along the width of the joint by translating the specimen along the x-direction in step increments of 5 mm. The oscillator in the waveguide is powered by a DC voltage source, and a diode detector located inside the waveguide couples the reflected energy to a digital voltmeter for measurement readings. A portable computer controls the stepping motor of the x-y scanner and records the measurements from the digital voltmeter.

In the near-field, the phase and amplitude of the energy reflected back to a microwave waveguide is characteristic of the dielectric properties and geometry of the specimen. In addition, the reflected energy is also characteristic of the waveguide frequency, standoff distance, and orientation [2]. Changes in the phase of the reflected energy due to defects or changes in the material properties are detected by the waveguide diode.

Three different microwave sensors incorporating open-ended, rectangular waveguides at K-band, at K-band with a small pyramidal horn, and at X-band were evaluated for inspection of the Deck Joint. The X-band is defined over the frequency range of 8.2-12.4 GHz, and the K-band is defined over the frequency range 18-26.5 GHz [3]. The dominant mode electric field in the waveguide is orientated parallel to the shortest dimension of the rectangular waveguide. During the inspections, the waveguide was positioned such that the electric field was orientated perpendicular to both the composite fibers and the scan direction. In all of the inspections, one edge of the waveguide was positioned flush against the composite joint producing a 0 mm standoff distance at the edge closest to the material.

RESULTS

The first step was to verify that the microwave signal could penetrate the Deck Joint and that sufficient energy could be reflected back from the void for detection by the waveguide diode. In order to determine if the microwave signal was penetrating the bond layer, a thin metal plate, approximately 1.5 mm thick and 30 mm wide, was inserted into the void. If the microwave signal completely penetrates the Deck Joint, then an abrupt change in the phase of the signal should occur at the metal plate boundaries. In addition, the signal should be phase-shifted by an observable amount in the presence of the metal plate. Alternatively, if the microwave signal is only partially penetrating the Deck Joint then no changes in phase of the reflected signal occur as a result of the metal plate.

Figure 2 shows the experimental inspection results of the Deck Joint with the metal plate located from 100-250 mm. The K-band waveguide was located 6.7 mm above the base of the bond joint. The figure contains a reference inspection (no metal plate) of the joint as scanned along the x-direction. The reference signal (dashed line in the figures) shows moderate variation of the detector voltage along the width of the specimen. The dashed line represents the inspection of the Deck Joint with no metal plate located in the void. Changes in fiber placement, resin thickness, and void thickness will cause slight variations in the reflected signal as seen by a 15 mV fluctuation in the dashed line.

The solid line in Figure 2 represents a second inspection of the Deck Joint with the metal plate located in the void from 100-250 mm. From 0 to 90 mm the solid line closely tracks the dashed line and little change in the reflected signal is observed as a result of the metal plate. After about 100 mm, the two signals diverge because of the change in the phase of the reflected microwave signal as a result of the metal plate. Once the metal plate is encountered,

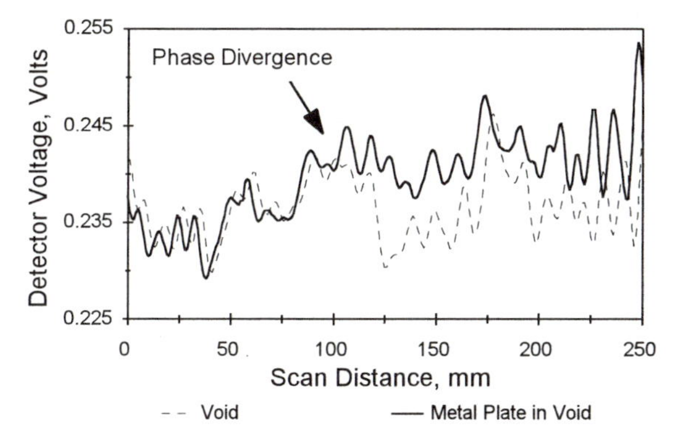

Figure 2. Microwave inspection of the Deck Joint with a metal plate located in the void from 100-250 mm (K-band, 0 mm standoff distance).

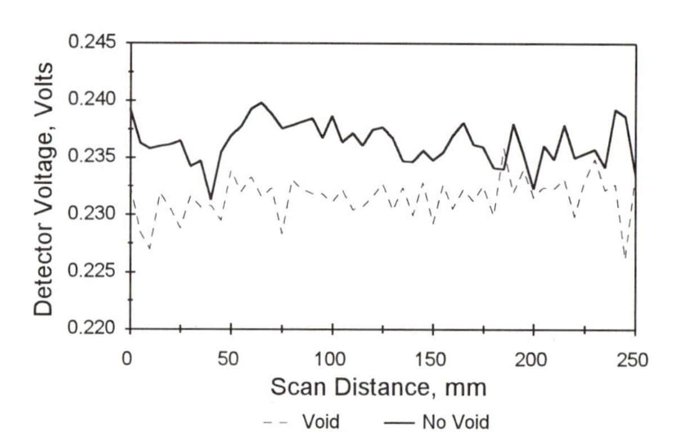

Figure 3. Microwave inspection of the Deck Joint with a K-band waveguide placed 6.4 cm above the joint base (0 mm standoff distance).

approximately a 5-10 mV increase in the diode detector voltage is observed. The results from this first experimental step clearly show that the microwave signal is sufficiently penetrating the laminate and core materials to the depth of the Deck Joint base.

The next experimental step was to perform microwave inspections of the joint using different waveguides and to compare the results. Three different waveguides were tested, one operating in the K-band, one operating in the K-band with a pyramidal horn, and one operating in the X-band. For these experiments, a scan was performed on the side of the triangular joint containing the void, side A (see Figure 1), and then a scan was performed on the opposite side of the triangular joint where no defect was present, side B.

The inspection result using the K-band waveguide is shown in Figure 3. The dashed line in Figure 3 is a plot of the diode detector voltage from a scan on the side of the joint containing the void, side A, and the solid line is a plot of the voltage from a scan on the opposite side of the joint containing no defect, side B. In both scans, the electric field of the waveguide was orientated perpendicular to the scan direction and the open end of the waveguide was placed 6.4 cm above the Deck Joint base. The recorded measurements

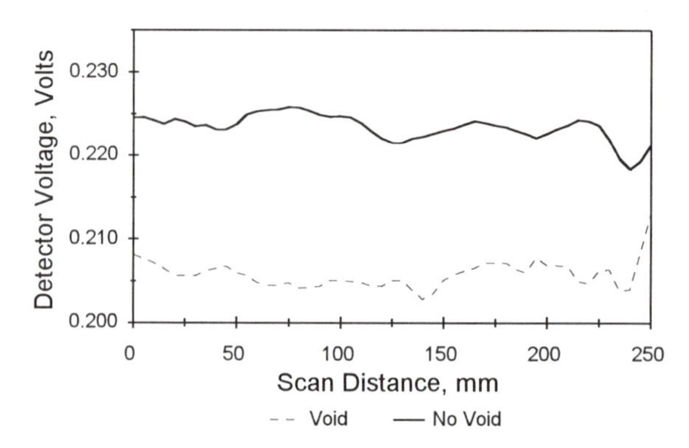

Figure 4. Microwave inspection of the Deck Joint with a K-band waveguide with a pyramidal horn placed 7.1 cm above the joint base (0 mm standoff distance).

show some variability in voltage along the length of the scan as a result of minor variations in the material, resin thickness, fiber location, and surface roughness. However, the signal

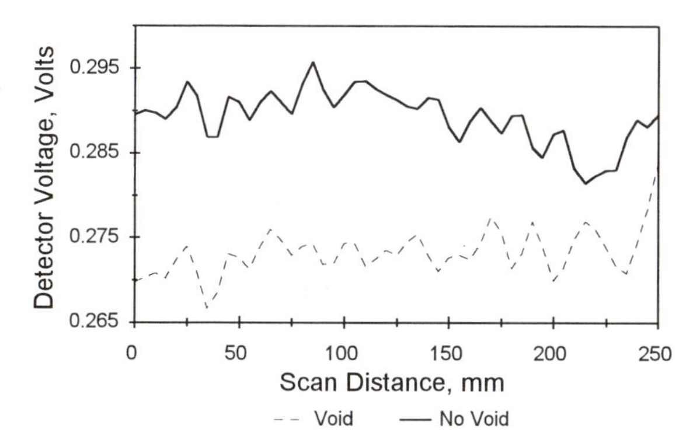

Figure 5. Microwave inspection of the Deck Joint with a X-band waveguide placed 6.7 cm above the joint base (0 mm standoff distance).

from the side with a void consistently shows a decrease in voltage, approximately 4 mV, from the side with no void.

The inspection result using the K-band waveguide with a pyramidal horn is shown in Figure 4. The pyramidal horn yields a smoother signal that is less sensitive to minor fluctuations in the material properties. The presence of the void causes a phase shift in the microwave signal, recorded at the diode detector and converted to a voltage, of approximately 12-22 mV, relative to the signal voltage with no void present. Adding the pyramidal horn to the open-ended waveguide improves the separation of the two signal and improves the capability to detect the void.

Figure 5 shows the inspection result using an X-band waveguide placed 6.7 cm above the joint base. As in the case of the K-band open-ended waveguide, the signals from the X-band exhibit a 5-7 mV variability along the length of the scan. However, at the lower frequency of X-band, the phase sensitivity of the microwave signal to the void appears similar to the sensitivity of the signal using the K-band waveguide with the horn. Both waveguide configurations produce a change in phase due to the void resulting in a 15-20 mV decrease in signal voltage at the diode detector.

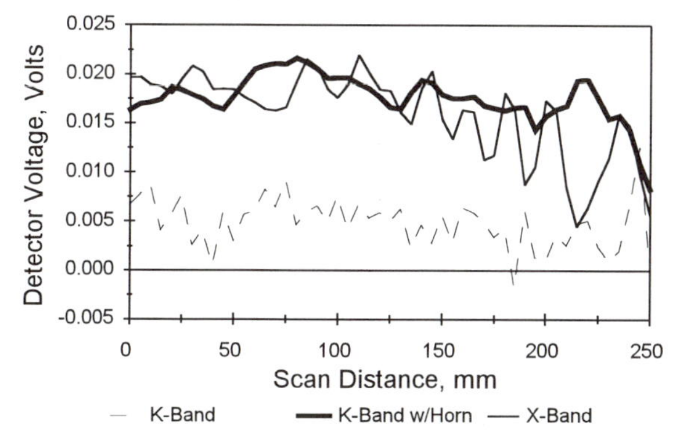

Figure 6. Summary of three inspections, (a) K-band positioned at 6.4 cm, (b) K-band with pyramidal horn positioned at 7.1 cm, and (c) X-band positioned at 6.7 cm, showing the difference in microwave signals, Side B with no void subtracted from Side A with a void (0 mm standoff distance).

Figure 6 presents a summary of the three inspections using the K-band waveguide, the K-band waveguide with a pyramidal horn, and the X-band waveguide. Each line in the figure represents the difference in the signal collected while scanning the side of the Deck Joint without a void and while scanning the opposing side of the Deck Joint containing a void. The side of Deck Joint with a void caused a consistent decrease in signal voltage, resulting in a positive difference in the two signals. The open-ended K-band waveguide had a signal difference around 5 mV, whereas the K-band waveguide with a horn caused an increased difference in the signals, around 18 mV. Both the K-band and X-band waveguide produced considerable variability in the voltage signal, whereas by adding the pyramidal horn to the K-band waveguide, the variability in the signals decreased and the difference in scans from opposite sides of the Deck Joint are more uniform. Overall, the K-band waveguide with a pyramidal horn yielded the best results.

SUMMARY

In summary, the experimental results clearly demonstrated that near-field microwaves can penetrate through the composite and foam structure of the Deck Joint for detection of voids and delaminations. Initial results indicate that the higher frequency of the K-band waveguide in conjunction with a pyramidal horn may provide best sensitivity for detection of defects. Further experimentation is necessary to determine to what extent the microwave signal will change due to changes in the thickness foam core versus signal changes resulting from the presence of defects in the material.

REFERENCES

1. J.J. DeLoach, and C.A. Lebowitz, unpublished results.
2. S. Bakhtiari, N. Qaddoumi, S.I. Ganchev, and R. Zoughi, "Microwave Noncontact Examination of Disbond and Thickness Variation in Stratified Composite Media," *IEEE Transactions on Microwave Theory and Techniques*, Vol. 42, No. 3, March 1994, pp. 389-395.
3. C.A. Balanis, *Antenna Theory Analysis and Design*, John Wiley & Sons, New York, NY, 1982.

ANALYSIS OF LOADING vs. MICROWAVE FATIGUE CRACK DETECTION SENSITIVITY USING OPEN-ENDED WAVEGUIDES*

Nasser Qaddoumi[1], Radin Mirshahi[1], Emarit Ranu[1], Vladimir Otashevich[1], Christian Huber[1], Philip Stepanek[1], Reza Zoughi[1] and J. David McColskey[2]

[1]Applied Microwave Nondestructive Testing Laboratory (*amntl*)
Electrical Engineering Department
Colorado State University
Ft. Collins, CO 80523

[2]National Institute of Standards and Technology (NIST)
Materials Reliability Division
325 Broadway
Boulder, CO 80303

INTRODUCTION

Detection of fatigue and stress induced hairline surface cracks is an important practical concern in a variety of metallic structures and in particular in steel bridges. There are several conventional nondestructive techniques used for this purpose[1]. Each of these techniques possesses certain advantages as well as some limitations. Microwave methods have also been used for surface crack detection in metals[2]. However, their utility have been quite limited thus far. In 1992 a novel microwave approach was introduced employing an open-ended rectangular waveguide for the detection and evaluation of surface cracks in metals[3-5]. This method has the potential for sizing a crack and determining crack tips very accurately even when the crack is filled with a dielectric material or when it is covered with a thin coating such as paint[6-7]. There have also been two electromagnetic models developed in conjunction with these efforts[1,8-9].

Although this approach has been extensively tested in the laboratory, its utility for stress induced fatigue cracks has only been demonstrated on a limited basis using fatigue specimens in the laboratory and by using the models. Therefore, the primary purpose of this study has been to investigate the potential of this approach by inspecting stress induced fatigue cracks under loading. In this way the applicability as well as any limitations associated with this technique may be fully understood. Consequently, stress induced fatigue crack specimens were produced at the National Institute of Standards and Technology (NIST) in Boulder Colorado. These specimens were then subjected to different loads rendering different crack openings or widths. This microwave approach was then used to detect the crack as a function of the impressed loading.

SPECIMEN PREPARATION

An A-36 steel specimen with a thickness of 12.7 mm (0.5") and with the dimensions shown in Figure 1 was used in these experiments. A small hole was drilled through the specimen and a sawcut starter notch was introduced on both sides of the hole as shown in Figure 2. The specimen was then fatigued at 8 Hz, using a closed loop servo-hydraulic fatigue machine, at a maximum stress of 103 Mpa (15 ksi) and a stress ratio of 0.05. Consequently, a tight fatigue crack was generated at the end of the notch whose width varied as a function of the distance away from the notch tip, as shown in Figure 3. The waveguide measurements were made at a point half way between the tip of the machined starter notch and the final crack tip. As the crack tip advances during fatigue, the surface of the specimen (at the crack tip) is in a state of plane stress, which influences the size of the plastic zone. The crack front typically tunnels along its path, with the specimen surface crack trailing the inner portion of the crack front. The uncracked ligament between the inner portion of the crack front and the specimen surface yields as the crack advances, creating a slight indentation along the crack length. This indentation depth varies with the size of the plastic zone. When the plastic zone is large, the indentation can be large, (analogous to necking in a tensile specimen). But when the plastic zone is small, as in fatigue of bridge structures, the indentation can be quite small. The indentation on this A-36 steel sample was very small, and could not accurately be measured without destroying the sample. An estimate of the indentation, using a replication technique, showed the depth to be less than 0.040 mm (0.0016"). Figure 4 shows the crack opening or width as a function of loading, measured halfway between the starter notch tip and the fatigue crack tip.

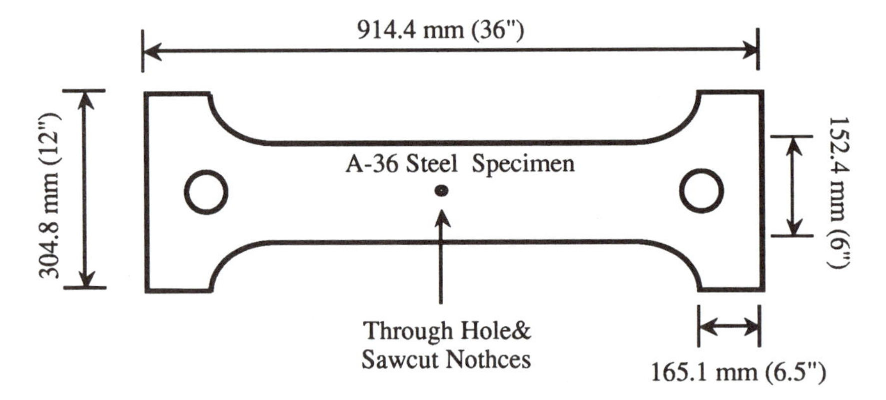

Figure 1: A-36 steel specimen used to generate a fatigue crack.

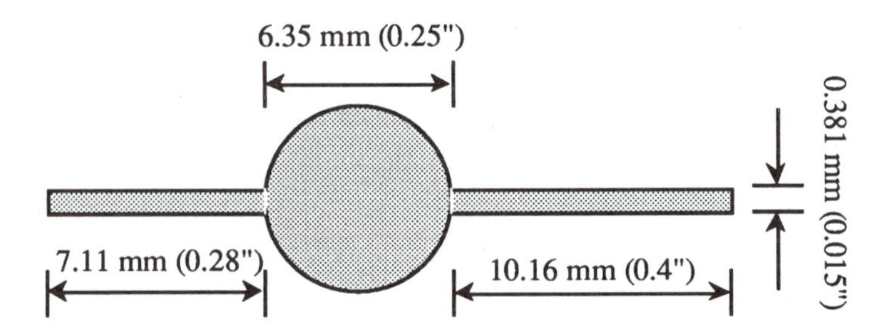

Figure 2: The dimensions of the through hole and the sawcut starter notch.

MEASUREMENT APPARATUS

A single-axis stepper motor driven platform was used on which the microwave measurement system was mounted. The entire apparatus was then mounted on the fatigue machine (at NIST). Using a computer controlled data acquisition software program,

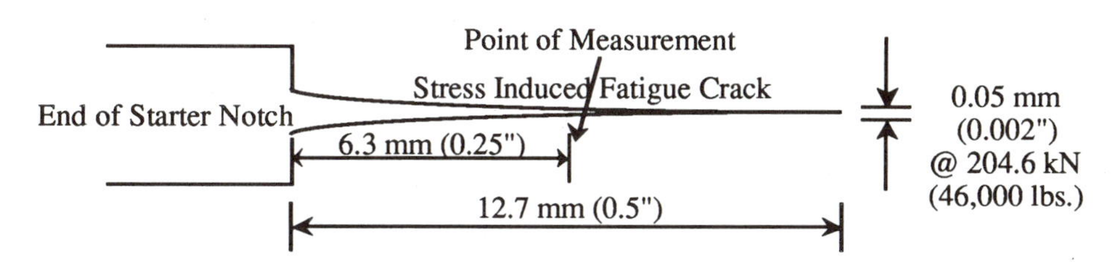

Figure 3: Dimensions of the stressed induced fatigue crack under 205 kN (46,000 lb.) of load.

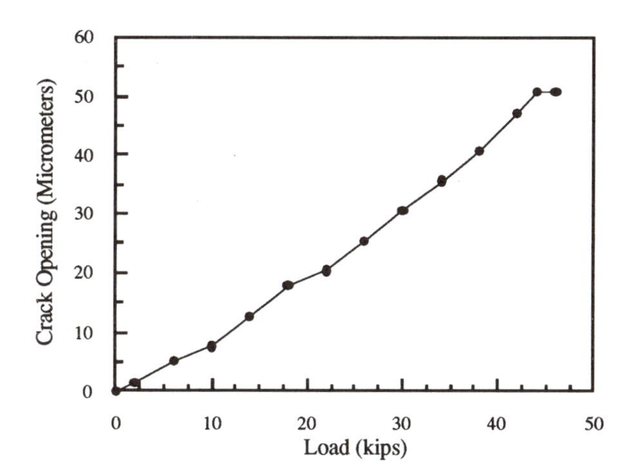

Figure 4: Crack opening vs. impressed loading.

developed at the Applied Microwave Nondestructive Testing Laboratory, the fatigue crack was scanned while under different loads. The microwave measurement system employed the dominant mode detection approach[3,8] at a frequency of 24 GHz (K-band) with aperture dimensions of 10.7 mm x 4.3 mm (0.42" x 0.17"), 35.05 GHz (Ka-band) with aperture dimensions of 7.11 mm x 3.56 mm (0.28" x 0.14") and 65 GHz (V-band) with aperture dimensions of 3.8 mm x 1.9 mm (0.148" x 0.074"). It must also be mentioned that hereon each figure must be studied individually and output voltage from one figure to the next are not directly related to each other. This is due to the fact that the amplifier gain used while making the measurements may have been different for each figure. In some cases an inverting amplifier was used and in some cases a non-inverting one. This was done to maximize the detector output voltage in each case.

It is important for the reader to have an idea of the properties or shape of the crack characteristic signal as obtained by scanning a crack with an open-ended waveguide[3-9]. Figure 5 shows the crack characteristic signal (output voltage vs. scanning distance) for the sawcut starter notch at 24 GHz. The notch represents a deep through crack. The first peak in the output signal indicates when the waveguide flange comes over the notch and a similar peak occurs when it leaves the notch. The signal in the middle, at a scanning distance of 17.5 mm, represents the notch. The nullwidth of this signal is closely equal to the narrow dimension of the waveguide plus the width of the notch, as expected[6]. Furthermore, since this is a deep notch, its crack characteristic signal is consistent with deep cracks investigated before[6]. Figure 6 shows the crack characteristic signal for the fatigue crack under two different loadings at a frequency of 24 GHz. The output signal obtained over an area devoid of a crack is also shown for comparison. Clearly, the presence of the crack is detected. Furthermore, the wider crack corresponding to the higher load level has a higher signal dynamic range. What is unexpected is the combination of the three peaks. The middle peak is once again consistent with a signal from a deep crack. The two adjacent peaks may be due to the indentation or shear lip along the crack edges caused by the fatiguing process.

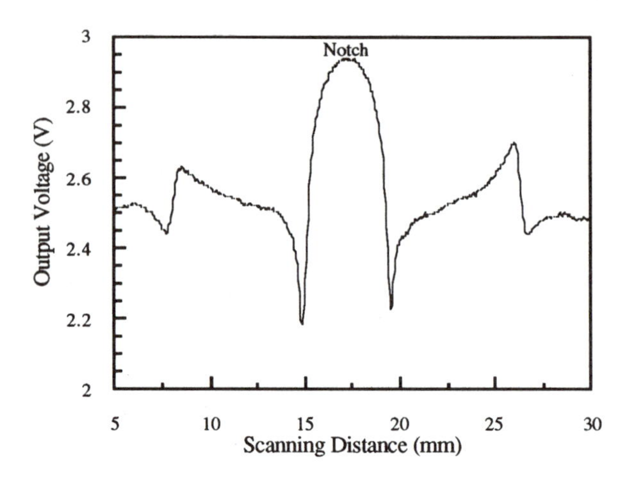

Figure 5: Crack characteristic signal due to the sawcut starter notch at 24 GHz.

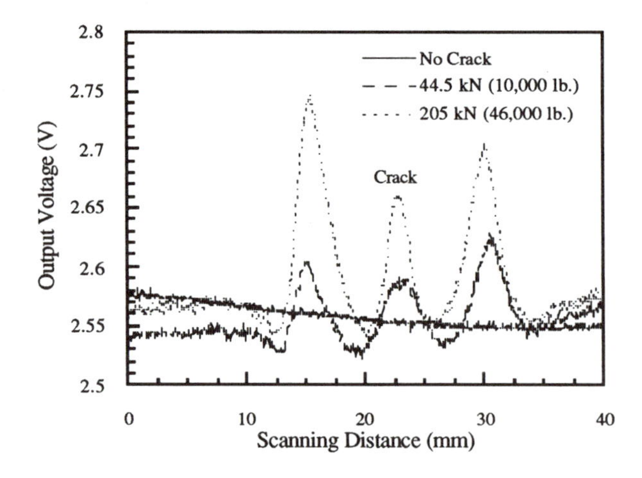

Figure 6: Crack characteristic signals for the fatigue crack under two different loads at 24 GHz.

However, the fact that their magnitudes are larger than that of the crack is somewhat puzzling and must be investigated further. Figure 7 shows similar results but at a frequency of 35.05 GHz. The middle two nulls correspond to the crack edges and the outer two nulls to the indentation. As before, higher loading causes a greater signal dynamic range, as expected since the crack opening is respectively larger for higher loading levels.

A given crack appears larger in width and depth at higher frequencies than at lower frequencies. Therefore, its crack characteristic signal should be more indicative of the presence of the crack at 65 GHz than at the two lower frequencies investigated thus far. Figure 8 shows the crack characteristic signal for the fatigue crack under two different load levels at a frequency of 65 GHz. The crack is distinctively detected in addition to the fact that consistent with previous results, higher loading causes higher output/detected signal dynamic range. Figure 9 shows a similar scan except for a large scanning distance to show the effect of the shear lip along the crack. Nevertheless, the crack is still detected (e.g. the deep null in the middle of the figure).

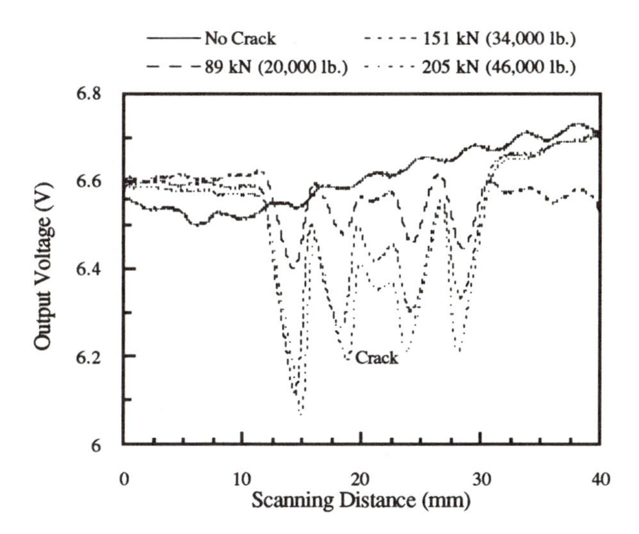

Figure 7: Crack characteristic signals for the fatigue crack under three different loads at 35.05 GHz.

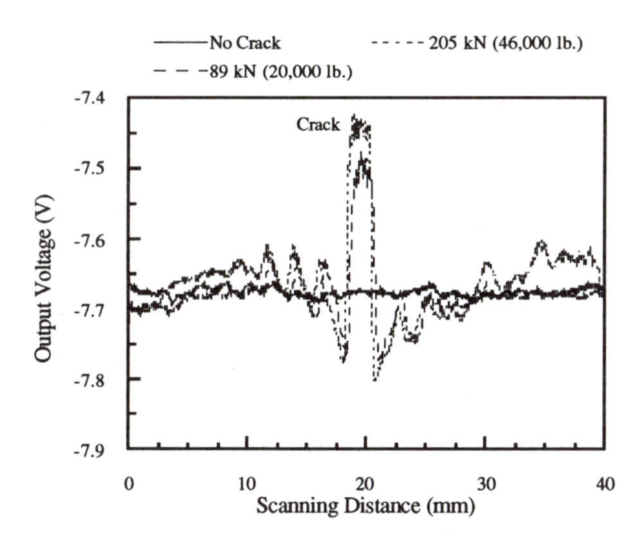

Figure 8: Crack characteristic signals for the fatigue crack under two different loads at 65 GHz.

SUMMARY

The potential of using open-ended rectangular waveguides in the microwave and millimeter wave region to detect tight fatigue cracks was demonstrated. Due to the manner by which the fatigue was generated some indentation along the crack edges were produced. The result showed that these indentations play a significant role in the overall behavior of the crack characteristic signal. This issue will be further studied in the months to come. Assuming that the crack produced for this study closely represents a real crack in a steel member, the results of these experiments are quite encouraging for the potential employment of this technique for the detection of real fatigue cracks in steel bridges and other metallic structures. This is particularly true when considering that the tight crack in this study had a maximum opening that is two to four orders of magnitude smaller than the operating wavelength. Clearly, more investigations must be conducted to completely understand the role of crack opening, shear lips and other features on the crack characteristic signal.

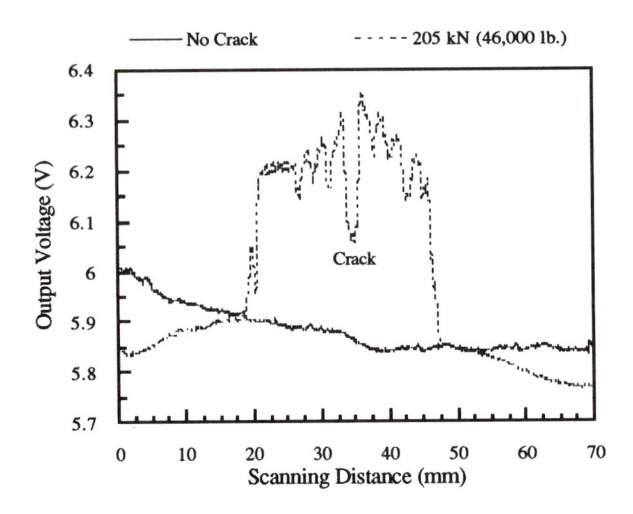

Figure 9: Crack characteristic signals for the fatigue crack under loading at 65 GHz for a large scanning distance.

Acknowledgment: This research has been funded by the Federal Highway Administration (FHWA) grant no. DTFH61-94-X-00023.

REFERENCES

1. Bovig, K.G., "NDE Handbook," *Teknisk Forlag A/S (Danish Tech. Press)*, 1989.
2. Bahr, A. J., "Microwave Nondestructive Testing Methods," Volume I, Gordon and Breach Science Publishers, 1982.
3. Yeh, C. and R. Zoughi, "A Novel Microwave Method for Detection of Long Surface Cracks in Metals," *IEEE Transactions on Instrumentation and Measurement*, vol. 43, no. 5, pp. 719-725, October, 1994.
4. Yeh, C., E. Ranu and R. Zoughi, "A Novel Microwave Method for Surface Crack Detection Using Higher Order Waveguide Modes," *Materials Evaluation*, vol. 52, no. 6, pp. 676-681, June, 1994.
5. Yeh, C. and R. Zoughi, "Microwave Detection of Finite Surface Cracks in Metals Using Rectangular Waveguide Sensors," *Research in Nondestructive Evaluation*, vol. 6, no. 1. pp. 35-55, 1994.
6. Yeh, C. and R. Zoughi, "Sizing Technique for Surface Cracks in Metals," *Materials Evaluation*, vol. 53, no. 4, pp. 496-501, April, 1995.
7. Ganchev, S., R. Zoughi, C. Huber, R. Runser and E. Ranu, "Microwave Method for Locating Surface Crack Tips in Metals," *Materials Evaluation*, vol. 54, no. 5, pp. 598-603, May, 1996.
8. Huber, C., H. Abiri, S. Ganchev and R. Zoughi, "Analysis of the Crack Characteristic Signal Using a Generalized Scattering Matrix Representation," *IEEE Transactions on Microwave Theory and Technique*, vol. 45, no. 4, pp. 477-484, April, 1997.
9. Huber, C., H. Abiri, S. Ganchev and R. Zoughi, "Modeling of Surface Hairline Crack Detection in Metals Under Coatings Using Open-Ended Rectangular Waveguides," To appear in the *IEEE Transactions on Microwave Theory and Techniques*, 1997.

NONDESTRUCTIVE CHARACTERIZATION OF MATERIALS: APPLICATIONS FOR HUMANITARIAN DEMINING

Gary W. Carriveau, Ph.D.

Senior Scientist
Science Applications International Corporation
4161 Campus Point Court
San Diego, California 92121 USA

INTRODUCTION

The threat to mankind due to landmines is extensive and diverse. Humanitarian demining (removal of landmines outside of the time of military operations) is an occupation that contains great risks. This paper is intended to review the problem, the costs associated with solving the problem, an overview of current demining methods, and a brief review of some nondestructive characterization methods that may aid in improving humanitarian demining activities.

THE PROBLEM

The presence of landmines may represent one of the most threatening form of pollution faced by mankind. These devices, which have been used extensively in all military conflicts since World War 1, have many different sizes and shapes and are designed for the destruction of vehicles and for killing and injuring people.

There are two general classifications of landmines: anti-tank (AT) mines and anti-personnel (AP) mines. AT mines are designed to destroy tanks and other large vehicles. They are relatively large and weight more than 5 pounds and are most often buried. Anti-personnel mines are much smaller and are designed to kill or maim people at random. They are most often scattered on or near the ground surface and may be as small as a cigarette package and weigh several ounces. AP mines are produced in a wide variety of shapes and colors (over 700 different types); some even look like toys for children.

Table 1 contains a review of the comparison of demining operations in support of military operations and humanitarian demining. It is quite obvious that there are very little similarities between the two types of operations.

Nondestructive Characterization of Material VIII
Edited by Robert E. Green Jr., Plenum Press, New York, 1998

Anti-personnel mines are increasingly used as a terrorist tool, being laid anywhere to cause unrest and restrict access. Although AP mines can kill a person, they are mainly intended to maim; in battle a wounded person requires more attention and assistance.

Table 1. Comparison Between Humanitarian and Military Demining Operations

Humanitarian	Military
Anti-personnel mines	Anti-tank and anti-personnel mines
Single or low density	Densely mined areas
Total demining	Cleared lanes and gaps
At an acceptable time	At any time
No risk is tolerated	Casualties are expected
Placed any time in the past; little or no documentation	Recently laid, often maps are used
No self-destroying mechanism	"Smart" mines
Demining technology has not change in over 50 years	Modern technology

During military operations AP mines are used to protect mine fields against clearance, to obstruct and disorient the enemy, and to reduce access to various areas requiring protection.

AP mines may be characterized by four methods of detonation. These are: pressure, trip wire, electronic, and remote control. Many AP mines in current use can be set off by more than one initiating process. The AP mines that are presently laid have no self-destroying or self-neutralizing mechanism (they are "dumb" mines).

Approximately 125 million (125,000,000) land mines are located in 60 countries around the world. These are summarized in Table 2.

In addition to those already in place, there are approximately 150 million (150,000,000) landmines stockpiled for potential use, along with a constant production of new mines. From these stocks, somewhere between 2 and 3 million (2,000,000 to 3,000,000) additional mines are placed each year.

Annually, over 25,000 people, especially women and children are killed, injured or maimed. This equates to about one person every twenty minutes, every day of the year, around the world.

THE COST

The cost to produce and place an anti-personnel land mine is less than $5. The cost to locate and remove an AP mine is between $500 and $1,000 per mine.

Table 2. World-wide Pollution of Landmines

Continent/Zone	Countries Affected	Mines (in millions)
Africa	16	44
Middle East	9	26
East Asia	6	23
South Asia	4	25
Europe	13	7
Latin America	8	1
	60	**$\geq$ 126**

Currently, the estimated cost for demining activities is approximately $50,000 per square kilometer cleared. This cost is excessive to most of the countries where the problem exists. For example, in Cambodia, it is estimated that the entire national economy could be consumed over the next five years for demining purposes alone.

The existence of mines greatly reduces the economic recovery process in affected areas because they are often used near agricultural fields, water supplies, and area roads and paths.

CURRENT DEMINING SITUATION

Approximately 100,000 land mines are located (and sometimes) destroyed each year. The reservation about their final disposition is intended to indicate that mines that have been found and removed are often added to stockpiles for future use. There are also stories concerning the open marketing of demined AP mines that are sometimes used by local farmers to protect their fields

For each mine removed, many times more new mines (factors of ten) are put into place. Reports from the United Nations indicate that while 80,000 mines were located and destroyed in 1993, approximately 2,500,000 new mines were laid. Considering the rate of continued mine emplacement, it is clear that current demining methods places the humanitarian deminer in an impossible task to ever catch up. It would require nearly a 100-fold improvement in demining efficiency to effectively reduce the number of mines in place by the year 2005.

This increase cannot occur using today's humanitarian demining technology, which has changed very little in over 50 years. Methods used today rely on three major steps. These are:

1. Visual detection using the human eye
2. Detection of metal parts using hand-held metal detectors
3. Location of the mine by means of probing the ground with a sharp wooden or plastic stick.

Sometimes these steps are supplemented by the use of specially trained "sniffer" dogs which sense the characteristic odors from the mine with their nose and indicate to the dog handler the possible mine location.

The increase in humanitarian demining effectiveness cannot occur by simply multiplying the current efforts. Current humanitarian demining practices are not safe or cost effective. The United Nations reports that for every 500 mines that are removed, an average of 1 deminer is killed and two are wounded. An increase in the number of demining teams using current methods would cause an unacceptable increase in killed or wounded deminers. In addition, it is very difficult to imagine where additional funding could be found to support more teams.

The frustration in continued use of accepted present day mine detection methods may be discerned in the following quote by U. S. Army Colonel Robert Greenwalt, "Today, highly trained, scared soldiers use all of their senses, augmented with a coin detector and pointed stick".

APPLICABLE NONDESTRUCTIVE CHARACTERIZATION TECHNOLOGIES

To locate and identify a single mine, several characteristic parameters of the mine (and mine laying process) may be used. These include:

- detection of components of the mine itself: metal, plastic, explosive
- emission of vapor from the explosive and/or other characteristic vapors
- the size, shape, and burial depth of the mine
- the physical difference of the mine in respect to the surrounding (i.e., electrical properties, magnetic properties, electromagnetic properties, etc.)
- changes in the surroundings brought about in placing the mine.

Effective detection and validation of these parameters are complicated by the characteristics of the soil surrounding the mine. Factors that influence mine detection include: the soil water content, soil type and texture, uniformity of the surrounding soil matrix, the magnetic properties of the soil, as well as the type and thickness of the covering vegetation and difficulties with rough terrain.

Nondestructive characterization technologies that may offer potential (or are already being used in a very limited degree) for humanitarian demining include:

- acoustic methods - acoustic and ultrasonic sensors may be used in searching for mines in ponds or flooded rice paddies where existing water provides the necessary energy couplant.
- electromagnetic induction instruments - This class of sensor uses a current in a coil to produce a magnetic field, these generate eddy currents in conducting metals within the field. The induced eddy currents produce their own magnetic fields that can be measured by the induction sensor. These sensors, the most common type of metal or "coin" detector, locate ferrous and other metal objects. They are often hand-held and work best when very near the metal object.
- livestock - it may be argued that this approach is not nondestructive in that livestock, such as sheep or goats, are herded into areas thought to contain mines, where they are sacrificed when they "detect and eliminate" the mines.
- magnetometer/gradiometer - these instruments measure the earth's magnetic field in the inspected area looking for perturbations created by ferrous metal mine components. The drawback in this class of instruments is that only ferrous objects can be located and ferrous soils reduce their effectiveness.
- nuclear based methods - these methods rely on the use of nuclear-based effects. They include the use of single-sided nuclear quadruple resonance, particle backscatter (both x-rays and neutrons), and the detection of nitrogen in the high energy explosives through neutron capture - gamma reactions.
- optical methods - optical sensors in the visible and IR ranges are often useful in detecting a disturbance in either the soil covering the mine or heat flow in the region surrounding the mine. Overgrowth vegetation as well as rough terrain and rocks reduces the use of these systems. Video technology enhances their use.

- radar - these detection systems generally fall into two categories: down looking (ground penetrating radar) and forward looking. Both rely on the comparison of the transmitted electromagnetic energy with the received, reflected energy, to characterize the reflecting object. They can be used to locate both metal and plastic mine components.
- vapor detection - specially trained dogs have been used in mine detection for decades. In addition, there are effort to produce an "electronic nose" using advanced technology from analytical chemistry. An additional area of development is in training and/or biomodification of insects to enhance their ability to detect mines from the characteristic chemical emanations. The use of honey bees, fruit flies, and cockroaches has been investigate.

This brief overview shows that there exists a wide variety of mature technology that can be used in humanitarian mine detection. Because of the variety of materials that require detection and the range of sensitivities found in these methods, it suggests that a combination of sensors may best serve the purpose for mine detection.

An important adjunct to the detection and characterization techniques list above is available in the Global Positioning System (GPS). This satellite-based position/navigation sensing system can provide critical information on the location of potential and actual mines as well as serve as an indispensable mapping tool in areas cleared of mine. Measured positions accurate to within a range of less than one-foot are possible with today's technology and with near-term improvements accuracy to within a few inches are anticipated.

HOW NONDESTRUCTIVE CHARACTERIZATION CAN HELP

Many nondestructive evaluation and characterization technologies that were outlined in the previous section are well developed for other applications and offer real possibilities for humanitarian demining. Some are currently being used in military demining and efforts are being made to develop and produce useful hardware specifically for humanitarian demining. Because financial support is quite limited, this research and development is very slow.

To date, the physical parameters that are measured to the locate a mine are most often interrogated one at a time, with little effort being made to correlate information from different sensors. Improved methods of detection and verification should provide information from several of the characteristic parameters together and in real-time. All of this information can be integrated/correlated with accurate GPS position/location data.

Several considerations are important when thinking of technology development for humanitarian demining. These include (but are not limited to):

- effort must obviously be placed on instruments that can acquire data above the surface of the ground being investigated; they must be able to get close to the surface but not too close and they must not use any mechanism which may detonate a mine
- AP mines should be detected by identifying and verifying the presence of one or more of the major constituents of the mine, i.e. metal, plastic, and high explosives - verification through the detection of more than one of the complimentary mine materials will greatly reduce the false positive rate and speed up actual mine removal
- data fusion, the use of data from different nondestructive characterization methods (such as information on mine materials, shape, position, burial depth, etc.) can be used to enhance the probability of detection and reduction of the false positive rate for detection of AP mines

- primary importance must be placed on rugged, battery-operated, hand held detection systems which can be used in very difficult terrain with overgrown vegetation
- there should be special emphasis on low cost, highly reliable, easy to maintain equipment for use in adverse environmental conditions

It is clear that there is no easy solution to the humanitarian demining problem. The detection of many types of mines based on their metal content alone is extremely difficult (and impossible in some cases). Furthermore, false signals from metal objects other than in mines (background clutter from shrapnel, empty shell casings, other metal fragments, etc.) creates many false positive indications which makes the conventional approach to mine clearance very slow and unsafe.

To develop new technology for humanitarian demining the scientists/engineers that conceive and develop new methods <u>must</u> obtain guidance from the deminer in the field so that equipment that is actually needed and usable can be produced. A high technology approach that does not address all of the real-world problems simply contributes a flawed solution.

MONITORING OF ATTENUATION DURING PHASE TRANSFORMATIONS IN STEEL USING LASER-ULTRASONICS

Marc Dubois,[1] André Moreau,[1] Matthias Militzer,[2] and Jean F. Bussière[1]

[1]Industrial Materials Institute, National Research Council Canada
75 Mortagne Blvd., Boucherville, Québec, Canada, J4B 6Y4
[2]The Centre for Metallurgical Process Engineering, The University of British Columbia, Vancouver, British Columbia, Canada, V6T 1Z4

INTRODUCTION

Ferrite grain size strongly influences the mechanical properties of steels. In hot rolled steels, the ferrite grain size is determined by the austenite microstructure during the finish rolling, and by the rate of cooling through the transformation temperature range. A sensor to monitor phase transformations and grain sizes in strips during hot rolling would allow to control the process parameters, optimize the ferrite grain size, and better achieve the desired mechanical properties. Unfortunately, such a sensor is presently not available in hot strip mills.

Currently, processing conditions in mills are set according to empirical knowledge, computer modeling, and laboratory simulations. In the laboratory, the standard technique to monitor phase transformations is dilatometry. This technique, however, provides only a quantitative measure of the overall phase decomposition and laborious metallographic techniques are required to determine further microstructural details.

Ultrasound has been known for many years to be an excellent method to characterize steel microstructure.[1,2] However, ultrasonic measurements at temperatures of the austenite-to-ferrite transformation in low-carbon steels are not easily obtained using conventional ultrasonic transducers. Laser-ultrasonics, a technique based on the generation of ultrasonic waves by a pulsed laser and on their detection by a laser interferometer, is a truly remote technique[3] (standoff distances of order 1 m) and works efficiently at high temperatures.[4-6]

Papadakis *et al.*[7] used conventional ultrasonic transducers and a momentary contact technique to measure ultrasonic attenuation and velocity during the ferrite-to-austenite phase transformation in an AISI 52100 steel. Scruby and Moss[5] also measured ultrasonic attenuation and velocity during the same phase transformation in a low-carbon steel using laser-ultrasonics. However, the development of the laser-ultrasonics in recent years allows to obtain new ultrasonic data at high temperatures with a significantly improved accuracy.

This paper presents laser-ultrasonic measurements of ultrasonic attenuation in a hot rolled A36 steel at temperatures between 500°C and 1100°C. The sudden variations of ultrasonic attenuation observed in this temperature range are related to microstructural changes caused by the austenite-to-ferrite phase transformation.

Nondestructive Characterization of Material VIII
Edited by Robert E. Green Jr., Plenum Press, New York, 1998

Ultrasonic Scattering Theory

Ultrasonic attenuation is caused by various microstructural parameters but is, for most metals, primarily due to grain scattering. The relationship between attenuation and grain size depends on the ratio of the acoustic wavelength, λ, to the average of some measure of grain size, D. Generally, three regimes are considered:[8]

$$\text{Rayleigh regime } (\lambda \gg D) \qquad \alpha = K_r D^3 f^4$$

$$\text{Stochastic regime } (\lambda \approx D) \qquad \alpha = K_s D f^2$$

$$\text{Diffusion regime } (\lambda \ll D) \qquad \alpha = K_d / D,$$

where α is the attenuation coefficient, K_r, K_s, and K_d are material constants, and f is frequency ($f = V/\lambda$, where V is the acoustic velocity). In this paper, the ratio of acoustic wavelength to grain size is such that the scattering regime lies between Rayleigh and stochastic. Therefore, the ultrasonic attenuation is expected to increase with grain size.

Ultrasonic attenuation has not yet been studied in detail during phase transformations, especially in the case of alloyed metallic materials like steel.[9] Furthermore, the effect of differences in the elastic constants between austenite and ferrite on ultrasonic attenuation during the phase transformation cannot be calculated because literature values of the elastic constants of austenite are only available for one temperature, 1150°C,[10] which is 240°C above the equilibrium transformation temperature in pure iron (910°C). Elastic constants of pure ferritic iron are nonetheless known at all temperatures below the equilibrium ferrite-to-austenite transformation temperature.[11]

In the Rayleigh regime, ultrasonic attenuation in cubic crystals is proportional to the square of the factor (C_{11}-C_{12}-2 C_{44}),[8] sometimes called anisotropy factor, where C_{ij} are the elastic constants of the single crystal. The values of this factor are -171.4 GPa in ferrite at 900°C[11] and -122 GPa in austenite at 1150°C.[10] The difference between the two values is less than 30%. This difference is small when compared to the observed variations in ultrasonic attenuation. Therefore, the effect of the elastic constant differences between austenite and ferrite will be neglected in a first approximation and ultrasonic attenuation variations during phase transformations will be interpreted hereafter only as grain size variations.

EXPERIMENTAL

Laser-ultrasonic measurements were made on A36 steel samples 25x25 mm^2 x 1 mm. The samples were machined from as-received hot-rolled material (transfer bars) provided by the Gary Works of US Steel. Ultrasonic measurements were performed near the center of the sample. Table 1 gives the chemical composition of the A36 steel samples.

Table 1. Chemical composition (wt%) of the A36 steel samples.

Steel	C	Mn	P	S	Si	Cu	Ni	Cr	Al	N
A36	0.17	0.74	0.009	0.008	0.012	0.016	0.010	0.019	0.040	0.0047

The experiments were conducted under an atmosphere of pure argon in a radiant furnace equipped with windows transparent to the laser radiation wavelengths. The sample temperature was measured with a thermocouple spot-welded approximately 5 mm from the edge of the laser detection spot.

Ultrasonic waves were generated using an excimer laser at an optical wavelength of 248 nm (ultraviolet). Pulse duration was 6 ns, pulse energy was 250 mJ, and the laser-beam was focused on the sample into a nearly uniform rectangular spot of approximately 4x6 mm^2. The optical power density was high enough to vaporize some material at the sample surface. However, no measurable change of sample thickness was observed after experiments involving thousands of light pulses.

Ultrasonic displacements were detected on the opposite surface of the sample with a laser interferometer based on a 3 kW long-pulse Nd:YAG laser and a confocal Fabry-Pérot interferometer operating in the reflection mode.[12] For surface displacements much smaller than the 1.064-μm wavelength of the detection laser and for ultrasonic frequencies higher than 5 MHz,[13] the electronic output of the interferometer was approximately proportional to the surface displacement. The detection laser beam was focused on the sample into a uniform disk approximately 5 mm in diameter.

Generation and detection spots were carefully aligned before each experiment. However, slight misalignments would have had almost no effect due to the large laser beam sizes with respect to sample thickness. The acoustic near-field parameter, $\lambda z / a^2$, where λ is the acoustic wavelength, z is the distance traveled by the ultrasonic wave, and a is the generation laser spot radius; was 0.13 for the first two longitudinal echoes at a frequency of 15 MHz. Therefore, the ultrasonic waves are essentially plane waves and no diffraction correction is required.

Figure 1 shows a typical single-shot signal obtained from an A36 steel sample at 1200°C. The power spectra ratio of the first two echoes divided by twice the thickness provided the ultrasonic attenuation as a function of frequency.

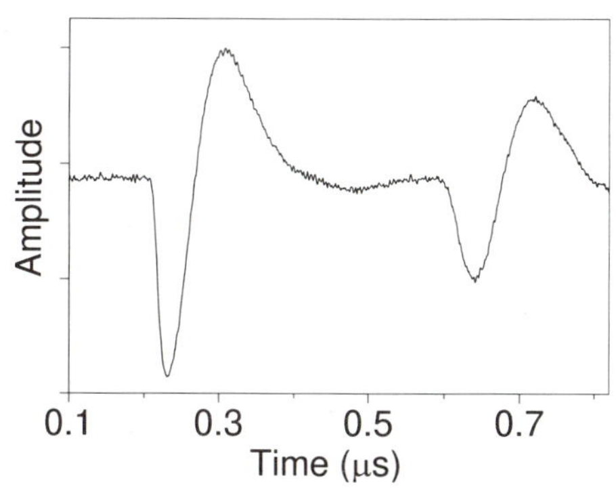

Figure 1. Ultrasonic single-shot signal obtained from an A36 steel sample at 1200°C.

Dilatometry measurements were also made on tubular A36 specimens (50 mm long with an inner diameter of 8 mm and a wall thickness of 1 mm) to quantify, for comparison, the austenite-to-ferrite transformation kinetics. These measurements were made on a Gleeble 3500 TS thermal simulator which reproduced the heating and cooling conditions of the furnace.

RESULTS AND DISCUSSION

Laser-ultrasonic attenuation at 15 MHz was monitored while A36 steel samples were heated from room temperature to 1000 and 1100°C at a rate of 5°C/s and cooled, after 10 and 15-minute holding periods, by turning off the furnace. The samples cooled at a rate of approximately 1°C/s at 700°C. The attenuation is presented as a function of temperature in

Figures 2a and 2b where the austenite fraction during cooling and heating is also indicated.

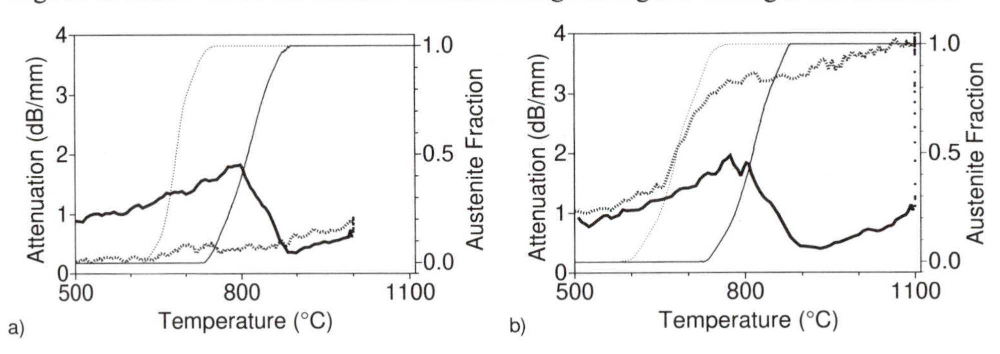

Figure 2. Attenuation measurements at 15 MHz in A36 steel samples during heating (thick solid line), cooling (thick dashed line) and isothermal treatment at a) 1000°C b) 1100°C (bold dots). A moving average of 3 was used to smooth the attenuation data. The austenite fractions during cooling (thin dashed lines) and heating (thin solid lines), as measured by dilatometry, are also indicated.

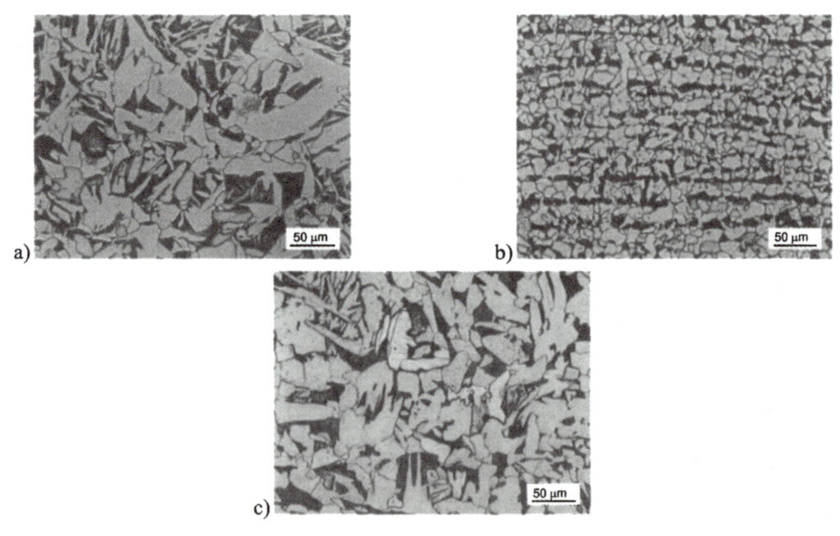

Figure 3. Optical micrographs of A36 steel samples: a) as-received and isothermally treated at b) 1000°C, and c) 1100°C. (Nital 4%).

During heating an A36 sample to 1000°C (Figure 2a), the ultrasonic attenuation increased slowly from 1 dB/mm at 500°C to almost 2 dB/mm at ~800°C, dropped to 0.3 dB/mm between ~800°C and ~900°C, which corresponds approximately to the transformation temperature range,[*] and increased again with temperature above ~900°C to reach 0.7 dB/mm at 1000°C. While temperature was maintained constant at 1000°C for 15 min, the ultrasonic attenuation increased slightly to 0.9 dB/mm. During furnace cooling (at approximately

[*]During heating at 5°C/s using the radiant furnace, the sample temperature was not uniform and the indicated temperature is slightly higher than the temperature measured during the dilatometry tests using the Gleeble 3500 TS. During cooling, however, the temperature was uniform in the radiant furnace and the measurements obtained using the radiant furnace and the Gleeble are in agreement.

1°C/s), the attenuation decreased monotonically from 0.9 to 0.2 dB/mm at 500°C, showing no special feature during the transformation. The attenuation measured during cooling at 500°C is considerably lower than the attenuation measured at the same temperature during heating.

During heating an A36 sample to 1100°C (Figure 2b), the ultrasonic attenuation followed approximately the same behavior than during heating to 1000°C, to reach a value of 1 dB/mm at 1100°C. While the temperature was maintained at 1100°C for 10 min, however, the attenuation increased considerably to 4 dB/mm. During cooling, the attenuation decreased to 3 dB/mm between 1100°C and 750°C. During the phase transformation, the ultrasonic attenuation decreased more rapidly to a value of ~1.5 dB/mm. When the phase transformation was completed, attenuation decreased approximately at the same rate than before phase transformation to reach ~1 dB/mm at 500°C, an attenuation similar to the value measured at the same temperature while heating.

These results are interpreted as follows. During heating, the elastic constants of ferrite vary with temperature[11] and cause a monotonically increase in ultrasonic attenuation between 500°C and 800°C. In the phase transformation range, austenite nucleates and forms a finer microstructure than the prior ferritic structure. The smaller austenite grains cause a decrease in ultrasonic attenuation. When a sample was heated to 1000°C for 15 min (Figure 2a), the austenite grain size did not increase significantly because AlN precipitates pinned the austenite grain boundaries.[14,15] During cooling back through the phase transformation region, ferrite nucleated and formed grains having approximately the same size as the prior austenite grains. Consequently, no sudden change in attenuation is associated with this phase transformation. The resulting ferrite microstructure is finer than it was before thermal processing, as confirmed by comparing the microstructures shown in Figures 3a and 3b. However, when a sample is heated to 1100°C for 10 min, significant austenite grain growth occurs[15] which results in a marked increase in attenuation (Figure 2b). The transformation during cooling then produces a ferrite microstructure which is finer than the prior austenite grains but comparable to the as-received microstructure, as confirmed by the micrographs of Figure 3. Consequently, the attenuation dropped back to levels similar to those observed initially during heating.

CONCLUSION

The ferrite-to-austenite transformation in a low-carbon steel was monitored using laser-ultrasonics. Sudden variations of ultrasonic attenuation were observed at transformation temperatures and were interpreted in terms of nucleation and growth of the new phase. Dilatometry and standard metallographic observations support these interpretations.

Ultrasonic attenuation appears mainly sensitive to the grain size difference between phases. This information is obtained in real-time and, as a complement to dilatometry, may contribute to facilitate greatly laboratory studies on phase transformations. Finally, this new technique may constitute, in the future, a new tool to obtain data in industrial environments on phase transformations during hot rolling.

ACKNOWLEDGMENTS

The financial support received from the American Iron and Steel Institute (AISI) and the US Department of Energy (DOE) is gratefully acknowledged.

REFERENCES

1. E. P. Papadakis, Ultrasonic attenuation and velocity in three transformation products in steel, *J. Appl. Phys.*, 35:1474 (1964).
2. F. Bergner and K. Popp, Mechanisms of ultrasonic attenuation in a bainitic low-alloy steel, *Scripta Met. Mat.,* 24:1357 (1990).
3. C. B. Scruby and L. E. Drain, *Laser-Ultrasonics: Techniques and Applications,* Adam Hilger, Bristol (1990).
4. J.-P. Monchalin, R. Héon, J. F. Bussière, D. Pascale, and B. Farahbakhsh, "Laser-Ultrasonic Determination of Elastic Consants at Ambient and Elevated Temperatures" in *Nondestructive Characterization of Materials II,* Plenum Publishing Corp., J.F. Bussière, J.P. Monchalin, C.O. Ruud, R.E Green, ed., New-York (1987).
5. C.B. Scruby and B.C. Moss, "Non-Contact Ultrasonic Measurements on Steel at Elevated Temperatures", *NDT&E International,* 24:177 (1993).
6. O.N. Senkov, M. Dubois, and J.J. Jonas, Elastic moduly of titanium-hydrogen alloys in the temperature range 20°C to 1100°C, *Metall. Mater. Trans. A,* 27A:3963 (1996).
7. E. P. Papadakis, L. C. Lynnworth, K. A. Fowler, and E. H Carnevale, Ultrasonic attenuation and velocity in hot specimens by the momentary contact method with pressure coupling and some results on steel to 1200°C,*J. Acoust. Soc. Am.*, 52:850 (1972).
8. E. P. Papadakis Revised grain-scattering formulas and tables *J. Acoust. Soc. Am.,* 37:703 (1965).
9. C.W. Garland, Ultrasonic investigation of phase transition and critical points, in *Physical Acoustics, vol. 7,* W.P. Mason and R.N Thurston, ed., Academic Press, New-York (1970).
10. J. Zarestky and C. Stassis, Lattice Dynamics of γ-Fe, *Phys. Rev. B,* 35:4500 (1987).
11. D.J. Dever, Temperature dependence of the elastic constants in α-iron single crystals: relationsship to spin order and diffusion anomalies, *J. Appl. Phys.,* 43:3293 (1972).
12. J.-P. Monchalin, R. Héon, P. Bouchard, C. Padioleau, Broadband optical detection of ultrasound by optical sideband strippoing with a confocal Fabry-Perot, *Appl. Phys. Lett.,* 55:1612 (1989).
13. A. Moreau and M. Lord, High Frequency Laser Ultrasonics, in this issue of *Nondestructive Characterization of Materials VIII* (1997).
14. F. G. Wilson, T. Gladman, Aluminium nitride in steel, *Int. Mat. Rev.,* 33:221 (1988).
15. M. Militzer, A. Giumelli, E.B. Hawbolt and T.R. Meadowcroft, Austenite grain growth kinetics in Al-killed plain carbon steels, *Metall. Mater. Trans.,* 27A:3399 (1996).

RAPID MICROSTRUCTURE ASSESSMENT IN ROLLED STEEL PRODUCTS USING LASER-ULTRASONICS

Marc Dubois and Jean F. Bussière

National Research Council Canada, Industrial Materials Institute
75 Mortagne, Boucherville, Québec, Canada, J4B 6Y4

INTRODUCTION

Metallography is probably the preferred metallurgists' method to characterize microstructures. Unfortunately, metallographic techniques are time-consuming and destructive. For these reasons, quite often, only a few micrographs are used to characterize large specimens and uniformity is assumed. This assumption may lead, in some cases, to inaccurate interpretations of mechanical properties in terms of microstructures. A non-destructive tool is therefore required to rapidly characterize microstructures.

Ultrasound has been known for many years to be a good method to characterize steel microstructure.[1] In steel, ultrasonic attenuation is generally sensitive to grain size, and ultrasonic velocities are sensitive to texture. However, conventional piezoelectric transducers suffer some disadvantages: limited bandwidth, tedious alignment, contact required with the sample, and necessity of an acoustic couplant. These inconveniences limit the use of piezoelectric transducers for rapid and easy measurements of microstructures.

Laser-ultrasonics, a technique that uses a laser to generate ultrasonic waves and another laser combined with an interferometer to detect ultrasonic displacements, does not require any contact with the sample.[2,3] Measurements are therefore rapid, easy, reliable and even possible on moving samples.

In this paper, the laser-ultrasonic characterization of hot-rolled and cold-rolled steel samples is presented. Ultrasonic velocity and attenuation were measured over the whole widths of three hot-rolled steel strips of different thicknesses, and ultrasonic attenuation was measured in thin cold-rolled steel sheets annealed in different conditions. The results obtained with the laser-ultrasonic system are compared with quantitative metallographic evaluations and hardness measurements.

ULTRASONIC ATTENUATION AND VELOCITY

Ultrasonic attenuation is caused by various microstructural parameters but, for most metals, is primarily due to grain scattering.[4] The relationship between attenuation and grain

Nondestructive Characterization of Material VIII
Edited by Robert E. Green Jr., Plenum Press, New York, 1998

size depends on the ratio of the acoustic wavelength, λ, to the average of some measure of grain size, D. Generally, three regimes are considered:[4]

$$\text{Rayleigh regime } (\lambda \gg D) \qquad \alpha = K_r D^3 f^4$$
$$\text{Stochastic regime } (\lambda \approx D) \qquad \alpha = K_s D f^2$$
$$\text{Diffusion regime } (\lambda \ll D) \qquad \alpha = K_d / D,$$

where α is the attenuation coefficient, K_r, K_s, and K_d are material constants, and f is frequency ($f = V/\lambda$, where V is the acoustic velocity). In this paper, the ratio of acoustic wavelength to grain size is such that the scattering regime lies between Rayleigh and stochastic. Therefore, the ultrasonic attenuation is expected to increase with grain size.

Ultrasonic velocity, on the other hand, is sensitive to the crystallographic orientation distribution of grains (texture). Texture may be expressed as a crystallographic orientation distribution function (CODF) describing the probability of finding one grain in a given orientation. The function is written as a series expansion, and the expansion coefficients are called the orientation distribution coefficients (ODCs). Therefore, the ODCs completely specify the CODF. Once the ODCs are known, (using x-ray diffraction or other techniques) ultrasonic velocities may be calculated in any direction. Conversely, a measurement of various ultrasonic velocities may be used to evaluate the ODCs up to the fourth order.[5]

EXPERIMENTAL

Three hot-rolled full-width steel strips were obtained directly from steel plants. These hot-rolled samples were made of plain-carbon commercial-quality steel having a carbon concentration between 0.05 and 0.10%. Strip A had a nominal thickness of 2.6 mm and a width of 107 cm, strip B had a nominal thickness of 4.9 mm and a width of 174 cm, and strip C had a nominal thickness of 7.9 mm and a width of 110 cm. Thicknesses were measured with a micrometer with an accuracy of ± 1 μm. Grain sizes were measured according to the ASTM-E112-88 standard using the Heyn procedure.

Table 1. Description of the cold-rolled sheets.

Sample set	Annealing soak temperature (°C)	Annealing soak time (s)	Hardness (R30T)	Volume fraction recrystallized (%)
1	600	15	79.1	8
2	600	30	74.1	12
3	600	90	66.6	55
4	625	30	65	77
5	675	30	61.9	100

Cold-rolled samples were plain carbon steel sheets with a carbon concentration of 0.12%. These sheets had received a cold reduction of 84% for a final nominal thickness of 0.33 mm. 20 sheets were annealed in five different conditions to produce four nominally identical samples for each thermal treatment. One sheet for each thermal treatment was used to metallographically evaluate the volume fraction recrystallized and to measure hardness. The thermal treatment given to each sample set, the hardness, and the volume fraction recrystallized are indicated in Table 1.

Ultrasonic waves were generated using an excimer laser at an optical wavelength of 248 nm (ultraviolet). Pulse duration was 6 ns, pulse energy was 250 mJ, and the laser-beam was slightly focused to an almost uniform square having an area of approximately 5x5 mm^2. The optical power density was high enough to vaporize a thin oil layer applied to the sample surface before laser-ultrasonic measurements.

Ultrasonic displacements were detected on the opposite surface of the sample using a laser interferometer based on a 3 kW long-pulse Nd:YAG laser and a confocal Fabry-Pérot interferometer operating in the reflection mode.[6] For surface displacements small with respect to the 1.064-μm wavelength of the detection-laser and for ultrasonic frequencies higher than 5 MHz, the electronic output of the interferometer was proportional to the surface displacement. The detection laser beam was focused to a uniform disk approximately 5 mm in diameter.

The delay between the first two longitudinal echoes was calculated using an autocorrelation and interpolation procedure that yielded an accuracy better than the digitizing delay of 1 ns.[7] Fourier transforms on the first and second longitudinal echoes were made, and the ratio of these power spectra divided by twice the thickness provided the ultrasonic attenuation as a function of frequency.

RESULTS AND DISCUSSION

Hot-rolled Strips

Thickness was measured using a micrometer along the width of each strip every 1 cm near edges and every 5 cm elsewhere. Laser-ultrasonic measurements were performed at all points where thickness had been previously measured.

Assuming a constant ultrasonic velocity, thickness variations with respect to the center of each strip were calculated using the ultrasonic delays. Figure 1 compares, as a function of position and for each hot-rolled strip, thickness variations measured using a micrometer (true thickness) with thickness variations estimated using ultrasonic delays (ultrasonic thickness).

For strip A, the ultrasonic and true thicknesses are in agreement over the whole strip width. The strip thickness decreases monotonously by approximately 50 μm from the center to the edges.

For strip B, the ultrasonic and true thicknesses are in agreement in the center of the strip, but at the edges, the ultrasonic thickness is 80 μm larger than the true thickness. The true thickness drops by 200 μm at the edges. Some thickness variations are also observed in the center: at a distance of approximately 25 cm from both edges, the strip is slightly thicker than at the center.

For strip C, the ultrasonic and true thicknesses are in agreement in the center of the strip but an important discrepancy is observed at the edges: ultrasonic thickness increases sligthly whereas true thickness decreased of approximately 225 μm. The thickness is approximately constant in the center.

For strips B and C, important variations of the ultrasonic velocities explain the discrepancies observed at the edges between the true and ultrasonic thicknesses. Figure 2 presents, for the three strips and as a function of position, the ultrasonic velocity calculated using true thickness and ultrasonic delays.

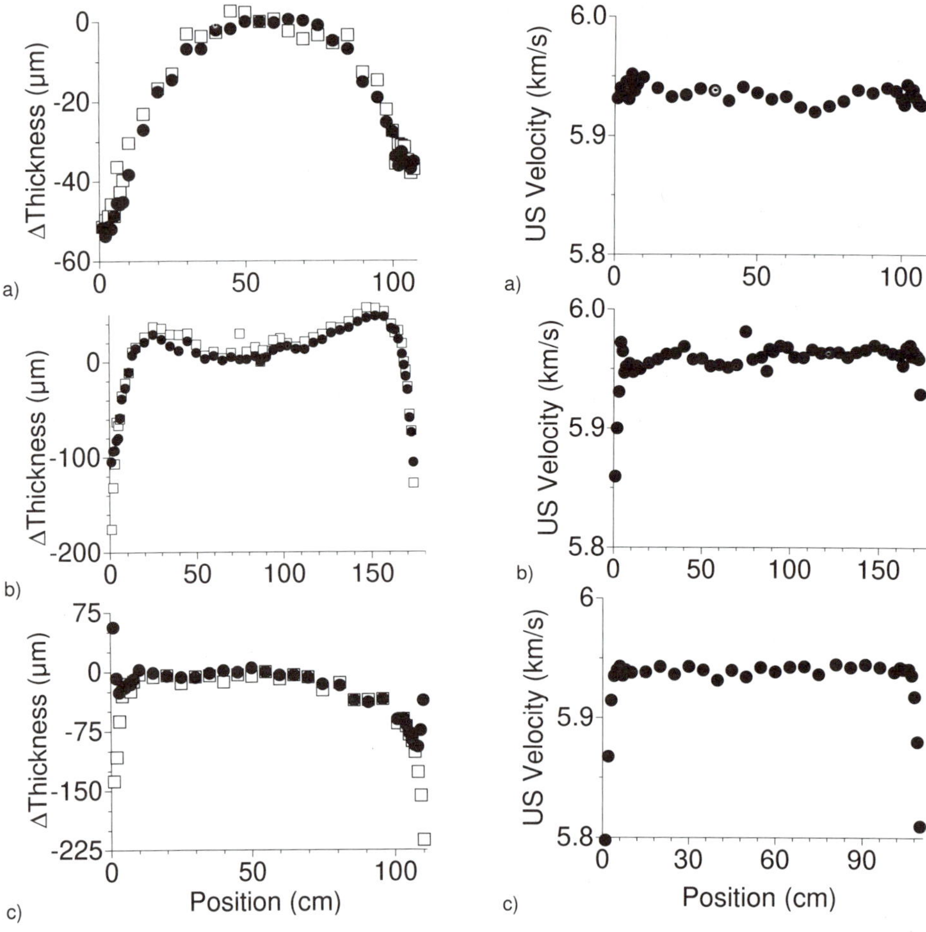

Figure 1. Thickness variations with respect to the center measured with a micrometer (□), and estimated with the ultrasonic delay assuming a constant velocity (●) as a function of position for hot-rolled a) strip A, b) strip B and c) strip C.

Figure 2. Velocity as a function of position in hot-rolled a) strip A, b) strip B and c) strip C.

Figure 2 shows that the longitudinal velocity is approximately constant along the whole width of strip A. For strips B and C, 100 m/s and 150 m/s drops in longitudinal velocity are observed at the edges. These velocity drops may be due to a texture variation caused by a difference in thermomechanical treatments between edges and center during hot rolling.[8]

For the three strips, the ultrasonic attenuation as a function of frequency was calculated from the ultrasonic signals, and grain sizes were measured metallographically at some points. These two parameters are presented in Figure 3. Variations of the ultrasonic attenuation (at 50 MHz for strip A, 40 MHz for strip B and 30 MHz for strip C) along the strip widths follow the grain size variations reasonably well. Figure 3 also shows that ultrasonic attenuations and grain sizes are smaller at the edges than at the center for strip A, approximately constant for strip B and larger at the edges than at the center for strip C. Once again, hot processing conditions probably explain the differences observed between edges and center.

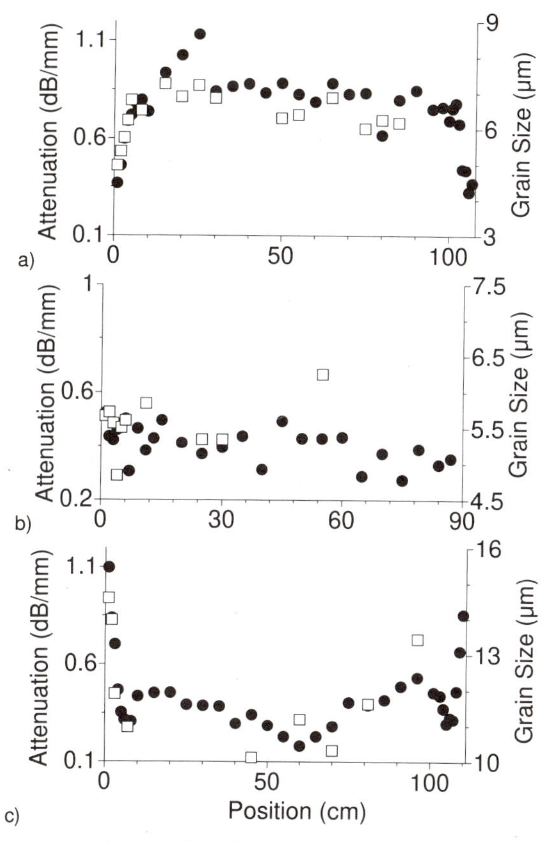

Figure 3. Grain size (□) and ultrasonic attenuation (●) as a function of position in hot-rolled a) strip A, b) strip B and c) strip C.

Cold-rolled Sheets

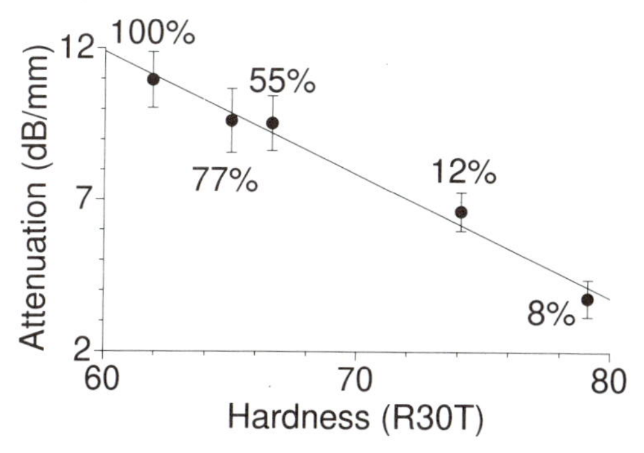

Figure 4. Ultrasonic attenuation at 180 MHz in cold-rolled sheets as a function of hardness. The volume fraction recrystallized, which was evaluated metallographically, is also indicated.

Ultrasonic attenuation as a function of frequency was measured once in each cold-rolled sheet. The average attenuation at 180 MHz for each sample set is plotted in Figure 4 as a function of hardness. Error bars indicate the difference between the highest and lowest attenuation measured in each set. The 180-MHz frequency corresponds to the frequency at which the relative error bars were smallest.

Figure 4 shows a linear relationship between ultrasonic attenuation at 180 MHz and hardness. Therefore, ultrasonic attenuation could be used to evaluate hardness measurements using laser-ultrasonics on cold-rolled steel sheets.

CONCLUSION

Using a single laser-ultrasonic system, ultrasonic attenuation and velocity were measured as a function of position in three hot-rolled strips, and ultrasonic attenuation was measured in thin cold-rolled sheets annealed in different conditions.

For the hot-rolled strips, the ultrasonic attenuation indicated significant grain size variations along the strip widths. These grain size variations were confirmed by metallographic measurements. Ultrasonic evaluation of the thicknesses, using simple ultrasonic delay measurements and a constant velocity, worked well in the center of the strips but texture variations prevented to use this technique to measure thickness variations near edges.

For the cold-rolled sheets, the ultrasonic attenuation at 180 MHz correlates linearly with hardness. Laser-ultrasonics could therefore evaluate hardness on moving cold-rolled sheets.

In summary, a laser-ultrasonic system is capable to rapidly assess microstructure uniformity in hot-rolled strips and in cold-rolled sheets. Such a system could be used for the real-time monitoring of microstructure during processes such as hot rolling and annealing.

REFERENCES

1. E.P. Papadakis, Ultrasonic attenuation and velocity in three transformation products in steel, *J. Appl. Phys.,* 35:1474 (1964).
2. J.P. Monchalin, Progress towards the application of laser-ultrasonic in industry, in: *Review of Progress in Quantitative Nondestructive Evaluation Vol. 12,* D.O. Thompson and D.E. Chimenti, ed., Plenum Press, New-York (1993).
3. C. B. Scruby and L.E. Drain, Laser-Ultrasonics: Techniques and Applications, Adam Hilger, Bristol (1990).
4. E.P. Papadakis, Revised grain-scattering formulas and tables, *J. Acoust. Soc. Am.,* 37:703 (1965).
5. C.M. Sayers, Ultrasonic velocities in anisotropic polycrystalline aggregates, *J. Phys. D: Appl. Phys.,* 15:2157 (1982).
6. J.-P. Monchalin, R. Héon, P. Bouchard, C. Padioleau, Broadband optical detection of ultrasound by optical sideband stripping with a confocal Fabry-Perot, *Appl. Phys. Lett.,* 55:1612 (1989).
7. J.-D. Aussel, J.-P. Monchalin, Precision laser-ultrasonic velocity measurement and elastic constant evaluation, *Ultrasonics*, 27:165 (1989).
8. M.P. Butron-Guillen and J.J. Jonas, Effect of Finishing temperature on hot band textures in and IF steel, *ISIJ Int.,* 36:1 (1996).

COMPENSATED LASER-BASED ULTRASONIC RECEIVER FOR INDUSTRIAL APPLICATIONS

G. J. Dunning, D. M. Pepper, M. P. Chiao, P.V. Mitchell* and T.R. O'Meara

Hughes Research Laboratories
3011 Malibu Canyon Road
Malibu, California, USA

*Melles Griot, Inc.
1770 Kettering Street
Irvine, CA, USA

Keywords: Laser based ultrasound, Photo induced-emf, composites, nondestructive.

ABSTRACT

Cost effective real-time process control can potentially give the competitive edge to companies from such diverse areas as aerospace, energy, automotive and microelectronics. Improvements in operating efficiency and production yield while simultaneously reducing the material and labor requirements can positively effect long-term company performance. We will examine different manufacturing arenas where closed-loop, in-process control; could be implemented using laser-based ultrasound (LBU). We will present data which demonstrates the applicability of LBU for composite cure monitoring and process control.

INTRODUCTION

Conventional ultrasonic methods can be employed to diagnose many materials, in that their acoustic properties are typically functions of the parameters to be ascertained. Present techniques, such as liquid immersion, jet-spray approaches, air-coupled and direct transducer contact may be of limited use in many process-control applications, including those involving vacuums, high temperatures, plasmas, and workpieces with highly structured and complex topography.

LBU is an emerging remote nondestructive diagnostic technique which can be incorporated in adverse manufacturing environments. Laser-based ultrasound [1] offers a means for long-standoff-distance ultrasonic inspection without destroying the workpiece and without physically contact. Optical probing enables high bandwidth sensing at precise locations on the part with the capability of rapid reconfigurability. The part or process being evaluated by LBU is diagnosed remotely by laser excitation of ultrasonic waves. The ultrasound generated in the part is used to probe the physical properties of the region and a second laser beam reflected from the part surface is then detected to monitor the ultrasonic displacements.

Although laser-based ultrasound is well known as a potential remote inspection tool, it has yet to make a major impact in the manufacturing community. For example, in order to inspect rough-cut parts while undergoing relative platform motion, the receiver requires real-time compensation for both speckle and dynamic beam wander. Fabry-Perot techniques [2] and single-speckle interferometry have been demonstrated, but require active stabilization (in the former case) or multiple-averaging (in the latter case). We have demonstrated a compact ultrasonic laser receiver, which employs an all-optical wavefront compensation technique called nonsteady-state photo-induced emf [3]. Our detection element consists of a crystal of GaAs:Cr, which functions as an adaptive photodetector, with minimal post electronic processing and tracking [4]. A schematic showing the experimental layout for both the excitation and reception of laser-based ultrasound is shown in Figure 1.

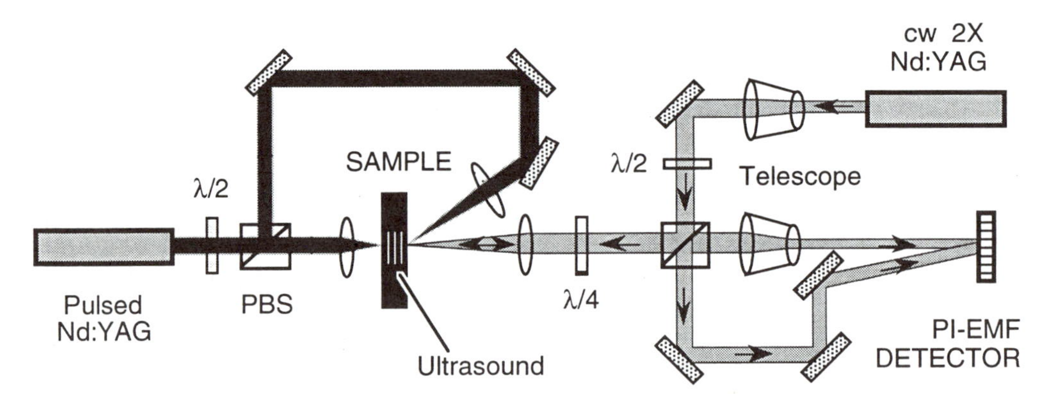

Figure 1. Schematic of Laser Ultrasonic Evaluation System

The laser ultrasonic receiver includes the cw, doubled YAG laser and most of the beam delivery optics and hardware to the right of the sample. The laser output is used to generate both a plane wave reference beam and a probe beam. The probe beam is brought to a focus and directed to the interrogation region of the sample. The beam reflected from the part is then directed to the PI-EMF detector. At the detector surface the reflected probe interferes with the reference beam producing an interference fringe pattern. In the steady state the photo excited carriers form a spatially periodic and stationary space charge field mimicking the fringe pattern. When the ultrasound or any other longitudinal surface displacement modulates the probe beam, the interference fringes move relative to the initial stationary space charge field grating and induces a net current or photo-emf. For displacements comparable to the response time of the detector material, the space charge gratings track the interference fringe motion producing no emf. However, if the intensity pattern moves much faster than the material tracking response time, typical of ultrasonic displacements, the space charge gratings cannot track the motion of the intensity pattern across the material. Therefore, the output current as a function of time from the detector is proportional to the displacement of the part. The upper frequency response may be as high as 100 MHz and is limited by the carrier recombination time of the material, the grating fringe spacing and signal-to-noise considerations.

The laser ultrasonic exciter includes the pulsed laser (Nd:YAG, 1.06 mm, 7 ns, 10 Hz) and all of the optical components used to direct the beam to the sample location. The pulsed laser can be configured to excite the ultrasound in the part using one of two geometries. The reflection mode of is defined when the pulse impinges on the part from the same side as the receiver while ultrasonic excitation originating from the opposite side from the receiver is the transmission mode.

COMPOSITE CURE MONITORING

One potential application of laser based ultrasound would be in is the process control of composite production [5,6,7]. It is necessary to monitor and control the process because of the variability caused by inhomogeneities in the starting materials and deviations of the process variables, e.g., pressure and temperature. Typically the manufacturer's recommended cure cycles for organic matrix composites are often longer than necessary in order to compensate for the variability. Therefore, there is considerable interest in reducing the cycle time by incorporating an appropriate *in situ* sensor which can be used to determine state of the cure and specifically the completion of the cure. It would be desirable to monitor one or more material parameter that are highly correlated to the cure state of the composite. There are several potential avenues for obtaining the relevant data during the process. One technique would rely on the changes in the time of flight of ultrasound in the material as a function of the process. The time of flight data could be measured without disturbing the process by launching and receiving the ultrasound through the composite mold. Alternatively a sensor embedded in the composite during the cure cycle may also be used to monitor the properties.

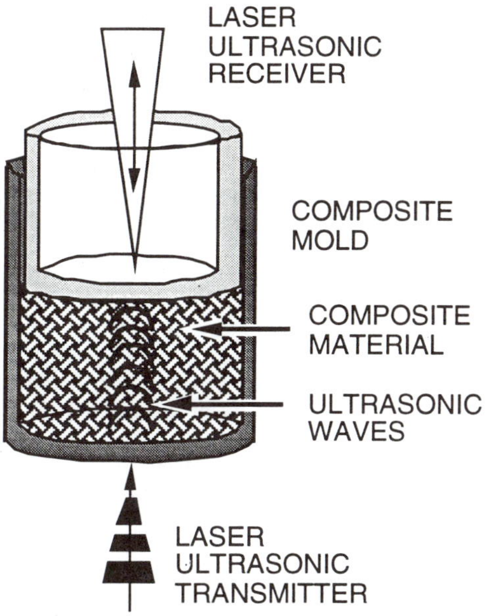

Figure 2. Schematic diagram of the composite mold incorporating in situ diagnostics.

We performed a series of experiments designed to demonstrate the feasibility of both of these approaches. In the first set of experiments we constructed a specially designed composite mold which could accommodate ultrasound launched and received by either high temperature piezoelectric transducers or laser based ultrasound. A schematic of the mold and the geometry for launching and receiving the ultrasound is shown in Figure 2. The ultrasound is launched at the exterior surface of the outer mold assembly and received at the exterior surface of the inner mold. Depending upon the state of the composite the resultant ultrasonic signal is modified. A photograph of a mold which can use either PZTs or lasers to launch and receive the ultrasound and a composite sample is shown in Figure 3. We conducted tests monitoring the curing process in a laminate press of woven graphite fiber with Hexcel F650, Hexcel F652, and graphite unidirectional tape with YLA RS3 cyanate ester. For benchmarking in-situ monitoring we used high-temperature PZT's (2.5 and 10 MHz) and an automated data acquisition system. The signals from the ultrasound included both transmit/receive and transceiver (pulse-echo) modes of operation. The data was simultaneously acquired from the PZTs along with the process parameters of time, pressure and temperature. A graph of the ultrasonic time of flight versus process time (solid line corresponding to the left-hand Y-axis) is shown in Figure 4.

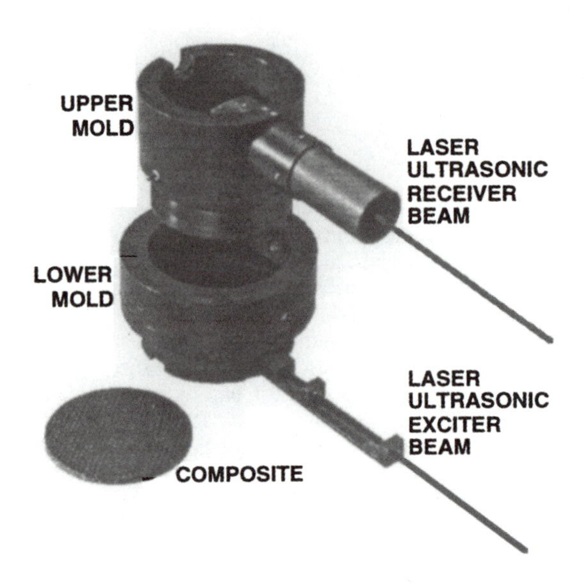

Figure 3. Photograph of mold assembly anc composite

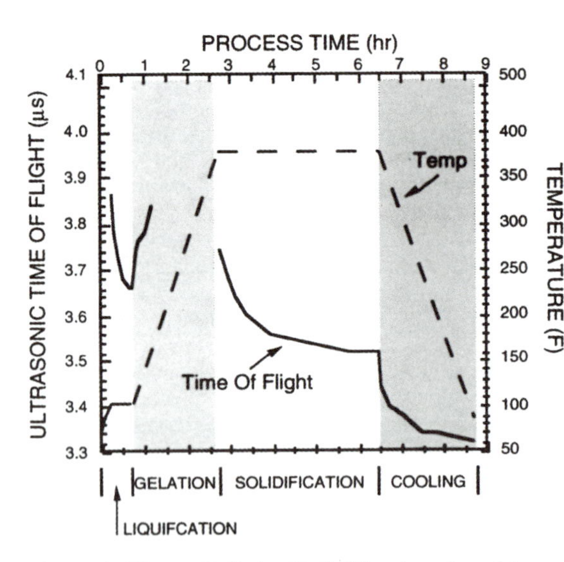

Figure 4. Time of Flight (Solid lines) and mold temperature (dashed line) versus process time.

The laminate press temperature (dashed line corresponding to the right Y-axis) versus process time is overlaid in Figure 4. Interpreting the plot of the time of flight (ToF), one sees that initially there is no ultrasound transmitted because the prepreg material is in a powdered form. As the material liquefies the ToF is seen to first decrease and then increase as a function of time/temperature. Next there is a period of time when no ultrasound is transmitted corresponding to the case when the composite is in a gelatinous state and cannot support the high frequencies associated with ultrasound. As the composite cools, the ToF is seen to decrease. It appears to change very little in during the time from approximately 5.5 to 6.5 hours in the process. If the ToF can be directly correlated to the state of the composite, then during the manufacture process the cool down process may be initiated 1 hour earlier.

Another technique for monitoring the composite process in situ involves detecting the propagation of ultrasound transmitted through fibers (optically opaque) which conduct ultrasound. For these experiments we used Silicon carbide fibers (SCS-6) which were embedded in the middle of the 3"x3" composite prepreg material. A schematic diagram showing how the "ultrasonic" fiber can be used as an in situ diagnostic is shown in Figure 5. A pulsed ultrasonic excitation laser is focused onto one end of the fiber and the ultrasound which is generated propagates the length of the fiber. The frequency and amplitude of the ultrasound transmitted to the receiving end of the fiber depends upon the state of the surrounding composite material inside of the mold. Depending upon the specific state of the composite, different amounts of energy can be coupled out of the fiber as evanescent waves into the surrounding material.

We conducted a series of experiments to demonstrate the applicability of the photo induced-emf receiver for this technique. The receiver probe beam was focused on one end of the fiber while the pulsed laser ultrasonic exciter was focused on the other end. A 7 ns pulsed Nd:YAG laser was used to excite a series of compressional waves in a 477 mm length of SCS-6 fiber.

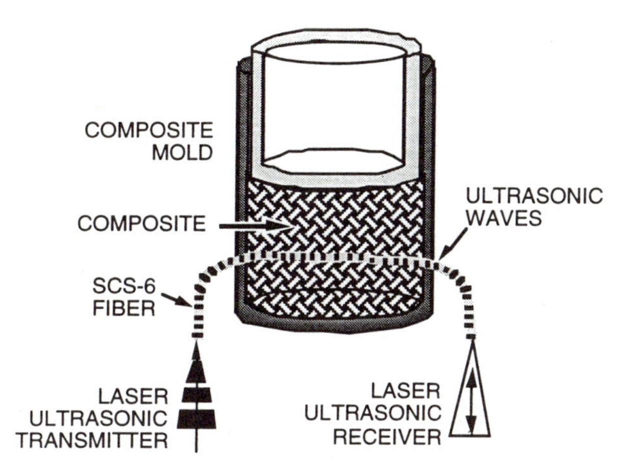

Figure 5. Composite mold with in-situ laser-based ultrasonic diagnostic.

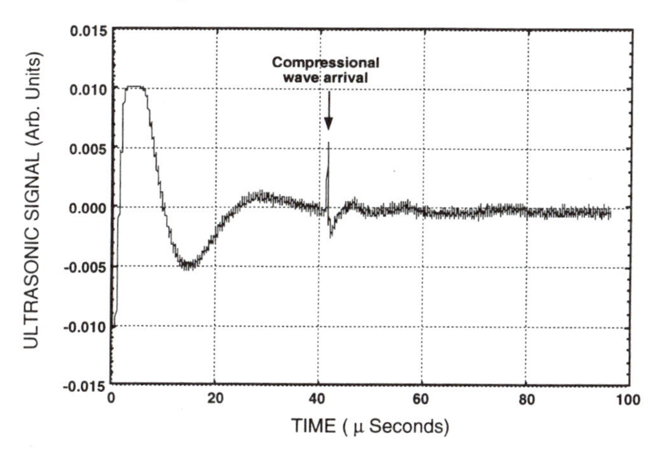

Figure 6. Ultrasonic signal transmitted in SCS-6 fiber in free space.

The first experiments were conducted with the fiber suspended in free space in order to obtain a baseline measurement and then repeated with the fiber embedded in the composite. An example of the received signal for the free-space case is shown in Figure 6. After an initial transient due to the exciter laser, the first compressional wave is seen to arrive 41.4 μsec after the excitation pulse and is relatively narrow pulse. A similar fiber (1.045 cm) was then embedded in an RS--300/T300 unidirectional tape 12 ply composite. The composite was then processed using a typical temperature and pressure time cycle.

After the composite solidified, the SCS-6 fiber was probed using the same laser excitation energy and receiver power. The resulting ultrasonic signal amplitude and frequency content were substantially altered when compared to the case of the fiber in free space. The received signal is shown in Figure 7 on the next page. Not only was the peak amplitude of the compressional wave significantly decreased but also the full width at half maximum was broadened. After an initial transient due to the exciter laser, the first compressional wave arrives 275 μsec after the excitation pulse. After this pulse there were additional features not seen in the free space case. All of these modifications of the transmitted waveform are due to the presence of the composite. While these results are encouraging, it is apparent that additional work is needed to quantify the correlation between the received signals and the cure state. It would be important to perform parametric characterizations of the received signal as function of the process time, temperature and pressure.

CONCLUSIONS AND FUTURE WORK

In conclusion, we have demonstrated techniques applicable for manufacturing process control using laser based ultrasound receivers incorporating photoinduced-emf detectors. Specifically we explored two ways to remotely monitor the processing of composite materials. The first approach used ultrasound generated and received on the mold surface and transmitted through the enclosed composite material. The second method relied on fibers embedded in the composite which guided the ultrasound. In the latter case, the possibility exists for sensing the cure state using the actual fibers which are used in the matrix. The preliminary data obtained using these two methods is encouraging and further work could yield a cost effective method for composite process control. In other related experiments demonstrating the applicability of

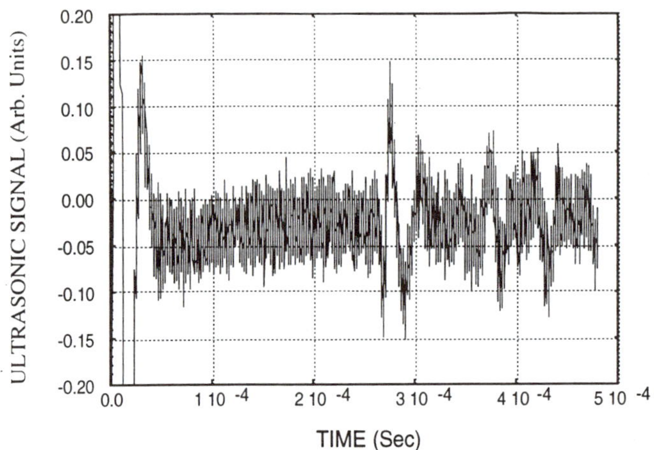

Figure 7. Ultrasonic signal transmitted through SCS-6 fiber embedded in composite.

the PI-EMF sensors to manufacturing, we have monitored parts moving with a velocity in excess of 2m/sec tangent to the receiver probe beam. This rapid response has allowed us to routinely measure the thickness, hardness and temperature of moving parts.

In other work we are investigating the performance properties of the receiver when multiple PI-EMF sensors are combined into detector arrays. This concept is economically possible because the solid state nature of the PI-EMF detector lends itself to forming monolithic structures. The sensors can easily be integrated into one and two dimensional arrays. Computer simulations of the point spread spatial responses under various conditions show that when used in conjunction with a short pulsed exciter laser, arrays can provide simultaneous high spatial and temporal resolution. In the future we plan to construct arrays and experimentally verify the computer simulations.

ACKNOWLEDGMENTS

The authors wish to thank D. Bohmeyer, S. Bourgholtzer, K. Fuller, R. Harold, R. Lohr, and D. Saboe for their excellent technical assistance and are grateful to J. Brown, R. Kent, and J. Wysocki for helpful technical discussions. This work was supported in part by DARPA and the Hughes Research Laboratories.

REFERENCES

1. See, for example, C.B. Scruby and L.E. Drain, *Laser Ultrasonics: Techniques and Applications* (Adam Hilgar Press, Bristol, 1990).
2. J.-P. Monchalin, IEEE **UFFC-33**, 485 (1986);
   J.-P. Monchalin, Rev. of Prog. in Quant. Nondest. Eval. **12,** D. Thompson and D. Chimenti, Eds. (Plenum Press, New York, 1993), pp. 495.
3. M.P. Petrov, S.I. Stepanov, and G.S. Trofimov, Sov. Tech. Phys. Lett. **12**, 379 (1986); I.A. Sokolov and S.I. Stepanov, J. Opt. Soc. Am. **B10**, 1483 (1993).
4. P.V. Mitchell, G.J. Dunning, T.R. O'Meara, M.B. Klein and D.M. Pepper, Rev. of Prog. In Quant. Nondest. Eval. **15B**, D. Thompson & D. Chimenti, Eds. (Plenum Press, N.Y., 1996); D.M. Pepper, G.J. Dunning, P.V. Mitchell, S.W. McCahon, M.B. Klein, and T.R. O'Meara, SPIE Proc. **2703**, 91 (1996).
5. A. McKie, R. Addison Jr and T. Liao, IEEE Ultrasonics Symposium (1992), **2**, p. 641.
6. M.M Ohn, A. Davis, K. Liu and RM. Measures, SPIE Proc. **1798**, 134 (1992).
7. R.M. Kent and M.J Ruddell, JOM **48**, no. 9, p. 32 (1996).

INDUSTRIAL DEMONSTRATIONS OF LASER-ULTRASONICS BASED PROCESS CONTROL, UTILIZING THE TEXTRON LASERWAVE™ ANALYZER

P. Kotidis and D. Klimek

Textron Systems Corporation
Wilmington, MA, USA

INTRODUCTION

In industrial applications, workpiece temperature has traditionally been measured by two methods, thermocouples and pyrometers. The recent industry trend towards continuous processing, such as continuous annealing and casting, has significantly limited the applicability of thermocouples, because they require physical contact with the moving workpieces. When fixed, atmosphere thermocouples are used in enclosed furnaces, they only measure the ambient temperature, which can be quite different from that of the actual moving workpiece. In that case, the process needs to be slowed down in order to "soak" the workpiece and guarantee thermal equilibrium with the furnace atmosphere, leading to reduced productivity and higher thermal losses. However, thermocouples are still being used in non-critical situations due to their significantly lower cost.

Given the limitations of the thermocouples, the industry has widely adopted the use of pyrometers, which measure the radiation emitted by the workpiece and can convert the reading to actual workpiece surface temperature. The critical limitation of the pyrometers is their dependence on target emissivity. For an ideal "black body" the value of emissivity is equal to one, but for almost any practical target, it is less than one and varies with temperature and radiation wavelength. In addition, for targets such as aluminum, the emissivity is very low and hence slight errors in the estimated values can lead to substantial errors in temperature readings. To complicate matters further, reflections from furnace walls and interference with furnace gases and sight port windows can lead to serious inaccuracies. Multicolor and laser-compensated pyrometers have been proposed as alternative technologies to compensate for the emissivity variations, but their operation is very complex and requires elaborate installations and sophisticated, application specific algorithms. Another limitation of the pyrometers is their inability to provide any indication of the internal temperature of the targets. In fact, for such measurements, the only method available is the use of embedded thermocouples, but that can only be done when the target is stationary. No practical method exists today to measure internal temperature of moving targets.

Laser ultrasonics has been proposed as an alternative method to measure target temperature.[1,2] Due to its fundamental operating principle, this technology can provide emissivity independent surface and internal temperature measurements for both stationary and moving targets. The LaserWave™ Analyzer, developed by Textron Systems, is based on this technology and hence it has attracted significant interest from the industrial community. Its additional capability to provide direct measurement of other target properties beyond temperature, such as grain size, hardness, flaws, etc., has enhanced its competitive advantage.[3] A serious drawback of the laser ultrasonics technology has always been its "R&D image" and the concern by the industry that it is only applicable to laboratory applications. The LaserWave™ Development Team at Textron Systems has focused its attention towards the demonstration of the industrial viability of the LaserWave™ technology and products through industrial Field Tests. This paper presents the results of two such Field Tests at large, full-scale facilities, a Continuous Aluminum Annealing Strip Line and an Oil Refinery. (In order to comply with the host sites requests, the names of the facilities and critical, competition sensitive process parameters have been withheld.)

OPERATING PRINCIPLE

The LaserWave™ Analyzer operating principle relies on the generation and detection of ultrasonic waves using lasers.[4] Depending on the application, a variety of lasers can be used for that purpose.[5] For the tests described in this paper, a 10-ns pulse-width Nd:YAG laser at 532 nm was employed to generate the ultrasonic waves. Its power was varied and properly controlled in order to avoid target damage. The waves were detected by a modified Michelson, polarizing interferometer, driven by a 100 mW CW monolithic semiconductor diode laser operating at 850 nm. This laser was chosen because of its low cost, rugged design and long coherence length. A critical advantage of the LaserWave™ system is that it requires no compensation on the reference leg of the interferometer. The optical layout is sized properly to receive and utilize a single speckle. Proprietary signal processing and pattern recognition algorithms have allowed the system to operate with diffusive targets and achieve sufficient signal-to-noise ratio for these measurements.

DETECTION SYSTEM DESIGN OPTIONS (Fabry-Perot, Michelson, etc.)

The LaserWave™ Analyzer incorporates a rugged design concept, which has allowed it to become a viable commercial product. Several other concepts were considered for this product, including the Fabry-Perot interferometer.[6] The polarizing Michelson design was chosen over the Fabry-Perot, because of the following reasons: (a) Compactness - Due to the fundamental physics involved, the sensitivity of the Fabry-Perot interferometer is controlled by the length of its cavity, which for practical applications cannot be less than a few feet, hence forcing the overall system dimensions to be considerably larger. For detection lasers with long coherence length, a corresponding Michelson interferometer can be, and has been, manufactured with overall dimensions of 2x2x4 inches; (b) Low Cost - The simplicity of the Michelson design eliminates the need for complicated electronic stabilization networks and high reflectivity, complex-shaped cavity mirrors; (c) Rugged Design - The Michelson concept is far superior to other designs due to its insensitivity to environmental conditions and its minimum requirements for components adjustments and moving parts; (d) Applications - Due to its ability to collect many speckles, the Fabry-Perot system is very efficient. However, for most industrial applications, where size, cost and ruggedness are the dominant design parameters, such a system is an "overkill." Depending on the application, if the proper signal processing is developed, the Michelson interferometer can provide comparable performance with slightly slower time response. Such performance has been demonstrated by the results shown in this paper. It should be pointed out that under laboratory conditions, a Michelson interferometer can be more sensitive than a Fabry-Perot interferometer.[7,8]

TESTS AT A CONTINUOUS ALUMINUM ANNEALING STRIP LINE (CAASL)

The objective of these tests was to demonstrate LaserWave™ Analyzer's capability to measure the temperature of moving aluminum strips with sufficient accuracy (± 10°F) under realistic field conditions. The customer needed a direct, emissivity independent measurement of aluminum strip temperature in order to: (a) approach the upper limit of line speed for maximum productivity, (b) detect possible over- or under-heating of the strip due to operational malfunctions and, hence, reduce scrap and avoid product liability issues.

The Facility was a Continuous Aluminum Annealing Strip Line (CAASL) and was jointly selected by Textron Systems and Surface Combustion, Inc., a leading US furnace manufacturer and our collaborator in this effort. Part of these tests were funded by the Department of Energy. The 600-ft. long CAASL was operated on a 2-3 shifts a day schedule and processed several types of aluminum alloys for aircraft and automotive applications. Strip speeds ranged up to a maximum of 200 ft./min. and maximum strip thickness was from 0.250" at approximately 5 ft. wide. The aluminum coils were loaded at the front end of the furnace and were automatically welded ("stitched") together for continuous operation.

A schematic illustration of the CAASL is shown in Fig. 1 with the LaserWave™ Analyzer installed. The unit was placed at the top of the furnace, approximately 35 ft. off the ground level, where special ports were available for optical access to the strip and was remotely operated from the Control Console, located at the "catwalk," approximately 25 ft. off the ground level. During the tests the ambient temperature at the top of the furnace, where the unit was installed, would routinely reach temperatures as high as 140°F. Therefore, use of a chiller with a closed-loop heat exchanger became necessary to avoid overheating of the lasers. As shown in Fig. 1, the standoff distance was approximately 5.5 ft. from the front window of the LaserWave™ Analyzer and the overall focal length of the collection lens was 7 ft. The beams were directed into the furnace through a 2" diam., 0.25" thick sapphire window. Special precautions were taken to avoid thermal gradients on the window and hence potential beam misalignments. Sapphire was selected for this application due to its high thermal conductivity and, hence, ability to eliminate thermal gradients. A line generation-point detection configuration was chosen and the beams were properly oriented to minimize the effect of the "flapping" aluminum strip. Typical movement of the strip was approximately ± 0.5" and it was easily handled by the LaserWave™ Analyzer.

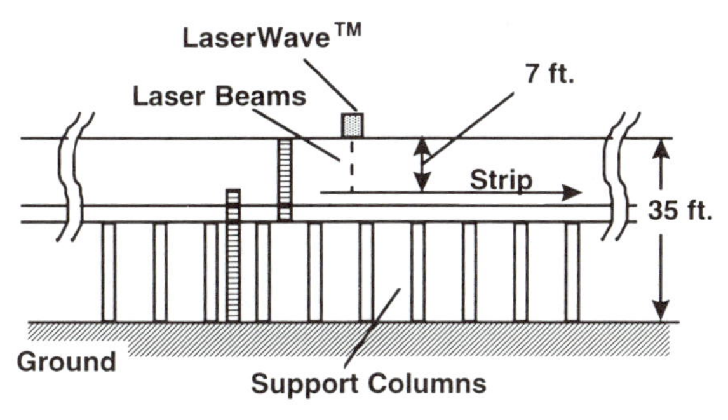

Fig. 1. Sketch of the CAASL

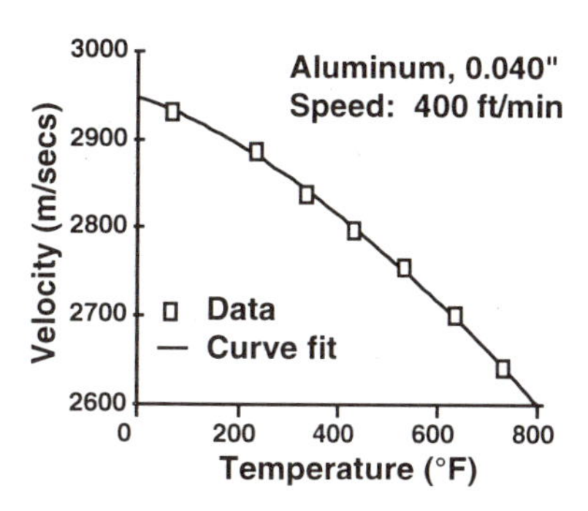

Fig. 2. Calibration curve with a moving target

In preparation for the field tests, several validation and calibration experiments were conducted. Figure 2 shows a typical calibration curve for a representative 0.040" thick aluminum sample moving at 400 ft./min. Operation in the thermoelastic regime, i.e., no target damage, was demonstrated and confirmed by further analysis of the targets at the host site's Metallurgical Laboratory. A library of calibration curves for various aluminum alloys, thicknesses and speeds, representative of the anticipated operating parameters at the field tests, was established. Several installation-related problems were overcome, including misalignment of the installed viewport, lack of adjustability in the mounting arrangement, ambient noise and temperature, electrical/ electromagnetic interference from main power feed and large-scale motors located close to the LaserWave™ installation. A test matrix was formulated with several aluminum alloys varying in thickness, furnace temperatures up to 950°F and varying strip speeds up to 200 ft./min. Extensive data were collected over a period of three months, in order to ensure the validity of the readings and the reliability of the LaserWave™ system. The built-in calibration curves, like the one shown in Fig. 2, were utilized to convert the time-of-flight readings to temperature.

Figure 3 shows a typical set of data, i.e., LaserWave™ temperature readings vs. time of the day. The furnace atmosphere thermocouple reading was steady at the furnace set point. As described below, this data set illustrated the importance of the LaserWave™ readings and its advantage over the existing temperature measuring methods. As shown in Fig. 3, the strip speed was intentionally varied in order to detect overheating conditions. As expected, the temperature read by the LaserWave™ Analyzer is inversely proportional to the speed of the line and the correlation is very good. At time point A, when the speed was reduced, an overheating condition was detected by the LaserWave™ system, which might have caused defects in that portion of the coil (about 400 ft.). Without such readings, this malfunction would not have been detected and the coil would have been shipped. Line speed variations are quite common and can be caused by many factors, such as power surges, "brown-outs," etc. Even if the line slowdown were detected in time, the whole coil would have to be rejected due to the uncertainty in the furnace thermocouple readings. The LaserWave™ Analyzer, on the other hand, provides the capability to precisely locate and mark the defective portion and save the rest of the coil. If this overheating were not detected, defects might have appeared farther down in the fabricating process during manufacturing of final products, such as aircraft skin, automotive components, etc., with severe cost and liability implications for the producer of the raw material.

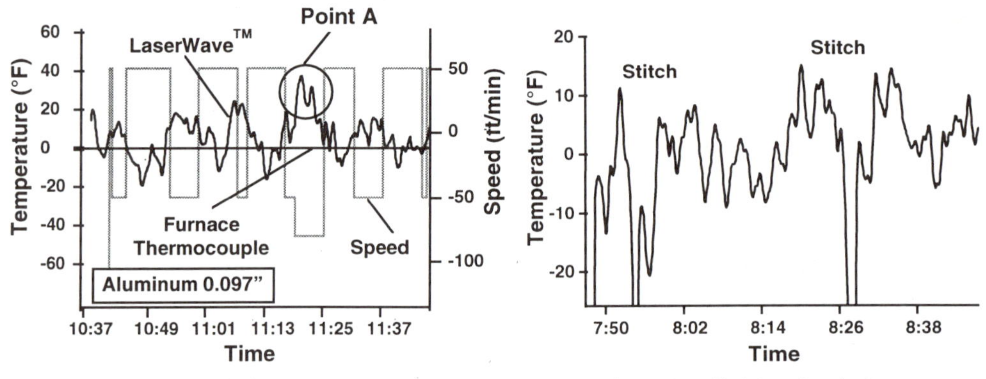

Fig. 3. Temperature & line speed measurements at CAASL

Fig. 4. Detailed data for single coil at CAASL

The accuracy of the measurement and the effect of variations within a specific alloy type were also validated during these tests. Figure 4 shows the readings from a complete coil, which started at the "stitch" indicator at 7:55 am and ended at the other "stitch" indicator at 8:30 am. The line speed was kept constant (5200 ft. long coil). The "stitch" area caused an erroneous reading due to the distorted surface, but it provided an independent indicator of the line speed and a convenient way to locate the data during analysis. The peak-to-peak variation of the temperature readings was about ± 12°F. However, it would not be appropriate to assign all the uncertainty to instrument error, since the strip temperature might have actually varied. If the "low frequency" modulation of the temperature readings is eliminated, then, as shown in Fig. 4, temperature accuracy of ± 6°F can be claimed. Figure 4 illustrates another important observation. At about 8:30 am, a new coil passed in front of the LaserWave™ system, which was expected to be of the same thickness and alloy type. After detecting the "stitch," the unit recovered and, using the same calibration curve, it predicted the right temperature with the same accuracy.

TESTS AT AN OIL REFINERY

The objective of these tests was to demonstrate the LaserWave™ Analyzer's capability to measure the temperature of refinery process tubes with sufficient accuracy (± 40°F) under realistic field conditions. The customer needed such direct, emissivity independent measurement in order to avoid overheating and dangerous explosions, caused by coke deposits on the inside wall of the process tubes. Pyrometry is unable to provide the information, because of radiation interference from the refractory walls and burners and emissivity changes due to scale deposits on the surface of the tubes.

The Facility selected was a large, full-scale Oil Refinery, and the tests were supported by both the host site and Textron Systems. One of the refinery's 40 ft. high Charge Furnaces was utilized for these tests. The tests took place during the summer months (avg. temp. 95°F, hum. 80%) and the unit was placed outdoors. A small, portable air-conditioning unit (standard option on the LaserWave™ product) was used to cool the electronics inside the LaserWave™ unit. A sketch showing a top view cross section of the furnace is shown in Fig. 5. The furnace utilizes natural gas for fuel and the burners are pointing upwards. Heat is transferred from the furnace atmosphere to the fluid (e.g., ethylene) being processed inside the tubes. Reduced heat transfer, due to coke deposits inside the tubes, can cause local overheating of the tube wall and explosions. Wall tube temperatures can be as high as 1800°F and the pressure inside the tubes can be hundreds of atmospheres.

In order to validate the data, an instrumented target probe was utilized to simulate the surface of the tubes (see detail in Fig. 5). Cooling air can be supplied through hollow passages and several thermocouples were embedded to provide independent temperature readings. Surfaces A and B were shaped accordingly to simulate the curvature of the actual process tube. As shown in Fig. 5, the LaserWave™ beams were directed to the target probe through one of the sight ports, and the target probe was mounted through a neighboring port. The test location was approximately 20 ft. off the ground on an existing platform. Plant air was used to control and vary the temperature of the probe. It is important to point out that the laser beams passed through the vertically-oriented flames from one of the nearby burners inside the furnace. No apparent signal degradation was observed due to that arrangement. Because of the extreme heat radiating from the furnace sight port, a heat shield was placed in front of the LaserWave™ unit.

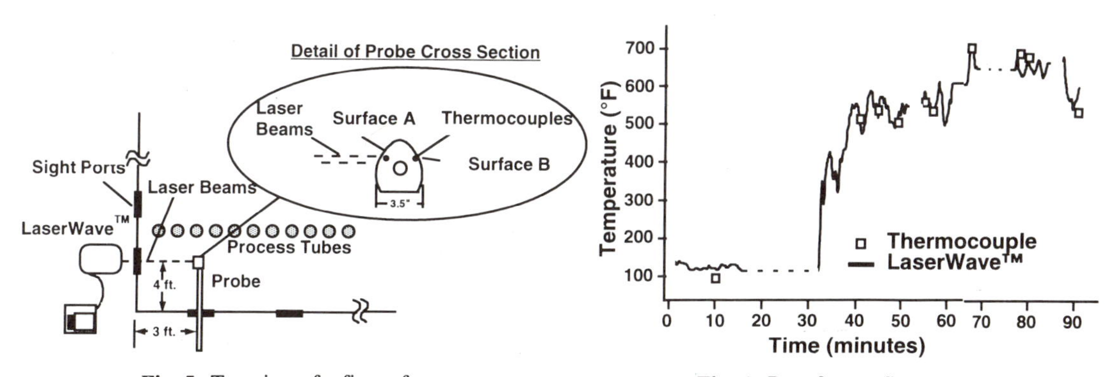

Fig. 5. Top view of refinery furnace

Fig. 6. Data from refinery tests

Following standard procedure, the unit was first calibrated at Textron's facilities and then shipped to the refinery with the calibration curves stored in the computer. A test plan was developed to determine the range of temperatures and operating conditions to be tested. Since the tests were performed on a "live" furnace, the field test team was instructed not to exceed a few hours per test in order to avoid process interruptions. Host site engineers were present at all times to record the data and assist in the experiments. The built-in calibrations were used to convert time-of-flight readings to temperature readings. Figure 6 presents a representative sample of the data acquired during these tests. As soon as the sight port was opened, the target probe was also inserted into the furnace from the other sight port (see Fig. 5). After a short heating time, the embedded thermocouple and the LaserWave™ readings were within a few °F. The sight port was repeatedly closed and opened in order to demonstrate the system's ability to recover and read the correct temperature. Flat, dashed lines in Fig. 6 indicate "no readings," i.e., sight port closed. It should be pointed out that the furnace was designed to maintain negative pressure, which implied that each time the sight port was opened, a jet of air would flow into the furnace and impinge onto the target probe, causing a cooling effect. In order to test the reliability of the LaserWave™ Analyzer, no adjustments or calibrations were allowed between readings throughout these tests. Temperature readings accuracy was estimated to be better than ± 20°F under all operating conditions. This error includes the effect of thermal currents and flames inside the furnace and around the open sight port, vibrations on the steel-mesh platform, high air temperature and humidity, vibrations of the target probe, oxidation and scale formation on the target probe and ambient temperature fluctuations.

CONCLUSIONS

The LaserWave™ field unit has successfully met the objectives of two sufficiently different field tests. Rugged and reliable operation has been achieved under realistic industrial conditions. Temperature accuracies well within customer specifications have been demonstrated. This pioneering demonstration illustrated for the first time that an inexpensive, laser ultrasonic instrument can provide sufficient accuracy and exhibit convincing performance under field conditions. The success of these tests has shaken the previously established notion that diffuse targets and industrial conditions require sophisticated systems, like the Fabry-Perot, and very powerful and expensive lasers. It is to the benefit of the laser-ultrasonics community to intensify its efforts towards the development of even more rugged, less expensive products in order to meet the needs of the ultimate customer, the industrial process control community. Textron Systems is continuing the development of such products and has identified an enormous number of new promising applications.

ACKNOWLEDGMENTS

This effort was partially supported by the U.S. Department of Energy, Office of Industrial Technologies.

REFERENCES

1. U. S. DoE, Final Report, Phase 1, "Development and Evaluation of a Workpiece Temperature Analyzer for Industrial Furnaces," DOE/ID/12830-1 (May 1990).
2. U. S. DoE, Final Report, Phase 1-A, "Development and Evaluation of a Workpiece Temperature Analyzer for Industrial Furnaces," DOE/ID/12830-2 (November 1991).
3. P. Kotidis and J.A. Woodroffe, "On-Line Noncontact Inspection of Composites Using the Textron Laser Ultrasonic System," Proceedings of 6th European Conference on Nondestructive Testing, Nice, France (1994).
4. C.B. Scruby and L.E. Drain, "Laser Ultrasonics: Techniques and Applications," Adam Hilger, Bristol (1990).
5. P. Kotidis, et al., "Development and Evaluation of a Workpiece Analyzer for Industrial Furnaces," Nondestructive Characterization of Materials VI, Plenum Press (1994).
6. J.P. Monchalin, "Optical detection of ultrasound," *IEEE Trans. on Ultrasonics, Ferroelectrics and Frequency Control* 33:485 (1986).
7. J.W. Wagner and J.B. Spicer, "Theoretical noise-limited sensitivity of classical interferometry," *J. Opt. Soc. Am.* B/Vol. 4, 8:1316 (1987).
8. U.S. DoE, Final Report, "Development and Evaluation of a Workpiece Temperature Analyzer for Industrial Furnaces," G.W. Roman, J.W. Berthold, DOE/ID/12875-3 (April 1991).

ADVANCES IN ULTRASONIC INSPECTION METHODS FOR PIM PROCESS MONITORING AND CONTROL

Joseph L. Rose,[1] Randall M. German,[2] and Derrick D. Hongerholt[1]

[1] 114 Hallowell Building
[2] 118 Research Building West
University Park, PA 16802

INTRODUCTION

Previous research efforts concentrating on proof of principle studies indicate that ultrasonic nondestructive evaluation (NDE) can be applied to process monitoring of a Powder Injection Molding (PIM) system [1, 2]. This research resulted in successful detection of several injection molding problems such as flashing, short shot, weld line defects, and jetting. Past techniques, however, require that the sensors be positioned so that the ultrasonic wave, bulk or guided, directly interacts with the defect. Recent work has shown that the data collected from an ultrasonic pulse-echo technique may be used as inputs into a neural network classification and simplex control algorithms for successful PIM process monitoring and control [3, 4]. This technique is sensitive to changes in the boundary conditions at the mold/specimen interface that result from the presence of various specimen defects. The current work investigates the use of ultrasonic data collected from a thru-transmission technique for PIM process monitoring and control. An overview of the data acquisition and analysis system is discussed along with data collection and preprocessing methods. In addition, results obtained from a simplex PIM process variable tuning algorithm experiment are presented. Adjustments in the PIM process variables are made based on information obtained from thru-transmission amplitude profiles.

EXPERIMENTAL APPARATUS

The hardware selected for digitizing the ultrasonic signatures is an eight-bit, 100 MHz analog-to-digital converter produced by GAGE Applied Science Incorporated. This instrument is based on a plug-in board for an AT-BUS PC that can be controlled under the LabWindows environment using device drivers from GAGE. This board was selected because it allows real time data collection at sufficient analog-to-digital conversion rates to prevent ultrasonic frequency aliasing during inspection. In addition, the eight-bit converter provides sufficient dynamic range to properly represent the range of signal amplitudes dealt with.

A high power RF amplifier, produced by RITEC Incorporated, has been selected as the excitation source generator. The RITEC system provides high voltage input to the transducer to assist in forcing energy into and through the material. Equipment of this nature is necessary since PIM materials in their green state are very attenuative and dispersive to ultrasonic frequency sound waves. Another benefit of this instrument is its capability of being controlled *via* GPIB. This option, developed specifically for this project, permits the instrument to be controlled from the LabWindows programming environment. Instrument drivers were developed under the LabWindows programming environment for the RITEC system. The RITEC system also operates as a receiving device with software controllable gain and impedance. In addition, a variety of high and low pass analog filters are available to precondition the received signal. A schematic of the inspection system is included in Fig. 1a. The PC, with integrated data

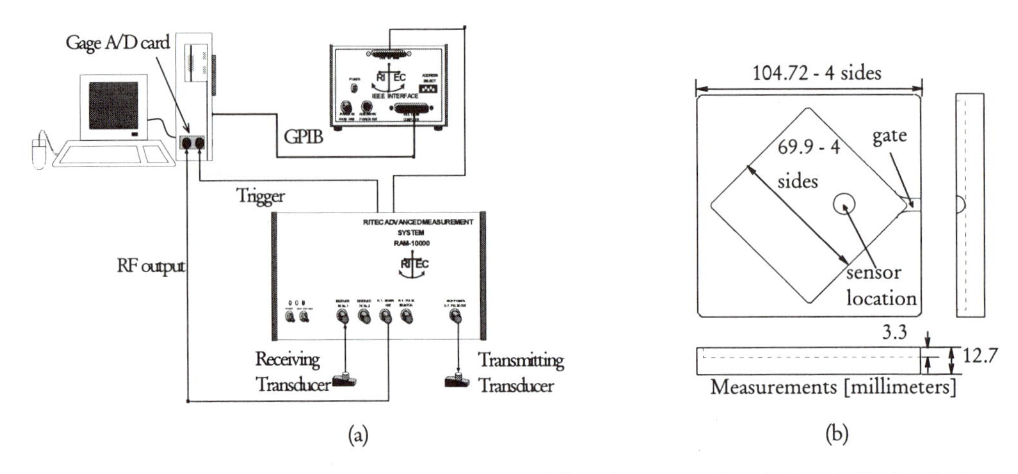

Figure 1. (a) A schematic diagram of the RITEC inspection system and the PC based Gage A/D card. Control of both devices is accomplished using National Instruments LabWindows for DOS software. (b) A schematic showing the geometry and dimensions for one side of the mold insert used to produce the plate specimen. The mating side of the insert is a flat surface. The location of the gate, where material enters the die cavity, is indicated.

acquisition card, and the RITEC system are shown. The LabWindows software resides on the PC, and RITEC system control is accomplished through GPIB.

The mold insert for the specimens molded during the experiments is described by the schematic in Fig. 1b. The specimen is a square plate with the dimensions indicated in Fig. 1b. The molding material is a carbonyl iron feedstock commonly used in the PIM industry. Two 6.4 mm diameter piezocomposite transducers, one attached to each side of the mold insert, are used. One functions as a sender and the other a receiver. The transducers are mounted, facing each other, to the back side of each half of the mold insert. The position of the sensors, relative to the mold insert geometry, is shown in Fig. 1b. The center frequency of each transducer is 2.25 MHz. The sending transducer is excited with a one cycle pulse of 2.25 MHz.

DATA PREPROCESSING

Ultrasonic thru-transmission signals collected at different times during the molding cycle are included in Fig. 2. The velocity and amplitude of the wave changes rapidly as the material cools from a viscous liquid melt to a solid. No signal is present immediately after the mold closes, before injection begins. The cavity at this time is filled with air and no ultrasonic energy passes through. Fig. 2a is the signal collected six seconds after the mold closes. Material has been pushed into the mold and at this point is a viscous melt. As the material starts to cool and solidify, the velocity and amplitude of the wave traveling through the specimen increase. After sufficient cooling, the specimen solidifies and the velocity and amplitude increase more (Fig. 2b). The data collected forty-two seconds after injection begins, Fig. 2c, indicates the same. The ultrasonic waveforms are processed in a real-time fashion. Preprocessing involves computing the maximum amplitude of the signal, in the time window shown in Fig. 2, as a function of time after injection begins. The resulting amplitude versus time profile is used as the feature source used in the process control algorithm.

PROCESS CONTROL STRATEGY

Ultrasonic sensors show promise in defect detection and classification when monitoring the PIM manufacturing process. Implementing these sensors into a closed-loop control system has enormous potential. In this section, a simplex method is introduced that can automatically tune the PIM manufacturing parameters based on *in-situ* ultrasonic sensor readings. Earlier works from Box that have used this method extend back to 1955 [5]. In this work, Box called the method Evolutionary Operation. He applied the method to increase industrial productivity. At this time, however, the algorithm was not automated and was performed with charts and hand calculations. In 1962, Spendley reviewed the method of evolutionary operation and came up with a method to make the procedure automated [6]. This was

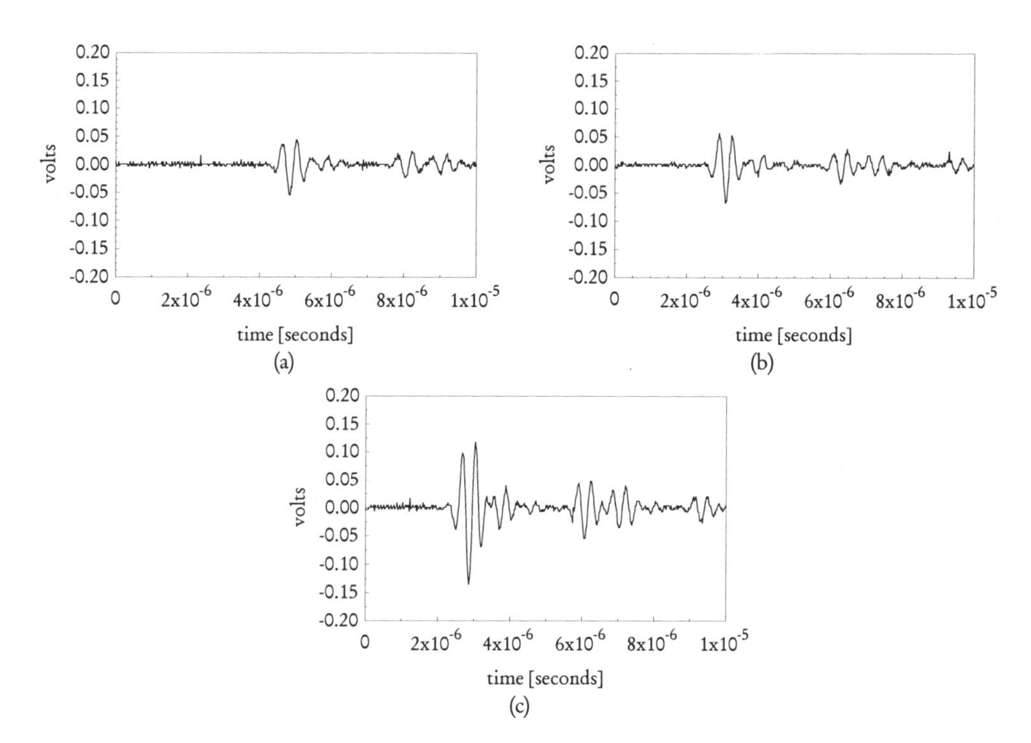

Figure 2. Thru-transmission ultrasonic signals collected during the injection molding process. (a) The signal collected six seconds after injection begins. (b) The signal collected seventeen seconds after injection begins. (c) The signal collected forty-two seconds after injection begins.

necessary at the time since Box had not defined such rules as when and in what direction to move the operating conditions in the operating condition space. The equations resulting from the work of Spendley are precisely the same as those presented in this section. At this time, the idea of using the method for mathematical function minimization was also presented. In 1965, Nedler and Mead applied the method to such problems [7]. The method finally lent itself to solving the inverse problem of fitting curves to data [8]. This is the application that the simplex method is commonly known for today in the NDE arena. The work in this thesis applies the simplex method to manufacturing related problems, more precisely, to tuning PIM manufacturing parameters. Potential applications for this system include the initial setup of optimal manufacturing conditions and also tracking of the optimum conditions *in-situ* during long term operation.

In general, consideration is given to the minimization of some unknown system, S, of n variables, V_1, V_2, V_3, ..., V_n, with constraints, C_1^l:C_1^h, C_2^l:C_2^h, C_3^l:C_3^h, ..., C_n^l:C_n^h. The system is unknown in the respect that the system response is not defined by a known function or functions. Each constraint defines the limits of the variable with matching subscript. The superscripts l and h correspond to low and high parameter limits, respectively. The output of the system for a particular set of variables is defined by the vector O, and is represented by sensor readings that describe the quality of the system product, the system efficiency, or the measured system response. For the PIM process, this vector is defined by any combination of *in-situ* ultrasonic, hydraulic, pressure, and temperature sensor readings.

A simplex is formed in $n+1$-dimensional space and is defined by $n+1$ vertices, P_0, P_1, P_2, ..., P_n, P_{n+1}, all of which are vectors of length $n+1$ describing the coordinates of the vertex in space. The extra dimension is the response surface, and the extra coordinate represents the response defined by the squared residual sum, $\sum R^2$. The value for $\sum R^2$ at each vertex is computed from the following relation (Eq. 1),

$$\sum R^2 = \sum_{i=1}^{np} \left(O_{(optimal)i}{}^2 - O_i^2 \right) \tag{1}$$

where $O_{(optimal)}$ is the vector containing the optimal sensor readings, and np is the number of points being evaluated. Eq. 1 is the condition that quantifies how well the sensor readings match the ideal sensor readings. Lower values of $\sum R^2$ indicate a closer match.

The vertices with the best (lowest) and worst (highest) response are P_{low} and P_{high}, respectively. The definitions of the best and worst vertices evolve from the fact that when minimizing the error in the residuals, a low response is considered to be better. A low response for the residual would correlate to a feature vector that matches the optimum feature vector closely. If the feature vector collected from a specimen matched the optimum feature vector exactly, the residual error would be zero. The worst vertex would be defined as the vertex corresponding to the feature vector, and specimen, that is farthest from the optimum feature vector. The center of the simplex is P_{center} and is calculated using all of the points making up the simplex except P_{high}. This is performed by adding all of the individual coordinates from the points and dividing by n, since only n points are used. The worst vertex, P_{high}, is excluded since it is the vertex that is replaced.

There are three underlying operations that allow the simplex to update and therefore optimize S. These are reflection, expansion, and contraction. Reflection of the worst vertex in the simplex is the first operation performed in the procedure. The governing relation, Eq. 2, behind reflection is,

$$P_{ref} = (1+\phi)P_{center} - \phi P_{high}, \tag{2}$$

where P_{ref} is the reflected vertex and ϕ is the reflection coefficient that can take on any value greater than zero. A reflection coefficient of unity creates a specular reflection of the worst vertex, referenced to P_{center}. If the response at P_{ref} is better than (less than) or equal to the response at all the vertices in the simplex, P_{ref} is accepted (replaces P_{high}) and the expansion operation is performed on the new P_{high}. This allows the simplex to expand in a direction that is more desirable than the worst direction of the previous simplex. If the response at P_{ref} is worse than (greater than) the response at P_{low} but better than the response at P_{high}, P_{high} is replaced by P_{ref} and the termination criterion is checked.

Expansion occurs, as mentioned above, if the response at P_{ref} is better than or equal to the response at all the vertices in the simplex. The governing relation, Eq. 3, behind expansion is,

$$P_{exp} = \gamma P_{high} + (1-\gamma)P_{center}, \tag{3}$$

where P_{exp} is the expanded vertex, and γ is the expansion coefficient that can take on any value greater than unity. The expanded vertex replaces P_{high} if its response is better than or equal to the response at all the vertices in the simplex including P_{ref}, if not P_{ref} replaces P_{high}. After P_{high} is replaced by either the expanded or reflected vertex, the termination criterion is checked. Expansion allows the simplex to move in a potentially desirable direction in order to find the global minimum.

Contraction occurs if the response at the reflected vertex is worse than the response of all vertices in the current simplex. The contracted vertex, P_{con}, is determined from the relation in Eq. 4,

$$P_{con} = \beta P_{high} + (1-\beta)P_{center}, \tag{4}$$

where β is defined as the contraction coefficient and can take on values between zero and unity. If the response at P_{con} is better than or equal to the response at all vertices in the current simplex, P_{con} replaces P_{high}. If the response at P_{con} is worse than the response at any vertex in the current simplex, all vertices, except P_{low}, are contracted towards P_{low}. By contracting the vertices, the simplex geometry shrinks in space and concentrates on searching for the global minimum near P_{low}.

The basic operations of a two dimensional simplex (BWV) are graphically displayed in Fig. 3a. The new vertex is computed from Eqs. 2-4. The new vertex is substituted for the worst vertex (higher ΣR^2 values). By following this procedure the simplex moves to lower ΣR^2 values. If the new vertex is expanded, W is replaced by E. The distance, d, between the center of BV and E now becomes γd. A contracted vertex results in d becoming βd, the new vertex then moves to C. Finally, the reflected vertex, R, becomes the mirror image of W with respect to BV. A schematic that shows the interaction between the PIM process, the simplex tuning algorithm, the data collection algorithm, and the feature extraction algorithm is included as Fig. 3b. The desired sensor readings and the initial system parameters are provided by the system operator before the tuning algorithm begins.

EXPERIMENTAL RESULTS

This section discusses the results from an experiment that uses the simplex tuning algorithm to tune the PIM process variables. Three PIM process variables are dynamically adjusted based on the simplex tuning algorithm suggestions. The parameters that are adjusted are the packing pressure profile (bar), the packing cycle time (seconds), and the injection velocity profile (mm/sec). These are three of the most critical parameters for molding this particular feedstock and specimen. The total number of machine variables that the parameters affected are twenty-one (ten for the packing pressure profile, ten for the injection velocity profile, and one for the packing cycle time). The total packing pressure profile is represented by ten points; each point is associated with a time frame of 0.1*V_2. The profile is then defined

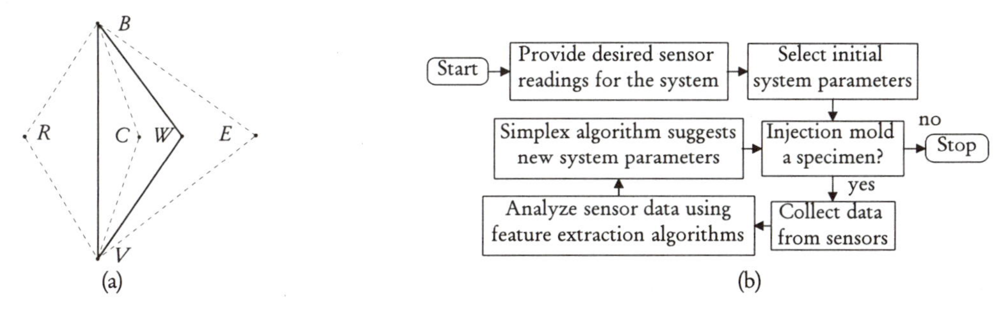

Figure 3. (a) A schematic describing the basic operations of a two dimensional simplex (*BWV*). The operations of reflection (*R*), expansion (*E*), and contraction (*C*) on the worst vertex (*W*) are graphically demonstrated. (b) A schematic diagram explaining how the simplex algorithm, data collection algorithm, and feature extraction algorithm communicate.

to be V_1, 0.95*V_1, 0.90*V_1, 0.80*V_1, 0.75*V_1, 0.70*V_1, 0.60*V_1, 0.50*V_1, 0.325*V_1, and 0.125*V_1. The injection velocity profile is adjusted by selecting velocity values for a particular injection velocity profile, then moving the range that the injection profile is applied. The beginning position of the range is defined as the parameter V_3 (mm). Other critical parameters are the die temperature and the barrel temperature. The die temperature is not selected since there is no reliable method, with the current equipment, of controlling this parameter over a wide range. The die temperature is therefore kept at room temperature. The barrel temperature is not controlled due to time limitations. It requires a considerable amount of time for the barrel temperature to stabilize after adjustment.

A total of forty-three specimens are manufactured during the experiment. The experiment is stopped when it is apparent that the simplex algorithm has converged to an error surface minima. Since several different combinations of parameters can result in a good specimen, several minima will appear in the feature space error surface. The simplex algorithm converges to the closest minima that satisfies $O_{optimal}$, Fig. 4. The complete profile is used to represent this vector. The parameter values for collecting the data used as $O_{optimal}$ are 38 bar, 20 seconds, and 55 mm. Specimens produced under these molding conditions show no signs of defects. The initial parameter values for V_1, V_2, and V_3 are 20 bar, 10 seconds, and 60 mm, respectively. The initial system starting parameters are selected within safe machine operating limits. These values are reasonable guesses based on operator molding experience. The constraints, C_1^l:C_1^h, C_2^l:C_2^h, and C_3^l:C_3^h, are set to values that are define the range of normal process operation for this particular mold geometry. Figs. 5a-5b are thru-transmission profiles collected during the experiment. Initial manufacturing attempts produce specimens with cracking. Cracking results from insufficient packing pressure during molding. The profiles collected while molding the first eight specimens, Fig. 5a, have ultrasonic amplitudes associated with them that are considerably different than the optimal feature vector shown in Fig. 4. Continued attempts using the parameters suggested by the tuning algorithm result in specimens that show no signs of cracking or other defects. The profiles collected while molding the last eight specimens are very similar to the profile in Fig. 4. The final values for V_1, V_2, and V_3 are 41 bar, 11 seconds, and 61 mm, respectively.

SUMMARY AND CONCLUSIONS

An *in-situ* ultrasonic sensor based control algorithm is developed for the PIM process. The control algorithm, based on the simplex methodology, is used to tune the PIM manufacturing variables, *in-situ*, to achieve increased specimen quality. The algorithm relies on having suitable sensor readings that discriminate good specimens from defective specimens. An optimum feature vector is selected to represent the PIM manufacturing process when it is operating under conditions that produce defect free specimens. The optimum feature vector selected is the ultrasonic amplitude profile. The algorithm begins by having the user select a set of manufacturing variables that they think will get the process close to the optimum operating conditions. From that point on, the simplex algorithm controls which variables will be used to manufacture future specimens. The sensor readings collected from the future specimens are compared to the optimum sensor readings using the least squares method as a closeness criteria. The simplex algorithm moves through the process variable space and finds a set of variables that result in sensor readings that closely match the optimum sensor reading vector. An experiments is conducted where three (twenty-one manufacturing variables) process parameters are tuned. During the experiment, specimen quality increased while the simplex algorithm was in control of the process.

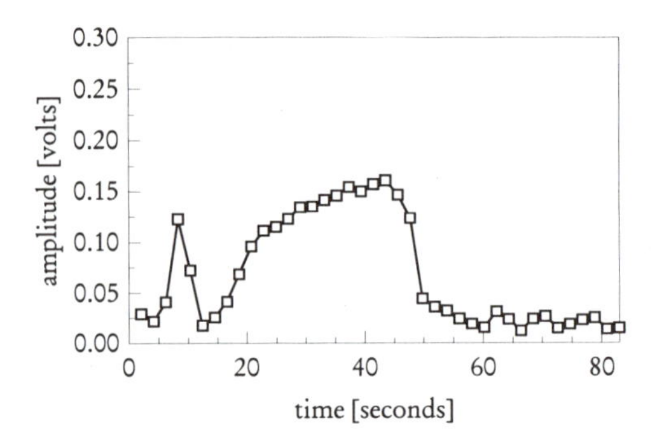

Figure 4. The thru-transmission amplitude profile collected with the manufacturing variables set to values that produced a defect free plate specimen. The profile is defined to be the optimal feature vector for the two parameter process control experiment.

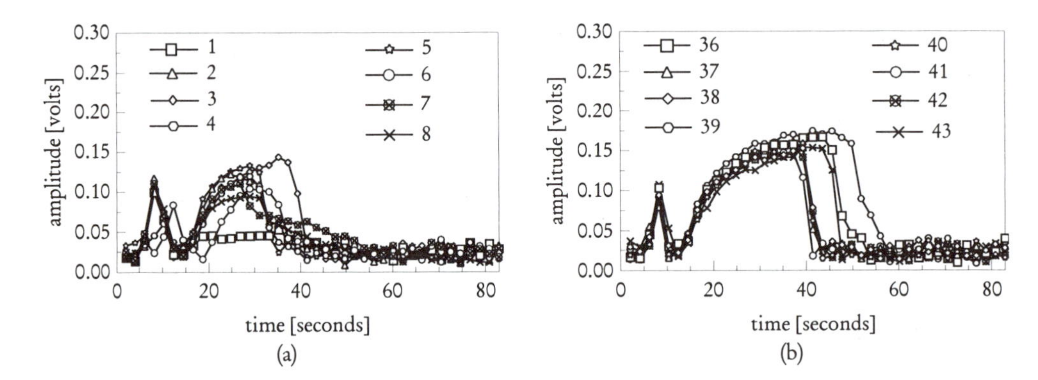

Figure 5. (a) The thru-transmission amplitude profiles that are collected for plate specimens 1 - 8. Specimen 1 resulted in severe short shot. The other specimens contained varying degrees of cracking due to insufficient packing pressure. Specimen 3, which was molded using the highest packing pressure out of the five specimens, showed the least amount of cracking. (b) The thru-transmission amplitude profiles that are collected for plate specimens 36 - 43. None of the specimens showed visible signs of cracking or flash. The process parameters were the same for seven of the specimens. Specimen 39 had a slightly higher packing pressure profile than the other seven.

ACKNOWLEDGMENTS

This work is supported by Grant #DMI9408878 from the National Science Foundation. Fellowship resources were also received by Derrick D. Hongerholt from a Department of Energy Integrated Manufacturing Doctoral Fellowship.

REFERENCES

1. S. M. Menon, K. F. Hens, R. M. German, and J. L. Rose, "*Ultrasonic Sensors for Powder Injection Molding*," Advances in Powder Metallurgy and Particulate Materials, Vol. 4, (1994), pp. 71-84.
2. J. L. Rose, S. Menon, R. Yi and R. M. German, "*A Guided Wave Resonance Matching Technique for In-Situ Monitoring of Powder Metal Injection Molded Products*," Proceedings QNDE Conference, Seattle, WA, July - August (1995), pp. 2053-2060.
3. D. D. Hongerholt, J. L. Rose, and R. M. German, "*Using Ultrasonic Sensors for Powder Injection Molding*," *JOM*, Vol. 48, No. 9, September (1996), pp. 24-28.
4. D. D. Hongerholt, J. L. Rose, and R. M. German, "*A Self-Tuning Ultrasonic Sensor Based Powder Injection Molding Process,*" Accepted for publication by NDT&E International, December, (1996).
5. G. E. P. Box, "*Evolutionary Operation: A Method for Increasing Industrial Productivity*," Paper presented to the International Conference of Statistical Quality Control, Paris, July 1955. Reproduced in Applied Statistics, Vol. 6, (1957), pp. 3-22.
6. W. Spendley, G. R. Hext, and F. R. Himsworth, "*Sequential Application of Simplex Designs in Optimization and Evolutionary Operation*," Technometrics, Vol. 4, No. 4, November (1962), pp. 441-461.
7. J. A. Nedler and R. Mead, "*A Simplex Method for Function Minimization*," The Computer Journal, Vol. 7, (1965), pp. 308-313.
8. S. Caceci and W. P. Cacheris, "*Fitting Curves to Data*," Byte, May (1984), pp. 340-362.

ANISOTROPIC ELECTRIC CONDUCTIVITIES IN GRAPHITE FIBER REINFORCED COMPOSITES

John M. Liu,[1] and Susan N. Vernon,[2]

[1]Carderock Div. Naval Surface Warfare Center
West Bethesda, Md. 20817-5700
[2]AMRON Corp.
Arlington, Va. 22202

INTRODUCTION

As a result of manufacturing processes, there is a strong tendency for the atomic basal planes in graphite to aligned preferentially along the fiber axis. Such fibers are anisotropic in their electrical properties. In composite laminates containing these fibers, the electrical properties are dependent on the precursors from which the fibers are made (e.g. pitch or PAN based), and upon the heat treatment the fibers and the composite have been subjected to after manufacturing.

Vernon[1-3] reported a technique to evaluate the electrical resistivity of conductors based on the excitation of eddy currents, and the measured probe geometry and the frequency dependence of the probe impedance. In this paper, we extend this approach to the case of anisotropic composites. The input electromagnetic field was made to polarized preferentially in one direction on the surface of the composite. As this direction was changed, the resulting probe impedance changed following a cosine square dependence, known to be appropriate for a second order conductivity tensor.

In the following we will briefly describe the bases of our technique, followed by examples of measured orientation dependence of resistivity. In addition, results for two and three layer laminates will be shown to illustrate the possibility of using our technique for the detection of fiber misalignment in laminates.

PRINCIPLES OF MEASUREMENT TECHNIQUE

Single sided eddy current measurement were made using an impedance analyzer. A ferrite core was used to concentrate the electromagnetic field generated by a multiturn coil. Some basic features of the impedance of the probe with and without the presence of the material are illustrated in Fig. 1a, for frequencies ranging from 125 KHz to 8000 KHz. A normalized impedance diagram of the same data is shown in Figure 1b, together with the definition of the lift-off angle(Θ), expressed as the ratio of the inductive to the resistive component of the probe impedance. Vernon's technique was based on the following relationship,

$$X_n/R_n = A + B\ r\ (f/\rho)^{1/2}$$

where X_n is the normalized inductance, R_n is the normalized resistance, r is the effective probe radius, f is the frequency, ρ is the appropriate component of the resistivity, A and B are constants for a particular probe.

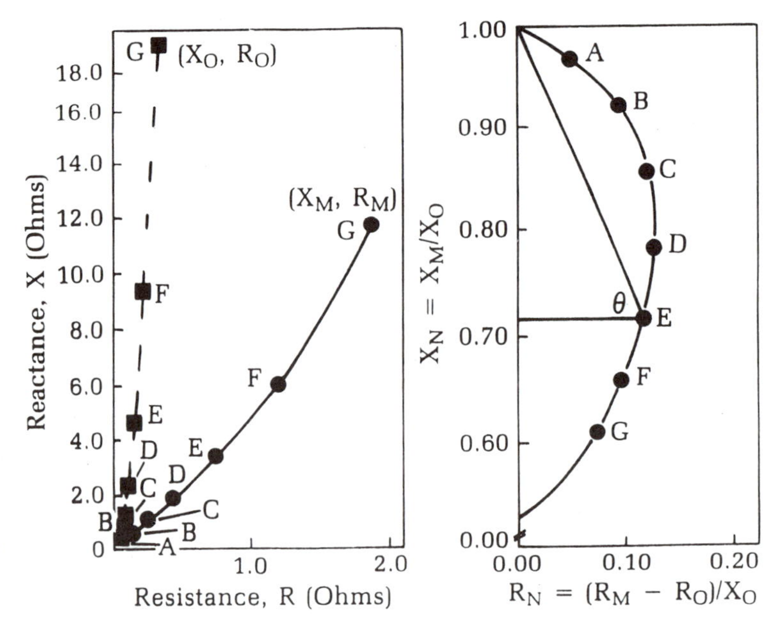

Figure 1. (a) Reactive and resistive components of the impedance of a 0.44 inch O.D. eddy current probe in air (dotted curve) and on a carbon/carbon composite (resistivity of 842 μΩcm). The frequencies in kilohertz are: A, 125; B, 250; C, 500; D, 1000; E, 2000; F, 4000; and G, 8000. (b) Normalized impedance diagram for data in (a).

An eddy current probe exciting a linearly polarized field to generate currents in one direction was constructed using a pair of line sources wound around a cup core having a length much larger than its width. Examples of such probes are shown in Fig. 2.

RESULTS AND DISCUSSION

In Fig. 3 the electrical resistivity is shown as a function of the angle between the fiber direction and the eddy current path in a uniaxial composite panel. It is noted that the resistivity in the fiber direction is about a tenth of that transverse to this direction. In addition, the orientation dependence is reasonably well described by the expected behavior of a second order tensor.

Figure 2. Examples of NSWC designed eddy current probes for nondestructive evaluation of composites. The large one is a ferrite-core probe approximating two line sources. In the smaller one, the coil windings are opposing each other for differential measurements.

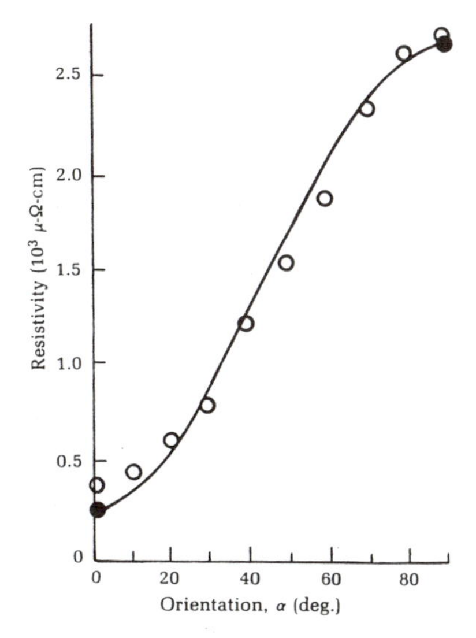

Figure 3. Variations of the electrical resistivity as the direction of eddy current flow changes from the fiber direction (∝ = 0) to the transverse direction (∝ = 90). o, measured data. The continuous curve shows the expected variation of a second order tensor, based on D.C. contact measurements at ∝ = 0 and ∝ = 90.

For laminates it is sometimes possible to assess properties one layer at a time. Since the depth of penetration of the electromagnetic field can be controlled by changing the frequency, our techniques allow for the evaluation of electrical resistivity and its anisotropy, as a function of the stacking sequence in a laminate. An example of such an application is shown in Fig. 4A for a stack of two uniaxial layers of graphite composite in which the fiber direction of the top layer is orthogonal to that in the bottom layer. It is seen that at low frequency both the top and the bottom layers interacted with the electromagnetic field, exhibiting an apparent 4-fold symmetric resistivity for the

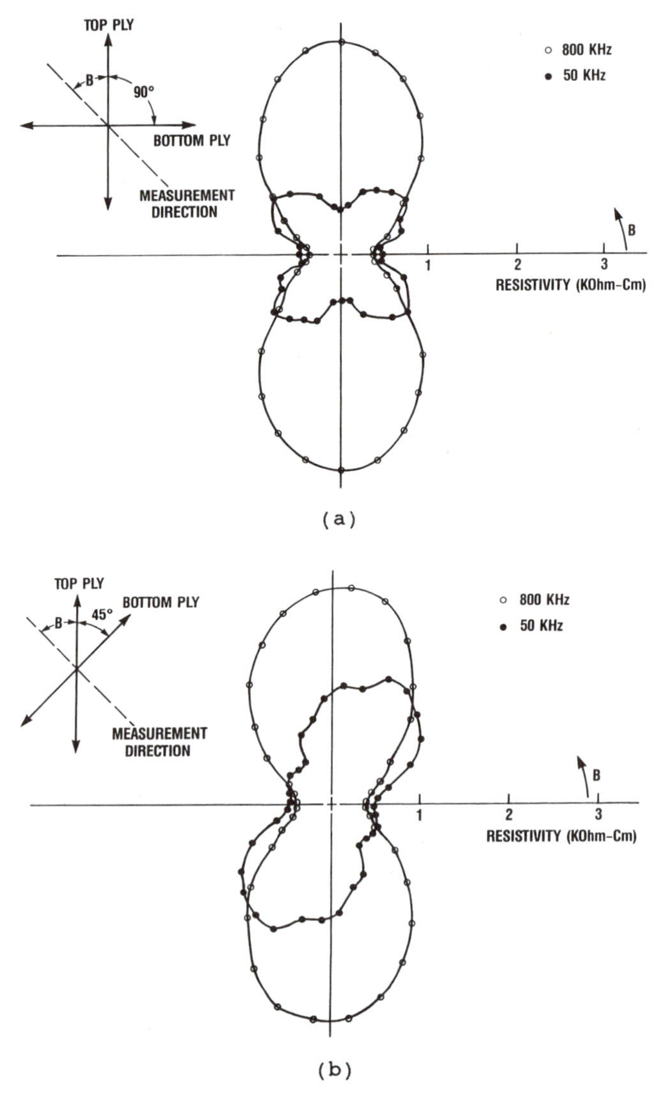

Figure 4. Polar plot of the variations of electrical resistivity in a two-layer, graphite composite laminate, as the direction of the input electric field varies on the surface of the laminate. (a) The fiber directions in the top layer and the bottom layer are at $\alpha = 0$ and $\alpha = 90$, respectively. (b) same as in (a) except the fiber directions were 45 deg. apart between the top and the bottom layer.

laminate as a whole. On the other hand, when the frequency was such that only the top layer interacted with the input field, a 2-fold symmetric pattern characteristic of a uniaxially reinforced top layer was observed. Results for an experiment in which the top and bottom uniaxial layers were 45 degrees apart are plotted in Fig. 4B. Again, the assessment of the two layers is seen to be possible by changing the frequency.

Another example of anisotropic resistivity is shown in Fig. 5. Here three uniaxial layers were stacked up as shown in the insets. When the fiber directions of the stack were aligned in a symmetric manner with respect to one another, the resulting effective resistivity was found to be symmetric as the probe was rotated over the stack, as shown in Fig. 5A. When the stacking was no longer symmetric, the resulting effective resistivity was also found to be asymmetric, as shown in Fig. 5B. Thus this technique can be exploited for the detection of fiber misalignment.

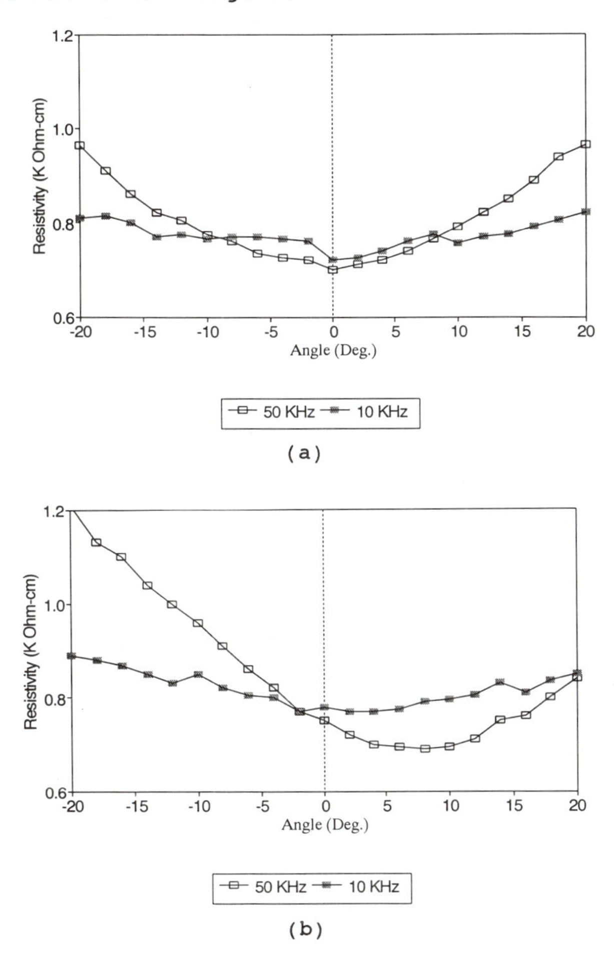

Figure 5. Orientation dependence of the resistivity in a three layer stack as the direction of the input electric field varies on the surface of the laminate. (a) the fiber direction in the top and the bottom layer is 90 deg. with respect to that for the middle layer (symmetric alignment), and (b) the fiber direction of the top and bottom layers are 82 and 98 degrees with respect to that of the middle layer (asymmetric alignment).

CONCLUSIONS

We have presented data on the anisotropic resistivity in graphite fiber composites. Taking advantage of the frequency dependence of the skin effect, we show examples of the orientation dependence of the resistivity for different layers in a stack of uniaxial layers. The possibility of detecting fiber misalignment in a laminate is indicated.

REFERENCES

1. Vernon, S. N. "Eddy Current Resistivity Model", *Materials Evaluation*, Vol. 46, No. 12, 1988, pp. 1581-1587.

2. Vernon, S. N. and Gammell, P. M., Patent #4922201: "Eddy Current Method for Measuring Electrical Resistivity and Device for Providing Accurate Phase Selection", 1 May 1990.

3. Vernon, S. N. and Liu, J. M., "Eddy Current Probe Design for Anisotropic Composites", *Materials Evaluation*, Vol. 50, No. 1, 1992, pp. 36-41.

ACOUSTIC CHARACTERIZATION OF MORPHOLOGICALLY TEXTURED SHORT-FIBER COMPOSITES: ESTIMATION OF PHYSICAL AND MECHANICAL PROPERTIES†

Martin L. Dunn[1] and Hassel Ledbetter[2]

[1] Center for Acoustics, Mechanics, and Materials, Department of Mechanical Engineering, University of Colorado, Boulder, Colorado 80309

[2] Materials Science and Engineering Laboratory, National Institute of Standards and Technology, Boulder, Colorado 80303

INTRODUCTION

Short-fiber composites are increasingly being used in a wide range of applications spanning the automotive, computer, communications, aerospace, and home-appliance industries to name a few. They are most often fabricated in the form of short glass or ceramic fibers embedded in a polymer, metal, or ceramic matrix. It is widely accepted that the microstructure of the composite plays a tremendous role on its overall physical and mechanical properties. For a prescribed fiber/matrix system, perhaps the most significant microstructural variable in a short-fiber composite is the fiber-orientation distribution; it dominates the aspect ratio effect. It dictates the overall symmetry of the composite and it strongly affects almost all macroscopic material properties: stiffness, strength, thermal expansion, thermal conductivity, dielectric constant, plastic-deformation behavior, processing-induced residual stresses, and fracture toughness. The fiber orientation distribution is predominately controlled by processing and forming operations.

There are numerous ways to describe the orientation distribution of short fibers in composite materials; however, a comprehensive review is not presented here. Interested readers may find details in numerous studies[1-14] and the many references contained therein. Regardless of the way the orientation distribution function (ODF) is described mathematically, the most challenging problem to date is its measurement. Currently, techniques for the measurement of the orientation distribution of short-fiber composites consist of two general approaches: (i) measurement by x-ray or neutron diffraction in the same way that crystallite orientations are measured in textured polycrystals;[15] (ii) measurement by a repeated slice-polish-image procedure.[7] The former is quite accurate but only applicable to composites with crystalline fibers (glass does not diffract), and there must be a unique relationship between the crystalline axes and the fiber axis. In particular, the technique can not be used when the fibers are polycrystalline. The technique is also expensive, requiring careful specimen preparation and x-ray or neutron diffraction facilities. The latter technique is based on the repeated polishing and imaging of a cross-section of the composite. Ideally the sectioned fibers appear as ellipses on the cross section. After measurement of the elliptical shape of numerous fibers at numerous cross sections, the orientation distribution of the fibers can be inferred. This method has numerous drawbacks including the fact that great care must be taken in surface preparation, it is destructive, and it is time-consuming. Nevertheless, it is the state of the art.

---

† Partial contribution by U. S. Government, noncopyrightable.

Nondestructive Characterization of Material VIII
Edited by Robert E. Green Jr., Plenum Press, New York, 1998

We have recently proposed an alternative approach to characterize the ODF of short-fiber composites. The basic idea is that we couple an experimentally validated micromechanics model that describes the effect of the fiber ODF on the overall elastic moduli of the composite, with measured elastic moduli (acoustic velocities) in an inverse manner to infer the fiber ODF.[16] The objective of this study is to take the fiber ODF so-obtained, and use it in conjunction with micromechanics models of other physical and mechanical properties to predict those properties. Specifically, we will use our acoustic characterization to predict thermal expansion, the elastic-plastic behavior, and the thermal conductivity of short-fiber composites. Thus, based on acoustic measurements, we hope to predict other physical and mechanical properties of the composite.

ACOUSTIC CHARACTERIZATION OF THE FIBER ODF

Micromechanics Model

Consider a two-phase matrix-based composite where quantities associated with the matrix and the reinforcing fibers are denoted by 0 and 1 respectively. The fibers are randomly distributed, but their orientation distribution is arbitrary and can be described by a suitable orientation distribution function (ODF). The orientation of a given fiber is specified uniquely by the three standard Euler angles θ, ψ, and φ relating two coordinate systems: a fiber coordinate system, $x_1'x_2'x_3'$, and a sample coordinate system, $x_1x_2x_3$. The orientation distribution of the fibers can be described by the function $w(\xi=\cos\theta,\psi,\varphi)$ that describes the probability of finding a fiber oriented at angles $w(\theta,\psi,\varphi)$. To express the ODF mathematically, we adopt Roe's[17] method and expand $w(\xi,\psi,\varphi)$ in a series of generalized Legendre functions $Z_{lmn}(\xi)$:

$$w(\xi,\psi,\phi)=\sum_{l=0}^{\infty}\sum_{m=-l}^{l}\sum_{n=-l}^{l}W_{lmn}Z_{lmn}(\xi)e^{-im\psi}e^{-in\phi} \quad (1)$$

In eq. (1) W_{lmn} are the texture coefficients, and, if known, completely specify the ODF. It turns out that when the ODF is used in the averaging of elastic constants, which are fourth-rank tensors, only the coefficients W_{lmn} with $l\leq4$ enter the problem, and the existence of symmetry further reduces the number of independent W_{lmn}. For second-rank tensorial properties, only the coefficients W_{lmn} with $l\leq2$ enter the problem. This is a result of the truncation theorem for averaging with weight function.[10]

We recently developed a micromechanics model to predict the elastic moduli of the morphologically textured composite. It results in an expression for the effective moduli given by

$$C_{ijkl}=C_{ijkl}^{0}+cT_{ijklmnop}(C_{mnrs}^{1}-C_{mnrs}^{0})A_{rsop} \quad (2)$$

where

$$A_{rsop}=\left[I_{rsop}+cS_{rsij}{C_{ijmn}^{0}}^{-1}(C_{mnop}^{1}-C_{mnop}^{0})\right]^{-1} \quad (3)$$

Here C_{ijkl}^{0} and C_{ijkl}^{1} are the elastic stiffness tensors of the matrix and fibers, c is the fiber volume fraction, and A_{ijkl} is the strain concentration tensor that relates the average strain in the reinforcement phase to that compatible with uniform displacement boundary conditions. I_{ijkl} is the fourth-order identity tensor, S_{ijkl} is Eshelby's[18] tensor, which is a function only of the shape (aspect ratio) of the reinforcement and the elastic constants of the matrix and has been tabulated by Mura.[19] $T_{ijklmnop}$ is the eighth-order texture tensor that is a function only of the ODF and thus only of W_{lmn}. Complete details regarding our model, along with simplified equations for special cases, are given elsewhere.[16,20] Finally, we note that the elastic constants are connected to acoustic velocities and the material density in the standard manner via solution of the Christoffel equation.

Acoustic Measurements and Inverse Procedure to Obtain W_{lmn}

We used a MHz-frequency pulse-echo-overlap technique to measure longitudinal and shear ultrasonic velocities in the $x_1=x_2$ and x_3 directions in cubical specimens machined from extruded composite rods. Measured velocities were combined with the measured mass density to determine the corresponding elastic constant: $C = \rho v^2$. This gives the on-diagonal elastic constants C_{11}, C_{33}, C_{44}, and $C_{66} = (C_{11}-C_{12})/2$. The remaining elastic constant C_{13} must be obtained by other means. In this study we obtained C_{13} by acoustic resonance spectroscopy. The ultrasonic wavelengths used in our measurements are $\approx$1300 μm and 650 μm for longitudinal and shear waves, respectively. They are much longer than the fiber dimensions for the SiC/Al composites that will be discussed later (length $\approx$ 20 μm, diameter $\approx$ 1.5 μm). We used the measured velocities (elastic constants) with our micromechanics model in an inverse manner to solve for the unknown W_{lmn}. Specifically, we make more velocity measurements than unknown W_{lmn}, and solve for the unknown W_{lmn} in a least-squares sense. Additional details are given by Dunn and Ledbetter.[16]

Prediction of Properties in Terms of W_{lmn}

We have developed a series of micromechanics models for the thermal expansion, elastic-plastic behavior, and thermal conductivity of morphologically textured short fiber composites. Details are presented elsewhere,[21-23] but it is noted that the micromechanical framework is similar to that for the elastic constants described in the previous section. In all cases, the effects of morphological texture are described by the W_{lmn} coefficients of the fiber ODF. We can then use the acoustic W_{lmn} obtained in the manner discussed in the previous section to predict these properties. Such predictions, along with comparisons with experimental results, are described in the following section.

APPLICATION TO METAL- AND POLYMER-MATRIX COMPOSITES

We applied the above-described approach to a series of SiC/6061-Al short-fiber metal-matrix composites. The SiC/Al composites consist of an Al matrix containing short SiC monocrystal fibers. They were extruded into long circular rods. The SiC fibers have an average aspect ratio of 7.4 and are oriented in an axisymmetric distribution about the extrusion axis x_3. The fiber concentrations are c = 17.7 and 27.9 volume percent. The elastic constants and density of the Al , SiC, and the SiC/Al composites were measured and are tabulated in Dunn and Ledbetter.[16] The W_{lmn} obtained by inversion of measured ultrasonic velocities are $W_{200} = 0.0223$ and $W_{400} = 0.0302$ for the c = 0.177 composite and $W_{200} = 0.0188$ and $W_{400} = 0.0266$ for the c = 0.279 composite. We imposed a condition of transverse isotropy on the solution and thus all other W_{lmn} are zero. Previously we showed that the acoustic W_{lmn} agree well with independent measurements by neutron diffraction, and that the computed elastic constants are in excellent agreement with measurements.[16] In addition, the C_{13} predicted with the acoustic W_{lmn} are in excellent agreement with values measured by acoustic resonance spectroscopy for both volume fractions.[16]

In Fig. 1 we show measured axial (along the x_3-direction) thermal expansion for both the matrix and the c = 0.279 composite. These measurements were made using a dilatometer technique with a pure alumina rod as a standard. Also shown is the predicted linear thermal expansion for this composite obtained using our analytical model with the acoustic W_{lmn} for $l \leq 2$ (only W_{200}). Input to the model are the elastic moduli and thermal-expansion coefficients of both phases, the fiber aspect ratio and concentration, and the W_{lmn}. We used the measured SiC thermal-expansion coefficients of Li and Bradt[24] as input for the calculations. Agreement between predictions and measurements is excellent. Results for the transverse (in the x_1=x_2-direction) thermal expansion are similar, as are those for the c = 0.177 composite. Complete details regarding both the thermal-expansion measurements and modeling are presented elsewhere.[21]

In Fig. 2 we show the measured axial stress-strain curve for both the Al matrix and the c = 0.177 composite. The elastic response of the SiC fibers is also shown, as is the predicted stress-strain curve for the composite obtained using our analytical model with the acoustic W_{lmn}. Input to the model are the elastic moduli of both phases, the stress-strain curve of the matrix, the fiber aspect ratio and concentration, and the W_{lmn} for $l \leq 4$. The agreement between predictions and measurements is good. Complete details regarding both the stress-strain measurements and modeling are presented elsewhere.[22]

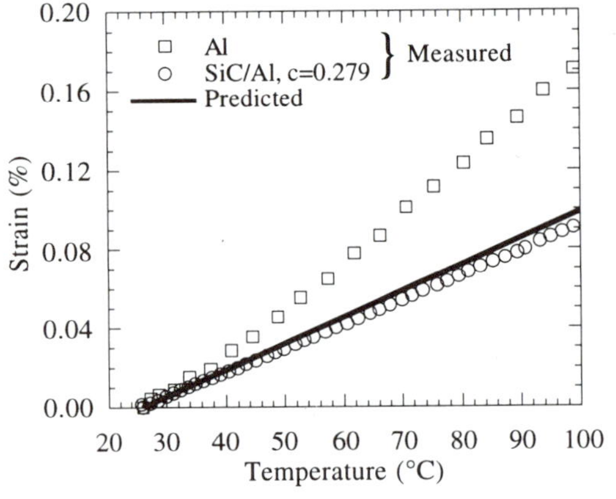

Figure 1. Measured and predicted thermal expansion of the c = 0.279 SiC/Al composite.

Figure 2. Measured and predicted axial stress-strain curves of the c = 0.1779 SiC/Al composite.

We now apply our approach to an injection-molded glass/polyphenylene sulfide (PPS) composite that was studied by Choy et al.[25] Specifically, they used an immersion technique to measure the entire set of orthotropic elastic constants for two specimens; one extracted from the middle of the injection-molded panel, and a second extracted from a region near the surface of the panel. They also measured the anisotropic thermal conductivity of these specimens using a flash-radiometry method. The fiber volume fraction is 0.264 and the aspect ratio is 17. Young's modulus, Poisson's ratio, and the thermal conductivity of the PPS matrix are 4.0 GPa, 0.4, and 2.0 mW/cm K, respectively. Those for the glass fibers are 76.0 GPa, 0.25, and 10.4 mW/cm K, respectively.

Table 1. Measured and predicted (using the acoustic W_{lmn}) elastic constants (in GPa) of the glass/PPS composites.

	Measured		Predicted	
	Surface	Middle	Surface	Middle
C_{11}	24.7	21.2	24.5	21.0
C_{22}	12.1	13.0	11.9	12.8
C_{33}	11.5	12.0	11.3	11.8
C_{44}	2.4	2.65	2.00	2.26
C_{55}	2.52	2.95	2.11	2.56
C_{66}	3.12	3.46	2.72	3.06
C_{23}	7.1	7.1	7.9	7.6
C_{13}	7.1	6.9	6.0	6.8
C_{12}	7.5	7.2	6.0	6.9

We used these properties along with the measured on-diagonal elastic constants (C_{11}, C_{22}, C_{33}, C_{44}, C_{55}, C_{66}) given in Table 1 to compute the fiber ODF coefficients W_{lmn} of both the surface and middle specimens. The orthotropic symmetry of the sample results in five non-zero W_{lmn}. A useful representation of the ODF is a pole figure. In the resent context, this is essentially a density plot of fiber orientation vs. volume fraction. Figure 3 shows the acoustic W_{lmn} and pole figures that were constructed using the acoustic W_{lmn}. The ability of the acoustic pole figures to display the dominant behavior of the ODF is evident, although complete details are clearly not represented because only the W_{lmn} for $l \leq 4$ are obtainable from acoustic velocities. In Table 1 we show the measured elastic constants[25] along with the on-diagonal elastic constants computed from the acoustic W_{lmn} and our micromechanics model. We also show predicted off-diagonal elastic constants (C_{23}, C_{14}, C_{12}). They are in good agreement with measured values.

Finally, we predicted the anisotropic thermal conductivity of the glass/PPS composites using our analytical model with the acoustic W_{lmn} for $l \leq 2$ (W_{200} and W_{220}). Input to the model are the thermal conductivities of both phases, the fiber aspect ratio and concentration, and the W_{lmn}. Predictions are given in Table 2 along with the measurements of Choy et al.[25] for both the surface layer and middle layer. Good agreement exists between measurements and predictions. Complete details regarding the modeling will be presented elsewhere.[23]

	Surface	Middle
W_{200}	-0.025	-0.019
W_{220}	0.025	0.017
W_{400}	0.026	0.016
W_{420}	-0.022	-0.012
W_{430}	0.025	0.014

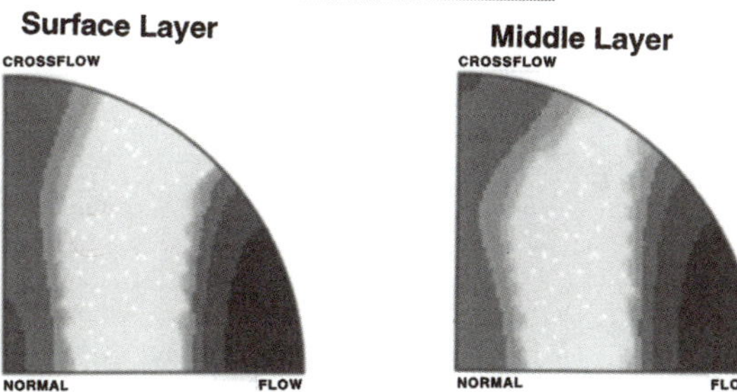

Figure 3. Acoustic W_{lmn} and acoustic pole figures for the glass/PPS composites.

Table 2. Measured and predicted thermal conductivity (in mW/cm K) of the glass/PPS composites.

	Measured		Predicted	
	Surface	Middle	Surface	Middle
K_1	4.08	3.99	4.22	3.94
K_2	3.24	3.22	2.91	3.06
K_3	2.79	3.18	2.78	2.91

CONCLUSIONS

We recently proposed an approach to characterize the orientation distribution function (ODF) of short fibers in two-phase composite materials, based on a coupling of micromechanical modeling and acoustic velocity measurements. Our acoustic approach can not completely characterize the ODF, but delivers information regarding the lower-order terms of the ODF when expanded in a series of generalized spherical harmonics. We have previously demonstrated the utility of the approach for SiC/Al and Al_2O_3/Al short-fiber composites.[16,26] In this study we showed that the lower-order terms of the ODF expansion are sufficient to characterize a large number of other physical and mechanical properties of the composite, when coupled with appropriate models. Specifically, we showed that good agreement is obtained between measurements and predictions for the effective thermal expansion, thermal conductivity, and elastoplastic behavior of SiC/Al and glass/PPS short-fiber composites. The former exhibit transversely isotropic symmetry and thus two independent nonzero W_{lmn} exist, while the latter exhibit orthotropic symmetry and thus five independent nonzero W_{lmn} exist. Thus, based on acoustic measurements, we were able to accurately predict other physical and mechanical properties of the composite. Complete details regarding the application of the acoustic approach to characterize the fiber ODF will be presented elsewhere.[23]

REFERENCES

1. Takao, Y., Chou, T. W., and Taya, M., Effective longitudinal Young's modulus of misoriented short fiber composites, *J. Appl. Mech.* 49:536 (1982).
2. Zhao, Y. H., Tandon, G. P., and Weng, G. J., Elastic moduli of a class of porous materials, *Acta Mechanica*. 76:105 (1989).
3. Taya, M., Dunn, M. L., Derby, B., and Walker, J., Thermal residual stress in a two-dimensional in-plane misoriented short fiber composite, *Applied Mechanics Reviews*. 43:S294 (1990).
4. Advani, S. G. and Tucker, C. L., The use of tensors to describe and predict fiber orientation in short fiber composites, *Journal of Rheology*. 31:751 (1987).
5. Advani, S. G. and Tucker, C. L., A numerical simulation of short fiber orientation in compression molding, *Polymer Composites*. 11:164 (1990).
6. Camacho, C. W., Tucker, C. L., Yalvac, S., Mcgee, R. L., Stiffness and thermal expansion predictions for hybrid short fiber composites, *Polymer Composites*. 11:229 (1990).
7. Bay, R. S. and Tucker, C. L., Stereological measurement and error estimates for three-dimensional fiber orientation, *Polymer Engineering and Science*. 32:240 (1992).
8. Gong, X., Hine, P. J., Duckett, R. A., Ward, I. M., February, The elastic properties of random-in-plane short fiber reinforced polymer composites, *Polymer Composites*. 15:74 (1994).
9. Toll, S. and Andersson, P. O., Microstructural characterization of injection molded composites using image analysis, *Composites*. 22:298 (1991).
10. Ferrari, M. and Johnson, G. C., The effective elasticities of short fiber composites with arbitrary orientation distribution, *Mechanics of Materials*. 8:67 (1989).
11. Wetherhold, R. C., Scott, P. D., Prediction of thermoelastic properties in short-fiber composites using image analysis techniques, *Composites Science and Technology*. 37:393 (1990).
12. Ferrari, M., Asymmetry and the high concentration limit of the Mori-Tanaka effective medium theory, *Mechanics of Materials*. 11:251 (1991).
13. Sayers, C. M., Elastic Anisotropy of short-fibre reinforced composites, *Int. J. Solids Structures*. 29:2933 (1992).
14. Munson-McGee, S. H. and McCullough, R. L., Orientation parameters for the specification of effective properties of heterogeneous materials, *Polymer Engineering and Science*. 34:361 (1994).
15. Schierding, R. G., Measurement of whisker orientation in composites by x-ray diffraction, *J. Comp. Mater*. 2:449 (1968).
16. Dunn, M. L. and Ledbetter, H., Estimation of the orientation distribution of short-fiber composites using ultrasonic velocities, *J. Acous. Soc. Am*. 99:283 (1995).
17. Roe, R. J., Description of the crystallite orientation in polycrystalline materials: III. General solution to pole figure inversion, *J. Appl. Phys*. 36:2024 (1965).
18. Eshelby, J.D., The determination of the elastic field of an ellipsoidal inclusion, and related problems, *Proc. R. Soc. Lon*. A241:376 (1957).
19. Mura, T. Micromechanics of Defects in Solids, Second ed., Martinus Nijhoff Publisher (1987).
20. Dunn, M. L., Ledbetter, H., Heyliger, P. R. , and Choi, C. S., Elastic constants of textured short fiber composites, *J. Mech. Phys. Solids*. 44:1509 (1996).
21. Dunn, M. L., Ledbetter, H., and Li, Z., Thermal expansion of morphologically-textured short fiber composites, submitted for publication.
22. Dunn, M. L. and Ledbetter, H., Elastic-plastic behavior of textured short fiber composites, *Acta Materialia*. 45:3327 (1997).
23. Dunn, M. L. and Ledbetter, H., Acoustic characterization of the orientation distribution of short fiber composites, to be submitted for publication.
24. Li, Z. and Bradt, R. C., Thermal expansion and thermal expansion anisotropy of SiC polytypes, *J. Am. Ceram. Soc*. 70:445 (1987).
25. Choy, C. L., Leung, W. P., Kowk, K. W., and Lau, F. P., Elastic moduli and thermal conductivity of injection-molded short-fiber-reinforced thermoplastics, *Polymer Composites*. 13:69 (1992).
26. Dunn, M. L. and Ledbetter, H., Ultrasonic characterization of the orientation distribution of short fiber composites," in *Proc. 27th SAMPE Technical Conference*, R. J. Martinez, et al., eds., 150 (1995).

LASER SCATTERING DETECTION OF MACHINING-INDUCED DAMAGE IN Si_3N_4 COMPONENTS

J. G. Sun, M. H. Haselkorn,* and W. A. Ellingson

Energy Technology Division
Argonne National Laboratory
Argonne, IL 60439

*Caterpillar Inc.
Mossville, IL 61552

INTRODUCTION

Silicon nitride (Si_3N_4) ceramics are considered the primary materials of choice to replace metals in many structural applications because of their desirable mechanical and physical properties, such as high stiffness, corrosion and wear resistance, and greater thermal stability. For such applications, the most critical portions of a ceramic component, i.e., those under the greatest stress during operation, are the surface or near-surface (usually to depths of <200 μm) regions. The most common types of defects in these critical regions are mechanical, e.g., cracks, spalls, inclusions, voids, etc., and can be induced by either machining or operation.

Machining of engineering ceramic surfaces is conducted by various material removal processes. During the machining, material directly in the path of an abrasive particle encounters high stresses and temperatures and, as a result, is broken and/or deformed. Material adjacent to the abrasive particle is placed in compression. After the particle passes, this material rebounds and causes tensile stresses that lead to the formation of radial, lateral, and longitudinal cracks. Neither the radial nor the lateral cracks are thought to significantly reduce the strength of the ceramic. Longitudinal (sometimes called median) cracks are parallel to the direction of grinding and perpendicular to the surface and are therefore thought to cause the greatest strength reduction.

Because Si_3N_4 ceramics are partially translucent to visible (and IR) light, we developed an elastic optical scattering method for detecting surface and near-surface defects in Si_3N_4 ceramics. For many Si_3N_4 (and other) ceramics, the optical penetration depth is usually >100 μm in the visible spectrum, depending on grain size, second-phase composition, and material absorption (Ellingson et al., 1993). Thus, elastic optical scattering can be used as a noncontact, nondestructive method for detecting surface and near-surface defects in Si_3N_4 ceramics (Steckenrider and Ellingson, 1994; Ellingson and Brada, 1995; Sun et al., 1997). This technique has been used to analyze diamond-ground and rotary-ultrasonic-machined Si_3N_4 specimens subjected to various machining conditions. The scatter results were identified to show the characteristics of the machining-induced damage, which were then correlated with machining conditions. The results were further substantiated through examination of surface layers by photomicroscopy and SEM analysis.

Nondestructive Characterization of Material VIII
Edited by Robert E. Green Jr., Plenum Press, New York, 1998

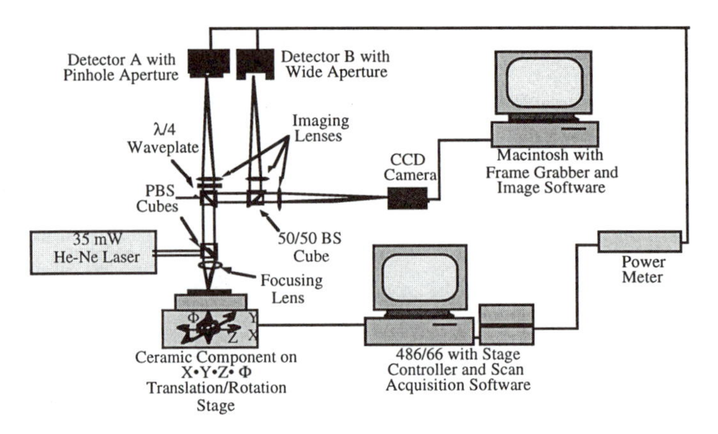

Figure 1. Setup for laser scattering inspection system.

ELASTIC OPTICAL SCATTERING SYSTEM

The experimental arrangement of the optical scattering system is illustrated in Fig. 1. A vertically polarized laser beam is directed through a polarizing beam-splitter (PBS) cube onto the specimen surface. Light reflected from the component surface will not undergo change in polarization unless the surface is extremely rough; therefore, all surface-scattered light will again be reflected in the PBS and directed back toward the laser. However, any light that is scattered from the subsurface material undergoes several reflections and refractions at the grain boundaries; each of these serves to alter the polarization of the light. The net effect of this behavior is to randomize the polarization of the subsurface scattered light and make the scatter completely diffuse. Thus, half of the subsurface-scattered light will also be reflected by the PBS and directed back to the laser. However, the other half of the subsurface-scattered light will be transmitted by the PBS into the detection train. The back-scattered light that passes through the surface-illuminating PBS is incident on a second PBS, through which it also passes. It is then directed through a quarter-wave plate, imaged by a positive lens onto a polished stainless steel pinhole aperture ($\approx$100 μm in diameter), and recorded by Detector A. Any light that is scattered from the subsurface directly beneath the incident spot passes through the aperture and onto Detector A. The remaining light that is scattered from the area around the illuminating spot is reflected back through the lens and quarter-wave plate. In this case, its polarization has been rotated to horizontal, and it is reflected by the detecting PBS and directed to a 50/50 beam splitter. One side of the 50/50 beam splitter is imaged by a positive lens onto Detector B, while the other side is imaged onto a CCD array to monitor the scattering surface.

The total back-scattered intensity can be measured by monitoring the sum of the outputs of Detectors A and B, i.e., A + B. This sum will be most indicative of lateral defects. As the laser illumination is rastered across the specimen surface, these sum values are assembled into a gray-scale image of the surface, hereafter called the "sum" image. However, if the ratio of outputs from Detectors B and A, i.e., B/A, is computed, we obtain an indication of the degree of lateral spread of the subsurface scatter. This value is primarily sensitive to median defects. Again, as the specimen is scanned, these ratio values are assembled into a gray-scale image called the "ratio" image. Most real defects will have some orientation between median and lateral, and will therefore provide an indication in each image, although one orientation will often dominate the other.

VICKERS INDENT SPECIMEN

To evaluate the sensitivity of the system to median-type defects, which are more likely to be induced during machining, median cracks were generated with a Vickers indenter in an NT164 flexure bar. The specimen was then subjected to inspection by laser scatter and by advanced dye penetrant methods. The dye penetrant image and the laser scatter sum and ratio images are shown in Fig. 2.

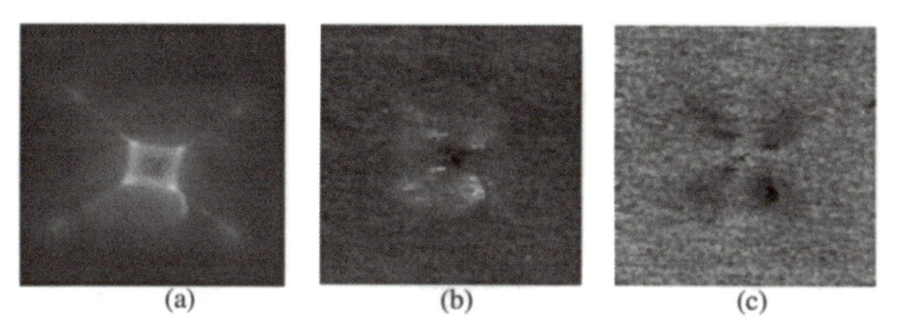

Figure 2. Images of Vickers indent showing (a) dye penetrant image in 0.5 mm square, and elastic optical scattering (b) sum, and (c) ratio images in 1 mm squares.

Several characteristic features are clearly evident from Fig. 2. The elastic optical sum image (b) shows two types of cracks: lateral cone-type cracks (indicated by a brighter halo around the indent), which emanate beneath the surface from the indent; and median cracks extending from the corners of the indent, with the crack in the lower right being the most severe, because the cracks are not perfectly straight and exhibit a lateral component. The presence of both of these features is supported by the dye penetrant image. In addition, the actual surface indent is visible in the sum image as a darker region near the indent center. By comparison, the elastic optical ratio image (c) shows no indication of lateral defects and is almost completely insensitive to the actual surface indent itself. Rather, it indicates only the presence of the median cracks that extend from the corners of the indent.

ROTARY-ULTRASONIC-MACHINED SPECIMENS

Rotary ultrasonic machining (RUM) is a hybrid machining process that combines the material-removal mechanisms of diamond grinding and ultrasonic machining, resulting in a higher material removal rate (MRR) than that obtainable by either diamond grinding or ultrasonic machining alone. One approach of RUM to surface machining is illustrated in Fig. 3 (Pei et al. 1995). The material removal in RUM of ceramic materials was attributed to the brittle fracture of the machining process. Many machining parameters affect the process performance. The applied static pressure has a great effect on MRR. As the static pressure increases, MRR increases to a maximum value and then decreases. Increase of the vibration amplitude also initially increases the MRR, but decreases the MRR after passing a maximum. Similarly, MRR increases as the diamond grit size (diameter) increases and then reduces as the grit size is larger. Other parameters, such as vibrating frequency, tool rotation speed, feed rate, may also have significant effects on the MRR.

Four rotary-ultrasonic-machined specimens were analyzed; machining conditions are listed in Table 1. Specimens HP38AAA and HP38BAA were machined with higher MRR corresponding to higher forces, and specimens HP38BAB and HP38BBB were machined with lower MRR corresponding to lower forces. There are also changes in feed rate, grit size, and vibrational amplitude for machining of these specimens. An optical photomicrograph of one of the specimens (HP38AAA) is shown in Fig. 4a, showing dark spots and white regions on the specimen surface. The dark spots can be easily identified as contaminants. The white regions, however, indicate excessive light scattering. To identify the surface microstructure, the surface was examined with a scanning electronic microscope (SEM). The surface is generally dense with uniformly distributed grains. Scattered abnormal regions can be found, and a SEM image of one of these regions in shown in Fig. 4b. It is evident that there is a loose surface region from the lower right to the center of the image, with a cracklike line extending to the upper left corner of the image. This region is about 150 μm in length. We suspect that this may be a region that was crushed by a machining grit; the region shows excessive light scattering, as indicated in Fig. 4a.

These rotary ultrasonic machined specimens were analyzed by elastic optical scattering for machining-induced subsurface defects. Figures 5a-d show the ratio and sum images of these specimens. The resolution of all the images was 15 μm and the specimen

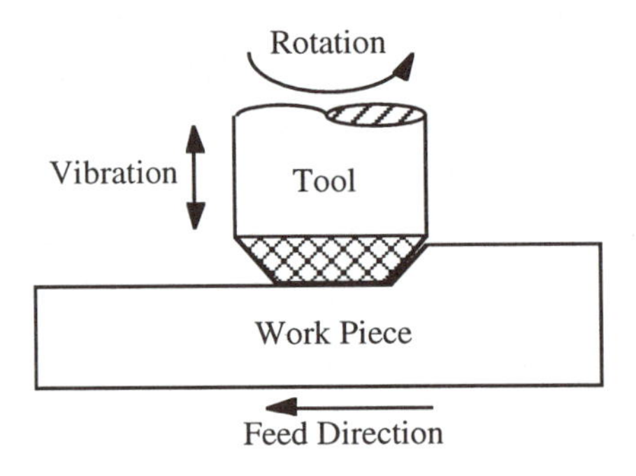

Figure 3. Illustration of rotary ultrasonic machining system.

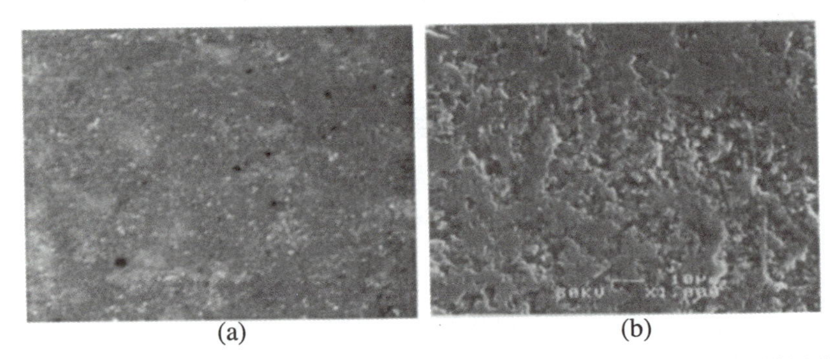

Figure 4. (a) Photomicrograph and (b) SEM image of the surface of specimen HP38AAA.

Table 1. Machining conditions of rotary-ultrasonic-machined specimens

Specimen ID	MRR (10^{-3} mm^3/s)	Max. Force (N)	Feed Rate (BLU/s)	Grit Size (μm)	Amplitude (μm)	R_a (μm)
HP38AAA	108	168	100	60/80	33	0.11
HP38BAA	108	236	100	170/200	33	0.21
HP38BAB	54	101	50	170/200	33	0.06
HP38BBB	54	132	50	170/200	23	0.09

image size was 2.5 mm x 2.5 mm. The prominent features in the images are dark spots (or regions) in the ratio images and brighter spots in the sum images. As discussed above for the Vickers indent specimens, these features represent surface/subsurface defects or damage in these regions. This observation can be further substantiated by examining the patterns of the spots. It is evident from the images that the spots are not randomly distributed, but are clearly arranged in circular patterns with various radius. This circular distribution pattern exactly matches the RUM process illustrated in Fig. 3. Therefore, we conclude that the dark and bright regions in the ratio and sum images represent regions damaged by machining.

Once the direct relationship is established between the spotted regions in the scatter images and the machining-induced damage, the machining damage under different machining conditions may be studied from the scatter data. Specimen HP38AAA was machined with a smaller grit size and shows abundant but relatively smaller damaged regions (Fig. 5a). For specimen HP38BAA (Fig. 5b), which was machined with the highest pressure, the damaged regions are close to each other along circular lines. At least two shades of gray scales are evident in the damaged regions. Spots with the lighter shade are relatively abundant in the images of specimens HP38BAB and HP38BBB (Figs. 5c-d), and this may be attributed to the lower feed and MRR rates.

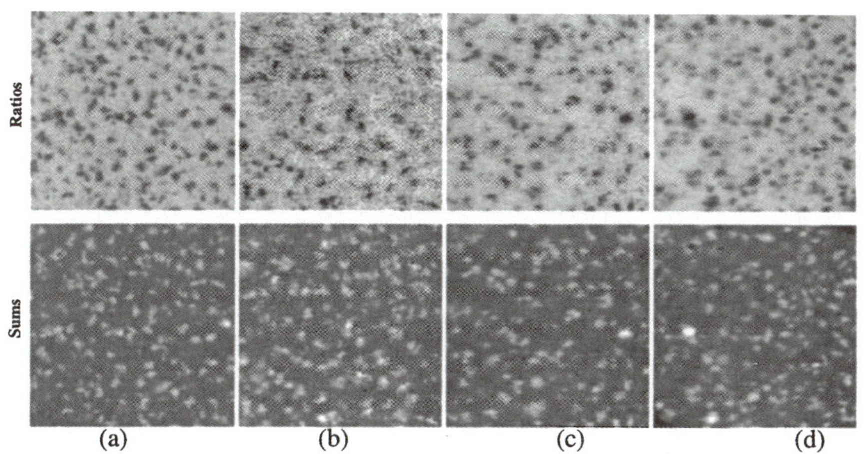

Figure 5. Elastic optical scattering ratio and sum images of specimens (a) HP38AAA, (b) HP38BAA, (c) HP38BAB, and (d) HP38BBB.

Table 2. Machining condition for diamond-ground GS44 specimens

Specimen ID	Diamond Bond	Depth of Cut (in.)	Feed Rate (in./s)
GS44AAA	soft	0.001	40
GS44AAB	soft	0.002	20
GS44BAA	hard	0.001	40

DIAMOND-GROUND SPECIMENS

Three diamond-ground GS44 specimens were analyzed, and machining conditions for these specimens are listed in Table 2. The machining parameters to be examined are diamond bond type and depth of cut (the latter corresponds to feed-rate variation). Figures 6a-c show the elastic optical scattering ratio and sum images of the these specimens. The scanned surface area was 1.8 x 1.8 mm and resolution was 10 μm. The machining direction is vertical in these images. In the sum images, the white spots represent surface regions with excessive light scattering due to surface/subsurface machining-induced damage. Correspondingly, the damaged regions are shown as dark spots in the ratio images. As described above, the ratio and sum images are sensitive, respectively, to median and lateral cracks in the specimen subsurface.

Two types of machining-induced damage are visible in these images. First, machining marks are represented by vertical lines (darker lines in ratio images and whiter lines in sum images); these marks are most likely median cracks generated when the grinding particles pass through the specimen surfaces. Second, individual damaged regions, or individual spots, are distributed throughout the specimen subsurfaces. The scattering images in Figs. 6a-c show different patterns for these specimens that can be correlated with machining conditions. For specimen GS44AAA, which was machined with a soft diamond bond and smaller cut depth, many damaged spots appear on the surface/subsurface while the machine marks show only occasionally, as seen in Fig. 6a. When the cut depth increases, as for specimen GS44AAB, the damage was almost exclusively represented as machining marks, as shown in Fig. 6b. The specimen GS44ABB was machined with a hard diamond bond at the lower cut depth, and its damage pattern in Fig. 6c seems to lie between the other two machining conditions, except that the damaged spots are larger than those in Fig. 6a.

CONCLUSIONS

A novel, noncontact, nondestructive elastic optical scattering technique has been developed to detect machining-induced damage in the surfaces and subsurfaces of ceramic

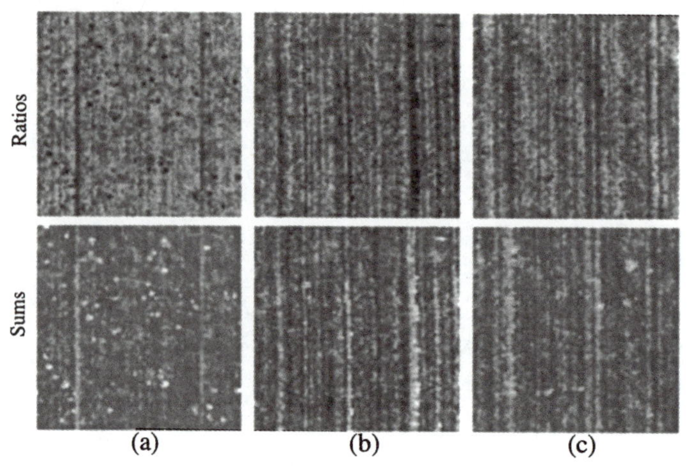

Figure 6. Elastic optical scattering ratio and sum images of diamond-ground specimens (a) GS44AAA, (b) GS44AAB, and (c) GS44BAA.

materials. The technique is based on the unique property of ceramics to partially transmit light into their subsurfaces, combined with polarization techniques to separate the effects of surface and subsurface defects.

This technique was used to detect machining-induced damage in Si_3N_4 ceramic specimens that were machined with diamond-ground and rotary-ultrasonic machining under various machining conditions. The results indicate that the laser scattering technique may detect and identify various types of surface/subsurface machining-induced damage that are critical to component strength and lifetime. Damage that is determined from the scatter data can be correlated with the machining conditions. Thus, the laser scattering method holds promise for automated inspection and qualification of ceramic components.

ACKNOWLEDGMENT

This work was sponsored by the U.S. Department of Energy, Assistant Secretary for Energy Efficiency and Renewable Energy, Office of Transportation Technologies, as part of the Heavy Vehicle Propulsion System Materials Program, under Contract DE-AC05-96OR22464 with Lockheed Martin Energy Research Corp.

REFERENCES

Ellingson, W. A., Ayaz, D. M., Brada, M. P., and O'Connell, W., in *Machining of Advanced Materials*, NIST Special Publication 847, Washington, DC (1993), p. 147.

Ellingson, W. A., and Brada, M. P., “Optical Method and Apparatus for Detection of Surface and Near-Subsurface Defects in Dense Ceramics,” U.S. Patent #5,426,506 (issued June 20, 1995).

Pei, Z. J., Ferreira, P. M., Kapoor, S. G., and Haselkorn, M., “Rotary Ultrasonic Machining for Face Milling of Ceramics,” Int. J. Machine Tools and Manufacture, Vol. 35, No. 7, pp. 1033-1046 (1995).

Steckenrider, J. S., and Ellingson, W. A., "Surface and Subsurface Defect Detection in Si_3N_4 Components by Laser Scattering," *1993 Review of Progress in Quantitative NDE*, Vol. **13**, D. O. Thompson and D. E. Chimenti, eds., Plenum Press, New York (1994), pp. 1645-1652.

Sun, J. G., Ellingson, W. A., Steckenrider, J. S., and Ahuja, S., “Application of Optical Scattering Methods to Detect Damage in Ceramics,” in *Machining of Ceramics and Composites*, NIST Special Publication, Washington, DC (in press, 1997).

NONDESTRUCTIVE CHARACTERIZATION OF THE NUCLEATION AND EARLY VERTICAL BRIDGMAN CRYSTAL GROWTH OF $Cd_{1-x}Zn_xTe$

B.W. Choi and H.N.G. Wadley

Department of Materials Science and Engineering
University of Virginia
Charlottesville, VA 22903

INTRODUCTION

$Cd_{1-x}Zn_xTe$ alloys (where 0.03<x<0.05) used as substrates for infrared focal plane arrays detectors [1,2] are typically grown by the unseeded directional solidification of a melt via either a vertical or horizontal Bridgman process [1,3,4]. In a vertical Bridgman process, the initiation of the growth process appears to be a crucial step controlling subsequent bulk ingot quality[1,3,4,5]. A number of experimental and modeling effort have sought to correlate the controllable growth parameters with resulting characteristics of the solidified ingot such as degree of single crystallinity, dislocation density and solute segregation etc.[6,7]. However, an inadequate understanding of melt undercooling and the detailed thermal conditions at the ampoule tip have seriously hampered the design of the initial solid nucleation process.

In-situ non-contact eddy current methods sense the electrical conductivity at elevated temperature[8] and provide an approach for monitoring the solid/liquid interface position and shape during the crystal growth for a variety of semiconductors[9,10]. It has recently been shown that the CdTe system is reasonably well suited for this sensor approach because the electrical conductivity of the liquid is 4-5 times higher than that of the solid at the melting point [8,11]. In this experiment, an eddy current sensor situated in the cone area region of an ingot has been non-invasively integrated into a commercial scale 17 zone vertical Bridgman furnace and used to detect the formation of the solid phase thereby yielding new insights about nucleation and early growth characteristics of vertical Bridgman grown $Cd_{0.955}Zn_{0.045}Te$ ingots.

FEM SIMULATION AND SENSOR DESIGN

An electromagnetic finite element model can be used to relate the multifrequency response of eddy current sensor designs to the position of a solid/liquid interface located near the tip of an ampoule. The method used here essentially solved for the magnetic vector potential, $\boldsymbol{A}(r, z)$ where r and z are the radial and the axial coordinates using a commercial electromagnetic FEM code[12]. The magnetic vector potential obtained could then be directly used to obtain the sensor's electrical impedance as a function of test frequency. The modeled problem consisted of a $Cd_{0.955}Zn_{0.045}Te$ sample contained within a conically shaped non conducting crucible. Since the model geometry was axisymmetric, a two dimensional FEM analysis could be used. For modeling purposes, the region of interest (the coil, the sample, and the intervening space) were divided into elements whose size were determined by skin depth considerations at the highest test frequencies. Figure 1 shows the geometry of the sensor sample in the *r-z* plane in which a six turn primary coil was used to excite an electromagnetic field and a four turn pickup coil then detected the perturbation of the primary coil's field resulting from the presence of the conducting sample. Figure 1 also shows the calculated 500 kHz magnetic vector potential field for a flat solid/liquid interface located at a height, *h*, of 19.1 mm from the cone tip.

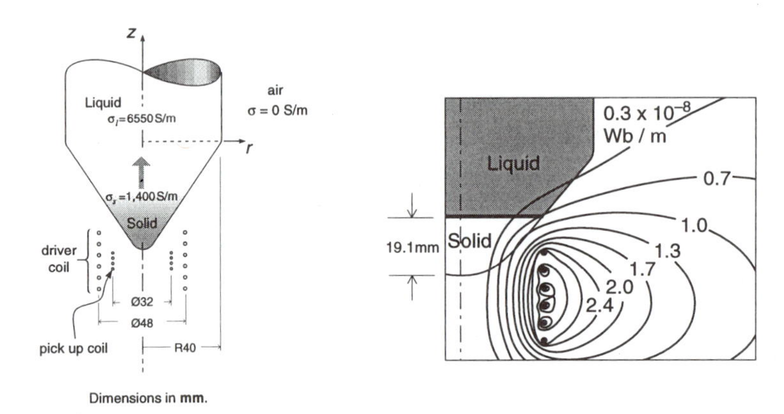

Figure 1 The eddy current sensor and sample geometry, and the resultant 500 kHz magnetic vector potential contours near the cone area for an interface height of 19.1mm at 500 kHz.

To simulate the response of the sensor during the onset of nucleation and subsequent propagation of the solid/liquid interface, a series of FEM calculations were performed with eleven different locations of a solid/liquid interface in the conical region of the cylindrical 80 mm diameter sample. In this case, input values of the conductivity for a liquid and a solid phase were 6550 s/m and 1400 s/m for the sample, respectively[8]. Figure 2 shows the calculated normalized imaginary impedance curves at four different frequencies as function of interface position from the cone tip (*h*=0mm) to the cone shoulder (*h*=45mm) for a flat solid/liquid interface.

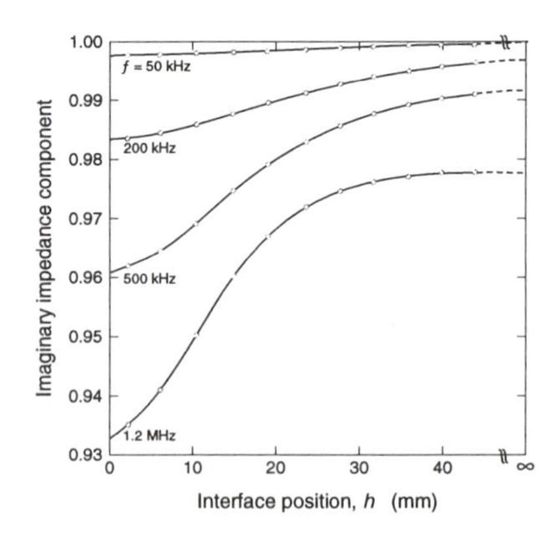

Figure 2 The sensors calculated imaginary impedance component variation with interface position at four frequencies.

In a completely melted condition prior to the onset of nucleation, the imaginary impedance component has a value determined by the liquid's electrical conductivity. As the solid volume increases during solidification (i.e. as the interface moves upwards), the imaginary impedance increases at all frequencies and continues to converge asymptotically to the value of the homogeneous solid.

The largest variations in response to interface position are exhibited at a test frequencies between 500kHz and 1.2MHz where both the eddy current density and the skin depth are large. This frequency range is therefore the most desirable for sensing the nucleation and early growth of $Cd_{1-x}Zn_xTe$.

EXPERIMENTAL PROCEDURES

Sensor Installation and Measurement Methodology

A two-coil "encircling" eddy current sensor, with a design similar to that analyzed above, was placed near the ampoule tip region of a commercial scale 17 zone vertical Bridgman furnace, as shown in Figure 3. To minimize disturbances to the crystal growth thermal environment, the sensors coils were wound on the pre-existing concentric alumina tubes which normally used to support the conically shaped ampoule inside the furnace. A 1.02mm platinum wire was then used to wind a 38.1 mm long primary coil while a 0.25mm platinum wire was used to wind a 12.7 mm long secondary coil.

For the measurement of two coil multifrequency eddy current sensors[8,9,10], a continuous signal was supplied to the primary coil by the variable frequency oscillator with Impedance Gain/Phase Analyzer (HP4194A). Frequency dependent gain (g) measurements for the two coil system were obtained by recording the ratios of the voltage induced across the secondary coil with the voltage drop across a 1 ohm low inductance precision resistor in the primary circuit. The phase difference (ϕ) between these two voltage signals was also monitored. The sensor's gain/phase response was normalized by that of the empty sensor. These empty sensor measurements were made at the growth temperature and gave reference empty coil gain (g_0) and phase (ϕ_0) measurements at each test frequency. During a growth experiment, gain/phase data was collected at test frequencies between 50 kHz and 5 MHz. Since the translation rate of the furnace was less than 2mm/hour, the data was collected and downloaded to a personal computer once every 5 to 10 minutes throughout growth.

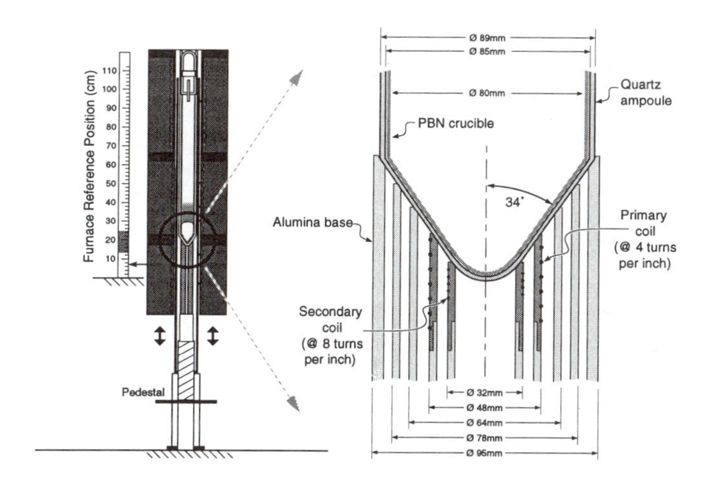

Figure 3 Schematic diagrams of 17 zone vertical Bridgman furnace and eddy current sensor located at the ampoule tip.

The real and imaginary components of the normalized impedance (Z) in the presence of a sample are given by:

$$Re(Z) = \left(\frac{g}{g_0}\right)\sin(\phi - \phi_0)$$

$$Im(Z) = \left(\frac{g}{g_0}\right)\cos(\phi - \phi_0)$$

Growth Experiments

The axial temperature profile of a stationary empty ampoule in the cone area was measured prior to two growth runs. The temperature gradient where solidification was initiated was about 8.6°C/cm.

Two growth runs were monitored to observe melting, nucleation and initial growth with the nucleation sensor. The charge of about 3.3 kg was placed in a 80mm inner diameter conical pyrolitic boron nitride crucible and then sealed in an 85mm diameter quartz ampoule under 10^{-6} torr pressure. A first growth run compounded the constituents and thus allowed the monitoring of precompounding as well as subsequent solidification of the fully melted charge. The position of the furnace with respect to the stationary ampoule was conveniently referenced by a pointer attached to the furnace, Figure 3. The pointer position for the start of the first run was 7.4 cm. The same charge was used in a second run, but with a raised furnace at a pointer position of 9.3 cm to start. Normally, growth of previously compounded material is accomplished from a cooler region of the furnace and the follow-up second run explored the consequence of this using the remelted charge.

RESULTS AND INTERPRETATION

For the first run, the ampoule tip temperature was about 10°C below the melting point and about 20°C below for the second run. This resulted in significantly different melting behaviors in the tip region. In the first run, complete melting of the charge occurred before the main growth run, whereas in the second run, the eddy current sensor clearly indicated that melting was incomplete.

Compounding / Complete Melting (first) Run

Figure 4(a) shows the normalized imaginary impedance as a function of time during the first experiment beginning with furnace heating. The shaded areas in Figure 4 represent periods during which the furnace was held stationary.

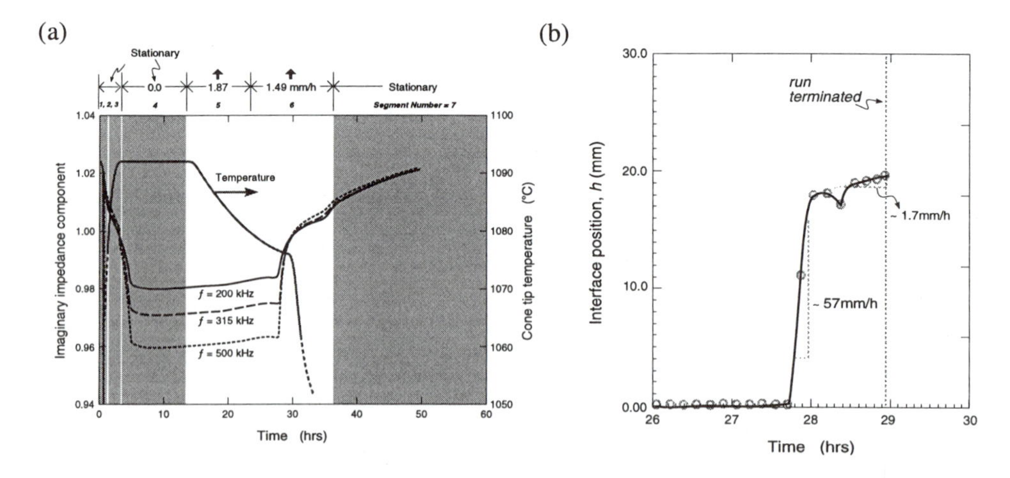

Figure 4 (a) Measured impedance response at three frequencies as functions of time during the CZT0322 run and the estimated cone tip temperature, and (b) The solid/liquid interface position, h, as a function of time during initial nucleation and growth of solid.

A sharp drop in impedance occurred at 1.1 hours, and appears to be melting of the elemental Zn and Cd as the temperature approached 700°C. These elements were added to precompounded equiatomic CdTe to bring the target composition to $Cd_{0.955}Zn_{0.045}Te$. As each element dissolved to eventually form solid with the $Cd_{0.955}Zn_{0.04}5Te$ composition, the impedance returned towards its "empty" value as the metallic components were compounded. This solid has a very low conductivity at temperatures less than 700°C. The solid state mixing process proceeded as the furnace temperature was increased to a set point of

1150 °C (at the end of segment #3). Towards the end of this stage, the impedance began to drop sharply, consistent with the beginning of melting (and the associated rise in conductivity) near the ampoule tip. On entering segment #4, the cone tip approached a temperature of 1090°C, while the cone shoulder approached 1105°C. The melting temperature for this composition ingot is 1098°C. The continued drop in impedance indicates that it required approximately 2 hours during segment #4 for the charged material to melt completely to the cone tip. The normalized impedance curve at the end of segment #4 compared well with the FEM results for an entirely liquid sample. After this melting transient, a small increase in the imaginary component was observed during the remainder of segment #4.

During segment #5, the furnace was translated upwards for 10 hours (from a starting pointer position of 7.4 cm) at a rate of 1.87mm/hour. The temperature of the cone tip decreased during this segment while the eddy current data showed a very slight rise consistent with the retention of a supercooled liquid whose conductivity was slightly decreasing. At the end of segment #5, the supercooling was estimated to be about 18°C while the melt conductivity was approximately 6,100 mhos/m. Continued cooling in the cone area occurred during segment #6 as the furnace was translated at a rate of 1.49 mm/hr. At a process time of 27.8 hours, an abrupt return of the imaginary impedance towards its null value was seen at all test frequencies. The furnace position at this moment was 9.9cm and the estimated supercooling at the tip of the ampoule was estimated to be about 20°C for this composition alloy. Examination of the normalized impedance curve immediately after the sharp rise in impedance indicated an interface shift that increased with test frequency and so this abrupt impedance change corresponded to nucleation of solid at the cone tip. For the location of the solid/liquid interface, the imaginary impedance data was compared with the FEM result shown in Figure 2. Figure 4(b) clearly shows that nucleation from a homogeneous liquid abruptly occurred at a process time of 27.8 hours. The unstable solidification event caused the interface position to move upwards about 19 mm from the cone tip over a period of 20 minutes. The solidification rate during this nucleation was therefore estimated to be 57 mm/h, in contrast to a furnace translation rate of only 1.49 mm/h.

Remelted (second) Run

The second run were conducted using the sample from first run but with a furnace starting position closer to that normally used for precompounded growth runs. Figure 5(a) shows that the variation of the imaginary impedance component with process time is quite different to that of the first run.

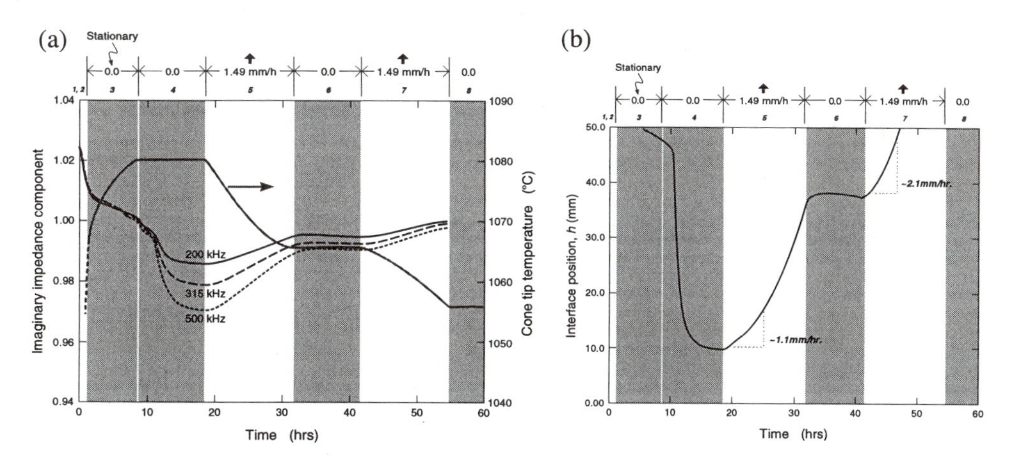

Figure 5 (a) Measured imaginary component impedance response for run CZT0405. The estimated cone tip temperature is also shown for comparison. (b) Interface position area during the early remelting/solidification stage of run CZT0405.

During the first 20 hours of the process period, no abrupt variations in imaginary component were observed. Instead, the imaginary component gradually decreased with process time as the samples electrical conductivity gradually increased. Note that the imaginary impedance component was still decreasing at the end of segment #4, indicating that the sample had not reached a quiescent state prior to the start of furnace translation. Melt back was seen to have occurred progressively with process time during segment #4. However, complete melting was not achieved. Using the FEM result of Figure 2, the interface position, h, was determined from the measured imaginary impedance data in Figure 5(b). It appeared to have arrested 9.6 mm above the cone tip at the beginning of the segment #5. As the furnace was raised (in segment #5), the interface propagated upwards away from the cone tip. The initial solidification velocity was estimated to be 1.1mm/hr compared to a furnace translation rate was 1.49 mm/h. As the process continued beyond 25 hrs, the rate of interface movement appeared to increase to about that of the furnace. It is interesting to note that this process sequence has usually been thought to have caused complete melting and to have induced nucleation from an undercooled melt. The results shown in Figure 5 show clearly that this does not occur.

SUMMARY

A two coil "encircling" eddy current sensor for monitoring the initial stages of crystal growth has been designed and integrated into a 17 zone commercial vertical Bridgman furnace. A electromagnetic finite element model was used to convert sensor impedance data to a solid/liquid interface's position so that solid nucleation and the growth velocity could both be measured. The sensor was used to determine the degree of melting, the extent of liquid supercooling, the moment of solid nucleation and the initial growth velocity from fully melted and partially remelted ingots. It reveals large supercoolings and unstable nucleation/ growth from insitu compounded melts. The normal growth process conditions used for precompounded charges are found to result in incomplete melt back, no supercooling and a growth velocity slightly less than that of the furnace translation rate. These observations indicate that this two coil "encircling" eddy current sensor may be helpful for optimizing the growth process conditions of current processes as well as initiating the seeded solidification, and introducing new feedback control strategies into vertical Bridgman growth processes.

ACKNOWLEDGMENTS

This work has been performed as a part of the research of the Infrared Materials Producibility Program conducted by a consortium that includes Johnson Matthey Electronics, Texas Instruments, II-VI Inc., Loral, the University of Minnesota and the University of Virginia. We are grateful for the many helpful discussions with our colleagues in these organizations and inparticular to Kumar Dharmasena at UVa, Pok-kai Lio at Texas Instruments, and to Brent Bollong, Art Socha, Daniel Bakken and the staff of JME for their assistance in preparing the samples. The consortium work has been supported by ARPA/CMO under contract MDA972-91-C-0046 monitored by Raymond Balcerak.

REFERENCES

1. S. Sen, W.H. Konkel, S.J. Tighe, L.G. Bland, S.R. Sharma and R.E. Taylor, J. Crystal Growth, 86, (1988), 111.
2. F. Butler, F.P. Doty, B. Apotovsky, S.J. Friesenhahn, C. Lingren, in Semiconductors for room temperature Radiation Detector Applications Eds R.B. James, T.E. Schlesinger, P. Siffert, F. Franks, Mat. Res. Soc., (1993), 502.
3. S. Sen and J.E. Stannard, "Improved Quality of Bulk II-VI Substrates for HgCdTe and HgZnTe Epitaxy," *Materials Research Society Symposium Proceedings*, 302 (1993), 391-401.
4. M. Pfeiffer and M. Muhlberg, J. Crystal Growth, 118, (1992), 269.
5. P. Rudolph and M. Muhlberg, Materials Science and Engineering B16, (1993), 8.
6. T. Wu and W. R. Wilcox, J. Crystal Growth, 48, (1980), 416.
7. S. Brandon, and J. J. Derby, J. Crystal Growth, 110, (1991), 481.
8. H.N.G. Wadley and B.W. Choi, J. Crystal Growth, 172, (1997), 323.
9. H. N. G. Wadley and K. P. Dharmasena, J. Crystal Growth 172 (1997), 313.
10. K. P. Dharmasena and H. N. G. Wadley, J. Crystal Growth, 172 (1997), 337.
11. V. M. Glazov, S.N. Chizhevskaya, and N.N. Glagoleva, Liquid Semiconductors, Plenum Press (1969).
12. Ed. B.E. MacNeal, MacNeal-Schwendler Corporation, MSC/EMAS Modelling Guide, 1991.

MULTI-CHANNEL PYROMETRY AND THERMOGRAPHY - A NON-CONTACT METHOD FOR THE DETERMINATION OF THE SPECTRAL EMISSIVITY OF HIGH-TEMPERATURE MATERIALS

T. Vetterlein, G. Walle and N. Meyendorf

Fraunhofer-Institut für Zerstörungsfreie Prüfverfahren (IZFP)
66123 Saarbrücken, Germany

Keywords: high-temperature materials, emissivity, pyrometry, thermography, multi-channel techniques

Materials for space technology and also for other high-temperature applications have to fulfil specific requirements, e.g. a definite behaviour of spectral emission coefficient is necessary. The aim of the experiments was to measure the spectral emissivity of the material for temperatures above 1500 °C. The solution of this task was a noncontact multi-channel pyrometric and thermographic technique which is described below.

INTRODUCTION

The contactless temperature measurement is based on the fact, that all materials emit electromagnetic radiation. The spectrum of this radiation is depending on the temperature as well as on the surface properties of the material. The body with the highest emission coefficient is the „black body" (emission coefficient $\varepsilon = 1{,}0$). Compared to the black body all components and material surfaces emit less radiation at the same temperature. The temperature measured by pyrometers is always lower than the real temperature because all pyrometers are calibrated on black body radiation.

In the case of technical materials the emission coefficient can become strongly dependent on the wavelength (spectral emission coefficient $\varepsilon = \varepsilon(\lambda)$). A variation of the surface properties of the material caused by oxidation or a change of the surface roughness may change the spectrum during a measurement. Especially at high temperatures, where chemical and physical processes occur very fast, account should be taken of these events during the measurement.

Generally applicable connections of the dependency of the emission coefficient upon temperature and wavelength are not available yet. One possibility is, to measure and compare the radiation densities in several wavelength ranges with the data measured in a „teach-in“ session. The multi-channel thermography system observes and records the differences between the intensities of radiation in the several spectral channels, emitted from the sample surface. Out of this all, a quantitative dependency of the emission coefficient on the wavelength can be determined. If these measurements are carried out with well-known materials whose emission coefficient varies depending on the wavelength in a characteristic way, this spectral dependence can be used to characterise the surface properties of the material.

STATE OF THE ART

For the contactless temperature measurement exists basically the problem of distinction between temperature and emission coefficient. Beside that, other disturbances such as reflected radiation or emission by surrounding media have to be taken into account.

For the determination of the emission coefficient of component surfaces a semidestructive method using a drillhole is available. This technique uses a drillhole with a depth-to-diameter relation of 3:1 drilled into the component. The radiation emitted by the hole is then compared to the radiation intensity emitted by the surface. Due to the fact that a drillhole with a depth-to-diameter relation of 3:1 is a very good approximation of a black body radiation source, a measurement of the emission coefficient can be carried out via comparison of intensities [1].

Colour pyrometry is a method for contactless temperature measurement, in which the influence of the emission coefficient is negligible. This technique uses the comparison of two different radiation densities measured at two different wavelengths to obtain the temperature.

Two concepts of multi-channel pyrometry are in use. One with a filter displacement system and one using a multi-wavelength analysing detector [2, 4, 6, 8].

Scanning IR cameras give an intensity distribution of the emitted infrared radiation with spatial resolution. Inhomogeneities within the IR image can be caused by different temperatures or varying emission coefficients. For the determination of the moisture content of buildings, IR cameras with filter displacement systems are in use. Here, two different ranges of wavelength will be compared. It is the range from 3 - 5 µm and the range from 8 - 12 µm[3, 5].

At high temperatures the surface properties of components or materials may vary in location and time. Due to this, location- and time-depending variations of the emission coefficients are possible which can not be detected by focusing pyrometers.

SET-UP OF A MULTI-CHANNEL THERMOGRAPHY SYSTEM

A multi-channel thermography system was set up. Fig. 1 shows the shematic set-up of this device. Here, a parallel arrangement of two IR cameras observes an area of 50 mm x 50 mm of the surface of the object in discrete wavelength from 800 nm to 9.5µm. The IR radiation emitted by the object is reflected by a scan mirror and detected by the IR cameras via a filter positioning system and two mirrors.

After rotating the filter system and positioning the scan mirror on position 1 the first image will be taken. After positioning the scanner on camera 2 the second picture will be taken. For the acquisition of six single images at different wavelength this process has to be repeated three times. The total time for one run is 500 ms.

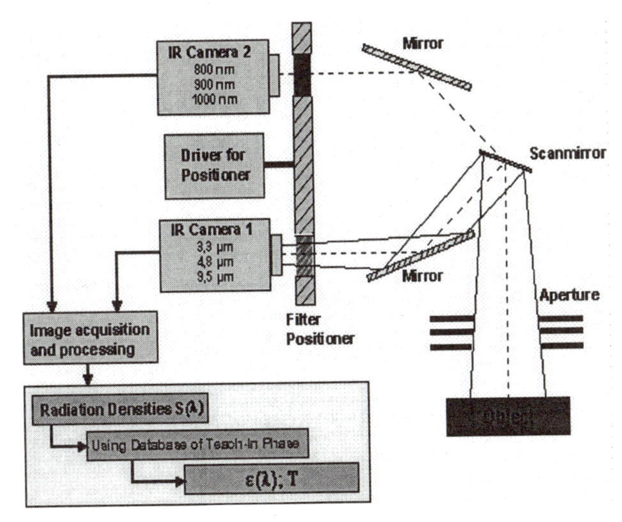

Figure. 1 Setup of the multichannel thermography system

EXPERIMENTS

The infrared camera used is working in the two atmospheric windows. That means in the short wave region from 3-5 µm and in the long wave region from 8-12 µm. Five small band pass filters were selected for spectral emissivity measurements in these radiation bands as follows: 3.3 µm, 3.8 µm, 4.8 µm, 8.5 µm, and 9.5 µm. The bandwidth of each filter is five percent of the specified wavelength.

Using these filters, calibration measurements were carried out at a black body radiator from 800 °C to 1700 °C. Based on these measurements calibration curves were obtained, allowing to evaluate the spectral radiation density by measuring the grey value of the thermographic image with each infrared filter.

The calibration curves were obtained as follows:

Using the respective filter curve and Planck's radiation law the radiation density transmitted through each filter was calculated for the relevant temperature region. Combining the calculations of the radiation densities with the calibration curve one can generate a relation between the radiation densities and grey scale values of the IR camera.

A laser heating technique based on a CO_2-laser with a maximum power of 1500 W was used for the heating of the specimen. A ring-shaped beam mode instead of a Gaussian laser beam mode was applied in order to achieve a more homogeneous heating of the specimen. The specimen positioned on three tungsten needles was heated with a laser beam orthogonally radiating onto the specimen surface. The infrared camera was arranged in a distance of about 40 cm from the specimens. The filter positioning system is placed in front of the infrared camera. This camera together with the filter positioning system was used in order to measure the radiation densities at the various wavelengths. Another infrared camera equipped with a telephoto lens of a spatial resolution of 0.1 mm is used for the temperature measurement. This infrared camera is positioned in such a way that a temperature measurement in the radial bore hole of the cylindrical specimen can be realised. The bore hole has a diameter of 2 mm and a length of about 7 mm so that it is a good approximation for a black body radiator. Numerical calculations according to the instationary heat diffusion theory are used to correct the surface temperature due to a temperature gradient in the test specimen.

During the heating procedure of the specimens the laser power (in the cw-mode) was increased slowly starting up at a low value. In this way a moderate heating speed could be realised avoiding thermal induced stresses inside the specimens. Using different beam powers various surface temperatures from approximately 1000°C to approx. 1700°C were generated at the specimen surface. The spectral emissivities in this temperature range are presented in Fig. 2 for SiC and in Fig. 3 for $MoSi_2$.

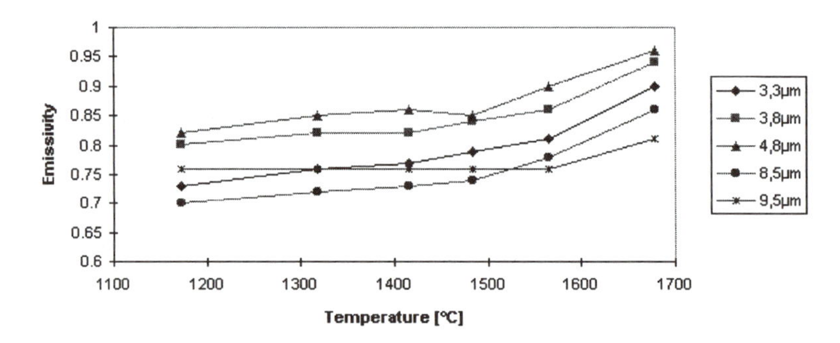

Figure. 2 Emissivity as a function of the temperature for SiC

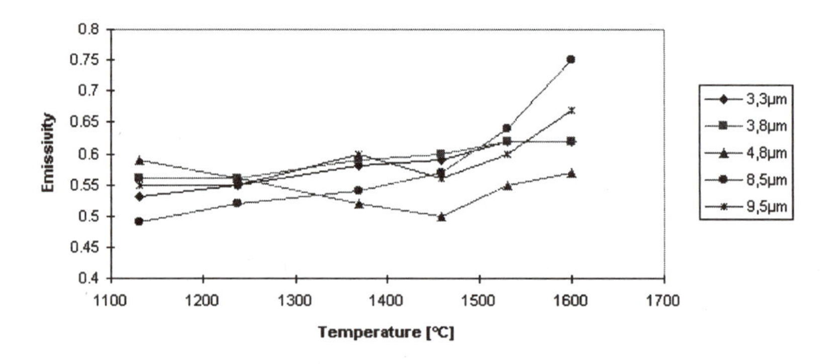

Figure. 3 Emissivity as a function of the temperature for $MoSi_2$

METHODS FOR AN ESTIMATION OF TEMPERATURES AND SPECTRAL EMISSIVITIES

Teach-in technique

For all relevant temperatures T_j the spectral radiation densities $S_j(\lambda_i)$ are measured for the various specimens. For all temperatures the spectral radiation density data are stored in a data bank together with the estimated surface temperatures (T_j) and the spectral emissivity values calculated from it (E_j, λ_i). So data sets as follows are produced:

$S_j(\lambda_1)$, $S_j(\lambda_2) \ldots S_j(\lambda_5)$; $\varepsilon_j(\lambda_1)$; $\varepsilon_j(\lambda_2) \ldots \varepsilon_j(\lambda_5)$; T_j

for $j = 1, \ldots n$; n: number of temperatures

In this way by training one can find for each material, temperature and surface state a characteristic spectral radiation density behaviour, similar to a fingerprint.

During the measuring phase it is then possible to estimate the spectral emissivities and the temperature by comparing the actual measured radiation density values (using a least square algorithm) with the data-bank values won during the teach-in phase.

Two different high-temperature materials -SiC-graphite and $MoSi_2$ were examined.

First teach-in measurements with the SiC-graphite specimen as well as with the MoSi specimen were carried out and the radiation density data together with the estimated surface temperatures were stored in a data-bank.

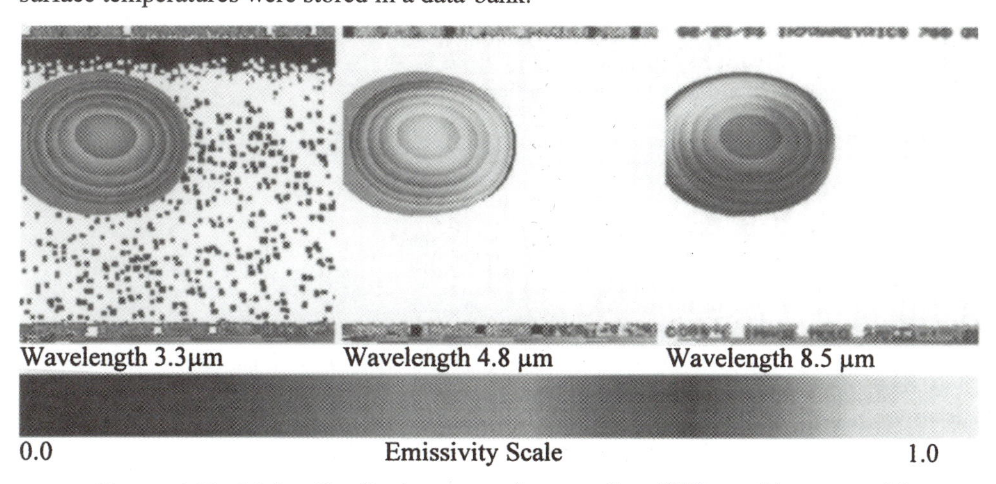

Figure 4. Emisivity distribution at specimen surface (SiC-graphite, example)

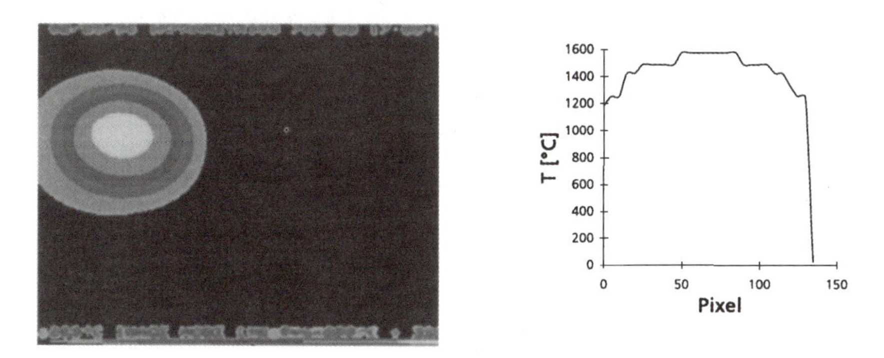

Figure 5. Temperature distribution at specimen surface (SiC-graphite, example)

Some time later the measurement was repeated with the SiC-graphite specimen at various laser beam powers and the spectral radiation density images were evaluated. Using the least square algorithm and the taught radiation density values stored in the data bank the spectral emissivities Fig. 4 as well as the temperatures Fig. 5 were evaluated pixel by pixel of the image. The results showed that the material was always identified correctly. The measurement, calculation and presentation of such spatially resolved spectral emissivity and temperature images is very fast. It takes about two seconds for one surface state.

SUMMARY

- The aim of the project was to measure the spectral emissivities for high-temperature materials at elevated temperatures.
- The multi channel thermographic system has been equipped and tested in the lab for two materials: SiC and $MoSi_2$.
- Spectral emissivities in the wavelength range 3 μm-10 μm as a function of the temperature show a dependence upon wavelength and temperature for these materials.
- It was shown that by a teach-in technique temperature and emissivities of a multi-channel image can be calculated pixel by pixel. The time for such a measurement is approx. 2 sec, so that this technique can be used to study fast processes where surface properties and emissivities vary as a function of time.

REFERENCES

1. Bergmann-Schäfer: *Experimentalphysik* - Optik, Band 3, 8. Auflage, Verlag De Gruyter, Berlin, New York 1987
2. C. Schiewe: *Mehrkanalpyrometrische Temperaturmessung* , VDI-Bericht Nr. 982, 1992, S. 119-125
3. G. Schickert: *Infrared thermography as a possible tool to detect damaged areas in buildings, Durability of building materials 3*, No. 2, 1985 , S. 87 - 99
4. U. Kienitz: *Pyrometrisches Meßverfahren und Mehrkanalpyrometer*, Offenlegungsschrift DE 3611634 A1, 1987
5. H. Wiggenhauser: *Verbesserte Aussagefähigkeit der Infrarotthermographie durch digitale Bildverarbeitung*, Berichtsband DGZfP-Jahrestagung, Luzern, 1991
6. J. Fissler: *Möglichkeiten und Grenzen der Mehrwellenlängenpyrometrie als emissionsgradunabhängiges Temperaturmessverfahren*, Dissertation Uni Stuttgart, 1990
7. G. Walle, N. Meyendorf, G. Dobmann, U. Netzelmann: *Impuls-Video-Thermographie ein leistungsfähiges Verfahren zur zerstörungsfreien Prüfung neuer Werkstoffe*, Berichtsband DGZfP-Jahrestagung, Garmisch- Partenkirchen, 1993
8. V. Norkus u.a.: *Multispektralsensor*, DE-Offenlegunsschrift P 4133481-7., 1992

A COMPARISON OF DIELECTRIC AND ULTRASONIC CURE MONITORING OF ADVANCED COMPOSITES

David D. Shepard, Kim R. Smith, and David C. Maurer
Micromet Instruments, Inc.
7 Wells Avenue
Newton Centre, MA 02159

SUMMARY

Dielectric measurements are routinely used for the in-process cure monitoring of thermosetting resins and composites. Measurements of the ultrasonic sound speed can also be used as a cure monitoring technique. New developments in ultrasonic sensor technology now enable ultrasonic measurements to be easily made in press, autoclave, and RTM processes. The ultrasonic sensors can withstand repeated thermal cycling up to 260° C and are mounted within the mold such that they do not contact the part. This paper compares dielectric and ultrasonic cure monitoring data collected simultaneously during the compression molding of epoxy composites. During both isothermal and ramp and hold cure cycles, a very good correlation was obtained between the electrical resistivity (as measured through dielectric measurements) and the bulk sound speed. The ultrasonic system also successfully monitored the cure of phenolic molding compound, which can not be measured by the dielectric technique. The relative advantages and disadvantages of each technique are discussed.

1. INTRODUCTION

Dielectric measurements are widely used for the cure monitoring of thermosetting resins and composites. The ability to implant disposable dielectric sensors within a part or install permanent dielectric sensors in a mold wall or tool makes dielectric cure monitoring a very versatile in-process cure monitoring technique. Commercial dielectric cure monitoring systems, comprising of an electronics package, sensors, cabling and software are available which enable routine use in research, development, and production.

Measurements of the ultrasonic sound speed of thermosetting resins and composites can also be used for in-process cure monitoring. Previously, use of ultrasonic measurements as a cure monitoring

Nondestructive Characterization of Material VIII
Edited by Robert E. Green Jr., Plenum Press, New York, 1998

technique had been hindered by poor reliability of ultrasonic transducers under typical processing conditions, especially thermal cycling to elevated temperatures. A new self-contained, in-mold, robust ultrasonic sensor has been developed. This robust sensor is capable of withstanding hundreds of thermal cycles up to 260°C. Ultrasonic sound speed measurements may now be routinely made during the processing of thermosets using this new sensor technology. An electronics instrumentation and software package has also been developed for the specific application of making a complete ultrasonic cure monitoring system.

2. DIELECTRIC MEASUREMENT TECHNIQUE

Dielectric cure monitoring involves measuring the changes in the electrical properties of a thermosetting resin or composite to study changes in its viscosity and cure state. A set of electrodes is placed in direct contact with the material and an AC signal of frequency between 0.1 Hz and 100 kHz is applied to one electrode. The application of this alternating voltage causes a signal of the same frequency to be induced in the second electrode. The change in the amplitude and the phase angle shift relative to the applied voltage are measured and used to calculate the dielectric properties of the material.

Two dielectric properties, permittivity and loss factor, result from dipole motion and ionic conduction. During a typical polymerization, permittivity changes are mainly a function of dipole motion. The loss factor is influenced by both dipole motion and ionic conduction. The dipole motion contribution is relatively small throughout typical polymerization processes so that the ionic conduction will often dominate the loss factor. Since the contribution of the ionic conduction to the loss factor is inversely related to frequency, it can be made to dominate the loss factor if measurements are made at a low enough frequency. In a typical dielectric cure monitoring experiment, a number of frequencies are measured and the ionic conductivity extracted from the multi-frequency loss factor data.

Since ionic conduction is essentially a measure of the mobility of ions in the material, it is inversely related to viscosity prior to gelation and rigidity after gelation. The inverse of ionic conductivity, the resistivity, is therefore directly related to viscosity/rigidity changes. Previous work has shown a strong correlation exists between the change in resistivity and actual polymer viscosity during the reaction of thermosetting resins as measured with a rheometer (1). The increase in the resistivity has been shown to correlate with the increase in glass transition temperature of epoxy resins during isothermal cure and is very sensitive to the end of cure (2,3).

Figure 1 shows typical log resistivity and temperature data during the compression molding of an epoxy-graphite prepreg. The initial decrease in resistivity corresponds to the decrease in viscosity as the resin

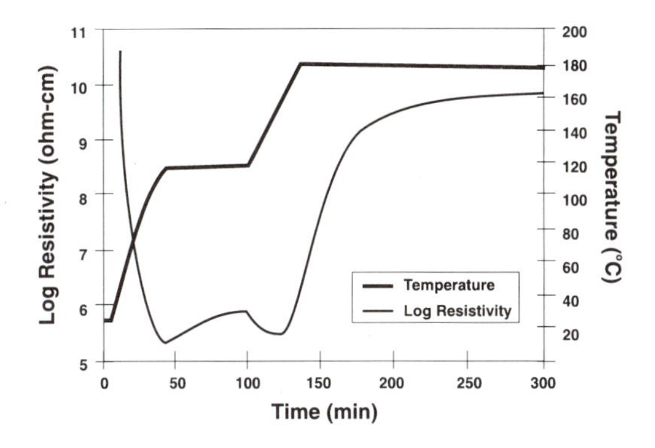

Figure 1: Typical dielectric cure monitoring data of graphite-epoxy prepeg

is heated to the initial 116° C hold temperature. During this hold period, the resistivity increases slowly as the resin viscosity increases due to the loss of volatiles and/or reaction of the resin system. As the temperature is ramped to the final 177°C hold temperature, the resistivity begins to decrease and reaches a minimum. After this point the curing reaction proceeds rapidly as can be seen through the rapid increase in the resistivity. Gradually the rate of increase in the resistivity slows as the rate of reaction slows. The near zero slope of the resistivity after 300 minutes shows that the curing reaction is near completion. Dielectric cure monitoring data is thus able to monitor the entire cure cycle.

3. ULTRASONIC MEASUREMENT TECHNIQUE

Ultrasonic cure monitoring uses changes in the sound speed of a material to monitor changes in viscosity and cure state. Ultrasonic transducers (sensors) are mounted co-linearly on opposites sides of a mold or tool and perform a through-transmission measurement. The ultrasonic measurement system uses piezoelectric elements to generate an acoustic wave. High voltage pulses excite the piezoelectric element in the

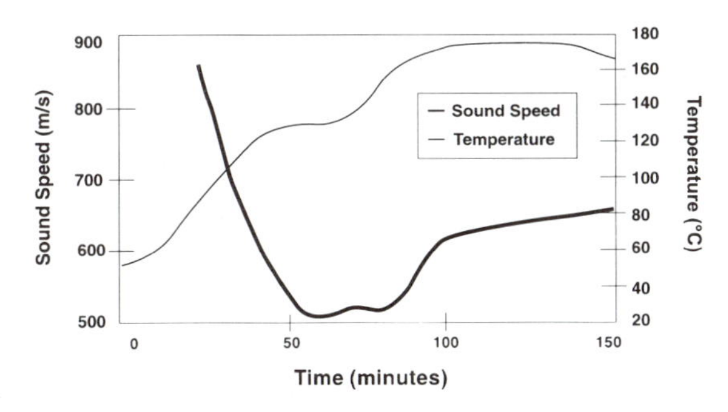

Figure 2. The change in ultrasonic sound speed and temperature of an epoxy-graphite fiber prepreg during a compression molding process

ultrasonic transducer causing this element to oscillate at 5 MHz. This creates the acoustic wave that propagates at a material specific velocity. Whenever the wave reaches a boundary some of the wave is transmitted through the boundary into the second material, and some of the wave is reflected back. When an acoustic wave reaches the piezoelectric element, the element produces a voltage which is the sensor signal. Measurements of sound velocity by through-transmission require two in-line ultrasonic transducers. One is used to generate the acoustic pulse while the opposing transducer records the arrival time of the acoustic wave. The ultrasonic transducers acts as both a transmitter and receiver. The transmitted wave is detected by the transducer in the opposite mold half. The reflected wave from the mold/material interface is detected by the transmitting transducer. It is necessary to detect both of these waveforms to be able to resolve the sound speed of the material under test and not have contributions due to the mold. The excitation and detection function is alternated between the two transducers. By subtracting the reflected waveform peak time from the transmitted waveform peak time for both directions and averaging the two resultant times, the total time of flight for the mold is subtracted out leaving only the transit time of the material under test. The part thickness is monitored by the system in order to provide accurate distance measurements between the opposing sensors. The ultrasonic sound speed is then calculated from the time and distance information. The transmitted signal attenuation is also measured.

The change in sound speed has been correlated with the change in the viscosity and cure state of advanced composites (4,5). Heating the prepreg decreases the material's viscosity and the acoustic wave velocity is decreased. As the material cures, the molecular structure is constrained and supports a higher acoustic wave velocity.

The change in ultrasonic sound speed of an epoxy-graphite fiber prepreg during a compression molding process is plotted in Figure 2. The initial decrease in sound speed coincides with the decrease in viscosity as the prepreg increases in temperature. A broad minimum is seen as the temperature cycle enters a 121 °C hold period. As the temperature is increased to 177 °C, the sound speed increases as curing proceeds. The rate of increase in sound speed then slows as the rate of cure slows and the reaction nears completion.

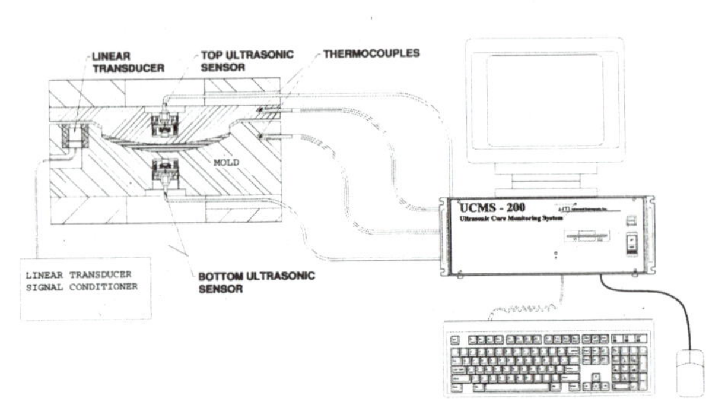

Figure 3. A schematic representation of the UCMS-200 installed in a compression molding process

4. NEW ULTRASONIC MEASUREMENT SYSTEM

The ultrasonic cure monitoring data was collected with the newly developed UCMS-200 Ultrasonic Cure Monitoring System manufactured by Micromet Instruments, Inc. The UCMS-200 consists of sensors and cabling, a rack-mountable electronics package, and a Windows-based LabVIEW software package. A computer is integrated into the electronics package complete with monitor, keyboard, mouse, and disk drives. Figure 3 shows a schematic of the ultrasonic measurement system installed in a compression molding process. An ultrasonic sensor set consists of two sensors located on opposite sides of the mold.

The system can measure the ultrasonic sound speed and attenuation at two different locations using two sets of ultrasonic sensors. Part thickness is monitored using up to four capacitive non-contact displacement probes or linear voltage displacement transducers (LVDT). Accurate measurement of the part thickness is important in accurately calculating the sound speed, especially in processes where the part thickness changes during the cure cycle. Temperature can be measured from two thermocouples. The LabVIEW based software package enables easy operation of the system and allows the display screen to be customized by the user. Algorithms can be easily written by the user to perform special functions and closed loop control of the process.

5. COMPARISON OF DIELECTRIC AND ULTRASONIC CURE MONITORING DATA

Figure 4 compares the electrical resistivity measured by a Micromet Instruments Eumetric System III Microdielectrometer to the ultrasonic sound speed during the non-isothermal cure of an epoxy-fiberglass prepreg. The sound speed and resistivity data were measured simultaneously in a compression mold. The mold temperature was ramped at 2.5°C/min. and reached the final cure temperature of 150°C at approximately 60 minutes. The data is scaled so that the relative increase in log resistivity and sound speed from their respective minimums during cure are equivalent. During this portion of the cure, when the

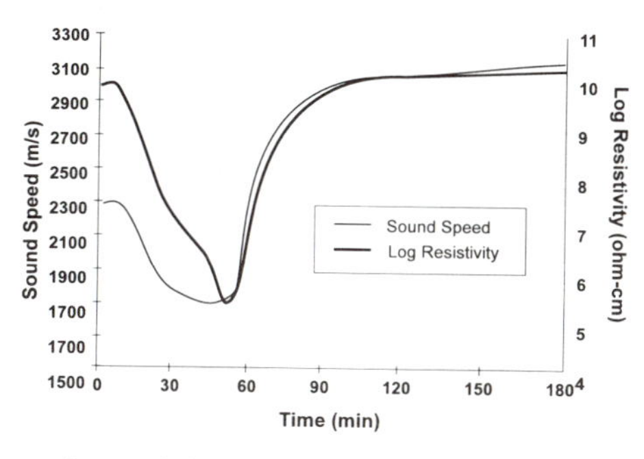

Figure 4. A comparison of the ultrasonic sound speed to the log resistivity during the non-isothermal cure of an epoxy-fiberglass prepreg

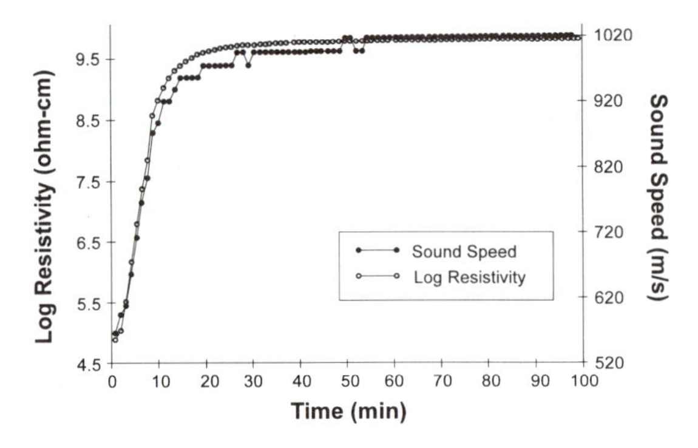

Figure 5. Isothermal cure of an epoxy fiberglass prepreg

temperature is virtually isothermal, the sound speed and log resistivity have an excellent correlation. However, during the heating ramp, the log resistivity shows a greater relative decrease and a small difference in shape than the sound speed. This is because the sound speed of the liquid resin does not change as greatly as a function of temperature as does the mobility of an ion.

The excellent correlation of the sound speed to the resistivity under isothermal conditions can be studied more closely in Figure 5 which compares the data from the two techniques during the 150°C isothermal cure of a 2.3 mm thick epoxy-fiberglass prepreg. Once again, the sound speed and log resistivity data were measured simultaneously in a compression mold The step increments in the sound speed data are due to the limit of resolution of the time of flight through such a thin sample and the lack of any filtering or averaging being performed on the data. On thicker samples, the measured time increments are a smaller percentage of the time of flight so that the resolution will increase and the sensitivity of the data to the end of cure will improve. In Figure 6 the sound speed is plotted as a function of log resistivity. A near linear relationship is observed until the sound speed increases above a level of 900 m/s. After this point, the sound speed appears to be more sensitive to the end of cure but the quantification of the sound speed data and the

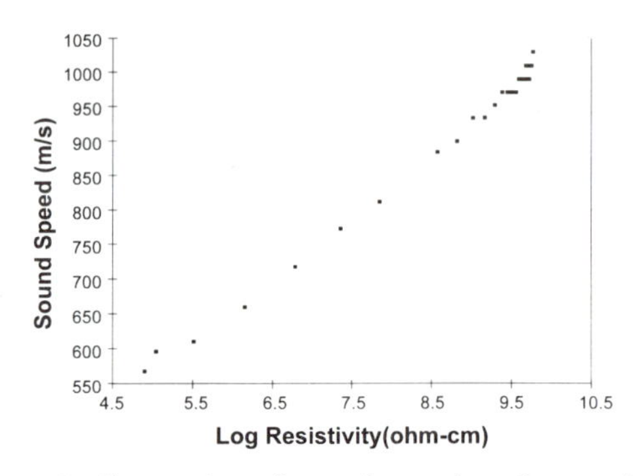

Figure 6. Cross-plot of sound speed vs. Log resistivity

lack of data from other resin systems prohibits a valid assessment of the relative sensitivity of the two measurement techniques to the end of cure.

6. COMPARISON OF DIELECTRIC AND ULTRASONIC TECHNIQUES

There are advantages and disadvantages to both dielectric and ultrasonic cure monitoring techniques. The decision as to which technique is the most appropriate is dependent upon the application and its requirements. The following can be used as a guide to selecting the appropriate technique.

A) Advantages of Dielectric Cure Monitoring:

1) Dielectric cure monitoring is easily implemented into almost any process. Disposable sensors can be easily implanted into a part which enables a simple, cost-effective means for obtaining cure monitoring data. Measurements can be routinely made in ovens, presses, autoclaves, RTM molds, or in small laboratory samples without machining a mold to accommodate a sensor.

2) Dielectric sensors can measure a localized area so that the curing behavior at a specific location can be observed. It also allows measurements on samples with thicknesses as small as one micron.

3) Multiple locations may be monitored in a single run by placing multiple sensors into a part. These sensors can be placed in any location so that any specific location in the part can be monitored.

4) Permanent sensors located in a mold or tool can make measurements using only a single sensor. This requires only a single hole to be machined into a mold and no caul plate is required in an autoclave tool.

5) Sensors are available which can withstand up to 400°C.

B) Disadvantages of Dielectric Cure Monitoring:

1) Dielectric measurements can be affected by reaction by-products. These by-products may contain ionic species which mask the curing reaction.

2) Sensors must be imbedded in the part or located on the surface of the part. This causes a defect in the part or a witness mark to remain on the part which may be unacceptable in a production application. In addition, for graphite filled materials, a filter must be placed over the sensor to prevent shorting of the electrodes.

3) Implantable sensors located in the part add cost and additional cabling and labor in laying up the part.

4) True bulk measurements cannot be made on parts over approximately 1.5 cm in thickness. On graphite composites, true bulk measurements can not be made at all.

C) Advantages of Ultrasonic Cure Monitoring:

1. The ultrasonic sensors are permanently located within the mold or tool and do not contact the part. This provides a non-intrusive measurement with no incremental sensor cost or additional lay-up time.

2) The system makes a true bulk measurement even on thick parts of over 5 cm in thickness (the actual thickness limitation is material dependent). Bulk measurements can be made on graphite composites.

3) The sound speed can be used to calculate the actual modulus of the sample.

4) The sound speed is not affected by reaction by-products which makes it useful for resin systems where dielectric cure monitoring is not successful. For example, phenolic based materials generate ammonia or water during cure which, due to their conductive nature, counteract the normal increase in resistivity due to cross-linking. Figure 7 plots the sound speed during the compression molding of a Rogers RX-640 phenolic molding compound at 172°C. The sound speed data closely resembles typical dielectric cure monitoring data collected on non-phenolic molding compounds, such as polyester bulk molding compound (7).

D) Disadvantages of Ultrasonic Cure Monitoring:

1) Holes must be carefully machined on both halves of the mold. For autoclaves, or other processes that employ only a single tool, a caul plate with sensor must be located on the top of the part (5).

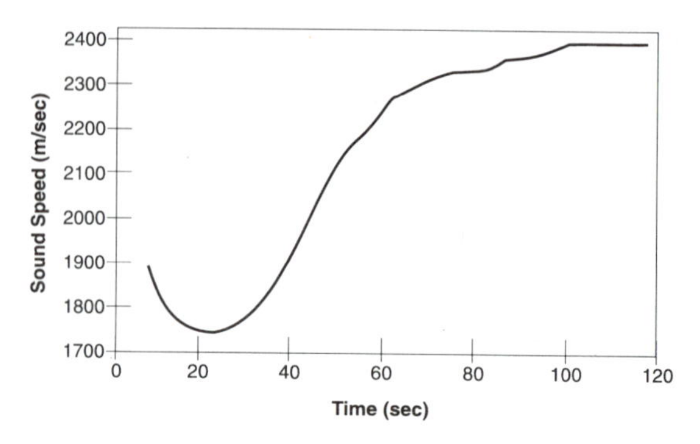

Figure 7. Sound speed during compression molding of Rogers RX-640 phenolic molding compound

2) Measurements can only be made at the location of the permanently mounted sensors.

3) The part must have a significant thickness (at least 2 mm), be homogeneous, and cannot have high damping properties (such as a foam). The thinner the part, the poorer the resolution. If the part is non- homogeneous, spurious reflections will prevent good measurements. If the part has high damping properties, there may be no transmission of the sound wave.

4) The part must not contract away from either mold wall during processing. Loss of contact to the mold wall will result in loss of signal.

5) The upper temperature limit of the ultrasonic sensors is 260°C.

6) For accurate sound speed measurements the part thickness must be measured. This is especially important when the part thickness will change during processing.

7. CONCLUSIONS

Complete dielectric and ultrasonic cure monitoring system are commercially available. Both dielectric and ultrasonic cure monitoring can provide in-process measurements of changes in viscosity and cure state of thermosetting resins and composites. Localized or bulk dielectric measurements can be made with disposable, implantable sensors or with a single permanent sensor mounted in a mold or tool. Ultrasonic measurements are made with permanent sensors located in both halves of a mold. A good correlation is observed between the electrical resistivity and the ultrasonic sound speed. Dielectric measurement advantages are ease of use, ability to measure at multiple locations, ability to measure thin samples, and sensors are commercially available for use at temperatures up to 400° C. Ultrasonic measurements are made with robust sensors permanently mounted in a mold or tool which leaves no sensor in the part and no witness mark. Ultrasonic measurements are unaffected by reaction by-products (they can successfully monitor phenolic materials), they require no further incremental costs for process monitoring, and further information, such as bulk modulus, may be extracted from the sound speed. The choice of the most appropriate cure monitoring technique depends upon the specific application.

8. REFERENCES

1) J. Gotro, NATAS Conference, 19, (1990), 523

2) D.R. Day and D.D. Shepard, NATAS Conference, 16, (1987), 52

3) D.R. Day, D.D. Shepard and K.J. Craven, International SAMPE Technical Conference, 22, (1990), 724

4) W. Veronesi, 41st Sagamore Conference, (1994.)

5) C.A. Lebowitz, P.J. Biermann, J.H. Cranmer, and L.M. Brown, SPIE-The International Society for Optical Engineering Conference, 2948, (1996), 72

6) D.D. Shepard and T.A. Senturia, American Laboratory, 25, (Nov. 1993), 38

NONINVASIVE MEASUREMENT OF ACOUSTIC PROPERTIES OF FLUIDS USING AN ULTRASONIC INTERFEROMETRY TECHNIQUE

W. Han, D. N. Sinha, K. N. Springer and D. C. Lizon

Electronic and Electrochemical Materials & Devices Group, MS D429
Los Alamos National Laboratory
Los Alamos, NM 87545

INTRODUCTION

Ultrasonic propagation provides a useful approach for extracting physicochemical properties of fluids. Traditional ultrasonic techniques for the measurement of sound attenuation and sound speed in fluids primarily include resonance reverberation, pulse-echo, and ultrasonic interferometry. Of these techniques, ultrasonic interferometry stands out for its ability to provide frequency-dependent measurements and for its ability to work with small liquid samples[1-3]. Eggers and Kaatze[1] give a recent comprehensive review on this subject. Unfortunately, this technique requires the transducers to be in direct contact with the test fluid, thereby limiting the applicability of this technique primarily to precision laboratory fluid characterization. For industrial and other practical applications, it is important to be able to characterize fluids in containers without requiring the transducers to be in direct contact with the fluid inside. Such noninvasive fluid characterization was recently demonstrated by Sinha et al.[4] who adapted the ultrasonic interferometry technique for noninvasive measurements to identify chemicals inside sealed metal containers.

The purpose of this paper is to describe the analytical approach behind this noninvasive adaptation of the technique and show how physical properties, such as sound speed, sound attenuation, and density can be determined with reasonable accuracy even when the measurements are made with transducers external to the container. Two types of containers, commercial rectangular optical glass cells and cylindrical steel shells were used in our studies. The measurements provide an ultrasonic resonance (interference) spectrum characteristic of the fluid and the container, over a frequency range 1-12 MHz. For reliable determination of sound attenuation and liquid density, two analytical approaches are derived from a theoretical model that is based on normal acoustic wave propagation through multiple-layered media[5]. One method is based on frequency-dependent relative strengths of minima and maxima (peaks) of the measured spectra, and the other is based on the resonance half-power bandwidths. Good agreement of extracted physical parameters with literature values demonstrates the potential of this noninvasive ultrasonic interferometry technique for industrial applications as well as in basic research.

THEORETICAL ANALYSIS

A noninvasive ultrasonic interferometry measurement can be described in terms of planar ultrasonic wave transmission and reflection through a multiple-layered system consisting of the test fluid sandwiched between symmetric layers of transducer crystal, wear plate, coupling gel, and cell (container) wall. This multi-layer model formulation is an extension of the well-known theory of planar wave propagation through one layer[6] (a liquid layer embedded between two solid layers of infinite dimension) and the details are presented elsewhere[5]. In the present case, the sound intensity transmission coefficients are observed experimentally as a spectrum of repeated peaks (see Fig. 2) that depend on the physical properties (acoustic impedance, longitudinal sound speed, and sound attenuation) and geometry (thickness) of each layer. With proper analysis, these physical properties can be extracted from the observed spectrum.

Although the full theory including all layers can be used to derive the physical properties of the liquid, we have found that by properly selecting the measurement frequency range to avoid resonance contributions from the walls, a much simpler approach is possible. Essentially, the problem can be reduced to a basic one-layer model making the analysis significantly simpler. Small errors introduced by this simplification are discussed later in the results section. The intensity transmission coefficient T_I for the simplified case[5] of a fluid layer of path-length L, attenuation coefficient α_L ($\alpha_L L<<1$), and sound speed c_L between two identical wall boundaries reduces to

$$T_I = \frac{1}{(1+\frac{1}{2}\sigma\,\alpha_L\, L)^2 + \frac{\sigma^2-4}{4} Sin^2 \frac{\omega}{c_L} L}. \tag{1}$$

Here, $\sigma = z_w/z_L + z_L/z_w$, ω is the angular frequency, z_w and z_L are acoustic impedance of wall and liquid, respectively. For most fluids inside an elastic cell, $\sigma \approx z_w/z_L$.

T_I in Eq. (1) is a periodic function of ω. The ratio of the transmission coefficient mimima $T_{I,min}$ and maxima $T_{I,max}$ can be expressed in terms of σ and α_L as

$$\left[\frac{1}{(T_{I,\min}/T_{I,\max})} - 1\right]^{-0.5} = \frac{2}{\sigma} + L\,\alpha(f^2). \tag{2}$$

Eq. (2) shows that both α_L and σ can be determined from a linear fit of the data of the transmission ratio factor TRF=$[(T_{I,min}/T_{I,max})^{-1} - 1]^{-0.5}$ vs. f^2 over a wide frequency range.

Another traditional method is to determine the fluid attenuation coefficient from the half-power bandwidth of observed resonance peaks. From Eq. (1), one can derive an inverse solution for half-power bandwidth δf in terms of acoustic properties in the fluid[5] as

$$\delta f = \frac{2\,c_L}{\pi\,\sigma\,L} + \frac{c_L\,\alpha_L(f^2)}{\pi}. \tag{3}$$

The second term is the contribution from liquid sound absorption and is identical to the solution obtained in a previous resonator theory[1-3] for transducers in direct contact with the test liquid. The first term is independent of frequency and depends on σ, c_L, and L. This results from the reflection loss at the wall-liquid interface and can be used to determine liquid density if the acoustic impedance of the wall is known.

In practical measurements, however, other loss mechanisms, such as diffraction loss and losses due to the transducer backing, tangential mode generation in the container, etc., although small in magnitude, do also contribute to the measured bandwidth and their effects need to be included. The diffraction loss is subtracted from the measured δf using the expression of Labhardt and Schwarz[3]. Other losses can be accounted for by using a reference liquid of known acoustic properties[1-3]. Therefore, the attenuation coefficient of the test liquid α_L can be written as

$$\alpha_L = \frac{\pi}{c_L}(\delta f - \delta f_0) - \frac{\pi}{c_L}(\delta f_r - \delta f_{r0}) + \frac{c_r}{c_L}\alpha_r . \tag{4}$$

Where, δf is the measured half-power bandwidth of the test liquid at any selected frequency f and δf_o is the extrapolated width at $f=0$. Subscript "r" refers to the reference liquid. From Eq.(4), the attenuation coefficient factor $\alpha_0 = (\alpha_L / f^2)10^{17}$ np s^2/cm can be determined with good accuracy. The density of the test liquid can be determined from

$$\rho_L = \delta f_{0c} \frac{\pi c_w \rho_w L}{2 c_L^2} . \tag{5}$$

Here, δf_{0c} is the extrapolated peak-width value (at $f^2 = 0$) that is corrected for miscellaneous small losses by using a reference liquid. Sound speed c_L is related to the frequency difference between any two consecutive resonance peaks $\Delta f = f_{n+1} - f_n$ and, can be determined independently by $c_L = 2\,\Delta f\, L$. However, experiments and model predictions[5] find that Δf oscillates within 1% of the mean value over a frequency range due to effects of wall resonance modulation, a characteristic behavior of the noninvasive measurement. Thus, c_L is obtained using the averaged frequency difference $<\Delta f>$

$$c_L = 2\,\mathrm{L} < \Delta \mathrm{f} > . \tag{6}$$

EXPERIMENTAL SETUP

The ultrasonic interferometer shown in Fig.1 consists a cell filled with a liquid to be tested, a Digital Synthesizer and Analyzer (NEEL Electronics, CA) PC plug-in-card (DSA Board), two disk-shaped piezoelectric transducers, a personal computer, and a voltage amplifier. The DSA board contains all electronics for sine-wave sweep signal generation, and signal detection and processing. The sweep frequency range available is 1 kHz - 15 MHz. The sweep rate can be varied from 1 - 800 frequency steps per second. With a resolution of 0.1 Hz over the entire frequency range, the system provides a frequency response directly in real-time. Typical excitation voltage levels used are 1-10 V pk-pk. Broadband piezoelectric transducers (Panametrics Corp.,

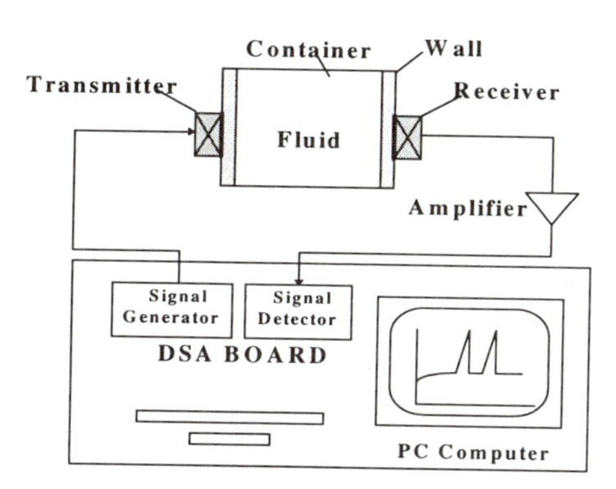

Figure 1. Experimental set up of a swept-frequency ultrasonic interferometer

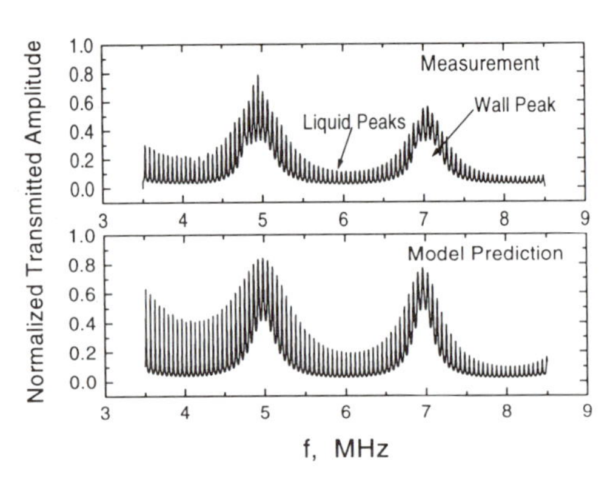

Figure 2. Comparison of measured interferometry spectrum with multi-layered model prediction

MA) with center frequency of 5 or 10 MHz are placed on directly opposite sides of the cell. One transducer is used as the excitation source and is driven by the DSA board to set up standing waves in the resonator cavity of the cell at certain frequencies. The second transducer detects these standing waves as resonance peaks, and its output is fed to the DSA board for analysis following signal amplification. The signal analysis involves a heterodyne mixing technique followed by signal rectification, envelope detection, and 14-bit digitization. The digitized amplitude spectrum is then analyzed in the PC to determine sound speed, sound attenuation, and density of the sample liquid by the methods described above. The containers used were optically polished rectangular glass cells (Starna Cells, Inc., CA) of liquid pathlength 1.0 cm and wall thickness 0.125 cm, and cylindrical stainless steel shells with inner diameter 5.27 cm and wall thickness 0.225 cm.

RESULTS AND DISCUSSION

For measurements made through the container walls, the observed interference spectrum is a superimposition of the interference pattern due to the liquid itself, and that due to the wall. An example is shown in Fig. 2 for isopropanol in a glass cell where the closely spaced liquid peaks are modulated by the underlying wall peaks. The predicted normalized transmission spectrum using the multi-layered model[5] is in good agreement with the measured data as seen in Fig. 2. The respective peak separations are mainly dependent on the pathlengths in the liquid and the wall in addition to the sound speeds in the respective media according to Eq. (6). However, in the vicinity

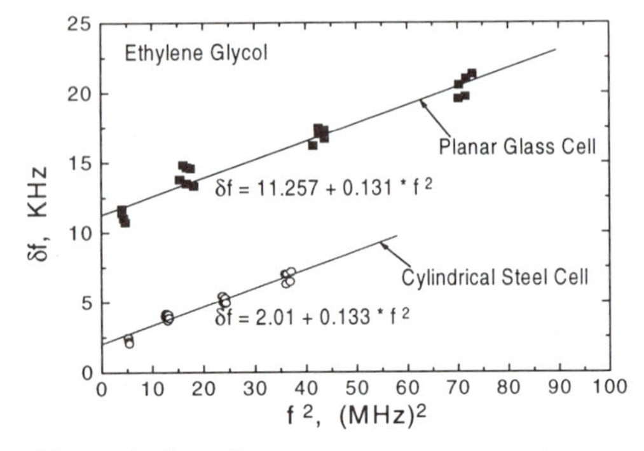

Figure 3. Interferometry measurements on planar glass cell and cylindrical steel shell filled with ethylene glycol

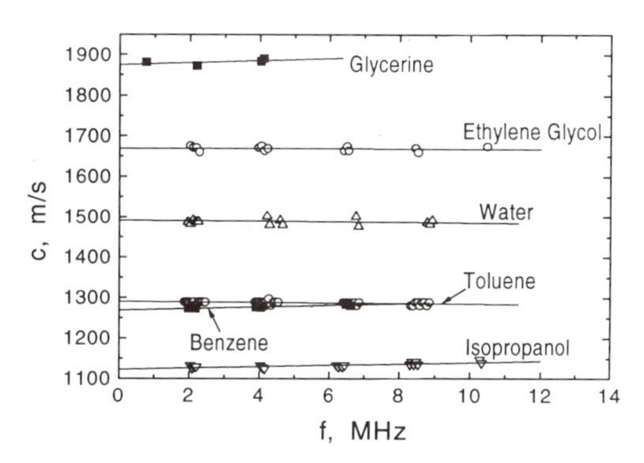

Figure 4. Sound speed of six liquids determined from interferometry measurements

of the wall peaks (e.g., 5 and 7 MHz), both liquid peak spacing and peak width are affected. The peak spacing gets progressively narrower and the peak width gets broader as a wall peak is approached from either side. The system behaves as a coupled oscillator and the liquid peak spacing gets narrower because of frequency pulling effect of the wall. Moreover, sound transmission increases and reaches a peak at the wall resonance frequency. This increases the reflection loss at the wall-liquid interface and broadens the liquid resonance peak width. Any measurement made further away from the wall peaks (e.g., near 4, 6, and 8 MHz) can provide appropriate measurements for the liquid, and we thus routinely use our analytical solutions (see Eqs. 4-6) to extract data from these frequency regions.

Although the theoretical model discussed is a 1-D model and therefore applicable to flat, parallel geometry, we have found that this model is equally applicable to cylindrical geometry. This is illustrated in Fig. 3 where peak-width measurements from two types of containers (planar glass cell and cylindrical shell) filled with the same liquid (ethylene glycol) are presented as a function of f^2. A planar-to-concave thin Plexiglas disk was used to couple the flat transducers to the curved cylinder walls. Besides providing better coupling, this disk reduces diffraction loss by focusing the sound beam, and also reduces shear wave generation that goes around the circumference of the cylinder. The measured attenuation coefficients of the liquid are found to be nearly identical with α_o being 246 and 254 np s^2/cm for the planar glass cell and cylindrical shell, respectively. The different intercepts reflect the different acoustic impedance ratios z_w/z_L and the correspondingly different amount of energy loss contribution to the resonance peak width: a higher impedance ratio signifies lower loss. Therefore, the steel shell shows a lower value of the intercept

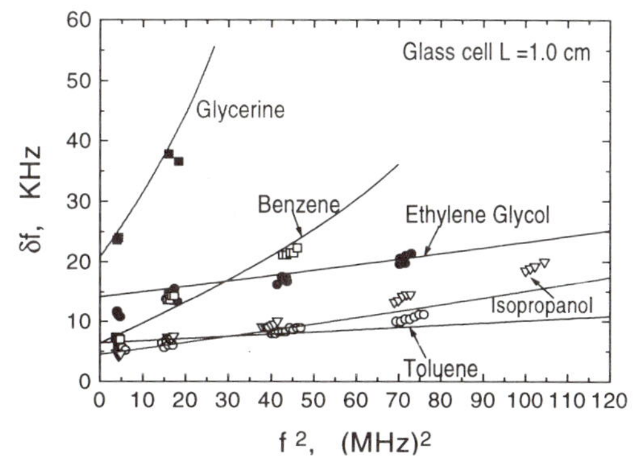

Figure 5. Comparison with predictions of half-power width δf vs. f^2 for five liquids

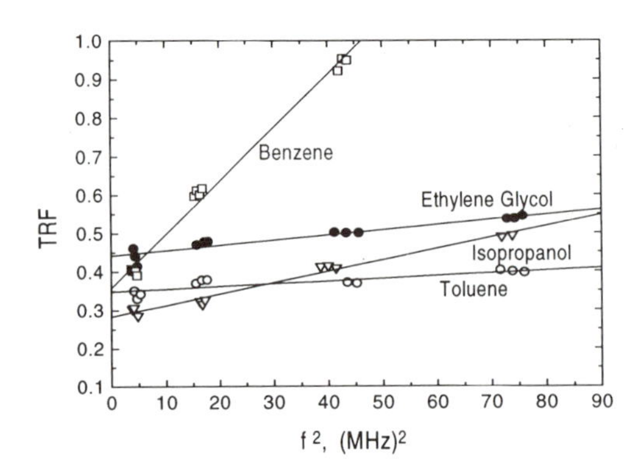

Figure 6. Determination of attenuation coefficients of four liquids in glass cells using the transmission ratio approach

than the glass cell. These results agree with the analytical approach presented in this paper and demonstrate the container geometry and material independence of the interferometry technique for liquid characterization.

Sound speed can be quite accurately determined by Eq. (6). Fig. 4 shows the sound speed measurements for glycerin, ethylene glycol, water, toluene, benzene, and isopropanol in a glass cell over a frequency rage of 1-12 MHz. For the above liquids, the maximum standard deviation relative to the mean values of sound speeds is 0.59%. The scatter (actually an oscillation) in the sound speed data is not due to measurement error but primarily due to the wall resonance modulation effects as discussed earlier. This oscillation is also seen in predictions from our theoretical model.

Measurements in a glass cell of half-power width δf of the liquid peaks are shown in Fig. 5 for glycerin, benzene, isopropanol, ethylene glycol, and toluene at 26.8°C. Because of its low attenuation coefficient, toluene is selected as reference liquid to subtract out the various small loss contributions discussed earlier, that cannot be otherwise accounted for by our 1-D model. Also shown are the predictions using Eq. (3). The measured values of α_o using Eq.(4) for the former four liquids are 1530, 778, 406, and 191.4 nps^2/cm, as compared to the reported values[7] at 20°C of 1500, 860, 270, and 130 nps^2/cm. It is difficult to attribute any particular explanation as to why the agreement between the measured and the reported values appear to be appreciably better for the high-attenuation liquids (glycerin and benzene) than the low-attenuation liquids (isopropanol and ethylene glycol) because of the different temperature used in measurements than the literature value and the lack of data on the temperature dependence of sound attenuation for these liquids. .

Figure 6 presents the determination of sound attenuation using the transmission ratio approach (see Eq. (2)), for four liquids as used previously for the δf analysis. As predicted by Eq. (4), all data show a strongly linear relationship over the entire frequency range. The values of α_0, determined directly from the slopes are 1380, 133, 294, and 67.5 nps^2/cm for benzene, ethylene glycol, isopropanol, and toluene, respectively. Except for benzene, the agreements of the other three liquids with the reported values (130, 270, and 82 nps^2/cm) are very good and better than the δf method previously discussed.

It is normally difficult to determine liquid density reliably using any noninvasive technique. However, the interferometry approach appears to provide a simple way to determine this important physical parameter. Using the peak-width approach (Eq.(5)), we obtained densities for several liquids and found that the relative deviation of measured density from the literature value is in the range of ±(2-10)%. This may be quite acceptable in many practical applications. Again, if toluene is used as the reference liquid, the values of measured liquid density using the transmission ratio approach, are 0.918, 1.068, and 0.566 g/cm^3, compared to the reported values[7] of 0.88, 1.108, and 0.79 g/cm^3 for benzene, ethylene glycol, and isopropanol. Both approaches provide equivalent accuracy in the sound attenuation and liquid density. However, the transmission ratio approach is easier to implement and appears to be somewhat more robust than the peak-width approach.

CONCLUSION

It is shown that the ultrasonic interferometry technique and the analysis approach developed here are capable of noninvasively determining sound speed, sound attenuation coefficient, and density of liquids. The accuracy of the technique is not dependent on container geometry or materials properties. The theoretical model for one-dimensional wave propagation through multi-layered media provides good agreement with experiment. Two analytical methods developed yield good results in the practical determination of acoustic properties in fluids. Swept-frequency ultrasonic interferometry is a versatile and adaptable noninvasive fluid characterization technique, and has a wide range of potential applications in industry and other nondestructive evaluation applications.

ACKNOWLEDGMENT

This work was supported by the Office of the Assistant to the Secretary of Defense, Counterproliferation Program, and by the Defense Special Weapons Agency.

REFERENCES

1. F. Eggers and U. Kaatze, "Broad-band ultrasonic measurement techniques for liquids", Meas. Sci. Technol., **7**, 1-19 (1996).
2. F. Eggers and Th. Funck, " Ultrasonic measurements with milliliter liquid samples in the 0.5-100 MHz range", Rev. Sci. Instrum., **44**, 969-977 (1973).
3. A. Labhardt and G. Schwarz, "A high resolution and low volume ultrasonic resonator method for fast chemical relaxation measurements", Ber. Bunsenges. Phys. Chem., **80**, 83-92 (1976).
4. D. N. Sinha, B. W. Anthony and D. C. Lizon, "Swept frequency acoustic interferometry technique for chemical weapons verification and monitoring", Third Intl. Conference On-Site Analysis, Houston, Texas, January 22-25, 1995, in press.
5. W. Han, D.N. Sinha, K.N. Springer and D.C. Lizon, "An ultrasonic interferometry method for noninvasive determination of acoustic properties in liquids", J. Acoust. Soc. Am., submitted for publication.
6. L.E. Kinsler, A.R. Frey, A.B. Coppens and J.V. Sanders, *Fundamentals of Acoustics*, Chapter 6, Third Edition, John Wiley & Son (1982).
7. L.W. Anson and R.C. Chiver, "Thermal effects in the attenuation of ultrasound in dilute suspensions for low values of acoustic radius", Ultrasonics, **28**, 16-26 (1990).

CHARACTERIZATION OF FIBER-WAVINESS IN COMPOSITE SPECIMENS USING DEEP LINE-FOCUS ACOUSTIC MICROSCOPY

W. Sachse[1], K. Y. Kim[1], D. Xiang[2] and N. N. Hsu[2]

[1] Department of Theoretical and Applied Mechanics
Cornell University, Ithaca, NY - 14853 USA

[2] Ultrasonics Group, Automated Production Technology Division
National Institute of Standards and Technology (NIST)
Gaithersburg, MD - 20899 USA

INTRODUCTION

A line-focus transducer used as the transduction element in an acoustic microscope forms the basis of a powerful materials characterization tool. When such a transducer is excited with rf-burst signals the transducer output voltage V exhibits strong amplitude variations which are related to the transducer's defocus distance z, that is, the distance between the transducer's focal point and the sample surface. These amplitude variations result from the interference between the leaky surface wave and the direct-reflected wave from the surface of the specimen. Analysis of such $V(z)$ curves permits determination of the wavespeed and attenuation of the surface wave which is the basis of the materials surface characterization measurement. Developed by Chubachi and Kushibiki [1] the line-focus beam has been used by a number of investigators to detect and characterize material anisotropy and stresses. Using a small aperture and high *f*-number lens as well as high-frequency excitations, the system is capable of high spatial resolution on a specimen. It forms the basis of an *acoustic microprobe* for determining near-surface material properties. Scanning the transducer permits the mapping of material properties over a region of the specimen, c. f. [2]–[6].

A large aperture, lensless, line-focus transducer was recently developed by the Ultrasonics Group at NIST.[7]–[9] Possessing a focal length of 25.4 mm and an aperture of 28.2 mm this transducer has a low *f*-number of 0.9. It is fabricated using a polyvinylidene fluoride (PVDF) piezoelectric film which is backed by a tungsten powder-loaded epoxy. The transducer is capable of broadband, short-pulse operation and has a center frequency of approximately 10 MHz. These characteristics make the transducer ideal for time-resolved, polarization-sensitive ultrasonic measurements on materials.

Nondestructive Characterization of Material VIII
Edited by Robert E. Green Jr., Plenum Press, New York, 1998

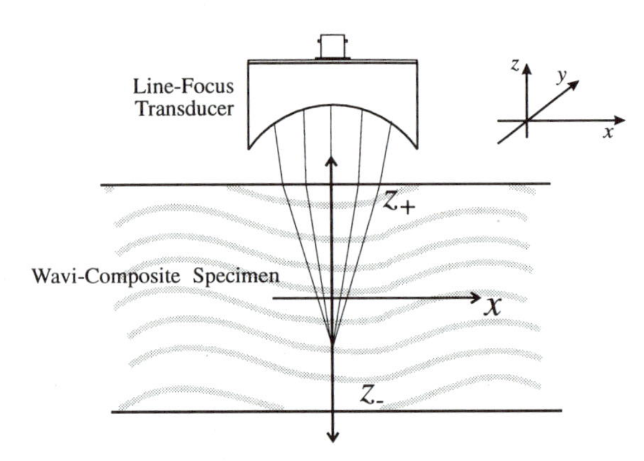

Figure 1: Schematic drawing of a wavi-composite and the ultrasonic testing configuration (deep focus configuration).

Operation of such a transducer has been demonstrated by measuring the wave arrivals and their amplitudes as a function of defocus distance in isotropic as well as anisotropic materials.[9] This data permits determination of a material's near-surface sound velocity and attenuation. Other applications have included the characterization of rough and porous ceramic coatings from measurement of the leaky surface wave velocity.[10]

In this paper we describe the use of such a system to detect and to characterize a composite material possessing waviness. We present results which are obtained when the focus of the transducer is near the front surface of the specimen. But further, we shall determine the sound field obtained when the transducer is focused on a region deep in the interior of a specimen.

SPECIMEN AND TEST PROCEDURE

A specimen possessing strong waviness was fabricated using graphite-fiber prepregs which were laid up unidirectionally, ten plies at a time, in an aluminum mold. The two-piece mold possessed mating sinusoidal surfaces with a spatial period of 40 mm and an amplitude of 2 mm. The completed layup, consisting of approximately 180 plies, was placed in a vacuum bag and then into an oven for curing. The density of the section of the cured composite specimen was 1524 kg/m^3 . The two opposite sides of the wavi-composite slab were machined flat and parallel. The specimen was subsequently polished to give a plate 15.93 mm thick.

We denote the direction along the thickness dimension as the z-axis and the x-axis is aligned with the average fiber direction. The specimen and its testing configuration are shown schematically in Fig. 1. Scans of the transducer were carried out along both the $x-$ as well as the $z-$directions as ultrasonic waveform data was acquired. A two-axis scanning system, operating under control of the waveform acquisition program running on a PC, was used for data acquisition.

The transducer was connected to a broadband ultrasonic system in which a short-duration pulse is used as excitation signal. Two commercially available systems were

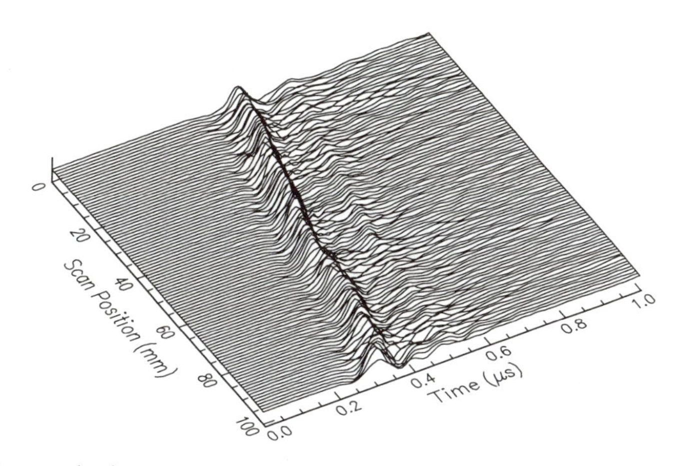

Figure 2: Scan image $v(x,t)$ at $z = z_+$ of sub-front-surface signals of a wavi-composite specimen.

used. One was a conventional 0.1 to 25 MHz bandwidth system and the other was a system whose bandwidth extended from .001 to 200 MHz. The signals were captured at up to a 450 MHz sampling rate, using an 8-bit digitizing oscilloscope which had capability for waveform processing and specifically, signal averaging to achieve a higher effective bit rate. Typically 1000 signals were averaged before the transducer was moved to the next position in a scan. A so-called *scan-image* is obtained by stacking the detected time-domain waveforms together. The axes of such an image are the transducer scan position, time and signal amplitude, in volts. In contrast to conventional $V(z)$ data we denote the measured ultrasonic time-domain waveform data by lower case v.

MEASUREMENTS AND MEASUREMENT RESULTS

Front Surface Scans, $v(x,t)$ at $z = z_+$

It is not surprising that when the transducer was focused on the front surface of the composite specimen, the largest signal was detected. Scans were then carried out as in a conventional line-focus acoustic microscope in which the transducer was slightly defocused below the front surface. We denote the near front surface defocus distance by z_+. The scans were along the x–direction and extended a distance of 90 mm to include more than two cycles of specimen waviness. A typical scan image of the near front surface signals is shown in Fig. 2. The variations in in first signal arrival-time and amplitude appear to be a function of surface roughness and apparently to a lesser extent, the local fiber orientation to the surface. The small amplitude signal following the first-arrival is probably the leaky wave but the variations in these small-amplitude signals do not appear to be sensitive to the sub-surface material waviness. It is clear that images obtained with the transducer slightly defocused below the front surface of the specimen will not readily provide information about microstructural features in the interior of a thick specimen.

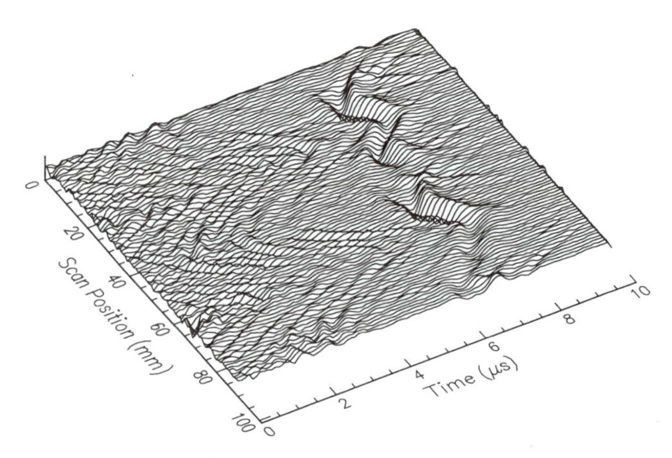

Figure 3: Scan image $v(x,t)$ at $z = z_-$ of *deep-focus* signals of a wavi-composite specimen.

Deep-focus Scans, $v(x,t)$ at $z = z_-$

By focusing the transducer deep in the interior of the specimen, at z_-, about 1–2 mm from the rear surface, we obtain the so-called *deep focus* configuration of the system. When the transducer has its line of focus oriented perpendicular to the fiber direction (aligned along the y–direction) and it is scanned in this configuration in the x–direction, the scan image shown in Fig. 3 is obtained.

It is seen that the rear-surface echo extends over a longer time period, and is correspondingly comprised of lower frequencies than are in the front surface specular reflection signal shown in Fig. 2. The surface roughness of the specimen does not seem to significantly influence the arrival-time and amplitude of the deep-focus signal. Further, the alignment of the specimen relative to the scanning transducer is not as critical as in conventional acoustic microscopy measurements.

It is immediately obvious that the deep-focus signal clearly reflects the variations of microstructure of a wavi-composite specimen. That is, the signal amplitude and arrival-times exhibit the periodicity of the waviness. In regions where the fiber waviness is closest to the transducer or farthest from the rear surface, the signal amplitude is reduced. The signal appears to become *de-focused*. In contrast, in regions where the fiber waviness is farthest from the transducer or nearest to the rear surface, the signal amplitude is largest and it arrives at long arrival times. The signal appears to become *focused*. Further, in regions of the scan image which arrive prior to the arrival of the deep-focus signal, there is also a structure reflecting the characteristics of the waviness, albeit more weakly than the deep-focus signal.

Measurements were also made in which the line of focus of the transducer was oriented parallel to the fiber direction (along the x–direction) and the transducer was set at deep-focus. In that case, the periodic structure visible in Fig. 3 is not visible. It becomes increasingly apparent in the scan images as the transducer is oriented to bring its line of focus to right angle to the fiber direction.

$v(z,t)$ Scans

In order to permit quantitative investigation of signal arrivals in a wavi-composite it is essential that the sequence of wave arrivals from the specimen be clearly identified. The time-resolution of signals which this film transducer possesses can be used to advantage if the transducer is scanned in the z–direction (direction normal to the specimen surface) at a particular location on the specimen from the maximum defocus distance to a point at which the focal spot is exterior to the specimen. Fig. 4 is one example scan. It is seen that the evolution of signal arrivals with z–scan position provides a powerful means for identifying many of the arrivals in the detected signals.

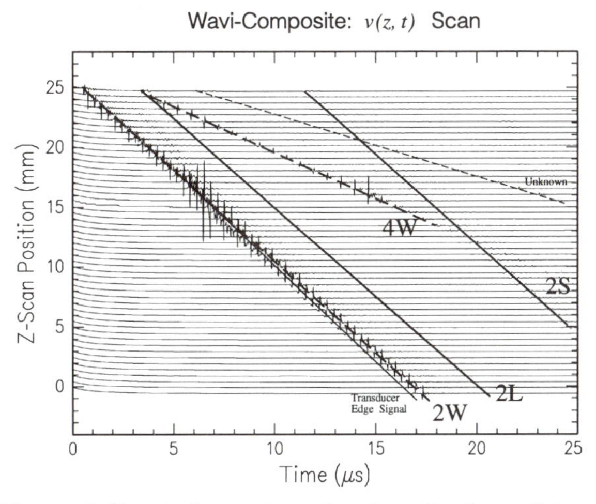

Figure 4: Signals detected as a function of z–focus distance.

The most obvious and clearly identified signals have been labelled in Fig. 4. There are the '2W' and '4W' signals which refer to signals corresponding to the waves which are specularly reflected once or twice, respectively, from the specimen surface directly under the line-focus transducer. Also clearly identified are the '2L' signal arrivals. These are the bulk longitudinal waves propagating at approximately 4.5 mm/μs down and back through the composite plate. Similarly, the signals arriving at the line labelled '2S' correspond to the bulk shear wave signals propagating at approximately 1.5 mm/μs one round-trip through the composite plate.

Visible also are some surprises and unknowns. Clearly identifiable are signals which emanate from the edges of the line-focus transducer. When the focus line is exterior to the specimen, these small signals arrive prior to the specularly reflected signal in a predictable way. The trace of the signal arrivals labelled 'Unknown' may be the result of the wave being propagated back and forth in the water between the transducer and specimen, corresponding to the '6W' signal. But this requires further investigation. The small signal amplitudes following immediately after the arrival of the '2W' signal when the transducer is positioned closer to the specimen than its focal length include the leaky wave signal.

The $v(z,t)$ scans clearly provide a powerful means for identifying many of the signals present in a transducer waveform. Further, by analyzing these images it is possible to clearly determine the wavespeed and attenuation of particular wave modes propagating in the specimen and surrounding fluid and from such data, the moduli and density of the composite specimen.[6]

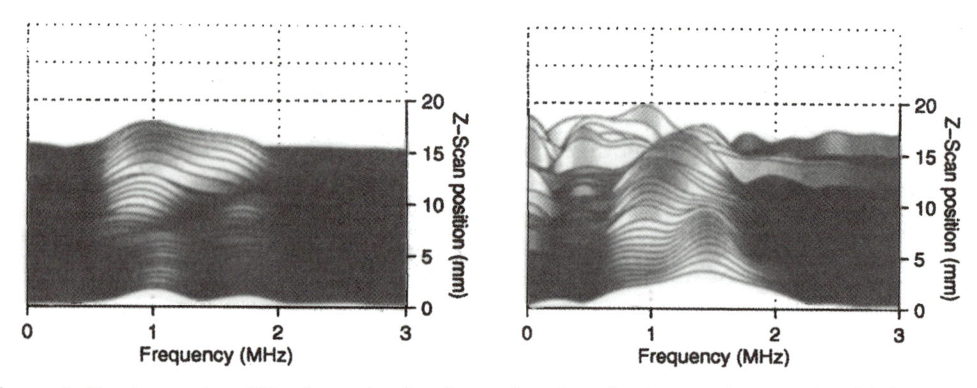

Figure 5: Fourier spectra of the deep-echo signals as a function of z in a wavi-composite. (a) At scan position 3 mm, (b) At 35 mm.

$v(z, x, t)$ Scans

An investigation of the heterogenous microstructure of a material such as a wavi-composite requires, in principle, that the wavespeeds and density of the material is determined for all points in the interior of a specimen. In some cases, it may be assumed that the material heterogeneity is only a function of x and y and it is constant in the thickness dimension. In that case, $v(z,t)$ scans collected at various x, y coordinates of the specimen are needed, i. e. $v(z,x,y,t)$ scans. In this paper we show the simpler case of $v(z,t)$ data collected along a second scan made along just one direction of the specimen, say the x–direction, that is, $v(z,x,t)$ scans.

It is a bit problematical on how to best present such data. What is required is a 4-dimensional representation of the values of x, z, t and v. The data shown in Figs. 2 and 3 are simply slices of such 4-dimensional data sets. An even more restricted view is obtained when one attempts to use conventional ultrasonic A- or B-scans or the $V(z)$ data of conventional acoustic microscopy measurements to characterize material heterogeneities. For some materials characterization applications it is unnecessary to collect a full 4-dimensional data set – or the 5-dimensional data set if the scans also include motions along the y–direction.

If data collection time is not an issue, it is possible to collect all the data and to extract from it specific features. For the case of the wavi-composite, we show in Fig. 5 the result that is obtained when one Fourier-transforms the windowed '2L' waveform as a function of defocus distance z, for two different x positions along a scan. While we draw no conclusions from the details of these spectra, we simply show that with today's powerful and fast signal processing software it is possible to rapidly extract slices of time-, frequency- or position-information from multi-dimensional ultrasonic data arrays.

CONCLUSIONS

We have described in this paper the application of the broadband ultrasonic, lensless, line-focus transducer with large aperture and low *f*-number as a critical element in an ultrasonic microscope which can be used to detect and characterize fiber waviness in a thick composite plate. We have shown that the waveforms obtained when the focus of the transducer is near the front surface of the specimen are insensitive to the waviness. But when the transducer is focused on a region deep inside a specimen, the waviness of a composite specimen is detected and can be clearly imaged provided that the line of focus of the transducer is co-linear with the waviness of the specimen. We show that the collection of $v(z,t)$ data aids in an unambiguous identification of wave arrivals. Waveform acquisition with the line-focus transducer scanned along two simultaneous scan directions has also been shown. The resultant multi-dimensional ultrasonic data arrays can be processed to extract slices of ultrasonic waveform information as a function of time (or frequency) and position.

ACKNOWLEDGEMENTS

The measurements on the wavi-composite was supported by the Office of Naval Research under Contract #N00014–95–1–0429. We thank P. Petrina of Cornell University for fabricating the wavi-composite specimen. The work related to the design, fabrication and testing of the line focus transducers was carried out at the National Institute of Standards and Technology. The data shown in Fig. 5 was processed by S. Holland at Cornell University.

REFERENCES

1. J. Kushibiki and N. Chubachi, "Material characterization by line-focus-beam acoustic microscope", *IEEE Trans. Sonics and Ultras.*, **SU-32**, 189–212 (1985).
2. G. A. D. Briggs, *Acoustic Microscopy*, Clarendon Press, Oxford (1992).
3. D. R. Weglein, "Acoustic microscopy applied to SAW dispersion and film thickness measurement", *IEEE Trans. Sonics and Ultras.*, **SU-27**, 82–86 (1980).
4. K. K. Liang, G. S. Kino and B. T. Khuri-Yakub, "Material characterization by the inversion of *V(z)*", *IEEE Trans. Sonics and Ultras.*, **SU-32**, 213–224 (1985).
5. T. Mihara and M. Obata, "Elastic constant measurement by using line-focus-beam acoustic microscope", *Experimental Mechanics*, **32**, 817–821 (1987).
6. Y.-C. Lee, J. O. Kim and J. D. Achenbach, "Acoustic microscopy measurement of elastic constants and mass density", *IEEE Trans. Ultras. Ferroelect. Freq. Control*, **42**, 2534–264 (1995).
7. N. N. Hsu, D. Xiang, S. E. Fick and G. V. Blessing, "Time and polarization resolved ultrasonic measurements using a lensless, line-focus transducer", *1995 IEEE Ultrasonics Symposium*, IEEE, New York (1996), pp. 867–871.
8. D. Xiang, N. N. Hsu and G. V. Blessing, "The design, construction and application of a large aperture lens-less line-focus PVDF transducer", *Ultrasonics*, **34**, 641–647 (1996).
9. D. Xiang, N. N. Hsu and G. V. Blessing, "Materials characterization by a time-resolved and polarization-sensitive ultrasonic technique", in *Review of Progress in Quantitative Nondestructive Evaluation*, Vol. 15, D. O. Thompson and D. E. Chimenti, Eds., Plenum Press, New York (1996), pp. 1431–1438.
10. D. Xiang, N. N. Hsu and G. V. Blessing, "Ultrasonic evaluation of rough and porous ceramic coatings with a dual-element large aperture lensless line-focus transducer", in *Review of Progress in Quantitative Nondestructive Evaluation*, Vol. 16, D. O. Thompson and D. E. Chimenti, Eds., Plenum Press, New York (1997), pp. 1563–1570.

NEW X-RAY REFRACTOGRAPHY FOR NONDESTRUCTIVE EVALUATION OF ADVANCED MATERIALS

Manfred P.Hentschel, Derk Ekenhorst,
Karl-Wolfram Harbich, Axel Lange and Jörg Schors

Federal Institute for Materials Research and Testing
BAM-VIII.32
D-12200 Berlin, Germany

X-RAY REFRACTION

The Nondestructive Characterization of high performance composites, ceramics and other Advanced Materials can be difficult. Anisotropy, heterogeneity and complex shapes reduce the performance of traditional nondestructive techniques, which have been optimized for isotropic single phase materials, preferably for metals.

The effect of X-ray refraction provides unconventional small angle X-ray scattering (SAXS) techniques which have been developed and applied to meet the actual demand for improved nondestructive characterization of Advanced Materials. X-ray refraction reveals the inner surface and interface concentrations of nanometer dimensions due to the short X-ray wavelength near 10^{-4} µm. Sub-micron particle, crack and pore sizes are easily determined by "X-ray Refractometry" without destroying the structure by cutting or polishing for microscopic techniques.

Beyond this analytical potential for (integral) analysis, spatial resolution can be achieved, when the sample is scanned across a narrow X-ray beam. This is possible within relatively short time, as the scattered intensity at very small angles of few minutes of arc is much higher than in conventional wide angle X-ray scattering (WAXS). In this case we have "X-ray Refraction Topography".

PHYSICS AND INSTRUMENTAL

The physics of X-ray refraction is quite similar to the well known refraction of light by optical lenses and prisms, which is governed by Snell`s law. However a major difference from optics is the deflection at very small angles, as the refractive index n of X-rays in matter is nearly one:[1]

Nondestructive Characterization of Material VIII
Edited by Robert E. Green Jr., Plenum Press, New York, 1998

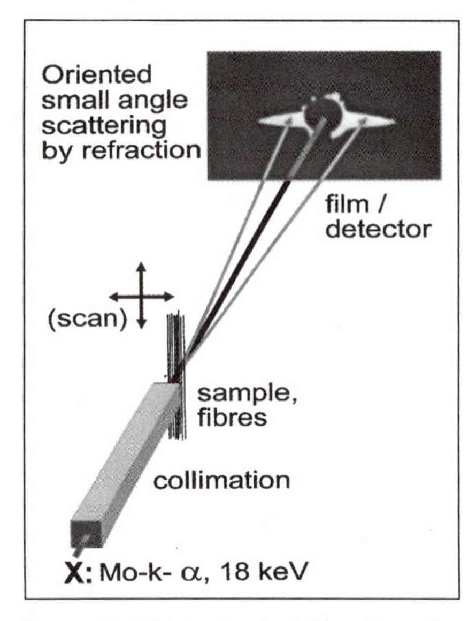

Figure 1. Effect of oriented small angle scattering by refraction of glass fibers

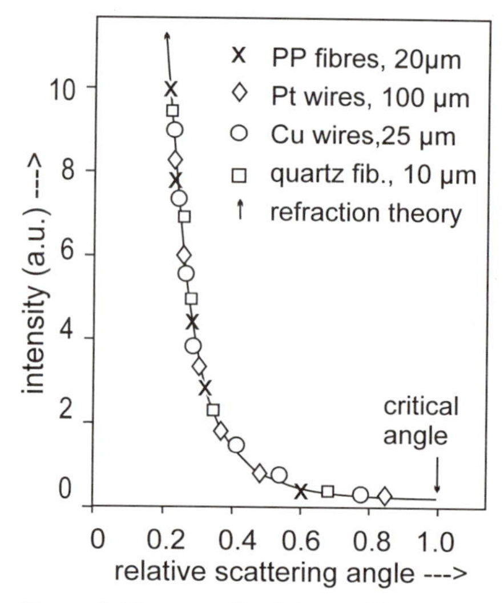

Figure 2. The normalized shape of the angular intensity distribution of cylindrical objects

$$n = 1 - \varepsilon \quad (\varepsilon \sim \rho * \lambda^2, \ \varepsilon \simeq 10^5 \text{ for glass/8keV radiation}) \tag{1}$$

ε is the real part of the complex index of refraction, ρ the electron density and λ the X-ray wavelength. With $n < 1$ the converging effect of convex lenses changes to divergence in case of X-rays. Figure 1 demonstrates the effect of small angle scattering by refraction of cylindrical lenses: A bundle of 15 μm glass fibers as used for composites (GFRP) deflects collimated parallel X-rays within several minutes of arc. The oriented intensity distribution is collected by X-ray film and the straight (primary) beam is omitted by a beam stop. Monochromatic radiation below 20 keV is applied like in crystallography, which is relatively soft for NDT purposes.

The shape of the intensity distribution of cylindrical objects is always the same even for very different materials, if the scattering angle is normalized to the "critical angle" θ_C of total reflection (Figure 2). This parameter depends only on the refractive index: $\theta^2_C = 2\varepsilon$.

Figure 3 demonstrates the X-ray deflection at circular objects (sections of fibers or spheres) by refraction and (very few) total reflection. In fibers and spherical particles the deflection of X-rays occurs twice, when entering and when leaving the object (Figure 3,

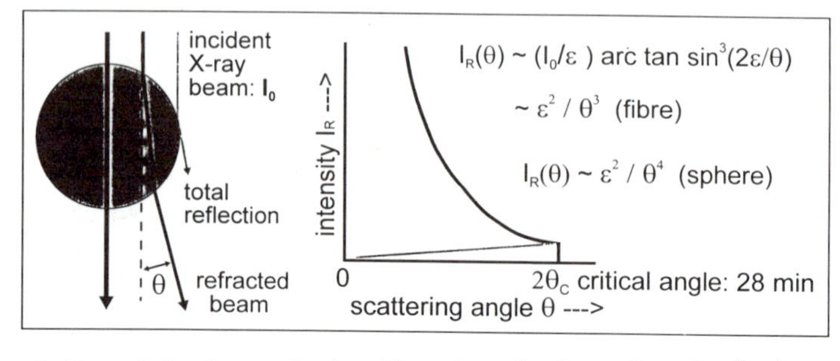

Figure 3. X-ray deflection at circular objects by refraction and total reflection; angular intensity distribution for fibers and spheres

left). The intensity of the deflected X-rays falls down near zero at the critical angle of total reflection (Figure 1, right). A cross section of 10^{-3} of the fiber diameter contributes to the detectable intensity above typically 2 minutes of arc. The effect of total reflection of X-rays occurs as well at the angle of grazing incidence but only 10^{-6} of the diameter is involved and therefore negligible. But well oriented planar surfaces can produce strong reflection Based on Snell's Law the angular intensity distribution has been calculated and approximated for cylindrical fibers and spheres, as illustrated by Figure 3. The refracted intensity of a cylinder without absorption effects is[2]

$$I^*_R(\theta) = \frac{I_0 2R}{\varepsilon} \sin^3\left(\arctan \frac{2\varepsilon}{\theta}\right) \propto \frac{I_0 R \varepsilon^2}{\theta^3} \tag{2}$$

R is the radius of the fiber, I_0 the intensity of the incident X-ray, ε the real part of the complex refractive index n and θ is the scattering.

The SAXS instrumentation is relatively simple but sometimes delicate in (thermo-) mechanical stability (Figure 4). It requires a fine structure X-ray generator, a (commercial) small angle X-ray camera (collimator C_S), X-ray refraction detector D_R and a sample manipulator S_C. An absorption detector D_A looks at the scattering foil F_S in order to monitor sample absorption and beam stability. For practical measurements the refraction slit remains at a fixed scattering θ. The additional refraction intensity of a refracting object can be measured according to[3]

$$I_R^*(\theta) = I_R(\theta) - I_A(\theta) = C\, d\, I_A(\theta), \text{ with } C = k\, S/V \tag{3}$$

The refraction factor C is proportional to the inner surface density (surface/volume ratio) S/V and the sample thickness is d. $I_A(\theta)$ is the scattering background under sample absorption and k is a constant of the apparatus, determined by a reference sample of known specific surface..

The conventional understanding of "continuous" small angle X-ray scattering (SAXS) is governed by the interpretation of diffraction effects. Apart from Guinier's theory for separated particles Porod[4] explains diffraction of densely packed colloids similar to Eq. (3). However both deal with particles two orders of magnitude smaller. A simple proof for

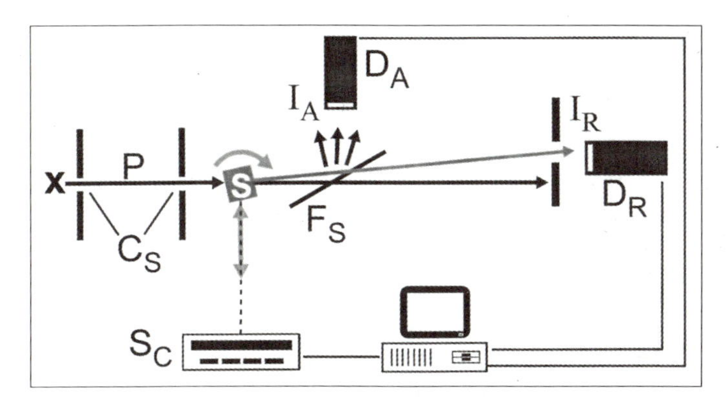

Figure 4. SAXS instrumentation: primary beam P, collimator C_S, sample S, scattering foil F_S, X-ray refraction detector D_R, absorption detector D_A and a sample manipulator S_C.

the refraction effect at large objects can be found by scanning a fiber through a narrow X-ray beam and collecting the intensity at each position (Figure 5). Even focussing by pores is possible. The behavior is exactly the same as in an optical experiment. Any diffraction effect would result in a symmetric intensity above background level.

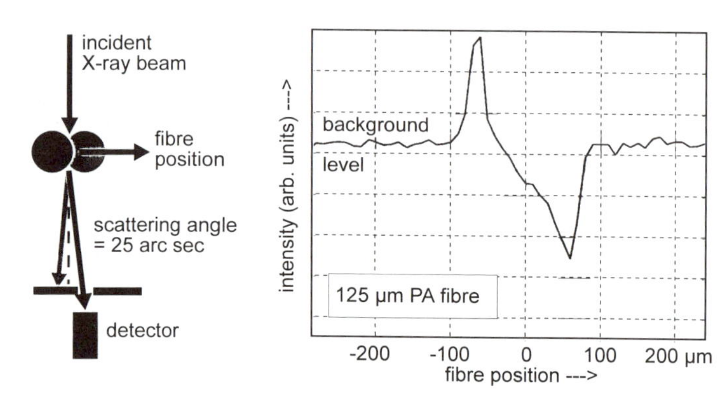

Figure 5. X-ray refraction by a 125 µm polymer fiber is demonstrated by scanning a fiber through a narrow X-ray beam and collecting the intensity at each position.

ANALYTICAL APPLICATIONS

Average specific surfaces can be determined by "stationary" X-ray Refractometry. Such kind of analytical investigations can be useful in the field of new materials, when cutting or polishing has to be omitted. In Figure 6 the specific surface densities of selected nonmetallic materials are compared. The values are relatively small in case of very "porous" materials as the pores are very large, which reduces the surface to volume ratio. The plotted surface values are taken from the refraction factors C, corrected for the different densities of the materials: C/ρ^2 (see Eqs. (1), (2)). In case of composites like paper and carbon fiber reinforced plastics (CFRP) the refraction value C is a composition of the refraction at inner surfaces of each component and at the interfaces. In case of the CFRP the po-

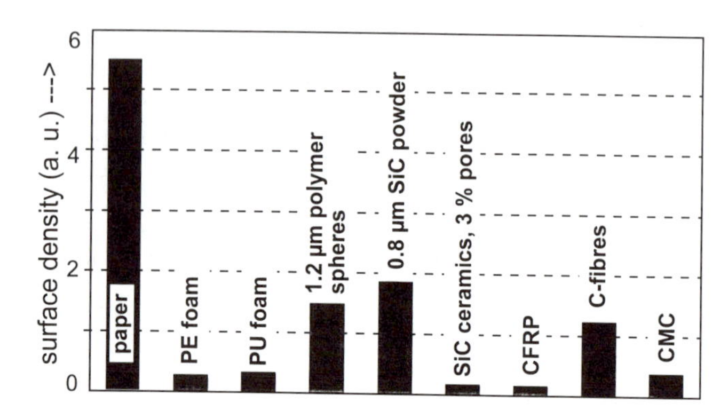

Figure 6. Comparison of inner surface densities of selected nonmetallic materials

rosity is very low, therefor only the interface contributes to the signal. The measuring time is usually a few seconds or less, if 1% error is accepted. The short measurement time allows scanning for spatial resolution or statistical evaluations.

The determination of the pore size distribution in glass ceramics by X-ray refraction results in diameters which correspond to the chordlength distribution in microscopic analysis. The mean values of the diameters are identical within ±3%. The measurements are performed with Mo-k-α-radiation at different positions of a 1,4 mm ceramics plate, sintered at 850 °C (Figure 7).[5] Further pore size measurements on SiC and Al_2O_3 ceramics by X-ray refractometry reveal good agreement with other techniques, especially with high pressure Mercury intrusion.

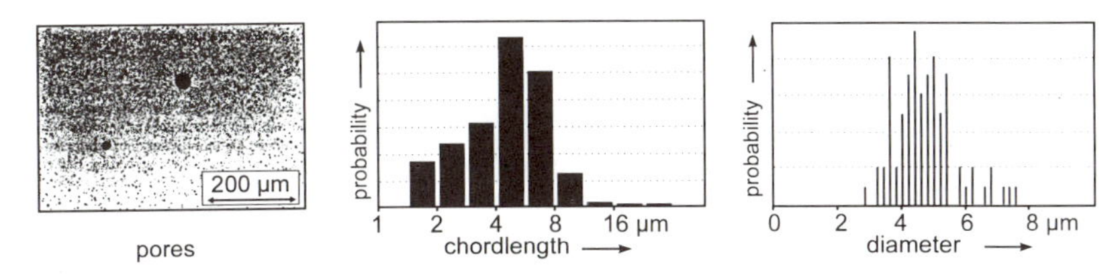

Figure 7. Pores in glass ceramics: micrograph, optical chordlength analysis of pores and pore diameter probability by X-ray Refraction

The measurement of the crack density in light weight materials can be performed by X-ray refractometry as well. The knowledge of the crack development is believed to play the key role in all long-term material behavior. Figure 8 correlates the residual shear strength of CFRP to the average inner surface of cracks created by aging treatment at 150 °C, 180 °C, 200 °C up to 10,000 h. The investigation compares epoxy and BMI matrix systems for high temperature applications in supersonic aviation. Although BMI has a high strength at the beginning, it falls below epoxy at the end of the aging treatment. The results explain clearly the dependence of the shear strength on the crack density. The slope defines an aging module which can be regarded as a new materials parameter.[6]

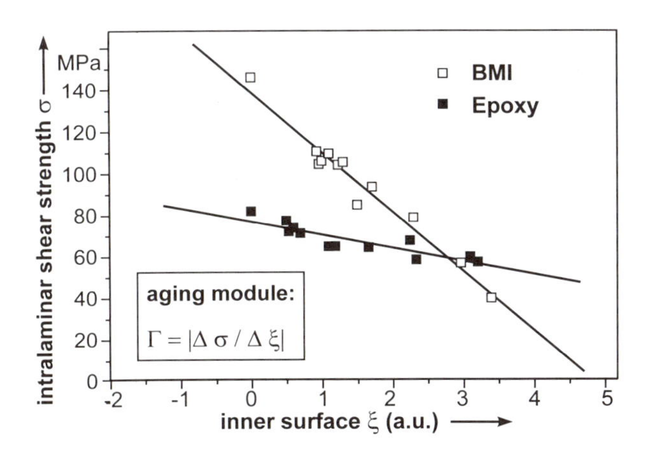

Figure 8. Correlation of the residual shear strength of CFRP to the average inner surface of cracks created by aging treatment at 150 °C, 180 °C, 200 °C for 1,000, 3,000, 5,000, 10,000 h.

Single fiber debonding in composites is not measurable - except by X-ray refractometry (although some attempts of pulling off individual fibers under the microscope have high artistic value). It is a central parameter of composites characterization. The basic principle can be understood by the optical analogue: compare the focussing properties of a lens (fiber) in air and in a liquid (fiber in matrix)! The refraction effect is lower in the second case. The density difference between fiber and matrix determines the X-ray scattering effect as well.

A model composite has been made in order to demonstrate the refraction behavior of a debonded and a bonded 140 μm sapphire fiber in wax matrix (Figure 9, left). The upper ray crosses the bonded fiber matrix interface causing a small amount of deflected intensity. At the debonded fiber and at the matrix surfaces (lower ray) much more X-rays are deflected, as the larger density difference between the materials and air corresponds to a higher index of refraction.

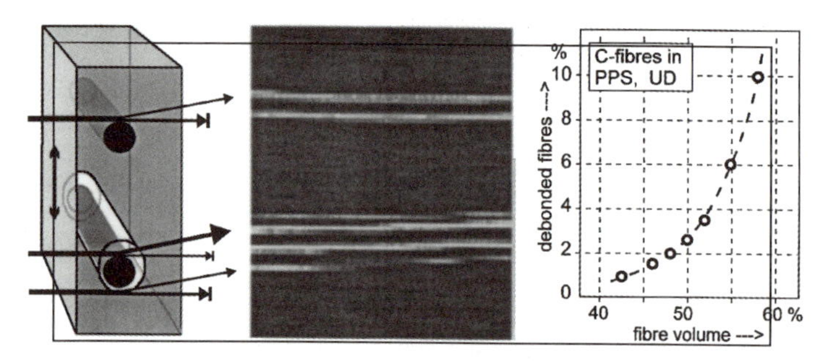

Figure 9. Model of X-ray refraction at interfaces of bonded and debonded fibers of a composite, X-ray topography of model, investigation of single fiber debonding at different fiber volume ratios.

The middle of Figure 9 shows the resulting intensity distribution of a refraction scan of the model composite. The wax channel is clearly separated from the fiber surface. The bonded fiber is much less contrasted. A practical measurement of the fraction of debonded fibers in a real thermoplastic C-fiber composite is given on the right. There is a nonlinear dependence of debonding on the fiber volume fraction. This can be explained by the very viscous thermoplastic matrix, which is hindered to penetrate between densely packed fibers during melt impregnation processing. Formulas for the calculation of individual or collective fiber debonding have been given.[3]

REFRACTOGRAPHY (REFRACTION TOPOGRAPHY)

Scanning X-ray refraction localizes the projection of inner surface concentrations or individual edges of surfaces and interfaces such as sub-micrometer pores or cracks. The spatial resolution can be better than 10 μm, although this is not the main advantage of refraction techniques, as the signal level itself contains the information about inner surfaces.

An example of Refraction Topography is given in Figure 10. It images the crack pattern in unidirectional CFRP after aging of 10,000 h at 180°C (epoxy) and 200°C (BMI). Although the average inner surface densities are the same (see Figure 8) the additional spatial information shows a difference in the type of crack distribution. The differences in shear strength can be understood by the different fractal behavior of cracks. The type of cracks is of course a mixture of cracks at the fiber /matrix interface (single fiber debonding) and matrix cracks. Figure 10 shows as well the directions of crack.

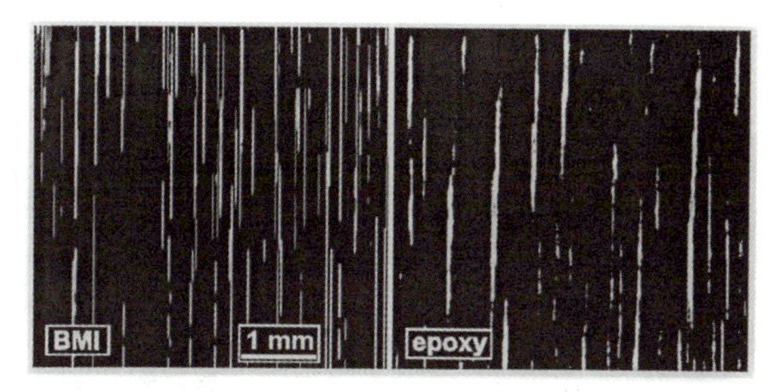

Figure 10. X-ray Refraction Topographs of crack patterns in CFRP after aging of 10,000 h at 180°C (epoxy) and 200°C (BMI).

Another problem of CFRP characterization relates to impact damages. Ultrasound C-scans resolve delaminations created by impact very well, but the single fiber debonding area, which develops at lower loads, is only detectable by X-ray refraction topography. In Figure 11 seven impact areas are imaged at 1 mm resolution. The reduction of details compared to Figure 10 is compensated by 100 times faster measurements (10 mm^2/s). (The three bright capitals are not impacted, simply pencil written [graphite scattering].)

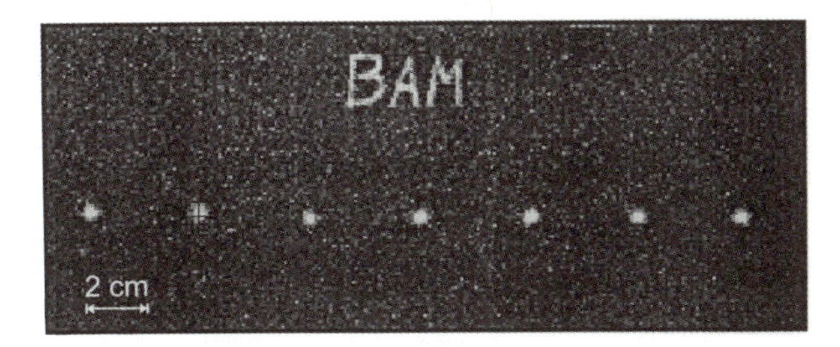

Figure 11. Large area X-ray Refraction Topography of seven impact areas at 1 mm resolution

X-RAY REFRACTION COMPUTER TOMOGRAPHY

Although two-dimensional Refraction Topography provides an effective new probe for analysing meso-structures of all kind of heterogeneous materials, it is sometimes interesting to have section images of transversal resolution as known from X-ray computer tomography in order to overcome the overlap of details by projection effects. Figure 12 demonstrates the feasibility of X-ray Refraction Computer Tomography: The sample micrograph shows a 3 by 3mm bar of phenolic resin CFRP laminate, which is a standard precursor in C/C and C/SiC CMC processing.

The computer tomography experiment is carried out by 18 keV single beam scanning in a Kratky camera according to Figure 4. Linear scans are performed for 360 angular positions, Fourier filtered for linear smearing on a PC and added up in an image file (filtered back projection). The reconstruction of detector signals I_A shows a quite homogeneous density of the conventional (absorption) computer tomographic image (Figure 12, center). The final refraction image reveals the spatial interface/inner surface distribution free of absorption effects. The typical layer and crack structure of the micrograph can be recognized by a nondestructive technique.

Figure 12. X-ray refraction Computer-Tomography of CFRP laminate: micrograph, left; conventional absorption tomography, middle; interface tomography, right.

X-ray refraction techniques combine analytical capabilities of sub-micrometer structure detection with the requirements of nondestructive full volume characterization. X-ray refraction therefore might help faster materials development, better understanding of meso structures and partly replace micro analysis and mechanical testing in advanced materials science.

REFERENCES

1. A. H. Compton, S.K. Allison. *X-ray in Theory and Experiment,* Macmillan and Co. Ltd., London (1935)
2. M.P. Hentschel, R. Hosemann, A. Lange, B. Uther, R. Brückner, Röntgenkleinwinkelbrechung an Metalldrähten, Glasfäden und hartelastischem Polypropylen, *Acta Cryst.* A 43:506 (1987)
3. M.P. Hentschel, K.-W. Harbich, A. Lange, Nondestructive evaluation of single fiber debonding in composites by X-ray refraction, *NDT&E International* 27:275 (1994)
4. G. Porod, Die Röntgenkleinwinkelstreuung von dichtgepackten kolloidalen Systemen, I. Teil, *Kolloid.-Z.* 124:83 (1951)
5. U. Mücke, K.-W. Harbich, T. Rabe, Determination of pore sizes in sintered ceramics using image analysis and X-ray refraction, *cfi/Ber.DKG* 74:95 (1997)
6. D. Ekenhorst, M.P. Hentschel, A. Lange, J. Schors, X-ray refraction: a new nondestructive evaluation method for analyzing the interface of composites in the nanometer range, in: *Proceedings of the Tenth International Conference on Composites Materials, Whistler, B.C., Canada, Aug.14-18th, 1995,* 5: 413, A. Poursartip, K. Street, eds., Woodhead Publishing Ltd., Cambridge (1995)

CCD-CAMERAS FOR X-RAY INVESTIGATIONS

F. Fandrich[1,2], R. Köhler[1], F. Jenichen[3]

[1] Humboldt-Universität Berlin
Institut für Physik, Arbeitgruppe Röntgenbeugung an Schichtsystemen
Hausvogteiplatz 5-7
D-10117 Berlin

[2] now c/o:
Universität Dortmund
Lehrstuhl für Qualitätswesen
D-44221 Dortmund

[3] Proscan GmbH
Kaspar-Kindl-Weg 10
D-86929 Penzing

ABSTRACT

In many NDT methods based on x-ray investigations the use of electronic x-ray detectors with lateral resolution ("sensor arrays") is helpful, delivering the result immediately after measurement as digitized data. CCD camera based systems are well suited to be x-ray detectors of high quality. This contribution focuses on so-called slow-scan CCD-camera systems used for measurements with high accuracy. The general setup of CCD camera systems and the pecularities for x-ray sensitivity are explained, followed by examples of the application of CCD-cameras for double crystal x-ray topography at 8 keV with high lateral resolution and low intensities.

INTRODUCTION

Several mechanisms of interaction between x-rays and matter are used for nondestructive characterization methods. Some very important ones are: absorption for radiographic applications, diffraction and scattering for investigations of crystal structures, phase, stress and surface analysis and fluorescence for elemental analysis and astrophysics.

For imaging and non-imaging methods, electronic imaging systems allow for a fast availability of digitized data or can even be applied for on-line techniques. For non-imaging methods, which deliver e. g. fluorescence spectra, scattering or diffraction patterns, imaging systems (often as a one-dimensional type) can be used as position sensitive detectors for the simultaneous registration of a larger part of the spectrum or pattern.

Suitable imaging detector system have to be designed according to the special experimental demands: energy of the x-rays, field of view, lateral resolution, dynamic range, frequency and accuracy of the measurements. Besides applications mainly focused to on-line control with no high requirements to resolution, sensitivity and accuracy, which are often based on standard (video) systems, this contribution shall only deal with so-called "slow-scan" CCD camera systems with high digitization accuracy (e. g. where the picture elements of the CCD array are used as localized small detectors). Standard and scientific CCD cameras can be made sensitive for x-rays in two manners: using a phosphor[1,2,3] or an open detector[4,5].

The main benefits of CCD cameras are a high lateral resolution and a linear intensity measurement; for a good resolution, however, the field of view is small. Compared with image plates[6], which have a much larger field of view, the CCD camera based systems provide an on-line read-out and a better resolution in the range of 10-20 μm.

PRINCIPLE OF OPERATION

Operation of a CCD-camera

Photons impinging the CCD ("charge-coupled device") produce electron-hole pairs in the silicon. A mean energy of about the three-fold energy gap is necesssary to produce one electron-hole-pair (about 3.6 eV in silicon)[7]. By use of an internal field at a p-n-junction the electrons and holes are separated, the electrons are collected and then transported by applying clock voltages along the path of transportation to the read-out node where the charge is converted to a voltage and pre-amplified. The operation of the CCD sensor and the read-out is controlled by a timing scheme.

During the operation of the slow-scan CCD-camera, there is an exposure period (when the read-out is stopped) and the read-out period, triggered by an external event or by a pacer clock (when the exposure should be stopped, if the exposure time is shorter or similar to the read-out time, to avoid artefacts by "smearing".)

To detect small signals the exposure time can be prolonged for an on-chip accumulation. For this kind of accumulation the dynamic range is limited by the sum of thermal and read-out noise in relation to the full-well capacity of the pixels. For a low thermal noise, the CCD can be cooled by a Peltier element or so-called MPP-mode CCDs (even operating at room temperature) should be used. A decreased read-out frequency reduces the read-out noise. The read-out frequency is limited by the AD-converter. Modern camera controllers allow for a 14 bits AD conversion with a frequency of 2 MHz, lower digitization depth can be obtained faster. If the picture elements of the CCD sensor shall be used as small

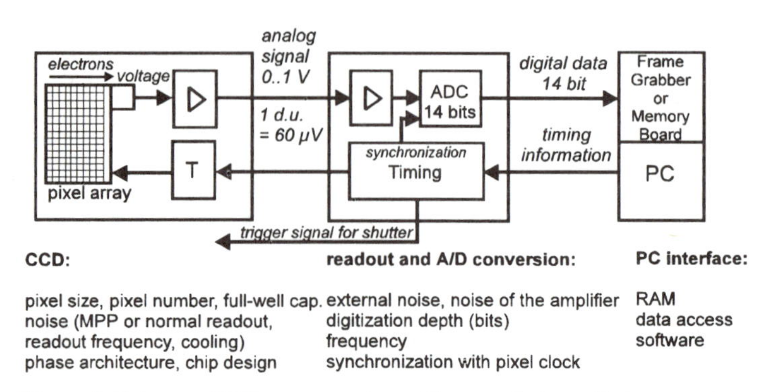

Figure 1. Setup o a CCD camera system with some design criteria

individual detectors in a laterally resolving detector, the AD-conversion has to be synchronized with the read-out ("pixel-synchronized"). For a high digitization depth one digital unit corresponds to a very small voltage step, to reduce additional noise by external elctromagnetic fields (e. g. x-ray generators operating at about 20 kHz), a well shielded cable should be used or the AD-converter should be as close as possible at the detector.

The digital data, which are much less sensitive to external electromagetic disturbance, are read via the ports or additional hardware into the PC (frame grabber or interface board supplied by the producer of the camera) (Figure 1).

X-ray Sensitivity Using a Phosphor Screen

For standard CCD sensors have a highest sensitivity in the spectral range of visible light, phosphor material[8,9,10,11] can be used to convert the incident radiation to that spectral range. The absorption, efficiency of conversion and emission spectrum is determined by the chemical composition and grain size of the phosphor (energy conversion efficiencies of few to about 15 %), the presence of an additional mirror coating and the deposition technique. With this method also radiation of high energy (> 20 keV) can be detected by increasing the thickness of the phosphor screen. Standard phosphor material consists of polycrystalline powder: a part of the light produced in the phosphor is scattered or absorbed before reaching the next optical element, limiting the resolution (which scales with the thickness) and efficiency. Therefore for low intensity measurements which make use of 8 keV (Cu-Kα) x-ray photon counting, an image intensifier is required, which is attached between the phosphor screen and CCD, Figure 2 (left). Modern image intensifiers often consist of one- or two-stage multi-channel plates (MCP), also limiting the resolution (see chapter "Setup and Performance"). In most cases the image intensifier can be used as a shutter.

The light-optical coupling can be maintained with high efficiency using fiber optics, which also allows (de-) magnification, depending on the field of view and resolution to be achieved, by use of so-called tapers. CCDs of 1024 x 1024 pixels have typically a size of 2 x 2 cm^2. A field of view in the order of 10 x 10 cm^2 can be achieved using a demagnifiying 5:1 taper optics. The resolution is then limited by the fiber diameter at the primary side, which can be calculated from the ratio of demagnification and the fiber diameter on the secondary side (the latter is typically about 5 μm). Fiber optics can induce an artifical pattern ("chicken wire"). Taper optics additionally induce a distortion of the image.

In summary this setup has a high flexibility. Based on the image intensifier, which can be uesd as shutter, the quantum efficiency (ratio of incident and detected x-ray quanta) is high. The CCD sensor will not suffer from radiation damage because of the preceeding phospor and glass. This setup, consisting of several components, however, impairs the image quality, increases the costs and requires trade-offs between lateral resolution, field of view and efficiency (selection of phosphor, image intensifier, and taper).

X-ray Sensitivity by Direct Exposure

When the glass window in front of the CCD is removed, incident x-ray photons can directly impinge the sensor and produce electron-hole pairs in the same mechanism as light photons, Figure 2 (right). The amount of charge, however, is much higher (which scales with the x-ray energy) and the absorption length is much larger compared with visible light (e. g. about 70 μm in silicon for 8 keV x-ray photons). For the standard CCD architecture is not prepared for this, not all charge[12,13] is collected, which results in a low quantum efficiency of few percent (for energies > 20 keV the use of standard CCDs becomes unreasonable). There are special CCD designs with a thicker sensitive layer ("depletion layer"), but these are special products at higher costs. Furthermore radiation damage[5,14,15] has to be expected especially for use in the intense polychromatic beam. Certain chip architectures ("burried-channel", "virtual-phase") are less sensitive for radiation damage. For this setup only mechanical shutters can be used. It is not possible to adapt the field of view or the sensitivity as for the indirect system. Advantages are that this one-component system is considerable cheaper and can provide a good resolution (where some numerical operations may be necessary) without distortions. The direct method is also suited for spectroscopic applications[16].

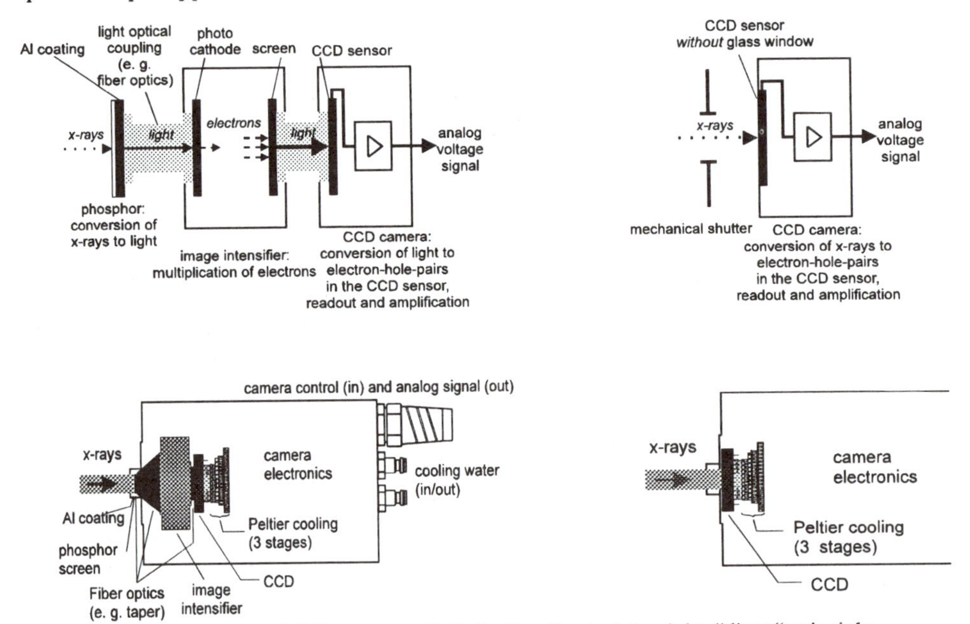

Figure 2. X-ray sensitivity of CCD-cameras. Left: "indirect" principle, right: "direct" principle

APPLICATION: X-RAY TOPOGRAPHY WITH CCD-CAMERAS

Experiment

For single-crystalline material, images based on the diffraction of collimated monochromatic x-rays (x-ray topography[17,18]) at the lattice planes show irregularities in the crystalline structure. A high sensitivity adapted to highly perfect material and a high lateral resolution of this technique is achieved by the so-called double-crystal setup, where the x-rays (8 keV, Cu-Kα) are collimated by a reflection at a preceeding nearly perfect crystal, which decreases the divergence to some seconds of arc. A lateral resolution down to few micrometers can be achieved, pecularities of the technique used here are described in [19]. Besides photographic emulsion, CCD-cameras have been used for this application. Figures 3 and 5 show x-ray topographs of epitaxial grown islands of silicon-germanium on silicon and of silicon-germanium bulk material.

CCD-Camera Setup and Performance

Indirect system:

- CCD with 1024 x 1024 pixels, 19 x 19 μm^2, 3-stage Peltier cooling
- 10 μm thick P43 (Gd_2O_2S:Tb) phosphor with thin aluminum coating
- one-stage MCP image intensifier with figer-optical windows (35 lp/mm @ 3 % MTF)
- fiber optical coupling with magnification of 2.1:1 to achieve a better resolution.

Direct system :

- 1317 x 1035 pixels (1024^2 used), 6.8 x 6.8 μm^2, MPP-mode, cooling not necessary
- mechanical shutter (50 µm steel blade)

The data of the 14 bits AD-conversion at 1 MHz maximum have been read by an advanced frame grabber. For the accumulation of the x-ray images a numerical procedure based on photon counting has been used, which is essential for a quantitative measurement with high lateral resolution. With this procedure and above described setup, the following performance has been achieved: The indirect system has a much better quantum efficiency of about 80 %, a field of view of 8 x 8 mm^2 and a resolution of better than 20 µm. The resolution of the direct system is about 7 µm (cf. Figure 4) with a quantum efficiency of 11 % and a field of view of 7 x 7 mm^2.

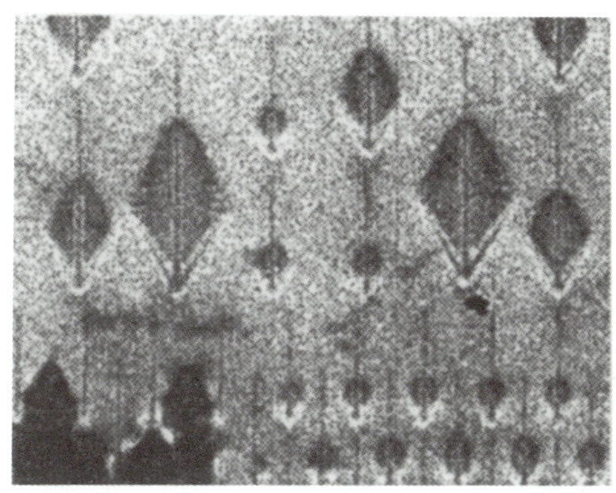
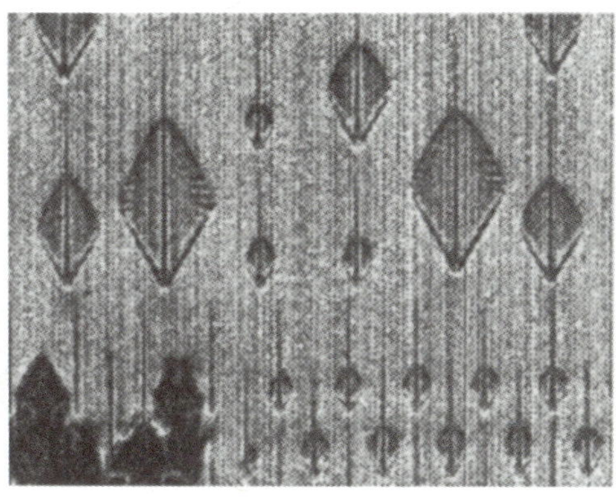
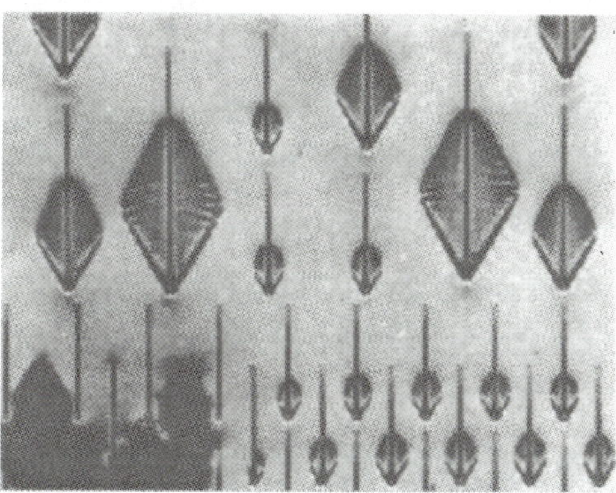

Figure 3.
X-ray topograph of GeSi islands on Si (1 1 1) with nominal 0.8 % germanium, grown by liquid-phase epitaxy (samples provided by MPI Stuttgart),

double-crystal setup, Cu-Kα_1 , (4 4 0) reflection

upper left: indirect camera system (1:15 h)
upper right: direct camera system (4 h)

left: nuclear plate (high-resolution photographic emulsion on glass) (3,5 h)

1 mm

SUMMARY

CCD camera based detectors are suitable electronic x-ray detectors. The setup of slow-scan CCD camera systems for quantitative measurements has been explained and several design criteria have been given. The systems described have been optimized for high lateral resolution and quantitative measurements using a numerical accumulation. For other experimental demands the setup or accumulation procedure should be adapted.

The indirect system is rather flexible, has a high quantum efficiency and the image intensifier can be used as a shutter, but reqires a number of trade-offs when selecting the components. The invariable direct system is available at lower costs, delivers a better image quality (here a numerical accumulation was used) but at a much lower quantum efficiency. It can suffer from radiation damage, a mechanical shutter could be necessary.

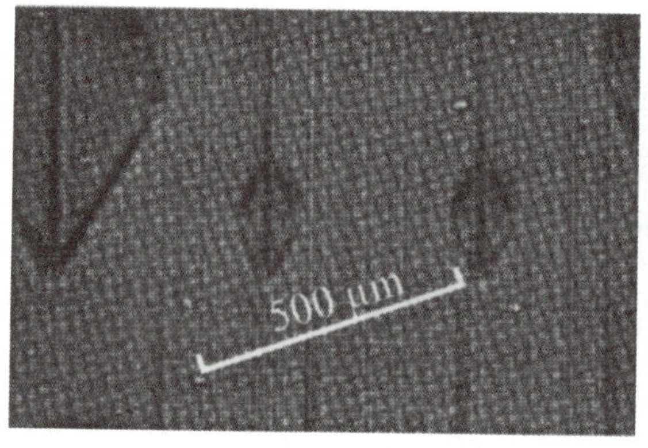

Figure 4. Cut of a topograph with a grid (mesh 1000: 25 μm period, 7.4 μm width of bridges) in front of the camera, sub-pixel algorithm used, direct system, 16 h

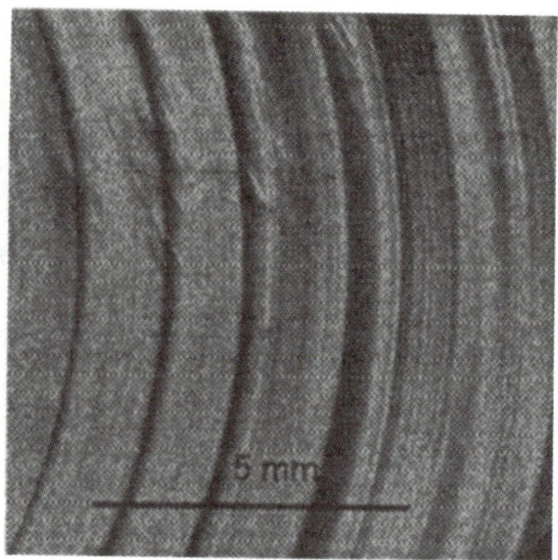

Figure 5. X-ray topograph of GeSi (1 1 1) bulk material with nominal 5.3% germanium, double crystal setup, (4 4 0) reflection (sample provided by IKZ Berlin), cut of a sample revealing inhomogenities in the germanium percentage (concentric "striations"), indirect system, 1 h

ACKNOWLEDGEMENTS

This project was supported by the former German Bundesministerium für Forschung und Technologie (BMFT) under contract number 01 M 2929. The samples have been provided by the epitaxy group of the Max-Planck-Institute für Festkörperforschung, Stuttgart, E. Bauser and A. Gutjahr, and the Institut für Kristallzüchtung, Berlin, which is gratefully acknowledged.

REFERENCES

1. E. F. Eikenberry, M. W. Tate, A. L. Belmonte, J. L. Lowrance, D. Bilderback, and S. M. Gruner, *IEEE Trans. Nucl. Sci.* 38:110 (1991)
2. A. Koch, Nucl. Instr. Meth. *Phys. Res. A* 348:654 (1994)
3. D. O'Mara, W. Phillips, M. Stanton, D. Saroff, I. Naday, E. M. Westbrook, *Proc. SPIE* 1656:450 (1992)
4. H. Tsunemi, S. Kawai, K. Hayashida, *Jap. J. Appl. Phys.* 30:1299 (1991)
5. R. Clarke, *Nucl. Instr. Meth. Phys. Res. A* 347:529 (1994)
6. Y. Amemiya, *J. Synchrotron Rad.* 2:13 (1995)
7. G. F. Knoll, *Radiation Detection and Measurement* (1989)
8. S. M. Gruner, S. L. Barna, M. E. Wall, M. W. Tate, and E. F. Eikenberry, *Proc. SPIE* 2009:98 (1993)
9. J. H. Chappel and S. S. Murray, *Nucl. Instr. Meth. Phys. Res.* 221.159 (1984)
10. J. P. Moy, A. Koch, and M. B. Nielsen, *Nucl. Instr. Meth. Phys. Res. A* 326:581 (1993)
11. M. Gurvic, *Röntgenleuchtstoffe und Röntgenlumineszenzbildwandler*, Akademische Verlagsgesellschaft Geest & Portig K.-G., Leipzig (1988)
12. D. H. Lumb, E. G. Chowanietz, and A. Wells, *Opt. Eng.* 26:766 (1987)
13. G. R. Hopkinson, *Opt. Eng.* 26:283 (1987)
14. B. G. Maggorian and N. M. Allinson, *Nucl. Instr. Meth. Phys. Res. A* 273:599 (1988)
15. N. M. Allinson, D. W. E. Allsopp, J. A. Quayle, and B. G. Magorrian, *Nucl. Instr. Meth. Phys. Res. A* 310:267 (1991)
16. G. A. Luppino, N. M. Ceglio, J. P. Doty, G. R. Ricker, and J. V. Vallerga, *Opt. Eng.* 26:283 (1987)
17. A. R. Lang in "International Tables of Crystallography", Vol. C, Ed. A. J. C. Wilson, Kluwer Acad. Publ., Dordrecht/Boston/London 81992), p. 113-123
18. R. Köhler, *Appl. Phys. A* 58:149 (1994)
19. B. Jenichen, R. Köhler, W. Möhling, *J. Phys. E: Sci. Instrum.* 21:1062 (1988)

DEVELOPMENT OF X-RAY DIFFRACTION METHODS TO EXAMINE SINGLE CRYSTAL TURBINE BLADES

Kirsten G. Lipetzky and Robert E. Green, Jr.

Center for Nondestructive Evaluation
The Johns Hopkins University
Baltimore, Maryland 21218

Paul J. Zombo

Westinghouse Electric Corporation
Turbine Materials/Power Generation Technology Division
Orlando, Florida 32826

INTRODUCTION

Technological advances in recent years have led to the development of single crystal, nickel based alloys for turbine blade applications. Among the problems encountered with single crystal turbine blades is the determination of the overall crystalline perfection of the final blades. The existing method relies upon chemical etching and visual inspection. While the problems associated with visual inspection are self evident, there are intricacies associated with the etch process. Following visual inspection of the etched blades, those which meet with approval are then "unetched." That is, a mechanical peening process is used to remove the shiny finish produced from the etch. The final step in the crystal perfection/orientation process for single crystal turbine blade inspection is to determine the crystallographic orientation of a "good" blade by performing Laue x-ray diffraction, in the back-reflection configuration, at one point on the blade. It is interesting to note that none of the inspection processes above can determine the crystalline perfection of the interior of the blades; as will be discussed below, this is a legitimate concern.

The present research focuses on nondestructive inspection techniques, based on x-ray diffraction, which would eliminate the need for chemical etching and visual examination. One such technique which has been used successfully in previous research[1-2] is termed asymmetric crystal topography (ACT); it is an x-ray diffraction based method which permits imaging of a large portion of a single crystal turbine blade at one time. Incorporation of an x-ray sensitive electro-optical detector permits this to be done in real-time. This paper describes the details of the ACT system, shows several topographic images illustrating the utility of the system, and discusses how this technique and other conventional x-ray techniques have been applied to the inspection of nickel based alloy, single crystal specimens.

EXPERIMENTAL PROCEDURE

The term topography literally means "to describe a place" (topos = place, graphein = to write) and the x-ray diffraction information which is obtained by asymmetric crystal topography may be bulk (transmission) or surface (back-reflection) in nature. The topographic technique is based on Bragg diffraction from a periodic crystal and is extremely sensitive to imperfections and strains in the crystal, since any alteration to the interplanar spacing of the crystal will effect a corresponding change in the Bragg condition. In the ACT set-up used in this investigation, a slit collimated, white radiation x-ray source was incident upon a high quality asymmetrically cut silicon crystal. This crystal served as both a monochromator and a beam expander, resulting in a x-ray beam of approximate dimensions 1 ½ inches high by ½ inch wide. Specimens of interest were placed in the path of this monochromated and expanded x-ray beam and diffraction information was detected using an image intensifier with fluorescent screen faceplate. Due to the relatively weak x-ray source available in the laboratory (copper target tube operated at 50 kV, 32 mA), as well as to the thickness of the specimens examined, the back-reflection mode was utilized in this research. A schematic of the ACT system showing positions of the highly perfect, asymmetrically cut silicon first crystal (monochromator and beam expander), the second crystal (specimen under investigation), and the x-ray image intensifier (for direct real-time viewing of topographic images) is shown in Figure 1.

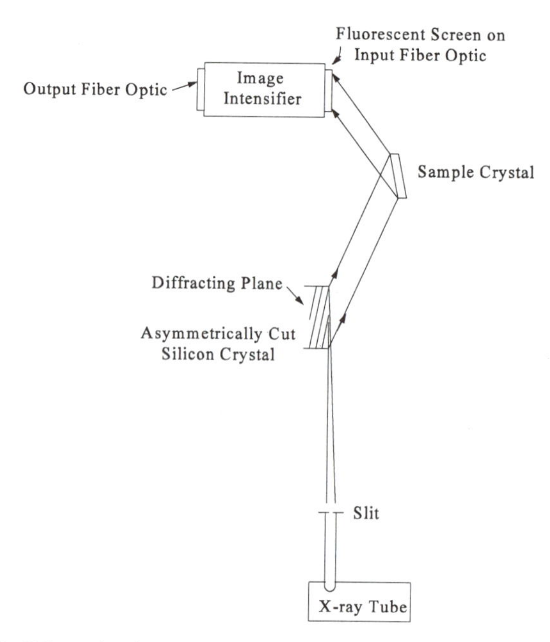

Figure 1. Schematic of asymmetric crystal topography system in back-reflection mode.

In the ACT technique each individual topographic image is essentially a large Laue "spot" generated by diffraction from a particular set of "parallel" lattice planes covering a large area of the crystal. The x-ray beam incident on the specimen illuminates a large area unlike conventional Laue pin-hole techniques and, because of the special beam expanding monochromizing silicon crystal, this large incident beam experiences minimal divergence.

The desirability of obtaining x-ray diffraction images with extremely short exposure times has long been realized. Several review papers have been presented which describe real-time imaging of x-ray topographs.[3-4] The method used in this research is called the indirect method in that the x-ray topographic image is converted into a visible light image by a fluorescent screen. This visible light image travels through the coherent fiber optic input faceplate of a low-light-level image intensifier tube and is converted into an electron image at the photocathode. The electrons constituting this image are accelerated by an electrostatic field and focused by a conical electron lens onto the output phosphor screen where they are converted into a stronger visible image. By cascading individual image tube stages using coherent fiber optic coupling, light gains as high as several million can be obtained. The current system possesses enough gain to permit viewing of individual photons. The visible output image from the last stage is sufficiently bright so that is can be viewed directly, recorded on film, or displayed on a closed circuit television monitor and videotaped. The topographs obtained from nickel based alloy, single crystal turbine blades were first viewed in real-time at the output of the image intensifier tube, recorded on video-tape, or permanently recorded on a sheet of dental x-ray film placed in a holder attached in front of the image intensifier tube imaging detector. It should be pointed out, however, that crystallographic microstructural misorientation of portions of the turbine blades were clearly visible in real-time at the output of the image tube.

EXPERIMENTAL RESULTS

Figure 2 shows a schematic of the ACT experimental arrangement used for inspection of a single crystal, nickel based alloy turbine blade. Figure 3 shows ACT images obtained from different regions along the airfoil of the turbine blade. In Figure 3 (a) the platform of the turbine blade is partially imaged (right hand side of the figure). While in Figure 3 (b) an elliptical region containing no diffraction information is apparent along the right hand side of the topographic image. This ellipse occurred because the incident x-ray beam was blocked by the platform section of the turbine blade and as a result, diffraction could not take place along the airfoil in this region. In both topographic images, (a) and (b), substantial subgrain structure can be observed and is indicative of the relatively poor quality of the crystal. In a "perfect" crystal, the contrast arising from diffraction would be uniform in intensity.

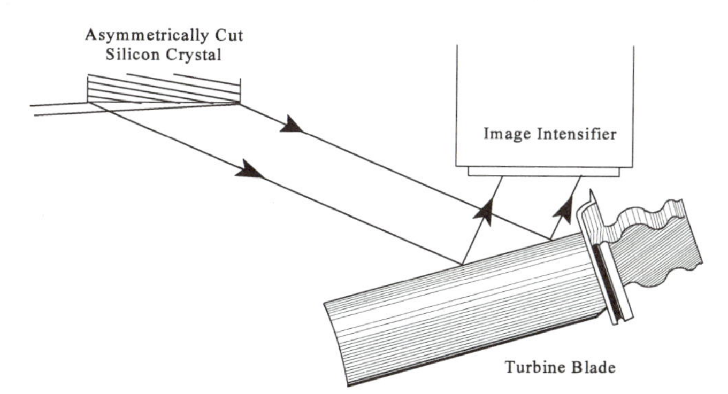

Figure 2. Schematic of ACT experimental arrangement used for inspection of a single crystal, nickel based alloy turbine blade.

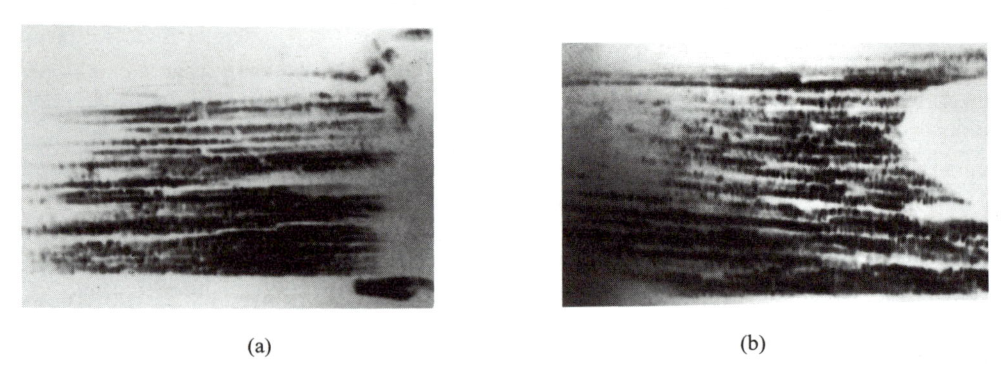

Figure 3. ACT images of different regions along the airfoil of a single crystal, nickel based alloy turbine blade.

Visual inspection of a different etched single crystal turbine blade sample revealed that it in fact contained a secondary crystal on the surface of the blade. As described above, the ACT system was used to examine this sample. Figure 4 (a) shows a schematic drawing of the turbine blade sample with secondary crystal which was examined. The "bowtie" which appears on the sample was made from a piece of fluorescent tape and it was used as a fiduciary mark. The marker wasn't necessary for the x-ray topography study, but it was beneficial in subsequent testing where Laue x-ray diffraction pictures in the back-reflection configuration were obtained from the three regions indicated by black dots. Figures 4 (b), (c) and (d) are topographic images obtained from (b), a region on the primary crystal (main portion of the turbine blade) which surrounds the secondary (stray) crystal, from (c), the secondary (stray) crystal and from (d), a montage of the two topographs. Laue x-ray diffraction analysis revealed the misorientation of the secondary crystal with respect to the main crystal to be 8 degrees.

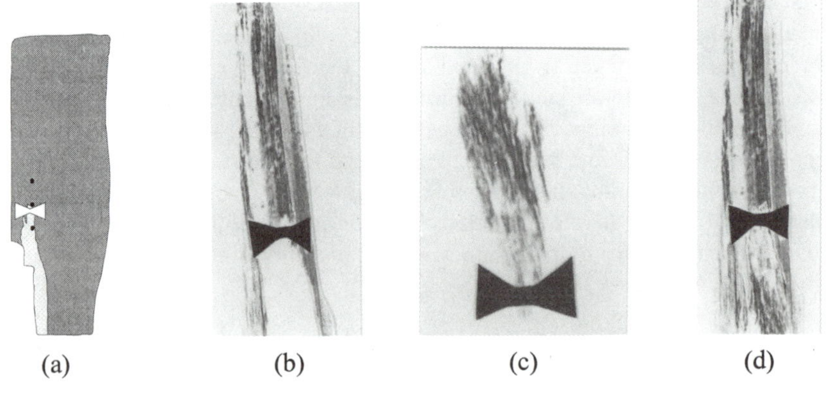

Figure 4. (a) Schematic of a turbine blade sample which contains a secondary crystal on the surface of the blade and resulting x-ray topographic images obtained from (b) a region on the primary crystal (main portion of the turbine blade) which surrounds the secondary (stray) crystal, from (c), the secondary (stray) crystal and from (d), a montage of the two topographs shown in (b) and (c).

In all of the cases described above, the turbine blade samples had undergone the etch process or the etch/peening process. In order to determine if the ACT technique might be applied to samples removed directly from the mold, a piece of single crystal, nickel based alloy gate (excess material from the casting process) was examined. Figure 5 (a) shows a schematic drawing of the gate piece which was inspected. Three different regions, labeled A, B and C were examined and in Figure 5 (b), (c) and (d) the topographic images obtained from these areas are shown respectively. Results clearly show that ACT can be performed on unetched single crystal, nickel based alloy parts.

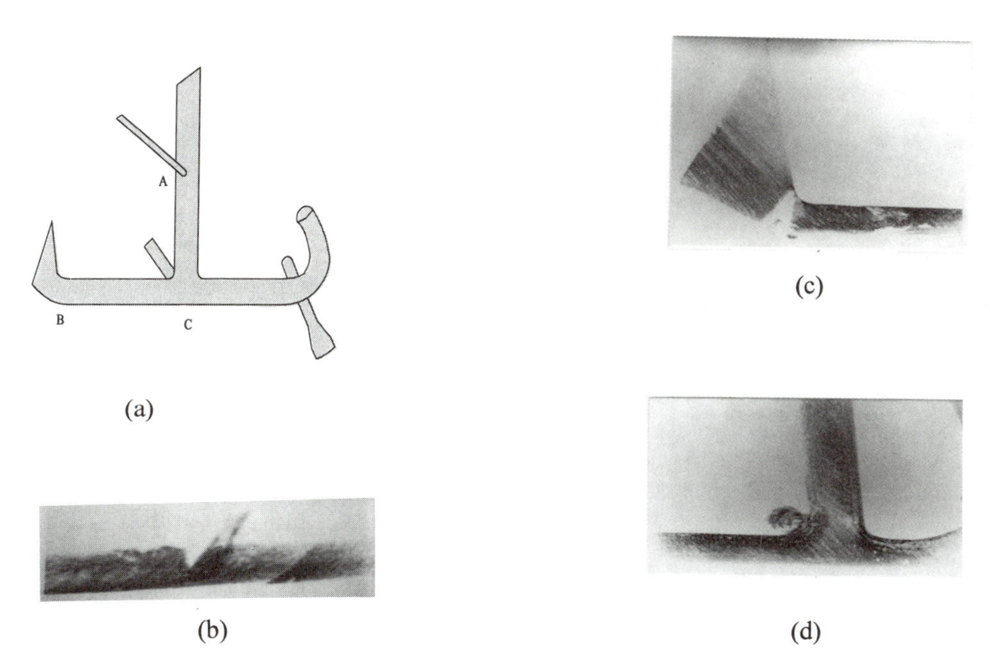

Figure 5. (a) Schematic drawing of an unetched piece of single crystal, nickel based alloy gate (excess material from the casting process) which was examined using ACT. X-ray topographic images which were obtained from the gate material at (b) position A, (c) position B, and (d) position C.

While past research conducted by the present investigators has focused on the exterior crystalline perfection of single crystal, nickel based alloy turbine blades, concern has also been expressed as to the possibility of secondary crystal growth occurring within the cooling chambers of the blade core. For this reason a sectioned turbine blade was analyzed to determine whether further attention should be given to this issue. Figure 6 (a) shows a schematic diagram of the sectioned single crystal, nickel based alloy turbine blade which was examined and Figure 6 (b) is a topographic image of a large portion of the sectioned turbine blade. As can be seen, the topographic image does not show the entire portion of the blade on which the incident x-ray beam impinged. This is because the entire blade is not a single crystal. In fact, by changing the Bragg angle, the misoriented portion of the turbine blade can be made to appear, as is shown in Figure 6 (c). The subgrain structure, which was observed in the topographic images of Figures 3, 4 and 5 is once again present. It should be noted that the size of these subgrains is approximately the same size as the area of the blade that would be covered by an incident x-ray beam when using a conventional Laue pin-hole collimating system (another technique which is commonly used for determination of crystalline orientation/perfection of turbine blades). This vividly illustrates the short-comings of pin-hole collimating systems for inspection of turbine blades. Furthermore, it is evident from this example, that the potential for secondary crystal growth occurring within the cooling chambers of the blade core is a legitimate concern.

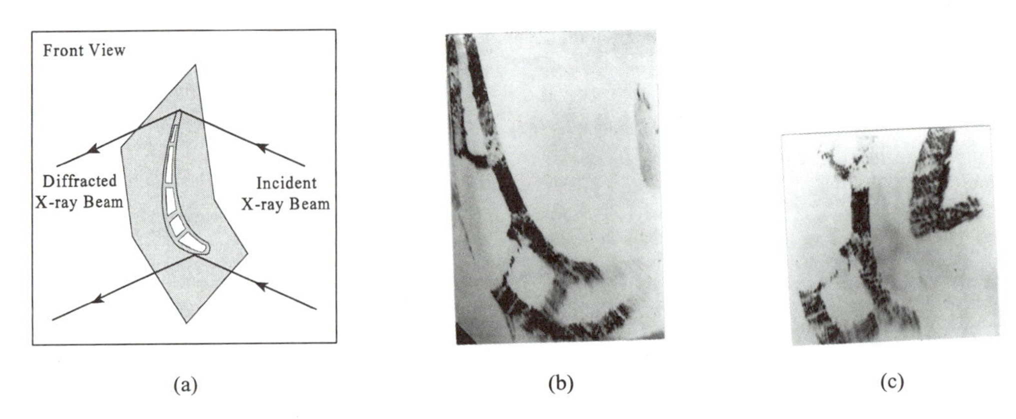

Figure 6. Schematic of a sectioned single crystal, nickel based alloy turbine blade (a) and corresponding x-ray topographic images (b) and (c) which were obtained using different Bragg angles for diffraction.

Since it is impossible to section turbine blades which are to be placed in service, a technique needs to be developed that would enable an inspector to analyze the internal crystallographic orientation of a turbine blade. The obvious approach to this problem is to use transmission x-ray diffraction techniques; in practice, however, this is not a trivial matter. The nickel based alloy, utilized in the production of single crystal turbine blades, tends to be a good absorber of the soft x-rays used in diffraction methods. In order to best evaluate different transmission diffraction techniques which might be utilized and to determine the limitations of each of these techniques, a panel was made from the same nickel based alloy used in the production of single crystal turbine blades. The dimensions of the panel length and width was 4 inches square. The thickness of the panel was designed to consist of varying step heights. Furthermore, holes were drilled through the entire length of one of the 4 inch dimensions of the panel. A schematic of the panel appears in Figure 7. It should be mentioned that the panel was machined from a single crystal and as a consequence should be a single crystal as well, although this hasn't been fully evaluated to date. In addition to determining what thicknesses might be characterized using x-ray diffraction transmission techniques, it was also desirable to evaluate whether or not diffraction from a secondary crystal could be detected. For this reason both single crystal and polycrystalline rods were supplied which had dimensions slightly smaller than the drilled hole diameters.

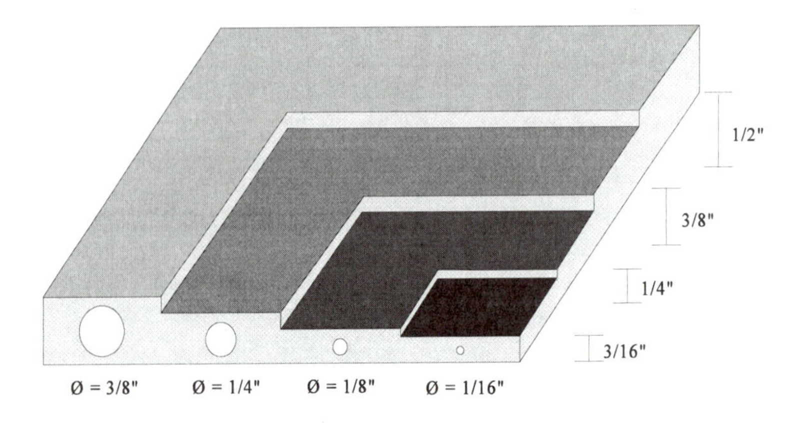

Figure 7. Schematic of the single crystal, nickel based alloy test panel.

Conventional methods of using Laue x-ray diffraction in the transmission mode (Cu target tube operated at 50 kV and 32 mA) proved unsuccessful for obtaining diffraction information from the single crystal, nickel based alloy test panel. Although not commonly practiced, it is possible to get diffraction information from a x-ray radiographic system and as a consequence it was decided that this route might hold the most promise for obtaining transmission x-ray diffraction information from the single crystal, nickel based alloy panel. A collimating system was machined out of lead and when placed in a Faxitron radiographic unit (W target tube, 115 kV, 3 mA capability) the normally conically divergent x-ray beam was "focused" to a line of x-rays of 3 inches by 1/16 inch dimension. The collimator and hence x-ray "line source" was placed in direct contact with the single crystal alloy test panel. Polaroid type 57 film was placed in a film cassette with a fluorescent screen and this was positioned in the radiographic unit on the side of the sample opposite the x-ray source at a distance of approximately 4 centimeters. Exposure times ranged from approximately 15 minutes up to 4 ½ hours for specimen thicknesses of 1/8 inch to 1/4 inch respectively. A typical x-ray diffraction pattern which was acquired from the stepped test panel is shown in Figure 8. In the particular diffraction pattern shown the nominal step thicknesses (the step dimension minus the drilled hole dimension) were, from left to right: 3/16 inch (1/4 inch - 1/16 inch), 1/8 inch (1/4 inch - 1/8 inch) and 1/8 inch (3/8 inch - 1/4 inch). Note that while the nominal thickness is the same for the two 1/8 inch regions, the region on the far right contains a greater amount of "thin" area and as a result the diffraction spots from this region appear larger.

Figure 8. Typical x-ray diffraction pattern obtained from the single crystal, nickel based alloy test panel. The pattern was obtained using a Faxitron radiographic unit (115 kV, 3 mA) which was line collimated (dimensions: 3 inches by 1/16 inch).

Because it had been established that it was possible to obtain diffraction information from the single crystal, nickel based alloy stepped test panel, it was next necessary to determine if diffraction from a secondary crystal could be detected. In order to do this, the line collimator was again used with the test panel under the same test conditions (115 kV, 3 mA). The only significant difference in the secondary crystal experiments was that the region of the test panel which was examined only consisted of one step and not three at a time as was done previously. (This was accomplished by using lead to downsize the length of the line collimator to 1 inch; compared to the 3 inch dimension described above.) Furthermore, diffraction information was first acquired from a given step while the drilled hole was empty. A second exposure was then obtained with the hole being filled by one of the single crystal rods. Figure 9 shows the resulting diffraction patterns acquired from such an experiment. In Figure 9 (a) the nominal thickness of the specimen is 1/8 inch (1/4 inch step - 1/8 inch empty hole). In Figure 9 (b) the smallest diameter single crystal rod (1/16 inch) was placed in the hole for a total sample thickness of 3/16 inch (1/8 inch + 1/16 inch). Exposure times were 30 minutes and 4 hours for diffraction pattern (a) and (b), respectively. Note the additional diffraction spots present in Figure 9 (b).

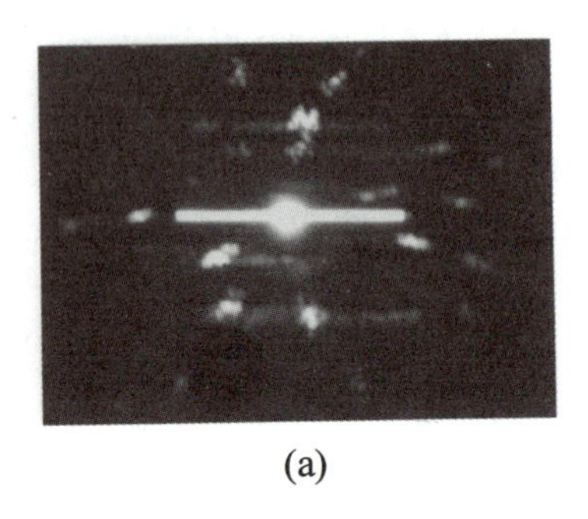

(a)

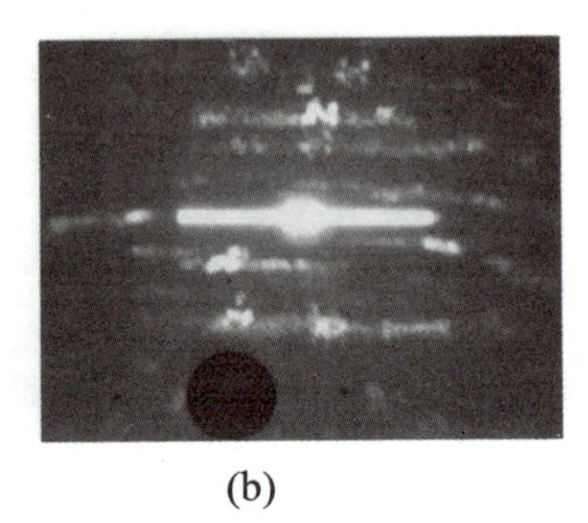

(b)

Figure 9. X-ray diffraction pattern obtained from the single crystal, nickel based alloy test panel; nominal specimen thickness is (a) 1/8 inch with an empty hole and (b) 3/16 inch with the hole filled with the smallest diameter (1/16 inch) single crystal rod. Note the additional diffraction spots present in (b).

CONCLUSIONS

Asymmetric crystal topography (ACT) in the reflection configuration can detect and image the presence of secondary crystals on the surface of "single" crystal turbine blades. Furthermore, when ACT is used in conjunction with back-reflection Laue x-ray diffraction, the misorientation of a secondary crystal can be precisely determined. Results obtained from unetched, single crystal nickel based alloy gate material (excess material from the single crystal casting process) demonstrates the potential application that ACT will be able to detect the presence of secondary crystals on turbine blades even in the unetched condition.

Transmission x-ray diffraction patterns have been obtained from a single crystal, nickel based alloy test panel up to a thickness of 1/4 inch using x-rays generated from a conventional radiographic unit (115 kV, 3 mA). Perhaps of greatest significance is the fact that it is possible to detect the presence of a secondary crystal in the transmission mode using x-rays generated from a conventional radiographic unit.

REFERENCES

1. Kirsten A. Green and Robert E. Green, Jr., "Application of x-ray topography to improved nondestructive inspection of single crystal turbine blades," *Proceedings of the 16th Symposium on NDE* (Southwest Research Institute, San Antonio, Texas), 13 (1987).
2. Kirsten A. Green and Robert E. Green, Jr., "Application of x-ray topography for nondestructive inspection of industrial materials," *Review in Progress in Quantitative Nondestructive Evaluation*, 13, 571 (1994).
3. R.E. Green, Jr., "Direct display of x-ray topographic images," in *Advances in X-ray Analysis*, 20, 221, H.F. McMurdie et al., eds., Plenum Press, New York (1977).
4. J.M. Winter and R.E. Green, "Rapid imaging of x-ray topographs," in *Applications of X-ray Topographic Methods to Materials Science*, 45, S. Weissmann, F. Balibar and J.-F. Petroff, eds., Plenum Press, New York (1984).

WHITE BEAM TRANSMISSION TOPOGRAPHY OF NICKEL-BASED ALLOY SINGLE CRYSTAL TURBINE BLADES USING SYNCHROTRON RADIATION

John M. Winter, Jr.[1], Robert E. Green, Jr.[1], and George Strabel[2]

[1]Center for Nondestructive Evaluation
The Johns Hopkins University
Baltimore, MD., 21218

[2]Howmet Corporation
Whitehall, MI, 49461

INTRODUCTION

X-ray diffraction in metals and alloys involves only low energy x-rays because of the typical interatomic spacings encountered in single crystals of these materials. Since these low energy diffracted x-rays are readily absorbed in alloys involving elements of higher atomic numbers, a diffraction imaging technique in transmission is not necessarily trivial for these materials. Several nickel-based alloy single crystal turbine blades have been examined using the x-ray diffraction technique known as "White Beam Transmission Topography", (WBTT), in combination with the extremely high flux of low energy x-rays available from the National Synchrotron Light Source at Brookhaven National Laboratory. The technique is described, and diffraction images are presented and discussed for several specific cases.

PRINCIPLES OF WHITE BEAM TRANSMISSION TOPOGRAPHY

White Beam Transmission Topography can be viewed as a variation of the classical Laue x-ray method used for orienting single crystals. In this classical method, an x-ray beam from a laboratory generator is collimated to about a millimeter diameter to achieve a beam with nominally parallel rays incident on the stationary single crystal specimen. Since this incident beam contains a continuous spectrum of all wavelengths up to perhaps 40 or 50 Kev, each set of lattice planes in the single crystal with its own specific interplanar spacing and its own specific orientation to the incident beam can select the single wavelength from the incident beam which will allow it to satisfy the Bragg condition for diffraction. For the Laue transmission geometry, the resulting diffracted beam will be directed at some specific off-axis angle downbeam from the single crystal. In this classical geometry, a sheet of film placed downbeam from the specimen and orthagonal to the axis of the incident beam will record the pattern of diffracted Laue spots. Generally, the diffracted Laue spot is kept small, thus enhancing the precision of crystallographic orientation deduced from their pattern.

If the film and the specimen were translated synchronously together across the incident x-ray beam, then a continuous series of contiguous volume elements in the specimen would be probed by the beam, and a corresponding series of adjacent and overlapping Laue spots would be recorded on the film. In this way, the original Laue spot could be expanded into a Laue image. Imperfections such as missing atoms in the lattice planes would cause density differences at various locations in the image, and in that way provide a contrast mechanism for displaying shades of gray to represent various defects in the perfection of the lattice array. This resulting Laue image is properly termed a topograph taken in transmission. It could be generated by scanning the single crystal specimen and the film synchronously across the incident beam, as described above, or by using a beam of much larger cross-section, or by both. Fortunately, a beam of larger cross-section which still is composed of essentially parallel rays is available at the National Synchrotron Light Source located at Brookhaven National Laboratory at Upton, New York. The work described here was conducted at Beamline X19C at that facility. It has a beam cross-section about 1.0" wide and about 0.25" high. The source generates a continuous spectrum (similar to a Bremsstralung spectrum) with a high energy cut-off at 20 kev. It is distinguished by a flux which is about 10^5 times larger than that of a conventional laboratory generator. This feature enables transmission topographs through specimens ordinarily too highly absorbing of x-rays to allow such studies with conventional laboratory generators.

TOPOGRAPHS OF SINGLE CRYSTAL TURBINE BLADES

Figure 1 is a white beam transmission topograph of a single crystal turbine blade grown from a nickel-based superalloy. The blade was oriented with its long axis horizontal and in the same plane as the width direction of the incident x-ray beam. The incident beam illuminated the concave side of the blade. The axis of the incident beam made approximately a 45 degree angle with the long axis of the blade, (with both axes in the same horizontal plane). The topograph was recorded on a 4"x5" sheet of Polaroid Type 57 film placed perpendicular to the axis of the incident beam and 7.5" downbeam from the blade. This film is not particularly sensitive to x-rays, and no intensifying screen was used to enhance its sensitivity. Other film types with larger dynamic range (number of shades of gray) and finer spatial resolution could be expected to yield images with more detail. Relative x-ray doses acquired by the film were tracked by arbitrary units which were determined by the product of exposure time and the nominal electron beam current in orbit in the synchrotron at the time of exposure. This beam current is proportional to the x-ray flux generated at any particular beamline at any given time. Figure 1 was the result of a 15 second exposure at a time when the synchrotron beam current was 216 ma. Hence the relative dose for fig. 1 is 54 ma-minutes. The rectangular image slightly above center (with the small black notches on its lower edge) is a radiograph of the illuminated section of the blade. It is caused by the residual non-diffracted transmitted beam. The other four images are Laue images generated by specific and different sets of crystallographic planes in the single crystal. Each of these images contains a unique record of the defects in its own particular set of lattice planes. Type 57 film lacks the dynamic range to properly resolve these details, but provides a suitable initial proof-of-feasibility.

Figure 2 is another topograph of the same blade in the same orientation with respect to the incident beam, except the film was 24" from the blade and positioned to catch the Laue

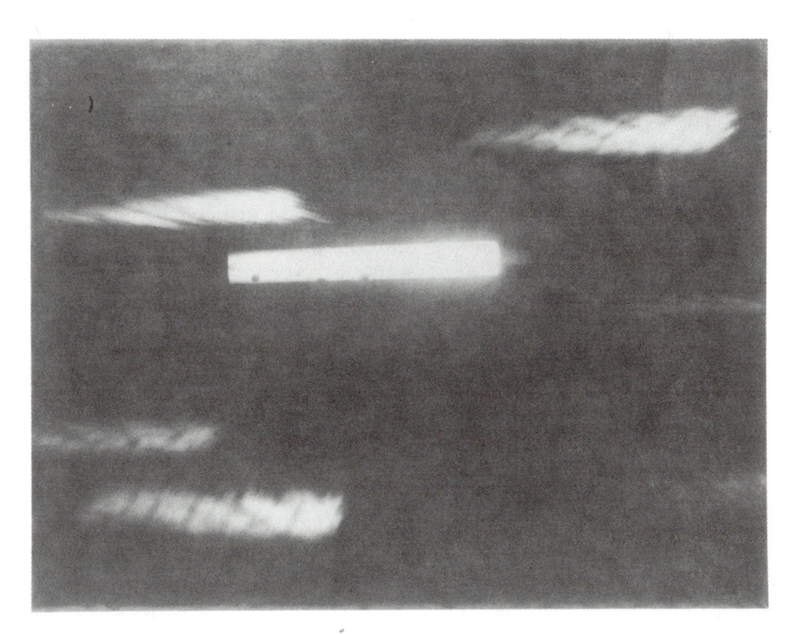

Figure 1. White beam transmission topograph of turbine blade with 7.5" between blade and film

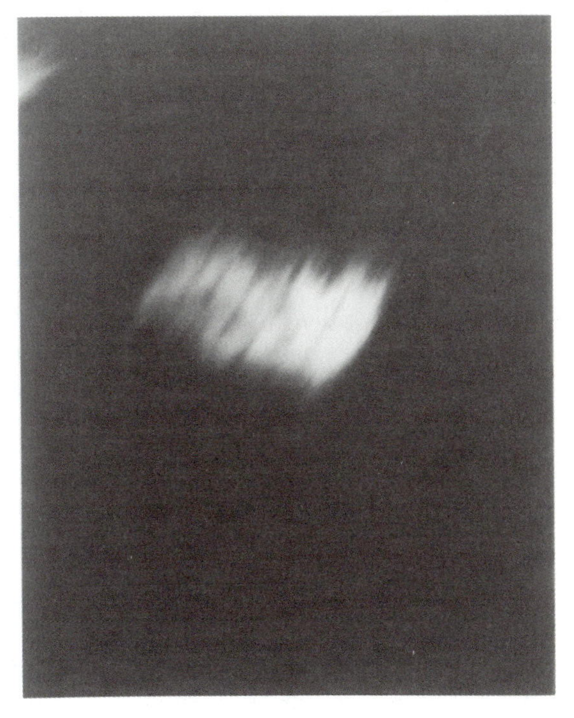

Figure 2. Topograph of same bladeas fig. 1, except with 24" between blade and film

image which was the lowest (and at the left of center) of those shown in fig. 1. The overall length of the image is essentially the same, but the height is much greater. It has been speculated that this could be related to the convex curvature of the surface of the blade in the vertical direction where the x-rays exit, but the authors have no quantitative explanation at this time. The exposure conditions for fig. 2 were 60 seconds when the beam current was 225 ma. Hence the relative dose was 225 ma-minutes, or almost 4.2 times the dose of fig. 1. The height of the Laue image in fig. 2 appears to be close to 4 times the height of the same image in fig. 1, even though the exact heights are difficult to define in these images. The factor of 4.2 might be understood as only the increase in dose required to expose four times the area of film to the same level of contrast.

Figure 3 is a topograph of a different single crystal blade than the one examined in the previous figures. Again, the blade was oriented with the axis of the blade in a horizontal plane, but unlike fig. 1 and fig. 2, the axis of the blade was oriented perpendicular to the axis of the incident beam. Another difference was that the incident beam was collimated by passing through an aperture in a lead block. The aperture was about 0.25" high by about 0.6" wide. The incident beam was again incident on the convex side of the blade. Six sheets of Type 57 Polaroid film were positioned downbeam with suitable overlapping so there were no gaps in coverage by areas of film surfaces with emulsion on them. After exposure and developing, the film was reassembled in the montage shown in fig. 3 by positioning each of the six films in regard to each other exactly as they were originally. The exposure time was 20 seconds with a beam current of 316 ma for a relative dose of 105 ma-minutes. The blade-to-specimen spacing was 7.0". The dark vertical and horizontal bars in the montage are locations where additional layers of paper laid over the emulsion as a result of overlapping the film packets. The darkening demonstrates the capacity of a few layers of paper to absorb the low energy incoherently scattered x-rays present.

The residual incident beam passing through the blade causes the bright oval image slightly above and to the left of the center of the montage. This is the radiograph of the section of the blade illuminated by the incident beam. It is well overexposed. Other spots, smaller and fainter, are the Laue images from the from the same volume of blade which is probed by the radiographic image. These images would probably be larger and more completely filled out had the relative dose been larger.

Figure 4 shows a topograph of the same blade as examined in fig. 3, except in a different location. The blade was moved horizontally in the beam to a new location where a small stray crystal had nucleated and grown, embedded in the single crystal matrix. The stray was visible on the concave surface of the blade (the exit surface for the transmitted x-rays). Other than the horizontal translation, the orientation of the blade in the incident beam was exactly the same as in fig. 3. The extra Laue images are from the stray crystal with its different crystallographic orientation than the single crystal matrix. In this particular case, the deployment and spacing was the same as for fig. 3. The exposure time was 25 seconds at a beamline current of 286 ma for a relative dose of 119 ma-minutes.

Although fig. 3 and fig. 4 illustrate the capability for white beam transmission topography to locate strays within a single crystal blade matrix, it must be noted that the stray in this case was located on the exit side of the blade. Since the diffracted x-rays which form the Laue image are very low in energy (compared to typical industrial or medical radiographic x-rays), a question arises as to the resolution of strays embedded further from the exit surface of the blade. The nickel-based superalloys used for these blades are highly absorbing for these soft x-rays. As a quantitative indication of how highly absorbing these alloys are, a separate experiment using the synchrotron x-rays showed it took a relative dose of about 1200 ma-minutes to exhibit Laue images in transmission through 0.130" of alloy CMSX-4.

Figure 3. Topograph of a different blade using collimator

Figure 4. Same blade and geometry as for fig. 3, except illuminating a stray crystal embedded in the single crystal matrix

DISCUSSION

The topographs shown here demonstrate the feasibility of acquiring Laue images in transmission from single crystal turbine blades made of nickel-based superalloys with reasonably short exposure times. Previous work[1] has established that a great deal more information about the defect structure can be obtained by proceeeding to the next step, that is, scanning the Laue image continuously along the single crystal with an appropriate adjustment in scan rate to obtain an optimum dynamic range of the resulting gray scale for whatever image recording medium is employed. It was also established in previous work that a real time x-ray imaging system can acquire scanned images comparable to ones on film, only with many orders of magnitude lower relative dose[2]. It was speculated that with proper system refinement, the relative dose for real time scanning might even be reduced to fluxes comparable to those available from conventional x-ray generators. It would seem rational to pursue this direction with the topography of nickel-based alloy single crystal turbine blades.

ACKNOWLEDGMENT

This work was performed at Beamline X19C at the National Synchrotron Light Source which is a facility located at Brookhaven National Laboratory, Upton, New York, and which derives support from the U.S. Department of Energy.

REFERENCES

1. J.M. Winter, Jr., R.E. Green, Jr., and K.A. Green, "Application of synchrotron and flash x-ray topography to improved processing of electronic materials", in *Advances in X-Ray Analysis*, C.S. Barrett, *et al*, eds., **35**:239, Plenum Press, New York, (1992).

2. J.M. Winter, Jr. and R.E. Green, Jr.,"Real time synchrotron topography using a CID array camera witih digital image acquisition and processing", *Advances in X-Ray Analysis*, P. Predecki, *et al*, eds., **38**:215, Plenum Press, New York, (1995).

NON-DESTRUCTIVE EVALUATION OF PLASTIC STRAIN IN DEFORMED LAYER USING X-RAY DIFFRACTION

Mitsuaki Katoh, Kazumasa Nishio, and Tomiko Yamaguchi[1]

[1]Department of Materials Science and Technology
Kyushu Institute of Technology
Kitakyushu, Fukuoka, JAPAN 804

INTRODUCTION

When metallic materials receive mechanical machining, polishing and grinding, plastically deformed layers are developed near the surfaces. Besides, such layers are developed by both thermal cutting and water-jet cutting[1]. Thickness of deformed layers have been measured using a scanning acoustic microscope[2]. X-ray diffraction technique, which is sensitive to structures, is suitable for estimating the degree of plastic deformations in the deformed layers. We have already reported that the distribution of plastic strain in the deformation layer, which is experimentally measured using X-ray diffraction technique, can be represented by an exponential function[3]. In this case, it is necessary to consider the phenomenon that the intensity of diffracted X-ray from a some depth from the specimen surface decreases because of the absorption of X-ray. In addition, it is reported that X-ray diffraction intensity curve (referred to as intensity curve) is accurately approximated by Gaussian functions[4].

In this report, we perform some simulation for estimating the distribution of plastic strains in the deformed layers. That is, we prepare some aluminum foils (thickness of a foil:0.012mm) with different plasic strains, whose plastic strains are changed by annealing cold rolled foils at different temperatures. We can prepare some distribution of plastic strain by putting these foils on top each other. The plastic strain obtained experimentally agrees well with those calculated.

MATERIALS USED AND EXPERIMENTAL PROCEDURES

Materials used in this study are commercially pure aluminum sheet A1100-O (thickness:1mm) whose purity is about 99.2 mass%, 1N30-H18 aluminum foil (thickness: 0.012mm) whose purity is about 99.3 mass% and Al-Mg commercial alloy A5083-O (thickness:2 and 10mm) which contains 4.55 mass% Mg.

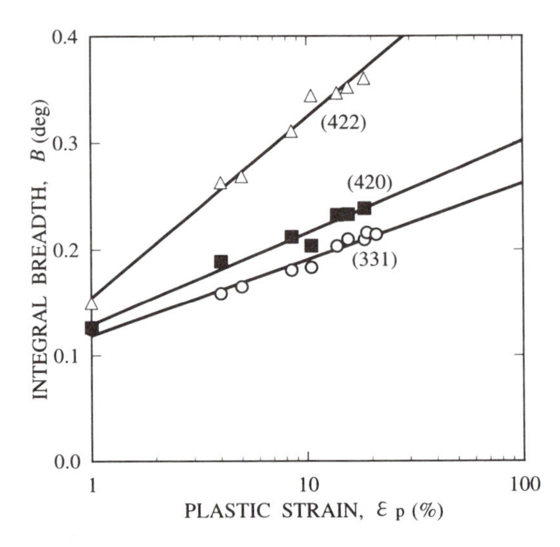

Figure 1. Relation between integral breadth and plastic strain for A1100 aluminum

The specimens, which had been beforehand annealed at 350℃ for one hour, were tensile strained between about 1 and 20% at room temperature by an Instron type tensile test machine. Then, the relation between integral breadth of an intensity curve, which is defined as the quotient of the area surrounded by the intensity curve divided by the maximum intensity, and platic strain was obtained using CuK α radiation.

Aluminum foil as rolled was annealed at several temperatures between 100 and 350℃ for one hour. Then, the integral breadth was obtained from each aluminum foil.

The strain distribution near the specimen surface which is obtained by cutting A5083-O using water jet is also estimated.

RELATION BETWEEN INTEGRAL BREADTH AND PLASTIC STRAIN

An intensity curve is composed of both a curve due to K α_1 radiation and a curve due to K α_2 radiation. In this study, all intensity curves experimentally obtained are separated using Gaussian functions and we use only the curves due to K α_1 radiation. When an intensity curve is approximated by a Gaussian function given by Eq.(1),

$$G(x) = q\exp(-p^2x^2) \tag{1}$$

the area S (equal to integral intensity of X-ray) surrounded by Eq.(1) is given by $S=q\sqrt{\pi}/p$ and integral breadth B is given by $B=\sqrt{\pi}/p$. Since the area S is constant even if the plastic strain is changed, we can obtain the Gaussian function when we know the integral breadth.

Fig.1 shows the relation between integral breadth and plastic strain in A1100 aluminum. The integral breadth linearly increases with the increase in the logarithm of plastic strain. In Fig.1, the results for diffraction planes of (422), (420) and (331) are shown. For the same plastic strain, the integral breadths of (422) are the largest. We can estimate the plastic strain by measuring the integral breadth using the relation shown in Fig.1.

CHANGE OF PLASTIC STRAIN IN FOIL BY ANNEALING

Fig.2 shows the relation between plastic strain and annealing temperature for aluminum foil and we get the relation independent of diffraction planes. The plastic strain as rolled is as large as 0.80 and this decreases a little when annealed at 100℃. Then, this linearly decreases between 100 and 200℃. When the annealing temperature is higher than 250℃ plastic strain is nearly zero. This shows that recovery occurs due to annealing. Hence, we can get some strain distributions by using these foils with different plastic strains.

COMPARISON OF PLASTIC STRAINS EXPERIMENTALLY OBTAINED WITH THOSE CALCULATED

When there is some distribution of plastic strain in the depth direction, the integral breadth also has some distribution. In this case, it is necessary to consider the contribution from each depth from the specimen surface to the integral intensity measured. Let the intensity from the thin layer Δx_i, which is located at the depth x_i from the specimen surface, be I_{xi}. Then, I_{xi} is given by Eq.(2) [5],

$$I_{xi} = \frac{I_0 ab}{2\mu}\left[\exp\left(-\frac{2\mu x_i}{\sin\theta}\right) - \exp\left\{-\frac{2\mu(x_i + \Delta x_i)}{\sin\theta}\right\}\right] \tag{2}$$

where I_0:intensity of incident X-ray, a:volume fraction of crystals oriented in the direction reflecting the incident X-ray, b:ratio of the intensity diffracted from a crystal of unit volume to the incident X-ray. Consider a thin layer of 0.012mm in thickness and assume that there is no change in the integral intensity in this region. Then, we can calculate the ratio R of the integral intensity due to the thin layer which is located at any depth less than the penetration depth of X-ray to that due to the surface thin layer. This shows the contribution of any thin layer to the integral intensity measured.

X-ray used in the X-ray diffraction technique has a penetration depth. This is defined as the depth from the specimen surface at which the ratio of the integral intensity of X-ray diffracted from the depth x from the specimen to the total X-ray intensity obtained from the

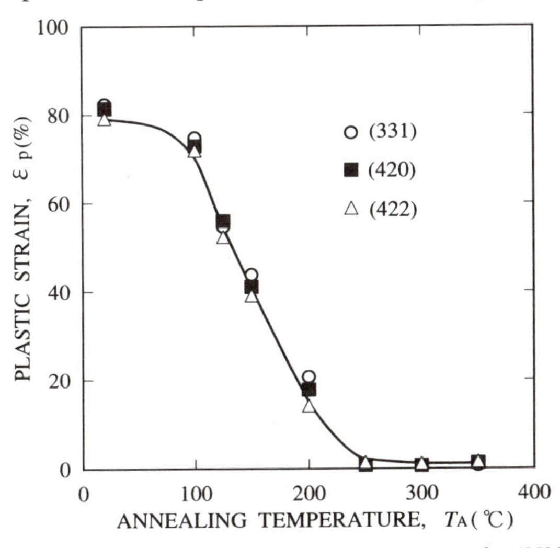

Figure 2. Relation between plastic strain and annealing temperature for 1N30-H18 aluminum foil

Table 1. The ratio R of integral intensity at arbitrary depth from the surface to that on the surface for different diffraction planes

Depth, x (mm)	Ratio, R		
	(422)	(420)	(331)
0.012	1.000	1.000	1.000
0.024	0.763	0.682	0.675
0.036	0.497	0.465	0.456
0.048	0.351	0.317	0.308
0.060	0.247	0.216	0.208
0.072	0.174	0.148	0.140

specimen having the infinite depth (on the order of one tenth mm) is equal to 0.95. Since the linear absorption coefficient of aluminum is 13.56/mm when we use CuKα radiation, the penetration depths of (422), (420) and (331) are 0.101, 0.094 and 0.092mm, respectively.

Table 1 shows the ratio R of integral intensity at arbitrary depth from the surface to that on the surface for (422), (420) and (331) when six sheets of aluminum foil are laminated. Consider the case when we laminate two sheets of aluminum foil whose strains are ε_p and ε_o. When we irradiate X-ray against this set of foil, the intensity curve obtained is as denoted synthesized ($\varepsilon_{oc}+\varepsilon_p$) in Fig.3. In Fig.3, the intensity curve denoted by annealed (ε_o) is given by Eq.(3).

$$G_0(x)=q_0\exp(-p_0^2x^2) \tag{3}$$

Actually, however, this intensity curve should be corrected to Eq.(4) when we use (422),

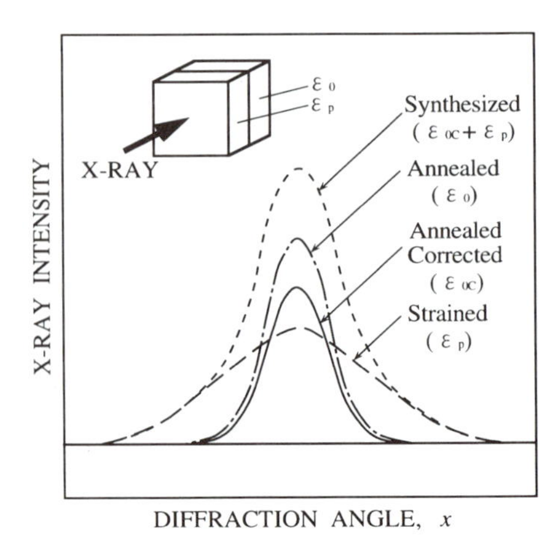

Figure 3. Method for obtaining synthesized intensity curve from a specimen with different plastic strain

Table 2. Comparison of plastic strain measured with those calculated

Diffraction Plane	Strain Measured	Strain Estimated
(331)	0.405	0.392
(420)	0.377	0.367
(422)	0.350	0.362

$$G_{0c}(x)=0.763q_0\exp(-p_0^2x^2) \tag{4}$$

because the intensity curve is absorbed by the surface layer. When $G_p(x)=q_p\exp(-p_p^2x^2)$ represents the intensity curve denoted by strained (ε_p), we can obtain the intensity curve denoted by synthesized ($\varepsilon_{0c}+\varepsilon_p$) by adding $G_{0c}(x)$ and $G_p(x)$. Then, the integral breadth of this is given by $1.763S/(0.763q_0+q_p)$, where S is the area surrounded by $G_0(x)$ and/or $G_p(x)$. When the number of laminated sheets increases, we can obtain the integral breadth of the synthesized intensity curve by the similar procedure mentioned above. Needless to say, this procedure is effective for the thickness less than the penetration depth of X-ray for each diffraction plane.

We get the relation between the integral breadths measured and those calculated when we laminate six sheets of foil in the order of as rolled, and annealed at 100, 125, 150, 200 and 250℃. Table 2 shows the comparison between the plastic strain which is obtained using the relation between the integral breadth and the plastic strain shown in Fig.1. For each diffraction plane, there is a good correspondence between the plastic strain measured and that calculated. The strain obtained is the largest in (331) and the smallest in (422) due to the difference of the values of R in Table 1.

METHOD FOR ESTIMATING STRAIN DISTRIBUTION

According to the results experimentally obtained, the plastic strain distribution in the deformed layer is approximated by an exponential function. Hence, the simulation is performed to obtain the correspondence between the assumed plastic strains, whose equation is as given by Eq.(5), and those estimated following the procedure as already mentioned.

$$\varepsilon_p=a\exp(-bx) \tag{5}$$

Fig.4 shows the result when we assume that a=0.20, and b=15.0 and 59.9. Diffraction planes of (511), (422), (311), (220) and (200) are used and the material is A5083-O whose linear coefficient is 13.26/mm. The solid line and the broken line are the plastic strain distributions for $\varepsilon_{p1}=0.20\exp(-59.9x)$ and $\varepsilon_{p2}=0.20\exp(-15.0x)$, respectively. The horizontal coordinates of the results estimated are plotted at X-ray penetration depth of each diffraction plane. Though there are differences between them, the important point in Fig.4 is that we can get the same plastic strain on the surface by extrapolating the estimated resuts to $x=0.00$mm as that initially assumed.

Since we can estimate the plastic strain that is the value of a in Eq.(5) on the surface the next procedure we should do is to select the most appropriate values of b in Eq.(5). The data we can experimentally obtain are those shown by circles and triangles in Fig.4. Therefore,

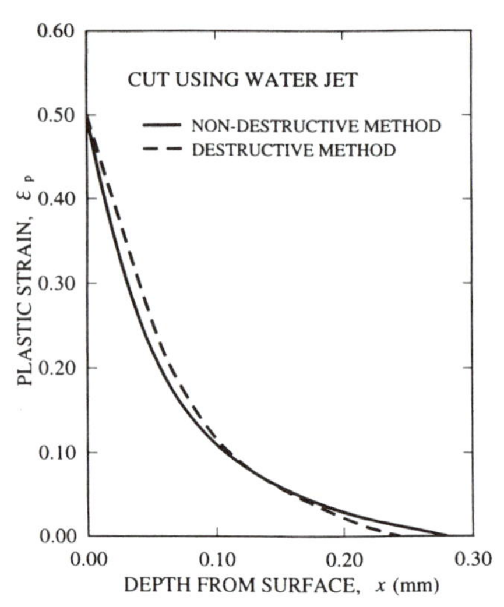

Figure 4. Correspondence of the distributions of plastic strains assumed with those simulated

all we have to do is obtain each data corresponding to those shown in Fig.4 by assuming arbitrary values of b in Eq.(5). If we can obtain the value of b which has the least error between the plastic strains experimentally obtained and those calculated, this is what we want to obtain.

Through this procedure, the plastic strain distributions of A5083 (thickness: 6mm) which are cut using water jet are estimated and these agree with those obtained by removing the surface thin layer little by little (destructive method).

CONCLUSIONS

(1) We propose the method for estimating the plastic strain distributions in the deformed layer due to maching and/or cutting using X-ray diffraction technique by considering the X-ray penetration depth.
(2) The effectiveness of this method is made clear by the experiment using laminated thin aluminum foils whose plastic strains are changed by annealing temperatures.
(3) The plastic strain distributions of A5083 which are cut using water jet are estimated by non-destructive method and these agree with those obtained by removing the surface thin layer little by little.

The aluminum foil was supplied by Showa Aluminum Co. The authors would like to express our sincere gratitude for their kindness.

REFERENCES

1. S.Matsui, H.Matsumura, Y.Ikemoto, H.Shimizu, T.Ochi, and K.Tsujita, Development of water jet cutting and its application, *Welding Technique*, 36:89(1988) (in Japanese).
2. K.Ishikawa, Peeling a thin layer, *Boundary*, 2:30(1988) (in Japanese).
3. S.Mukae, M.Katoh, K.Nishio, and M.Abe, X-ray study on deformation layer of aluminum alloy A5083 of cutting surface using water jet, *Quarterly Journal of J. W. S.*, 10:280(1992) (in Japanese).
4. M.Kurita, Statistical study on X-ray stress measurement using three-points parabolic and three-points Gaussian curve approximations, Trans. J.S,M.E., 43: 964(1977) (in Japanese).
5. B.D.Cullity(translated by G.Matsumura), *Elements of X-ray Diffraction*, Agune(1980) (in Japanese).

NONDESTRUCTIVE TESTING OF CERAMIC AUTOMOTIVE VALVES

U. Netzelmann[1], H. Reiter[1], Y. Shi[2], J. Wang[1], and M. Maisl[1]

[1]Fraunhofer-Institute for Nondestructive Testing (FhG-IZFP)
University, Bldg. 37, 66123 Saarbrücken, Germany
[2]Beijing Institute of Aeronautical Materials
P. O. Box 81, Beijing 100095, P. R. China

INTRODUCTION

Ceramic components exhibit an attractive combination of advantageous properties, e. g. low density, good temperature and corrosion resistance. Therefore, they have a high potential for application in combustion engines. Valves made out of sintered silicon nitride are of primary interest. The ceramic valves promise fuel savings and a reduction of noise emission. For a widespread use of such valves, mass production and quality control have to be mastered reliably. Here, nondestructive testing (NDT) can yield a significant contribution. In this article, a survey on application of various NDT techniques for testing volume and surface of valves is given.

DYE PENETRANT TESTING

Fluorescent dye penetrant testing allows to search nearly completely the surface of ceramic valves for open cracks. Indications for defects were found at many valves taken from production, e. g. on the valve heads, on the shafts and at the notch areas. Different surface finish in certain regions of the valve, in particular in the horn-like transition area between head and shaft (in the following called "transition region"), may easily cause background indications making detection of the defects more difficult. A disadvantage of dye penetrant testing is that cracks not open to the surface can not be detected.

HIGH-FREQUENCY ULTRASOUND TESTING

For detection of critical defects with dimensions of 50-100 µm, a high-frequency ultrasound equipment for the frequency range 10-200 MHz is employed[1]. Using broadband focusing polymer foil probes (80 MHz, Krautkrämer, Hürth), the short ultrasound pulses necessary are excited and detected. With this technique, the head region of the valves is tested for volume defects by C-scanning. Fig. 1 (left part) shows a defect in the head region of a valve which was found by this technique.

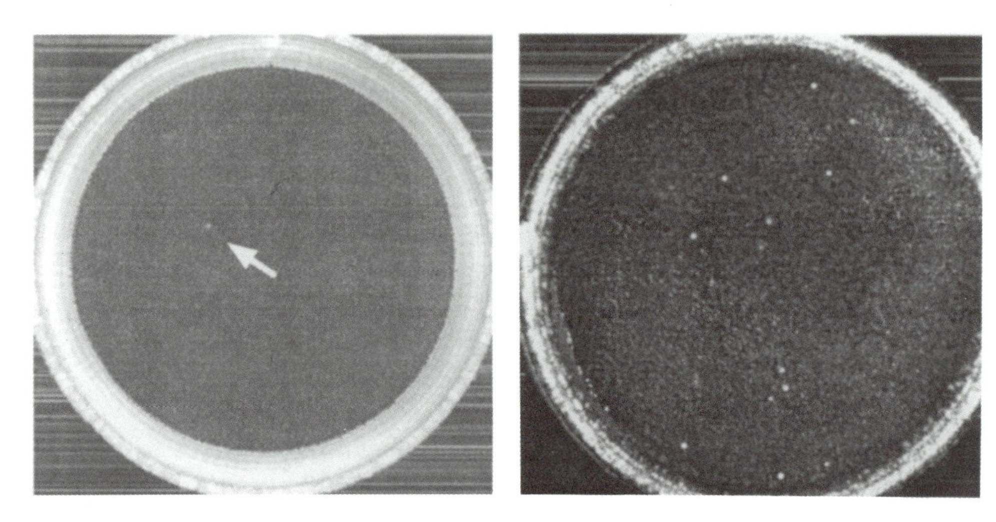

Figure 1. Left side: C-scan of the head region of a Si_3N_4-ceramic valve showing a "natural" defect in the volume (arrow), image size 28 mm x 28 mm. Right side: C-scan of the head region of a test valve containing test defects with about 100 μm diameter (image size 32 mm x 32 mm)

It is difficult to set a time window for selection of the depth of interest because of the irregular shape of the valve head. A compromise between background signals from sound scattered at the inclined outer surfaces on one hand and unnecessarily reduced depth range on the other hand is necessary. A way out of this situation is a volume measurement by recording all backscattered signals, followed by analysis using a volume visualization system[2]. Reduction of detection sensitivity in depths beyond the focal point can be compensated by applying synthetic aperture focusing techniques[3]. As the smallest detectable defects are scattering in the Rayleigh regime, a clear statement on the size of the defect can not be easily obtained. Therefore, this point was investigated in more detail by using test valves (valve blanks). Diamond grains with known dimensions were mixed into the ceramic powder. During sintering, the diamonds were burning out partly, leaving defects of defined size[4]. For such valves it could be shown that using longitudinal waves in normal incidence, 200 μm and 100 μm defects were easily detected, and even 50 μm defects could be detected with somewhat more effort, in spite of relatively rough surfaces (Fig. 1, right part). Such defects can not be detected with higher signal amplitude when mode-converted transverse waves are used, even considering their more favorable ratio of wavelength to defect size[5]. Usually it is difficult to obtain information about the type of defect for the very small sizes[6].

Volume testing of the shaft region is a challenge for ultrasound testing, as refraction of sound at the interfaces is very strong and anisotropic because of the high speed of sound (10.5 mm/μs longitudinal) and the low radii of the shaft. Up to now, only larger defects could be detected in the shaft.

Another application of high-frequency ultrasound is the detection of surface cracks by Rayleigh wave techniques. Using the same probes as for volume testing, the sample is insonified under the critical angle for excitation of a leaky Rayleigh wave. In spite of strong damping, the wave can be traced over a distance of a few mm. This technique is particularly applied to the valve seat surface (Fig. 2). A B-scan with the rotation angle around the long axis of the valve as the spatial coordinate is performed to test this critically loaded area of the valve. In the same way, the shaft surface can be scanned for surface defects. In contrast to dye penetrant testing, cracks not open to the surface can be detected.

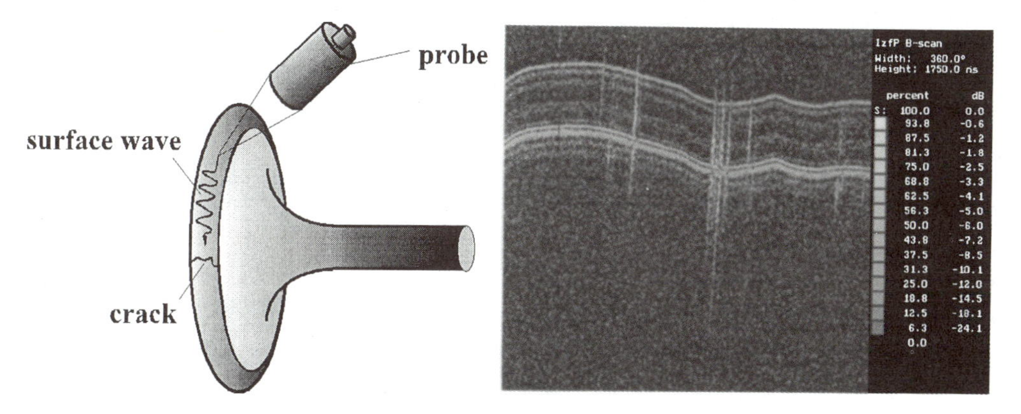

Figure 2. Left: Scheme of leaky Rayleigh wave testing of the valve seat. Right: Test result obtained on a valve. The rotation angle (360°) is the horizontal coordinate, ultrasound time-of-flight is the vertical coordinate. The surface defects appear as nearly vertical lines.

X-RAY TESTING

Microfocus X-ray testing is particularly applied in regions, which are difficult to inspect by ultrasound because of their complicated geometry. Examples are the transition region and the notch area.

A first technique, which is used to inspect the valve head region, is digital radiography. The X-ray cone-beam, which is subject to different attenuation is first converted into visible light, then video-imaged and digitized. The raw image is evaluated by digital image processing. A typical image resolution is 512 x 512 pixels with a pitch of 10 μm. By subtraction and edge enhancement algorithms small defects are emphasized. In Fig. 3 (left part), an image with defects of 200 μm size in the head region of the test valves mentioned above is shown. A complete inspection of the plate requires to measure many of such images in an automated raster scan.

Other X-ray techniques are the 2D- and 3D-computed tomography (CT). They are particularly used for the shaft and the transition region. They also work with pixel sizes of 10 μm. A line detector or a detector array are employed[7]. Due to the large number of projections to be collected, the time consumption of these techniques is relatively high. Fig. 3 (right part) shows a reconstruction result from the transition region of a valve. Other than for digital radiography, the defect position can be reconstructed in three spatial coordinates.

Compared with the other techniques described here, a good characterization of the defect is possible. Pores and metallic inclusions (the latter are often surrounded by reaction zones in ceramics) can be well distinguished. Density variations can be detected as well[8].

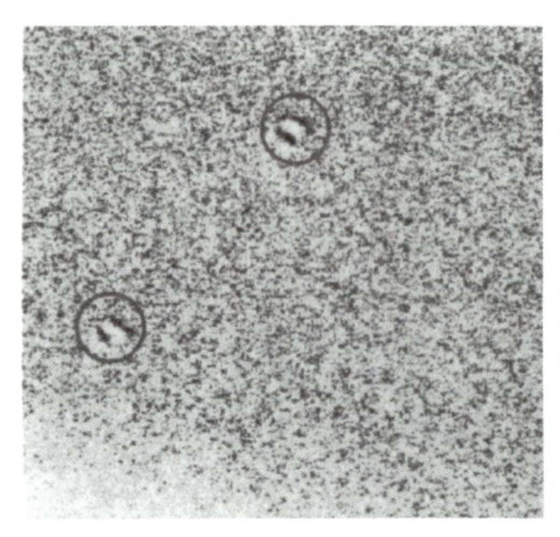

Figure 3. Left: Digital radiography showing two test defects in the valve head. Image size 5 mm x 5 mm. Right: Visualization of a pore detected by 3D-CT in the transition region between valve head and shaft.

RESONANT TESTING

Acoustic resonant testing is a well known technique for fast quality control of components and has recently come again into the focus of interest[9,10]. Its basic principle is to induce a vibration of the component in one or several of its eigenmodes by using pulsed or periodic mechanical excitation. The resulting oscillation is analyzed. Defects in the component become visible as a change of the resonant spectrum (position of resonance lines, width and amplitude of the lines, line splitting for components with symmetries).

In the setup used for the present experiments, the valve is supported by piezoelectric transducers at two points (Fig. 4). One transducer is operated as transmitter, the other as receiver. Two approaches were followed for excitation and detection. In the first, the transmitter is excited by a sine wave and the acoustic signal transmitted through the valve is detected using a lock-in amplifier. The frequency of the sine wave is slowly sweeped, allowing to record amplitude and phase of the spectrum over a wide frequency range.

The second approach uses an excitation by a step-function. The resulting resonant signal is recorded over the full decay time using a transient digitizer with large storage depth. The signal is averaged over many repetitions of the experiment and then Fourier transformed over 50000 or more points.

As expected, the result of both approaches was equivalent, but the time-domain technique turned out to be faster and more flexible. The experiments performed up to now suffered from the poor reproducibility of the mechanical coupling of the vibration to and from the component. In particular the low-frequency bending modes showed significant variations in the resonant frequencies.

The resonance lines of the valves inspected exhibit Q-values in the range of 1000 to 3000. Therefore, a high detection sensitivity can be expected. On the other hand, it is questionable, if defect sizes of 50 to 100 μm are detectable by this technique. Finite element calculations to study the influence of cracks in disk-shaped samples show, [11] that crack lengths of about 10 % of the radius are required to cause relative frequency changes in the order of 10^{-4}. Such changes have to be detected with the background of tolerable variations of density, shape and mass already present in the production lot. Experimentally, by weighing a lot of ceramic valves a standard deviation of the mass of 37.4 mg was determined. If this mass is concentrated in a single point of the valve, it is represented by a pore of 2.8 mm in diameter, when one uses the density of silicon nitride.

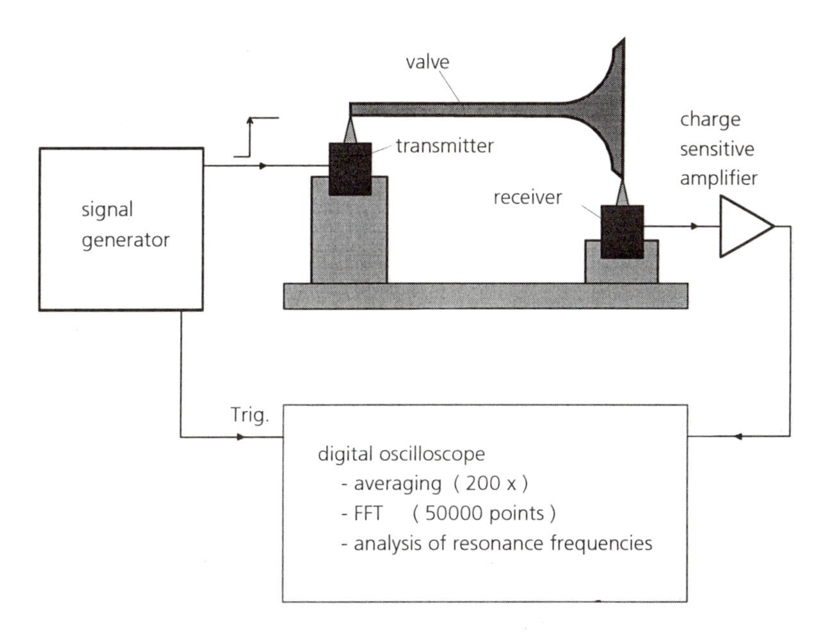

Figure 4. Setup used for resonant testing of ceramic automotive valves.

CONCLUSION

Table 1 summarizes some major results obtained with the testing techniques discussed above. The testing times are referring to setups as they are realized at present in FhG-IZFP. They certainly could be reduced for testing in the production site.

The optimum technique, which would be able to detect smallest defects in all regions of the sample in a fast way and with highest sensitivity, does not exist. In some cases the numbers for detection limits are still unproved due to lack of suitable test defects.

The present investigations performed in the framework of research programs were aiming at an utmost complete testing for volume and surface defects. Such testing will not be feasible for series components due to economic limitations. From the viewpoint of NDT, an improved specification of the defect size to be detected as a function of the position in the component is necessary. This requires furthermore a close cooperation of developers in construction, production and application.

Acknowledgments

This work was supported by the BMBF. The authors thank H. A. Lindner, Cremer Forschungsinstitut, Rödenthal, for supply of test samples.

Table 1. Summary of main results obtained for ceramic valve testing

	ultrasound volume testing	ultrasound surface wave testing	resonant testing	X-ray digital radiography	X-ray computed tomography	dye penetrant testing
testing time	4 min	4 min	0.5 min	30 min	60 min	10 min
detection limit (valve head)	50-100 μm	20 μm	unknown	100 μm	150 μm	50 μm (open crack)
valve seat (surface)	-	30 μm (?)	unknown	100 μm	150 μm	50 μm (open crack)
detection limit shaft	100-200 μm (?)		cracks: 1 mm (?)	100 μm	50 μm	50 μm (open crack)
defect characteri-zation	moderate	moderate	poor	good	very good	moderate

REFERENCES

1. S. Pangraz, H. Simon, R. Herzer, and W. Arnold, 'Non-destructive evaluation of engineering ceramics by high-frequency acoustic techniques', in: *Acoustical Imaging, Vol. 18*, H. Lee and G. Wade, eds., (Plenum Press, New-York 1991), p. 189
2. W. Arnold, R. Herzer und U. Netzelmann, 'Visualisierung von Ultraschall-Volumendaten mit moderner Computergrafik', *DGzfP Berichtsband* **28**, Jahrestagung 6.-8.5.1991, Luzern, (Deutsche Gesellschaft für zerstörungsfreie Prüfung, Berlin 1991), p. 287
3. M. Maisl, P. Kreier, W. Müller und U. Netzelmann, 'Rekonstruktion von Fehlstellen in Keramik mittels 3D-Röntgen CT und Hochfrequenz-Ultraschall - ein Vergleich', in: *DGZfP Berichtsband* **47**, (DGZfP, Berlin 1995), p. 171
4. B. Caspers, J. Hennicke, H.-A. Lindner, M. Maisl, U. Netzelmann und H. Reiter, 'Zerstörungsfreie Prüfung von Maschinenbau-Komponenten aus Siliziumnitrid am Beispiel des Kfz-Ventils', *Fortschrittsberichte der DKG* **11**, 29 (1996)
5. Y. Shi, J. Hennicke, and U. Netzelmann, 'High-frequency ultrasound detection of small defects in Si_3N_4 ceramic', *Nondestr. Testing and Evaluation* **13**, 139 (1997)
6. U. Netzelmann, H. Stolz, W. Arnold and G. Giunta, 'Defect sizing in ceramic materials by high-frequency ultrasound techniques', in: *Acoustical imaging* **20**, Y. Wei, B. Gu (eds.), (Plenum Press, New-York 1993), p. 303
7. J. Buck, M. Maisl, H. Reiter, Èntwicklung eines schnellen Rekonstruktionsverfahrens für die 3D-Computertomographie', in: Zerstörungsfreie Materialprüfung, DGZfP Berichtsband 47.2, (DGZfP, Berlin 1995), p. 477
8. H. Reiter, 'Zerstörungsfreie Prüfung zur Qualitätssicherung keramischer Bauteile', Fortschrittsberichte der DKG **11**, 9 (1996)
9. J. J. Schwarz, G. W. Rhodes, 'Resonance inspection for quality control', *Rev. Progr. QNDE* **15**, D. O. Thompson and D. E. Chimenti (eds.), (Plenum Press, New-York 1996), p. 2265
10. U. Schlengermann, W. Hansen, 'Das Resonanzverfahren - die Antwort auf neue industrielle Forderungen an die Qualitätssicherung', *DGZfP Jahrestagung* Dresden 5.-7.5.1997, Vortrag 55
11. H. A. Lindner, J. Hennicke, B. Caspers, H. Feuer, and I. Petzenhauser, 'Production-oriented non-destructive testing of structural ceramic components: vibration analysis', *cfi/Ber DKG* **70**, 294 (1993)

ADVANCEMENT IN NONDESTRUCTIVE INVESTIGATION OF LIQUID YAG AT VERY HIGH TEMPERATURE BY SYNCHROTRON RADIATION

Claude Landron[1,2], Xavier Launay[1], Jean-Pierre Coutures[1], Marc Gailhanou[2] and Michel Gramond[3]

[1]Centre de Recherches sur la Physique des Hautes Températures,45071 Orléans cedex 2, France.
[2]Laboratoire pour l'Utilisation du Rayonnement Electomagnétique, 91405 Orsay cedex France.
[3]Centrale Paris, 92290 Chatenay France.

Keywords : Liquid, YAG, High-Temperature, Synchrotron Radiation, Structure.

ABSTRACT

Synchrotron radiation experiments have been performed on liquid YAG, yttrium aluminum garnet $Y_3Al_5O_{12}$, (melting point : 1950°C), as well as on solid YAG and YGG, yttrium gallium garnet $Y_3Ga_5O_{12}$, to obtain structural information at high temperature. The development, construction and performance of a new cell for X-ray Absorption Spectrometry for in-situ high temperature analysis of solid and liquid materials in the presence of a CO_2 laser radiation is described. The sample is heated up to the liquid phase by using a continuous wave laser associated with levitation by a gas jet. We have investigated short and long range order at the atomic scale by performing combined X-ray absorption and X-ray diffraction real time measurements.

INTRODUCTION

The structural aspects of liquid oxides at high temperature are still poorly understood[1,2]. The characterization at the atomic scale of the liquid state is a challenge that can be met from X-ray studies which is now possible with the recent development of very intense synchrotron radiation sources [3,4]. Nevertheless, the structural investigation of liquids at high temperature is not easy to undertake due the difficulties encountered in high temperature experiments deriving from the geometry of the heating system, the consequences of the contact of the sample with the walls of the container and the measurement of the temperature as well as the measurement of the signal of interest. Nowadays, the development of new intense X-ray sources, such as the storage rings of the third generation, has made possible the high temperature investigation of disordered materials[5].

In this paper, we report the first results related to the development of an experimental device designed for high temperature structural measurements in contactless conditions. The device is suitable for X-ray absorption (XAS) as well as for X-ray diffraction (XRD) studies, giving complementary structural information. XAS is mainly sensitive to the local surrounding of atoms in condensed matter whereas XRD probes structures with long range

Nondestructive Characterization of Material VIII
Edited by Robert E. Green Jr., Plenum Press, New York, 1998

order[6]. This system based on aerodynamic levitation and laser heating presents the important advantage of providing a free space around the sample for various detectors[7,8]. Another advantage is that the heating is produced by the irradiation of a focused laser beam, avoiding the heating by the sample surrounding, the target reaching very rapidly the desired temperature. A 100 W continuous wave CO_2 laser allows to reach on many refractory oxide samples, with a spherical shape diameter of 3 mm, temperatures higher than 2000°C.

Our preliminary measurements have been performed on YAG and YGG compounds. In order to improve the production of YAG single crystals which are used as hosts for laser devices, information are needed on the structure of liquid YAG at temperatures higher than 2000°C. We have synthesised the starting powders by a conventional sol-gel process and millimetre sized spheres have been shaped by using an aerodynamic levitatorand a 500 W CO_2 laser, XAS and XRD experiments have been performed at the LURE (Orsay, France) synchrotron radiation facility.

X-RAY ABSORPTION AND HIGH TEMPERATURE

It is usual to describe disordered materials at the atomic scale by determining the one-dimensional radial distribution function which is the probability for an atom to be located at a given distance of the scatterer atom. This information of short range order in liquids replace the long range description given by X-ray diffraction on crystalline materials. The XAS data[9] correspond to the fractional cross section $\chi(k) = \{\mu(k)-\mu_0(k)\}/\mu_0(k)$ of the absorption coefficient, as a function of the wave number k of the photoelectron, on the higher energy side of the absorption edge of a given element in a compound. The oscillations observed in the absorption cross section of a photoelectron results from the interference between the wave function associated to an outgoing electron emitted during the photoemission process and the part of this wave function reflected by the neighboring atoms. Eisenberger and Brown[10] calculated the EXAFS oscillations in k-space for an atom imbedded in a medium with a pair distribution function g(R). They found :

$$k\chi(k) = -|F(k)| \int_0^{+\infty} [1/kR^2]\, g(R) \exp(-2R/\lambda) \sin\{2kR+\phi(k)\}\, dR \tag{1}$$

where λ is the mean free path of the photoelectron, R is the distance between the absorber atom and the shell of interest, F(k) and $\phi(k)$ are the backscattered amplitude and the phase shift respectively encountered by the photoelectrons. EXAFS depends only on the local structure by the fact that this length is short. The knowledge of the near surrounding of an atom is particularly useful in the case of the study of disordered materials. Eq(1) was expanded with the result :

$$k\chi(k) = -|F(k)|\ |X(k)|\ \sin\{2kR+\phi(k)+\mathrm{Arg}X(k)\} \tag{2}$$

$$\text{where : } X(k) = \int_{-\infty}^{+\infty} g(x)(1+\frac{x}{R})^{-2} e^{-2x/\lambda(k)} e^{2ikx} dx \tag{3}$$

The separation of the shells derives from the determination of the radial distribution function around the excited atom. The determination of the coordination sphere is obtained by isolating the first peak after extraction of the modulus of the complex Fourier transform g(R) of the oscillating functions $\chi(k)$. The radial distribution function g(R) describes the variation in the distances of neighbouring atoms from a given central one. By increasing the temperature of the sample, anharmonic contributions to the vibration of the atoms have to

Eisenberger and Brown[10] have developed a cumulant expansion model which is useful for a correct treatment of highly disordered systems such as liquid oxides at high temperature.

EXPERIMENTAL

The heating of a sample in contactless conditions presents important advantages due to the fact that no thermal or atomic diffusion processes occur on the walls of the container[11]. The cooling of the sample is controlled up to high speed quenching conditions resulting from the large ratio surface/volume of the sample levitated in a gas jet. This analysis cell which has been developed can operate for experiments from room temperature up to 2300°C under various gas conditions. The heating is produced by continuous CO_2 laser. The power of the 100W CO_2 laser is stabilized by a loop control system. An advantage of the laser heating system is that the equilibrium temperature of the sample is rapidly reached. The thermal exchange rate is high because of the large ratio surface/volume of a millimetre sphere. Another benefit is the fast cooling of the sample in aerodynamic conditions after shutting the laser. During the radiative regime the cooling of the sample can be faster than 1000°C/s. We have recorded spectra in the fluorescence mode with new large area silicon photodiodes. The position of the detectors has been adjusted to achieve optimum fluorescence signal emitted by the sample and to avoid self-absorption. The photodiode supports were water cooled. Kapton windows, covered by a graphite film, have been used to transmit X-rays and to absorb visible light. For security reasons, the CO_2 laser beam is displayed by a diode pointer which emits a visible laser light coaligned with the CO_2 laser beam. It is obvious that the temperature in contactless experiments cannot be measured by thermocouple. We have measured the temperature of the sample with an optical pyrometer operating at a wavelength of 0.85 μm in the (800°C - 3000°C) range.

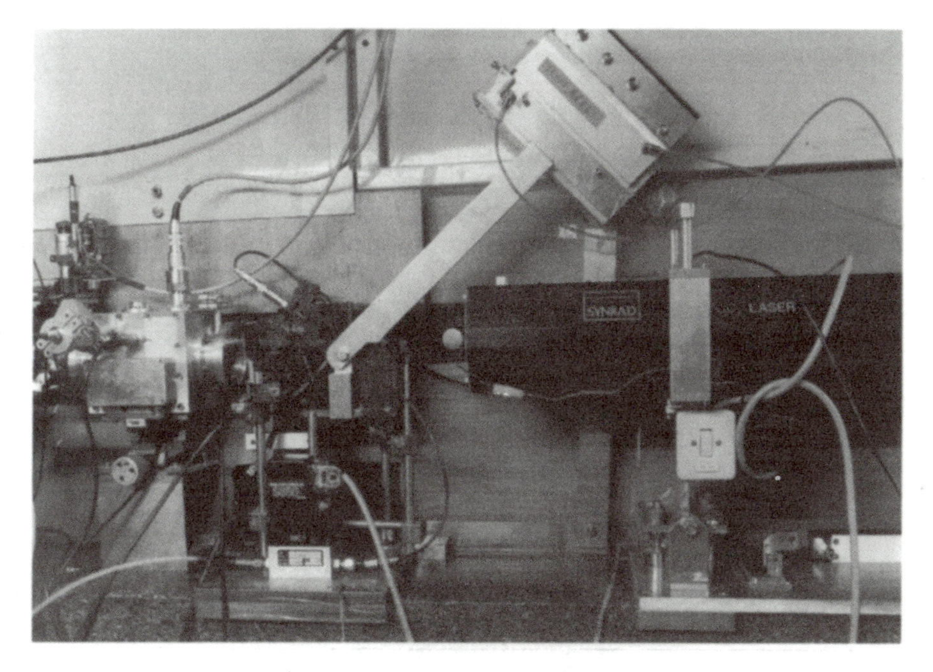

Figure 1 : View of the analysis cell which has been used at LURE for the combined X-ray absorption and diffraction experiments on samples in contactless conditions heated by a laser beam.

X-ray absorption experiments have been performed with the device represented in the Fig.1. The absorption spectra were recorded on the EXAFS IV station at LURE, on the Y K-edge (16950 - 18000 eV energy range) and on the Ga K-edge (10300 - 11300 eV energy range). The synchrotron radiation was provided by the 1.85 GeV storage ring of DCI. The positron intensity was 300 mA at the beginning of the run with a life time of 59 hours. The white X-ray beam was monochromatized by a double Bragg reflection of Si (311) crystals for the yttrium K-edge and Si (111) for the gallium K-edge. Diffraction patterns have been recorded with a linear detector giving long range structural information.

RESULTS

The analysis of EXAFS spectra was conducted by using the conventional Fourier transform method after applying a smoothing of the raw EXAFS data and a normalization to the edge jump[12]. E_0 was chosen by deriving the experimental spectra.

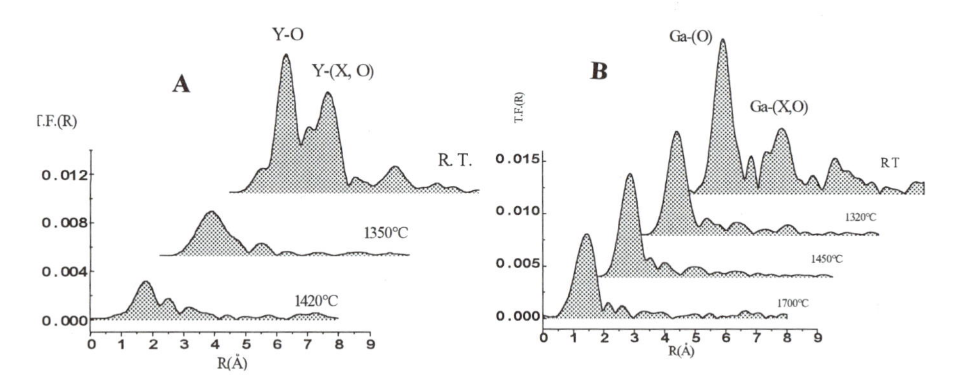

Figure 2 : Fourier transform moduli at the yttrium K-edge (A), and at the Gallium K-edge (B) of the YGG solid sample $Y_3Ga_5O_{12}$ at different temperatures.

Each experimental fractional cross section $\chi(k)$ was converted to momentum k-space. The $\chi(k)$ were Fourier transformed into RDF curves. Then the first shell was backtransformed.

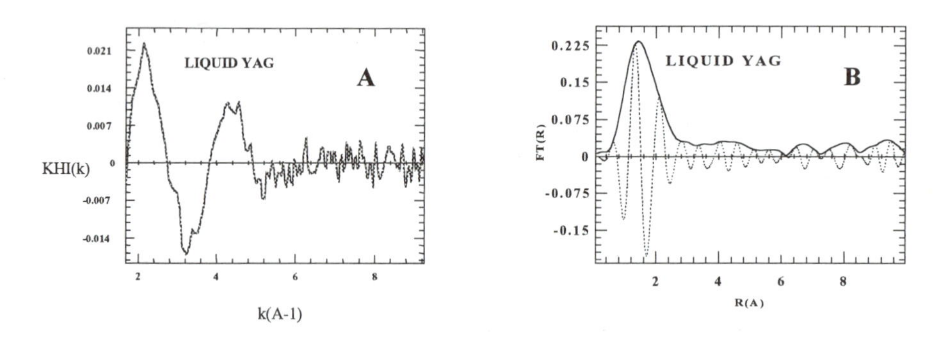

Figure 3 : Normalized, background subtracted, EXAFS spectra (A), modulus (full line) and imaginary part (dotted line) of the Fourier transform of the experimental signal (B) at the yttrium K-edge, for a levitated liquid YAG sample $Y_3Al_5O_{12}$ at 2200°C.

The fits of the filtered EXAFS spectra produces a set of structural data that includes the mean interatomic distance d(Ga-O), the number of neighbors N(O) and the Debye-Waller factors. Fig.2 presents the absolute value of the Fourier Transform of the $\chi(k)$ recorded above the yttrium K-edge (at Room Temperature, 1350°C, 420°C) and above the gallium K-edge (at Room Temperature, 1320°C, 1450°C, 1700°C) for crystallized $Y_3Ga_5O_{12}$

A representative plot of the experimental oscillations above the yttrium K-edge of liquid YAG, in contactless conditions, at high temperature, is represented in Fig. 3A. The diameter of the liquid drop was about 1 mm. Fig. 3B present the Fourier transform of the previous signal. We note a single peak in the RDF, showing a well defined oxygen surrounding of yttrium atoms.

Since XRD and XAS are complementary techniques, we have also performed XRD experiments with the same analysis chamber. A sequence of heat treatments in flowing argon has been given to a YAG sample. X-ray diffraction patterns have been simultaneously recorded. At room temperature, YAG exhibits the garnet structure. This structure is cubic, space group Ia3d, with eight formulas $Y_3Al_5O_{12}$ per unit cell. For one unit formula, Y ions are dodecahedrally surrounded. $Y_3Al_5O_{12}$ has an unit cell of parameter a = 1.2010 nm. Each yttrium is surrounded by four oxygen atoms at d(Y-O) = 0.2303 nm and four oxygen atoms at d(Y-O) = 0.2432 nm. Aluminium atoms are located in two types of sites : the first is six-fold coordinated with d(Al-O) = 0.1937 nm and the second is tetrahedrally coordinadated with d(Al-O) = 0.1760 nm. The successive XRD patterns obtained from this YAG spherical specimen are presented in Fig. 4. Each spectrum acquisition required 1 min. The series of XRD diagrams has been collected while heating the YAG from room temperature to 2083°C, at this temperature $Y_3Al_5O_{12}$ is liquid. By decreasing the power of the laser, we observe in our experimental conditions that supercooled liquid $Y_3Al_5O_{12}$ exists up to 1910°C.

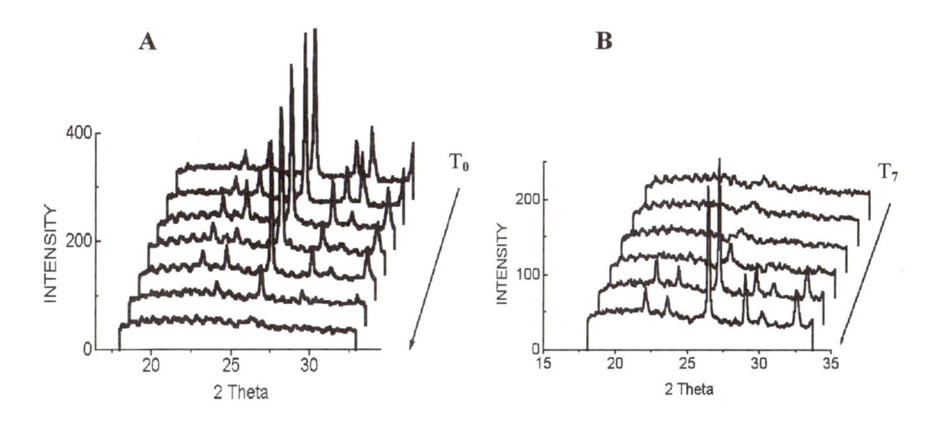

Figure 4 :X-ray diffraction pattern obtained during the heating (A) and the cooling (B) of a levitated YAG spherical sample at various temperatures :
A) T_0 = RT, T_1 = RT, T_2 = 1430°C, T_3 = 1520°C, T_4 = 1750°C, T_5 = 1850°C, T_6 = 2020°C,
B) T_7 = 2083°C, T_8 = 2020°C, T_9 = 1910°C, T_{10} = 1700°C, T_{11} = RT, T_{12} = RT,

CONCLUSION

In-situ techniques are increasingly developed in order to extend the knowledge of material-properties in extreme conditions in term of structure. Our aim in this study was to perform a nondestructive investigation of the structural properties of liquid YAG at a very high temperature by using synchrotron radiation. We note that the high temperature experiments on levitated samples are only in their early stage. They are very promising for the future especially in combination with other X-ray techniques. We have shown the good capability of the X-ray absorption spectrometry combined with X-ray diffraction for estimating the order at short and long distance of high temperature solids and liquids. The structure of pure YAG and YGG samples was studied under various temperature conditions with a special interest related to the modification of the yttrium coordination sphere.

Acknowledgements

The authors wish to gratefully thank the staff of LURE for providing synchrotron radiation and associated facilities during our dedicated runs. This paper is part of the "Synchrotron Radiation and High-Temperature" project funded by a "XIème plan Etat-Région Centre" contract.

REFERENCES

1 I. Egry, G. Lohofer, E. Gorges and G. Jacobs, J. Phys. : Condens. Matter, 8 : 9363 (1996)
2 S. Ansell, S. Krishman, J. K. Weber, J. J. Felten, P. C. Nordine, M. A. Beno, D. L. Price and M. L. Saboungi, Phys. Rev. Lett., 78 : 464 (1997)
3 A. Filliponi and A. Di Chicco, Nucl. Inst. Meth. Phys. Res. B 93 : 302 (1994)
4 A. Filliponi and A. Di Chicco, Phys. Rev. B, 51 : 12322 (1995)
5 F.Farges, G.Fiquet, D.Andrault and J.P.Itié, Physica; B : 208&209, 263 (1996)
6 D. C. Koningsberger, "General principle of analysis", In X-ray Absorption, (J. D. Winefordner ed.), John Willey & Sons, New-York. (1989)
7 P. F. Paradis, F. Babin and J. M. Gagne, Rev. Sci. Instrum. 67 : 262 (1996)
8 K.R.Weber, D.S.Hampton, D.R.Merkley, C.A.Rey, M.M.Zatarski and P.C.Nordine, Rev. Sci. Instrum. 65 : 456 (1984)
9 B. K. Teo, "EXAFS : basic principles and data analysis", Inorg. Chem. Conc., Vol. 9, Springer Verlag, Berlin, (1986)
10 P. Eisenberger and G. S. Brown : Solid St. Comm., 29 : 481 (1979)
11 P. C. Nordine, Ceram. Bull., 70 : 71 (1991)
12 International Workshops on Standards and Criteria in XAFS, Physica B, 158 : 701 (1989)

MULTI-ENERGY RADIOSCOPY - AN X-RAY TECHNIQUE FOR MATERIALS CHARACTERIZATION

N. Meyendorf[1], G. Walle[1], H. Reiter[1], M. Maisl[1], H. Bruns[1]
A. Hilbig[2], F. Heindörfer[2], R. Pohle[2], S. Ehlers[2]

[1]Fraunhofer Institut für Zerstörungsfreie Prüfverfahren
Universität, Geb. 37
66123 Saarbrücken, Germany
[2]Otto-von-Guericke-Universität
Magdeburg, Germany

INTRODUCTION

The interaction of X-rays with matter was already analyzed to a high extent by 1926, as the patents by Simon[1,2] and Danin[3] demonstrated possibilities for the production of color X-ray images utilizing energy, atomic number and density dependence. Medical diagnostics applied and developed these ideas first.

Beyer[4] first described practical applications of color radiography for nondestructive materials testing purposes in 1962. The papers by Linke[5] in 1965, and Richter and Linke[6] in 1966 describe the basics and applications of color radiography in industrial materials testing.

Most multi-energy techniques are generally based on the application of computed tomography. Some methods use monochromatic synchrotron radiation, or energy-selective detectors. Information about the composition of material can be derived from the resulting images due to extremely high apparative efforts (e.g.,[7,8,9]). The aim of the work, presented here, was to develop an imaging technique for the characterization of different materials by means of a radioscopic system using a X-ray tube as radiation source.

In this kind of "shadow projection" the influence of the extinction coefficient interferes with the irradiated material thickness according to the law of extinction. Thus, the correlation of radioscopic images obtained at different accelerating voltages does not directly point to a change of the extinction coefficient. To overcome this problem an image linearization and a normalization procedure is necessary

Nondestructive Characterization of Material VIII
Edited by Robert E. Green Jr., Plenum Press, New York, 1998

NORMALIZATION METHOD

After linearization and adjustment the quotient of the image matrices obtained at two different accelerating voltages represents a matrix with its elements being the ratios of extinction coefficients. As the extinction coefficients of many different materials are well-known in terms of radiation energy (e.g.,[10,11]) accelerating voltages can be determined by simple simulation calculations in which the quotients of the extinction coefficients of the materials under investigation have different values. This is important as it is impossible to achieve the same quotient for different layer thicknesses because of the shift in average photon energies of the X-ray spectrum during penetration of the material. However, the investigations presented here proved that a change in the effective atomic number in the path of the rays of $\Delta Z_{eff} \geq 3$ can produce a sufficient clustering of the color values in the image matrix in the color feature space. With the help of simple classificators these clusters can be used for qualitative materials characterization.

A more qualitative approach is the normalization to a reference material. In this method the atomic composition (effective atomic number) of a known material, e.g. of the main component of the material under investigation, is chosen as reference in order to present differing components by means of color values and color saturations. An equidistant step wedge of characteristic thickness according to the test specimens under investigation is produced from the reference material. This step wedge is irradiated together with the test specimens at three different accelerating voltages. The efficiency of the X-ray tube determines the minimum and the maximum accelerating voltages. It is helpful to have informations concerning the compositions of the specimens when choosing the mean accelerating voltage for an optimization by means of simulation calculations. The three gray scale value images obtained by irradiation are referred to in the following as the red, green, and blue color separations of a multi-plane color image. The image obtained at the minimum accelerating voltage can be taken as the **r**ed color separation, the image obtained at the maximum accelerating voltage as the **b**lue color separation and the image obtained at the third accelerating voltage as the **g**reen color separation. The dependence of the extinction coefficients on the atomic number and energy leads to different contrasts in these images which will produce a real-color image when correlated. This real-color image still does, however, neither provide any qualitative nor quantitative information for materials characterization.

First, the gray values GW_i of the step wedge made from the reference material have to be measured for the three radioscopic images. The following functions can be formed with the help of these measuring values and the known thicknesses (d) of steps of the step wedge: $GW_i(\text{measurement}) = f_i(d)$ with $i = (U_1, U_2, U_3)$.

$f_i(d)$ are non-linear regression functions. A transformation functions $f_{T,i}$ map the functions $f_i(d)$ on a selectable performance function $f_Z(d)$ which should be the same for all individual images. $f_i(d) \cdot f_{T,i} = f_Z(d)$

If one selects a straight line for $f_Z(d)$ image adjustment and linearization are achieved in a single step.

Using this method in all three images (color separations) for the reference material the same gray-scale values are produced which characterizes it "gray" in the multi-plane image. If the effective atomic numbers of test specimen differ from that of the reference material due to the changed extinction behavior different gray-scale values will occur in the transformed individual images and therefore, distinctive color values and saturations will occur in the true color image.

Due to the above-mentioned allocation of the individual images to the color channels a higher effective atomic number (higher value of the extinction coefficient) compared to the reference material leads to the highest values in the blue color separation. These regions turn out blue in the true color image. Image regions with smaller effective atomic numbers compared to the reference material will be red.

A possible result of the calibration may also be different performance functions for the individual images. However, it seems to be more elegant to change from the RGB (red, green, blue) model to the HSV model (hue, saturation, value) and to carry out a concerted analysis and processing of color value and saturation.

For this model can be demonstrated that hue (color angle in the ground plane) is the appropriate parameter for materials characterization for the investigations carried out. Color saturation is more or less additional information. Saturation lower than 5% point to a small change of $\Delta Z_{eff} \leq 3$. In the color feature space defined by lightness, hue and saturation image spots (pixels) with the same effective atomic number will form clusters. Differences in material composition can be separated by means of cluster analysis or classification.

EXPERIMENTAL RESULTS

Step wedge images

An arrangement of three step wedges made from steel, tin, and aluminum was irradiated. The irradiation was carried out at accelerating voltages of 81, 100, and 160 kV (corresponding to the mean energies of 62, 77, and 123 keV). The step wedges were mounted to a 5mm steel base plate which was placed on the same side of the experimental set-up as the X-ray sensor and provided a sufficient filtering of scattered radiation for all accelerating voltages. Reference material for the normalization was steel. In the experimental set-up an additional 3 mm aluminum prefiltering was carried out leading to a narrowed irradiation spectrum.

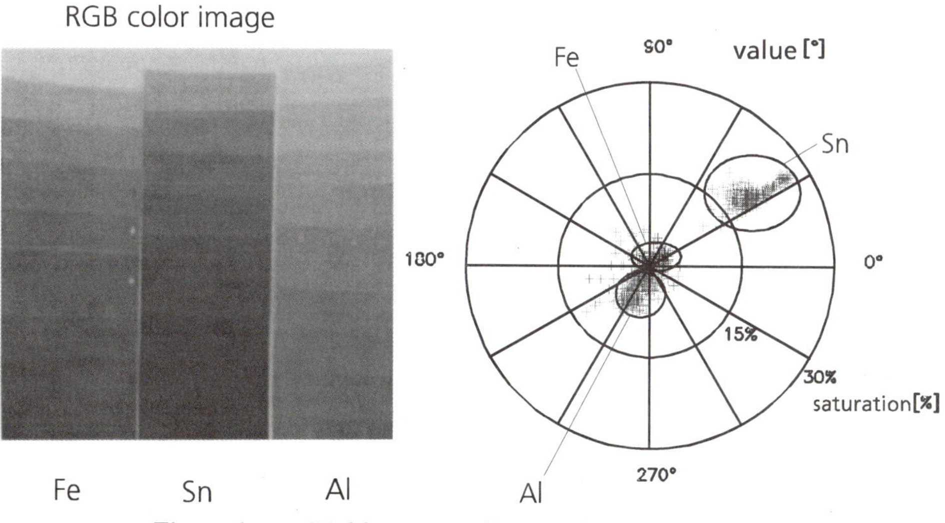

Figure 1. Multi-energy radioscopy image of step wedges

Investigation of a layered system

In the polar diagram in figure 1 the angle corresponds to the color value and the position of each measuring point (radii) corresponds to the color saturation. Iron as reference material shows only weak color saturations for different color values. The minor deviations from zero (i.e., no color = gray) in the polar diagram are caused by image noise and digitalization errors. All measured values of the tin step wedge are distributed at 30° (which corresponds to the color blue in the color image). The variations of the color saturation is caused by the different irradiated tin step wedge thicknesses. The color image of aluminum step wedge has a reddish-yellow color which corresponds to an angle of 240° in the polar diagram. In the investigations carried out here, the color value allows an identification of the material and the color saturation gives information on the thickness of the irradiated material.

The specimen investigated was a one-layer titanium deposition welding onto a copper layer which was explosive plated onto a steel plate. Titanium is used as coating material for platings because of its good strength, low specific weight, and good corrosion resistance. In order to produce the titanium coating a welding method has to be applied which causes an alloy with the interlayer because of the much higher melting point of titanium (1670°C) in comparison to copper (1083°C). The degree and homogeneity of the formed alloy has an impact on the corrosion resistance. The degree of the alloy can be derived from the thickness of the remaining copper layer. The homogeneity of the copper interlayer was evaluated by means of multi-energy radioscopy.

The radioscopic images were produced with the above-described pre-filtering at accelerating voltages of 97, 120, and 160 kV. The normalization functions were calculated from auxiliary images in which a copper step wedge mounted to the base material was imaged with the same irradiation parameters. The normalization was carried out by means of the method described above in order to achieve a linear correlation between the gray-scale value and the thickness of the copper layer.

Figure 2 shows measuring tracks. The color values of these tracks are shown in the following polar diagrams (The image position along the measuring tracks corresponds to the position on the radii, the color value corresponds to their angle). The column value 0 is on the left-hand side for the horizontal measuring tracks. The line value 0 is at the bottom of the vertical measuring track.

The measuring track 3a corresponds to the base material and shows actually good clustering at the color angles of 200-205° (Fig. 3a). The 3b measuring track was assigned to the region of the specimen with almost no influences on the copper layer. The expected scattering of the values in the color plane is, however, limited to the angles between 210-30° (Fig. 3b). Here, the normalization to the layer structure can be seen clearly.

The measuring track 3c shows the changes in welding direction. As long as there is enough welding deposit a relatively stable cluster can be found in the polar diagram at an average value of 23° (Fig. 3c). This is interrupted starting with line 188. The strong variations in color indicate a slightly increased copper content. The deposition welding ends in line 200. Here, the transition to the unwelded copper layer starts, followed by the transition to the base material. This is very nicely demonstrated by the color values.

The specimen was cut following the measuring track 3c in order to check the results predicted by the polar diagrams. The polished sections support the qualitative predictions (Fig. 4).

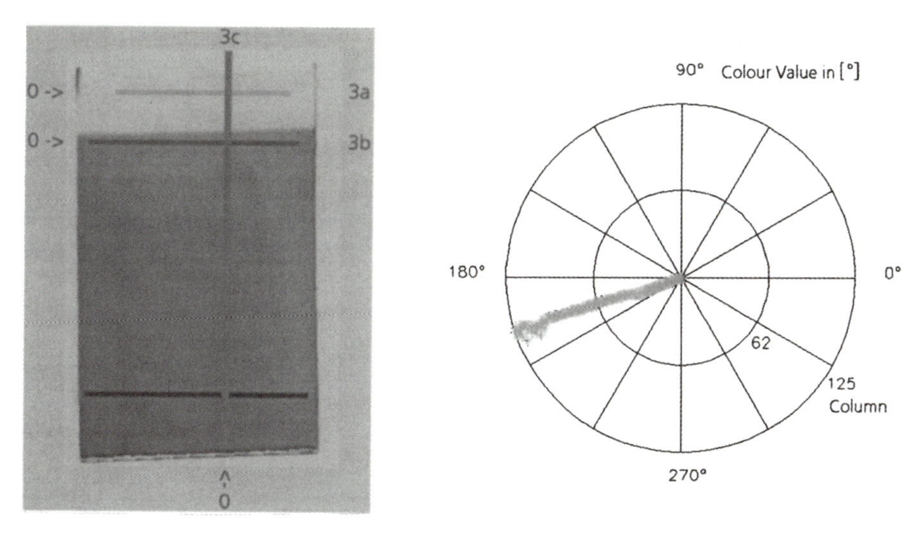

Figure 2. Colored measuring Tracks

Figure 3a. Measuring Track (3a)

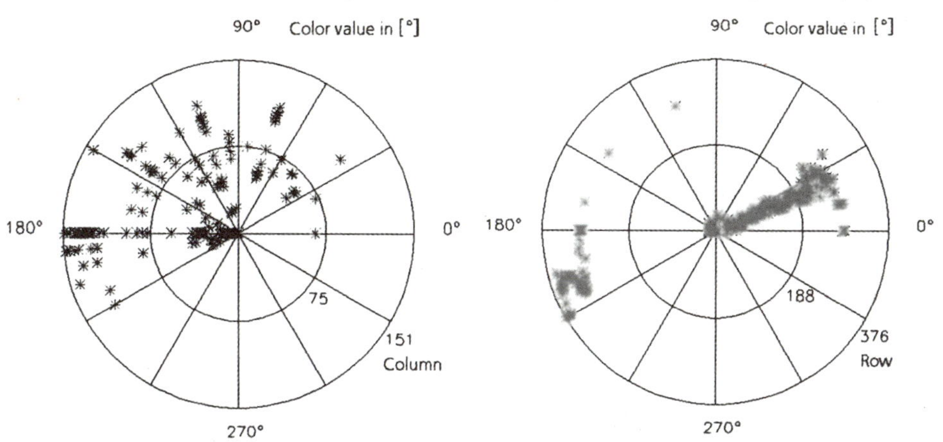

Figure 3b. Measuring Track (3b)

Figure 3c. Measuring Track (3c)

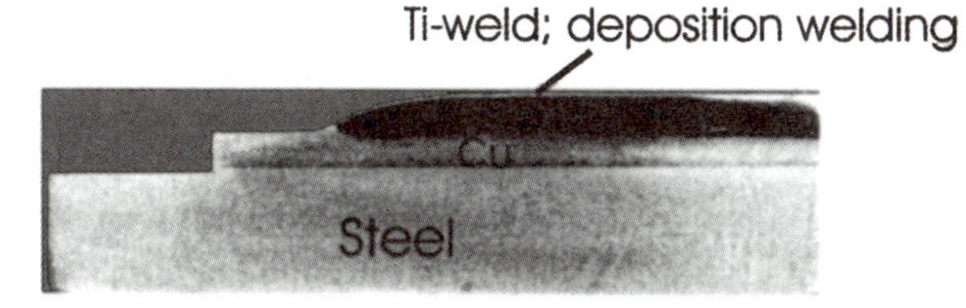

Figure 4. Deposition Welding

SUMMARY AND OUTLOOK

Multi-energy radioscopy with normalization to a reference material is basically an easy-to-realize method with no special requirements to the equipment. Investigations as described above can be done with any radioscopic system. It is possible to identify a material by the comparison of two radioscopic images obtained at different accelerating voltages. In order to evaluate composite materials and to characterize material layers which are covered by a surrounding medium three irradiation energies are necessary. Therefore, the images obtained at three accelerating voltages should be imaged as color radioscopic images and evaluated by means of color image processing. Experiments and simulation calculations showed that this technique also allows the identification of materials in layered systems.

An important result of simulation calculations is the fact that in many cases signal noise and digitalization errors are critical due to the little differences in the gray-scale values of two normalized images. Therefore, low-noise images, i.e. the application of high integration times and an improved gray-scale value resolution (12-16 bit), are necessary.

The practical advantage of this technique is the testing of components which can't be accessed with conventional analytical methods. This is especially a problem for the characterization of inner layers in layered structures. Multi-Energy Radioscopy can also be applied to composite materials.

Acknowledgements

The authors would like to thank the „Deutsche Forschungsgemeinschaft" (German Research Society) for funding the presented work.

REFERENCES

1. F. Simon. Verfahren zur fotografischen Aufnahme von Röntgenstrahlen unter Umsetzung der Härteunterschiede in sichtbare Töne und Aufnahmematerial, *DRP* 442 807 (1926).
2. F. Simon. Improvements relating to X-ray Photography, *Great. Brit. Pat. 276 678 (1927).*
3. L. Danin. Röntgenbild und Verfahren zu seiner Herstellung, *DRP 437 507 (1926).*
4. N. N. Beyer. New Techniques in Radiography, *Proceedings of the Third Annual Symposium on Nondestructive Testing of Aircraft of the Missile Components*, San Antonio, Texas (USA) 1962.
5. D. Linke. Beitrag zur Anwendungsmöglichkeit der Color-Radiografie in der industriellen Werkstoffprüfung, *Belegarbeit der TH Magdeburg*, Institut für Werkstoffkunde und Werkstoffprüfung.
6. Richter und D. Linke, 18., 19. Mitteilung der Forschungsgruppe für Zerstörungsfreie Werkstoffprüfung im VEB Schwermaschinenbau „Karl Liebknecht" Magdeburg, *die Technik* 7, 8 and 9 (1966).
7. Engler, Friedman, Armstrong, Determination of material composition using dual energy computed tomography on a medical scanner, *Proc. of Indust. Computed Tomo. Topical*, ASNT (1989).
8. R. Leukel. Mikro-Computer-Tomographie mit Dual-Energy-Rekonstruktion, *Dissertation*, Saarbrücken (1994).
9. J.H. Kinney, M.C. Nichols, X-Ray tomographic microscopy (XTM) using synchrotron radiation, *Review of Materials Science* 22 (1992)
10. G.L. Clark. *The Encyclopedia of X-Rays and Gamma-Rays*, Reinhold Publishing Corporation, New York; Chapmann & Hall Ltd., London (1963).
11. http://physics.nist.gov/PhysRefData/XrayMassCoef/

IN-SITU, POSITIVE MATERIALS IDENTIFICATION AT ROOM AND ELEVATED TEMPERATURES WITH MODERN, PORTABLE, OPTICAL EMISSION AND X-RAY FLUORESCENCE ANALYZERS

Stanislaw Piorek, Jyrki Ojanpera, Ewa Piorek, and James R. Pasmore

Metorex Inc.
Princeton, New Jersey

ABSTRACT

Spectrometric techniques utilizing the phenomena of optical emission (OE) and X-ray fluorescence (XRF) have been used for years for QA/QC purposes by metals manufacturing and reprocessing industries. The advent of portable, OE and XRF based alloy analyzers - stimulated by the development in electronics and computer technology - made possible in-situ grade identification and analysis of metals of construction either before their use or in post-factum verification. The advantages of portable OE and XRF alloy analyzers were quickly discovered by petroleum refining, nuclear, aviation and power generating industries, in which a 100% positive materials identification, (PMI), is routinely required.

In this paper we describe the operation principles of these state-of-the-art analyzers, and discuss their performance. We also report on the extension of useability of portable OE and XRF alloy analyzers for the identification and analysis of metallic objects operating at elevated temperatures. Specifically, it will be shown that with only a minor modification to its probe, the x-ray analyzer can identify in 5 sec the grade of an alloy heated up to 500°C (932°F). Similarly, a portable, OE-based analyzer can accurately assay the hot element of construction after only a minor extension of the preburn time.

By extending the analytical capabilities of the portable alloy analyzer to measure objects operating at up to 500°C, it is now possible to perform the PMI on working installations without the necessity for disruptive and costly downtime.

INTRODUCTION

A 100 percent Positive Materials Identification (PMI) is currently a mandatory requirement in many industries, which range from those critical to life and environment, such as nuclear and aviation, to those which are driven often by just pure economics as in

Nondestructive Characterization of Material VIII
Edited by Robert E. Green Jr., Plenum Press, New York, 1998

the case of the metal scrap industry. The PMI is most often associated with alloys, as these are still the predominant materials of construction.

Many properties of alloys, such as mechanical strength, corrosion resistance, etc., depend on their chemical makeup. Steel mills have been providing the heat analyses of their products, and for years this has been the sufficient assurance that the alloy meets the requirements of the job for which it has been designed. Continuous progress in chemical, automotive, semiconductor and aviation industries resulted in the demand for new materials capable of operating at higher temperatures, hostile environments, and often at the extremities of absolute zero temperature or high pressure. A new discipline of materials engineering emerged as a tool to conceive and design the materials (alloys) with predetermined properties. As the result of this we now have more different materials (alloys) than just twenty or thirty years ago. All of this creates even greater need for fast, in-situ, and reliable PMI.

Both X-ray fluorescence spectrometry and optical emission spectrometry stand on their own as analytical methods for alloy analysis. However, they also complement each other. While X-ray fluorecscence is excellent in the analysis of alloying elements in high temperature alloys and stainless steels, it cannot analyze carbon in carbon and low alloy steels. On the other hand, the optical emission spectrometry has no problem with carbon analysis in carbon and low alloy steels, but does have difficulty in transforming a high temperature alloy into a plasma, a prerequisite for optical emission analysis.

ALLOY IDENTIFICATION - APPROACH TO THE PROBLEM

Alloys are best characterized by their chemical composition. The examination of the specifications of thousands of alloys reveals that from about 50 different elements involved in the alloying process, any given alloy contains only 10 to 20 elements. Out of this number, only about 10 are responsible for the main characteristics of any given alloy. Thus, the number of elements to be analyzed in order to identify an alloy is somewhere between 10 to 15.

Present instrumentation usually employs two approaches to alloy identification. The first which is more traditional and straightforward, is based on performing accurate quantification of alloying elements and then comparing the result with tables of specifications listing chemical compositions along with alloy designations (trade names). This mode of identification is often called *identification via a grade table*. The ruggedness of this method greatly depends on the accuracy of analysis, which requires longer rather than shorter measurement time per sample. This is especially true for XRF whose precision improves twofold for each fourfold increase of measurement time.

The second approach is based on the concept of *pattern recognition* in which spectral features of the unknown alloy are compared for a match with previously stored features of known reference alloys. The spectral features are usually spectral intensities from individual elements in the measured sample. The multidimensional vector of intensities from the unknown alloy sample is compared for colinearity with similar vectors generated for known alloy standards, and stored in the analyzer memory (library). The alloy reference whose intensity vector is the most similar to that of unknown is selected as a match. An example of the identification formula, which has been successfully used for the last ten years in the X-MET™ family of portable XRF analyzers is given below[1]. As can be seen from it, the value of the calculated comparison parameter, t, can also be interpreted as the squared, normalized distance between two points, x and r, in a multidimensional space. The acceptance criteria for parameter t, which are solely based on probability and statistic, can be adjusted by the operator to the required confidence level.

$$t_r = \sum_{i=1}^{n} \frac{(I_{ix} - I_{ir})^2}{\sigma^2(I_{ix}) + \sigma^2(I_{ir})}$$

where:
- I_{ix} and I_{ir} are intensities of the i-th element of the unknown, x, and the r-th reference, respectively;
- $\sigma(I_{ix})$ and $\sigma(I_{ix})$ are the standard deviations of these intensities;
- n is the number of intensities (elements) measured in the sample.

When using the XRF analyzer, identification via pattern recognition is much faster than via a grade table. Usually, the first can be completed within 5 sec, while the identification via grade table typically requires about 60 to 100 sec measurement time per sample. This distinction does not apply to OES instruments.

A similar identification algorithm is also used in modern portable optical emission spectrometers.

The advantage of speed and no need for knowledge of alloy specifications in the identification via pattern recognition is somewhat compromised by the absolute need of possession of physical samples (standards) of various alloys; the task not always feasible.

PMI WITH PORTABLE X-RAY ANALYZER

Initial applications of portable XRF analyzers were in mining and prospecting[2]. The progress in electronics (on-board memory, microprocessors) made possible their wider acceptance and use for alloy identification and analysis. Since then, thousands of these analyzers have been sold making alloy sorting and analysis truly a "flagship" application for field portable XRF analyzers.

A modern, portable x-ray analyzer for alloy analysis, consists of a hand-held probe connected to an electronics unit. The probe contains x-ray source(s), a detector, and a means of a reproducible presentation of the sample for measurement. The electronics unit accepts the signal from the probe, processes it, and displays the results. It also contains power supplies (rechargeable batteries) and interfaces for communication with the operator and the peripheral devices. Alternatively, in the most recent designs, the electronics unit may be replaced by a dedicated multichannel analyzer plug-in board in a portable, battery-operated computer[3]. A modern, portable XRF analyzer features not only empirical calibration software, but also the so called "standardless" calibration software, based on the fundamental parameters approach, which is usually included with the analyzers featuring semiconductor detectors[4].

A small, rugged, sealed radioisotope capsule, emitting either x-rays or low energy gamma rays, is a preferred source of primary radiation for portable instruments. The typical radioisotopes used are Fe-55, Cd-109, and Am-241, with strengths of up to 40, 20, and 30 mCi, respectively. An essential part of the probe is its x-ray detector. This can be a high resolution, gas filled proportional detector, or a semiconductor detector for even better energy resolution. The better the energy resolution, the better the analytical performance of the instrument. Not a trivial matter for a portable instrument is its weight and size, because for obvious reasons, the portable analyzer should not challenge the operator "gravitationally". This and other requirements pertaining to portable XRF instruments have been recently addressed within the forum of the International Standardization Organization[5].

Table 1 shows typical performance data for a modern, contemporary field-portable XRF analyzer operating in assay mode, while Table 2 illustrates performance of the analyzer in pattern recognition mode for alloy identification.

Table 1. Standard performance data for a typical, contemporary, commercially available Portable XRF Alloy Analyzer (X-MET^{tm}, Models 880, 960)

Alloy Group	Ti	Cr	Mn	Fe	Co	Ni	Cu	Zn	Nb/Mo	Sn	Pb
Low Alloy	0.01	0.04	0.1	0.25	0.25	0.10	0.05	0.10	0.006	0.15	0.15
Steels	0.02	0.1	0.2	0.5		0.25	0.15		0.01		
Stainless	0.015	0.20	0.10	0.20	0.20	0.20	0.06	0.20	0.01	0.3	0.05
Steels	0.03	0.30	0.20	0.30		0.30	0.10		0.03		0.30
Ni/Co	0.10	0.20	0.10	0.12	0.10	0.20	0.05	0.30	0.02	0.3	0.15
Alloys		0.50	0.30	0.50	0.50	0.50	0.30		0.08		
Cu-based											
Alloys,	0.02	0.1	0.02	0.02	0.05	0.05	0.15	0.07	0.01	0.008	0.20
Brass/Bronze			0.06	0.06		0.08	0.40	0.30		0.20	0.30
Aluminum	0.02	0.05	0.10	0.05	0.05	0.04	0.05	0.06	0.003	0.005	0.01
Alloys		0.20		0.10					0.005	0.20	0.02
Titanium	0.4	0.10	0.10	0.06	0.05	0.05	0.02	0.02	.008	0.005	0.01
Alloys											

Notes: The values listed are typical precision ranges in percent absolute for total assay time per sample not longer then 60 sec. The differences are due to the analyzer model and/or to the radioisotopes used.

Table 2. Performance of a portable X-ray analyzer in alloy identification.

Alloy Group	Identification Results (% feasible)
Ni/Co Alloys	100
Cu Alloys	90 - 100; 95 - 100
SS and HiTemp Alloys	90 - 100; 100
Cr/Mo Steels	95 - 100; 100
Low Alloy Steels	65 - 80; 90 - 100
Ti Alloys	95 - 100
Al Alloys	90 - 100; 95 - 100

Note: If two results are given, the first refers to gas-filled detector while the second to semiconductor detector.

Figure 1. X-MET^{tm} probe shown with high temperature shield installed.

Measurements at Temperatures up to 500°C

There has always been a definite need in the petroleum industry for the ability to identify or assay metallic components of an industrial installation without the need for a costly shutdown. The difficulty of measurement stems from the fact that refinery installations usually operate at elevated temperatures, frequently up to 500°C. Surprisingly, the obstacle of elevated temperature can be overcome with a very simple heat dissipating adapter mounted over the standard probe, as shown in Figure 1.

A simple shield made of 0.5 mm thick steel is mounted over the probe aperture. This helps to dissipate the heat during measurement when the probe is in direct contact with the measured object. The sides of the shield which come in direct contact with the probe are perforated to hinder the heat flow to the probe. In addition a plastic film window covering the probe aperture is aluminized to mirror finish to reflect the heat emanating form the analyzed object. Up to a 60 sec. long measurement can be performed on the hot object. The probe cools down sufficiently by convection between consecutive measurements if they are taken at about 30 sec. apart. If shorter measurements are performed, such as 5 sec in identification mode, the probe cools down in just a few seconds. No statistically significant degradation in quality of the results has been observed due to the elevated temperature of the sample.

PMI WITH PORTABLE OPTICAL EMISSION SPECTROMETER

The excellent analytical sensitivity and accuracy offered by laboratory optical emission spectrometers is achieved at the cost of their large size and mass. This is because most of these spectrometers have their optics constructed around a number of photomultipliers (PM for short) as detecting elements. Since even the smallest PM device is about half inch in diameter, this imposes dimensional restrictions on the spectrometer optics. Consequently, the smallest spectrometer optics cannot be reduced below about 50 cm. This has been one of the two major roadblock in the successful development of portable optical emission spectrometers. The other one was the fact that the measured sample had to be positioned right at the entrance slit of the optics. The radical change came after flexible glass fiber optic bundles became a commercial reality. Now, a small pistol-like probe could be used to strike the arc between the electrode and sample, and the light of the plasma could be carried via a fiber optic bundle to the spectrometer resting on a dolly[6]. However, even that solution had its problems because the UV light of the plasma caused premature deterioration of the glass fiber optic. The situation improved diametrally with a radically new spectrometer design known under the name of ARCMET™. The new design represents a total departure from the traditional concept of transmitting the light via fiber optic bundle from probe to bulky spectrometer. The whole optics of the new spectrometer is now contained inside the probe. There are no signal losses since there is no need to transmit the light trough the fiber optics. The speed of the fiber optics is utilized solely to transmit to the main computer already digitized spectral data. This feat has been made possible by using as a detector a 2048 element, 2 inches long, linear photodiode array (PDA). Due to the small size of the PDA detector, the focal length of the spectrometer could be reduced to 15 cm, which allowed it to fit inside the probe clamshell[7]. The new spectrometer is designed to operate in argon atmosphere, which allows carbon to be analyzed from its direct line of 193 nm .

Another exceptionally useful feature of the new design is the ability of the PDA element to record the continuum of spectrum within the range from 170 to 340 nm, rather than the limited number of discrete, preselected wavelengths, as is the case with the photomultiplier

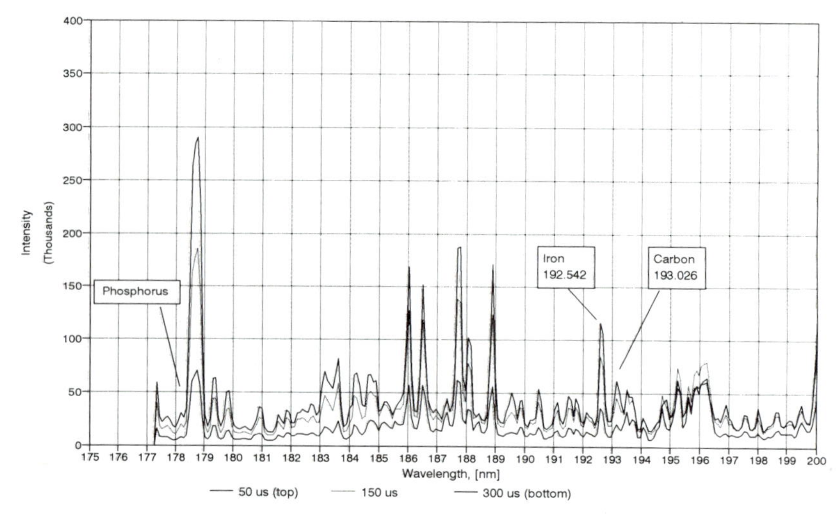

Figure 2. Fragments of emission spectra of sample of carbon steel C1117 shown for different duration of the burn pulse.

based spectrometers. This feature allows the operator to select analytical lines as well as proper preburn and burn time in a matter of minutes, as opposed to lengthy guesswork and a tedious daily line profiling characteristic of PM based spectrometers. Figure 2 illustrates several spectra of low alloy steel collected from the same sample, the only difference being burn time.

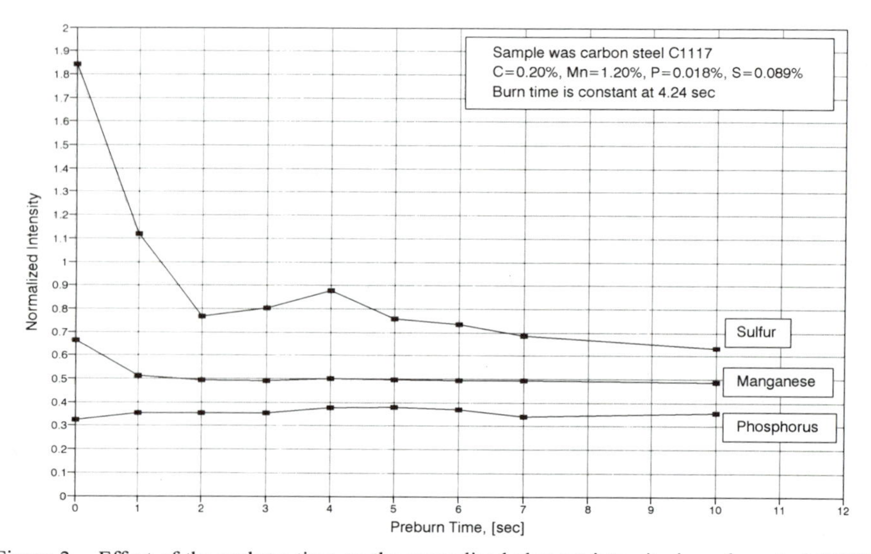

Figure 3. Effect of the preburn time on the normalized element intensity in carbon steel C1117

It is easy to see how the burn time affects the intensity of elemental lines and based on that select proper length of the burn. Figure 3 shows the intensities of sulphur, manganese and phosphorus, all normalized to iron, plotted as function of preburn time. Based on this plot we can say that preburn time of 7 sec. or longer should be selected as a working parameter to assure repeatable burn conditions from sample to sample. Again, such analysis can be performed in a matter of minutes because the operator has immediate access to the full spectra and can select via instrument software these elemental lines which offer the best analytical performance. This lack of constrains on the selection of analytical and reference lines is perhaps the most valuable feature of the design.

The performance of the analyzer described above is summarized in Table 3. The ARCMET™ in a standard configuration can measure hot surfaces up to 200°C. Hotter surfaces require a special adapter made of stainless steel. Deeper grinding of sample surface is usually required accompanied by an extended to 15 sec. preburn time to cut through the oxidation layer. At 25°C ambient temperature maximum one measurement per 6 min. can be made to allow the probe to cool down. Typically, depending on the element, the repeatability of measurement will be worse by a factor of two to seven when compared to room temperature results.

Table 3. Performance of a portable optical emission analyzer at different object temperatures (ARCMET, Model 900 and 930)

	Flat sample, at 25°C		Flat sample, at 400°C		Tubular sample, at 400°C	
Element	**Aver.**	**Stdev**	**Aver.**	**Stdev**	**Aver.**	**Stdev**
Fe	70.32	0.02762	70.55	0.22129	70.37	0.33912
C	0.037	0.00168	0.050	0.01348	0.043	0.00259
Si	0.32	0.00806	0.30	0.02000	0.32	0.01164
Mn	1.42	0.00530	1.44	0.00894	1.88	0.02707
Cr	18.76	0.05653	18.56	0.47486	19.33	0.34115
Ni	8.33	0.05850	8.25	0.51551	7.25	0.18319
Mo	0.17	0.00374	0.18	0.00657	0.23	0.00727
Cu	0.21	0.00211	0.21	0.00954	0.15	0.00272
Ti	0.003	0.00022	0.0053	0.00246	0.010	0.00120
Nb	0.03	0.00309	0.027	0.01811	0.038	0.00539
Al	0.068	0.00125	0.072	0.00488	0.0078	0.00013
V	0.050	0.00090	0.048	0.00917	0.083	0.00379
W	0.088	0.00068	0.083	0.00845	0.036	0.00443
Co	0.18	0.00144	0.20	0.02762	0.21	0.01405
S	0.011	0.00043	0.018	0.00538	0.031	0.00324
P	0.018	0.00305	0.012	0.00848	0.024	0.01020

Notes: - All data above are in % absolute.
- Stdev is determined on a series of at least 3 measurements
- Sample measured was stainless steel **AISI 304**.

CONCLUSIONS

On-site, in-situ PMI for alloys is now possible with use of the portable XRF and OES instrumentation, which offers performance often comparable to that of the laboratory systems. In addition, the on-site, multielemental analysis of alloy can be completed in 5 to 60 sec. per sample, and at the cost per sample of less than one dollar. Portable instrumentation is also capable of analyzing the metal objects at elevated temperatures. Elevated temperatures promote corrosion of material, which occurs when two dissimilar, incompatible with each other alloys are mistakenly joined together. This feature is particularly useful at petrochemical plants because it allows inspection of working installations without a costly downtime.

Perhaps the *Titanic* disaster would never have happened if optical emission spectrometers had been available then.

REFERENCES

1. S. Piorek - "Modern alloy analysis and identification with a portable X-ray analyzer", in Adv. in X-Ray Anal., Vol. 32, pp 239 - 250, Plenum Press, New York, 1989.

2. "Radioisotope X-Ray Fluorescence Spectrometry" - Report of a Panel held in Vienna, Austria, May 13-17, 1968, Edited by C.G. Clayton, Technical Report Series No. 115, IAEA, Vienna, 1970.

3. S. Piorek, J.R. Pasmore - "Standardless, in-situ analysis of metallic contaminants in the natural environment with a PC-based, high resolution portable X-ray analyzer", Proc. of the 3rd Int. Symp. on Field Screening Methods for Hazardous Wastes and Toxic Chemicals, Las Vegas, Feb. 24-26, 1993, pp. 1135-1151.

4. P.F. Berry, G.R. Voots - "On-site verification of alloy materials with a new field-portable XRF analyzer based on a high-resolution HgI_2 semiconductor X-ray detector", in Non-Destructive Testing (Proceedings of the 12th World Conference on Non-Destructive Testing, Amsterdam, The Netherlands, April 23-28, 1989), Boogaard J. and Van Dijk G.M., editors, Elsevier Science Publishers, Amsterdam, 1989, pp. 737- 742.

5. M.S. Simmons, S.E. Chappell - "Performance requirements for field applications of portable and transportable X-ray fluorescence spectrometers", Proc. of the 3rd Int. Symp. on Field Screening Methods for Hazardous Wastes and Toxic Chemicals, Las Vegas, Feb. 24-26, 1993, pp. 1152-1161.

6. K. Slickers - "Automatic Atomic Emission Spectroscopy", Sec. Edition, Bruhlsche Universitatsdruckerei, D-35334 Giessen, P.B. 10 04 51, Germany, 1993.

7. M-L. Jarvinen, H. Katajamaki, J. Koskinen, J. Ojanpera, S. Piorek S. - "On-site analysis of steels", presented as a poster at "The Pittcon '91", Pittsburgh Conference, Chicago, March 4 - 8, 1991.

PROBLEM OF CHARACTERIZATION OF VACUUM VESSEL SHELL FOR SPHERICAL TOKAMAK GLOBUS-M AT DIFFERENT STAGES OF ITS FABRICATION

German P. Gardymov,[1] Vasilly K. Gusev,[2] Naum Ya. Dvorkin,[1]
Viktor M. Komarov,[3] Evgeni G. Kuzmin,[3] Vjacheslav V. Mikov,[1]
Vladimir B. Minaev,[2] Vladimir I. Nikolaev,[2] Alexander N. Novokhatsky,[2]
Klara A. Podushnikova,[2] Igor E. Sakharov,[4] Nikolai V. Sakharov,[2]
Sergei V. Shatalin,[4] Vitaly V. Shpeizman[2]

[1] Northern Plant, S.Petersburg, Russia
[2] A.F.Ioffe Physico-Technical Institute, S.Petersburg, Russia
[3] D.V.Efremov Scientific Research Institute of Electrophysical Apparatus, S.Petersburg, Russia
[4] State Technical University, S.Petersburg, Russia

INTRODUCTION

Spherical tokamak shows considerable promise for the peaceful uses of thermonuclear energy in future. A number of spherical tokamak projects are developed in different research centers (MAST project at Culham Laboratory, UK, NSTX at Princeton University, USA, ETE, Brazil, GLOBUS-M, Russia etc.). Spherical tokamak GLOBUS-M[1] is under construction now at A.F.Ioffe Physico-Technical Institute, S.Petersburg. Its basic parameters are: plasma major radius 0.36m, plasma minor radius 0.24m, plasma current ≤0.5MA, toroidal magnetic field ≤0.62T. In all the projects the vacuum vessel is one of the most complex and critical part of the device. The overall dimensions of the shell are: diameter -1.96m, height - 1.85m. The large volume of the vessel, high requirements necessary for its final shape and sizes, for the mechanical properties of its material, as well as for magnetic ones generate a need for development of the special technological procedure of the vacuum vessel shell fabrication and methods of material state characterization and shape control at the different stages of the shell manufacturing. For their realization, a number of novel technological facilities was designed and made at the state enterprise "Leningradsky Severny Zavod (Northern Plant)", Sankt-Petersburg, Russia. The developed procedure of spherical tokamak vacuum vessel shell fabrication and the control of its shape, sizes and material state is an example of application of nondestructive characterization methods to the real design of large weight and complex configuration.

This project is fulfilled in the framework of the conversion program and is funded by the International Science and Technology Center.

PECULIARITIES OF ENGINEERING DECISIONS, USED AT FABRICATION OF TOKAMAK GLOBUS-M VACUUM VESSEL

Stainless steel 12Cr18Ni10Ti and 08Cr18Ni10Ti was chosen as a material for manufacturing of the vacuum vessel (Fig.1).

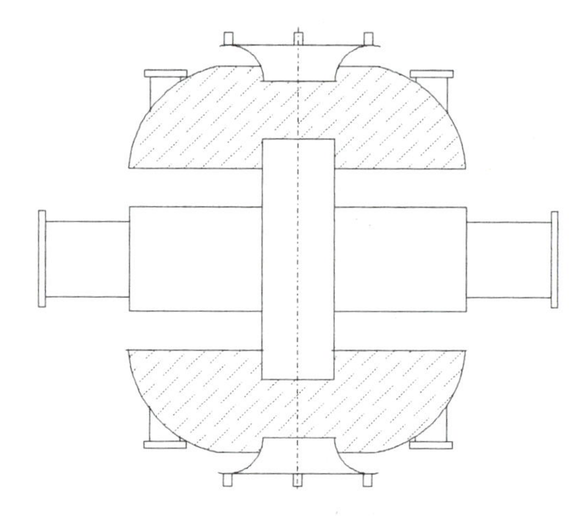

Figure 1. A general view of the tokamak GLOBUS-M vacuum vessel shell

The vacuum vessel is an all-welded shell, strengthened by a central belt with 14 mm thickness of a wall. A central belt joins two hemispheres by thickness of 3 mm. The hemispheres through upper and lower cups are connected with the central cylinder. The hemispheres have "near-pure-compression" shape relatively the atmosphere pressure. The variable thickness of cup walls provides their connection both with hemispheres of 3 mm wall thickness, and with the central cylinder, of 2 mm wall. Diagnostic ports and hatches for access to inner elements are located on periphery of the chamber. All elements of the chamber are connected by welding. The welds should provide vacuum 10^{-8} torr inside the vessel. The deviation from theoretical profile has to be no more than a half of material thickness. As the length of welds is great, and their form is rather complex, the automatic and semiautomatic welding in environment of argon is used in 90% cases. The control of results of machining and welding was implemented by three-coordinate measuring machines with the accuracy 0.25 microns. The quality of welded seams was checked up by X-ray testing. The vacuum control of connections was made by a helium detector. The strength control of shell included the loading up to overpressure 140 kPa during 5 minutes in an armored tank with a water. As a final test, the shell is loaded by overpressure 200 kPa in autoclave.

Manufacturing of the cups

The manufacturing of cup (see Fig.2) is a process of the deep drawing in a special die on press of double action. The drawing factor is 0.3, that exceeds a permissible degree of deformation for the given material for one operation. The total process of a cup molding was carried out for 12 steps. After the seventh step the heating up to 1050^0C within 1 hour with the cooling in air was made. Thus with the purpose of obtaining the smooth change of thickness of a wall from 3.0 mm on an external edge down to 2.0 mm on an internal edge the operation of drawing was performed by a non-traditional method, namely: a combination of drawing with forced ironing of a wall.

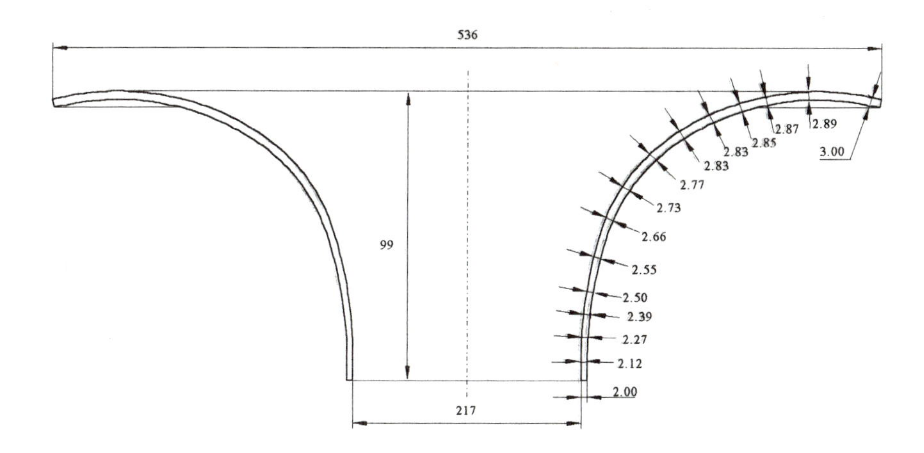

Fig. 2. The varying thickness of a cup.

Then the cups were welded with the cylinder. This unit was undergone by the certification for the geometrical sizes, quality of welding, mechanical and vacuum properties.

Drawing of hemispheres

For hemisphere deep drawing a special die was designed and made. A polyethylene film of 70 ...80 microns thickness and the special oil was applied as a technological lubrication at drawing. The drawing factor at fabrication of a hemisphere was ~0.79. Drawing was made for 25 ...30 steps up to depth 330±5 mm. The depth of drawing for one step was 8 ...12 mm. After achievement of drawing depth more than 140 mm, local surface ripples of 6 ...12 mm altitude took place due to increase of a gap between a punch and matrix. To remove it the correction of the form of a hemisphere was made by the method of template rolling of that part of a surface where the ripples were formed.

Certification of hemispheres after drawing

The certification parameters of hemispheres after drawing were following:

- Thickness of a wall in a radius direction
- Internal diameter of a cylindrical part
- Deviation of the real contours of a hemisphere from theoretical one.

The thickness of a wall was measured with an error of measurement of 0.01 mm. For all hemispheres the minimum wall thickness in their upper parts was 2.75 ..2.80 mm, that corresponded to the requirements of the design documentation.

The internal diameter of hemispheres varied from 1263.0 to 1263.3 mm.

The measurement of deviation of the true form of contours of hemispheres from the theoretical one was carried out by 3-D measuring machine of model MISTRAL made in DEA (Italy). The accuracy of measurements was better than 0.25 microns. The results of certification have demonstrated the conformity of fabricated hemispheres to the requirements of the design documentation.

In addition a state of material of hemispheres was checked up. The results of revision of magnetic properties showed that the magnetic phase appeared in all parts of the hemisphere, especially in zones, where correction of the profile of the hemisphere by rolling was made. The measurement of magnetization was carried out by two methods. The first of them was nondestructive, but indirect. In this case the force of attraction of a calibrated magnet was measured. In the second one the plates were cut from the various parts of hemisphere and were placed into external magnetic field, varied in the range 0 - 3.5T. Measurements performed with Hall's gauge indicated significant, up to 30%, increase of external magnetic field near the plate edges without visible saturation of additional magnetic induction, produced by the material even up to a maximum external field. X-ray structural analysis have showed that the content of martensitic phase after deep drawing reached more than 50%. It appeared that the small amount of magnetic phase arose as a result of any mechanical treatment.

For magnetization removal, the hemispheres were subjected to heat treatment in the following mode:

- Heating up to temperature 950 ^{0}C during 30 minutes
- Cooling on air

In the process of preparation to heat treatment the analysis of the dependence of material strength on the annealing temperature was studied (Fig.3).

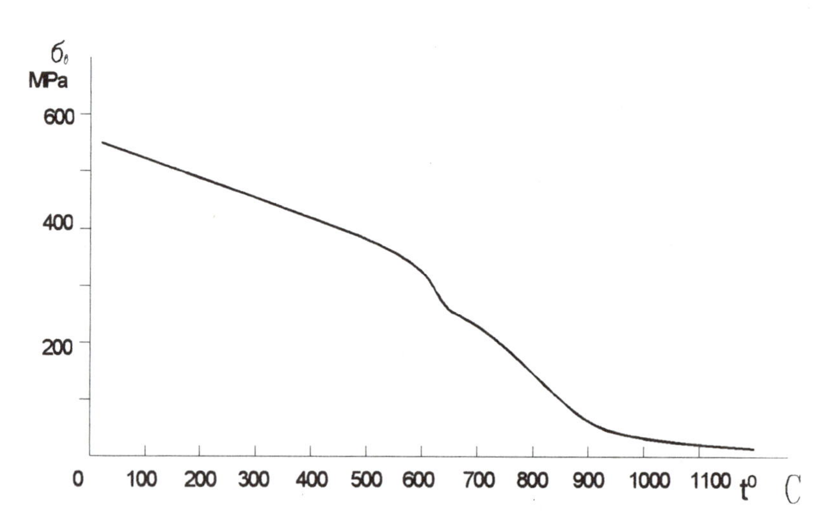

Fig.3. Strength dependence on the annealing temperature

It is seen from Fig.3, that at temperature 950^{0}C the value of ultimate tensile stress decreases down to 43 MPa. At the same time tests revealed that heating of small samples of the material at this temperature completely removed magnetization in external magnetic field. To provide the shape preservation it was important to make the temperature gradient minimal. For temperature control a lot of thermocouples was used. Scheme of their arrangement on a hemisphere at heat treatment is shown on Fig.4.

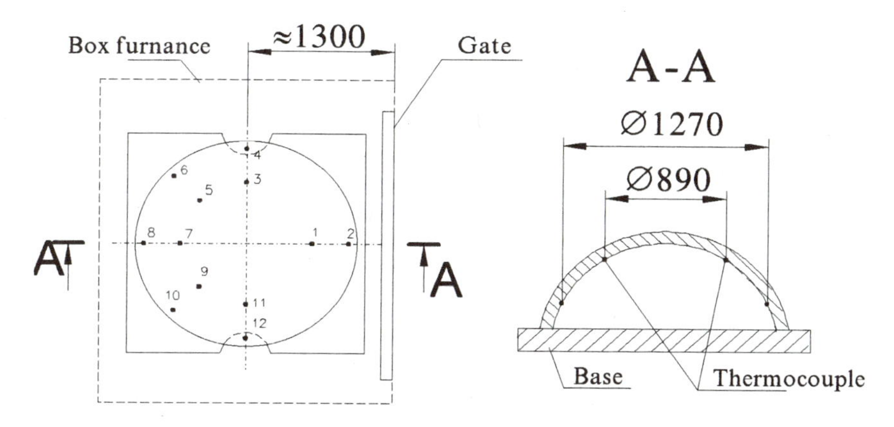

Fig.4. Scheme of thermocouples arrangement on a hemisphere at annealing

The certificated hemispheres were moved to the specialized working places for leak tests at overpressure 100 kPa. After these tests the hemispheres passed the final machining before welding with ports, cups and the belt. The ports of the complex form were placed as required with gaps in welded connections no more than 0.1 ..0.2 mm. After machining the hemispheres were certificated by measuring with the three-coordinate machine and then were undergone to electropolishing and processing by boron carbide.

Manufacturing of flanges

For manufacturing of flanges and ports it was necessary to develop a series of the complex technological methods, connected mainly with a used type of vacuum seal. The scheme of vacuum seal is shown in Fig. 5.

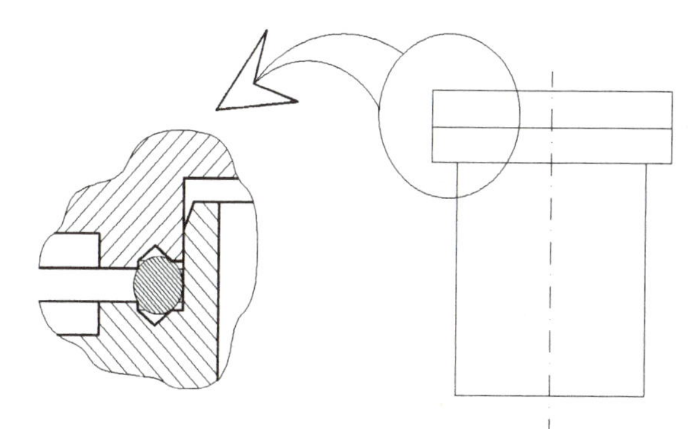

Fig. 5. The scheme of vacuum seal.

The vacuum sealing is provided by copper or aluminum gusset between the edges of flange grooves. This type of sealing is usually used for flanges of circular shape. In GLOBUS-M vacuum vessel the same principle of sealing was implemented for flanges of nonstandard configuration (close to rectangular). In this case high precision of manufacturing was provided by the use of multi-coordinate machine.

Manufacturing of a belt

The belt machining was made in a few stages:
- Preliminary processing.
- Diagnostic ports welding in special fixture;
- Test of all welding connections of a belt by a method of vacuum operation;

Fine processing of belt edges to agree its dimensions with those of hemisphere.

Assembly of the vacuum vessel shell

The manufacturing of the vacuum vessel shell included the following stages:

1. Welding of the upper hemisphere with a belt on the special multicoordinate installation in a semiautomatic mode.
2. Welding of the lower hemisphere in a similar way.
3. Certification of geometrical parameters of obtained assembly unit.
4. Machining of this assembly unit for the installation of the central core of 536 mm in diameter.
5. Certification of geometrical parameters after machining for determination of final length of the core.
6. X-ray control of welded connections.
7. Welding of the second cup with the core.
8. Cutting off the technological extra- edges of the core.
9. Mounting the core and shell in the special multicoordinate installation and their welding.
10. Certification of geometrical parameters of the shell.
11. X-ray control of welded connections.
12. Leak test.
13. Test with overpressure in autoclave.

SUMMARY

1. Technological process of GLOBUS-M vacuum vessel shell manufacturing was developed. For vacuum vessel manufacturing forty five units of novel technological equipment were designed and manufactured. Five special devices for automatic welding were made and reconstructed.
2. Special facilities were created for assembly and tests of the vacuum vessel and its parts.
3. Elements and units of the vessel passed vacuum and mechanical tests.
4. Magnetization of a stainless steel occurred in the process of drawing and mechanical treatment was studied and eliminated by means of thermal treatment.
5. Precise measurements of vacuum vessel details and units were performed. The deflection of dome shape from designed one don't exceed ± 1.5 mm.
6. Full scale model of the vacuum vessel was manufactured. The model was subjected testing under extra pressure of 2 atm., accompanied by strain measurements and thermo-cyclic testing. At present the real vacuum vessel for the spherical tokamak GLOBUS-M is manufactured.

REFERENCES

1. V.E.Golant, V.K.Gusev, V.V.D'yachenko et al., "Basic peculiarities of GLOBUS-M spherical tokamak project", *Proc. of 16th Conf. on Fusion Energy, IAEA-CN-64/GP-15*, Montreal, Canada (1996).

MEASUREMENT OF MATERIAL PROPERTIES FOR EARLY DETECTION OF FATIGUE DAMAGE IN HIGHWAY BRIDGE STEELS

Richard A. Livingston[1], Ward Johnson[2], David McColskey[2], Christopher Fortunko[2] and George Alers[2]

[1]Exploratory Research Team, HNR-2
Federal Highway Administration
McLean, VA 22101
[2]Materials Reliability Division
National Institute of Standards and Technology
Boulder, CO 80303

INTRODUCTION

A recent workshop sponsored by the Federal Highway Administration (FHWA) into research needs for structures concluded that fatigue cracking is the most critical problem for the near future (FHWA, 1992). As a result, a major research thrust has concerned NDT methods to detect fatigue cracks. However, it would be extremely desirable to identify steel that is a high risk for cracking before the cracks themselves appear. A related problem is that fatigue crack prediction is based on the knowledge of the number of fatigue cracks that the steel has experienced. However, for most existing bridges, a complete fatigue cycle history is not available. Consequently there is a need for nondestructive methods that can characterize the fatigue state of steel.

Several nondestructive linear and non-linear ultrasonic methods as well as magnetic techniques have been proposed to analyze the fatigue state of bridge steel in order to predict or detect the onset of fatigue cracking (Teller, 1993). However, it is not clear in all cases that the method will sense a significant change in the relevant material property. This may be due to insufficient sensitivity of the method or to a lack of a significant change in the property itself. To evaluate the latter possibility requires the collection of data on the changes in relevant material properties as a function of fatigue damage using the most sensitive methods available.

Under a collaborative program between FHWA and the National Institute of Standards and Technology (NIST), these data are being obtained by monitoring the material properties as a function of fatigue cycles. Table I lists the methods employed, the parameters employed and the possible microstructural characteristics that can cause changes in these parameters.

Table I: Materials Properties for Characterization of Aging Steel

MEASUREMENT TECHNIQUE	MEASURED PARAMETER	MICROSTRUCTURAL CHARACTERISTIC
Resonant Ultrasound Spectroscopy	Elastic Constants, Q	Texture, Voids, Grain size
Nonlinear Acoustics	Higher-order Moduli	Strain, Dislocations
Trapped-mode Ultra-sonic Resonance	Q, Velocity	Dislocations,
Magnetostrictive Transduction	Magnetostrictive Coefficient, Damping	Magnetic Domain Pinning
Neutron Diffraction	Atomic Spacing, Peakshape	Residual Stress, texture
Positron Annihilation	γ-ray Peakshape	Dislocations, microvoids

WORK PLAN

The FHWA provides samples of steel that have been in actual service . These are obtained from obsolete bridges that are being replaced. The samples are characterized by standard test procedures including chemical analysis, Charpy toughness test and grain structure (Nelson, 1996). The data for these tests are included in the FHWA Historical Steel Data Base which can be found on the Internet at: www.tfhrc.gov

The steel samples are then transferred to NIST at Boulder,Colorado, where fatigue test specimens are machined from this steel in a special cylindrical shape that permits trapped torsional mode measurements. During the fatigue cycling, the specimens are monitored for changes in ultrasonic resonance frequency and damping. Periodically, specimens will be removed and cubical specimens cut for measurement of elastic constants, nonlinear acoustic response and other variables listed in Table I. Neutron diffraction measurements for texture and residual stress will be done at NIST in Gaithersburg, Maryland. Finally, positron annihilation methods will be applied by FHWA.

TRAPPED TORSIONAL MODE MEASUREMENT

A major part of the program consists of the nondestructive measurement of the torsional modulus and damping while the specimen is subjected to uniaxial cyclic loads. The torsional vibration modes are excited by pulses applied to the specimen by cylindrical electromagnetic-acoustic transducers (EMATs). As shown in Fig. 1, the cylindrical steel specimens are machined with a slightly enlarged radius in its mid length. This shape allows trapping of the torsional vibrational energy so that measurements of the damping will not be affected by the end grips.

The theory of torsional mode trapping is discussed in detail by Johnson et al. (1996). Briefly, it depends on the torsional wave angular frequency, ω, the axial wavenumber, κ, and a dimensionless constant, η_o, which are related by:

$$\kappa^2 = \eta_o^2\left[\left(\frac{\omega}{\omega_o}\right)^2 - 1\right] \tag{1}$$

where ω_o is the cutoff frequency, equal to the ratio: $\eta_o v_s / a$. The parameter a is the radius of the cylinder and v_s is the plane wave shear velocity of the steel. For frequencies above cutoff, κ is real, and, for frequencies below it, κ is purely imaginary.

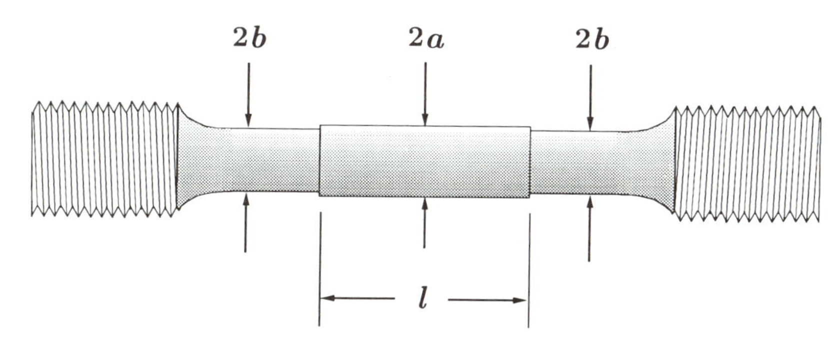

Figure1: Schematic of Trapped Torsional Mode Test Specimen

The condition for torsional wave trapping is that κ be real in the trap region and imaginary everywhere else. This leads to the constraint on the driving frequency: ($\omega_o \leq \omega < \omega_o'$) where ω_o and ω_o' are the cutoff frequencies in the trap region and outside the region, respectively. The continuity condition across the step in radius leads to:

$$\frac{b\kappa}{a|\kappa'|}\tan\left(\frac{l}{2a}\kappa'\right) = 1 \tag{2}$$

where a and b are the radii of the trap region and the region external to the trap respectively, and l is the length of the trap (Fig. 1).

Coupling to the trapped resonant torsional modes is accomplished with an unusual noncontacting electromagnetic-acoustic transducer or EMAT (Johnson and Alers, 1997). Two permanent magnets provide a static field that is essentially perpendicular to the axis of the cylinder in the trapped region. A rectangular spiral coil, which acts as both a transmitter and receiver, is wound flat and then bent round the cylinder such that the wires are oriented in the axial direciton under each magnet. A gated amplifier generates a 1-5 ms driving tone burst from a continuous sine wave, and a phase-sensitive receiver enables calculation of the log decrement damping factor, δ, and the resonant frequency from the ring-down following the tone burst (Johnson, 1996).

LINEAR AND NONLINEAR ULTRASONIC PROPERTIES

During the accumulation of fatigue damage, the trapped torsional mode measurements are expected to indicate when significant changes in the microstructure have occurred. At these times, the cyclic loading schedule will be terminated and the central portion of the specimen shown in Fig. 1 will be cut out to produce a cylindrical sample that has a microstructure characteristic of material that is part way through its fatigue life. For this shape of sample, high resolution ultrasonic elastic constant and attenuation measurements can be made using a resonant ultrasonic technique described elsewhere (Heyliger et al., 1993).

This technique yields all the elements of the elastic modulus tensor and hence detects any anisotropy that may be present or being developed in the material. It also allows the Q factor of various modes to be measured and, thus, allows changes in internal friction to be monitored. By attaching a lithium niobate transducer to one circular face of the cylinder, a well-characterized ultrasonic tone burst can be launched along the axis of the the cylinder. When this ultrasonic signal reaches the opposite face of the cylinder, a carefully designed laser interferometer will measure the surface displacements at both the frequency applied to the transmitter and its second harmonic. This provides enough information to calculate the nonlinearity parameter, β, of the material (Hurley and Fortunko, 1997). Since this parameter depends on the anharmonic properties of the interatomic forces as well as on dislocation and microcrack substructures, it is expected to be a sensitive indicator of the development of fatigue damage.

MAGNETOSTRICTION

Since the material being studied is a ferromagnetic steel being driven into mechanical resonance by an EMAT, it is possible to collect information on the amplitude of the induced vibrations as a function an applied magnetic field. For certain orientations of the field relative to the conductors in the EMAT coils, this amplitude of vibration i.e. the efficiency of the EMAT, can be related to the magnetostrictive properties of the steel (Thompson, 1978). Various features of the efficiency vs field curves can be related to the coercive force and to the presence of residual stresses (Igarashi et al., 1997) which in turn may indicate microstructural changes produced by fatigue damage.

NEUTRON DIFFRACTION

Like X-ray diffraction, neutron diffraction makes of use of Bragg's Law to measure interatomic spacing in crystal structures (Holden, 1996). However, neutrons can penetrate into materials like steel to provide a three-dimensional scan of the structure. Shifts in the spacing of diffraction peaks indicate residual stresses, while microstructural texture is measured by peak shape and by preferred orientation of crystal axes. A triple-axis diffractometer optimized for stress mea-surements, with a 1 mm spatial resolution has been installed at NIST in Gaithersburg (Prask and Brand, 1997). This will used to measure the fatigued steel specimens.

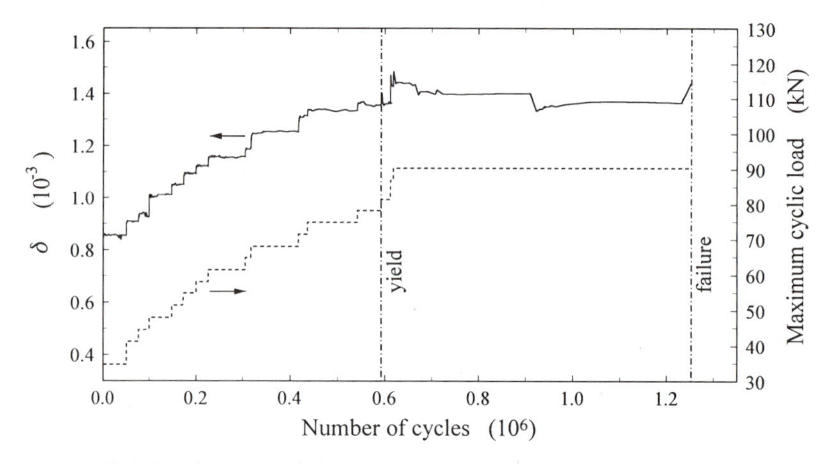

FIGURE 2: Positron Annihilation Peak (0.511 MeV) Peakshape Analysis

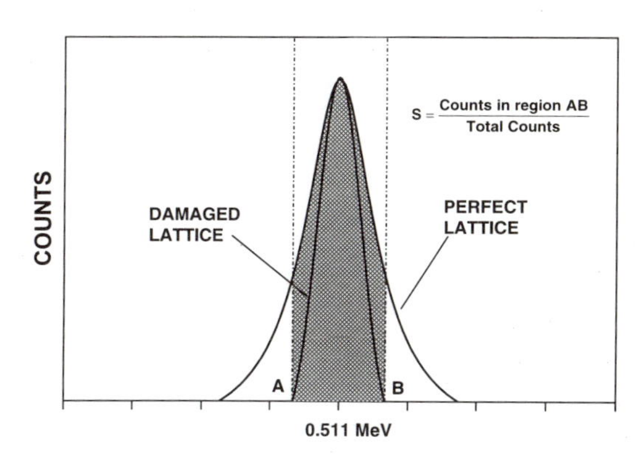

FIGURE 3: Plot of Damping vs Fatigue Cycles

POSITRON ANNIHILATION

In this technique, positrons emitted from a source such as ^{22}Na are used to irradiate the electronsin a material. When these antimatter particles interact with electrons, they annihilate each other. Their masses are converted to pure energy in the form of a pair of 0.511 MeV γ- rays. However, there is actually a distribution of energies around this peak, and the shape of this distribution is a function of the fatigue state of the steel (Uchida et al., 1992) . As shown in Fig. 3, the perfect lattice of the undamaged steel has a wider distribution than damaged steel. Thus by using the ratio of the area AB to the total area of the peak, a ratio defined as S, it is possible to develop a measure of damage.

PRELIMINARY RESULTS

An extensive fatigue test was conducted on A-36 material. The design of this experiment was unique and provided unprecedented accuracy in the measurement of ultrasonic damping and resonant frequency as a function of load and number of fatigue cycles. The sample was submitted to tension-tension load-controlled fatigue with the maximum cyclic stress increased in steps from 50% to 110% of the monotonic yield strength over a total of 1.2 million cycles until the sample failed. Each increase in the load caused the ultrasonic damping to systematically increase over several thousand cycles and, then, level off to a constant value (Fig. 3)

Since dislocation densities and pinning are known to strongly affect ultrasonic damping in metals, the observed changes in damping are assumed to result from the changes in the dislocation structure. Further observed changes in damping during cycling at the highest load may be related to crack initiation and growth. Perhaps the most intriguing result from the standpoint of nondestructive testing is that measurement of damping as a function of static stress showed the development of a pronounced peak after higher cyclic loads. This feature may be directly related to the embrittlement (strain hardening) of the material. Similar peaks in aluminum have been interpreted in terms of the evolution of tightly knit dislocation structures (Kung and Zhu, 1994)

REFERENCES

FHWA [Federal Highway Administration] in: *Proceedings: Conference on Nondestructive Evaluation of Bridges, FHWA-RD-93-040A*, National Technical Information Service, Springfield VA (1993).

Heyliger,P., Jilani, A., Ledbetter, H., Leisure, R.G. and Wong, C.L., 1993, Elastic constants of isotropic cylinders using resonant ultrasound, *J. Acous. Soc. Am* .Pt 1:1482.

Holden, T.M., 1996, The determination of macrostresses (type-I stresses) and microstresses (type II stresses) by neutron diffraction, in: *Neutron Scattering in Materials Science II*, D.A.Neumann, T.P. Russell and B.J. Wuensch. Eds., Materials Research Society, Pittsburgh PA.

Hurley, D. and Fortunko, C., 1997, Determination of the nonlinear ultrasonic parameter β using a Michelson interferometer, *Meas. Sci. Tech.* 8:634.

Igarashi, B., Alers, G. and Purtscher, 1997, Magnetostrictive EMAT efficiency as an NDE tool, in: *Proc. 8th Int. Conference on Nondestructive Characterization of Materials*, B. Djordjorvic, ed. Plenum, NY (this volume).

Johnson, w. and Alers, G., 1997, Force measurement using vibrational spectroscopy, *Rev. Sci. Instrum.* 68:102.

Johnson, W., Auld, B.A. and Segal, E., 1996, Trapped torsional modes in solid cylinders, *J. Acoust. Soc. Am.* 100(1):285.

Johnson, W., 1996, Ultrasonic resonance of metallic spheres at elevated temperatures, *J. de Phys. IV*, 6:C8-849.

Kung, G. and Zhu, Z., 1994, A study of dislocation movement during push-pull fatigue by ultrasonic attenuation, *Phys.Stat.Sol.(a)*142:357.

Nelson, G. , 1996, Development of an Historic Bridge Steel Database, Upublished Masters Thesis, University of Wisconsin, Madison WI.

Prask, H. and Brand, P., 1997, Residual stress determination by means of neutron diffraction, in: *International Conference: Neutrons in Research and Industry, SPIE 2867*, G. Vourvopoulos, ed., SPIE, Bellingham, WA.

Teller, C.M., *1993*,The State-of-the-Art in Nondestructive Evaluation of Steel Bridges, in: *Proceedings: Conference on Nondestructive Evaluation of Bridges FHWA-RD-93-040A*, National Technical Information Service, Springfield VA (1993).

Thompson, R.B., 1978, New configurations for the electromagnetic generation of SH waves in ferromagnetic materials, in: *Proc. 1978 Ultrasonics Symp. IEEE Cat. No 78CH1344-ISU*, 374.

ULTRASONIC MEASUREMENT OF STRESS IN BRIDGES*

A.V. Clark,[1] P.A. Fuchs,[2] M.G. Lozev,[3] D. Gallagher,[1] C.S. Hehman[1]

[1]National Institute of Standards and Technology, Boulder, CO 80303
[2]Federal Highway Administration, McLean, VA 22101
[3]Virginia Transportation Research Council, Charlottesville, VA.

INTRODUCTION

Of the approximately one half million bridges in the United States about 40% are either in need of upgrade or have structural deficiencies [1]. Since it is impractical to replace all deficient bridges, means to quantify their safety and prioritize repairs are needed. Ultrasonic measurements of stress can be useful in evaluation of bridge safety. Here the acoustoelastic effect, (the change in sound speed due to stress) is used.

We report here on ultrasonic measurements of static stress in field tests on two bridges. On the first bridge we measured the residual stress due to a thermal treatment to camber the girders. In the second test we measured the stress in an integral backwall bridge which has no expansion joints.

We also conducted proof-of-concept experiments to determine stress in pin and hanger connections. These are used to suspend interior spans and also to accommodate thermal expansion. They are placed near expansion joints; salts from deicing can wash down into the connections. Corrosion products can lock up the pins and cause large stress buildup in the connections.

We also report on field measurements of time varying stresses (live loads) due to vehicle traffic. We used the time-of-flight of Rayleigh waves propagating along the flanges of the I-beams. Here we wish to determine the stress range and the number of cycles of stress to evaluate the remaining fatigue life.

THEORY

Acoustic Birefringence Method

Consider a state of plane stress with stresses referred to a coordinate system parallel and perpendicular to the rolling direction of the I-beams. For a typical steel grade (A-36 used in

Nondestructive Characterization of Material VIII
Edited by Robert E. Green Jr., Plenum Press, New York, 1998

bridge construction, in the absence of stress there will be a faster SH-wave velocity for waves polarized along the rolling direction; the slower wave is polarized in the transverse direction.

We define the birefringence B by $B = (V_F - V_S)/V$ where F represents the fast wave, S represents the slow wave and V = average velocity. Assume the rolling and transverse directions are principal stress directions (which is the case for our measurements). Then the relation between B and the stresses is [2]

$$B = B_0 + m(\sigma_{yy} - \sigma_{xx}) \tag{1}$$

where B_0 is the unstressed birefringence, m = stress acoustic constant. For A-36 steel, m = 0.92 (10^{-5})/MPa. Here σ_{xx} is the normal stress along the rolling direction, σ_{yy} is the normal stress in the transverse direction.

Bridge girders typically have σ_{xx} due to bending moment, M, and a shear stress σ_{xy} due to shear force P

$$\sigma_{xx} = \frac{My}{I} \; ; \quad \sigma_{xy} = P\left(\frac{h^2}{4} - y^2\right)/2I \tag{2}$$

where y = vertical distance from the girder centerline, h = beam depth, I = area moment of inertia. In general $\sigma_{yy} = 0$ except in the immediate vicinity of bearings which support the girders. For most applications we wish to determine the bending stress σ_{xx} since it typically acts as a crack driving stress for fracture and fatigue.

Rayleigh Wave Method

We used this method for determining dynamic variations in σ_{xx} induced by truck traffic. We mounted transmitting and receiving transducers on the bridge girders. As the girder flexes there will be a strain $\varepsilon_{xx} = E^{-1}\sigma_{xx}$ where E = Young's Modulus. Hence the transducer spacing will vary with stress, causing time-of-flight (TOF) changes. In addition we will have a velocity change due to stress so the normalized change in TOF is [3] $\Delta t/t = \epsilon_{xx} - \Delta V/V$ where the acoustic pathlength is along the girder axis (in the x direction). For steel the acoustoelastic effect $\Delta V/V$ is about one-tenth of the strain effect for Rayleigh waves. We have performed experiments for common steel girders to determine the calibration factor F: $\Delta t/t = F\,\epsilon_{xx}$. For A-36 steel, F ≅ unity.

Fatigue Life Prediction

One purpose of Rayleigh wave measurements is to estimate the remaining fatigue life. Fatigue cracks in bridges typically initiate at welded details. The relation between fatigue life, characterized by the number of cycles, N_f to failure, and S_{re} (stress range) is given by [4] $N_f = AS_{re}^{-n}$. Here A is a constant depending upon design of the detail, n = 3 for steel.

Vehicles cause varying amplitude stress ranges, S_{ri}. It is conventional to use Miner's rule to convert stress ranges S_{ri} to equivalent constant amplitude S_{re} [5]

$$S_{re} = \left[\sum_i \frac{n_i S_{ri}^3}{N}\right]^{1/3} \tag{3}$$

Here N is the total number of cycles measured. A histogram is constructed from the data where n_i is the number of counts in the ith interval.

FIELD TESTS

Measurement of Residual Stress

The first bridge tested was a simply supported structure. We made measurements of birefringence on the vertical centerline of the outermost girder. Thermal treatment had been applied to the girders to give an upward camber to compensates for the deadload of the deck. We wanted to determine residual stress due to cambering.

We used a commercial unit which measures the phase of the wave. By symmetry the shear stress $\sigma_{xy} = 0$ on the vertical centerline, and $\sigma_{yy} \cong 0$ since we are away from the bearings. Therefore from equation 1

$$\sigma_{xx} = \frac{-(B - B_0)}{m} \quad . \tag{4}$$

The birefringence is obtained from the phase measurements $B = (P_s - P_f)/P$ where P_s is phase in the slow wave, P_f = phase in the fast wave.

To determine B_0 we used the fact that there is no net axial force acting on the girders, therefore from the preceding equation we have

$$B_0 = \frac{\int\int B dA}{A} \tag{5}$$

where the integral is over the girder cross section of area A.

With B_0 determined, we obtained σ_{xx} from equation 4. σ_{xx} has components due to dead load and cambering. We calculated the first from standard bridge design procedures [8] and subtracted it to obtain the residual stress distribution.

Measurement of Thermal Stress in Integral Backwall Bridge

The integral backwall bridge is a novel design which has no expansion joints. It consists of 10 girders which rest on fixed bearings on a central pier, on sliding bearings at the abutments. Concrete is poured over the girder ends locking them together in a rigid cap (integral backwall). Soil is then backfilled at the backwalls. As the girders expand the backwall compresses the soil.

At the time of erection thermocouples and strain gages were installed on two girders near the backwalls; pressure sensors were also buried in the backfilled soil. In normal operation the thermocouples record approximately sinusoidal day- night temperature variations. This is mimiced by strain gage readings on the girders and by pressure gage readings. At the time of our field test strain gage readings on two girders at the west side of the bridge were reading large compressive stresses (near yield).

We made birefringence measurements on the bottom flanges of the girders at both west and east abutments. On the bottom flanges σ_{xy} is 0 (see equation 2). Measurements were made away from the bearings so σ_{yy} is also 0; hence σ_{xx} is given by equation 8.

For each girder we subtract data at the east side (which is operating properly) from west side data. This subtraction removes in the effect of any deadload stresses from our readings since the bridge is symmetrical about the center pier. Then from equation 4

$$\sigma_{xx}^{W} - \sigma_{xx}^{E} = \frac{B^{E} - B^{W}}{m} \tag{6}$$

where W and E denote quantities at the west and east sides.

In normal operation the left hand side of the above equation is small since the stresses due to thermal expansion are almost the same at opposite sides. Our data indicated a difference of 54 MPa at the time of the test. This is considerably below the value of approximately 300 MPa indicated by the strain gage at the west side, so we infer the bridge is safe.

At this point we suspected that strain gage electronics were experiencing d.c. shifts. We recommended that these strain gage electronics be replaced. When this was done new strain gage measurements were 12 MPa [6] compared to 54 MPa from the ultrasonic results.

We also made measurements in the girder webs. We again subtracted data at the east side from the west side and found that the result could be represented by a constant (axial) component plus a linear component (due to bending). We calculated the resulting bending moment M at the west abutment and used standard structural analysis techniques to calculate the bending moment as a function of length along the girder. From this we determined the rotation of the backwall at the west end. Our data indicated a rotation of about 0.5° which was confirmed by visual observation [6].

Liveload Measurements

We made liveload measurements on the I-64 bridge using Rayleigh wave EMATs installed on the bottom flange of a girder at midspan. Strain gages were mounted on the flange to compare with the ultrasonic results. Traffic was closed and a test vehicle was driven over the span at controlled speeds.

The received Rayleigh wave is digitized and stored in computer memory. A threshold technique is used to determine changes in TOF. 50 waveforms are averaged to improve signal-to-noise. For a pulse repetition rate of 400 Hz, this results in a response time of 1/8 second. Since the lowest modes of vibration of typical bridges have resonant frequency of the order of a few Hz, this response time is adequate for most practical applications.

Typical EMAT and strain gage responses are shown in Figure 1. There is a peak stress level of about 14 MPa (2 ksi). Such low stress levels are common for well designed bridges.

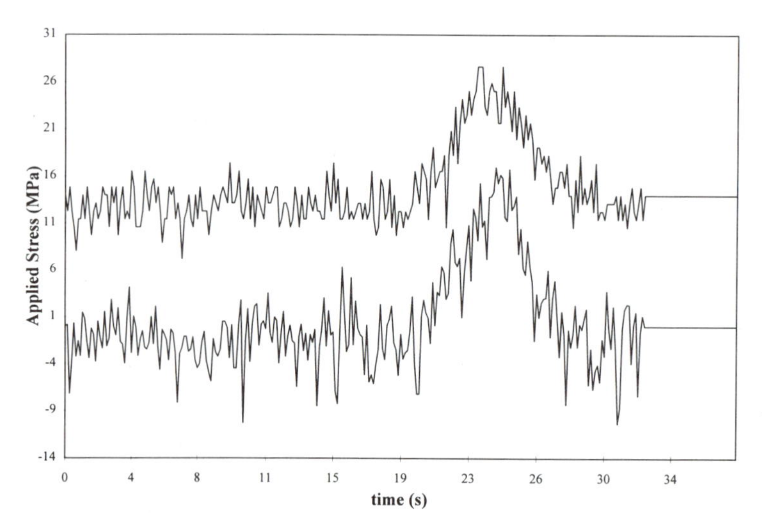

Figure 1. Typical EMAT (lower trace) and strain gage (upper trace) data for vehicle passing over I-64 bridge.

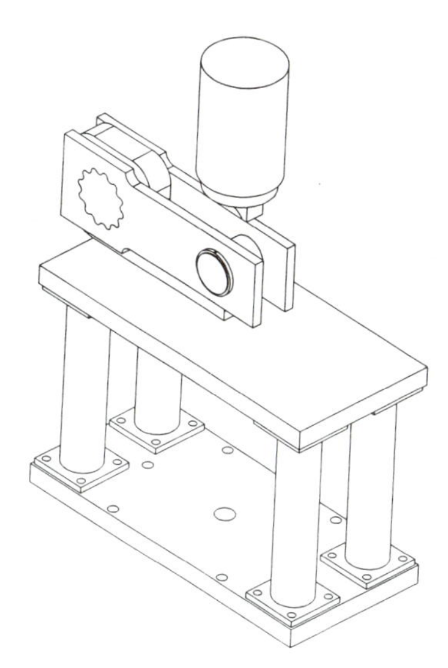

Figure 2. Apparatus for simulation of locked up pin and hanger connection. Note 12-tooth spline at left-hand side of hangers and loading machine piston (shown retracted) on right-hand side.

The noise level is approximately 7 MPa and 3.5 MPa respectively for EMAT and strain gage data. Because stress levels are low and because the calibration factor F is approximately unity we must resolve TOF changes of the order of 2 parts per million.

Three runs were made at each speed and S_{re} was determined for each run. Data was processed with no filtering; using a 5-point filter; rejecting data below noise level of strain gages. The average discrepancy (for total of 15 runs) between EMAT and strain gage data was 20% for no filtering. With 5-point filtering, the discrepancy was reduced to 14%. Rejecting counts below the noise level of strain gages gave a discrepancy of 11%. Filtering improves the agreement between strain gage and EMAT data as does rejecting data below the noise threshold.

Pin and Hanger Connections

We designed and constructed an apparatus to simulate lockup in pin and hanger connections, see Figure 2. A 12-tooth spline models a locked up top pin by preventing rotation of the ends of a pair of hangers. The hangers are connected by a free pin at the opposite ends. The effect of expansion of the bridge (which causes horizontal shear force P acting at both pins) is simulated by loading by the piston of a mechanical testing machine.

The load P causes a bending moment M in the hanger; M = Px, where x = distance from the pin. The bending moment causes bending stress σ_{xx} = My/I. To characterize hanger stress we measure B at opposite sides at the same value of x; from equation 4

$$\frac{B^{l} - B^{r}}{m} = \sigma_{xx}^{r} - \sigma_{xx}^{l} \tag{7}$$

where l and r denote left and right sides of the hanger. If the connection is good (no bending stress) then $B^l - B^r = 0$. For bad connections the difference is nonzero.

In equation 7 we assume that B_0 is constant at both left and right locations. In general the steel is not completely homogeneous; if we measure B_0 at several locations we will obtain an average value and a standard deviation, δB_0. This causes a measurement uncertainty equivalent to $\delta B_0/m$.

We performed measurements to simulate three other scenarios: 1) continuous monitoring from a known state; 2) intermittent monitoring from a known state; 3) measurements with no a prior knowledge. In scenario 2 we left our transducer in situ as the load was changed and measured the change in birefringence. In scenario 3 B_0 was measured in an initial state, then the transducer was removed and the load changed. The transducer was then replaced, and the difference $B - B_0$ was measured.

We measured $B^l - B^r$ at several x locations for no load. We took an average value and calculated $\sigma_{xx}^l - \sigma_{xx}^r$. Because of variations in B_0 measurements indicated a stress state equivalent to 12 MPa.

For scenario 1 we have good agreement between strain gage and ultrasonic results (to within about 5 microstrain). For scenario 2 the agreement was within ±25 microstrain (±5 MPa). The increased discrepancy is attributed to repeatability in replacing the transducer at the measurement location. For scenario 3 the discrepancy between strain gage and ultrasonic results was ±55 microstrain (11 MPa) which is due in large part to variations in B_0.

CONCLUSION

We performed ultrasonic measurement of static loading on two bridges. On the first bridge, we first determined the combined state of deadload plus residual stress due to cambering and then subtracted the deadload stress (calculated from standard design procedures). In the second we determined the thermal stress due to expansion of an integral backwall bridge. By subtracting data from the good abutment from that at the bad abutment we removed the effect of deadload stresses; the difference is due to any anomalies from thermal expansion. Our measurements indicated that the bridge was well within the safe range. Upon replacement of suspect strain gage electronics, subsequent strain gage measurements confirmed these results.

Field measurements were also performed to determine liveload stress. In general strain gage and ultrasonic measurements were in good agreement, especially if means such as 5-point filtering or rejecting data below the strain gage noise measurement were used.

We have constructed a pin and hanger facility. Proof of concept measurements were done to demonstrate the feasibility of ultrasonic measurement to determine state of stress in pin and hanger connections.

REFERENCES

1. U.S. Department of Transportation Report to Congress Tech. Dep. Senate Document 99-6, U.S. Gov't Printing Office, Washington, DC 1985.

2. K. Okada, 1980, J. Acoustic Soc. Jpn (E) Vol. 1, No. 3, p. 193.

3. A.V. Clark, P. Fuchs and S.R. Schaps, 1995, J. Nondest. Eval., Vol 14, No. 3, p 83.

4. G.R. Cudney, Michigan Dept. of State Highways, Research , Dep. No. R-638.

5. P.B. Keating. J.M. Kulichi, D.R. Mertz and C.O. Hess, June 1990, Participant Notebook, Nat. Highway Inst. Course, 13049.

6. F. Hoppe, Va. Trans. Res. Council, Charlottesville, VA, private communication.

DEVELOPMENT OF SCANNING HIGH-T_C SQUID-BASED INSTRUMENT FOR NONDESTRUCTIVE EVALUATION IN A MAGNETICALLY NOISY ENVIRONMENT

Nilesh Tralshawala, James R. Claycomb, Hsiao-Mei Cho, and John H. Miller, Jr.

Texas Center for Superconductivity at the University of Houston
Houston, Texas 77204-5932

INTRODUCTION

Safety of space shuttle and aircraft requires airframe inspections for damage from widespread fatigue cracks and corrosion in *underlayers*. Ultrasonic techniques* of any kind can not detect cracks and corrosion in the underlayers. Although thermal wave imaging techniques[1] have reported some success in this regard, eddy current methods still dominate airframe inspections. This is because of their ability to penetrate multiple layers, with or without gaps (bonded or disbonded layers), and to identify both fatigue cracks and material loss from corrosion. They are also inexpensive, non-contacting, and versatile. However, the conventional eddy-current technique, aside from being labor intensive,† is unable to detect defects much more than a skin depth below the surface at a given frequency. Low frequency (less than 500 Hz) eddy current excitations are required to detect deeper subsurface flaws (1 cm or deeper in aluminum alloys). The sensitivity of conventional pickup coils typically employed imposes a lower limit on the frequency of operation (generally a few tens of kilohertz) and thus an upper limit on the skin depth, typically a few millimeters in aluminum. The Superconducting QUantum Interference Device (SQUID) is the only magnetic field sensor that offers high enough sensitivity in this frequency range.‡

---

*For example, a discussion with M. Choquet of National Research Council, Canada, at this conference; and M. Choquet, et al, "Detection of Corrosion under Fasteners by Laser-Ultrasonics," elsewhere in these proceedings.

†This problem can be alleviated by using magneto-optic eddy current imaging techniques[2]. Another solution is a mechanized robotic scanning setup designed to scan the whole aircraft using multiple small area surface scans by the Center for Nondestructive Evaluation at the Iowa State University (e.g. see *http://www.cnde.iastate.edu/develop.html*).

‡A SQUID with a 1 mm size loop can resolve magnetic flux of 10 $m\Phi_0/\sqrt{Hz}$, uniformly from 1 Hz to 10 KHz. A conventional eddy current coil of 500 turns, operating at 100 Hz, resolves a flux of 10^5 $\Phi_0/\sqrt{Hz}$. A quantum of magnetic flux, Φ_0, is 2.07 × 10^{-15} Webers.[3]

SQUIDS FOR EDDY CURRENT TESTING

There are two main obstacles to widespread use of SQUIDs in practical eddy current testing systems. Firstly, the SQUIDs must be operated at cryogenic temperatures, using cryostats and liquid cryogens. Recent advances in high transition temperature (high-T_c) superconductivity and cryocooler technology has enabled the use of SQUIDs with a substantially reduced refrigeration cost and ease of operation[4]. Secondly, the unparalleled sensitivity that makes SQUIDs attractive for deep subsurface flaw detection also renders them unusable outside of magnetically shielded enclosures. In the conventional low-T_c SQUID systems, one can construct niobium-wire-wound or monolithic second-order gradiometers that cancel out uniform and first gradient noise fields[5]. One has to then deal with the encumberance of cooling down to 4K. A disadvantage of using high-T_c SQUIDs (HTSQUIDs) is that there is no technology for making wire-wound gradiometers, and it is very difficult to make efficient monolithic gradiometers.[§] One of the most popular solutions to this problem involves using three SQUIDs to setup an electronic gradiometer[7]. This technique has two drawbacks: (1) one of the individual SQUIDs must be operated in the presence of ambient magnetic field noise, and hence must have extremely high dynamic range (as high as 140 dB) and linearity; (2) the other two SQUIDs need to be exactly matched pairs, otherwise subtle phase shifts affect electronic subtraction adversely[8].

Apart from ambient magnetic noise, in eddy current testing one encounters an additional unwanted magnetic field source. This is the primary magnetic field of the eddy current drive coil. One practical NDT approach employed so far, uses differential eddy current drive coils. This setup reduces primary eddy current field, but the individual HTSQUIDs are still in a noisy environment.[9]

Our approach to this problem is to employ a combination of small, cylindrical superconducting and mu-metal shields to reduce the ambient electromagnetic field noise at the HTSQUIDs and permit their operation outside of costly and impractical magnetically shielded rooms. Moreover, this also enables the HTSQUID instrument to be scanned, rather than the sample. An excitation solenoid induces eddy currents in the object under test. The geometry of the shields is selected, not only to expel the excitation and ambient noise fields from the HTSQUID, but also to focus the eddy currents into a narrow region underneath the probe. The design thus enhances spatial resolution and minimizes undesirable edge effects. Our tests demonstrate that the instrument is able to image simulated cracks and pits below isolated multiple layers (ultrasonic techniques do not work in this case) of aluminum with a combined thickness of at least 5 cm.[10]

EDDY CURRENT PHENOMENA

In order to better understand the need for using SQUIDs in eddy current testing, we will start out with basics of low frequency eddy current phenomena. In eddy current testing, time varying magnetic fields are produced by an excitation coil close to a conducting sample. These time varying magnetic fields induce electric fields according to Maxwell's equation,

$$\frac{1}{\mu}\vec{\nabla} \times \vec{E}_1 = -\frac{d}{dt}\vec{H}_1.$$

[§]Recently, a new technique has been invented at the University of California–Berkeley, that allows fabrication of efficient monolithic gradiometers that can, in principle, perform even better than low-T_c gradiometers.[6]

Here $\vec{H}_1$ is the primary magnetic field produced by the excitation coil. The electric field induced by $\vec{H}_1$ will produce a "first order" eddy current, $\vec{J}_1$, in the conducting medium as given by Ohm's law, $\vec{J}_1 = \sigma \vec{E}_1$. This current, in turn, gives rise to a "secondary" magnetic field, $\vec{H}_2$, according to the equation $\vec{\nabla} \times \vec{H}_2 = \vec{J}_1$, where $\vec{H}_2$ is the secondary magnetic field produced by $\vec{J}_1$. For the case of time harmonic fields (fields that vary as $e^{j\omega t}$), $\vec{H}_2$ will be $\pi/2$ out of phase with the primary excitation field, $\vec{H}_1$ (as $\vec{E}_1 = -d\vec{A}_1/dt$ leads to $\vec{J}_1 = -j\omega\sigma\vec{A}_1$, $\vec{J}_1$ being the primary eddy current). Now the time variation of the secondary magnetic field gives rise to a higher order electric field,

$$\frac{1}{\mu}\vec{\nabla} \times \vec{E}_2 = -\frac{d}{dt}\vec{H}_2.$$

This electric field gives rise to a secondary eddy current $\vec{J}_2 = \sigma\vec{E}_2$. The secondary eddy current is $\pi/2$ out of phase with the primary eddy current and gives rise to a magnetic field that is π out of phase with the excitation coil current. This magnetic field induces an even higher order eddy current that is π out of phase with the primary eddy current, and so on. As we will show in the next subsection, this has a huge practical implication for eddy current testing systems. The net result is an infinite series of eddy currents $\vec{J}_s$ giving rise to a magnetic field $\vec{H}_e$, where

$$\vec{\nabla} \times \vec{H}_e = \sum_{s=0}^{\infty} \vec{J}_s, \tag{1}$$

where $\vec{J}_0$ represents the excitation coil current. The in-phase and quadrature response of a SQUID magnetometer is then obtained from the field produced by the even- and odd-order eddy currents, respectively.

Circular Excitation Coil Above a Conducting Plate

Now we are ready to look at exact solution for a practical case depicted in Fig. 1. We solve equation (1) using cylindrical symmetry to bring out important features of the problem. We essentially modify the analysis of Dodd and Deeds[11], for the problem at hand, i.e., calculating magnetic field produced at the location of the SQUID magnetometer, due to eddy currents induced in object under test.

The magnetic field is derived from the vector potential $\vec{\nabla} \times \vec{A} = \mu\vec{H}$. In the Coulomb gauge, $\left(\vec{\nabla} \cdot \vec{A} = 0\right)$, the vector potential satisfies Poisson's equation $\vec{\nabla}^2\vec{A} = -\mu\vec{J}$. In axial symmetry, $\vec{A} = A(r,z)\hat{\phi}$, and thus we get

$$\left(\vec{\nabla}^2\vec{A}\right)_\phi = \frac{\partial^2 A_\phi}{\partial r^2} + \frac{1}{r}\frac{\partial A_\phi}{\partial r} - \frac{A_\phi}{r^2} + \frac{\partial^2 A_\phi}{\partial z^2} = \nabla^2 A_\phi - \frac{A_\phi}{r^2} = -\mu_o J.$$

Solving this equation by separation of variables in the four regions of Fig. 1, we get

$$A_\phi^1(r,z) = \int_o^\infty B_1 \cdot e^{-\alpha \cdot z} \cdot J_1(\alpha r) d\alpha \qquad z \geq \ell$$

$$A_\phi^2(r,z) = \int_o^\infty (B_2 \cdot e^{-\alpha \cdot z} + C_2 \cdot e^{\alpha \cdot z}) \cdot J_1(\alpha r) d\alpha \qquad 0 \leq z \leq \ell$$

$$A_\phi^3(r,z) = \int_o^\infty (B_3 \cdot e^{\alpha_1 \cdot z} + C_3 \cdot e^{-\alpha_1 \cdot z}) \cdot J_1(\alpha r) d\alpha \qquad -d \leq z \leq 0$$

$$A_\phi^4(r,z) = \int_o^\infty B_4 \cdot e^{\alpha \cdot z} \cdot J_1(\alpha r) d\alpha \qquad z \leq -d$$

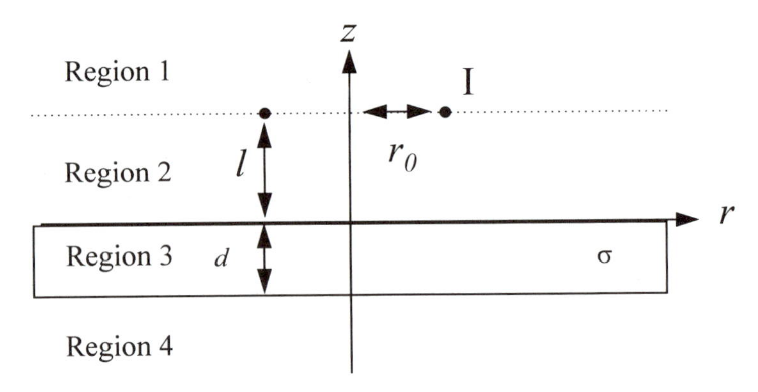

Figure 1. Cross sectional view of a circular excitation coil of radius r_0, carrying a current I, at distance ℓ above a conducting plate of thickness d.

where $J_1(\alpha r)$ is the first order Bessel function. The parameter α is the separation constant with dimensions of inverse length and $\alpha_1 = \sqrt{\alpha^2 + j\omega\mu\sigma}$. Since there are no restrictions on α in performing the separation of variables, this parameter must be summed over continuously in constructing a solution for the vector potential. The constants $B_1, B_2, B_3, B_4, C_2$and C_3 are obtained by requiring continuity of E_ϕ at $z = 0$, $z = l$, and $z = -d$, and continuity of H_r at $z = 0$ and $z = -d$, and the discontinuity in H_r equal to $\mu_o J$ at the position of the excitation coil. This gives six equations with six unknowns. Solving these equations and back substituting yields expressions for the vector potential in the four regions. In the region where the SQUID magnetometer is located (region 1 in Fig. 1),

$$A_\phi^1(r,z) = \frac{\mu_o I r_o}{2} \int_o^\infty J_1(\alpha r) J_1(\alpha r_o) e^{-\alpha \cdot z} \times \tag{2}$$

$$\left\{ e^{\alpha \cdot \ell} + e^{-\alpha \cdot \ell} \frac{\alpha - \alpha_1}{\alpha + \alpha_1} + e^{-\alpha \cdot \ell} \frac{\alpha - \alpha_1}{\alpha + \alpha_1} \cdot \frac{4\alpha \cdot \alpha_1}{\left[(\alpha - \alpha_1)^2 - (\alpha + \alpha_1)^2 \cdot e^{\alpha_1 \cdot d} \right]} \right\} d\alpha.$$

In this expression, the factor $\frac{\alpha - \alpha_1}{\alpha + \alpha_1}$ can be thought of as a "reflection factor" that is zero in the absence of a conducting medium and is -1 for the case of infinite conductivity (zero skin depth)[12]. The first term in the curly brackets corresponds to the vector potential of a current loop in free space. This is the only term that remains for the case of zero conductivity (as reflection factors are now zero). The second term corresponds to the change in vector potential of a current loop above a semi-infinite conducting half space. The third term represents the change in vector potential of a plate of finite thickness. It can be seen that as the plate thickness d becomes infinite the third term vanishes. Equation (2) can be used to calculate magnetic field measured by the SQUID magnetometer, using the formula

$$B_z(r,z) = \frac{1}{r} \frac{\partial}{\partial r} \left(r\, A_\phi \right).$$

The variation in B_z due to small changes in d can then be related to the signature of material loss due to corrosion. Note that the first term in equation (2) does not contain any useful information but will affect the sensitivity of measurement and hence must be eliminated.[10]

Similarly, we obtain the vector potential in the conducting plate (region 3 in Fig. 1),

$$A_\phi^3(r,z) = \frac{\mu_o I r_o}{2} \int_o^\infty J_1(\alpha r) J_1(\alpha r_o) e^{-\alpha \cdot \ell} \times \tag{3}$$

$$\left\{ \begin{array}{l} e^{\alpha_1 z} \left(\frac{2\alpha}{\alpha+\alpha_1} - \frac{2\alpha(\alpha-\alpha_1)^2}{(\alpha+\alpha_1)\left[(\alpha-\alpha_1)^2-(\alpha+\alpha_1)^2 \cdot e^{2\alpha_1 \cdot d}\right]} \right) \\ +e^{-\alpha_1 z} \left(\frac{2\alpha(\alpha-\alpha_1)}{\left[(\alpha-\alpha_1)^2-(\alpha+\alpha_1)^2 \cdot e^{2\alpha_1 \cdot d}\right]} \right) \end{array} \right\} d\alpha.$$

The eddy current in this region is given by $-j\omega\sigma A_\phi^3(r,z)$. When $d \to \infty$, the expression for the eddy current in the conducting half space is obtained. A plot of eddy current contours is shown in Fig. 2a. The most important feature of which is the sign change

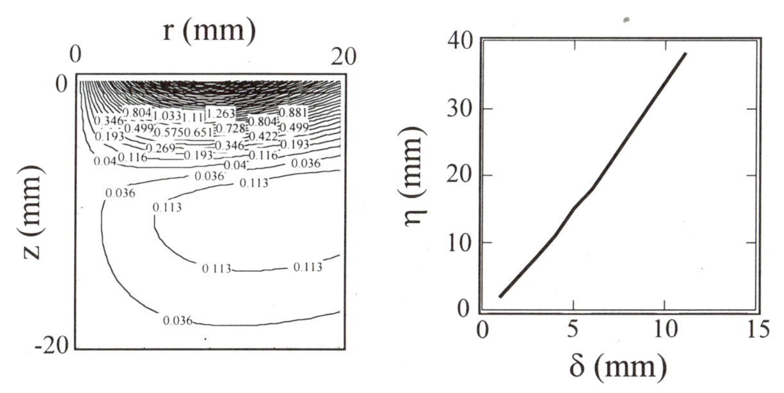

Figure 2. Eddy current phenomena: (a) Quadrature phase eddy current density (A/m^2) contours in a conducting plate with $\sigma = 10^7 S/m$, directly underneath a coil carrying 1 mA current at 1000 Hz. Regions of positive and negative current flow are clearly visible. Thus, a defect in the deeper portion will give rise to a signal that has opposite sign and $1/10^{th}$ the magnitude of a defect signal from the shallower region. (b) Depth at which the quadrature phase eddy current changes sign, η, as a function of skin depth, δ, for plate in (a).

exhibited by induced eddy currents at a certain depth. The depth at which this sign change occurs is directly related to the operating frequency, and hence the skin depth (Fig. 2b). This shows a way of differentiating between defects that are underneath each other, but at different depths[13]. This is very important in practical aircraft structures where one is required to locate corrosion or fatigue cracks underneath fasteners that in general mask the signatures of the underlying defects.¶

¶For example, issues raised by W. Bisle of Daimler-Benz Aerospace Airbus Gmbh, during this conference regarding practical difficulties in locating flaws in aircraft structures. The ultrasonic techniques fail because of insufficient coupling through fasteners and disbonded layers, and conventional eddy current techniques just do not have enough resolution to isolate a signature of a flaw from that of a fastener.

SUMMARY

We show that it is possible, in principle, to tackle one of the most vexing problems in airframe safety testing, namely, the subsurface flaws hidden under sharp features in the structure. Our proposed technique uses the phenomenon of a sign change in quadrature-phase eddy currents at a characteristic depth (see Fig. 2). The resulting sign change in the magnetic field signature of a flaw (hidden underneath a fastener) will let us distinguish it from the signature due to that fastener. This signal, though, is at least an order of magnitude smaller and hence requires a very sensitive sensor. This can not be resolved by conventional eddy current techniques and the ultrasonic techniques fail due to insufficient coupling through the fasteners. As nothing else seems to work in solving this problem, we make a very strong case for the use of SQUID magnetometers in tackling this problem. We believe that we have overcome one of the major problems associated with the use of HTSQUIDs, namely, being able to operate them in a practical environment.[10] Also, with the advent of new refrigeration technologies,[4] the cooling requirements of HTSQUIDs will be easily met in the near future.

Acknowledgments

This work was supported by NASA Grant No. NAG 9-905, by the State of Texas through the Texas Center for Superconductivity and the Advanced Research Program, and by the Robert A. Welch Foundation.

REFERENCES

1. R. L. Thomas and L. D. Favro, From acoustic microscopy to thermal wave imaging, *MRS Bull.,* 21:47 (1996).
2. G. L. Fitzpatrick et al, Magneto-optic/eddy current imaging of aging aircraft: a new NDI technique, *Mat. Eval.,* 1402 (1993).
3. W. Podney, Electromagnetic microscope for deep, pulsed, eddy current evaluation of airframes, Preprint.
4. C. Heiden, Pulse tube refrigerators: a cooling option for high-T_c SQUIDs, in: *SQUID Sensors: Fundamentals, Fabrication and Applications,* H. Weinstock, ed., Kluwer Academic Publishers, Amsterdam (1996).
5. J. Clarke, SQUID concepts and systems, in: *Superconducting Electronics,* H. Weinstock and M. Nisenoff, eds., Springer-Verlag, Berlin (1989).
6. E. Dantsker et al, High-T_c superconducting gradiometer with a long baseline asymmetric flux transformer, to be published in *Appl. Phys. Lett.* (1997).
7. R. H. Koch et al, Three SQUID gradiometer, *Appl. Phys. Lett.,* 63:403 (1993).
8. A. Matlashov et al, Phase shifts in dc SQUID readout electronics, Preprint.
9. H.-J. Krause et al, Mobile HTS SQUID system for eddy current testing of aircraft, preprint submitted to *Rev. of Prog. in Quant. Eval.,* vol. 16, Plenum, NY (1997).
10. N. Tralshawala et al, A practical SQUID instrument for nondestructive testing, to be published in *Appl. Phys. Lett.* (1997).
11. C. V. Dodd and W. E. Deeds, Analytical solution to eddy current probe coil problems, *J. of Appl. Phys.,* 39:2829 (1968).
12. A. J. M. Zaman, A Theoretical Investigation of Eddy Current Response to Subsurface Flaws for Application in Nondestructive Evaluation, Ph. D. Thesis, Department of Electrical Engineering, University of Houston (1981).
13. N. Tralshawala et al, Phase shift phenomena in deep subsurface eddy current measurements, Preprint.

A DETECTION METHOD FOR FINE PLASTIC DEFORMATIONS OF STEEL USING A HIGH PERFORMANCE MAGNETIC GRADIOMETER

Hiroshi Yamakawa[1], Noboru Ishikawa[1], Kazuo Chinone[2], Satoshi Nakayama[2], Akikazu Odawara[2], and Naoko Kasai[3]

[1]Shimizu Corporation
Tokyo, Japan
[2]Seiko Instruments Inc.
Chiba, Japan
[3]Electrotechnical Laboratory
Ibaraki, Japan

INTRODUCTION

It is important to design buildings and civil engineering structures to take into account earthquakes in Japan. When a big earthquake occurs, some structural members are damaged with present designs absorbing the energy of earthquakes. It is important when evaluating the seismic capability, to evaluate absorption capabilities of the structural members, but the nondestructive evaluation methods for absorption capabilities of structural members are not well understood.

For the first step in developing the nondestructive evaluation methods of structural members of building and civil engineering structures, spontaneous magnetic fields of a damaged hot-rolled steel were measured using a SQUID (Superconducting Quantum Interference Devices) gradiometer. This paper describes the detection method for plastic deformations and the possibility of quantitative evaluation of steels' degradations by magnetic measurement.

EXPERIMENTAL SYSTEM

A schematic diagram of our experimental system is shown in Figure 1. The system is composed of an x-y scanning stage assembly made of non-magnetic materials, a SQUID sensor, electronics and a computer for control.

A SQUID device is equipped with a concentric second-order pick-up coil which is integrated on the Si-chip (3 mm × 3 mm)[1]. The outline of the pick-up coil is shown in Figure 2. It consists of two coils which are connected inversely on the same plane. The diameter of the outer coil is 2 mm and that of the inner coil is 1 mm. It is effective to canceling out

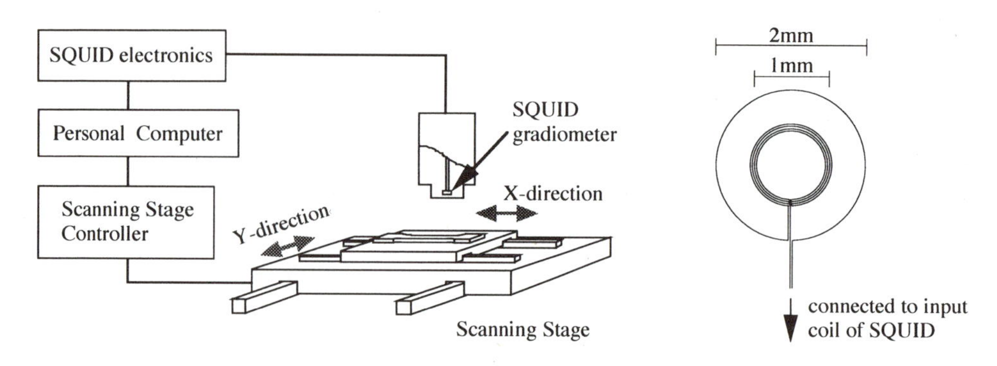

Figure 1. SQUID NDE System.

Figure 2. Pick-up coil of SQUID.

ambient magnetic noises. The SQUID was driven with flux locked loop (FLL) operation.

TEST PIECE

A test piece was made of hot rolled steel plate for welded structures. The chemical and mechanical requirements prescribed in Table 1 is specified in Japanese Industrial Standards (JIS) G 3106. It was machined to a shape shown in Figure 3 and demagnetized by heating for 2 hours in hydrogen gas atmosphere between 780℃ and 800℃ after the machining.

A mechanical tensile load was applied to the test piece by a hydraulic testing machine. Deformation was measured using a pair of displacement gage adhered to the side faces. The gage length is 100 mm. The load was applied as a function of stress in elastic deformation area and plastic deformation area during strain-hardening and as a function of displacement in plastic deformation area before strain-hardening. Each of the spatial second differentials of magnetic field was measured after removing the load. Hysteresis of tensile stress and strain applied to the test piece is shown in Figure 4.

MEASUREMENT

The test piece mounted on the scanning stage was scanned in a magnetically shielded room. The pick-up coil was 4.5 mm above the surface of the test piece. The pick-up coil was set parallel to the surface and detected the normal component of the magnetic field generated from the test piece. The scanning stage moved to the X-direction for 100 mm at a rate of 10 mm per second, and moved to the Y-direction for 1 mm after moving to the X-direction. In this way, the scanning stage moved continuously in a boustrophedonic fashion. The scanned area was 100 mm by 70 mm. Data was sampled every 1 mm over a square 2-dimensional grid.

Table 1. Tensile Requirements and Chemical Requirements of JIS G 3106.

Tensile strength and Yield point, MPa		Composition, %				
Tensile strength	Yield point min.	Carbon, max.	Silicon, max	Manganese, max	Phosphorus, max	Sulfur, max
490 to 610	325	0.20	0.55	1.60	0.035	0.035

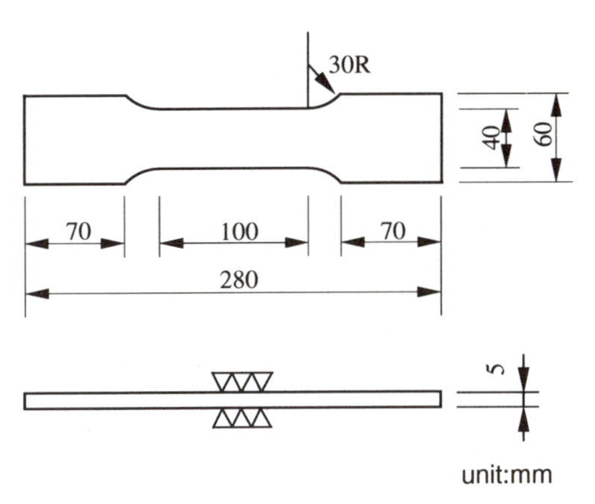

Figure 3. Test piece.

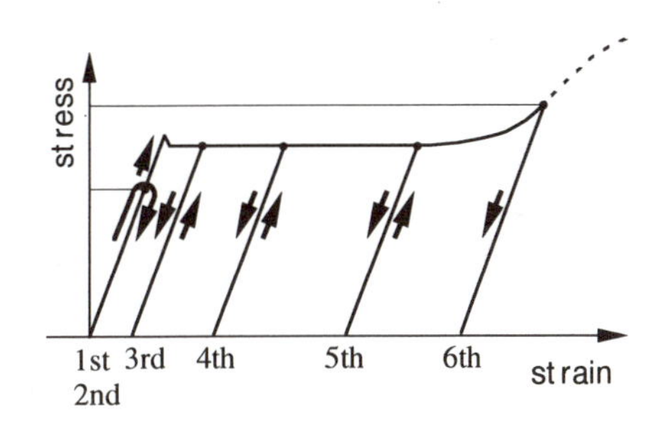

Figure 4. Measuring points.

We didn't apply any artificial magnetic fields to the test piece during this measurement.

RESULTS

Drawing contours of magnetic signals detected by the SQUID system, the area of narrow space of contour lines corresponds to the Lüders band which is observed on the surface of the test piece deformed plastically[2]. A pair of examples is shown in Figure 5.

At a point (X,Y), we calculated the difference between the maximum and minimum value in the area of X-2 mm to X+2 mm and Y-2 mm to Y+2 mm. We call it the "gradient" of the magnetic signal at the point (X,Y). Shaded contour graphs of the "gradient" of the magnetic signal are shown in Figure 6. The dark shade indicates a large value of the "gradient" and the light shade indicates a small one in Figure 6. The large value of the "gradient" corresponds to the location which the Lüders bands are observed approximately. Therefore plastic strains are able to be found in measuring the "gradient" of the magnetic signals.

For the quantitative evaluation of the plastic deformation, we reviewed distributions of the "gradients" in each state. As the characteristics of the distributions, we calculated mean, maximum and minimum values, and standard deviations omitting the area arround edges. These are shown in Table 2.

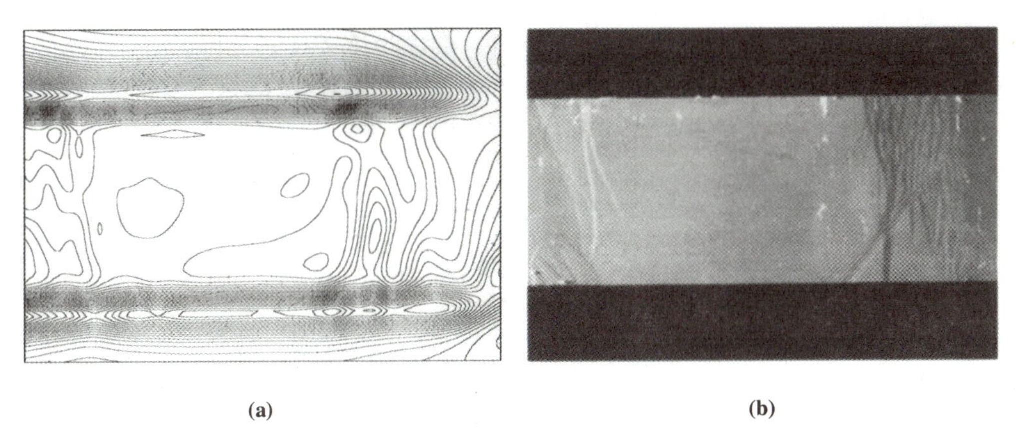

(a) **(b)**

Figure 5. The test piece which is deformed plastically (the residual strain is 0.5%); (a) contour of magnetic signals and (b) photograph (The stripes at both ends of the test piece in (b) are the Lüders band.) (a) and (b) are same scale.

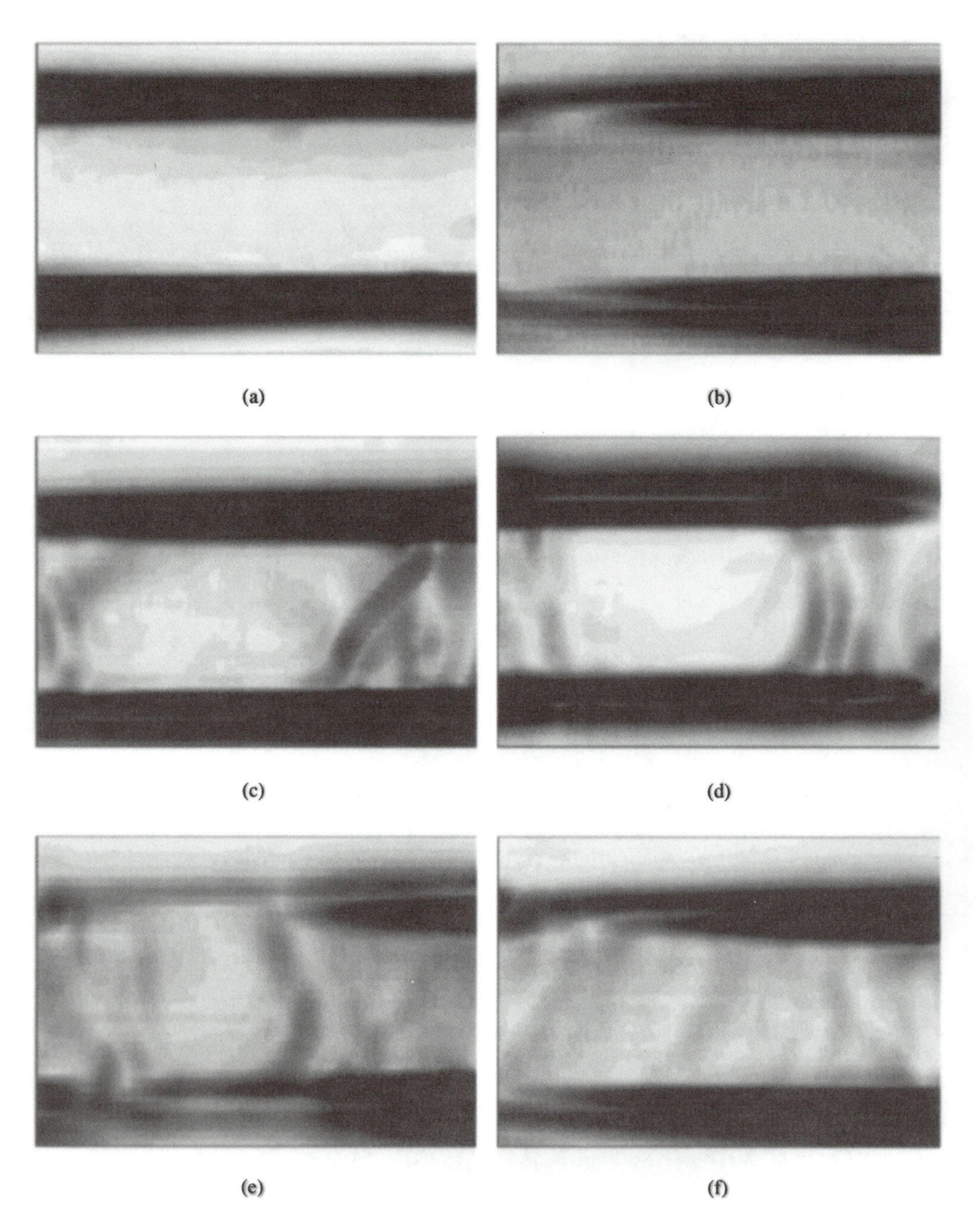

Figure 6. The shaded contours of the "gradients"; (a) initial state, (b) loaded with elastic stress, (c), (d), (e), and (f) deformed plastically. (The residual plastic strains of (c), (d), (e), and (f) are 0.2%, 0.5%, 0.8%, and 1.5% respectively.) The dark shade means a large value of the "gradient" and the light shade means a small one.

The mean values of each states after loading is larger than that of initial state. But objectively speaking, each state is not distinguished by the mean values. The maximum values of initial state and elastic loaded state are smaller than those of plastically deformed states and the value is decreasing as the strain becomes large. The difference of the maximum value and the minimum value and the standard deviations show the same tendency as the maximum value. The results can be reasoned as follows. If a steel is loaded with a stress over the upper

Table 2. Mean values, maximum values, minimum values, distribution ranges, and standard deviations of the "gradients".

	Statistics of "Gradients", 10^{-18}Wb					
	initial	elastic	0.2% strain	0.5% strain	0.8% strain	1.5% strain
mean	0.20	0.50	0.50	0.31	0.59	0.39
max.	0.59	0.96	1.58	1.32	1.26	1.19
min.	0.06	0.30	0.13	0.02	0.22	0.10
max.-min.	0.53	0.66	1.45	1.30	1.04	1.09
std. dev.	0.08	0.13	0.31	0.25	0.20	0.16

yield point, it is deformed locally plastically. The dislocations are generated in the plastic deformed area and internal stresses in the neighborhood of the dislocations induce anisotropic magnetic field[3,4]. As a strain becomes large, the plastic deformed area is spread. Gathering the anisotropic magnetic field to the random directions, the apparent magnetic field may be reduced.

It is considered that this result suggests the possibility of the quantitative evaluation for steel degradation with measuring spontaneous magnetic fields.

CONCLUSIONS

To develop a nondestructive evaluation method of degradation of steels, we measured spontaneous magnetic fields of a hot-rolled steel which had been loaded with tensile stress using a SQUID gradiometer with a concentric second-order pick-up coil. We calculated differences between the maximum and minimum value in an area of 4mm by 4mm around each measuring point and defined them as "gradients". We analyzed the "gradient" and obtained the results as follows:

1. The value of "gradient" is comparatively small and uniform before being deformed plastically and it becomes large at Lüders band after being deformed plastically. This shows the "gradient" is able to detect localized plastic deformations.
2. As the deformation becomes large, the "gradient" becomes small and its distribution narrows. It suggests the "gradient" may be able to evaluate the plastic strain quantitatively.

REFERENCES

1. T. Morooka et al., Developement of integrated direct current superconducting quantum interference device gradiometer for nondestructive evaluation, in: *Jpn. J. Appl. Phys.* , 35: L486-L488 Part2,No.4B,(1996),
2. N. Kasai et. al., Nondestructive detection of dislocation in dteel dsing a SQUID gradiometer, in:*IEEE trans. on Appl. Supercon.*,5. (Preparation of Publication)
3. A. Seeger et. al.,Die Einmündung in die ferromagnetische Sättigung-I, in: *Phys. Chem. Solids*, 12: 298. (1960)
4. H. Kronmüller et. al., Die Einmündung in die ferromagnetische Sättigung-II, in: *Phys. Chem. Solids*, 18: 93. (1961)

DISTRIBUTED REAL-TIME SYSTEM FOR ACOUSTIC EMISSION WAVEFORM CAPTURE

Jeffrey D. Gentry

Digital Wave Corporation
11234-A East Caley Avenue
Englewood, CO 80111

Keywords: Acoustic Emission, Waveform Capture, System Architecture, Digital Signal Processing

INTRODUCTION

Acoustic Emission (AE) is a transient elastic stress wave propagating in a material that can be detected using displacement sensitive transducers. Characterizing the source of the AE through analysis of the wave modes which are excited is a recent development (called Modal AE) which has many advantages in materials characterization.

Digitizing these AE waveforms presents a unique challenge in data acquisition. AE signals are transient in nature, have broad frequency content (20 kHz to over 2.0 MHz), and can occur randomly among a large number of input channels. A Modal AE system must detect and digitize AE waveforms on one or more input channels, record the arrival time at each channel with enough resolution for accurate source location, move the waveform from each channel to permanent storage, and be ready for another completely new signal within a few microseconds.

This paper presents a distributed real-time system architecture for capturing wide band AE waveforms. This system is currently under development for the Federal Highway Administration in conjunction with the National Institute for Standards and Technology in Boulder, Colorado, for characterizing AE in bridge steels.

The distributed architecture allows multiple processors to operate simultaneously, permitting a large number of input channels, and provides real-time digital signal processing (DSP) for rejecting extraneous noise and characterizing the AE source. Other advantages of the distributed architecture include increased flexibility in configuring systems with different components and increased reliability obtained by allowing a single component to fail while the rest of the system continues to operate normally.

The improvements in performance of a system using a distributed architecture are quantified through direct comparison with a system which has the same number of input channels and the same processing power but which uses a centralized architecture. The benefits of this technology for material characterization are then discussed.

BACKGROUND AND MOTIVATION

The declining state of the nation's infrastructure has been a growing concern for several years. In particular, the problem of aging steel highway bridges has been well documented in numerous studies[1,2]. Bridge structures being used past their design life, increasing highway traffic, and lack of maintenance are problems which are typically cited as the causes of many bridges being rated as structurally deficient. Without adequate funding to replace all of these bridges, improving nondestructive inspection (NDI) techniques to verify structural integrity has become a high priority for both state and federal transportation departments. One NDI method which offers unique capabilities for steel bridge inspection is acoustic emission (AE).

ADVANTAGES OF ACOUSTIC EMISSION

Unlike other NDI techniques which interrogate a small area of the structure and require that a probe be placed in the immediate vicinity of a flaw to detect it, AE allows sensors to be placed some distance from the flaw in somewhat arbitrary locations. If a flaw exists in the structure, and grows as the structure is loaded, the resulting release of energy excites a stress wave which propagates away from the source. This propagating stress wave can be detected with the remotely located sensors, and the time of arrival at each sensor can be used to locate the source.

As promising as AE technology is, the current methods used for collecting and analyzing data often leave the inspector with a great deal of doubt as to the exact nature of the AE data and its relevance to the structural integrity of the bridge. The problems with current AE data collection and analysis methods include: (1) using a narrow band, resonant response sensor which distorts the wide band frequency characteristics of the propagating wave; (2) using narrow band filters with nonlinear phase response which further distort the analog signal; (3) using a threshold dependent analog circuit to define a few signal parameters such as rise time, peak amplitude, duration above threshold, and the area under the enveloped signal; and (4) basing the analysis of structural integrity on the values of these parameters.

The most significant problem with this approach is that it has absolutely no theoretical basis. The analysis of parameters, which are measured from a signal highly influenced by the sensors and analog signal conditioning, is purely empirical. Given a large number of specimens with the same sources and the same geometry, the parameters should correlate with different damage mechanisms. However, this correlation only occurs when the test environment is free of extraneous noise.

The advantages of digitizing and storing the entire AE waveform for analysis have been recognized for many years. Even with resonant sensors, the information in the full waveform can provide much more information than a few parameters. Using wide band, high fidelity sensors with careful analog signal conditioning and high speed A/D boards provides waveform signals which are very accurate measurements of far field surface displacement and can be theoretically predicted[3].

Because of the theoretical foundation of Modal AE analysis, interpretation of AE waveform data does not rely on an empirical correlation, and results obtained in structures of different geometry can be compared by accounting for the differences in wave propagation.

Collecting and analyzing the full AE waveform requires more effort than the traditional parameter based approach. In addition to the wide band, high fidelity and low noise requirements, acquiring the entire waveform requires from 2 to 16 kilobytes of data (depending on the sample size and time window) for each waveform from each channel. Compare this with 20 - 30 bytes to store a handful of floating point parameter values. Furthermore, analysis of the waveform data requires a great deal of understanding about source functions and wave propagation. Only recently have automated routines been developed which eliminate tedious manual interrogation of each waveform signal.

SYSTEM THROUGHPUT REQUIREMENTS

The system requirements specified by the Federal Highway Administration include analog to digital conversion at 10 Megahertz sampling rate and 12 bits of vertical resolution on each of twelve channels. The number of samples per channel for each defined event must be adjustable up to at least 8,192 points. The amount of pretrigger, or the number of points in memory prior to a trigger event that are kept as part of the signal, must be adjustable up to 4,096 points. The dead time, or time between the end of digitizing one signal and rearming the A/D for another trigger event, should be no greater than 500 microseconds with sample sizes of 8,192 points. Larger sample sizes could result in proportionally larger dead times. The dead time specification is applicable to each of the 12 channels for the entire test duration up to 20,000 events per channel.

The computation of required system throughput and storage for the specified AE waveform acquisition system is shown below.

$$\frac{12 bits / sample}{8 bits / byte} x 819.2 \mu s / \sec . x 10 MHz = 12{,}288 bytes / waveform \tag{1}$$

$$\frac{1}{819.2 \mu s + 500 \mu s} x 1 x 10^{6} \mu s / \sec = 758 events / \sec / channel \tag{2}$$

In other words, sampling 8,192 samples at 10 MHz results in an event time of 819.2 microseconds. If there is a dead time of 500 microseconds after each event the total time from the beginning of one event to the next is 1.3192 micro seconds. This results in 758 events that can be stored per second on any given channel.

The total system throughput can then be computed as follows:

$$\frac{12{,}288 bytes}{waveform} x \frac{758 events / \sec}{channel} x 2 channels = 106.6 Mbytes / \sec \tag{3}$$

A system that provides adequate throughput and accounts for the random nature of AE activity has unique requirements that cannot be met with existing waveform capture instruments.

DISTRIBUTED REAL TIME ARCHITECTURE

One architecture that provides several potential advantages for handling random transient events such as AE while providing the necessary throughput, is a distributed system. A distributed system has several processes running on different processors working toward specific system requirements. Advantages of a distributed system include better performance from parallel processing, increased resource availability, improved reliability because the system can be designed to recover from failures, and increased flexibility since processors can be added or removed to adjust the size of the system for varying requirements. Figure 1 below shows an example of a distributed architecture.

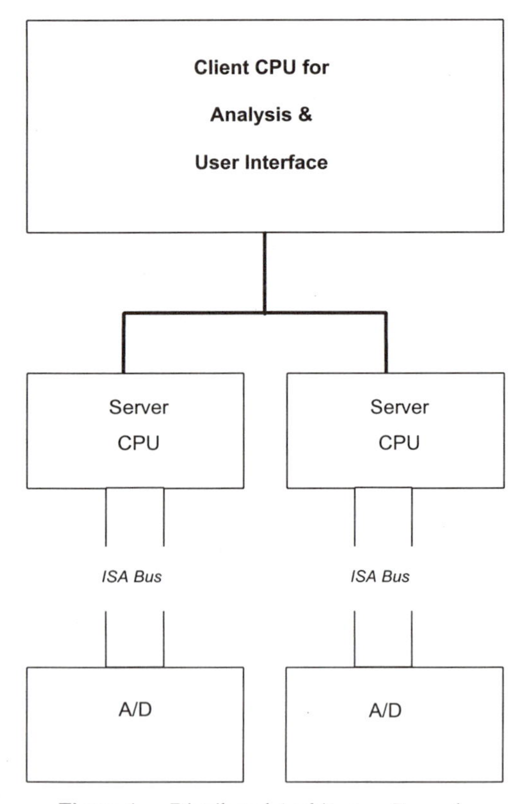

Figure 1. Distributed Architecture Example

A real-time system is characterized by the need for timing constraints[4]. In terms of AE waveform acquisition, the system is constantly sampling the physical condition of various structural locations by looking at the analog voltage produced by transducers at those locations. The sampled data is processed and the status of a display is updated. All three processes must be performed within specified times which are the timing constraints of the system.

Critical timing aspects for capturing and storing AE waveforms are determining accurate wave arrival times at each sensor and knowing how much time elapses between the end of one signal and the start of the next signal.

PROPOSED DESIGN SOLUTION AND FEASIBILITY DEMONSTRATION

To meet the requirements of the FHWA SBIR solicitation, Digital Wave developed the design of a distributed real time system. The system was designed around the PC architecture to provide a commercially viable solution that would take full advantage of the emerging trends in the PC market such as full-motion video, advanced 3-D graphics, integrated stereo sound, and virtual reality which require large block data transfers to be accomplished expeditiously.

The distributed architecture design, with a CPU and PCI bus dedicated to a small number of channels, is shown in Figure 2 below.

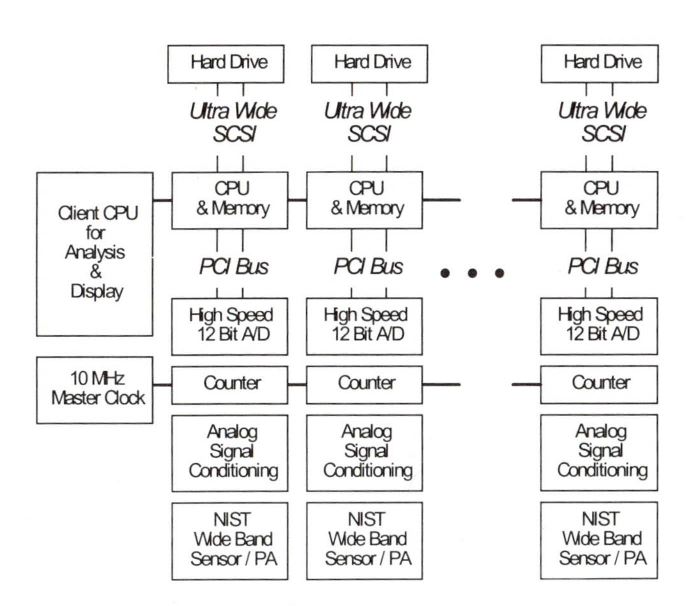

Figure 2. Distributed Real Time Architecture for AE Waveform Capture

This system design results in unprecedented performance and allows the benefits of using high fidelity, wide band AE sensors and high speed waveform capture to be realized. While the system was designed to improve inspection techniques for steel bridges, commercial viability is maintained by using the latest PC technologies.

To demonstrate the feasibility of such a design, Digital Wave developed a Windows 95 version of waveform acquisition and display software using a distributed approach. A dedicated server application communicates with the high speed A/D board and moves the waveform data to permanent storage. A separate client application periodically polls the server application for waveform data. If the server application has moved waveform data since the last request from the client, the client is given the latest waveform to analyze and display. The client and server applications can run on dedicated CPUs or on a single CPU using Windows 95 preemptive multitasking capabilities. The communication between the two applications is accomplished via a standard TCP/IP network connection.

RESULTS

Testing consisted of supplying a simulated AE signal with an arbitrary function generator at 200 events per second to each of four channels of high speed A/D. The A/D boards digitized the signal at 30 MHz with 12 bits of vertical resolution. Each individual waveform consisted of 1,024 samples before the A/D was rearmed. A comparison of a single CPU system running a both client and server applications with a dual CPU system with dedicated processors for each application is shown in Figure 3.

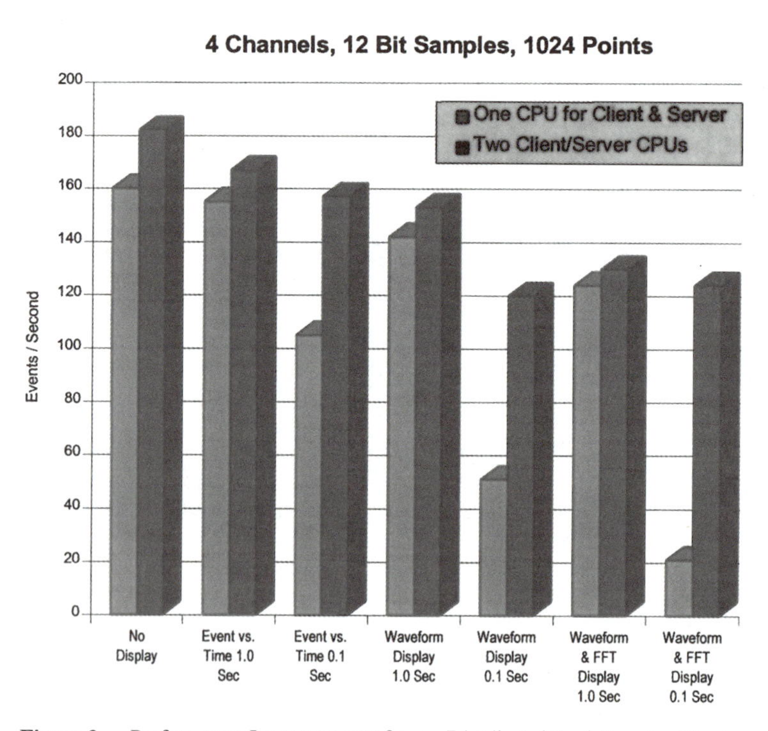

Figure 3. Performance Improvements from a Distributed Real-time Architecture

The performance improvements of the distributed architecture become apparent as the analysis and display requirements increase. With no display or analysis the distributed system only provides a 15% performance increase (1490 to 1310 kbytes/sec) compared with a 590% increase (1014 to 172 kbytes/sec) if FFTs are performed and both waveforms and FFTs are displayed 10 times a second.

CONCLUSIONS

Modal AE offers significant advantages to traditional parameter based AE approach. These include a strong theoretical foundation that allows superior source identification and location. These advantages come at a cost in system data handling requirements.

A distributed real time architecture offers many critical features for developing large Modal AE systems. These include known timing constraints, parallel processing, parallel bandwidth, flexible system configurations and superior maintainability.

During Phase I of a FHWA sponsored SBIR, Digital Wave demonstrated the feasibility of using a distributed real time architecture for capturing acoustic emission waveforms. The results show that the performance improvements from 15% to 590% with performance improvements dependent upon processing and display requirements.

References

1. A. Ghorbanpoor, "Acoustic emission characterization of fatigue crack growth in highway bridge components," Ph.D. dissertation, University of Maryland, College Park, MD, 1985.

2. M., Sison, J.C. Duke, Jr., C. Clemena, M.G. Lozev, " Acoustic emission: a tool for the bridge engineer," *Materials Evaluation*, 54(8):888-900, (1996).

3. K. Graff, *Elastic Wave Motion in Solids*, Ohio State University Press, Columbus, OH, 1975.

4. J. Tsai, et al., *Distributed Real-Time Systems*, John Wiley & Sons, Inc. 1996, pp. 4-5.

ACOUSTIC EMISSION METHOD FOR PARTIAL DISCHARGES MEASUREMENTS OF ELECTRIC POWER COMPONENTS

J. Sikula[1], B. Koktavy[1], P. Vasina[1], Z. Weber[1], M. Korenska[1], L. Pazdera[1], T. Lokajicek[2], F. Matejka[3]

[1] Technical University of Brno, Zizkova 17
602 00 Brno, Czech Republic
[2] Geophysical Institute, Academy of Science of the Czech Rep., Bocni 1401
141 31 Praha 4, Czech Republic
[3] Skoda Factory Research Centre
Plzen, Czech Republic

INTRODUCTION

Nowadays, prediction and assessment of reliability of high power components are important problem[1,2]. Any method which can give us some information about quality and reliability of high power insulators should be studied in detail and outgoing results of the study will cause component lifetime to increase. Exact location of defect can save the time of maintenance and can find out positions of next potential defects[3].

Acoustic emission testing method has been used to localised partial discharges in electric power components. This measurement and investigation technique has some unique features: one of them - partial discharges which are possible source of acoustic emission could be localised in insulation layers due to the difference time of acoustic and electric signal propagation[4].

Partial discharges in high voltage components are created by electric field in gas bubbles inside of insulating layers. Sources of partial discharges are usually in high resistance region on insulating surface or in bulk. Our method is based on the measurements of current pulses arising in the electrical circuit created by series combination of machine conducting parts and the insulating systems. The quality and reliability depends on density of bubbles which are unwillingly implanted in insulation layer. Acoustic emission allows us to detect the partial discharges density and defect position.

The shape of commutative curve of overshoot number of acoustic emission signal depends on the load voltage. Delay of the acoustic signals according to electric discharge impulse depends on the velocity of propagation of acoustic waves in insulator and metal. On the interface one can observe different mechanical waves.

EXPERIMENTAL

Study of partial discharges by the acoustic emission measurements have been performed. The frequency of evoked random-switching (hereinafter overshoots) signals has been used as a characterisation quantity. Measurements have been carried out in laboratory at Technical University of Brno (measurements on sample of high voltage bar with RELANEX insulation) and at Skoda factory (measurements on the stator of high voltage generator).

The acoustic emission signal measurement set-up is shown in Fig. 1. Sample-under-test have been biased up to 10 kV by the special DC high voltage source. The signals, generated into RELANEX insulation layer, have been registered by B&K 8303 wide-band piezoelectric detector. Afterwards, these electric signals were fed to Acoustic emission signal analyzer and amplified in frequency range from 10 kHz up to 1 MHz. Than, these signals have been divided into two independent ways. The first way was fed into counter BM 640, where frequency of signals pulses above a threshold voltage level were counted. The second one monitored signal in time domain by the digital plotter DL 912.

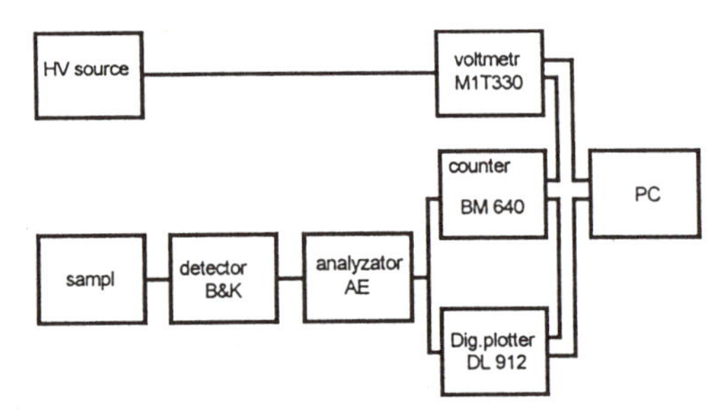

Fig. 1 Experimental measurement set-up.

Experiment has been powered by PC AT 486 equipped by a standard IEEE-488 plug-in card. Measurements have been carried out on the iron sample with diameters of 12x36x1000 mm. This sample was overlaid by the RELANEX insulating strip No. 45750 (provided by Elektroizola Tabor) of width 1.9 mm. Annealing was processed in the pressing machine for certain period of two hours at temperature of 165° C. However, we have prepared conductive measurement electrode, in the middle of the sample with length of 100 mm, performed by the semi-conductive paint SIB 643. Wide-band piezoelectric detector B&K 3303 have been fixed in the distance of 75 mm from the conductive electrode border.

RESULTS

Measurement results show (see Fig. 2.), that shape of commutative curve of the acoustic emission signal depends on the slope of the high voltage ramp. Plot shows the remarkable increase of the signal overshoots within the time interval 10/15 s. These signals of the acoustic emission are probably evoked even when voltage reached level of the 10 kV (curve U). The next region has an exponential increase of number of signal overshoots. The time interval since 1080 up to 1095 s shown rapid increase of the acoustic signal overshoots, where bias was dropping from 10 kV up to zero value. Till the end of measurement exponential increase of the signal overshoots one can observed. When we have avoided intervals where the bias voltage has change, the shape of commutative curve expressed by simple exponential formula in form

$$N = 277.6 \cdot \exp(8.733 \cdot 10^{-4} \cdot t) \tag{1}$$

where N is the number of acoustic emission signal overshoots and t represents time.

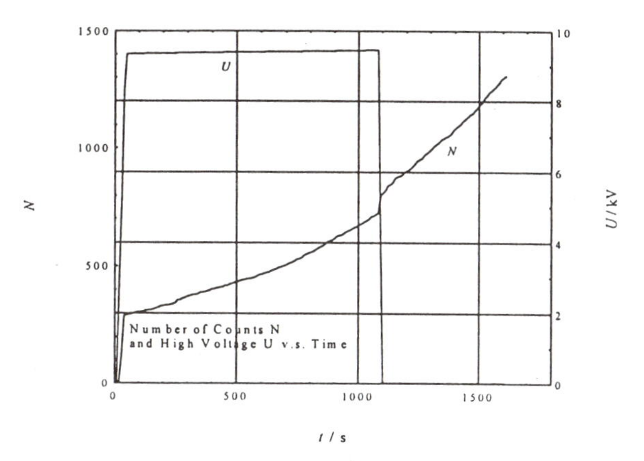

Fig. 2. Bias voltage and commutative curve of the acoustic emission signal vs. time.

Fig. 3. shows the acoustic emission signal envelopes registered in time domain by the digital plotter of signal transitions DL 912. The time record in Fig. 3a represents signal record at the bias voltage of 7 kV and Fig. 3b. at the bias 9.5 kV, respective. Comparison both of them yields shorter threshold time (time up to pulses appear) for the higher bias. Hence, at bias 9.5 kV threshold time was 22 μs and for 7.5 kV one can express from Fig. 3b. threshold time equal to 61 μs.

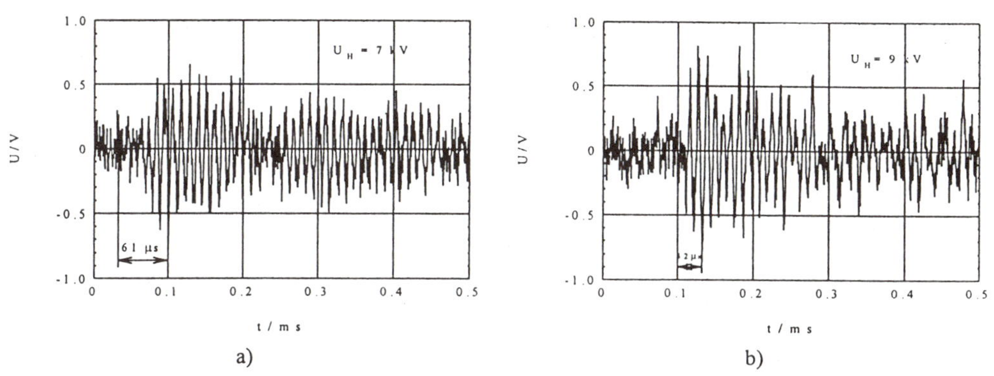

a) b)

Fig. 3 Acoustic emission signal record in time domain, at different supply bias of 7 kV and 9.5 kV, respective.

Together with acoustic signal emission measurements we have provided localisation of these sources into RELANEX insulation layer. Especially, at this kind of measurements we have used analyser of signal acoustic emission Locan 320, which allowed us to measure signals of acoustic emission from more (up to 6 channels) piezoelectric detectors sampling at the same time. Consequently, in our study we have used 4 similar channels equipped by detectors B&K 8313 (resonance frequency 250 kHz) to localised the defect position. The positions of the detectors mounted and acoustic emission sources co-ordinates on` the laboratory iron sample is depicted in Fig. 4. (crosses No. 1.-4. consider detector positions in mm. according to defects location in sample represented by vertical solid lines).

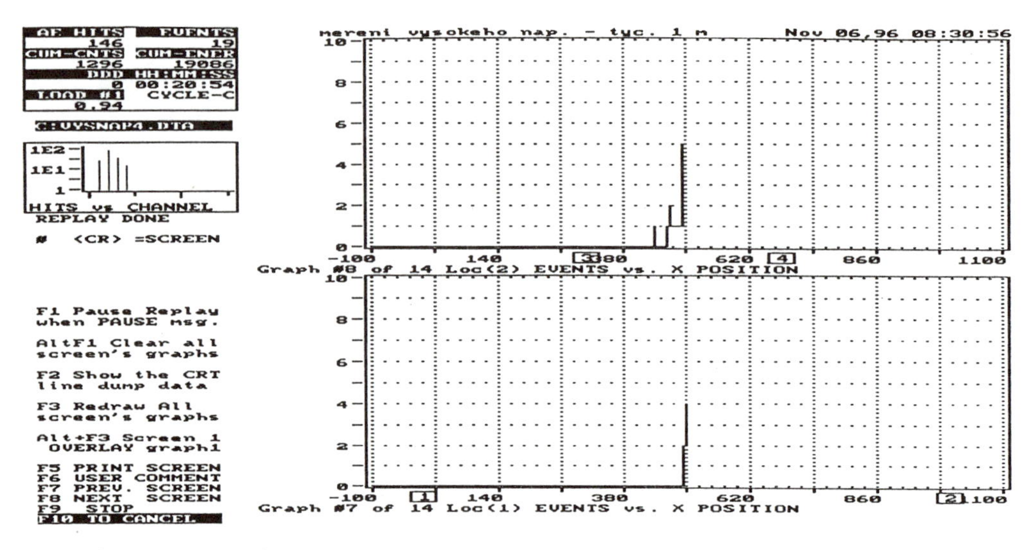

Fig. 4. Detectors and defect positions in sample.

CONCLUSION

Measurements of the acoustic emission signals and its comparison with electromagnetic signals detected at different position of tested sample can give us some useful information about number of defects and determine exact position in sample. Shape of commutation curve of acoustic signal pulses increased exponentially in time (Fig. 2.). The time record expressed shorter threshold time (time up to pulses appear) for the higher bias. For example, at bias 9.5 kV threshold time was 22 μs and for 7.5 kV we expressed threshold time equal to 61 μs.

ACKNOWLEDGEMENT

This work have been supported within Grant agency of the Czech Republic and grant No. 103/97/0899 „Electromagnetic emission and its application in the civil engineering and geophysics“.

REFERENCES

1. J. Šikula, B. Koktavý, P. Vašina, Z. Weber, M. Kořenská a T. Lokajíček: Electromagnetic and Acoustic Emission from Solids, Proc. of 22nd European Conference on Acoustic Emission Testing, Aberdeen, UK, 1996.
2. J. Sikula, A. Touboul: Noise spectroscopy, diagnostics and reliability of electronic devices. Proc. of the 12th Int. Conf on Noise in Phys. Systems, , Aug. 1993, pp. 206 - 211, 1993, (ed. P. Handel and A.L.Chung) St.Luis,USA
3. T. Lokajíček, and J. Vlk, 1994: Interface Card for Multichannel Processing of Acoustic Emission Signals, Progress in Acoustic Emission VII, Proceedings of 12th International Acoustic Emission Symposium, Sapporo Oct. 17 - 20, 577 - 582.
4. K. R. McCall, 1994: Theoretical Study of Nonlinear Elastic Wave Propagation, J. Geoph. Res., B, 99, 2.

THE RECENT RESULTS OF COMPUTER PROCESSING OF DIGITIZED SIGNALS GAINED DURING FLAW DETECTION TYPE CHECKING OF STEEL CABLES

Michal Lesnak,[1] Oldřích Lesňák[2]

[1]Institute of Physics, Technical University, Ostrava,
[2]Research Mining Institute, Inc., Ostrava
Czech Republic

INTRODUCTION

The basic knowledge gained during the computer processing of digitized signals from flaw checking of steel cables has been published in the paper "Evaluation of Digitized Signals from Steel Cable Flaw Detector" by O. Lesňák during the Seventh International Symposium on Nondestructive Characterization of Materials, held in Prague in 1995,[1,2,4,5,6].

In the paper, some recent knowledge primarily on the development of methods characterizing the technical state of the checked rope and verified by practical tests has been presented.

The above mentioned results have been verified on flaw detectors equipped with coil gauges, Hall´s probes as well as sets equipped not only with coil gauges but also with Hall´s probes. There were used the following flaw detecting devices [3]:

- MID-1, the flaw detecting device equipped with coil sensors (4 channels),
- MID-3, the flaw detecting device equipped with coil sensors (2 channels),
- MID-5HVS, the flaw detecting device equipped with Hall´s probes (2 channels),and
- MID-6HVS, the flaw detecting device equipped with coil sensors and Hall´s probes (2 channels).

The digitized flow detecting apparatus consists of

- measuring head with optic-electronic incremental rotational gauge,
- indication module and
- computer.

The paper deals on data gained by means of flow-detecting devices having two gauges, i. e.

- defects gauge (fatigue defects) and
- corrosion gauge (corrosion).

All both gauges are equipped by Hall's probes or coil gauges and measured data from both gauges are also displayed separately in two channels,[2] i.e.

- channel of defects and
- channel of corrosion.

As mentioned above, the processing of the digitized signal is provided in two stages, i.e.

- baseline visualization methods and
- evaluation methods characterizing the state of the steel cable.

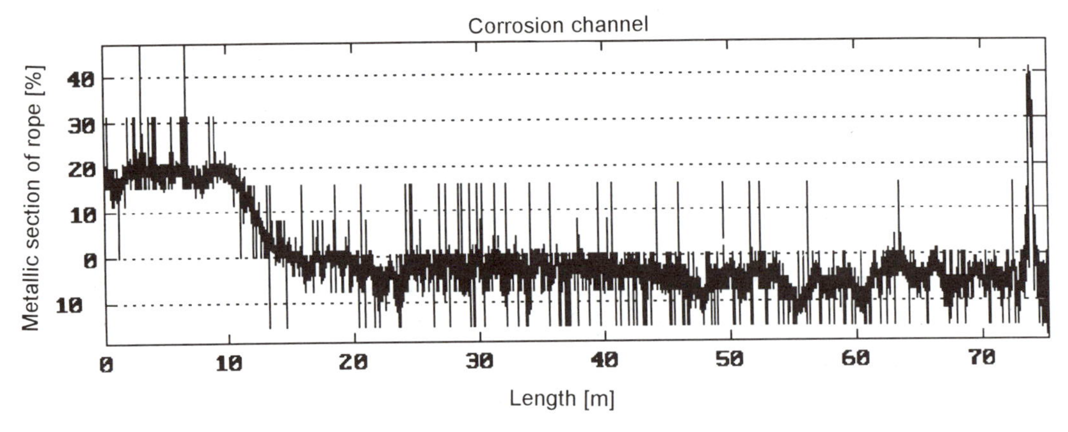

Figure 1. Original record

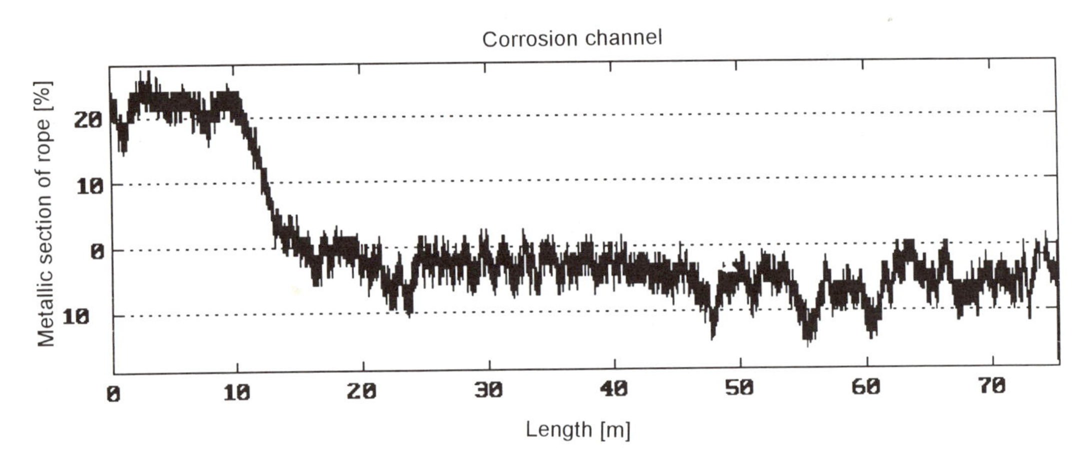

Figure 2. After processing

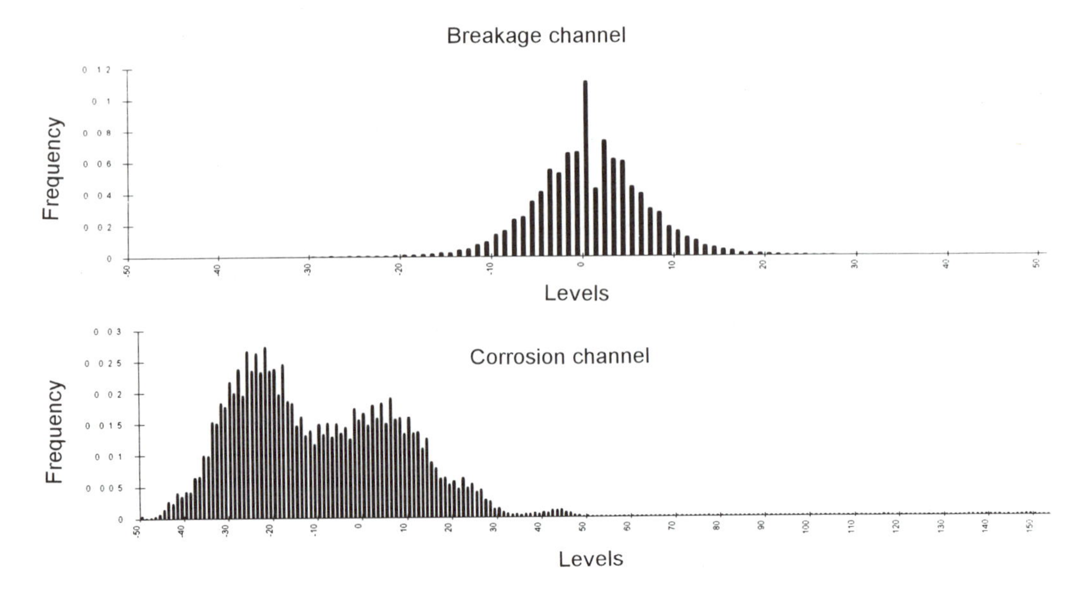

Figure 3. Histogram related to worn steel mine rope

EXPERIMENTAL

Baseline visualization methods deal with raw data only, and the evaluation is done by the specialist, who performs the testing. This is the first thing we can do on the place of testing.

The used software allows us to browse both channels simultaneously (possibility of comparison of corrosion and breakages) or separately. After observing the whole-length record in one graph, experienced specialist can choose some specific parts of the rope and display them with appropriate scaling (zoom), which is independent for both axes (signal amplitude - y-axis, the position on the rope - x-axis). The detailed inspection of chosen places can lead to decision whether there is a real defect on the rope, or the signal change was produced by some external influence (shocks in the power network and so on).

The main purpose of the defectoscopic measurement is to evaluate the condition of the rope. If we use only the simple visualization described previously, we rely mainly on the specialist's experience (human factor). The using of mathematical methods enables us :

to avoid (or at least to reduce) the influence of human factor,

to obtain some additional information from the measured data.

The main purpose of all evaluation methods is to determine the technical condition of the steel rope more precisely than the specialists can do, when they evaluate the record looking at the „raw“ data only. Generally, there is no universal method for every situation and one has to employ various mathematical methods, each of them designed for one specific task.

The most frequently used evaluation methods are :

- preprocessing,
- evaluation of statistical data,
- histograms,
- frequency analysis,
- input signal filtration,
- correlation,
- wavelet analysis,
- evaluation of the worst location on the rope.

Preprocessing

The preprocessed signals could be used with good effect in course of records evaluation directly in case of steel rope checking. The preprocessing method enables also definitely to avoid the pseudo defects, e.g. shocks in the power network, etc.

Figure 1. shows a record featured by a great number of pseudo defects caused by the power network shocks. On Figure 2. you can see the same record after preprocessing.

Histogram

Histogram is statistical tool for numeric expression of technical condition of the rope. It can graphically express the frequency of respectively levels of monitored rope measured data. The determination attribute is the width of the histogram. The more frequent and higher changes (peaks) in the output signal, the bigger histogram width. The changes in the signal are result of the changes in crossection of the rope. This holds for both channels.

The histogram in Figure 3. is related to the worn steel mine rope.

Frequency Analysis

The information about the defects is in most cases mixed (if not completely hidden) with the noise. To design optimal filters, it is essential to perform the frequency analysis of recorded data. The analysis refers mainly to fatigue fractures of wires and deep nicks recorded during the measurement in time domain. The purpose is to identify the response of individual faults in the frequency domain.

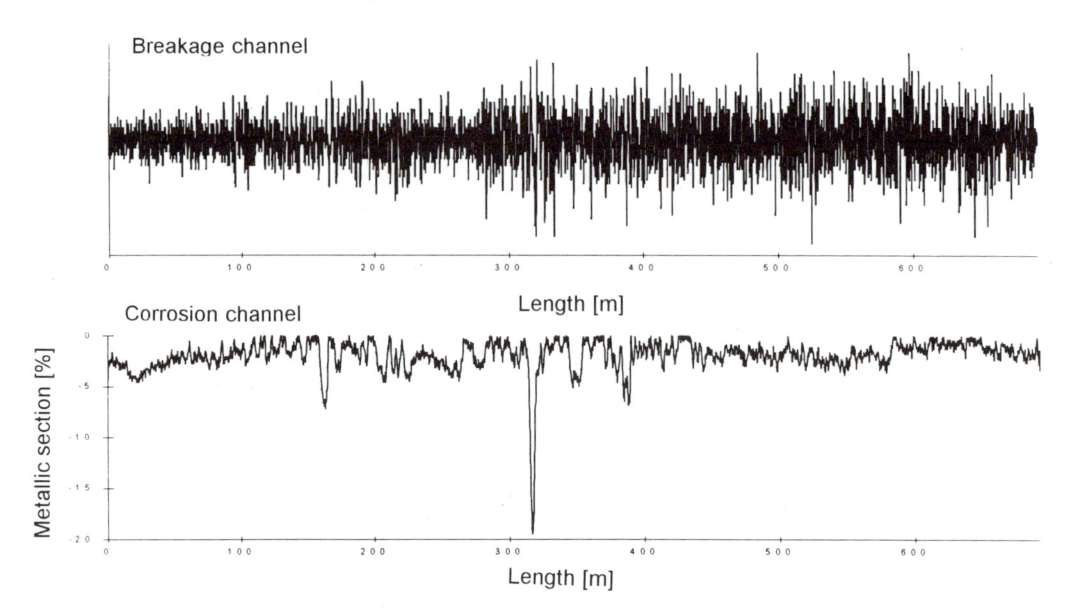

Figure 4. Original record

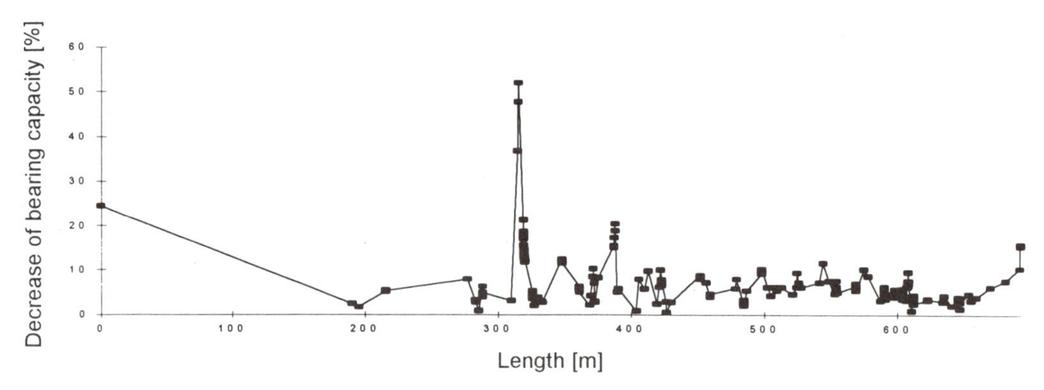

Figure 5. Determination of the worst place on the rope

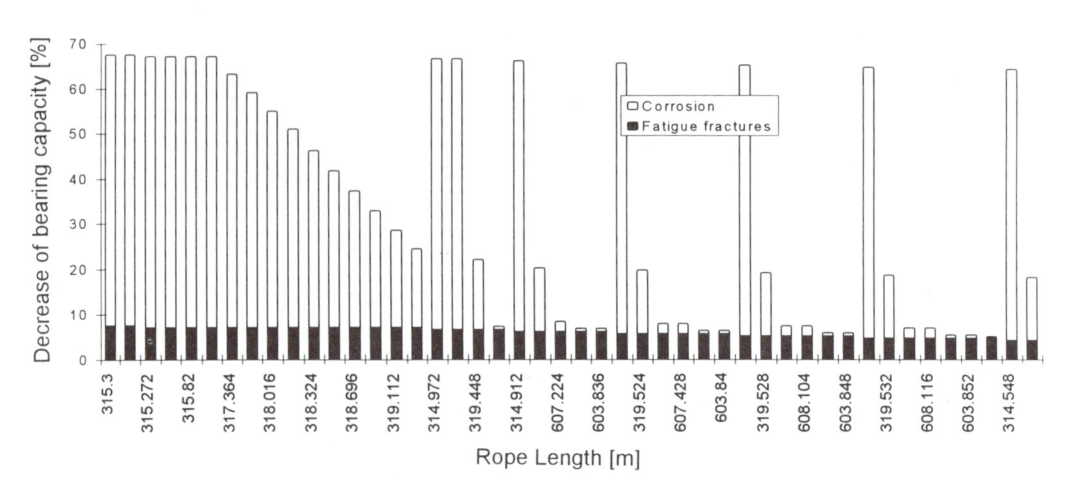

Figure 6. The worst spot on the rope, total decrease of bearing capacity acc. to fatigue fractures of wires and corrosion

The time dependent measured signal is expressed as a sequence of digital samples in frequency domain, i.e. to the decomposition in to the elementary harmonic signals characterized by three parameters - amplitude, frequency and initial phase.

Input Signal Filtration

In order to suppress the undesirable noise and obtain more clear shape of the record, we can use various type of filters - analog filters, digital filters and filter based on the principle of Dolby system. The purpose is not only to suppress the noise (frequency filters), but to get better response from wire breakages and/or deep notches (Dolby system). After the filtration is possible to distinguish the breakages more clearly and effectively.

Autocorrelation

Autocorrelation is the other way, how to get rid of unwanted noise. Generally, the autocorrelation (in time domain) is the relationship between the signal and its time shifted version. If the signal n(t) is sampled every Δ seconds, its autocorretion is done by following expression :

$$N(t) = \frac{1}{M}\sum_{j=1}^{M} n(j\Delta)\cdot n(j\Delta+\tau) \tag{1}$$

where τ is time delay and M is the autocorrelation interval.

In the case of our defectoscopic measurement, the linear shift in the longitudinal length of the rope is used instead of the time delay τ. This shift depends on the density of sampling - used values are 1, 2 or 4 mm. The shift used in the presented records is one sampling value and with respect to applied digitalization density, it is equal to 4 mm.

Wavelet Analysis

This up-to-data signal processing method can be a convenient alternative to well known frequency analysis. In comparison with it, wavelet analysis can give high resolution for low frequencies in the time domain and for high frequencies in frequency domain. Continual wavelet transformation can be defined by following integral transformation :

$$WT_X(\tau,s) = \frac{1}{\sqrt{|s|}}\int_{-\infty}^{\infty} x(t)\psi\left(\frac{t-\tau}{s}\right)dt \tag{2}$$

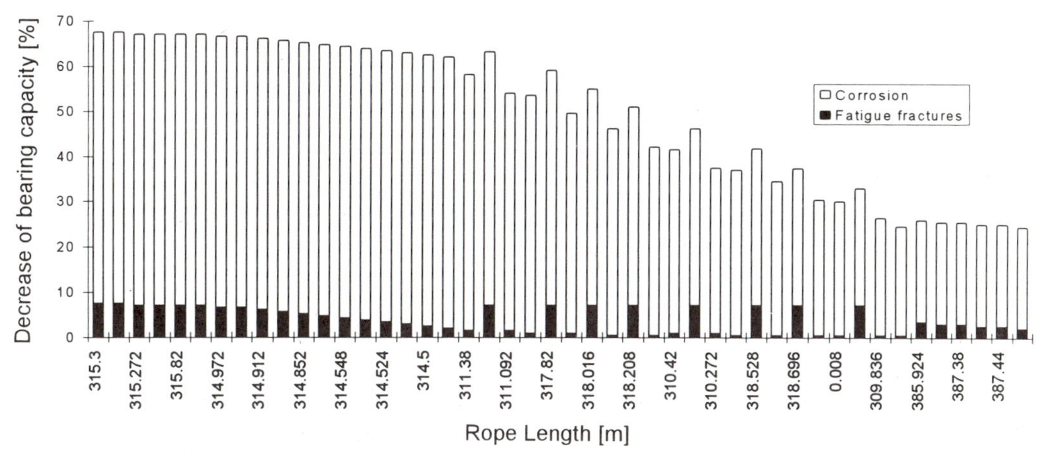

Figure 7. The worst spot on the rope, total decrease of bearing capacity acc. to corrosion and fatigue fractures of wires

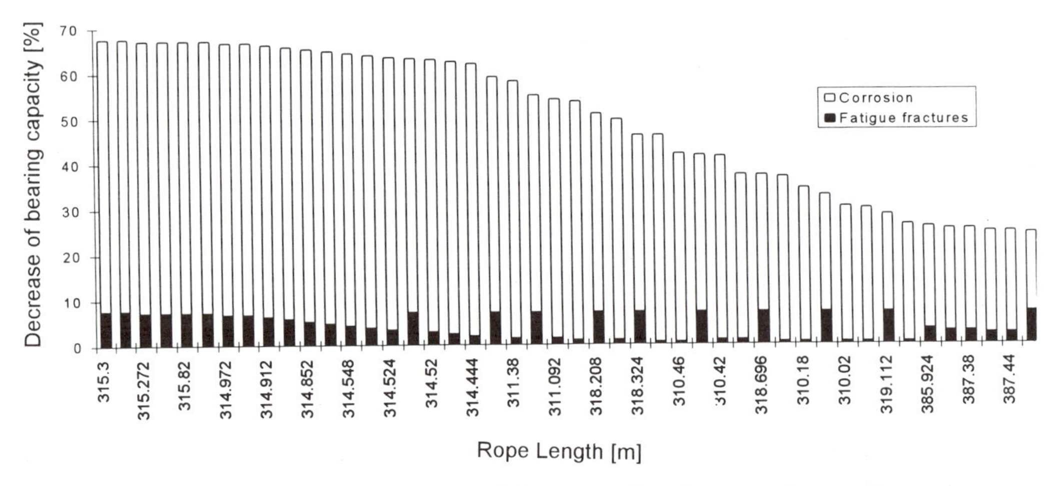

Figure 8. The worst spot on the rope, total decrease of bearing capacity acc. To maximum decrease of bearing capacity

The signal filter based on wavelet analysis can be easily modified for various tasks. In our case the filter was designed in order to conserve the amplitudes of high peaks (the response from surface defects) . Such a filter is not so easy to design using the method of frequency analysis.

Evaluation of the Worst Location on the Rope

The purpose is to specify the position with maximal reduction of bearing capacity of the rope. The bearing capacity is affected by the simultaneous influence of breakages and corrosion. First we evaluate the bearing capacity reduction caused by corrosion and by breakages and then we sum both results for every appropriate length interval. The length position, at which the sum reaches its maximal value is the worst location on the rope. Result original record in Figure 4. Is in Figures 5, 6, 7 and 8.

CONCLUSION

The paper incorporates a number of improvements made on the basis of the experience with the analysis of the output signals from magnetic defectoscopes. The major addition is the application of the new methods in the signal processing. On this base we can observe the new quality in the signal interpretation. It has been shown that the presented process enables to specify the changes of the rope crossection with accuracy level better than 0.5 %. A more detailed discussion is planned for a future paper.

REFERENCES

1. Rabiner L.R., Gold B. : Theory and Application of Digital Signál Process, Prentice-Hall, Inc. Englewood Cliffs, New Jersey 1975,
2. M. Lesňák : Využití výpočetní techniky při zpracování digitalizovaných dat z defektoskopických kontrol ocelových lan, VŠB - TU Ostrava 1994
3. J. Boraška, J. Hulín, O. Lesňák : Ocelová lana, Vydavatelstvo technickej a ekonomickej literatury Bratislava, 1982
4. S. K. Mitra, J.F.Kaiser : Digital Signal Process, John Wiley and Sons, Inc. New York, 1993,
5. H. Havlová, M. Vošvrda : Analýza etalonového signálu, Zpráva ÚTIA Praha č.1565, 1987
6. O. Lesňák : Evaluation of Digitized Signals from a Steel Cables Flaw Detector, Seventh International Symposium on Characterization of Materials, Prague 1995, Czech Republic.

DETECTION OF CRACK POSITION BY AE AND EME EFFECTS IN SOLIDS

T. Lokajíček[1], J. Šikula[2], and P. Vašina[2]

[1]Geophysical Institute, Academy of Sciences of the Czech Republic, Boční II/1401, 141 31 Praha 4 - Spořilov, Czech Republic, E-mail: TL@IG.CAS.CZ
[2]Technical University of Brno, Physics Department, Žižkova 17, 602 00 Brno, Czech Republic

INTRODUCTION

When a stress is applied to the rocks, due to its inner structure and stress concentrator orientation acoustic emission can be observed. The acoustic emission could be accompanied by electromagnetic field generation. during the process of microcracks generation, the weakening of mechanical bonds at he faces of the cracks gives arise to the origin of the electric charge between the crack faces. This results in dipole or more complicated system of electric charge origin. The faces of the crack could behave as an electric capacitor. The electric dipole system could be source of electrostatic and electromagnetic field. The electrostatic field results mostly from the electric charge redistribution and electromagnetic field could be caused by non-harmonic vibrations of electric charge. The electric voltage across the plates depends both on the relaxation constant of the dipole system and on the vibrations of the microcrack faces and dislocations.

THEORY

When a crack is produced a certain portion of the mechanical energy is converted to that of vibrations. The time dependent displacement of the faces results in an AC component whose frequency corresponds to that of mechanical vibrations of the specimen.

Measurable quantity is the electric voltage $u(t)$ on parallel plate capacitor C_m. In the case when the crack appears the induced electric charge Q_i creates on the plates of capacitor C_m voltage $u(t)$ given by

$$u(t) = \frac{Q_i}{C} \tag{1}$$

where C is the capacity of the parallel plate capacitor C_m (see Figure1).

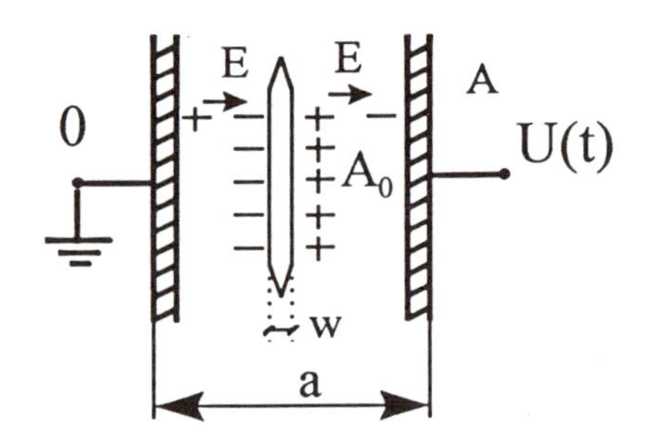

Figure. 1 Schematic diagram of electric set-up

We will suppose that, the crack walls, when appear, are charged by electric charge $Q_o(t)$. The dipole moment $p(t)$ is given by the electric charge $Q_o(t)$ and the crack with w and holds:

$$p(t) = Q_o(t).w. \tag{2}$$

The induced charge Q_i on the parallel plate capacitor is related to the electric charge on the elementary capacitor given by cracs walls:

$$Q_i = NQ_o = Np(t)/w, \tag{3}$$

where N is the electric charge reduction on the parallel plate capacitor C_m.

The crack appearance is accompanied by the mechanical vibration. Due to this effect the crack walls are vibrating and crack width w is a function of time

$$w(t) = w_o + \sum_n A_n \sin(\omega_n t) = w_o(1 - x(t)) \tag{4}$$

where A_n, ω_n are amplitude and angular frequency of vibrating spectrum, $x(t)$ is relative displacement given by

$$x(t) = \frac{-1}{w_o} \sum_n A_n \sin(\omega_n t) \tag{5}$$

The electric voltage $u(t)$ on capacitor C_m is then using (1) to (4)

$$u(t) = \frac{N.p(t).a}{\varepsilon A.w_o(1 - x(t))}. \tag{6}$$

Electrical voltage measured on the plate capacitor C_m is time dependent due to vibration of crack walls and dipole moment $p(t)$ time dependence.

The electrical conductivity of measured sample is discharging the elementary capacitor representing crack and we suppose that the dipole moment $p(t)$ is given by

$$p(t) = p_o \exp(-t/\tau), \tag{7}$$

where τ is the relaxation constant.

Then the voltage

$$u(t) = U_o(t) + U_m(t). \tag{8}$$

For $x \ll 1$ we have

$$U_o(t) = \frac{Np(t)a}{\varepsilon A w_o} = U_o e^{-t/\tau} \tag{9}$$

and

$$U_m(t) = U_o e^{-t/\tau}.x(t). \tag{10}$$

In Figure 2 the time dependence of $U_o(t)$ and $U_m(t)$ is shown. Both quantities are dependent on electric dipole moment created by the crack and follow the exponential dependence with relaxation constant $\tau = 24$ μs.

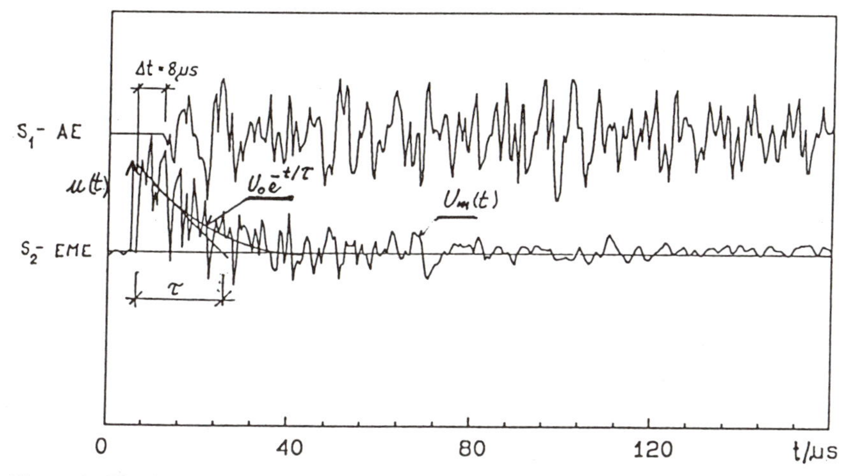

Figure 2. The time dependence of U_o (t) and U_m (t), s_1 - AE signal, s_2 - EME signal for the same event, Δt = 8 μs is time shift of AE signal with respect to EME signal.

EXPERIMENT

The granite samples dimension of 4 x 4 x 1 cm^3 were used in the experiment. The samples were loaded with a constant stress rate up to their failure. Two conducting plates were fixed to the specimen symmetrically making up a plate capacitor whose dielectric was the material under the test (see Figure. 3). The acoustic emission pick-up was fixed on the edge of the sample to record acoustic emission signals (AE) during the loading of the sample. The voltage changes - caused by electro-magnetic emission (EME) arising across the capacitor were amplified by a low noise preamplifier. To eliminate ambient influence, the sample and preamplifier were shielded electrically and magnetically.

Time dependence of recorded signals is depicted in Figure 4, but the initial part of the EME signal is opposite polarity. It seems that the configuration of the source - dipole orientation - is reversed too. In these cases the charge configuration of dipole moment is oriented perpendicularly to the capacitor plates.

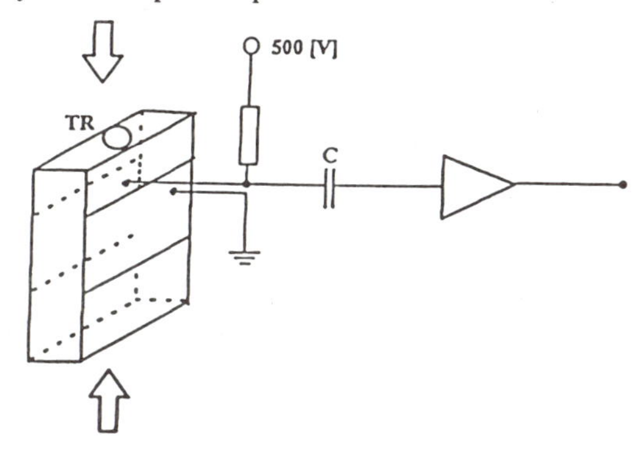

Figure 3. Schematic diagram of experimental set-up

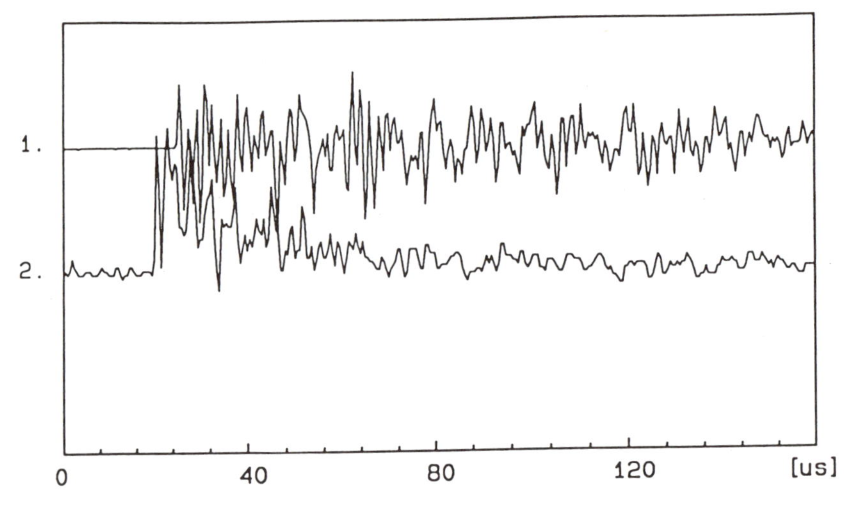

Figure 4. Acoustic and electromagnetic emission signals

During the measurement also no initial DC step component was measured. This phenomena could be explained by the change of dipole to the quadrupole configuration on the crack face, or by the orientation of the dipole moment - parallel with the capacitor plates.

DISCUSSION:

The EME signal precedes the AE signal by the time difference ranging from 2 - 10 microseconds. According to the dipole moment occurrence during AE radiation and its orientation - DC step function was observed. This effect is followed by a electric charge relaxation with a time constant between 20 to 30 microseconds.

When a crack is produced, a certain amount of the mechanical energy is connected to vibrations. The time dependent displacement of the crack faces results in an AC component whose frequency corresponds to that of mechanical vibrations of the specimen.

While the amplitude of the mechanical vibrations (as evidenced by AE signals) decreases very slowly, the high frequency oscillations of EME signal (its amplitude) decreases faster then the AE signal - by one order faster. Due to this fact the measured oscillations cannot be due to the capacitor plate displacement induced by the sample face vibrations.

To decide whether the observed effects were due to the electromagnetic waves or a quasi-stationary electromagnetic field - a wire flat coil antenna was used to detect electromagnetic waves. This experiment proved that a quasi-stationary electromagnetic field was observed rather then an electromagnetic wave.

CONCLUSIONS

When a granite samples are disintegrated mechanically, electric charges are generated at the faces of the crack; arising due to the stress concentration at the grain boundaries or already existing cracks. These charges constitute a dipole or quadrupole electrostatic system. The electric voltage magnitude depends on the charge distribution in the sample. Due to the non-zero electric conductivity of the sample the discharging phenomena can be observed, what is manifested by a relaxation phenomena (see Figs. 2 and 3). In our experiment the relaxation constant was from 20 to 30 microseconds. It can be concluded that generation of microcracks is accompanied not only by acoustic emission, but also by the generation of electric charge, which can be measured by means of a capacitance antenna. In our opinion, charge redistribution, rather then electromagnetic emission can be observed.

ACKNOWLEDGEMENT

Portion of this work was supported by the Grant Agency of the Czech Republic, Grants No. 102/94/0858 „Fluctuation Phenomena in a Metal-Insulator-Metal Structures", No. 205/95/0263 „Experimental investigation of Seismo-electromagnetic effects" and No. 103/97/0899 „Electromagnetic emission and its application in the civil engineering and geophysics.

REFERENCES

[1] Hanson, D.R., Rowell, G.A., Electromagnetic Radiation from Rock Failure, *Report of investigations 8594*, United States Department of the Interior, 1981.

[2] Park, S.K., Johnston, M.J.S., Madden, T.R., Morgan, F.D., Morrison, H.F., Electromagnetic Precursors to Earthquakes in the ULF Band: *A review of Observations and Merchanisms*, 1992.

[3] Bella, F., Biagi, P.F., Caputo, M., Della Monica, G., Ermini, A., Plastino, W., Sgrigna, V., Zilpimiani, D., *Il nuovo cimento*, 18 (1995), 19 - 32.

[4] Hayakawa, M., Tomizawa, I., Ohta, K., Shimakura, S., Fujinawa, Y., Takahashi, K., Yoshino, T., *Physics of the Earth and Planetary Interiors*, 77 (1995) 127 - 135.

ANALYSIS OF ACOUSTIC EMISSION FROM ROCK SAMPLES LOADED TO LONG-TERM STRENGTH LIMIT

V. Rudajev,[1] J. Vilhelm,[1] and T. Lokajíček[2]

[1]Institute of Rock Structure and Mechanics, Academy of Sciences of the Czech Republic, V Holešovičkách 41, 180 00 Prague 8, Czech Republic
[2]Geophysical Institute, Academy of Sciences of the Czech Republic, Boční II/1401, 141 31 Prague 4, Czech Republic

INTRODUCTION

As shown in the previous papers by Rudajev,[1,2] the parameters of the statistical distribution of the acoustic emission (AE) are not random, although the origin of the acoustic signal is random in time. In the above papers it was shown that the parameters of distribution cohere with the strain stage of the loaded samples. The object of this paper is the analysis of the statistical distribution of AE, of the relation of their parameters to the process of fracturing, and mainly determining the probability of total fracture.

The experimental study was carried out on rock samples under laboratory conditions. The importance and difficulty of the experiments consist in the reaching the long-term strength of rock samples, i. e. in reaching such a value of the acting pressure at which without additional increasing of load spontaneous total fracturing of the sample is observed. The granite samples were loaded at a constant stress rate. On the basis of the behaviour of the AE activity (very high activity) the loading of the sample was stopped and was held at the constant acting stress. AE was recorded during the whole loading of the sample - during the linear increase of the acting stress, as well as after reaching its constant stress level - the limit of long-term strength and also during the time interval close to the total rupture of the sample.

EXPERIMENT

The rectangular granite samples, dimensions 40 x 40 x 80 mm were, loaded with uniaxially acting stress in the direction of the longer axis of the sample. The loading rate was constant at 21 kPa/s. When the limit of the long-term strength of the sample was reached, the loading of the sample was stopped and kept at the constant load level (Figure 1c). The

Nondestructive Characterization of Material VIII
Edited by Robert E. Green Jr., Plenum Press, New York, 1998

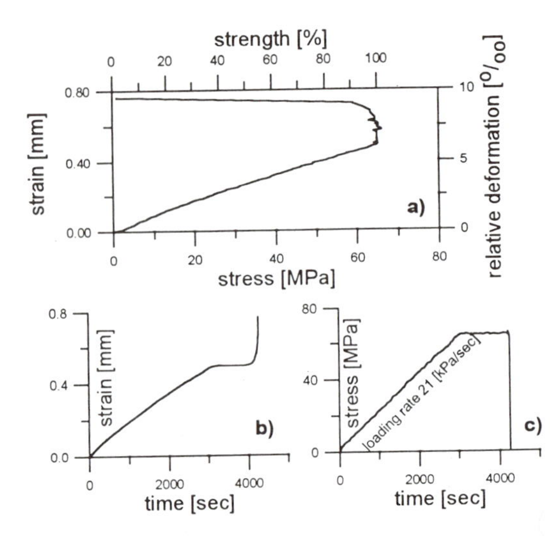

Figure 1. a) stress-strain curve, b) time-stress curve, c) strain- time curve.

limit of the long-term strength of the sample was determined on the basis of a series of experiments on identical granite samples based on the analysis of AE activity. The criterion was determined from the limit of the sample strength and expected AE activity.

The loading of the sample was measured continually. The total deformation of the sample was measured by an LVDT transducer with a resolution of 1 μm (Figure 1). The acoustic emission occurring during sample loading was recorded by one piezoceramic sensor WD-PAC located in the centre part of the lateral surface of the sample. The AE signals were processed by means of an SF41 electronic device, which could be used for continuous 4-channel recording of the acoustic emission. The measuring system was realised as an interface card for an IBM/PC,[3].

METHOD OF AE ANALYSIS

The data describing the course of loading and AE activity were compared from the point of view of finding possible significant anomalies. From this point of view the dependence of the AE activity on time was determined. This dependence is shown in Figure 2. The number of AE events can be described very well by an exponential law (dot-dashed line) in the form:

$$\Sigma N(t) = 283.58 * 10^{(0.00063 * t)} \quad (1).$$

Since the development and state of material loading is not known during the monitoring of AE under field conditions (for instance, mine-induced seismicity) a priori, it is usual to analyse the acoustic emission in a discrete time interval. For the interpretation of the results obtained in situ and their comparison with laboratory experiments, four discrete time intervals were determined (denoted in Figure 2 as I, II, III, IV), in which the cumulative number of AE events can be well approximated by the line. Table 1 shows the values of slope "s" of the linear regression lines versus time and percentage of the sample strength limit.

This change of slope "s" is more significant in the time interval close to the long-term strength than the mean value of the centred exponential function (1).

Table1. Analysis of AE series in the course of loading

Interval	I	II	III	IV
time [sec.]	200 - 400	1400 - 1600	2000 - 2200	2600 - 2800
%	7 - 12	46 - 51	65 - 70	84 - 90
slope "s"	1.39	2.78	6.14	24.92

After reaching the pressure of 65.4 MPa the loading of the sample was stopped and kept at the same acting level. The total deformation in this interval is shown in Figure 1.

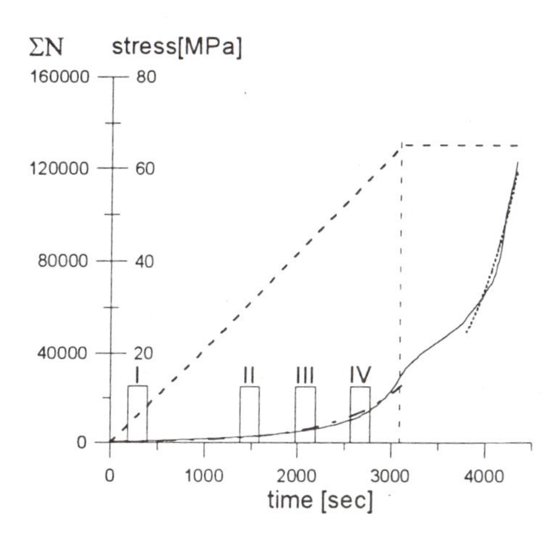

Figure 2. Cumulative number of AE events - full line.

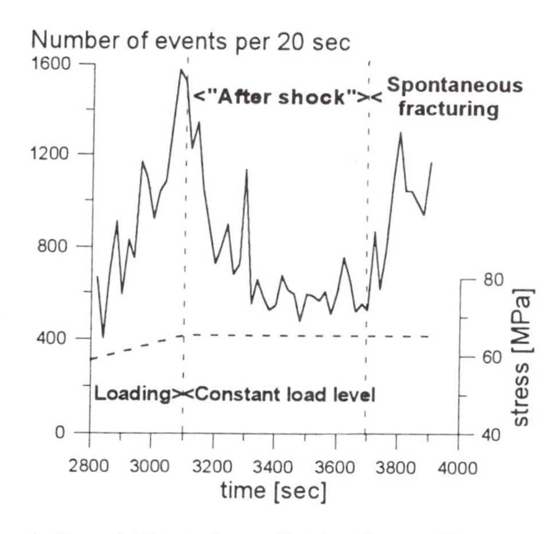

Figure 3. Description of AE rate immediately after reaching constant loading level.

The overall behaviour of the AE activity as displayed in Figure 2, can be described as follows:

Immediately after the loading was stopped, the decay of the acoustic rate with time could be observed, which is reflected by the inflexion of ΣN at the time equal to 3100 s and by the curve of $\Sigma N=f(t)$ changing from convex (1) to concave. The next point of inflection is at the time equal to 3700 s, where the concave dependence is changed to convex. Starting at this point in time a rapid increase of acoustic activity is observed. This phenomenon is observed even if no deformation changes are observed (Figure 1). The AE activity just before reaching the maximum stress state (dashed line - Figure 2) and immediately after is shown in Figure 3. In the time interval between the two points of inflection the variation of the acoustic activity with time can be described by the well-known Omori-Utsu law, which is valid for aftershock sequences :

$$N = N_0 \, (t - \text{const.})^{-0.29} \qquad (2),$$

where N_0 is the number of acoustic events at the time when the loading of the sample was stopped, and const. = 3100 s.

At the end of loading, i.e. just before the total rupturing of the sample, an exponential increase of AE activity is observed. The approximated exponential shape in the time interval 3800 - 4200 s is depicted by a dotted line. The exponential dependence has the form

$$\Sigma N(t) = N_{3800} + 2061.88 * 10^{(0.0031 * (t - 3800))} \qquad (3),$$

where $N_{3800} = \Sigma N(3800) = 55920$. The exponential dependence of the acoustic rate on time describes the fracturing process only phenomenologically. Another very important feature is also the occurrence of non-accidentally AE events in time, which does not follow from the acoustic rate data. The mutual interaction of the occurence of AE events was analysed by autocorrelation analysis of the acoustic event series. The experimental approach enables the analysis in the time domain to be apply very well. The autocorrelation method, which was used for processing the AE series, is described in detail in, [2].

Since non-steady phenomena were subjected to the analysis, when also the mean value of the number of AE events per time unit varies with time, the last stage of the sample loading (time interval 3700 - 4300 s) was divided into subintervals. Three subintervals of the same duration 40 s, were subjected to the analysis. The comparability of the results obtained in each of the individual subintervals was ensured by selecting elementary intervals of such duration that the mean value of the AE events was constant and equal to 10 events/sec.

To verify the plausibility of the precursor autocorrelation characteristics also the time intervals (2500 - 2590 s, 3000 - 3040 s) in the course of sample loading were subjected to the same analysis. In these time intervals the acting stress was equal to the 80% and 90% of the total strength of the sample. Autocorrelation analysis was also applied to the time interval 3200 - 3240 s, on the aftershock sequence, which was observed immediately after the constant level of acting stress (long-term limit of strength) has been reached.

DISCUSSION

The results of the autocorrelation analysis are depicted in Figure 4. In all time intervals the numerical decay of the autocorrelation coefficients can be fitted well by linear function, irrespective of the manner of loading or level of acting stress. The basic parameters of this decay can be described as:

a) the value of autocorrelation coefficient $R(1)$, $[R(k) = C(k)/C(0)]$,

b) the slope dR/dt of the fitted linear function,

c) the length ρ of the time interval in which the fitted linear function crosses the zero value,

d) the value of the correlation coefficient r^2 of the fitted straight line,

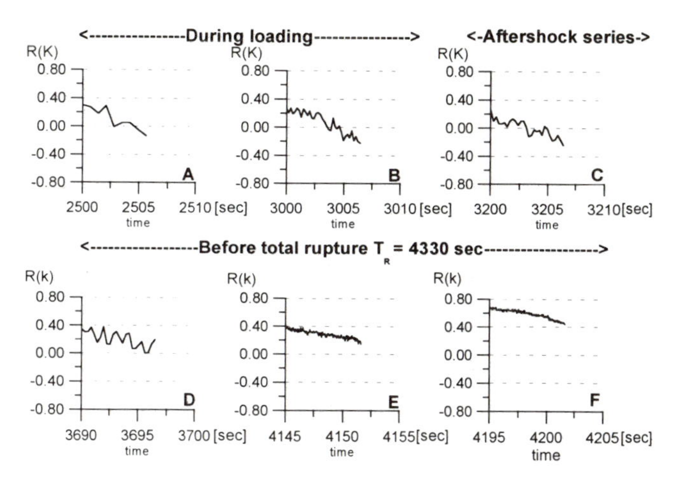

Figure 4. Autocorrelation functions for different loading and fracturing states.

e) the value of the null covariance coefficient C(0).

All parameters are summarised in Table 2. The fitted lines are unambiguously determined by parameters R(1) and slope dR/dt. Parameter ρ can be calculated as $\rho=R(1)/(dR/dt)$. Value r^2 is the deviation computed from coefficient of correlation determined from experimental data and the expected values given by linear decay in the form $R(t)=R(1) - dR/dt * t$. The last row of Table 2 shows the value of the autocovariance coefficient C(0) for t=0, i. e. at the beginning of each time interval. Although the values of this coefficient vary in the individual time subintervals (during the increasing loading of the sample, as well as during the time of constant acting load), their changes do not show any precursor character.

Table 2. Summary of autocorrelation analysis.

	During loading		Aftershock series	Before total rupture		
	A	B	C	D	E	F
time [s]	2500-2510	3000-3010	3200-3210	3690-3700	4145-4155	4195-4205
R(1)	0.26	0.30	0.25	0.34	0.42	0.68
dR/dt	-0.076	-0.076	-0.053	-0.038	-0.028	-0.032
ρ [s]	2.8	3.5	3.2	5.6	more than 10	more than 10
r^2	0.83	0.86	0.74	0.45	0.88	0.92
C(0)	20.2	12.6	14.7	14.2	9.37	16.3

The linear fit of the number of AE events in time for different time subintervals has a reasonable foundation as it is evidenced by the small change of the coefficients of correlation, which vary from 0.74 - 0.92. The exception is only subinterval D, which is in the transient part between the aftershock sequence and the acoustic emission activity before total fracture. This time subinterval D (3690 - 3700 s; compare Figure 4) physically coincides with reaching the long-term strength of the material.

On the contrary, the parameters given in the first three lines of Table 2 - R(1); dR/dt; ρ - exhibit a systematic dependence on the stress state and the process of fracturing, and thus can be considered as a precursor of total fracturing. In time subintervals before reaching the long-term strength limit, parameter R(1) displays a low value (R(1) < 0.3) and the slope, which characterises the decay of the autocorrelation coefficients in time - dR/dt, displays a high value. This indicates a higher contingency of the occurrence of AE events in time. On the contrary, just before total fracturing, the mutual interaction between the origin of AE events can be observed (due to the redistribution of the local stress in the sample loaded to the limiting value). This phenomenon is pronounced by the increase of the value of autocorrelation coefficient R(1) together with decay of the values of the subsequent autocorrelation coefficients. The changes of both the parameters are characterised by the time duration of positive autocorrelation ρ, i. e. by time T, for which R(T)>0. This time duration is several times longer once the limit of long-term strength is reached than during the beginning of loading.

The changes of the autocorrelation coefficient values normalised to the number of AE events as well as the changes of the linear decay slope according to the acting stress level were also determined during previous experiments made on sandstone samples,[2]. The increase of parameters R(1) and ρ was also observed during another regime of loading, i.e. during sample loading at a constant stress rate up to total fracture. From this point of view the prognostic features of the autocorrelation parameters can be regarded as independent of the loading regime (constant increasing of load or reaching the long-term strength limit and maintaining the same acting load) as well as independent of the rock type.

CONCLUSION

The results of the investigation of rock sample fracturing, mainly the study of the precursors of total fracturing, proved the worthiness of the AE activity interpretation. It was shown that the variation and values of autocorrelation coefficients incorporate significant information about the stress state or about the stage of rock disintegration. The origin of the total fracturing can be evaluated on the basis of quantitative data. It can be assumed that the above-mentioned laboratory results can be used to study mine-induced seismicity, where the acoustic emission from the rock body due to the accumulation of strain energy is monitored, and thus also these areas could be inclined to rockburst origin.

Acknowledgements

A portion of this work was supported by The Grant Agency of the Czech Republic, grants Nos. 205/95/0263, 105/96/1065 and by The Ministry of Education of the Czech Republic (project PG97330).

REFERENCES

[1] V. Rudajev, J. Vilhelm, and J. Kozák. Simulation of aftershock series from rock samples. *Int. J. Rock. Mech. Min. Sci. Geomech. Abst.* 31, 253 - 260 (1994).

[2] V. Rudajev, J. Vilhelm, J. Kozák, and T. Lokajíček. Statistical precursors of instability of loaded rock samples based on acoustic emission, *Int. J. Rock. Mech. Min. Sci. Geomech. Abstr.* 33, 743 - 748 (1996).

[3] T. Lokajíček, and J. Vlk. Interface card for multichannel processing of acoustic emission signals, in: *Progress in Acoustic Emission VII*, Proceedings of The 12th International Acoustic Emission Symposium, Sapporo, October 17-20, (1994).

NEUTRON DIFFRACTION AND ULTRASONIC SOUNDING: A TOOL FOR SOLID BODY INVESTIGATION.

T. Lokajíček,[1] Z. Pros,[1] K. Klíma,[1] A.N.Nikitin,[2] and T.I.Ivankina[3]

[1]Geophysical Institute, Academy of Sciences of the Czech Republic, Boční II/1401, 141 31 Prague 4, Czech Republic
[2]Joint Institute for Nuclear Research, 141980 Dubna, Russia
[3]Teacher's Training University, 300026, Tula, Russia

INTRODUCTION

Texture analysis of coarse grained polyphase rock samples can be presently determined by using neutron diffraction with a position sensitive detector and the time-of-flight diffractometry. A pulsed source of neutron was used. The results show that neutron diffraction provides quantitative texture data and has several advantages over conventional X-ray diffraction. Most important is the high transmission for most materials, the measuring pole figures on large coarse-grained samples and the measuring all pole figures with high accuracy (no corrections) simultaneously. The method is especially useful for geological samples, which are often composites of low symmetry constituents and therefore characterised by complex diffraction patterns with numerous partly and completely overlapping Bragg reflections. Theoretically, the textures of all mineral phases may be determined even from such complicated experimental data. The advantage of the neutron diffraction on spherical samples is that the scattering takes place in the whole sample volume. Consequently, the average texture of the whole sample is obtained.

The aim of this paper is the study of olivine spherical sample diameter 50 mm ± 0.01 mm and the comparison of its P-wave anisotropy expressed by isolines with its texture determined by means of neutron diffraction.

EXPERIMENTAL

The anisotropy of P-waves has mostly been studied in three mutually perpendicular directions, exceptionally also in more directions on samples in the shape of polyhedrons [Giesel, 1963; Babuška, 1966]. The study of the anisotropy of P-wave velocities on spherical samples [Pros, Babuška, 1967; Thill, et al, 1969] enables the velocity of elastic waves to be measured in any direction with the same accuracy. Later, measuring equipment was developed to enable elastic anisotropy to be studied under conditions of high

Nondestructive Characterization of Material VIII
Edited by Robert E. Green Jr., Plenum Press, New York, 1998

hydrostatic pressure up to 400 MPa [Pros, 1977, Arts, 1993]. The spherical shape of the sample enables us to investigate the spatial distribution of P-wave velocities for any pressure value, and this can then be used to derive the changes of velocity with pressure for any direction in space [Pros, 1977].

Thus the influence of the matrix and of the "soft" part of the rock (pore space, contacts between grains and populations of micro-cracks) on the resultant (observed) anisotropy can be distinguished.

For a detailed investigation of the anisotropy, the measurements have to be carried out in a sufficiently dense net of directions. The net of measuring directions is established by dividing the geographical co-ordinates λ and ϕ regularly in steps of 15° [Pros, 1977], thus creating 132 independent measuring directions.

Texture analysis of olivine sample can be determined by a neutron diffraction technique with a position sensitive detector and the time of flight diffractometr [Walther et al., 1990]

The principal objective of texture analysis is obtaining information on the distribution of the orientation of crystallites in the investigated samples. The traditional method for representing the preferred orientation is the pole figure (PF) (polar diagram), i.e. the stereographic projection of the normals of the reflecting planes (hkl).

The pole figures are constructed from a rather large number of neutron diffraction spectra recorded in samples with different positions stipulated by the angles θ and λ. A special texture goniometer, enabling the sample to be rotated in a beam of neutrons about different axes, is used for this purpose. A specific spectrum corresponds to each individual position of the sample. By determining the integral intensity of one reflection from this spectrum, we obtain the value of the pole density for a single point in the pole figure corresponing to angles θ and ϕ [Nikitin et al, 1995; Walther et al, 1993].

By rotating the sample about the axis lying in the scattering plane, the goniometer is able to scan the angle κ for 50 points along the full circumference at intervals of 7.2^{0}; this angle corresponds to azimuth angle ϕ in the pole figure. The direction of this vector is determined by the difference of two vectors: the vector coinciding with the direction of the incident wave and the vector fixing the direction from the sample to the detector. Polar angle θ is the angle between the scattering vector and the sample rotation vector.

The measuring system detectors are designed to record 7 diffracted rays simultaneously. This detector configuration can be used to realize a system with an angular step of 7.2^{0} by 7.2^{0} which is sufficiently dense for describing the sample texture.

RESULTS

The P-wave velocities isolines determined at different values of confining pressure are depicted in Fig. 1. The isolines are expressed as an equal area projection to the lower hemisphere. The isolines at pressure value equal to 5 MPa display complex shape and the minimum velocity value is 5.7 km/s and maximum velocity is 6.7 km/s. At pressure value equal to 50 MPa the minimum velocity value is 7.1 km/s and maximum velocity is 7.7 km/s.

With increasing hydrostatic pressure, due to the closing of grain boundaries and microcracks, the isolines pattern becomes more clear and also the velocity increases. At confining pressure equal to 400 MPa the isolines are smooth and reflect preferred orientation of crystals close to orthorombic symmetry. The velocity ranges from 7.6 km/s for minimum velocity and 8.3 km/s for maximum velocity.

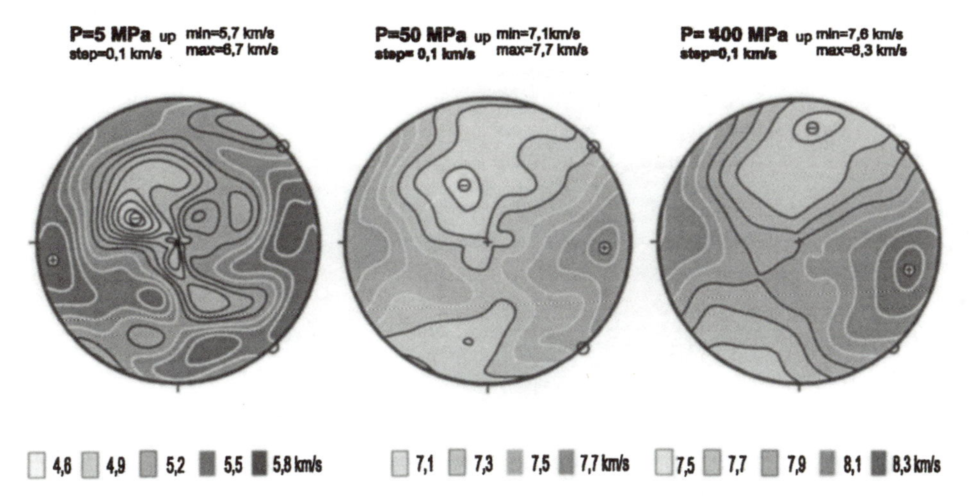

Figure 1.P-wave velocity isolines determined for 5 MPa, 50 MPa and 400 MPa confining pressures. + denotes direction of maximum velocity, - denotes direction of minimum velocity.

The information about the texture can be obtained from the set of diffracted spectra, where peak position indicates the pole figure and area under the peak indicates the pole intensity. The indexing of the spectra is based on the theoretical spectra computed from the structural parameters of olivine crystal and Bragg diffraction law. The texture components provides to reconstruct pole figures which cannot be measured directly, as (001), (100) and (010) pole figures. Such a pole figures coincide with the main crystallographic axes of olivine and with the main axes of elastic tensor too. Figure 2 shows the pole figures determined for the main crystallographic axis. The isolines are also expressed as an equal area projection to the lower hemisphere. The dark colour denotes the maximum intensity of pole figures.

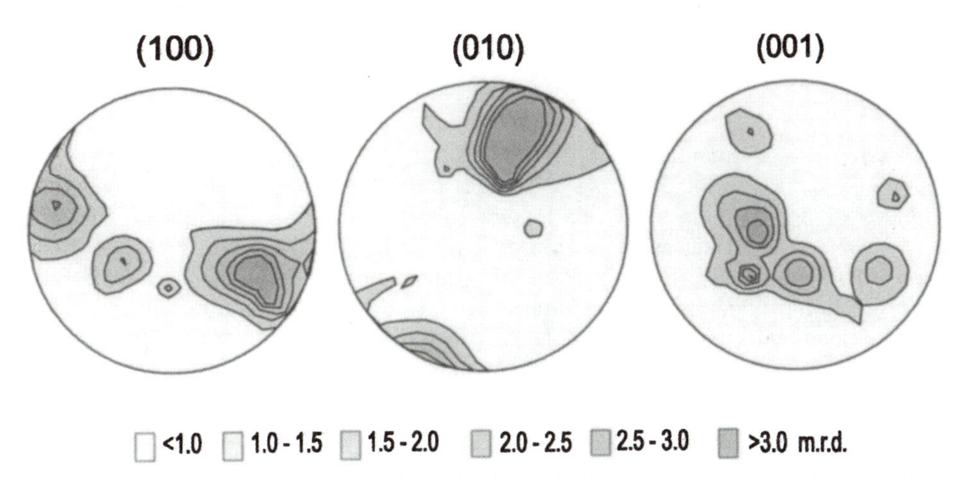

Figure 2. Reconstructed pole figures for (001), (100) and (010) orientation. The pole figure isolines are expressed in multiple random distribution values (m.r.d).

Figure 3 shows the pole figures derived by means of neutron diffraction for (100) and (010) orientation - left part of the Figure. This orientation coincides with the maximum and minimum velocity determined for olivine single crystal - central part of the Figure (Kumazawa and Anderson, 1969). Right part of the Figure shows P-wave anisotropy determined by ultrasonic sounding at confining pressure 400 MPa.

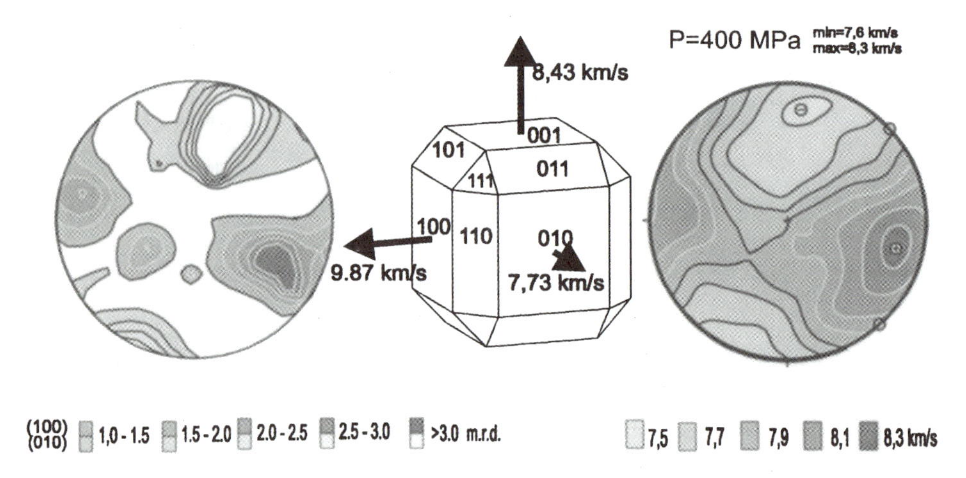

Figure 3. Comparison of pole figures derived from neutron diffraction for (100) and (010) orientation (pole figures are composed to one picture) with P-wave anisotropy determined for 400 MPa by ultrasonic sounding.

At such an acting hydrostatic pressure most of grain boundaries and microcracks are closed and thus the resultant isolines describe the P-wave anisotropy of the matrix of the sample. It can be seen from this picture that position of maximum velocity axis (100) - coincide very well with the maximum velocity direction denoted by + determined by means of ultrasonic sounding. Also the position of minimum velocity axis (010) coincide very well with P-wave minimum velocity direction (-). Comparing texture data with P-wave anisotropy determined by ultrasonic sounding can be find very good correlation between results obtained by quite an independent experimental approaches.

CONCLUSIONS

The ultrasonic sounding of olivine sample gives P-wave isolines picture, which describe the anisotropy of the sample under the study. The neutron diffraction gives the pole figures of crystal main axis. The texture analysis by means of neutron diffraction shows the preferred orientation of olivine grains, which control the anisotropy of elastic properties. The orientation of axis (010), where the olivine crystal has minimum P-wave velocity direction - coincide with minimum velocity direction determined by ultrasonic sounding of spherical sample. The orientation of axis (100), where the olivine crystal has maximum P-wave velocity direction - coincide with maximum velocity direction determined by ultrasonic sounding.

The results of neutron diffraction and ultrasonic sounding of polycrystalline sample of olivine, expressed by the map of isolines of preferred orientation of olivine crystal axes and map of P-wave velocity isolines, display very similar pattern, which reflects the texture of the sample. More over the determined texture display that olivinic xenolith preserved certain part of its original texture in spite of very complex geological process during its exhumation.

ACKNOWLEDGEMENT

A portion of this work was supported by Grant Agency of the Czech Republic, Nos. 205/97/0905, 205/95/0263, 205/94/1556 and 103/93/2201 and by the Grant Agency of the Academy of Sciences of the Czech Republic, Grant No. A3012603.

REFERENCES

Arts, R., 1993, A study of general anisotropic elasticity in rocks by wave propagation, -Theoretical and experimental aspects, *PhD Thesis*, The University Pierre et Marie Curie, Paris VI.

Babuška V., 1966, Velocity of compressional waves and anisotropy of some igneous and metamorphic rocks, *Travaux Inst. Geophys. Acad. Tchecosl. Sci,* 223: 275.

Giesel W., 1963, Elastische Anisotropie in tektonisch verformten Sedimentgesteinen, *Geophys. Prosp*, 11: 423.

Kumazawa M., and Anderson O. L., 1969, Elastic Moduli, Pressure Derivatives and Temperature Derivatives of Single-Crystal Olivine and Single-Crystal Forsterite, *J. Geophys. Res*., 74: 5961.

Pros Z., 1977, Investigation of anisotropy of elastic properties of rocks on spherical samples at high hydrostatic pressure. In: High Pressure and Temperature Studies of Physical Properties of rocks and Minerals, *Naukova Dumka*, Kijev , 56.

Pros Z., and Babuška V., 1967, A method for investigating the elastic anisotropy on spherical rock samples, *Zeitschrift für Geophysik,* 33: 289.

Thill R. E., Willard R. J., and Bur T. R., 1969, Correlation of longitudinal velocity variation with rock fabric, *J. Geophys. Res*., 74: 4897.

Walther K., Kurtasov S. F., Nikitin A. N., and Torina E. G., 1993, Modelling deformation textures in high-temperature guartz, *Fizika Zemli, Izv.RAN*, N 6: 45.

Walther K., Nikitin A.N., Shermengor T.D., and Yakovlev V. B., 1993, Determination of effective electroelastic constants of textured polycrystal quartz, *Fizika Zemli, Izv. RAN*, N 6: 83.

Walther, K., Isakov, N.N., Nikitin, A.N., Ullemeyer, K., and Heinitz, J., 1990, Research of the structure of geomaterials by the diffraction method using a high-resolution neutron spectrometer at the I.M.Frank Neutron Physics Laboratory of Joint Institute for Nuclear Research, *Fizika Zemli*, 6: 37.

DEVELOPMENT OF AN EPITHERMAL NEUTRON DETECTOR FOR NON-DESTRUCTIVE MEASUREMENT OF CONCRETE HYDRATION

Richard A. Livingston and Habeeb Saleh

Exploratory Research Team, HNR-2
Federal Highway Administration
McLean, VA 22101

INTRODUCTION

Portland cement concrete technology depends on the reactions between water and the constituents of the Portland cement powder, which are primarily tricalcium silicate, dicalcium silicate and tricalcium aluminate. The main reaction, involving tricalcium silicate, is given by:

$$3CaO \bullet SiO_2 + zH_2O \rightarrow xCa \bullet SiO_2 \bullet y(OH) \bullet nH_2O + (3-x)Ca(OH)_2 \qquad \textbf{(1)}$$

where: $z = n+2y+(3\text{-}x)$.

The first phase is on the reaction products side is not crystalline, but instead is a gel, usually symbolized by C-S-H. It is the development of this gel that develops the stiffness and strength of the concrete. The reaction proceeds relatively slowly for the first few hours after mixing. It then accelerates dramatically, so that the reaction is roughly fifty percent complete on the order of one day, which thus determines the setting of the concrete. Thereafter, the reaction slows down again significantly, and may require as much as a year to reach completion. The reaction of dicalcium silicate is similar, although slower. In contrast, the tricalcium aluminate reaction proceeds rapidly, and it is essentially complete in less than an hour (Christianson and Lehman, 1984).

Since the degree of hydration (the ratio of unreacted to the total water content) is a direct measure of the completion of the concrete hardening reaction, it could be used to determine the development of strength in concrete structures. A nondestructive method for measuring this would lead to improved quality of Portland cement concrete construction and reduced construction time schedules. Neutron scattering can provide very sensitive measurements of hydrogen-bearing materials, and therefore this research focuses on the application of this to Portland cement hydration. There is a precedent for the use of neutron-based instruments in the construction field in the form of the neutron moisture gauge (ASTM,1978), although this method measures total water content, not free water content.

QUASI-ELASTIC NEUTRON SCATTERING MEASUREMENTS

Neutrons traveling in a material at near thermal equilibrium velocities can scatter quasi-elastically from free water molecules; i.e. the neutrons gain or lose kinetic energy in collisions with freely moving water molecules. The resulting Doppler shift in the neutron wavelength can be measured in the laboratory(Livingston et al.,1995). In this approach the energy distribution of a

Nondestructive Characterization of Material VIII
Edited by Robert E. Green Jr., Plenum Press, New York, 1998

scattered neutron beam is measured as a function of the scattering wave vector, Q, and the Doppler wavelength shift, ω, or the energy equivalent, $\hbar\omega$. This type of measurement requires instruments that can measure both the angle and the velocity of the scattered neutrons.

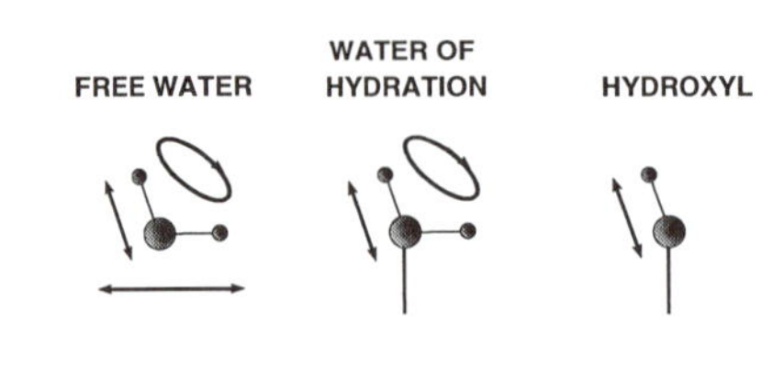

Figure 1: Modes of Motion of Water

Inspection of Equation 1 reveals that the water can be present in three different states: freely moving water, water bound in the gel structure, and hydroxyl (OH) groups. Each of these three states can be characterized by its degrees of freedom of motion, as illustrated in Fig. 1. Free water can translate, as well as rotate around the oxygen atom, and vibrate along the O-H bonds. Water molecules bound in the gel structure cannot translate, although they retain rotational and vibrational motions. Finally, the hydroxyl groups can only vibrate. Each of these motions has a characteristic set of frequencies, and thus the wavelength of the scattered neutron will depend on the state of water molecule it scatters off.

From first principles, it can be shown that the distribution of energies of neutrons scattering from free water has a Lorentzian shape while the distribution for scattering off bound water and other elements has a Gaussian shape. Then the neutron scattering data, reduced to the form of the incoherent scattering function,$S_{inc}(Q,\omega)$, can be fitted to the function:

$$S_{inc}(Q,\omega) = \frac{A}{\sigma\sqrt{2\pi}}\exp\left[-\frac{1}{2}\left(\frac{\omega}{\sigma}\right)^2\right] + \frac{B}{\pi}\frac{DQ^2}{(DQ^2)^2 + \omega^2} \tag{2}$$

where A and B are constant coefficients, D is the diffusion constant of water, and σ is the standard deviation of the Gaussian function, which is determined by the instrument resolution function. The values of A and B are directly proportional to the number densities in the specimen of the bound and free water molecules respectively. Consequently this provides a simple index for the degree of hydration in the concrete (Harris et al. 1974). The free water index

$$FWI = \frac{B}{A+B}$$

One advantage of this index is that it is independent of the absolute amount of water in the concrete. Another is that it is insensitive to other constituents of the concrete because hydrogen dominates the scattering cross-section.

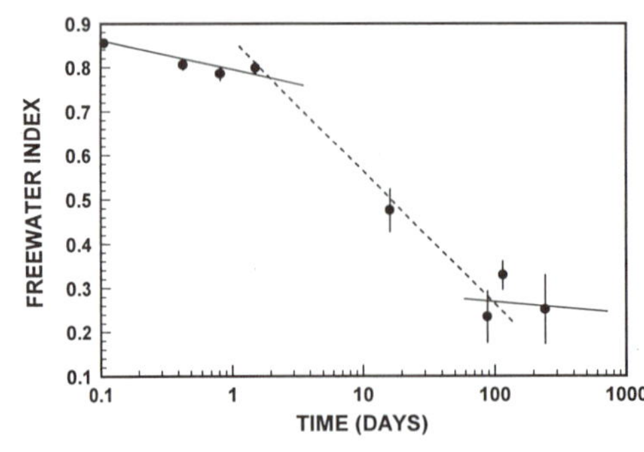

Figure 2: Plot of Hydration Reaction for Portland Cement Mortar (Redrawn from Livingston et al., 1995)

The change in the free water index over time for an actual Portland cement mortar sample is illustrated in Fig. 2. In this plot, lines have been drawn to highlight differences in the rate of hydration. At early ages, less than one day, the reaction proceeds relatively slowly. The dashed line indicates a significant acceleration in the rate, which is followed by a third phase which has a slower rate . This is consistent with the general hydration process described above. and thus demonstrates that it is possible to use neutron scattering measurements to monitor the reaction progress. With this technique, Livingston et al., (1996) have studied the reaction of tricalcium silicate under isothermal conditions to obtain detailed kinetic rate laws.

NEUTRON ENERGY SPECTROSCOPY

Despite the advantages of the neutron scattering method, it can only be performed at specialized nuclear facilities where monochromatic neutron beams, and instruments to measure energy and momentum at very high resolution area available. Also, it is necessary to restrict the thickness of the sample so that the neutron undergoes a single scattering event in the material, which effectively limits the thickness of Portland cement samples to less than a millimeter. Consequently, it cannot be applied to concrete, because the presence of the coarse aggregates requires a minimum thickness on the order of centimeters.

As an alternative, there is an indirect method of sensing the Doppler effect that could be applied to large concrete specimens, and does not require specialized facilities. This is based on measuring the energy distribution of neutrons in thermal equilibrium with the material (Livingston, 1997). In the absence of free water, the neutron energy spectrum would have a Maxwellian distribution around a value of 0.025 eV, but in the presence of free water, the distribution would also show a tail at higher energies in the epithermal region of 0.1 to 1 eV (LaMarsh, 1966). This tail is the result of the Doppler shift accumulated over multiple scattering events.

The design of the system includes a low-intensity radioisotope neutron source, typically ^{252}Cf. which is placed in proximity to the concrete specimen to be measured. The neutrons enter the concrete and scatter off its constituents, producing a flux of neutron back out of the concrete. The magnitudes of neutron flux in thermal and the epithermal regions are detected. The ratio of the two is then an index of the state of hydration of the water in the concrete. However, unlike the quasi-elastic method described above, there is no simple theoretical model that can be applied. This means that the magnitude of the epithermal flux shift as a function of the progress of the hydration reaction has to be determined empirically.

EPITHERMAL DETECTOR DESIGN

For this method, it is necessary to have an energy-sensitive neutron detector that can operate in the energy range 0.025 eV $<E<$ 1 eV. Unfortunately, the nuclear physics involved rule out any proportional type of detection in this energy range (Tsoulfanidas, 1983). The resolution of energy sensitive neutron detectors is typically in the range of 700 keV.

Since it is not possible to make the required neutron energy measurements with a single detector, a method using multiple detectors with different energy detection efficiencies is used. One detector is sensitive only to thermal neutrons. Another is sensitive to both. The difference in the total number of neutrons detected between the two is then a measure of the epithermal flux. Detectors with the specified energy efficiencies can be made from lithium glass sensors.

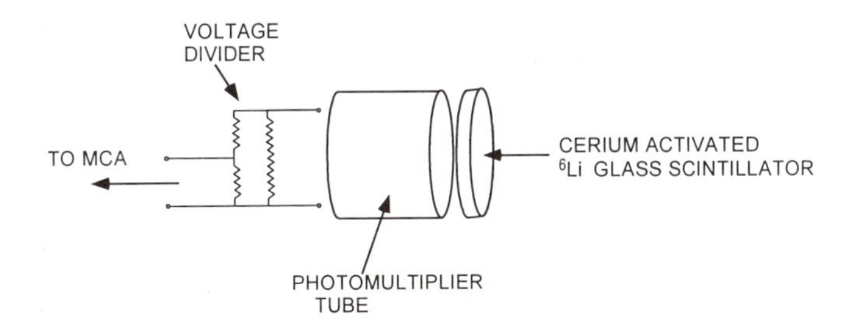

Figure 3: Schematic Diagram of Lithium Glass Neutron Detector.

Lithium glass neutron detectors are scintillation devices. As shown in the schematic diagram in Fig. 3, it consists of a glass scintillator disk, a photomultiplier tube, and a voltage divider circuit. The neutrons are detected in the glass by the capture reaction : ${}^{6}Li + n \rightarrow \alpha + {}^{3}He$. The two charged particles thus produced interact with the glass to produce photons which then can be detected by photoelectric sensors.

The detection efficiency is a function of the glass thickness, x, and the attenuation coefficient, μ_{th}, where the subscript indicates that the coefficient is evaluated at the thermal neutron energy, 0.025 eV. The fraction of the thermal neutron flux entering the glass captured is:

$$f = 1 - \exp(-\mu_{th}x) \qquad \textbf{(4)}$$

The Li glass scintillator system detects essentially all the neutrons that are captured (Wraight et al., 1965).

The attenuation coefficient is a function of the microscopic absorption cross section, σ_a. For a light element like Li, σ_a varies inversely with the neutron velocity, or equivalently with the square root of the neutron energy. Hence:

$$\mu(E) = \sqrt{\frac{E_{th}}{E}\mu_{th}}$$

Thus for a fixed thickness of glass, the fraction detected as a function of energy is:

$$f(E) = 1 - \exp\left[-\mu_{th}\left(\sqrt{\frac{E_{th}}{E}}\right)x\right] \qquad \textbf{(6)}$$

This defines a lowpass filter.

If two detectors with different thickness, x_1 and x_2, respectively, are used to measure the same neutron flux, the difference in the fractions detected is:

$$f_1(E) - f_2(E) = \exp\left[-\mu_{th}\left(\sqrt{\frac{E_{th}}{E}}\right)x_2\right] - \exp\left[-\mu_{th}\left(\sqrt{\frac{E_{th}}{E}}\right)x_1\right] \qquad \textbf{(7)}$$

This describes a bandpass filter. Therefore, by proper selection of x_1 and x_2 it is possible to have a detector array that is tuned for epithermal neutron energies. For standard Li- glass type GS20, with an enrichment factor for ${}^{6}Li$ of 95%, μ_{th} is 14.5 cm^{-1}. Thus thicknesses of 2 mm and 4 mm were chosen for x_1 and x_2 respectively. A third detector with a thickness of 12.7 mm was also fabricated to measure the neutron range beyond 1 eV. The energy response curves for the three detectors are plotted in Fig. 4. It can be seen that all three have the same efficiency at 0.025 eV. However, the 2 mm detector's efficiency rolls off faster with increasing energy than the other two. There is thus a significance difference in response over the epithermal region.

DETECTOR CALIBRATION

To evaluate the actual efficiency of each detector, it is necessary to have a monochromatic neutron beam that can be varied over the energy range of interest. This was provided by the linear accelerator at the Gaerttner Laboratory at Rensselaer Polytechnic Institute in Troy, NY. In this facility,

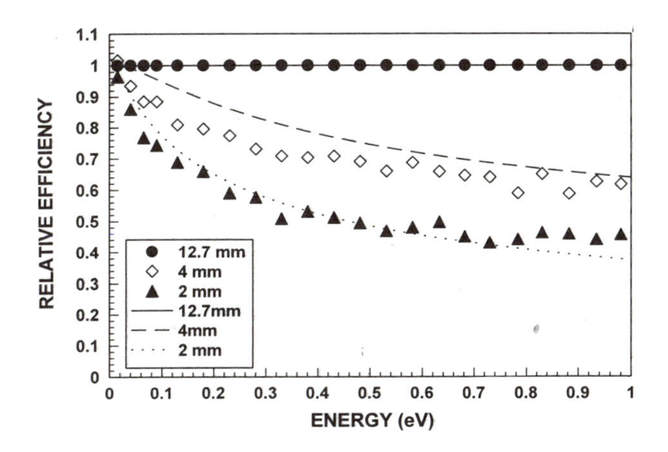

Figure 4: Glass Neutron Detector Efficiency Normalized to 12.7 mm Sensor.

a pulsed beam of electrons traveling at relativistic velocities generates a pulse of photo neutrons in a tantalum target. The detectors were located at a station 25 m from the source. The neutron energies were determined by time-of-flight.

Preliminary results are plotted in Fig. 4 for comparison with the model response curves. It can be seen that the points generally follow the predicted rolloff curves. This demonstrates that it is feasible to make an epithermal neutron energy measurement system using a tuned array of Li glass sensors. However, the actual values of the data depart from the predictions. This suggests that either the thicknesses or the ^{6}Li enrichment factor, or both, differed from the original specifications.

CONCLUSIONS

The degree of hydration of Portland cement paste can be measured nondestructively by quasi-elastic neutron scattering at reactor-based facilities. However, it is not feasible to use this method directly on concrete specimens. Therefore, an approximate technique has been developed that measures the ratio of the epithermal to the thermal neutron flux. The feasibility of using a tuned array of Li glass sensors to measure neutron energies in the epithermal region has been demonstrated . Remaining research includes the determination of the sensitivity of the system to measurement of the degree of hydration on actual concrete test cylinders.

ACKNOWLEDGMENTS

The authors gratefully acknowledge Dan Neumann of the National Institute of Standards and Technology, and Robert Block and Peter Brand of the Gartner Laboratory for their work on the calibration of the Li glass detectors.

REFERENCES

ASTM, 1978, Standard Test Method for Moisture Content of Soil and Soil-Aggregate in Place Nuclear Methods (Shallow Depth), in: *Annual Book of ASTM Standards-Part 19*, American Society for Testing and Materials, Philadelphia PA, D 3017.

Christensen, A.N and Lehmann, M.S., 1984, Rate of reactions between D_2O and Ca_x Al_y O_z , *J. Sol.State.Chem.* 51: 196

Harris, D.H,. Windsor, C.G. and Lawrence,C.D., 1974, Free and bound water in cement pastes, *Mag. Concrete Res.*, **26**[87] 65.

LaMarsh, J.R., 1966, *Nuclear Reactor Theory*, Addison-Wesley, Reading, MA.

Livingston, R.A., 1997, *Nondestructive Measurement of Portland Cement Hydration by Epithermal Neutron Flux Shift* , patent application.

Livingston, R.A., Neumann, D.A., Allen A. and Rush, J.J., 1995, Application of neutron scattering methods to cementitious materials, in: *Neutron Scattering in Materials Science, II, Materials Research Society Proceedings Vol. 376,* D.A. Neumann, T.P. Russell and B.J. Wuensch eds., Materials Research Society, Pittsburgh, PA

Livingston, R.A., Neumann, D.A., FitzGerald, S. and Rush, J.J., 1997, Quasi-elastic neutron scattering study of the hydration of tricalcium silicate, in: *International Conference: Neutrons in Research and Industry*, G. Vourvopolous, ed., SPIE, Bellingham, WA. 2867:148.

Tsoulfanidis, N., 1983, *Measurement and Detection of Radiation*, Hemisphere Publishing Corp. NY.

Wraight, L.A., Harris, D.H.C. and Egelstaff, P.A., 1965, Improvements in thermal neutron scintillation detectors for time-of-flight studies, *Nucl. Inst. Meth.* 33:181.

ULTRASONIC ASSESSMENT OF DAMAGE IN CONCRETE UNDER CYCLIC COMPRESSION

Z. Radakovic[1], K. Willam[1] and L. J. Bond[2]

[1]CEAE Department
University of Colorado at Boulder
Boulder, CO 80309-0428
[2]Denver Research Institute
University of Denver
Denver, CO 80208

INTRODUCTION

The main objective of the present study is to evaluate progressive damage in concrete using ultrasonic wave transmission[1,2,3,4,5,6]. The paper presents results of ultrasonic experiments on concrete prisms that were cyclically loaded in uniaxial compression while recording longitudinal and shear wave velocities during loading. Furthermore, after each load cycle, wave velocities in at least five different directions were recorded and used to determine progressive damage within the format of transverse isotropy. Even though concrete exhibits anisotropic behavior right from the beginning of loading, the degree of anisotropy is much greater in the post-peak softening region when the axial strength deteriorates progressively. The results show good correlation between mechanical and ultrasonic measurements of degradation in terms of transversely isotropic elastic stiffness properties. The investigation extends previous ultrasonic studies by the authors[7,8] in steel and concrete from isotropic to anisotropic damage assessments.

EXPERIMENTAL SETUP

Concrete specimens of dimensions 4 x 4 x 8 in^3 were prepared according to the mix proportions $C : S : G = 1 : 2.63 : 2.14$, using a water-cement ratio of $W/C = 0.65$, and a maximum agregate size of $d_a \leq 3/8\, in$. The specimen surfaces were polished, and the contact faces with the loading platens were greased to reduce interface friction and minimize boundary constraints. The specimens were loaded in the 110 kip MTS load frame of the Structures and Materials Laboratory of the University of Colorado, Boulder. For the external displacement control, feedback signals for the control loop were provided by averaging signals of two external LVDT-s that were attached between the upper and lower loading plate to measure axial

Nondestructive Characterization of Material VIII
Edited by Robert E. Green Jr., Plenum Press, New York, 1998

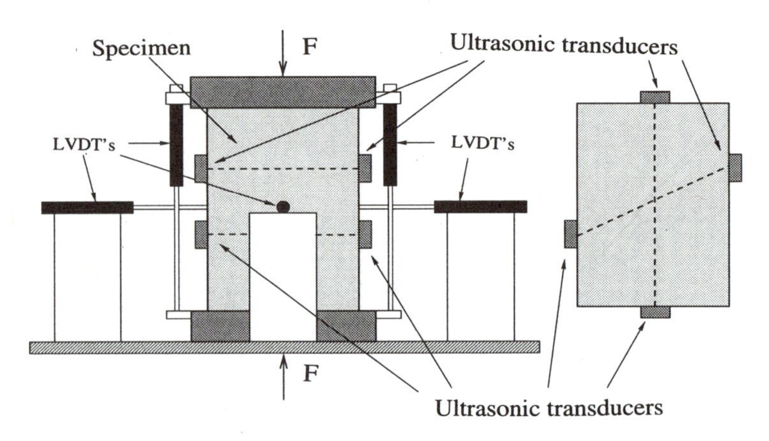

Figure 1: Test Setup of the Compression Test on Prismatic Concrete Specimen

deformation. The transducers were 200 hrdc LVDT-s with a ±0.2 in (5 mm) nominal linear range and 15 V/in (0.59 V/mm) sensitivity. For measuring lateral deformations, four 100 hrdc LVDT-s with $d_t = \pm 0.1$ in (2.5 mm) nominal linear range and 54 V/in (2.13 V/mm) sensitivity were used. For the acquisition of test data the Labview software package was used.

The ultrasonic equipment consisted of a pulse generator, a digital oscilloscope and two pairs of transducers. The ultrasonic signal was generated by Panametrics Pulse Generator 5052.PRX with one channel input and output. The digital oscilloscope was a Nicolet PRO50, 50 MHz oscilloscope with 100 MHz band-width and 8 bit vertical resolution. Both pairs of piezoelastic transducers had a diameter of the contact surface of 1 in (25 mm) with central frequency of 250 kHz for the shear wave transducers and 500 kHz for longitudinal wave transducers. The overall test setup and specimen layout are depicted in Figure 1.

The axial and lateral stress/strain diagrams in Figure 2 illustrate the cyclic response as a function of axial loading. Due to internal friction and granular action large lateral dilatation develops in the post-peak regime primarily due to axial splitting, whereby tension is plotted with a positive sign.

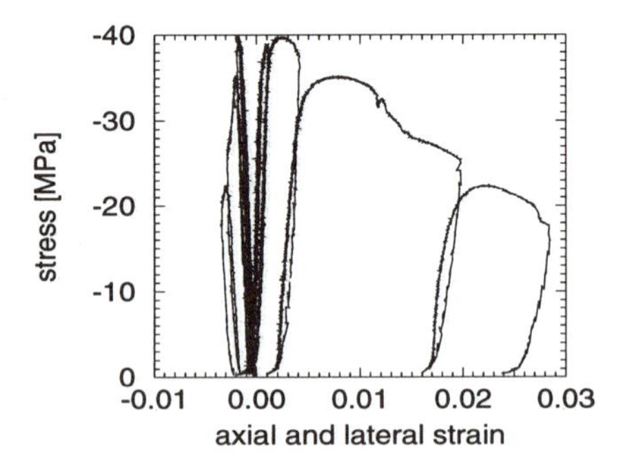

Figure 2: Cyclic Response History - Axial Stress versus Axial and Lateral Strain

ULTRASONIC DAMAGE MEASUREMENTS

In the simplest case of scalar damage[9], the uniaxial stress-strain relationship for elastic damage may be written in the form of the nonlinear secant expression

$$\sigma_z = E_s \, \epsilon_z \tag{1}$$

where σ_z and ϵ_z denote the nominal values of axial stress and strain, and where E_s designates the secant stiffness modulus of the damaged material. Measuring damage in terms of degrading elastic stiffness properties, the stress-strain ratio $\tilde{E} = \frac{\sigma_z}{\epsilon_z}$ provides a simple damage indicator, whereby d defines damage in terms of the ratio between the damaged and undamaged elastic moduli, i.e.

$$d = 1 - \frac{\tilde{E}}{E_0} \tag{2}$$

Damage may be measured directly in terms of the velocities of ultrasonic signals that are transmitted through undamaged and damaged concrete specimens. Assuming isotropic behavior the elastic longitudinal and shear wave velocities are related to the elastic properties as follows:

$$v_l^2 = \frac{E_0}{\rho_0}\frac{1-\nu_0}{(1+\nu_0)(1-2\nu_0)}, \qquad v_t^2 = \frac{E_0}{\rho_0}\frac{1}{2(1+\nu_0)} \tag{3}$$

where ρ_0 is the mass density and ν_0 is Poisson's ratio of the undamaged elastic specimen. Assuming damage to behave isotropic, the longitudinal wave speed in the loaded specimen is proportional to

$$\tilde{v}_l^2 = \frac{\tilde{E}}{\tilde{\rho}}\frac{1-\tilde{\nu}}{(1+\tilde{\nu})(1-2\tilde{\nu})} \tag{4}$$

For loading up to 70 % of the peak stress it is fairly realistic to assume that $\tilde{\nu} \approx \nu_0$ and $\tilde{\rho} \approx \rho_0$ since damage consists mainly of diffuse microcracks emanating from the interfaces of the two-phase concrete composite. In this case, the level of damage may be determined directly from the ratio of longitudinal wave speeds in the damaged and undamaged specimens, i.e.

$$d = 1 - \frac{\tilde{v}_l^2}{v_l^2} \tag{5}$$

This expression[9] provides a very simple estimate of damage by means of ultrasonic velocity measurements which holds for linear isotropic behavior as long as $\tilde{\rho} = \rho_0$ and $\tilde{\nu} = \nu_0$. Figure 3

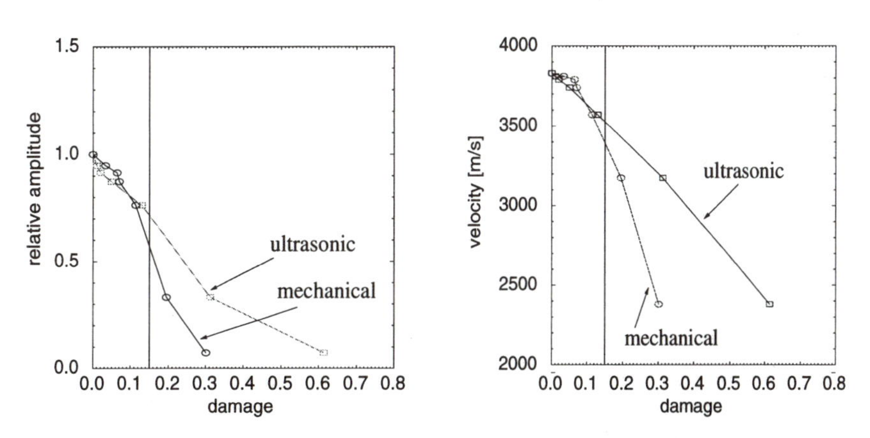

Figure 3: Reduction of Relative Wave Amplitude with Increasing Damage

depicts the variation of wave amplitude and velocity with the level of damage that is measured by mechanical and ultrasonic means (Equations 2 and 5). The agreement between those two damage measures is very close in the pre-peak region, but the correlation in the post-peak region turns rapidly very poor after the maximum stress level has been exceeded. Thereby, it might surprise that the mechanical measurement of axial stiffness degradation compares well with the ultrasonic measurement of stiffness degradation derived from longitudinal wave transmission data in the transverse direction.

Ultrasonic Evaluation of Transversely Isotropic Damage

The assumption of isotropic damage under axial loads led to significant discrepancies of the damage measure in the post-peak region. To account for the effects of stress induced anisotropy, concrete was considered to be transversely isotropic with the principal axis parallel to the load direction and a plane of isotropy perpendicular to the direction of load as shown in Figure 4.

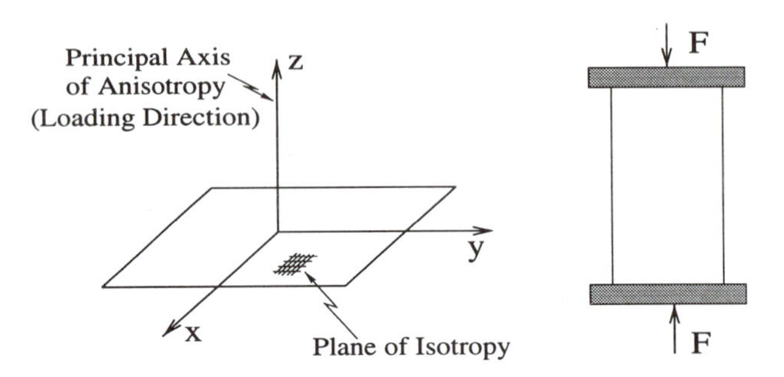

Figure 4: Principal Axes for Transversely Isotropic Material

Upon completion of the load cycle, arrival times of shear and longitudinal wave propagation were measured parallel, perpendicular (both directions) and at an angle of 45 degrees with the load direction, and concomitant velocities were calculated. The specimen was weighed at the beginning of the test and the density was assumed to remain constant. Knowing the mass density and the relationships between the five elastic moduli and the wave velocities, the elastic stiffness matrix of transverse isotropy may be cast in matrix form as

$$\boldsymbol{E}_s = \begin{bmatrix} \frac{E(-E'+E\nu'^2)}{(1+\nu)(E'(\nu-1)+2E\nu'^2)} & \frac{-E(E'\nu+E\nu'^2)}{(1+\nu)(E'(\nu-1)+2E\nu'^2)} & \frac{-EE'\nu'}{E'(\nu-1)+2E\nu'^2} & 0 & 0 & 0 \\ \frac{-E(E'\nu+E\nu'^2)}{(1+\nu)(E'(\nu-1)+2E\nu'^2)} & \frac{E(-E'+E\nu'^2)}{(1+\nu)(E'(\nu-1)+2E\nu'^2)} & \frac{-EE'\nu'}{E'(\nu-1)+2E\nu'^2} & 0 & 0 & 0 \\ \frac{-EE'\nu'}{E'(\nu-1)+2E\nu'^2} & \frac{-EE'\nu'}{E'(\nu-1)+2E\nu'^2} & \frac{E'^2(\nu-1)}{E'(\nu-1)+2E\nu'^2} & 0 & 0 & 0 \\ 0 & 0 & 0 & G & 0 & 0 \\ 0 & 0 & 0 & 0 & G' & 0 \\ 0 & 0 & 0 & 0 & 0 & G' \end{bmatrix} \quad (6)$$

where E, ν and G denote Young's modulus, Poisson's ratio and the shear modulus in the plane of isotropy, respectively, while E', ν' and G' are the moduli in planes normal to the plane of isotropy. In other words, ν characterizes the cross effect in the plane of isotropy due to a stress in the same plane, while ν' governs the cross effect between the plane of isotropy and the direction normal to it. Note that the five independent constants E, ν, E', ν' and G' reduce to the standard isotropic form of linear elasticity when $E = E'$, $\nu = \nu'$ and $G = G' = \frac{E}{2(1+\nu)}$. The secant stiffness of transverse isotropy in Equation 6 can be expressed in terms of the compression wave velocities

$$v_1 = \sqrt{\frac{E(-E' + E\nu'^2)}{\rho(1+\nu)(E'(\nu-1) + 2E\nu'^2)}} \qquad v_2 = \sqrt{\frac{E'^2(\nu-1)}{\rho E'(\nu-1) + 2E\nu'^2}} \quad (7)$$

and the shear wave velocities

$$v_3 = \sqrt{\frac{G}{\rho}} \qquad v_4 = \sqrt{\frac{G'}{\rho}}, \quad (8)$$

as

$$
\boldsymbol{E}_s = \rho \begin{bmatrix} v_1^2 & v_1^2 - 2v_4^2 & s & 0 & 0 & 0 \\ v_1^2 - 2v_4^2 & v_1^2 & s & 0 & 0 & 0 \\ s & s & v_2^2 & 0 & 0 & 0 \\ 0 & 0 & 0 & v_3^2 & 0 & 0 \\ 0 & 0 & 0 & 0 & v_4^2 & 0 \\ 0 & 0 & 0 & 0 & 0 & v_4^2 \end{bmatrix}, \tag{9}
$$

where ρ denotes the density of the solid, and $s = 2\nu'(v_1^2 - v_4^2)$.
With reference to Figure 4, v_1 represents the longitudinal wave velocity in the x-direction, v_2 is the longitudinal wave velocity in the z-direction, v_3 is the velocity of the shear wave propagating in the x-direction with the particle motion in the $x - y$ plane, and v_4 is the velocity of the shear wave propagating in the z-direction polarized in any vertical plane.

To evaluate the five elastic moduli of transverse isotropy from ultrasonic measurements, s is considered to be unknown with all other quantities (v_1, v_2, v_3 and v_4) being measured. To determine s, the acoustic tensor of the damaged secant stiffness $\boldsymbol{E}_s$ is evaluated as $\boldsymbol{Q}_s = \boldsymbol{N} \cdot \boldsymbol{E}_s \cdot \boldsymbol{N}$ where $\boldsymbol{N}$ designates the direction of wave propagation. Upon measuring v_1, v_2, v_3, v_4 and the velocity in the fifth direction denoted as v_*, the eigenvalues of $\boldsymbol{Q}_s$ are found to express v_* as a function of the four previously measured velocities as well as s. Solving for s, the five moduli of transverse isotropy are fully established. Figure 5 depicts the variation of the five moduli during one of the tests on normal strength concrete.

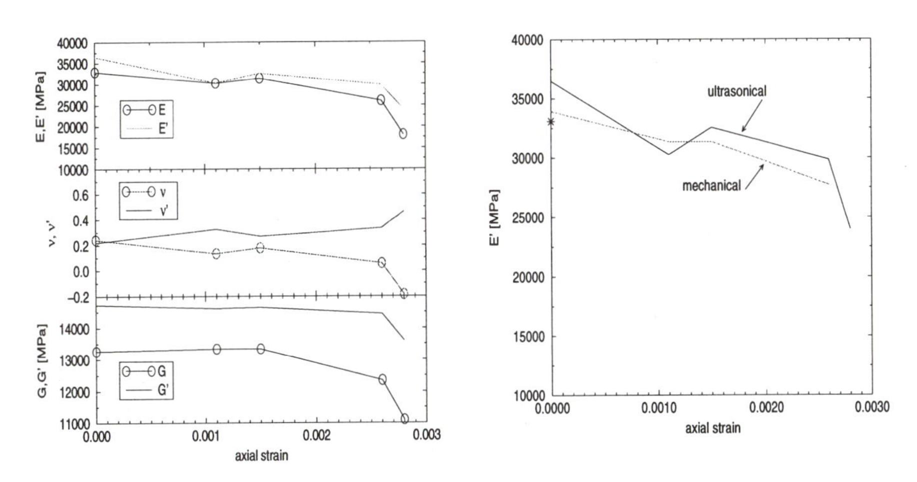

Figure 5: Change of Elastic Constants due to Damage

All secant moduli exhibit significant reduction during progressive damage, except for ν'. This Poisson's ratio, which characterizes the cross-effect between the plane of isotropy and deformation perpendicular to it, increases indeed due to lateral dilatation during axial compression, which was confirmed by mechanical measurements. Since the decrease of the isotropic in-plane Poisson's ratio ν was not measured mechanically, it can not be compared with the ultrasonic value. However, the negative values of Poisson's ratio, though possible in theory, are probably caused by slight errors in the evaluation of the off-diagonal terms of the elastic stiffness matrix. Figure 5 on the right depicts the variation of E' that was measured ultrasonically and mechanically. The ultrasonic measurement of E' based on transverse isotropy did provide good agreement among ultrasonic and mechanically measured E'-values in both, the pre- as well as the post-peak regions. The asterisk on the ordinate of the plot on the right-hand-side indicates the value of E' evaluated according to the proposed ASTM relation: $E = 57000\sqrt{f_c'}$, where f_c' denotes the maximum compressive stress.

Finally, Figure 6. shows the phase velocity plots at five different stages of compressive load cycles. Each phase velocity plot depicts the variation of one longitudinal and two shear wave velocities. The polar distance from the origin to a point on the curve (which traces a circle, in the isotropic case) determines the magnitude of the velocity while the angle with the horizontal axis indicates the direction of wave propagation with respect to the x-axis (Figure 4). The three 'circular tracings' on the left-hand-side depict the phase velocity changes in the pre-peak region of damage, the first of which refers to the initial, 'undamaged' stage at zero loading. The unloaded concrete specimen behaves nearly isotropic with longitudinal and shear wave velocities that are approximately the same in all directions. The two shear velocity diagrams, which almost coincide with each other at the beginning, depict SH and SV velocities (the propagation and polarization direction are in the horizontal plane of isotropy in the first case and the propagation and polarization direction are in the vertical plane perpendicular to the plane of isotropy in the second case). This might surprise in the light of Figure 5 which shows on the left-hand-side significant differences between E and E', ν and

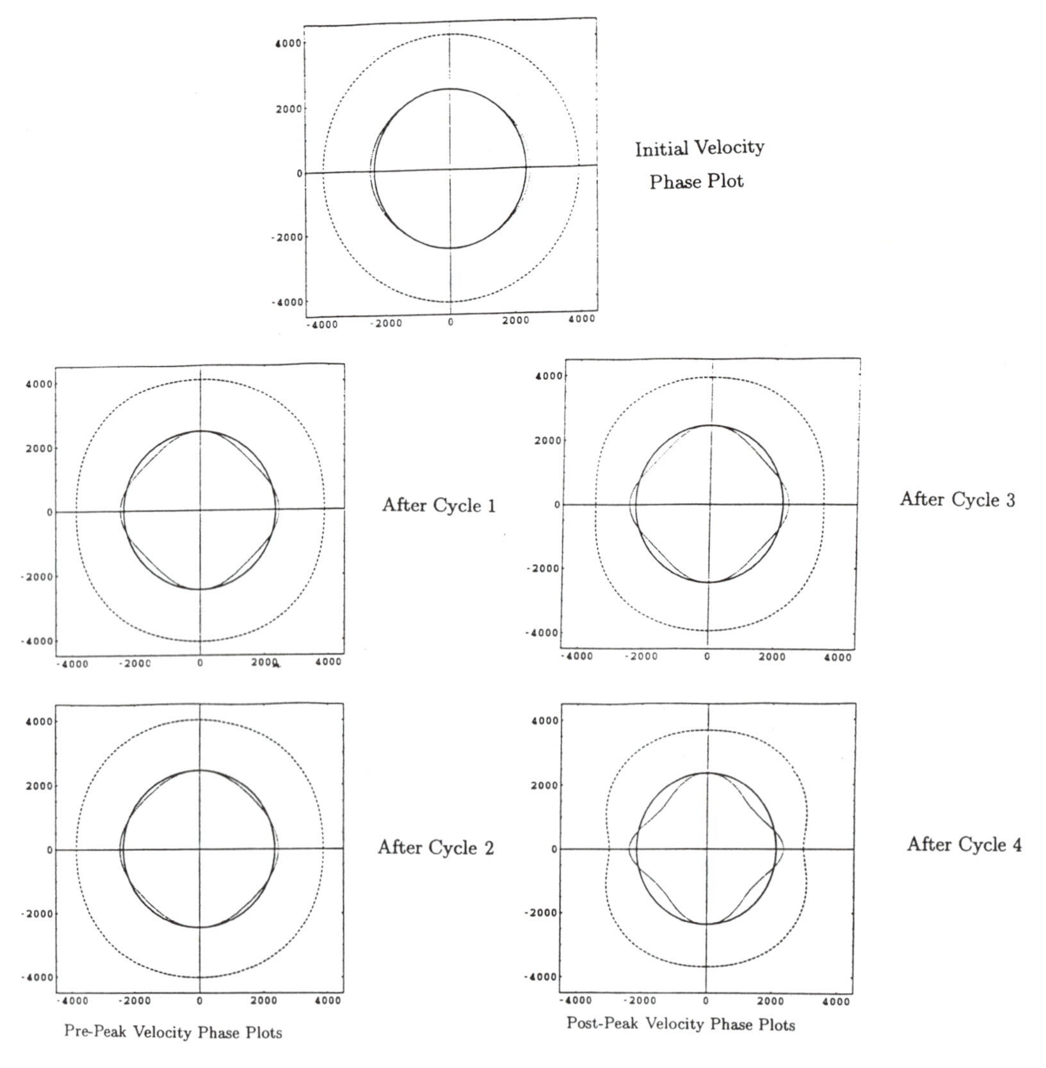

Figure 6: Pre-Peak and Post-Peak Polar Plots of Phase Velocity Diagrams, Initial and Pre-Peak Cycles (lhs) and Post-peak Cycles (rhs)

ν' and G and G' even in the initial stage at zero axial strain. However these differences do not show-up in the phase velocity plots of Figure 6. The results of the first two load cycles indicate that damage progresses in an isotropic manner in the pre-peak region with nearly equal reduction of wave velocities in all directions. This is not true, however for the post-peak load cycles shown on the right-hand-side of Figure 6 after cycles 3 and 4, where anisotropic damage is much more pronounced. The maximum reduction of the wave velocity occurs when the longitudinal wave propagates perpendicular to the direction of axial loading along which splitting cracks develop, which are quite visible in the post-peak region. The smallest eigenvalue of the acoustic tensor, though, is the SV shear wave velocity which attains a minimum at 45 degrees with respect to the direction of axial loading. This suggests formation of axial tension cracks in the brittle concrete material along 45 degree diagonals. They correspond to the pseudo-frictional failure mode which is normally observed in the shape of pyramidal failure surfaces near the loading platens in the case of frictional restraint.

CONCLUSIONS

The results of the ultrasonic studies provide quantitative information to evaluate anisotropic degradation of the elastic properties to be combined with mechanical measurements of plastic deformation and strength. The evaluation of stiffness and strength properties through a combination of ultrasonic damage and plastic softening measurements provides a novel experimental platform to characterize deterioration processes in concrete materials and cementitious particle composites in general.

ACKNOWLEDGMENTS

The authors wish to thank Dr. J. Scalzi for his liaison efforts and acknowledge the partial support of NSF under grant CMS-9634923 to the University of Colorado Boulder.

REFERENCES

1. Berthauld, Y., (1991),"Damage Measurements in Concrete via an Ultrasonic Technique. Part I: Experiment", Cement and Concrete Research, Vol. 21, 73-82
2. Berthauld, Y., (1991),"Damage Measurements in Concrete via an Ultrasonic Technique. Part II: Modeling", Cement and Concrete Research, Vol. 21, 219-228
3. Malhotra, V.M. and Carino, N.J., Eds., (1991),"CRC Handbook on Nondestructive Testing of Concrete", CRC Press.
4. Popovics, Sandor and Popovics, John S., (1991), "Effect of Stresses on the Ultrasonic Pulse Velocity in Concrete", Materials and Structures, Vol. 24., 15-23.
5. Popovics, S., Rose, J.L. and Popovics, J.S. , (1990), "The Behavior of Ultrasonic Pulses in Concrete", Cement and Concrete Research, Vol. 20., 259-270.
6. Teodoru, G.V.M., (1994), "Nondestructive testing of concrete from the research to the use in practice", International advances in Nondestructive Testing, Vol.17, Edited by Warren J.McGonnagle, Gordon and Breach Science Publishers, 117-137.
7. Radakovic, Z., Willam, K. and Bond, L, (1995), "Ultrasonic Detection of Plastic Yielding in Steel Specimens", Proc. Review of Progress in Quantitative NDE, Plenum Press, New York, Vol. 14, Ed. D.O. Thompson and D.E. Chimenti, pp. 1593-1600.
8. Radakovic, Z., Willam, K. and Bond, L, (1996), "Ultrasonic Assessment of Concrete in Uniaxial Compression", Proceedings 3rd Conference NDE of Civil Structures and Materials, Sep. 9-11 1996, Boulder Colorado, 455-469.
9. Lemaitre, J. and Chaboche, J.L., (1989), Mechanics of solid materials, Cambridge University Press.

ASSESSMENT OF ACOUSTIC TRAVEL TIME TOMOGRAPHY (ATTT) AT BARKER DAM

William F. Kepler[1], Leonard J. Bond[2,3], Dan M. Frangopol[3]

[1] U. S. Bureau of Reclamation, Denver, Colorado
[2] University of Denver Research Institute, Denver, Colorado
[3] University of Colorado at Boulder, Boulder, Colorado

INTRODUCTION

The primary focus of an evaluation of a large concrete dam is to determine the structure's ability to withstand a major earthquake or flooding conditions. The dam's reaction depends on the mechanical strength developed throughout the complex discontinuous mass of the structure. Construction techniques combined with normal loading conditions ensure that most defects in a large concrete dam will develop in horizontal planes along construction joints, also known as lift-lines. A dam's survival under extreme loading conditions often depends on the strength developed across these horizontal construction joints.

The traditional method for determining the strength of a concrete dam is to extract large diameter core samples drilled from the top of the dam down to the foundation, which are then destructively tested to determine strength and elastic modulus. These data are then used in a finite element model to simulate responses of the structure to various loading conditions. Depending on the size of the structure, two to four drill holes are cored down to the foundation, often a distance of more than 90 meters (300 feet). Large diameter cores, 250 to 300 mm (10 to 12 inches), are required to provide representative samples. Extracting this amount of large diameter concrete cores is very expensive. In addition, although a coring program of this magnitude is considered sufficient, it only samples a very small percentage of the dam volume, typically less than 0.1%. This procedure cannot therefore be expected to find most local anomalies, such as regions of disbonded lift-lines or cracks. A new testing procedure is required that will provide a more thorough evaluation of the physical properties of the dam, and is less expensive than a full coring program.

ACOUSTIC TRAVEL TIME TOMOGRAPHY

A novel approach is employed which combines aspects of ultrasonic nondestructive testing of small concrete structures, nondestructive evaluation of interfaces, and shallow seismic surveying, in "Acoustic Travel Time Tomography," ATTT. It is proposed that ATTT will be implemented together with a reduced coring program [1]. Acoustic testing can provide reliable estimates of the modulus of elasticity and compressive strength of hardened concrete. ATTT uses a sparse array of receivers, and an impulse source. The procedures determine both local and global bulk modulus and strength values of a structure, and are also be able to detect cracks, voids, and anomalies within the structure. ATTT will increase the percentage of the dam volume inspected, and do it at a reduced cost when compared to conventional methods.

For older dams it is expected that "weak zones" in lift-lines will have developed into complete delaminations. The inspection problem then simplifies to one of the detection and mapping of cracked or delaminated zones. This inspection can be performed using an array of receivers and an impulse source. An acoustic pulse will not pass through a major delamination, in effect causing a shadow zone. The only waves that will be received immediately below the delamination will be scattered pulses and tip-diffracted waves, both of which will have clearly identifiable travel time delays when compared with the expected direct pulse. When the relative positions of the source and receivers are varied, and ray analysis employed, the arrival times can be used in a tomographic inversion scheme to estimate crack penetration through the thickness of the dam [2].

A sparse array of receivers can be used to detect the presence of anomalous zones [3]. Once detected and located, the defective zones in the lift-lines can then be characterized using receivers in an increasingly dense array to increase the number of data points, and hence the resolution. Obviously the denser the data point grid, the more sensitive the data but the more costly the testing and modeling of the dam will be.

It is proposed to estimate the physical properties of a large concrete dam by correlating the results of acoustic tomography with destructive strength testing. Nondestructive measurements will be made in the frequency range between 1 to 50 kHz, depending on the size and attenuation characteristics of the structure and the source employed. Data will initially be taken using a sparse array of receivers spaced every 1.5 meters (five feet) down the side of the dam. These arrays will be up to 90 meters (300 feet) long and set in lines spaced every three meters (10 feet) horizontally.

As the distance between source and receiver increases, the arrival time is increasingly delayed. The velocity can be determined by dividing the arrival time by the distance between source and receiver. By correlating the velocity with destructive strength tests, an estimate for the strength and modulus of elasticity can be determined using a well-established relationship. The results of this testing can help determine the best locations to drill core that will provide the most data for critical parts of the structure.

WORK TO DATE

The current program is divided into three phases; Phase I - *Laboratory Work*, and Phase II - *Testing a Section of a Large Dam* have been completed. In Phase III we will evaluate the reliability of the testing procedure. Phase I - Laboratory Work has been presented previously [1].

Phase II - Testing a Section of a Large Dam - Barker Dam Studies

To evaluate the testing procedure at full scale, the next phase of the study was to exam a section of a large concrete dam. A portion of Barker Dam, above Boulder Colorado, containing known horizontal cracked lift-lines was selected. Barker Dam was chosen because its location, ease of access, simple geometry, and because it's design and construction are similar to those of structures that we will be examining in the future.

Barker Dam, is a cyclopean mass concrete gravity dam originally constructed in 1909. It is 53.3 meters (175 feet) high, and 220 meters (720 feet) wide [4]. The thickness across the top is 6.7 meters (22 feet), across the bottom it is 34.4 meters (113 feet). As was typical of the time period, the concrete was placed in lifts that ranged in thickness from 0.9 to 1.2 meters (three to four feet). The concrete had a nominal maximum size aggregate of 63.5 mm (2 ½ inches), with the addition of plums, which are large rocks weighing anywhere from 10 to 140 kilograms (25 to 300 pounds). Joint preparation between lifts was minimal, consisting of sweeping with a broom and placing a low strength grout of cement and water on top of the lift prior to placing the subsequent lift. This method of joint preparation almost always ensures poor bond between lifts.

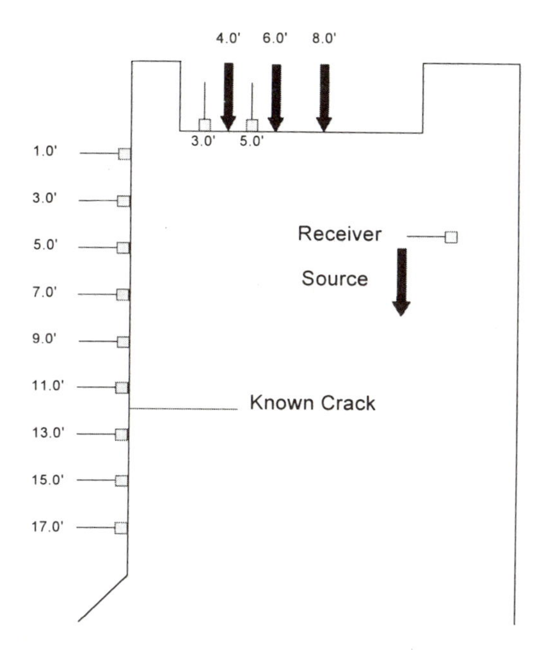

Figure 1. Cross section of Barker Dam showing 1996 testing setup

In this phase, a 5 channel transducer system developed in Phase I was expanded to 20 channels. The sensors used were one-inch diameter piezoelectric PZT type 5a disks with micro-dot connectors, mounted on copper foil. The disks were encapsulated in copper to provide electrical shielding. Each disk was attached to a source follower to match the impedance of the coax cable. The coax cable from each sensor was attached to a 20-channel data acquisition system with LeCroy 6810 digitizer modules. The system can sample upto 5 million samples per second, with a capacity of 128 thousand samples per channel. The data acquisition system was attached to a computer with a 9600 baud modem.

The top 6 meters (20 feet) of each block was tested along five vertical lines. The vertical lines were spaced 3 meters (10 feet) apart, starting at 1.2 meters (4 feet) from the southern edge of the block. This portion of the dam has a rectangular shape, which will ease data reduction efforts. The acoustic receivers were placed in pairs along a vertical line on the downstream face of Barker Dam by members of Reclamation's Climb Team.

The sensor pairs were spaced on approximate two-foot centers, as shown in Figure 1. The sensors were attached to the dam with petroleum jelly, then the leads of each sensor were hot-glued to the dam. Additional sensors were placed on the top of the dam, one at 0.9 meters (3 feet), and one at 1.5 meters (5 feet) from the downstream face. The wires from each sensor were bundled together with the wires from the other sensors so that the sensors would hang down at the appropriate intervals. The wire bundle then was connected to a data acquisition system in the testing van.

The impact source consisted of an 3.6 kilogram (8-pound) sledge hammer with an accelerometer attached to one end of the head. The hammer struck a 100-mm (four-inch) square metal plate, 25 mm (one inch) thick. The location of the plate was typically placed at 1.2 meters, 1.8 meters, and 2.4 meters (four-feet, six-feet, and eight-feet) from the downstream face, an additional repeat impact was typically made at 2.4 meters (8-feet)[6].

Table 1. Comparison between known horizontal crack locations and the tomographic estimate of the crack locations in the top 5 m (17 ft) of Block 4 of Barker Dam. The vertical distance from the top of the dam in meters (ft) are given.

Block 4-14 North				Block 4-24 North			
Known Crack location		Tomographic estimate of crack location		Known Crack location		Tomographic estimate of crack location	
0.61	(2.0)	0.91	(3.0)	0.46	(1.5)	0.55	(1.8)
1.37	(4.5)	Not apparent		1.37	(4.5)	1.40	(4.6)
2.56	(8.4)	2.52	(8.3)	2.56	(8.4)	2.62	(8.6)
3.29	(10.8)	3.29	(10.8)	3.29	(10.8)	3.20	(10.5)
4.42	(14.5)	4.42	(14.5)	4.39	(14.4)	4.39	(14.4)

Data analysis.

Over 1800 data traces were gathered at Barker Dam. The data analysis consisted of the initial wave form analysis, picking arrival times. Then tomographic images were created from the arrival times and the distances between the source and receiver locations. The resulting tomograph is shown in Figure 2.

Comparison of Acoustic Travel Time Tomographs results and known conditions

The downstream face of Block 4 of Barker dam has 5 horizontal cracks in the top 5 m (17 ft). The crack locations are shown in Table 1.

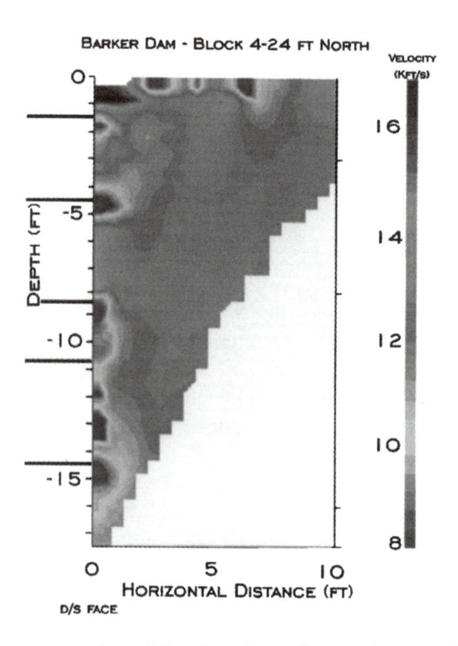

Figure 2. Tomography of Barker Dam from 1996 testing program. Showing known locations of horizontal cracks on the left and tomographic results

Figure 3. Ray paths for 1997 Barker Dam testing program

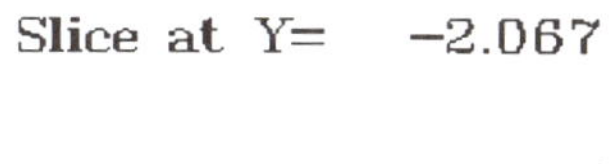

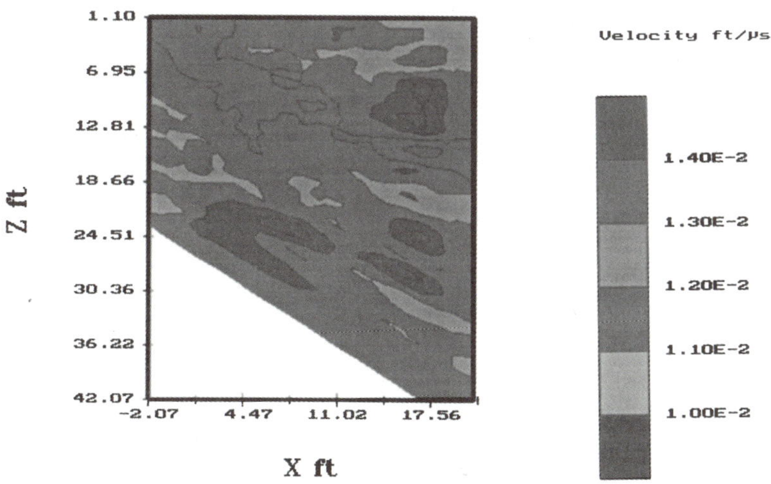

Figure 4. Preliminary tomography for 1997 Barker Dam testing program

Phase III - Reliability testing

In 1997 we returned to Barker Dam to determine the reliability of ATTT. We tested a single cross-section of the dam in Block 4, between the previous tests. This time, we placed 20 receivers on the downstream face, spaced at one foot intervals. We then used a sledgehammer as an impact source on the top of the dam, and on the upstream face. The spacing of the impact locations was every two feet, however, we did have one foot spacing on the upstream face on the top 20 feet. The ray paths of this testing are shown in Figure 3. At this time only a preliminary tomograph has been produced, see Figure 4. The tomographic program used was "3DTOM", which can run on a PC, rather than the "full seismic" program that was used in Phase II [7]. The results of the initial analysis look promising.

Later this year we plan on returning to Barker Dam to extract concrete core. This will be compared to the ATTT test results. In addition, we will cast a large test block in the laboratory to develop probability of detection curves for the testing procedure.

CONCLUSIONS

Acoustic Travel Time Tomography is a novel testing procedure which can characterize the physical properties of the unjointed concrete, and map lift-line delaminations. The sonic test results can be correlated with destructive tests to determine bulk strength and modulus. Such tests can be used to locate changes in the physical properties of the unjointed concrete within the dam. Using ATTT, lift-lines can be characterized and anomalies mapped. This testing procedure provides a more comprehensive data set for use in the finite element model at a significantly reduced cost than is currently possible.

REFERENCES

1. Kepler, W. F., and Bond, L. J. 1995. "Improved Assessment of Concrete Dams Using Sonic Tomography." In *Review of Progress in Quantitative Nondestructive Evaluation.* Ed. D. O. Thompson and D. Chimenti Vol 15. Plenum, New York. pp 1807-1814.
2. Schuller, M. P, and Atkinson, R. H. 1995. "Evaluation of Concrete Using Acoustic Tomography." In *Review of Progress in Quantitative Nondestructive Evaluation*. Ed. D. O. Thompson and D. Chimenti. Vol 14. Plenum, New York. pp 2215 - 2222.
3. Dines, K. A. and Lytle, R. J. "Computerized Geophysical Tomography, " *Proceedings of the IEEE*, Vol 67, No. 7 July 1979.
4. Davis, R. E., Jansen, C., and Neelands, W. T. 1948. "Restoration of Barker Dam." *Journal of the American Concrete Institute.* Vol 44. pp 633- 668.
5. Kepler, W. F. "Improved Assessment of Concrete Dams Using Acoustic Travel Time Tomography." Ph.D. Dissertation. University of Colorado at Boulder. In Preparation.
6. Keiswetter, D. A., and Steeples, D. W. "A Field Investigation of Source Parameters for the Sledgehammer. *Geophysics*. Vol 60, pp 1051 - 1057.
7. Jackson, M.J., and Tweeton, D.R. 1996. "3DTOM: Three-Dimensional Geophysical Tomography." U. S. Bureau of Mines, Report of Investigations 9617.

A NONLINEAR ACOUSTIC TECHNIQUE FOR CRACK AND CORROSION DETECTION IN REINFORCED CONCRETE

Dimitri M. Donskoy,[1] Keith Ferroni,[2] Alexander Sutin,[1] and Keith Sheppard[2]

[1]Davidson Laboratory
[2]Material Science Department
Stevens Institute of Technology
Hoboken, NJ 07030, USA

INTRODUCTION

This investigation is a preliminary study on the possible implementation of a novel nonlinear vibro-acoustic nondestructive testing technique to low corrosion rate concrete-steel systems. The method is extremely sensitive to the presence of integrity reducing flaws such as steel-concrete debondings and cracks developed as a result of volume expansions associated with corrosion products. The acoustical method was compared with tradition electrochemical techniques such as Linear Polarization and Electrochemical Impedance Spectroscopy so as to observe acoustical behavior changes in active and passive systems. The nonlinear vibro-acoustic method showed a significant increase in nonlinear interaction for low corrosion rates. These results suggest that the proposed concept is indeed capable of detecting the early stage of degradation.

Concrete is a porous non-homogeneous solid containing an alkaline pore solution of about 12.5 pH.[1] The high pH provides a passivating or protective environment for the reinforcing steel embedded within the concrete matrix. Often in service, chloride ions from deicing salts or sea water along with moisture and oxygen, can penetrate a concrete-steel system via capillary action through an extensive pore network. Once chloride ions come in contact with the steel, the passivated system will locally deteriorate. Preferred sites on the metal surface (e.g., local flaws, compositional inhomogeneities or thin portions of the protective oxide film) can begin to corrode in the form of pits covered by corrosion product. The conditions inside the pit are quite different from the environment at the metal surface. The presence of a pit can create localized oxygen as well as possible corrosion products and pH differentials at the metal surface. Reduction reactions of oxygen as well as possible corrosion products[2] occur outside the pit while iron oxidation and hydrolysis of chloride ions occur within the pit. As the oxidation of iron ($Fe \rightarrow Fe^{+2} + 2e^{-}$) increases inside the pit, the migration of chloride ions through the porous corrosion product cap is accelerated due to electrostatic attractions. The pH within the pit continues to decrease thereby increasing the oxidation process. The corrosion mechanism is therefore an auto-catalytic process. If allowed to continue, the pits can become quite extensive on the metal surface (general corrosion). Associated with this accumulation of corrosion product is a substantial increase in volume, approximately four times the volume of iron consumed during corrosion. This volumetric increase can create stress cracks in the surrounding concrete matrix thereby increasing the number and size of migratory paths for moisture, oxygen and chloride ions. At first, the cracks are very small, on the order of microns. However, as the corrosion product volume becomes more substantial, macro flaws such as concrete spalling can occur. At this time, the concrete-steel interface is permanently destroyed and the reinforcement can no longer provide the necessary tensile properties required by most concrete structures.

The nondestructive detection of reinforcement corrosion is of significant, practical importance for *in-situ* monitoring of concrete structures, i.e., parking decks, bridge decks, etc. If the steel-concrete interface is destroyed, the structural integrity can be sacrificed. Early detection can allow for timely preventive measure. Because the degradation is an electrochemical process, often electrochemical techniques can be utilized as sensory devices. However, electrochemical techniques are not without limitations; for example, current confinement problems during polarization techniques[3] and long testing times[1] have limited the employment of such methods. Other alternative techniques, such as acoustics, may provide viable prospects for corrosion sensor devices. To date, most acoustical materials testing techniques utilize linear interaction of sound with a medium. Often changes in sound velocity, resonant frequencies and elastic properties of materials can be correlated with material flaws. However, conventional linear acoustic methods of detecting degradation are limited by low sensitivity to flaws in heterogeneous materials such as concrete. Nonlinear vibro-acoustic technique used in the present study avoids many of the problems of conventional methods and has the potential for detecting corrosion and other structural integrity reducing flaws at an early stage.

A recently developed nonlinear nondestructive testing methods utilizing a low frequency vibration coupled with a high frequency ultrasound has shown promise as a sensitive diagnostic tool for the selective detection of cracks.[4,5] This technique could potentially be used to detect corrosion product, volume expansion cracks, and debondings in steel-reinforced concrete structures. By applying a low frequency (Ω) vibration to a material, any crack-like defect, if present, will change in contact area in its interface: increasing the area in the compression phase and decreasing it in the tensile phase. Because of this, the phase and the amplitude of the transmitted higher frequency (ω) ultrasonic probe wave varies accordingly, leading to the modulation of this probe signal. In the frequency domain the result of this modulation manifests itself as sidebands with the combination frequencies $\omega + \Omega$ and $\omega - \Omega$. The presence and the amplitude of the sidebands can be used to determine the presence and degree of degradation.

EXPERIMENTAL PROCEDURE

Sample Preparation

Type III concrete (water/cement ratio of 0.68) and 30.5 cm long pieces of #8 reinforcing bar were used in the construction of 15 x 15 x 36 cm specimen for corrosion and acoustical testing. The steel-reinforcing bar was wire brushed to remove pre-existing corrosion products and to provide a reproducible surface finish. The concrete was poured into wooden pre-forms and vibrated to ensure homogeneity. Forty-eight hours after pouring, the samples were removed from the pre-forms and moistened to enhance hydration. Testing of the samples occurred after a curing period of 21 days. The sample had a single piece of reinforcement embedded 25 cm into the test concrete block (with a minimum of 5 cm concrete cover).

To enhance corrosion, 1 N NaCl solutions were applied to the block. The solution was periodically added to a level just below the entrance of the reinforcement into the block. The solution level would fluctuate from this maximum level to a completely dry state. This wetting schedule allowed for a cycle in moisture content at the steel-concrete interface.

Electrochemical Procedures

Linear polarization and Electrochemical Impedance Spectroscopy were employed on the sample to obtain the polarization resistance of the metal. The sample was three electrode corrosion cells where the working electrode was polarized by an external counter electrode made from a steel foil. Corrosion potential was measured between an external saturated calomel reference electrode (SCE) and the encased working electrode.

Linear polarization was conducted on the specimen to obtain the apparent resistance which is inversely proportional to the corrosion current density and the corrosion rate. The metal was polarized ±10 mV from the corrosion or equilibrium potential. A scan rate of 0.1 $mV \cdot sec^{-1}$ was used in this investigation and has been shown to agree well with gravimetric results in highly passivated concrete systems.[6]

Due to the large ohmic drop associated with the concrete matrix, Electrochemical Impedance Spectrometry was also employed. A sinusoidal current signal with a 10 mV amplitude and the frequency ranged from 100 μHz to 100 kHz was applied to the encased steel. From a Nyquist plot, the solution resistance was determined. In some cases the resistance of the metal was graphically determined for comparison with the linear polarization technique. However, due to the high impedance of the concrete-steel system, low frequency information was not obtainable or was considered not useful for corrosion rate determination, in most cases. The solution resistance was always obtained to compensate for the large ohmic drops associated with concrete.

Acoustic Procedures

A nonlinear vibro-acoustical method was employed on samples. Two piezoelectric transducers were bonded to the surface of the concrete with epoxy. One transducer functioned as the transmitter of a 28.5 kHz high frequency sinusoidal signal. The second crystal was the receiver and was positioned on the opposite side of the specimen An additional vibration source applied a 250 Hz low frequency signal. A lead anchor was inserted into a pre-drilled hole to provide adequate bonding and therefore good transmission of the vibration signal to the concrete. Both the low and high frequency signals were controlled by two synthesized function generators which were externally amplified. A dual channel Fast Fourier Transformation Analyzer (FFT analyzer) was used to acquire and process all data.

Prior to each acoustical test, the frequency response was measured by applying a noise signal with a maximum frequency of 100 kHz. The resonances of the test block acted as a calibration tool to observe any linear changes in the material such as attenuation or shift in resonance. In addition to measuring the frequency response, the intensity of the low frequency vibration was measured for several settings prior to each test. This also served as a calibration tool to monitor any changes in the setup or concrete block which might effect the degree of excitation. The same setting were used then during the tests to study dependence of the modulation on amplitude of the low frequency excitation.

RESULTS AND DISCUSSION

Electrochemical Results

The test block was subjected to a daily wetting schedule of 1N NaCl solution for 180 consecutive days. A significant decrease in the corrosion potential occurred during this testing duration, indicative of reinforcement depassivation due to chloride contamination, see Fig.1(a). The initial corrosion potential, - 0.09 V_{SCE} (+0.151$_{SHE}$), and the high concrete pH provided a passive environment for steel. This early state of depassivation in steel systems exposed to chloride solutions has been observed in other investigations.[7] After 25 testing days, the continued introduction of chloride ions to the metal-concrete interface resulted in a notable decrease in the corrosion potential to approximately -0.430 V_{SCE} (-0.188 V_{SHE}). Potential measurements remained at this low corrosion potential for the duration of the test. It was speculated that in the presence of chloride ions, the passive oxide film locally depassivated, forming pits. Conditions inside the pits have been shown to be different from that of the bulk material, i.e., a much lower pH (5 to 1.5) due to hydrolysis of chloride ions. During depassivation, the pit concentration on the metal surface can become quite significant forming macro corrosion cells. Corrosion potentials from -0.181 to -0.641 V_{SCE} (+0.006 0 -0.40 V_{SHE}) have also been reported in other studies for general corrosion conditions.[7] In addition, similar decreases in corrosion potential were observed by Avila-Mendoza, et. al, for concrete steel samples.[8]

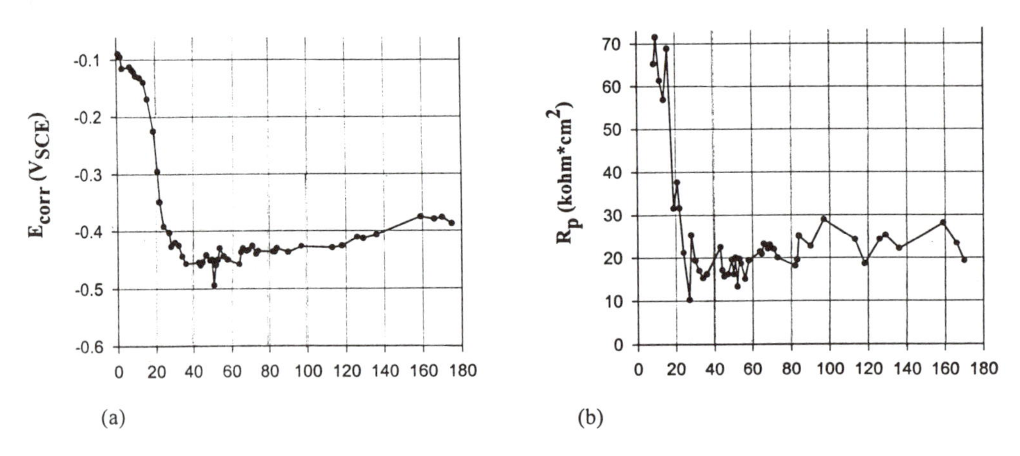

Figure 1. Corrosion potential (a) and polarization resistance (b) vs. testing days

Associated with the potential drop of the test block was a decrease in the polarization resistance, see Fig.1(b). The integrated average polarization resistance was 24.9 $k\Omega \cdot cm^2$ for the testing duration corresponding to a 1.0 $\mu A \cdot cm^{-2}$ current density and a dissolution rate of 12 $\mu m \cdot yr^{-1}$. The maximum current density observed was approximately 3 $\mu A \cdot cm^{-2}$.

The low corrosion kinetics were a result of the significant concrete cover encasing the steel reinforcement ($\approx$ 5 cm) which limited the permeation of moisture, chloride contaminants and oxygen to the steel-concrete interface. It was initially proposed that the majority of corrosion occurred at the entrance of the reinforcement into the concrete test block, where solution and oxygen permeation rates were maximized. This was latter confirmed after removing the steel from the concrete block at the completion of the test. The range of degradation on the steel reinforcement (quite extensive in some areas) extended approximately 4 cm along the bar length into the concrete test block.

The electrochemical impedance spectroscopy results also alluded to the depassivation of the encased steel. The ability of this technique to accurately measure the apparent resistance was limited (low frequency results) for the low corrosion rates observed in this investigation. However, the technique was capable of clearly distinguishing the difference between a passive and active system.

Acoustical Results

The concrete test block was tested with the nonlinear vibro-acoustic technique at various stages of corrosion. Initially the samples exhibited no modulation of the high frequency signal prior to solution application for the testing frequency of 28.5 kHz and modulation frequency of 250 Hz. The initial power spectrum shown in Fig.2(a) displayed a linear response with no side peak formation. The second acoustical test was conducted after 37 exposure days. The power spectrum shown in Fig.2(b) exhibits significant sideband formation (28.5 ± 0.25 kHz) around the high frequency signal, indicative of nonlinear acoustical interaction created by the presence of flaws in the block. Increase in the nonlinear response of a material was characterized by the degree of modulation (ΔdB), defined as the difference between the high frequency and the sideband intensities. The test block exhibited an increase in modulation Δ of +13 dB after initiation of corrosion. At the end of the corrosion testing period the final measurement exhibited no definitive change in modulation from the last measurement.

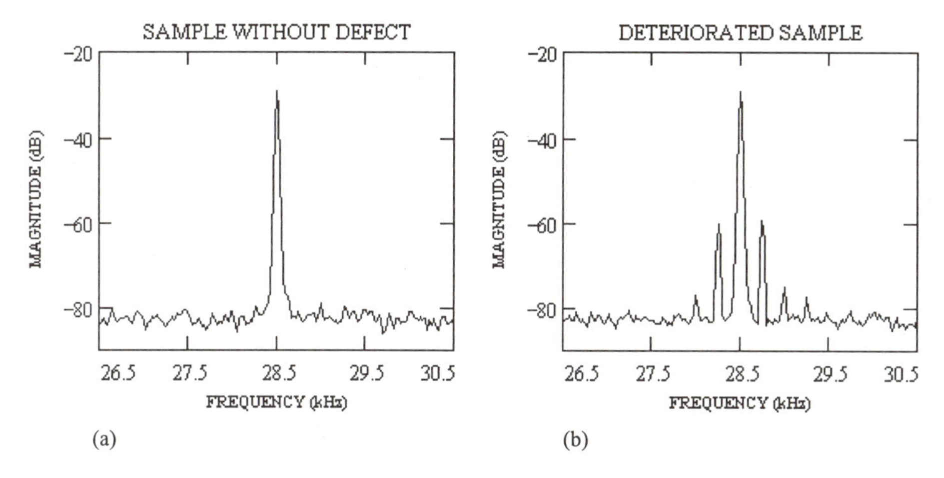

Figure 1. Spectra of acoustic signal before initiation of corrosion (a) and after 37 testing days

It was observed that the excitation frequencies between 20 to 30 kHz and the low frequency vibrations of 250 Hz had very low attenuation in the concrete specimens. Therefore, it was reasonable to assume that multiple reflections of the input signals had occurred and provided significant excitation of the concrete such that all regions were probed. Interaction at the region of degraded concrete-steel interface was very likely and supported the acoustical findings. Calibration of the high and low frequency response for the test block indicated minimal shift in resonances.

The nonlinear response of the concrete-steel system was modeled as a quadratic polynomial. The measured relationship between the degree of modulation and the low frequency amplitude did show a proportional relationship with a slope close to unity in accordance with theoretical model. In addition, the extent of modulation showed no dependence on the amplitude of the high frequency as expected.

Comparison of Electrochemical and Acoustical Results

The change from passive to active conditions occurred quickly for the test block. Fig.1 clearly shows initiation of corrosion after only 25 testing days. The acoustic test displayed a substantial difference in power spectra during the same testing period. The acoustical transition from a linear to non-linear system agrees well with the electrochemical changes (passive $\rightarrow$ active state) observed from LP, EIS and corrosion potential measurements. Physical observation of the steel reinforcement confirmed corrosion product formation, the extent of which ranging 4 cm along the length of the reinforcement. After the initial increase in corrosion kinetics, the polarization resistance remained constant, most likely indicating that the area of degradation was no longer increasing in size due to kinetic reaction limitations. The final acoustical test exhibited no distinguishing change in modulation during this period, consistent with the corrosion kinetics of the testing period.

CONCLUSIONS

The acoustical results exhibit an increase in the degree of nonlinear interaction as the polarization resistance of the steel reinforcement decreases. This result coupled with the theoretical stress development and physical examination of the steel reinforcement suggests a dependence between the increase in acoustical-nonlinear response and the increase in corrosion. The corrosion of the steel in concrete is theorized to behave as a debonding at the steel-concrete interface, i.e., a nonlinear element. From the preliminary results, it appears that the extent of the degradation area controls the degree of modulation. As the area of interfacial degradation increases, the change in nonlinear response also increases. The nonlinear acoustical results exhibit high sensitivity for the detection of steel-concrete interface degradation at its earliest stages, well before visible cracking in the surrounding concrete. Therefore this technique is a possible early detection technique for corrosion degradation in concrete structures. Nonlinear acoustics are shown to be easily applied to heterogeneous materials such as concrete and can provide quick information about the integrity of the reinforcement-concrete interface.

Further investigations, where the corrosion kinetics are accelerated and the degradation is more uniform, may provide supporting results and help to quantify the dependence between degradation area and degree of modulation.

REFERENCES

1. S. Berke, *Concrete, Corrosion Tests and Standards - ASTM Manual Series*, American Society for Testing and Materials, Philadelphia, PA, 331-337 (1995).
2. M. Stratmann, K. Bohnenkamp, and H.J. Engell, An electrochemical study of phase transitions in rust layers, *Corrosion Science*, 23(9):969-985 (1983).
3. S. Feliu, J.A. Gonzalez, and C. Andrade, Effect of current distribution on concrete rate measurements in reinforced concrete, *Corrosion*, 51(1):79-86 (1995).
4. A.M. Sutin, Nonlinear acoustic non-destructive testing of cracks. *Nonlinear Acoustics in Perspective*, 14th-Intern. Symp. on Nonlinear Acoustics Ed. R.J. Wei, Najing University Press, China, (1996).
5. A.M. Sutin and D.M. Donskoy, Nonlinear vibro-acoustic nondestructive testing, *Eight International Symp. on Nondestructive Characterization of Materials*, Boulder, CO (1997).
6. J.A. Gonzalez, A. Molina, M.L. Escudero, and C. Andrade, Errors in the electrochemical evaluation of very small corrosion rates — II. Other electrochemical techniques applied to corrosion of steel in concrete, *Corrosion Science*, 25(7):519-530 (1985).
7. V. Novokshchenov, Brittle fractures of pre-stressed bridge steel exposed to chloride-bearing environments caused by corrosion-generated hydrogen, *Corrosion*, 50(6):477-485 (1994).
8. J. Avila-Mendoza, J. M. Flores, and U. C. Castillo, Effect of superficial oxides on corrosion of steel reinforcement embedded in concrete, *Corrosion*, 50(11):879-885 (1994).

EVALUATION OF WOOD PRODUCTS BASED ON ELASTIC WAVES

M. L. Peterson

Mechanical Engineering Department
Colorado State University
Fort Collins, CO 80523

INTRODUCTION

While timber is ubiquitous as a building material, it is common to associate it with lightweight residential construction. However, in a number of areas of the United States and in a number of applications timber construction is an important part of the infrastructure. Based on the National Bridge Inventory (NBI) 41,743 timber bridges are included in the inventory with an additional 42,102 steel bridges with timber decks. The number of timber bridges is a significant percentage of the total bridge inventory in a number of states as well. Colorado, Nebraska, Montana and North Dakota are the four western states with 20% or more of the bridge inventory built with timber main spans. In the southern portion of the country Louisiana is notable with 42.1% of the bridge inventory built of timber with Arkansas, Alabama and Mississippi with 20% to 35% of the bridge inventory with timber main spans [1].

Of the number of total number of timber bridges, over 47% are classified as structurally deficient in the NBI. Based on the methodology of the NBI, deficiency of these bridges is based on visual inspection and may overstate the significance of visual defects. In spite of the reduced load ratings of many of these bridges, the rural locations of the bridges increases the likelihood that overloaded trucks use the bridges on a regular basis. Thus, the lack of failures may indicate excessive conservatism in the evaluation of some of the structures. A new array of tools is required for improved in-situ evaluation of the timber members as well as the construction details of the bridges. Currently only a limited number of alternatives exist. A review of the currently available alternatives and the technology associated with those alternatives can form a basis for considering a couple of obvious and less obvious new approaches. However, the application of this technology can only be realized if field test technology is available for in-situ evaluation of timber structures.

CURRENT TECHNOLOGY

NDT of timber has the potential to make use of elastic wave methods as well as electromagnetic wave methods. A recent review of NDT of wood members in structures focused primarily on elastic wave methods of assessing the integrity of the material [2]. The use of elastic wave methods in general appears to be dominant in the NDT of wood although potential exists for using electromagnetic methods as well. Both radar and microwave have the potential for imaging of timber, however in the large sections currently under consideration the impact of wood preservatives on the dielectric of the material would be a concern. In general however the NDT of timber at the present time relies on the use of elastic wave methods from either impact or piezoelectric signal sources. Moisture meters are required to eliminate the significant impact of moisture content on the ultrasonic wave velocity. Finally, the alternative technologies remain to be investigated in the future based on a need and an ability to be used in this application.

Hammers and Pulsers

Elastic wave methods can either be passive, using acoustic emission, or active using either impact or piezoelectric excitation. Analysis methods which have been employed for the received signal are either velocity measurements, estimates of attenuation or descriptions of the shape of the received signal. For the current discussion acoustic emission will not be considered. A number of investigators have used AE techniques successfully in timber [3]. However, on the large sections used in many of these applications and for the low operational loads which are suitable for in-service inspection, AE is not expected to be a significant factor except to facilitate proof loading of a structure. The most basic measurement of elastic waves is the velocity of propagation in the media.

Velocity Measurements

By far the most common type of measurement made for timber sections is the measurement of pulse velocity. Pulse velocity appears on the surface to be a fundamentally simple measurement to perform and an number of instruments are currently on the market which can be easily used for field testing. Except for differences in the bandwidth of the excitation signal, excitation by hammer strike or piezoelectric signal is similar. The signal used in either case has significant bandwidth. The typical method used for determining the arrival time of the signal is to use a voltage threshold. When the received signal exceeds a particular voltage threshold, the signal is determined to have been received. Both of the field units on the market currently use this type of approach.

The use of velocity measurements for any characterization of timber beyond a measure of the modulus of elasticity is not well supported. Even as a measurement of the modulus, the use of a velocity threshold is suspect. As will be discussed further, the assumption of one-dimensional wave propagation is not well supported by either experiment or by theory. An accurate measurement of the wave velocity in a timber section must consider not only the issue of the development of an appropriate measure of the arrival time, but must also be prepared to consider significant geometrical dispersion.

A far more questionable claim is often made for the pulse velocity measurements. The claim is that the strength of the specimen is correlated to the velocity of an elastic wave in the specimen. While arguably, strength is the most important property in many applications, no fundamental relationship exists between strength and the velocity of propagation of a disturbance. This lack of correlation is evident in statistical measures of the correlation between bending strength and pulse velocity with one author giving a coefficient of correlation (r) between bending strength and pulse velocity ranging from approximately 0.42 to 0.52 [4]

Moisture Meters

All of the elastic wave methods have a commonly accepted limitation, the measurements are quite sensitive to the moisture content of the material. This sensitivity is evident in attenuation as well as velocity measurements. This sensitivity is shown in results for which the distribution of the group velocity of an ultrasonic waves are clearly separated for the dry and wet timber [7]. This situation has been addressed in some commercial inspection instruments by inclusion of a moisture meter into the pulse velocity measuring instrument [5]. Current moisture meters are typically simple resistance meters which have a limited range of operation for the moisture content and which do not measure the moisture content a significant distance into the sample. A potentially important alternative method of measuring the average moisture content through the sample would use microwave methods. Initial work on this approach showed promise although the complexity and cost of the required equipment may have been restrictive [6]. Additional work in this area has recently been performed which used low cost instrumentation which had been developed for other work and which showed good sensitivity to the moisture content of the wood [7]. By using the reflected microwave power, this instrument has the potential to be actively tuned to penetrate half way into the sample thus producing an average moisture content through the thickness.

In general, moisture content measures are well established and a number of the relationships are well known. Opportunities and needs remain however to facilitate the use of such measures to produce a more complete picture of the moisture content in a sample so that other elastic wave measures may be used more reliably. A more complete understanding of the physics of this interaction are needed to allow the elastic wave methods to develop into a robust evaluation strategy.

ISSUES IN TRADITIONAL PULSE MEASUREMENT

A number of important issues in associated with the use of elastic waves in wood have not been traditionally addressed in the inspection of wood. Because of the complexity and variability of the material, there has been a reluctance to attempt to apply many of the basic principles of elastic wave propagation which can help to illuminate the measurements which are being performed. While a full evaluation of such an anisotropic material is a significant undertaking, a number of important points can be made regarding the expected shape of received signals. If the received signals are considered from a perspective other than one-dimensional wave propagation, then a clear pattern of a complex but solvable problem emerges. It remains, however, to develop material models and to apply these models at a suitable level of sophistication. For a number of the geometries and section sizes which are of interest it is critical to consider not only material dispersion but also geometrical dispersion and, in some cases, the propagation of multiple waveguide modes in the sample.

Dispersive signal

Recognizing the existence of dispersion in a waveguide signal is possible based simply on the wavelength to diameter ratio of the specimen or from simple observation of a received experimental sample. Dispersion may be directly observed in the time domain signal as shown in Figure 1 where higher frequency components appear to arrive earlier than the slower oscillations evident later in the time record. The signal shown has propagated through a 28 cm diameter section of a telephone pole. Excitation of the disturbance was from a modal hammer struck directly on the wood surface using a brass hammer tip. Direct impact contact between the hammer and the wood was chosen to provide a relatively low frequency input signal which would be most likely to result in a received signal which looked more like the one dimensional problem considered by other authors. In the received signal which resulted, it is not immediately evident that more than one waveguide mode is present. However, for a single dispersive waveguide mode, the reason for early arrival of the high frequencies seen in Figure 1 is not clear. If more than a

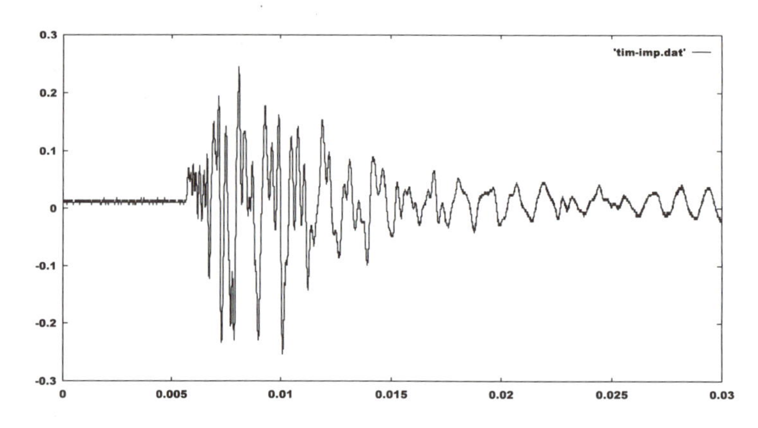

Figure 1. Received impact excited signal propagated through a 28 cm. diameter cylindrical timber section.

single mode propagates then the early high frequency arrivals may be propagating as a different mode with a higher group velocity which makes this characteristic of the signal more understandable.

The degree of dispersion in a received signal is more evident from the change in the spectrum as a function of time. A graphical representation of the change in the spectrum with respect to time is seen in a scalogram. The scalogram is produced from a wavelet decomposition of the received signal.

Figure 2 shows a scalogram produced from the received impact signal in Figure 1. The wavelet used is a Morlet wavelet with standard parameter settings [8]. The change in the location of the peaks of the spectrum with time provides a graphical view of the dispersion in the signal. What is most evident in the scalogram of diameter for the radial dimensions to be neglected. As a result dispersion would need to be considered as well as the possibility of propagation of more than one mode.

Beyond the possibility of dispersion occurring as a result of geometry of the specimen, more than one waveguide mode can propagate if the spectrum of the input signal

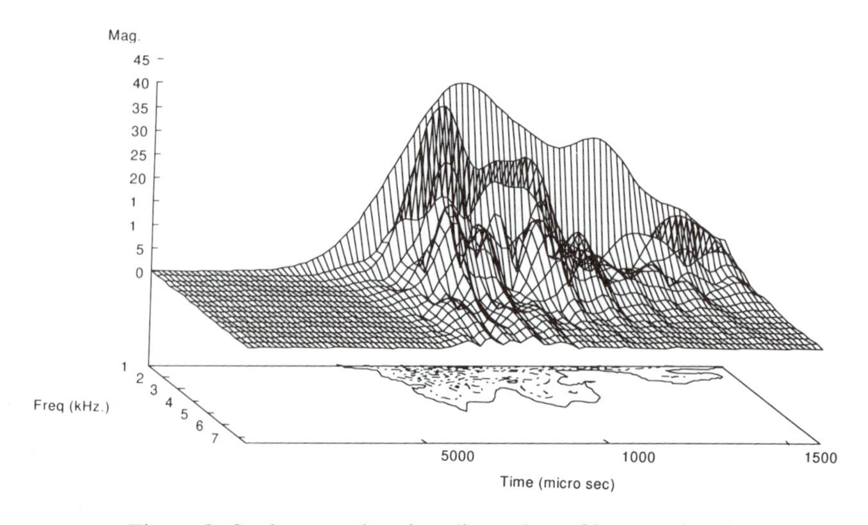

Figure 2: Scalogram showing dispersion of impact signal.

contains frequency components which are greater than the cut-off frequency for the second waveguide mode. A frequency of 20 kHz. in the telephone pole considered above corresponds to more than one wavelength in diameter. The description of waveguides which are on the order of one wavelength in diameter is particularly problematical because the modes have relatively similar group velocities and thus overlap in time for waveguides of modest length. This would explain the received signal shown in Figure 1. The problem with similar group velocities does improve until the signal become clearly separated in time which clearly occurs when the frequency is increased by a factor of ten. At these frequencies a clearly distinct early arrival frequency can be used for evaluation although significant dispersion is evident in all of the higher modes. This technique is well established for other types of applications.

FUTURE DIRECTIONS

It appears from the discussion of elastic waves in timber that several issues remain to be addressed both by the practicing and the research community. Most importantly, the robustness of the measurement of elastic wave parameters is in question because of the significant dispersion which can be expected in many of the received signals. If the sample cross section and phase velocity is consistent the reliable results will be produced. However, since the primary interest is in in-situ measurements of old wood elements for which decay and other damage may be present, these assumptions do not seem to be appropriate. The property of the elastic waves which seems most promising is the measurement of attenuation. Measurements of changes in attenuation with frequency overcomes one of the major difficulties with attenuation measurements, the need to maintain consistent coupling. As long as surface roughness is present which contains some lengths scales on the order of the wavelengths of interest, coupling variation will continue to impact the relative attenuation measurement. However by deconvolving the input signal to the sample the reflection coefficient may be used to determine the expected output signal.

The strength of the attenuation approach, regardless of the associated difficulties, can also be thought of in an historical context. From the perspective of timber testing, an original assumption led to the use of elastic waves for the characterization of timber. In 1959 Jayne's hypothesis was proposed which suggested that vibrational characteristics of wood could be used as an index of the quality of the material. It is useful to reconsider what was proposed in this seminal paper. Jayne's hypothesis states that "Energy storage and dissipation properties of wood are controlled by the same mechanisms that determine the mechanical behavior of such materials" [9]. Thus, Jayne's hypothesis never proposes the physically questionable concept that wave velocity is related to the strength of the material. This damping ratio may be related to the attenuation measured and may be brought back to consideration into the constitutive behavior of wood.

CONCLUSIONS

It appears that the state of wood NDE, while well developed, is in need of some additional consideration of the principles which underlie many measurements. The complexity of a natural anisotropic material makes it possible to become lost in the mechanics of sample to sample variation so that the unifying principles of material evaluation are lost. In many cases common measures of elastic wave properties such as velocity are incorrectly used to determine strength categories of wood. This is not supported by the physics of the situation and can lead to a loss of credibility for the techniques. However, the promise of elastic wave methods can be met in timber by careful application of some of the important measurement parameters such as those used in acousto-ultrasonic testing. However, the existing literature in the area shows the promise of the approach and the knowledge of basic wood science necessary to at least to begin to understand these measurements of the condition of timber structural elements.

REFERENCES

[1] U.S.Department of Transportation, FHWA, 1986, "Seventh Annual Report to the Congress on Highway Bridge Replacement and Rehabilitation Program".

[2] Ross, Robert J. and Pellerin, Roy F., 1994, "Nondestructive Testing for Assessing Wood Members in Structures", Forest Product Laboratory, General Technical Report FPL-GTR-70.

[3] Beall, F.C., 1987, "Fundamentals of acoustic emission and acousto-ultrasonics, in *Proc. 6'th Non-Destructive Testing of Wood Symposium*, Washington State University, Pullman, WA.

[4] Boström, Lars, 1994, "A Comparison between four different timber strength grading machines", Proc. 9'th International Symposium on Nondestructive Testing of Wood, Forest Products Society, Madison WI.

[5] Boström, Lars, 1994, "A Comparison between four different timber strength grading machines", Proc. 9'th International Symposium on Nondestructive Testing of Wood, Forest Products Society, Madison WI.

[6] James, William L., You-Hsin Yen and Ray J. King, 1985, "A Microwave Method for Measuring Moisture Content, Density and Grain Angle of Wood", Forest Products Laboratory Research Note FPL-0250, March 1985.

[7] Peterson, M.L., D. Maas, C. Mittlestadt, S. Srinath and R. Zoughi, Preliminary Results for a Multi-Sensor Non-Destructive Test of Timber Strength, to appear in *Review of Progress in Quantitative Non-Destructive Evaluation* (Plenum Press).

[8] M.L. Peterson, 1994, "A Signal Processing Technique for Measurement of Multi-Mode Waveguide Signals: An Application to Monitoring of Reaction Bonding in Silicon Nitride", *Research in Nondestructive Evaluation*, Springer Verlag New York Inc., 5, p. 239-256.

[9] Jayne, B.A., 1959, "Vibrational Properties of Wood as Indices of Quality", Forest Products Journal, 9(11), 413-416.

NUMERICAL MODELING OF ELASTIC WAVE PROPAGATION IN RANDOM PARTICULATE COMPOSITES

Frank Schubert and Bernd Koehler

Fraunhofer-Institute Non-Destructive Testing (IzfP)
Branch Lab Dresden (EADQ)
Kruegerstrasse 22
D-01326 Dresden (Germany)
e-mail: schubert@eadq.izfp.fhg.de

INTRODUCTION

Elastic wave propagation in a random particulate composite is a very complex phenomenon. In such a material, the application of ultrasonic nondestructive testing methods is difficult due to the multiple scattering processes, the strong backscattering from the aggregates, the frequency dependent attenuation and dispersion of the coherent wave fields and the mode conversions between pressure, shear and Rayleigh waves. Therefore, the interpretation of the received signals is complicated and the signal to noise ratio is low. In order to improve the applicability of pulse-echo and impact-echo testing methods and to optimize inverse reconstruction techniques, it is necessary to study the process of wave propagation systematically.

Multiple scattering theories[1,2] yield good results for the mean wave field in low-packed composites including spherical inclusions. They are in a good agreement with experimental studies by Kinra et al.[3] The most important physical phenomenon is the resonant scattering of the particles in the effective medium. The resonance modes can be easily observed in the attenuation and dispersion spectra. Their intensity depends on the mismatch of elastic stiffness between matrix and particles and that of density as well.

Up to now, the applicability of multiple scattering theories is limited to simple scatterer arrangements with a low packing density. Therefore, for highly packed composites with complex scatterer geometries, e.g. concrete, it is more suitable to use numerical modeling techniques. In this paper, we used the Elastodynamic Finite Integration Technique (EFIT) which was originally developed by Fellinger et al.[4]

The FIT-discretization of the integral form of the basic equations of linear elasticity, Hooke's law and the equation of motion, leads to a staggered grid formulation and produces a very stable and efficient numerical code. Working with different material cells and taking into account appropriate boundary conditions allows to realize arbitrary scatterers (elastic, rigid or cavity).

Nondestructive Characterization of Material VIII
Edited by Robert E. Green Jr., Plenum Press, New York, 1998

Because of the complexity of the modeled composites and for reasons of computer capacity, we carried out two-dimensional simulations, that means the realization of a plane strain propagation process.

WAVE PROPAGATION IN LOW-PACKED COMPOSITES

Figure 1 (left side) shows a two-dimensional composite model including circular scatterers. The acoustic parameters of matrix and particles are in accordance with the parameters of a cement matrix ($c_P = 3950$ m/s, $c_S = 2250$ m/s, $\rho = 2050$ kg/m³) with gravel / sand as aggregate ($c_P = 4180$ m/s, $c_S = 2475$ m/s, $\rho = 2610$ kg/m³). The discretization of the quadratic EFIT grid was $\Delta x \approx 55$ µm. Since the overall dimensions of the synthetic composite sample were 100 x 50 mm, nearly 1800 x 900 grid points were required.

Figure 1 (right side) shows an EFIT wavefront snaphot at time t = 11.5 µs after a normal force impact at the bottom side of the sample. The duration of the impact was 1 µs. The temporal disretization amounted to $\Delta t = 5$ ns in 2300 time steps. The elastic wave field is displayed as a gray scale image representing the absolute value of the particle displacement velocity vector. One can identify the longitudinal or pressure wave (P-wave), the transverse or shear wave (S-wave), the two surface or Rayleigh waves and moreover the two head waves which are also shear waves.

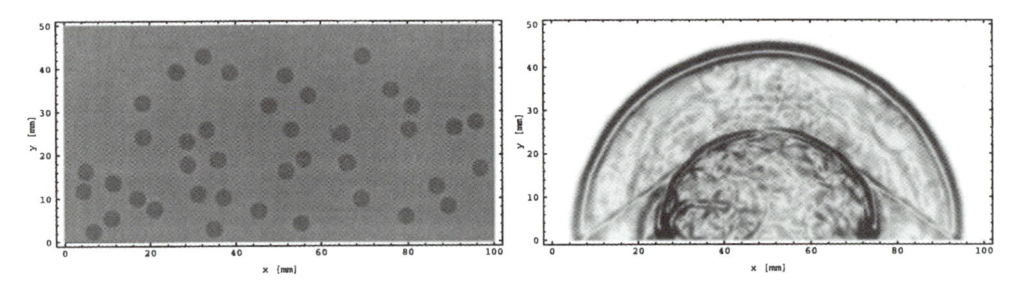

Figure 1. 2D composite model (left side) and EFIT-$|\underline{v}|$-snapshot after normal force excitation (right side).

Based on the composite model in Figure 1, the frequency dependent attenuation and dispersion of P- and S-wave was determined using the Point Source / Point Receiver (PS/PR)-technique which was first introduced by Sachse and Kim for attenuation measurements.[5] More details about the EFIT-PS/PR procedure used in the simulations can be found in another paper.[6]

Figure 2 shows the attenuation and dispersion of the P- and S-wave in the frequency range from f = 0 to f ≈ 3 MHz. The attenuation curve of the P-wave (left side, top) is characterized by an remarkable up and down caused by different particle resonances. The phase velocity (left side, middle) shows a minimum at low frequencies followed by an abrupt rise to a higher level. This trend in the wave speed diagramm is in a good qualitative agreement with multiple scattering theories.[1,2] It can also be noted, that the positions of the attenuation maxima coincide with the maxima of the group velocity (left side, bottom) while the phase velocity shows a significant rise at these frequencies. This fact demonstrates the qualitative validity of local Kramers-Kronig relations and is also in accordance with the results of multiple scattering theories.

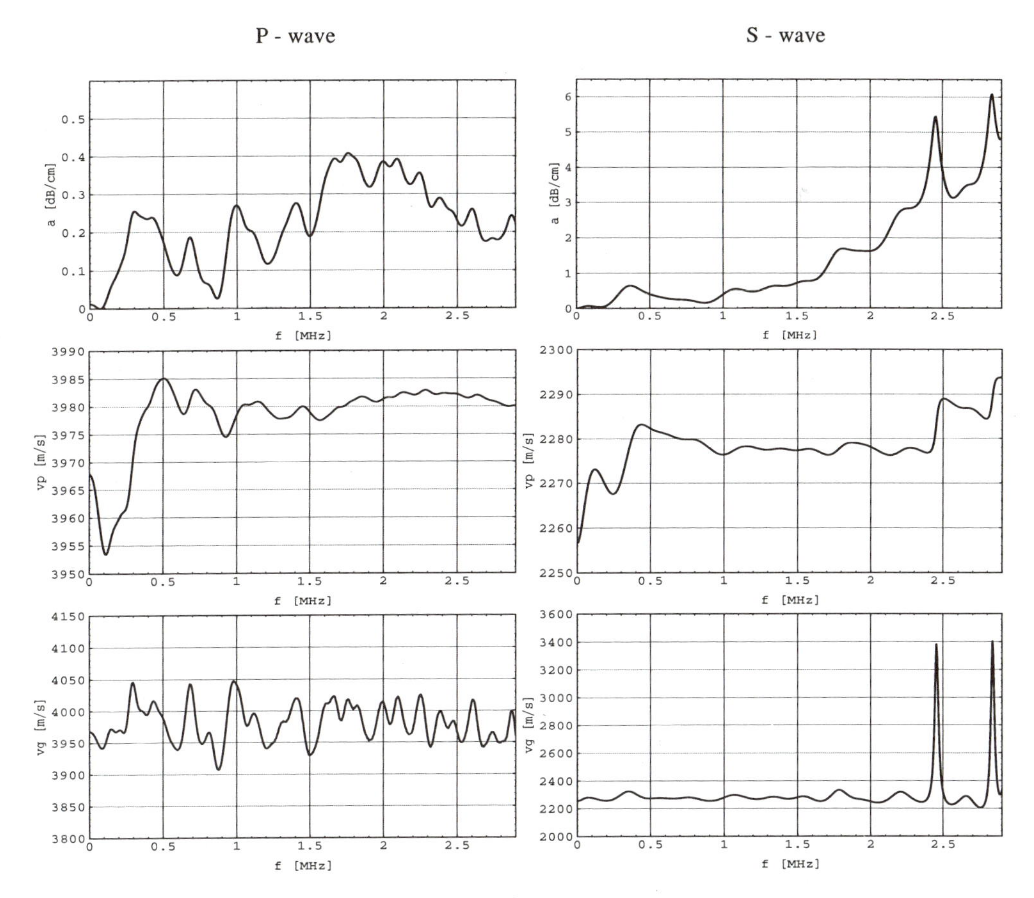

Figure 2. Attenuation (top), phase velocity (middle) and group velocity (bottom) of P- and S-wave.

All statements mentioned above for the P-wave are also valid for the S-wave (Figure 2, right side). At the end of the spectrum, two remarkable resonances can be observed. The physical reasons for this behaviour are still unknown, but the phenomenon is very similar to the discontinuities in wave speed found in an experiment by Kinra et al. for a steel/PMMA composite.[7]

WAVE PROPAGATION IN HIGH-PACKED COMPOSITES

If the volume fraction of the scatterers increases to 30 % or more, the results of multiple scattering theories are not very similar to experimental results.[2] Therefore, for highly packed composites like concrete only numerical simulations can be used.

Figure 3 shows two models of a synthetic concrete specimen according to the grading curve A8 with porosity (left side) and without porosity (right side). About 73 % of standard aggregates, e.g. gravel and sand, are embedded in a cement matrix. The location, the orientation and the shape of the scatterers are randomly distributed. Density and ultrasound velocities vary Gaussian-distributed around the given mean value.

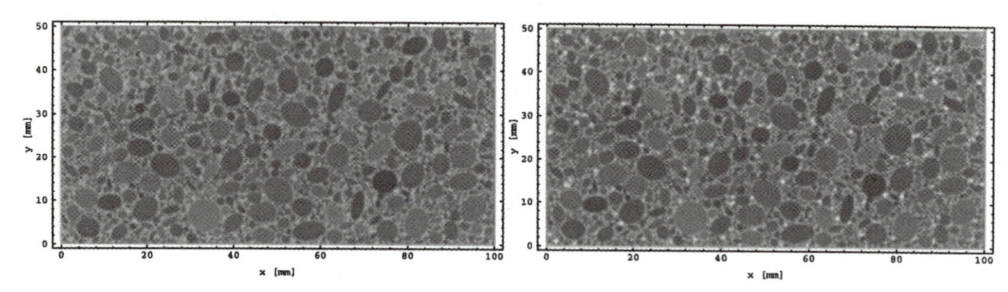

Figure 3. 2D concrete models (grading curve A8) with porosity (right side) and without porosity (left side).

Based on the concrete models above, parametric studies were carried out systematically. They revealed that the maximum aggregate size, the elastic properties of the aggregates and especially the cement porosity have a significant effect on attenuation and dispersion.[6] As an example, Figure 4 shows EFIT wave propagation snapshots in the concrete model with and without pores (cement porosity 5 %).

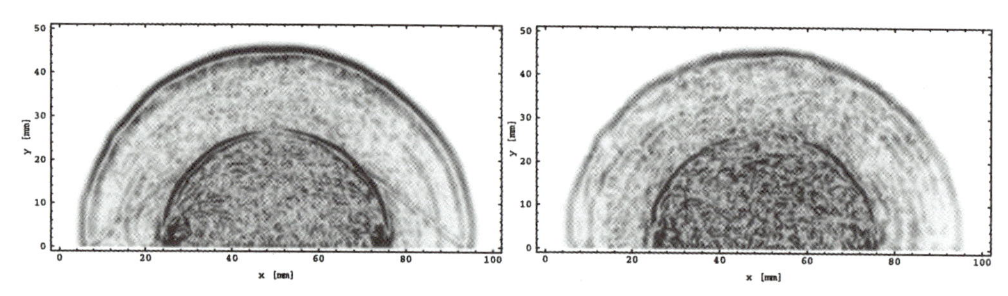

Figure 4. EFIT-|$\underline{v}$|-snapshots of the wave propagation after normal force excitation in the concrete model with porosity (right side) and without porosity (left side).

The differences between the two models are obvious. In the porous concrete model, the primary wave fronts are more attenuated. The secondary scattering reflexes are higher, especially behind the S-wave. The head waves are no longer visible. Figure 5 gives the attenuation of both, P- and S-wave, in the two different models. The results were again obtained by applying the EFIT-PS/PR technique.

The first caracteristic rise in the S-wave attenuation shifts to lower frequencies when the pores are added. A similar shift in the P-wave attenuation cannot be observed. There might be a small shift to higher frequencies. Generally for both wave modes, the attenuation becomes higher at a fixed frequency. It is important to point out, that the particle resonances in the porous concrete model are more significant than in the model without pores.

The appearance of particle resonances in concrete as a result of the simulations is confirmed by experimental PS/PR measurements by Kim, Lee, Kim[8] and Landis, Shah[9] at concrete and mortar slabs. Their results also show significant resonance phenomena in the P-wave attenuation curve.

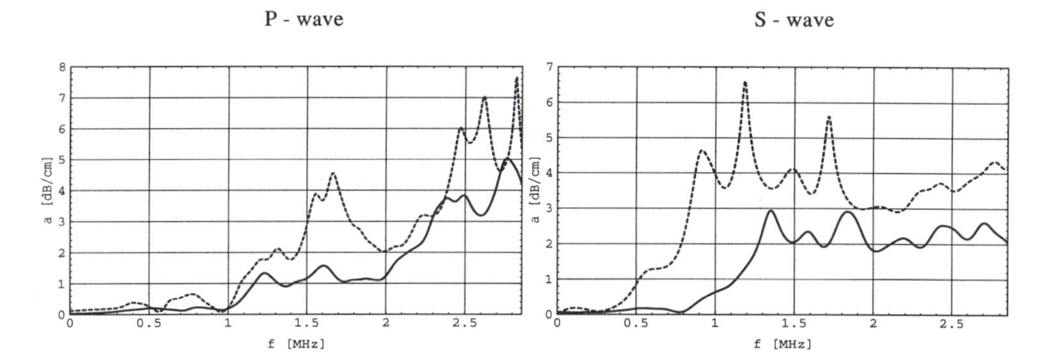

Figure 5. Attenuation of P- and S-wave in a concrete model with pores (- - - -) and without pores (——).

To demonstrate the dominant effect of porosity on the applicability of ultrasonic NDT methods in concrete, a pulse-echo simulation in different media was carried out. Figure 6 shows the synthetic test specimen (60 x 20 cm) including an elliptical cavity. A longitudinal wave transducer (5 cm aperture) was placed at x = 25 cm at the bottom side of the specimen. The center frequency of the input signal was $f_C = 200$ kHz. The simulation was done in an ideal homogeneous medium and in a concrete medium with and without pores. The discretization parameters were $\Delta x = 375$ µm (1600 x 533 grid points) and $\Delta t \approx 48$ ns (2900 time steps).

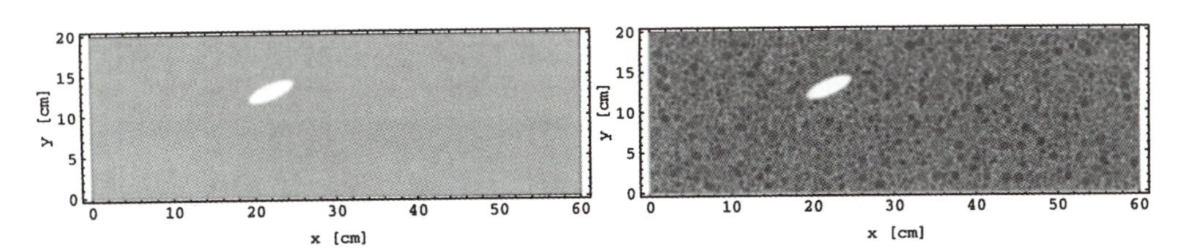

Figure 6. Pulse-echo testing geometry with an elliptical cavity in a homogeneous medium (left side) and in a concrete medium (grading curve A16, right side).

The Figures 7-9 display snapshots of the wave propagation in the three different models. Figure 10 shows the corresponding A-scans detected at the transducer position by averaging the velocity signal over all discrete grid points inside the transducer aperture. The interaction of the dominant P-wavefront with the inclusion and the backwall, respectively, produces a reflected P-wave and a mode-converted S-wave.

In the ideal homogeneous medium (Figure 7), the reflected wave fronts are clearly visible. In the A-Scan (Figure 10a), the echos from the backwall (at t ≈ 100 µm) and from the cavity (at t ≈ 60 µm) can be identified. In the concrete medium without porosity (Figure 8), the wavefronts are more attenuated and a lot of ''scatterer noise'' appears in the time signal. But nevertheless, the echos can still be related to the backwall and the inclusion (Figure 10b). After adding the pores into the concrete model (Figure 9), the testing conditions become worse. The attenuation is very strong and the elastic wave energy is smeared over a large region. Therefore, the echos from the backwall and the cavity in the A-scan can no longer be identified (Figure 10c).

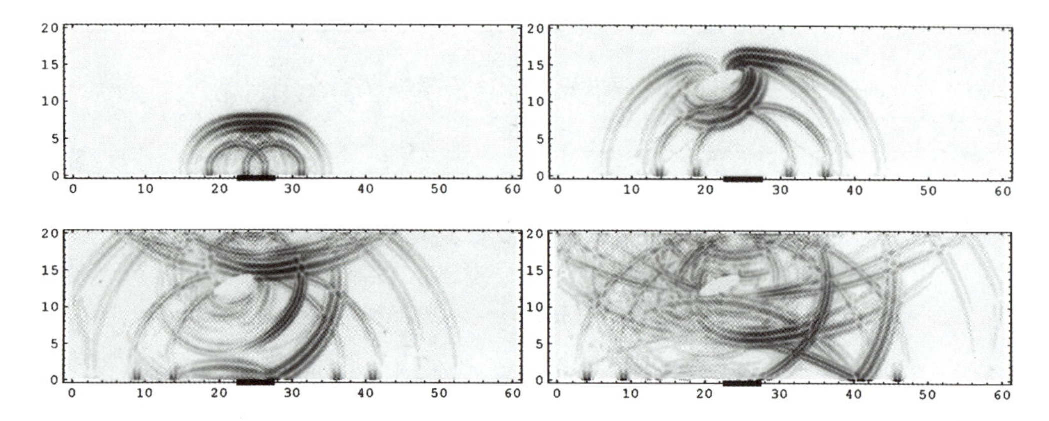

Figure 7. Pulse-echo simulation in an ideal homogeneous medium including an elliptical cavity.

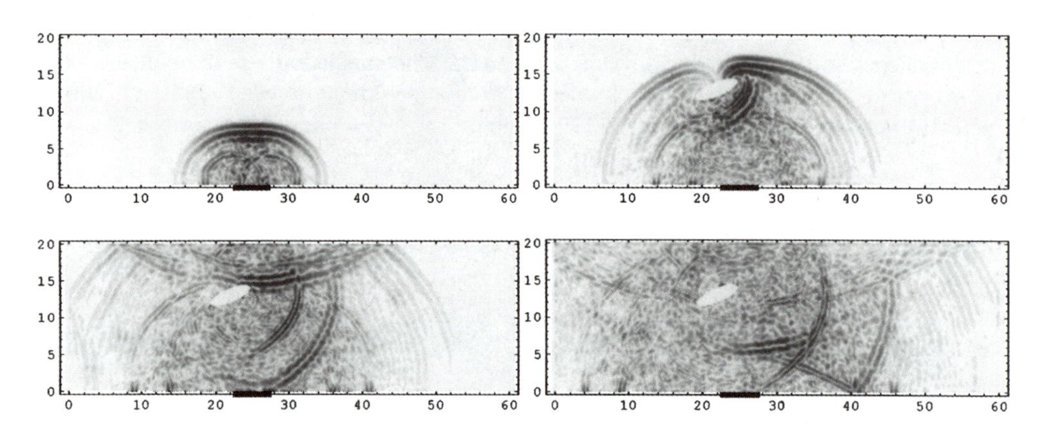

Figure 8. Pulse-echo simulation in a concrete model (grading curve A16) without porosity.

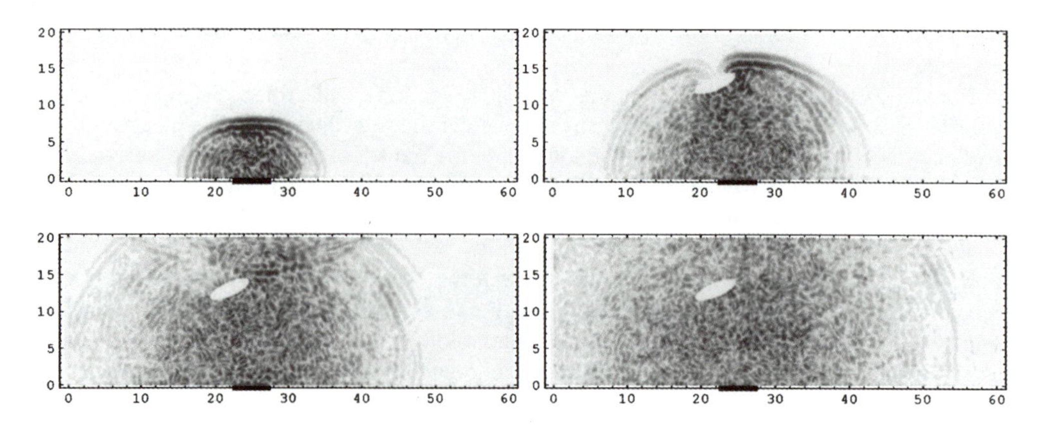

Figure 9. Pulse-echo simulation in a concrete model including 5 % cement porosity.

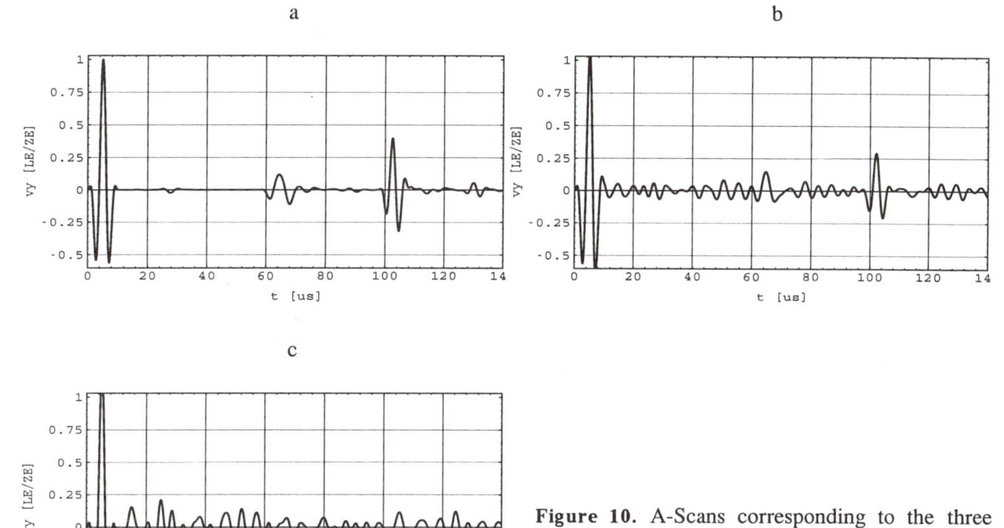

Figure 10. A-Scans corresponding to the three different pulse-echo testing situations (a: homogeneous medium, b: concrete model without pores, c: concrete model with pores).

CONCLUSIONS AND OUTLOOK

In this paper, it was demonstrated how numerical modeling techniques like EFIT can be used to investigate the elastic wave propagation in low- and high-packed composite materials systematically. For the first time, the appearance of particle resonances and the Kramers-Kronig relationships between attenuation and dispersion could have been shown by a numerical simulation. The results for low-packed composites are in a good qualitative agreement with multiple scattering theories. Therefore, in further studies the simulations can be used to prove multiple scattering models, especially their two-dimensional versions.

The simulations in conrete as a very complex and high-packed composite material also showed significant particle resonances confirming experimental results of PS/PR attenuation measurements. The comparison of porous and nonporous concrete models revealed the dominant effect of cement porosity on the applicability of ultrasonic nondestructive testing methods. In the future, the concrete model must be designed in a more realistic way. Although the most likely mechanism for the attenuation is the scattering of the waveforms at the aggregates and the pores, further research has to take into account the attenuation due to absorption (which was neglected till now) as well as mechanisms concerning the transition zone around the cement-aggregate interface (e.g. nonlinear effects, imperfect adhesion). This will be the subject of further publications.

ACKNOWLEDGMENTS

The authors would like to thank the 'Deutsche Forschungsgemeinschaft' (DFG) for supporting the project "Ausbreitung von Ultraschall in Beton (Propagation of Ultrasonic Waves in Concrete)" (grant No Ko 1386/1-1).

REFERENCES

1. V.K. Varadan, Y. Ma, and V.V. Varadan, A multiple scattering theory for elastic wave propagation in discrete random media, *J. Acoust. Soc. Am.* 77: 375-385 (1985).
2. J.-H. Kim, J.-G. Ih, and B.-H. Lee, Dispersion of elastic waves in random particulate composites, *J. Acoust. Soc. Am.* 97 (3): 1380-1388 (1995).
3. V.K. Kinra, E. Ker, and S.K. Datta, Influence of particle resonances on wave propagation in a random particulate composite, *Mech. Res. Commun.* 9: 109-114 (1982).
4. P. Fellinger, R. Marklein, K. J. Langenberg, and S. Klaholz, Numerical modeling of elastic wave propagation and scattering with EFIT - elastodynamic finite integration technique, *Wave Motion* 21: 47-66 (1995).
5. W. Sachse and K.Y. Kim, Point source / point receiver materials testing, *Rev. Progr. Quant. NDE* 6A: 311-320 (1987).
6. F. Schubert and B. Köhler, Numerical Modeling of Ultrasonic Attenuation and Dispersion in Concrete - The Effect of Aggregates and Porosity, *Proceedings of the International Conference 'Non-Destructive Testing in Civil Engineering'*, Liverpool, 143-157 (1997).
7. V.K. Kinra, Dispersive wave propagation in random particulate composites, *Recent Advances in Composites in the U.S. and Japan*, ASTM STP 864: 309-325 (1985).
8. Y.H. Kim, S. Lee, and H.C. Kim, Attenuation and dispersion of elastic waves in multiphase materials, *J. Phys. D: Appl. Phys.* 24: 1722-1728 (1991).
9. E.N. Landis and S.P. Shah, Frequency-dependent stress wave attenuation in cement-based materials, *J. Engineer. Mech.* 121 (6): 737-743 (1995).

GAS-COUPLED ULTRASONICS FOR HIGH-PRESSURE PIPELINE INSPECTION*

Christopher M. Fortunko,† Raymond E. Schramm,† and Jerry L. Jackson‡

†National Institute of Standards and Technology
Boulder, Colorado

‡Southwest Research Institute
San Antonio, Texas

INTRODUCTION

Ultrasonic pigs (Fig. 1) frequently inspect large-diameter oil pipelines so there is considerable expertise for using the ambient liquid to couple sound into the steel walls. We set out to show feasibility and determine obstacles when using compressed gas as the sound couplant. The main obstruction to inspecting natural gas pipelines with ultrasonics is that the specific acoustic impedance of gas is many orders of magnitude smaller than that of oil and other liquids used as coupling agents. In oil, an impedance of about 1.5 MRayls allows reasonable energy transfer into steel at 46 MRayls. However, in methane the impedance is only about 300 Rayls at atmospheric pressure. In the past, it seemed that such a mismatch, along with sound attenuation in the gas, would preclude using the gas as the coupling agent. Instead, considerable resources went into alternative approaches: liquid-filled wheel probes,[1] electromagnetic-acoustic transducers (EMATs),[2] and liquid slugs.[3] While prototypes of high-speed, in-line inspection systems employing such principles now exist, all exhibit serious operational shortcomings that prevent widespread commercial exploitation. We addressed this problem with gas-coupled ultrasonics and took advantage of recent improvements in ultrasonic pulser, receiver, and piezoelectric-transducer technologies. High-pressure gas exhibits both little propagation loss and a greatly increased specific acoustic impedance. We have already demonstrated concept feasibility.[4] Our present focus here is some of the physical factors influencing practicality.

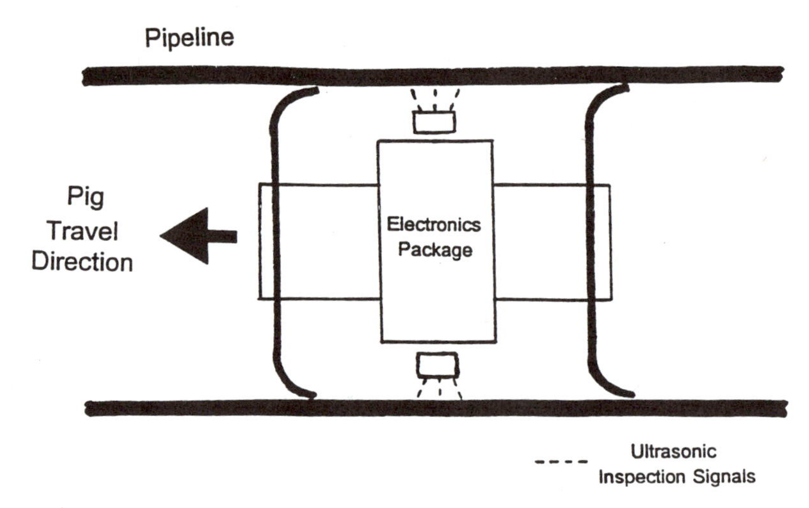

Figure 1. A pig is a platform carrying sensors, such as ultrasonic transducers. The flowing fluid moves the system through the pipe. The fluid (oil or compressed gas) in the pipe also couples the ultrasound into and out of the pipe wall.

APPROACH

The details of our experimental system showing the feasibility of the concept have been described before.[4-6] The key result of our earlier work was the possibility of using the same transducer inside the pipe for generating and detecting ultrasonic signals to reveal cracks and corrosion in the pipe wall (the same system used for inspecting oil-filled pipelines). The high-pressure gas environment transfers sufficient energy into the pipe wall; electronics with fast recovery and large dynamic range make it possible to analyze the back-wall reflection.

EXPERIMENTAL RESULTS

Figure 2 shows how the amplitude of the ultrasonic signal reflected from the front surface of a steel plate varies with the pressure of a nitrogen couplant. Even at relatively low pressures (~2 MPa), useful signals are possible.

We used normal-incidence ultrasonic signals to probe the pipe-wall thickness, and refracted signals detected crack-like features (Fig. 3). To generate a standard shear wave propagating inside the pipe wall at 45°, the incident beam was at approximately 4½° with respect to the surface normal. Our commercial flat transducers (2¼ MHz, 13 mm diameter) were designed for water immersion. However, we modified one transducer by completely removing the wear plate. This reduced much of the coherent ultrasonic clutter, now considered a main obstacle to the practicality of this approach.

We also used a focused transducer (same frequency and size, 100 mm focal length in water) for thickness measurement. Because of the large difference in sound velocity between water and gas, the focus in gas is markedly different, and it would be necessary to alter the design specifically for gas. This transducer was used to explore briefly the effect of surface coatings on the outside surface of the pipe wall.

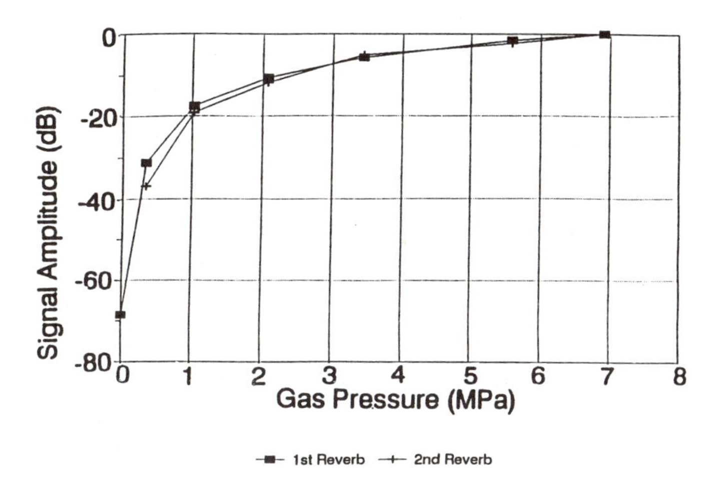

Figure 2. Relative amplitude of the front-surface reflection from a steel plate as the pressure of the N_2 atmosphere decreased to atmospheric. Frequency = 2¼ MHz; transducer-plate standoff = 34 mm; normal incidence.

The data presented here came from two types of steel specimens cut from pipelines with a diameter of 508 mm:.

1. Wall thickness of 17½ mm. This plate had two artificial flaws along the axis on the outer wall (Fig. 3).
   a. A milled slot , 5 mm wide, 38 mm long, 50% of wall thickness.
   b. A fatigue crack, grown to 50% of wall thickness, welded shut, weld cap ground off.
2. Wall thickness of 6½ mm. One specimen had no coating, while a second had an outside tar coating ~2 mm thick.

Figure 4 shows two normal-incidence circumferential scans across the milled slot in the thick plate. In the original scan (Fig. 4a), the rms front-surface roughness was about 11 μm, so the average peak-to-peak distance of the irregularities was about 31 μm. This is a significant fraction of $\lambda/2$ ($\lambda = 150$ μm in the gas), where destructive interference occurs; diffuse scattering also generates cancellations across the face of a phase-sensitive detector. The result is a low signal-to-noise (S/N) ratio. After the surface was ground to an rms roughness of 0.5 μm, a condition more typical of the inside wall of a pipeline, the signal is much cleaner (Fig. 4b). The low S/N hid the slot reflections, even with additional

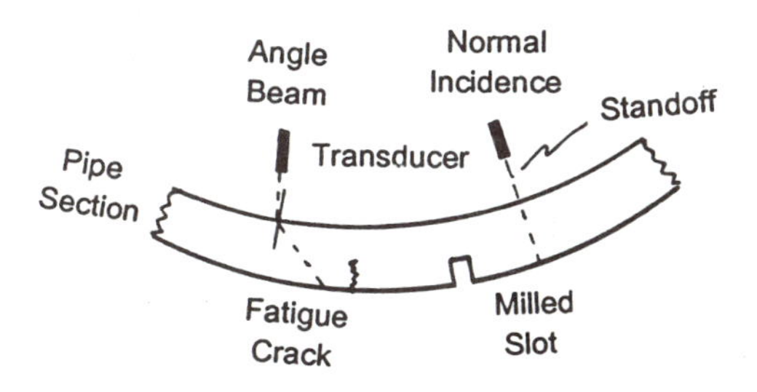

Figure 3. In this work, the transducer was on the inside of the pipe and the defects on the outside wall. The transducer was scanned around the circumference to sweep the beam across the axial flaws.

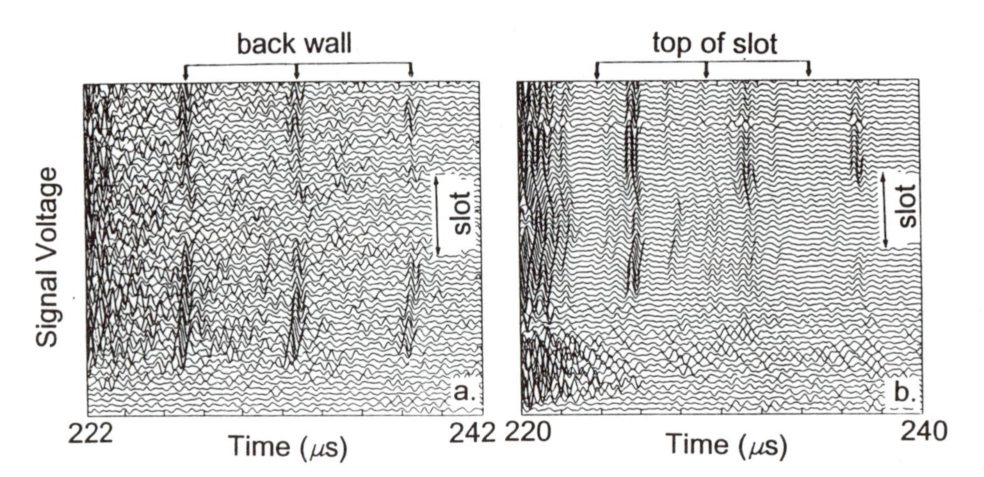

Figure 4. These plots show the relative signal voltage over time. The transducer was scanned circumferentially along the inside of the pipe section, normal to the milled slot on the outside or back wall. The wall thickness was 17½ mm, N_2 pressure was 6.8 MPa. The standoff of the flat transducer (faceplate removed) was 40 mm, and the beam was at normal incidence. This shows 60 signals collected at 1 mm intervals as the transducer moved across the slot. The horizontal time scale begins at the end of the first front-face echo. The signals seen here are echoes from the back wall and from the top of the slot.

a. Front-surface rms roughness was 11 μm. Each trace is the average of 16 signals.

b. After grinding, surface roughness was 0.5 μm. Each trace is the average of 8 signals. (Scale factor used in plotting was 75% of that used in a.)

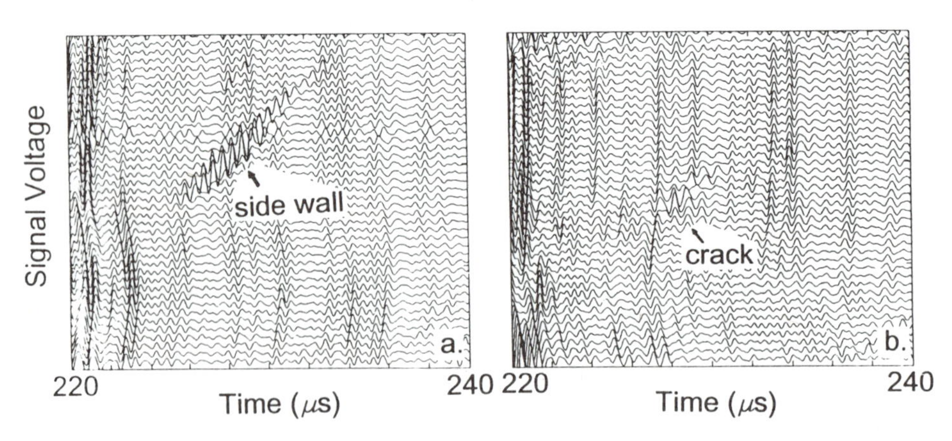

Figure 5. Same experimental configuration as in Fig. 4, except that the beam was a 45° shear wave to detect planar flaws normal to the surfaces. Vertical and horizontal scales as in Fig. 3, but there were only 40 steps of 1 mm each. The diagonal pattern is the result of the beam scanning the height of the crack: the longer the pattern, the deeper the crack. Each trace is the average of 8 signals.

a. Reflections from the smooth side wall of the milled slot (idealized crack).

b. Reflections from the tightly closed, rough fatigue crack.

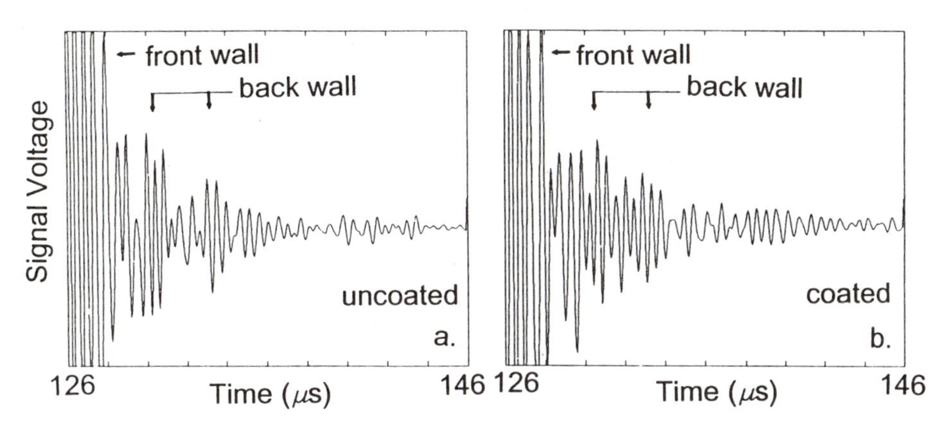

Figure 6. Through-thickness reflections from the back wall of the 6½ mm pipe section. This was from the focused transducer at a 20 mm standoff and 6.8 MPa N_2 pressure. Each trace is the average of eight signals.

a. Bare-metal specimen. Back-wall signals are distinguishable even in the thin plate.

b. Tar-coated outside wall. Echoes are less discrete due to absorption by and reverberations within the coating.

signal averaging. The high S/N revealed additional echoes at 3 μs intervals (the round trip through the full thickness took 6 μs). Surface roughness can have a significant effect on the signal quality.[7, 8]

In Fig. 5, the probe was a 45° shear beam swept across both the slot and the fatigue crack. The smooth side wall of the slot stands out just above the middle of Fig. 5a. Again, surface roughness appears to reduce echo strength, but the crack is still evident in the middle of Fig. 5b. In general, front-surface roughness will have a greater effect for gas-coupled systems than for liquid-coupled systems, because the sound propagates more slowly in methane (466 m/s) than in oil or water (~1500 m/s). The effect of the outer wall, however, will not depend on the couplant used.

Figure 6 shows single-position signals in thin-wall pipes. The external coating sometimes used for corrosion protection does make the back-wall echoes less distinct (and more difficult to analyze). This increases the importance of eliminating internal transducer reverberations that can lower the system sensitivity. The limitations of the system will be determined by the operating frequency, the bandwidth, and the ability to eliminate coherent backgrounds. We are now addressing such issues.

SUMMARY

Pressurized gas is a feasible ultrasonic coupling agent for inspecting natural gas pipelines. We believe it is possible to use a pulse-echo, gas-coupled ultrasonic system to detect flaws and measure thickness in pipelines. We found practical differences between gas-coupled and liquid-coupled approaches as a result of the differences in the speed of sound. It will be necessary to overcome other obstacles as well, specifically, transducers to minimize coherent backgrounds, transmit/receive electronics to increase S/N, and system optimization to determine the best inspection frequencies and other operational parameters. Future work will focus on such issues. Also, we are working closely with representatives of the pipeline-inspection industry to develop a testbed system. This will refine our understanding of the physical effects governing gas-coupled inspection configurations. Quantitative models will be necessary for system optimization and development of engineering specifications.

ACKNOWLEDGMENTS

This research was supported by the Gas Research Institute and A. Teitsma was technical monitor. We are grateful to J. D. McColskey, R. L. Santoyo, W. P. Dubé, and M. C. Renken of NIST for the design and construction of the pressure vessel and as well as the software for motion control and data collection. J. Crane of SwRI designed and built the transducer scanning mechanism.

REFERENCES

1. C. R. Ward and A. S. Mann, *Eighth Symposium on Line Pipe Research*, Houston, Texas, Sept. 26-29, 1993, paper no. 21.
2. G. A. Alers and L. R. Burns, *Mats. Eval.* **45**, 1184 (1987).
3. J. A. de Raad and J. v. d. Ent, *Proceedings Twelfth World Conference on Non-Destructive Testing*, eds. J. Boogaard and G. M. van Dijk (Elsevier B. V., Amsterdam, 1989), p. 156.
4. C. M. Fortunko, W. P. Dubé, and J. D. McColskey, *Proceedings, IEEE 1993 Ultrasonics Symposium*, Vol. 2, eds. M. Levy and B. R. McAvoy (IEEE, NY, 1993), p. 667.
5. J. D. McColskey, W. P. Dubé, C. M. Fortunko, R. E. Schramm, M. C. Renken, C. M. Teller II, and G. M. Light, U.S. Patent 5,587,534, "Wall Thickness and flaw detection apparatus and method for gas pipelines," (Dec. 24, 1996).
6. C. M. Fortunko, R. E. Schramm, C. M. Teller, G. M. Light, J. D. McColskey, W. P. Dubé, and M.C. Renken, *Proceedings, Review of Progress in Quantitative Nondestructive Evaluation*, Vol. 14a, eds. D. O. Thompson and D. E. Chimenti (Plenum Press, NY, 1995), p. 951.
7. K. W. Hollman, M. R. Holland, J. G. Miller, P. B. Nagy, and J. H. Rose, *J. Acoust. Soc. A.* **100**, 832 1996.
8. J. H. Rose, M. Bilgen, and P. B. Nagy, *J. Acoust. Soc. Am.* **95**, 3242 1994.

ELASTIC MODULUS AND DAMPING OF CONCRETE ELEMENTS

Richard Kohoutek

Department of Civil and Mining Engineering
University of Wollongong
Wollongong 2522
AUSTRALIA

INTRODUCTION

An important property in the analytical investigation of vibration of structures is damping of the material. In particular, for the analytical model of hysteretic damping, the determination of the properties of dynamic elastic modulus and loss factor is important. These properties are normally estimated from the response of a structure around its resonance frequency. The resonance condition leads typically to large amplitudes of vibrations, consequently large amplitudes of stresses, and likely to some distortion of results if applied in the nonresonant part of a spectrum of the structure. Therefore, the application of results from these measurements to forced vibrations outside of resonant frequencies, preferred situation in civil engineering structures, is of limited value.

A forced vibration of a simple beam for asphalt by Zaveri and Olsen (1972) was adopted here and improved. Similar to asphalt samples, there is a minimum size of a specimen which needs to be tested for concrete. Damping properties of a concrete mix, which consists: agreggates, cement, water and voids is further modified by other factors such as reinforcement, stress induced by presstressing, and additives.

The improvement of analytical model was achieved by using dynamic deformation method for a simply supported beam. The advantage of the method is removal of the upper limit for a tested frequency, which is the first natural frequency of the tested specimen. The method presented here makes it possible to measure both properties, the elastic modulus and the loss factor for a frequency above as well as below the first and higher natural frequencies of the specimen. Furthermore, additional influences such as the mass of the core of the shaker and the accelerometer mass acting at midspan; shear deformations and rotary inertia of the beam; and overhang on both ends can be also included in the measured modulus and the loss factor.

To consider the need for a range of frequencies, it must kept in mind that the first frequency of a building or a bridge is within the first decades from zero, where the exciting frequency induced by machinery is 50 or 60 Hz, depending on the country, but also below as well as above this value, because of the need of gear boxes when rotating speeds of motors are unsuited to various needs in buildings. Hence, the range of frequency of interest is broad and not only related to the resonant frequency of a structure. In most engineering structures the resonance is intentionally avoided.

PROBLEM FORMULATION

The theoretical model is a simply supported beam. It is assumed that the sinusoidal load applied will develop only small amplitudes of deformation, which will be within the linear range of the material of the specimen. Only the bending stress is considered to contribute to the deflection, which is furthermore small in respect to the dimensions of the beam. The bending of such a beam may be then described by an ordinary differential equation of fourth order and the dynamic load may be expressed in a matrix using frequency functions as presented in Kohoutek (1985).

The most common types of damping are *viscous, dry friction* and *hysteretic*. The model of damping used here is that of *hysteretic* damping, where the stress-strain relationship is that the stress (harmonic force) leads the strain (deformation) by a constant angle, φ.

An harmonic strain, $\in = \in_o \sin \varphi t$, where the induced stress is $\sigma = \sigma_o \sin(\theta t + \varphi)$. Hence, an harmonic stress is $\sigma = \sigma_o \cos \varphi \sin \theta t + \sigma_o \sin \varphi \cos \theta t$ which is also $\sigma = \sigma_o \cos \varphi \sin \theta t + \sigma_o \sin \varphi \sin (\theta t + \pi/2)$. The last expression can be rewritten with use of a complex variable as:

$$\sigma = \sigma_o \cos \varphi \sin \theta t + i \sigma_o \sin \varphi \sin \theta t. \tag{1}$$

Hence a complex modulus E is,

$$E = \sigma/\in = \sigma_o/\in_o \cos \varphi + i \sigma_o/\in_o \sin \varphi = E' + E'', \tag{2}$$

where the first component is storage modulus and the second is loss modulus. The loss factor η is a measure of hysteretic damping, $\eta = \tan \varphi = E''/E'$. The relationships above lead to the angular difference between applied force $P(t) = P_o \sin \theta t$ and the displacement v_1 by a phase angle φ as derived thus.

Consider a beam loaded by a single periodic force $P(t) = P_o \sin \theta t$ in the midspan as shown in Figure 1. Three different models are considered here, a) simply supported beam on ideal hinges, b) semi-rigidly supported beam on marked with a cross inside of the circle, and c) semi-rigidly supported beam with overhang.

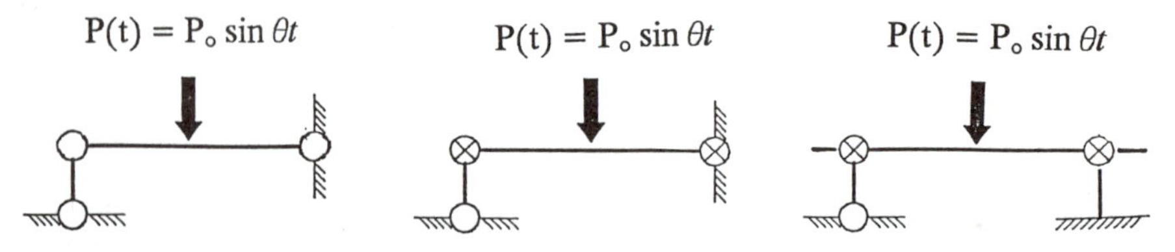

a) Ideal hinges b) Semi-rigid supports c) Semi-rigid supports with overhang

FIGURE 1. Three models of simply supported beam subjected to dynamic load.

The equilibrium of the dynamic moment and the dynamic shear force at the point under the load must be maintained for any value of t, where the details can be found in Kohoutek (1985). The system for the amplitudes of ξ_1 and v_1, for any frequency of the load, is

$$\begin{aligned} a_{11} \xi_1 + a_{12} v_1 &= 0 \\ a_{21} \xi_1 + a_{22} v_1 &= P(t) \end{aligned} \tag{3}$$

and the solutions can be found exactly, in this case decoupled. The solution vector has the first component $\xi_1 = 0$ for any θ, and taking the amplitudes only, the second component v_1 (real part) is:

$$v_1 = \frac{4\,P\,l^3\,((-1+\exp(-2\lambda))\cos\lambda + (1+\exp(-2\lambda))\sin\lambda)}{\lambda^3\,E'\,I\,(1+\exp(-2\lambda))\cos\lambda} \tag{4}$$

and subsequently taking a ratio of the amplitudes P/v_1 in combination of the relationship $E = E'(1 + i\,\eta)$,

$$P/v_1 = \frac{4\,E'\,I\lambda^3\,(1+\exp(-2\lambda))\cos\lambda + \eta\,4EI\lambda^3\,(1+\exp(-2\lambda))\cos\lambda}{l^3\,((-1+\exp(-2\lambda))\cos\lambda + (1+\exp(-2\lambda))\sin\lambda)}. \tag{5}$$

The loss factor $\eta = \tan\varphi$ and the modulus of elasticity can be calculated from the equation (5). Because the modulus is also included on the right hand side of the equation in λ, the iterative process used is numerical. However, the equation (5) is the exact solution of the differential equation for flexural deformations of a beam. The solution for the basic support conditions are summarized in Kohoutek (1985). Where the rotary inertia and shear deformations can be included by using frequency functions of Kohoutek (1985). The result of application to equation (4), when including the influences of shear deformations and rotary inertia on the total deflection v_1

$$v_1 = \frac{4\,P\,\lambda^3(aa_o + dd_o)\{d_o(1+\exp(-2d))\sin a - a_o(1-\exp(-2d))\cos a\}}{E'\,I\,a_o d_o(aa_o + dd_o)^2(1+\exp(-2d))\cos a} \tag{6}$$

and similar expressions can be derived for the ratio of P/v_1, followed by an iterative evaluation of E, thus including the influence of shear deformations and rotary inertia.

Influence of Support Conditions and Overhang

It is very difficult to create ideal support conditions, especially for a structure dynamically loaded. The dynamic performance is very sensitive to the support conditions, and when this influence is neglected it could lead to large errors in the estimate of a modulus.

Formulae have been developed to accommodate a variation of the support conditions and the above process for ideal hinge supports can be carried out numerically for semi-rigid supports. The same method can be used by creating a stiffness matrix for any other tested structure using Kohoutek (1985), also inclusive of an influence of rotary inertia and shear deformations of a specimen.

TESTING SET-UP

Beams were subjected to a sinusoidal force of varying frequency applied in the mid-span by an electromagnetic shaker. The tests were performed by selecting particular frequency and adjusting the displacement to be approximately constant. This led inevitably to a variation in the amplitude of the dynamic force, measured by a piezoelectric load cell. An accelerometer with a charge amplifier was the second channel into the Spectrum Analyzer to record a displacement and acceleration. The set-up allows comparison of four channels: force, displacement, acceleration, and strain to be correlated and spectra on each and other functions taken.

PROPERTIES OF CONCRETE BEAMS

The experimental programme is made up of eight concrete beams. The mix proportions of the constituents and reinforcement follows in Table 1.

TABLE 1. Mix proportions and reinforcement of concrete beams

Beam #	Cement kg/m^3	Water kg/m^3	Fine A. kg/m^3	Coarse A. kg/m^3	Density kg/m^3	Comp.S. MPa	W/C	Slump mm
B1-8	340	170	1060	785	2235	42.8	0.5	80

where the *coarse aggregate* was a mixed blend of 20mm, 14mm and 7mm irregular shaped crushed basalt. The *fine aggregate* was 70% river sand (coarse) and 30% beach sand (fine).
Cement used was *slag cement blend* used in the mix was 65% Portland, type A, complying with the Australian Standard and 35% granulated blast furnace slag.

TABLE 2. Reinforcement and measured properties of concrete beams

Beam #	Reinf. %	Stirrups *mm*	Modulus* *GPa*	Damping** *Log. Decr.*	Nat.freq. 1 *Hz*	Modulus† *GPa*	*Nat.freq. 2* *Hz*	*Nat.freq. 3* *Hz*
B1	-	-	30	0.0641	45.31	35.6	152.65	254.69
B2	0.17	-	30	0.0699	47.29	38.9	137.46	279.40
B3	0.42	-	30	0.0756	42.64	31.6	136.05	277.20
B4	1.11	120	30	0.0763	42.29	31.1	150.52	260.51
B5	1.11	70	30	0.0824	43.08	32.2	132.34	294.49
B6	1.57	120	30	0.0704	43.60	33.0	136.92	290.97
B7	1.57	70	30	0.0720	43.05	32.2	135.16	287.55
B8	3.1	70	30	0.0890	42.04	30.7	131.96	288.71

* calculated according to AS1012
** calculated by free decay test

Nat.freq.1,2,3 measured on hinges
† calc. from the first nat. freq.

MODULUS OF ELASTICITY AND HYSTERETIC DAMPING OF CONCRETE BEAMS

A theory presented in section above was applied to the concrete beams. The values of natural frequencies shown in Table 2 were measured initially. However, the beams were loaded by shaker over the range of frequencies (from 0 to 200 Hz), apart from doing a modal analysis (impact tests). This loading history reduced the beams natural frequency, sometimes substantially (5-6 Hz). Furthermore, spectra taken on different channels (deflection, acceleration and strain) were also different, particularly at higher frequencies. Therefore, calculations of modulus based on those frequencies will be influenced accordingly.

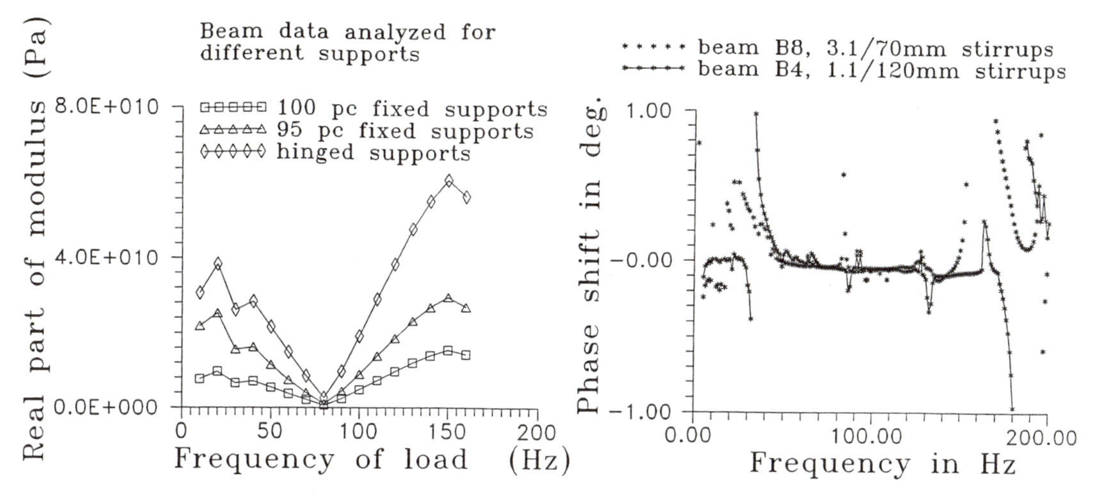

FIGURE 2. Dependency of the real part of modulus on the support.

FIGURE 3. Dependency of the phase shift on the frequency.

Specific Damping Capacity, used because of large value of φ, is defined as a ratio of an energy absorbed by the material against the energy supplied. This is an area of ellipse generated by the phase shift φ on two coordinates, a force and displacement; over the area of amplitudes of a force and displacement.

The area of an ellipse, for amplitudes of a force P_o and a displacement v_o, based on equation (1) is:

$$A_1 = P_o \, v_o \, \pi \sin \varphi, \tag{7}$$

and the energy supplied is $A_2 = P_o \, v_o$. Taking a ratio of A_1 / A_2 leads to the expression for a Specific Damping Capacity:

$$\psi = \pi \sin \varphi. \tag{8}$$

The test arrangement, in particular the support conditions, have also marked influence on the results. This is shown in Figure 2, where the same measured data were analyzed for the three support conditions. Completely fixed supports, 95% fixity at the supports, and ideal hinges result in completely different modulus. It is noted, that the small change at the fixity has considerably larger effect on the result than the same change at a hinge (not shown here). Therefore in interpretation of data, some attention should be paid to creation known boundary conditions, which will be used correctly in the analysis. The beams tested here were supported on almost ideal hinges.

CONCLUSIONS

Several main points can be made from this round of tests on beams made from plain and reinforced concrete. The advantages of present test arrangement is in testing the global properties of concrete, on a relatively large size of specimen. Furthermore, the stress distribution within the tested structure is similar to the real structural elements and includes real size aggregates and its properties. However, the disadvantage is the result, which clearly includes several mechanisms of damping without defining the individual contributions of each of three mechanism.

1) The modulus of elasticity for plain concrete is frequency dependent over the range of tested frequencies, in particular: below, above and at the first natural frequency of the specimen.

2) The Specific Damping Capacity measured by a phase shift is also frequency dependent.

3) It is possible, as reported earlier (1992), to vary damping properties of concrete at will by changing the composition of mix, therefore introducing additional design parameter for those structures, where such a property is required.

4) The amount of reinforcement does not appear to have large influence on damping properties outside of resonance region, however, this might be a function of load and needs further investigation.

5) There is a substantial shift in natural frequencies due to the loading, which indicate change of properties, generally softenning of structural elements. The change could be used to investigate (nondestructive test) global history of loading on progressive deterioration of strutural elements.

All main points of the present study need further investigation with a different concrete mixes and possibly some variations in support conditions. However, the method of testing presented makes this forced vibration possible for any beam structures and materials. The solution presented is an exact in terms of the differential equation for a beam.

ACKNOWLEDGEMENT

Several my students helped with the data collection. The facilities of the Department of Civil and Mining Engineering of University of Wollongong were used.

REFERENCES

Zaveri, K. and Olsen, H.P. (1972). Measurement of elastic modulus and loss factor of asphalt, *Technical Review No.4,* Bruel & Kjaer, pp. 3-15.

Kohoutek, R, (1985). Analysis of beams and frames, Chapter 4 in *Analysis and Design of Foundations for Vibrations,* pp. 99-156; P. Moore, ed., 512pp, 1985, published by A.A.Balkema.

Kohoutek, R., Bazant, Z. P. (1991). Damping properties of concrete beam, *ASTM International Symposium on* M^3D : Mechanics and Mechanisms of Materials Damping, Baltimore, March 13-15, poster session.

Kohoutek, R. (1992). Damping of concrete beams of different mix design, *Proceedings of Materials Week'92,* 2-5 November, Conference at Hyatt Regency, Chicago, pp95-102.

Lazan, B.J. (1968). *Damping of Materials and Members in Structural Mechanics*, Pergamon Press.

Myklestad, N.O. (1952) The concept of complex damping, *J. of Applied Mechanics*, pp. 284-286.

NONDESTRUCTIVE EVALUATION OF CONNECTION STIFFNESS

Richard Kohoutek

Department of Civil and Mining Engineering
University of Wollongong
Wollongong 2522
AUSTRALIA

INTRODUCTION

This paper is concerned with the total dynamic behaviour of steel structures considering semi-rigid connections. Due to the dynamic character of the load analysed, it leads primarily to serviceability performance but also to fatigue and finally static design considerations.

Our research provides supportive information used in making current progress towards better utilization of a material used in designs. The applications of our research will be in the area of fatigue for dynamically loaded structures as well as excessive deformations and other service performance problems.

PROBLEM FORMULATION

Behaviour of Connection

We will differentiate, in agreement with literature and the terminology adopted by AISC, between a connection and a joint. The joint comprises the detailing of an assembly connecting typically a beam and column. A connection includes some parts of a beam and column in addition to the joint. It is the connection which is the subject of this research. Our interest is in the performance of the connection and the structure, rather than in the sizing of a bolt or a weld. This is not to deny the effect of joint details, which may exert a large influence on the performance of the connection. This joint influence will, of course, be included in the behaviour of the connection, together with the adjoining parts of the beam and column which are subjected to local peaks of moments and shear forces.

Common frame analysis assumes, sometimes implicitly, that the connection is infinitely rigid and consequently carries the moment or shear forces to the adjoining bars in their full amplitude. The fact that this is not so was recognised very early (1930 Baker), and attempts to make the analysis more relevant to the actual behaviour have been made ever since [11]. This is certainly true of static analysis and it was recently extended to the dynamic behaviour of frames.

The rigid and hinged connections are clearly misnomers because *all* real connections are flexible to some degree. This is especially true of dynamic behaviour where a hinge creation is virtually impossible to achieve even in the laboratory [14].

Nondestructive Characterization of Material VIII
Edited by Robert E. Green Jr., Plenum Press, New York, 1998

The connection behaviour has a major influence on the performance of a bar; be it a column, loaded predominantly by an axial force, or a beam, subjected mainly to bending moment and shear force. There is an intrinsic relationship between the static and dynamic behaviour of a bar, not discussed here because of space limitations, which allows findings of the dynamic behaviour to be utilized also in the static analysis. The connection's performance influences natural frequencies, the dynamic as well as static moment distribution in frames, static stability through effective length of bars and the critical loads, as will be shown below.

For demonstration purposes in this paper a simple structure is selected, but using the same method, expansion to analyses of frames can be readily made.

DYNAMIC ANALYSIS

The structure under consideration is in Figure 1 below.

The structure can be modelled as a beam with two semi-rigid connections at the end supports *i* and *j*. The analytical model allows an arbitrary value for semi-rigidity to be entered where the coefficient of rigidity $\Gamma(i,j)$ is in the range of *0* for a hinge and *1.0* for a complete fixity.

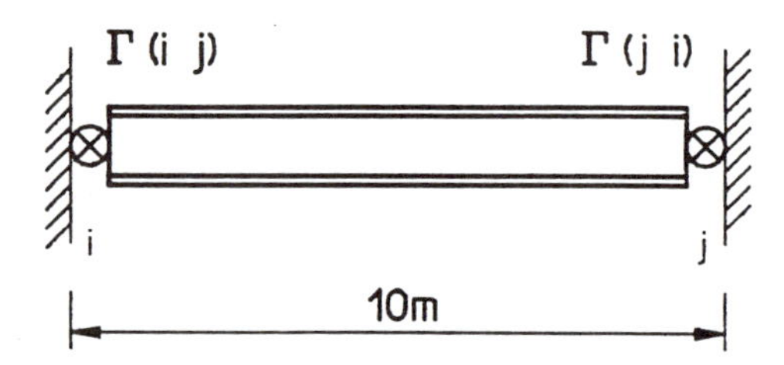

Figure 1. Beam with semi-rigid connections.

If the frequency of the oscillation coincides with the natural frequency of a beam, resonance will occur and large amplitudes of vibrations can be expected on the real structure. Consequently, serviceability problem or, in time, fatigue can be expected because of the large forces generated by the beam's inertia. Therefore, the analysis of the beam to determine such frequency, which is related to the physical properties, is of interest. The amplitude of the dynamic load is of secondary importance because vibration without damping is considered here.

If we analyse the beam with the assumption of uniformly distributed mass and flexural stiffness, the results for three different cross sections in [24] are given in Table 1 below. As can be seen from Table 1, the connection parameter Γ makes a substantial variation of several design parameters, such as critical load due to the effective length and the first natural frequency. The second column of the first natural frequencies includes the effect of shear deformations and rotary inertia on a flexural deformations of the beam. Within each section, where the cross section parameter is constant, the type of connection at ends makes some tuning of desired property possible. A designer can influence the selection of suitable natural frequency, effective length, or critical load by a choice of the suitable connection at the ends together with other traditional design parameters such as the cross section and the material of the beam.

Beam - Influence of Shear Deformations

Comparison of values for the first natural frequency in Table 1 shows considerable variation due to the influence of the shear deformations and rotary inertia on the flexural stiffness of a beam. This influence will be now investigated further.

Table 1. Variation of natural frequencies and effective lengths related to support conditions. Sectional properties of three alternatives are as in [24]. The first column for the natural frequencies frequency does NOT include the influence of shear deformations and rotary inertia, where the second column includes those influences. The actual length for this calculation is assumed to be 10 m.

Steel grade Alternative No.	Load frequency 10 Hz		20 Hz		30 Hz		40 Hz		50 Hz		Support ends conditions Fixed = 1.0
AS 1204	1.03	1.03	1.14	1.16	1.41	1.49	2.39	2.92	15.66	4.19	Fixed/fixed
Grade 250	1.32	1.34	1.52	1.75	2.42	4.87	23.46	1.75	1.05	0.36	Γ=0.75 /0.75
	1.64	1.67	2.21	2.78	12.14	5.48	1.42	0.63	0.36	0.11	Γ=0.5 /0.5
# 1	2.01	2.05	3.82	5.04	3.36	1.96	0.58	0.37	0.11	1.27	Γ=0.25 /0.25
	2.44	2.46	11.77	13.91	1.36	1.27	0.28	0.25	2.05	3.57	Hinged/hinged
AS 1204	1.04	1.04	1.16	1.17	1.49	1.57	2.98	3.94	4.05	2.25	Fixed/fixed
Grade 350	1.32	1.35	1.56	1.84	2.79	8.16	5.40	1.20	0.74	0.25	Γ=0.75 /0.75
	1.65	1.69	2.36	3.11	152.	3.17	1.02	0.46	0.24	4.11	Γ=0.5 /0.5
# 2	2.04	2.08	4.51	6.64	2.27	1.44	0.42	0.26	4.25	3.76	Γ=0.25 /0.25
	2.51	2.52	34.27	64.39	1.03	0.97	0.19	0.17	6.64	7.91	Hinged/hinged
Bisalloy 80	1.04	1.04	1.18	1.20	1.59	1.69	4.11	6.42	2.16	1.45	Fixed/fixed
	1.32	1.36	1.62	1.93	3.37	33.19	2.86	0.87	0.54	0.16	Γ=0.75 /0.75
	1.66	1.72	2.54	3.56	9.41	2.13	0.75	0.34	0.16	1.19	Γ=0.5 /0.5
# 3	2.08	2.12	5.63	10.16	1.65	1.09	0.31	0.17	9.30	7.93	Γ=0.25 /0.25
	2.58	2.59	31.38	21.65	0.80	0.76	0.11	0.10	0.10	0.11	Hinged/hinged

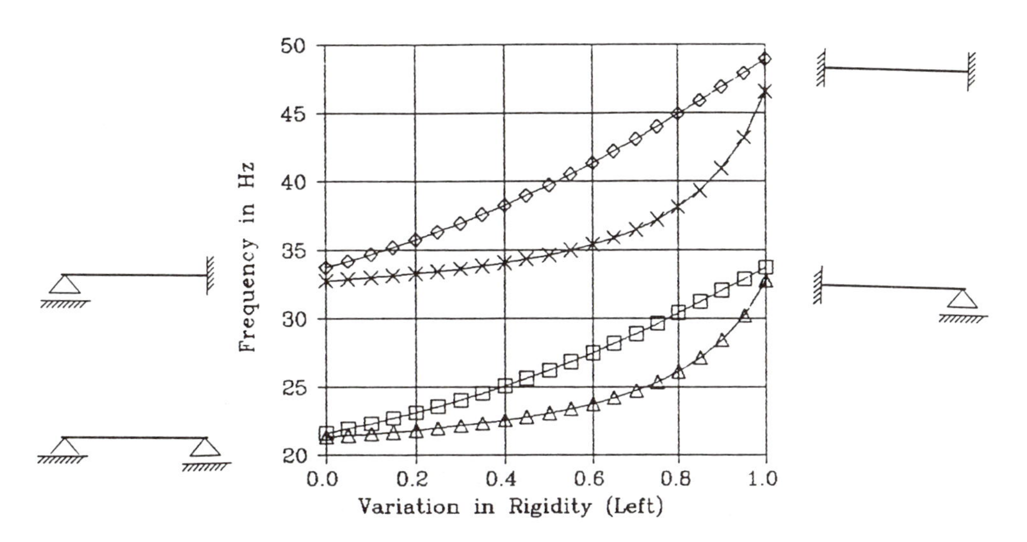

Figure 2 Variation of the first natural frequency of the beam from Figure 1, made of Grade 250 steel. The top curve is without shear deformations and rotary inertia, the bottom curve includes those influences.

The frequency on the left, lower (about 22 Hz) is the simply supported beam, where the right, top corner (about 33 Hz) is the result for the propped cantilever; full fixity at the left support, and hinge at the right support. This is also starting point on the left; propped cantilever, completely fixed at the right, hinge on the left. Variation of this left support produces two upper curves, with the fixed/fixed beam at the top right (about 47 Hz).

OUTLINE OF TESTING PROCEDURE

Behaviour of connection

The general idea of the dynamic testing is based on the variation of natural frequency of a beam subjected to some exciting force. Vibration induced can be analysed experimentally with a Spectrum Analyzer for the natural frequencies of a structure. Since the mathematical model of the beam includes inertia and mass distribution which is continuous, and the model is very closely modelling a real structure, any variation in the natural frequency is attributed to the connection.

A simple structure fabricated and tested for natural frequencies was chosen to be a simple T-connection. The structure is supported in a manner of closely approximating hinged/hinged beam, where the rigidity of the supports is measured dynamically. Specifically, a variation in the first natural frequency is used to deduce the rigidity of supports. A specimen is then excited in the bending mode to activate connection contribution and the second natural frequency is determined experimentally. The matching second natural frequency is then determined analytically by varying rigidity index.

The method has proved to be very sensitive, but with good repeatability, where we have dismantled the whole testing supports and assembled the specimen again. Furthermore, the method was used previously used on open sections reported elsewhere (Kohoutek, 1991a), also with very good results.

Test program in progress consists of eleven T-connections (to date, but more are in preparation) with the properties tabulated in Table 2 below.

Table 2. Sizes of chord and branch of tested joints.

Joint designation	Chord size(mm)	Branch size(mm)
T1/3	219.1 x 8.2	219.1 x 4.0
T2/3	219.1 x 8.2	114.3 x 6.0
T3/3	219.1 x 8.2	114.3 x 4.8
T4/3	219.1 x 8.2	60.3 x 3.9
T5/3	219.1 x 6.4	219.1 x 4.8
T6/3	219.1 x 6.4	114.3 x 6.0
T7/3	219.1 x 6.4	114.3 x 4.8
T8/3	219.1 x 6.4	60.3 x 3.9
T9/3	219.1 x 4.8	219.1 x 4.8
T10/3	219.1 x 4.8	114.3 x 4.8
T11/3	219.1 x 4.8	60.3 x 3.9

The specimens are made from steel pipes readily available in Australia. Subsequently, there is a limited choice of diameters and thicknesses which could be used. A limited survey of non dimensional parameters β, γ, τ was made before the test to ascertain applicability of test pieces to the real structures, with the nomenclature at the end of this paper.

PARAMETRIC RELATIONSHIP FOR Γ

Several possible relationships were investigated using statistical methods. First, the proposed relationship is linear in the form:

$$\Gamma = b_o + b_1\beta_i + b_2\gamma_i + b_3\tau_i + Error_i \tag{1}$$

where Γ is an index of rigidity, b_i, $i = 1,3$, are unknown constants, the un observable *Error* terms are assumed to have independent $N(0,\sigma^2)$ distributions, and β, γ, τ are non dimensional parameters. It was found that the model gave an R^2 value of *92.5%*, or *adjusted* R^2 value of *89.3%*.

The output from the first model suggested that γ was not of great value in explaining the variability in the rigidity values, so a second model was fitted in which only β and τ were the explanatory variables. This model gave $R^2 = 91.3\ \%$ and *adjusted* $R^2 = 89.2\ \%$, suggesting that the simpler model would be virtually as good in explaining the variability in rigidity.

However, the plot of the *residuals* against β shown in Figure 3, suggested that it might be more appropriate to include β^2 in the model. When the four explanatory variables β, γ, τ,and β^2 were fitted, it was τ which appeared to be superfluous. The residuals were obtained from the model which fitted rigidity as a function of β and τ.

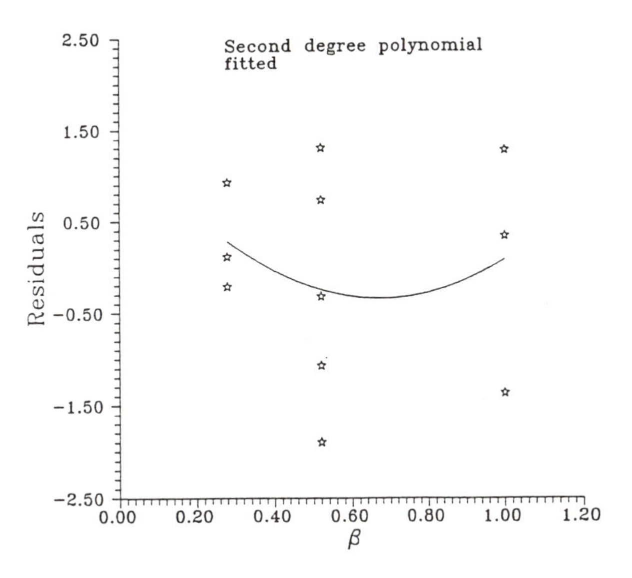

Figure 3. Relationship of residuals and β. The curve fitted is the second degree polynomial.

Finally, the proposed relationship for the rigidity index Γ and geometrical properties can be reformulated as:

$$\Gamma = 1.307 - 0.819\,\beta - 0.0130\,\gamma + 0.488\,\beta^2 \tag{2}$$

where all the coefficients are significantly different from zero, $R^2 = 98.4\ \%$ and *adjusted* $R^2 = 97.7\ \%$.

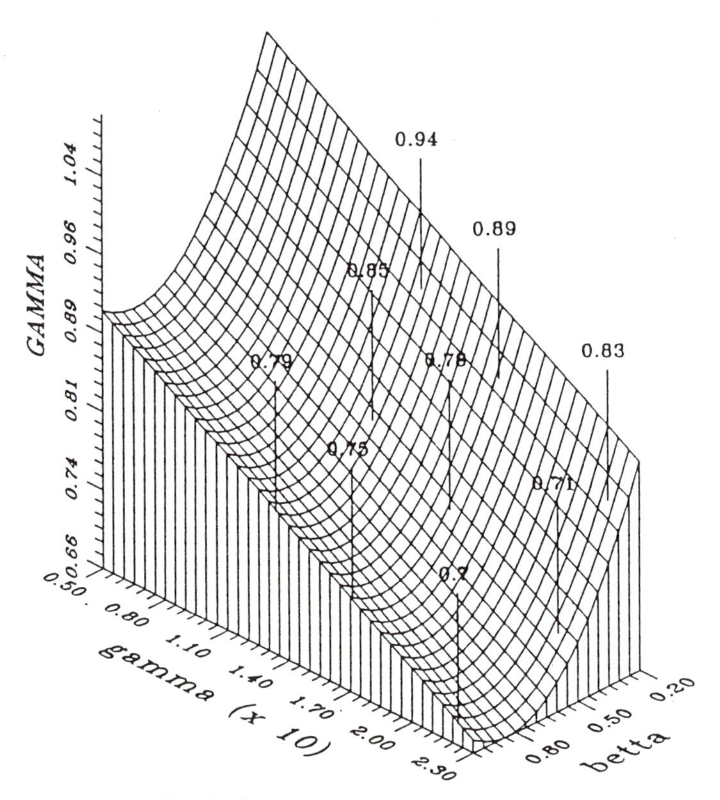

Figure 3. Model adopted {eq (2)}, with the measured values of data shown.

This model was found to be the best, because it is relatively simple, and reasonably consistent with the predicted models found in each of the subsets of the data.

CONCLUSIONS

Frequency measurement can be used to determine rigidity as a single characteristic of a tubular connection. Because the natural frequencies are measured, they include nearly all physical properties of the connection. Hence, testing prototypes capture the most contributing factors for the actual behaviour of a structure. Therefore, the final analysis of a jacket or generally any other structure can be done by first testing the fabricated connections and then apply the results in the analyses for moment distribution, static stability, and dynamic analyses.

It is possible to formulate rigidity indices as function of connection parameters β, γ, β^2 and a parametric formulae was derived to estimate the rigidities directly from the geometrical properties. For the tested connections the relationship is quadratic in β.

However, the range of tested connections should be further enlarged and further studied, work now in progress.

ACKNOWLEDGEMENT

The project is supported by the University of Wollongong, Australian Research Council, and Tubemakers Pty Ltd; their co-operation is gratefully acknowledged. The author wish to thank Dr K.G.Russell from the Department of Mathematics for his help with the statistical analysis of data. The tests were carried out by my former PhD student and contribution of Dr I.Hoshyari is gratefully acknowledged. The facilities of the Department of Civil and Mining Engineering of University of Wollongong were used.

REFERENCES

1. Ackroyd, M. H., (1987). Simplified frame design of type PR construction, *Amer.ISC, Eng. J.*, Vol. 24, No. 4, pp 141-147.
2. Bjorhovde, R. et al., (1988). *Connections in Steel Structures,* Elsevier Science Publishers.
3. Chen, W.F., (1987). *Joint Flexibility in Steel Frames,* Elsevier Science Publishers.
4. Bouwkamp J.G., Hollings J.P., Maison B.F., Row D.G., (1980). Effects of joint flexibility on the response of offshore towers, *Proceedings of the 12th Annual Offshore Technology Conference*, Texas, paper 3901.
5. Behrens M., Kohoutek R., (1990). Dynamic of beam with semi-rigid joints, Part II-Experimental evaluation, *Proceedings of Australian Vibration and Noise Conference*; held at Monash University 18-20 September, pp 344-8.
6. Ellingwood, B. (1989). Serviceability guidelines for steel structures, *Amer.ISC, Eng. J.,* Vol. 26, No. 1, pp 1-8.
7. Firt, V., (1974). Stabilita a kmitani konstrukci s netuhymi spoji, (Stability and Vibrations of Structures with Semi-Rigid Joints). Academia-Praha.
8. Hogan, T.J., Firkins, A., Thomas, I.R. (1985). Standardized Structural Connection, *Aust.ISC,* (first edition).
9. Howlett, J.H. et al. (1981). Joints in structural steelwork, *Proceedings of the International Conference held at Teesside Polytechnic,* Cleveland 6-9 April 1981. Pentech Press.
10. Koch S. P., (1989). Field measurements of jacket member structural properties, *Proceedings of the 21th Annual Offshore Technology Conference*, Texas, pp 569-576.
11. Kohoutek, R., (1983). Computational improvements of the dynamic deformation method, *Proceedings of CTAC - 83*, August, held at Sydney University; published by North-Holland. R.A. Anderson, ed., pp 769- 784.
12. Kohoutek, R., (1984). The dynamic instability of the bar under tensile periodic force, *Transactions of Canadian Society for Mechanical Engineers*, Vol.8, pp 1-5.
13. Kohoutek, R., (1985a). Dynamic analyses of frames with semi-rigid joints, I.E. Aust. *Proceedings of Metal Structures Conference*, held at Monash University, Melbourne, 23-24 May, L. Pham, ed., pp 7-10.
14. Kohoutek, R., (1985b). *Analysis of Beams and Frames*, Chapter 4 in Analysis and Design of Foundations for Vibrations, pp 99-156; P. Moore, ed., 512pp, 1985, published by A.A.Balkema.
15. Kohoutek R., (1990). Dynamic of beam with semi-rigid joints, Part I-analytical model, *Proceedings of Australian Vibration and Noise Conference*; held at Monash University 18-20 September, pp 339-343.
16. Kohoutek, R., (1991a). Dynamic tests of semi-rigid connections, *Proceedings of The Second International Workshop on Connections*, Invited paper, Pittsburgh, April 10-12, 9pp.
17. Kohoutek, R., (1991b). Analysis of frames for stability, *Proceedings of The Second International Workshop on Connections*, Invited paper, Pittsburgh, April 10-12, 9pp.
18. Kohoutek, R. and Hoshyari, I., (1991a). Dynamics of tubular semi-rigid joints, *Proceedings for Offshore Mechanics and Artic Engineering*, Stavanger, Norway, June 23-28, V. I pp 579-585.
19. Kohoutek, R. and Hoshyari, I., (1991b). Flexibility of tubular joints in offshore structures, *The First International Offshore and Polar Engineering Conference*, Edinburgh, United Kingdom, August 11-15, pp 61-66.

20. Kohoutek, R. (1988). Serviceability performance of steel joints, *Aust.ISC, Steel Construction,* Vol. 22, No. 3.

21. Lay, M.G., (1973). Effective length of crane columns, *Aust.ISC, Steel construction,* Vol. 7 No. 2.

22. Leon, R. T. et al. (1987). Semi-rigid composite steel frames, *Amer.ISC, Eng. J.,* Vol. 24 No. 4, pp 147-155.

23. Pham, L., Mansell, D. S.(1985). Survey of research needs for metal structures, *I.E. Aust. Proceedings of Metal Structures Conference,* held at Monash University, Melbourne, 23-24 May, L. Pham, ed., pp 172-177.

24. Rosier, G.A., Croll, J.E. (1987). High strength quenched and tempered steels in structures, *Aust.ISC, Steel construction,* Vol. 21 No. 3.

25. Segal, A. and Barnuch, M. (1980). A nondestructive dynamic method for the determination of the critical load of elastic columns, Experimental Mechanics, pp 285-288.

26. Sweet, A.L., Genin, J. and Mlakar, P.F. (1977). Determination of column-buckling criteria using vibratory data, Experimental Mechanics, pp 385-391.

27. Ueda Y., Rashed S. M. H. and Nakacho K., (1986). Flexibility and yield strength of joints in analysis of tubular offshore structures, *Proceedings of the Fifth International Offshore Mechanical and Arctic Engineering*, Symposium, pp 293-301.

28. Underwater Engineering Group, (1985). Design of tubular joints for offshore structures, *UEG Publication*, UR33.

29. Wardenier J., (1982). *Hollow Section Joints,* Delft University Press.

Nomenclature

D	....	chord diameter
d	....	branch diameter
L	....	chord length
l	....	length of the beam
T	....	chord thickness
t	....	branch thickness
Γ	....	rigidity index
α	....	$2L/D$
β	....	d/D
γ	....	$D/2T$
τ	....	t/T

RENEWING ORIGINAL SHAPE OF ACOUSTIC EMISSION SIGNALS CHANGED WHILE PROPAGATION IN A MATERIAL

O.V.Abramov, O.M.Gradov,

Laboratory of Ultrasound Technology,
N.S.Kurnakov Institute of General and Inorganic Chemistry of Russian Academy of Sciences, Leninsky prosp.,31, Moscow, 117907, Russia.,

Tel.: (095)-955-48-38, Fax: (095)-954-12-79.

INTRODUCTION

The development of the acoustic testing system of a new generation supposes to use the non-distorted acoustic information taken from the surface of the tested object. To achieve this aim it is necessary to have as wide-band transducers as new methods of a signal treatment which permit to rebuild the original impulse form taking place near the source of the generation of acoustic emission (AE) signals.

Among the many reasons causing the spectral changes of the acoustic impulse there is of the universal kind independent on the recording instruments properties the mechanism connected with the dissipate processes in a material. It's acting is conditioned by the dependence of the ultrasound waves decrement of damping on the frequency that defines the spectral changes of the signal while propagating from the source of it's generation. The theoretical investigations are based on the system of equations of an elasticity and of a thermal conductivity.

The constructive method of the practical calculation of the acoustic oscillations damping rates is developed for deriving the expressions establishing the connection between the damping length of the ultrasound waves and parameters of the viscosity and thermo-conductivity of the material where they propagate. The kind of the frequency dependence of the damping rate is found to be defined by the type and structure of the material and also by the frequency range of the signal spectrum. It is created the algorithm of rebuilding the original shape of the acoustic impulse which the last had near the source of it's appearance. The algorithm is based on the use of the method of the AE source localization that permits to define the distance which the signals went up to each of recording instruments. With help of measured values of the signal shape by each of the transducers and also knowing the distances gone by the impulse before it's registration is it possible to define the damping rate of each frequency line in the signal spectrum, that permits in one's turn to find the frequency dependence of the decrement and to derive the values of the thermo-physical parameters of the material. The knowledge of the dampingrate gives the possibility to rebuild the original spectrum of the acoustic impulse with help of measurements made by each recording instruments at different points of the tested material surface where they are placed.

The spectral analyze of the rebuilt AE signal together with measurements of the test impulse characteristics permits to define the physical and mechanical parameters of the material by the methods of the NDT. The algorithm of the damping rates calculation of the ultrasound oscillations may be useful not only for rebuilding of the acoustic waves spectrum but also for many another applications and investigations where such waves are used.

BASIC RELATIONSHIPS

We start with the equation for elastic deformations taking into account the inhomogeneity of a heated material and its viscosity [1-3]

$$\rho \ddot{\mathbf{u}} = \frac{E}{2(1+\upsilon)}\Delta\mathbf{u} + \frac{E}{2(1+\upsilon)(1-2\upsilon)}\text{graddiv}\,\mathbf{u} + \eta\,\Delta\dot{\mathbf{u}} + (\zeta + \frac{1}{3}\eta)\ \text{graddiv}\,\dot{\mathbf{u}} - K\alpha\ \text{grad}\ T \tag{1}$$

Here ρ is the density, E and ν are the module of elasticity and Poison rate respectively, $\mathbf{u}$ is the vector of elastic displacements, $\dot{\mathbf{u}} = \partial\mathbf{u}/\partial t$, ζ and η are rates of viscosity, α is the rate of temperature expansion, K is the module of thorough compression, T is the temperature of the material.

Another equation describes the heat conductivity of the materials [4]

$$C_V\dot{T} + \frac{C_P - C_V}{\alpha}\text{div}\,\dot{\mathbf{u}} = \kappa\Delta T \tag{2}$$

where C_V and C_P are thermal capacities for the constant volume and constant pressure respectively, κ is the rate of thermal conductivity.

For longitudinal oscillations when rot $\mathbf{u}$ = 0 the equation (1) takes the following form

$$\rho \ddot{\mathbf{u}} = \rho c_l^2 \text{graddiv}\,\mathbf{u} + (\zeta + \frac{4}{3}\eta)\ \text{graddiv}\,\dot{\mathbf{u}} - K\alpha\ \text{grad}\ T \tag{3}$$

Here $c_l = \sqrt{\dfrac{E(1-\upsilon)}{\rho(1+\upsilon)(1-2\upsilon)}}$ is the velocity of longitudinal oscillations. Because the ultrasonic oscillations are of the adiabatic type it is possible to use the known expression [4] for K

$$K = \rho c_l\ (1 - \frac{4}{3}\frac{c_t^2}{c_l^2})\frac{C_V}{C_P} \tag{4}$$

Here $c_t = \sqrt{\dfrac{E}{\rho(1+\upsilon)}}$ is the velocity of transverse elastic oscillations.

For adiabatic oscillations we can represent the temperature as a sum of the basic value T_0 and additional terms δT_0 and δT_1 connected with the wave propagation, i.e.

$$T = T_0 + \delta T_0 + \delta T_1 \tag{5}$$

For grad δT_0 we can derive from (2) in the first approximation

$$\text{grad } \delta T_0 = -\frac{T_0 \alpha K}{C_V} \text{graddiv} \vec{u} \tag{6}$$

Here we used the thermodynamic equality

$$C_P - C_V = T\alpha^2 K \tag{7}$$

For the value δT_1 we get from (2) the following equation

$$C_V \delta \ddot{T}_0 = \kappa \text{ div}\left(-\frac{T_0 \alpha K}{C_V} \text{graddiv } \dot{\mathbf{u}} \right) \tag{8}$$

$$\delta \ddot{T}_1 = c_l^2 \Delta \delta T_1 \tag{9}$$

It gives

$$\text{graddiv} \delta T_1 = -\frac{\kappa T_0 \alpha K}{C_V^2 c_l^2} \text{graddiv } \dot{\mathbf{u}} \tag{10}$$

With help of (6), (10) and (4) we can rewrite (3) in the form

$$\rho \ddot{\mathbf{u}} = \rho c_l^2 (1 + \gamma_0) \text{graddiv} \mathbf{u} + \gamma \text{ graddiv } \dot{\mathbf{u}} \tag{11}$$

where

$$\gamma_0 = \rho c_l^2 \left(1 - \frac{4}{3} \frac{c_t^2}{c_l^2} \right) \frac{C_V T_0 \alpha^2}{C_P^2}, \qquad \gamma = \zeta + \frac{4}{3}\eta + \frac{\kappa}{C_V} \rho \gamma_0 .$$

Because $\gamma_0 << 1$ we can neglect the value γ_0 in the first term of the right side of (11). But in the second term of the right side of (11) the value γ_0 must be taken into account.

The equation (11) is sufficient to solve the problem of rebuilding the original shape of acoustic signal propagating from the source to the point of the observation and registration. We study this problem for the simplest one dimensional case.

Let us consider the longitudinal ultrasonic wave propagating along x-axis from the point $x = 0$ in the direction of positive values of x to the point of the observation and registration $x = L$ (in this case $\mathbf{u}(\mathbf{r},t) = (u(x,t), 0, 0)$) where the known measured function is $u(x=L,t)$ and the task is to find $u(x=0,t)$. So, such a case represents the boundary problem that may be solved with help of the Furrier transformation of a time dependence

$$u(x,t)=\frac{1}{2\pi}\int_{-\infty}^{+\infty} d\omega e^{-i\omega t}\ u(x,\omega),\quad u(x,\omega)=\int_{-\infty}^{+\infty} dt\ e^{i\omega t}\ u(x,t), \tag{12}$$

and the Laplace transformation of a space distribution

$$u(x,t)=\frac{1}{2\pi}\int_{-\infty+i\sigma}^{+\infty+i\sigma} dke^{ikx}\ u(k,t),\quad u(k,t)=\int_{0}^{\infty} dxke^{-ikx}\ u(x,t) \tag{13}$$

The application of (12) and (13) to (11) gives for u(ω,k) the following expression:

$$u(k,\omega)=-i\frac{\left.\frac{\partial u(x,\omega)}{\partial x}\right|_{x=0}+ku(x=0,\omega)}{k^2-k_0^2\left(1+i\frac{\gamma\omega}{\rho c_l^2}\right)},\qquad k_0=\frac{\omega}{c_l} \tag{14}$$

In the case of a stress-free source when $\partial u/\partial x|_{x=0}=0$ we can write for the spectral function of the elastic signal at the point x = 0

$$u(x=0,\omega)=u(x=L,\omega)\exp(i\omega t_0+\omega^2/\Omega^2) \tag{15}$$

where $\Omega=\sqrt{\frac{\gamma}{2\rho c_l^2}}$, $t_0=L/c_l$.

The direct transformation of (15) with help of (12) to receive u(x=0,t) is out of a practical interest because of the term $\exp\{\omega^2/\Omega^2\}$ that grows exponentially with increasing of ω. It means that small mistake of measuring u(x=L, ω) gives the essential wrong distortion of shape rebuilding of u(x=0,t).

Another way to receive the real form of the original signal u(x=0,t) consists of the solution of the integral equation that may be derived from (15) with help of (12)

$$u(x=L,t+t_0)=\frac{\Omega}{4\sqrt{\pi}}\int_{-\infty}^{+\infty} dt'\ u(x=0,t')\ \exp\left(-\frac{(t-t')^2\Omega^2}{4}\right) \tag{16}$$

Here t = 0 corresponds to the moment of the appearance of the acoustic signal at the point x = 0 where the ultrasonic source is situated. As we can see from (16) the signal appears at the point x = L some times later in the moment $t=t_0$ because the time t_0 is required for it to overcome the distance L.

In practice, we measure the signal at the point x = L in some fixed moments of the time $t=t_i+t_0$ where i = 1, 2, 3, ... As ussually, the meassure is carried out automatically through the equal interval of time defined by possibilities of the detector and digital transformer of the acoustic signal. Because of that the solution of (16) may be written in the form suitable for a practical use

$$u(x=0,t-t_i)=2\sum_{n=0}^{\infty}\frac{H_n(0.5\Omega(t-t_i))}{n!\Omega^n}\left.\frac{\partial^n u(x=1,\tau)}{\partial\tau^n}\right|_{\tau=t+t_0} \tag{17}$$

Here H_n (x) is the Hermit polynom [5] of the argument x and we assume that $|t - t_i| << t_i$.

The result (17) for longitudinal waves may be derived by the similar way also for transverse waves. In this case the following expression must be used instead of γ written in (11)

$$\gamma = \eta$$

Such a simple expression takes place because of the known independence of the transverse waves propagation on the heat conductivity.

CONCLUSIONS

The problem of a rehabilitation of an original shape of acoustic signals propagating from a source through a dissipate medium to a point of observation and registration is solved in the form suitable for a practical use. The general formula derived is appropriate for every kind of oscillations with corresponding corrections concerning the wave velocity and space damping rate.

The direct application of these results described above must be began from measuring the distance between the source of acoustic signals and the point of the registration. It is possible with help of methods of the AE localization. Then the time dependence of the signal intensity must be defined by measuring the AE amplitude a few times with given interval in the time. Further investigations consists of the direct application of the general formula giving the result awaited.

REFERENCES

1. H.M.Westergaard. *Theory of Elasticity and Plasticity*. Harvard University Press, Cambridge, (1952).
2. I. S. Sokolnikoff. *Mathematical Theory of Elasticity*, McGraw - Hill Book Company, New York, (1956).
3. Y.C. Fung. *Foundation of Solid Mechanics*, Prentice - Hall, New Jersey, (1965).
4. L.D. Landau, E.M.Lifshits. *Theory of Elasticity, Nauka*, Moscow (1987) (Russia).
5. G.A. Korn, T.M.Korn. Mathematical Handbook, McGraw - Hill Book Company, New York, (1968).

NONDESTRUCTIVE EVALUATION OF THE RESIDUAL LIFE OF STEEL-BELTED RADIAL TRUCK TIRES

Henrique L.M. dos Reis and Kris A. Warmann
Department of General Engineering
University of Illinois at Urbana-Champaign
Urbana, Illinois 61801, U.S.A.

INTRODUCTION

Underinflated (80% or less than recommended operating cold inflation pressures or runflat) radial truck tires can be subjected to steel cord fatigue damage in the upper sidewall area caused by overflexing of the tire. When the tire is being serviced, often during the retreading or during remounting and inflation steps, weakened cords may break with rupture of the tire upper side wall with potential catastrophic consequences such as loss of life. This failure mode of a truck tire is usually called the *Zipper Break Mode*. Following the American Retreaders Association (ARA), *"Zipper"* is a term used in industry to describe a rupture that occurs in the sidewall flex area of a medium or light duty truck's steel belted radial tires. The rupture normally ranges from 12 inches (30.5 cm) to 36 inches (91.4 cm) around the radius of the tire, releasing an air pressure wave that can cause serious injury or even death. Some nondestructive testing devices, such as the ones based upon radioscopy, may show (but not always) this hazard condition if the cords are already broken. However, this approach is expensive and, if the cords are not yet broken, radioscopy does not evaluate the residual strength of the tire side walls, which has the potential to lead to a very serious safety hazard condition.

The structural integrity of tire casings varies so widely that no single casing can be considered to be representative of an entire group. Nondestructive testing and evaluation provides the means by which all tire casings could be tested for structural integrity. Using nondestructive testing and evaluation, each tire casing can either be accepted as is, accepted conditionally upon repair and further testing, or rejected outright. Most of the research and development of nondestructive techniques for the evaluation of pneumatic tires has been conducted during the last half of this century. Almost all of the developed techniques have been in four main areas: x-ray, holographic and shearography, infrared, and ultrasonic testing. All of these techniques have drawbacks, and while these approaches can detect some delaminations in the tire, they can not evaluate the remaining useful life of the tire casing or of a new tire[1-5]. A good literature review on the current nondestructive testing and evaluation techniques applied to tires is presented by Reis and Golko[6]. Clearly, a new nondestructive testing and evaluation approach is needed[6-7].

Analytical ultrasonics implies the measurement of material microstructure and associated factors that govern mechanical properties and dynamic response. It goes beyond flaw detection, flaw imaging and defect characterization and includes assessing the inherent properties of material environments in which the flaws reside. Acousto-ultrasonics is an analytical ultrasonic NDE technique which measures the relative efficiency of energy transmission in the specimen[8]. An ultrasonic pulse is injected with a transmitting transducer mounted on the surface of the specimen. A larger amount of damage (i.e., flaws, changes in the microstructure, etc.) in the specimen produces a higher signal attenuation, resulting in lower stress-wave-factor (SWF) readings. Traditionally, the

Nondestructive Characterization of Material VIII
Edited by Robert E. Green Jr., Plenum Press, New York, 1998

SWF has been evaluated as the number of oscillations higher than a chosen threshold in the ring down oscillations in the output signal from the receiving transducer. The stress-wave-factor does not yet have a standard definition. In this study, a stress-wave-factor is any stress wave parameter in any domain, such as the time and frequency domains, that help to characterize the acousto-ultrasonic signal. A good review of analytical ultrasonics in materials research and testing is given by Duke[8].

Using Acousto-ultrasonics, a tire with seeded defects[1-2] and a tire with one hundred road miles of use[3] were investigated by Reis and Warmann. Using the same methodology, Reis and Warmann[4] also presented and discussed the results of a blind study using seven different tires of several manufactures and construction. Furthermore, the technical specifications of a prototype instrument to evaluate the damage in radial truck tires are also presented and discussed by Reis and Golko[5-6]. Here, the results of an experimental investigation to study the possibility of using the acousto-ultrasonic approach to evaluate the residual life of a steel belted radial truck tire is presented.

EXPERIMENTAL PROCEDURE

To demonstrate the feasibility of using the acousto-ultrasonic approach for on-line inspection, eight Goodyear radial truck tires (model G167, size 11R24.5) were tested with different levels of controlled damage. Each tire was broken in by running it against a flywheel for approximately 800 miles (1287 km) under normal operating conditions. In addition to the break-in period, the tires were run through a test designed to fatigue the ply cords in the upper sidewall. In this test, the tires were run in an underinflated and overdeflected state. The tires were divided into a control group and an experimental group. The control group consisted of two tires that had been run the 800 miles (1287 km) break-in period but had not been run on the fatigue test. These two tires were provided in order to establish a baseline for comparison against the experimental group. The amount of damage, i.e., number of fatigue miles in addition to the break-in miles for each tire used in this study, is listed in Table 1.

Table 1. Tires used in this study.

Tire Number	Number of Underinflated Miles (km)
1	0
2	0
3	105 (169)
4	180 (290)
5	300 (483)
6	354 (555)
7	520 (837)
8	560 (901)

Each of the experimental tires was x-rayed in order to locate regions of ply cord damage. The x-rays taken of tire number three showed no visible signs of ply cord fatigue damage anywhere on the tire. Tire number four contains four regions with ply cord anomalies -- two on the nonserial and two on the serial side of the tire. The first region of tire number four has about 1.5 in (38 mm) of ply cords with obvious fretting, but none are completely severed; the second region has a slightly irregular ply cord spacing; the third region has about two inches of fretted cords with some completely severed, and the fourth region has a slight amount of fretting evident on only a few ply cords. Tire number five shows no visible damage using x-rays. Tire number six did not show any damage using x-rays. However, five indications of possible weakened areas were detected using infrared thermography during the fatigue test; these five areas were also detected using the acousto-ultrasonic approach as discussed by Reis and Warmann[1], which shows that acousto-ultrasonic approach is capable of detecting damage before it becomes visible with x-ray inspection. The x-ray

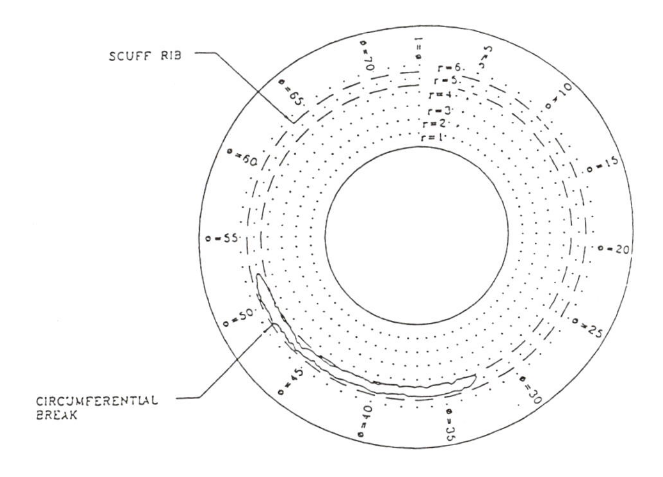

Figure 1. Typical coordinate system affixed to each tire sidewall

results of tire number seven, indicated one area approximately two inches (51 mm) long on the nonserial side of the tire with cords that had started to buckle but were not frayed; this region was clearly identified with the acousto-ultrasonic approach along with a second region that appeared to be weakening. The x-ray results also identified another damaged region on the serial side of this tire which contains approximately one inch (25.4 mm) of frayed cords. Tire number eight contains two regions with severe damage; the first has many fretted and severed cords and the second contains many moderately fretted ply cords. A complete description of these tires including radiographs of the damaged areas is presented and discussed by Reis and Warmann[1].

As shown in Figure 1, a polar coordinate system was affixed to each sidewall of each tire to enable accurate transducer positioning. The radial dimension was marked in distances of inches from the inside diameter of the tire, while the angular dimension was marked in five degree increments for a total of seventy-two angular locations. The tire was sampled by changing the angular position while keeping the radial position of the transducers fixed. A polar grid was affixed to the tire using "r" as the radial distance and "ϕ" as the angular position. Each tire was tested by

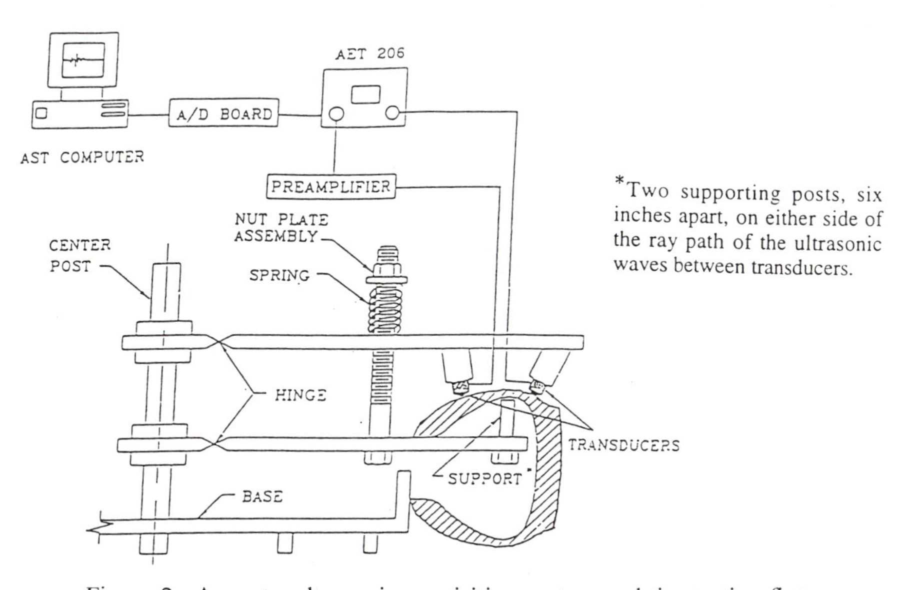

Figure 2. Acousto-ultrasonic acquisition system and tire testing fixture

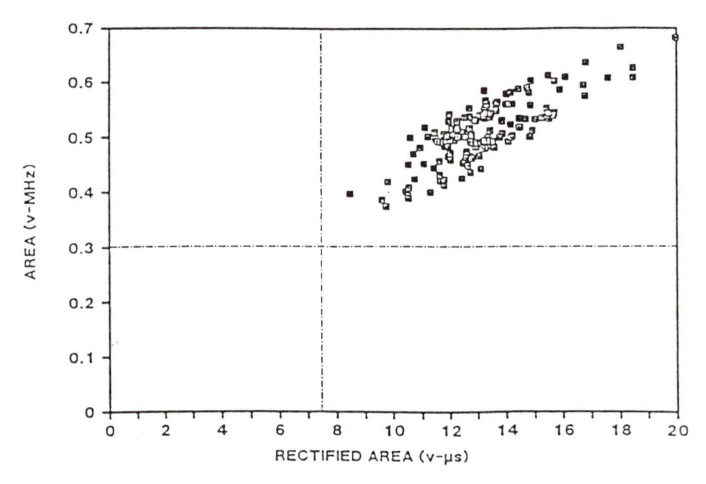

Figure 3. Area under the power spectral density curve vs. the rectified area in the time domain for tire number one

collecting five averaged waveforms at each of the seventy-two angular positions for a total of three hundred and sixty averaged waveforms per tire side wall. In Figure 1, the circumferential break on the tire sidewall represents a typical tire failure induced by steel ply cord fatigue damage, i.e., *"Zipper"* mode of failure.

Acousto-ultrasonic waveforms were collected from the tire using the data acquisition system shown in Figure 2. The system consisted of a portable Acoustic Emission Technology (AET) model 206 Portable Acousto-Ultrasonic Stress Wave Analyzer, preamplifier, two transducers, and an AST portable computer equipped with a Sonix STR*825 analog-to-digital converter board. Both the transmitting and the receiving transducers were a broad-band AET model FC-500 with a flat sensitivity of approximately -85 dB (relative to 1V/μbar) from 0.1 MHz to 3 MHz. Each transducer was fitted with a waveguide, reducing its contact surface to a circle with a diameter of 0.25 in (6.4 mm). The transmitting transducer was excited by a -250 V pulse generated by the AET 206 AU unit, which was set at a pulsing rate of 500 pulses per second. The output of the receiving transducer was amplified by 60 dB by the AET model 160B preamplifier containing a filter with a passband between 30 kHz and 2 MHz. The output signal from the preamplifier was amplified another 65 dB in the AET 206 for a total gain of 125 dB. The analog output from the AET 206 was digitized by the Sonix STR*825 analog-to-digital converter using a sampling rate of 1.563 MHz. To reduce the noise level, at each location sixteen digitized signals were averaged to create a single digitized waveform which was stored on a floppy disk for later analysis. Five averaged acousto-ultrasonic waveforms were collected at each of the five degrees increment locations around the tire for a total of three hundred and sixty waveforms. Because the ply cord fatigue damage was known to occur at about the same radial distance as the scuff rib, the pulsing transducer was placed approximately 0.25 in (6 mm) above the scuff rib for a total transducer separation distance of 3 in (76 mm).

EXPERIMENTAL RESULTS AND DISCUSSION

Each of the eight tires was tested and evaluated using the methodology described by Reis and Warmann[1-3]. In order to rank the tires from the least damaged to the most damaged, see Table 1, a combined stress-wave-factor (SWF) was used. This combined stress-wave factor was calculated using the frequency of maximum amplitude, rectified area, signal energy, maximum amplitude in the time domain, and the area under the spectral density curve[1-3]. This five-dimensional damage decision space was used to quantify the damage state of each tire; as the tire wears and damage accumulates, the cluster of data points moves closer to the origin as it is shown in Figures 3 and 4. In order to quantify the state-of-damage, the distance between the centroid of the averaged seventy-two data points and the origin of the damage decision space was calculated and used as a combined stress-wave-factor. Prior to calculating the combined SWF, each of the SWF was normalized with

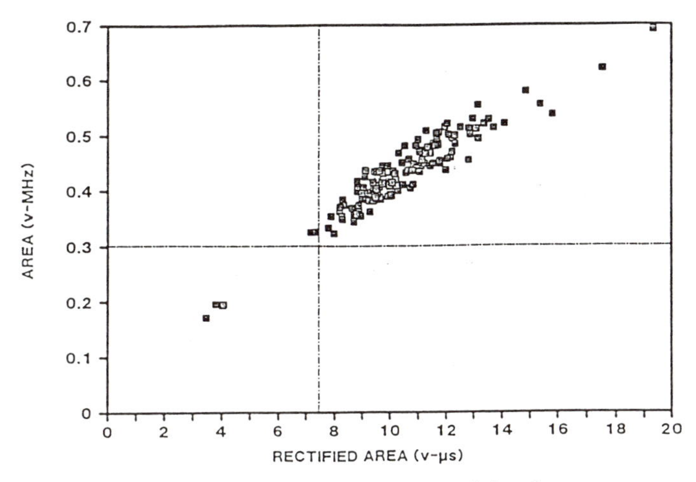

Figure 4. Area under the power spectral density curve vs. the rectified area in the time domain for tire number four

respect to the threshold level of that particular SWF; this normalization gives equal weight to each SWF rather than biasing the centroid toward the SWF with the largest absolute value. For each stress-wave-factor, the corresponding threshold level was obtained from the base line measurements of a new tire.

Figure 3 shows the area under the power spectral density curve plotted against the rectified area in the time domain for tire number one, i.e., a tire that was broken-in but has no damage caused by being subjected to run underinflated. Figure 4 shows the area under the power spectral density curve plotted against the rectified area in the time domain for tire number four. In both Figures 3 and 4, the threshold lines, i.e., the dash-dot lines, separate the damaged from the non-damaged areas. Figure 3 shows a normal distribution of a tire, while Figure 4 shows how the centroid of the data points is used to estimate the remaining life of the tire and to evaluate if local areas of the tire are severely damaged, i.e., have crossed the threshold lines.

Figure 5 shows the normalized centroid distance averaged over the whole tire as a function of the mileage on the tires. The linear regression results give an R value of 0.871 which indicates a relatively high correlation between the average centroid distance and the number of mileage in the tire. Figure 6 shows the normalized centroid distance averaged over each side, i.e., the serial (S)

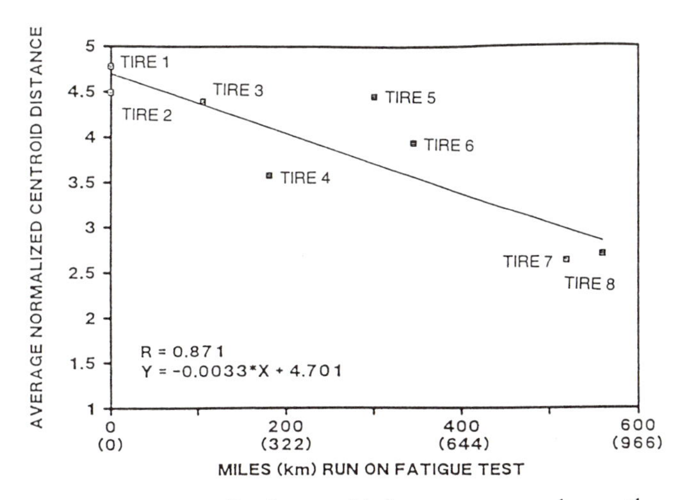

Figure 5. Normalized centroid distance averaged over the whole tire as a function of the tire mileage

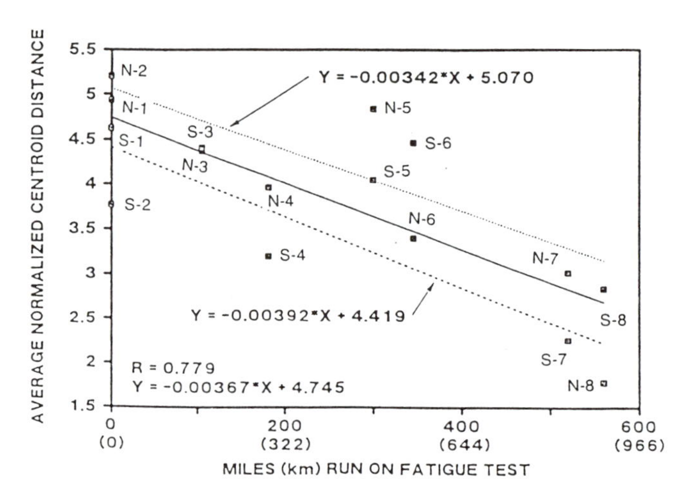

Figure 6. Normalized centroid distance averaged over each side of the tires as a function of the tire mileage

and nonserial (N) side of the tires. The solid line shows the best fit line through all of the data points. The dotted line is the best fit line through only the highest data points from each tire and the dashed line is the best fit line through only the lowest data points from each tire. The slope of the best fit line through only the lowest point on each tire is slightly higher than the slope of the best fit line through the highest point on each tire. This supports the idea that one side of the tire tends to accumulate damage at a faster rate than the other side of the tire.

CONCLUDING REMARKS

The experimental results show that the acousto-ultrasonic technique could be used to identify and evaluate the severity of local defects as well as estimate the residual life of a tire casing. Local defects were located by setting minimum acceptable threshold levels for the five stress wave factors; if any SWF value fell below the predetermined threshold level, the tire would be discarded. The severity of the defects was determined by two different methods. The first method was based solely on the magnitude of the SWF at a given location; lower SWF values indicated an increase in the severity of the defects. The second method takes into consideration the relative size of the indication in addition to the magnitude of the SWF. The ability of the AU approach to quantify the severity of the damage is a distinct advantage of this approach over other methods such as radiography. By quantifying the severity of defects, some less severe defects could be easily repaired rather than scrapping the tire casing. The residual life of the tire casing was evaluated by using a combination of five stress-wave-factors. This combined SWF had a correlation of 0.871 when it was regressed against the number of miles that the tires had been run. Furthermore, some tires showed considerable difference in their SWF readings on opposite sides of the same tire which could potentially be used as a sign of a tire that is susceptible to an early failure.

ACKNOWLEDGMENTS

The authors are very grateful to Dr. Thomas G. Ebbott of the Goodyear Tire and Rubber Company in Akron, Ohio, for his support.

REFERENCES

1. H.L.M. dos Reis, and K.A. Warmann, "Nondestructive Evaluation of Damage in Steel Belted Radial Truck Tires," Technical Report UILU ENG 91-3003, University of Illinois, Urbana, Illinois, 1991.
2. H.L.M. dos Reis, K.A. Warmann, "Nondestructive evaluation of damage in steel belted radial tires using acousto-ultrasonics," *Journal of Acoustic Emission,* Vol. 11, No. 3, 1993, pp. 107-115.

3. H.L.M. dos Reis, K.A. Warmann, "Acousto-ultrasonic nondestructive evaluation of fatigue damage in steel belted radial tires," INSIGHT -- *Non-Destructive Testing and Condition Monitoring,"* Vol. 36, No. 12, 1994, pp. 958-963.
4. H.L.M. dos Reis, Warmann, K.A., "Acousto-ultrasonic damage evaluation in steel belted radial tires," *Nondestructive Characterization of Materials YII*, A.L. Bartos, R.E. Green, and C.O. Ruud, Eds., Transtec Publications Ltd., Zurich, Switzerland, 1996, pp. 671-678.
5. H.L.M. dos Reis, and P.J. Golko, "A prototype instrument to evaluate fatigue damage in the side walls of steel belted radial truck tires," To appear in Tire Science and Technology, The Tire Society, Akron, OH, 1997.
6. H.L.M. dos Reis, and P. J. Golko, "Prototype Acousto-Ultrasonic Machine for Nondestructive Damage Evaluation/Characterization of Steel Belted Radial Tires," Technical Report UILU ENG 97-3001, University of Illinois, Urbana, Illinois, 1997.
7. "Proceedings of the Fourth Symposium on Nondestructive Testing of Tires," P.E.J. Vogel, ed., Army Materials and Mechanics Research Center, Watertown, Mass., 1978.
8. "Acousto-Ultrasonics--Theory and Application," J.C. Duke, Jr., ed., Plenum Press, New York, N. Y., 1989.

APPLICATION OF THE PSEUDO WIGNER-VILLE DISTRIBUTION TO THE MEASUREMENT OF THE DISPERSION OF LAMB MODES IN GRAPHITE/ EPOXY PLATES

W. H. Prosser[1], M. D. Seale[2], and B. T. Smith[3]

[1]MS 231
NASA Langley Research Center
Hampton, VA 23681-0001

[2]Wheeling Jesuit University
316 Washington Ave.
Wheeling, WV 26003

[3]Norfolk Academy
1585 Wesleyan Drive
Norfolk, VA 23502

INTRODUCTION

Acoustic waves propagate in thin plates as guided or Lamb modes. The velocities of these modes are dispersive in that they depend not only on the material elastic properties and density, but also on the frequency. Accurate characterization of Lamb wave dispersion is important in many acoustic based nondestructive evaluation techniques. It is necessary for ultrasonic measurements in thin plates to determine elastic properties and for flaw detection and localization. In acoustic emission (AE) testing, if not taken into account, highly dispersive Lamb mode propagation can lead to large errors in source location[1].

In this study, the pseudo Wigner-Ville distribution (PWVD) was used for measurement of group velocity dispersion of Lamb waves in a unidirectional graphite/epoxy (AS4/3502) laminate. The PWVD is one of a number of transforms which provide a time-frequency representation of a digitized time series. Broad band acoustic waves were generated by a pencil lead fracture (Hsu-Neilsen source) and were detected with broad band ultrasonic transducers. The arrival times for the lowest order symmetric (S_0) and antisymmetric (A_0) Lamb modes were determined from measurements of the time at which the respective peak amplitudes occurred in the PWVD. Measurements were made at several source-to-detector distances and a least squares fit used to calculate the velocity. Results are presented for propagation along, and perpendicular to, the fiber direction. Theoretical dispersion curves were also calculated and a comparison between theory and experiment demonstrates good agreement.

The PWVD and other time-frequency analyses offer two advantages for dispersion measurements in comparison to other techniques. First, only a single measurement of a broad band signal is required to determine the velocity over a wide range of frequencies. Narrow band measurement techniques such as continuous wave resonant or tone burst pulse methods only provide measurements at a single frequency and require a large number of measurements to map out dispersion curves. The Fourier phase method is a velocity measurement technique that overcomes this problem using a broad band excitation[7,8]. The phase difference, as a function of frequency, between signals recorded at different distances of propagation is used to calculate dispersion. A difficulty arises in calculating the true phase which must be "unwrapped" to remove discontinuities when the calculated phase exceeds its $\pm\pi$ limits. A further problem with this and the previously mentioned narrow band techniques occurs when multiple propagating modes and/or reflections are superimposed in the time domain signal.

The second advantage of using time frequency analysis for dispersion measurement is that often the multiple modes or reflections can be separated in time-frequency space and such signals can still be analyzed. The reflection coefficient method can also analyze multiple mode signals and has been successfully used in Lamb wave dispersion measurements[2-4]. However, measurements must be repeated at multiple reflection angles and the plate must be immersed. The two-dimensional Fourier transform method[5] can be used on signals containing multiple modes. In this approach, signals are detected at closely spaced intervals. A two-dimensional Fourier transform is then applied which is used for phase velocity dispersion calculations. A large number of measurements are needed to avoid aliasing when Fourier transforming from the spatial domain. Other spectral estimation techniques such as Prony's method can be used to reduce this requirement for such high spatial sampling[6].

THEORY

The general equation of a time-frequency distribution, w(t,ω) for signal s(t) is given by

$$w(t, \omega) = \frac{1}{2\pi}\iiint e^{-i\theta t - i\tau\omega - i\theta u}\phi(\theta, \tau)s^*\left(u - \frac{\tau}{2}\right)s\left(u + \frac{\tau}{2}\right)du\, d\tau\, d\theta \qquad \text{Eq. 1}$$

where the integrals are evaluated from $-\infty$ to ∞. In this equation, s* is the complex conjugate and $\phi(\theta,\tau)$ is an arbitrary function known as the kernel. Time and frequency are represented by t and ω respectively[9]. For the Wigner distribution[10-11], the kernel function has a value of 1. Substituting and applying to a sampled signal yields

$$w(m\Delta t, k\Delta\omega) = 2\Delta t \sum_{n=0}^{2N-1} s[(m+n)\Delta t]s^*[(m-n)\Delta t]e^{\frac{-i2\pi nk}{2N}} \qquad \text{Eq. 2}$$

where Δt is the sampling interval and $\Delta\omega = \pi/(2N\Delta t)$. The Wigner distribution has several characteristics which limit its usefulness. The first is a higher sampling requirement to avoid aliasing. The sampling frequency must be four times that of the highest frequency content of the signal as opposed to the usual Nyquist criteria of only twice the highest frequency[12]. Additionally, the Wigner distribution produces complicated and unexpected results when more than one frequency component is contained in a signal. The Wigner distribution may contain signal amplitude at frequencies and times not actually contained in the original time signal. This artifact is caused by interference consisting of cross terms in the distribution from the multiple frequency components. It can make interpretation of results very difficult. The final undesirable characteristic of the Wigner distribution is that it may have negative values. These do not have physical meaning and most often occur as a result of this interference.

An approach to overcome the more restrictive sampling requirement was formulated by Ville[13]. The Wigner-Ville distribution is calculated using analytic signal where the imaginary component is the Hilbert transform of the original time signal. The requirement to avoid aliasing is then reduced to that of the Nyquist criteria. An intuitive rational for this approach is based on the fact that a single sample of the analytic signal provides two effective samples (the real and imaginary parts) of the original signal[12]. Convolving a smoothing window with the distribution is used to minimize the effects of interference terms in the distribution and eliminate negative values. Smoothing, which results in the pseudo Wigner-Ville distribution, emphasizes deterministic components and reduces those due to interference[11].

EXPERIMENT

Dispersion measurements were performed on a 16 ply unidirectional plate of AS4/3502 graphite/epoxy. The nominal plate thickness was 2.26 mm with lateral dimensions of 0.508 m. along the fiber direction (0 degree direction) and 0.381 m. along the 90 degree direction. Signals were generated by fracturing a 0.5 mm. diameter pencil lead on the surface of the plate (Hsu-Neislen source). This source mechanism produces broad band, transient acoustic waves and is often used to simulate acoustic emission.

The acoustic waves were detected by a 3.5 MHz ultrasonic sensor (Panametrics V182) which has a diameter of 1.27 cm. This sensor is heavily damped. It was operated far off resonance in detecting these signals which had maximum frequency contents below 500 kHz. At these frequencies, this sensor provides a flat frequency, displacement sensitive response. The signals were digitized at a sampling frequency of 10 MHz with 12 bit vertical resolution (Digital Wave Corporation F4012). The 2048 point waveforms were padded with 1024 zeros both in front of, and after the signal to increase the frequency resolution of the PWVD.

To improve accuracy, seven measurements were taken at different distances of propagation in 1.27 cm. increments over a range of 8.89 cm. to 16.51 cm. A least squares fit of arrival time versus distance was used to calculate the velocity. The source position was kept fixed for all measurements and the receiver was moved to different positions. Waveform acquisition was triggered by a narrow band, 150 KHz resonant sensor (Physical Acoustics Corporation R15) positioned adjacent to the source.

The original Fortran source code for the PWVD calculation from Jeon and Shin [11] was modified for these dispersion measurements. The maximum number of points in the input signal and the maximum number of points in the calculated distribution were increased to provide enhanced time and frequency resolution. Peak detection routines were added to determine the arrival times of the two Lamb modes. A time separating the latest arrival of the S_0 mode and earliest arrival of the A_0 mode was determined the time domain signals and input into the program. At each frequency, the peak in the distribution before this input time value were used as the basis for the S_0 arrival times. The distribution peak after the input time value was used for the A_0 arrival time. A seven point cubic spline fit was used to improve the resolution in determining the peak amplitudes (i.e. arrival times) of the two modes.

The PWVD results were displayed as grey-scale images. The grey level at a given x and y point in the image represents the amplitude of the distribution at a particular time and frequency. There was a large amplitude difference between the two modes. To more clearly show the two modes in the same image, the values of the distribution were compressed by taking the 4'th root which allowed adequate visualization of the data.

RESULTS AND DISCUSSION

Images of the PWVD along with the corresponding time domain signals for two distances of propagation are shown in Figure 1. The large dispersion of the A_0 mode is clearly seen in both the time domain signals and the PWVD images. Higher frequencies arrive at earlier times in this mode and the signal changes shape as it propagates longer distances. Lit-

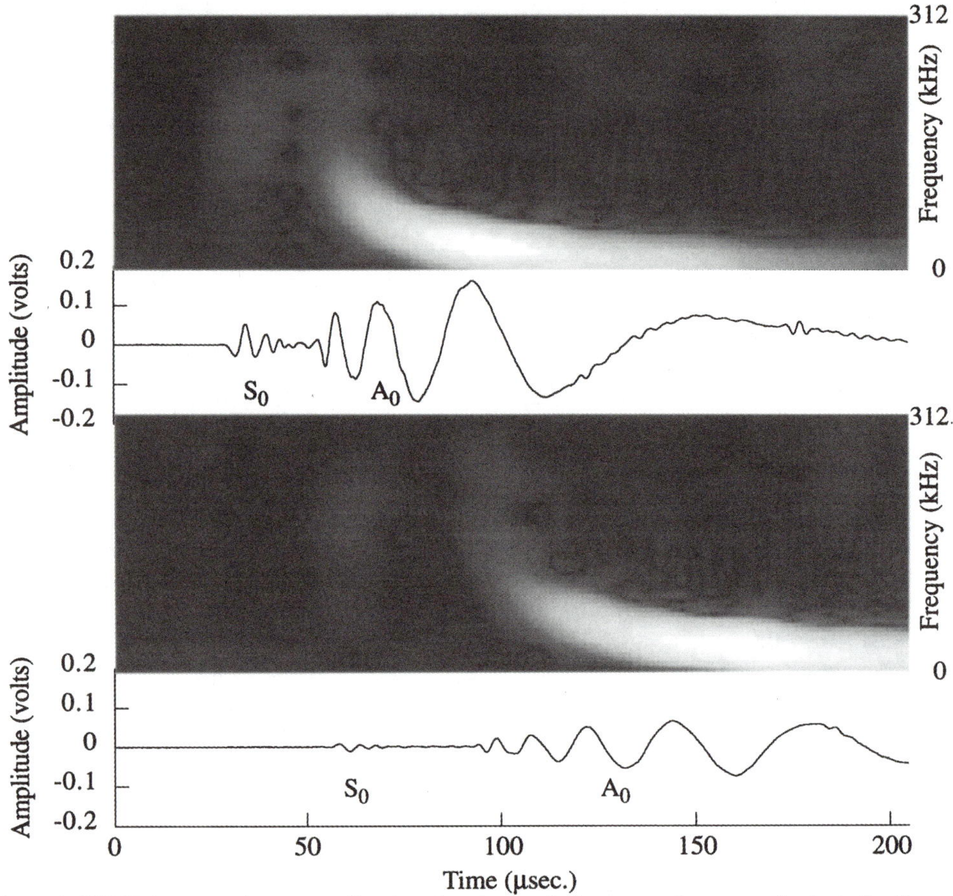

Figure 1. PWVD images and corresponding time domain signals for propagation perpendicular to the fiber direction at distances of 8.89 cm. (upper) and 15.24 cm. (lower).

tle dispersion is seen in the S_0 mode.

The measured dispersion results were compared with theoretical group velocity curves based on a finite element model[14,15]. Nominal elastic properties for a single lamina were obtained from the manufacturer and used in this calculation. These values are shown in Table 1 where the coordinate reference system is such that the 1 axis is along the fiber direction. The nominal density value of 1550 kg/m^3 was used.

The measured Lamb mode dispersion results for propagation perpendicular to the fiber direction are shown in Figure 2. The uncertainties in the measured velocity values as determined from the uncertainty of the slope in the least squares fit are displayed as error bars on the measured data. Comparison with theoretical curves show good agreement, particularly for the A_0 mode. The measured values for the S_0 mode are slightly higher than predicted. Similar results were obtained in phase velocity measurements in this material in an earlier study[8]. A possible explanation for this slight discrepancy is that the material properties obtained from the manufacturer and used in the theoretical calculations are somewhat different from those of the actual material. Material property variations are not uncommon in graphite/epoxy and may be caused by fiber volume variations, differences in cure processing conditions, and variations in resin chemistry.

The results for propagation along the fiber direction are shown in Figure 3. Comparison with theoretical curves shows good agreement for the A_0 mode. The measured values of S_0 mode are slightly lower than predicted along this direction. These results are again consistent

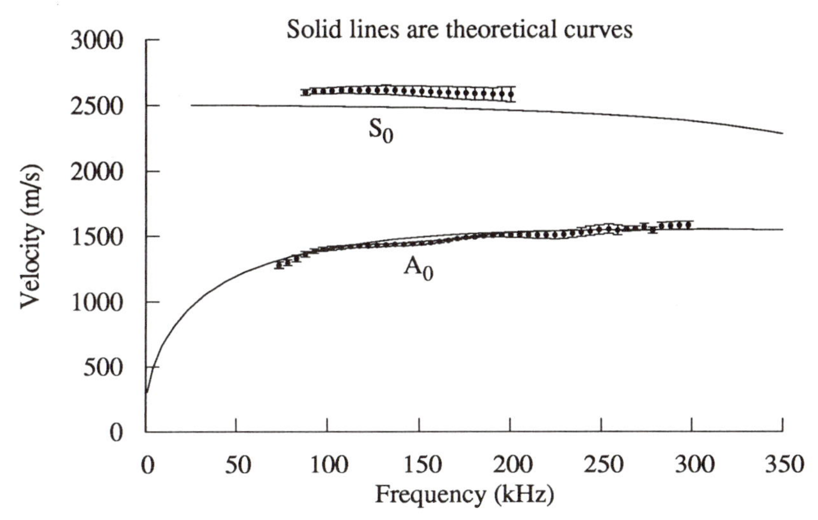

Figure 2. Measured Lamb mode group velocities along 90 degree propagation direction and comparison with theoretical predictions.

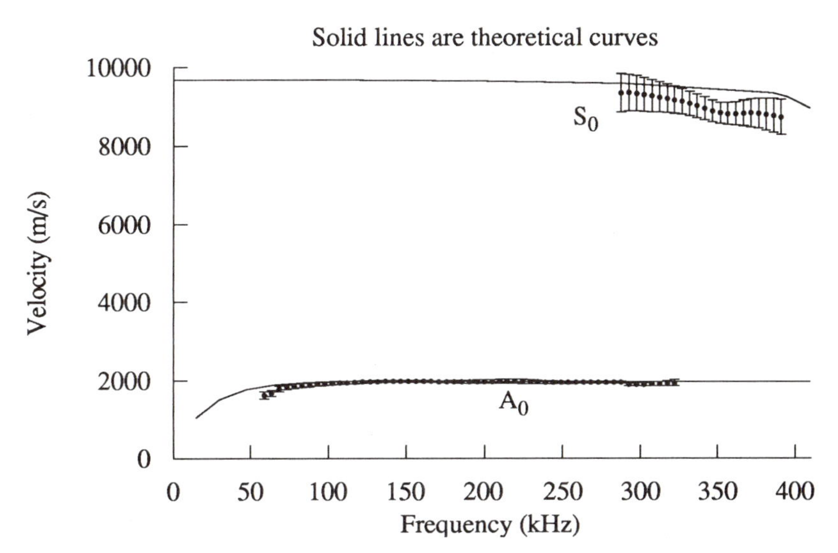

Figure 3. Measured Lamb mode group velocities along 0 degree propagation direction and comparison with theoretical predictions

with the previous phase velocity measurements.

In summary, a Pseudo Wigner-Ville distribution analysis has been used to measure the group velocity dispersion of Lamb waves in a unidirectional graphite/epoxy laminate. One advantage of this method is that the dispersion over a wide range of frequencies can be determined from a single measurement. Multiple measurements were used in this study to provide increased accuracy. Another advantage of time-frequency methods is that they can be used for signals which contain multiple modes. Results are presented for measurements both along, and perpendicular to the fiber direction. Comparison with theoretical predictions shows good agreement. The slight discrepancies between theory and experiment for the S_0 mode velocities are consistent with previous phase velocity measurements in this material.

Table 1. Nominal elastic moduli of AS4/3502 used in theoretical dispersion calculations.

Elastic Modulus	(GPa)	Elastic Modulus	(GPa)
c_{11}	147.1	c_{22}	10.59
c_{12}	4.11	c_{55}	5.97
c_{23}	3.09		

REFERENCES

1. S. M. Ziola and M. R. Gorman, Source location in thin plates using cross-correlation, *JASA* 90:2551-2556 (1991).
2. A. K. Mal, C. -C. Yin, and Y. Bar-Cohen, The influence of material dissipation and imperfect bonding on acoustic wave reflection from layered solids, *Review of Progress in Quantitative Nondestructive Evaluation* 7B:927-934 (1988).
3. D. E. Chimenti and A. H. Nayfeh, Leaky lamb waves in fibrous composite laminates, *J. Appl. Phys.* 58:4531-4538 (1985).
4. K. Balasubramaniam and J. L. Rose, Physically based dispersion curve feature analysis in the NDE of composites, *Res. in NDE* 3:41-67 (1991).
5. D. Alleyne and P. Cawley, A two-dimensional fourier transform method for the measurement of propagating multimode signals, *JASA* **89:**1159-1168 (1991).
6. S. M. Kay and S. L. Marple, Jr., Spectrum analysis - a modern perspective, *Proceedings of the IEEE* 69:1380-1419 (1981).
7. W. Sachse and Y. H. Pao, On the determination of phase and group velocities of dispersive waves in solids, *J. Appl. Phys.* **49:**4320-4327 (1978).
8. W. H. Prosser and M. R. Gorman, Plate mode velocities in graphite/epoxy plates, *JASA* 96:902-907 (1994).
9. L. Cohen, Time-frequency distributions - a review, *Proceedings of the IEEE* 77:941-981 (1989).
10. E. Wigner, On the quantum correction for thermodynamic equilibrium, *Physics Review* 40:749-759 (1932).
11. J. Jeon and Y. S. Shin, Pseudo Wigner-Ville Distribution, computer program and its applications to time-frequency domain problems, *Naval Postgraduate School Report NPS-ME-93-002* (1993).
12. B. Boashsah, Note on the use of the Wigner Distribution for time-frequency signal analysis, *IEEE Transactions on Acoustics, Speech, and Signal Processing* 36:1518-1521 (1988).
13. J. Ville, Theorie et applications de la notion de signal analytique, *Cables et Transmission* 2:61-74 (1948).
14. S. B. Dong and K. H. Huang, Edge vibrations in laminated composite plates, *J. Appl. Mechs.* 52:433-438 (1985).
15. S. K. Datta, A. H. Shah, and W. Karunasena, Wave Propagation in Composite Media and Material Characterization, *Elastic Waves and Ultrasonic Nondestructive Evaluation,* S.K. Datta, J.D. Achenbach, and Y.S. Rajapakse, eds., Elsevier Science Publishers B.V., North-Holland (1990).

STOCHASTIC MICROFRACTURE PROCESS ANALYSIS OF SiC PARTICLE DISPERSED GLASS COMPOSITES BY ACOUSTIC EMISSION

Manabu Enoki, Hiroki Fujita, and Teruo Kishi

Research Center for Advanced Science and Technology
The University of Tokyo
4-6-1 Komaba, Meguro-ku, Tokyo 153, Japan

INTRODUCTION

Fracture of ceramics is essentially brittle and a probabilistic phenomenon. Parameters for fracture such as fracture strength, fracture location and fracture time should be treated statictically or stochastically, and scattering and volume effect of fracture strength have been explained by Weibull distribution which is a main statistic distribution[1]. However, the fracture phenomenon which controls the final strength and fracture is intrinsically cumulative and should be treated stochastically[2]. An integrated result of microfractures before the final fracture demonstrates the final strength of final toughness. The final strength has been treated statistically, but the probabilistic investigations for microfracture have not been reported.

Acoustic emission (AE) is a phenomenon due to microfractures before the final fracture and many studies have been reported to understand the fracture process of various materials. It was reported that some ceramics were affected by environment and demonstrated the stress rate dependence on strength [3]. However, the most of these studies focused on the final strength only and the microfracture process was not so well considered.

In this paper, we consider two models for microfracture; microfracture occurs independently before the final fracture, and a crack propagates from a microfracture as an origin. We obtain the probability distribution functions for microfracture stress and for microfracture location. The SiC particle dispersed borosilicte glass composites were used in experiments. The four point bending tests were carried out in air by using the specimens of 3 by 4 by 40 mm. Acoustic emission sensors were attached at the both ends of specimen and AE waveforms of two channels were recorded. One dimensional locations of AE signals were analyzed. Microfracture process of composites was analyzed by these models. The analysis of microfracture stress and location from AE data demonstrates that the increase of volume fraction of SiC particle changes the microfracture process of the materials. The relation between strength and microfracture process has been discussed from the analysed results.

MODELS FOR MICROFRACTURE PROCESS

It is demonstrated by the acoustic emission results that microfracture occurs before the final fracture even in ceramics like brittle materials, not sudden fracture[4]. We consider two models for microfracture, (1) microfracture occurs independently before the final fracture, (2) a crack propagates from a microfracture as an origin. These models are ideal ones and real microfracture process can be considered to be a mixed model of these two models.

Independent microfracture model

We assume that microfracture occurs independently before the final fracture ant each microfracture is subject to the Weibull distribution[5, 6]. The probability density function for microfracture, $f_1(\sigma_m, x)$ can be represented as,

$$f_1(\sigma_m, x) = \exp\left\{-\int\left(\frac{\sigma}{\sigma_0}\right)^m dx\right\}\frac{\partial}{\partial\sigma_m}\left(\frac{\sigma}{\sigma_0}\right)^m \tag{1}$$

where σ_m is the reference stress, x is location, m is a scale parameter of Weibull distribution, σ_m is a shape parameter of Weibull distribution. The probability density functions for microfracture stress and microfracture location can be represented from the integrals of equation (1). If we consider the stress field during the four point bending, the probability distribution functions $G_1(\sigma_m)$ for microfracture stress and $H_1(x)$ for microfracture location is given by[9],

$$G_1(\sigma_m) = 1 - \exp\left\{-2b\left(x_1 - \frac{mx_0}{m+1}\right)\left(\frac{\sigma_m}{\sigma_0}\right)^m\right\} \tag{2}$$

$$H_1(x) = \begin{cases} \dfrac{x_0}{2c}\left(\dfrac{x}{x_0}\right)^{m+1} & (0 \le x \le x_0) \\ \dfrac{1}{2c}\{x_0 + (m+1)(x - x_0)\} & (x_0 \le x \le 2x_1 - x_0) \\ \dfrac{1}{2c}\left\{2c - x_0 + \dfrac{x_1^{\,m+1} - (2x_1 - x)^{m+1}}{x_0^{\,m}}\right\} & (2x_1 - x_0 \le x \le 2x_1) \end{cases} \tag{3}$$

where $2(x_1 - x_0)$ means the upper span, $2x_1$ means the lower span and σ_m is the maximum tensile stress.

Crack propagation model

We consider that a crack propagates from an origin at a stress, then the probability distribution functions for microfracture stress and location can be represented as, respectively,

$$G_2(\sigma_m) = u(\sigma_m - \sigma_f) \tag{4}$$

$$H_2(x) = u(x - x_f) \tag{5}$$

where σ_f is a fracture stress, x_f is a fracture location, u() means the Heaviside unit function.

EXPERIMETAL PROCEDURE

In this study $PbO-SiO_2-B_2O_3-Al_2O_3$ glass was chosen as matrix glass and SiC was chosen as dispersed ceramics particle. Because thermal expansion constants of $PbO-SiO_2-B_2O_3-Al_2O_3$ glass and SiC are almost same and the difference of elastic modulus is large. The average size of SiC particle were about 8μm and 50μm and the volume fraction of SiC particle were 0, 10, 20 and 30%. The glass powder and SiC particle were mixed by ball-milling in methanol and were dried in air. The powder was sintered by hot pressing under the following conditions. The hot pressing temperature was 630°C that was 30°C higher than softening point and the pressure was 25.5MPa, the sintering time was 30 minutes in argon gas atmosphere. The sintered samples was performed X-ray diffraction (XRD) analysis and density was measured by Archimedes method and elastic modulus by ultrasonic method.

Samples were cut to 3×4×40 mm specimens and were polished by 1μm diamond slurry. The bending strength was measured according to JIS of four-point bending tests. The cross head speed was 0.5 mm/min., and upper span was 10 mm, lower span was 30 mm. Both bending strength test and fracture toughness test were performed in air, in order to adjust the effect of stress corrosion, we controlled the humidity of experimental room from 50% to 65%. Fracture surfaces after four-point bending tests were observed by scanning electron microscope (SEM). AE measurements were performed during both strength tests and fracture toughness tests. The AE sensors used in this study was wide range response type (M304A, Fuji Ceramics, Japan) and the AE measuring system was DCM120E (JT, Japan), respectively. The AE sensors were attached to both ends of the specimens. Figure 1 shows the AE measuring system and measuring condition. The one dimensional AE location was carried out by solving the equations for the arrival time difference of AE signals.

RESULTS

Four point bending strength

The bending strength as a function of volume fraction of particle was measured. The bending strength increased with volume fraction of particle. The bending strength as a function of particle size was also measured. The bending strength decreased with increasing of particle size.

The flat fracture surface that was typical fracture of glass and holes that was the evidence of crack deflection or bowing by dispersion of SiC particle were clearly observed. The fracture origin was also observed clearly in the sample of low volume fraction. With the increase of the volume fraction of SiC particle the fracture surface become rough. We can not distinguish the part of glass fracture from the part of crack deflection or bowing in the sample of high volume fraction because the fracture surface was like that of polycrystal materials.

Locations of microfracture

The one-dimensional source location with two channel AE system was carried out. The location of each source event is determined by measuring the differences in the wave arrival time between two transducers. We can represent the general equation for source location and the least-square method can be used to solve this equation for the one dimensional source location. The results of one-dimensional location and stress level of AE generation in figures. Figure 2 shows the one dimensional AE locations during four-point bending tests. The horizontal axis means longitudinal direction of the bending specimen. The zero point represents the end of the lower span. The vertical axis means the stress of AE

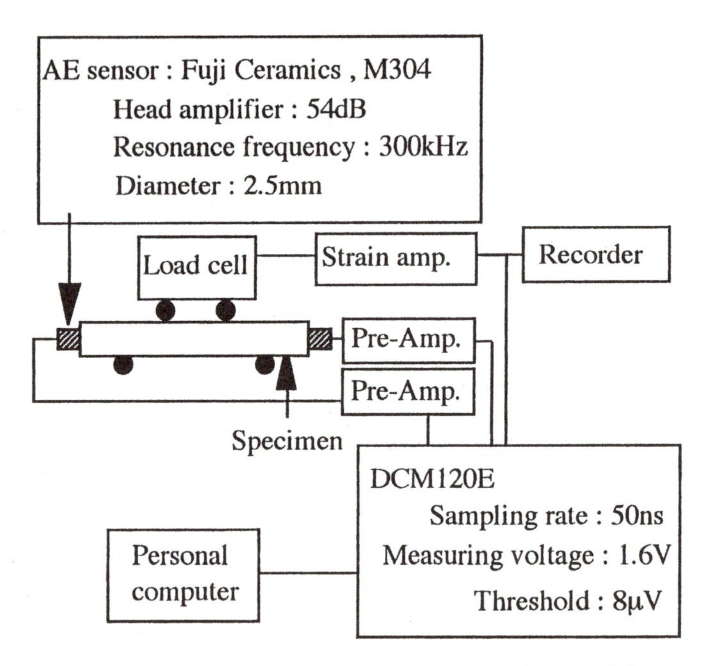

Figure 1 AE measuring system and measuring condition.

generation. Figure shows the AE behaviors of the glass, composites reinforced with 10vol%, 20vol% and 30vol% SiC particle, respectively. The total number of AE events was very few among all materials. For example, there were two events in glass and one of them was the signal from final fracture. There were about thirty events even about in composite reinforced 30vol% SiC particle. So we can consider the composite in this study also revealed an unstable fracture that a crack propagates in a moment as soon as it occurs.

Analysis by microfracture process model

The experimental results of relations between micro-fracture stress and microfracture location were analyzed by equations (2) - (5). Figure 3 shows the probability distributions of microfracture stress and microfracture location. The median rank method was used to plot the data. The results of low volume fraction materials is corresponding to the crack propagation model and in these materials fracture occurs imidately after the first microcrack generates. The results of high volume fraction materials can be considered to be a mixture type of microfracture process of two models.

DISCUSSION

About glass composite, main mechanisms of strengthening are due to surface compressive residual stress and due to load transfer or load sharing. Strengthening by surface compressive residual stress is the mechanism that residual stress at nearly surface of materials, which were introduced by physical or chemical methods, close the crack and disturb crack growth. Strengthening by load transfer is the mechanism that ceramics fibers or particle which have higher elastic modulus than glass matrix share and decrease the stress added to glass matrix. In this study materials were not performed any special treatment for surface compressive residual stress. And there was little thermal residual stress and no crystallization

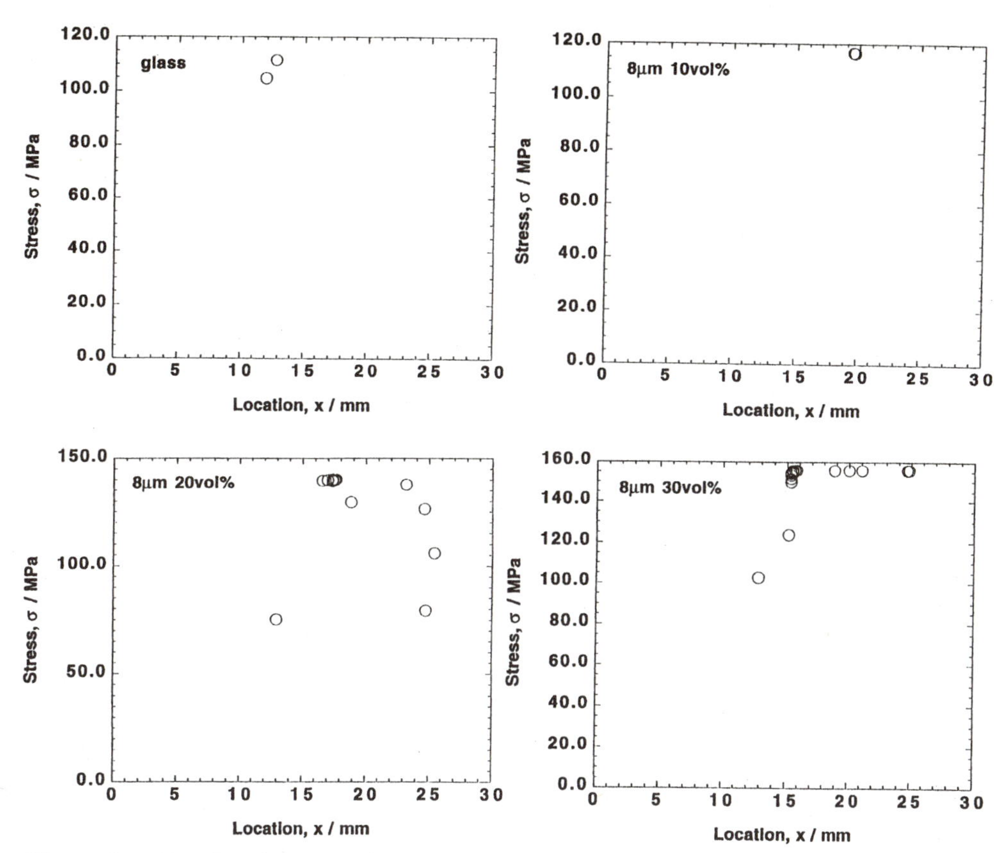

Figure 2 Results of the one dimensional AE locations during four-point bending tests.

in this composite. The composites included the same initial defects because all materials in the present study were almost full density and composites had the same size particle, about 8μm. The total number of AE events increased with increasing of volume fraction of SiC, but the number is few. Fracture origin was observed clearly. About bending test we consider that the specimen with low volume fraction of particle failures as soon as critical cracks generate. So we can think the microcracks were less effective and the elastic modulus was effective for strengthening. The results of location of AE signals in Figure 3 demonstrate that microfracture process in the high volume fraction materials is different from that in low volume fraction materials. The microfracture process in low volume fraction materials could be considered as the crack propagation microfracture model because a crack propagates in restricted area. On the other hand, the microfracture process in the high volume fraction materials seems to be the mixed model because microfracture occurs randomly in location firstly and a crack propagates in restricted area. So we can fit the line according to the independent microfracture model in the first part of the plots for microfracture stress and draw a perpendicular line according to the crack propagation model. This analytical method could also give an excellent explanation for the experimental data.

CONCLUSIONS

(1) The glass composites reinforced by SiC particle with good densification were produced by hot pressing controlling the mismatch of thermal expansion constant between particle and matrix.

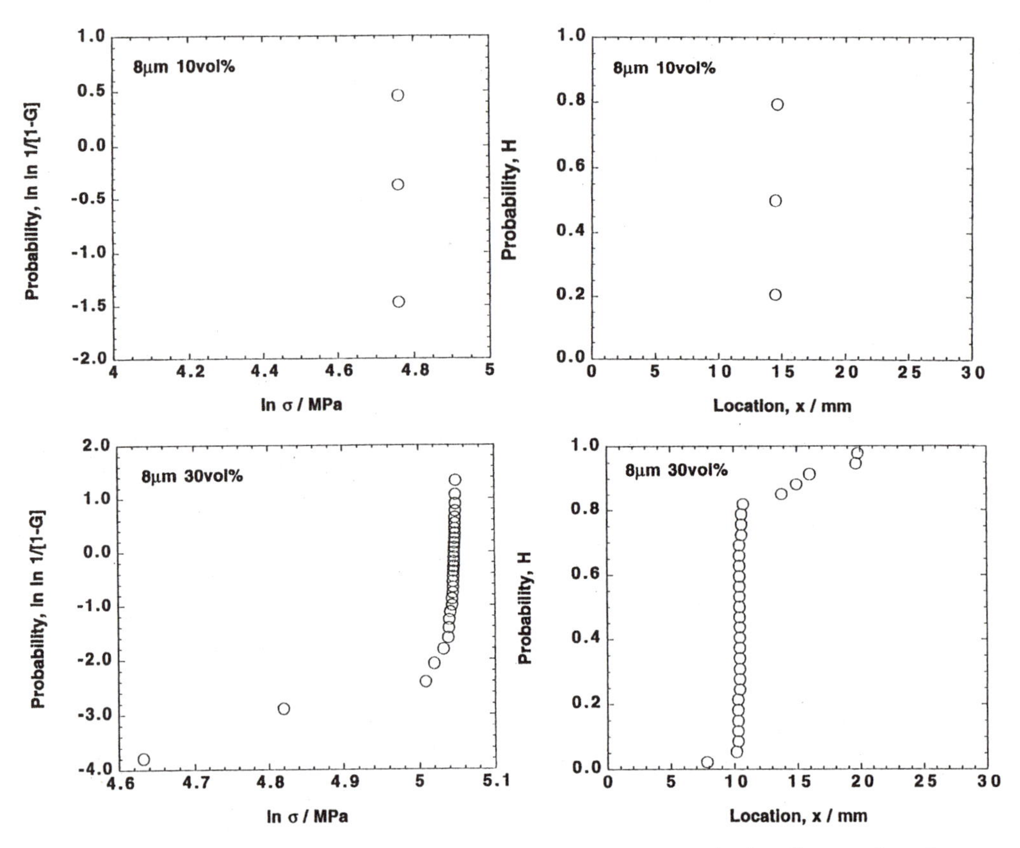

Figure 3 Probability distributions of microfracture stress and microfracture location.

(2) The bending strength of the materials were measured by four-point bending test. The strength of glass composites is significantly dependent on the volume fraction of particle and the increase of volume fraction increases the strength.

(3) The analysis of microfracture stress and location from AE data demonstrates that the increase of volume fraction changes the microfracture process of the materials from the crack propagation type to mixed microfracture type.

(4) The increase of strength can explained by the increase of elastic modulus in the low volume fraction materials. On the other hand, the increase of the strength can be considered to be due to both the increase of elastic modulus and the effect of microfracture in the high volume fraction materials.

REFERENCES

1. W. Weibull, Ingevetenskakad Hadl., No. 151, Stockholm (1939).
2. Y. Yokobori, The Strength of Materials, Iwanami Press, Tokyo (1974). (in Japanese)
3. S. M. Wiederhorn and L. H. Bolz, J. Ceram. Soc. Am., 53 (1974) 543.
4. A. G. Evans and K. T. Faber, J. Am. Ceram. Soc., 67 (1984) 255.
5. H. L. Oh and I. Finnie, Int. J. Fracture, 6 (1970) 287.
6. M. Enoki, Y. Utoh and T. Kishi, Mater. Sci. Eng., A176 (1994) 289.
7. K. Kageyama et al. J. Japan Inst. Metals, 59 (1995) 140. (in Japanese)
8. M. Enoki and T. Kishi, Int. J. Fracture, 38 (1988) 295.
9. M. Enoki, H. Fujita and T. Kishi, Mater. Trans. JIM, 37 (1997) 399.

PREDICTIVE MEASUREMENT USING AE: FRACTURE AND LIFETIME OF Al-ALLOYS

X. Lu[1], W. Sachse[1] and I. Grabec[2]

[1] Department of Theoretical and Applied Mechanics
Cornell University, Ithaca, NY – 14853 USA

[2] Faculty of Mechanical Engineering
University of Ljubljana
1000 Ljubljana SLOVENIA

INTRODUCTION

Living beings are remarkable in their ability to use their senses and to learn from their experiences in order to predict the future performance of a system. Such beings are an example of a synergistic measurement system which possesses capability for *sensing, processing, modeling* and *forecasting*.[1] A synergistic *ultrasonic* measurement system is one which is based on the union of sensing and processing of active (UT) or passive (AE) ultrasonic signals, and the subsequent modeling, forecasting and possibly even controlling the condition of a system or a machine with respect to its performance. In the future, such ultrasonic systems will likely be used in non-destructive testing, materials characterization, structural integrity monitoring and process control applications. The development of synergistic measurement systems is described in a recent monograph.[1]

In this paper, we explore the use of a synergistic system which relies on data measured during the initial portion of a fatigue-loaded, thin-plate, pre-cracked Al-alloy specimen to forecast its subsequent crack growth and lifetime. Here, *sensing* refers to the detection of AE waveform data on the specimen which can be used to predict the fatigue performance of the material. Specifically, we seek a reliable on-line, AE-based, crack sizing measurement procedure. *Processing* refers to the amplification, filtering and modification of signals to enhance their information content and to recover missing information. Here we are referring to evaluating the signals from active sources of emission and relating these to the growth of a crack. *Modeling* refers to a concise description of the fatigue-fracture phenomenon which will permit predicting as well as forecasting missing information. *Forecasting* refers to the prediction of the future evolution of a crack in a specimen or structure from past measurements of the growing crack.

APPROACH

Central to a synergistic measurement system is a neural-like, adaptive signal processing procedure which permits modeling the non-linear relationships between detected signals and the characteristics of a source, the condition or properties of a specimen. The procedure which we utilize here, relies on a statistical treatment of measured data to generate an empirical modeler of the natural law describing the phenomena. Such an *automatic modeler* [2] is based on a self-organized, optimal preservation of empirical information that utilizes the principle of maximum entropy of information and an optimal, associative estimation of missing information resembling a non-parametric regression. The approach corresponds, in part, to a neural network based on a set of radial basis functions or a 3-layer perceptron.[1]

We denote s–components of the measured sensor signals as $S = (x_1, x_2, \ldots, x_s)$. These may be functions of time or frequency, $S(t)$ or $S(\omega)$. The corresponding material or property descriptors are written as $P = (x_{s+1}, x_{s+2}, \ldots, x_{s+P})$. Thus, the concatenated signal description of a material is expressed by the data vector

$$X = S \oplus P = (x_1, x_2, \ldots, x_s, x_{s+1}, \ldots, x_{s+P}) \tag{1}$$

To learn from examples, one collects the data vectors $X_1, X_2, \ldots, X_N$ during a series of training measurements. These data vectors form the basis of the *memory* of the modeler. The formation of this memory corresponds to *adaptation* of the system. For processes in which there is a continuous set of measured data, as, for example, continuous AE data, one represents the data by a fixed, finite set of representative *prototype data* vectors which are selected by a self-organization procedure.[1] Once the memory has been developed, an optimal estimation of the material property characteristics $\hat{P}_O(S)$ from the measured sensor data is obtained via *multi-dimensional, non-parametric regression*. This is the *analysis* mode of the modeler. Specifically,

$$\hat{P}_O(S) = \int P \cdot f(P|S)\, \mathrm{d}P \implies \sum_{n=1}^{N} C_n P_n \tag{2}$$

where the *measure of similarity* is expressed by the coefficients

$$C_n = \frac{g\,(S - S_n)}{\sum_{n=1}^{N} g\,(S - S_n)} \tag{3}$$

Here, g represents the Gaussian functions which are formed from the data measured during learning, S_n, and during an subsequent, actual experiment, S.

Our goal here, is to train the automatic modeler to model the fatigue crack phenomenon so that it can subsequently be used to predict the crack growth and hence the lifetime of a specimen. In this paper we first consider the use of acoustic emission measurements to sense and to monitor the crack growth in aircraft Al-alloys loaded under tension and torsion mixed-mode fatigue loading.

When there is only one crack in a specimen, fatigue damage is usually described in terms of the length of the crack and its growth rate as a function of fatigue cycle count. One of the standard procedures for determining crack size and monitoring its growth is the use of direct optical measurements. Such a determination relies on a calibrated measuring microscope. However, an alternative procedure for sizing cracks is sought – one which would permit the monitoring of cracks remotely and in an actual structural component while it is in-service.

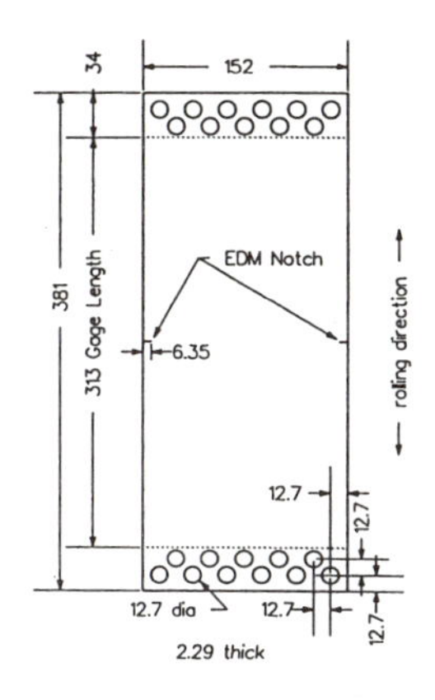

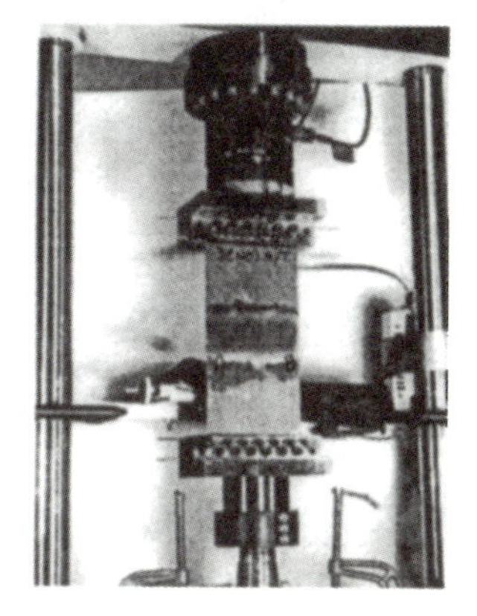

Figure 1: (a) Tension/torsion specimen geometry; (b) Mechanical testing configuration (from Ref. 3)

In the past, a number of crack sizing measurement techniques have been proposed and demonstrated. These include techniques based on eddy current, electrical resistance, and several based on ultrasonics. The use of crack tip diffracted signals in active ultrasonic measurements to determine the position of the crack tip is described elsewhere.[4, 5] The advantages of AE measurements as the basis of a continuous, on-line structural monitoring system are often cited. It is recognized that the complexities associated with the propagation of transient elastic waves through a bounded medium and their detection with a piezoelectric transducer complicate the analysis and the interpretation of the detected AE signal. But just as a living being is able to use its sense of hearing to quickly identify sources of sound, we explore here whether it is possible to use a minimum of AE data as input to a synergistic measurement system.

MEASUREMENTS AND MEASUREMENT RESULTS

It is recognized that the pressurization of an aircraft fuselage results in both tensile and transverse (out-of-plane tearing) stresses on a crack near a lap joint.[3] Crack tip tensile stresses arise from the hoop stress in the fuselage skin while the out-of-plane tearing stresses arise from the internal pressure in the fuselage. To study the fatigue crack growth under both tensile and transverse shear stresses, Zehnder and Viz [3] tested double-edge notched specimens of 2024-T3 Al-alloys such as that shown in Fig. 1(a) under constant amplitude cyclic tensile and torsional loadings using the mechanical testing system shown in Fig. 1(b). The work presented here will include fatigue studies as these and, in addition, measurements on 7076-T6 Al-alloy as well as single-edge notch specimens (SEN).

The direct optical measurements which were used to measure and monitor a growing crack are based on a traveling microscope which was capable of resolving the position of the crack tip to within about 0.001 in (25 μm). In this study the growth of the crack was measured approximately every 0.020 in. (0.5 mm) and the fatigue cycle count noted.

Simultaneously, we sought to determine the crack length using acoustic emission techniques. As our goal is an on-line monitoring system, we exclude quantitative AE approaches which rely on the time-consuming signal processing to recover the micro-mechanical details of the source. A description of the simple AE measurements we have used instead are given in the following sub-section.

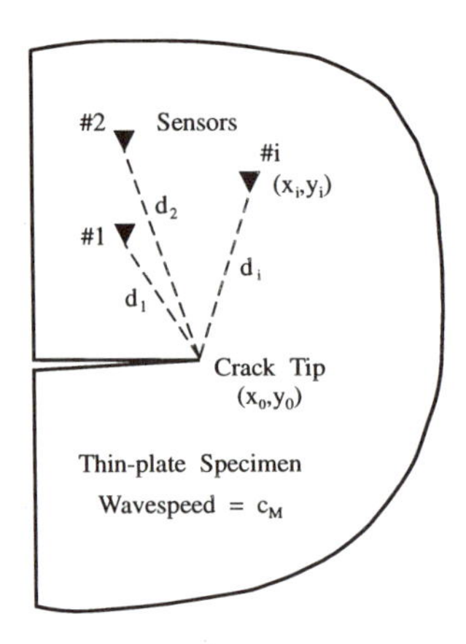

Figure 2: AE source/sensor location geometry.

Monitoring Crack Growth via AE

There are two AE measurement procedures which have been used to characterize the evolution of damage in a stressed material. One is based on counting the number of events above a detection threshold, emanating from the fracture region of the specimen. The assumption made is that the number of emissions is directly related to the growth of the crack. The second approach relies on AE source location data to determine the position of an event which is assumed to coincide with the position of the crack front and hence, when monitored over time, provides information about the propagation of the crack through the specimen. We explored both approaches.

Our measurement system was capable of recording an AE waveform parameter as well as the entire emission waveform. For either measurement, the AE signals were detected with an array of four, small-aperture (active element 1.3 mm diameter) sensors mounted around the starting notch as shown in Fig. 2. The signals were amplified by +60 dB using low-noise preamplifiers whose bandwidth extended from 20 kHz to 2 MHz. The AE event count was determined by a counter connected to the output of the trigger detection circuit of a transient recorder. This system was capable of recording a maximum event rate of about 500 events/s which is far in excess of the event rate encountered in the Al-alloys tested here. Entire waveforms could also be recorded using four channels of 10-bit waveform recorders which were typically operated at a sampling rate of 30 MHz. The digitized waveforms were transferred to a PC for processing and storage.

To determine whether the detected AE signals emanated from a source at the crack tip, the following procedure was used. The location of each sensor comprising the sensor array is known, that is (x_i, y_i). When an event is recorded, the arrival-time of the same dominant feature in each detected waveform, that is, t_1, $t_2, \ldots t_i$ is determined. Using these, the arrival-time relative to one of the signals is formed. This is denoted as Δt_i. Since the experiments are carried out in a laboratory setting, we simultaneously determine the actual location of the crack tip, (x_0, y_0) optically. We assume that the plate specimen is sufficiently thin so that the ray path between the crack tip to each sensor can be approximated by

$$d_i = \sqrt{(x_i - x_0)^2 + (y_i - y_0)^2} \quad . \tag{4}$$

The path length Δd_i is defined as the path length relative to that ray path corresponding to the signal between the source of emission in the crack tip and the sensor that was chosen as the reference signal in forming the Δt_i. Next, the "speed of propagation" of

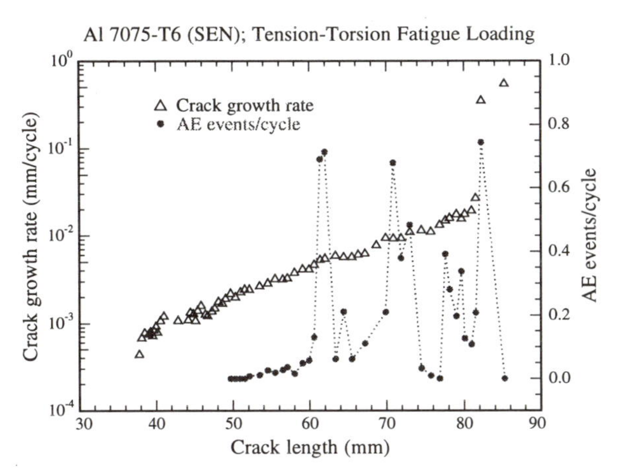

Figure 3: Relationship between optically measured crack growth rate and AE events/cycle (detected at one sensor) and the crack length.

the dominant feature is determined in each waveform. This is given by

$$c_M^i = \frac{\Delta d_i}{\Delta t_i} \quad . \tag{5}$$

If $c_M^1 \approx c_M^2 \approx c_M^i$, then the assumed crack location at the crack tip is verified to be correct. When the measured time delays resulted in an inconsistency in the speed for the equivalent feature in the waveforms detected at each sensor, this indicated that the source of the acoustic emission was not at the crack tip.

AE Measurement Results

We show in Fig. 3 the correlation between the optically measured crack growth rate and the AE event rate as a function of the crack length of a single-edge notched specimen of Al 7075-T6. There is clearly no obvious relationship between the optically measured crack growth and AE event rates. While it appears that transitions in the optically measured growth rate are accompanied by bursts of higher event rates, it does not appear to be possible to recover the crack length from the measured AE event rate. It is noted that the measured number of AE events depends strongly on the detailed characteristics of the AE measurement system (trigger level, window duration, hold-off, etc.). For this reason it is doubtful that data of a single AE waveform parameter as this, collected on Al-alloy thin plate specimens, can be used as input data to an automatic modeler for obtaining a reliable prediction of future performance of the material.

We next sought to determine whether direct determination of crack growth activity at the crack tip might be used to follow the growth of the crack in a sample undergoing fatigue loading. Fig. 4 shows the signals from an event detected at each of the sensors comprising the measurement array and the identified arrival times. Also shown in this figure are the results of following the procedure outlined in the previous subsection.

While Fig. 4 is an example in which AE signal arrival data could be used to correctly locate the position of the crack tip, there were many events for which the verification procedure led to the conclusion that the source was not at the crack tip. Further, many events were detected in which the signals detected at one or more sensors exhibited no clear, large amplitude wave arrival. An example is shown in Fig. 5. Such signals were sometimes observed at reduced load amplitudes when there was no visible crack growth

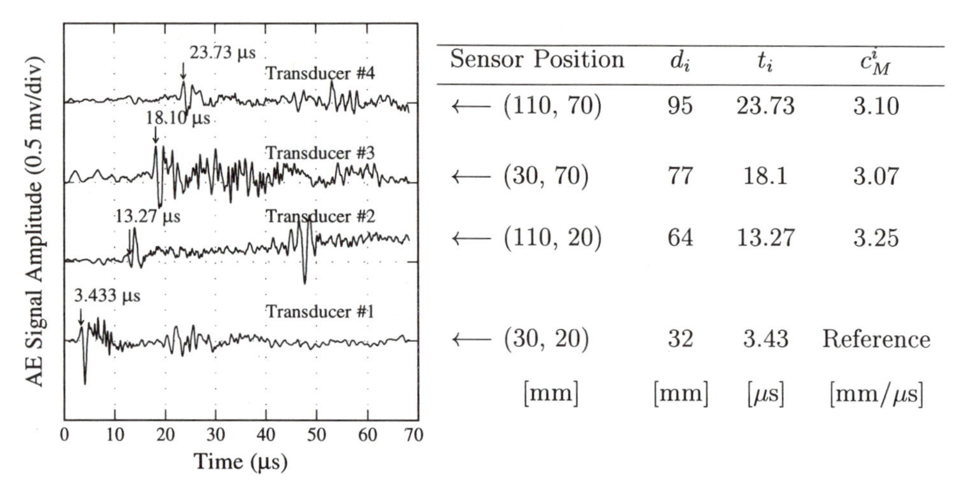

Sensor Position	d_i	t_i	c^i_M
⟵ (110, 70)	95	23.73	3.10
⟵ (30, 70)	77	18.1	3.07
⟵ (110, 20)	64	13.27	3.25
⟵ (30, 20)	32	3.43	Reference
[mm]	[mm]	[μs]	[mm/μs]

Figure 4: Verification of AE source location. (a) AE waveforms detected at four sensors; (b) Observation: The crack location could have been determined from measurement of the arrival of the waveform feature.

detected in the microscope. From signals as those shown in Fig. 5 in which no clear first wave arrival can be identified, it is impossible to determine the location of the source and hence the position of the crack front is impossible to determine with any certainty. It is likely that the source of such events is either a micro-crack forming at a site away from the crack tip or that these signals result from crack face rubbing frictional effects. The presence of such sources will clearly limit the usefulness of AE source location data to form the basis of the input data for training the memory of an automatic modeler.

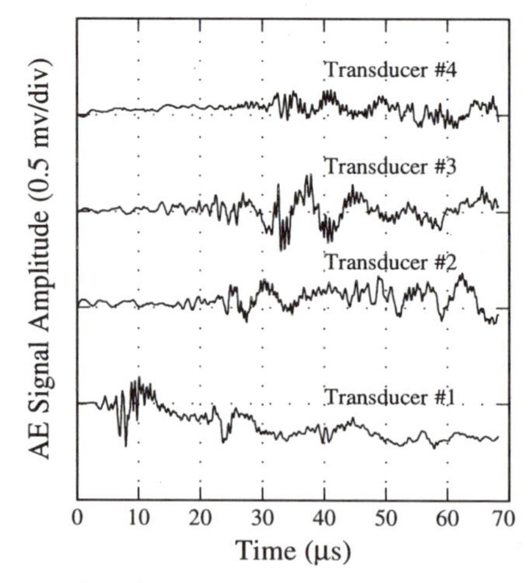

Figure 5: AE waveforms detected at four sensors from an event accompanied by no visible crack growth.

FATIGUE MODELING BASED ON MEASURED CRACK LENGTH

A critical question is whether the premise that a synergistic system can be adapted and subsequently used to predict fatigue crack growth is true. For this, we require precise crack growth data while a specimen is undergoing fatigue. The compilation of the crack growth data measured optically on a number of single-edge-notch specimens of 2024-T3

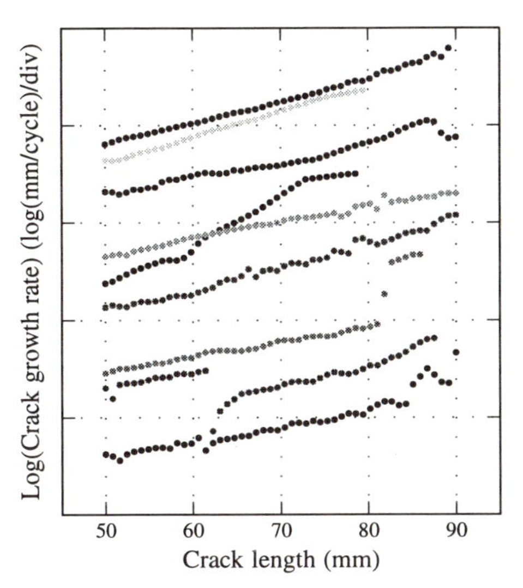

Figure 6: Crack growth data on 2024-T3 and 7075-T6 fatigued Al-alloy specimens which was used to develop the modeler *memory*.

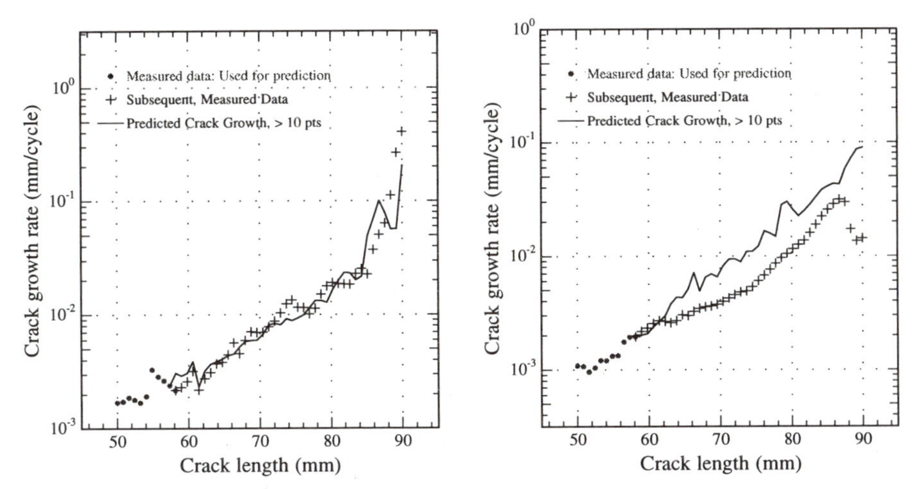

Figure 7: Modeler prediction of crack growth data in fatigue-loaded Al-alloy specimens. (a) 7075-T6 showing excellent prediction; (b) 2024-T3 showing poor prediction.

and 7075-T6 Al-alloys in the testing machine is shown in Fig. 6. This data was collected in tension-tension, torsion-torsion fatigue tests under various maximum and minimum applied loads and torques. This data constitutes the *memory* of the modeler.

Once the memory of the modeler has been developed, it can be used to recover missing information in the data vectors (i. e. Eq. (1)) which are presented to it. Here we wish to recover components of the missing property descriptors, P, which correspond to predicting expected crack growth. In the *analysis* mode, Eq. (2) was used with only the first ten, initial crack length data values as input to the modeler from which the remaining evolution of the crack growth was predicted. We show in Figs. 7(a)-(b) the results of two blind experiments.

These results confirm others which show that reliable prediction of the crack growth appears to be only possible when the input data reasonably closely corresponds to data already in the memory. The result of the 7075-T6 Al-alloy specimen which is shown in Fig. 7(a) exhibits good agreement between the predicted and the subsequently measured crack growth curve. It happens that the first ten points of the measured data used for prediction as well as the subsequent, measured data closely resemble one of the curves in the memory which was generated under similar loading conditions but had a different initial crack length. In that case, the predictive capability of the automatic modeler based on only the first few points in a crack-growth curve seems to be quite good.

In contrast, the crack growth data for the 2024-T3 Al-alloy specimen differs significantly from each of the curves comprising the memory because the test shown in Fig. 7(b) was carried out under loading conditions which differed significantly from those used to produce the memory. It appears that under these loading conditions the modeler cannot properly predict the crack growth. It may need to be trained with additional data, such as loading information, that also controls the crack growth, in order to improve its predictive capabilities. Exactly which data or parameters are the controlling factors in such crack growth and thus need also to be input to the modeler is not yet known and needs to be investigated.

CONCLUSIONS

We have described in this paper the use of acoustic emission waveform parameters to characterize the size and growth of a crack in a fatigue-loaded, thin-plate specimen of Al-alloy. The results show that the AE data does not reliably correlate with the characteristics of the crack. When direct optical measurements are used to determine the crack size, then only a few data need to be input to an automatic modeler to predict the entire subsequent growth of the crack. Good prediction of actual crack growth rate is obtained however only when the input data used to make the prediction closely resembles that which already resides in the memory. This work shows that AE data may not always be a suitably reliable indicator of crack growth and hence it is likely that we must rely on other crack length measurement techniques as the source of the input data to the modeler.

ACKNOWLEDGMENTS

This work was supported by the Air Force Office of Scientific Research under Contract #F49620–95–1–0383. We appreciate the use of the mechanical testing machine in Prof Zehnder's lab. Use of the facilities of the Materials Science Center at Cornell University which is funded by the National Science Foundation is also acknowledged.

REFERENCES

1. I. Grabec and W. Sachse, *Synergetics of Measurement, Prediction and Control*, Springer-Verlag, Heidelberg (1997).
2. I. Grabec and W. Sachse, "Automatic modeling of physical phenomena: Application to ultrasonic data", *J. Appl. Phys.*, **69**(1), 6233–6244 (1991).
3. M. J. Viz and A. T. Zehnder, "Fatigue crack growth in 2024-T3 Aluminum under tensile and transverse shear stresses", in *FAA/NASA International Symposium on Advanced Structural Integrity Methods for Airframe Durability and Damage Tolerance*, NASA Conference Publication 3274, Part 2, C. E. Harris, Ed., NASA, Washington, D. C. (September 1994), pp. 891–909.
4. W. Sachse and S. Golan, "The scattering of elastic pulses and the non-destructive evaluation of materials", in *Elastic Waves and Non-destructive Testing of Materials*, AMD-Vol. 29, Y. H. Pao, ed., American Society of Mechanical Engineers, New York (1978), pp. 11–31.
5. X. Lu and W. Sachse, "Self-calibrating active ultrasonic technique for crack tip location" MSC Report #8239, Materials Science Center, Cornell University, Ithaca, NY (August 1997).

AE CHARACTERIZATION OF MICROFRACTURE PROCESS IN CERAMIC MATERIALS

Shuichi Wakayama and Byung-Nam Kim

Department of Mechanical Engineering, Tokyo Metropolitan University
1-1 Minami-Ohsawa, Hachioji-shi, Tokyo 192-03, JAPAN

INTRODUCTION

Whisker reinforced ceramic composites have been investigated[1-4] because of their high toughness, due to the toughening mechanisms of the crack deflection[1], crack bowing[2] and especially crack bridging[3] by whiskers. But further progress may be restricted by the fact that the parameters obtained experimentally are macroscopic but the parameters included in theoretical models are microscopic.

On the other hand, the authors have studied the toughening mechanisms and fracture behavior in ceramic materials using acoustic emission technique[4,5], because of its ability to detect the nucleation times and locations of microcracks in the material. In particular, an experimental technique to evaluate the critical stress for maincrack formation has been developed[5], using acoustic emission monitoring and a fluorescent dye penetrant observation of fracture process during bending test of the monolithic alumina.

In this study, two kinds of alumina matrix composites reinforced with different whisker (silicon carbide and alumina) were fabricated. The microfracture process during four point bending tests were evaluated by acoustic emission technique. The critical stress of microcracking in matrix of the composites was determined experimentally. Consequently, the influences of the whisker arrangement and the combination of whisker/matrix materials on the fracture behavior and the critical stress were investigated.

TESTING PROCEDURES

Preparation of Materials

In this study, three types of materials were prepared, i.e. monolithic alumina and alumina reinforced with silicon carbide or alumina whisker. The volume fractions of whisker

Nondestructive Characterization of Material VIII
Edited by Robert E. Green Jr., Plenum Press, New York, 1998

were 10 and 20 % in both composites. The materials were sintered by the hot pressing in Ar gas with suitable sintering conditions (1600 - 1800 °C, 20 - 40 MPa) for each materials, then the relative densities of obtained materials were higher than 99 %.

Bending Tests

Specimens were cut from sintered discs by diamond saw to the dimensions of 3 × 4 × 40 mm. Bending tests were carried out using the Instron-type tensile testing machine in air with upper and lower span of 10 and 30 mm.

Cutting direction of samples was selected so that the hot press axis was parallel (P direction) or normal (N direction) to the surface where tensile stress was applied during bending tests. The whiskers in the P direction samples were parallel to the tensile axis, where the toughening mechanism due to crack bridging caused by whisker was effective, while the orientations of those in the N samples were 2-dimensionally random. Both P and N direction samples were available for silicon carbide whisker reinforced alumina (SiC/Al_2O_3[P] and SiC/Al_2O_3[N]), but other materials were only P direction samples (Al_2O_3 and Al_2O_3/Al_2O_3[P]).

AE Measurement

The AE measuring system used in this study is shown in Figure 1, schematically. Two piezo-electric elements were used directly as AE sensors and attached on both ends of the specimen. In this study, AE source locations were calculated from the difference of arrival times between two sensors and used for the removal of mechanical noise. Since the accuracy of AE source location strongly depends on the equilibrium of sensitivities of sensors, the sensitivity of two sensors, especially those equilibrium, were calibrated carefully using pencil lead breaking as a simulated source, before each testing.

Since it is well known that the AE activity of ceramic materials is quite low, the minimization of noise level of the system is indispensable for the AE measurement of such materials[4,5]. In this study, the connection between sensor and pre-amplifier was modified the same way as differential type transducers, in which the electromagnetic noise can be automatically canceled between the opposite phase signal cables. Consequently, the noise level at the input terminal of pre-amplifier was decreased to 14 μV, and then the threshold level was selected as 18 μV. AE signals were measured by the AE analyzer with load signals and sent to a personal computer through the RS-232C interface to be analyzed.

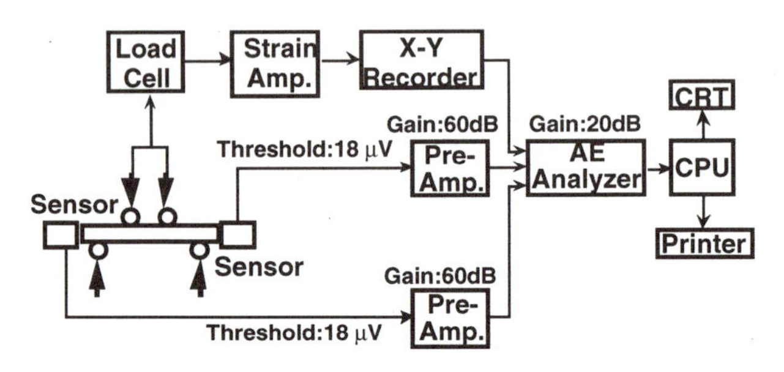

Figure 1. AE measuring system used in this study.

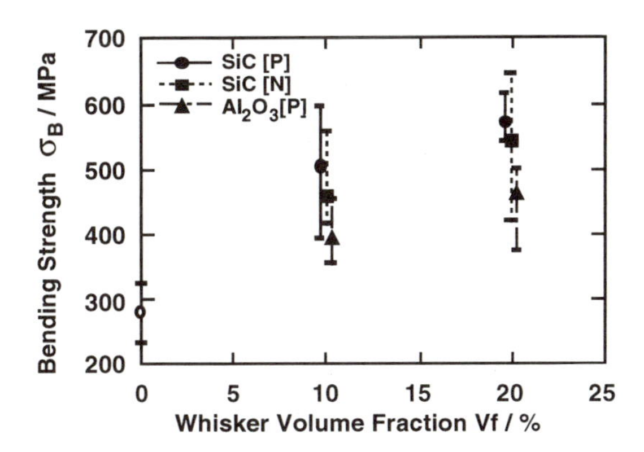

Figure 2. Bending strength of materials. Symbols indicate the average values; "O" for monolithic alumina, "●" and "■" for SiC whisker reinforced Al_2O_3 for P direction (SiC/Al_2O_3[P]) and N direction (SiC/Al_2O_3[N]), and "▲" for Al_2O_3 whisker reinforced Al_2O_3 for P direction (Al_2O_3/Al_2O_3[P]), respectively.

RESULTS

Bending Strength

Figure 2 shows the bending strength, σ_B, of materials. It can be understood by the figure that the strength of SiC whisker reinforced composites (SiC/Al_2O_3[P] and [N]) is larger than Al_2O_3 reinforced composites (Al_2O_3/Al_2O_3[P]). These results can be explained by the fact that pull out of whiskers was rarely observed on the fracture surface of Al_2O_3/Al_2O_3[P] although they were very apparent in SiC/Al_2O_3[P] and SiC/Al_2O_3[N].

On the other hand, it is interesting that the strength of SiC/Al_2O_3[P] is larger than SiC/Al_2O_3[N]. The whiskers in SiC/Al_2O_3[P] are normal to the crack growth direction during bending tests. Therefore it can be concluded that these results were caused by the arrangement of SiC whiskers which contribute the enhancement in the crack growth resistance by crack bridgings.

AE Generation Behavior

The typical results of AE generation patterns during bending tests are shown in Figure 3, for (a) monolithic alumina, (b) 20 % Vf silicon carbide whisker reinforced alumina (SiC/Al_2O_3[P]) and (c) 20 % Vf alumina whisker reinforced alumina (Al_2O_3/Al_2O_3[P]). It can be observed in Figure 3 (a) that cumulative AE events increases rapidly at about 350 s. It has been already made clear by the authors[5] that the apparent bending stress, σ_C, at that point corresponds to the critical stress for maincrack formation, investigating the dependence of the critical stress on AE threshold level, AE source locations and the observation of fracture process using fluorescent dye penetrant method. Those points can be also determined in the AE generation patterns of SiC/Al_2O_3[P] at about 250 s and Al_2O_3/Al_2O_3[P] at about 570 s. The apparent stresses at those points were determined to be the critical stress for maincrack formation in alumina matrix.

The critical stress, σ_C, of monolithic alumina is 230 MPa and close to the bending strength, σ_B. σ_C of SiC/Al_2O_3[P] is 240 MPa and less than 50 % of σ_B. It is important that the σ_C of SiC/Al_2O_3[P] is equivalent to that of monolithic material. On the contrary, the σ_C

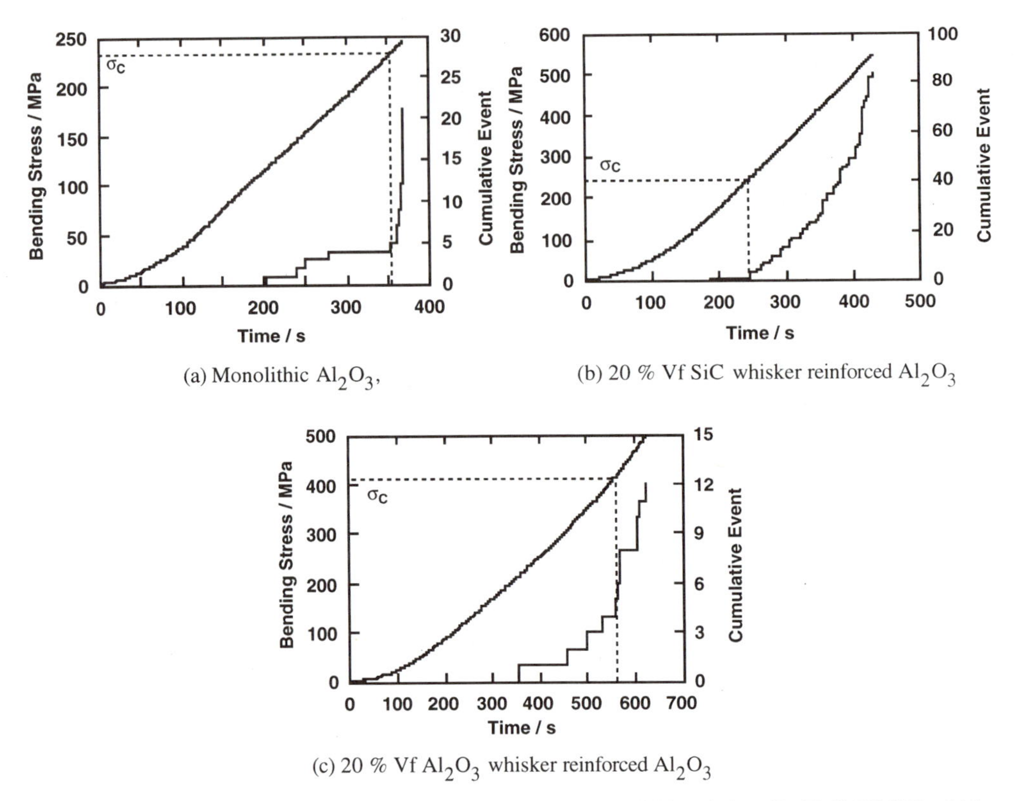

Figure 3. Bending stress and AE generation pattern of (a) Monolithic Al_2O_3, (b) 20 % Vf SiC whisker reinforced Al_2O_3 (SiC/Al_2O_3[P]) and (c) 20 % Vf Al_2O_3 whisker reinforced Al_2O_3 (Al_2O_3/Al_2O_3[P]), respectively.

of Al_2O_3/Al_2O_3[P] is 410 MPa and approximately 80 % of σ_B. It is much larger than those of monolithic alumina and SiC reinforced composite.

DISCUSSIONS

Microcracking in Matrix of Whisker Reinforced Composites

In this study, maincrack formation in matrix of silicon carbide and alumina whisker reinforced alumina composites were detected and the critical stress, σ_C, was determined by the acoustic emission technique. From the view points of both structural and material design of ceramic composites, it is important to investigate their characteristics.

The relationships between σ_C and volume fraction of whisker are shown in Figure 4. It is understood that the value of σ_C is hardly influenced by the arrangement of whiskers. On the other hand, the value of σ_C for Al_2O_3/Al_2O_3[P] is larger than SiC/Al_2O_3[P]. The differences are 80 MPa at a Vf of 10 % and 70 MPa at 20 %, respectively, and they are much larger than the differences between SiC/Al_2O_3[P] and SiC/Al_2O_3[N].

The value of σ_C measured in this study is the macroscopic critical stress for maincrack formation. The microscopic criteria for microcrack initiation was introduced by Laws et al.[6] as

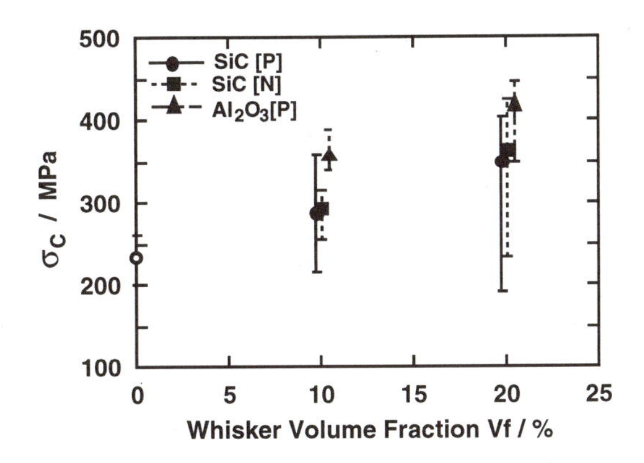

Figure 4. Increase in σ_C with the increase in Vf.

$$l_c \cdot (\sigma_{yy}^2 + \sigma_{xy}^2) = b \cdot K_c^2 \tag{1}$$

where l_c is the critical crack length, K_c is the microscopic fracture toughness, b is a constant and σ_{yy} and σ_{xy} are local tensile and shear stress, respectively. Since the monolithic alumina and the matrices of SiC/Al_2O_3[P] and SiC/Al_2O_3[N] were sintered from pure alumina powder without any additives, the differences of l_c and K_c between the materials are quite small. Consequently, it is concluded that the criteria for microcracking as well as maincrack formation is dominantly controlled by local stress, σ_{yy} and σ_{xy}, which are superposition of the applied stress and the residual stress due to the thermal mismatch between the whiskers and the matrix. Because the thermal residual stress of SiC/Al_2O_3[P] and SiC/Al_2O_3[N] is considered as equivalent, the effect of whisker arrangement on σ_C was quite small. And the value of σ_C for Al_2O_3/Al_2O_3[P] was larger than SiC/Al_2O_3[P], because the residual stress in Al_2O_3/Al_2O_3[P] is smaller than SiC/Al_2O_3[P].

Microfracture Process in Whisker Reinforced Composites

It can be seen from Figures 2 and 4 that the differences between σ_B and σ_C strongly depends on material system; 40 MPa for monolithic alumina, 230 MPa for SiC/Al_2O_3[P], 180 MPa for SiC/Al_2O_3[N] and 40 MPa for Al_2O_3/Al_2O_3[P]. The differences of fracture toughness measured by IF method between composites and monolithic material were 1.4 - 2.5 MPa√m for SiC/Al_2O_3[P], 1.7 - 1.8 MPa√m for SiC/Al_2O_3[N], and 0.7 - 0.8 MPa√m for Al_2O_3/Al_2O_3[P]. On the other hand, the increases in σ_B - σ_C are 190 MPa for SiC/Al_2O_3[P], 140 MPa for SiC/Al_2O_3[N], and 0 MPa for Al_2O_3/Al_2O_3[P]. It is understood from these results that σ_B - σ_C strongly depends on the increase in fracture toughness. The toughening of ceramic composites resulting from crack bridging of whisker, dK^w, was estimated by Becher et al.[3] as

$$dK^w = \sigma_f^w \left(\frac{Vf \cdot r}{6(1 - \nu^2)} \cdot \frac{E^c}{E^w} \cdot \frac{G^m}{G^i} \right)^{1/2} \tag{2}$$

where, σ_f^w, r and E^w are the strength, radius and elastic modulus of whisker, ν and E^c are Poisson's ratio and elastic modulus of the composite, and G^m and G^i are the critical strain energy release rates of matrix and interface, respectively. The value of G^i of Al_2O_3/Al_2O_3 is considered larger than SiC/Al_2O_3, because the materials of the whiskers and the matrix are same. Therefore, the toughening of Al_2O_3/Al_2O_3 is poorer than that of SiC/Al_2O_3. Although precise measurement of R - curve of the materials is needed, it can be concluded that the fracture process after σ_C is dominated by the R - curve behavior of the material.

It is understood that the lifetime after the maincrack formation in silicon carbide whisker composites are much larger than those in alumina whisker composites. Consequently, it can be concluded that reinforcing silicon carbide whisker is effective for the enhancement of crack growth resistance while alumina whisker contributes the increase in crack initiation resistance.

CONCLUSIONS

In this study, monolithic alumina, silicon carbide reinforced alumina and alumina whisker reinforced alumina were sintered by hot pressing, and fracture toughness and bending strength of the materials were measured. Furthermore, microfracture process during four point bending tests was evaluated by AE technique.

The critical stress, σ_C, for microcracking, especially the maincrack formation, in the matrix was determined from AE generation behavior. The value of σ_C is not influenced by the arrangement of whiskers in SiC whisker reinforced Al_2O_3. However it strongly depends on the combination of whisker and matrix materials; σ_C of Al_2O_3 whisker reinforced Al_2O_3 was larger than that of SiC whisker reinforced Al_2O_3. These results suggest that the microcracking in the matrix of whisker reinforced composites is controlled by the residual stress due to the thermal mismatch between the whiskers and the matrix.

The fracture process after the maincrack formation during bending test yields the R-curve behavior of the composites. Consequently, the contributions of the whiskers to the strengthening and toughening were understood.

REFERENCES

1. K. T. Faber and A. G. Evans, Acta Metall., 31:565(1983).
2. A.G. Evans, Phil. Mag., 26:1327(1972).
3. P.R. Becher, C.H. Hsueh, P. Angelini and T.N. Tiegs, J. Am. Ceram. Soc.,71:1050 (1988).
4. T. Kishi, S. Wakayama and S. Kohara, Fracture Mechanics of Ceramics, Vol. 8, R. C. Bradt, A. G. Evans, D. P. H. Hasselman and F. F. Lange, ed., Plenum Press, New York, 85(1985).
5. S. Wakayama and H. Nishimura, Fracture Mechanics of Ceramics, Vol. 10, R. C. Bradt, D. P. H. Hasselman, M. Sakai and V. Ya. Shevchenco, ed., Plenum Press, New York, 59(1985).
6. N. Laws and J.C. Lee, J. Mech. Phys. Solids, 37:603(1989).

CHARACTERIZATION OF FIBER FRACTURE VIA QUANTITATIVE ACOUSTIC EMISSION

David J. Sypeck and Haydn N.G. Wadley

School of Engineering and Applied Science
University of Virginia
Charlottesville, Virginia 22903

INTRODUCTION

Fiber fragmentation tests[1,2,3] can be used for in situ recovery of fiber-matrix interface shear strengths and/or fiber fracture spatial and strength distributions. Often neglected with the tests however, is important quantitative information (e.g. the crack volume and its history) contained in the elastic waves (acoustic emission) released during fracture. The final crack volume is related to the fiber fracture stress and the average sliding stress at the fiber-matrix interface. Thus, quantitative acoustic emission analysis provides an alternative approach for in situ recovery of these important fracture characteristics. Furthermore, knowledge of the crack volume history can help with the study of transition from the cumulative to noncumulative failure mode in composites since this transition is likely connected to the rate of stress redistribution in the fracture wake.

ACOUSTIC EMISSION FUNDAMENTALS

To model the motion caused by abrupt failure processes in elastic bodies, Burridge and Knopoff[4] treated the failure as an expanding dislocation loop and expressed spatial and material characteristics of the defect in terms of a distribution of "equivalent" body forces. A single point-like source can be modeled by a source moment tensor[4,5]

$$M_{ij} = c_{ijkl}[u_k]\Sigma_l \tag{1}$$

where i and j indicate the direction and separation of body force dipoles. c_{ijkl} is the elastic constant tensor, $[u_k]$ is the displacement discontinuity across the defect in the k-th direction and Σ_l is the defect surface area projected onto a plane having a normal in the l-th direction.

The time-dependent displacement in the i-th direction, $u_i(\boldsymbol{x}, t)$, at location, $\boldsymbol{x}$, and time, t, caused by a source centered at $(\boldsymbol{x}', t')$ is obtained by a convolution

$$u_i(\boldsymbol{x}, t) = M_{jk}\int_0^t G_{ij,k}(\boldsymbol{x};\boldsymbol{x}', t-t')S(t')dt' \tag{2}$$

Nondestructive Characterization of Material VIII
Edited by Robert E. Green Jr., Plenum Press, New York, 1998

where $S(t)$ is the source time-dependence (e.g. the normalized crack volume history) and $G_{ij,k}(\boldsymbol{x};\boldsymbol{x}', t-t')$ is the spatial derivative of the dynamic elastic Green's tensor[5]. Each component of the Green's tensor represents the displacement at $(\boldsymbol{x}, t)$ in the i-th direction due to a unit strength impulsive body force dipole concentrated at $(\boldsymbol{x}', t')$, acting in the j-th direction, with separation in the k-th direction.

For a unidirectional composite under tensile load in the fiber direction, mode I cracking of the fiber is usually accompanied by mode II shear at the fiber-matrix interface. Static equilibrium considerations necessitate that sliding must occur along a length, $l = r_f T/2\tau_s$, where r_f is the fiber radius, τ_s is the average interface sliding stress and T is the fiber fracture stress and the remote fiber stress thereafter. The crack opening is approximately

$$\Delta = \frac{r_f T^2}{2E_f \tau_s} \tag{3}$$

where E_f is the Young's modulus of the fiber. Suppose the normal to the fracture surface is oriented in the x_1 direction. Because far field contributions to the moment tensor from shear at the fiber-matrix interface cancel due to symmetry, Equations (1) and (3) give

$$M_{ij} = \begin{bmatrix} \lambda_f + 2\mu_f & 0 & 0 \\ 0 & \lambda_f & 0 \\ 0 & 0 & \lambda_f \end{bmatrix} \cdot \frac{\pi r_f^3 T^2}{2E_f \tau_s} \tag{4}$$

where λ_f and μ_f are the Lamé elastic constants of the fiber. Observing Equation (4), we see that the magnitude of the moment tensor is scaled by the final crack volume, $\pi r_f^3 T^2/2E_f \tau_s$.

Solutions to the wave equation for Heaviside or unit ramp excitation are available for the infinite linear elastic isotropic plate[6]. Wave dispersion, attenuation and the presence of the fiber are neglected. In terms of the unit ramp tensor, $G^R_{ij,k}(t-t')$, the time-dependent surface displacement normal to the plate (i.e. the acoustic emission signal) has the form

$$u_3(t) = M_{jk} \int_0^t G^R_{3j,k}(t-t') \ddot{S}(t') dt' \tag{5}$$

for $\dot{S}(0) = 0$ where $\boldsymbol{x}$ and $\boldsymbol{x}'$ have been omitted for simplicity. Combining Equations (4) and (5) we find

$$u_3(t) = \frac{\pi r_f^3 T^2 \int_0^t [(\lambda_f + 2\mu_f) G^R_{31,1}(t-t') + \lambda_f G^R_{32,2}(t-t') + \lambda_f G^R_{33,3}(t-t')] \ddot{S}(t') dt'}{2E_f \tau_s} \tag{6}$$

EXPERIMENTAL

A thin groove was scribed in a Ti-6Al-4V (wt.%) plate and a single SCS-6 fiber (Textron Specialty Materials, Lowell, MA) was placed in it. An unscribed, but otherwise identical section of plate was placed over the fiber and the plates were electron beam welded in vacuum around their edges to encase the fiber. The sample was hot isostatically pressed for 90 minutes at 100 MPa and 900°C for microscopically complete consolidation. This procedure was performed identically to produce a neat (fiberless) sample. Both samples were machined to a dogbone geometry (with the fiber located at the center of the plate) with a 95 mm long, 50 mm wide and 5 mm thick gauge section, Fig. 1. A 300 kN capacity electromechanical testing machine with serrated face wedge action grips was used to apply tensile load at a constant crosshead rate of 0.1 mm/min at 25°C. Load was mesured using a strip chart recorder.

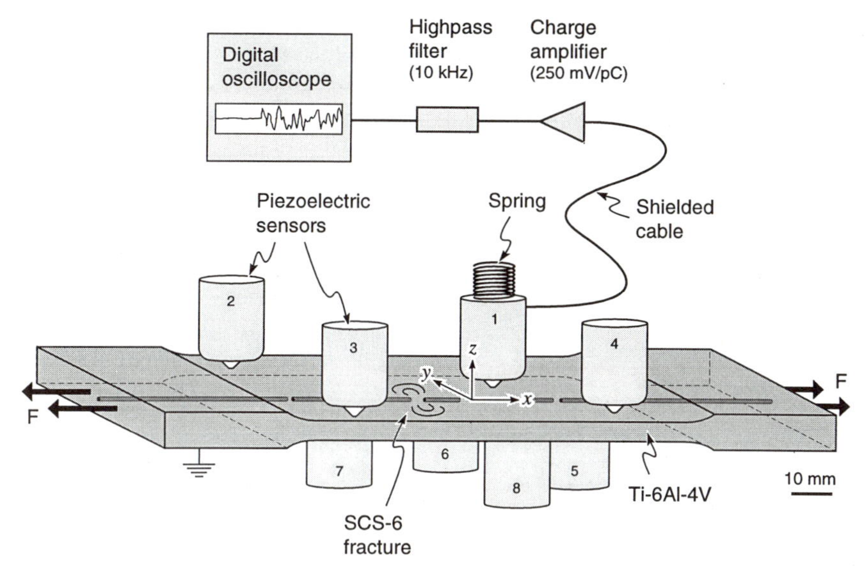

Figure 1. Experimental arrangement for quantitative analysis of fiber fragmentation.

To measure the acoustic emission activity, eight miniature piezoelectric sensors were spring loaded to the plate surface (see Fig. 1) for measurement of displacement normal to it. Sensors were based on the broad band design[7] of the National Institute of Standards and Technology (NIST). When calibrated (on steel using the same model 250 mV/pC charge amplifier) at NIST[8], a displacement response, Fig. 2, which compared favorably to that of the NIST design was achieved but with a smaller element backing. The maximum displacement sensitivity was 45 dB (RE 1 V/μm) or 178 V/μm at 450 kHz and decreased to 30 dB or 32 V/μm at 2 MHz. Signals were recorded at 1 nsec per data point with 10.5 bits voltage resolution for a 15 MHz upper frequency limit.

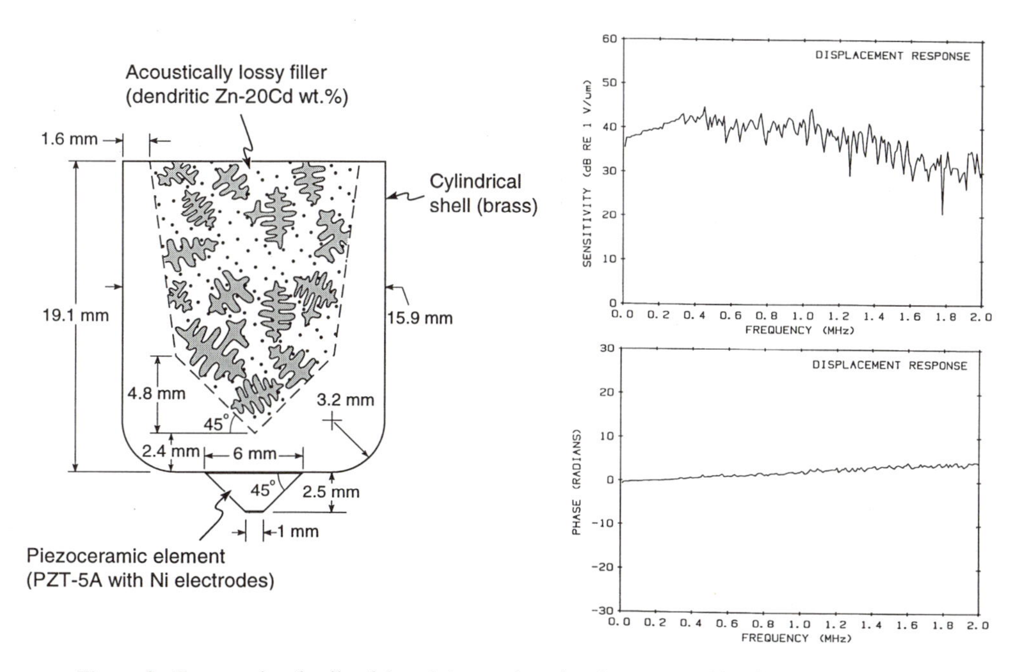

Figure 2. Construction details of the miniature piezoelectric sensor and its displacement response.

ACOUSTIC EMISSION ACTIVITY

During the initial stages of loading, both the neat and single fiber samples experienced a few weak, low frequency emissions. As loading progressed beyond the elastic regime, more than 35 additional strong, high frequency emissions were measured for the single fiber sample. The first of these occurred at a plate tensile stress, σ, of 773 MPa and the last at 852 MPa. Acoustic emission signals recorded during testing of the single fiber sample at $\sigma = 788$ MPa are shown in Fig. 3.

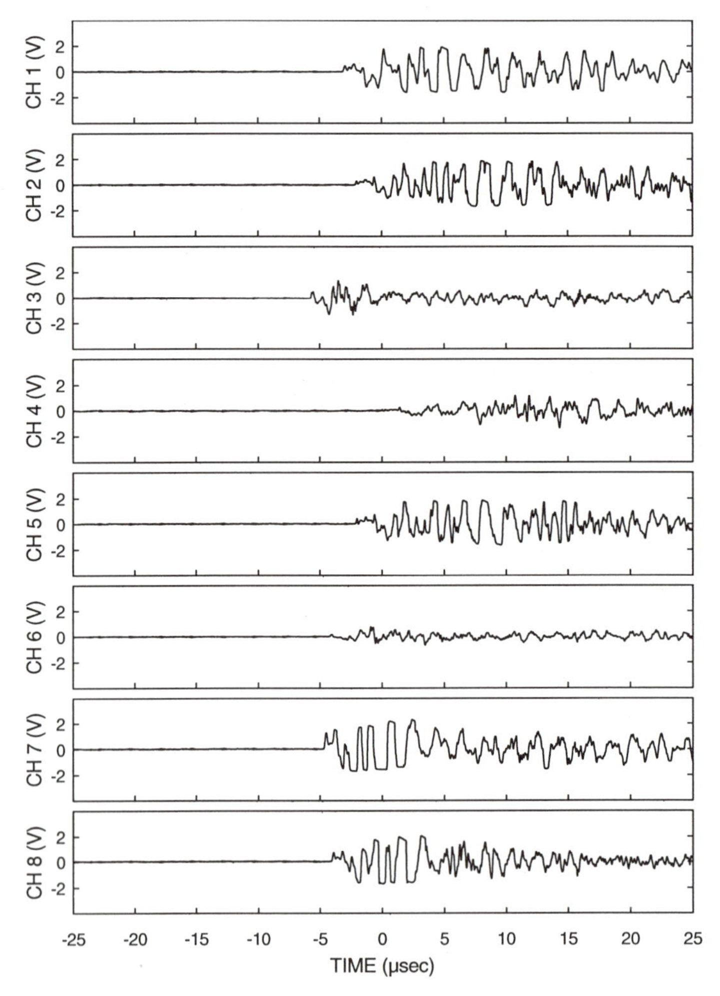

Figure 3. Acoustic emission signals for a fracture event which occurred at a plate stress of $\sigma = 788$ MPa.

FRACTURE SITE LOCATION

If x', y' and z' is the unknown location of the fracture, the i-th receiver located at x_i, y_i and z_i will first experience a signal when

$$(x' - x_i)^2 + (y' - y_i)^2 + (z' - z_i)^2 = (c_p t_i)^2 \tag{7}$$

where t_i is the travel time for the first compression wave to reach the i-th receiver and c_p is its velocity. Let t_0 be the travel time for the wave to reach the closest receiver and Δt_i be the travel time difference between that receiver and the i-th receiver. Substituting $t_i = t_0 + \Delta t_i$ into Equation (7) and subtracting any i-th equation from any j-th to generates one linear equation. For η receivers, there are a maximum $N = \eta(\eta - 1)/2$ unique subtractions.

Travel time differences, Δt_i, were found within ~20 nsec accuracy by observing when the AE signal magnitude first exceeded the background noise. The $N = 28$ equations were solved using a least squares algorithm with $c_p = 5.9$ mm/µsec for Ti-6Al-4V. The returned fracture location was $x' = -6.6$ mm, $y' = 0.0$ mm, $z' = -2.3$ mm and $t_0 = 1.73$ µsec. This location was consistent with the centerline of the plate and the known fiber location. Note that accuracy in z' is problematic for thin plates because the straight line distance between the source and a receiver 'far'' from it is not substantially affected by changes in z'.

CRACK VOLUME HISTORY

Sensor 5 was oriented ~4.7° with respect to the x_1 axis of the selected fracture event. This small rotation was disregarded and Equation (6) was used directly. Infinite plate unit ramp responses[6] were computed for the source located at plate center (i.e. $z' = 0$) with a radial (x_1 - x_2 plane) source-receiver distance of 31.9 mm. For Ti-6Al-4V, the shear wave velocity was 3.1 mm/µsec, the shear modulus was 43 GPa and again $c_p = 5.9$ mm/µsec. 500 data points beyond the first wave arrival and a 5 nsec step size allowed reasonable comparison with the measured AE signal while avoiding reflections from the edges of the plate (not considered in the infinite plate model).

Equation (6) was evaluated (in terms of T^2/τ_s) using $r_f = 70$ µm, $E_f = 400$ GPa and a fiber Poisson's ratio of 0.14 (find $\lambda_f = 68$ GPa and $\mu_f = 175$ GPa), Fig. 4. Through trial and error involving convolution (discrete) of $\ddot{S}(t)$ with the appropriate unit ramp responses, good agreement (normalized) between the measured and modeled acoustic emission signal (see Fig. 4) was obtained with $\ddot{S}(t) = \pm 44.4$ 1/µsec^2 and a 0.3 µsec rise-time. Integrating this function twice with respect to time, a crack volume history, $S(t)$, having the shape of a symmetrical parabolic ramp was recovered. The shape of this function is thought to correctly portray the physics of the fracture in the sense that it accelerates (i.e. nucleation and growth) and then decelerates to reach a final static value (i.e. arrest).

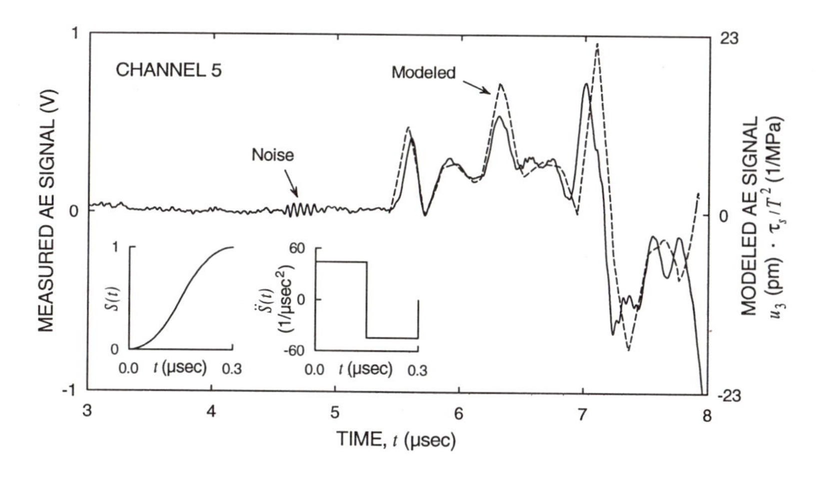

Figure 4. Comparison between the measured and the modeled acoustic emission signal.

INTERFACE SLIDING STRESS

At the time of the selected fracture event, minimal plasticity had occurred. For small fiber volume fraction composites in the elastic regime, the remote fiber stress can be approximated by

$$T = \left[\frac{\sigma}{E_m} + (\alpha_m - \alpha_f)\Delta T\right]E_f \qquad (8)$$

where E_m is the Young's modulus of the matrix, α_m and α_f are the matrix and fiber linear expansion coefficients and ΔT is the processing temperature change. For Ti-6Al-4V, E_m = 113 GPa and $\alpha_m = 8.5 \times 10^{-6}$ 1/°C. For SCS-6, $\alpha_f = 4.8 \times 10^{-6}$ 1/°C. Using σ = 788 MPa and ΔT = –875 °C, the fracture stress is estimated to be T = 1494 MPa.

Observing Fig. 4, we see that the measured AE signal was dominated by sinusoids having frequencies in the neighborhood of 3 MHz (the expected upper limit for the modeled signal is approximately 1/rise-time = 3.3 MHz) which is higher than the range of frequencies (10 kHz - 2 MHz) which our piezoelectric sensors had been calibrated for (see Fig. 2). To estimate the 3 MHz sensitivity of our sensors, we note that beyond 2 MHz, the "roll off" rate for sensors of this type typically increases to about 15 dB (RE 1 V/μm) / MHz[7]. Since the 2 MHz sensitivity of our sensors was 30 dB, the 3 MHz sensitivity should be about 15 dB or 5.6 V/μm. With this value and T = 1494 MPa, nearly absolute agreement between the measured and the modeled AE signal (see Fig. 4) is then obtained with an average interface sliding stress of τ_s = 287 MPa. We note that recovery of T for a known τ_s presents another possibility. This analysis would proceed in a similar fashion.

ACKNOWLEDGEMENTS

We gratefully acknowledge the help of N. Hsu, F. Breckenridge and T. Proctor. Funding for various aspects of this work was provided by an ARPA-URI grant (contract No. N0014-86-K0753) managed by S. Fishman and W. Coblenz and a NASA-DARPA program (NAGW-1692) directed by W. Barker and D. Brewer.

REFERENCES

1. R.B. Clough. F.S. Biancaniello, H.N.G. Wadley and U.R. Kattner, Fiber and interface fracture in single-crystal aluminum / SiC fiber composites, *Met. Trans.* 21A:2747 (1990).
2. A. Manor and R.B. Clough, In-situ determination of fiber strength and segment length in composites by means of acoustic emission, *Comp. Sci. and Tech.* 45:73 (1992).
3. W. Sachse, A.N. Netravali and A.R. Baker, An enhanced, acoustic emission-based, single-fiber-composite test", *J. NDE* 11:251 (1992).
4. R. Burridge and L. Knopoff, Body force equivalents for seismic dislocations, *Bull. Seis. Soc. Am.* 54(6):1875 (1964).
5. K. Aki and P.G. Richards, *Quantitative Seismology Theory and Methods*, Volumes I and II, W.H. Freeman and Company, San Francisco (1980).
6. N.N. Hsu, Dynamic Green's functions of an infinite plate - a computer program, NBSIR 85-3234, National Institute of Standards and Technology, Gaithersburg (1985).
7. T.M. Proctor Jr., More recent improvements on the NBS conical transducer, *J. Acoustic Emission,* 5(4):134 (1986).
8. N.N. Hsu and F.R. Breckenridge, Characterization and calibration of acoustic emission sensors, *Mat. Eval.* 39:60 (1981).

ACOUSTIC EMISSION CHARACTERISTICS DURING RING BURST TEST OF FW-FRP MULTI-PLY COMPOSITE

Akihiro Horide, Shuichi Wakayama and Masanori Kawahara

Department of Mechanical Engineering, Tokyo Metropolitan University
1-1 Minami-Ohsawa, Hachioji-shi, Tokyo 192-03, Japan

INTRODUCTION

The fracture process in fiber reinforced plastic (FRP) composite materials under stress consists of several kinds of microfractures such as fiber breakage, matrix cracking, delamination, etc. Since each microfracture has specific mechanical features, each emits an AE signal with particular parameters. Therefore, many AE studies[1,2,3,4] have been carried out including authors'[5].

In this paper, the cancellation of fiber waviness as a fracture mechanism in filament wound (FW) FRP composite layers is investigated. During FW fabrication, the wound outer layer binds the inner layers, and the inner layers are compressed in a circumferential direction, forming fiber waviness.

Ring burst tests [6] were carried out in order to evaluate the influence of fiber waviness on the burst pressure and the fracture behavior. The evaluation of the fracture behavior inside FRP layers was done by direct observation using a high magnification video (HMV) system. The fracture behavior of FW-FRP composites, especially cancellation of fiber waviness, was characterized by a combination of AE analysis and observation using the HMV system. Then, an experimental model of the fracture process considering the effect of cancellation of fiber waviness on the burst pressure was proposed in order to predict the actual burst pressure.

EXPERIMENTAL PROCEDURE

Ring Burst Test

In this study, the ring burst test developed by the authors was used to investigate the strength and the fracture behavior of composite rings under internal pressure. The internal pressure was generated by means of inserting a tapered rod into a 12 piece steel ring. The specimen was then pressurized through a rubber ring located between the specimen and

Table 1. Results of winding tension measurement. 1st ply designates the innermost layer

Ply Number		1st	2nd	3rd	4th	5th	6th	7th
Average Tension [N]	Type I Increasing Tension	16.0	16.8	17.4	18.1	18.4	18.9	19.2
	Type D Decreasing Tension	16.3	15.2	14.0	12.6	11.5	10.6	8.56

the 12 piece ring. In order to determine the relationship between the corresponding internal pressure and the rod's axial load, preparatory experiments were carried out. The fracture behavior of cancellation of fiber waviness was observed by the acoustic emission technique and HMV system until the ultimate fracture of the composite ring.

Preparation of Specimens for Ring Burst Test

The ring specimens were made from E-glass fiber (diameter: 16μm, roving tensile strength: 1.9GPa) and epoxy resin. The fiber roving with the epoxy was wound in the hoop direction on an Al tube (A 6061-O, outer diameter: 100mm, thickness: 3mm) using the filament winding method, where tension in the fibers was controlled as the load was increased (Type I) or decreased (Type D) gradually. The tension in the fibers was measured as shown in Table 1 by a load cell attached between the resin bath and the feed eye. After curing in an oven to obtain the ring type specimens, the FRP wound Al tubes were cut to 10mm in width.

AE Measurements

A schematic diagram of the AE measuring system is shown in Figure 1. The total gain was 60dB (20dB in the main amplifier and 40dB in the pre-amplifier), and the threshold level was 42dB(i.e. 128μV at the input terminal of the pre-amplifier). AE sensors with a resonant frequency of 500kHz (PAC: PICO) were used in this study. These sensors were attached to the specimen surface, and the distance between the sensors was 60mm. The AE signals, load and strain data were measured by an AE system (PAC: LOCAN). Then the data was sent to a personal computer (EPSON: PC486NAS) through an RS-232C interface in order to analyze it in detail using the computer.

RESULTS AND DISCUSSIONS

Results of Observation using High Magnification Video System

It was observed using the HMV system that fiber waviness moved in a radial direction and was cancelled from the outer to inner layers. In addition, the cancellation

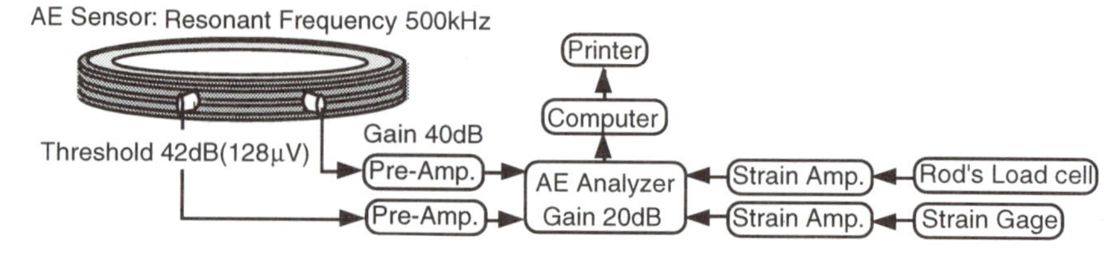

Figure 1. Schematic diagram of AE measuring system

behavior was more noticeable in Type I than in Type D specimens. The time of cancellation of fiber waviness in each ply was determed by the HMV system.

Characterization of Cancellation of Fiber Waviness by AE Analysis

In order to understand the relationship between AE amplitude and the cancellation of fiber waviness, AE amplitude distributions were investigated. AE amplitude distributions of Type I and Type D specimens are shown in Figure 2 (a) and (b), respectively. In Type I, which has some fiber waviness, the peak point is found around 63dB. On the other hand, in Type D, which has less few fiber waviness, the peak point is found around 47dB. Consequently, it is understood that higher amplitude AE can be related to the fracture behavior of cancellation of fiber waviness.

The typical AE event count rate, defined as the AE event counts for 10 s, for Type I specimen is shown in Figure 3. During a sudden increase of A, any fracture behavior or cancellation of fiber waviness could not be observed by the HMV system. The amplitude distributions before and during a sudden increase of A are shown in Figure 4 (a) and (b), respectively. By comparing these figures, it can be noted that although the total number of AE event count for 10 s in (a) are approximately twice as much as those in (b), the distribution shapes are similar.

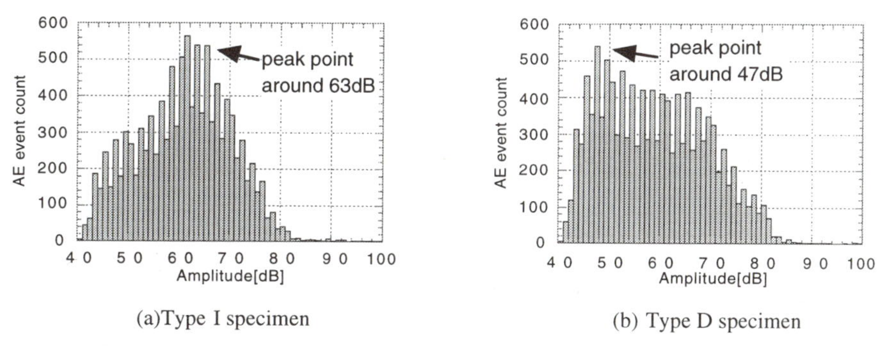

(a)Type I specimen (b) Type D specimen

Figure 2. Typical amplitude distributions during ring burst test

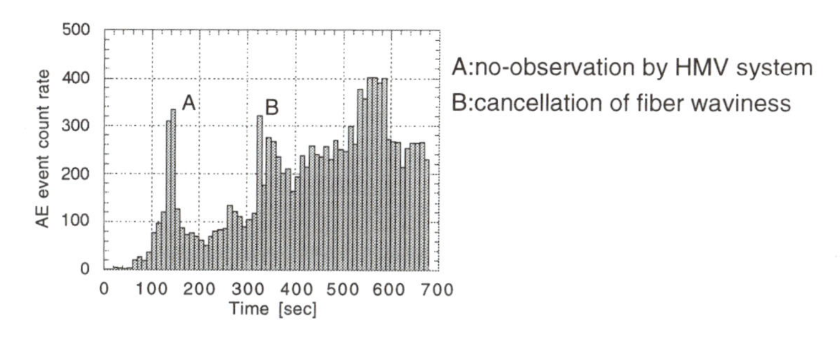

Figure 3. Typical AE event count rate for 10 sec. of Type I specimen

On the other hand, with a sudden increase of B, the cancellation of fiber waviness in the second layer was observed by the HMV system. The amplitude distributions before and during a sudden increase of B are shown in Figure 5 (a) and (b), respectively. The amplitude distribution before B has a peak point around 50dB and that during B has a peak point around 60-70dB.

Considering these results, it is understood that the cancellation of fiber waviness can be related to AE with an amplitude greater than 60dB. The following parameter is suggested to distinguish the effective AE signal from the others.

$$\text{AE event count rate ratio} = \frac{\text{AE event count rate for 10 [s] (Amplitude} \geq \text{60dB)}}{\text{AE event count rate for 10 [s] (Amplitude} < \text{60dB)}} \quad (1)$$

Figure 6 shows the typical behavior of the AE event count rate ratio and strain in the outermost layer. In this figure, A, B, C, and D correspond to the time of cancellation of fiber waviness in the 4th, 3rd, 2nd and 1st ply observed by the HMV system. It can be concluded that the above parameter may be applied to distinguish the cancellation of fiber waviness from the other fracture behaviors. Considering the mechanical features of fracture behavior in FRP composites, the defined ratio may be concerned with the interfacial fracture between fiber and matrix and interlaminar friction.

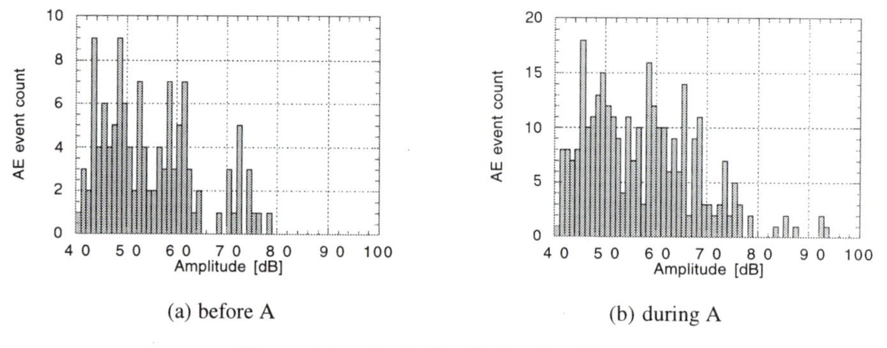

(a) before A (b) during A

Figure 4. Amplitude distributions for 10 sec.

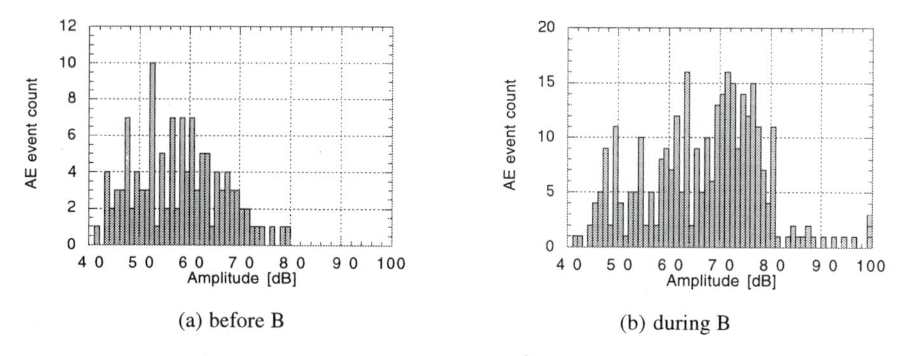

(a) before B (b) during B

Figure 5. Amplitude distributions for 10 sec.

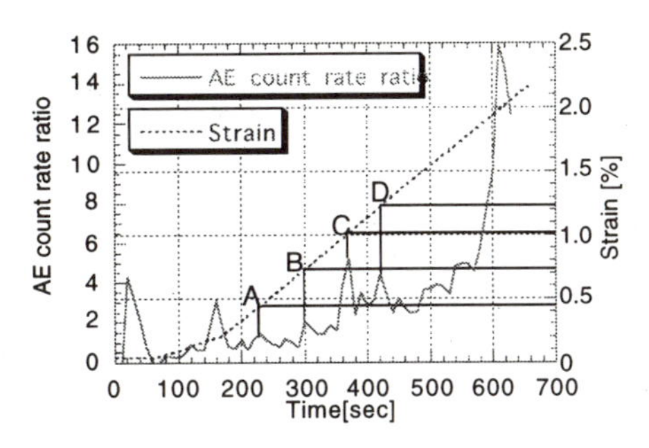

Figure 6. AE event count rate ratio of 5 ply type I specimen

Prediction of Reduction of Burst Pressure Due to Fiber Waviness

Under the rule of mixtures, strain in all ply is assumed homogeneous during the entire test. In the actual FRP composite, however, upon the fracture of this outermost ply, the strain in each ply that has fiber waviness may be lower than that in the outermost ply. This reduction of burst strain in each ply decreases the burst stress of the ply and consequently the overall burst pressure.

In this study, the hoop strain in the outermost ply at the cancellation of fiber waviness in the inner ply could be determined by the AE analysis and the HMV system. The burst pressure of each specimen (eg. 5-ply specimen) with fiber waviness can be estimated using the following procedure:

1) The reduction of i-th ply (i=1, 2, 3, 4) hoop strain just after the cancellation of fiber waviness is determined by the 5th ply hoop strain, $\varepsilon_{CN,i}$, and the i-th ply hoop elongation changed by the cancellation of fiber waviness, ε_i. Using Hooke's law, the reduction of i-th hoop stress at the cancellation of fiber waviness, $\Delta\sigma_i$, is determined as follows:

$$\Delta\sigma_i = E\left(\varepsilon_{CN,i} - \varepsilon_i'\right) = E\left\{\varepsilon_{CN,i}\left(\varepsilon_{CN,i} + 1\right) / \left(\pi + \varepsilon_{CN,i}\right)\right\} \tag{2}$$

2) If the reduction, $\Delta\sigma_i$, remains until the ultimate fracture of the 5th ply, the i-th ultimate stress at the fracture of the 5th ply, $\sigma_{Bi,}$ is given by:

$$\sigma_{Bi} = \sigma_{B5} - \Delta\sigma_i = E\left\{\varepsilon_{B5} - \varepsilon_{CN,i}\left(\varepsilon_{CN,i} + 1\right) / \left(\pi + \varepsilon_{CN,i}\right)\right\} \tag{3}$$

where σ_{B5}, E and ε_{B5} are 5th ply burst stress, Young's modulus and 5th ply burst strain, respectively.

3) Considering each ply as a thin-walled cylinder, the ultimate pressure of each ply is determined from the stress, $\sigma_{Bi,}$ and the dimension of the ply. Superimposing all pressures of FRP and Al ring at the burst of the 5th ply, the total burst pressure is given by:

$$P = P_{Al} + \sum_{i=1}^{N} \sigma_{Bi} \cdot t_{fiber} / r_i \qquad (N=5) \tag{4}$$

where P_{Al}, t_{fiber} and r_i are the internal pressure loaded on the Al ring at the burst of the 5th ply, fiber thickness and radius of i-th FRP layer, respectively.

Table 2. Comparison between experimental results and predicted values

Specimen Type	Type I with fiber waviness		Type D with little fiber waviness	
No. of ply	5	7	5	7
Experimental Burst Pressure [MPa]	35.9	47.4	39.1	52.5
Model 1(Rule of mixtures) [MPa]	39.7	53.4	39.7	53.4
Experimental Model [MPa]	36.8	45.9	38.4	49.3

The results of estimated burst pressure are summarized in Table 2. Comparing the results of the rule of mixtures and the present model with experimental results, it is obvious that the present model provides better prediction of burst pressure than does the rule of mixtures. In particular, this is noticeable in Type I specimens with fiber waviness. Using the experimental model considering cancellation of fiber waviness, the actual burst pressure of FW-FRP composites was predicted.

CONCLUSIONS

The fracture process during ring burst tests of FW-FRP composites was evaluated by AE techniques with HMV observations. From these results, the following conclusions were obtained:

1)The cancellation of fiber waviness was related to high amplitude AE events. The AE event count rate ratio was proposed to detect the cancellation of fiber waviness.

2)The improvement of burst pressure and the restraint of occurrence of fiber waviness in FW-FRP composites are achieved by controlling the winding tension as the load is decreased gradually.

3)An experimental model considering the fiber waviness was proposed using the outermost hoop strain at the cancellation of fiber waviness determined by AE technique. The burst pressure predicted by this model showed good agreement with the experimental results.

REFERENCES

1. Fuwa, M., Bunsell, A. R. and Harris, B., *J. Mat Sci., JMTSA*, 10(12): 2062 (1975).
2. Komai, K., Minoshima, K. and Shibutani, *JSME International Journal*, Series I, Vol.34, No.3: 381 (1991).
3. Suzuki, M., Nakanishi, H., Iwamoto, M., Jinen, E., Maekawa, Z., Mori, A. and Sun, F., *J. Trans. Jpn. Soc. Mech. Eng.*, (in Japanese), Vol.53, No.492, A:1459 (1987).
4. Sun, F., Suzuki, M., Nakanishi, H., Iwamoto, M.and Jinen, E., *J. Jpn. Soc. Mat. Sci.*, (in Japanese), Vol. 37, No.416: 517 (1988).
5. A. Horide, A. Wakayama, and M. Kawahara, Progress in Acoustic Emission VIII:28 (1996)
6. O. Fujishima, S. Wakayama, and M. Kawahara, *Progress in Acoustic Emission* VII:463 (1994)

THE STUDY OF PVDF ACOUSTIC EMISSION SENSOR

Zhengxian Wang*, Luming Li, Minsheng Wu, Hua Hou

Mech. Eng. Dept.
Tsinghua University
Beijing, China 100084

ABSTRACT

The newly developed piezoelectric material PVDF has the advantages of high piezoelectric conversion ability, nice pliability and low impedance matching with various kinds of substances. It has a promising future in AE sensor production. Series experiments were designed to study some possible influence factors on the signal receive sensitivity and frequency response of the sensor, including the thickness and area of the membrane, the polar connection and backing materials. The result is hoped to be useful for the further application of PVDF AE sensor.

INTRODUCTION

Acoustic Emission(AE) sensor is an important component to obtain acoustic signals, which is extremely crucial for NDE in acquiring dynamic characteristic parameters and analyzing the frequency, power and sound field distribution of ultrasonic signals. It is normally composed of shell, protective membrane, piezoelectric component, resistance block and high-frequency cable. Usually an AE sensor requires high signal-detect sensitivity, high response velocity, strong anti-disturb ability and wide range of dynamic frequency response. PZT is one of the most frequently used piezoelectric material for AE sensor. However, it has strong stiffness and relatively high manufacture expense. New materials to overcome such shortcomings are expected for industrial application. PVDF (Polyvinylidene fluoride), which has a repeat unit of CH2-CF2, is the represent of newly used piezoelectricity high polymer films in AE sensor. In 1969, the strong piezoelectricity and thermoelectricity of PVDF were discovered. Then other advantages as good pliability, low impedance matching with human tissue and chemical rusting of PVDF have been

* Corresponse should be made to: Zhengxian Wang, Mech.Eng. Dept. University of Maryland Baltimore County, Baltimore MD 21250

Table 1. Property Parameters of PVDF

property parameter	value	property parameter	value
c (m/s)	2.260	d_{33} (10^{-12}C/N)	35
pc (10^6kg/m2s)	1.78	k_{33}	~0.2
c^k_{11} (10^{10}N/m^2)	0.3	k_{31}	0.117
s^k_{11} (10^{-12}m^2/N)	330	k_t	~0.19
s^E_{33} (10^{-12}m^2/N)	320	$\varepsilon^\tau_{33}/\varepsilon_0$	13
e (C/m^2)	0.06	Q_c	3.5
h (10^8V/m)	5.3	Q_m	~10
g_{31} (10^{-3}V·m/N)	200	N_{3t} (Hz·m)	1.200
g_{33} (10^{-3}V·m/N)	330	ρ (10^3kg/m^3)	1.75~1.80
d_{31}(10^{-12}C/N)	23	ρ (μΩ·m)	2×10^{18}
Tm(°C)	158-197	Tc (°C)	120

Table 2. Comparison of some piezoelectric material

material	PVDF	PVF	PZT	$BaTiO_3$	crystal
ε	13	5	1200	1700	4.5
d_{31} (10^{-12}C/N)	20	1	110	78	2
g_{31} (10^{-3}V·m/N)	174	20	11	5	50
k_{33}	0.1		0.31	0.21	0.09
ρ (10^3g/cm^3)	1.78	1.38	7.5	5.7	2.65
Z (10^6kg/m^2)	2.7		25	25	15.1

reported. Some typical properties of PVDF and its comparison with some other piezoelectric materials are shown in table 1 and 2.[1,2,5,8]

From table 2 we can see that PVDF has the highest g_{31} value among all the materials listed, which means it may has the highest piezoelectric sensitivity. It also has a wide frequency response range, from 0.1 to over 10^7 Hz, which makes it suitable for AE sensor for receiving ultrasonic signal. Relative to ceramics, the low acoustic impedance of PVDF makes excellent mechanical match to water and biological systems possible. PVDF has good pliability. It can be processed by normal polymer techniques so that large area detectors or detectors with curve surface can be readily made.[8] Moreover, PVDF has relatively higher mechanical resonance sharpness, so we can obtain short pulse vibration. The uniform in thickness make the batch process possible to get stable and reliable transducer and can also acquire better frequency response.

Nevertheless, the utilization of AE sensor with PVDF on the industrial detection has not been widely used. Many aspects, which may influence its use in practical process need further exploration to provide instruction for the manufacture of PVDF AE sensor. Therefore, we designed this experiment to study several influence factors, i.e., the thickness and area of PVDF membrane, different backing material and the way of polar connection. Based on this experiment, we tried to setup a PVDF sensor. The comparison with this sensor to traditional PZT sensor favored the promising application of PVDF AE sensor in industry.

EXPERIMENT

Subjects

PVDF film (d_{33} : 25 PC/N, K_{33} :10-14%, ε 12~13 (1KHz), $\rho 10^3$ (Ω.cm)) with two type of thickness: 25μm and 110μm. The film has aluminum-plated layers as electrodes.

Two kinds of simulated acoustic emission source -- electric spark and breaking pencil were used.[1]

An A/D sampling card (F902) was used. Its sampling speed is from 156.3K/s to 20M/s. The input voltage is from 0.2V to 4.0V, resolving power is 8 bit. Because acoustic signal is a sudden signal, in this experiment A/D card used positive edge trigger function and stored the signal from an advance period of time.

Procedure

During experiment PVDF membrane was firmly fixed under pressure on the workpiece using the specific apparatus, AE signals were provided by simulated AE source and absorbed by PVDF membrane. Then we adjusted the analogue signal and used A/D conversation to get the data figure of the signal. In the experiment, the backing material used were polytetrafluoroethylene(PTFE) and polymethyl methacrylte. Further comparison and analysis were performed based on the data and the FFT (Fast Fourier Transfer) result .

RESULTS AND ANALYSIS

1. The influence of membrane thickness on signal receive. Used 25μm and 110μm round PVDF film separately. Shows in Figure 1 and 2.

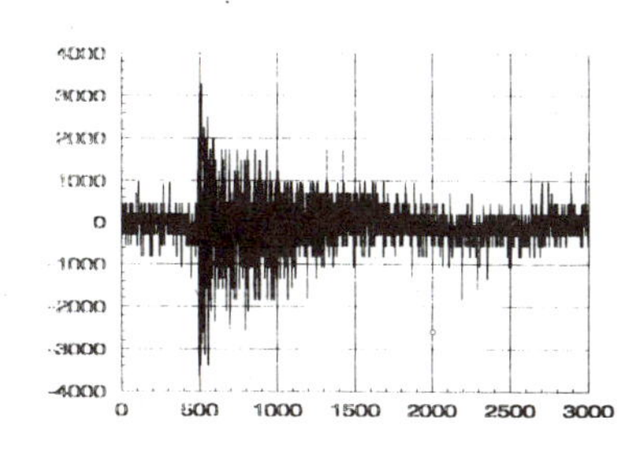

×1/156K (S)

Figure 1. Simulated acoustic signal received by 25μm PVDF film

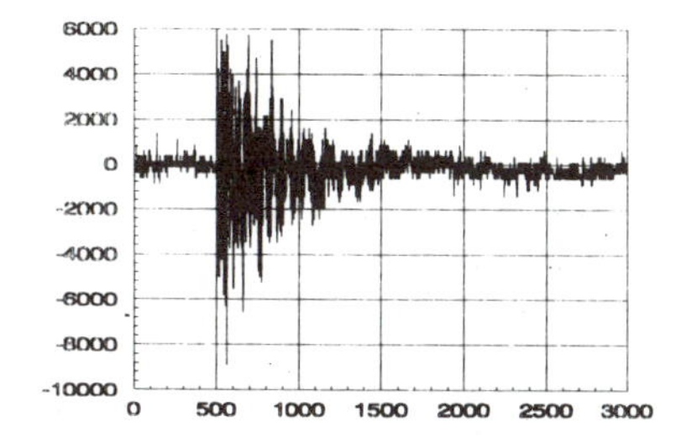

×1/156K (S)

Figure 2. Simulated acoustic signal received by 110μm PVDF film

The result showed that the amplitude of signal received by 25μm PVDF film was less than 50% of that of 110μm film. That is, the thinner the film, the worse the receiving sensitivity. The FFT result was shown in Figure 3 and 4. No significant central frequency change with different film thickness was found. But we still can see the thicker film has better low frequency responce.

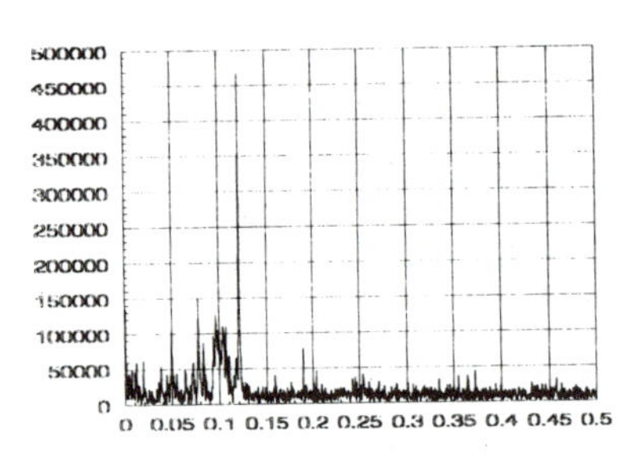

×156.3K (Hz)

Figure 3. Signal frequency analysis of 25μm PVDF film

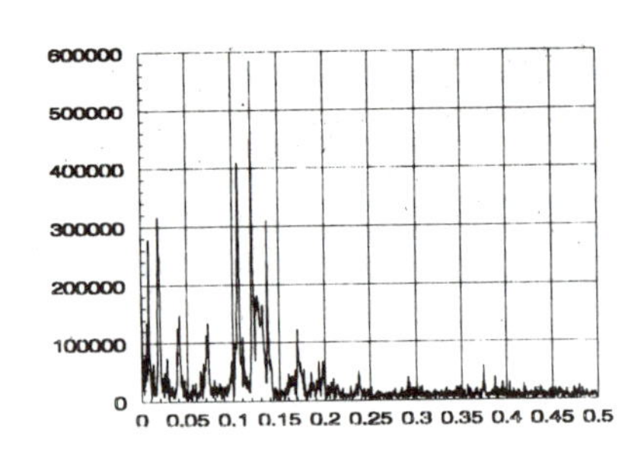

×156.3K (Hz)

Figure 4. Signal frequency analysis of 110μm PVDF film

2. The influence of different area of the film on signal receive

To test how film area affects the receiving sensitivity, two PVDF films had been applied. One was round with 20 mm in diameter and the other was rectangle as 5×10 mm, and both had thickness as 110 μm. The result was shown in figure 2 and 5:

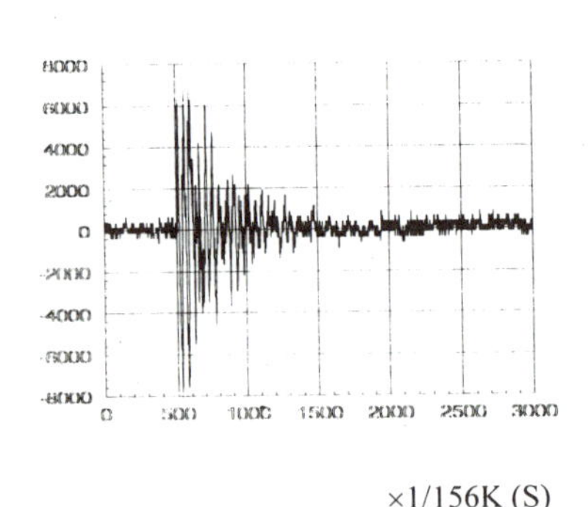

×1/156K (S)

Figure 5. Simulated acoustic signal received by rectangle 110μm PVDF film

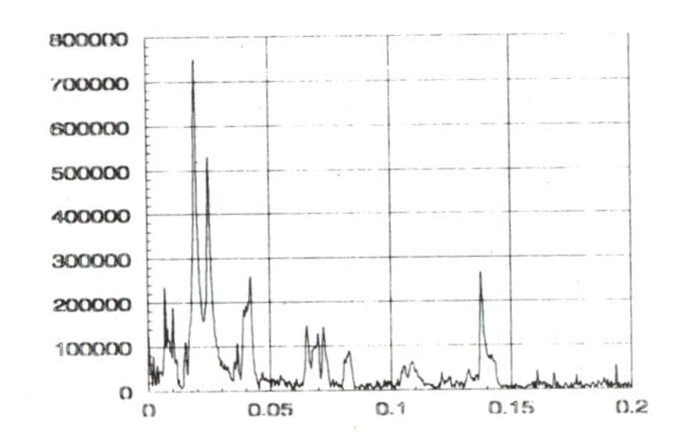

×15 (MHz)

Figure 6. Signal frequency analysis of rectangle 110μm PVDF film

From the result, we can see that the area of the film mainly affects the equational capacitor, but has little influence on the signal receiving sensitivity. In fact, the area of the film may only need to maintain its equation capacitor to be several times greater than that of signal transmission cable, then the receiving signal will not be emerged by the disturbance of the cable.

One difference between PVDF sensor and traditional PZT sensor is that the PZT sensor uses column piezoelectric piece and it vibrates as longitudinal thickness model. For PVDF film, when the signal frequency received is less than 1 MHz, the film vibrates as transverse winding model, which means that it is not required to use round film in the sensor. Additionally, to ensure uniform pressure on the film, it will be more preferable to chose narrow rectangle film which has a smaller area.

3. The influence of polar connection of the film on signal receive

PVDF film has different polar on its upper and lower sides, so experiment was designed to test its polar influence with the signal cable connected to both sides separately. Shown in figure 7 and 8:

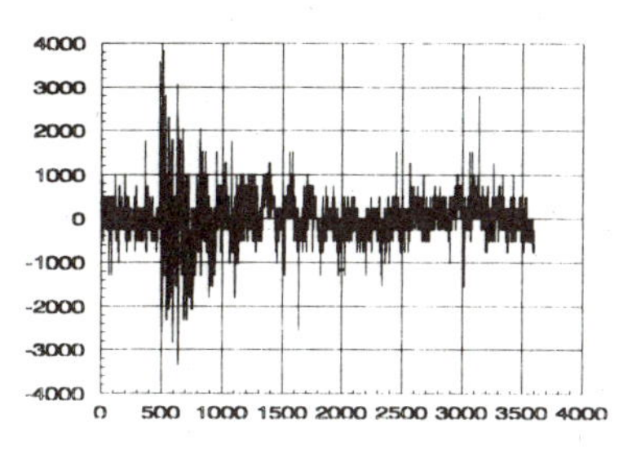

×1/156K (S)

Figure 7. Simulated acoustic signal received by negative polar connection of 110μm PVDF film

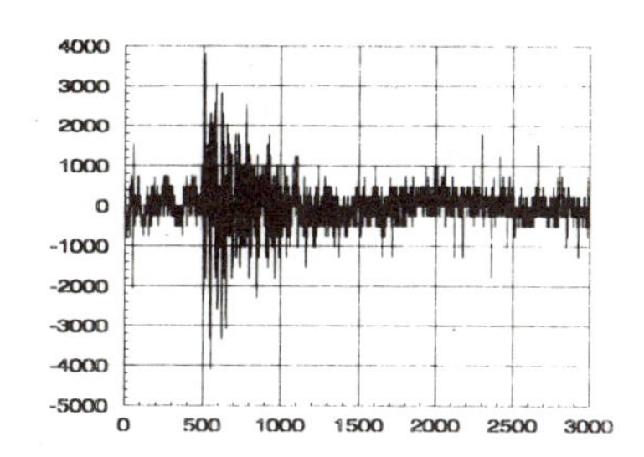

×1/156K (S)

Figure 8. Simulated acoustic signal received by positive polar connection of 110μm PVDF film

The experiment result showed that there was no significant difference between the two conditions, which means that the change of polar will not have great influence on signal receiving sensitivity.

4. The influence of differential connection on signal receive

Differential connection can reduce offset disturbance signal and enforce isolation effect. In this experiment, we folded the PVDF film. The signal cable was led from the central of the film and the other two top and bottom sides were grounded. Shown in figure 9 and 10:

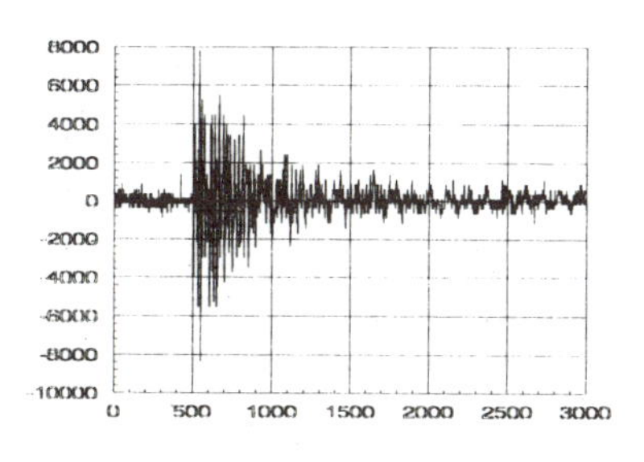

×1/156K (S)

Figure 9. Simulation acoustic signal received by folded 110μm PVDF film

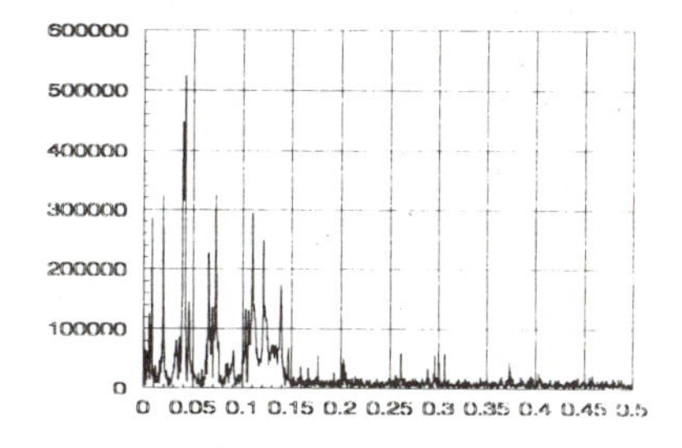

×156.3K (Hz)

Figure 10. Signal frequency analysis of folded 110μm PVDF film

The result showed that the signal received by the folded film was about the same as single layer film, but the disturbance signal was smaller than the latter one. This means that double film connected in differential function has a better common-model noise restrain ratio. The frequency analysis showed that, compared to single layer film, the central frequency of signal received by folded film shifted toward the acoustic frequency .

5. The influence of backing material on signal receive

In the experiment, two kinds of materials were selected as the backing. One was PEFT and the other was polymethyl methacrylate. The result showed no significant difference between these two materials.

CONCLUSIONS

This paper have conducted series experiments testing the possible influence factors on the characteristics of PVDF AE sensor. From the experiment we can see that:

1. Sensitivity

Physical differences , i.e. the membrane area, single or double layer and the way of polar connection will not have significant influence on PVDF membrane's sensitivity on signal receive.

Different backing materials can have some influence on its censoring without much significance.

The thickness of the membrane has a great influence on the sensitivity, the thicker the membrane, the higher the sensitivity. Sensitivity is also influenced by the acoustic coupling among the PVDF membrane, the protective membrane and the backing material. The better the coupling is, the higher the sensitivity is.

2. Range of frequency response:

PVDF has wider frequency response range than PZT. Different thickness of the membrane won't significant influence the central response frequency. When the membrane is folded the frequency band shift toward acoustic frequency. Also, the central frequency response shift toward acoustic frequency with smaller membrane area.

3. Application

PVDF piezoelectric membrane is a promising material for acoustic emission sensor. It can has similar sensitivity to PZT under appropriate equipment. Then, with flexibility, wide frequency range and low economical expense, it is supposed to have better application in industry.

REFERENCES

1. Weiheng Huang, *The Improvement of High Technology High Polymer Material*, Chemical Industry Press, China (1994)
2. Zhenming Yuan, *Acoustic Technology and Its Application*, Mechanical Industrial Press, China (1984)
3. Habeger et al, Using neoprene-faced PVDF transducers. J. Acoustic Soc. Am. 4:84 (1988)
4. Habeger CC and Wink WA, *Development of a Double-element Pulse echo PVDF Transducer*, Ultrasonic 1990, V28:52(1990)
5. Petra Hammes, *Infrares Matrix Sensor Using PVDF on Silicon*, Delf University Press Strvinweg 1 2628 CN Delft, the Netherlands. (1994)
6. Drouiillard TF et al, *Industrial Use of Acoustics Emission for NDT*, Monitoring Structural integrity by acoustic emission. ASTM STP 571(1957)
7. Lovinfer AJ, *Development In Crystalline Polymers*. Applied Science, London, 195 (1982).
8. P T Moseley and A J Crocker. *Sensor Materials*. IOP Publishing Ltd. (1996)

MICROSTRUCTURE AND TEXTURE INFLUENCES ON ULTRASONIC QUANTITIES FOR WELDING STRESS ANALYSIS

Uwe Arenz, Eckhardt Schneider

Fraunhofer Institut Zerstörungsfreie Prüfverfahren IZFP, Saarbrücken, Germany

INTRODUCTION

Light metals and especially aluminum alloys are of increasing interest for applications which have been dominated by steel. Applications of aluminum alloys for automotive bodies, train waggons and ship constructions are already under way. Among all joining techniques, welding is still a widely applied one and it is known that the residual stress states due to welding strongly influence the dynamic and static behavior of a welded component. But in contrast to steel welding, there are only limited possibilities to relieve the welding stresses in Al-components by post weld treatments. Hence, a nondestructive technique which enables a fast and locus continuous stress analysis would be a very helpful tool to optimize the welding parameters and the post weld treatments. Ultrasonic techniques to evaluate stress states are nondestructive ,fast and easy to apply and they are already in industrial use for some specific applications. Whether or not and with what accuracy the ultrasonic techniques enable the evaluation of stress states in and around Al welds depends heavily on the microstructural change in the area of interest and its influence on the material dependent quantities needed for the quantitive stress analysis.

OBJECTIVE

Based on the principal equations, describing the influence of elastic strain states on the ultrasonic velocities, the relative change of ultrasonic velocities can be expressed in terms of principal stresses /1/:

$$\frac{v_{ii} - v_L}{v_L} = \frac{t_L - t_{ii}}{t_{ii}} = \frac{A}{C} \cdot \sigma_i + \frac{B}{C}\left(\sigma_j + \sigma_k\right) \tag{1}$$

$$\frac{v_{ij} - v_T}{v_T} = \frac{t_T - t_{ij}}{t_{ij}} = \frac{D}{K} \cdot \sigma_i + \frac{E}{K} \cdot \sigma_j + \frac{F}{K} \cdot \sigma_k \tag{2}$$

$$\frac{v_{ij} - v_{ik}}{v_{ik}} = \frac{t_{ik} - t_{ij}}{t_{ij}} = S\,(\sigma_j - \sigma_k) \tag{3}$$

The first index of the velocity v_{ij} describes the propagation direction, the second gives the polarisation direction of the ultrasounic wave. The relative change of velocities can be formulated into the relative change of the ultrasonic time-of-flight t_{ij} if the pathlength of the wave remains unchanged. σ_i , σ_j and σ_k are the principal stresses. The influence of each of the three principal stresses is weighted by factors A till F, K and S which are specific functions of the material dependent second order elastic constants Youngs- and shear moduli and of the third order elastic constants l, m and n. It can be seen from the equations that the accuracy of the elastic constants has a significant influence on the accuracy of the stress analysis.

The objective is to evaluate the second and third order elastic constants using samples cut from a welded plate in such a way that different microstructural states are propagated by the ultrasonic waves. Using samples, cut parallel and perpendicularly to the principal directions of the texture in the plate and in the weld, the texture influence on the elastic constants is to evaluate. With these elastic constants, the factors A till F, K and S are to calculate and to be used to evaluate the stress states in an second plate made from the same material welded in the same way as the one used to cut the samples /2/.

SAMPLES AND CHARATERIZATION OF MICROSTRUCTURE

An Al alloy and a welding technique which are already widely used are made available for the experimental investigations by an industrial partner. The aluminium sheets of the hardenable ISO grade designated alloy AlCu6Mn (equivalent to Al 2219) are 440 x 230 x 4,5 mm^3 in size. The weld feed material is AlMg4,5Mn. The sheet is welded by TIG (DCVP) and preheated to 150 °C. The welding velocitiy is aproximatly 20-30 mm/min.

The main characteristics of microstructure in the base material (BM), in the heat affect zone (HAZ) and in the weld seam (WS) are shown in micrographs in Fig.1 and summarized in Table 1.

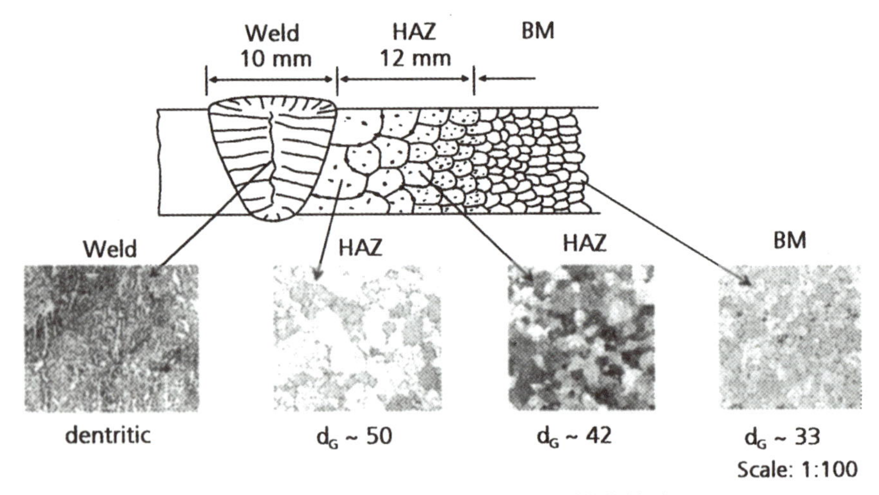

Figure 1: Micrographs and sketch of the welded Al-plates

Table 1. Characterization of microstructure.

Weld Seam	Heat Affected Zone	Base Material
dendritic	larger grains	fine grains
welding slack inclusions	precipitation of copper	
	$A^G = 2000\ \mu m^2$	$A^G = 1100\ \mu m^2$
	d^G = 40 till 50 μm	d^G = 30 till 35 μm
width: 10 mm	width: 12 mm	

d^G is the grain diameter and A^G is the grain area evaluated using the area method.

INFLUENCE OF MICROSTRUCTURE ON ELASTIC CONSTANTS

The velocities of longitudinal and shear waves in samples of about 10 x 10 mm^2 cut from different positions in the BM, HAZ and WS are determined. Using these velocities and the density of the material, the second order elastic constants are calculated. The samples are cut to the mentioned size in order to relieve the residual stress state. Table 2 gives mean values for the different microstructural states.

Table 2. Second order elastic constants.

	Yougs Modulus [GPa]	Shear Moduls [GPa]	Poisson Ratio
Base Material	72,5 ± 1,6 %	27,0 ± 1,1 %	0,341 ± 1,5 %
Heat Affected Zone	71,0 ± 1,6 %	26,4 ± 1,1 %	0,341 ± 1,5 %
Weld Seam	71,5 ± 1,6 %	26,7 ± 1,1 %	0,343 ± 1,5 %

As expected, the changes of the second order elastic constants are only within the error bars. The softening effect in the area around the weld can only be seen if locus continuous measurements are done under constant measuring conditions /2/.

A common technique to evaluate the third order elastic constants is the measurement of the relative change of ultrasonic velocities or times-of-flight as function of the elastic strain of a sample. Figure 2 shows a schematic of the experimental procedure and the measuring result. The solid and dashed lines indicate the influence of the difference in microstructure in the the heat affect zone and in the base material. The slopes of the measured lines also called acousto-elastic constants (AEC) are functions of the second and third order elastic constants. Using the second order constants, evaluated in the before mentioned way, the third order constants are evaluated.

Using samples of the base material, from the heat affected zone and from the weld and performing the measurements after the position of the ultrasonic probes have been systematically changed, the above mentioned measurements were performed. The calculated third order elastic constants (TOEC) l, m, n are shown in Figure 3 as function of the distance from the centerline of the weld.

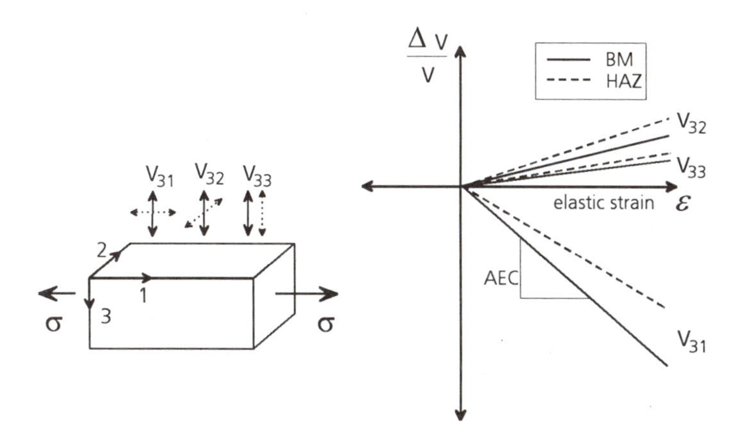

Figure 2. Schematic of the experimental procedure to determine the third order elastic constants.

As to be seen, the microstructural change with the distance from the weld has no significant influence on the elastic constant n. Although the overall change of the individual values of the constant m is only within the error bars, there is a systematic change in the area of the weld. The same result holds for the constant l also.

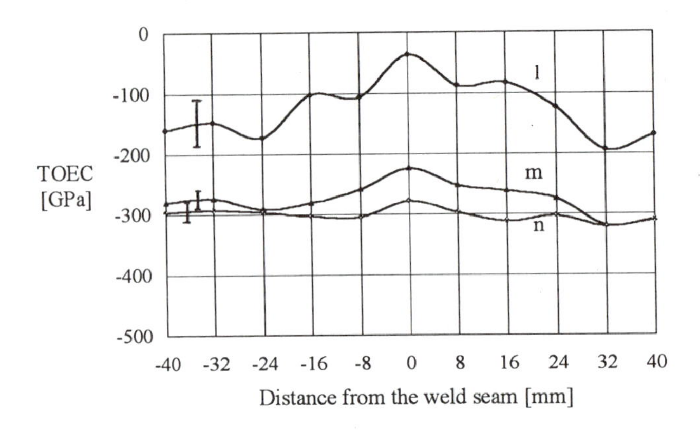

Figure 3. Changes of the third order elastic constants (TOEC) with the distance from the centerline of the weld.

INFLUENCE OF TEXTURE ON ELASTIC CONSTANTS

The easiest but sufficient way to characterize texture by ultrasonic means is the measurement of the ultrasonic time-of-flight t as function of the polarization direction of a linear polarized shear wave. Using the before mentioned 10 x 10 mm^2 sized samples, the shear wave polarization was at first parallel to the rolling direction (RD) of the plate, then perpendicular to it (transverse direction TD) or parallel and perpendicular to the welding direction, respectively. The relative change of $(t_{RD} - t_{TD})/t_{TD}$ (called texture grade) evaluated for the stress free samples is a measure for the texture of the samples. Table 3 gives the mean values of the texture grades and of the shear moduli determined for all samples cut from comparable positions.

Table 3. Texture grades and second order constants of samples cut from the weld seam, from the heat affected zone and from the base material.

	$(t_{RD} - t_{TD})/t_{TD}$ [10^{-3}]	G_{RD} [GPa]	G_{TD} [GPa]
WS	1.0 ±0.1	27.0 ±1%	27.0 ±1%
HAZ	3.5 ±0.1	25.9	25.7
BM	1.5 ±0.1	26.6	26.5

It can be seen that there is about the same texture in the base material and in the weld seam, whereas there is a stronger texture in the heat affected zone. Texture causing a $(t_{RD} - t_{TD})/t_{TD}$ value of 3.5 10^{-3} can be considered as small and of negligable influence as it has been confirmed: the shear moduli change only within the measuring error. Hence, there is also no texture caused change of the Youngs modulus. Also the third order constants were found to be the same for samples cut parallel and perpendicular to the rolling and the welding direction, respectively.

For comparison reasons, similar investigations were performed using samples of a welded plate of Al MgSi 0,7 with a texture grade of about 9 10^{-3}. The shear modulus for the rolling direction G_{RD} is found to be 28 GPa and the modulus G_{TD} for the transverse direction is about 26.6 GPa. The texture also causes a change in the third order elastic constants. The strain dependence of the velocity of the shear wave polarized parallel to the applied tensile load is shown in Figure 4 in order to demonstrate the texture influence. The experiment is done using a tensile test sample with its length direction parallel to the rolling direction (solid line) and using a sample with its lenght direction perpendicular to the rolling direction (dashed line).

The difference in the slopes found in case of the samples with a texture grade of about 2.5 10^{-3} is in the error bar. The influence of texture in the samples with a texture grade of about 9 10^{-3} on the slopes is significant.

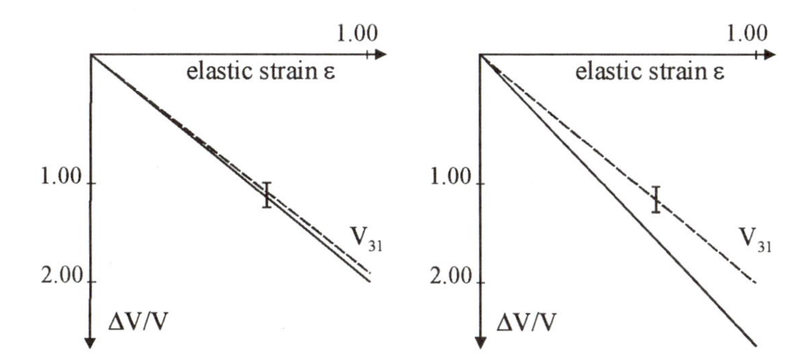

Figure 4. Relative change of shear wave velocity as function of the elastic strain of samples cut parallel (solid line) and perpendicular (dashed line) to the rolling direction of a slightly (left diagram) and a stronger (right diagram) textured Al-plate.

From these investigations it is concluded, that the texture influence on the second and third order elastic constants is negligible if the texture grade is smaller than about 3 10^{-3}. But the texture caused direction dependency of the ultrasonic velocities has to be taken into account.

ULTRASONIC EVALUATION OF STRESS STATES

Based on the described results, the stress state in a second Al-plate of the same alloy, welded in the same way as the previous one, was evaluated using ultrasonic techniques. In order to facilitate the data aquisition, the weld reinforcement was grinded. The Figure 5 shows the evaluated stress parallel to the weld σ_{II} along two traces perpendicular to the seam. The advantage of the ultrasonic stress analysis is the possibility to map the stresses in areas of interest as to be seen in Figure 6. The Figure shows the stress acting perpendicular to the weld seam.

The comparison of the ultrasonic results with these, evaluated using the drilling hole techique was very encouraging. With one exception the agreement of the stress values for both, the stresses parallel and perpendicular to the weld was within ± 20 MPa.

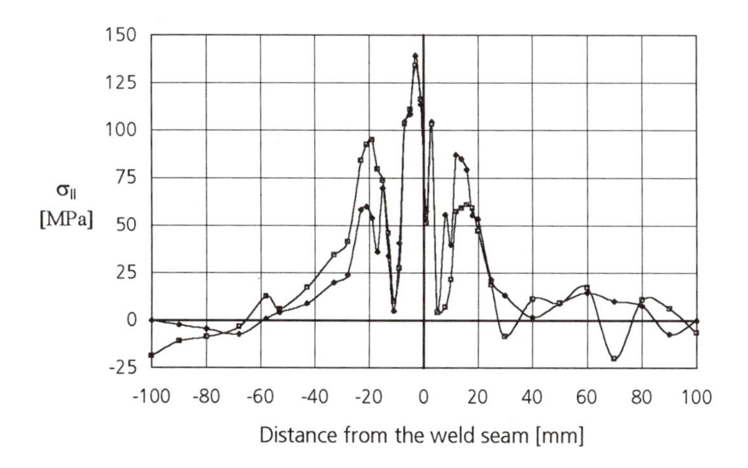

Figure 5: Stress parallel to the weld seam along two traces prependicular to the seam.

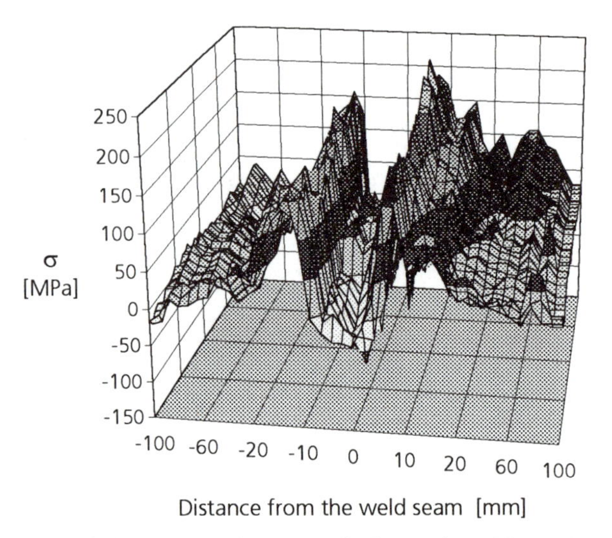

Figure 6. Map of the stress acting prependiculare to the weld seam in an welded Al-plate.

CONCLUSION

The ultrasonic evaluation of the stress states in welded structures of Al alloy AlCu6Mn is possible since the microstructural changes are not of significant influence on the elastic quantities needed for the quantitative stress analysis. Using one set of material dependend quantities, the stresses in the base material and in the heat affected zone can be evaluated. A second set of material dependent quantities is recommended for the stress analysis in the weld seam. That requires also the machining of the weld reinforcement.

Since the ultrasonic technique allows a fast and easy to apply stress analysis, it might be a helpful tool to optimize welding parameters and to localize stress inhomogeneities in order to support the post weld treatments.

REFERENCES

1. E. Schneider, Ultrasonic Techniques, *Structural and Residual Stress Analysis by Nondestructive Methods,* V. Hauk (ed) Elsevier Amsterdam, will be published in Dezember (1997).

2. U. Arenz, Gefügeabhängigkeit der materialspezifischen Kenngrößen zur Ermittlung der Schweißeigenspannungen in Aluminiumschweißnähten mittels Ultraschall, *Diplomarbeit,* Universität des Saarlandes, Fraunhofer Institut Zerstörungsfreie Prüfverfahren IZFP (1996).

3. N.M. Mourik, E. Schneider, K. Salama, Ultrasonic Characterization of Microstructural States in Aluminum Welds Depending on the Welding Parameters, *Nondestructive Characterization of Materials VIII*, R.G. Green, Jr. (ed) (1997)

RESIDUAL STRESS MEASUREMENTS IN FRONT OF A CRACK TIP WITH HIGH SPATIAL RESOLUTION BY USING BARKHAUSEN MICROSCOPY

Iris Altpeter[1], Gerd Dobmann[1], Norbert Meyendorf[1], Horst Blumenauer[2], Dirk Horn[2], Martin Krempe[2]

[1]Fraunhofer Institut für Zerstörungsfreie Prüfverfahren
66123 Saarbrücken, Germany
[2]Otto-von-Guericke Universität
39106 Magdeburg, Germany

INTRODUCTION

Upon loading a cracked specimen or a component, a plastic zone emerges in front of the crack tip. If an unloading takes place before the crack growth, residual stresses of first and higher order build up which influence the subsequent crack growth upon reloading. Of practical importance is the so-called WP (warm-prestress) effect which includes the possibility of increasing the brittle-fracture resistance of a pressure vessel by pre-straining. The detection of the local residual stress state in the plastically deformed zone at the crack tip by using high-resolution methods is thus an important prerequisite for the understanding of the fracture processes with broader aim of improving the damage tolerance of structural materials. Whereas there have been several reports on the application of X-ray measurement techniques,[1-4] up to now only limited experience is available concerning the using of micromagnetic effects for the determination of local residual stress states in crack tip region.[5] It is the aim of this paper to verify the methodical basis for the determination of residual stress states in crack tip region.

METHODS

Barkhausen microscopy

The ferromagnetic domain structure as well as the Bloch-wall movements are strongly influenced by macro- and micro-residual stress state.[6] This effect is used to evince the residual stresses in components by using the magnetic Barkhausen noise. The Barkhausen noise refers to electric impulses induced in an inductive sensor mainly by irreversible 180° Bloch-wall jumps during the advanced by continuous cycling through the hysteresis curve. For the residual stress measurement purpose maximum amplitude of

Nondestructive Characterization of Material VIII
Edited by Robert E. Green Jr., Plenum Press, New York, 1998

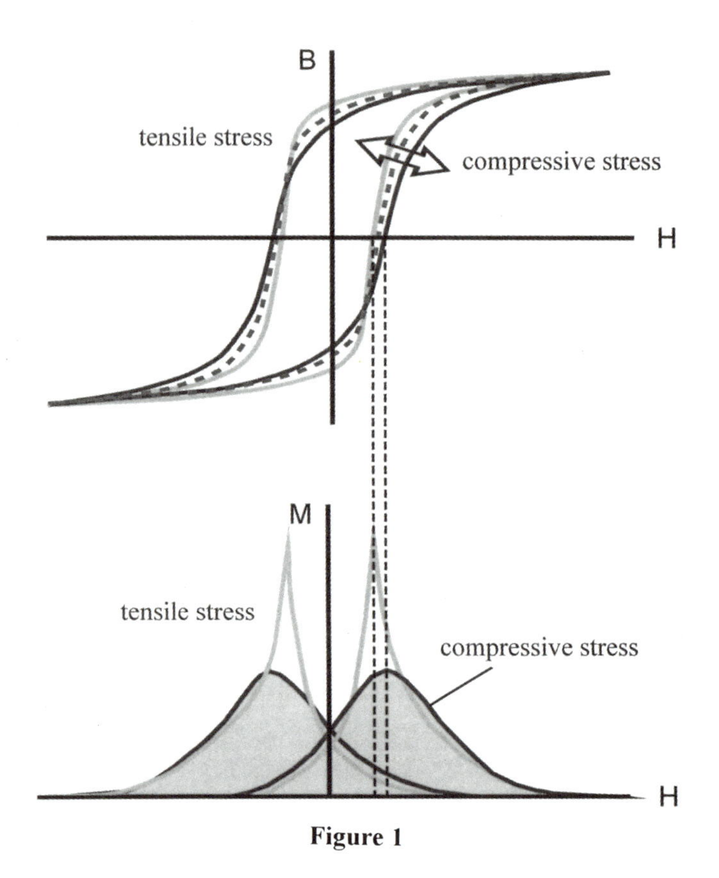

Figure 1

the rectified Barkhausen-noise events M_{MAX}, as well as the coercive field strength H_{CM} derived from the magnetic Barkhausen noise may be used (figure 1).[7] Under the influence of residual and loading stresses due to tension and pressure, the measuring parameters change in a characteristic way for every steel quality as figure 1 shows schematically. A quantitative micromagnetic residual stress measurement requires the calibration of the procedure with a corresponding set of calibration specimen. The calibration is performed by means of X-ray determined residual-stress values in the tensile test.

Corresponding to the demands for a high lateral resolution, the BEMI (Barkhausen and Eddy Current Microscopy)[8]-method for the assessment of Barkhausen noise in the crack tip region was used. Figure 2 shows schematically the conception of the monitoring Barkhausen and eddy current microscope. The test rig consists of the microscopy unit, the motor control, a PC for the image processing, file and the documentation module. The microscopy unit consists of a portal device with free unwinding cable with a x-y-z manipulation unit which controls the sensor over the specimen magnetized by an electromagnet. By software controlling the miniaturized inductive sensor can be positioned in all directions with an accuracy of 1 µm. Measurements on microprofiled test specimens show a detection limit better than 5 µm as well as a resolution of about 10 µm, depending on sensor type and specimen material.[9] Advantage of the micromagnetic determination of residual stresses is the high lateral resolution in comparison to the X-ray procedure as well as the short measuring time (of approx. 1 sec per measuring point).

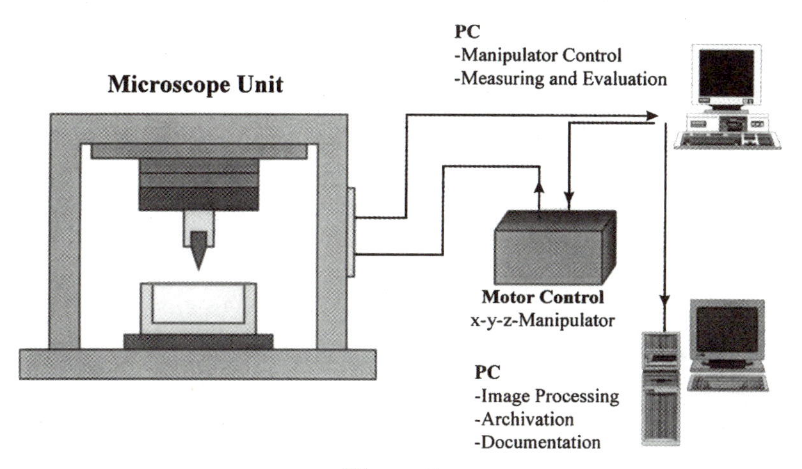

Figure 2

MEASUREMENT RESULTS

Local residual stress measurements are of practical significance for understanding material damage processes. A fracture mechanic evaluation at crack tips always requires assumptions about stress state. Therefore, it was undertaken the measurement of the residual stress field in front of the crack tip by using the Barkhausen microscopy. The used material is: a pressure vessel steel 10 MnMoNi 5-5 with a typical bainitic microstructure. Charpy-V-Notch specimens were produced and a fatigue crack of 3 mm length was initiated under cyclic loading. With a pendulum impact testing machine, a part of the specimens were subsequently pre-strained at 300 °C so that crack tip blunting or stable crack propagation growth occurs.

A slice with a thickness of 3 mm was cut out of the middle of each impact bending Charpy-V specimen. This specimen segment was grounded and polished. Thereafter, a layer of 0.1 mm thickness of the specimen surface in the crack tip region (10 x 10 mm) was removed by electrolytic atching in order to eliminate residual stresses due to previous treatment. H_{CM} and M_{MAX} distributions were recorded of the not prestressed Charpy-V-Notch specimens.

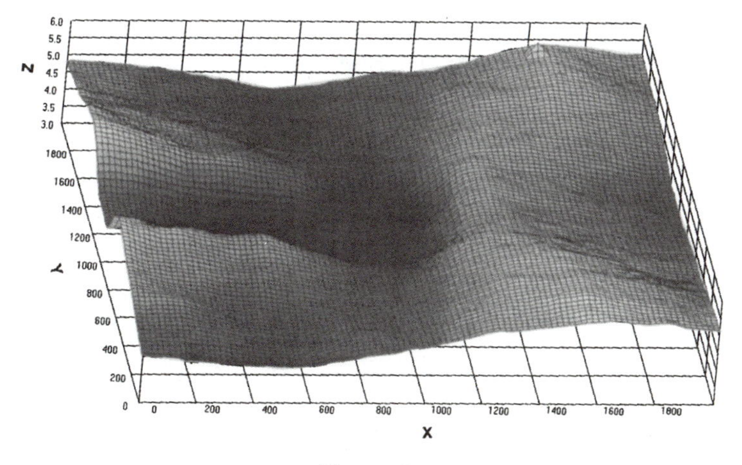

Figure 3

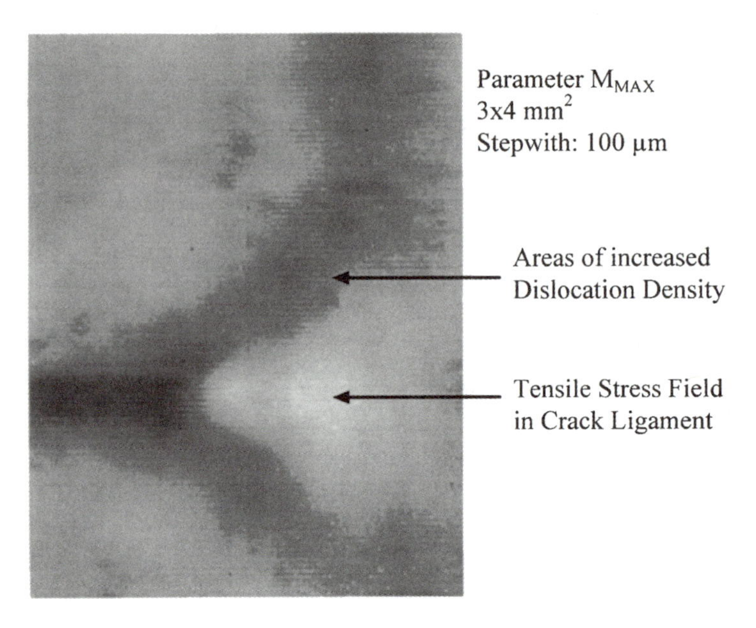

Figure 4

Figure 3 is an image of the Barkhausen noise amplitude at the crack tip. The lower noise amplitude at the left side of the image corresponds to the end of the crack. Remarkable are the higher amplitudes in the right half, shaped like a bone. This can be interpreted, with caution, as the stress in front of the crack tip. The differently loaded specimens show zones of increased dislocation density (45° angle from the crack tip) in the crack tip region and a drop-shaped tensile stress field in front of the crack tip (figure 4). With increasing load, the bone-shaped residual stress field is more and more overlapped by the zone of plastic deformation.

Further investigations have been done at the prestressed CT-specimens. The measured parameters M_{MAX}, H_{CM}, and HWB_{75} (the width at 75% of M_{MAX}) were recorded according to the BEMI method. Residual stress values obtained by X-ray measurements were used for the calibration of the micromagnetic measuring parameters. Micromagnetic and X-ray measured parameters were correlated by means of neural network analysis with the network being taught before by a training specimen set.

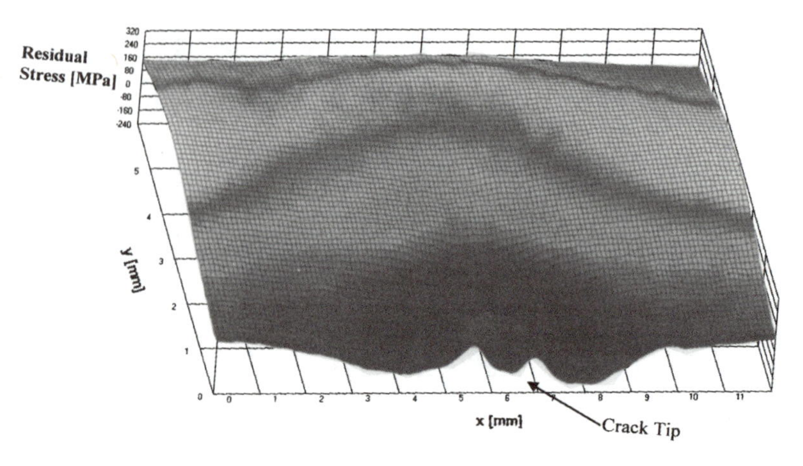

Figure 5

A good correlation was found between residual stress values obtained with micromagnetic and X-ray measurements respectively. The residual stress distribution obtained with micromagnetic measurements is presented in figure 5.

Half-circle-shaped compressive residual stresses were measured in the crack tip region (diameter 2 mm). Ahead of the crack tip, compressive stresses change more and more into tensile residual stresses having their maximum in a distance of 5 mm from the crack tip.

CONCLUSIONS

A good correlation was found between residual stress values at the crack tip obtained with micromagnetic and X-ray measurements; respectively. For the first time quantitative results on residual stress propagation and distribution of plastic deformations in front of a crack tip of pre-stressed specimens were obtained nondestructively, with high local resolution, and within a very short measuring times. However, the measuring results presented here have to be supported statistically by further research with a larger population of specimens.

REFERENCES

1. E. Welsch. *Einfluß rißspitzennaher Eigenspannungen auf die Ausbreitung von Ermüdungsrissen*, Dissertation, Universität Karlsruhe (1985).
2. H. Blumenauer and U. Räcke, Residual stresses on fatigue crack growth behavior of case-hardening steels, in: *Residual Stress*, V. Hauk, ed., DGM Informationsgesellschaft Verlag (1992).
3. M. Krempe and H. Blumenauer. *Röntgenographische Eigenspannungsmessungen an Rißspitzen*, 9. Sommerkurs Werkstofftechnik Uni Magdeburg, (1996).
4. M. Krempe. *Werkstoffmechanisches Verhalten von postulierten Anrissen in druckführenden Komponenten mit vorbeanspruchter Rißspitze bei Belastung infolge rascher Abkühlvorgänge*, Zwischenbericht zum Forschungsvorhaben des BMBF, Förderkennzeichen 1500988, (1996).
5. I. Altpeter, I. Detemple, R. Kern and H. Blumenauer. *Hochauflösende zerstörungsfreie Ermittlung lokaler Gefüge- und Eigenspannungszustände mit Hilfe der Wirbelstrom- und Barkhausenmikroskopie*, DGM-DVM-Tagung Werkstoffprüfung Bad Nauheim 341-349 (1995).
6. E. Kneller. *Ferromagnetismus*. Springer Verlag, Berlin (1962).
7. I. Altpeter and W. Theiner, Spannungsmessung mit magnetischen Effekten, in: *Handbuch für experimentelle Spannungsanalyse*, C. Rohrbach, ed., VDI-Verlag, Düsseldorf (1990).
8. I. Altpeter and W. Theiner. *Vorrichtung zum ortsaufgelösten, zerstörungsfreien Untersuchen des magnetischen Barkhausenrauschens*, DE Patent Nr. 235387 (1994)
9. J. Bender, Barkhausen noise and eddy current microscope (BEMI), *Rev. of Progress in Quantitative Nondestructive Evaluation* 16 (1997).

NONDESTRUCTIVE STRESS MEASUREMENT IN STEEL USING MAGNETOSTRICTION

Tomohiro Yamasaki and Masahiko Hirao

Faculty of Engineering Science
Osaka University
Toyonaka, Osaka 560, JAPAN

INTRODUCTION

Nondestructive measurement of residual stress plays an important role in assuring the structural safety. Several methods, such as ultrasonics and X-ray diffraction, have been studied to improve the accuracy of the stress evaluation. Among these methods, the magnetic method can be easily applied to ferromagnetic materials. In the magnetic stress measurement, we use the magnetic properties which depend on the stress.[1] The stress dependence is based on the domain realignment induced by the stress. When the ferromagnetic material is magnetized, the dimensional change, called the magnetostriction, appears.[2] The external magnetic field causes the domain realignment, which results in both magnetization and magnetostriction. The stress also restructures the domains. Then the magnetic properties shows the stress dependence. For a low carbon steel, the magnetostriction increases as the magnetization, and starts to decrease when the field becomes strong, showing the maximum value. In this study, we apply the maximum magnetostriction[3] to the nondestructive stress measurement.

STRESS DEPENDENCE OF MAGNETOSTRICTION

Individual grains of polycrystalline steel are divided into many magnetic domains. Each domain is magnetized parallel to one of the crystallographic axes <100> and is slightly elongated in the magnetization direction. In the demagnetized state, the domain distribution is random. External magnetic field expands the domains magnetized parallel to the field at the sacrifice of the other domains as illustrated in Fig. 1. As a result of the realignment of the spontaneous elongation, the dimensional change appears in the magnetization direction as a whole specimen. When the domain reorientation is almost completed by the strong field, the rotation of the domain magnetization starts to occur. Then the domain magnetization direction is no longer parallel to the crystallographic axis, which induces the decrease in the magnetostriction. The magnetostriction shows the maximum value, which depends on the domain structure in the demagnetized state.

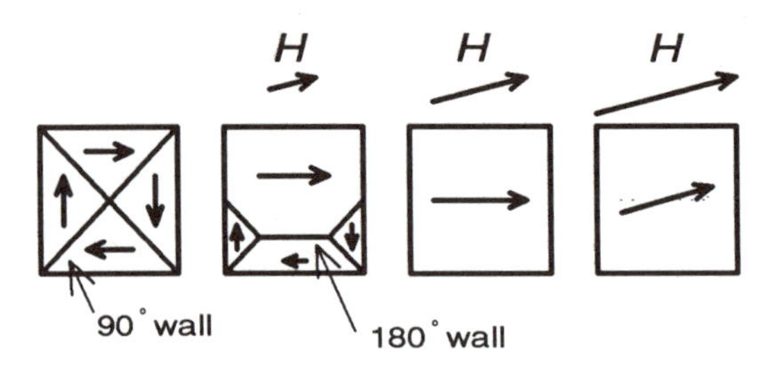

Figure 1. Domain wall movement induced by external magnetic field.

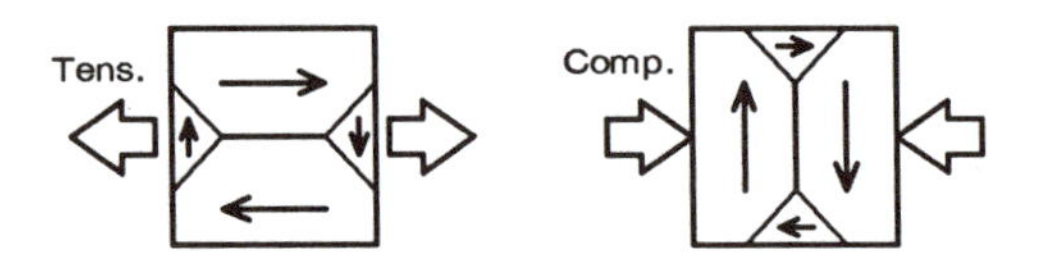

Figure 2. Domain realignment induced by stress.

If the stress is applied, the 90° domain walls move so as to expand the domains magnetized either parallel to the tensile stress or perpendicular to the compressive stress, while 180° walls remain still, as shown in Fig. 2. The resultant magnetostriction appears without accompanying any increase in the magnetization. The stress-induced domain realignment in the demagnetized state alters the maximum magnetostriction. It can be predicted that the tensile stress decreases the maximum magnetostriction while the compression increases it for the magnetization parallel to the stress. On the contrary, the maximum magnetostriction will increase with tension and decrease with compression, if the magnetization is normal to the stress.

UNIAXIAL LOADING TEST

Experimental Procedure

We prepared both tensile and compressive specimens of JIS-SS400 low carbon steel. The tensile specimens are 400 mm long with 39 mm width and the compressive specimens are 39 mm squared, both being 11 mm thick. To investigate the effect of the magnetic anisotropy introduced in the manufacturing process, two types of specimens were machined from as-rolled plate. The specimens being loaded parallel to the rolling direction are called specimens R, and those being normal are called specimens T. We also prepared the specimens from the annealed plate, which are loaded parallel to the rolling direction and are called specimens HR.

Figure 3 shows the experimental setup for the loading test. We measured the magnetostriction, while increasing the magnetic field strength stepwise by adjusting DC driving current to an electromagnet. To detect fractional changes in the strain, semiconductor strain gauges were used. Biaxial gauges measured the magnetostriction in both directions parallel and perpendicular to the field. Two gauges were attached on both surface of the specimen to cancel the effect of the bending. The magnetic field was applied either parallel or perpendicular to the uniaxial applied stress. Prior to the magnetostriction measurement at each stage of the loading, we demagnetized the specimen by decaying AC field to minimize the effect of the magnetic hysteresis.

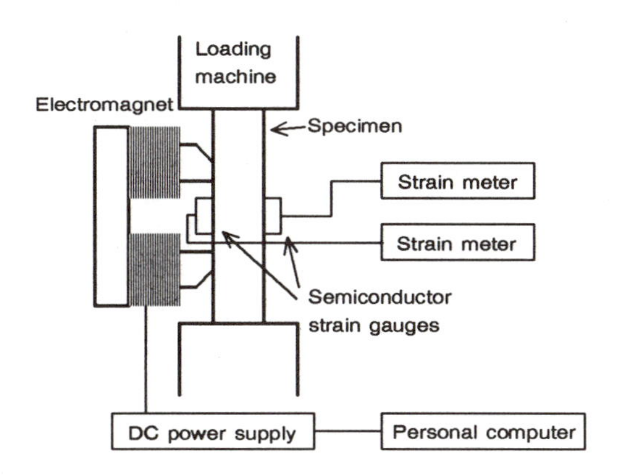

Figure 3. Experimental setup.

Results and Discussion

Figure 4 shows the magnetostriction curves of specimen R for the magnetization parallel to the tensile stress. As predicted, the maximum magnetostriction in the magnetization direction decreases with the tensile stress. Under strong tension, the magnetostriction decreases monotonically and the maximum disappears. It is because the tensile stress realigns the domains to such an extent that the 180°-wall movement and the rotation magnetization are dominant even in the demagnetized state. Both magnitude and stress dependence of magnetostriction normal to the magnetization direction are about half of those in the magnetization direction. This ensures that what we measure is the Joule magnetostriction which keeps the volume of the specimen unchanged. In this study, we focus on the magnetostriction in the magnetization direction. Figure 5 shows the stress dependence of the magnetostriction for the magnetization parallel to the compressive stress. The compressive stress increases the magnetostriction. As shown in Fig. 6, for the magnetization normal to the stress, the tensile stress increases the maximum magnetostriction, while the compressive stress decreases it. However, the maximum

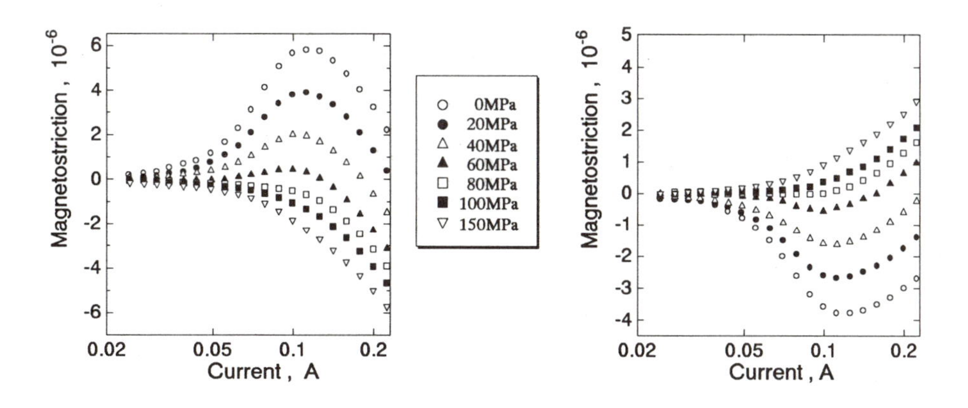

(a) Parallel to magnetization (b) Normal to magnetization.

Figure 4. Magnetostriction curves of specimen R for magnetization parallel to tensile stress.

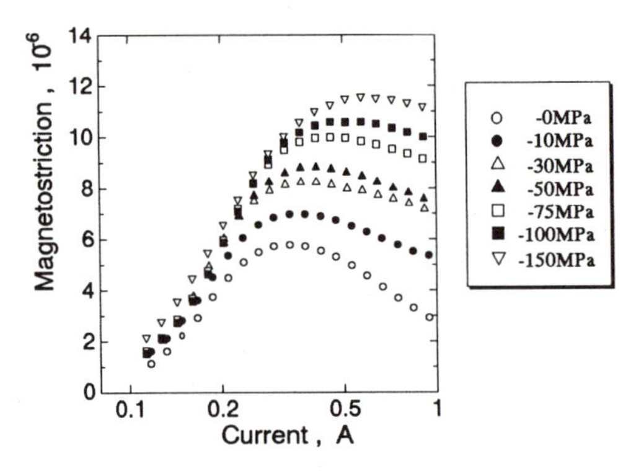

Figure 5. Magnetostriction curves of specimen R for magnetization parallel to compressive stress.

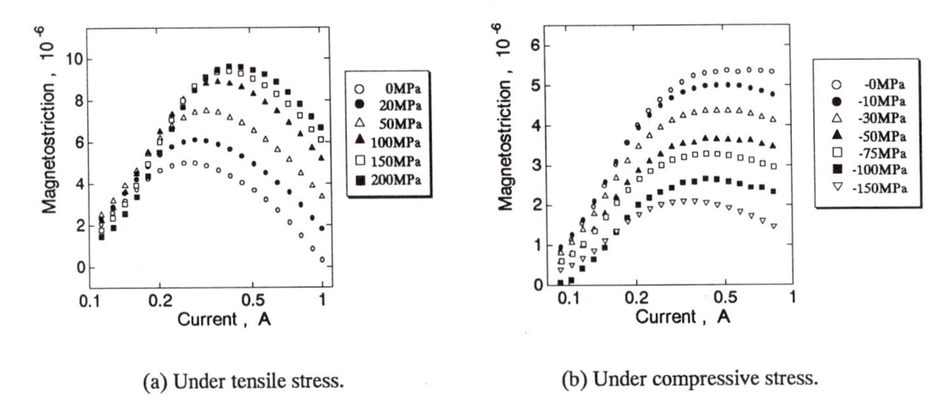

(a) Under tensile stress.

(b) Under compressive stress.

Figure 6. Magnetostriction curves of specimen R for magnetization normal to stress.

appears even under a strong compression. Because the domains magnetized in the plane normal to the compressive stress expands, the 90°-walls bounding the domains parallel and normal to the field remains.

Figure 7 summarizes the stress dependence of the maximum magnetostriction for specimens R,T and HR. Comparison of the results of specimens R and T reveals that the effect of the magnetic anisotropy is negligible. However, the effect of the microstructural change by the annealing is not negligible.

NONDESTRUCTIVE RESIDUAL STRESS MEASUREMENT

Using the results obtained in the loading test as master curves, we measured the residual stresses in butt welded steel plates nondestructively. We prepared the butt welded plates from the same steel plate as the loading test specimens. The dimensions, along with the measuring points, are shown in Fig. 8. At each measuring point, the magnetostriction was measured in the

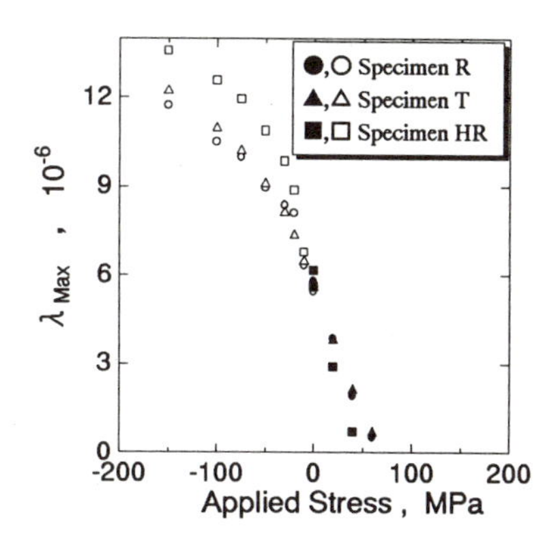

(a) Magnetization parallel to stress.

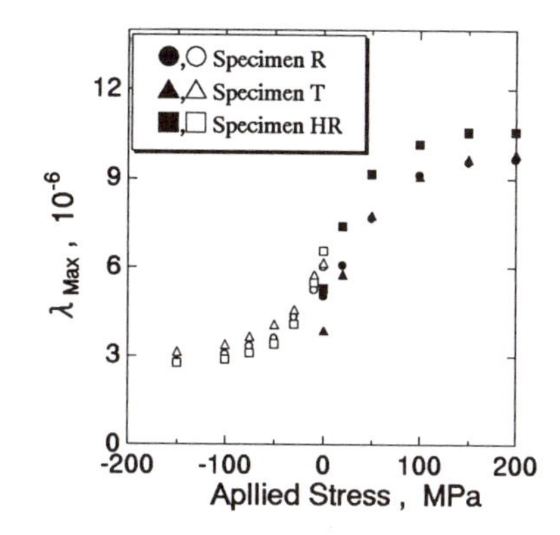

(b) Magnetization normal to stress.

Figure 7. Stress dependence of maximum magnetostriction.

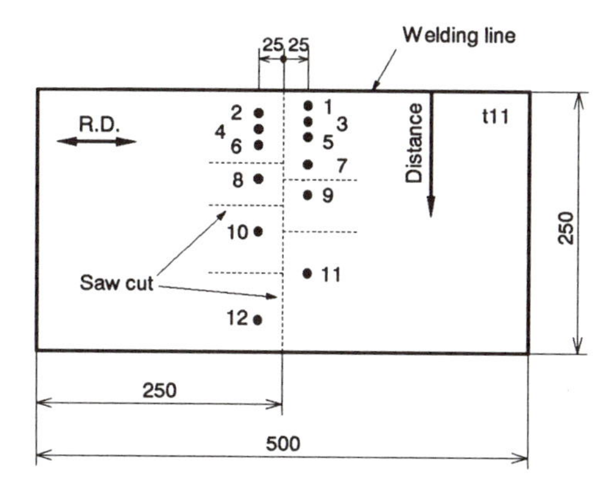

Figure 8. Butt-welded steel plate.

directions both parallel and normal to the welding line, because these directions coincides with the principal stress directions.

After the measurement of the maximum magnetostriction, the residual stresses were evaluated as follows. We assumed that the residual stress is uniaxial either parallel or normal to the welding line. For example, at measuring point 7, the maxima were 9.34×10^{-6} and 3.69×10^{-6} for the magnetization parallel and normal to the welding line, respectively. Assuming that the stress is parallel to the welding line, the stress candidates are -53MPa and -56MPa, as illustrated in Fig. 9. In the same manner, they will be 130MPa and 22MPa, if the stress is normal to the welding line. Then the stress is predicted to be -55MPa and parallel to the welding line, because the difference between two stress candidates are the smaller in this case. The evaluated stresses are plotted in Fig. 10 together with the strain gauge measurements. The magnetostrictive method predicted that the stress component parallel to the welding line, σ_1, is dominant at all measuring points. Comparing with the strain gauge measurements, the accuracy of the magnetostrictive method is revealed to be better than 50MPa. The error appearing near the welding line can be attributed to the microstructural change, which was introduced during the heating process.

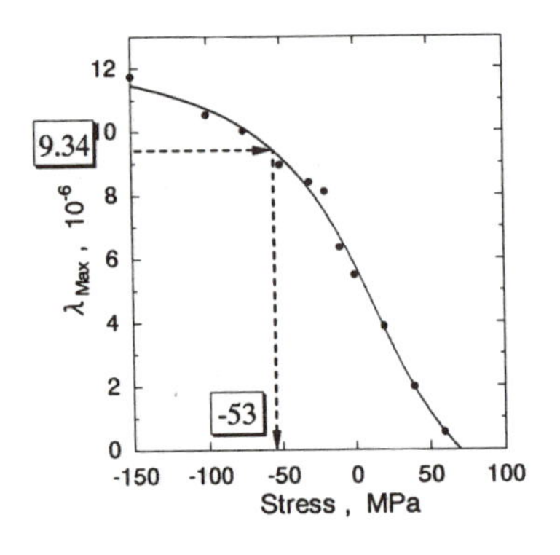

(a) Magnetization parallel to stress.

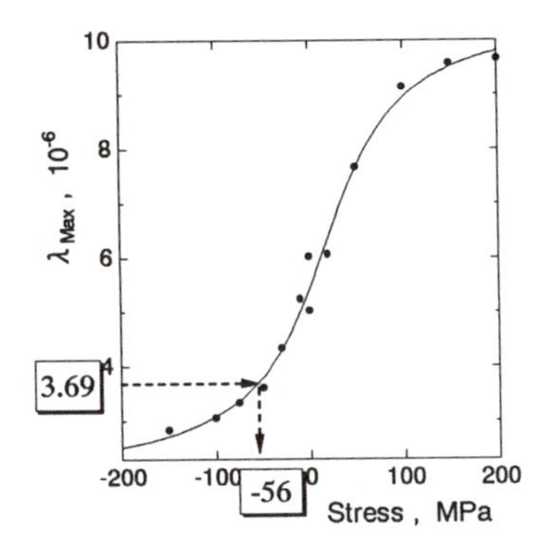

(b) Magnetization normal to stress.

Figure 9. Master curves for stress parallel to rolling direction.

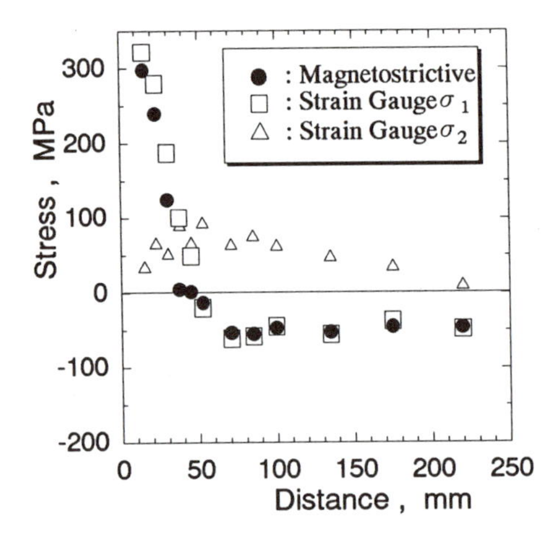

Figure 10. Results of residual stress measurements.

CONCLUSION

The stress dependence of the maximum magnetostriction was applied to the residual stress measurement in the butt welded steel plates. The accuracy was as good as 50MPa. Considering that the method is completely nondestructive, the magnetostrictive method is useful in the residual stress measurement.

REFERENCES

1. D.L. Jiles, Review of magnetic methods for nondestructive evaluation, *NDT Intl.*, **21**-5:311 (1988) .
2. R.M. Bozorth, *Ferromagnetism*, Van Nostrand, New York, (1951).
3. T. Yamasaki, S. Yamamoto, and M. Hirao, Effect of applied stresses on magnetostriction of low carbon steel, *NDT&E Intl.* **29**-5:263 (1996).

ELECTROMAGNETIC ACOUSTIC SPECTROSCOPY IN THE BOLT HEAD FOR EVALUATING THE AXIAL STRESS

H. Ogi and M. Hirao
Graduate School of Engineering Science
Osaka University
Toyonaka Osaka 560, Japan

INTRODUCTION

Nondestructive technique for evaluating the bolt's axial stresses has long been an important subject in order to ensure the safety and the reliability of structures. The torque wrench method is most widely used, but it often presents unfavorable results due to the friction in the bolt threads and between the nut and the work. In place of this method, several ultrasonic techniques have been studied [1-4]. Responding to the load, the transit time of the ultrasonic wave along the length changes due to the increase of the propagation distance and the acoustoelastic effect. The measurement of the transit time can give the axial stress. Most of the previous techniques, however, need some coupling materials to provide the acoustic coupling to the bolt head, which requires special skill and inhibits an easy and quick measurement. Because there exists a large number of bolts in a structure, we need more practical technique. The authors presented a couplant free method using electromagnetic acoustic transducers (EMATs) [5]. Although the phase shift of the shear wave showed a good linearity to the axial stress, the requirement for the initial phase value which is easily changed by the temperature and the bolt length is still a problem in practice.

In this paper, another EMAT technique is proposed, which is free from the temperature and the bolt length. The technique relies on the electromagnetic acoustic resonance [6] in the bolt head. An EMAT is specially developed to generate and detect the axially polarized shear wave reverberating within the bolt head. A superheterodyne spectrometer system excites the EMAT by the high-power rf bursts and measures the amplitude spectrum of the axial shear wave. Several resonant peaks are selected for the stress measurement by comparing with the reference spectrum obtained with a hexagonal rod of the same dimension. Each peak displays the resonant mode, which corresponds to the shear waves propagating in a specific region in the head; at the lower mode, the shear wave deformation is concentrated near the outer surface region, and at a higher mode, it penetrates inward. When the bolt is

Nondestructive Characterization of Material VIII
Edited by Robert E. Green Jr., Plenum Press, New York, 1998

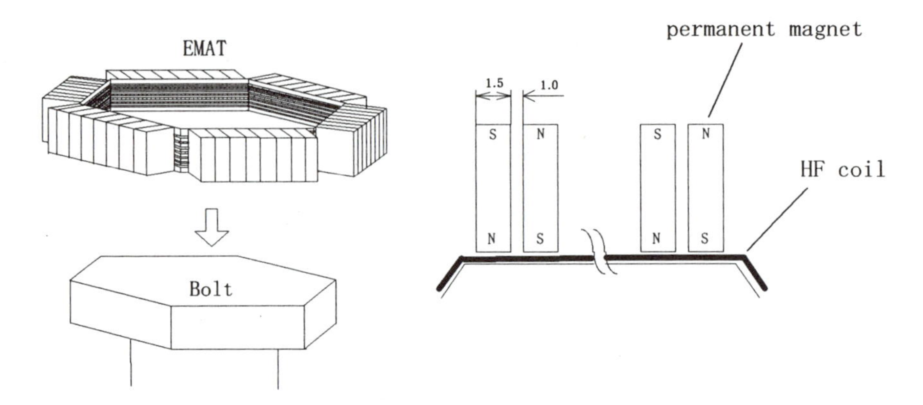

Figure 1 The EMAT configuration for generating and receiving the shear wave propagating in the circumferential direction with the axial polarization.

tightened, the resonant frequency shifts depending on the stress in each region of vibration. The stress field varies in a complicated manner in the head and the resonant frequencies indicate the stress representing the different regions. Change of the resonant frequency ratio thus indicates the magnitude of the whole stress field proportional to the axial stress, being independent of the temperature and the bolt's size and material. The technique is inherently free from the effect of the bolt length because the measurement is done only in the head.

EMAT AND MEASUREMENT SYSTEM

Figure 1 shows the EMAT configuration for the generation and reception of the axial shear resonance in the head of a bolt. It consists of thin permanent magnets arrayed to serve the alternate bias field on the side faces, and the copper wire solenoidally wound in a the hexagonal shape. When the driving current is applied to the coil, the shearing force in the axial direction arises periodically and generates the shear wave propagating to the circumferential direction with the axial polarization.

The superheterodyne-EMAT system [6] is used for measuring the resonant frequencies as shown in Fig.2. The rf-burst signal of 100μs duration drives the EMAT. The same EMAT receives the highly overlapped echoes. The received signals are fed to the quadrature phase sensitive detectors and the in-phase and out-of-phase outputs are extracted. The analog integrators are used to integrate these signals and the amplitude spectrum is determined by the root of the sum of the squares of the integrator outputs. The resonant spectrum is obtained by sweeping the frequency of the input burst signal and getting the amplitude spectrum as a function of the frequency.

RESONANT MODES OF THE AXIAL SHEAR WAVE

When the sample is a cylindrical rod, the resonant mode of the axial shear wave is easily determined by solving the frequency equation [7,8]. Calculation of the deformation distribution at the resonant modes concludes that at a lower resonant mode, the shear wave propagates near the outer surface region and as the mode becomes higher, the oscillating region penetrates to inside. We consider that the similar phenomena occurs in the case of the hexagonal rod, although the reso-

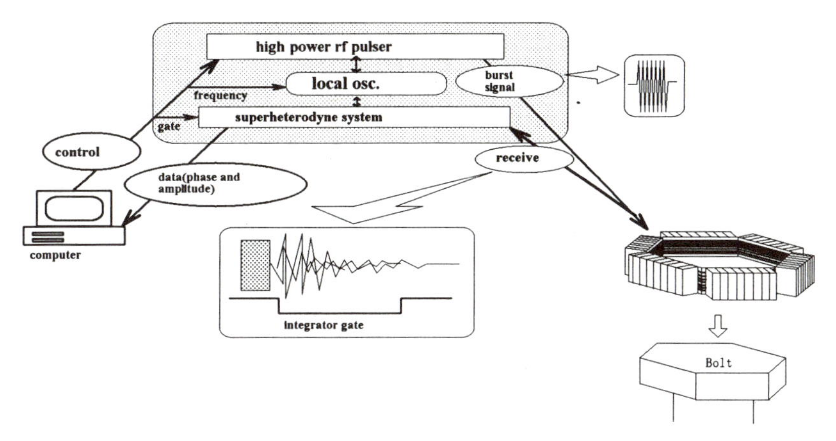

Figure 2. Superheterodyne-EMAT system for acquiring the resonant spectrum.

nant modes will not be explicitly determined with a simple frequency equation. Figure 3 presents the measured resonant spectra for the head of the M24-bolt (JIS-S45C) and the steel hexagonal rod with 20mm sides, which has nominally the same shape with the bolt head. The spectrum for the hexagonal rod is constructed purely by the axial shear wave because the length of the rod is long enough (250mm) and no mode conversion occurs at the side surfaces. But, the spectrum in the bolt head may contain other modes as well as the axial shear modes due to the mode conversion. We are only interested in the axial shear modes, and selected several peaks, from f_a to f_g, referring to the hexagonal spectrum pattern. These peaks don't interface with other resonant peaks when shifted by the stress application.

For searching the probing region of a resonant mode, a two-dimensional FEM calculation is done. The cross section perpendicular to the axial direction is divided into 13,824 regular triangle elements of 0.41mm sides. The driving line-sources, which act along the axial direction, are placed on the sides of the hexagon at the same geometry with the permanent magnet array. The deformation normal to the

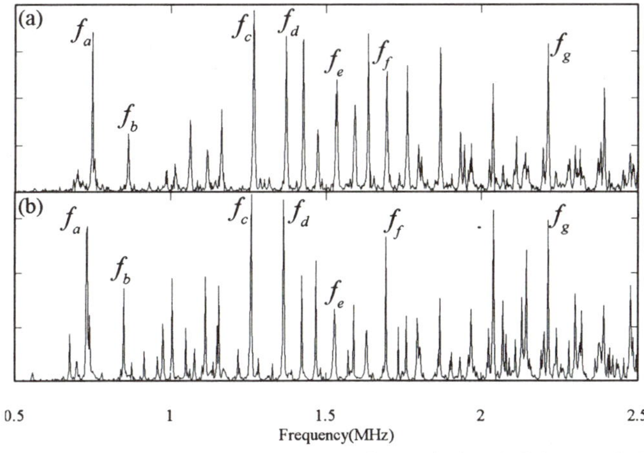

Figure 3. Measurement of EMAR spectra for (a) the head of the M24-bolt and (b) the hexagonal rod.

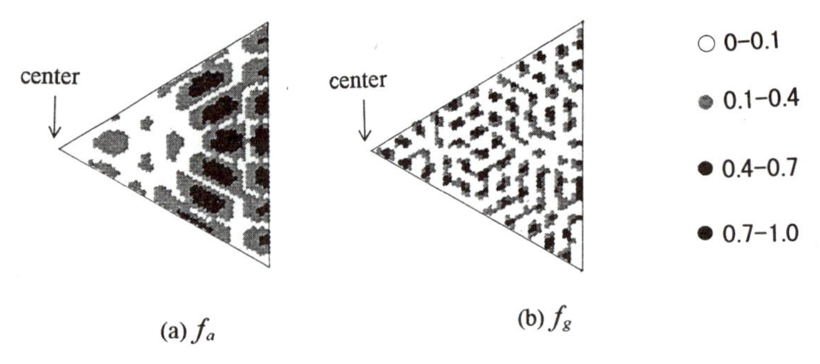

Figure 4. Distribution of the amplitude for (a) the low mode (f_a) and (b) the high mode (f_g).

area is only considered. The frequency-domain response is obtained by sweeping the frequency and calculating the resonant amplitude near the source region, which should be proportional to the measured resonant intensity. The calculated spectrum were favorably compared with the measurements of Fig.3(b), but it involved several resonant frequencies which were absent in the measurement, indicating that all modes are not detected. Figure 4 presents the amplitude distribution at the two frequencies corresponding to f_a and f_g in Fig.3. As expected, the amplitude of the lower mode are forced to concentrate near the outer region, while the higher mode oscillates in the whole region of the head.

RESONANT FREQUENCY SHIFT DUE TO AXIAL LOADING

Generally, the *n*th resonant frequency (f_n) is approximately expressed by $f_n=\eta_n c/L$, where η_n is the coefficient specific to the mode, c the velocity, and L the representative length of the sample. The resonant frequency shifts with the applied stress because of the acoustoelasticity, which establishes the linear dependence of the velocity on the stress:

$$\frac{f_n^{(\sigma)}-f_n^{(0)}}{f_n^{(0)}}=C\sigma, \tag{1}$$

where $f_n^{(\sigma)}$ and $f_n^{(0)}$ express the resonant frequencies at the stress σ and at the stress free state, respectively. C is an acoustoelastic constant. When a bolt is tightened, a stress profile arises in the head of the bolt. We denote f_L for the lower resonant mode whose propagation region is concentrated near the outer surface and σ_L for the averaged stress over this region; and f_H for the higher mode and σ_H for the averaged stress over the region of the higher mode. Taking the ratio $f_L^{(\sigma)}/f_H^{(\sigma)}$ and considering $1>>|C\sigma|$, Eq.(1) is reduced to

$$\frac{f_L^{(\sigma)}}{f_H^{(\sigma)}}=\frac{f_L^{(0)}}{f_H^{(0)}}\left(1+C_L\sigma_L-C_H\sigma_H\right), \tag{2}$$

with the acoustoelastic constants of C_L and C_H. Assuming that σ_L and σ_H are proportional to the axial stress σ_{axis}, we have

$$\frac{f_L^{(\sigma)}}{f_H^{(\sigma)}} = \frac{f_L^{(0)}}{f_H^{(0)}} + C_{L/H}\sigma_{axis}, \tag{3}$$

where $C_{L/H}$ is another acoustoelastic constant related to the two modes. It should be noted that the first term of the right hand side in Eq.(3) depends only on the coefficients η_L and η_H which two modes are involved, being free from the material, temperature, and initial bolt dimension. Actually, we measured $f_a^{(0)}/f_g^{(0)}$ for three hexagonal rods made of the low carbon steel, an aluminum alloy, and an austenitic stainless steel. Their values were between 0.3370-0.3381.

EVALUATION OF THE AXIAL STRESS

Four steel bolts of the same material (JIS-SCM435) were used for the measurement. They were two bolts of 185mm long (L_1 and L_2) and two of 155mm long (S_1 and S_2). The axial stress was monitored by strain gages attached on the surface of the cylindrical part. We placed the EMAT on the head and tightened the nut by torque wrench. Figure 5 shows the frequency shift of eight modes obtained from the bolt L_1. We find that f_a is most noticeably shifted by the stress and as the mode becomes higher order, the resonant frequencies tend to be less sensitive to the stress. The stress profile in the head region is considerably complex and it is unclear which stress component or which combination is responsible for the resonant frequency shift. But, it is reasonable to consider that all the stress components are continuously varied from the inner to the outer region and some of them are changed their signs. This leads to the prediction that the averaged stresses over the whole region are smaller than those in the outer region, resulting in the observation in Fig.5.

Figure 6 plots f_a/f_g against the axial stress obtained from the strain gauges. We calibrated $C_{L/H}=2.58\times10^{-5}$ MPa^{-1} for the four bolts. The measurement error

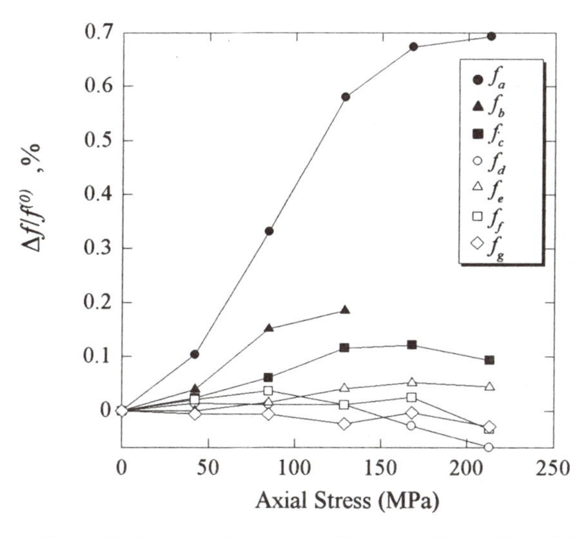

Figure 5. Resonant frequency shifts responding to the axial stress.

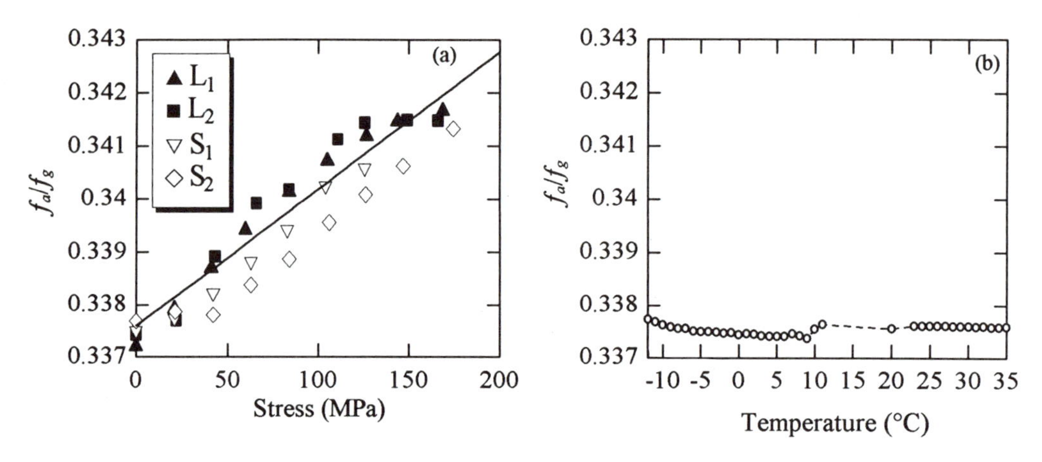

Figure 5. Responses of the frequency ratio to (a) the axial stress and (b) temperature without stress.

caused by the scattering of the initial value is estimated to be within 15MPa and that by the data scattering is within 25MPa. We also show the temperature effect of the measurement in Fig.6 (b) in the same vertical scale with Fig.6 (a). The measurement fluctuation due to the temperature change is only 8MPa for the range from -12° to 35°.

CONCLUSION

The EMAR method presented here is capable of evaluating the bolt-axial stress without the information of the temperature and the bolt length. The axial shear wave spectrum contains a series of resonant peaks and the individual modes have the unique ribration regions in the bolt head. The axial stress changes the resonant frequencies depending on the stress profile in the each region. But, other factors such as the temperature shift them at the same rate and their effects are minimized by taking the ratio of the two resonant frequencies. This method was revealed to provide the axial stress for the M24-bolts with 25-MPa error band, and the error caused by the temperature effect is estimated less than 8MPa. Further work should involve the investigation of the initial value ($f_L^{(0)}/f_H^{(0)}$) and the acoustoelastic constant ($C_{L/H}$) for a large number of bolts of different materials and different forming processes.

REFERENCES

1. J. S. Heyman, Exper. Mech., **17**: 183 (1977).
2. J. S. Heyman and E. J. Chern, J. Test. Eval., **10**: 202 (1982).
3. S. G. Joshi and R. G. Pathare, Ultrasonics, **22-6**:270 (1984).
4. S. M. Zhu, J. Lu, M. Z. Xiao, Y. G. Wang, and M. A. Wei, J. Phys. IV, Colloq. (France), Journal de Physique IV (Colloque), **2**: 923 (1992).
5. H. Ogi, M. Hirao, and H. Fukuoka, Proc. of the 1st US-JAPAN Symp. on Advan. in NDT, 37 (1996).
6. M. Hirao, H. Ogi, and H. Fukuoka, Rev. Sci. Instrum., **64**: 3198 (1993).
7. W. L. Johnson, B. A. Auld, and G. A. Alers, Review of Progres in QNDE, Vol. 13, eds. D. O. Thompson and D. E. Chimenti (Plenum, New York, 1994), p.1603.
8. H. Ogi, M. Hirao, and K. Minoura, J. Appl. Phys., **81**: 3677 (1997).

DETECTION OF RESIDUAL STRESSES AND NODULAR GROWTH IN THIN FERROMAGNETIC LAYERS WITH BARKHAUSEN AND ACOUSTIC MICROSCOPY

Iris Altpeter[1], Gerd Dobmann[1], Silvia Faßbender[1], Jochen Hoffmann[1], Jane Johnson[1], Norbert Meyendorf[1] and Wolfgang Nichtl-Pecher[2]

[1]Fraunhofer Institut Zerstörungsfreie Prüfverfahren IZFP
66123 Saarbrücken, Germany
[2]Exabyte Magnetics GmbH
90429 Nürnberg, Germany

INTRODUCTION

In order to develop and optimize thin ferromagnetic layers which are used in combined write-read heads it is necessary to evaluate the residual stress state and defects in microstructure of these layers nondestructively. Residual stresses in ferromagnetic thin layers are due to non-optimized process conditions, undesired phase transitions or insufficient ductile adaptation to the substrate. The ratio of signal and noise and the sensitivity of the write-read heads are influenced negatively by residual stresses. The bad ductile adaptation between the substrate and the ferromagnetic layer leads to insufficient workability of the write-read heads.

The **B**arkhausen Noise and **E**ddy Current **Mi**croscope (BEMI) developed by the IZFP Saarbrücken enables fast and high-resolution measurement of magnetic parameters. These parameters are recorded for later data processing. Mechanical properties, especially residual stress state and hardness state of the specimen can be obtained from these data by means of appropriate calibration[1,2]. The **S**canning **A**coustic **M**icroscope (SAM) was used for the detection and characterization of nodular growth.

MEASURING METHODS

Barkhausen microscopy provides the possibility to detect high-resolution Barkhausen noise signals and enables line and area imaging of Barkhausen noise signals of ferromagnetic specimens with the help of appropriate sensor and positioning devices. Therefore, the specimen has to be excited by a variation of the magnetic field in time. During the magnetic hysteresis is passed through the Barkhausen noise occurs. An adjustable magnetization yoke guarantees the homogeneous magnetization of the specimen tangential to the measuring surface. The tangential field strength of the specimen is measured with a Hall probe. The high-frequency Barkhausen noise signal is detected by a miniaturized inductive sensor. This sensor is a magnetic circuit consisting of a small yoke with a very small air gap of approx. 0.1µm. The Barkhausen noise signal measured by this sensor at each measuring point will first be amplified and filtered. The option of an adjustable amplification and various selectable filters provides the adaptation to various testing tasks.

The following measuring parameters can be derived from magnetic Barkhausen noise: M_{MAX} - maximum of the noise amplitude and H_{CM} - value of external field at which the maximum occurs. The computer-controlled positioning unit allows the sensor to find selected measuring points or to produce meandric measuring scans. The integrated three-axes positioning system has a positioning accuracy of approximately 1µm for all directions. Additionally to the Barkhausen noise sensor a second Hall probe can be integrated in order to measure the distribution of the tangential field strength on the surface.

By means of acoustic microscopy near-surface regions of materials can be characterized. In SAM, the specimen surface is scanned by a focused ultrasonic beam[3,4]. The reflection amplitude of each measuring point is recorded, converted into gray scale values and finally into an image. A resolution of 1.5 µm is obtained at 1 GHz. Large aperture angles of the acoustic lens excite surface waves (so-called Rayleigh waves) with a penetration depth of wavelength λ (approx. 3-6 µm at 1 GHz). This enables a high-sensitivity detection of surface cracks.

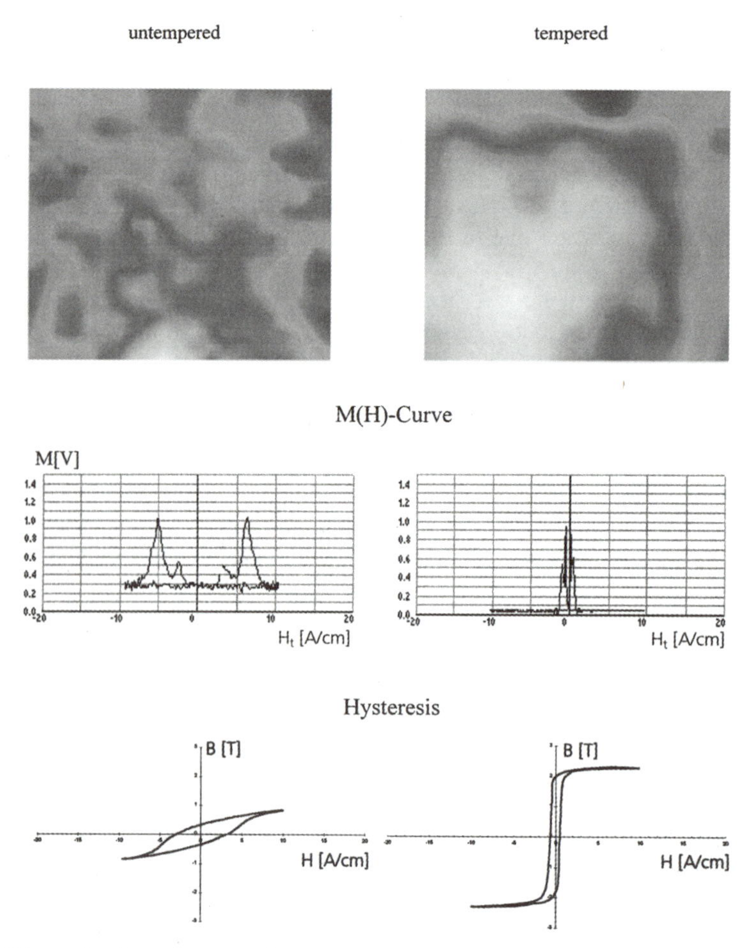

Figure 1. Comparison of untempered and tempered thin Sendust-layers

RESULTS

The specimens under investigation were Sendust layers sputtered onto a ceramic substrate. The specimens differ in layer thickness and applied heat treatment. Sendust is a FeSiAl alloy having ideal soft magnetic properties, especially after annealing. Figure 1 shows the difference between tempered and untempered layers. Characteristic and reproducible inhomogeneities can be seen in the different Barkhausen noise C-scans (Figure 1., upper part). These inhomogeneities correlate with the residual stress distribution, which will be demonstrated later. Moreover can be seen, that tempered specimens are more homogeneous than untempered specimens of the same geometry. This is mainly due to the different order of the lattice structure before and after tempering (550°C/1h). Sendust occurs untempered basicly in the unordered α phase which changes into the ordered DO_3 phase after tempering. In tempered specimens the observed inhomogeneities decrease corresponding to the layer thickness. This observation indicates: the larger the distance of the Sendust layer from the substrate the more independent, and thus more homogeneous this layer can be grow. There are huge differences in the measuring parameters for one scan which can be up to 20% of the corresponding average value.

Barkhausen noise profile curves (Figure 1., central part) are a fast way of detecting whether the specimen under investigation is a tempered or untempered thin layer. Whereas tempered specimens have comparatively high M_{MAX} noise signals in a very narrow region close to H_{CM}, untempered specimens show noise activity in a much wider region of the external field values. The higher H_{CM} values of the untempered specimens indicate that they are not as soft magnetic as the tempered specimens. I.e., only tempered thin layers show the ideal magnetic behavior desired for layer materials in inductive sensors: abrupt magnetic reversal, low coercivity field strength, high remanence. Hystereses (Figure 1., lower part) measured by means of a differerential coil device show a very good correlation with the corresponding noise profile curves.

The magnetic and magnetoelastic measurements at different texture states of different steel grades show that all dynamic magnetic and magnetoelastic measuring parameters depend on texture and stress state. In order to separate the two influences two measuring parameters independent of one another have to be applied. Thus, the measuring parameters M_{MAX} (maximum of the Barkhausen noise amplitude) and H_{CM} (position of the maximum in the magnetic field) were evaluated. In the testing task described here, i.e. the determination of the residual stress state in ferromagnetic layers, the parameters influencing the magnetic parameters are residual stress state, texture and layer thickness. I.e., for the solution of this task at least three independent measuring parameters have to be applied. The BEMI provides not only the measurement of M_{MAX} and H_{CM} but also signal full-width at 75%, 50% and 25% of the noise maximum.

By means of a X-ray residual stress analysis it was possible to obtain several reference stress values for each specimen. The used X-ray diffractometer work with an X-ray beam with a diameter of about 100 μm. Figure 2. presents an M_{MAX} area scan for a tempered specimen with a layer thickness of 2 μm in comparison to an X-ray area scan. The X-ray area scan is based on 25 measuring points with interpolated scanning increments of 0.5 mm. The magnetic image consists of 400 measuring points with increments of 400 μm. Measuring time for the X-ray measurements was approximately 5 hours per measuring point, for Barkhausen noise measurements only about 1 second. A good correlation was found between the two results. The high M_{MAX} values are caused by

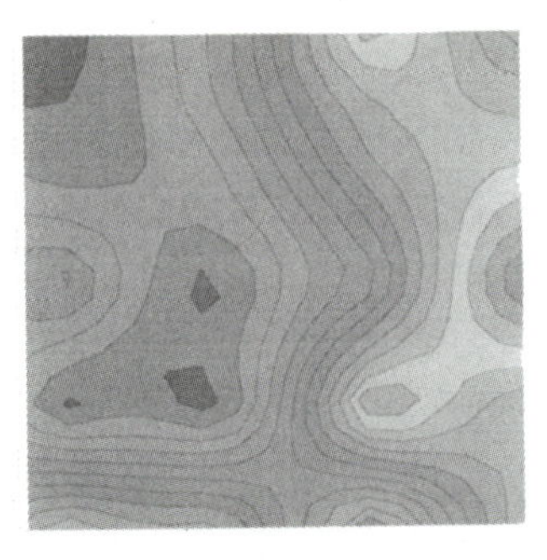

X-rax residual stress measurement:

5x5 measuring points, interpolated
stepwidth: 0.5 mm
X-ray-beam diameter: 100-200 µm

Residual stress range: 310 MPa - 400 MPa

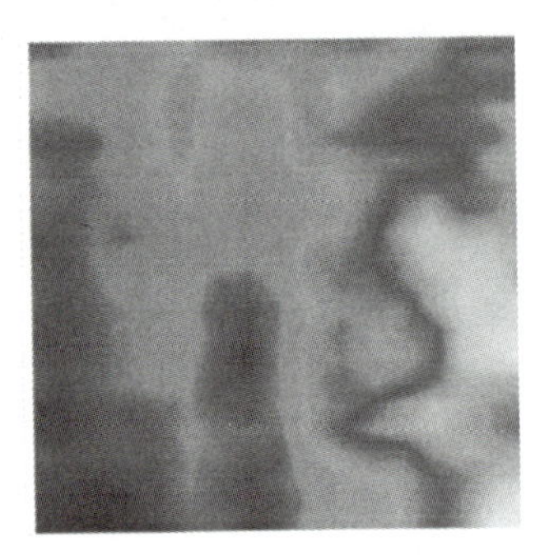

BEMI-Measurement - Parameter M_{MAX}:

20x20 measuring points
stepwidth: 100 µm

value range: 1.26 V - 2.84 V

Figure 2. Residual stress distribution in thin Sendust-layers

high tensile residual stress values, the low M_{MAX} values by low tensile residual stress values.

The results obtained with the acoustic microscope about nodular growth in the layers are used as pre-examination results. The same specimen were measured with the BEMI and the obtained Barkhausen noise measurements will be interpreted by means of the acoustic images. The aim of the measurements is the detection of nodular growth in intermediate layers by Barkhausen microscopy only.

Several different thin-layered specimens were measured by means of acoustic microscopy (SAM = **S**canning **A**coustic **M**icroscopy). Representative results were obtained for a set of specimens consisting of a ferritic layer material and a ferritic substrate.

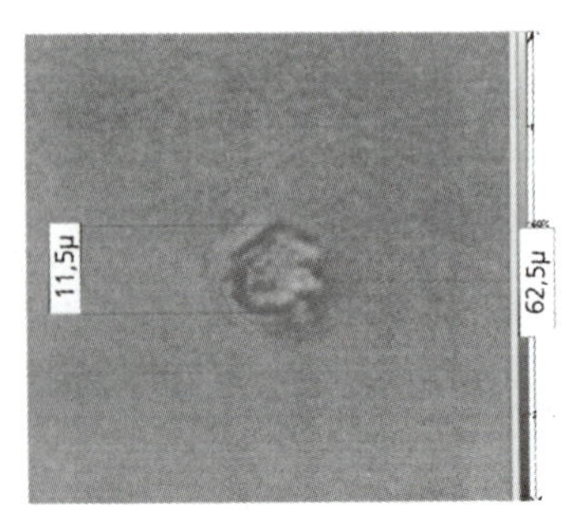

Figure 3a. Nodular growth in thin ferrite-layer

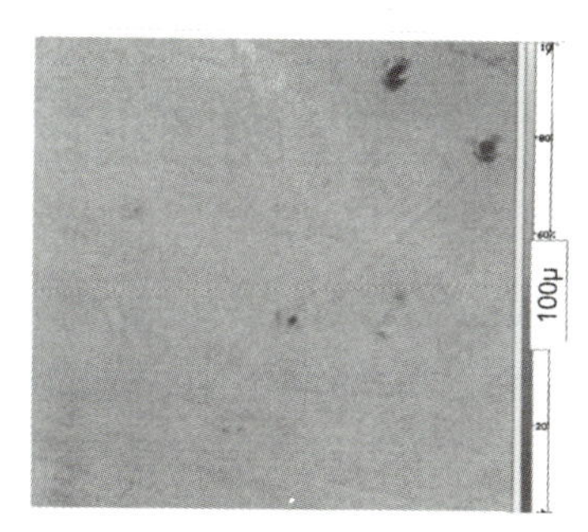

Figure 3b. Holes with different ranges in thin ferrite-layer

Two different kinds of voids can be seen on the acoustomicroscopic images. The first kind are circular structures with dimensions of 5-30 µm which can be identified as nodular growth (Figure 3a.). These structures can mainly be found in specimen with additional surface treatment. There is a discontinuity between the nodular growth areas and the surrounding material which cannot be passed by the acoustic wave. Therefore, a strong contrast is produced at the borders of the nodular growth areas. Thus, these defects can be detected and characterized easily.
The second kind are circular structures with a dimension of 1-3 µm (Figure 3b.). A lot more of these defects could be found than nodular growth areas. These defects can be indentified as holes in the layer material.

CONCLUSION

The measurements described here showed that Barkhausen microscopy allows qualitative and quantitative determination of residual stresses in ferromagnetic thin layers. Moreover, the BEMI enables fast qualitative imaging with high local resolution of residual stress distributions in specimens. A good correlation was found between residual stress values obtained by magnetic and X-ray measuremens. Reference residual stress values can be obtained by means of four-point-curvature instead of X-ray measurement. Thus, the calibration would no longer depend on the X-ray measurements bearing the big disadvantage of lower resolution and immense time efforts. Calibration by means of tensile tests is impossible because of the brittleness of the ceramic substrate. A coupling of micromagnetic and reference measuring values can be achieved by means of training with a neural network. With the help of acoustic microscopy growth defects in thin layers could be detected and characterized. Thus, new knowledge about the structure and properties of these defects were obtained.

Acknowlegement

The research work was sponsored by the German Ministry for Education and Research under the project number 03N8003.

REFERENCES

1. I. Altpeter, J. Bender, J. Hoffmann, M. Kopp, D. Rouget: Barkhausenrausch- und Wirbelstrommikroskopie zur Werkstoffcharakterisierung und Eigenspannungsmessung im µm-Bereich, DGzfP-Jahrestagung, Mai 1996, Dresden

2. I. Altpeter, S. Faßbender, U. Netzelmann, W. Nichtl-Pecher: Hochauflösende Materialcharakterisierung mit mikroskopischen Techniken, DGzfP-Jahrestagung, Oktober 1996, Jena
3. A. Briggs: 'An Introduction to Scanning Acoustic Microscopy', Microscopy Handbooks, **12**, Oxford University Press (1985).
4. S. Hirsekorn, S. Pangraz, G. Weides and W. Arnold: Appl. Phys. Lett. **67**, 745 (1995).

FATIGUE CHARACTERIZATION OF AISI 321 AUSTENITIC STEEL BY MEANS OF HTC-SQUID

Marco Lang,[1] Hans-Jürgen Bassler,[2] and Jane Johnson[1]

[1]Fraunhofer Institut Zerstörungsfreie Prüfverfahren IZFP
66123 Saarbrücken, Germany
[2]Lehrstuhl für Werkstoffkunde, Universität Kaiserslautern
67663 Kaiserslautern, Germany

INTRODUCTION

Austenitic steel of the grade AISI 321 (German Grade 1.4541) is often used in power station and plant constructions. The evaluation of the fatigue damage and thus the remaining lifetime of this and related austenitic materials is a task of enormous practical relevance. In spite of intensive research work it is still impossible to evaluate the fatigue damage of materials nondestructively in order to determine the remaining lifetime. AISI 321 austenitic steel forms martensite due to quasi-static and cyclic loading. This presupposes the exceeding of the threshold value of cumulated plastic strain. The main aim is to determine the fatigue damage of austenitic steel by characterizing the martensitic structure with the help of the SQUID measuring technique. Several specimen batches were evaluated and thereby the load amplitudes and the test temperature were varied (room temperature and 300°C).

CYCLIC DEFORMATION BEHAVIOUR OF AISI 321 AUSTENITIC STEEL

According to its good mechanical and technological properties, i.e. high toughness, as well as its corrosion resistance the test material AISI 321 (German grade 1.4541) is widely used, e.g. in the chemical industry and in power plant construction. The magnetizable phase fraction, i.e. δ-Ferrite, of the test material was reduced to less than 1 vol.% by annealing in the initial state. Sufficient amounts of mechanical energy due to plastic deformation lead to phase transformation from fcc austenite without diffusion to tetragonal or bcc ferromagnetic α'-martensite[1].

Nondestructive Characterization of Material VIII
Edited by Robert E. Green Jr., Plenum Press, New York, 1998

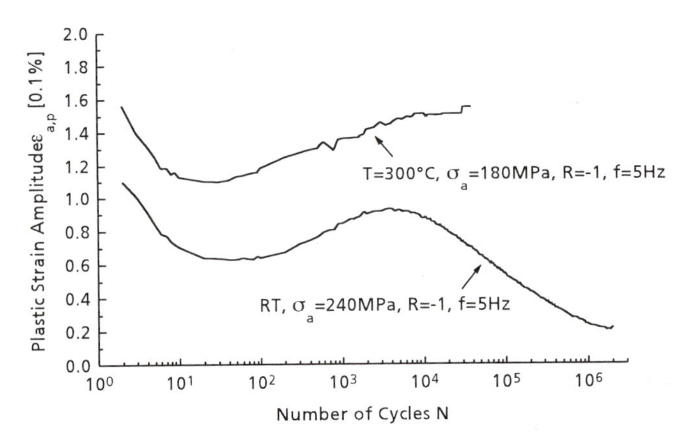

Figure 1. Both at room temperature and at T=300°C AISI 321 austenitic steel shows hardening within the first tens of cycles superseded by a softening. A significant secondary hardening only shows for experiments at room temperature.

Cyclic deformation curves for AISI 321 show for both room temperature and T=300°C hardening up to a few tens of cycles followed by softening that is only slight in case of T=300°C. For room temperature a secondary hardening that correlates with the beginning of the martensite formation[2] starts at about 4×10^3 cycles (Figures 1 and 2). Figure 1 shows the typical development of the plastic strain amplitude $\varepsilon_{a,\,p}$ of this test material for a stress amplitude σ_a=240MPa at room temperature and a stress amplitude σ_a=180MPa at T=300°C.

PRINCIPLES OF THE INVOLVED SQUID-TECHNIQUE

In the first paragraph the correlation of fatigue in AISI 321 austenitic steel and the formation of martensite has been shown. As the martensitic volume fractions are especially low for in-service-temperatures of about 300°C highly senisitive measuring systems are necessary. Systems on the basis of HTC-SQUIDs (High Temperature Super Conducting QUantum Interference Devices) combine a higher sensitivity than for example fluxgate-sensors with a higher practicability than Low-Temperature SQUIDs cooled with liquid Helium. Apart from portable cryostats for liquid Nitrogen also portable cooling devices such as Joule Thomson coolers or Stirling engines are available for HTC-SQUIDs. Here both a SQUID-magnetometer in a magnetically shielded box and a SQUID-system with pick-up-coils were used. The principles of the systems including hard- and software are given by Krause et.al.[3].

RESULTS OF THE SQUID-MEASUREMENTS AND THEIR DISCUSSION

The results discussed in this paper mainly refer to two sets of samples representing high-cycle-fatigue damage HCF at room temperature (HCF-I, σ_a=240MPa, R=-1) and at 300°C (HCF-II, σ_a=180MPa, R=-1). The samples were investigated in defined states of magnetization. The cylindrical samples representing the actually damaged material were cut of the center of fatigue specimens. Figure 2 shows the remanent magnetizations as a function of the number of cycles. As expected the flux densities are much higher for the

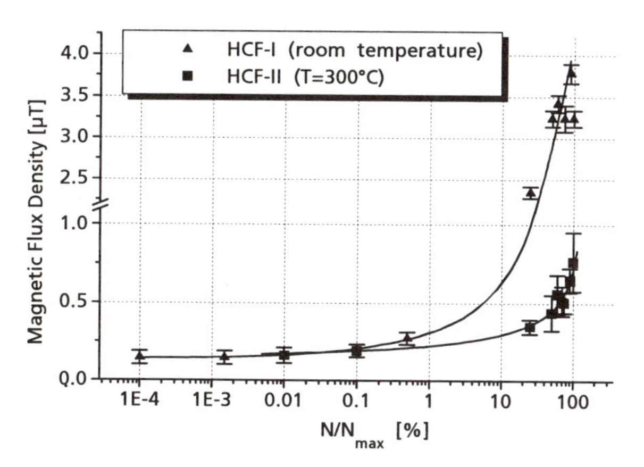

Figure 2. Magnetic Flux Density of the cylindrical HCF-I-specimens (room temperature) as a function of the N/N_{max} with $N_{max}=2\times10^6$ for HCF-I and $N_{max}=N_f=37209$ for HCF-II.

room temperature samples than for the samples fatigued at 300°C due to the much higher volume fractions of martensite at room temperature compared to 300°C[4]. Both sets of samples show an exponential growth of the remanent magnetic field with the number of cycles. Similar results were found for a comparable austenitic steel by Otaka et.al.[5] with a LTSL-SQUID-System (cooled with He_{liquid}).

Neutron depolarization measurements proved a linear increase of the mean size of the interacting magnetic phase parts with the martensitic volume fraction[6]. This leads to an exponential rise of the magnetic remanent field with the martensitic volume fraction.

Compared to those for room temperature the results for the 300°C specimens show a higher variation. This is due to two main reasons. First, the maximum number of cycles limited by specimen failure, and the total amount of martensite are lower than at room temperature. This leads in a first step to an increase in the relative variation of the measured values. For 12 samples with 22300 cycles the values for the remanent magnetization varied around a mean value of $293.5\Phi_0$ with a standard deviation of $65\Phi_0$. Second the number of cycles to fracture N_f varied from 30359 to 45787. As the N/N_f-

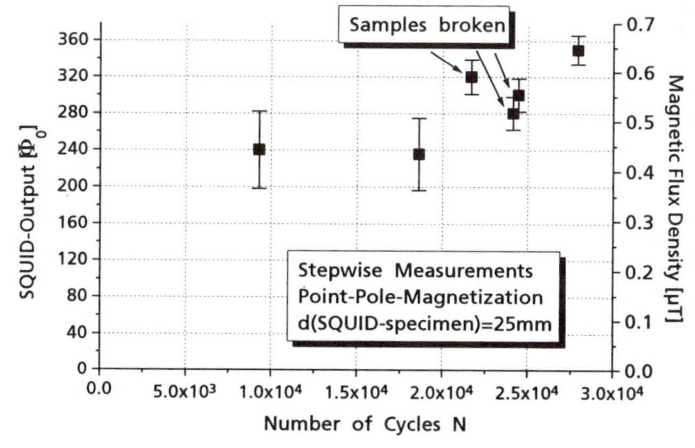

Figure 3. First experiments with stepwise fatigued and tested specimens show an increase in the measured magnetic flux density near the end of the lifetime. For a more deatiled interpretation further experiments with optimized excitating methods have to be carried out.

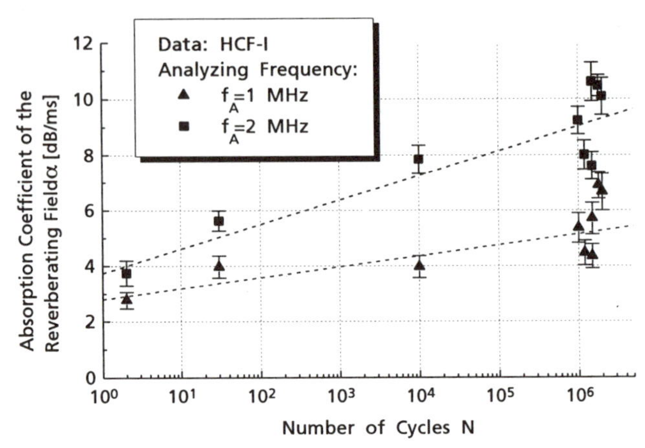

Figure 4. Ultrasound absorption coefficient of the reverberating field as a function of the number of cycles for the HCF-I-samples (room temperature, σ_a=240 Mpa, R=-1).

values in Figure 2 refer to a mean N_f=37209 this also effects the correlation of the measured field values to the fatigue level N/N_f.

Also stepwise experiments were carried out at complete fatigue samples (T=300°C, σ_a=180MPa, R=-1). The samples were fatigued for a certain number of cycles, measured with SQUID, fatigued, measured etc.. The magnetization was realized perpendicular to the the axis of the specimens as a point-pole-magnetization with a NdFeBr permanent magnet. Although this magnetization method was not optimized for this application and only few points could be gained it can be pointed out that there is an increase in the measured remanent magnetic field within the last ca. 20% of the specimens' lifetimes (Figure 3).

ALTERNATIVE NONDESTRUCTIVE EVALUATION METHODS

Apart from the SQUID-measurements other measurements with several, especially nondestructive, techniques, e.g. REM, TEM, Neutron Diffraction and Laser Ultrasonics as well as Scanning Acoustic Microscopy (SAM) were carried out.

The absorption of ultrasound, i.e. the dissipation of sound energy, reveals information about the interaction of ultrasonic waves with structural defects in the material. For the two-phase material austenite/martensite the absorption is dominated by both the high dislocation and grain boundary density. The ultrasonic waves were excitated by Laser-pulses (Nd-YAG-Laser, λ=1064nm, pulse energy 2J), the movement of the specimen's surface was observed with a Laser-interferometer. The absorption coefficient of the reverberating field shows a correlation with the number of cycles (Figure 4). Compared to the SQUID-results this correlation is much less significant.

For flat surfaces (after electropolishing) the contrast differences in the SAM-images represent different elastic properties[7] (Figure 5). For a grain size exceeding ca. 50µm it is possible to measure the sound velocity within single grains with the scanning acoustic microscope using leaky surface or Rayleigh-waves (Figure 6). This can be used to identify phases within a multi-phase material. For the austenitic matrix a sound velocity v_{aust}=3628±28m/s was determined in comparison to a sound velocity v_{mart}=3781±34m/s

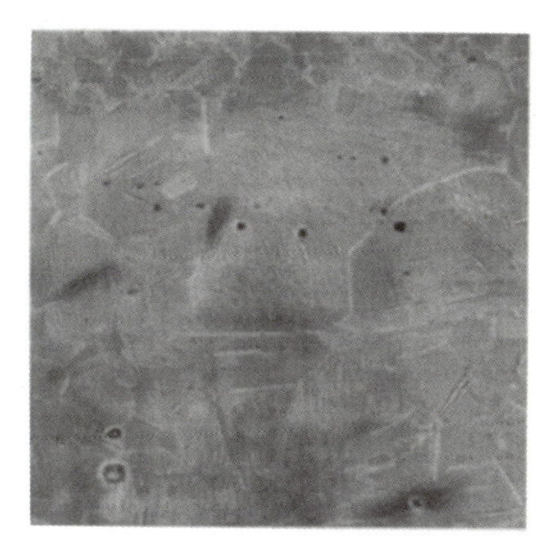

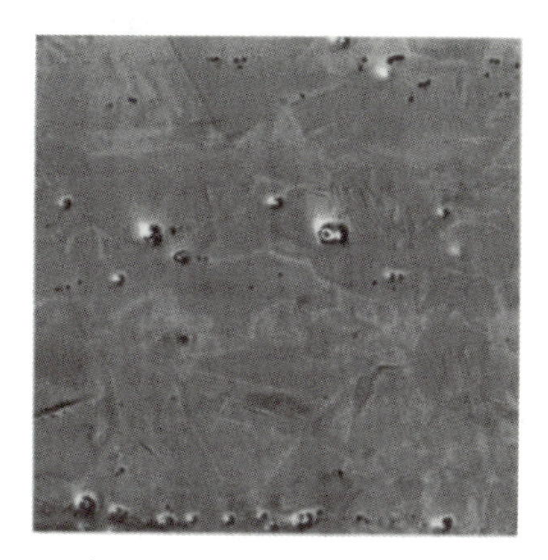

Figure 5. Acoustic images (312×312µm) of one HCF-I after 10^6 cycles, σ_a=240MPa, R=-1, RT (left) and one HCF-II specimen after 28000 cycles (75%N/N_f), σ_a=180MPa, R=-1, T=300°C (right).

for the martensitic grains. The higher velocity for the martensitic phase is due to the higher strength and hardness and thus higher elastic moduli.

CONCLUSIONS AND OUTLOOK

For AISI 321 austenitic steel the starting point of microcracks leading to macrocracks and finally to failure has to be expected at the locations of highest martensite concentrations. Information about the ferromagnetic martensitic phase fraction gives access to a possible way to determine the state of early fatigue damage of the material and reveals the risk for local failure by cracking. For realistic service temperatures of about 300°C, the fraction of martensite even up to failure is rather low, making highly sensitive measuring techniques such as SQUIDs compulsory. Having shown the general potential of SQUIDs to characterize fatigue damage in austenitic steels future work will focus on the development of a testing system for measurements of in

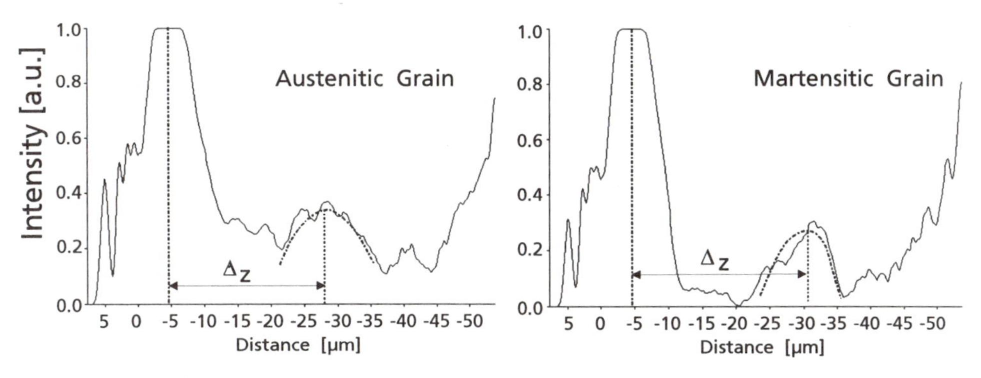

Figure 6. For grains of sufficient size the measurement of V(z) curves within single austenitic (left) and martensitic (right) grains is possible. The higher Δz-value for martensite represents the higher Rayleigh wave velocity due to higher elastic moduli.

service components. This testing system will not only focus on the measurement of the remanent magnetic field but also on dynamic measuring parameters such as the incremental permeability. This development will be based on the results on the use of HTSL-SQUID-systems in eddy current testing[8].

The magnetic behaviour of ferromagnetic material is highly influenced by stresses, also extremly localized residual stresses due to structural defects for example. In other applications this dependency is used to determine these stresses quantitatively[9]. Although X-ray and neutron diffraction stress measurements on the fatigue samples have been carried out it will be necessary to further focus on the analysis of the stress-impact on the magnetic properties of the ferromagnetic phase. This will improve the nondestructive testing method to be developed.

ACKNOLEDGEMENT

We like to thank the German Ministry for Education, Research, Science and Technology for funding these investigations (13 N 6358).

REFERENCES

1. Bayerlein, M.; Christ, H.-J.; Mughrabi, H., Plasticity induced martensitic transformation during cyclic deformation of AISI 304L stainless steel, Mater. Sci. Eng., vol. 114A, 1989, L11-L16
2. Lang, M., Bassler, H.-J., Untersuchungen zur Charakterisierung der Werkstoffermüdung in austenitischen Werkstoffen unter Anwendung von HTSL-SQUID, DGZfP-Berichtsband 55, Berlin 1996, pp. 76-82
3. Krause, H.-J. et. al., HTS-SQUID Magnetometer with digital feedback control for NDE Applications, Review of Progress in Nondestructive Evaluation, Vol 16A, pp. 2129-2135, Plenum Press, New York, 1997
4. Schreiber, J., Dobmann, G., Lang, M., Eifler, D., Bassler, H.-J., Gampe, U., Ehrlich, R., Charakterisierung von Werkstoffermüdung in austenitischen Werkstoffen unter Anwendung von HTSL-SQUID, 5. Statusseminar Supraleitung und Tieftemperaturphysik, Köln, 1996
5. Otaka, M., Enomoto, K., Hyashi, M., Sakata, S., Shimizu, S., Detection of Fatigue Damage in Stainless Steel Using a SQUID Sensor, Residual Stress and Integrity with NDE, ASME 1994, PVP-Vol. 276, NDE-Vol. 12
6. Wagner, V., private communication
7. Briggs, A., An introduction to Scanning Acoustic Microscopy, Microscopy Handbooks, 12, Oxford University Press, 1985
8. Krause, H.-J., Junger, M. et. al., Mobile HTS-SQUID System for Eddy Current Testing of Aircraft, Review of Progress in Nondestructive Evaluation, Vol 16A, pp. 1053-1060, Plenum Press, New York, 1997
9. Altpeter, I., Kern, R., Lang, M., Quantitative Evaluation of Thermally Induced Residual Stresses in White Cast Iron and Steels with different Cementite Morphologies, Review of Progress in Nondestructive Evaluation, Vol 16B, pp. 1649-1653, Plenum Press, New York, 1997

NONDESTRUCTIVE CHARACTERIZATION OF MATERIALS AT THE NIST RESEARCH REACTOR

H. J. Prask

NIST Center for Neutron Research
National Institute of Standards and Technology
Gaithersburg, MD 20899

INTRODUCTION

In the last three decades, neutrons have become an essential probe of the properties of matter, particularly condensed matter. This usefulness arises from the intrinsic properties of the neutron, which can be summarized as follows[1]:

- Determination of both atomic positions and motions can be made in solids and liquids because thermal neutrons have appropriate de Broglie wavelengths and energies.
- The neutron has a magnetic moment so that magnetic structure and magnetic excitations can be characterized.
- Neutrons penetrate condensed matter generally about three orders-of-magnitude deeper than xrays for wavelengths in the diffraction regime.
- Neutrons are sensitive to different isotopes of the same element in scattering so that, for example, hydrogen (which has a very high scattering cross section) is easily distinguishable from deuterium.
- Neutron absorption by a nucleus is isotope-sensitive so that chemical analysis is possible.

In recent years, there has been a significant increase in the use of "cold" neutrons (wavelength > 0.4 nm; energy < 0.005 eV) for a broad variety of studies[1]. Since the critical angle for total reflection increases with neutron wavelength, the increased total reflection at interfaces, which cold neutrons provide, makes it feasible to build neutron guides, which can be used to transport intense beams over many tens of meters. Increased spatial and energy resolution, and the ability to characterize larger structures (up to ~500 nm) are possible with cold beams.

The NIST Research Reactor (NBSR) has been operational since 1969; however, since about 1990 capabilities for materials characterization with neutrons have increased dramatically. Facilities at the NBSR include a liquid-hyrogen cold source and cold neutron guide hall, and state-of-the-art thermal-beam instruments. These constitute the NIST Center for Neutron Research -- a national user facility, unique in this country.

At present, the instrumentation for materials characterization at the NIST CNR includes

Nondestructive Characterization of Material VIII
Edited by Robert E. Green Jr., Plenum Press, New York, 1998

three thermal and one cold neutron "triple-axis" spectrometers (primarily for inelastic scattering); one 32-detector powder diffractometer; a new diffractometer for residual stress, texture, and single-crystal structure studies; two cold neutron reflectometers; two 30 m and one 8 m small angle neutron scattering instruments; a medium resolution time-of-flight spectrometer; one thermal- and one cold-neutron depth-profiling instrument; one thermal- and one cold-neutron prompt-gamma activation analysis instrument; a cold-neutron interferometer; a neutron optics facility; and a cold-neutron station for fundamental neutron physics studies. Many of the instruments have polarized beam capabilities for magnetic property studies. Instrumental and radiochemical neutron activation analysis and neutron radiography facilities are also available. Under development are three cold-neutron high-resolution inelastic scattering spectrometers and two thermal-beam instruments.

Together, the number and capabilities of the thermal- and cold-neutron instruments place the NIST Research Reactor among the front ranks of neutron centers in the world.

MATERIALS RESEARCH EXAMPLES

Numerous applications of neutrons to materials characterization problems have been described previously in this conference series[2,3]. The following examples illustrate new or significantly improved capabilities in various materials research efforts at the NIST Center for Neutron Research.

Phase-Analysis by Powder Neutron Diffraction[4]

The phase analysis of yttria-stabilized zirconia (YSZ) powders used as feedstock for the preparation of ceramic thermal barrier coatings has been reported previously[5]. The addition of yttria (Y_2O_3) to zirconia (ZrO_2) at a mass fraction of about 8 % Y_20_3 stabilizes the tetragonal phase relative to the low-temperature monoclinic phase (so-called partially stabilized zirconia); higher amounts of yttria stabilize the cubic phase (fully stabilized zirconia). Plasma spraying is a rapid solidification process that yields a phase composition of the resulting ceramic coating that is different from that of the feedstock powder, and often results in the formation of metastable phases.

Industrial experience with thermal barrier coatings applications has shown that tight control of the coating phase composition is imperative for optimal coating lifetime and survival. Larger amounts of the monoclinic phase in the deposits are considered undesirable, as these transform on heating into the tetragonal phase. Since this transformation is accompanied by a large volume change, stresses are generated in the deposits leading to premature failure. Coatings made of fully stabilized zirconia (cubic phase) have shorter lifetimes and generally inferior properties compared to partially stabilized zirconia coatings.

The zirconia phase analysis is challenging. Whereas the monoclinic phase is easily identified and quantified using x-ray patterns, the tetragonal:cubic ratio is much more dffficult since the cubic peaks are all coincident with tetragonal peaks. Previous work has shown that the cubic phase content is often underestimated[5]. Neutron Rietveld refinement provides a superior (albeit inconvenient for the industrial analyst) technique. In addition to yielding more data at higher scattering angle to help differentiate the tetragonal and cubic phases, the neutrons penetrate the entire coating sample and thus give results for the bulk material and make unimportant such x-ray analysis considerations as surface roughness and texture, preferred orientation, and gradations of phase composition through the thickness of the sample.

Deposits made of Sylvania SX233 powder were manufactured at SUNY Stony Brook. The deposits of about 5 mm thickness were sprayed on a steel substrate covered by an Al layer deposited by wire arc spraying. The Al layer was later dissolved using HCl to obtain free-standing deposits. Samples were cut into 10x10x5mm^3 sections using a diamond saw and annealed for one hour in ambient atmosphere at varying temperatures prior to phase analysis.

The results of the neutron Rietveld phase analysis are given in Table 1. As can be seen for the samples with a spray distance of 90 mm, little change occurs in the phase composition with heating until 1400°C, when the composition reverts to the thermodynamically stable

Table 1. Phase content in mass fraction % of SX233 plasma-sprayed and annealed YSZ coatings.

Spray Distance		% Monoclinic	% Tetragonal	% Cubic	% $YO_{1.5}$ (tetragonal phase)
Powder	--	25±1[a]	49±3	26±3	8.3±0.1
65 mm	as sprayed	3	60	37	8.8
145 mm	as sprayed	2	71	27	9.0
90 mm	as sprayed	2	74	24	8.8
90 mm	1100 °C, 1 h	2	72	26	7.5
90 mm	1200 °C, 1 h	3	72	25	7.2
90 mm	1300 °C, 1 h	3	70	26	7.1
90 mm	1400 °C, 1 h	3	60	37	6.2
65 mm	1400 °C, 1 h	4	59	37	5.9

[a] Standard uncertainties.

composition, with an increase in the relative amount of the cubic phase as predicted by the phase diagram. For the material sprayed at 65 mm, the as-sprayed phase composition is close to the thermodynamically stable composition, and little change occurs with annealing. In addition, it is possible to calculate the yttria content of the tetragonal phase from the tetragonal lattice parameters. For the materials sprayed at both 65 mm and 90 mm, there is some loss of yttria from the tetragonal phase with annealing even at 1100 °C to 1300 °C, but the loss is accelerated at 1400 °C. Further work is planned on samples prepared under varying plasma spray and heat treatment conditions.

Triaxial Residual Stress Measurements Using Neutron Diffraction

Several techniques are available for the nondestructive determination of residual stress, e.g. x-ray diffraction, eddy current, magnetic methods, and ultrasonics. Of these, the best established for quantitative characterization is x-ray diffraction which is, generally, a surface probe. Neutron diffraction closely parallels x-ray diffraction in methodology and analytical formalism. However, because neutrons interact with nuclei and x-rays with electrons, neutrons are typically about a thousand times more penetrating than x-rays in the wavelength range for diffraction ($0.07 \leq \lambda \leq 0.4$ nm). In addition, different elements exhibit significantly different relative scattering powers for neutrons and x-rays. Highly absorbing materials such as cadmium can be used to collimate the beam such that strains can be measured--and stresses inferrred--in gauge volumes a few millimeters in size. In recent years the technique has become quite well-established with a variety of applications described in recent review articles and conferences (see, e.g., ref. 6, and literature cited). Very recently a new, dedicated, state-of-the-art diffractometer for residual stress determination has been installed at the NIST reactor. With it, the power of the technique is shown by the results of recent measurements on steel tubes[7].

In this work, two mild steel tubes, of outside diameter 168 mm, wall thickness 8 mm and length 1000 mm, were subjected to four-point bends using a force of ≈580 kN. Tube I is seamless, tube II has a weld parallel to the cylindrical axis. After bending, the central 260 mm, together with end pieces, were cut out of the tubes for neutron diffraction measurements. Using a 2x2x2 mm^3 gauge volume, strains were measured in the mid-plane of the 260 mm piece, along radii 22.5° apart. Because of the non-circular cross-section produced by bending,

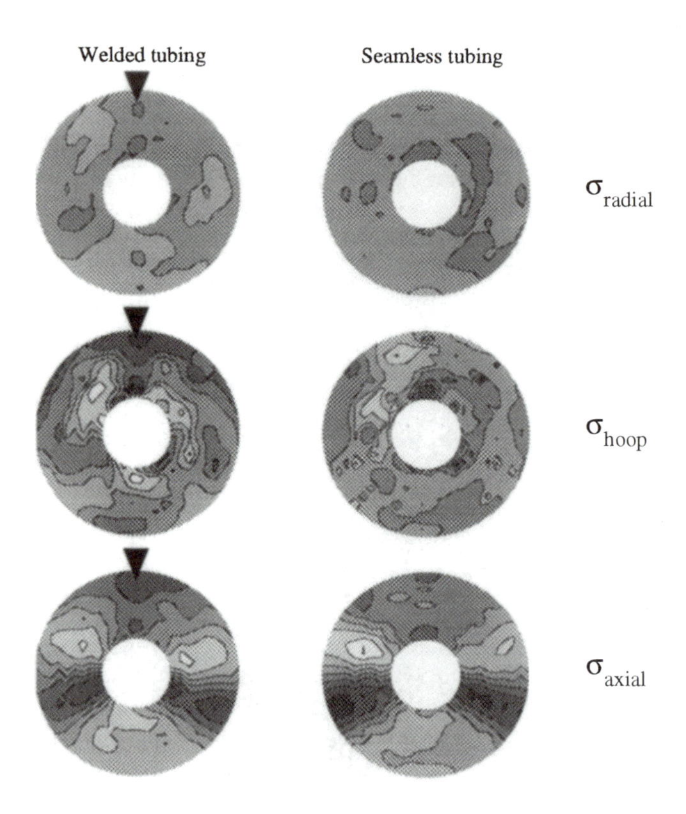

Figure 1. Neutron diffraction residual stress measurement results for a bent welded tube (left three plots) and a bent seamless tube (right three plots). The weld location is indicated by the filled triangles. The stress level is proportional to the grey scale such that white corresponds to -400 MPa and black to 400 MPa. For presentation purposes, the wall thickness is not drawn to scale. The force on the central sections was applied from the top in this figure.

steps of 0.2 mm were utilized along each radius from outside the OD to inside the ID. Measurements were then correlated according to the observed intensity[8]. The d_0 was obtained from the unbent tube portion. Residual stresses were inferred from the measured strains by utilizing Hooke's Law and the appropriate diffraction elastic constants.

In Figure 1 are shown residual stresses through the thickness in the central plane of each tube. These results for welded and unwelded tubes closely resemble one another and can be qualitatively predicted from the classic derivation of residual stress development in a plastically bent bar. Characteristic of this residual stress pattern is the very sharp transition of sign of the axial residual stress at approximately +90 and -90 degrees from the weld/top position. The stress in the radial direction is essentially equal to zero throughout the wall thickness and for all values around the circumference. This is commensurate with the notion that the radial stress has to be equal to zero at both the inside and outside diameter surfaces. The hoop stress component is the only component that shows a significant difference between the welded and the seamless tubing. The welded tubing shows a transition from tensile to compression going from the outer to the inner radius for most of the half tube section at the side where the weld is located ($-90° < \phi < 90°$). The seamless tube does not exhibit this behavior. Additional measurements in the weld region in both the bent and unbent welded tube

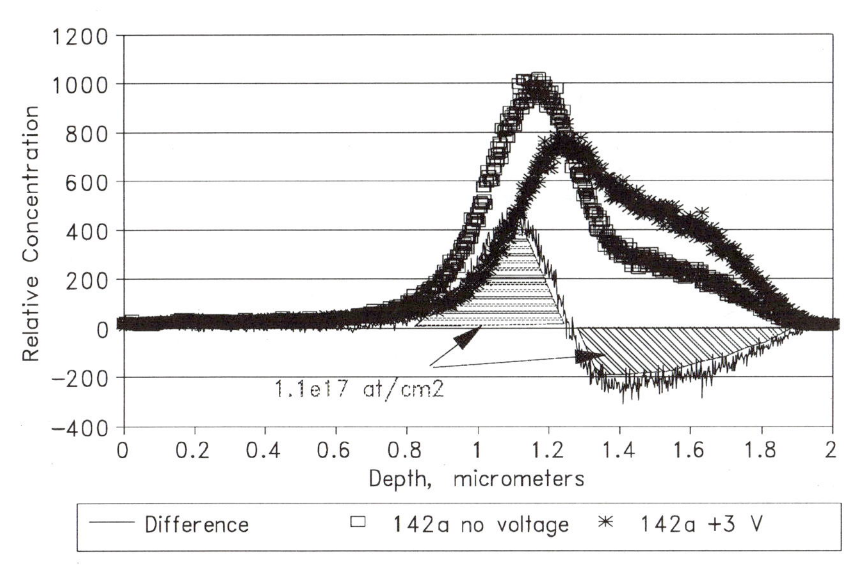

Figure 2. Lithium profile at two different voltages with difference spectrum.

samples show the same behavior for the hoop stress: tensile stress at the outer radius, compressive stress at the inside, while the radial and axial stresses are essentially zero in the unbent tube. Based on those measurements it appears that the behavior of the hoop stress in the vicinity of the weld is caused by the welding process alone.

New Cold Neutron Depth-Profiling (CNDP) Applications

Manufacturers of high technology devices need to know both thickness and composition of surface layers on their devices. Two recent investigations have included nitrogen concentration measurements in TiN[9] and lithium profile measurements in multilayers for electrochromic devices[10]. Both of these measurements presented challenges for the newly-configured CNDP instrument. The latter measurements were particularly challenging in that the lithium mobility was being assessed in active devices.

Depth profiling of nitrogen is based on the measurements of the proton from the $^{14}N(n,p)^{14}C$ reaction. The nitrogen measurements were the first quantitative evaluation of nitrogen concentrations since the instrument was modified. The advantages of the new configuration were apparent in the higher signal rate and the lower background rate compared with nitrogen measurements made at the previous instrument location. The nitrogen profiles obtained for the titanium nitride films showed a clear correlation with the conditions of the film making process. Future plans call for increasing resolution and further reducing the background. This will allow measurements on films that are both thinner and of lower nitrogen concentration.

The lithium profiles are based on the measurement of the energy of alpha particles from the $^{6}Li(n,\alpha)^{3}H$ reaction. The energy of the detected particle provides a direct measurement of the depth of the originating lithium nucleus. In this case, *in situ* measurements were taken with different bias voltages on the film layers. The bias causes the lithium to migrate between different layers and changes the optical transparency of the film.

The lithium profile of a multilayer under two different voltage conditions as well as the resulting difference spectrum is shown in Figure 2. The cross-hatched areas give the total amount of lithium moved by the voltage change. Future plans call for making simultaneous optical transmission measurements and lithium profile measurements.

SUMMARY

Neutron techniques can provide important, often unique, information for a broad spectrum of materials science and engineering problems. In the present paper we have attempted to illustrate the diversity of types of information and applications for which cold and thermal neutron methods are currently being utilized.

ACKNOWLEDGMENTS

The author is indebted to J. Stalick, G. Lamaze, H. Chen-Mayer and P. Brand for helpful discussions.

REFERENCES

1. See *NIST Journal of Research* 98: #1 (1993) for comprehensive review articles on cold neutron techniques for materials research.
2. H. J. Prask., "Materials characterization with cold neutrons," in *Nondestructive Characterization of Materials VI*, R. E. Green, Jr., K. J. Kozaczek, and C.O. Ruud, pp. 773-780 (1994).
3. H. J. Prask., "Materials characterization with cold neutrons," in *Nondestructive Characterization of Materials VII*, A.L. Bartos, R.E. Green, Jr. and C.O. Ruud, eds., *Matls. Sci. Forum* 210-3:711 (1996).
4. J. Ilavsky and J. K. Stalick, "Phase composition of plasma-sprayed YSZ," in *Reactor Radiation Technical Activities 1996*, L. Clutter, ed., NISTIR 5594 pp. 29-31 (1997); and Proc. of United Thermal Spray Conf. (Indianapolis, IN), September, 1997.
5. D. N. Argyriou and C. J. Howard, "Re-investigation of yttria-tetragonal zirconia polycrystal (Y-TZP) by neutron powder diffraction -- a cautionary tale," *J. Appl. Cryst.* 28:206 (1995).
6. H. J. Prask and P. C. Brand, "Residual stress determination by means of neutron diffraction," in *Neutrons in Research and Industry, Proc. of Intn'l. Conf. on Neutrons in Research and Industry*, ed. G. Vourvopoulos, *SPIE Proceedings Series*, 2867:106 (1997).
7. P. C. Brand et al., "Residual Stresses in Bent Steel Tubes," *Proc. of 5th International Conf. on Residual Stress*, Linköping, Sweden, June, 1997.
8. P. C. Brand and H. J. Prask,"New methods for the alignment of instrumentation for residual stress measurements by means of neutron diffraction," *J. Appl. Cryst.* 27:164 (1994).
9. G. P. Lamaze, H. Chen-Mayer, J. K. Langland and R. G. Downing, "Neutron depth profiling with the new NIST cold source," *Surf. & Inter. Anal.*, in press.
10. G. P. Lamaze, H. Chen-Mayer and M. Badding, "In situ measurement of lithium in thin -film electrochromic coatings," *Surf. & Inter. Anal.*, in press.

AN ULTRASONIC COMB TRANSDUCER FOR GUIDED WAVE MODE SELECTION IN MATERIALS CHARACTERIZATION

J. L. Rose, S. P. Pelts, J. N. Barshinger and M. J. Quarry

Department of Engineering Science and Mechanics
The Pennsylvania State University
114 Hallowell Building
University Park, PA 16802

INTRODUCTION

It is proposed to use guided waves for material and defect characterization analysis. Rather than use normal beam ultrasonic techniques that move point by point over a structure, guided waves are more global in nature allowing examinations of regions over 2-3 cm on up to 4-5 m with excellent sensitivity. One of the most important areas of study is therefore focused on transducer design and mode selection analysis. Phase and group velocity dispersion curve computation required in guided wave analysis, has taken place for decades. The ability to use these curves from a Nondestructive Evaluation point of view in materials characterization, however, is just now provoking great interest. Every point on the dispersion curve with a different phase velocity and frequency value has different performance characteristics with respect to wave penetration power and defect detection and/or material characterization sensitivity analysis. Sensitivity is a function of wave structure distribution across the thickness. It is therefore critical to select specific points for inspection and to have the ability to isolate the particular mode and frequency for optimal inspection performance. It is therefore proposed to utilize a specially designed comb type transducer with an excellent capability of mode selection and control.

Utilization of a small multi-element comb type ultrasonic transducer is therefore proposed for guided wave mode control in Nondestructive Evaluation. Theoretical methods are developed and experimental results are presented for guided wave generation and mode control with this efficient and versatile novel comb type ultrasonic transducer. Proper excitation and probe design is crucial in mode selection and isolation. The comb transducer generates waves that are influenced by such parameters as number of elements, spacing between elements, dimensions, pulsing sequence, and pressure distribution. The excited elastic field depends on the excitation frequency, structure

Nondestructive Characterization of Material VIII
Edited by Robert E. Green Jr., Plenum Press, New York, 1998

thickness, and elastic properties. Techniques are studied to optimize the applied loading and the comb transducer design parameters so that only the modes most sensitive to particular material characteristics can be generated. Complete understanding of the comb transducer parameters and their impact on the elastic field allows us to generate efficiently higher order modes as well as low value phase velocity modes valuable in composite material characterization. Theory of the comb transducer and a few experimental feasibility studies will be presented.

THE COMB TRANSDUCER

Utilization of a multi-element comb type ultrasonic transducer is proposed as an inspection tool employing multi-mode waves for improved probability of detection and defect classification. One of the first applications of a comb type transducer in the NDE field was discussed in Victorov[1], and some features of wave excitation by a comb type were studied[2,3]. It is well known that for a fixed frequency of time harmonic excitation in a linearly elastic plate that there exists a finite number of propagation modes and an infinite number of evanescent modes, with their own displacement and stress distribution patterns through the thickness. Since the effect of the evanescent modes decays exponentially away from the transducer, only the propagating modes make a significant contribution to the field at some point far from the transducer. From an NDE point of view, some propagation modes are more sensitive to particular material characteristics. It is advantageous therefore, to modify the applied loading distribution so that only modes with significant amplitude would be excited. The influence of the ultrasonic comb transducer parameters on the amplitudes of the excited modes is investigated.

Figure 1 illustrates the comb transducer principle. The structure of the comb transducer consists of a group of equally spaced parallel elements (teeth or legs) in the front face. The physical principle of the comb transducer is the periodical vibration of each tooth in phase at the same frequency to generate guided waves of the wavelength, λ, equal to the spacing, ΔS, with the said vibration frequency, f. The guided waves include waves propagating along plates and waves propagating along material surfaces, called surface waves. The guided waves generated with a comb transducer propagates in directions along which the transducer elements are arranged. For the relationship between the comb transducer element spacing and the guided wave mode[4].

There are many benefits of a comb transducer in material characterization and NDE applications. Conventional ultrasonic probes may not be suitable because of

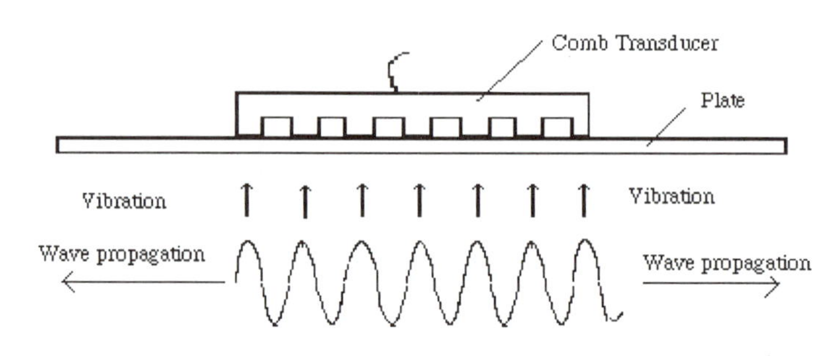

Figure 1. Comb transducer principle

overall size and numbers required to achieve certain test modes. Some of the principal benefits of a comb transducer are summarized.

Benefits of a comb transducer

1. Generate guided waves in plates or pipes in a variety of materials, even in materials of low phase velocities where the angle beam technique does not work.
2. Can be designed to generate guided waves at numerous multi-mode and frequency desired points on the dispersion curves; therefore, the best wave mode with a suitable wave structure can be selected and generated for any particular problem. Optimal sensitivity to certain defects as well as penetration capability can therefore be obtained.
3. More efficient penetration potential, no angle beam highly attenuative shoes or wedges are required; hence no loss in Plexiglas shoes or exiting interfaces.
4. Higher frequency mode generation is possible for increased resolution; and lower frequency for increased penetration. Piezocomposite broad band frequency elements are easily utilized.
5. Smaller size transducers are possible. The low profile comb transducer can be mounted on components that calls for smaller size because of poor access. This is particularly appropriate in smart material applications for embedded probes in a structure. Wire telemetry potential is also possible.
6. Suitable for high temperature plate or pipe inspection using a comb transducer consisting of a longitudinal wave transducer and possibly a metal comb shoe.
7. Pure modes can be generated since all of the energy can be in phase as it enters a curved or multi-layered structure.

MULTI-MODE INSPECTION MOTIVATION

The benefits of multi-mode inspection are associated with opportunities of achieving different penetration powers and sensitivities to different structural configurations or defect types and locations respectively. This is accomplished by adjusting wave structure across the thickness of a structure. For example, energy concentration on the outer surface is more sensitive to tiny defects on the outer surface or greater penetration of an immersed product is achieved by minimizing out of plane displacement on the outer surface. It is therefore important to examine many points on a dispersion curve. Each point has different performance characteristics associated with penetration power, energy leakage, and sensitivity to certain defects.

Let us now consider another excellent example where multi-mode inspection and wave resonance tuning plays an important role. Suppose we are using guided waves in a tubing experiment for defect screening purposes. We can consider distance, frequency, and phase velocity perturbation to find a defect. Any of the tuning mechanisms could find a defect, some even different defects. Any defects found would be correct[5,6].

Many other examples of multi-mode inspection can be presented. See [7] for lap splice and tear strap inspection in the aging aircraft industry. Also see [8,9] for an outer surface crack detection experiment. Improved sensitivity by changing modes for surface crack detection becomes possible. Energy concentration on the outer surface helps

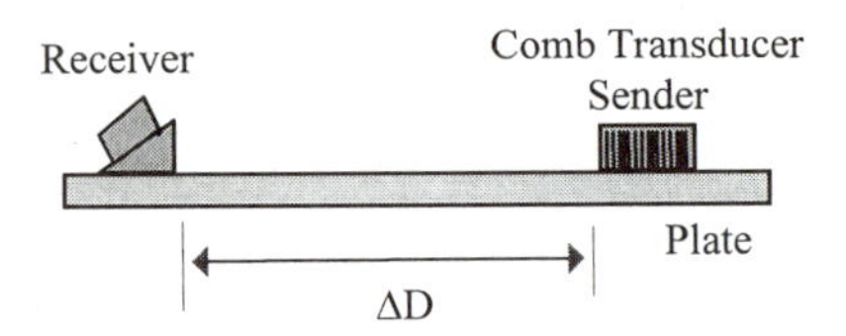

Figure 2. Transducer setup (Set at 3.00" and 6.00" between transducers).

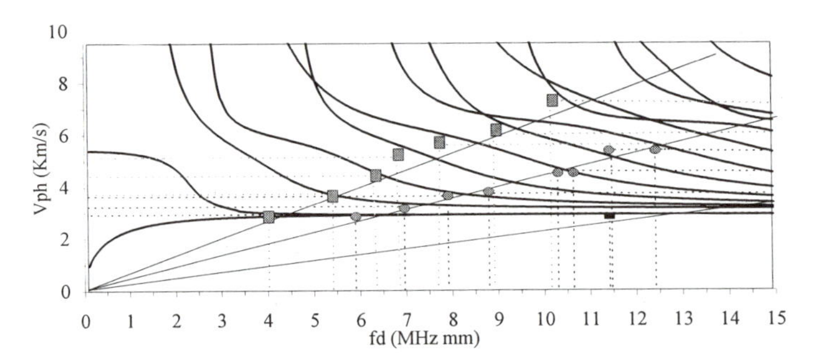

Figure 3. Generated guided wave modes in phase velocity (Vph) and group velocity (Vg) dispersion curve diagrams for an aluminum plate (V_L=6.3 Km/s, V_T=3.1 Km/s).

considerably, hence the Al mode at 2.5MHz becomes much more sensitive than So at 1 MHz..

Note that the wedge technique activates a horizontal line in the dispersion curve for a particular phase velocity according to Snell's law. The comb transducer activates a sloped line from the origin which can be swept by adjusting element spacing or properly applying time delays from one element to the next.

EXPERIMENTAL RESULTS

To demonstrate the capability of the comb transducer for guided wave generation, experiments were conducted using a 5.4 MHz broad band flat comb transducer on a 3mm thick aluminum plate. Twenty- elements (single spacing), ten-elements (double spacing), and seven-elements (triple spacing) of a comb transducer were used as senders and variable angle beam transducers were used as receivers for a through transmission setup as shown in Figure 5. A tone burst system was used for data acquisition. Function generator output was 0.5 Volts. Twenty cycle pulses were amplified to 167 volts to the transducer. Received guided wave RF signals were recorded with a LeCroy 9310 digital oscilloscope and IBM computer.

According to the phase and group velocity dispersion curves, Lamb wave modes were generated by adjusting the function generator excitation frequency and the receiver transducer angle to obtain optimum wave mode signals. The group velocities of each mode were measured by moving the receiver 3" from the sender. The modes were confirmed by comparing with both phase and group velocity dispersion curves. As expected, Lamb wave modes were generated that were very close to the intersection points between the comb straight line.

Guided waves were successfully generated in pipes as well[10,11]. In order to illustrate the wave resonance tuning technique to locate a defect, let us consider a curved comb transducer configuration for a 6" diameter steel pipe. Excellent

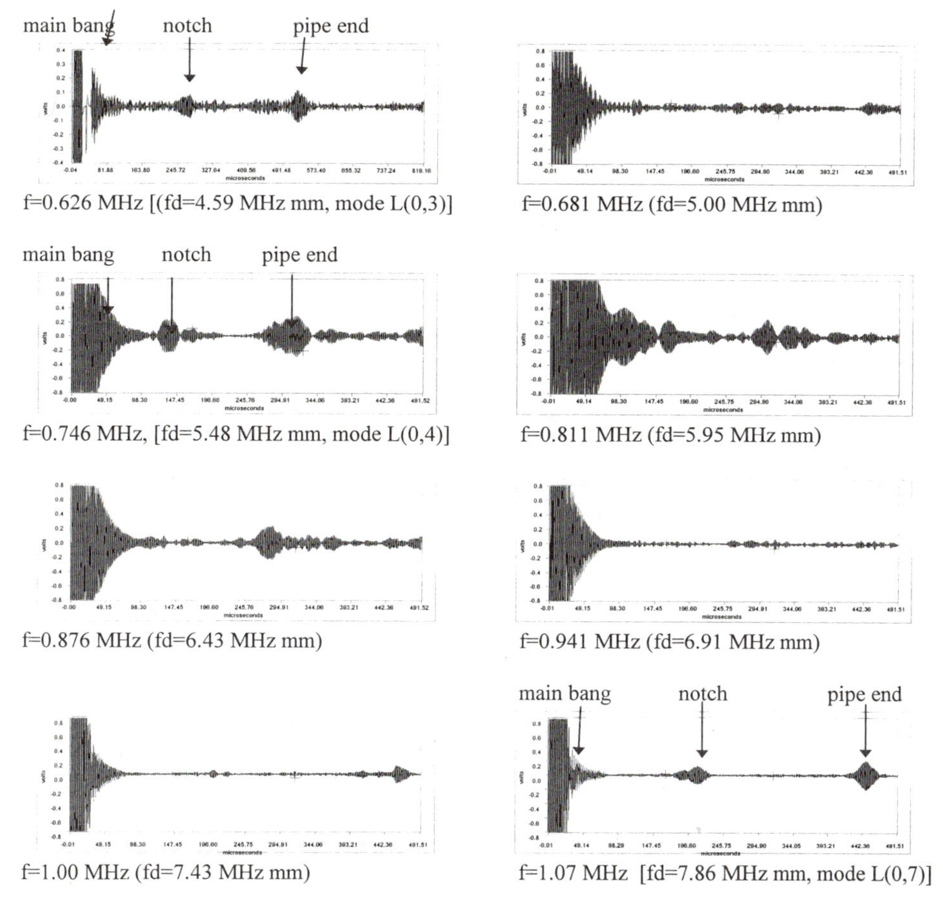

Figure 4. Guide waves generated in a 6" steel pipe with a comb transducer at different frequencies. Data taken from [6]. Note the frequency tuning and excellent defect detection results at .626 MHz, .746 MHz, and 1.07 MHz.

experimental results were obtained with the curved comb transducer[11]. The received RF signals are presented in Figure 4.

Another experiment on the selection of element numbers is presented here. In a pulse-echo set-up, a comb transducer with .005 inch element width and a .06 inch spacing was used to analyze the influence of changing the number of elements on the modes produced. The modes were produced in a 2.25 mm thick aluminum plate. Activation of the modes lies on the line with a slope of the spacing divided by the thickness. The influence of changing the number of elements was analyzed experimentally by producing S0, A1, and S2 modes at the proper positions on the dispersion curves. Increasing the number of elements in all cases led to improved S/N ratio.

CONCLUDING REMARKS

1. The linear array comb transducer shows great potential in guided wave inspection. Excellent penetration power and sensitivity is possible because of superior mode

control and choice compared to fixed angle beam methods of guided wave generation.

2. Design optimization of the linear array comb transducer is underway in the selection of element width, spacing, numbers, frequency, frequency bandwidth, and reception techniques.
3. The practical value of multi-mode inspection for improved defect detection, sizing, and classification is enhanced by comb transducer technology.

ACKNOWLEDGEMENT

Thanks are given to Dr. Vinod S. Agarwala of the Naval Warfare Center, Aircraft Division, Paxtuxent River, MD for technical support of this project and to the Office of Naval Research for financial support. Thanks are also given to EPRI and Jack Spanner Jr. for initiating efforts on this subject for piping and tubing.

REFERENCES

1. I. A Victorov, Rayleigh and Lamb Waves: Physical Theory and Applications, Plenum Press, 1967.
2. J. J. Ditri., J. L. Rose., A. Pilarski, "Generation of guided waves in hollow cylinders by wedge and comb type transducers," *Review of Progress in Quantitative Nondestructive Evaluation* (Ed. by D. O. Thompson and D. E. Chimenti), vol. 12, 1993, pp. 211-218.
3. T. Demol., P. Blanquet, and C. Delaberre, "Lamb wave generation using a flat multi-element array device," *1995 IEEE Ultrasonic Symposium Proceedings*, pp. 791-794.
4. Rose, J.L., "An Ultrasonic Comb Transducer for Smart Materials," Far East & Pacific Rim Symposium on Smart Materials, Structures & MEMS," Bangalore, India, Dec. 11-14, 1996.
5. Rose, J.L., Rajana, K., Hansch, M.K.T., "Ultrasonic Guided Waves for NDE of Adhesively Bonded Structures", Journal of Adhesion, Vol. 50, pp. 71-82, 1995
6. Shin, H.J., Quarry, M.J., Rose, J.L., "Non-axisymmetric Guided Waves for Tubing Inspection," QNDE Conference, Brunswick, Maine, July 28-Aug. 2, 1996
7. Rose, J. L., Barshinger, J. N., "Development of Guided Waves for Adhesive Bond Inspection," 41st International SAMPE Symposium & Exhibition, Anaheim, CA, March 24-28, 1996.
8. Cho, Y., J. L. Rose "An Elastodynamic Hybrid Boundary Element Study for Elastic Guided Wave Interactions with a Surface Breaking Defect", accepted by International Journal of Solids and Structures, October 1995.
9. Cho, Y., D. D. Hongerholt, J. L. Rose "Lamb Wave Scattering Analysis for Reflector Characterization, IEEE Transactions on Ultrasonics, Ferroelectrics, and Frequency Control, Vol. 44, No. 1, pg. 44-52, January 1997
10. J. L. Rose and D. Jiao, "Ultrasonic Guided wave NDE for piping," *The First ASNT U. S.-Japan Symposium on Advances in NDT*, Kahulu, Hawaii, June 24-28,1996.
11. J. L. Rose, D. Jiao, and J. Spanner,. Jr., "Ultrasonic Guided Wave NDE for Piping", *Materials Evaluation*, 54, 11, 1310-1313, Nov.1996.
12. Rose, J.L., S. Pelts, M. Quarry, "A Comb Transducer Model for Guided Wave NDE", to be presented at Ultrasonics International, July 1-4, 1997, Amsterdam.
13. Pelts, S.P., J.P. Cysyk, J.L. Rose, "The Boundary Element Method for Crack Sizing Potential with Guided Waves", submitted to JASA.

Characterization of the Microstructure of Laser-Hardened Carbon Steels by Means of Positron Lifetime Measurements and Micromagnetics

B. Somieski,[1]* N. Meyendorf,[2] R. Kern,[2] and R. Krause-Rehberg[3]

[1] Oak Ridge National Laboratory, CASD, Oak Ridge, TN 37831-6142, U.S.A.
[2] Fraunhofer-Institut für zerstörunsfreie Prüfverfahren, D-66123 Saarbrücken, Germany
[3] Universität Halle, Fachbereich Physik, D-06099 Halle (S), Germany

Keywords: Positron annihilation, micromagnetics, carbon steels, laser-hardening, microstructure, crystal lattice defects, martensite

ABSTRACT

Pure carbon steels containing different amounts of carbon were laser-irradiated. This hardening process, caused by martensitic transformation due to high self-quenching rates, created microdefects that were revealed by positron lifetime spectroscopy and micromagnetic techniques. The dependence of the microstructure upon the laser-deposited energy is presented. Changes in the defect structure are discussed as a function of the material. The stability of the produced microdefects is analyzed by isochronal annealing. The opportunity to obtain more detailed information about the material behavior by comparing the results of the positron lifetime spectroscopy and the micromagnetic spectroscopy is discussed.

1 INTRODUCTION

Thermalized positrons can be trapped in a crystal by lattice defects of open volume like dislocations, vacancies or microvoids [1]. The annihilation rate in the defect, which is determined by the electronic structure of the defect, is specific for each single kind of defect. Different positron annihilation rates or lifetimes, the inverse of the annihilation rates, in a lifetime spectrum provide information about different kinds of defects in the sample. The most

* corresponding author; e-mail: somieskib@ornl.gov

sensitive parameter in the positron lifetime spectroscopy is the average positron lifetime that can be considered as an integral value of the defect density.

A steel surface can be hardened by heat-treating with a high-powered CO_2-laser beam that scans the surface. The subsequent surface properties, i.e., hardness, hardening depth or residual stresses, depend on the hardening parameters, i.e., laser power, velocity of workpiece or energy density. After fast heating up of the surface beyond the austenizing temperature by the local, surface-near concentration of heat in the laser spot, martensite transformation is caused by a rapid cooling due to the self-quenching of the material.

A martensite transformation improves the strength and wear-resistance of highly stressed surfaces of components. The variation of the mechanical technological parameters in the component's surface is correlated with a modification of the microstructure and the defect structure. To clarify these dependencies that are important for the functional behavior of components, the technique of positron lifetime spectroscopy is an effective but not well known technique. This testing method has been applied to the study of defects such as vacancies or dislocations for about 20 years. It offers the possibility of indicating lattice defects of atomic size immediately.

In the following the application possibilities of positron lifetime spectroscopy will be demonstrated on laser-hardened microstructure states. The obtained results are analyzed in comparison with Barkhausen noise measurements, performed at the same samples at the same position. Because the positron lifetime spectroscopy is not affected by internal stresses, stacking faults, impurities, or interstitials, the results of the micromagnetic technique that is widely being used for the study of laser-hardened work pieces [2] can be analyzed in more detail by considering the positron lifetime results.

2 EXPERIMENTAL

Three kinds of carbon steel (Ck55, Ck60, and Ck75) containing different amounts of carbon (0.55 wt.%, 0.60 wt.%, and 0.75 wt.% respectively) were prepared. For comparison, 5N pure iron was taken into account also. The steel samples were annealed for 60 min at 1100°C before the laser irradiation was applied. Samples of pure iron, Ck55, and Ck75 had thicknesses of 2 mm; samples of Ck60 were 4 mm thick for the material comparison, cf. Figure 1, and 5 mm thick for the annealing experiments, cf. Figures 2 and 3.

Laser irradiation was carried out with a CO_2 laser of 1.3 kW DC. The laser diameter was 1 cm. The specimens were irradiated locally with varying irradiation times from 0.5 to 6 s.

30 μCi ^{22}NaCl was used as positron source. For the experiments comparing different steels, cf. Figure 1, the source was covered by a reference material (zinc-doped GaAs). The annealing experiments were performed by using two identically prepared samples. Details of source, source correction, stability of the numerical analysis, and obtaining the resolution function are described elsewhere [3, 4].

A standard fast-fast-coincidence positron lifetime system [5] with a resolution of about 250 ps was used. The spectra were analyzed with the program LIFESPECFIT [6] as described elsewhere [3]. All positron lifetime experiments were carried out at room temperature. The material volume analyzed by positron lifetime spectroscopy is, considering the effective source size, 2 mm x 2 mm x 0.1 mm.

Micromagnetic measurements are carried out with the 3MA-testing unit built in the IZFP [7, 8]. These measurements were done at exactly the same place as the positron lifetime measurements. The analysis frequency was 0.4 MHz that provides an average depth of information of roughly 0.05 mm [9]. Because the average penetration depth of the positrons is 33 μm in iron and steel [10], we are sure to get the information from the same depth with both methods.

3 RESULTS AND DISCUSSION

3.1 Influence of the Material

Figure 1 shows the average positron lifetime as a function of laser irradiation time. Obviously, no effect is found in pure iron because, owing to the lack of carbon atoms, no martensitic transformation can occur and the quenched-in excess vacancies anneal out below room temperature [11]. The slight increase in τ_{av} might be attributed to some dislocation loops created by excess vacancies.

Comparing the behavior of Ck55 and Ck75, no influence of the material is found; both curves are the same. In addition, the decomposition of the spectra shows very similar results [12].

After laser irradiation of 0.5 s, some defects, remaining in spite of the previous annealing treatment, annealed out in Ck55 and Ck75, indicated by the decrease of the average positron lifetime. No martensitic transformation occurred yet. After irradiation of 1 s, τ_{av} is increased by the generation of defects during the laser hardening process.

The different curve, obtained from Ck60, is caused by the different thickness of the material. Caused by the less efficient heat transfer into the underlayer compared with the heat transfer in the sample itself, in Ck55 and Ck75 (2 mm thickness) the volume influenced by the laser irradiation was smaller than in Ck60 (4 mm thickness). Thus, the amount of deposited energy, necessary for annealing of remaining defects as well as complete transformation bcc → fcc, is smaller in the thinner samples.

After further irradiation, until an irradiation time of 2.5 s, τ_{av} does not show a significant increase. Therefore, we conclude the defect structure of all three materials to be similar.

In the decomposition of the spectra we found two components, having values of 140 ps ... 150 ps and 200 ps ... 300 ps, attributed by their lifetime values to dislocations and small vacancy clusters, respectively [8]. The dislocations to adjust the different lattices of martensite and the remains of the austenite are mostly screw dislocations [13]. This finding is in good agreement with our results, considering screw and edge dislocations to have lifetime values of 143 ps and 165 ps, respectively [14].

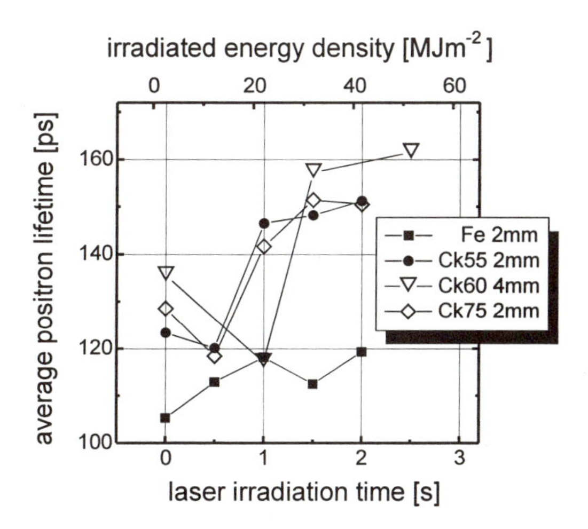

Figure 1 Average positron lifetime in carbon steels as a function of laser irradiation. Lines are to guide the eye.

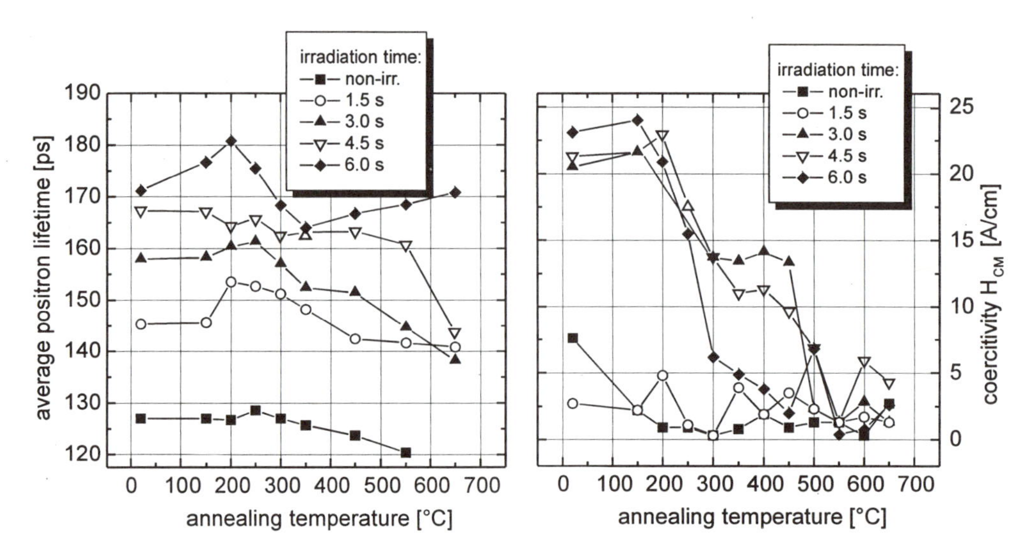

Figure 2 Average positron lifetime in laser-irradiated CK60 as a function of annealing temperature.
Lines are to guide the eye.

Figure 3 Coercitivity, derived from magnetic field position of max. Barkhausen noise amplitude; same specimens as in Figure 2.
Lines are to guide the eye.

3.2 Effect of Annealing on the Relaxation of Lattice Defects

In Figure 2 the effect of isochronal annealing treatment on the average positron lifetime is presented for Ck60 on a set of circular hardening spots performed with different laser irradiation times. The irradiation times from 1.5 to 6 s are compared with the non-irradiated microstructure state.

The first measuring value of each curve represents the hardening state after laser irradiation. Considering the results presented in section 3.1, the minimum laser irradiation time was chosen to provide hardening. It can be seen that the mean positron lifetime increases monotonously with the irradiation time. That is in agreement with the results obtained on the materials presented in Figure 1. Owing to increased sample thickness, cf. Section 3.1, that provides faster quenching rates, the concentration of vacancy clusters is slightly higher [12]. Therefore, the average positron lifetime is also slightly increased.

Annealing up to 200°C causes some changes in the defect structure. Especially, the quenched-in microvoids disappear at this temperature [12]. After that heat treatment all samples exhibit spectra that contain only a single component, i.e., all positrons are trapped by one kind of defect, which is believed to be dislocations. The samples irradiated for 1.5 s and 3 s show similar behavior. The average positron lifetime decreases during the annealing to a value of approximately 140 ps. Because edge and screw dislocations exhibit different lifetime values of 165 ps and 142 ps respectively [14], this decrease in τ_{av} can be attributed to a change in the density ratio of these types of dislocations. Compared with those two samples the one after 4.5 s laser irradiation shows a more stable configuration of dislocations that does not change very much up to 550°C. After the annealing at 650°C this sample exhibits a defect structure similar to the former ones.

The annealing behavior of the 6 s laser-irradiated material is different. The spectra obtained from this sample are more complicated. Especially the increase in τ_{av} from 350°C to 650°C is not well understood yet. This problem will be discussed in more detail elsewhere [12].

Figure 3 shows the results of the micromagnetic investigations of the same set of specimens. The increase of the coercitivity H_{CM} derived from the maximum of the Barkhausen noise amplitude, compared with the non-irradiated base material from about 3 A/cm to more than 20 A/cm, can be attributed to the formation of martensite that is mechanically and magnetically a much harder microstructure state. The material laser-irradiated for 3, 4.5, and 6 s behaves similarly in the temperature range from about 200°C's to 300°C. The H_{CM} shows the strongest decrease here. From 300°C to 450°C, H_{CM} remains nearly constant in the samples laser-irradiated for 3 and 4.5 s. It decreases further after annealing for higher temperatures. This behavior is not reflected in the positron lifetime measurements. Therefore, we assume that this effect is caused entirely by the relief of internal stress. In contrast with this, the sample irradiated for 6 s is, considering the coercitivity H_{CM}, almost completely annealed out after the heat treatment of 300°C. It is not well understood yet which process stabilizes the internal stress in the less-irradiated steel. It might be that during the longer irradiation a more homogeneous austenite is produced that results in less internal stress after the martensite transformation.

The Barkhausen noise amplitude M_{MAX}, measured at the position of H_{CM}, shows approximately the opposite temperature variation of H_{CM}. This is a well-known fact that is usually found for a variation of dislocation density or changing residual stress states [7, 15]. Increasing dislocation density leads to increasing H_{CM} values (decreasing M_{MAX}-values); increasing residual stress leads to decreasing H_{CM} values (decreasing M_{MAX}-values). This will be presented elsewhere [12].

The sample with irradiation of 1.5 s does not show any significant difference in H_{CM} from the unirradiated material. In [3] it is demonstrated that one can quantitatively analyze by positron lifetime spectroscopy the dislocation density in iron, at least in the range from 6×10^{8} cm^{-2} to 3×10^{10} cm^{-2}. Thus, we have to conclude that the average dislocation density in the investigated material is larger than 3×10^{10} cm^{-2}, because all positrons are trapped in dislocation. The generation of dislocations in this sample, obviously detected by the positron lifetime spectroscopy, is not reflected in the coercitivity. The reason for that might be that the value of the coercitivity H_{CM} was obtained from a Barkhausen-noise curve that contained two peaks, one from the martensite and one from partially annealed material. If the peak from the partially annealed material has a higher yield of M_{MAX}, it is used for the H_{CM} determination. Details will be discussed in [12].

Comparing the results of micromagnetics and positron lifetime measurements in Figures 2 and 3, even the overall shape of the temperature variation of coercitivity H_{CM} and positron lifetime is quite different. It is well established that variations of first order stresses have no influence upon the positron lifetime measurements. We have to conclude, therefore, that different material characteristics are reflected in both nondestructive testing methods.

The H_{cm} values for the martensitic state are quite similar for the different irradiation times, i.e., the saturation level is already reached after 3 s laser irradiation, whereas the average positron lifetime increases systematically with the irradiation time and seems to approach the saturation level after 4.5 s irradiation.

4 CONCLUSIONS

By means of positron lifetime spectroscopy, the microstructure of laser-hardened carbon steels was studied and the results were compared with those of micromagnetic methods. Both positron lifetime and Barkhausen noise were measured after laser-hardening and after several annealing steps during an isochronal heat treatment. The presented results prove that providing information about the microstructure of technical materials is also useful. The main measuring effects found in the laser-hardened materials are due to dislocations generated during the martensite transformation of the self-quenched steels. Significant changes in the magnetic quantities can be found in the temperature range up to 300°C. Comparing the positron lifetime spectroscopy and the coercitivity derived from Barkhausen noise after isochronal annealing steps, the results of both methods do not correlate.

The explanation is that the changes in the magnetic parameters in the discussed temperature range are caused mainly by the destruction of residual stresses whereas the positron annihilation is only sensitive to microstructural defects. This means that the use and comparison of both methods are necessary for a good material characterization.

Martensitic transformation, occurring in laser-irradiated carbon steels, creates dislocations as well as vacancy clusters. The positron lifetime parameters were found to be independent of the amount of carbon in the steel. After approximately 20 MJm^{-2} energy deposition the material was, as far as positron lifetime spectroscopy is concerned, completely martensitically transformed.

ACKNOWLEDGMENTS

The work was financially supported by the Deutsche Forschungsgemeinschaft.

We want to thank Lester D. Hulett, Oak Ridge National Laboratory, for helpfull discussions.

REFERENCES

1. R. N. West; in: *Positrons in Solids*, series: Topics in Current Physics (12), ed. by P. Hautojärvi, Springer, Berlin (1979) 89-144.
2. R. Kern, R. Meyer, W.A. Theiner, B. Valeske; Mat. Sci. Forum **210-213**, 687.
3. B. Somieski, T.E.M. Staab, R. Krause-Rehberg; Nucl. Instr. Meth. A **381** (1996) 128.
4. T.E.M. Staab, B. Somieski, R. Krause-Rehberg; ibid. 141.
5. P. Hautojärvi and A. Vehanen; in: *Positrons in Solids*, series: Topics in Current Physics (12), ed. by P. Hautojärvi, Springer, Berlin (1979) 1-23.
6. M. Puska; Program Lifespecfit, Teknillinen Korkeakoulu, Otaniemi (Finland) 1978.
7. G. Dobmann, W.A. Theiner, R. Becker; in: *Nondestructive Characterization of Materials*, ed. by P. Höller et al., Springer, Berlin (1989) 516.
8. W. A. Theiner, B. Reimringer, H. Kopp, M. Gessner; ibid. 699.
9. H. Heptner, H. Stroppe; *Magnetische und magnetinduktive Werkstoffprüfung*, Dt. Verlag f. Grundstoffindustrie, Leipzig, Germany (1972), p. 250.
10. W. Brandt, R. Paulin; Phys. Rev. B **15** (1977) 2511.
11. A. Vehanen, P. Hautojärvi, P. Johansson, J. Yli-Kauppila; Phys. Rev. B **25** (1982) 762.
12. B. Somieski, R. Krause-Rehberg, N. Meyendorf, R. Kern; to be publ. 1997.
13. P. Haasen; *Physikalische Metallkunde*, ch. **13**, Springer, Berlin (1984).
14. Y.-K. Park, J.T. Waber, C.L. Snead Jr., C.G. Park; in: *Positron Annihilation* ed. by P.C. Jain et al., World Scientific, Singapore (1985) 586.
15. W. A. Theiner, I. Altpeter; in: *New Procedures in Nondestructive Testing*, ed. by P. Höller et al., Springer, Berlin (1983) 575.

DETERMINATION OF MASS DENSITY AND ELASTIC CONSTANTS IN THE SURFACE LAYER WITH LEAKY SURFACE WAVES

Koichiro Kawashima, Ikuya Fujii and Naoki Takenouchi

Department of Mechanical Engineering
Nagoya Institute of Technology
Nagoya 466, Japan

INTRODUCTION

Various surface treatments such as cementation, nitriding, plating, sputtering, coating, etc., are used for preventing surface damage, corrosion or wear of structural or functional components. The physical and mechanical properties of the surface layers thus formed are not always identical with their bulk properties, therefore, we need to measure the surface properties, if possible, in-situ with some nondestructive methods.

The governing equation of the leaky Rayleigh wave[1] gives us the relation among the wave numbers of the leaky Rayleigh, transverse and longitudinal waves and mass density of water and material to be examined. Of course, the wave number of the leaky Rayleigh wave is complex, in which the real part determines the velocity and the imaginal does the attenuation of the wave.

By using an acoustic microscope, Lee et. al. measured the mass density and elastic constants of some glass and aluminum[2]. In their study, however, only leaky Rayleigh wave velocity was measured by the acoustic microscope. The longitudinal velocity was measured by a through-thickness transit-time measurement and the transverse one was estimated from those velocities with the equation of the leaky Rayleigh wave. Thus estimated density of aluminum deviated from the typical value.

The use of large aperture, broadband transducers made of PVDF polymer enables us to measure the velocities of the leaky Rayleigh, surface SV and creeping waves in time domain simultaneously[3,4,5]. If we measure accurately the attenuation of the leaky Rayleigh wave in addition to these velocities, we can determine directly the mass density of the surface layer with the Viktrov equation[1].

In this paper, the mass density as well as the elastic constants of fused quartz and aluminum alloy have been determined by time-domain measurement of the velocities of the leaky waves and the attenuation.

THEORY OF LEAKY RAYLEIGH WAVE

Viktrov[1] has derived the following equation of the leaky Rayleigh wave, which propagates through the solid/fluid interface.

Nondestructive Characterization of Material VIII
Edited by Robert E. Green Jr., Plenum Press, New York, 1998

$$4k^2qs-(k^2+s^2)^2=i\frac{\rho_W}{\rho}\frac{qk_t^4}{\sqrt{k_W^2-k^2}}$$
$$q^2=k^2-k_l^2,\quad s^2=k^2-k_t^2 \tag{1}$$

where, ρ is mass density and k, k_l and k_t denote the wave numbers of the leaky Rayleigh, longitudinal and transverse waves of solid, respectively. The suffix W stands for the adjacent fluid(water). The leaky Rayleigh wave propagates on the solid surface with leaking its energy into the adjacent fluid, therefore, the wave attenuates significantly. This means that k is complex, of which real part determines the velocity and imaginal part does the attenuation.

The mass density of water and longitudinal velocity in water are known as the function of temperature. With the above Equation, therefore, we can determine directly the density of the surface layer by measuring the velocities of the leaky Rayleigh, longitudinal and transverse waves of the solid as well as the attenuation of the leaky Rayleigh wave.

PRINCIPLE OF MEASURING SURFACE WAVE VELOCITY AND ITS ATTENUATION WITH PVDF SENSOR

We can distinguish the leaky Rayleigh, surface SV and creeping waves in time domain[3,4,5], by using large aperture, broadband PVDF transducer and selecting appropriate defocus distance, as shown in Figure 1. PVDF polymer transducers transmit and receive on the concave surface, therefore, it is free from the noise caused by the reflection at the surfaces of the focusing rod of the conventional focused PZT transducers.When we use a line-focused circular or point-focused spherical transducer, the time delay of each surface wave with respect to the specular reflection is given by as a function of the defocus distance z[6,7].

$$\Delta t_i=2z(1-\cos\theta_i)/V_W\equiv m_iz \tag{2}$$

where θ_i is the critical angle of each wave, and V_w is the velocity in water. With the Snell law, each surface velocity is given by

$$V_i=V_W[m_iV_W-(m_iV_W)^2/4]^{-1/2} \tag{3}$$

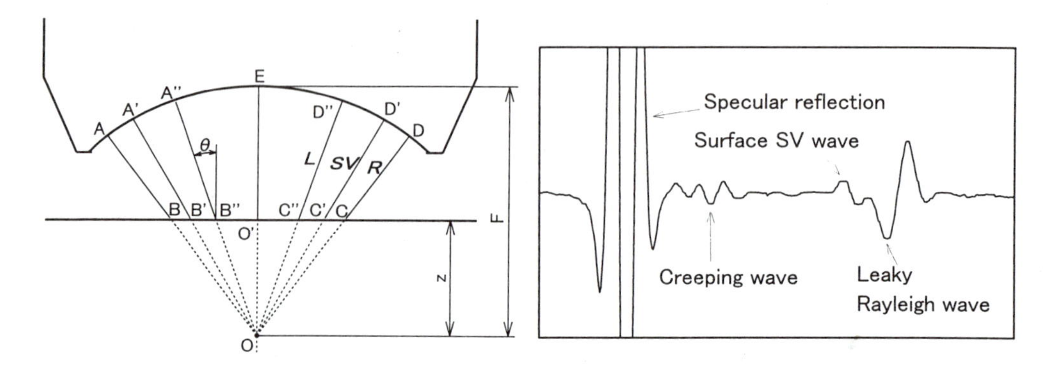

Figure 1. Principle of time-domain measurement of velocities of leaky surface waves.

By measuring of the time delay at various defocus distance, thus we can determine the velocities of the leaky Rayleigh, surface SV and creeping waves.

The amplitude of the leaky Rayleigh wave is given as a function of the defocus distance z, attenuation coefficient α_W in water and that γ at solid/water interface for the geometry shown in Fig.1.

$$A(z) = C\exp[-2\alpha_W(F - z/\cos\theta_R)]\exp(-2\gamma z\tan\theta_R) \tag{4}$$

where F and θ_R denote the focal length of the transducer and the critical angle of the leaky Rayleigh wave. From this equation, we obtain the attenuation coefficient.

$$\gamma = \frac{\alpha_W}{\sin\theta_R} - \frac{1}{2\tan\theta_R}\frac{d[lnA(z)]}{dz} \tag{5}$$

γ in this Equation is composed of the energy leakage into water and the absorption and scattering in the solid. However, the latter is far smaller than the former for common metals and glasses. Therefore we may assume the attenuation coefficient in Equation (5) to be the imaginary part of the wave number k.

APPARATUS FOR MEASURING VELOCITIES AND ATTENUATION OF SURFACE WAVES

The measurement system[9] used is schematically shown in Figure 2. The key components are the PVDF transducer, its alignment and sliding stage, pulser/receiver (Panametrics 5900), A/D converter board(Sonix STR8100) and personal computer. Two type of PVDF transducers of line–focus type are used: the nominal frequency 10 and 36MHz, the focal length 15 and 5mm. The angle of aperture and element width are 90° and 8mm for both transducer, respectively. The PVDF transducer is mounted on the swivel table which is rotated along mutuary orthogonal axes by stepping motors. The perpendicularity of the transducer to the sample surface is controlled within 0.02° with measurement of the received energy of the specular reflection from the surface. The accuracy of this alignment is crucial for precise measurement of the leaky surface wave velocities.

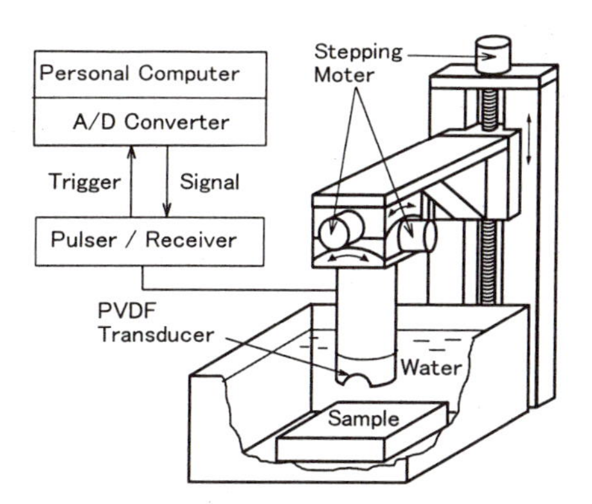

Figure 2. Experimental apparatus for measuring velocities and attenuating of leaky surface waves.

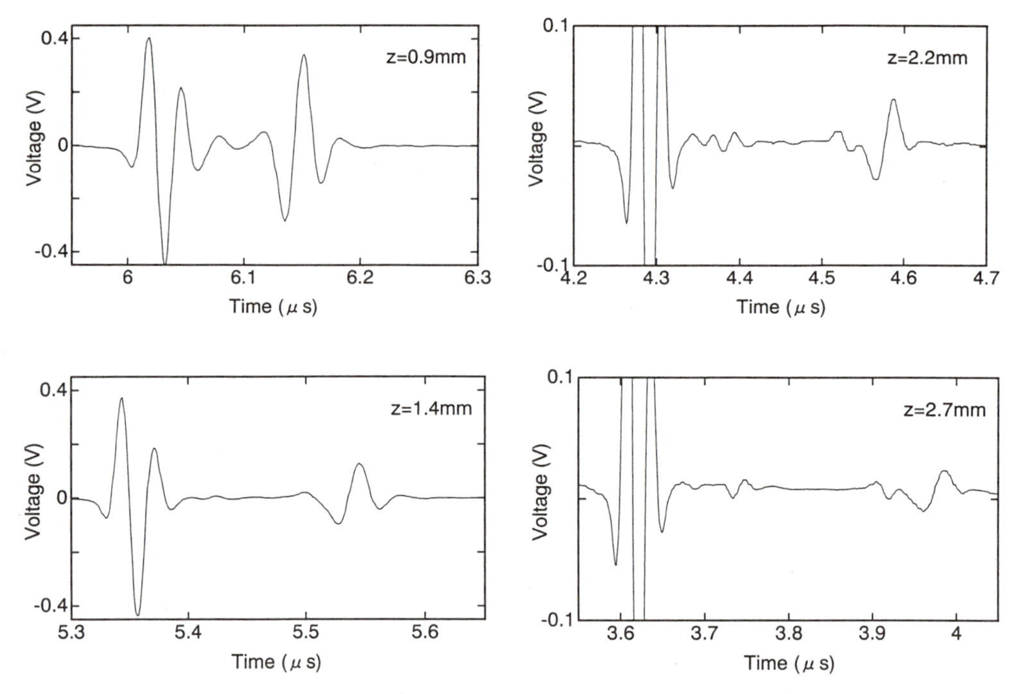

Figure 3. Wave signals of specular reflection and leaky surface waves at some defocus; fused quartz.

By the A/D converter, the waveforms of the specular reflection and the leaky waves are digitized at a sampling rate of 800MS/s. The digitized data are accumulated into the computer memory. By adding 256 waveforms, we obtain the clear waveform as shown in Figure 3. In this accumulation process, the trigger signal for the pulser should be synchronized with the sampling clock for the A/D converter.

The water temperature was kept 20°C during the measurement.

MEASURED VELOCITIES

We have measured the velocities of the leaky surface waves of fused quartz and 2017 aluminum alloy. The received waveforms of fused silica are shown for some defocus distance z in Figure 3. At small z, only leaky Rayleigh wave is observed. At large z, the SV wave is separated from the Rayleigh one and the creeping wave is identified between the specular reflection and the SV wave. The time delay of the leaky Rayleigh wave with respect to the specular reflection is nearly proportional to z and the leaky Rayleigh wave attenuates markedly with an increase in z.

Table 1. Measured velocities of leaky surface waves (m/s).

Material	Frequen -cy(MHz)	Leaky Rayleigh	Surface SV	Leaky creeping
Fused	10	3437	3778	5863
quartz	30	3429	3752	5933
2017	10	2943	3146	6359
Aluminum	30	2934	3138	6359

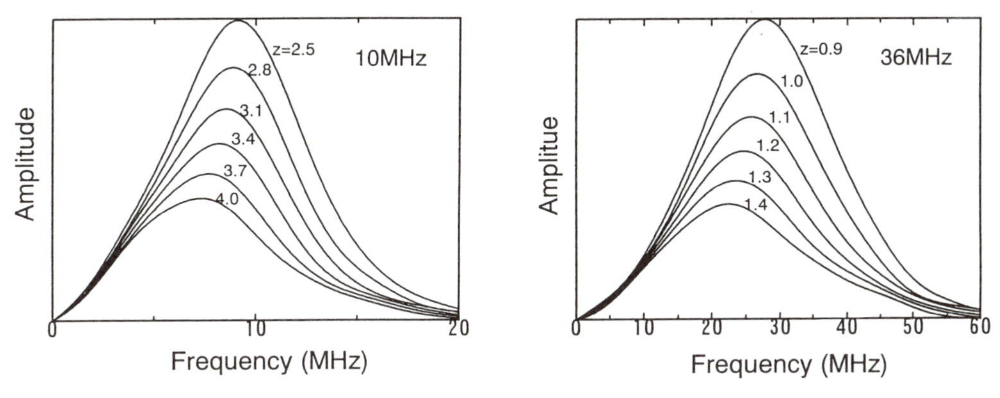

Figure 4. Amplitude spectra of fused quartz; left: 10MHz, right: 36MHz.

The time delay of each leaky surface wave with respect to the secular reflection was calculated by the cross-correlation method. With the slopes m_i in Equation (2) and Equation (3), we obtain the velocities of the leaky surface waves shown in Table 1. The measurement error of the velocities is less than 2m/s for the leaky Rayleigh and surface SV waves.

ATTENUATION OF LEAKY RAYLEIGH WAVE

The change in the amplitude of the leaky Rayleigh wave is given by Equation (4). The attenuation coefficient in the equation, of course, depend on the frequency of the received wave. Therefore the leaky Rayleigh wavelet was transformed into frequency domain. The amplitude spectra at several defocus distances are shown in Figure 4. By cross plot, we obtain the change in the amplitude against the defocus distance at each frequency. The attenuation coefficient calculated by Equation (5) are shown in Table 2. The theoretical energy leakage rate per wave length is calculated[10] as 0.25 and 0.18 for fused quartz and 2017 aluminum alloy, respectively. These are slightly lower than the measured due to the absorption and scattering in these solids.

MASS DENSITY AND ELASTIC CONSTANTS

The mass density estimated by Equation (1) is shown in Table 2 along with the Young modulus and modulus of rigidity, which are calculated with the density and velocities of longitudinal and transverse waves shown in Table 1 by the well known

Table 2. Measured density and elastic constants of fused quartz and 2017 aluminum.

Material	Frequen-cy (MHz)	Defocus range (mm)	Attenuation coefficient (/mm)	Density (10^3kg/m^3)	Young's modulus (GPa)	Rigidity modulus (GPa)
Fused	9.2	2.5–2.6	0.596	2.25	73.5	32.1
quartz	26.4	1.05–1.15	2.00	2.06	67.7	29.0
(Typical)				2.2	73.1	31.2
2017	9	2.0–3.0	0.560	2.72	71.7	26.9
Aluminum	26	0.8–1.0	1.62	2.73	72.0	26.9
(Typical)				2.79	71.5	26.7

formulas. The difference between the measured density and typical value is less than 7% for fused quartz and less than 3% for aluminum alloy. The error may result from that the absorption and scattering in the solid is also included in the attenuation coefficient measured.

CONCLUDING REMARKS

The precise measurement of the velocities and attenuation of the leaky surface waves was applied for measuring the mass density and elastic constants of the surface layer of fused quartz and 2017 aluminum. The stable measurement of the velocity and attenuation was performed by the combination of a digital signal processing and the use of PVDF transducers. For more precise measurement of density and elastic constants, we must separate the energy leakage into water from the absorption and/or scattering in the solid. Also we have to consider the change in the acoustic field. The proposed ultrasonic method provides effective means for in–situ measurement of mechanical properties of the surface layer which has been subjected to various surface treatment.

ACKNOWLEDGMENT

This work was supported in part by Grant–in–Aid for Scientific Research (B) 08455056 from Ministry of Education, Science, Sports and Culture

REFERENCES

1.I.A.Viktrov: *Rayleigh and Lamb Waves*, Plenum Press, New York(1967).
2.Y–C.Lee, J.O.Kim and J.D.Achenbach:Measurement of Elastic Constants and Mass Density by Acoustic Microscopy, *Proc. 1993 IEEE Ultrasonic Symp.*: 607(1993).
3.G.Congdard, F.Tady, M.H.Noroy and L.Paradis, Surface characterization of materials using Rayleigh velocity measurement in the broad–band mode, in *Review of Progress in QNDE*, Plenum Press,15:1605(1996).
4.D.Xian, N.N.Hsu and G.V.Blessing, The design, construction and application of a large aperture lens–less line focus PVDF transducer, *Ultrasonics*, 34:641(1996).
5.I.Fujii, K.Kawashima and N.Takenouchi, Measurement of leaky Rayleigh, creeping nd surface SV wave velocities with a line–focused PVDF transducer(in Japanese), *Trans. JSME Ser.A* 63, in press(1997).
6.K.Yamanaka, Surface acoustic wave measurements using an impulsive converging beam, *J.Appl.Phys.*, 54: 4323(1983).
7.J.C.Johnson and R.B.Thompson, The spatial resolution of Rayleigh wave, acoustoelastic measurement of stress, in *Review of Progress in QNDE*, Plenum Press, 12:2121(1993).
8.I.R.Smith and H.K.Wickramasinghe, SAW attenuation measurement in the acoustic microscope, *Electron. Lett.*, 18:955(1982).
9.K.Kawashima, I.Fujii, T.Sato and M.Okade: A digital signal processing for measuring leaky surface wave velocity with C–scan acoustic microscope, in *Review of Progress in QNDE*, Plenum Press, 15:2039(1996).
10.G.S.Kino, *Acoustic Waves; Devices, Imaging, and Analog Signal Processing*, Prentice–Hall, New York (1987).

Determination of Porosity in Fiber Reinforced Plastics by Computer Aided Ultrasonic Signal Analysis (CAMPUS)

Andreas Kück, Wolfgang J. Bisle, and Gustav Tober
Daimler-Benz Aerospace Airbus GmbH
Structure Technology and Test
D-28183Bremen, Germany

Introduction

With the increasing use of fiber reinforced plastics in many industrial applications like aerospace, naval and automotive construction, test techniques and especially the nondestructive test technique is of more and more importance. Defects like delaminations or voids introduced into the material during manufacturing, service or repair must be detected and quantified reliably. Detection of porosity especially in monolithic fibre reinforced plastic (FRP) structures is of great importance because a porosity contend higher than a certain limit (approximately 3 volume percent) decreases the mechanical strength of the FRP material. Numerous studies have been conducted in this domain in order to acquire a suitable nondestructive evaluation procedure for the inspection practice. One of the questions to be answered in this paper is: Up to which degree provides a specific spectral analysis of ultrasonic signals additional useful informations compared to the conventional signal height assessment ?

The present paper picks up a scatter model, that was assumed in order to estimate the porosity content in fiber reinforced plastics by measuring the frequency dependence of the acoustic attenuation. This assumption was made for the first time a few years ago. The paper describes the improvement of this model for practical application use. It is shown that interface effects and the basic attenuation of the intact material can to be taken into account by measurement techniques and calculations. Till now this has been neglected. Moreover, the computer program CAMPUS ("Computer Aided Measurement of Porosity by Ultrasonic Signal Analysis") is presented; it is specificly adapted to the requirements of these investigations.

Measurement Principle

The measurement of porosity by ultrasonic signal analysis is based on a model that is assumed for the simple scattering of ultrasonic waves on voids [1, 2, 3, 4, 5]. According to this model, the following equation is valid for a certain frequency range:

$$P_v = C \frac{d\alpha_{POR}}{df}$$

P_V: porosity [vol.-%]

C : material specific correlation factor [vol.-%•mm•MHz/dB]

$\frac{d\alpha_{POR}}{df}$: acoustic scatter dispersion (dB/(mm•MHz)]

α_{POR} : portion of acoustic attenuation coefficient caused by scattering on voids [dB/mm]

According to this scatter model, the correlation factor **C** and the relevant frequency range depend on material and shape, size, distribution and quantity of voids. The theoretically determined value **C** for the fiber reinforced material investigated here (Fibredux 913 C - 926 - F 35) is 7.5 [vol%•mm•MHz/dB] (spherical voids assumed). Moreover, $d\alpha_{Por}/df$ is a constant value because the acoustic attenuation coefficient $\alpha(f)$ is a linear function of frequency in the relevant range.

Knowing **C** the measurement of porosity is reduced to the task determination of the acoustic attenuation dispersion $d\alpha_{Por}/df$. Therefore, the amplitude spectra of signals with

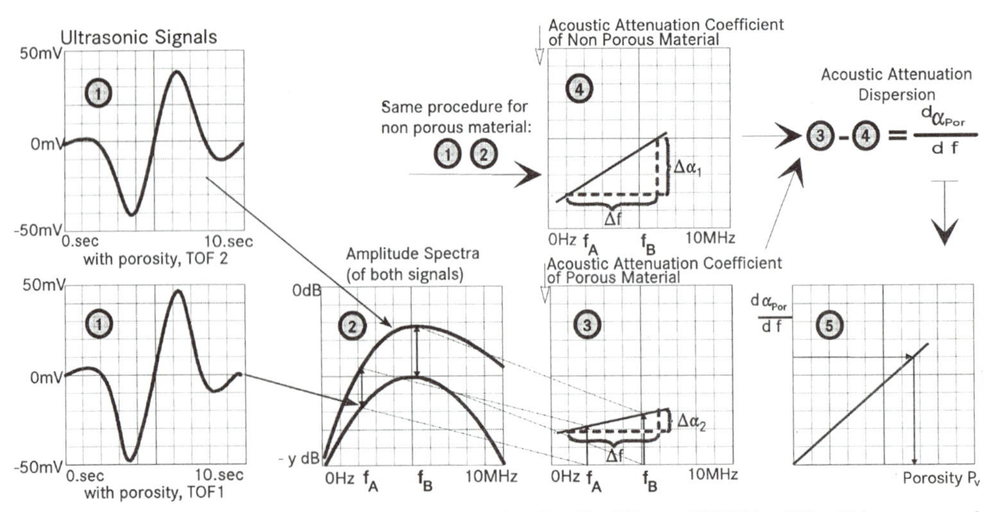

Fig. 1: Measurement principle of CAMPUS: (1) Two signals with different TOF (TT or PE) will be processed with FFT (2). The acoustic attenuation coefficient is in a certain range a linear function of the frequency (3). The gradient is calculated. The process is done either for plane material as reference (4) and for the porous material(3). Subtraction of both gradients is leading to the acoustic scatter dispersion (5)

the same features (either through transmission (TT) signal or back wall (BW) echo) of porous and intact area are calculated and computed with each other. The signal of the pore free area is necessary in order to compensate acoustic interface transitions and the basic attenuation of the intact material. $d\alpha_{Por}/df$ shall contain only the portion of acoustic scattering caused by voids. The amplitude spectra of the selected BW-echo or TT-signal are calculated using Fourier transformation and a simple rectangular window. Fig. 1 shows a scheme of the measurement principle.

Test Specimens

Two series of plane parallel monolithic test specimens made of CFRP fabric were investigated each containing both intact and porous areas (Fig. 2).

All the test specimens are made of the material Fibredux 913 C - 926 - 35 F. Their dimensions are 120 mm x 100 mm (length times width), their thickness' are 1.4, 2.1, 4.2 and 7.0 mm. In both series contain four test specimens for each of the four thicknesses. The two series differ by the porosity content. The porosity was determined afterwards by micrografic analysis:

Series 1: 1.3, 1.6 and 2.3 vol.-%
Series 2: 2.2, 2.6 and 2.8 vol-%

The porosity areas have diameters of 18 mm and were realized by strewing microencapsuled Freon between every internal layer during lamination by means of a punched template. In order to investigate the way of strewing Freon capsules on the acoustic scattering dispersion, a second test specimen series was backed. Inside these test specimen poros areas with quantities 2.2, 2.6 and 2.8 Vol% porosity were introduced, namely both using the punched template producing a cluster distribution and without

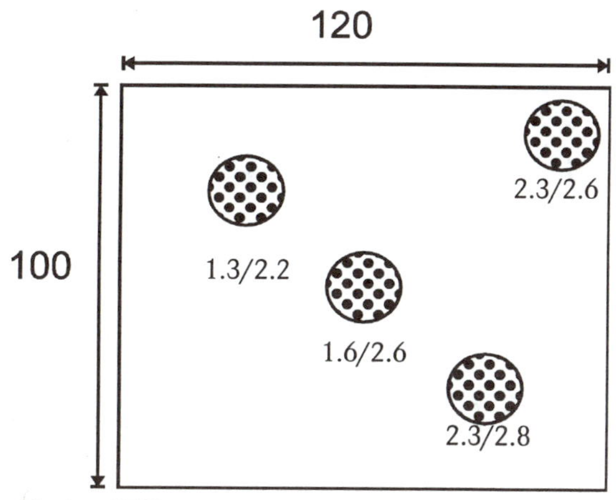

Technical data of test specimen:
CFRP, Fibredux 913C/926/F35
Dimension :120 x 100 mm
Thickness : 1.4, 2.1, 4.2, 7.0mm
Porosity :
Series 1: 1.3 Vol%, 1.6 Vol%, 2.3 Vol%
Series 2: 2.2 Vol%, 2.6 Vol%, 2.8 Vol%
Diameter of porosity area : 18mm
using microencapsuled Freon

Fig 2: CFRP testspecimen and technical data

template getting a uniform distribution. Each test specimen contains an additional porosity area for the microfractographic investigations (upper right corner).

Investigations

The test specimens were investigated by ultrasonic signal analysis, but also cross checked with micrographic and microfractograpic tests in order to determine acoustic scatter dispersions and porosities. The acoustic scatter dispersion of the pore free material was measured in order to find out up to which degree interface effects and the basic attenuation must be taken into account measuring the porosity content. The acoustic scatter dispersion of the pore free material was determined analysing two signal of different sound path in the intact material. The ultrasonic signals analysed were acquired with remote controlled through transmission as well as pulse echo techniques in an immersion tank but also by manual pulse echo contact technique . Conventional industrial inspection systems, test equipments and probes with nominal frequencies from 2 to 5 Mhz (for pulse echo techniques taking broadband probes) and transducer diameters from 6.25 to 10 mm were used (Fig. 3).

However, the signal analysis was performed by the in house developed computer program CAMPUS (Computer Aided Measurement of Porosity by Ultrasonic Signal Analysis"). CAMPUS was written on the base of the graphical programming language G using the programming environment LABVIEW® by NATIONAL Instruments, Inc.

CAMPUS executes all required computing operations automatically and provides the value of $d\alpha_{Por}/df$ both for a single position and for inspection areas (volume scans/ full waveform scans) digitally and graphically, too (Fig. 4).

After recording the ultrasonic data, sections from the test specimens were prepared for the determination of porosity by micrographic analysis. Moreover, the micrographic analysis served to determine the shape and the size of the voids, as the microfractographic investigations did as well.

Fig.3: Measurement built-up using conventional ultrasonic test equipment for manual inspection.

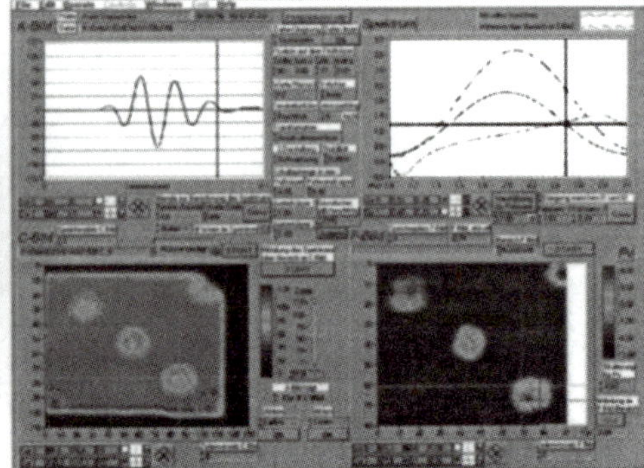

Fig. 4: Operating surface of CAMPUS

Results

The acoustic scatter dispersion of the intact material is 0.25 dB/(mm•MHz). The acoustic scatter dispersion of porous areas versus porosities (found by the micrographic section analysis) are shown in Fig. 5. This figure additionally shows the relation between P_V and $d\alpha_{Por}/df$ resulting from the assumed scatter model. This theoretical relation is drawn as dotted line.

Figure 5 shows that nearly all measured acoustic scatter dispersions can be fitted by a straight line with the slope of 0.2 dB/(mm•MHz•Vol%). From that, a material specific correlation factor of 5 (Vol%• mm•MHz)/dB. The signal record was free of interferences and is regarded as reliable.

More measurement results:

1. Measurements performed by manual pulse echo contact technique provided **C** = 4 (Vol%• mm•MHz)/dB, however, those measurements had a broader scatter band for $d\alpha_{Por}/df$
2. Comparing measurements without taking into account the influences of the intact material (basic attenuation) and interfaces resulted in different values of the acoustic scatter dsipersion.
3. The measurements performed on the test specimen containing uniformly strewed capsules (without using the punched template) provided similar acoustic scatter dispersions compared to porosity areas realized by the punched template.
4. The micrographic and the microfractographic analysis show that voids with porosity quantities < 7 Vol% have spherical shape with diameters of about 0.1 mm.

Summary

Detecting porosity in components made of fiber reinforced plastics in monolithic structure is of great importance because voids from a certain degree on (approximately 3 volume percent) decrease the mechanical strength of the material. Numerous studies have

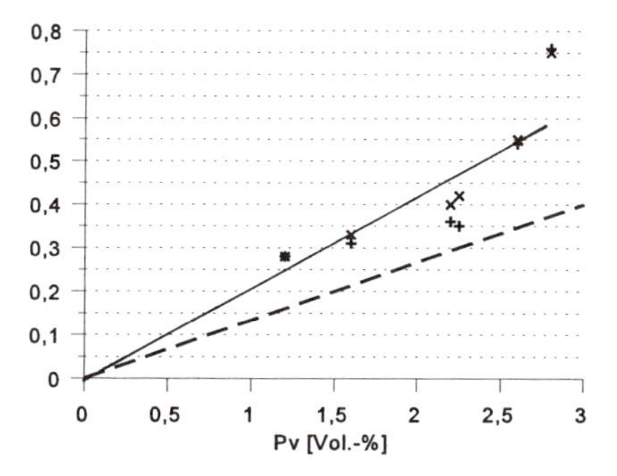

Fig.5: Thru transmission and pulse echo technique (remote controlled, probes with nominal frequencies of 2 MHz, relevant frequency range between 1.5 and 2.5 Mhz)

been conducted in this domain in order to acquire a suitable nondestructive evaluation procedure for the inspection practice. However, this paper presents a nondestructive evaluation procedure providing the possibility to determine porosity in fiber reinforced plastics with an accuracy of 0.5 volume percent points by ultrasonic signal analysis. It is based on the linear correlation between the frequency dependent acoustic attenuation dispersion and the porosity. The statement for determination of porosity resulting from that correlation was mentioned in the literature for the first time a few years ago and has been developed further to a test procedure that meets the requirements in the practical application. The realization of the procedure resulted in the computer program CAMPUS ("**C**omputer **A**ided **M**easurement of **P**orosity by **U**ltrasonic **S**ignal Analysis"). The procedure was verified on different porous components. The capability of the procedure could be validated. The method is superior to the conventional signal amplitude assessment for the determination of porosity since, because of the reproducibly measurable acoustic attenuation dispersion, the procedure is far less subject to influences resulting from the inspection system like coupling quality, acoustic interface transitions and choice of the ultrasonic probe. The most reliable measurement is performed by recording the signal by remote through transmission immersion technique. However, even manual pulse echo contact technique provides good results. The procedure is applicable to different inspection systems and to all materials that can be tested by ultrasonic.

References

Stone, Clarke "Ultrasonic attenuation as a measure of void content in carbonfibre reinforced plastics", NDT Int., Juni 1975, S. 137

Jones, Stone "Towards an ultrasonic-attenuation technique to measure void content in carbon-fibre composites", NDT Int., April 1976, S. 71

Martin, "Ultrasonic attenuation due to voids in fibre-reinforced plastics", NDT Int., Oktober 1976, S. 242

Hsu, Nair, "Evaluation of porosity in graphite-epoxy composite by frequency dependence of ultrasonic attenuation", Rev. of Prog. in QNDE, Vol. 6 B, Plenum Press NY 1987, S. 1185

David K. Hsu, "Ultrasonic measurements of porosity in woven graphite polymide composites", Rev. of Prog. in QNDE, Vol. 7 B, Plenum Press NY 1988, S. 1063

APPLICATION OF NEURAL NETWORKS IN AUTO-RECOGNITION OF EDDY CURRENT TESTING SIGNAL ON LINE

Shi Keren, Zhang Ping, and He Zhaohui

NDT Division
Department of Mechanical Engineering
Tsinghua University
Beijing 100084, China

INTRODUCTION

Eddy current testing is a very important non-destructive testing method. It has been used widely in nondestructive testing and characterization of conductive material. Eddy current testing has become an integral part of production and maintenance methods in many industries[1]. But there are many factors to affect the eddy current signal. Therefore, the results of eddy current testing is only qualitative and rough in the most practical testing. Furthermore, the final signal interpretation of eddy current testing is completed by human's eye in most cases, which makes the application of eddy current testing for automatic testing on-line to be limited. Therefore it is necessary to develop the automatic distinguished method of the eddy current testing signals on-line.

Artificial neural network have demonstrated their usefulness in functionally modeling processes involving many variables and complex quantitative reasoning, both linear and nonlinear, and even in the presence of noisy data[2]. Udpa.L etc. and H.Komastru etc. have researched the application of neural network to eddy current signal processing and got good results in 1991 and 1992[3][4].

This paper describes the application of a *feed-forward* neural network for intelligent identification of eddy current testing signals on line. The data for training and testing the neural network are from testing copper tubes and steel tubes with groove and hole defects, middle and lower carbon steel specimens, by a digital eddy current testing system. As the neural network is used for identifying on line, the architecture of the network is as simple as possible to reduce the identifying period.

How to choose the signal characteristic of eddy current testing as the input vector of the neural network is very important to get good identifying ratio of the network. Four types of the signal characteristic of the eddy current testing and their identifying ratios have been described in this paper.

The best result we have got is that the identifying ratio is 100%, identifying period

Nondestructive Characterization of Material VIII
Edited by Robert E. Green Jr., Plenum Press, New York, 1998

needs only 0.8ms on 486/66 computer. Thus, intelligent identification of eddy current testing signals on line can be achieved.

NEURAL NETWORK

Network Architecture

A neural network can be characterized by the type of rule that is used to compute the activation of units in it. There are more than one hundred models of architecture of the neural network now. The *feed-forward* neural network is a simple and widely applied model. The *feed-forward* neural network is one in which there are distinct layers of units, the input is the lowest layer, and the output is the highest layer, and all activation flows from lower to higher layer. Fig. 1 shows the architecture of unit *j* in layer *k*.

In Fig. 1, $y_j^{(k)}$ is the output of unit *j* in layer *k*. $y_1^{(k-1)}, y_2^{(k-1)}, \ldots, y_n^{(k-1)}$ are the input vector of unit *j* in layer k. $w_{1j}^{(k-1)}, w_{2j}^{(k-1)}, \ldots, w_{nj}^{(k-1)}$ are weight vector. $f_j^{(k)}(x)$ is the sigmoid activation of unit *j*. It is given by

$$f_j^{(k)}(x) = \frac{1}{1 - e^{-x}}.$$

That is:

$$y_j^{(k)} = f_j^{(k)}\left(\sum_{i=1}^{N_{k-1}} w_{ij}^{(k-1)} y_i^{(k-1)} - \theta_j^{(k)}\right), \qquad j = 1,2,\ldots,N_k; k = 1,2,\ldots,M.$$

Here $\theta_j^{(k)}$ is bias. N_k is the number of units in layer k, M is the number of the layer of the network.

Multi-layer perceptron is a typical *feed-forward* neural network. The architectures of the network used in this paper are seven models of multi-layer perceptron. The main model used in this paper includes three units in input layer, five units in hidden layer, one units in output layer (3 × 5 × 1), it is shown in Fig. 2.

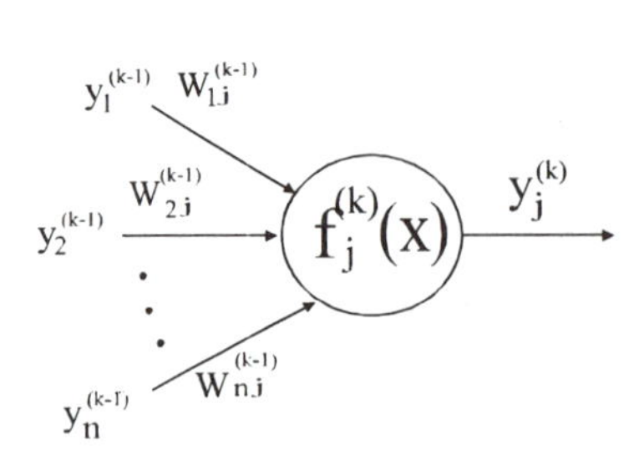

Figure. 1 Architecture of a note layers

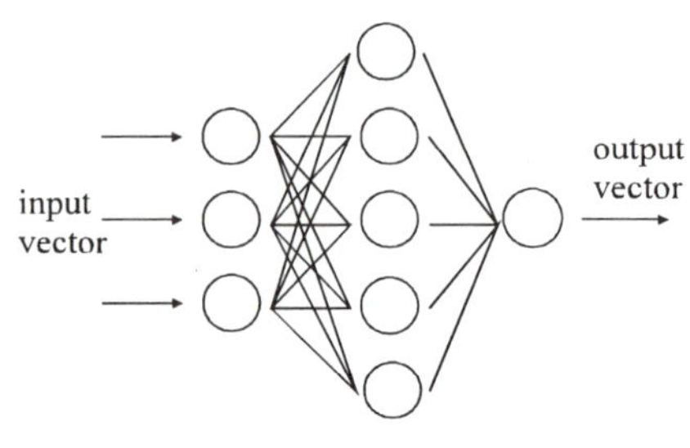

Figure 2. Multi-layer perceptron with three

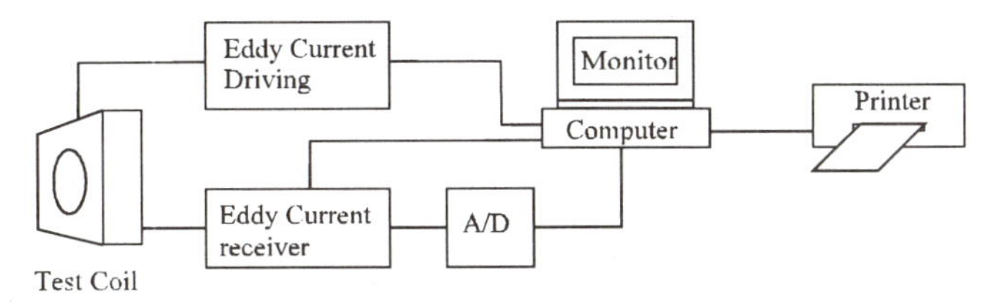

Figure 3. Digital eddy current system

Learning: Error Back-Propagation

Back-propagation is a supervised learning technique that compares the responses of the output units to the desired response, and readjust the weights in the network so that the next time when the same input is presented to the network, the network's response will closer to the desired response. Back-propagation is also called the *generalized delta rule.* This rule have been described in Some books of neural network[5].

DIGITAL EDDY CURRENT TESTING SYSTEM

The digital eddy current testing system used in this paper is consisted of test coil, eddy current driver, eddy current signal receiver and A/D converter, computer(Fig. 3)

The acquisition rate of the A/D converter is 40kHz. Both of the initial signal and reduced noise signal of the eddy current testing have been corrected when the testing object were tested.

SIGNAL CHARACTERISTIC OF EDDY CURRENT TESTING AND INPUT VECTOR OF THE NETWORK

As we know, the output vector is the map of the input vector in the neural network . Therefore, how to choose the input vector of the neural network is very important to get the useful output vector. Four types of signal characteristic of eddy current testing have been studied as the input vector of the network here and have got different results..

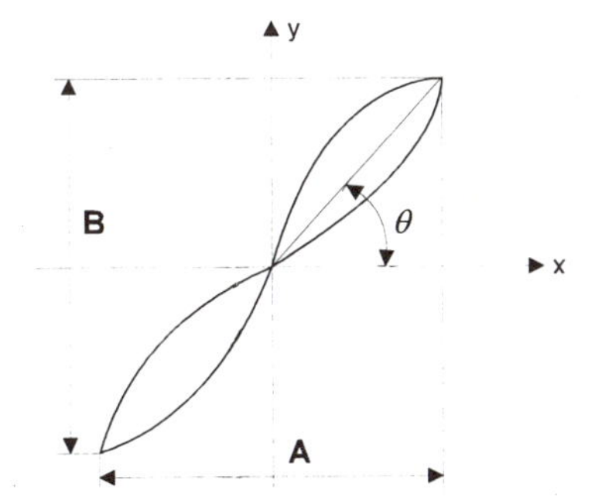

Figure 4. Resistance graph on the impedance plane

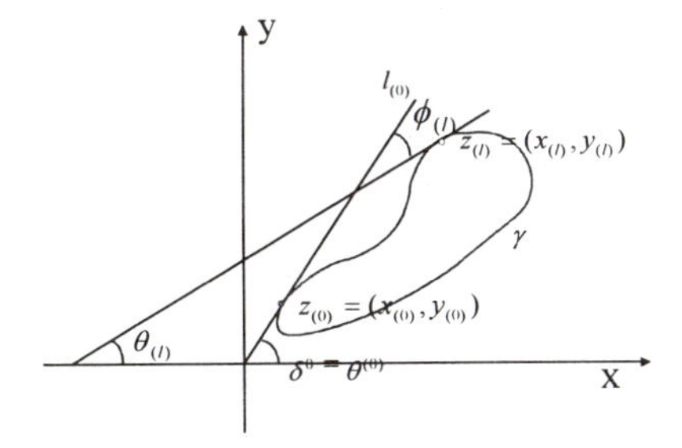

Figure 5. Fourier transforming the outline of the resistance graph of eddy current

$0.01F_0,...,0.01F_{29}$ of FFT of eddy current testing initial signal are selected as characteristic parameters forming the input vector of the network. One hundred specimens of steel pipes with groove or hole defects were used for training. Another one hundred specimens with groove or hole defects were used to test. The identifying ratios are shown in Table 2. In this case, the identifying ratio of using 30 × 18 × 10 × 1 architecture of the network is better than that of using 30 × 20 × 1 architecture of the network.

$0.01F_0,...,0.01F_{30}$ of FFT of eddy current testing initial signal for middle carbon steel and lower carbon steel are selected as characteristic parameters forming the input vector of the network. The identifying results were shown in Table 3.

Fourier Transforming Outline of Resistance Graph

In Fig. 5, γ is a sample of the outline of the resistance graph on impedance plane. The describing parameter is $z_{(l)} = (x_{(l)}, y_{(l)})$, here l is the arc length of γ, $0 \le l \le L$, here L is the total length of γ. $z_{(0)}$ is the initial point, $\delta_{(0)} = \theta_{(0)}$ is the included angle of tangent line $l_{(0)}$ at the initial point $z_{(0)}$. $\phi_{(l)}$ is the included angle of tangent line at the point $z_{(l)}$ and tangent line $l_{(0)}$. $z_{(l)}$ is the function of the arc length from initial point $z_{(0)}$ on curve. It

Table 1. Identifying results (for copper pipe; output>0.5 means groove defects , output≤0.5 means hole defects)

No.	A	B	θ	Output	Real defects
1	0.771	0.131	0.168	0.88	groove
2	0.792	0.164	0.204	0.85	groove
3	0.781	0.142	0.179	0.87	groove
4	0.759	0.153	0.199	0.87	groove
5	0.737	0.142	0.190	0.88	groove
6	0.939	0.667	0.617	0.16	hole
7	0.995	0.705	0.617	0.12	hole
8	0.978	0.688	0.613	0.13	hole
9	0.978	0.699	0.620	0.13	hole
10	0.989	0.683	0.604	0.14	hole

$A = A_i / A_{max}$, $B = B_i / B_{max}$, $\theta \in [0, 2\pi]$

Identifying ratio: groove 100%; hole 100%.
(Total is 50 specimens.)

Table 2. Identifying results for defects

Material	Architecture of the Network	Identifying ratio (%)	
		groove	hole
steel pipe	30 × 20 × 1	96.2	85.5
steel pipe	30 × 18 × 10 × 1	95.2	95.7
copper pipe	30 × 18 × 10 × 1	100	100

Table 3. Identifying results for material

Material	Architecture of the Network	Identifying ratio (%)
middle carbon steels	31 × 4 × 1	80
lower carbon steels	31 × 4 × 1	100

can describe the curve. That is:

$$\phi^*(t) = \phi(\frac{Lt}{2\pi}) + t, \quad \text{here} \quad \phi(\frac{Lt}{2\pi}) = \phi(l), \quad l = \frac{Lt}{2\pi}, \quad t \in [0,2\pi]$$

The Fourier series of $\phi^*(t)$ is given by :

$$\phi^*(t) = \mu_0 + \sum_{n=1}^{\infty} A_n \cos(nt - \alpha_n)$$

The (A_n, α_n) can be used as the characteristic to describe the outline of the curve γ.

The (A_n, α_n) (n=0,1,...,7) of the outline of the resistance graph on impedance plane have been chosen as the characteristic parameters forming the input vector of the network.

Characteristic Parameters of Resistance Graph

Three characteristic parameters of the resistance graph on impedance plane, horizontal peak-peak amplitude *A*, vertical peak-peak amplitude *B* and phase angel θ(0-2π), are used as the input vector of 3 × 5 × 1 perceptron (Fig. 4). Fifty specimens were used for training. Another fifty specimens were used to test. The ratios of identifying hole defects and groove defects all reach 100%. The results of identifying with 10 specimens of them are shown in Table 1.

Coefficients of Fourier Frequency Spectrum of Initial Eddy Testing Signal

For *n* sampling points, we have: $P = (p_0, p_1, \ldots, p_{n-1})$. Using FFT(Fast Fourier Transformation), the F(x) is given by:

Table 4. Identifying results for the input vector of the network from Fourier transforming parameters (A_n, α_n) (n=0,1,...,7) of the outline of the resistance graph on impedance plane.

Eddy current signal	Material	Architecture of the Network	Identifying ratio (%) groove	hole
without noise reduction processing	steel pipe	16 × 6 × 1	60.5	41.6
	copper pipe	16 × 32 × 1	60	50
after noise reduction processing	steel pipe	16 × 32 × 1	69	72
	copper pipe	16 × 12 × 1	50.6	65.6

$$F(x) = p_0 + p_1 x + p_2 x^2 + ... + p_{n-1} x^{n-1}$$

Then:

$$F_0 = p_0 + p_1 + ... + p_{n-1}$$

$$F_1 = p_0 + p_1 \omega + p_2 \omega^2 + ... + p_{n-1} \omega^{n-1}$$

$$F_2 = p_0 + p_1 \omega^2 + p_2 (\omega^2)^2 + ... + p_{n-1} (\omega^2)^{n-1}$$

.

.

.

$$F_{n-1} = p_0 + p_1 \omega^{n-1} + p_2 (\omega^{n-1})^2 + ... + p_{n-1} (\omega^{n-1})^{n-1}$$

Here $\omega = e^{-j\frac{2\pi}{n}}$, $n = 2^k$.

The 68 specimens were used for training. Another 132 specimens were used to test. The identifying ratios are shown in table 4. It shows that the identifying ratio is relatively low.

IDENTIFYING PERIOD

Using three characteristic parameters of the resistance graph on impedance plane as the input vector of $3 \times 5 \times 1$ perceptron, the identifying period needs only 0.8ms on 486/66 computer. It shows that the intelligent identification of the signals of eddy current testing on line by neural network can be achieved.

CONCLUSION

Artificial neural network is a very useful method for automatic signal identification of eddy current testing.

How to select the input vector of the neural network is very important for the identifying ratio of the network

In this case, using three characteristic parameters of the resistance graph on impedance plane as the input vector of the network is very successful. Because it can describe the characteristic of the eddy current signals and avoid the effect of the noise.

The intelligent identification of eddy current testing signals on line by neural network can be achieved.

REFERENCES

1. R.D.Shaffer, Eddy current testing, today and tomorrow, *Material Evaluation*, 52(1) :28 (1994)
2. Eric v. K. Hill, James L. Walker II, and Ginger H. Rowell, Burst Pressure Prediction in Graphite/Epoxy Pressure Vessels Using Neural Networks and Acoustic Emission Amplitude Data, *Material Evaluation*, 54(6) :744, (1996).
3. Udpa.L, and S.S.Udpa, Eddy current defect characterization using neural network, *Material Evaluation*, 49(1):34(1991).
4. H.Komastru, and Y.Matrumoto, Basic study on ECT data evaluation method with neural network, *Proceeding of Nondestructive Testing*, pp322-326, (1992)
5. David D. Morgan and Christopher L. Scofield, *Neural Networks and Speech Processing,* Kluwer Academic Publishers, Boston (1991)

INITIAL RESULTS OF APPLYING SINGLE TRANSDUCER THICKNESS-INDEPENDENT ULTRASONIC IMAGING TO TUBULAR STRUCTURES

Don J. Roth and Dorothy V. Carney

NASA Lewis Research Center, MS 6-1
21000 Brookpark Rd.
Cleveland, Ohio 44135
ph: 216-433-6017
fax: 216-433-8300
email: don.j.roth@lerc.nasa.gov

ABSTRACT

A single transducer ultrasonic imaging method that eliminates the effect of thickness variation in the image, and its application to hollow tubular/cylindrical structures, are described. The method, based on the measurement of ultrasonic velocity, thus isolates ultrasonic variations due to material microstructure. Its use can result in significant cost savings because the ultrasonic image, and consequently assessment of material and manufacturing quality, can be interpreted correctly without the need for precision thickness machining during nondestructive evaluation stages of material development. Additionally, it may be applicable to net shape parts having nonuniform thickness. The method has been commercialized via a cooperative agreement between NASA Lewis Research Center and Sonix, Inc.

BACKGROUND

It is the experience of these authors that ultrasonic velocity/time-of-flight imaging has proven to be more suited for quantitative characterization of microstructural *gradients* (such as those due to pore fraction, density, fiber fraction, and chemical composition variations) than has conventional ultrasonic c-scanning that maps waveform peak amplitude.[1–3] Variations in these microstructural factors can affect the uniformity of physical performance (mechanical [stiffness, strength], thermal [conductivity], and electrical [conductivity, superconducting transition temperature], etc.) of monolithic and composite components.[1,4–6] A weakness of conventional ultrasonic velocity/time-of-flight imaging (as well as ultrasonic peak amplitude c-scanning) is that the image shows the effects of thickness as well as microstructural variations unless the part is uniformly thick. This is easily observed from the equation for (pulse-echo) waveform time-of-flight (2τ)

$$2\tau = \frac{(2d)}{V} \qquad (1)$$

where d is the sample thickness and V is the velocity of ultrasound in the material. Therefore, interpretation of the image is difficult as thickness variation effects can mask or overemphasize the true microstructural variation portrayed in the image.

Nondestructive Characterization of Material VIII
Edited by Robert E. Green Jr., Plenum Press, New York, 1998

Thickness effects on time-of-flight can also be interpreted by rearranging eq. (1) to calculate velocity

$$V = \frac{(2d)}{(2\tau)} \tag{2}$$

such that velocity is inversely proportional to time-of-flight. Velocity and time-of-flight maps will be affected similarly (although inversely in terms of magnitude) by thickness variations, and velocity maps are used in this investigation to indicate time-of-flight variations.

THICKNESS-INDEPENDENT ULTRASONIC VELOCITY MEASUREMENTS

Several attempts to account for thickness variation effects in ultrasonic velocity measurements were noted in the literature. One study[7] used a two-transducer method whereby the transducers with the transducers located on opposite sides of the sample. Accurate thickness at each scan location was obtained via both transducers using the time-of-flights acquired from pulse-echo front surface reflections off both surfaces of the sample, and using the known constant velocity in water and the known distance between the two transducers. Additionally, time-of-flight from the first reflected pulse off the sample back surface was obtained by one of the transducers at each scan location. Knowing thickness (d) and time-of-flight (2τ) at each scan location, thickness-independent velocity (V) images were calculated according to eq. (2).

Several studies[8–11] described a single point ultrasonic velocity measurement method using a reflector plate located behind and separated from the sample, that does not require prior knowledge of sample thickness. The latter method was studied with success in prototypical scanning configurations for plate-like shapes,[12–14] and incorporated into a commercial scan system.[14,15] Figure 1 shows a schematic of the immersion pulse-echo testing set-up required to use this method and the resulting ultrasonic waveforms. The mathematical derivation for the method[14] results in ultrasonic velocity being calculated according to

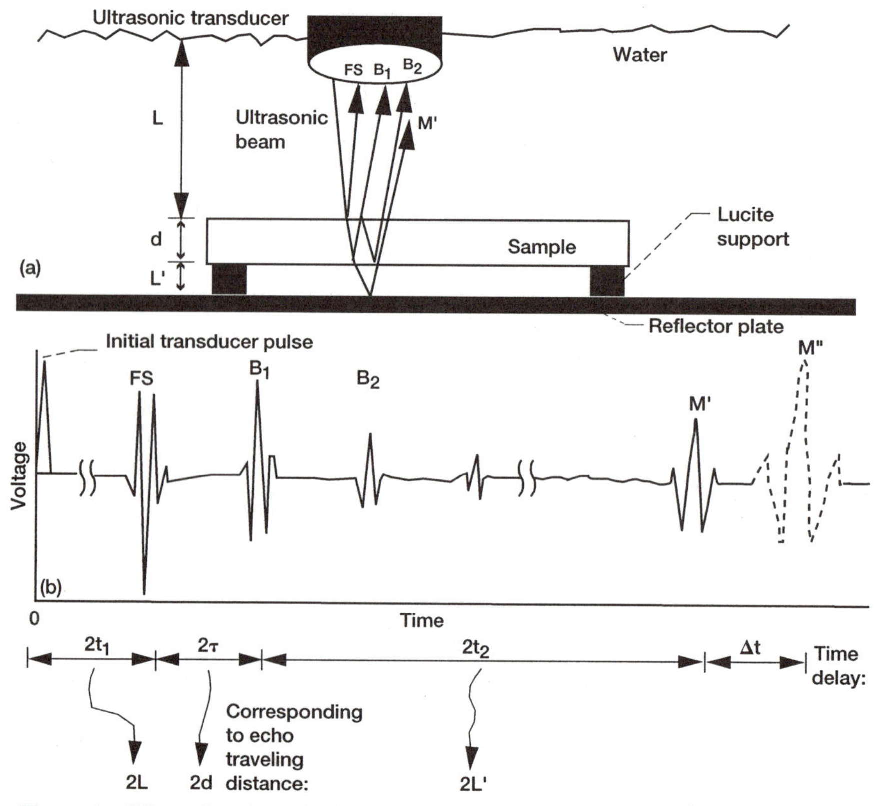

Figure 1.—Ultrasonic pulse-echo immersion testing. (a) Schematic of ultrasonic pulse-echo immersion testing. (b) Resulting waveforms.

$$V = c\left(\frac{\Delta t}{2\tau} + 1\right) \tag{3}$$

where c is water velocity, 2τ is the pulse-echo time delay between between a front and back surface echo or between two successive back surface echoes, and Δt is the pulse-echo time difference between the first echo off the reflector plate front surface with and without the sample present, respectively. Water velocity (c) is determined from known relations between water velocity and temperature[16] or by direct measurement using the time difference of ultrasonic wave travel between two transducer heights. This thickness-independent ultrasonic imaging method does not require prior knowledge of sample thickness as shown in eq. (3) and if engineered for scanning, the effect of thickness variation is eliminated in the resulting image. Precision and accuracy associated with this method are estimated at near 1 to 3 percent, respectively, for plate-like samples.[14] The thickness-independent methods noted here[7–15] all require access to both sides of the sample, i.e. there is not a single-sided technique available for scanning that will result in thickness-independence.

SCOPE OF INVESTIGATION

In this investigation, the single transducer reflector plate method shown in fig. 1 is applied to obtain thickness-independent ultrasonic images of mullite (alumina-silica compound) and silicon nitride ceramic tubular structures. Thickness-independent ultrasonic images for these materials are compared to conventional (apparent) velocity images (the latter being obtained using eq. (2)). The single transducer reflector plate method[8–15] was chosen over the two-transducer method[7] because of its simplicity and the fact that the some tubes under investigation at NASA for high temperature structural duty have inner regions too small to contain a perpendicularly-positioned second transducer but large enough to contain a reflector plate.

MATERIALS

The mullite tube was of nominal wall thickness 4.1 mm, 53 mm outside diameter, and 150 mm height. Five square 1 × 1 cm patches of depth 0.1, 0.2, 0.3, 0.4 and 0.5 mm were machined out of the exterior surface of the mullite tube to create thickness variation of up to 12 percent in the tube. The silicon nitride tubular material was of nominal wall thickness 7.5 mm, 48 mm outside diameter, and 150 mm height. It was manufactured unintentionally with a 0.8 mm thickness variation (~10 percent of total thickness) and currently unknown microstructural variation.

EXPERIMENTAL

The thickness-independent ultrasonic imaging method requires at least two scans to collect the necessary echoes required for calculation of time delays (fig. 1). The first of two scans is run with the sample in place and echoes B1 (first back surface echo), B2 (second back surface echo), and M' (first echo off reflector plate front surface with sample present) are collected. Following this, the sample is removed and a second scan is run to collect M" (first echo off reflector plate front surface with sample removed). In this investigation, three scans were used to obtain all echoes as required using the version of this method that is commercially-available.[15] B1and B2 are obtained in scan 1, M' is obtained in scan 2 and M" is obtained in scan 3. A computerized cross-correlation method was used to calculate precise time delays.[17] Broadband spherically-focused transducers of 20 and 10 MHz nominal center frequency were used to interrogate the mullite and silicon nitride tubes, respectively. Figure 2 shows photographs of the set-up for the scans. The tube was placed upon a motorized turntable assembly placed in the immersion scan tank. A machined stainless steel reflector plate was suspended from the scanner bridge and positioned internal to the tube ~1 cm from the inside wall. In this manner, the reflector plate remained stationary throughout the scan. Prior to the scan, the transducer was positioned perpendicularly to the tube front surface using lateral, gimbal and swivel adjustments to obtain highest front surface reflection near the scan start position. Then, the distance between the transducer and tube was adjusted starting from the focal length to that distance where tube back surface echoes and reflector echoes were highest. The scanning proceeded as follows. The turntable spun so that one scan line corresponded to a full rotation (360°) of the tube. Ultrasonic echoes were acquired at 1° increments. After the turntable completed a full rotation and thus returned to the scan starting position, the transducer was raised in height by 1 mm so that a new 360° line of data could be obtained. 250 and 500 MHz analog-to-digital sampling rates were used during data acquisition. Images were contrast-expanded to better reveal *global* material variation by replacing relatively high and/or low (abnormal) data values that occurred in very low numbers with specified maximum and minimum velocity value limits (extreme value filtering) or an average of nearest neighbors.

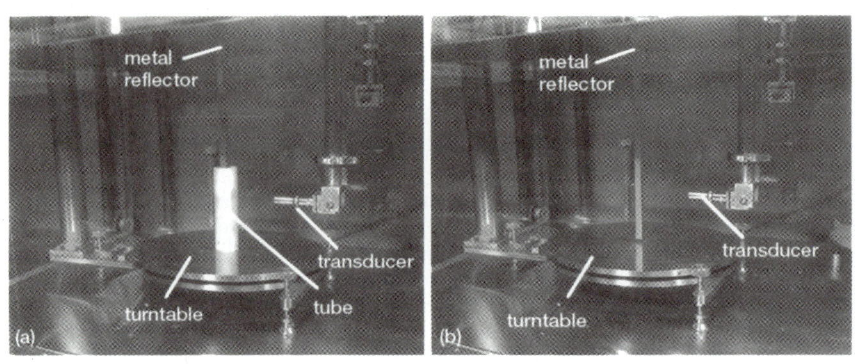

Figure 2.—Ultrasonic set-up using turntable scan for application to tubes. (a) Scan 1: sample present, collect B1, B2, M'. (b) Scan 2: sample removed, collect M".

RESULTS

Figures 3 to 6 show preliminary imaging results for the tubes. Figures 3 and 5 show image results in flat "unwrapped" form while figs. 3 and 5 show image results as "decaled" onto tubular models. Consider fig. 3 where unwrapped images of apparent and thickness-independent velocity for a 40 mm height slice of the mullite tube encompassing the machined-in patches are shown. Thirty discrete gray levels are used in the gray scale images. The apparent velocity images of figs. 3(a), 4(a) and (b) clearly shows the patches with velocity values increasing as material thickness decreases as expected based on eq. (2). (Remember that the value of thickness (d) input for the calculation of the apparent velocity image remains constant while time delay 2τ is decreasing as the material thickness decreases.) Patch edges appear "stepped" due to the coarse scan height increment (1 mm) employed. The total apparent velocity variation is on the order of 20 percent. The thickness-independent velocity images of figs. 3(b), 4(c) and (d) have undergone filtering to rid the images of most abnormal values present due to ultrasonic wave interference caused by the sharp edges of the patches.[16] These images show total velocity variation reduced to ~3 percent, and four of the five patches have nearly disappeared except for some remaining indications around the edges. The patch of ~depth 0.5 mm shows up as lower velocity than the surrounding material. This result may

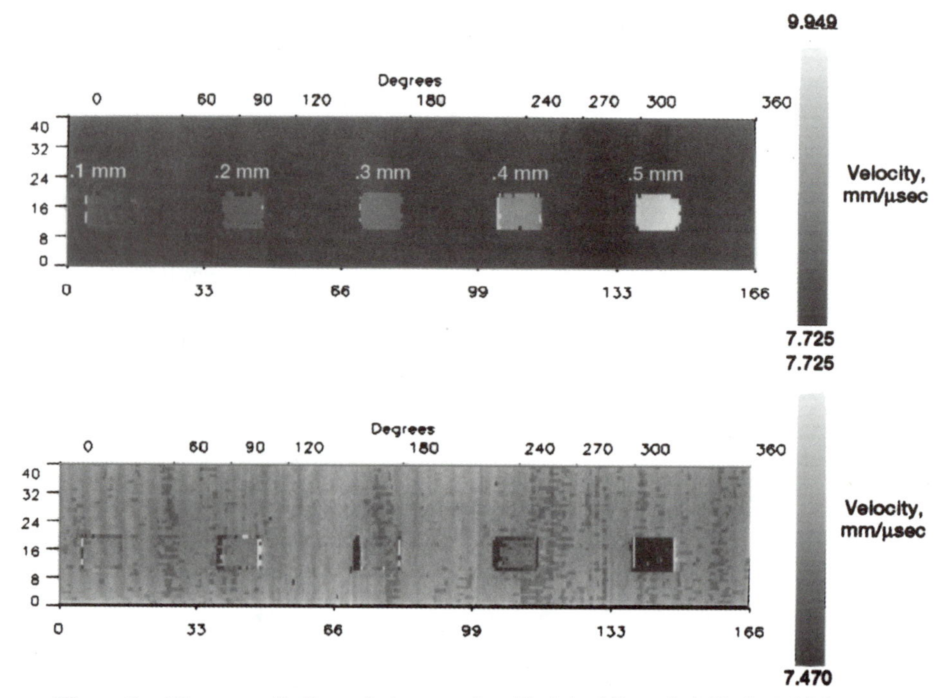

Figure 3.—"Unwrapped" ultrasonic images of mullite tube (40 mm height slice). (a) Apparent velocity. Text over patches indicates patch depth. (b) Thickness-independent velocity. (Some extréme value filtering performed to reduce abnormal values seen at patch edges).

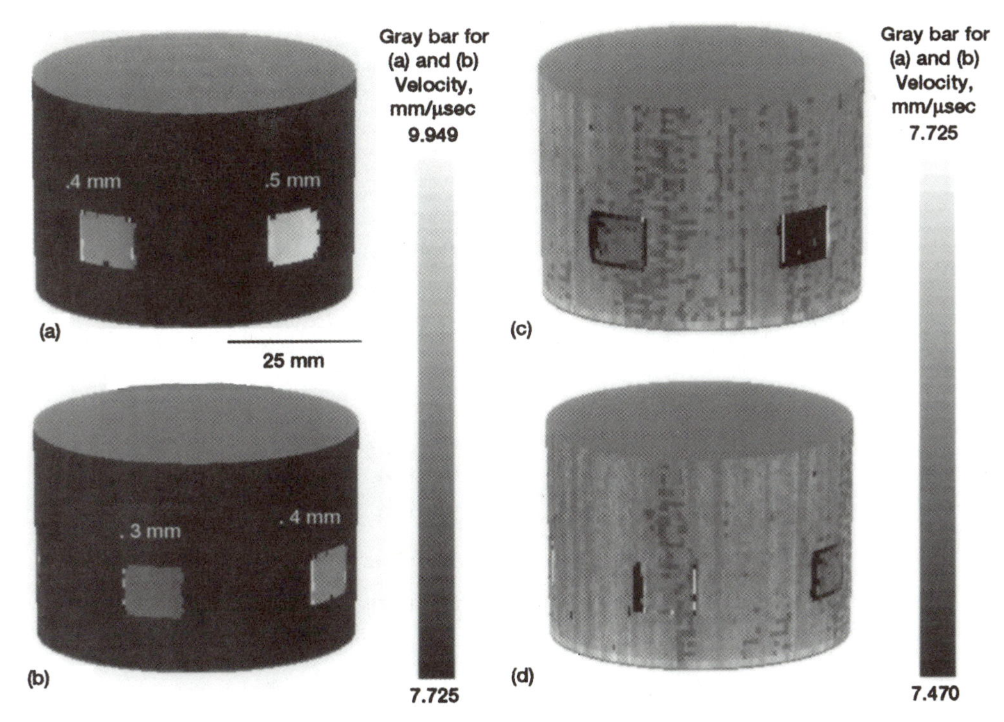

Figure 4.—Ultrasonic images of mullite tube (40 mm height slice) with ultrasonic info "decaled" onto cylinder. (a) Apparent velocity, initial orientation. Text over patches indicates approximate patch depth. (b) Apparent velocity, rotated 270 degrees with respect to initial orientation. Text over patches indicates patch depth. (c) Thickness-independent velocity, initial orientation. (d) Thickness independent velocity, rotated 270 degrees with respect to initial orientation.

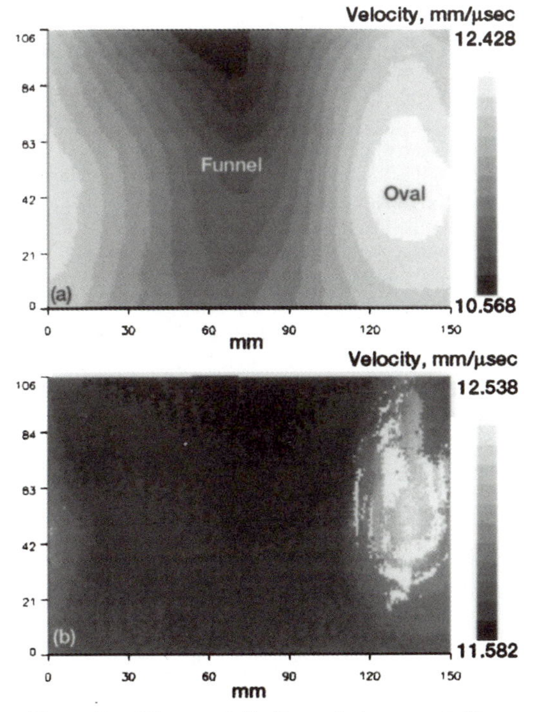

Figure 5.—"Unwrapped" ultrasonic Images of silicon nitride tube. (a) Apparent velocity. (b) Thickness-independent velocity.

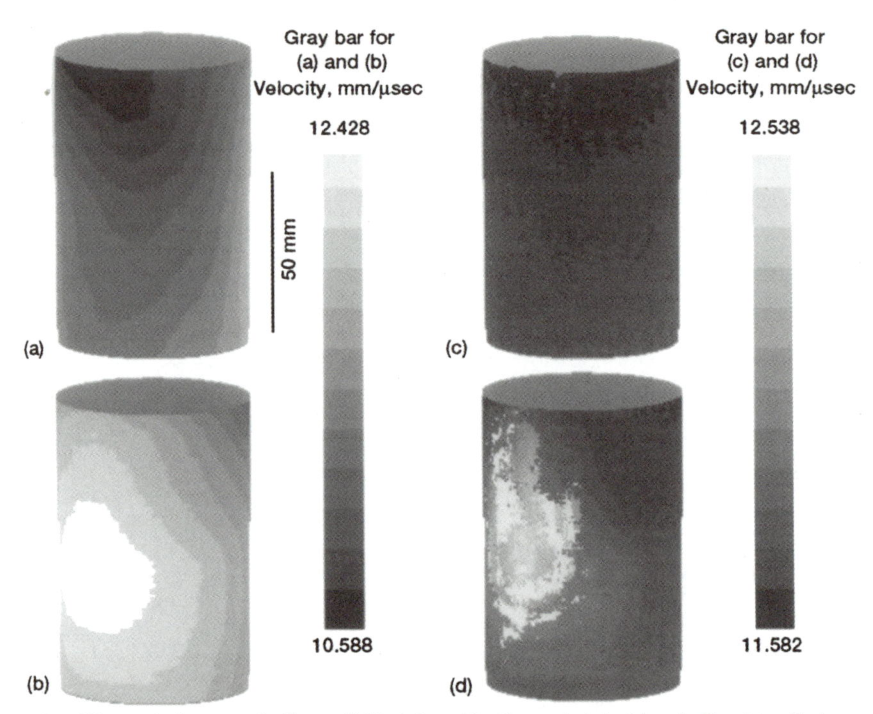

Figure 6.—Ultrasonic images of silicon nitride tube with ultrasonic info "decaled" onto cylinder model. (a) Apparent velocity, initial orientation. (b) Apparent velocity, rotated 180° with respect to initial orientation. (c) Thickness-independent velocity, initial orientation. (d) Thickness-independent velocity image, rotated 180° with respect to initial orientation.

be due to the previously-described edge effects, or refractive effects (due to slight deviations from perpendicular incidence of the ultrasonic wave on the tube) causing wave path length to the reflector plate to differ between scans with, and without, the tube present. Such refraction will result in time delays deviating from those expected if perfect perpendicularity existed, and velocity calculations based on these time delays will reflect this. The surrounding material with 255 level gray scale shows minor velocity versus position variation.

Figures 5 and 6 show preliminary imaging results for a 105 mm height of the silicon nitride tube having a 0.8 mm (10 percent) thickness variation and currently unknown microstructural variation. Twelve discrete gray levels are present in the gray scale images. A funnel-like feature containing concentric open contours showing decreasing velocity from tube bottom-to-top is indicated in the apparent velocity images of figs. 5(a) and 6(a). This feature disappears for the most part in the thickness-independent images of figs. 5(b) and 6(c) indicating it was primarily due to thickness variation. The apparent velocity image decal image of fig. 6(b) is rotated 180° (about the height axis) from the view shown in fig. 6(a). For this orientation, an oval-like feature is indicated. The oval-like feature contains concentric contours indicating different apparent velocity regions but overall the feature shows higher-than-average apparent velocity as compared to the surrounding region. In the thickness-independent image of fig. 6(d) (which is the same rotational orientation as that of the apparent velocity image decal shown in fig. 6(b), the overall oval-like feature, although somewhat diffused as compared to its indication in the apparent velocity image, does seem to be of slightly different velocity (and thus microstructure) as compared to the surrounding region. The contours seen in this region in the apparent velocity image are not seen as distinctly in the thickness-independent image indicating that some thickness variation is also present in this region of the tube. Further ultrasonic scanning trials and experimentation with different time delay calculation methods, as well as destructive testing with optical analysis and possibly x-ray computed tomography, are required to confirm these results.

CONCLUSION

The thickness-independent ultrasonic velocity images of the mullite tube region containing artificially-created thickness variations up to 500 mm in depth (highly visible in the apparent velocity image) showed only subtle indications of these thickness variations after extreme value filtering to remove abnormal values resulting from edge effects. For the silicon nitride tube containing natural thickness and microstructural variation, two major indications were revealed in the apparent velocity image results. Preliminary thickness-independent ultrasonic velocity image results indicated that one of these indications was due to thickness variation because it nearly disappeared in the thickness-independent image. The other indication was likely due to a combination of microstructural and thickness variation because the overall indication as seen in the apparent velocity image remained while internal details changed in the thickness-independent image. These results for the tubes indicate the increased degree of inspection accuracy that the thickness-independent ultrasonic imaging method can provide for hollow tube-/cylindrical-type structures.

REFERENCES

1. D.J. Roth, M.R. DeGuire, L.E. Dolhert, and A.F. Hepp, "Spatial Variations in a.c. susceptibility and microstructure for the $YBa_2Cu_3O_{7-x}$ superconductor and their correlation with room-temperature ultrasonic measurements," J. Mater. Res. 6[10] pp. 2041–2053 (1991).
2. D.J. Roth, et. al., Quantitative Mapping of Pore Fraction Variations in Silicon Nitride Using an Ultrasonic Contact Scan Technique. Res. Nondestr. Eval., Vol. 6, 1995, pp. 125–168.
3. D.J. Roth, G.Y. Baaklini, J.K. Sutter, J.R. Bodis, T.A. Leonhardt, and E.I. Crane, An NDE Approach For Characterizing Quality Problems in Polymer Matrix Composites. Proceedings of the 40th International SAMPE Symposium, May 8–11, 1995, pp. 288–299.
4. R.M. Christensen, Mechanics of Composites, John Wiley & Sons, 1979, pp. 47,51,52.
5. D.R. Flynn, Thermal Conductivity of Ceramics, Proceedings of Mechanical and Thermal Properties of Ceramics Symposium, April 1–2, 1968, pp. 92–93.
6. K. Bowles, et.al., Void Effects on the Interlaminar Shear Strength of Unidirectional Graphite-Fiber-Reinforced Composites. J. Comp. Mat., Vol. 26, No. 10, 1992, pp. 1487–1509.
7. J.J. Gruber, J.M. Smith, and R.H. Brockelman, Ultrasonic Velocity C-scans for Ceramic and Composite Material Characterization. Mater. Eval., Vol. 46, No. 1, 1988, pp. 90–96.
8. B.D. Sollish, Ultrasonic Velocity and thickness gage, United States Patent No. 4,056,970, Nov. 8, 1977.
9. L. Pichè, Ultrasonic velocity measurement for the determination of density in polyethylene. Polymer Engineering and Science, Vol. 24, No. 17, Mid-December 1984, pp. 1354–1358.
10. I.Y. Kuo, B. Hete, and K.K. Shung, A novel method for the measurement of acoustic speed. J. Acoust. Soc. Am. Vol. 88, No. 4, October 1992, pp. 1679–1682.
11. D.K. Hsu and M.S. Hughes, Simultaneous ultrasonic velocity and sample thickness measurement and application in composites. J. Acoust. Soc. Am. Vol. 92, No. 2, Pt. 1, August 1992, pp. 669–675.
12. V. Dayal, An Automated Simultaneous Measurement of Thickness and Wave Velocity by Ultrasound, Experimental Mechanics, September, 1992, Vol. 32, No. 2, pp. 197–202.
13. M.S. Hughes and D.K. Hsu, An automated algorithm for simultaneously producing velocity and thickness images, Ultrasonics 1994 Vol. 32, No. 1, pp. 31–37.
14. D.J. Roth, Single Transducer Ultrasonic Imaging Method That Eliminates the Effect of Plate Thickness Variation in the Image. NASA TM–107184, 1996.
15. D.J. Roth, et. al., Commercial Implementation of Ultrasonic Velocity Imaging Methods via Cooperative Agreement Between NASA - Lewis Research Center and Sonix, Inc. NASA TM–107138, 1996.
16. Nondestructive Testing Handbook, second edition, Volume 7 Ultrasonic Testing, eds. Birks, A.S., Green, R.E., and McIntire, P. American Society For Nondestructive Testing, 1991, pp. 227–230, 237.
17. D.R. Hull, H.E. Kautz, and A. Vary, Measurement of Ultrasonic Velocity Using Phase-Slope and Cross-Correlation Methods. Mater. Eval., Vol. 43, No. 11, 1985, pp. 1455–1460.

FLAW DETECTION IN STEEL WIRES BY ELECTROMAGNETIC ACOUSTIC TRANSDUCERS

Tomohiro Yamasaki,[1] Shingo Tamai,[2] and Masahiko Hirao[1]

[1] Faculty of Engineering Science
[2] Graduate School
Osaka University
Toyonaka, Osaka 560, JAPAN

INTRODUCTION

Since wires are used to suspend heavy components in the structure, such as bridges, the flaw detection method is strongly required to be established. One of the promising techniques is ultrasonics. Especially, use of the ultrasonic wave propagating in the axial direction makes the inspection for a long range accessible.[1]

Ferromagnetic materials show the dimensional change, called magnetostriction, when an external magnetic field is applied. For low carbon steels, the magnetostriction first increases with the magnetization, and then decreases under the strong field. If the magnetic field along the length of the steel wire changes rapidly, the consequent change in the magnetostriction can be the source of the longitudinal wave propagating along the wire. In this study, we made electromagnetic acoustic transducers (EMATs) for flaw detection in steel wires, which utilize the magnetostriction in generating and detecting the longitudinal wave. Dependence of the signal amplitude on the magnetic field strength was investigated for transmitter and receiver, respectively. The results showed that the use of the separate EMATs, each designed for transmitter or receiver, is recommended. The EMATs were then applied to the detection of artificial flaws.

ELECTROMAGNETIC ACOUSTIC TRANSDUCER

Magnetization process of polycrystalline steel starts with domain wall movement in the low field region, followed by rotation of domain magnetization under the strong field. The domain restructuring induces the fractional change in the dimension, called the magnetostriction. As illustrated in Fig. 1, the magnetostriction increases during the domain realignment process, and then starts to decrease when the rotation magnetization becomes dominant.

We made separate EMATs for transmitter and receiver, each of which consists of a coil wound around the wire and an electromagnet to give the static magnetic field along the length of the wire. Figure 2 sketches the structure schematically. When the input signal drives the

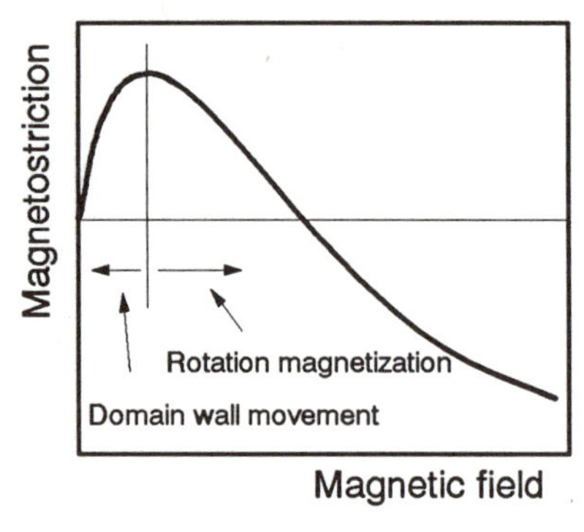

Figure 1. Magnetostriction curve of steel (schematic).

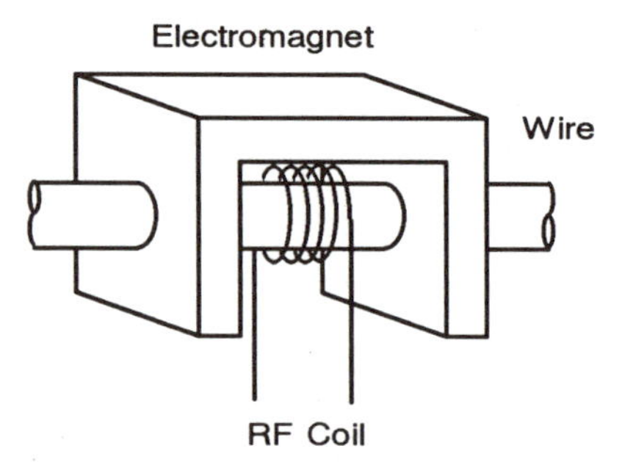

Figure 2. EMAT for steel wire.

transmitter coil, the effective magnetic field in the wire varies around the bias field. The resultant change in the magnetostriction launches the longitudinal wave propagating in the axial direction. Assuming that the magnetostrictive response for the dynamic field is the same as the static response, the amplitude will be proportional to the slope of the magnetostriction curve. When the wave arrives at the receiver, the accompanying stress varies the permeability of the wire. Then the flux density under the bias field changes, which can be picked up by the receiver coil.

TRANSDUCTION EFFICIENCY OF EMAT

Experimental Procedure

Figure 3 shows the experimental setup. We used the square pulse of 1 μsec width to excite the transmitter. After being propagated for 1m, the wave was received and amplified by the broadband receiver. The signal was then sampled by the storage oscilloscope, and was averaged 128 times to improve the signal to noise ratio. The bias field strength of each EMAT was varied by adjusting DC current to the electromagnet.

We prepared JIS-SS400 low carbon steel wires of 5mm diameter and 2m long. Using a flawless wire, we measured the dependence of the signal amplitude on the bias field strength for

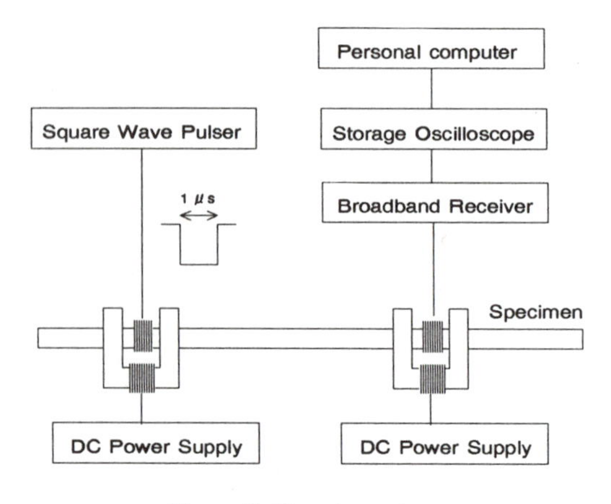

Figure 3. Experimental setup.

each of transmitter and receiver. After the demagnetization, the bias field strength was increased stepwise, while measuring the signal amplitude at each level of magnetization.

The magnetostriction curve of the wire was measured using the semiconductor strain gauges. The electromagnet of the EMAT applied the magnetic field. To minimize the effect of the bending, two gauges were attached on the surface axial symmetrically. At several levels of magnetization, we also measured the magnetic flux change in the wire induced by the static stress.

Results and Discussion

Figure 4 compares the magnetic field dependence of the amplitude with the magnetostriction curve of the wire. Magnetizing current is taken to be positive when the

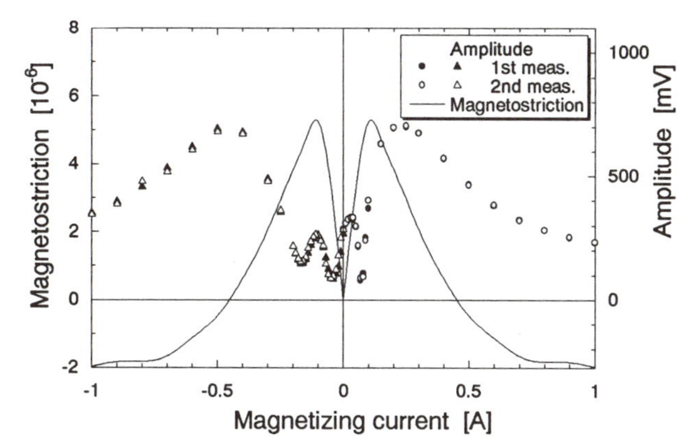

Figure 4. Comparison of magnetostriction and field dependence of generated signal amplitude.

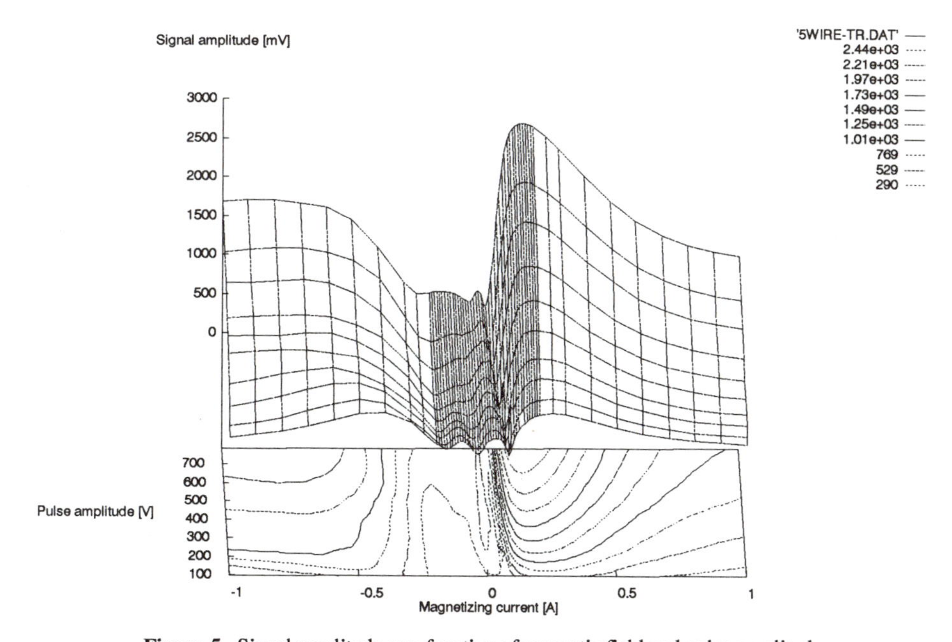

Figure 5. Signal amplitude as a function of magnetic field and pulse amplitude.

dynamic field is applied in the same direction as the static field. Neglecting the -0.04A offset, it is found that the amplitude depends on the slope of the magnetostriction curve. Especially, signal almost disappears when the magnetostriction shows the extreme value. Comparing with the rotation magnetization region appearing above 0.11A, the slope of the magnetostriction curve is steeper for the domain realignment process, while the amplitude is smaller. It reveals that the dynamic response of the magnetostriction differs from the static response. It may be due to the pinning of the domain walls by the dislocations or the precipitates. Figure 5 plots the signal amplitude against the magnetic field and the pulse amplitude. We found that the offset increases with the pulse amplitude, which can be explained as follows. Since we used the square wave pulse, the effective field below the transmitter shifts to one side and then returns back to the initial value. It can then be expected that the signal amplitude depends on the slope at the mean value of the effective field. The offset thus appears depending on the magnitude of the dynamic field, equivalently, the pulse amplitude. However, the -0.04A offset seems to be too large considering the experimental condition. The electromagnet consists of a 3500-turn coil and the core having much larger crosssectional area than that of the wire. While the skin effect may be one of the reasons, it is still left to be investigated.

Figure 6 shows the field dependence of the received signal amplitude. Even at the demagnetized state, that is, for zero magnetizing current, small signal was observed. It may be due to the effect of the residual magnetization and the magnetic flux leaked from the transmitter. Figure 7 plots the change in the flux density induced by the static stress. The flux was measured by using a 1000-turn coil and the flux meter before and after applying the static stress to the wire. Peaks are observed at almost the same magnetizing current in Figs. 6 and 7, which ensures that the EMAT receives the wave as the induction change caused by the stress of the wave.

As shown in Figs. 4 and 6, the optimum field strength differs from each other for transmitter and receiver. Thus the use of the separate EMATs for transmitter and receiver is recommended. Furthermore, the pulser is usually designed for 50 Ω impedance, which limits the turns of transmitter coil. However, the broadband receiver allows us to use a high impedance coil and we can make the receiver coil wound many turns to improve the sensitivity. The separate EMATs are also suitable for this reason. In this study, we used 29-turn and 100-turn coils for transmitter and receiver, respectively.

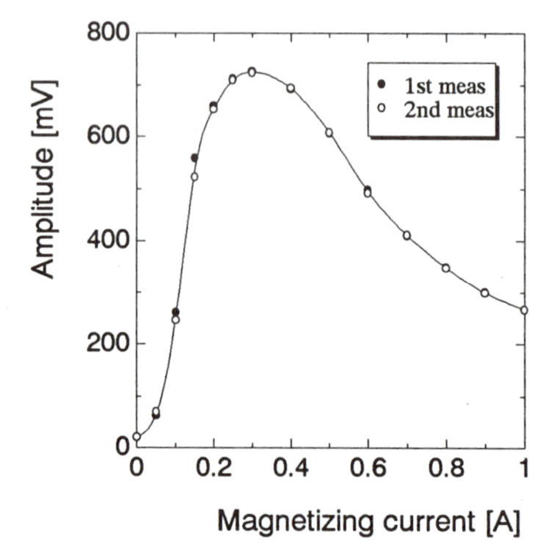

Figure 6. Magnetic field dependence of received signal amplitude.

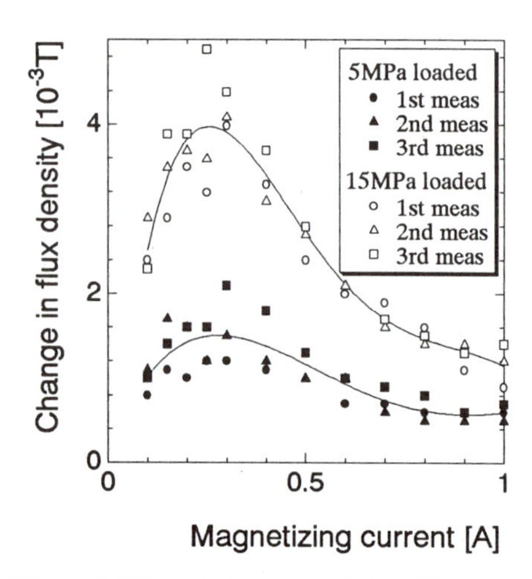

Figure 7. Magnetic flux change induced by stress.

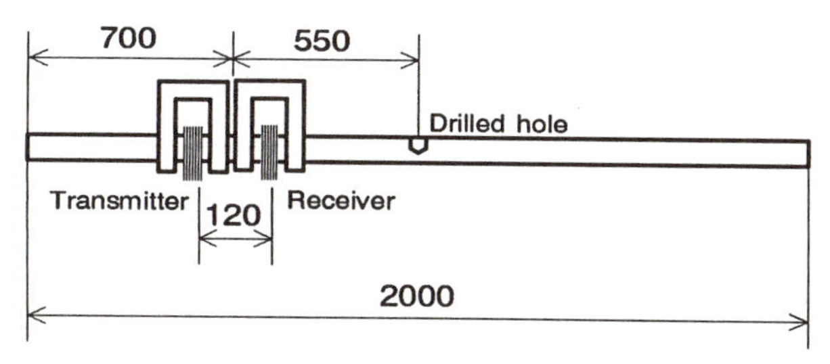

Figure 8. Location of EMATs and flaw.

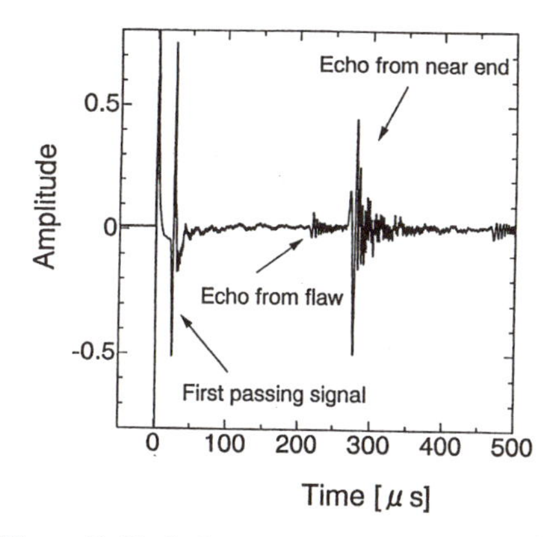

Figure 9. Typical waveform of separate EMATs.

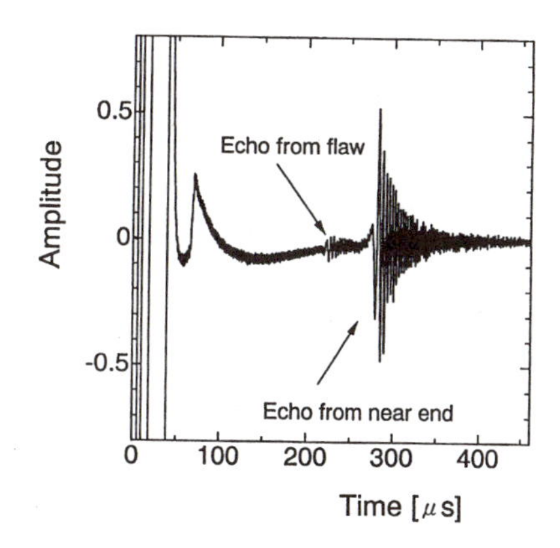

Figure 10. Waveform of single EMAT.

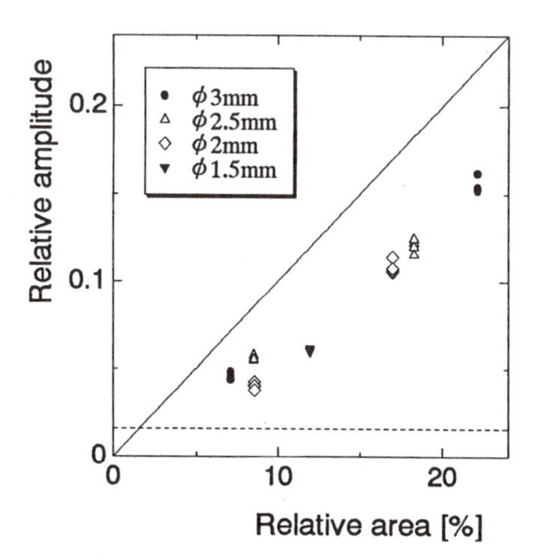

Figure 11. Amplitude of flaw echo.

FLAW DETECTION BY EMAT

As artificial flaws, we drilled holes on the wires. The location of the EMATs and the flaw is sketched in Fig. 8. Figure 9 shows the typical waveform. While the first passing signal shows the sharp shape, the wave is distorted as the propagation path increases. It can be attributed to the dispersion[2] and the frequency dependence of the attenuation. The waveform of single EMAT is also represented in Fig. 10. The transmitter EMAT was also used as the receiver. To prevent the input pulse from giving damage to the broadband receiver, we used a diplexer. Figures 9 and 10 show the normalized signal The absolute signal amplitude was larger for the separate EMATs configuration. Comparison of Figs. 9 and 10 reveals that the signal to noise ratio is improved by using the separate EMATs. For the single EMAT configuration, the leakage of input pulse cannot be avoided completely, which makes it difficult to detect a flaw located close to the EMAT.

Figure 11 plots the relative amplitude of the flaw echo against the relative area of the flaw. The amplitude of the flaw signal was normalized with that of the echo from the near end. The dotted line represents the noise level. If the relative amplitude equals to the relative area, the results will be plotted on the solid line. In this study, the curvature of the surface of artificial flaws is small compared with the wavelength, which reduces the echo amplitude. Since the signal amplitude is almost proportional to the flaw size, we can predict that the wave intensity distributes uniformly across the diameter of the wire. Assuming that the flaw surface is flat and normal to the axis of the wire, the results are expected to be more close to the solid line in Fig. 11.

CONCLUSION

The properties of the electromagnetic acoustic transducers for flaw detection in steel wires were investigated. It was found that the magnetostriction is the dominant mechanism and the optimum field strength exists for each of the transmitter and the receiver. The use of the separate EMATs is recommended to improve the flaw detectability.

REFERENCES

1. H. Kwun and C.M. Teller, Detection of fractured wires in steel cables using magnetostrictive sensors, *Mat. Eval.* 52:503 (1994).
2. H. Kolsky, *Stress Waves in Solids*, Dover, New York (1963).

NONDESTRUCTIVE CHARACTERIZATION OF THE THERMAL DIFFUSIVITY OF POOR CONDUCTORS BY A LOW COST, FLASH TECHNIQUE

John M. Liu,[1] Robert A. Brizzolara,[1] and Douglas N. Rose[2]
[1]Carderock Div. Naval Surface Warfare Center
West Bethesda, Md. 20817-5700
[2]AMSTA-RSA
U.S. Army TACOM
Warren, MI 48397-5000

INTRODUCTION

Numerous techniques and approaches exist for the determination of the thermal diffusivity through the thickness of materials. These include techniques based on photothermal radiometry, the detection of the acoustic wave excited by the heating beam, and the 'mirage' effect. These techniques typically incorporate a high power laser or an infrared camera, items that are relative expensive and not generally available in small-scale material evaluation facilities.

We report here a low-cost approach for the determination of through thickness thermal diffusivity in poor conductors. It is based on a pulse excitation using a photographic flash unit and temperature measurement using a commercially available, film-back thermocouple attached to the material. In the following, we will introduce the basic principles for converting the measured temperature rise to diffusivity, show the calibration using materials of known diffusivity, and discuss several experimental parameters affecting the precision and accuracy of the technique. We will present some data on composite materials of relatively low diffusivity to show the applicability of the apparatus.

THEORETICAL BACKGROUND

The thermal diffusivity is a quantity which is a measure of how quickly heat propagates through a material. It is defined by the expression:

$$\alpha = K/(C_p*\rho)$$

where α is the thermal diffusivity in m^2/sec, K is the thermal conductivity in W/m °K, C_p is the specific heat in J/kg °K, and ρ is the density in kg/m^3.

This apparatus measures the thermal diffusivity of a material using the flash technique.[1] This method requires recording the temperature of the rear surface of a material as a function of time after heating the front surface with an instantaneous energy pulse. The time needed for the rear surface to reach one-half its maximum temperature ($t_{1/2}$) can be used in a simple expression to determine the thermal diffusivity of the sample:

$$\alpha = 0.1388*L^2/t_{1/2}$$

where α is in mm^2/sec, L is the sample thickness in mm, and $t_{1/2}$ is in seconds. This expression assumes a one dimensional heat flow through the material with no complications from lateral material boundaries.

EXPERIMENTAL

A sketch of the apparatus lay-out is shown in Fig. 1. Some pertinent specifications of the equipment and experimental conditions are shown in Table I.
A photoflash unit was used as the energy source, and a film-back thermocouple was used to measure the rear-surface temperature of the sample. A digital voltmeter (1 μV resolution) was used to measure the thermocouple output. The whole experiment was controlled with a PC with a GPIB card. After the data-collection routine begins, the photoflash is fired after a user-specified time delay. This was done by sending a signal from the computer's parallel port to an external relay box.

The data-acquisition code possesses a number of capabilities. Among these are averaging, automatic wait between averaging runs for the sample to cool to a preset temperature, automatic photoflash-firing at a user-selected time, and data-plotting. The maximum data-acquisition rate of the system is dependent on the computer platform the data collection program is being run on. The nominal data-

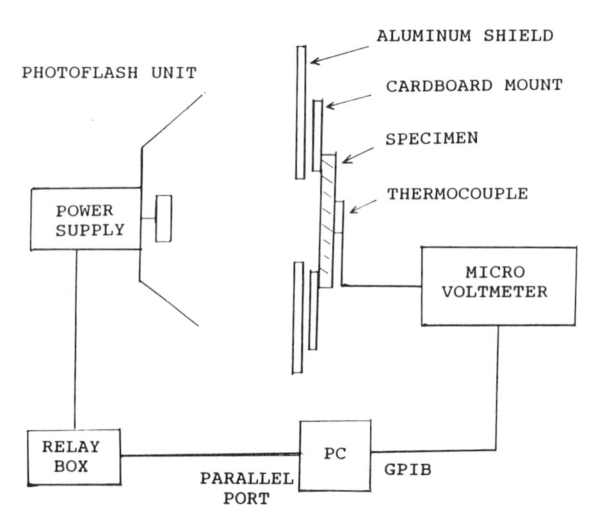

Figure 1. Sketch of experimental set-up.

TABLE I. Equipment specifications and experimental conditions.

source output	2900 BCPS
source pulse duration	< 1 msec
angles of illumination	45° x 60°
thermocouple surface area	5 x 20 mm
thermocouple response characteristics	10 - 20 msec
thermocouple sensitivity	40 μV/C @20 C
thermocouple backing material	polymer/glass
thermocouple couplant	heat sink compound
specimen surface area (exposed to source)	2.54 x 2.54 cm
specimen characteristics	flat, dark color
specimen thickness	1 - 5 mm

acquisition rate is determined by the settings on the voltmeter; however, at some point, the computer speed will limit the rate at which the data-acquisition program can run. The actual data acquisition rate was checked with a fast digital oscilloscope the result of which was compared to the number of data acquisition command stated in the computer program. The maximum, reliable data acquisition rate on the machine used was found to be approximately 120 points/sec.

The samples were mounted in one of two ways: by clamping between two rubber cushions or by fastening to a piece of cardboard with adhesive tape. The sample was placed about 3 to 5 cm from the flash unit. The front surface of refractory metal samples was painted black; the front surface of transparent samples was painted silver, then black. The black paint enhanced heat absorption and the silver paint prevented light from penetrating into the sample. An aluminum foil or cardboard shield was placed in front of the sample to prevent the input energy from striking the sides of the sample. The transparent plastic lens was removed from the flash unit to facilitate transmission of the broadband radiation to the sample. The flash unit was fired a known amount of time after the data collection computer program had started.

A fast digital oscilloscope was used to measure the time delay in firing the photoflash by measuring the time between the onset of the signal from the parallel port of the computer and the occurrence of the flash, as measured by a photodetector. This time delay was measured to be 2.3 milliseconds. This is insignificant. In addition, the time delay introduced by the "flash" command in the data-acquisition program was measured using the computer's clock, 12000 iterations of the data-acquisition loop and a data-acquisition rate (nominal) of 120 pts/sec. The time

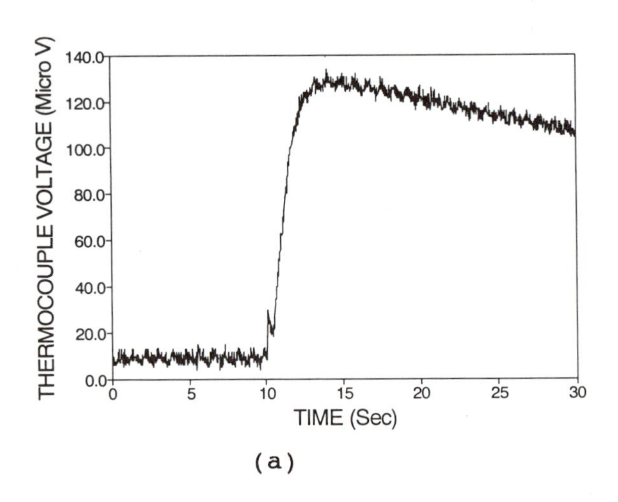

(a)

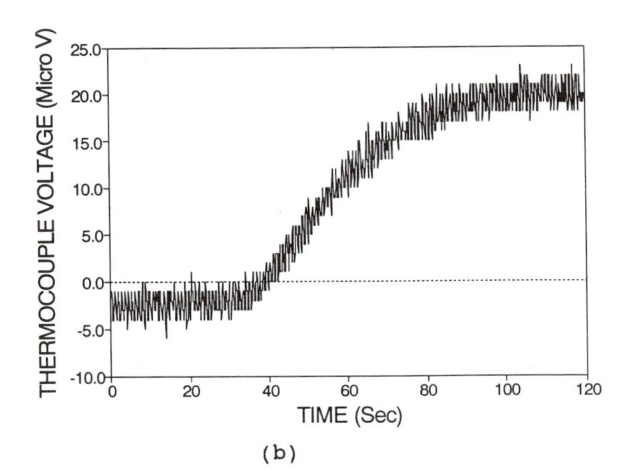

(b)

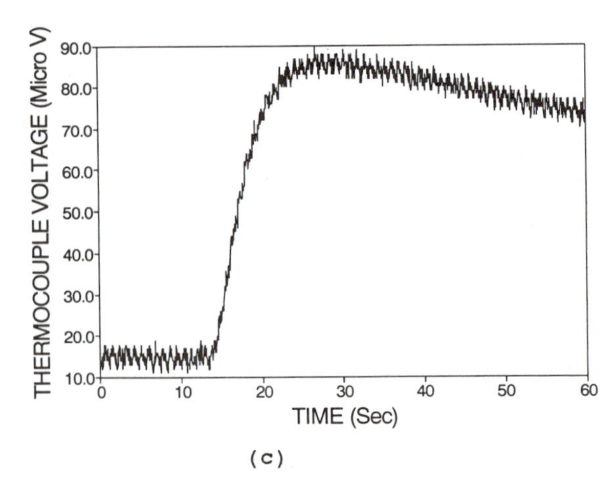

(c)

Figure 2. Typical temperature versus time trace for (a) lucite, 2.0 mm thick, (b) lucite, 4.0 mm thick, and (c) reinforced phenolic, 5.0 mm thick.

difference was found to be small enough so as not to be measurable.

RESULTS AND DISCUSSION

Some examples of temperature versus time traces recorded with our equipment will now be presented to illustrate the effects of experimental parameters on the measurements.

1. Sample Thickness: Figure 2A shows data for the 2.0 mm thick Lucite plate and Figure 2B for the 5.8 mm thick Lucite plate (a factor of 2.9 difference in thickness). The measured half rise times differ by a factor of 7.9, within 5% of the square of the thickness ratio - very good agreement. A typical trace for a 5 mm thick, reinforced phenolic composite is shown in Fig. 2C. Note that the time at which the flash unit was fired was many seconds after data acquisition had started, and is not shown in these traces. This time was set by the timing loop within the data acquisition computer program.

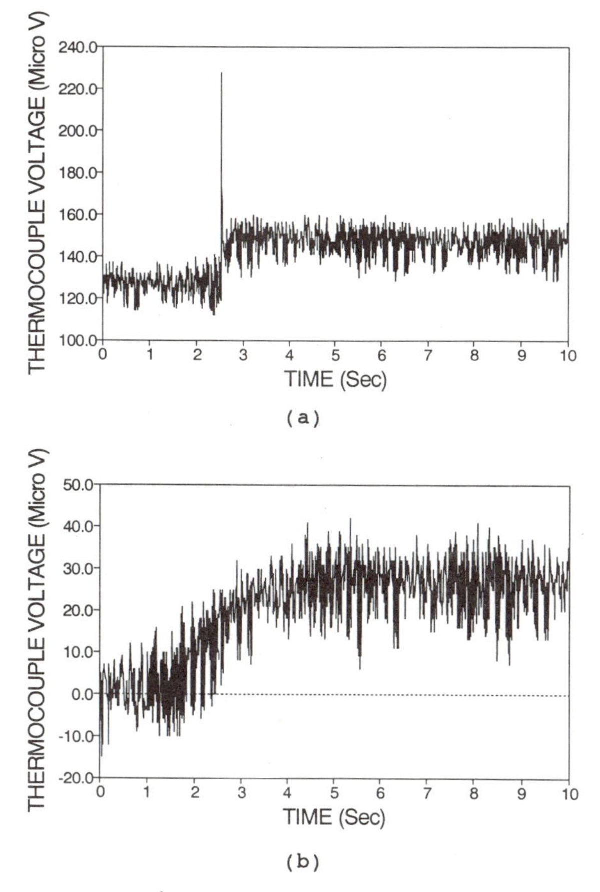

Figure 3. Temperature versus time trace for alumina; (a) with one layer of black paint, and (b) with 3 layers of black paint.

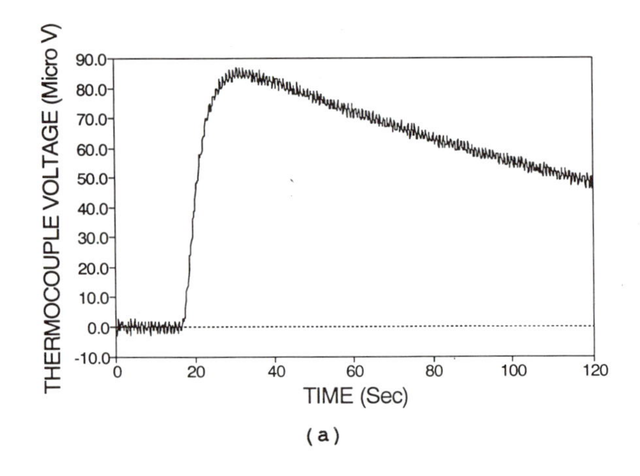

(a)

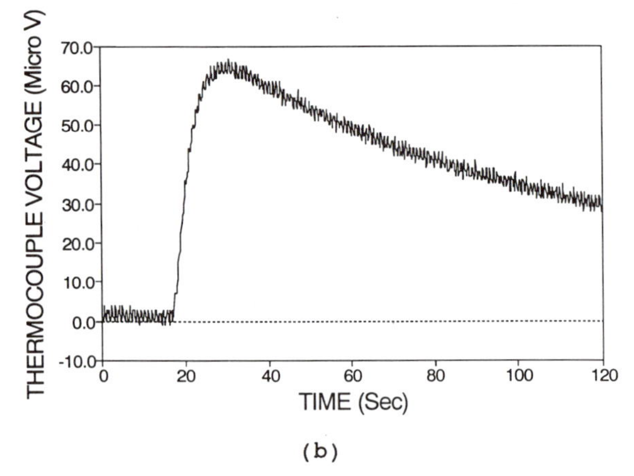

(b)

Figure 4. Effect of the size of the illumination opening on thermal response. Opening sizes are: (a) 2.54 x 2.54 cm, and (b) 1.27 x 1.27 cm.

2. Use of Paint on Front Surface: Figure 3A shows an alumina sample with a thin coating of black ink on the front surface. The spike in the data due to the instantaneous arrival of the light pulse is evident. Figure 3B shows the same sample with three coats of black ink. The spike has been eliminated.

3. Size of Hole in Shield: Figure 4A show data for Lucite with an aluminum shield with a opening of 2.54 x 2.54 cm. Figure 4B shows the result when the opening was changed to 1.27 x 1.27 cm. It can be seen that while the rise time is the same in both Figures, the smaller opening resulted in a maximum amplitude approximately 10% smaller, and an increase in the rate of fall-off of the rear surface temperature. These effects are presumably due to the non-uniform illumination of the front surface of the sample.

Examples of diffusivity measured with our apparatus for a number of materials are shown in the following Table.

TABLE II. Measured thermal diffusivity, α

MATERIAL	α_{ave} (mm^2/sec)	Std. Dev. (mm^2/sec)	α (known) (mm^2/sec)	comments
lucite	1.1	----	1.1	quick check
Al_2O_3	7.8	----	5 - 6	porosity unknown
graphite/ epoxy	0.53	0.07	0.4 - 0.6	chemistry unknown
reinforced phenolic A	0.41	0.03	0.42	chemistry unknown
reinforced phenolic B	0.29	0.05	0.29	chemistry unknown
reinforced phenolic C	0.17	0.05	(unknown)	chemistry unknown

CONCLUSIONS

A low cost apparatus and approach for thermal diffusivity measurement using the flash technique for poor conductors have been demonstrated. The cost of the equipment, excluding the PC and the voltmeter, was less than $200 US. After extraneous effects associated with the experimental set-up were minimized, diffusivities successfully measured ranged from approximately 0.2 to 6 mm^2/sec, using material specimen thickness larger than 1 mm.

REFERENCES

1. W. J. Parker, et al, J. Appl. Phys. 32, 1679 (1961).

RAPID, CONTACTLESS MEASUREMENT OF THERMAL DIFFUSIVITY

Zhong Ouyang, Li Wang, Feng Zhang, L.D. Favro, and R.L. Thomas

Department of Physics and Institute for Manufacturing Research
Wayne State University
Detroit, MI 48202

INTRODUCTION

We report a novel method for the rapid determination of the thermal diffusivity of materials. The technique utilizes pulsed heating in a geometrical pattern on the surface of the sample, with infrared imaging of the resultant temperature field on the surface as a function of time. By fitting the observed temperature fields to theoretical models, we extract the thermal diffusivities in three orthogonal directions. Knowledge of these diffusivities is important for the design and manufacture of advanced materials for aerospace structures, because of the high temperatures reached on leading edges of airfoils, etc., in supersonic flight conditions. This technique is applicable to a variety of new space-age materials, including ceramic-matrix composites, three-dimensionally woven polymer composites, and high-temperature alloys.

EXPERIMENTAL ARRANGEMENT

The configuration of the preliminary apparatus is shown schematically in Fig. 1. Here, the heat source is a flashlamp or flashlamps arranged to produce uniform illumination at the front surface of the slab-shaped material under investigation. A shadow of some geometrical shape (in this case a rectangular corner) is cast upon the surface by a screen placed just in front of the sample. The surface temperature of the sample, either on the illuminated surface, or (as shown in Fig. 1), on the opposite side, is monitored both spatially and temporally by means of a focal plane array (FPA) infrared (IR) camera. A computer controls the FPA, acquires the data, and synchronizes it with the flashlamps. Data acquisition occurs at a rate of 20M-pixels per sec, with 12-bit dynamic range. The FPA has 512x512 detector elements. The basic principle of the measurement is to measure the rate at which heat flows into the shadow region, as well as the rate at which the heat flows through the sample from the heated side across the slab to the rear of the sample. These data are then fit to theoretical expressions for the spatial and temporal dependencies of the temperature on the front (or rear) surface, to obtain the thermal characteristics of the material.

THEORY

An earlier method of determining thermal diffusivity [1], which more or less has become standard, also uses a flash of thermal energy applied to the front face of the sample,

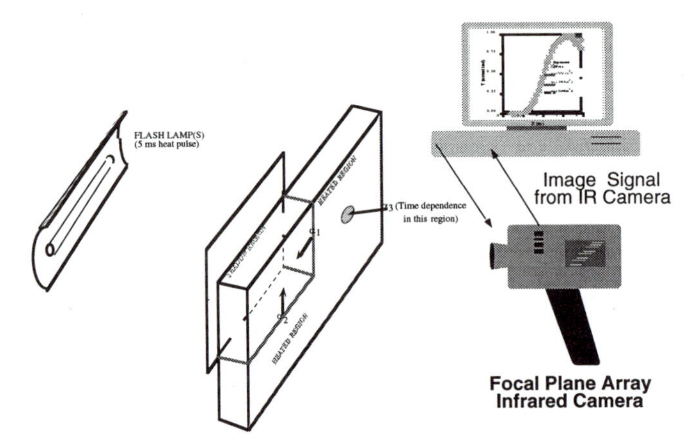

Fig. 1 Block diagram of the experimental apparatus.

with the resulting temperature variation on the back face being measured as a function of time. The standard technique only measures the flow normal to the surface. In that standard measurement, the temperature of the rear surface is recorded as a function of time, and once it reaches its maximum value, the heating record is examined to determine the time for which it had reached half that maximum value. The diffusivity is then determined from the single time determination, together with the measured thickness of the slab, using the formula $\alpha = \frac{0.139\ d^2}{t_{1/2}}$, where d is the sample thickness. Our method has the important advantage of providing diffusivity values in three distinct orthogonal directions, as well as considerably improved precision, which is gained by taking into account the full set of data. Equations 1 and 2 represent equivalent expressions for the front-surface temperature distribution, T_F, when a corner shadow is utilized (see Fig. 1), and Equations 3 and 4 represent the corresponding expressions for the rear-surface temperature distribution, T_R. Here, x and y are coordinates measured parallel to the two edges of the corner shadow, with their common origin at the corner, and α_1, α_2, and α_3 are the x-,y-, and z-components of the diffusivity, respectively. Similar expressions can be derived for other shadow geometries. These equations are the basis for the theoretical fits used to obtain the numerical values of the thermal properties of the materials to be investigated.

$$T_F = \left[1+\operatorname{erf}\left(\frac{x}{\sqrt{4\alpha_1 t}}\right)\right]\left[1+\operatorname{erf}\left(\frac{y}{\sqrt{4\alpha_2 t}}\right)\right]\frac{1}{\sqrt{4\pi\alpha_3 t}}\sum_{n=-\infty}^{\infty} e^{-\frac{(2n)^2 d^2}{4\alpha_3 t}} \quad \text{(Short times)} \tag{1}$$

$$T_F = \left[1+\operatorname{erf}\left(\frac{x}{\sqrt{4\alpha_1 t}}\right)\right]\left[1+\operatorname{erf}\left(\frac{y}{\sqrt{4\alpha_2 t}}\right)\right]\frac{1}{d}\sum_{n=-\infty}^{\infty} e^{-\frac{\alpha_3\pi^2 n^2}{d^2}t} \quad \text{(Long times)} \tag{2}$$

$$T_R = \left[1+\operatorname{erf}\left(\frac{x}{\sqrt{4\alpha_1 t}}\right)\right]\left[1+\operatorname{erf}\left(\frac{y}{\sqrt{4\alpha_2 t}}\right)\right]\frac{1}{\sqrt{4\pi\alpha_3 t}}\sum_{n=-\infty}^{\infty} e^{-\frac{(2n-1)^2 d^2}{4\alpha_3 t}} \quad \text{(Short times)} \tag{3}$$

$$T_R = \left[1+\operatorname{erf}\left(\frac{x}{\sqrt{4\alpha_1 t}}\right)\right]\left[1+\operatorname{erf}\left(\frac{y}{\sqrt{4\alpha_2 t}}\right)\right]\frac{1}{d}\sum_{n=-\infty}^{\infty} (-1)^n e^{-\frac{\alpha_3\pi^2 n^2}{d^2}t} \quad \text{(Long times)} \tag{4}$$

Fig. 2 Typical IR frame of T_R. The "shadow" region is in the upper left corner of the image, corresponding to the geometry pictured schematically in Fig. 1.

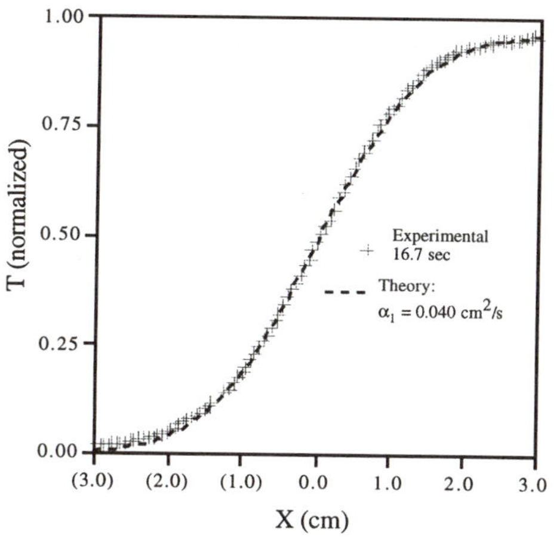

Fig. 3 Comparison of experiment and theory for diffusion along the fiber direction (lateral thermal diffusion).

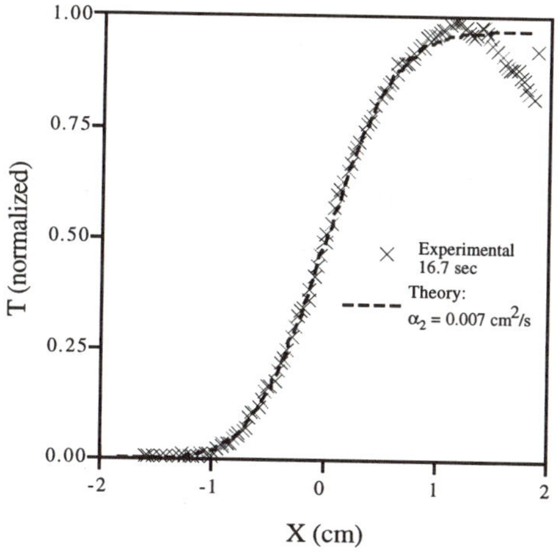

Fig. 4 Comparison of experiment and theory for diffusion perpendicular to the fiber direction (lateral thermal diffusion).

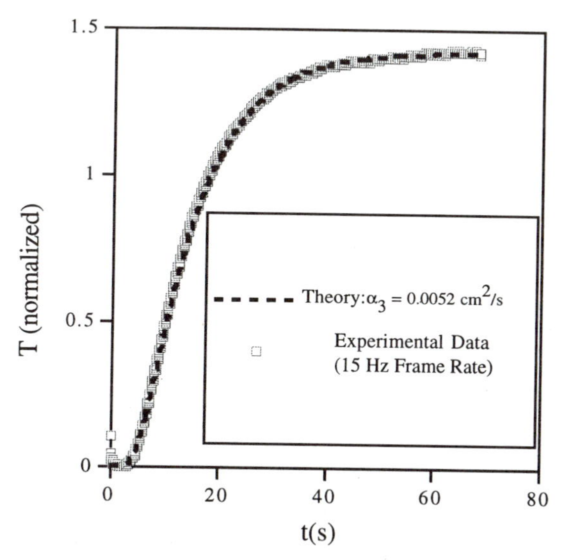

Fig. 5 Comparison of experiment and theory for diffusion perpendicular to the fiber direction (thermal diffusion through the plate).

EXPERIMENTAL RESULTS

A typical IR image frame of T_R is shown in Fig. 2. The "shadow" region is in the upper left corner of the image, corresponding to the geometry pictured schematically in Fig. 1. The blurring on the edge of the shadow in Fig. 2 results from lateral diffusion of the heat. The differences of the blurring for vertical and horizontal edges of the shadow are evidence for the in-plane anisotropy in the thermal properties of this sample. We have analyzed preliminary data from such a sample, which is a uniaxial carbon-fiber-reinforced composite sample. The analysis is displayed as theoretical and experimental curves in Figs. 3-5. The resulting three thermal diffusivities (0.040cm^2/s; 0.007cm^2/s; 0.0052cm^2/s) clearly show the anisotropy in all three dimensions for this preliminary test of the method on a polymer composite. In practice, we utilize not one such curve, but a family of curves for a large number of frame times (for lateral diffusion, as in Figs. 3,4), and a number of positions (for diffusion across the plate, as in Fig. 5) to obtain more precise estimates of the thermal properties of the material.

CONCLUSIONS

This method provides for rapid measurements of thermal diffusivities in three orthogonal directions. Theory and experiment are in good agreement, so that full use the large data set in a sequence of images allows a very accurate determination of the diffusivity tensor.

ACKNOWLEDGMENT

This work was supported by the Institute for Manufacturing Research, Wayne State University.

REFERENCE

1. W.J. Parker, *J. Appl. Phys.*, **32**: 1679 (1961).

EFFECTS OF TEXTURE ON PLASTIC ANISOTROPY IN SHEET METALS

Chi-Sing Man

Department of Mathematics
University of Kentucky
Lexington, Kentucky 40506

INTRODUCTION

Crystallographic texture determines the mechanical anisotropy of sheet metals and thus strongly influences their formability. For an alloy with a given composition, crystallographic texture is determined by the processing sequence employed to produce the end product. By suitably controlling thermomechanical processing parameters, a desirable blending of texture components could be obtained. The capability to monitor crystallographic texture on-line at various stages of the processing sequence is a prerequisite for controlling the texture of the end product.

Ultrasonic techniques offer a fast and inexpensive means for obtaining information on the texture of sheet metals on-line. By measuring in effect the texture coefficient W_{400}, ultrasonics has been used[1,2] successfully for monitoring the average plastic strain ratio $\bar{r}$ of steel sheets. Efforts[3,4] have also been made to use ultrasonics to characterize textures in aluminum alloys. As is well known, ultrasonics can deliver only very limited information on crystallographic texture. For the applications at hand, however, we are interested really not in crystallographic texture per se, but in its effect on formability of the sheet metal in question. Hence the bottom line is whether the information on texture that we could obtain from ultrasonics would suffice for our purpose. In this regard the success story on steel sheets serves as a good example. Earlier studies[5,6] suggest that the texture coefficients W_{400}, W_{420}, W_{440} would specify, to good approximation, the plastic anisotropy of cold-rolled steel sheets. Through the efforts of many researchers, this observation has ultimately led to the successful application of ultrasonics for on-line monitoring of the formability parameter $\bar{r}$ in steel sheets.

As the plastic flow of a sheet metal in forming operations is determined by its yield function, it is natural to ask how crystallographic texture, as described by the orientation distribution function w, would affect the yield function. Back in 1948 Hill[7] introduced a class of quadratic yield functions for describing the orthotropic plasticity of sheet metals. Hill's quadratic class of yield functions was widely adopted in the fifties and sixties for modelling the plastic anisotropy of steel sheets. On the assumption that the anisotropic part of the yield function depends linearly on the texture coefficients

W_{lmn} (which should be an adequate assumption if the sheet in question is weakly textured), it has been proved[8] recently that the principle of material frame-indifference in continuum mechanics entails the following theorem: For orthotropic aggregates of cubic crystallites, the anisotropic part of any yield function in Hill's quadratic class can depend on w only in the three texture coefficients W_{400}, W_{420}, W_{440}; moreover, this dependence is explicitly determined up to a material-dependent multiplicative factor. The preceding theorem leads to an hitherto unnoticed prediction, which is borne out by experimental data on low-carbon steel sheets.[9,10]

Hill's quadratic yield functions, however, are inadequate[11,12] for describing the plastic behavior of aluminum. This finding has prompted research efforts to develop non-quadratic yield functions. Hill[13] himself has lately introduced a "user-friendly theory" by adding cubic terms to his 1948 quadratic. On the other hand, a recent work of Wagner and Lücke[14] suggests that a yield function with seven texture coefficients (W_{4m0} for $m = 0, 2, 4$, and W_{6m0} for $m = 0, 2, 4, 6$) might suffice for characterizing the plastic anisotropy of aluminum alloys. Motivated by these developments, we shall add a general cubic term to Hill's quadratic and reduce the possible form of the resulting yield function f by appealing to the principle of material frame-indifference. Our work differs from Hill's in that we let the yield function f depend on the orientation distribution function (ODF) and, without going into any detailed micromechanical modelling, try to delineate this dependence as explicitly as possible. In this paper we restrict our attention to weakly-textured orthotropic sheets of cubic crystallites, and we assume that the anisotropic part of f depends linearly on the texture coefficients W_{lmn} ($l \geq 1$). Under these assumptions, the principle of material frame-indifference dictates that f depends on the ODF only in the aforementioned seven texture coefficients, and that the anisotropic part of f can be written as a sum of three terms, each of which is determined up to a material-dependent multiplicative constant. For want of space, we shall not exhibit these terms explicitly. Instead we shall examine, in some detail, the resulting formulae for three familiar r-values.

TEXTURE COEFFICIENTS AND R-VALUES: AN ELEMENTARY THEORY OF LINEAR CORRELATIONS

Let a homogeneous sample sheet of some cubic metal be given. Unless specified otherwise, we shall choose a spatial Cartesian coordinate system such that the 1-, 2-, and 3-direction are in the rolling direction (RD), the transverse direction (TD), and the normal direction (ND) of the sheet, respectively. We shall refer to this coordinate system as the standard coordinate system. We assume that the given sheet is orthotropic, and that the coordinate planes under the standard coordinate system are planes of orthotropic symmetry of the sheet in question.

Let $r(\Theta)$ be the plastic strain ratio in the direction in the plane of the sheet which makes an angle Θ with the 1-direction of the standard system. Let

$$\bar{r} = \frac{1}{4}(r(0) + r(\pi/2) + r(\pi/4) + r(3\pi/4)), \tag{1}$$

$$\Delta r = \frac{1}{2}(r(0) + r(\pi/2) - r(\pi/4) - r(3\pi/4)), \tag{2}$$

$$\delta r = r(0) - r(\pi/2). \tag{3}$$

Here, for later convenience, we have written $r(\pi/4) + r(3\pi/4)$ for the usual $2r(\pi/4)$ in Eqs. (1) and (2).

At the end of the next section, we shall list formulae which relate how $\overline{r}$, Δr and δr depend on the texture coefficients. Certain aspects of those formulae, however, can be derived by elementary arguments. To gain a deeper understanding of the formulae to come, we shall see in this section what we can derive from an elementary theory of linear correlations. To this end, we recall the following transformation formula for W_{lmn}:[8,15]

$$\widetilde{W}_{lmn} = \sqrt{\frac{2}{2l+1}} \sum_{p=-l}^{l} W_{lpn} Z_{lpm}(\cos\beta) e^{-ip\alpha} e^{-im\gamma}; \tag{4}$$

here W_{lmn} are the texture coefficients of a sample under a fixed spatial coordinate system (assuming that a fixed reference orientation has been chosen for the crystallites), $\widetilde{W}_{lmn}$ are the texture coefficients after the sample undergoes a rotation defined by the orthogonal matrix $\boldsymbol{Q}$ (with $\det \boldsymbol{Q} = 1$), and (α, β, γ) are the Euler angles pertaining to $\boldsymbol{Q}^{-1}$. (In defining Euler angles, we follow the convention[15] adopted by Roe.) Equation (4) can also be interpreted as the transformation formula which governs the texture coefficients for a passive rotation, i.e., the situation where the sample is fixed but the spatial coordinate system undergoes a rotation defined by the Euler angles (α, β, γ). We shall, however, use active rotations throughout our discussion.

Consider an experiment in which we determine δr for a batch of coupons (cut from homogeneous sample sheets) when they are undergoing plastic flow at a specified strain (e.g., at 15% strain). We assume that the sample sheets in question are orthotropic under the given coordinate system and are of the same cubic metal but may have different textures. As the sample sheets are anisotropic, the uniaxial flow stress σ in each coupon when δr is measured will depend on the angle Θ that the coupon axis makes with the rolling direction of the sheet. Suppose the sample sheets are weakly textured. Then we can write $\sigma = \sigma_0 +$ terms depending on the texture coefficients W_{lmn} $(l \geq 1)$, where σ_0 is the flow stress in the isotropic limit.

By definition of the anisotropic parameter δr, we know that if the coupon in question is isotropic (i.e., $W_{lmn} = 0$ for all $l \geq 1$), then $\delta r = 0$. As we consider only weakly-textured samples, we assume that there is a linear correlation between δr and the texture coefficients. For later comparison, here we assume further that we may drop all W_{lmn} with $l > 6$. Taking these assumptions as well as sample and crystal symmetry into account, we write

$$\delta r = c_1 W_{400} + c_2 W_{420} + c_3 W_{440} + c_4 W_{600} + c_5 W_{620} + c_6 W_{640} + c_7 W_{660}, \tag{5}$$

where c_i are parameters which depend on the material and possibly also on σ_0 but are independent of texture.

Let $\boldsymbol{Q}_1$ be the rotation with $\boldsymbol{Q}_1^{-1}$ defined by the Euler angles $(\alpha, \beta, \gamma) = (\pi/2, 0, 0)$. Consider another sample sheet, identical to the first one except that its texture is given by the orientation distribution function $\widetilde{w}$, where

$$\widetilde{w}(\boldsymbol{R}) = w(\boldsymbol{Q}_1^{-1}\boldsymbol{R}) \tag{6}$$

for each rotation $\boldsymbol{R}$. Physically we may interpret this second sheet as the original one rotated about the 3-axis clockwise by 90°. This second sheet remains orthotropic under the fixed spatial coordinate system. Hence we have for this second sheet

$$\widetilde{\delta r} = c_1 \widetilde{W}_{400} + c_2 \widetilde{W}_{420} + c_3 \widetilde{W}_{440} + c_4 \widetilde{W}_{600} + c_5 \widetilde{W}_{620} + c_6 \widetilde{W}_{640} + c_7 \widetilde{W}_{660}. \tag{7}$$

Note that we have the same coefficients c_i ($i = 1, ..., 7$) in Eqs. (5) and (7). From Eq. (3), we easily observe that $\widetilde{\delta r} = -\delta r$. Likewise, from the transformation formula (4), we obtain $\widetilde{W}_{660} = -W_{660}$, and $\widetilde{W}_{l00} = W_{l00}$, $\widetilde{W}_{l20} = -W_{l20}$, $\widetilde{W}_{l40} = W_{l40}$ for $l = 4, 6$. Substituting these expressions into Eq. (7) and comparing the resulting equation with Eq. (5), we conclude that $c_1 = c_3 = c_4 = c_6 = 0$ and

$$\delta r = c_2 W_{420} + c_5 W_{620} + c_7 W_{660}. \tag{8}$$

To obtain the corresponding reduced formulae for $\bar{r}$ and Δr, in addition to the rotation $\boldsymbol{Q}_1$ above we consider another (active) rotation of the sample by $\boldsymbol{Q}_2$ with $\boldsymbol{Q}_2^{-1}$ given by the Euler angles $(\alpha, \beta, \gamma) = (\pi/4, 0, 0)$. Under $\boldsymbol{Q}_2$, clearly $\widetilde{\bar{r}} = r$ and $\widetilde{\Delta r} = -\Delta r$ (cf. Eqs. (1) and (2)); on the other hand, the rotated sample will not be orthotropic with respect to the fixed spatial coordinate system. A slightly more sophisticated argument, but still entirely elementary and similar to the one above, will be necessary to derive the reduced formulae we seek (cf. Man[8] for an analysis of a similar situation). Here we omit the details and just report the results:

$$\bar{r} = 1 + a_1 W_{400} + a_4 W_{600}, \tag{9}$$

$$\Delta r = b_3 W_{440} + b_6 W_{640}, \tag{10}$$

where a_i and b_i ($i = 1, ..., 7$) are parameters similar to c_i.

A YIELD FUNCTION WHICH SHOWS EXPLICITLY THE EFFECTS OF SEVEN TEXTURE COEFFICIENTS

Let Dsym be the set of traceless symmetric tensors, and let w be the ODF. We consider yield functions f with w and the deviatoric stress $\boldsymbol{\sigma}$ as independent variables. The principle of material frame-indifference[16] dictates that f must satisfy[8] the identity

$$f(\boldsymbol{Q}\boldsymbol{\sigma}\boldsymbol{Q}^T, \tilde{w}) = f(\boldsymbol{\sigma}, w) \tag{11}$$

for each rotation $\boldsymbol{Q}$, each $\boldsymbol{\sigma}$ in Dsym, and each ODF w; here $\boldsymbol{Q}^T$ stands for the transpose of $\boldsymbol{Q}$, and $\tilde{w}$ is the rotated ODF defined by $\tilde{w}(\boldsymbol{R}) = w(\boldsymbol{Q}^T\boldsymbol{R})$ for each rotation $\boldsymbol{R}$.

Henceforth we restrict our attention to weakly-textured orthotropic sheets of cubic metals, and we adopt the standard coordinate system. We make the following assumptions:

1. The anisotropic part of f depends linearly on the texture coefficients W_{lmn} ($l \geq 1$). Unlike the elementary theory, here no assumption is made *a priori* on whether f should depend on those W_{lmn} with $l > 6$.

2. We can express the yield function as

$$f(\boldsymbol{\sigma}, w) = \boldsymbol{\sigma} \cdot \boldsymbol{C}(w)[\boldsymbol{\sigma}] + \boldsymbol{\sigma} \cdot \boldsymbol{D}(w)[\boldsymbol{\sigma}, \boldsymbol{\sigma}], \tag{12}$$

   where $\boldsymbol{C}(w)$ is a fourth-order tensor which enjoys the major symmetry and $\boldsymbol{D}(w)$ is a sixth-order tensor which satisfies $\boldsymbol{\sigma}_1 \cdot \boldsymbol{D}(w)[\boldsymbol{\sigma}_2, \boldsymbol{\sigma}_3] = \boldsymbol{\sigma}_{\tau(1)} \cdot \boldsymbol{D}(w)[\boldsymbol{\sigma}_{\tau(2)}, \boldsymbol{\sigma}_{\tau(3)}]$ for each permutation τ of $\{1, 2, 3\}$.

The yield function in Eq. (12) will fall in Hill's quadratic class if we drop the $\boldsymbol{D}$-term. For a given w, if we take f as a nonlinear function of $\boldsymbol{\sigma}$ and expand it by Taylor's formula, the quadratic and cubic terms will be of the form given by the $\boldsymbol{C}$-term and $\boldsymbol{D}$-term in Eq. (12), respectively.

Under the preceding assumptions, we can prove that f depends on w only in the seven texture coefficients W_{4m0} ($m = 0, 2, 4$) and W_{6m0} ($m = 0, 2, 4, 6$). By appealing to theorems in the theory of invariants and theory of group representations, we can further show that f can be cast in the following form:

$$
\begin{aligned}
2f(\boldsymbol{\sigma}, w) \;=\; & \frac{1}{Y_o^2}\Big(\frac{3}{2}\operatorname{tr}\boldsymbol{\sigma}^2 + \alpha\operatorname{tr}\boldsymbol{\sigma}^3 + \beta\,\boldsymbol{\sigma}\cdot\boldsymbol{\Phi}(W_{400}, W_{420}, W_{440})[\boldsymbol{\sigma}] \\
& + \gamma_1\,\boldsymbol{\sigma}\cdot\boldsymbol{\Psi}_1(W_{400}, W_{420}, W_{440})[\boldsymbol{\sigma}, \boldsymbol{\sigma}] \\
& + \gamma_2\,\boldsymbol{\sigma}\cdot\boldsymbol{\Psi}_2(W_{600}, W_{620}, W_{640}, W_{660})[\boldsymbol{\sigma}, \boldsymbol{\sigma}]\Big);
\end{aligned}
\tag{13}
$$

here $Y_o, \alpha, \beta, \gamma_1$ and γ_2 are material constants; $\boldsymbol{\Phi}$: Dsym $\to$ Dsym and $\boldsymbol{\Psi}_i$: Dsym $\times$ Dsym $\to$ Dsym ($i = 1, 2$) are explicitly determined fourth-order and sixth-order tensor functions of the listed texture coefficients, respectively.

Using the yield function given by Eq. (13) as the plastic potential in the flow rule, we deduce that the r-values defined in Eqs. (1)–(3) are given, correct to terms linear in the texture coefficients, by the formulae

$$\overline{r} = 1 - aW_{400} - bW_{600}, \tag{14}$$

$$\triangle r = \frac{2\sqrt{70}}{5}aW_{440} - \frac{2\sqrt{14}}{7}bW_{640}, \tag{15}$$

$$\delta r = \frac{4\sqrt{10}}{5}aW_{420} + b\left(\frac{34\sqrt{105}}{105}W_{620} - \frac{2\sqrt{231}}{7}W_{660}\right), \tag{16}$$

where

$$a = \frac{8\sqrt{2}\pi^2}{7}\beta + \frac{5}{24}\sigma_0\gamma_1, \qquad b = -\frac{7}{16}\sigma_0\gamma_2, \tag{17}$$

and σ_0 is the isotropic limit of the uniaxial flow stress when the r-values are measured.

Formulae (14)–(17) are consistent with Eqs. (8)–(10) derived under the elementary theory in the preceding section. The total number of unknown coefficients, however, is two in the present theory as compared with seven in the elementary theory. Indeed, by measuring empirically the coefficients of linear correlation, Eqs. (14) and (15) could serve as the basis for an experimental verification of whether the yield function f given in Eq. (13) could model the plastic behavior of a specific sheet metal. At the University of Kentucky we have done some preliminary experiments on copper in this regard.

If we drop the W_{6m0} terms in Eqs. (14)–(16), then these equations will become formally the same as those[9,10] that follow from a yield function in Hill's quadratic class.

CLOSING REMARKS

The conventional method of measuring directional dependence of ultrasonic wave speeds will directly deliver only the texture coefficients W_{400}, W_{420} and W_{440}. If the yield function given in Eq. (13) could adequately model the plastic behavior of aluminum alloys, evaluation of the texture coefficients W_{6m0} for $m = 0, 2, 4, 6$ should be on the agenda of researchers who strive to develop measurement systems for on-line monitoring of formability parameters of aluminum alloys. In this regard the work of Sakata et al.[17] on cold-rolled steel sheets might serve as a possible point of departure.

Acknowledgment

The research reported here was supported by a DEPSCoR grant from AFOSR (Grant No. F49620-94-1-0393).

REFERENCES

1. M. Hirao, H. Fukuoka, K. Fujisawa, and R. Murayama, On-line measurement of steel sheet r-value using magnetostrictive-type EMAT, *J. Nondest. Eval.* 12:27 (1993).
2. K. Kawashima, T. Hyoguchi, and T. Akagi, On-line measuremnt of plastic strain ratio of steel sheet using resonance mode EMAT, *J. Nondest. Eval.* 12:71 (1993).
3. A.J. Anderson, R.B. Thompson, R. Bolingbroke, and J.H. Root, Ultrasonic characterization of rolling and recrystallization textures in aluminum, *Textures and Microstructures* 26:39 (1996).
4. W. Lu, D. Hughes, and S. Min, Texture measurement using EMAT and laser ultrasonics, in: *ICOTOM-11*, 134 (1996).
5. C.A. Stickels and P.R. Mould, The use of Young's modulus for predicting plastic strain ratio of low-carbon steel sheets, *Metall. Trans.* 1:1303 (1970).
6. G.J. Davies, D.J. Goodwill, and J.S. Kallend, Elastic and plastic anisotropy in sheets of cubic metals, *Metall. Trans.* 3:1627 (1972).
7. R. Hill, A theory of yielding and plastic flow of anisotropic metals, *Proc. Roy. Soc. A* 193:281 (1948).
8. C.-S. Man, On the constitutive equations of some weakly-textured materials, *Arch. Rational Mech. Anal.*, to appear.
9. C.-S. Man, Elastic compliance and Hill's quadratic yield function for weakly orthotropic sheets of cubic metals, *Metall. Mater. Trans.* 25A:2835 (1994).
10. C.-S. Man and Q. Ao, Plastic strain ratio and texture coefficients in orthotropic sheets of cubic metals, in: *Review of Progress in Quantitative Nondestructive Evaluation*, volume 15B, D.O. Thompson and D.E. Chimenti, eds., Plenum, New York (1996), p. 1353.
11. J. Woodthorpe and R. Pearce, The anomalous behaviour of aluminum sheet under balanced biaxial tension, *Int. J. Mech. Sci.* 12:341 (1970).
12. F. Barlat, Crystallographic texture, anisotropic yield surfaces and forming limits of sheet metals, *Materials Science and Engineering* 91:55 (1987).
13. R. Hill, A user-friendly theory of orthotropic plasticity in sheet metals, *Int. J. Mech. Sci.* 35:19 (1993).
14. P. Wagner and K. Lücke, Quantitative correlation of texture and earing in Al-alloys, *Materials Science Forum* 157–162:2043 (1994).
15. R.-J. Roe, Description of crystallite orientation in polycrystalline materials. iii. general solution to pole figures, *J. Appl. Phys.* 36:2024 (1965).
16. C. Truesdell and W. Noll, *The Non-Linear Field Theories of Mechanics*, Second Edition, Springer, Berlin (1992).
17. K. Sakata, D. Daniel and J.J. Jonas, Estimation of 4th and 6th order ODF coefficients from elastic properties in cold rolled steel sheets, *Textures and Microstructures* 11:41 (1989).

APPLYING PHASE SENSITIVE MODULATED THERMOGRAPHY TO GROUND SECTIONS OF A HUMAN TOOTH

C. John,[2] D. Wu,[1] A. Salerno,[1] G. Busse[1] and C. Löst[3]

[1]Institute for Polymer Testing and Polymer Science,
IKP, University of Stuttgart, Pfaffenwaldring 32,
D - 70569 Stuttgart, Germany

[2]Ultrasonics & Biomedical Engineering Laboratory,
[3]School of Dental Medicine, ZMK-Dept. Conserv.,
UKT, University of Tübingen, Osianderstr. 2 - 8,
D - 72076 Tübingen, Germany

INTRODUCTION

Thermophysical properties and acoustical parameter pattern are of high interest for numerical modeling of heat transfer processes, transient stress analysis and predictability of wave propagation.[1-5] The determination of time characteristics [2, 3, 6] and the use of material properties, like density [5, 7-9], elasticity [3, 6, 8, 10, 11], moisture content [8, 12], specific heat [5, 7], speed of sound[3, 6, 8] and thermal conductivity [5, 13-16] are critical issues when considering small samples of inhomogeneous materials. Only a few investigations have been focused on more than a complete but single one-dimensional description of the physical properties of teeth. To avoid a damaging temperature rise in the living pulp inside a tooth determinations of the thermal diffusion were necessary.[14, 19, 20]

In this investigation a lock-in thermography technique was applied to ground sections of a human tooth to map the delay of a thermal wave as a function of location. Full two-dimensional reconstructions will be presented as phase images to emphasize the aspect of inhomogeneity. Local deviations from unique parameter values describing spectral absorption / reflection / transmission processes have been discussed [3, 5, 8, 17, 18, 21-24] and may influence a determination of one important thermophysical property: the thermal diffusivity.[1, 7, 12, 16, 24]

To predict the response of living tissues to applied heat [5, 25, 26] and temperature changes due to heat transients of new laser applications in dentistry [2, 4, 5, 22, 26] one has to determine the magnitude, the phase and the orientation of thermal waves propagating through oral structures, by thermal wave techniques [10, 16, 18, 21], thermography [23, 27-30] or a combination of both.[9, 17, 24]

Nondestructive Characterization of Material VIII
Edited by Robert E. Green Jr., Plenum Press, New York, 1998

MATERIALS AND METHODS

1. Infrared Imaging Systems

Infrared imaging systems visualize infrared radiation emitted by objects in the spectral region of interest (ROI). A variety of commercial products and new developments [23, 27-30] are able to capture two-dimensional thermal images in real time using infrared (IR) detectors (quantum detectors or microbolometers [24, 27]). For scientific applications mainly quantum detectors are used due to the better sensitivity and (until recently) the higher speed. Traditional IR cameras with one IR detector and an opto-mechanical scanning system are nowadays being replaced by focal-plane array (FPA) cameras [23, 24, 27, 28, 30] with an array of detectors and no moving parts in the electro-optical scanning system: the advantage is a better signal to noise ratio, with a noise equivalent temperature difference that can reach 0.01°C [23], and the ability to acquire images at a very high speed.

2. Thermography

Thermography is a rasterizing IR imaging approach and an established technique where the local thermal infrared emission power of a sample is monitored. IR emission of a body depends on temperature and surface emissivity coefficient following the well known Stefan-Boltzmann law,

$$W = \varepsilon \cdot B \cdot T^4 \qquad (1)$$

where W is the total radiation intensity, ε is the emissivity coefficient, $B = 5.667 \cdot 10^{-8} Wm^{-2}K^{-4}$ is the Stefan-Boltzmann constant, and T the absolute temperature. IR cameras provide thermal 2D images, which are related to IR emission of bodies, so that local temperatures can be remotely measured only if the emissivity coefficient is known. Infrared thermography uses IR cameras equipped with sensors that operate typically in the 3-5μm or in the 8-12μm waveband. For applications related to the human mouth at temperatures of between room and body temperature, the Wien's displacement law is providing the wavelength of maximum emittance λ_{max} to be,

$$\lambda_{max} = b / T \cong 10 \mu m \qquad (2)$$

where b is the Wien displacement constant $= 2.898 \cdot 10^{-3}$ mK. Because the object of IR testing is to accurately measure surface and subsurface temperature changes, local variations of emissivity across the surface of a (tooth) sample will cause false indications for inhomogeneous materials. Although all materials are continuously and simultaneously radiating and absorbing infrared energy, a local ROI with decreased emissivity (e.g. polished amalgam fillings) will show reduced radiation intensity, thereby falsely indicating a misleading local decrease in temperature. This problem may cause serious limitations to the use of infrared thermography as a rapid and contactless means for temperature measurements. But for nondestructive characterization of materials there is a more adequate and powerful technique, called thermal wave imaging, which is not suffering from surface features, including any unknown or varying emissivity coefficient.

3. Thermal Wave Imaging

Thermal wave imaging exploits the time dependent heat flow (governed by the Heat diffusion equation 3)) that occurs, when a time varying heat source is applied to an object with a uniform thermal conductivity κ [5,13-15], the density ρ [5,7,9] and the specific heat c [5,7],

$$\rho \cdot c \cdot \frac{\partial T}{\partial t} \cdot \kappa^{-1} = \alpha^{-1} \cdot \frac{\partial T}{\partial t} = \nabla^2 T \tag{3}$$

where T is the absolute temperature and $\alpha = \kappa/\rho c$ is the thermal diffusivity [7,12,24] of the specimen under test. The corresponding heat source is optimized for a specific application either to deposit the heat periodically or by a short excitation pulse in a remote way. When a modulated source with complex periodicity $e^{i\omega t}$ is applied on the surface of the sample, Eq. 3) reduces to the well known Helmholtz wave equation,

$$q = (1+i) \cdot \sqrt{\frac{\omega}{2 \cdot \alpha}} = (1+i) \cdot \mu^{-1} \qquad \nabla^2 T - q^2 \cdot T = 0 \tag{4}$$

where the decay length of a strong exponential damping term is called the thermal diffusion length μ[16-18], and q is the (complex) wave number of the periodic thermal wave which is propagating through the medium with a strongly frequency dependent velocity $\upsilon = (2 \cdot \alpha \cdot \omega)^{-1/2}$. The real part of the corresponding one-dimensional solution represents the actual physical temperature excursion which is approximated by Eq. 5):

$$T(x, t) = T_o \cdot e^{-x/\mu} \cdot \cos(\omega \cdot t - \omega \cdot x / \upsilon) \tag{5}$$

Different techniques can be used in order to generate and to detect thermal waves: among them 'photoacoustic imaging' and 'photothermal radiometry' allow for very sensitive point measurements using a chopped laser and a piezoelectric detection system [16,18], or an IR detector [10,17,21], respectively. The magnitude or the phase shift of thermal waves can be measured, both carrying information about material thermal properties, but the phase shift measurement has the advantage of being independent of local emissivity, and that the depth range is roughly double than that of the magnitude. [16-18] The depth range of a thermal wave depends inversely from the square root of the excitation frequency. However, raster scan techniques like photothermal radiometry suffer from being very slow, since for every point one has to wait at least one complete period, which can last several minutes.

4. Lock-in Thermography

'Lock-in Thermography', or 'Phase Sensitive Modulated Thermography' is a thermal wave imaging technique that combines the advantages of the previous IR techniques: a very low-frequency cw modulation provides phase angle information and a maximum depth range; an IR camera serves as a rapid scanning radiometer (Fig. 1).

The multiplexed read out data sequence of the conclusively sinusoidal modulation signal S_i of every pixel element i has to be monitored and stored at four (or even three) fixed points per modulation cycle (Fig. 2).

A nitrogen cooled thermographic scanner/IR camera was taken, whose sensors and systems performance for display, recording and analysis of thermal infrared radiation has been certified according to SS-EN ISO 9001. Prior to and after modulated light exposure (= heat application) thermal radiation measurements and weighings have been performed using additional equipment to non-simultaneously evaluate variations of moisture content.

A mathematical algorithm described by the formulas given in Eqs. 6) and 7) allows for

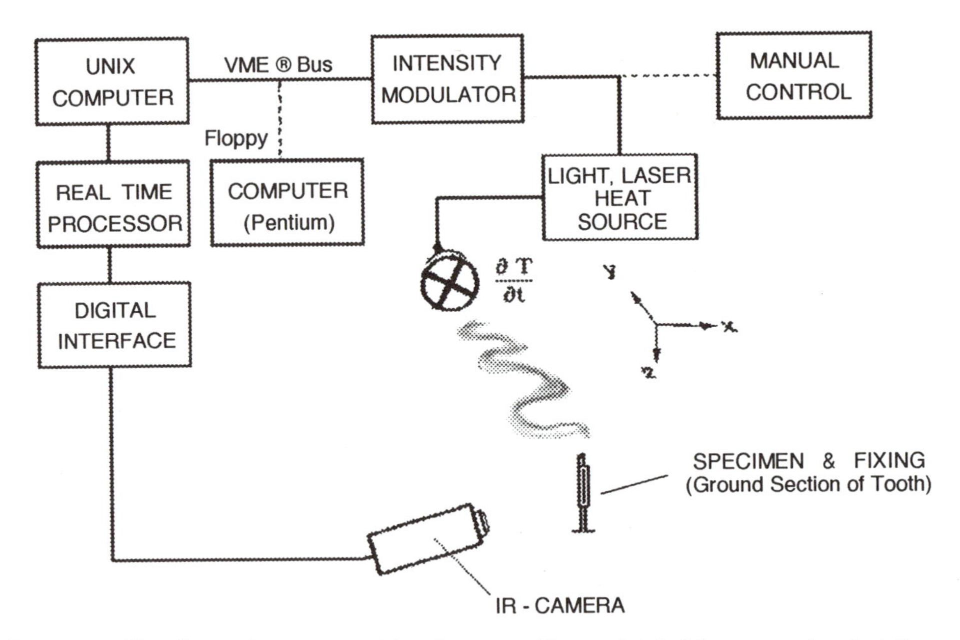

Figure 1. Experimental arrangement for phase sensitive modulated thermography. A stationary temperature field was generated in coplanar ground sections of a previously embedded tooth by means of a sinusoidal modulation of a light beam (f = 1 Hz ... 0.01 Hz). The lock-in system is based on an infrared camera (AGEMA 900®, Sweden), working in the 8-12 μm waveband.

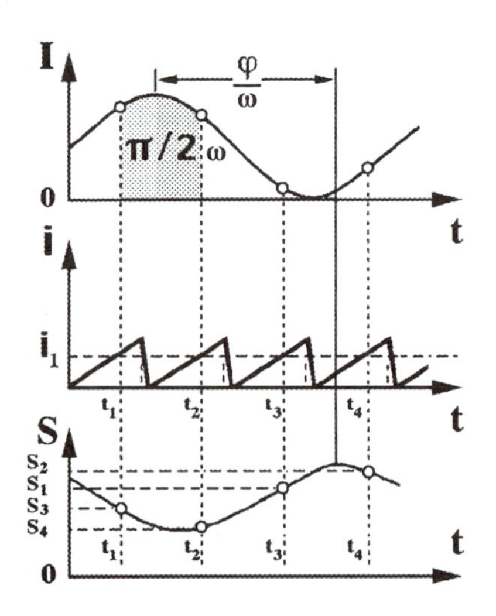

the automatic pixel by pixel reconstruction of a phase and a magnitude image. Although formally only four measurements are needed, actual data result from an averaging process aimed at increasing the signal to noise ratio.

$$\varphi = \text{arctg}\,(S_1 - S_3) / (S_2 - S_4) \quad (6)$$

$$M = \sqrt{(S_1 - S_3)^2 + (S_2 - S_4)^2} \quad (7)$$

Figure 2. Principle of the phase sensitive modulated thermography: The intensity of light **I**(t) is creating a thermal wave (curve S(t, x)), which has been reconstructed from four IR camera signal values S_1 to S_4 at pixel address i_1,- showing a phase shift φ relative to the modulation signal **I**(t).

CONCLUSIONS

Our first images reveal regions of inhomogeneous thermophysical characteristics in teeth. The phase images of coplanar hard dental tissues may be regarded as a 'map of thermal thickness'.

DISCUSSION

This was the first scientific approach to combine thermal wave lock-in measurements and the thermographic technique for dental research or medical element engineering &

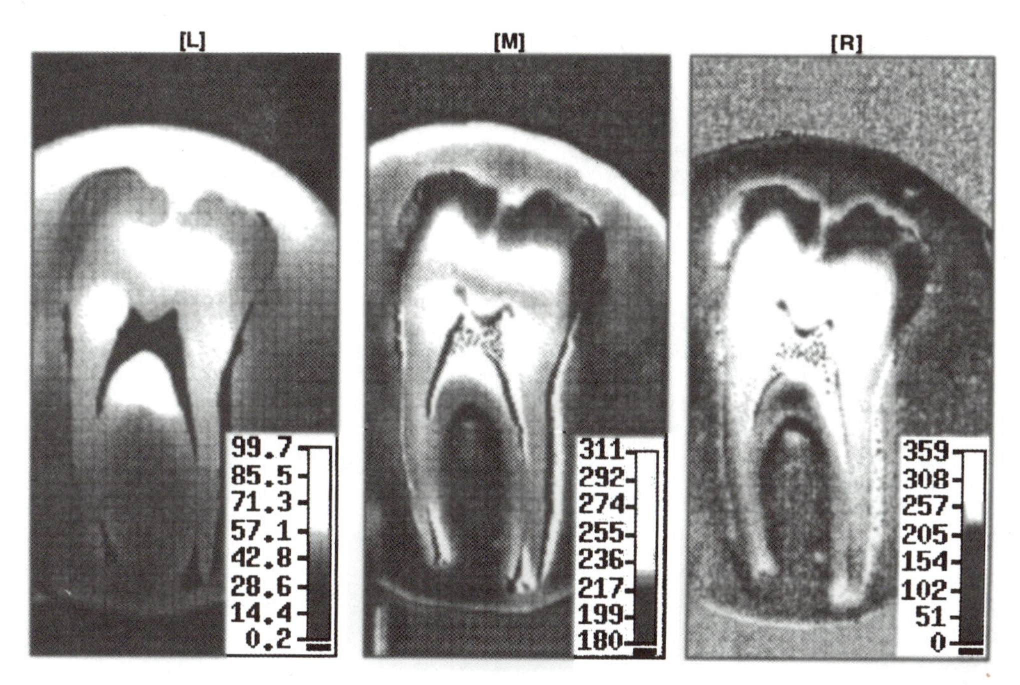

Figure 3. Phase sensitive modulated thermography images of an illuminated human tooth section. The magnitude image was calculated from Eq. 7) [L] ; in [M], [R] phase images (Eq. 6)) are capable to resolve different thermal structures with varying contrast for a reflection mode modulation frequency of 0.06 Hz [L, M] and 0.23 Hz [R], respectively. The enamel can be distinguished from the dentin.

simulation (MEES). However, the observable subsurface structures lead to promising aspects: a resonable explanation for the differing literature values, the controlled heating of tooth surfaces for *in vivo* sterilization[2] or heat effects on the living tissues inside healthy teeth [14], microbial *in vitro* studies which may be affected due to late re-contamination via random-reminiscent wash out effects into the storage medium[3,8], and the remainder of a few colony forming units of microorganisms inside the ineffectively sterilized root canal system [3].

RESULTS

Figure 3 shows a magnitude image with a slightly inhomogeneous sample illumination and two phase images of a ground sections of a resin embedded extracted human tooth. Thermal structures can be seen for different depth ranges (x2 in[M]) and gradual changes.

ACKNOWLEDGEMENT

The first author would like to express his gratitude to Mr. J. Stanullo for support and for getting into communication with each other.

REFERENCES

1. M. Braden: Heat conduction in teeth and the effect of lining materials. *J Dent Res* 43 (5-6): 315-322 (1964)
2. R. Hibst, K. Stock, R. Gall and U. Keller: Controlled tooth surface heating and sterilisation by Er:YAG laser radiation. *Proc SPIE* Vol. 2922: 119-126 (1996)

3. C. John, M. Brecx, B. Linder-Lais, W. Krüger, U. Faust and C. Löst: 2D-pattern of longitudinal sound velocity in endodontically treated teeth. *J Dent Res* 76 (5): 1109 (1997a)
4. D. Meyer and H.J. Foth: Treatment of hard dental tissue by very short CO_2 laser pulses. *Lasermed* 12: 58-66 (1996)
5. A. Sagi, A. Shitzer, A. Katzir and S. Akselrod: Heating of biological tissue by laser irradiation: theoretical model. *Opt Eng* 31 (7): 1417-1424 (1992)
6. C. John, B. Linder-Lais, M. Brecx, R. Weiger, W. Krüger and C. Löst: 2D-profiles of longitudinal sound velocity reveal inhomogeneity of human dentin. *J Dent Res* 75 (Special Issue): 197 (1996)
7. W.S. Brown, W.A. Dewey and H.R. Jacobs: Thermal properties of teeth. *J Dent Res* 49 (4): 752-755 (1970)
8. C. John, K. Vitsoroglou, C. Von Ohle and C. Löst: The loss of water for different storage conditions. *J Dent Res* 76 (IADR Abstracts): 340 (1997b)
9. D. Wu, G. Zenzinger, W. Karpen and G. Busse: Nondestructive inspection of turbine blades with lock-in thermography, in: *Nondestructive Characterization of Materials VII* (Vols. 210-213): 289-294, A.L. Bartos, R.E. Green, Jr. and C.O. Ruud, eds., Transtec Publications, Switzerland (1996)
10. F.C. Chen, U. Netzelmann, M. Disque and M. Kröning: High resolution photothermal imaging of metal matrix composite interface, in: *Nondestructive Characterization of Materials VII* (Vols. 210-213): 447-454, A.L. Bartos, R.E. Green, Jr. and C.O. Ruud, eds., Transtec Publications, Switzerland (1996)
11. D.C. Watts, O.M. El Mowafy and A.A. Grant: Temperature-dependence of compressive properties of human dentin. *J Dent Res* 66 (1): 29-32 (1987)
12. D.C. Watts and R. Smith: Thermal diffusivity in finite cylindrical specimens of dental cements. *J Dent Res* 60 (12): 1972-1976 (1981)
13. R.G. Craig and F.A. Peyton: Thermal conductivity of tooth structure, dental cements, and amalgam. *J Dent Res* 40 (5-6): 411-418 (1961)
14. V.F. Lisanti and H.A. Zander: Thermal conductivity of dentin. *J Dent Res* 29 (4): 493-497 (1950)
15. B.C. Soyenkoff and J.H. Okun: Thermal conductivity measurements of dental tissues with the aid of thermistors. *J Am Dent Assoc* 57 (7): 23-30 (1958)
16. G. Busse: Optoacoustic phase angle measurement for probing a metal. *Appl Phys Lett* 35 (10): 759-760 (1979)
17. G. Busse: Nondestructive evaluation of polymer materials. *NDT & E International* 27 (5): 253-264 (1994)
18. G. Busse and A. Rosencwaig: Subsurface imaging with photoacoustics. *Appl Phys Lett* 36 (10): 815-816 (1980)
19. F.A. Peyton: Temperature rise in teeth developed by rotating instruments. *J Am Dent Assoc* 50 (6): 629-632 (1955)
20. E.D. Voth, R.W. Phillips and M.L. Swartz: Thermal diffusion through amalgam and various liners. *J Dent Res* 45 (4): 1184-1190 (1966)
21. A. Brun, M. Marty, H. Roede, C. Gounelle and F. Giroux: Thermal-wave imaging: a non-destructive technique to characterize the electromigration on Al alloy, in: *Nondestructive Characterization of Materials VII* (Vols. 210-213): 309-316, A.L. Bartos, R.E. Green, Jr. and C.O. Ruud, eds., Transtec Publications, Switzerland (1996)
22. T.M. Odor, T.F. Watson, T.R. Pitt Ford and F. McDonald: Pattern of transmission of laser light in teeth. *Int Endod J* 29: 228-234 (1996)
23. M. Schulz: Wie Wärme sichtbar wird. Videokamera für den Infrarotbereich. *Forschung - Mitteilungen der DFG*, ISSN 0172-1518 (3): 22-32 (1996)
24. J. Stuckey, J.G. Sun and W.A. Ellingson: Rapid infrared characterization of thermal diffusivity in continuous fiber ceramic composite components, in: *Nondestructive Characterization of Materials VIII* (pp. see this Issue); R.E. Green, Jr., ed. Plenum Publishing, New York (1997)
25. L. Zach and G. Cohen: Pulp response to externally applied heat. *O Surg O Med O Pathol* 19 (4): 515-530 (1965)
26. D.M. Zezell, S.C.M. Cecchini, M. Pinotti and C.P. Eduardo: Temperature changes under Ho:YLF irradiation. *Proc SPIE* Vol. 2672: 34-39 (1996)
27. E.J. Lerner: Infrared array detectors create thermal images. *Laser Focus W* 32 (9): 105-111 (1996)
28. A. Owen: Surveillance cameras steal away the night. *Laser Focus W* 33 (2): 111-117 (1997)
29. L.A. Peach: Thermal imaging: Near-field microscopy maps semiconductors. *Laser Focus W* 33 (2): 38-40 (1997)
30. G. Russell: Thermal imager brings TV news out of the dark. *Laser Focus W* 32 (12): 145-146 (1996)

RECOVERING ECHO SIGNALS FOR *IN-VITRO* CHARACTERIZATION OF HARD DENTAL TISSUES

Christoph John

Ultrasonics & Biomedical Engineering Laboratory,
ZMK, University of Tübingen, Osianderstr. 2 - 8,
D - 72076 Tübingen, Germany

INTRODUCTION

To provide ultrasonic diagnosis for dentists has been sought for many years without success.[1] The missing link of ultrasonics and dentistry is both reasonable and questionable. The application of ultrasonic energy in the human mouth is difficult due to high reflectivities of oral multiboundaries between soft tissues, gaps and mineralized tissues. The elastic wave propagation in a porous and inhomogeneous natural composite, such as the anisotropic dentin core of a human tooth, is a very complex phenomenon. Teeth reveal a more individual pattern of known parameters and show larger variations of structure and composition than human soft tissues (medicine) or a commercial microfill composite (industry).[2] However, it appears appealing to replace radiographic examinations with ultrasonic examinations. When directing ultrasonic waves into organic but also crystalline materials of non-vital and/or living oral tissues one aims at recovering information from an acoustic parameter pattern related to the substructure.

Ultrasound has often been considered a progress in medical diagnostics. Very recent research activities include characterization of red blood cells,[3,4] the vessel wall,[5] the heart muscle,[6] the prostate and liver,[7] the cancellous bone[8] and uric acid stones,[9] - or applications in dermatology and ophthalmology.[10] New developments may contribute to establish layer-sensitive,[11] time-resolving,[5,12] precise,[13,14] safe,[15] and fast[10] high-resolution[16,17] ultrasonic (diagnostic) tools and techniques. With them, as with MRI and CT techniques, it is possible to create full data sets for the benefit of medical element engineering & simulation (MEES).[18] Using short pressure pulses, high-frequency fields and shock waves, the advantages of acoustic measurements are to determine the elastic properties in a nondestructive way and to avoid invasive *in vivo*-measurements of related parameters. Structural aspects become detectable by the pattern of the speed of sound,[1,11,18-21] if anisotropy,[22,23] inhomogeneity,[22,24] moaicity[12] and porosity[25] do not solely influence material's density and/or elasticity[23] but other physical properties, too.[22,26] The related acoustic parameters are of high interest for numerical modeling of transient stresses or elastic wave propagation by finite techniques (FEA, EFIT).[25]

Nondestructive Characterization of Material VIII
Edited by Robert E. Green Jr., Plenum Press, New York, 1998

AIM

The aim was to reconstruct images of ground sections of a human tooth by recovering echo signals of applied ultrasound, thereby mapping the delay of acoustic waves.

MATERIALS AND METHODS

1. Ultrasonic Imaging Systems

Ultrasonic imaging systems usually visualize a series of acquired rf-signals (radio frequency signals)[5, 7, 10] resulting from a scan. Post-processed data which may have been analyzed in the time and/or frequency domain can also lead to ultrasonic images.[6, 13, 14, 16, 21] In addition to commercially available (pulsed) Doppler systems, there are high resolution ultrasonic scanners which have been designed for medical real-time imaging.[6, 10] They provide users with a subset of A-Scan, B-, C- and M-Mode imaging capabilities at data acquisition rates of about 10 images per second and sampling rates between 50 MSa/s and 100 MSa/s. For scientific applications with more accurate A-Scans,[1, 3, 9, 13, 17, 21, 25] B-Scans,[6, 7] C-Scans,[13] D-Scans,[13, 16, 18, 20, 26] T-M-Scans [6] and precise industrial measurements,[11] instrumental setups of sophisticated components are used due to their increased flexibility, greater precision, higher resolution and the better sensitivity.[3-5, 9, 11-14, 17, 18, 20, 21, 26] A typical measurement arrangement consists of several high performance components, as can be shown up for our laboratory equipment (Fig. 1). Part of the arrangement is the world's leading digital sampling oscilloscope (DSO) which allows digitization of high frequencies, f, at a single shot acquisition sampling rate of about 100 times that one mentioned above.

2. Acoustic Waves

In a given medium, an ultrasonic wave causes particle vibrations along its path, and energy is transmitted at a finite velocity. Considering a propagating wave along the z-direction in a lossless system, the sum of the oscillating potential and kinetic energies is not varying with time, t. However, when a wave travels through (our) real media, like a gas (air), a liquid (saliva, saline solution, water) or a solid (cementum, dentin, enamel, gingiva, dental filling materials), the peak displacement u_0 of particles is not constant and the intensity I is reduced as a function of distance. The wavelength λ is the z-distance between consecutive particles where the displacement amplitudes are identical. The amplitude of a plane wave can be expressed by,

$$A(z) = A_0 e^{-\alpha \cdot z + \varsigma} \cdot \sin(k \cdot (c \cdot t - z)) \tag{1}$$

where A_0 is the peak value at $z = 0$ of a wave variable (like the particle velocity, $v = \partial u / \partial t$), α is the attenuation coefficient in [dB/m], ς is a quotient containing a factor of $20 \cdot \log_{10}(e)$, and $k = 2\pi/\lambda = \omega/c$ is the wave number; c is the velocity at which various types of waves are propagating. It is defined in the wave equation, Eq. 2), which relates the second differential of the particle displacement with respect to distance, to the acceleration of a (heavily/critically damped or simple) harmonic oscillator according to:

$$c^{-2} \cdot \frac{\partial^2 u}{\partial t^2} = \nabla^2 u \tag{2}$$

3. Ultrasonic Properties

Several wave types and acoustic fields can be distinguished. The spherical wave is the simplest form of a non-planar wave which radiates uniformly over a solid angle of 2π radians. At large distances, where the radius of the spherical surface is much larger than the uniform wavelength, the wavefront may again be considered to be plane over that region.The shape of the ultrasonic field of a particular transducer can be determined by application of the well known Huygen's principle. Two-dimensional examples of simulated field distributions have been displayed in the literature for pulse-echo field distribution measurements (contour plot)[17] and where the lateral resolution of a real transducer has been determined (colour coded map).[16] The elastodynamic finite integration technique (EFIT) and the point source / point receiver technique (PS/PR) were used to model the elastic wave propagation and the frequency-dependent attenuation of P- and S-waves in 2D synthetic composite media.[25] The kind of wave where the particles oscillate in a direction normal to the direction in which the wave travels is said to be a transverse or shear wave (S-wave). Plane waves may be shear-horizontally (SH) polarized (with displacement parallel to the free surface) or shear-vertically (SV) polarized. A generated S-wave is propagating at a shear wave velocity, c_s, described by,

$$c_s = \frac{1}{2}\sqrt{2}\cdot c_p\cdot\sqrt{(1-2\cdot\nu_o)/(1-\nu_o)} = \sqrt{E/(2\cdot\rho\cdot(1+\nu_o))} = \sqrt{\mu/\rho} \tag{3}$$

where c_p is the longitudinal wave velocity (P-wave), ν_o is the Poisson's ratio, E is the Young's modulus, ρ is the mass density, and μ is the transverse shear modulus given as one of the Lamé moduli.

As non-viscous fluids can't support shear stress no transverse waves occur in water. In this couplant the particles oscillate backwards and forwards in the same z-direction as that of the pressure wave travelling through the medium. Similarly, this wave travels through a material of interest at a longitudinal wave velocity, c_p, given by,

$$c_p = \sqrt{2}\cdot c_s\sqrt{(1-\nu_o)/(1-2\cdot\nu_o)} = \sqrt{E\cdot(1-\nu_o)/\rho\cdot(1+\nu_o)\cdot(1-2\cdot\nu_o)} \tag{4}$$

where both c_p and c_s are also related to the particle pressure and to the product of the corresponding wavelength and the frequency of the oscillation:

$$c_{...} = \lambda_{...}\cdot f = p_{...}/\rho\cdot v_{...} \tag{5}$$

Let us assume that (e.g. due to high frequencies and/or low elasticity-to-density ratios, ...) the wavelength is small in comparison with the dimensions of any interfering object/ medium and that a wave reaches a plane interface between two media 1 and 2. In the propagating wave there are no sudden discontinuities in either particle velocity v or particle pressure p. The wavefront coherence is being maintained for certain conditions until a critical angle for total reflexion is reached. As in optics the reflected angle α_r is equal to the incident angle α_i and the well known Snell's law can be applied:

$$\sin\alpha_i/c_1 = \sin\alpha_r/c_2 \tag{6}$$

4. Ultrasonic Measurement System

A semi-automatic measurement system (Fig. 1) has been arranged to measure sound velocities in hard dental tissues, because c is a relevant quantity of interest. It can be seen, that it is appearing in each of the given equations, Eqs. 1) - 7):

$$c_p = c_{H_2O} \cdot ((\Delta t_{Refl.} + \Delta t_{Subs.}) / \Delta t_{Refl.}) \qquad (7)$$

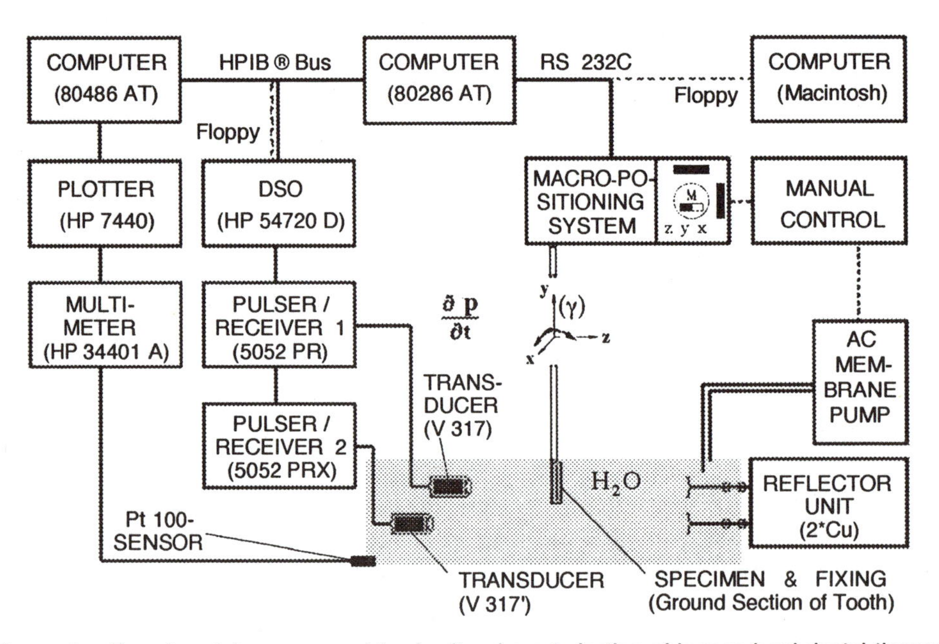

Figure 1. Experimental arrangement for *in-vitro* characterization of human hard dental tissues. Measurements were performed in a water basin. Coplanar ground sections of human teeth were scanned in a meander-like way using a x-y-z-table equipped with stepping motors. Transient high frequency pressure pulses were generated by means of spherically focused 20MHz-transducers connected to commercially available pulser/receivers. [4, 5, 11, 17*, 18-20]
In that way, ultrasonic energy was directed onto the front (and rear) of the specimen and two ideal reflectors. From the time delays of corresponding echo peaks one acoustical parameter has been determined at equidistant measuring points: The longitudinal sound velocity has been derived from A-Scans which had been sampled at single shot acquisition sampling rates of up to 4GSa/s. All data are based on *in vitro*-measurements of previously embedded teeth. The thickness of different ground sections varied between 0.6 mm and 1.8 mm. Samples were cut from the center of extracted teeth.

Sound velocities in teeth were calculated from Eq. 7) by determining the sonic speed in water, c_{H_2O} [6] and by combining [9, 13, 18-20, 24] two time-of-flight (TOF) measurements as defined by the reflection method (Refl.) [1, 17, 18, 20] with two consecutive TOFs which correspond to the substitution method (Subs.).[3*] In order to account for temporary temperature variations the resistance of a Pt 100 element has been measured simultaneously. Nonstationary transient pressure fields have been generated by commercially available transducers [16-20, 24] (Fig. 1). An algorithm [13, 26] allowed for avoiding any inaccuracies due to faulty caliper thickness data.

ACKNOWLEDGEMENT

The author would like to express his gratitude to Prof. Dr. C. Löst for his support.

RESULTS

The two-dimensional distribution of the longitudinal sound velocity in a longitudinally cutten tooth section (thickness Δz) is revealing a very inhomogeneous structure (Fig. 2,[R]):

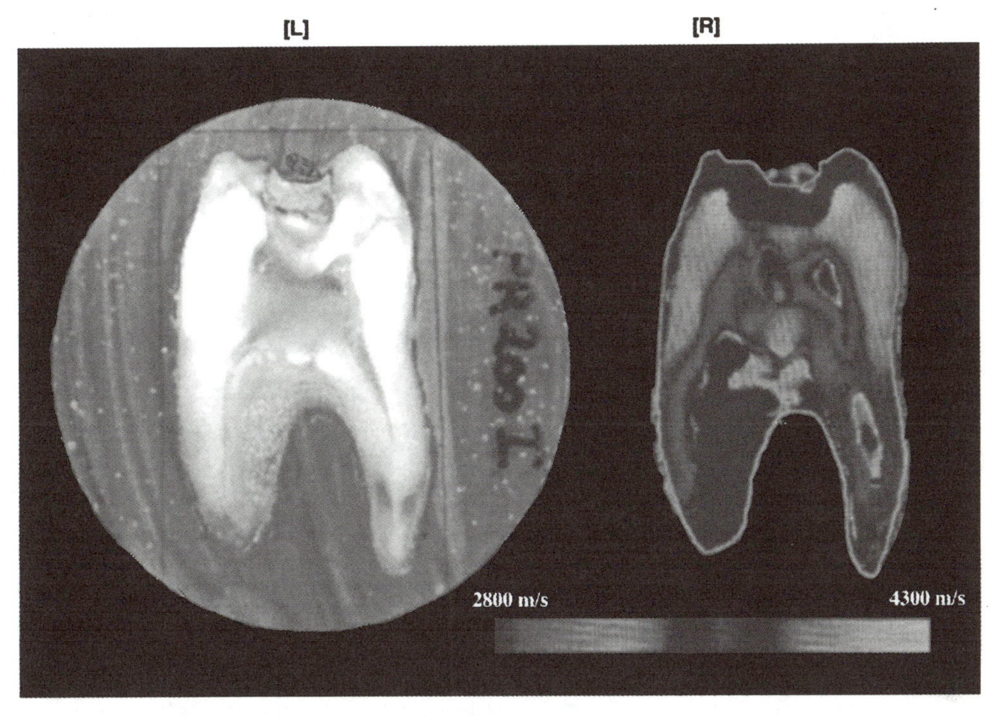

Figure 2. Original photograph [L](of the embedding resin, the tooth section through two roots & an amalgam filling [✎]) and corresponding D-Scan [R] of a previously vital molar of a 38 year old female. The two-dimensional distribution is displayed as a gray scale image representing 11605 values of the p-wave velocity within the range of 2800 m/s - 4300 m/s [Bar]. Brighter regions can be seen as a 'shoulder' of ~ 4000 m/s or lower 'ω'-shaped regions of ~ 3200 m/s. The x/y-plot's maximum, median and mean ± SD values of P-wave velocity were 4100 m/s, 3500 m/s and 3730 m/s ± 240 m/s, respectively. This window enables to automatically exclude almost the whole resin block and the enamel regions from being displayed.

CONCLUSIONS

The images reveal regions of inhomogeneous ultrasonic wave propagation characteristics. They may be regarded as a 'map of acoustic thickness'. All data represent a finite cross section area of thickness Δz for which the local variations of material properties were describable.

DISCUSSION

Except for the well established sonography of soft tissues, it is unrealistic, to assume any relatively constant sound velocity for future B-mode imaging approaches of hard dental tissues.

REFERENCES

1. F.E. Barber, S. Lee, and R.R. Lobene: Ultrasonic pulse-echo measurements in teeth. *Arch Oral Biol* 14: 745-760 (1969)
2. J.L. Ferracane, J.C. Mitchem, J.R. Condon, and R. Todd: Wear and marginal breakdown of composites with various degrees of cure. *J Dent Res* 76 (8): 1508-1516 (1997)
3. A. Sennaoui, M. Boynard, and C. Pautou: Characterization of red blood cell aggregate formation using an analytical model of the ultrasonic backscattering coefficient. *IEEE T Bio-Med Eng* 44 (7): 585-591 (1997)
4. S.-H. Wang, and K.K. Shung: An approach for measuring ultrasonic backscattering from biological tissues with focused transducers. *IEEE T Bio-Med Eng* 44 (7): 549-554 (1997)
5. C.L. de Korte, A.F.W. van der Steen, B.H.J. Dijkman, and C.T. Lancée: Performance of time delay estimation methods for small time shifts in ultrasonic signals. *Ultrasonics* 35 (4): 263-274 (1997)
6. H. Kanai, H. Hasegawa, N. Chubachi, Y. Koiwa, and M. Tanaka: Noninvasive evaluation of local myocardial thickening and its color-coded imaging. *IEEE T Ultrason Ferr, Freq Contr* 44 (4): 752-768 (1997)
7. F.L. Lizzi, E.J. Feleppa, M. Astor, and A. Kalisz: Statistics of ultrasonic spectral parameters for prostate and liver examinations. *IEEE T Ultrason Ferr, Freq Contr* 44 (4): 935-942 (1997)
8. P. Laugier, P. Giat, C. Chappard, Ch. Roux, and G. Berger: Ultrasonic backscatter coefficient in human cancellous bone in vivo. *Ultrasound Med Biol* 23 (S1): 135 (Abstract PIO 7406) (1997)
9. G. Pittomvils, J.P. Lafaut, H. Vandeursen, R. Boving, L. Baert and M. Wevers: Ultrasonic velocities of concentric laminated uric acid stones. *Ultrasonics* 34 (2-5): 571-574 (1996)
10. M. Berson, J.M. Grégoire, L. Vaillant, L. Colin, Ch. Yvon, F. Patat, F. Tranquart and L. Pourcelot: Real time high resolution microsonography. *Ultrasound Med Biol* 23 (S1): 131 (Abstract PIO 7003) (1997)
11. E. Biagi, A. Fort, and V. Vignoli: Guided acoustic wave propagation for porcelain coating characterization. *IEEE T Ultrason Ferr, Freq Contr* 44 (4): 909-916 (1997)
12. B. Bonello, B. Perrin, E. Romatet, and J.C. Jeannet: Application of the picosecond ultrasonic technique to the study of elastic and time-resolved thermal properties of materials. *Ultrasonics* 35 (3): 223-231 (1997)
13. M.S. Hughes, and D.K. Hsu: An automated algorithm for simultaneously producing velocity and thickness images. *Ultrasonics* 32 (1): 31-37 (1994)
14. C. Koch, G. Ludwig, and W. Molkenstruck: Calibration of an interferometric fiber tip sensor for ultrasound detection. *Ultrasonics* 35 (4): 297-303 (1997)
15. G. ter Haar, and the European Committee for Ultrasound Radiation Safety - The Watchdogs: Terms used in describing ultrasonic exposures. *Eur J Ultrasound* 5 (2): 127-130 (1997)
16. C. John, K.M. Irion, W. Nüssle, and C. Löst: Untersuchungen zur Auflösung zweidimensionaler Ultraschall-geschwindigkeitsprofile von humanen Zahnschliffen. *Schweiz Monatsschr Zahnmed* 104 (1): 25-30 (1994)
17. K. Raum, and W.D. O 'Brien: Pulse-echo field distribution measurement technique for high-frequency ultra-sound sources. *IEEE T Ultrason Ferr, Freq Contr* 44 (4): 810-815 (1997)
18. C. John, and C. Löst: Acoustic images of human teeth, in: *Comp Assist Radiol Excerpta Medica Int Congr Ser No. 1124:* xxvii, 902-908, 1089; H.U. Lemke, M.W. Vannier, K. Inamura, and A.G. Farman, eds. Elsevier Science B. V., Amsterdam, ISBN 0 444 82497 9 (1996)
19. C. John, M. Brecx, B. Linder-Lais, W. Krüger, U. Faust, and C. Löst: 2D-pattern of longitudinal sound velocity in endodontically treated teeth. *J Dent Res* 76 (5): 1109 (1997)
20. C. Löst, C. John, K.M. Irion, and W. Nüssle: Dentincharakterisierung mittels zweidimensionaler Ultraschall-geschwindigkeitsprofile. *Schweiz Monatsschr Zahnmed* 104 (1): 20-24 (1994)
21. L. Singher: Bond strength measurement by ultrasonic guided waves. *Ultrasonics* 35 (4): 305-315 (1997)
22. R.N. Klepfer, C.R. Johnson, and R.S. Macleod: The effects of inhomogeneities and anisotropies on electro-cardiographic fields: a 3-D finite-element study. *IEEE T Bio-Med Eng* 44 (8): 706-719 (1997)
23. F.A. Peyton, D.B. Mahler, and B. Hershenov: Physical properties of dentin. *J Dent Res* 31 (3): 366-370 (1952)
24. C. John, B. Linder-Lais, M. Brecx, R. Weiger, W. Krüger, and C. Löst: 2D-profiles of longitudinal sound velocity reveal inhomogeneity of human dentin. *J Dent Res* 75 (Special Issue): 197 (1996)
25. F. Schubert, and B. Koehler: Numerical modeling of elastic wave propagation in random particulate composites, in: *Nondestructive Characterization of Materials VIII* (pp. see this Issue); R.E. Green, Jr., ed. Plenum Publishing, New York (1997)
26. C. Löst, K.M. Irion, C. John, and W. Nüssle: Two-dimensional distribution of sound velocity in ground sections of dentin. *Endod Dent Traumatol* 8: 215-218 (1992)

INVERSION OF EDDY CURRENT DATA FOR RECOVERY OF THE ELECTROMAGNETIC PROPERTIES OF MATERIALS IN LAYERED FLAT AND TUBULAR PRODUCTS

C. González[1] and R. Martín[2]

[1]Departamento de Física Aplicada
Facultad de Ingeniería
[2]Departamento de Física
Facultad de Ciencias
Universidad Central de Venezuela
P.O. Box: 47533, Caracas 1041A, Venezuela

INTRODUCTION

Inversion of eddy current data is applied to the determination of the conductivity profile for metallic layered flat and tubular products. Material considered are stepwise continuous in their properties and test pieces are not very thick compared to the standard penetration depth.

The experimental assembly for flat products consists in a driving flat circular current coil, placed above and parallel to the surface of the product (Figure 1). The coil is driven by a low frequency harmonic source. A Hall effect detector is used for a radial scanning of axial and radial components of the magnetic field.

For tubular products an encircling driving coil is used. A Hall effect detector scans axially for axial and radial components of the magnetic field. Amplitude and phase are recorded.

Measurements are compared with numerical simulations corresponding to random conductivity distributions, subject to symmetry and other a priori restrictions. Comparison is performed introducing a cost function, which is the sum of the absolute values of the difference between the measured and the calculated fields. An optimizing algorithm is used to obtain the best fit between experimental and simulated data. The solution of the inversion procedure corresponds to the distribution of conductivity, which fits best experimental results.

The resolution of the procedure is established comparing the measured fields for a particular sample, with all the possible conductivity distributions fulfilling symmetry and other a priori restrictions. The particular discretization scheme is taken in other to keep the number of the configurations within a reasonable limit. The average conductivity is going to be important for the performance of the optimization algorithm.

Nondestructive Characterization of Material VIII
Edited by Robert E. Green Jr., Plenum Press, New York, 1998

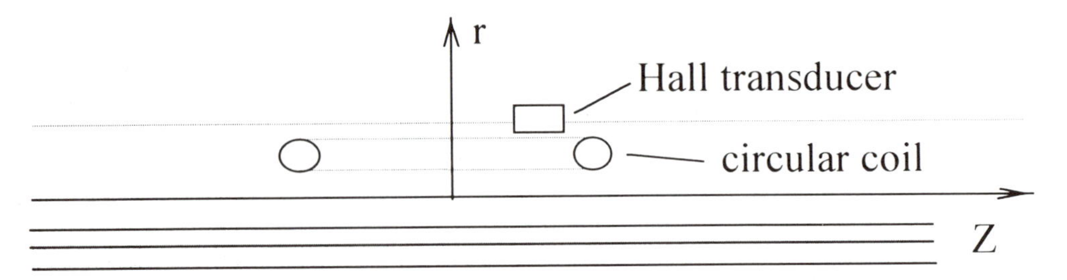

Figure 1. Stratified flat sample. The Hall detector scans radially from r = 0 to r = 2R, where R is the current loop radius. The Hall transducer is oriented for detection of -r- and -z- components of the magnetic field.

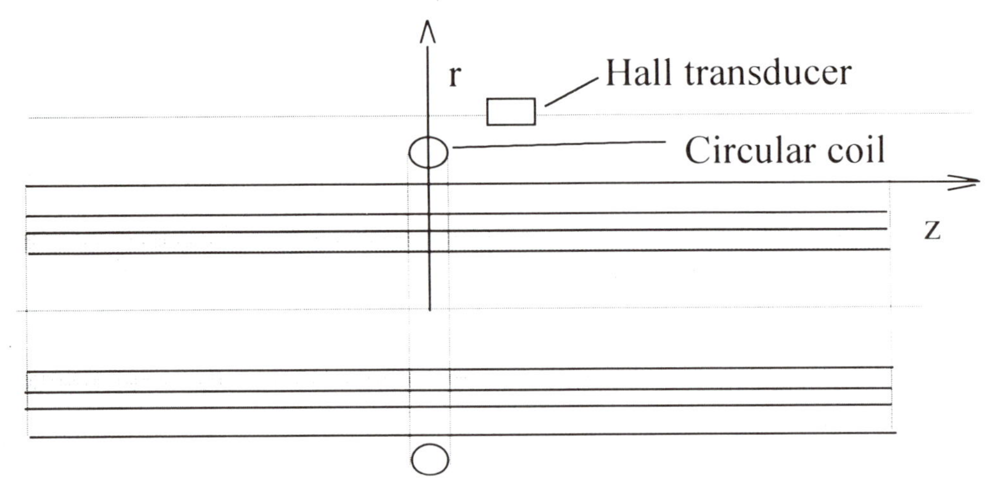

Figure 2. Multilayered tubular sample. The Hall detector scans radially from r = 0 to r = 2R, where R is the current loop radius. The Hall transducer is oriented for detection of -r- and -z- components of the magnetic field.

FORMULATION OF THE PROBLEM

The problem to be solved is the determination of the conductivity profile for layered conductive flat and tubular configurations. The developed approach is also applicable to measurement of the thickness of a conductive plate or tube wall thickness. Non conductive layers and interlayers thickness could be recovered without a precise knowledge of the conductivity.

Hall transducer output is processed by means of a cuadrature detector in order to obtain amplitude and phase for the components of the magnetic field in the radial and normal directions. With the implemented system, the conductivity profiles of metallic sheets (up to 3mm) have been estimated. The profile of tubular products with 30mm in diameter and with wall thickness of 3mm, have been also recovered. Frequency seting was within the range from 3 to 50 kHz.

Non magnetic materials like cooper, aluminum, lead; bronze and tin are used in these experiments, including non-conducting interlayers.

PHYSICAL BACKGROUND

Maxwell equations for the harmonic or quasi-stationary case with axial symmetry using cylindrical coordinates give rise to the following equation for the electric field [1,2].

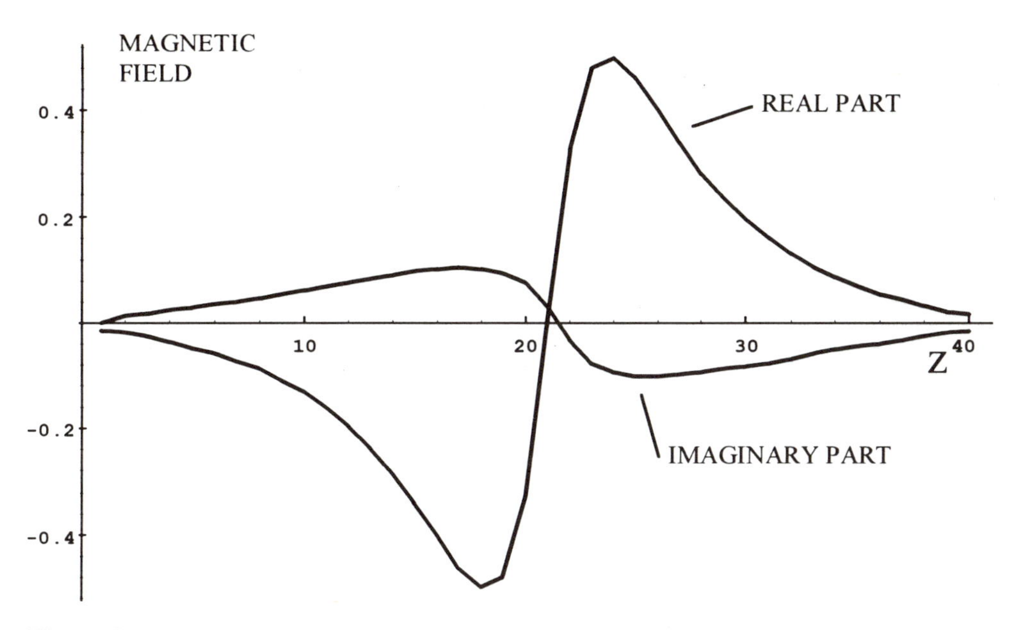

Figure 3. Real and imaginary parts for the radial component of the magnetic field for a coper tube with an encircling coil. Units are arbitrary.

$$\left(i\omega\mu \cdot \sigma(r,z) + \frac{1}{r^2} - \frac{\partial^2}{\partial r^2} - \frac{\partial^2}{\partial z^2} - \frac{1}{r}\frac{\partial}{\partial r} \right) E_\varphi(r,z) - \mu\omega i J_s = 0 \tag{1}$$

Where E_φ is the only non vanishing component and $\partial E_\varphi/\partial\varphi = 0$, ω is the angular frequency, σ the conductivity, μ the magnetic permeability, supposed to be nearly equal to that for air, and J_s is the current density in the coil.

Table 1. Profile corresponding to a model of 4 layers of conductive materials A, B, C and D, with conductivity 1.4, 2.8, 3.5, and 5.7 (10^5 S/m) respectively. Zero -0- stand for a non- conductive material. The Cost Function is calculated by comparison between de calculate and measured field values.

Profile	Cost Function	Average conductivity 10^5 S/m	Profile	Cost function	Average conductivity 10^5 S/m
CC00	9.465	1.74	BBAA	9.894	2.10
D0A0	9.492	1.77	BAC0	9.918	1.90
CAB0	9.503	1.90	AD0A	9.927	2.12
CBA0	9.506	1.90	D000	9.967	1.17
BCA0	9.525	1.90	CAA0	10.01	1.57
DA00	9.543	1.77	BCB0	10.02	2.27
CB0A	9.564	1.90	BB0B	10.03	2.12
CA0B	9.602	1.90	BD00	10.04	2.12
C0C0	9.648	1.74	AD00	10.05	1.77
BBB0	9.652	2.10	BABA	10.06	2.10
CAAA	9.681	1.92	BBC0	10.14	2.27
BC0A	9.696	1.90	CA0A	10.15	1.57
C0BA	9.721	1.90	ACB0	10.20	1.90
C0AB	9.819	1.90	B0D0	10.24	2.12
ADA0	9.847	2.12	BC00	10.26	1.45
CB00	9.891	1.57	C0B0	10.28	1.57

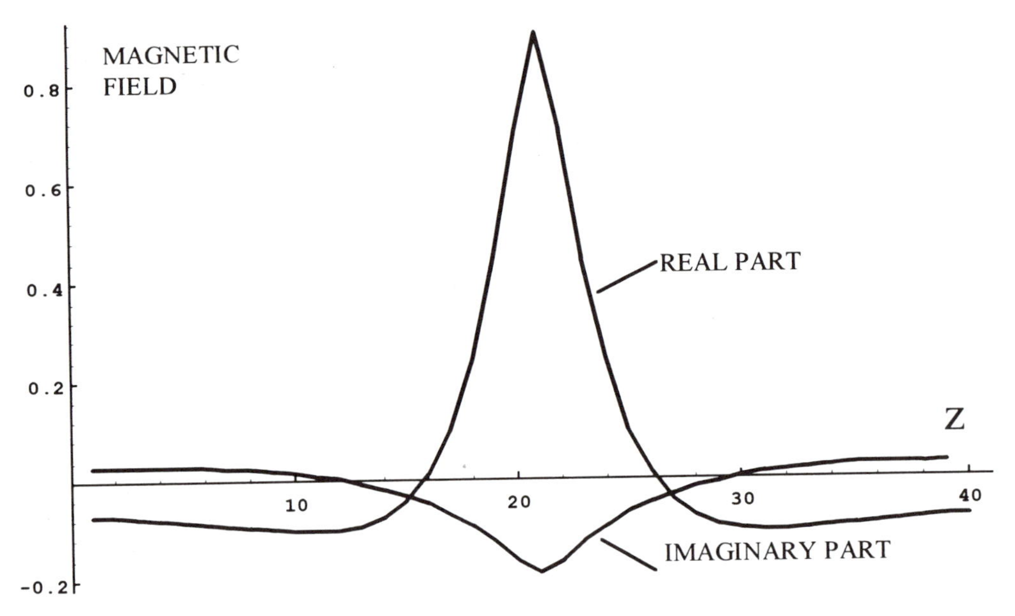

Figure 4. Real and imaginary parts for the axial component of the magnetic field for a coper tube with an encircling coil. Units are arbitrary.

Equation (1) has been solved analyticaly [1]. Finite elements as well as finite difference methods are also feasible. Figures 3 and 4 show the finite difference solution for the components of the magnetic field, for a tubular product. Empirical corrections have been introduced in the model in order to reproduce consistently the experimental results. Discretized region includes the radial interval $0<r<2R$ where R is the driving coil radius, under the consideration that measured field is nearly zero at that distance from the axis. The coil is placed at $z = 0$. Boundary conditions and the values for the field and its derivatives at $r = 0$ are readily obtained. Mesh parameter h is taken as $h = \delta/5$ where δ is the average standard skin depth [3,4].

EXAMPLE OF INVERSION PROCEDURE

The procedure considers a conductivity range, which is going from lead to cooper. The possibility of non-conducting layers (surface and internal) was also considered. Table 1 shows the cost function versus conductivity profile, in a model of 4 layers. Numerical simulation corresponding to a conductivity profile identical to the one of the measured sample has the lowest value for the cost function.

The software developed takes advantage of the *Simulated Annealing Method* [5] to optimize the search for the best fit between the measured field at the surface and the simulated one, which corresponds to a random generated conductivity profile. Generation is subjected to physical constrains related to *a priori* information, such as the range of allowed conductivity, possibility of non conducting surface and internal layers. This additional information allows a reduction in the number of degrees of freedom and also is going to reduce the number of required iterations. For example, if we know the overall thickness, the number of possible configurations for the conductivity layers is drastically reduced. Of course, if we know the conductivity, the thickness of a homogenous plate could be readily obtained.

For the case of an aluminum plate of thickness $2h$, an exhaustive search was implemented for the optimum profile for a model with 4 layers and conductivity close to aluminum. Possibility of non-conducting internal and surface layers was also considered. The cost function as a function of an arbitrary index, has many local minima values when all the allowed conductivity profile are generated. Our problem is to find the conductivity profile corresponding to a blobal minimum for the cost function. By sorting the values of the cost function, as shown in Table 1. The conductivity profiles with almost the lowest value have approximately the same average conductivity.

DISCUSSION AND CONCLUSION

The results have been obtained using a single frequency and from measurements of the field at a given distance from the surface. It is expected that scanning for several frequencies and distances from the surface may improve sensitivity and resolution, allowing industrial applications for profile determinations in conductive multilayer flat or tubular products. Applications to ferromagnetic materials could be implemented by magnetic saturation of the sample. For implementation of the algorithm we need to establish the dimension of the problem and choose variables in such a way that continuity between input and output should be guaranteed. Average conductivity proves to be an adequate parameter in order to vary the conductivity profiles.

Convergence of the method is quite slow and is strongly related to the variables chosen to generate conductivity profiles. In the experiments performed, the use of average conductivity as criteria for determination of closeness between two conductivity profiles was the best way found to improve convergence.

REFERENCES

1. E. E. Kriezis, Theodoros D. Tsiboukis, Stavros M. Panas and A. Tegopoulos: Eddy currents: Theory and applications, *Proceedings of the IEEE*, Vol 80, No. 10, Oct. 1992.
2. R. Albanese, G. Rubinacci: Formulation of the eddy-current Problem, *IEE Proceedings*, Vol 137, Pt. A, No. 1, Jan. 1990.
3. Riadh Zorgati, Bernard Duchene, Dominique Lesselier and Francis Pons: Eddy current testing of anomalies in conductive materials, Part I: quantitative imaging via diffraction tomography techniques, *IEEE Transactions on Magnetics*, Vol. 27, No. 6, Nov. 1991
4. Riadh Zorgati, Bernard Duchene, Dominique Lesselier and Francis Pons: Eddy current testing of anomalies in conductive materials, Part II: quantitative imaging via deterministic and stochastic inversion techniques, *Transactions on Magnetics*, Vol. 28, No. 3, May 1992.
5. S. Kirpatrick, C. D. Gelatt Jr., and M. P. Vecchi: Optimization by simulated annealing, *Science*, Vol. 220, Number 4598, May 1983.

ULTRASONIC LEAKY WAVE MEASUREMENTS FOR MATERIALS EVALUATION

Dan Xiang*, Nelson N. Hsu and Gerry V. Blessing

Automated Production Technology Division
National Institute of Standards and Technology
Gaithersburg, MD 20899

INTRODUCTION

Ultrasonic leaky waves at a liquid/solid interface have been utilized by many investigators to characterize the properties of materials. The leaky surface wave, for example, has been used in acoustic microscopy [1], especially in Line-Focus-Beam (LFB) microscopy [2], for many years. Nearly all leaky-surface-wave measurements by acoustic microscopy rely on the interpretation of a $V(z)$ curve, which is a record of the transducer's voltage V as a function of the defocusing distance z between the transducer's focal plane and the specimen surface while the transducer is operated in a tone-burst mode [3]. The focus of this $V(z)$ analysis is the interference of two principal acoustic components - the direct reflection and the leaky surface wave. One of the challenges of this technique is to extract, from the complex $V(z)$ interference patterns, the various leaky wave modes that may be simultaneously present in the material. Another challenge is the complexity of modeling and analyzing the $V(z)$ curve, especially involving the pupil function that must be determined for the microscope's particular lens [4].

In an effort to overcome these challenges, we had previously developed a simple time-and-polarization-resolved ultrasonic technique to generate and measure leaky waves on samples immersed in water [5, 6]. Using a specially designed large-aperture lensless line-focus polyvinylidene fluoride (PVDF) transducer [7] in conjunction with conventional ultrasonic pulse instrumentation, this technique features most of the attributes of LFB acoustic microscopy, but on a larger interrogated sample area (the order of millimeters). Based upon a simple methodology, the leaky surface wave velocity is readily determined from the arrival times of transient wave fronts. In this paper, we present measurements of various leaky wave modes such as the surface, longitudinal, shear and Lamb modes on various materials. We also outline and demonstrate a predictive algorithm for the transient

* NIST Guest Worker, The Johns Hopkins University, Baltimore, MD 21218

Nondestructive Characterization of Material VIII
Edited by Robert E. Green Jr., Plenum Press, New York, 1998

waveforms associated with this technique in order to obtain a better understanding of leaky-wave phenomena.

THEORY

Transient Waveform Simulation

A simplified algorithm for a theoretical simulation of the transient response of a lensless line-focus PVDF transducer defocused on an isotropic solid sample is outlined here. (The anisotropic case is discussed elsewhere in these proceedings [8].) Basically, the transducer's output $v(t,z)$ is a convolution of the transfer function $h(t,z)$ of the transducer-specimen system and the input source $s(t)$ where t is the time and z the defocusing distance,

$$v(t,z) = s(t) \otimes h(t,z). \tag{1}$$

The transfer function $h(t,z)$ is the Fourier transform of the system frequency response $H(f,z)$ which can be derived from the expression for $V(z)$ in acoustic microscopy [9]. Since our PVDF transducer is lensless, the complicated pupil function in the V(z) expression can be reasonably assumed to be unity [10]. Hence, the following simplified equation results

$$H(f,z) = \int_{-k_M}^{k_M} R(k_x)\exp(i2k_z z)\mathrm{d}k_x, \tag{2}$$

where k_x, k_z are the wave-number components in water ($k=2\pi f/c$), and k_M corresponds to k_x at the half-aperture angle of the transducer. $R(k_x)$ is the reflection coefficient at the liquid/solid interface. Therefore, the transfer function is simplified to an integration of the reflection coefficient at a liquid/solid interface. The input source $s(t)$ in equation (1) can be easily obtained by deconvolving the transfer function $h(t,0)$ with the experimentally recorded waveform $v(t,0)$ when the liquid/solid interface is at the focal point.

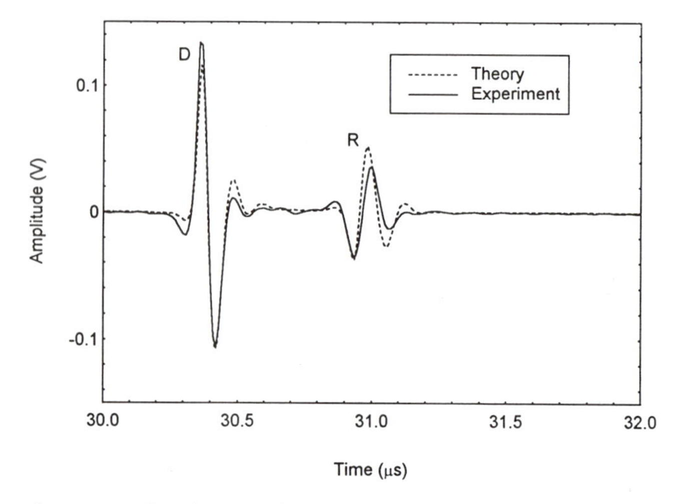

Fig. 1 A time waveform comparison between theory and experiment for an isotropic stainless steel sample at a defocusing distance of 3.0 mm.

Figure 1 presents a comparison of theoretical and experimental transient waveforms on an isotropic stainless steel sample at a defocusing distance of 3.0 mm. Good agreement is observed for the directly reflected wave, D, and the leaky surface wave, R, which are also clearly resolved in time.

Time-Resolved Leaky Wave Velocity Measurements

In a previous paper [6], we described in detail the time-resolved ultrasonic technique for leaky wave velocity measurements. This technique is based on a simple ray model which helps to establish the relationship between various wave arrivals and the defocusing distance. Since the arrival times of both the directly-reflected and the leaky waves are linear with defocusing distance, the leaky wave velocity V may be determined from the linear slope m of arrival time as a function of defocusing distance

$$V = [1/(V_W * m) - 1/(4 * m^2)]^{-1/2}, \qquad (3)$$

where the directly reflected wave is used as the time (zero) reference, and V_W is the wave velocity in water. The two-standard-deviation probable error of the leaky wave velocity measurements can be estimated with statistical methods.

EXPERIMENT

A prototype lensless line-focus PVDF transducer with a focal length of 25.4 mm, a half-aperture angle of 34°, a center frequency of 10 MHz and a 3 dB bandwidth of about 65%, is used in this work. A conventional ultrasonic pulser/receiver and a digital oscilloscope are the principal electronic system components. A vertical scanner controls the defocusing distance between transducer and sample. Both the data acquisition and scanner movement are computer controlled.

Leaky Surface Waves in Steel

A stainless steel sample was used for one set of measurements. The experimental waveforms for defocusing distances ranging from 2.0 mm to 10.0 mm, in steps of 0.2 mm, are shown as a “waterfall” plot in Fig. 2(a), where the directly reflected wave D is used as the time reference for convenience. A linear relationship between the time delay of the leaky surface waves R and the defocusing distance z is clearly observed. By simply selecting the positive and negative peaks of D and R respectively as the arrival-time markers, the arrival time as a function of defocusing distance can be plotted and fitted as shown in Fig. 2(b). From the slope of the linear-least-squares fit (m=5.254), and a pre-determined wave velocity in water (1485 m/s), the leaky surface wave velocity and its two-standard-deviation probable error for the stainless steel sample can be readily determined to equal (2897±3) m/s.

Leaky Longitudinal Waves in Plexiglass

A set of experimental waveforms, obtained on plexiglass at defocusing distances from 0.5 mm to 8.5 mm, are plotted in Fig. 3(a). For a low modulus material like plexiglass, a Snell’s law calculation reveals that the line-focus transducer’s half-aperture angle is much less than the Rayleigh critical angle. Therefore, it can be predicted that the leaky Rayleigh wave will not be observed. However, the leaky longitudinal wave (designated L) can be

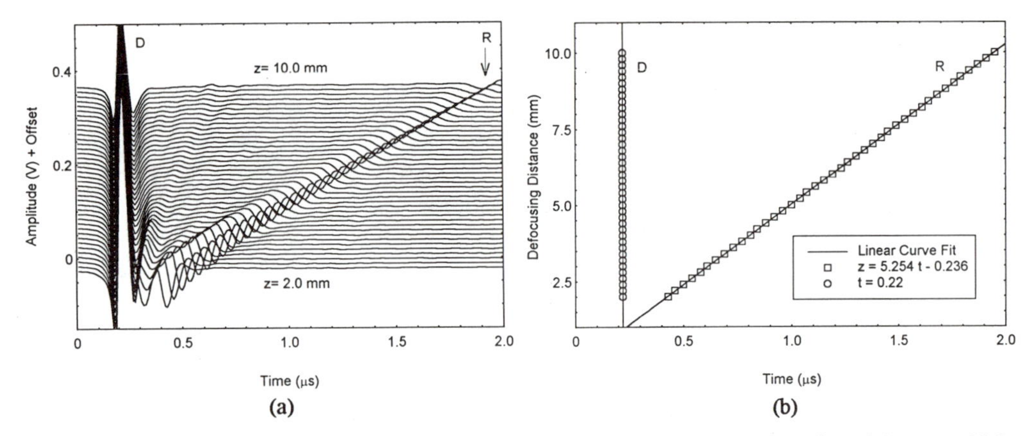

Fig. 2 (a) Experimental time waveforms on stainless steel for defocusing distances z from 2.0 mm to 10.0 mm, and (b) linear-least-squares fit of defocusing distance as a function of arrival time for the leaky surface wave R, where the directly reflected wave D is the time reference.

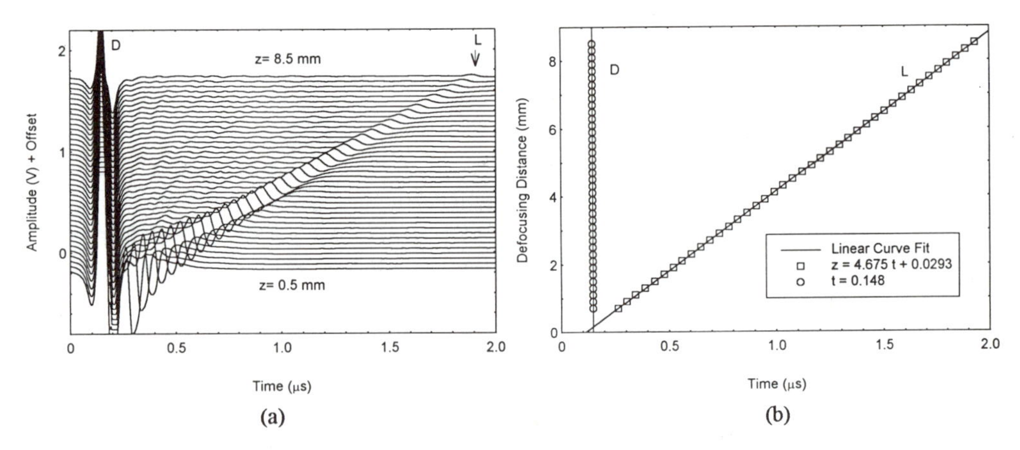

Fig. 3 (a) Experimental time waveforms on plexiglass for defocusing distances z from 0.5 mm to 8.5 mm, and (b) linear-least-squares fit of defocusing distance as a function of arrival time for the leaky longitudinal wave L, where the directly reflected wave D is the time reference.

clearly observed, with a time delay which is a linear function of the defocusing distance. Data extracted from these waveforms are plotted and fitted in Fig. 3(b). Again, we use the slope of the linear-least-squares fit (m=4.675) in equation (3) to calculate a value of (2746±1) m/s for the leaky longitudinal wave velocity for plexiglass.

Leaky Shear, Longitudinal, and Surface Waves in Aluminum

Experimental waveforms obtained on an aluminum sample at defocusing distances from 2.0 mm to 10.0 mm are plotted in Fig. 4(a). Based upon prior experience, the leaky surface wave R can be immediately recognized in the plot. The leaky longitudinal wave L can also be readily identified, and in addition a small amplitude leaky shear wave signal preceding the leaky surface wave can be resolved. Therefore, three leaky wave types possessing different slopes as a function of defocusing distance are simultaneously resolved in this one set of waveforms. Using these waveforms, we plot and fit the data in Fig. 4(b).

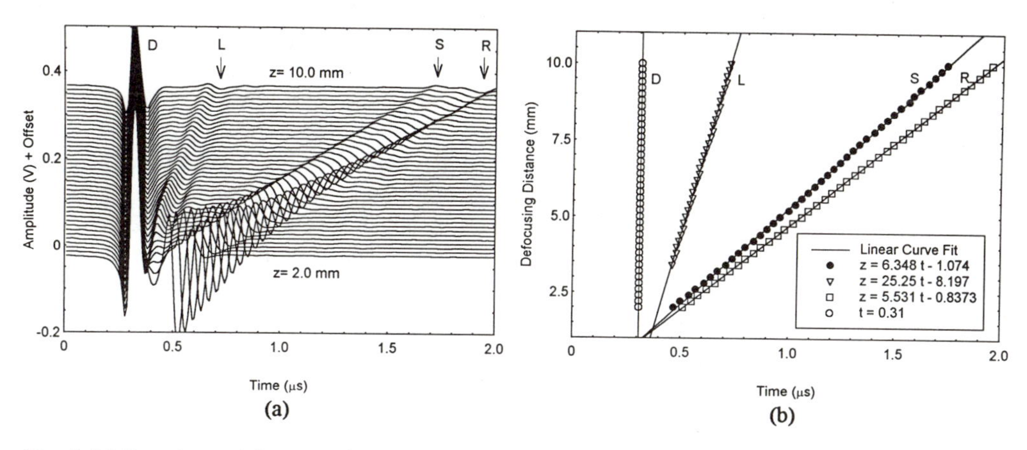

Fig. 4 (a) Experimental time waveforms on aluminum for defocusing distances z from 2.0 mm to 10.0 mm, and (b) linear-least-squares fits of defocusing distance as a function of arrival time for the leaky longitudinal wave L, shear wave S and surface wave R with the directly reflected wave D as the time reference.

From the respective slopes, we determine the velocities of leaky longitudinal, shear and surface waves in the aluminum sample to be (6170±40) m/s, (3164±8) m/s and (2967±4) m/s respectively.

Leaky Lamb Waves in Tape

For sufficiently thin or layered materials, leaky Lamb waves may be generated using the line-focus transducer. Here we use a videocassette recorder (VCR) tape as a demonstration sample. Experimental waveforms for defocusing distances from 0.0 mm to 3.0 mm in steps of 0.1 mm are plotted in Fig. 5(a). The leaky Lamb waves can be observed for defocusing distances of 1.0 mm or less. Careful observation reveals that most of these leaky Lamb modes, which depend on the sample thickness as well as the frequency, are not linear in time with defocusing distance. Their interpretation is relatively complicated and will be given elsewhere. One exception is the fundamental symmetric Lamb mode, identified as S_0 in Fig. 5(a). Data for this mode are plotted and fitted in Fig. 5(b). From the fitted slope (m=4.133), the wave velocity is determined to be (2600±20) m/s for the S_0 mode leaky Lamb wave in the VCR tape.

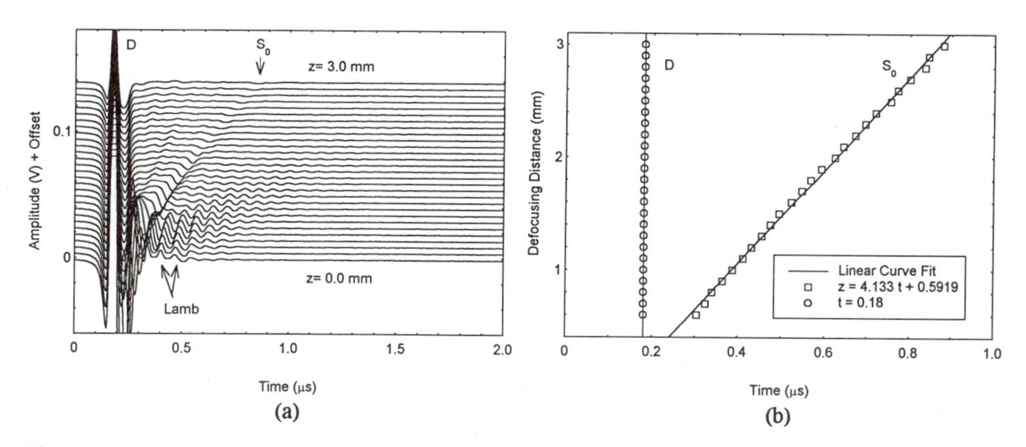

Fig. 5 (a) Experimental time waveforms from a VCR tape, and (b) linear-least-squares fit of defocusing distance as a function of arrival time for the fundamental-mode leaky Lamb wave S_0 with the directly reflected wave D as the time reference. Higher order leaky Lamb modes are also indicated in (a).

CONTACT MEASUREMENTS

The leaky wave measurements are corroborated by the following conventional contact measurement results: a surface wave velocity of 2882 m/s in stainless steel; a longitudinal wave velocity of 2744 m/s in plexiglass; and longitudinal, shear and surface wave velocities of 6402 m/s, 3166 m/s and 2965 m/s respectively in aluminum. We note that the difference is less than 0.6 %, with one exception. We also note that the leaky Lamb wave velocity in VCR tape falls within the range of values (2272 m/s - 2991 m/s) measured by Hirao et al. [11]. However, directional measurements on our VCR tape indicate an isotropy for Lamb wave propagation, in disagreement with the results from [11].

CONCLUSION

The time-and-polarization-resolved ultrasonic technique demonstrated here provides a simple methodology for leaky wave measurements on a variety of solid samples. The experimental results show that different (material-dependent) leaky waves can be identified from their time waveforms, and their velocities readily determined. In addition, a simplified formulation is shown to be a successful predictor of experimental waveforms. From the various leaky wave velocity measurements and theoretical simulations, characterization of a wide range of materials can be achieved.

REFERENCES

1. C. Quate, A. Atalar and H.K. Wickramasinghe, Acoustic microscopy with mechanical scanning - a review, *Proc. of IEEE*, 67:1092-1114 (1979)
2. J. Kushibiki and N. Chubachi, Material Characterization by line-focus-beam acoustic microscope, *IEEE Trans. Son. & Ultrason.* 32:189-212 (1985)
3. J. D . Achenbach, J.O. Kim and Y.C. Lee, Measuring thin-film elastic constants by line-focus acoustic microscopy, in: *Advances in Acoustic Microscopy*, Vol. 1, A. Briggs, ed., Plenum Press, New York, 153-207 (1995)
4. A. Briggs, *Acoustic Microscopy*, Clarendon Press, Oxford, (1992)
5. N.N. Hsu, D. Xiang and G.V. Blessing, Time and polarization resolved ultrasonic measurements using a lensless line-focus transducer, *Proc.1995 IEEE Ultras. Sym.*, 867-871 (1995)
6. D. Xiang, N.N. Hsu, and G.V. Blessing, Materials characterization by a time-resolved and polarization-sensitive ultrasonic technique, in: *Review of Progress in QNDE*, Vol. 15, D.O. Thompson and D.E. Chimenti, ed., Plenum Press, New York. 1431-1438 (1996)
7. D. Xiang, N.N. Hsu, and G.V. Blessing, The design, construction and application of a large aperture lens-less line-focus PVDF transducer, *Ultrasonics*, 34: 641-647 (1996)
8. N.N. Hsu, D. Xiang and G.V. Blessing, Time domain waveforms of a line-focus transducer probing anisotropic solids, (this volume)
9. W. Li and J.D. Achenbach, Measuring thin-film elastic constants by line-focus acoustic microscopy, *Proc.1995 IEEE Ultras. Sym.*, 883-892 (1995)
10. K. K. Liang, G.S. Kino and B.T. Khuri-Yakub, Material characterization by the inversion of V(z), *IEEE Trans. Son. & Ultrason.* 32:213-224 (1985)
11. M. Hirao and K. Yokota, Leaky Lamb wave along VCR magnetic tapes, in: *Review of Progress in QNDE*, Vol. 15, D.O. Thompson and D.E. Chimenti, ed., Plenum Press, New York. 239-246 (1996)

MODELING RAYLEIGH WAVE DISPERSION DUE TO DISTRIBUTIONS OF ONE AND TWO DIMENSIONAL MICRO-CRACKS: A REVIEW

Claudio Pecorari
Materials Department
Oxford University
Oxford OX1 3PH
England

INTRODUCTION

The characterization of surface and sub-surface damage has attracted the attention of NDE researchers for many years. Marshall et al.[1] used surface acoustic waves to investigate crack growth in sample with ground surfaces. They found that catastrophic failure is preceded by stable growth of the cracks because of the residual stress field generated by surface grinding. The cracks generated by grinding ceramic surfaces were described as being mostly parallel to the grinding direction, generally semi-elliptical in shape, and possibly overlapping each other. Khuri-Yakub and Clarke[2] reported that the compressive residual stress field due to machining extends from the surface into the bulk for about 20 % of the crack's depth, and causes the closure of the crack at its mouth. More recently, Scherer et al.[3] estimated crack densities of the order of 10^3 mm^{-2} on ground surfaces by using high-frequency acoustical images. Surface wave velocity on damaged surfaces was also measured and variations up to 5% were recorded. Finally, Tardy et al.[4] presented dispersion curves for a Rayleigh wave propagating over machined surfaces of ceramic components. In the frequency range between 6 MHz to 14 MHz, they observed a linear dependence of the wave velocity on the ultrasonic frequency.

Micro-cracks distributions are very complex systems, and to characterize them completely it is necessary to know many parameters. It is reasonable to expect, therefore, that several complementary techniques must be used to achieve such a goal. In view of the results described above, acoustic techniques are likely to play an important role in surface damage characterization. Towards this end, however, a lot remains to be done in order to extract quantitative information from experimental data. The purpose of this paper is to present a review of the models recently developed, which provide links between micro-crack distribution parameters and Rayleigh velocity. A comparison between experimental data published by Tardy et al.[4] and theoretical predictions will be presented.

Nondestructive Characterization of Material VIII
Edited by Robert E. Green Jr., Plenum Press, New York, 1998

THEORY

A distribution of one-dimensional cracks

Recently, a new heuristic model[5] to predict the dispersion of a Rayleigh wave propagating over a surface containing a distribution of one-dimensional, surface-breaking cracks has been presented. This model may be used to describe wave propagation on ground surfaces. In fact, the close proximity of the semielliptical cracks along the grinding direction[1] suggests that the collective elastic response of such a one-dimensional array of cracks can be modeled as that of a single one-dimensional crack. The model's heuristic nature is due to the representation of a real cracked surface by means of a fictitious layered system (Fig. 1). The elastic properties of the substrate are those of the uncracked host material, whereas those of the layer account for the effect of the micro-crack distribution. The basic assumption of the model is that the dispersion of a Rayleigh wave propagating over such a cracked surface may be simulated by that of the lowest order mode supported by the effective layered system. The modeling is developed in the low frequency limit, and under the constraint that the average distance between neighboring cracks is smaller than the Rayleigh wavelength. In addition, the interaction between neighboring cracks is neglected.

The strain field, $\vec{\varepsilon}(\vec{x})$, in the near-surface region of a system with a uniform distribution of one-dimensional cracks, which are parallel to the y-z plane and have the same depth, h, can be written as follows,

$$\vec{\varepsilon}(\vec{x}) = \vec{S}^{o}:\vec{\sigma}(\vec{x}) + \frac{1}{2}\sum_{r=1}^{N}\left[\vec{b}(\vec{x}_r)\hat{n}_r + \hat{n}_r\vec{b}(\vec{x}_r)\right]\delta(x - x_r)u(h). \tag{1}$$

In Eq.(1), $\vec{x} = (x,z)$, $\vec{S}^{o}$ is the compliance tensor of the uncracked material, $\vec{\sigma}(\vec{x})$ is the total stress field, $\vec{b}(\vec{x}_r)$ is the crack opening displacement (COD) of the r-th crack, $\hat{n}_r$ is the unit vector normal to the crack faces, and u(h) is the characteristic function of the r-th crack, i.e., u(h)=1 if $0<z<h$, and u(h)=0 otherwise. The delta function dependence of $\vec{\varepsilon}(\vec{x})$ on the variable x is to be ascribed to the displacement discontinuity localized on the crack face. The strain and stress fields also depend on the location of the N cracks. By averaging both sides of Eq. (1) first with respect to all the possible configurations of the crack distribution, and then over the range of depth $0<z<h$, Eq. (1) yields

$$\langle\vec{\varepsilon}(x)\rangle = S^{o}:\vec{\sigma}(x) + \frac{\nu h}{2}\left[\langle\vec{b}(x)\rangle\hat{n} + \langle\vec{b}(x)\rangle\hat{n}\right], \qquad \text{for } 0<z<h, \tag{2}$$

where ν is the crack density. By introducing the crack compliance tensor, B_{ij}, defined by

$$\langle b(x)\rangle_i = h\,B_{ij}\langle\vec{\sigma}(x)\rangle_{jk}\,n_k, \qquad i, j, k = 1, 2, 3 \tag{3}$$

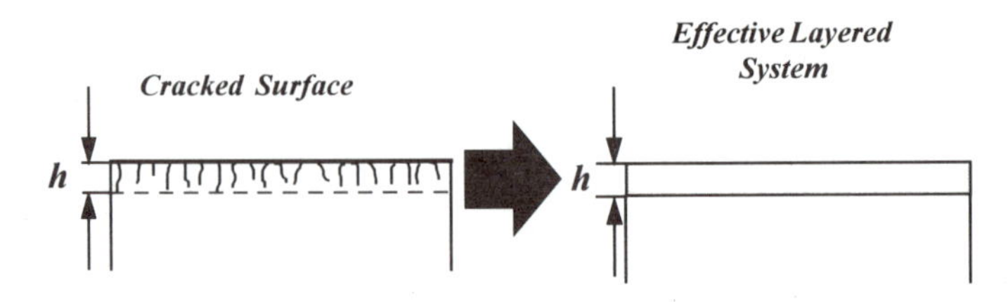

Figure 1. Cracked surface and its model consisting of a homogeneous, anisotropic layer on a substrate.

the ij-th component of $\vec{\varepsilon}(\vec{x})$ for 0<z<h becomes

$$\begin{aligned}\langle\vec{\varepsilon}(x)\rangle_{ij} &= S^{o}_{ijrs}\langle\vec{\sigma}(x)\rangle_{rs} + \frac{\nu h}{2}\left[B_{ir}n_s n_j + B_{jr}n_s n_i\right]\langle\vec{\sigma}(x)\rangle_{rs} \\ &= \left(S^{o}_{ijrs} + \Delta S_{ijrs}\right)\langle\vec{\sigma}(x)\rangle_{rs}. \end{aligned} \tag{4}$$

If a distribution of identical cracks which are parallel to the wave front of the propagating Rayleigh wave, that is, $\hat{n}=\hat{x}$, is considered, then the only non-zero elements of the extra compliance tensor ΔS_{IJ} are $\Delta S_{11}=\nu hB_1/2$, $\Delta S_{55}=\nu hB_3/2$, and $\Delta S_{66}=\nu hB_2/2$. B_1, B_2, and B_3 are the crack's normal and shear compliances. Here, Voigt's notation has been used. Finally, the layer's stiffness tensor is recovered by inverting the compliance tensor.

Figures (2) and (3) show the dependence of the normalized wave velocity versus the crack density for three different values of the crack depth, and versus the crack depth for three values of crack density, respectively. The frequency of the propagating wave is maintained constant. These results were obtained for alumina (Al_2O_3) assuming for the bulk velocity and mass density the following values: V_L=10822 m s^{-1}, V_T=6163 m s^{-1}, and ρ=3970 Kg m^{-3}. Both figures show phase velocity variations of the order of a few percentages, and, thus, easily measurable. The different dependencies of the wave velocity on the crack density and on the crack depth can be explained as follows. A change in crack density affects only the elastic properties of the compliant layer (see Eq. (4)), while a change in the crack depth affects the layer thickness as well. Therefore, the dispersion shown in Fig. (2) is entirely due to changes of the layer's mechanical properties, whereas that in Fig. (3) also includes geometrical effects. This model can also treat micro-cracks having distributed depths, and, under certain conditions, random orientation[5].

The effect of a compressive residual stress generated by surface grinding in the outermost part of the near-surface region on the velocity of a surface wave propagating over a cracked surface was also investigated[6]. The main effect of the residual stress is that of causing the closure of a surface-breaking crack at its mouth[1, 2]. Thus, the elastic response of such a crack can be reasonably approximated by that of a subsurface crack. With this assumption in mind, the real cracked surface can be modeled by a layered system consisting of two layers on a substrate. The elastic properties of the outermost layer and of the

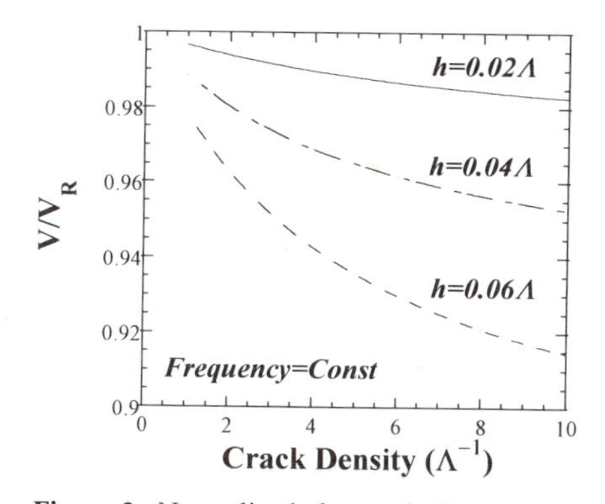

Figure 2. Normalized phase velocity versus normalized crack density: one-dimensional case.

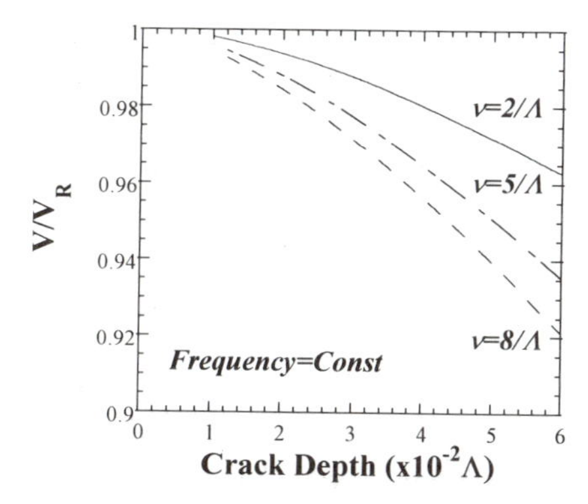

Figure 3. Normalized phase velocity versus normalized crack depth: one-dimensional case.

substrate are those on the host, uncracked material, whereas those of the intermediate layer account for the effect of the crack distribution. If the results of Khuri-Yakub and Clarke[2] are used to set the depth of the first layer equal to 20% of the crack depth, then this model predicts that the variation of the wave velocity due to a distribution of one-dimensional, open, surface-breaking cracks is reduced by the residual stress by about 70%.

A distribution of surface-breaking two-dimensional cracks

When surfaces of a higher quality are required, the finishing process continues after grinding by lapping the surface with finer and finer powders. Surfaces so treated contain distributions of two-dimensional micro-cracks which are randomly oriented and, generally, not in very close proximity to each other.

The model presented before can be adapted to describe the dispersion Rayleigh waves propagating over such surfaces[7]. The obvious difference with the previous case is the rather more cumbersome numerical evaluation of the COD. However, though at the cost of numerical accuracy, an approximating scheme can be used for a fast estimation of the COD. In this scheme, which was first adopted by Zhang and Achenbach[8], the surface of the crack is discretized into narrow strips normal to the surface of the sample and terminating at the crack front. The depth of each strip depends on its position within the crack face. The COD within each strip is calculated as if the strip were part of a one-dimensional crack with depth equal to that of the strip. The results so obtained overestimate the correct ones.

The components of the extra-compliance tensor are still formally given by Eq. (4), but, as a consequence of the two-dimensional nature of the cracks, the crack depth, h, is substituted by the crack area, Ω. For a uniform, random distribution of parallel cracks which are oriented normally to the direction of propagation of the surface wave, the only non-zero elements of the extra-compliance tensor are $\Delta S_{11} = \nu\Omega B_1/2$, $\Delta S_{55} = \nu\Omega B_3/2$, and $\Delta S_{66}=\nu\Omega B_2/2$. Figures (4) and (5) illustrate the dependence of the phase velocity versus the crack density and versus crack depth. These plots, which refer to a distribution of semielliptical cracks with an aspect ratio of 3 to 1, closely resemble those obtained in the one-dimensional case. The material was again alumina. It is worth noticing that even for distributions of two-dimensional cracks, velocity variations of the order of a few percentages

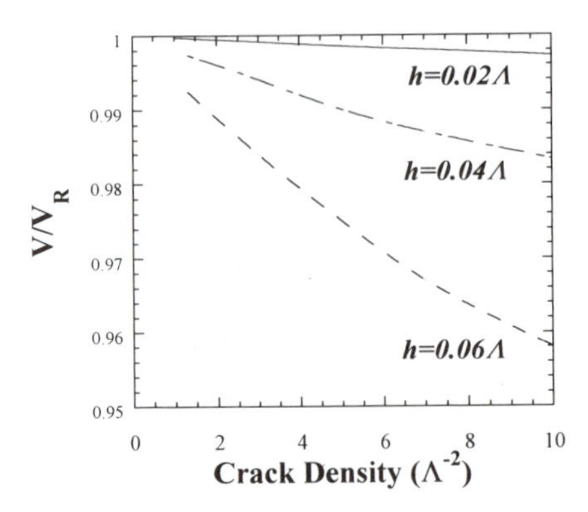

Figure 4. Normalized phase velocity versus normalized crack density: two-dimensional case.

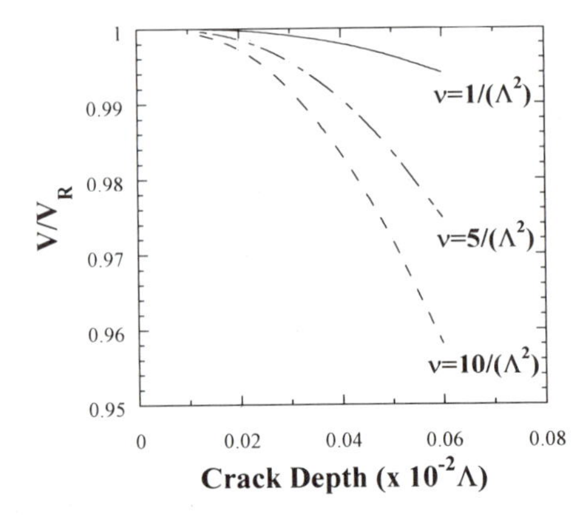

Figure 5. Normalized phase velocity versus normalized crack depth: two-dimensional case.

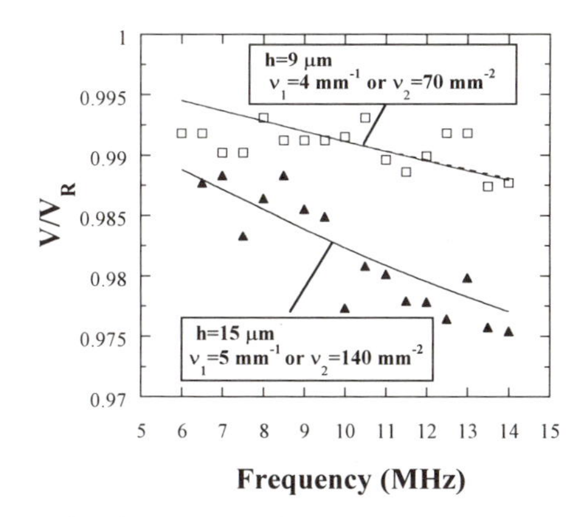

Figure 6. Comparison between experimental and theoretical dispersion curves

are predicted in agreement with the measured values reported by Scherer et al[3] and Tardy's et al[4].

THE MODELS AND EXPERIMENTAL RESULTS

In the previous section (see Eq. (4)), it has been shown that crack density, ν, and crack depth, h, contribute to the dispersion of a surface acoustic wave in different degrees despite the fact that they play the same role in determining the extra-compliance of the near-surface region. It is reasonable to expect, therefore, that such a property of the phase velocity may be exploited to obtain a simultaneous estimate of these two distribution parameters from acoustical data by minimizing, for example, a χ^2-like quantity in the two-dimensional spece (h, ν). However, the determination of one of the two parameters by means of another technique can greatly simplify the estimation of the second parameter from acoustical data, as shown next.

Tardy et al.[4] adopted a broad-band technique to investigate ground surfaces and to obtain the experimental data shown in Fig. (6). Here, the actual data have been normalized with respect to the value of the wave velocity measured on a sample which did not cause any significant dispersion within the range of frequency used in these measurements. They also directly measured the depth of the damage zones of the two samples and found them to be 20 μm and 12 μm, respectively. Assuming that these values represent the maximum depth of the damage, an effective average depth of 15 μm and 9 μm, i.e., about 70% of the maximum depth, can be used in the model.

Simulations were run with both models presented above, i.e., by assuming the actual crack distributions modeled by a distribution of one-dimensional, surface-breaking cracks, or, alternatively, by a distribution of semielliptical, surface-breaking cracks. In the second case, the possible proximity of neighboring cracks was included in the model by allowing for an unusually large aspect ratio of the semielliptical cracks, that is, 3 to 1. Figure (6) shows two sets of almost identical dispersion curves obtained with the two models. The curves relative to the more damaged surface were obtained assigning values of 5 mm^{-1} and 140 mm^{-2}, respectively, to the crack density, while those relative to the second surface were obtained assigning values of 4 mm^{-1} and 70 mm^{-2}, respectively, to the crack density. These simulations show that sets of parameters for both models can be found which yield identical dispersion curves. Thus, independent, possibly direct observation of the geometrical features of the crack distribution must decide between the two models.

CONCLUDING REMARKS

Three heuristic models have been presented which provide quantitative links between micro-crack distribution parameters and Rayleigh phase velocity under various surface conditions. No claim of mathematical rigor is put forward. For a model to be mathematically rigorous within the framework of an effective medium theory, it should include the non-local nature of the crack elastic behavior. The lack of mathematical rigor notwithstanding, the comparison of theoretical predictions with the available experimental results indicates that the models provide the correct dependence of the wave velocity from the ultrasonic frequency. It seems possible, therefore, to develop a numerical procedure which provides a way to estimate crack ditribution parameters simultaneously.

ACKNOWLEDGEMENTS

This work was funded by the EPSRC (Grant No. GR/L/27633). The author wishes to thank Dr. S.G. Roberts and Dr. G.A.D. Briggs for their support.

REFERENCES

1 D. B. Marshall, A. G. Evans, B. T. Khuri-Yakub, J. W. Tien, and G. S. Kino, The nature of machining damage in brittle materials, *Proc. R. Soc. Lond. A*, 385: 461 (1983).

2 B. T. Khuri-Yakub, L. R. Clarke, Effects of surface residual stress on crack behavior and fracture stress in ceramics," in *Review of Progress in QNDE*, 4B:1133, D. O. Thompson and D. E. Chimenti, ed., Plenum, New York, (1984).

3 V. Scherer, S.U. Fassbender, G. Weised, and W. Arnold, Microscopic characterization of machined high-performance ceramics using acoustical, optical and mechanical NDT techniques, *Proc. 99th Ann. Meeting Am. Ceram. Soc.* (1997), to be published.

4 F. Tardy, M.H. Noroy, L. Paradis, and J.C. Baboux, Material characterization by a Rayleigh wave velocity measurement, *Ultrasonic Testing*, (http://www.ultrasonic.de/backup/bup1_e.htm), Nov. (1996).

5 C. Pecorari, Modeling variations of Rayleigh wave velocity due to distributions of one-dimensional surface-breaking cracks," *J. Acoust. Soc. Am.* 100:1542 (1996).

6 C. Pecorari, On the effect of a residual stress field on the dispersion of a Rayleigh wave propagating on a cracked surface, submitted for publication.

7 C. Pecorari, Rayleigh wave dispersion due to a distribution of semielliptical surface-breaking cracks, submitted for publication.

8 C. Zhang and J.D. Achenbach, Dispersion and attenuation of surface waves due to distributed surface-breaking cracks, *J. Acoust. Soc. Am.* 88:1986 (1990).

SIZING OF 3-D SURFACE BREAKING FLAWS FROM THE DISTRIBUTION OF LEAKAGE FIELD

Dorian Minkov,[1] and Tetsuo Shoji[2]

[1] and [2] Research Institute for Fracture Technology, Tohoku University
Sendai 980, Japan
[1] on sabbatical leave from the Department of Electronic Engineering,
University of Natal, Dalbridge 4014 South Africa

INTRODUCTION

The electrical methods for sizing of surface breaking flaws in conductive materials are divided in two groups: DC methods and AC methods. Most of these methods are based on potential drop techniques where electrical current passes inside the sample, and information about the sizes of the flaw is extracted from measurements of the potential drop between two electrodes which contact the surface of the sample.

In the magnetic methods for sizing of surface breaking flaws, external magnetic field is applied to magnetic sample. The magnetic field lines are bent around the flaw which leads to leakage of magnetic flux out of the sample. The magnetic leakage flux is formulated mathematically by the finite elements method or the dipole model. The spatial distribution of the magnetic leakage flux is detected by a magnetometer such as a Hall generator[1] or magnetodiode. In some magnetic methods, the calculation of the sizes of flaws with the simplest parallelepiped shape is based on dependencies of some characteristic points of the distribution of the magnetic leakage flux intensity as a function of the width and the depth of the flaw[2]. Nishio et all[3] have proposed a method in which the influence of the size of the Hall element is eliminated by using a standard flaw with a known cross-section and performing of several Fourier transforms.

In this paper, a new electrical and magnetic method is proposed for sizing of surface flaws with a complex cross-section, by a regression for the distribution of the leakage field intensity. The method allows to distinguish the shape and the size of a flaw amongst different shapes and sizes of flaws which are inherent to the specimen, as well as to size precisely the flaw.

ANALYTICAL APPROACH

In the frame of the magnetic dipole model of a flaw, the magnetic field is distributed homogeneously in the part of the sample which is not disturbed by the flaw, and the repulsion of the magnetic flux lines from the flaw is represented by considering the flaw being filled homogeneously by magnetic dipoles with magnetic moments oriented in the same direction, which is opposite to the direction of the magnetic field. These magnetic dipoles generate magnetic field outside the sample which is equivalent to the leakage magnetic field. Analytical expressions about the components of the intensity of the leakage field $\mathbf{H_x}$ and $\mathbf{H_z}$ for a flaw with the simplest parallelepiped shape at a point with co-ordinates **x, y=0**, and **z** are given in [3] where the subscripts **x** and **z** of **H** indicate the components of the intensity of the leakage magnetic field along the axes **x** and **z** which are correspondingly perpendicular to the long axis of the flaw **y**, and to the surface of the specimen, while both $\mathbf{H_x}$ and $\mathbf{H_z}$ are functions of the parameter $\mathbf{c_m}$ which depends on the sample and the measurement conditions, and is proportional to the density of the magnetic charge at the flaw's wall.

By analogy, the behaviour of a flaw in the electrical models for sizing of flaws can be formulated assuming the flaw to be filled homogeneously by electrical domains with orientation opposite to the electrical field, provided that the electrical field is distributed homogeneously in the part of the sample which is not disturbed by the flaw, which could be true for DC current. In this case, the formulae for the distribution of the time averaged intensity of the leakage electrical field is described by equations similar to the equations about $\mathbf{H_x}$ and $\mathbf{H_z}$ for a flaw with the simplest parallelepiped shape, where $\mathbf{H_x}$ and $\mathbf{H_z}$ are replaced correspondingly by $\mathbf{E_x}$ and $\mathbf{E_z}$, and $\mathbf{c_m}$ by $\mathbf{c_e}$ where the parameter $\mathbf{c_e}$ depends on the experimental conditions and is proportional to the density of the electrical charge at the flaw's wall.

A flaw with an arbitrary shape can be represented as a set of flaws with the simplest parallelepiped shape. Such complex cross-section flaws can arise by surface scratching or around the welding on the surface of seamed tubes, and are correspondingly known as surface scratches and weld flaws. In such cases, the intensity of the leakage field in a given point above the sample is given by a sum of the leakage fields of the corresponding set of flaws with the simplest parallelepiped shape.

Let's assume that the intensity of the leakage field is measured in **N** points, and the value of e.g. the **x** component of the measured intensity of the leakage magnetic field in the point numbered **i** is $\mathbf{H_x^m(i)}$. The theoretical value of the **x** component of the intensity of the leakage magnetic field in the same point- $\mathbf{H_x^t(i)}$ can be determined using the equation for $\mathbf{H_x}$ for a particular flaw. The sizing of a flaw could be considered as a regression problem for minimisation of the root mean square error **RMS** between the measured distribution and the theoretical distribution of the intensity of leakage field, i.e. for computer predicted change of the shape and the sizes of the flaw used to generate the theoretical distribution, which leads to decreasing of the **RMS** error at every step of the minimisation. The minimisation procedure could end when the **RMS** error reaches some predetermined small value which would guarantee that the shape and the sizes of the flaw used at the last step of the minimisation procedure are almost identical to the shape and the sizes of the flaw for which the measurements are performed.

COMPUTED RESULTS

Computations are performed for four different flaws with the same length of **2l**=14 mm, and cross-sections which are invariable along the long axis of the flaw. The cross-sections of these flaws in the **xz** plane are shown in Fig.1, while Fig.1b, Fig.1c, and Fig.1d correspond to surface scratches or weld flaws, and Fig.1a corresponds to a crack flaw which satisfies the condition $\mathbf{d_1>10*2a_1}$. Case symbols are introduced for simplicity, with the numbers 1, 2, 3, and 4 being used for the flaws with cross-sections shown respectively in Fig.1a, Fig.1b, Fig.1c, and Fig.1d, while the letters **x** and **z** indicate the component of the leakage field measured, e.g. the case symbol **2z** shows that the flaw considered has the cross-section shown in Fig.1b, and that the measurements are performed for the **z** component of the leakage field.

It is assumed that:

- The length of the flaw is measured independently, i.e. the value of **l** is known. This assumption aims mainly at decreasing of the minimisation procedure time in comparison with the case when **l** is unknown.
- The flaws with complex cross-sections illustrated in Fig.1b, Fig.1c, and Fig.1d are represented by a set of five flaws with the simplest parallelepiped cross-section and the same width on each side of the inflection point of a complex flaw.
- The parameter in the equations for $\mathbf{H_x}$ and $\mathbf{H_z}$ has a value $\mathbf{c=c_m}=10^5$ A/m in all measurements performed which gives reasonable values for the components of the intensity of the leakage

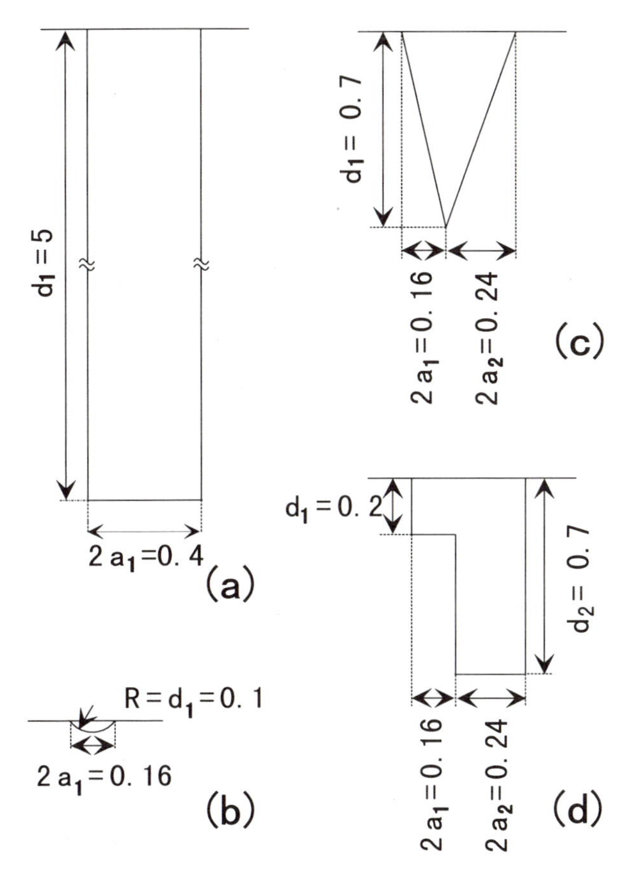

Figure 1. Cross-section in the **xz** plane of the flaws used in the computations. The dimensions are given in millimetres.

Table 1. RMS error after 100 minimisation steps for the regression between the measured distribution for the flaw from Fig.1c and the theoretical distributions for flaws with shapes shown in Fig.1 and known parameters. Measurements are performed for the **x** or the **z** component of the intensity of the leakage field for **y**=0 and **z**=10 mm, and the known flaw parameters are used as an initial approximation in the computation.

CASE	INITIAL APPROXIMATION					RMS ERROR AFTER 100 STEPS
	a_{1i} (mm)	a_{2i} (mm)	d_{1i} (mm)	d_{2i} (mm)	c_i*10^5 (A/m)	
1x	0.2		5		1	1.200
2x	0.08		0.1		1	4.980
3x	0.08	0.12	0.7		1	$1.9*10^{-6}$
4x	0.08	0.12	0.2	0.7	1	0.330
1z	0.2		5		1	17.292
2z	0.08		0.1		1	124.499
3z	0.08	0.12	0.7		1	$1.7*10^{-4}$
4z	0.08	0.12	0.2	0.7	1	2.102

magnetic field, while the measurements about the distribution of the intensity of the leakage field are simulated by replacing this value of **c** in the above mentioned equations.

- Measurements of the intensity of the leakage field are performed in **N** points positioned equidistantly along the **x** axis for **y**=0 at distances **z**=1 mm, and **z**=10 mm above the surface of the sample, and centred around the long axis of the flaw. Consequently, the minimisation of the **RMS** error between this measured distribution and the corresponding theoretical distribution would result in sizing of the cross-section of the flaw which is perpendicular to the long axis of the flaw.
- **c** is assumed to be unknown when theoretical data are generated, considering that **c** depends on the measurement conditions.

According to the assumptions made, the theoretical distribution of the leakage magnetic field depends on three flaw parameters for the flaws with cross-sections shown in Fig.1a, and Fig.1b, four parameters for the flaw from Fig.1c, and five parameters for the flaw from Fig.1d.

The software developed is written in **MATLAB**. The minimisation of the **RMS** error in respect to the above discussed flaw parameters which are unknown is performed using the built in MATLAB subprogram **fmins.m**. The minimisation procedure starts with introduction of some initial values for these parameters. At each step of the minimisation, the subprogram generates new flaw parameters for which the **RMS** error is smaller than its value at the previous step. In most cases, the minimisation procedure ends when the new flaw parameters which are generated by the subprogram are indiscernible from the flaw parameters generated at the previous several minimisation steps.

The aim of the following computations is to establish if the method proposed can distinguish the shape of a particular flaw. Correspondingly, it is assumed that a sample contains only flaws with the four shapes shown in Fig.1, and that the sizes of the flaws with these shapes are known. The

Table 2. RMS error after 100 minimisation steps for the regression between the measured distribution for the flaw from Fig.1c and the theoretical distributions for flaws with shape shown in Fig.1c and different known parameters. Measurement are performed for the **x** or the **z** component of the intensity of the leakage field for **y**=0 and **z**=10 mm, and the known flaw parameters are used as an initial approximation in the computation.

CASE	INITIAL APPROXIMATION				RMS ERROR AFTER 100 STEPS
	a_{1i} * 0.08 (mm)	a_{2i} * 0.12 (mm)	d_{1i} * 0.7 (mm)	c_i * 10^5 (A/m)	
3x	1.5	1	1	1	0.030
3x	1/1.5	1	1	1	0.012
3x	1	1.5	1	1	0.004
3x	1	1/1.5	1	1	0.003
3x	1	1	1.5	1	0.024
3x	1	1	1/1.5	1	0.007
3x	1	1	1	1.5	0.032
3x	1	1	1	1/1.5	0.017
3z	1.5	1	1	1	0.383
3z	1/1.5	1	1	1	0.221
3z	1	1.5	1	1	0.555
3z	1	1/1.5	1	1	0.416
3z	1	1	1.5	1	0.204
3z	1	1	1/1.5	1	0.374
3z	1	1	1	1.5	0.595
3z	1	1	1	1/1.5	0.397

studied flaw is considered to be the one shown in Fig.1c, and the measured distribution of the intensity of the leakage field along the **x** axis for **y**=0 and **z**=10 mm is simulated for field components **x** and **z**. Minimisation of the RMS error is performed between the theoretical distribution of the intensity of the leakage field for each of the flaws from Fig.1 and the above mentioned measured distribution. The initial approximation for generation of the theoretical distribution for each of the flaws from Fig.1 is the set of the corresponding known flaw parameters, which is denoted by **TSP**, keeping in mind that **TSP** depends on the flaw investigated. The value of the **RMS** error after only 100 minimisation steps is shown in Table 1 for the four flaws considered, and for the **x** and the **z** component of the leakage field. It is seen from Table 1 that the smallest **RMS** error is achieved when the measured distribution is compared with the theoretical distribution for the flaw from Fig.1c, for a fast minimisation procedure containing only 100 steps.

It has also to be verified whether the method can distinguish the sizes of a particular flaw. Correspondingly, it is considered that the sample contains only flaws with a given shape, but with known different sizes. The shape studied is the one from Fig.1c, the sizes of the particular flaw are the ones shown in Fig.1c, the measured distribution of the leakage field is simulated as described in the previous paragraph, the initial approximation for the theoretical distribution is the set **TSP** of the corresponding known flaw parameters which are different from the ones shown in Fig.1c, and the **RMS** error is shown in Table 2 for 100 minimisation steps for **y**=0 and **z**=10 mm and the **x** and the **z** component of the leakage field. The results from Table 2 show that the smallest **RMS** error is achieved when the measured distribution is compared with the theoretical distribution for the flaw with sizes given in Fig.1c.

The next point to be clarified is if the method can give the parameters of the cross-section of the flaw, and the parameter **c** which determines the density of the charge at the flaw walls provided that the shape of the flaw is known. All four shapes of flaws shown in Fig.1 are studied, and computations are performed for different measurement distance from the surface, i.e. for **z**=1 mm,

Table 3. Number of minimisation steps as an integer number of 500 for the regression between the measured distribution for all the flaws shown in Fig,1 and the corresponding theoretical distribution. Measurements are performed for the **x** or the **z** component of the intensity of the leakage field for **y**=0 and **z**=1 mm or **z**=10 mm, and the sets 3*__TSP__ and **TSP**/3 are used as initial approximations in the computations.

CASE	z (mm)	INITIAL APPROXI-MATION *1 (TSP)	NUMBER OF MINI-MISATION STEPS	CASE	z (mm)	INITIAL APPROXI-MATION *1 (TSP)	NUMBER OF MINI-MISATION STEPS
1x	1	3	2000	2z	1	3	2500
1x	1	1/3	500	2z	1	1/3	2000
1x	10	3	2500	2z	10	3	7000
1x	10	1/3	2000	2z	10	1/3	5000
2x	1	3	1500	3z	1	3	3000
2x	1	1/3	1000	3z	1	1/3	2000
2x	10	3	2500	3z	10	3	3500
2x	10	1/3	2000	3z	10	1/3	2000
3x	1	3	4000	4z	1	3	1500
3x	1	1/3	2500	4z	1	1/3	1000
3x	10	3	28000	4z	10	3	12000
3x	10	1/3	17000	4z	10	1/3	10500
4x	1	3	1500	5z	1	3	3000
4x	1	1/3	1000	5z	1	1/3	1000
4x	10	3	15000	5z	10	3	12000
4x	10	1/3	8500	5z	10	1/3	8000

and **z**=10 mm. Considering that the duration of the minimisation procedure depends strongly on the initial approximation, the computations are performed for initial approximations 3*__TSP__ and **TSP**/3 which differ significantly from **TSP.** The minimisation procedure ends when the new flaw parameters which are generated by the subprogram at a given step are indiscernible from the corresponding parameters within the last 50 minimisation steps. The number of minimisation steps which have been performed, as an integer number of 500, is shown in Table 3 for the different cases investigated. In all these cases, the computed results for the cross-section parameters of the flaw, and the parameter **c** are within 0.01% from the values of these parameters which have been used for simulation of the measurement distributions. It is seen from Table 3 that the minimisation process contains less steps for the smaller measurement distance of **z**=1 mm compared with **z**=10 mm in all cases considered. Also, less minimisation steps are required in all cases for an initial approximation **TSP**/3 in comparison with the corresponding initial approximation 3*__TSP__, but there is no distinct dependence of the number of minimisation steps as a function of the component **x** or **z** of the measured field used.

DISCUSSION

A method is proposed for sizing of 3-D surface breaking flaws. The method requires measurement of the distribution of the intensity of the leakage field at a constant distance above the surface of the sample in direction perpendicular to the long axis of the flaw.

The results from Table 1 show that the method can distinguish fast the shape of a particular flaw when the sample contains several flaws with different shapes, and the sizes of these flaws are known, in both cases when the **x** component and the **z** component of the leakage field are utilised.

It is seen from Table 2 that the method can also distinguish fast the sizes of a particular flaw when the sample contains several flaws with the same shape, but with different sizes, provided that these sizes are known, for either the **x** or the **z** component of the leakage field.

The sizes of the cross-section of the flaw in the plane **y**=0 and the density of the charge at the domain walls could be computed with a relative error of 0.01% according to the data from Table 3, when the shape of the flaw is known, independent of the initial approximation and the component of the leakage field used, when the measurement distance does not exceed 10 mm. The number of minimisation steps decreases, and correspondingly the speed of the minimisation increases, when the measurement distance decreases, and when the initial approximation is smaller than **TSP** rather than larger than **TSP**.

Further computations show that the sizing of flaws could be performed even when the flaw length **2l** is unknown which results in longer minimisation procedure compared with the case when the flaw length is measured independently. This means that a complete flaw sizing would be possible by using the proposed method for flaws with arbitrary shapes and the sizes. It turns out though that the accuracy of the flaw sizing decreases significantly with the increment of the measurement distance above 10 mm.

REFERENCES

1. W. Lord, W. and L. Srinivasan. *Review of Progress in Quantitative Nondestructive Evaluation*, Plenum Press, New York (1984)
2. M. Katoh, S. Mukae, K. Nishio, and A. Masumoto, Simulation for estimating breadth and depth of defect in magnetic leakage flux testing method, *Nondest Insp,* 33:59 (1984)
3. S. Mukae, M. Katoh, and K. Nishio, Investigation on quantification of defect and effect of factors affecting leakage flux density in magnetic leakage flux testing method, *Nondest Insp*, 37:885 (1988)

CHARACTERIZING THE SURFACE CRACK SIZE BY MAGNETIC FLUX LEAKAGE TESTING

Li Luming Zhang Jiajun

NDT Lab. Mechanical Engineering Dep.
Tsinghua University Beijing 100084 P.R.China

Abstract: The work of this paper is mainly on evaluation the surface crack of different depth and width just like natural defects. First, the effect of slot width to the leakage flux is carefully studied using finite element method(FEM.). Then the "Width model" about the influence of slot width is developed. Base on the research of the influence of width, a new quantitative testing method is advanced in order to size the surface cracks with different width and depth. The core of the new method is the consideration of the influence of width. The results of FEM. and the experiment show that: the characterizing result is very good after the correction of Byp-p (y component peak to peak value)by slot's width.

key words: magnetic flux leakage testing (MFL), depth evaluation, finite element method

INTRODUCTION

The research of MFL shows that the y component of leakage field peak to peak value Byp-p(or Hyp-p, Byp-p=Hyp-p* μ_o) has a linear relationship with the depth of cracks at same width, which becomes the fundament of crack depth evaluation. However, in actual situation, the crack width is usually not the same. And there are different opinions about the influence of crack width. With the enlargement of width, some believe that Byp-p turns to be small, some believe it to be large while others have other different opinions[1,2,3,4]. But they all had not interpreted the reason for such effect. Even more got the width influence into the depth characterization of crack. Therefore it is important and also useful to study the characterization of cracks with different width and depth. The work of this paper is focused on this respect, using slot to simulate crack in characterization.

THE EFFECT OF WIDTH

For specific cracks with same depth, it is assumed that the width ranges from infinite minor to infinite large. It's obviously that the crack should gradually disappear when the width changes to minor, and the leakage field of the defect should disappear too. On the other hand, if the width increases, finally to an infinite large one, the characteristic of the crack disappears as well as any actual leakage flux. Considering these two trends simultaneously, we can assume that during the change of the width, the Byp-p value at first increases, then after it reaches the peak point at a specific width, the value decreases along with the enlargement of width.

Results of FEM.

FEM. is the most effective method in the research of MFL, especially in studying the effect of width. First, it can be affected neither by the manufacturing capability and precision of manual defects nor by the randomness of natural cracks. Second, it can simulate almost freewill minute wide crack. And can get a satisfied result of the total leakage field by carefully adjusting the meshes, subdividing the meshes of the crack and its surrounding.

Fig.1 and 2 are the Byp-p values of 2D slots of different width at different lift-off(L) using FEM.. The magnetizing strength is 9000A/m, slots depth is 0.2mm. Material is 45# steel, anneal state.

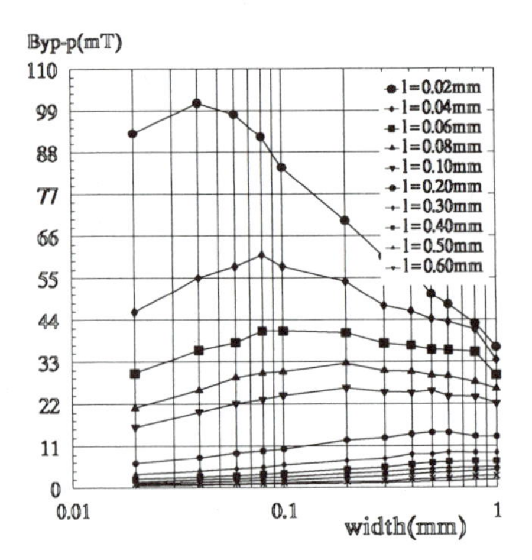

fig.1 Byp-p- widths of different lift-off

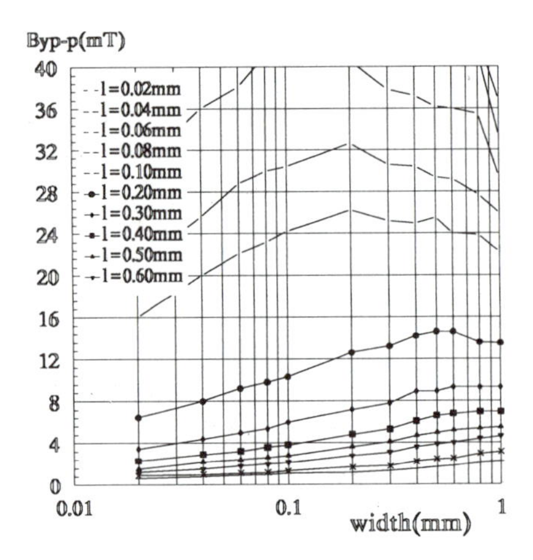

fig.2 the enlargement of fig.1

From the figures above we can see that :

(1) Byp-p values did not simply increase or decrease along with the changing of width, but increased at first and then decreased. It's just as same as the analysis of 2.1 above.

(2) The characteristic of increasing first and decreasing later is not total same at different lift-off. The prominent part is the turning points. In other words, the width of the max. Byp-p at a specific lift-off is firmly related with the lift-off.

The other very interesting phenomena of figure 1 and 2 is that the leakage field gets its maximal Byp-p value when the slot width is two times of the lift-off. Figure 3 shows the relation of width and lift-off at the Byp-p peak points. The line is the regressive result.

assume a factor of a ratio of slot width to width of turning point at a specific lift-off, to describe the effect of width relate to the magnetic leakage field Byp-p for the slots of a depth. Based on the analysis above, for the same deep slots, the width effect can be pointed out as follows :

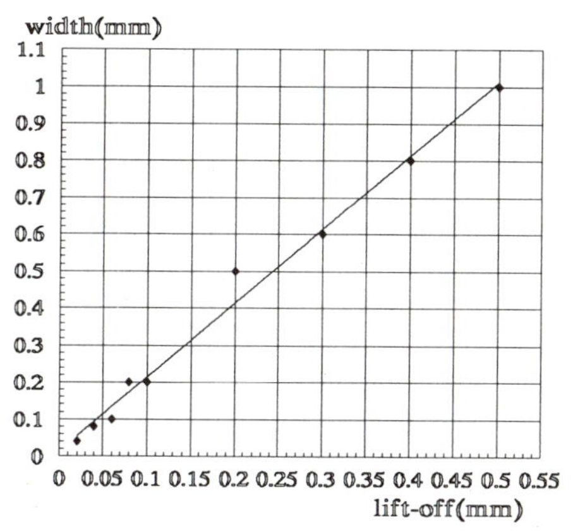

fig.3 positions of max. value with the lift-off

(1) Width will influence the distribution of the leakage field of slot.

(2) For the slots of one depth, if the lift-off is L, the leakage field will reach its Max. Byp-p when the width is 2L. The Max. Byp-p value can be expressed as Byp-pmax.

(3) the width effect can be expressed by the ratio k of sensor's lift-off to the slot width. The Byp-p of one slot, which width is w, can be got from the Byp-pmax at the same lift-off:

$$\text{Byp-p}=\text{Byp-pmax}*k \qquad (2)$$

when w<2L, $$k=\left(\frac{w}{2L}\right)^{\alpha} \qquad (3)$$

when w>=2L, $$k=\left(\frac{2L}{w}\right)^{\beta} \qquad (4)$$

(4) the values of α, β is related with the lift-off. The values can be got from the following table based on great amount of calculation.

Table 1 α, β value

lift-off(mm)	α	β
<0.05	0.1	0.2
0.05≤L<0.1	0.15	0.1
0.1≤L<0.15	0.2	0.1
0.15≤L<0.2	0.25	0.1
0.2≤L<0.3	0.3	0.1
0.3≤L<0.4	0.35	0.1
0.4≤L<0.6	0.4	0.1
0.6≤L<1.0	0.5	0.1

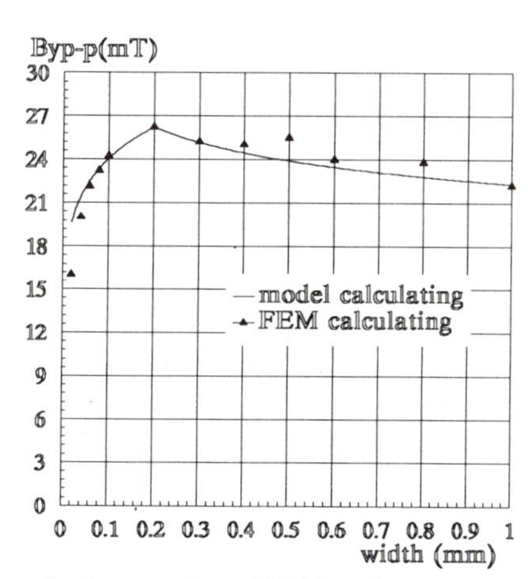

fig.4 comparison of FEM. and model results

the Width Model

Figure 4 is the comparison of FEM. and model calculating results at lift-off of 0.1mm. It shows they match perfectly. Moreover, the slots' calculating data of different depths, magnetic strength and practical testing also show almost the same result.

Why the width influences the leakage field in this way? The results above can be interrupted at a point of a view of the width effect to the magnetic force difference, magnetic resistance and distribution of flux. The detailed discussion will be given in other papers.

NEW DEPTH CHARACTERIZING RULE OF MFL

It's obviously that the real depth of a crack should not be characterized only by the Byp-p but without consideration of width effect. For example, a deep and narrow crack, dangerous to component, may be characterized as a little, safe one, running in the opposite direction of actual condition. Undoubtedly, the effect of width must be considered if the crack depth be evaluated actually. In other words, a new depth characterizing rule should be developed.

The new characterizing rule is:

(1) the surface slot can be characterized into two fundamental parameter: width and depth.

(2) the depth evaluation must based on the consideration of width. Only after the Byp-p modified by width, it can be used to characterize the slot depth.

Characterization of Width

The data of FEM. and experiential show that the y component of leakage field includes the information of slot width. The x direction distance of the two peaks of y component has a linear relation with crack width. This fixes with the work of Lord[2]. The definition of x direction distance is indicated in figure 4.

The relation of width, lift-off and Syp-p can be expressed as following formula:

$$\text{Syp-p} = L + w \qquad (5)$$

Figure 5 is the comparison of results by testing and formula (5).

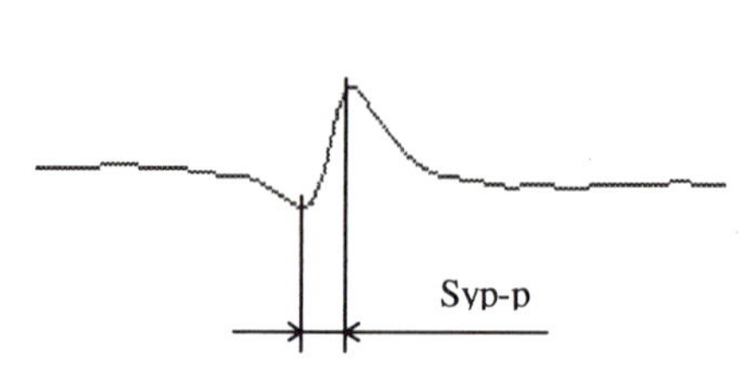

Fig.4 definition of Syp-p

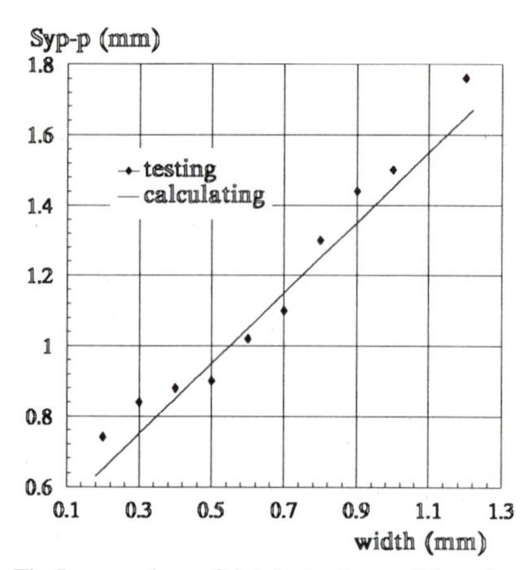

Fig.5 comparison of data by testing and formula

Modified Depth Characterization

How to modify the slot with different width is becoming the key point in the new depth characterizing method. The width model show the way. First, to modify the slots of different width into slot of a width. Second, to quantitative the depth of slots. But how to choose the width to be modified? It's known that the modifying factors are different at the two side of the peak point for a specific lift-off, L. Obviously, it's most appropriate to modify the Byp-p with width of peak point, 2L. So, the new method can be expressed in the following way after consideration of width effect.

(1) The Byp-p of slots with same width has a linear relation with depth at a specific magnetizing strength and lift-off. It can be expressed as:

$$\text{Byp-p} = c + b * d \tag{6}$$

which: c is count, changed along with lift-off, magnetizing strength, width etc.
b is linear factor, changed along with lift-off, magnetizing strength, width etc.
d is the depth of slot

(in practical testing, the linear formula 6 can always be got from manual defects of different depth but same width)

(2) to modify the linear formula with the width of two times of lift-off.

Suppose that the lift-off of sensor is L, the formula can be modified to one width of 2L. The method is:

Suppose that the width of manual defects is w:

If $w \leq 2L$, the formula is turned into: $$\text{Byp-p} = (c + b * d) / \left(\frac{w}{2L}\right)^{\alpha} \tag{7}$$

if $w > 2L$, the new formula is: $$\text{Byp-p} = (c + b * d) / \left(\frac{2L}{w}\right)^{\beta} \tag{8}$$

the values of $\alpha$and β can be chosen by table 1.

the formula 7 and 8 is the modified linear function.

(3) to modify the Byp-p of testing slot with width of 2L

suppose the width of testing slot is w_1(got by formula 5), and the y component peak to peak value is Byp-p_1,

if $w_1 \leq 2L$, Byp-p_1 modified to: $$\text{Byp-p}' = \text{Byp-p}_1 / \left(\frac{w_1}{2L}\right)^{\alpha} \tag{9}$$

if $w_1 > 2L$, Byp-p_1 modified to: $$\text{Byp-p}' = \text{Byp-p}_1 / \left(\frac{2L}{w_1}\right)^{\beta} \tag{10}$$

(4) to use Byp-p' and formula 7 or 8 to get the depth d of slot.

RESULTS OF NEW DEPTH CHARACTERIZING RULE

Result of FEM.

Table 2 is the FEM. result. The linear formula is got by the slots of 0.1mm width. The characterizing results only simply by the Byp-p and modified Byp-p are also compared in this table. The width of the slot being characterizing is 0.04mm. In this table, depth 1 is the result without modified and depth 2 is the modified one. Error 1 is the relative error of characterizing depth without modified and error 2 is the modified one.

Table 2 the characterizing result of 0.04mm wide, 0.2mm deep slot

lift-off(mm)	0.10	0.20	0.30	0.40	0.50	0.60
practical depth(mm)	0.20	0.20	0.20	0.20	0.20	0.20
depth 1(mm)	0.167	0.159	0.157	0.163	0.166	0.153
depth 2(mm)	0.206	0.208	0.198	0.209	0.218	0.199
error1(%)	16.5	20.5	21.5	18.5	17.0	23.5
error2(%)	-3.0	-4.0	1.0	-4.5	-9.0	0.5

It's clearly that the characterizing error with modified Byp-p is decreased greatly from the result of table 2.

Result of Experiment

Table 3 is characterizing result of manual slots. The characterizing linear formula is got from the wide slots of 0.13mm. The results of characterizing only simply by the Byp-p and modified Byp-p is also compared in this table. The definition of depths and errors are just as same as table 2.

Table 3 the characterizing result of manual slot

No. of manual slot	1	2	3	4	average error
practical depth(mm)	0.165	0.194	0.129	0.462	
width(mm)	0.247	0.654	0.181	0.401	
depth 1(mm)	0.208	0.366	0.138	0.774	
depth 2(mm)	0.154	0.178	0.118	0.483	
error 1(%)	26.06	88.63	6.78	67.53	47.25
error 2(%)	6.67	8.25	-8.01	4.55	6.87

The table 3 also show a good result for the manual slots of different width and depth. The average error decreases from 47.25% to 6.87%.

5. CONCLUSION

1 The width model developed in this paper reveals the effect of width and makes the fundament of characterization of slot of different width and depth.

2 The new depth characterization method based on consideration of width effect has solved the problem of slots' depth evaluation. And the new rule can also be used to the practical surface cracks evaluation. The work of this aspect will be given in other paper.

REFERENCES:

1. W.Lord, J.M.Bridge, W.Yen, Residual and active leakage field around defects in ferromagnetic materials, Materials Evaluation. July:47 (1978)
2. W.Lord and J.H.Hwang, Defect characterization from magnetic leakage Fields, British Journal of NDT. January :14 (1977).
3. F.Foerster , New findings in the filed of non-destructive magnetic leakage field inspection , NDT international. 19, No.1:3 (1986)
4. G.Dobmann, G.Walle and P.Hosller, magnetic leakage flux testing with probe: physical principle and restrictions for application. NDT International Vol 20 No.2:101 (1987)

TIME DOMAIN WAVEFORMS OF A LINE-FOCUS TRANSDUCER PROBING ANISOTROPIC SOLIDS

N. N. Hsu, D. Xiang*, and G. V. Blessing

Automated Production Technology Division
Manufacturing Engineering Laboratory
National Institute of Standards and Technology
Gaithersburg, MD 20899

INTRODUCTION

Ultrasonic wave velocity measurements can be used to evaluate material properties such as elasticity, texture (crystalline structure), surface roughness, coating thickness, porosity, and residual stress. These properties may vary as a function of material sample position and orientation. Scanning acoustic microscope systems equipped with line-focus transducers have been successfully used to measure many of these inhomogeneous and anisotropic properties. These systems use a technique that relies on the measurement of the reflected high frequency (~200 MHz) tone-burst echo amplitude, V, as a function of the defocus distance, z. Analysis of the interference minima in the $V(z)$ curve yields the surface wave velocity in a direction perpendicular to the focal line, with a resolution of tens of micrometers in the propagation direction. Both theory and instrumentation are well documented for this technique [1-3].

Recently we developed a low cost, large aperture, lensless line-focus ultrasonic transducer [4].The transducer can be used in a conventional ultrasonic scanning system, and is capable of performing most of the functions of the acoustic microscope mentioned above. Moreover, the wide bandwidth of the transducer allows it to be operated either in a short-pulse broadband mode or in a fixed-frequency narrowband mode, facilitating both waveform interpretation and system calibration. The center frequency of this transducer is about 10 MHz, and the spatial resolution is on the order of a millimeter in the propagation direction. For many crystals and composite materials, this lower spatial resolution is advantageous by providing the macroscopic material properties useful for engineering applications.

Since the focused transducer is lensless, its pupil function is easily defined for computational purposes. This allows for a relatively simple predictive analysis, and a straight forward interpretation of the measurement results. Examples of leaky surface wave

* NIST Guest Worker, The Johns Hopkins University, Baltimore, MD 21218

Nondestructive Characterization of Material VIII
Edited by Robert E. Green Jr., Plenum Press, New York, 1998

measurements using this transducer on a variety of isotropic materials are presented in a companion paper [5]. Here we present both theoretical predictions and experimental results for measurements on anisotropic materials such as single-crystal quartz. The same computational and experimental procedures can be used to obtain results on composite materials modeled as anisotropic layered structures.

THEORY

Time domain waveforms of a line-focus transducer probing a liquid/solid interface can be computed by integration of the time-domain Green's function. The integration is carried out twice: once over the curved transducer as the pressure source, and once as the detector. We have previously developed a computer program to simulate the time domain echo waveform of a line-focus transducer probing an isotropic solid through a liquid [6].

Cylindrical transducers are highly symmetric. The transducer-sample system has a simple frequency response that may be derived from angular spectra. This representation can serve as the theoretical basis for predicting the $V(z)$ curve for isotropic and anisotropic samples [7-10]. By taking into account the implicit frequency dependence of the $V(z)$ curve, the transfer function for a line-focus transducer coupled to a sample through a liquid can be derived. The time domain waveform can then be obtained by convolution of this frequency response function with the source waveform from the transducer [11]. The computed transducer output voltage, $v(t,\theta,z)$, may be represented by a convolution expression:

$$v(t,\theta,z) = s(t) \otimes h(t,\theta,z) \tag{1}$$

where $s(t)$ represents the source waveform determined by the characteristics of the transducer and the electronic pulser/receiver instrumentation, and $h(t,\theta,z)$ represents the transfer function of the transducer and sample. If we define the functions V, S, and H as the respective Fourier transforms of v, s, and h, then $V=SH$. Note that while the time functions v, s, and h are real, their Fourier transforms V, S, and H are complex.

Time Domain Waveform

The time domain waveform of a line-focus transducer probing an anisotropic sample is simply a convolution of Equation (1), which can first be carried out in the frequency domain, and then transformed to the time domain, i.e.

$$v(t,\theta,z) = \mathcal{FFT}^{-1}\{S(f)H(f,\theta,z)\} \tag{2}$$

where $\mathcal{FFT}^{-1}$ is the inverse fast Fourier transform.

Therefore, the transducer output waveform $v(t,\theta,z)$ can be synthesized in the frequency domain by a simple multiplication (frequency by frequency) of the source and response functions, followed by an inverse fast Fourier transformation. The synthesis requires a known band-limited source waveform $s(t)$, and a way to compute the frequency response function $h(t,\theta,z)$.

Transfer Function of a Line-Focus Transducer Probing an Anisotropic Solid

Shown in Fig. 1 is the test configuration of the transducer and sample in the designated coordinate system, where l is the focal distance. The frequency response function, H, of the transducer-sample system is simply

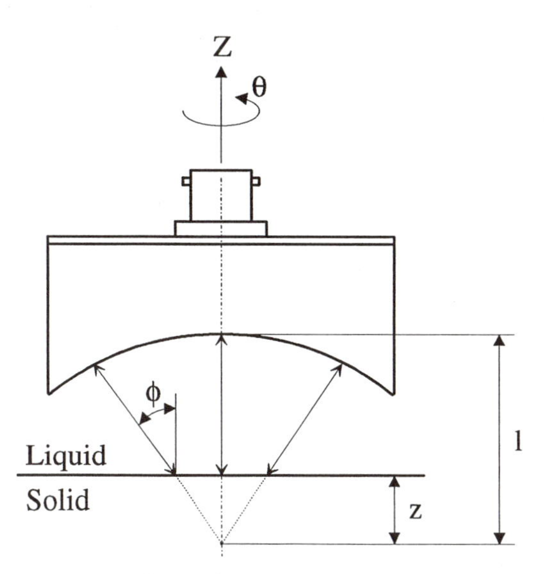

Fig. 1 Test configuration of a line-focus transducer probing a liquid-solid interface.

$$H(f,\theta,z) = \int_{-\alpha}^{\alpha} R(f,\theta,\phi) e^{i(4\pi f/c)z\cos\phi} f \cos\phi \mathrm{d}\phi \tag{3}$$

where f, θ, and z, are the frequency, rotation angle, and defocus distance respectively, c is the wave speed in the liquid, α is the half aspect angle of the line-focus transducer, and R is the reflectance function. Equation (3) is a modified version of *V(z)* as derived by many authors [7-10]. The frequency dependence is explicitly shown, and the pupil function is omitted since it is a constant over the integration variable ϕ.

The reflectance R must be known in order to perform a numerical integration of the right side of Equation (3). For materials modeled as isotropic, a half-space, a plate, or a layered half-space, formulas to compute R have been give by Brekhovskikh [12]. For anisotropic materials such as crystals, formulas to compute R have been derived in many of the references sited above for the *V(z)* computation. Additional derivations can be found in references [13-15]. In our work, we use the algorithm derived by Tsukahara [15] because it is general and simple.

Source Waveform

The source waveform can be determined experimentally by obtaining the echo amplitude from a thick sample of an isotropic material at the transducer focus ($z = 0$). The reflectance R for a thick isotropic material is independent of f and θ. From Equations (2) and (3), this output voltage is then

$$v(t,0) = \mathcal{F}^{-1}\{S(f) f \int_{-\alpha}^{\alpha} R(\phi)\cos\phi d\phi\} \tag{4}$$

Transforming to the frequency domain, we obtain

$$S(f) = a\mathcal{F}\{v(t,0)\} / f \tag{5}$$

where a is a constant which can be computed. The function *S(f)* is a function of the transducer and its pulser/receiver, and in general is independent of the sample being tested.

NUMERICAL COMPUTATIONS

The numerical computation of the time domain waveform is rather simple once the source waveform is determined. For a specific sample and test configuration, the reflectance function R is computed first. We employ a procedure developed by Tsukahara [15] for the computation of R for a layered solid structure composed in general of anisotropic materials. The frequency response function H is then computed by numerical integration of Equation (3). Then a convolution of the source and transfer functions in the frequency domain, followed by an inverse FFT, yields the time waveform $v(t)$.

As an example of the application of this procedure, we present results on a quartz crystal, an anisotropic material. Quartz crystals of various cuts are readily available, and their elastic constants are well known. Shown in Fig. 2 are the real and imaginary parts of the reflection coefficient for Z-cut quartz. The incident and reflected rays are in a plane containing the z-axis and the [100] direction. Equation (3) is used to determine H for a line-focus transducer with a half-aperture angle of 32.5 degrees, operating at a defocus distance of 11.5 mm. The integration of the reflection coefficient is shown in Fig. 3 for the magnitude of H. Finally, the simulated time waveform, $v(t)$, of the transducer output is shown in Fig. 4. The three wave pulses in the figure are the direct reflection, the pseudo-Rayleigh wave and the Rayleigh wave respectively.

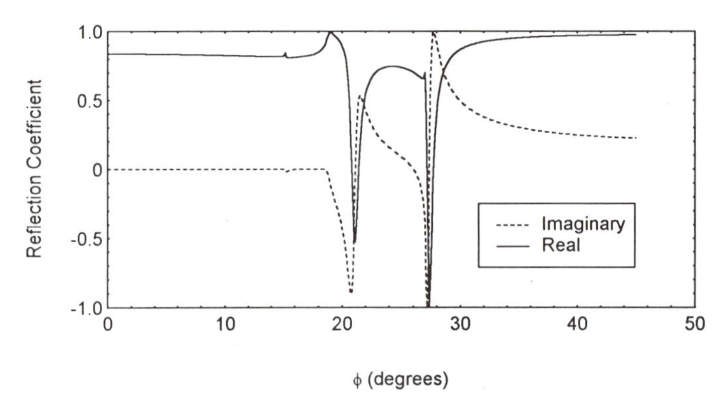

Fig. 2 Reflection coefficient R of a Z-cut quartz crystal submerged in water as a function of incident angle ϕ. Surface wave propagation in [100] direction.

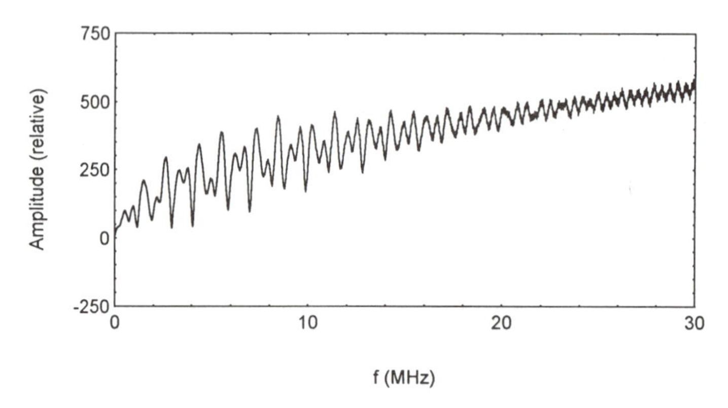

Fig. 3 Amplitude of transfer function $H(f)$ for Z-cut quartz, computed using Eq. (2) and the reflection coefficient shown in Fig. 2.

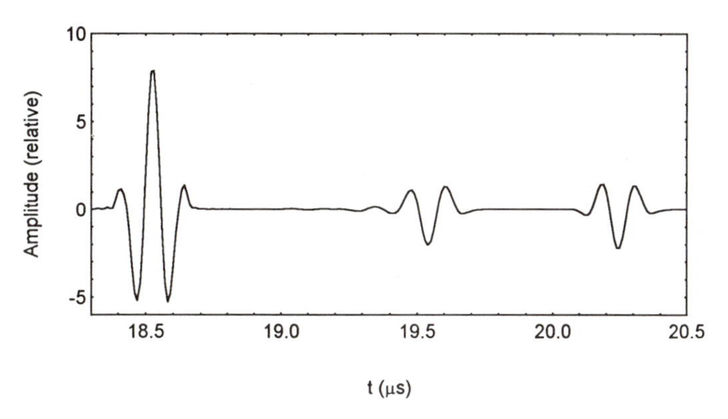

Fig. 4 Simulated echo waveform $v(t)$ for wave propagation in [100] direction in Z-cut quartz.

Simulated waveforms like that of Fig. 4 may be compared with experiment. A sequence of experimental waveforms of voltage vs time for anisotropic Z-cut quartz, obtained by rotating the transducer in discrete steps about the sample-surface normal, is presented in the waterfall plot of Fig. 5. (The experimental arrangement is described elsewhere [5].) A sequence of simulated waveforms are similarly presented in Fig.6. The agreement between experiment and theory is remarkable.

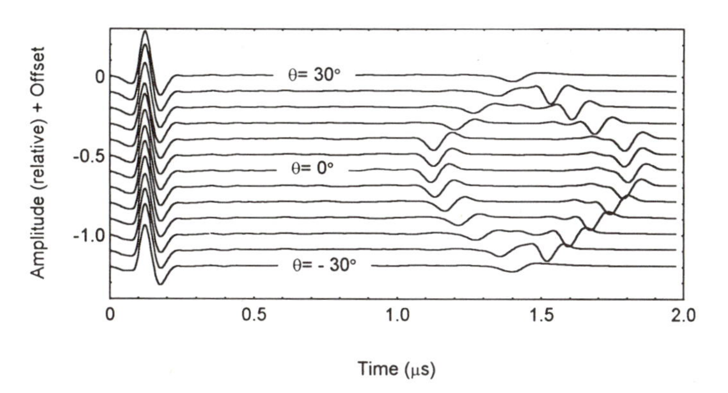

Fig. 5 Sequence of experimental waveforms as a function of rotation angle θ in Z-cut quartz, beginning at 30 degrees from the [100] crystalline direction.

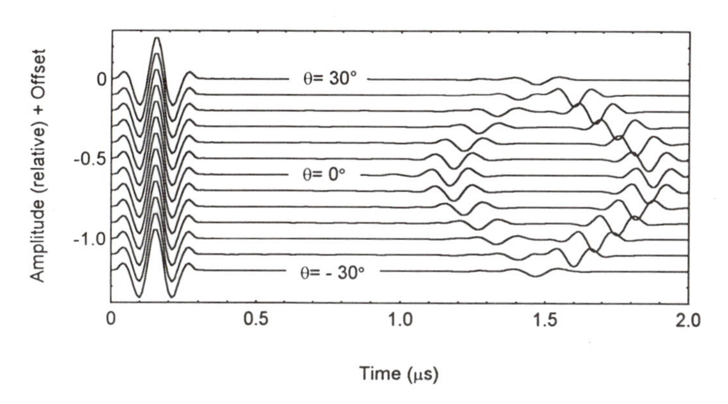

Fig. 6 Sequence of simulated waveforms, with same conditions as for Fig. 5.

CONCLUSIONS

The line-focus transducer which we have previously developed is capable of rotational scans for measuring orientation-dependent leaky surface waves. Because the transducer design is lensless, the theory for predicting the time waveforms can be greatly simplified. Furthermore, the agreement between experiment and theory is very good.

Acknowledgments

The authors are grateful to Dr. Yusuke Tsukahara of the Technical Research Institute, Toppan Printing Co. of Japan, for providing a Mathematica program used in this work to compute the acoustic reflection coefficients at an interface between a liquid and a layered half-space of arbitrary anisotropy.

REFERENCES

1. A. Briggs, *Acoustic Microscopy*, Clarendon Press, Oxford (1992)
2. J. Kushibiki and N. Chubachi, Materials characterization by line-focus beam acoustic microscope, *IEEE Trans. Sonics Ultrason.*, SU-32, 189-212 (1985)
3. C. Lee, J. O. Kim, and J. D. Achenbach, V(z) curves of layered anisotropic materials for the line-focus acoustic microscope, *J. Acoust. Soc. Am.* 94(2), 923-930 (1993)
4. D. Xiang, N. N. Hsu, and G. V. Blessing, The design, construction and application of a large aperture lens-less line-focus PVDF transducer, *Ultrasonics*, 34, 641-647 (1996)
5. D. Xiang, N. N. Hsu, and G. V. Blessing, Ultrasonic leaky wave measurements for materials evaluation, *Nondestructive Characterization of Materials* (this volume)
6. N. N. Hsu, D. Xiang, S. F. Fick and G. V. Blessing, Transient analysis of a line-focus transducer probing a liquid/solid interface, *Review of Progress in Quantitative Nondestructive Evaluation*, Vol. 15, 995-1002 (1996)
7. A. Atalar, An angular-spectrum approach to contrast in reflection acoustic microscopy, *J. Appl. Phys.* 49(10), 5130-5139 (1978)
8. K. K. Liang, G. S. Kino and B. T. Khuri-Yakub, Material characterizations by the inversion of V(z), *IEEE Trans. Son. & Ultrson.* 32, 213-224 (1985)
9. C.J. R. Sheppard and T. Wilson, Effects of high angles of convergence on V(z) in the scanning acoustic microscope, *Appl. Phys. Lett,* 38(11), 858-859 (1981)
10. M. G. Somekh, H. L. Bertoni, G. A. D. Briggs, and N. J. Burton, A two dimensional imaging theory of surface discontinuities with the scanning acoustic microscope, *Proc. R. Soc. London* Ser. A, 401, 29-51 (1985)
11. W. Li and J. D. Achenbach, Measuring thin-film elastic constants by line-focus acoustic microscopy, Proc. 1995 *IEEE Ultras. Sym.*, Vol. 2, 883-892 (1995)
12. L. M. Brekhovskikh, Waves in Layered Media, Second ed. Academic Press, New York. (1980)
13. D. E. Chimenti and A. H. Nayfeh, Ultrasonic reflection and guided waves in fluid-coupled composite laminates, *J. Nondestructive Evaluation,* 9 (2/3), 51-69 (1990)
14. A. H. Nayfeh, The general problem of elastic wave propagation in multilayered anisotropic media, *J. Acoust. Soc. Am.* 89(4), 1521-1531 (1991)
15. Y. Tsukahara, Acoustic reflection coefficients at an interface between liquid and layered half space with arbitrary anisotropy, *Proc. IEEE Ultrason. Symp.*, Vol. 2, 877-881 (1995)

RAPID INFRARED CHARACTERIZATION OF THERMAL DIFFUSIVITY IN CONTINUOUS FIBER CERAMIC COMPOSITE COMPONENTS

J. Stuckey, J. G. Sun,* and W. A. Ellingson

Energy Technology Division
Argonne National Laboratory
Argonne, IL 60439

INTRODUCTION

Continuous fiber ceramic composites (CFCCs) are currently being developed for a variety of high-temperature applications, including use in advanced heat engines due to their relatively high strength and toughness at high temperatures and their relatively low density. In the development of these material systems, quantification of the mechanical and thermal properties is necessary. Furthermore, it is necessary to detect flaws and defects that may result in subsequent failure of the part. The need for this inspection is critical in evaluating material system performance and reliability. In most instances, due to the high cost associated with material manufacturing, the testing must be nondestructive.

The critical issue in most CFCC applications is the distribution of thermal properties (conductivity, specific heat, and thermal diffusivity) within a component. To meet this need, we developed a system that uses infrared thermal imaging to provide "single-shot" full-field through-thickness measurement of thermal diffusivity distributions in a component (Ahuja et al., 1996). The system has been constantly improved to provide thermal diffusivity data of high resolution in the short periods of data acquisition and processing that are necessary for testing large full-scale components. The system uses an infrared focal plane array detector and associated electronics for imaging transient thermal events. A computer program has been written to obtain multiple thermal images and quickly calculate the thermal diffusivity. The model used in this program is based on a one-dimensional conductive heat transfer solution proposed by Parker et al. (1961). This system has been used to identify thickness variations in homogenous samples, and to detect flaws in CFCC materials due to delamination and density variations in test specimens. In addition, subscale CFCC components of nonplanar geometries have been inspected for manufacturing- and operation-induced variations in thermal properties.

THERMAL DIFFUSIVITY MEASUREMENT

To measure thermal diffusivity based on the Parker et al. (1961) method, the front surface of a sample is heated instantaneously. Heat conduction through the sample, which is related to thermal diffusivity of the material, must then be measured. Figure 1 shows the theoretically predicted back-surface temperature T as a function of time t and specimen thickness L according to the relationship

---

* To whom all correspondence should be sent.

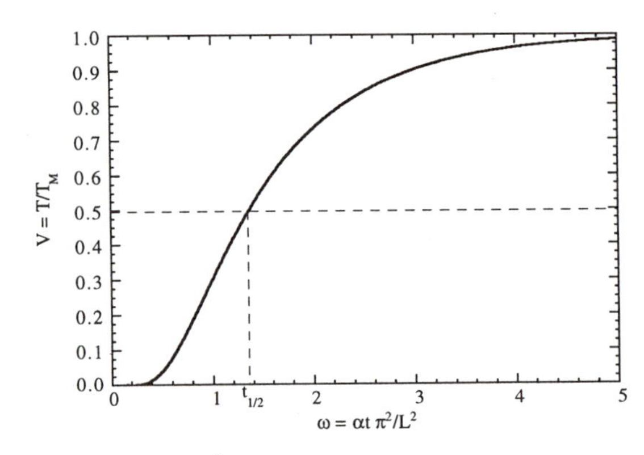

Figure 1. Theoretical prediction of back-surface temperature rise.

$$T(L,t) = \frac{Q}{\rho CL}\left[1 + 2\sum_{n=1}^{\infty}(-1)^n \exp\left(-\frac{n^2\pi^2}{L^2}\alpha t\right)\right], \tag{1}$$

where Q is the radiant energy incident on the front surface at $t = 0$, ρ is density, C is specific heat, and α is thermal diffusivity. Because the axes of Fig. 1 have been normalized (i.e., $V = T/T_M$ and $\omega = \alpha t\pi^2/L^2$, where T_M is the maximum front-surface temperature), the relationship is universal to all specimens. One method to determine the thermal diffusivity is the "half-rise-time" ($t_{1/2}$) method. When the back-surface temperature rise has reached half of its maximum, i.e., $V = 0.5$, we have $\omega = 1.37$. Thus, using the "half-rise-time" method,

$$\alpha = \frac{1.37L^2}{\pi^2 t_{1/2}}. \tag{2}$$

The experimental apparatus for the thermal imaging is illustrated in Fig. 2. The apparatus includes an IR camera consisting of a focal plane array of 256 x 256 InSb detectors, a 200 Mhz Pentium-based PC computer equipped with a digital frame grabber with DSP processor, a flash lamp system for the thermal impulse, a function generator for camera operation, and a dual timing trigger circuit for camera and external trigger control. An analog video system is used to monitor the experiments. For a typical measurement of a 256 x 256 array of diffusivity image, the processing time ranges from 8 to 20 min, depending on the number of frames taken.

To ensure the accuracy of the diffusivity measurement, the front-surface flash heating should occur at time zero with a short duration. The flash lamp discharge was measured to determine the flash characteristics. Figure 3 shows the measured flash intensity (dotted curve) as a function of frame number (or time) at a sampling rate of 1720 Hz. The measured curve was integrated in order to obtain the total energy of the flash pulse. The triangles superimposed on the graphs have the same area as the measured curve with their peak at the maximum flash intensity. The base of the triangles represents a characteristic time for the flash, i.e., 3.7 ms.

EXPERIMENTAL RESULTS

NIST Standard Specimen

To check the accuracy of the thermal diffusivity measurement, a NIST standard graphite specimen was used. Its thermal diffusivity value was reported to be between 0.74 and 0.76 cm^2/s (Taylor and Groot 1980; Chu et al., 1977). The specimen is 25.4 mm (1 in.) in diameter and its mean thickness is 5.87 mm. The sample was mounted in a sample holder

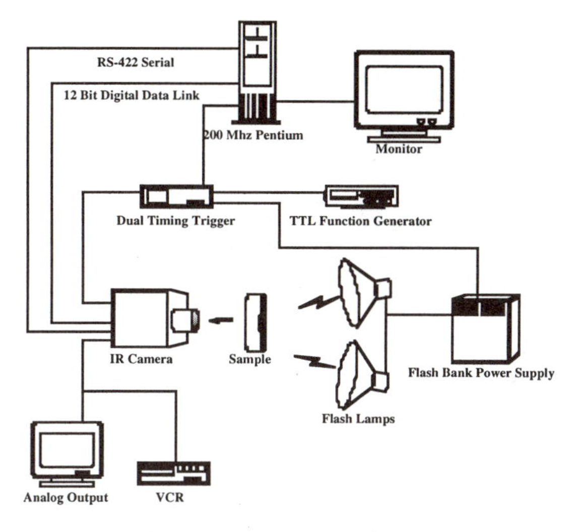

Figure 2. Experimental thermal imaging apparatus.

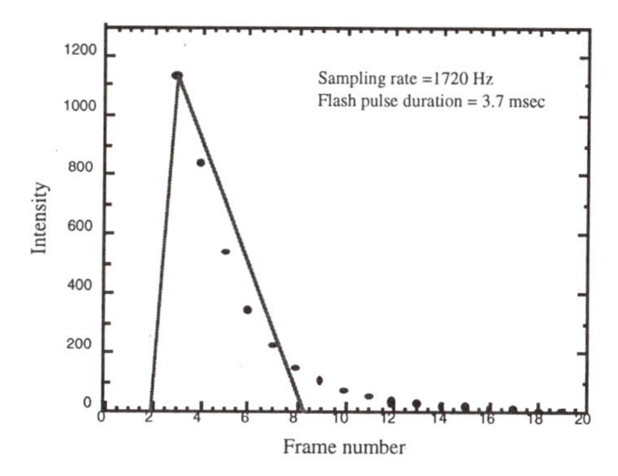

Figure 3. Flash pulse characteristics.

whose edges were carefully insulated to prevent radiation leaking through. A small piece of masking tape was placed on the sample to mark its geometry in subsequent images. The camera was set to an operation rate of 330 Hz. The resulting diffusivity image is shown in Fig. 4a, and a comparison of a single pixel history with the theoretical curve is shown in Fig. 4b. The image in Fig. 4a has a marked rectangular region in which the statistical results of the measured thermal diffusivity were determined. The mean diffusivity value was found to be 0.72 cm^2/s, with a standard deviation of 0.0267 cm^2/s. Therefore, the standard deviation is 3.7% of the mean value, which is greater than the ideal deviation reported in the ASTM standard of ±2%. This deviation can be attributed to a few sources of error. The largest source of error, however, may be in measurement of the specimen thickness.

Specimen thickness was measured at several locations on the specimen. Diffusivity values at these locations were also determined from the diffusivity image in Fig. 4a. Figure 5a shows the relationship between the diffusivity and the thickness square (L^2). It is evident that there is a correlation between specimen thickness and measured thermal diffusivity. Figure 5b is a contour plot of thickness variation in the specimen, obtained from the correlation. Each contour represents a thickness variation of 0.014 mm in the sample. The result here could be used to quickly map an object's thickness very accurately. The fact that diffusivity is a function of the square of the thickness makes small differences in thickness easier to detect.

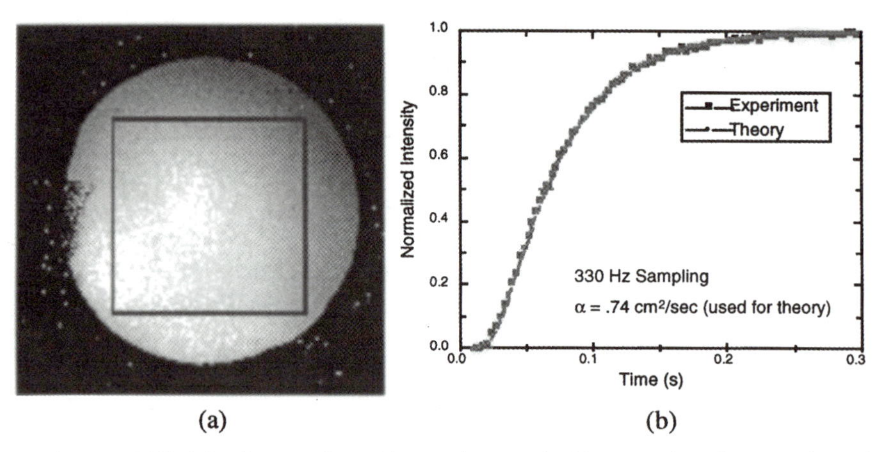

Figure 4. (a) Diffusivity image of graphite specimen and, (b) comparison between theoretical and experimental single-pixel history data.

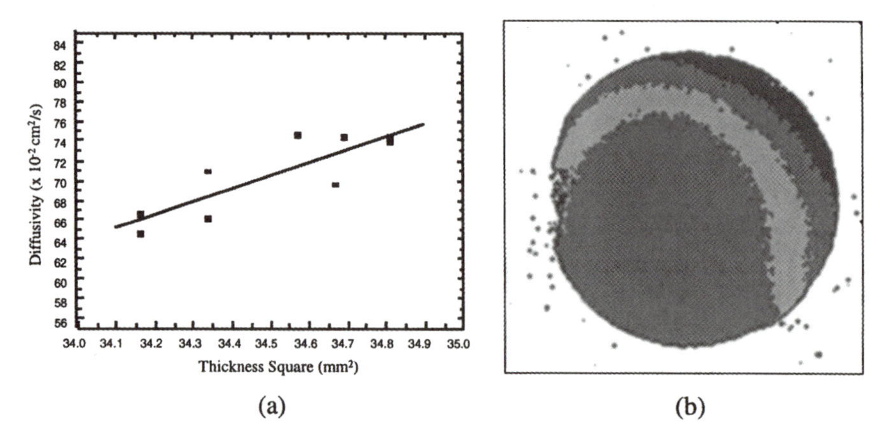

Figure 5. (a) Thickness squared vs. thermal diffusivity and (b) mapping of thickness variation (each contour represents 0.014 mm change in thickness) for NIST graphite standard.

CFCC Components

CFCC material systems consist of two or more phases. A typical CFCC material may be silicon carbide fibers surrounded by a silicon carbide matrix ($SiC_{(f)}/SiC$). The fibers may have different thermal properties from the matrix, causing the thermal diffusivity (macroscopic or "bulk") measurements to differ in each composite. The thermal diffusivity measurement can be taken parallel or perpendicular to the fiber orientation, and values obtained from these two geometries has been shown to vary up to 20% in a carbon/carbon composite (Heath and Winfree 1989). Differences in structure, fiber orientation, volume fiber ratios, and porosity (void, delamination) all contribute to the diffusivity measurement. As a result, the measured thermal diffusivity distribution may be used to infer the variation of the property in the components.

Thermal imaging was conducted on a CFCC standard component prepared by the Babcock and Wilcox (B&W) Corporation. The standard is a ring of Al_2O/Al_2O composite material that is 20.3 cm (8 in.) in diameter and 7.6 cm (3 in.) high and contains embedded defects of 20 x 20 mm, 12 x 12 mm, 8 x 8 mm, and 4 x 4 mm. The inside surface of the ring was painted black to reduce volumetric heating and emissivity effect from the white surface. The flash lamp was placed inside the ring, and a special cover was used to direct the flash radiation onto a section of the ring that corresponds to the area of interest. Figure 6 is a

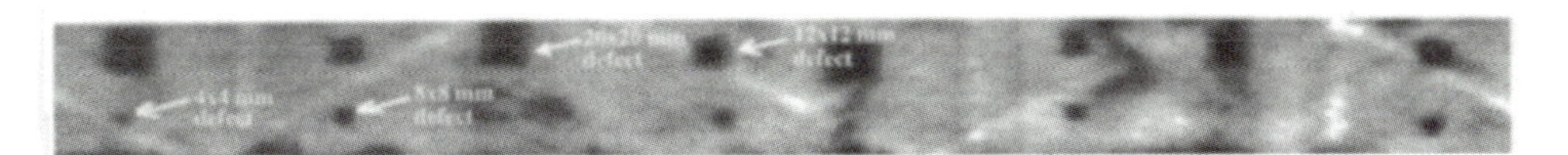

Figure 6. Thermal diffusivity image of B&W liner showing multiple defects as dark regions.

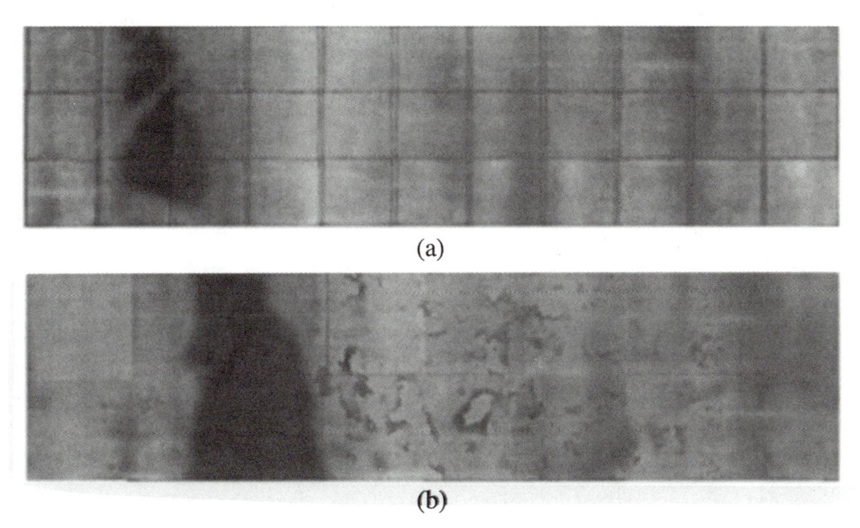

Figure 7. Thermal diffusivity images of SiC/SiC liner (a) pretest and (b) after ≈25 h test.

mosaic of 10 thermal diffusivity images taken for the entire ring. This thermal diffusivity mapping shows the defect regions as dark patches (low diffusivities) in the image. By comparing the image scale with the defects, the reported size of the defects was found to correlate with the size subtended by the image. This measurement is however subject to interpretation as to where the edges begin and end. An accurate edge-detection algorithm could be used to verify these results. Thickness variations were also detectable in the image. Regions of fiber overlap where the material was thicker are apparent in the image as lighter regions; a subsequent measurement finds this overlap region to be 2 mm.

Figures 7a and 7b show the thermal diffusivity images of a SiC/SiC subscale combustor liner before and after a test of ≈25 h in a high-temperature test rig, respectively. The liner is 20.3 cm (8 in.) in diameter and 20.3 cm (8 in.) high. The image in Fig. 7a was sectioned with horizontal and vertical lines marked on the surface. The image in Fig. 7b show some surface patterns due to the flaking of the surface coating. Both images show a large region with low thermal diffusivity, presumably a delaminated region. By comparing the two images, it is apparent that after the high temperature test, there is a significant increase in the area of the low-diffusivity region. However, the peripheral regions in which diffusivity had significantly declined after the test are faintly visible in the pretest image (Fig. 7a), showing as a relatively lighter gray scale around the dark region. Therefore, this result seems to indicate that a defect (delamination) may grow into regions that are not initially well formed.

CONCLUSIONS

A rapid infrared thermal imaging system has been developed. The processing time of a 256 x 256 array of diffusivity image is approximately 15 min depending on the number of frames taken. This fast processing time now allows for rapid inspection of large full-scale CFCC and other material components.

The thermal imaging system is shown to provide accurate measurements of thermal diffusivity for standard samples. Measured thermal diffusivity values showed correlation with specimen thickness variations. The system is effective for measuring manufacturing-

and operation-induced defects in CFCC components. Therefore, the thermal imaging method can be used for reliable characterization of CFCC materials.

ACKNOWLEDGMENT

The authors want to acknowledge Dr. R. M. Lueptow of Northwest University for useful discussions. This work was sponsored by the U.S. Department of Energy, Energy Efficiency and Renewable Energy, Office of Industrial Technologies, under Contract W-31-109-ENG-88.

REFERENCES

Ahuja, S., Ellingson, W. A., Steckenrider, J. S., and King, S., Thermal diffusivity imaging of continuous fiber ceramic composite materials and components, *Thermal Conductivity*, Vol. 23, Proceedings of the 23rd International Thermal Conductivity Conference, Technomic Publishing, 1996.

Chu, F. I., Taylor, R. E., and Donaldson, A. B., Flash diffusivity measurements at high temperatures by the axial heat flow method, *Proceedings of the Seventh Symposium on Thermophysical Properties*, ASME, 1977.

Heath, D. M., and Winfree, W. P., Thermal diffusivity measurements in carbon-carbon composites, *Review of Progress in quantitative nondestructive evaluation*, Vol. 8B, Plenum Press, 1989.

Parker, W. J., Jenkins, R. J., Butler, C. P., and Abbott, G. L., Flash method of determining thermal diffusivity, heat capacity, and thermal conductivity, J. Appl. Phys., 32:1679-1684, 1961.

Taylor, R. E., and Groot, H., Thermophysical properties of POCO graphite, *High Temp.-High Pres.*, 12:147-160, 1980.

RAMAN SPECTROSCOPY FOR IN-SITU CHARACTERISATION OF STEAM GENERATOR DEPOSITS

P.A. Rochefort, D.A. Guzonas, and C.W. Turner

AECL
Chalk River Laboratories
Chalk River, Ontario
Canada, K0J 1J0

INTRODUCTION

Fouling of the secondary-side of nuclear steam generators (SG) by corrosion products and other impurities that have been transported into the SG with the feedwater is a serious problem. The build-up of deposits on the SG tubes can lower the rate of heat transfer[1,2], and act as sites for the concentration of impurities leading to localised corrosion of the underlying substrate. When fouling is severe, the deposits must be removed by chemical and/or mechanical cleaning techniques. Information on the composition of the deposits is then required for the development of optimal cleaning conditions and procedures. The composition of the deposits also provides information on the chemistry conditions within the SG and the feedtrain[3,4].

Although samples of deposit can be removed from the steam generator for ex-situ characterisation, for example, after waterlancing, the quantity of deposit available is often limited, and the exact origin of the recovered deposit is not always known. Samples removed may also undergo chemical or physical surface alteration upon exposure to the atmosphere. An in-situ inspection technique capable of identifying the chemical composition of the compounds present in a deposit, and of giving a semi-quantitative measurement of their concentrations, would therefore provide valuable information.

Recent advances in filter and detector technologies coupled with the use of fibre optics make Raman spectroscopy a useful remote characterisation method of materials by vibrational spectroscopy. Fibre optics are used to transmit laser energy to the inspection area and scattered light back to the spectrometer. Remote Raman spectroscopy using fibre-optics is now being used in plant environments to characterise a variety of materials[5] . A recent review of the application of fibre optics for Raman spectroscopy can be found in a recent paper by Lewis and Griffiths[6].

We have constructed several fibre optic probes capable of measuring Raman spectra of secondary-side deposits, in an on-going program to develop and demonstrate fibre-optic Raman spectroscopy for in-situ deposit analysis.

Nondestructive Characterization of Material VIII
Edited by Robert E. Green Jr., Plenum Press, New York, 1998

EXPERIMENTAL

In Raman spectroscopy, a sample is irradiated by a monochromatic light source (typically from a laser) and the scattered light is collected and analysed by spectrophotometer. Most of the incident light is elastically (Rayleigh) scattered without any frequency change. However, a small portion of the light, typically 1 photon in 10^6, is inelastically (Raman) scattered by the material, the frequency shift of the light (normally measured in units of inverse centimetres, cm^{-1}) is proportional to the transition energy between vibrational energy levels of the molecules or crystal structures in the sample. The frequency, width, and relative intensities of the spectral peaks can be used to determine the identity and quantity of the compounds in the sample.

There are two major problems in obtaining Raman spectra of solid metal oxide samples using fibre optics. The first is preventing the Raleigh scattered light in the total scattered signal from entering the spectrometer so that the detection system is not swamped. The second is the removal of the silica (or glass) Raman spectra produced by the fibre as it transmits the laser energy to the sample and the Rayleigh scattered light from the inspection area. The principal bands of silica Raman spectrum occurs between about 100 to 600 cm^{-1}, the same region as the bands of most of the oxides found in SG deposits.

A demonstration Raman illumination and collection optical system that addresses these problems has been fabricated (see Figure 1). At its core, a holographic notch filter (Kaiser Optical HSPF-647-1.0) is used both as a reflective optical component for the incident laser light and as a notch filter for the Raman scattered light. Holographic notch filters reflect almost 100% of the design wavelength while transmitting more than 80 % of light with longer or shorter wavelength. Typical filter rejection bandwidth around the central wavelength is 350 cm^{-1} (10 nm).

The optical assembly is approximately 200 mm long and 40 mm in diameter. The input laser beam, steered directly from the laser as shown in the figure, or carried by an optical fibre, enters near the base of the assembly at right angles to the vertical optical axis. The beam is reflected up by an adjustable mirror onto the centre of the notch filter, which is set at 5 degrees off-normal to the optical axis. The position and angle of the mirror is adjusted so that the laser beam is reflected by the filter down the centre of the optical axis through a 0.4 numerical aperture (NA) long working distance microscope objective (Leitz H32X/0.60). When the laser light is transmitted by an optical fibre, the silica induced Raman spectrum is removed from the laser beam because the frequency shifted light is transmitted through the holographic filter.

A sample is placed at or near the focal point of the beam, and scattered light from the sample is collected and collimated by the objective and directed back up to the notch filter. Raleigh scattered light is reflected by the notch filter, whereas wavelength-shifted light is transmitted through the filter. The filtered light is concentrated by a condenser lens onto the end of a fibre optic bundle and transmitted to the spectrometer. The use of a single objective to both focus the laser beam and collect the scattered light significantly reduces the alignment complexity as compared to a two lens optical arrangement, one to focus the laser beam and one to collect the scattered light.

The fibre optic bundle consists of 7 step-index multi-mode fibres having 200 μm pure silica cores. The bundle is 3 meters long and terminated at each end with modified standard fibre optic couplers. At the collection end, the fibres are arranged in six-around-one close-packed geometry while at the other end, used as the source for the spectrometer the fibres are arranged in a close packed line. The output of the fibre bundle is focused on the spectrometer slits. Both ends are polished flat.

Raman spectra were measured using a SPEX 1000M single monochromator with a charge-coupled device (CCD) detection system. The detector integration times and the number of signal-averaged acquisitions were optimised to give the best signal-to-noise ratio for each sample. Typical values were 10 to 40 second integration times and 4 to 16 acquisitions. The 647.1 nm line from a Krypton ion laser was used for excitation; laser plasma lines were removed using a bandpass filter (10 nm bandwidth) centred at the laser frequency. The laser power incident on the samples was approximately 30 mW.

Raman spectra were acquired over the range 200-1600 cm^{-1} in several overlapping windows which were then spliced together. Data manipulations such as baseline correction and spectral splicing were carried out using a spectrum analysis software package (GRAMS/386, Galactic Industries Inc.)

RESULTS

The Raman spectra of the various iron oxide phases expected in these deposits have been well documented in the literature[7,8,9]. The Raman spectra of the three most common iron containing phases found in CANDU (CANada Deuterium Uranium) SG deposits, magnetite, hematite and nickel ferrite, are shown in Figure 2 over the spectral range of 150 to 840 cm^{-1}.

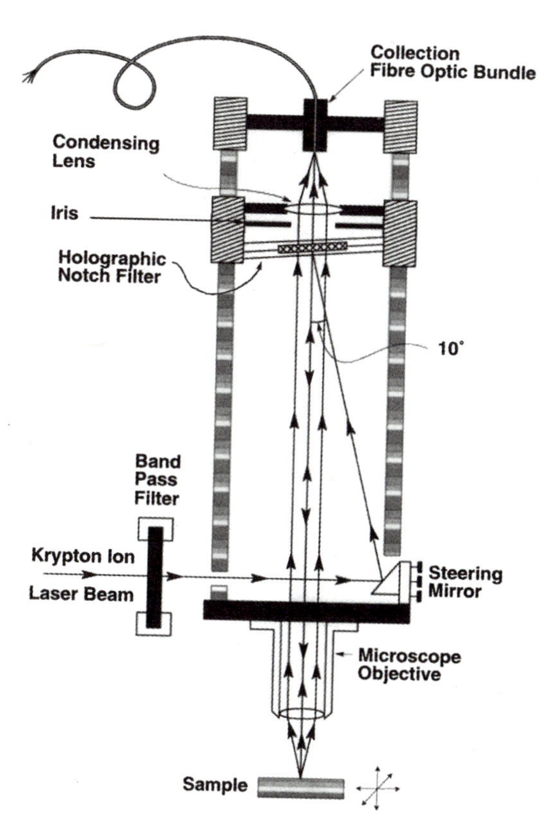

Figure 1: Schematic of the fibre optic Raman collection system

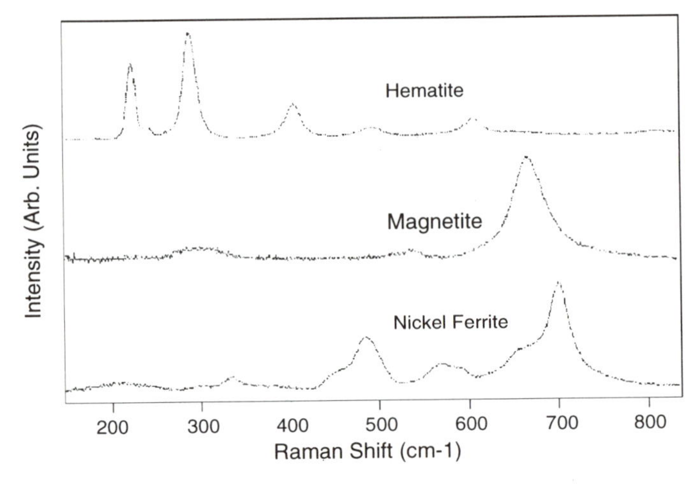

Figure 2: The Raman spectra of the three most common iron oxide phases found in the CANDU deposits.

In mixtures of particles of strongly absorbing oxides such as magnetite, Raman band intensities do not vary linearly with the concentration of the oxide phases present due to scattering and light absorption by the particles. Quantification of the composition of such a mixture therefore requires the use of suitable calibration mixtures. Calibration mixtures of magnetite and hematite as well as magnetite and nickel ferrite were prepared by mechanically mixing the pure powders to produce a homogeneous mixture. Raman band intensities were then plotted against concentration, and an empirical calibration curve fit to the data.

As the concentration of magnetite in the deposit samples increased, it became increasingly more difficult to obtain a Raman spectrum as a result of the strong absorption of the laser light by magnetite. Sample degradation often occurred unless care was taken to optimise the focus of the laser. This degradation was usually indicated by the appearance of strong hematite bands in the Raman spectrum after short exposure of the sample to the laser, a result of oxidation of the surface of the magnetite to hematite. A red coloured damage zone could often be observed visually at the spot where the laser beam had struck the sample. These factors made it much easier to obtain Raman spectra from the samples which contained low concentrations of magnetite. Recognisable spectra from samples containing large amounts of nickel ferrite or hematite could be obtained with an integration time of about one second, while samples consisting mainly of magnetite required integration times of about 10 seconds.

The three samples investigated had been removed from steam generators by waterlancing. Two were in the form of loose powders, and the third was a millimetre-sized flake. The samples had previously been examined by X-ray diffraction (XRD), and the first sample had been extensively characterised by Mossbauer spectroscopy and other analytical techniques.

Raman spectra of the three secondary-side deposits are shown in Figure 3, over the 200-800 cm^{-1} spectral region. The spectrum in Figure 3a contains a strong band envelope in the 640-710 cm^{-1} region characteristic of nickel ferrite (NF). The strongest magnetite band is also observed, at approximately 665 cm^{-1}, and is marked with an M in the figure. Although the nickel ferrite and magnetite bands overlap in this spectral region, the intensities of the magnetite and nickel ferrite peaks under the band contour can be used to estimate the relative amounts of magnetite and nickel ferrite. It was estimated that 65%

±5% of the total spinel phase present was nickel ferrite and 35% ±5% was magnetite. For comparison, the composition determined by X-ray and Inductively Coupled Plasma-Atomic Emmission Spectroscopy (ICP-AES) indicated 15% magnetite and 85% nickel ferrite, while Mossbauer spectroscopy suggested the composition is 44% magnetite and 47 % nickel ferrite, with the remainder being zinc ferrite. The Raman results are therefore intermediate between these two analyses.

The Raman spectrum of the deposit in Figure 3b clearly shows the presence of both hematite and magnetite. Based on reference mixtures of these two materials, the ratio of magnetite to hematite is estimated to be 5:2. Analysis by XRD gave an estimate of the magnetite to hematite ratio of 3:1, in good agreement with the Raman data. The Raman spectrum in Figure 3c, contains a band due to magnetite at 684 cm^{-1}, and also bands at 490, 634 cm^{-1}.

CONCLUSIONS

The work presented here demonstrates that good quality Raman spectra of steam generator secondary-side deposits can readily be obtained using a fibre optic based Raman spectrometer. The major chemical phases, other than metallic copper, can be readily identified, and a semi-quantitative estimate of the composition can be made, as verified by comparison with deposit analyses performed using other techniques. This work suggests that in-situ characterisation of secondary-side deposits using fibre-optic Raman spectroscopy is feasible.

Laser-induced decomposition was noted for samples containing large concentration of magnetite. The decomposition usually resulted in the formation of hematite, and an increase in the intensity of hematite bands with time was a good indicator for decomposition. This decomposition could be minimised by working at low laser powers

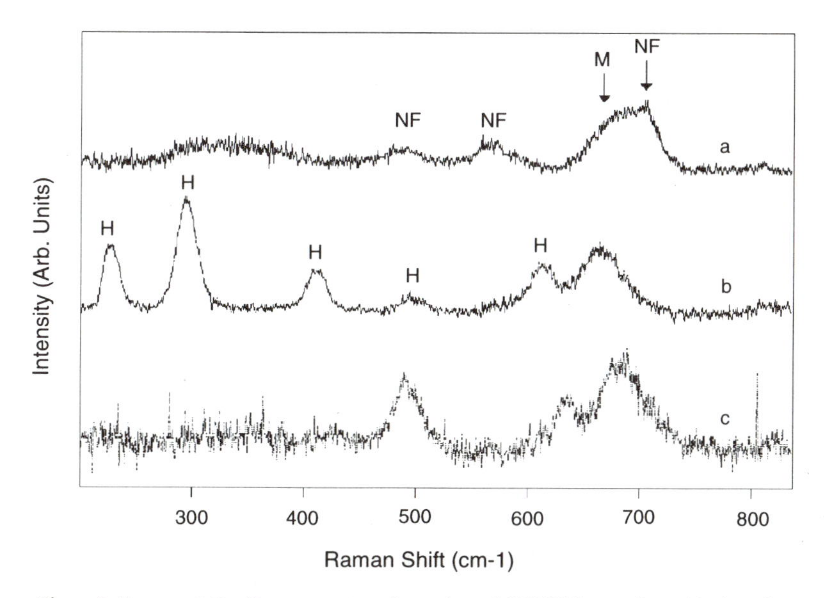

Figure 3: Representative Raman spectra of waterlanced CANDU secondary-side deposits

and using a slightly defocused laser. The presence of water with the samples did not adversely affect the spectra, but rather improved the quality of the spectra obtained from the powder samples by minimising laser-induced decomposition.

The spectra obtained from the flake samples clearly illustrate the heterogeneous nature of these deposits. Several Raman bands were found which could not be definitively assigned to a particular chemical species.

ACKNOWLEDGEMENTS

Funding for this work was provided by the CANDU Owners Group (COG), Technical Committee 19. The authors would like to thank COG for permission to publish this work. The authors would like to thank Bruce Nuclear Generating Station, Pickering Nuclear Generating Station, and Gentilly-2 Nuclear Generating Station for providing the samples studied in this work. The authors would also like to acknowledge J. Semmler for providing the XRD, ICP-AES, and Mossbauer analytical data for the sludge samples.

REFERENCES

1. T.F. Habib, P.A. Sherburne, J.F. Dunne, and C.L. Williams, Degradation in Ginna steam generator tube heat transfer due to secondary-side fouling, in: *Steam Generator Sludge Deposition in Recirculating and Once Through Steam Generator Upper Bundle and Support Plates*, NE-Vol. 8, ASME, 1992.
2. C.W. Turner and S.J. Klimas. Thermal resistance of boiler tube deposits at steam generator operating conditions, in: *Proceedings: Sludge Management Workshop*, May 10-12, 1994, Norfolk. Virginia. EPRI Report TR-104212.
3. W.A. Byers, P.J. Kuchirka, M. Rootham, and A.J. Baum. Can traces from the past predict the future?, in: *Electric Power Research Institute Surface Chemistry Workshop*, August 22-23, 1996, Myrtle Beach S.C.
4. C.W. Turner and K. Shamsuzzaman. The role of copper in sludge consolidation, in: *Proceedings: Sludge Management Workshop,* May 10-12, 1994, Norfolk. Virginia. EPRI Report TR-104212.
5. J. Andrews. Raman spectroscopy moves on-line, *Spectroscopy Europe*, 7, 8, (1995).
6. I.R Lewis, P.R. Griffiths, Raman spectroscopy with fiber-optic sampling, *Appl. Spectrosc.* 50, 12A (1996).
7. J. Gui, T.M. Devine. In-situ vibrational spectra of the passive film on iron in buffered borate solution, *Corrosion Science*, 32, 1105 (1991).
8. J. Dunwald and A. Otto. An investigation of phase transitions in rust layers using Raman spectroscopy, *Corrosion Science*, 29,1167 (1989).
9. R.J. Thibeau, C.W. Brown, and R.H. Heidesbach. Raman spectra of possible corrosion products of iron, *Appl. Spectrosc.*, 32, 532 (1978).

CHARACTERIZATION OF NEAR SURFACE MECHANICAL PROPERTIES OF ION-EXCHANGED GLASSES USING SURFACE BRILLOUIN SPECTROSCOPY.

Mar Puentes[1], John Bradshaw[2], Andrew Briggs[1], Oleg Kolosov[1], Keith Bowen[3] and Neil Loxley[3]

[1]Department of Materials
University of Oxford
Oxford, OX1 3PH
[2]Pilkington plc. Pilkington Technology Centre. Hall Lane Lathom Ormskirk Lancarshire England L40 5UF.
[3]Bede Scientific Instruments Ltd., Bowburn, Durham DH6 5AD, United Kindom

ABSTRACT

The technique of Surface Brillouin Spectroscopy is attracting considerable attention for the non-destructive characterization of near-surface properties that depend upon the elastic behaviour. These include surface strengthening, delamination in metallised layers and structural changes near surfaces. We here present a study of the chemical strengthening of glass. SBS measurements of surface acoustic wave (SAW) velocity in glasses strengthened by the exchange of K^+ for Na^+ ions were carried out using the Bede BriSc instrument.

The study of surface acoustic waves on a transparent material such as glass was made possible by depositing a thin layer of aluminium on the glass surface. At a working frequency of 20 GHz, the optimum thickness of the layer was found to be 30 nm. The replacement of Na^+ by K^+ ions that takes place during the treatment of the glass causes near-surface modifications of both density and elastic constants. An increase of 3.8% in density due to the replacement of light Na^+ ions by heavier K^+ ions at the surface could account only for a 1.9% decrease in the surface acoustic wave velocity. Residual stresses were estimated to have a negligible effect on surface acoustic wave velocity. At the same time a $2.8 \pm 0.4\%$ decrease in the surface acoustic wave velocity due to the ion-exchange process was observed by SBS. This result suggests that a significant part of the surface acoustic wave (SAW) velocity change should be attributed to variations in the elastic constants (approximately 1.8%) and agrees well with independent estimates of the Young's modulus (E) and shear modulus (C_{44}) changes caused by replacing the Na^+ ions by K^+ ions during the process.

INTRODUCTION

A general characteristic of brittle solids is that surface cracks can lead to local stress concentration and a fracture strength which is very much less than the intrinsic strength of the undamaged material. Glass is a typical example. Most of the successful methods for strengthening glass have involved the introduction of compressive stress in the surface layers to overcome the strength-impairing effects of surface cracks[1,2].

Nondestructive Characterization of Material VIII
Edited by Robert E. Green Jr., Plenum Press, New York, 1998

For this purpose a chemical strengthening method[3-5] is studied in this paper. In this process, small cations initially present in the glass (Na^+) are replaced by larger cations (K^+) from a molten salt bath. The compressive stress achieved in the strengthening process[6-12], depends on the nature of the exchanging ions[13,14], on the concentration of ions introduced into the glass[15-17], on the elastic properties of the glass[2] and on the other physical parameters, such as temperature[18-20] that characterize the replacement process.

Surface Brillouin spectroscopy (SBS) has already proved to be sensitive to changes in material properties such as elastic constants and density[21]. The goal of this study, therefore, is to apply SBS to characterize ion-exchanged glasses. Brillouin light scattering[22] is usually referred to as an inelastic scattering of an incident photon of wavevector $\vec{k}_i$ and frequency ω_i by an acoustic phonons of wavevector $\vec{q}$ and frequency Ω. In SBS the total energy and projection of momentum on the surface plane are conserved during the scattering process according to the following equations

$$\vec{k}_i - \vec{k}_s = \pm\vec{q} \tag{1}$$

where $\vec{k}_i$, $\vec{k}_s$ and $\vec{q}$ are the projections of the wavevectors of the incident and scattered photons and phonon, respectively on a surface, and

$$\hbar(\omega_i - \omega_s) = \hbar\Omega \tag{2}$$

where $\hbar\omega_i$ and $\hbar\omega_s$ are the energies of the incident and scattered photon and $\hbar\Omega$ is the energy of the acoustic phonon, respectively. Once the wavevector $\vec{q}$ is known, and the photon frequency shift ($\omega_i - \omega_s$) is measured by a spectrometer, the phase velocity of the surface phonon can be recovered from

$$v = \frac{2\pi\Omega\lambda}{2\sin\theta} \tag{3}$$

where λ, Ω and θ are the wavelength, the frequency shift measured in SBS and the angle of the incident light, respectively. Knowledge of the SAW velocity allows the estimation of the mechanical properties of the material.

Since the incident light wavelength is usually in the visual light spectrum and about 500-600 nm, the surface acoustic phonon wavelength detected with SBS is of the order of 300 nm. Changes in glass properties caused by the ion exchange process can be detected by changes in the phonon's wavelength. In particular, as the K^+ concentration is highest at the surface[23], surface acoustic phonons are expected to be sensitive to these surface composition changes.

In transparent materials the major contribution to Brillouin scattering is caused by the strain-induced modulation of the dielectric constant: this effect is known as the elastooptic effect[24]. For such materials mainly bulk phonons appear in the Brillouin spectra, and these cannot be used to characterize the glass surface. However, if the transparent material is coated with a thin (much less than the SAW wavelength) reflective layer, the interaction of light with surface acoustic phonons can be more easily detected with SBS since their amplitude is increased[25].

Sample Preparation and Experimental Set-up

Sample Preparation. The glass samples were prepared by Pilkington Technology Management Ltd., by melting primarily SiO_2, Na_2O, MgO and CaO. The approximate glass composition is presented in Table (1).

Table (1) Composition of the reference glass specimen (mol %).

SiO_2	Al_2O_3	Na_2O_3	K_2O	MgO	CaO	Fe_2O_3	SO_3
71.7	0.6	12.5	0.4	5.7	8.9	0.038	0.18

These glasses were polished on the side that was in contact with the tin bath of the float process. One of the samples (the original glass) was left in this state, while the other (the ion-exchanged glass), was immersed in a molten bath of KNO_3 (99.5%): H_2SiO_3 (0.5%) for 48 hours at 460° C.

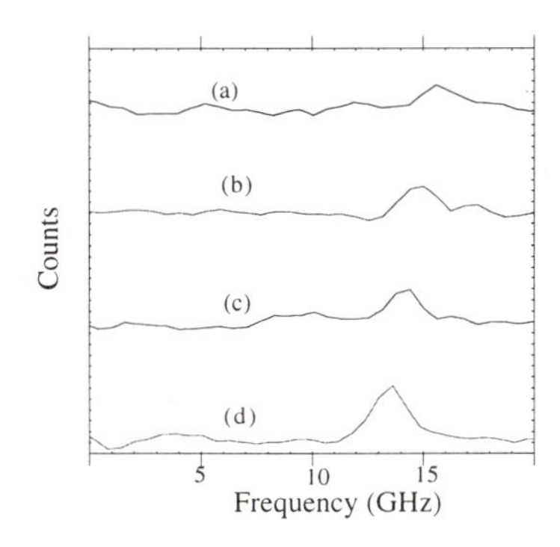

Figure 1. Brillouin scattering from Si (001) coated with different Al layer thicknesses: (a) 0 nm, (b) 20 nm, (c) 30 nm and (d) 40 nm.

Brillouin Spectrometer Set-up

Brillouin scattering measurements were performed in air, at room temperature, on a commercial BriSc SBS spectrometer[26] using a p-polarized light beam (a 532 nm single mode frequency doubled Nd-YAG laser). The incident light was focused onto the surface of the specimen and the backscattered light collected by a lens with a focal length of 50 mm. The frequency shift of the scattered light was measured using a Fabry Perot interferometer, characterized by a finesse of about 70 and a contrast ratio of about 109. The sampling time for the whole spectrum was typically 15 minutes, and the relative precision is about 0.4%. The incident and scattered light were polarized in the plane of incidence. Since the backscattering geometry was chosen, the component of the acoustic wavevector parallel to the surface is related to the optical wavevector of the incident light by the relationship

$$q = 2\vec{k}_i \sin\theta \tag{4}$$

with θ the angle of incidence.

RESULTS

Al-Coatings

Since the glass surfaces are transparent, a very thin coating of Al was deposited onto the glass surface in order to detect the surface phonons. The effect of the coating on the propagation of SAWs on glass was studied using the following procedure. The same glass sample was coated at different times with an Al layer of 10, 20, 30 and 40 nm in thickness. In order to check for possible

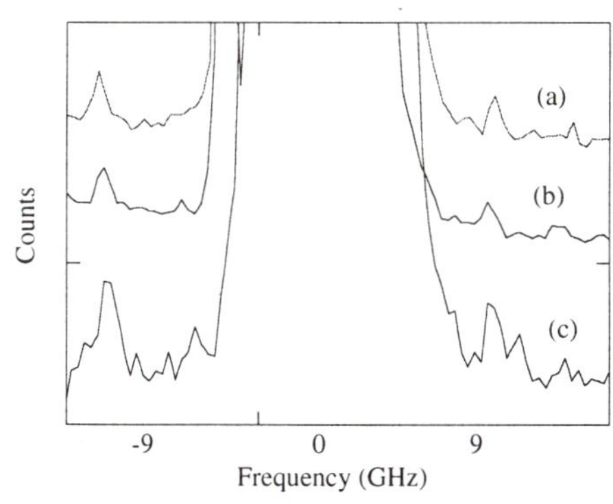

Figure 2. Brillouin scattering from a glass specimen coated with different Al layer thicknesses (a) 20 nm, (b) 30 nm and (d) 40 nm

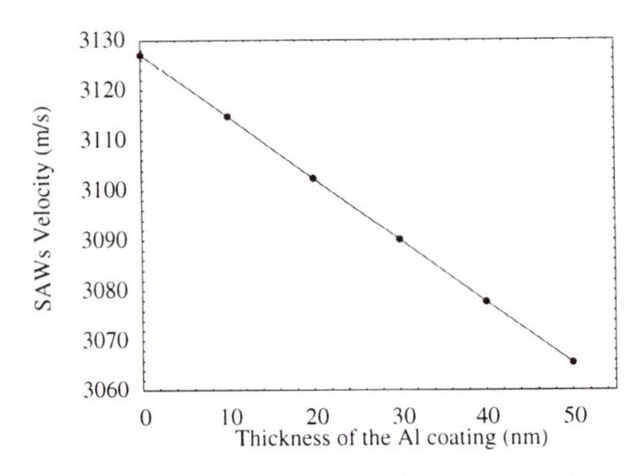

Figure 3. Calculated SAW velocities of Si (001) versus thicknesses of Al coatings

irregularities in the deposition of the Al a reference (or control) sample of known SAW velocity was simultaneously coated along with the glass specimen. The control sample used for this purpose was Si, which was chosen for the following two reasons. Firstly, its acoustic properties are well known and differ strongly from Al properties, therefore even small variation of Al layer properties could be easily determined by measuring the control sample. Secondly, because its surface is semi-opaque and, therefore surface phonons are always detected whether Si is coated with Al or not.

The Brillouin spectra in Figures 1 and 2 show the SAW peaks observed in the Al coated Si and glass samples. Since the glass surfaces are transparent, there is a critical thickness of the Al layer above which it is possible to detect the peaks associated with the SAWs due to increase of scattering signal. Fig. 2 shows the SAW peaks for glass coated with Al layer of 20, 30 and 40 nm thickness. As the SAW velocity of the Al is slightly lower than the SAW velocity of the glass, an increase in the thickness of the Al layer causes a decrease in the SAW velocity. This feature is observed in Fig. 2 where a frequency shift occurs as a consequence of an increase in the thickness of the Al layer. Also, it can be noticed that the thicker the Al-coating, the more intense and clear the SAW peaks. This effect is even more clear in Si, as can be seen by comparing Figures 1 and 2. Figures 3 and 4 show the calculated results of the effect of the Al-coating on SAW velocity as a function of the thickness of Al for Si and glass substrates, respectively. This study shows the optimum thickness of Al-coating for glass to be 30 nm. For this thickness value, the surface phonon peaks can clearly be seen in the spectra and the effect of the Al layer on the propagation of SAWs is kept to a minimum.

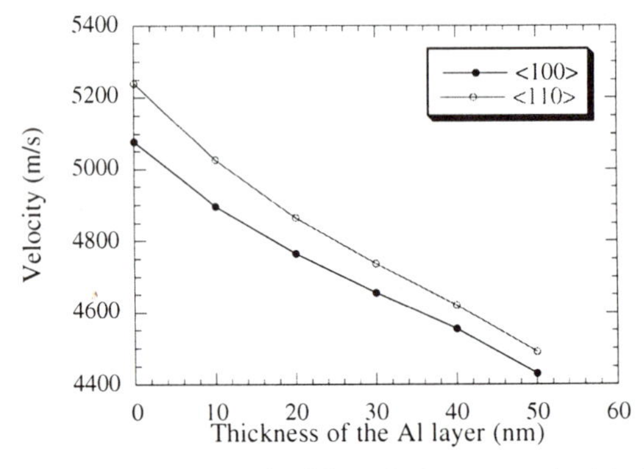

Figure 4. Calculated SAW velocities of glass coated with Al layer of different thickness

Surface Brillouin Spectroscopy (SBS)

SBS measurements of SAW velocities were performed for the ion-exchanged and the original glasses at the angle of incidence of 65° which corresponds to a SAW wavelength of about 300 nm. The experimental conditions were the same for both samples, and a set of measurements was taken at 5 points spaced by 0.1 mm along a straight line. The average SAW velocity obtained in these measurements was 3066 ± 25 m/s and 3154 ± 16 m/s for the ion-exchanged glass and the reference glass, respectively.

DISCUSSION

The 2.8% decrease of SAW velocity detected in the ion-exchanged glass using SBS can be attributed to the density ρ, Young's modulus E and the shear modulus C_{44} changes. For this purpose the SAW velocity can be approximated as[27]

$$V_R = V_S(a - b\sigma + c\sigma^2)^{-1} \tag{5}$$

where a, b and c are constants equal to 1.14418, 0.25771 and 0.12661, respectively, V_S is the shear velocity ($V_S = \sqrt{C_{44}/\rho}$) and σ is the Poisson's ratio ($\sigma = \frac{E}{2C_{44}} - 1$). Expressing V_R as a function of E, ρ and C_{44} as

$$V_R = \frac{\sqrt{\frac{C_{44}}{\rho}}}{a - b\left(\frac{E}{2C_{44}} - 1\right) + c\left(\frac{E}{2C_{44}} - 1\right)^2} \tag{6}$$

the relative variation of V_R due to a change in density and the elastic constants is then given by

$$\left(\frac{\Delta V_R}{V_R}\right)_{\rho,E,C_{44}} = -0.5 * \left(\frac{\Delta\rho}{\rho}\right) + 0.23 * \left(\frac{\Delta E}{E}\right) + 0.27 * \left(\frac{\Delta C_{44}}{C_{44}}\right) \tag{7}$$

Thus, an increase in density or a reduction in the Young's modulus or shear modulus due to the diffusion of the K^+ ions (heavier than Na^+ ions) into the glass are expected to yield an increase in the SAW velocity.

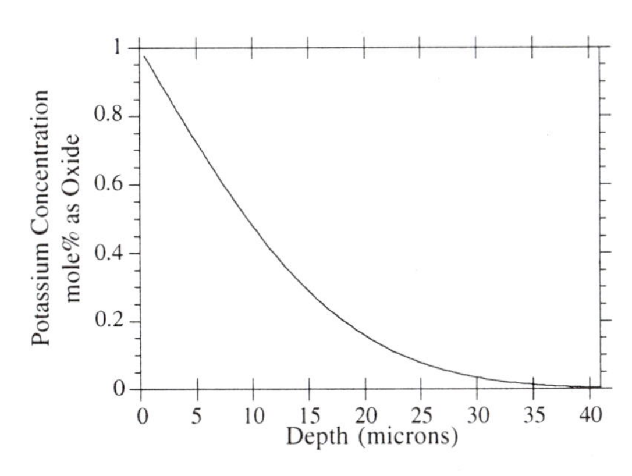

Figure 5. Calculated diffusion profile at 460°C for 48 hrs. The k concentration mole% as oxide is plotted versus the depth in microns

Using the diffusion profile given in Fig. 5[28], a 3.8% increase of density at the surface (within 300 nm depth) has been estimated for the ion-exchanged glass. The interdiffusion coefficient of this glass at 460° C is 5.8x10-16 m^2s^{-1}. The obtained increase of density accounts for only a 1.9% decrease in the SAW velocity. Therefore the 2.8% SAW velocity decrease observed by SBS cannot be explained only in terms of this increase in density.

A significant change in the elastic constants must therefore also take place to account for the decrease of the SAW velocities measured by SBS in the ion-exchanged glass. Kozlovskaya et al.[29] have calculated the change in the elastic constants due to changes in glass composition. This study has been used in this paper to account for the effect of replacing the sodium by potassium during the ion-exchange process on the elastic constants. This independent estimation shows a reduction in the Young's modulus and the shear modulus of 2.9% and 2.6%, respectively which accounts for a 1.4% increase in the SAW velocity. Therefore, the independently evaluated decreases in the SAW velocity of 1.9% and 1.4% due to density and elastic constants changes, respectively, account for a total change of 3.3%, which is in reasonable agreement with the experimental decrease of 2.8 ±0.4% measured by the SBS.

CONCLUSION

We have demonstrated the applicability of SAWs to detect changes in the near-surface mechanical properties introduced by the ion-exchange processes in glasses. The study of SAWs on transparent materials by means of SBS has been made possible by depositing a thin layer of Al onto the glass surface. The effect of the coating on the propagation of the SAW has been examined using different thicknesses of the Al layer.

Our results indicate that the velocity of the SAWs is significantly affected by variations of the density due to the exchange of K^+ ions for Na^+. However, of the 2.8±0.4% SAW velocity decrease observed by SBS, density increase accounts for only 1.9%. A significant change in the elastic constants must therefore also take place. Using estimates of the elastic constants variations due to the glass composition, a decrease of 1.4% has been found which agrees well with our experimental data.

ACKNOWLEGEMENT

Authors acknowledge a financial support and a fruitful collaboration within EPSRC-Link Nanotechnology project "Characterization of Surface Roughness and Sub-surface damage" particularly with Bede Scientific Instruments Ltd., and personally Max Robertson from Logitech Ltd., and Claudio Pecorari, University of Oxford.

REFERENCES

1. D.K. Hale, Nature 217: 1115 (1968).
2. R. Tandon, D.J. Green and R.F. Cook, J. Am. Ceram. Soc. 73 [9]: 2619 (1990).
3. H.A. Miska, Materials Engineering 83: 38 (1976).
4. F. Roger Bartholomew, Corning Inc., in: Engineered Materials Handbook, Ceramic and Glasses, Vol. 4, ed. by ASM international: 460 (1991).
5. H. M. Garfinkel, The Journal of Physical Chemistry 72 [12]: 4175 (1968).
6. I. W. Donald, M.J.C. Hill, B.L. Metcalfe, DJ. Bradley and A.D. Bye, Glass technol. 34 [3]: 114 (1993).
7. N.H. Ray, M. H. Stacey and S. J. Webster, Physics and Chemistry of Glasses 8 [1]: 30 (1967).
8. Martin E. Nordberg, Ellen L. Mochel, Harmon M. Garfinkel, and Joseph S. Olcott, J. Am. Ceram. Soc. 47 [5]: 215 (1964).
9. R. Tandon and D. J. Green, J. Am. Ceram. Soc. 71 [4]: C-192-C-193 (1988).
10. G.J. Fine and P.S. Danielson, Physics Chem. Glasses 29 [4]: 134 (1987).
11. David H. Roach and Alfred. Cooper, J. Am. Ceram. Soc. 53 [9]: 508 (1970).
12. J.E. Ritter and A.R. Cooper, Physics and chem. Glasses 4 [3]:76 (1963).
13. S.S. Kistler, J. Am. Ceram. Soc.45 [2]:59 (1962).
14. R.H. Doremus, J. Phys. Chem.: 2212 (1964).

15. Gary L. Mcvay and Delbert E. Day, J. Am. Ceram. Soc. 53 [9]: 508 (1970).
16. R. Terai, Physics Chem. Glasses 10 [4]: 146 (1969).
17. R. Hayami & R. Terai, Physics Chem. Glasses 13 [4]: 102 (1972).
18. J.S. Stroud, Glass technol. 29 [3]: 108 (1988).
19. Ajit Y. Sane, J. Am. Ceram. Soc. 70 [2]: 86 (1987).
20. E.E. Shaisha and A.R. Cooper, J. Am. Ceram. Soc. 64 [5]: 278 (1981).
21. L. Bassoli, F. Nizzoli and J. R. Sandercock, Physical Rev. B Vol. 34 n° 2: 1296 (1986).
22. M. Cardona, Ligth scattering in solids II, in: Topics in applied physics, V. 50, ed. M. Cardona and G. Guntherodt, Springer-Verlag Berlin (1982).
23. A.J. Burggraaf and J. Cornelissen, Physics Chem. Glasses 5 [5]:123 (1964).
24. A.M. Marvin, V. Bortolani and F. Nizzoli, J. Phys. C 13: 299 (1980).
25. H. Sussner, J. Pelous, M. Schmidt and R. Vacher, Solid State Commun. Vol. 36: 123 (1980).
26. Bede Scientific Instruments, Durham, UK.
27. G.A.D. Briggs, Acoustic Microscopy, Clarendon Press Oxford (1992).
28. J.M. Bradshaw and B. Taylor in: proc. VIIth Conference on Physics of Non-Crystalline Solids:471 (Taylor and Francis 1992).
29. E.I. Kozlovskaya, The structure of the Glass 2: 299 (1952).

NONLINEAR ULTRASONIC PROPERTIES OF AS-QUENCHED STEELS

D.C. Hurley, P.T. Purtscher, K.W. Hollman, and C.M. Fortunko

Materials Reliability Division
National Institute of Standards & Technology
325 Broadway
Boulder, CO 80303 USA

INTRODUCTION

We have investigated the effect of carbon content on the nonlinear ultrasonic parameter β and the longitudinal-wave velocity v_L in a series of martensitic steel specimens. The specimens were measured in the as-quenched state to insure that the carbon was present primarily as an interstitial in the martensite. Experimentally, β increased with increasing hardness, while v_L remained virtually the same for all specimens. These results indicate that β is sensitive to microstructural variations between the specimens, but v_L is not.

SPECIMEN PREPARATION

Three different types of steel stock (9310, 4320, and 4340) were used to fabricate the specimens for our experiments. The primary difference in composition between the steels was the measured mass percent of carbon: 0.10% (9310), 0.18% (4320), and 0.40% (4340). Specimens ranged in cross section from 30×30 to 60×60 mm and were approximately 19 mm thick. The specimens were prepared by "soaking" them at approximately 100 °C above their critical temperature, that is, the temperature at which ferrite first begins to form from austenite upon cooling. The soaking temperatures were 900 °C, 885 °C, and 870 °C for the 9310, 4320, and 4340 steels, respectively. The specimens were soaked for one hour and then rapidly quenched to room temperature by immersion in an agitated water bath with an approximate cooling rate of 100 °C/s. This process was intended to achieve a martensitic microstructure in which the chief difference was the variation in amount of interstitial carbon. Measurements were performed on the specimens in the resulting as-quenched state.

The Rockwell C hardness (HRC) was determined for each specimen. Measurements were made in several locations through a cross section of each specimen, and the average hardness was computed. Since individual values varied by only the measurement uncertainty

---

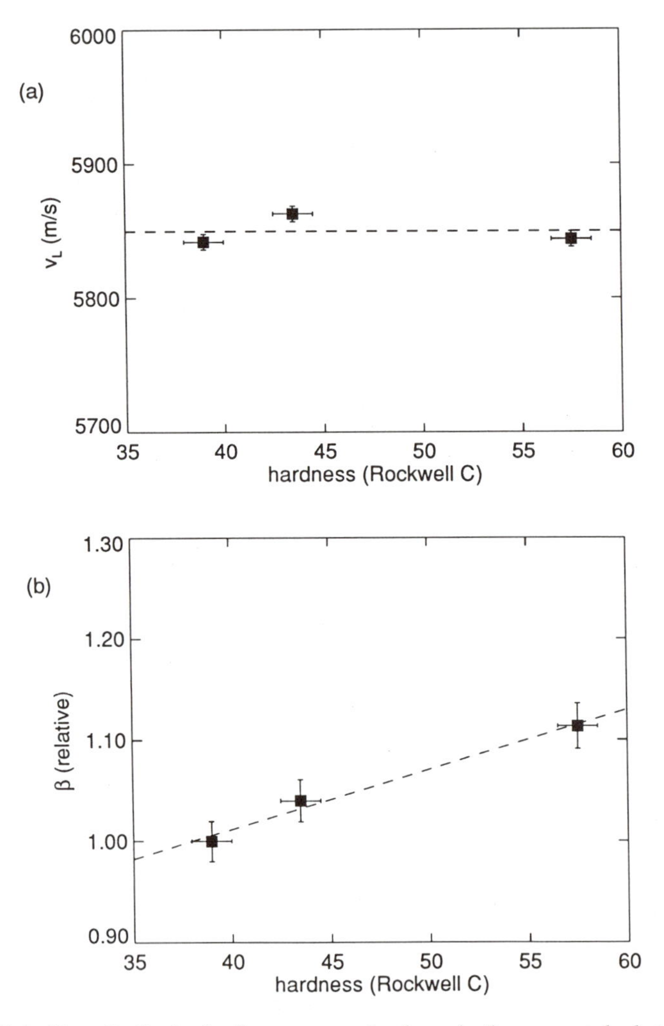

Figure 1. (a) Plot of longitudinal velocity v_L versus hardness in the as-quenched steel specimens. The broken line indicates the average velocity ($v_L = 5850$ m/s) of all three specimens. (b) Plot of the relative nonlinear ultrasonic parameter β versus hardness for the same steel specimens. The broken line is a linear least-squares fit to the data.

(±1 RHC), we concluded that the hardness of each specimen was uniform. The hardness values ranged from 39.0 RHC (9310 steel) to 57.5 RHC (4340 steel).

Specimen microstructure was examined using a light microscope (100X magnification). The micrographs revealed that all three specimens possessed a lath martensite structure, uniformly sized throughout each specimen. The average martensite packet size varied from specimen to specimen (approximately 5-20 μm). The micrographs indicated little or no evidence of retained austenite or second-phase formation.

ULTRASONIC PROPERTIES

Longitudinal Velocity

The longitudinal phase velocity v_L was measured in each of the samples using an immersion, pulse-echo superposition technique with an estimated experimental uncertainty of $\pm 0.1\%$. The measured values for v_L in the three as-quenched specimens are plotted versus hardness in Fig. 1(a). The figure reveals that v_L was nearly identical for all specimens. Minor differences in velocity slightly were measured, but no consistent trend with hardness was observed.

Nonlinear Ultrasonic Parameter

The value of the nonlinear ultrasonic parameter β for these specimens was obtained through harmonic-generation experiments. Finite-amplitude, longitudinal tonebursts at the fundamental frequency $\omega_0 = 2\pi \times 10$ MHz were generated by a piezoelectric transducer bonded to one side of the specimen. The waveform detected on the other side of the sample, which contained a component of amplitude A_1 at ω_0 as well as a component of amplitude A_2 at the second harmonic frequency $2\omega_0$, was detected using a near-infrared Michelson interferometer (laser wavelength $\lambda = 1064$ nm). Our interferometric detection techniques (which have been described in detail elsewhere[1]) allow the absolute ultrasonic displacements to be determined, since the displacements are related directly to the precisely known laser wavelength. The detected waveforms were digitally filtered to obtain the components A_1 and A_2.

Given the fundamental and second-harmonic amplitudes, the nonlinear parameter β was determined through the relationship[2]

$$|\beta| = \frac{8 v_L^2 A_2}{\omega_0^2 z A_1^2}, \tag{1}$$

where z is the sample thickness. Typically, the amplitude of A_2 was measured for several different amplitudes of A_1 using stepped attenuators. The slope of the line A_2 *vs.* A_1^2 was calculated and used with Eq. 1 to obtain the value for β in the specimen.[1] In these experiments, $A_1 \approx$ 2-6 nm.

Figure 1(b) shows the experimental results for β as a function of hardness, relative to β for the 9310 steel specimen. The estimated uncertainty for relative measurements is $\pm 2\%$. The graph shows that, unlike the longitudinal velocity, β varied markedly from specimen to specimen. Values for β were observed to increase monotonically with hardness or, equivalently, with carbon concentration. For example, a 12% increase from the 0.1% C to the 0.4% C specimen was measured. Additional experiments with specimens containing other carbon concentrations are planned to better quantify the relationship.

Table 1. Summary of experimental results for as-quenched steel specimens.

steel	mass % C	hardness (RHC)	v_L/v_L°	β/β°
9310	0.10	39 ± 1	1	1
4320	0.18	43.5 ± 1	1.004 ± 0.001	1.04 ± 0.02
4340	0.40	57.5 ± 1	1.000 ± 0.001	1.11 ± 0.02

DISCUSSION

Table 1 summarizes our experimental results. In the table, v_L and β for each specimen are expressed relative to the values for the 9310 steel specimen ($v_L^\circ = 5842 \pm 6$ m/s; $\beta^\circ = 5.48 \pm 0.33$). The table emphasizes the results of Fig. 1: β is much more sensitive to differences in microstructure between these specimens than v_L.

What are the possible causes for the increase in β? One microstructural difference between specimens is the extent of tetragonal distortion. With increasing carbon content, the martensite lattice becomes increasingly more tetragonal.[3] We have performed calculations based on previous theoretical work[4,5] to estimate the effect on v_L and β of a lattice strain ϵ induced by the tetragonal distortion. The predicted change in v_L is linear with ϵ, and approximately the same magnitude. Therefore, strains smaller than $\epsilon \sim 10^{-3}$ would change v_L by approximately the experimental uncertainty (or less), which is consistent with experimental results. Accurate calculations of the corresponding change in β require knowledge of the second-, third-, and *fourth*-order elastic stiffness moduli. Without more information about C_{1111} and C_{1112}, it is difficult to predict the magnitude of the effect. However, it appears likely that for strains less than $\epsilon \sim 10^{-3}$, the tetragonal distortion would increase β by less than one percent – much less than the experimental results.

Another microstructural feature of as-quenched steels which could affect nonlinear measurements is the large number of dislocations. The estimated dislocation density[3,6] for our specimens is $\sim$1-5$\times 10^{11}$/cm^2. Hikata and coworkers[7] showed that second-harmonic generation in single-crystal aluminum was sensitive to dislocations in combination with applied stress or internal stresses due to plastic deformation. It is possible that internal stresses present in the steel specimens could cause dislocations to generate additional second harmonics and thus affect β. We have made order-of-magnitude estimates based on the equations of Hikata *et al.*, using realistic values of dislocation and material parameters. The estimates indicate that dislocation-stress interactions could generate sufficient second harmonics to increase β by observed amounts.

Further evidence of a dislocation contribution to β is provided by the work of Kelly and Kehoe.[6] They demonstrated a linear relationship between the dislocation density and the carbon content in a series of as-quenched martensite specimens. Using this assumption, the equations of Hikata *et al.* predict increases in β of approximately the observed amount. Further work is underway to better understand the underlying mechanisms responsible for the experimental results.

CONCLUSIONS

We have evaluated the ultrasonic properties of as-quenched steel specimens in which the carbon content was varied. High-precision measurements revealed virtually no change in the longitudinal phase velocity v_L. In contrast, the nonlinear ultrasonic parameter β changed noticeably from specimen to specimen. A monotonic increase in β with increasing hardness over the range 39-57.5 HRC (0.1% - 0.4% C) was observed. It is currently believed that this behavior is due to large numbers of dislocations affected by internal stresses. We are now working to develop a systematic understanding of the microstructural mechanisms which account for the observed behavior.

ACKNOWLEDGMENTS

We thank H.M. Ledbetter for useful discussions, and R.L. Santoyo and J.D. McColskey for their assistance in sample preparation.

REFERENCES

1. D.C. Hurley and C.M. Fortunko, Determination of the nonlinear ultrasonic parameter β using a Michelson interferometer, *Meas. Sci. Tech.* 8:634 (1997).
2. M.A. Breazeale and J. Philip, Determination of third-order elastic constants from ultrasonic harmonic generation measurements, in: *Physical Acoustics Vol. XVII*, W. P. Mason and R. N. Thurston, eds., Academic, New York (1984).
3. R.E. Reed-Hill, *Physical Metallurgy Principles*, D. Van Nostrand Company, New York (1973).
4. R.N. Thurston and K. Brugger, Third-order elastic constants and the velocity of small amplitude elastic waves in homogeneously stressed media, *Phys. Rev.* 133:A1604 (1964).
5. J.H. Cantrell and W.T. Yost, Effect of precipitate coherency strains on acoustic harmonic generation, *J. Appl. Phys.* 81:2957 (1997).
6. P.M. Kelly and M. Kehoe, The role of dislocations and interstitial solutes on the stength of ferrous martensite, in *New Aspects of Martensitic Transformation, Suppl. Trans. Japan. Inst. Metals* 17:399 (1976).
7. A. Hikata, B.B. Chick, and C. Elbaum, Dislocation contribution to the second harmonic generation of ultrasonic waves, *J. Appl. Phys.* 36:229 (1965).

INDEX

Absorption, 455
Acoustic attenuation dispersion, 713
Acoustic characterization, 359
Acoustic emission, 499, 507, 517, 523, 595, 609, 615, 629, 635, 641, 647
Acoustic energy reflection, 183
Acoustic methods, 151
Acoustic microscope, 677
Acoustic microscopy, 401, 775
Acoustic nonlinearity, 183
Acoustics, 133, 555
Acousto-elastic constants, 229, 653
Acousto-ultrasonics, 239, 601
Active laser thermography, 263
Adhesion strength, 125
AE source location, 621
Aftershock series, 523
Aging aircraft, 73
Aging aircraft inspection, 487
Air-coupled transducer, 211
Air-coupled ultrasonics, 239
Al-alloys, 653
Alloy identification, 461
Aluminum, 27, 145
Amplitude, 595
Anisotropic, 747
Anisotropy, 117, 353
Arc welding, 257
Artificial neural networks, 719
Atomic number, 455
Attenuation, 145, 707
Attenuation coefficient, 393
Austenite, 683
Austenitic steel, 469
Autocorrelation coefficients, 523
Automatic identification, 719
Automatic modeling, 621

Barkhausen microscopy, 659
Barkhausen noise, 677
Biomedical ultrasonics, 763
Bolt, 671
Bond strength, 183
Bridges, 481
Brillouin spectroscopy, 817
Brinell hardness, 175
Brittle cracking, 757
Brittleness, 157
Buried ultrasound sources, 47

Calibrating measurements, 47
Capacitance transducer, 517
Carbon steel, 493, 701
Case depth, 211
Case depth in steel, 211
Casting, 217
CCD-camera, 417
Ceramic
  composites, 629
  pore size, 409
Ceramics, 443, 629
Charge redistribution, 517
Coatings, 27
Cold neutrons, 689
Comb type ultrasonic transducer, 695
Composite material, 91, 401
Composites, 59, 85, 305, 335, 359, 567, 609, 635, 641, 747
Compression molding, 383
Computation, 787
Concrete, 541, 549, 567, 581
Conductivity profile recovery, 769
Confining pressure, 529
Continuous fiber ceramic composites (CFCCs), 805
Copper alloys, 189
Corrosion, 73, 555, 575
Crack depth, 793
Crack detection, 409, 575
Crack tip, 659
Crack width, 793
Cracks, 133, 555, 781
Creep damage, 139
Crystal lattice defects, 701
Cure monitoring, 383
Cutting force dynamics, 205
Cylinders, 725

D-scan, 763
Dam, 549
Damage, 541, 601

Damping, 145, 581
Dead zone, 235
Decrement, 595
Deep drawability, 175
Defects, 133, 443
Deformation, 469
Deformation rate spectra, 157
Deformed layer, 437
Delamination, 601
Density, 393, 707
Dentin, 757, 763
Dielectric
  analysis, 383
  characterization, 291
  properties, 297
Diffraction plane, 437
Diffusion bonds, 183
Diffusion-bonding, 245
Diffusivity, 747, 757
Digital signal processing, 235, 499
Directional solidification, 217
Directivity patterns, 105
Disbond, 305
Dislocation hysteresis, 169
Dislocations, 145, 493
Dispersion, 609, 781
Dispersion compensation, 67
Dissipation, 595
Dynamic modulus, 581
Dynamic moduli, 277

Eddy current
  inspection, 487
  inversion, 769
  microscope, 677
  sensing, 371
  techniques, 353
  testing, 719
Efficiency, 417
Elastic anisotropy, 175, 529
Elastic constants, 229, 707
Elastic optical scattering, 365
Elastic stiffness, 85
Elastic waves, 561, 567
Electric conductivity, 353
Electric partial discharges, 507
Electromagnetic Acoustic Resonance (EMAR), 139
Electromagnetic AcousticTransducer (EMAT), 39, 139, 197, 671, 733
Electromagnetic effects, 517
Electromagnetic finite element modeling, 371
Electromagnetic techniques, 269
Emissivity, 377
Epithermal neutron, 535
Experimental evaluation, 587

Fatigue, 683
Fatigue crack, 311
Fatigue cracking, 475
Fatigue of Al-alloys, 621
Ferritic alloy steels, 163
Fiber fragmentation, 635
Fiber waviness, 401, 641
Fibre/matrix debonding, 409
Fibre reinforced plastics (FRP), 713
Field-portable XRF analyzer, 461
Filament winding, 641
Finite element mode (FEM), 671
Finite element method (FEM), 793
Flash techniques, 739
Flaw detection, 733
Fluids, 393
Focused imaging, 67
Fracture prediction, 621
Free water, 535
Free-space technique, 297
Frequency chirp, 1

Gas-coupled ultrasonics, 575
Generation and detection of ultrasound, 7
Glass composites, 615
Grain size, 329
Graphite composites, 353
Guided waves, 695

Hall effect, 511
Hard dental tissues, 757, 763
Hardness, 825
Heterogeneities, 117
High frequency, 27
High frequency ultrasound, 443
High pressure gas, 575
High strength, low alloy (HSLA) steel, 197
High temperature applications, 805
High temperature materials, 377
High temperatures, 33, 449
High voltage, 507
Humanitarian demining, 317

Imaging, 79, 431
Immersion ultrasonics, 239
In-line process control, 269
In-situ PMI, 461
Industrial, 341
Inelastic mechanical processes, 151
Infrared, 747
Infrared microscopy, 263
Infrared thermal imaging, 805
Inspection, 423
Insulators, 507
Integral breadth, 437
Interferometer, 13, 341
Interferometry, 79
Internal friction, 163, 169
Intravascular imaging, 105
Ion-exchanged glass, 817
IR-imaging, 757

Lamb wave testing, 235
Lamb waves, 67, 609, 775

Landmine detection, 317
Landmine material identification, 317
Lankford coefficient, 175
Lap joints, 73, 125
Laser, 39, 59, 91
  generation of acoustic waves, 7
  generation source, 7
  hardening, 701
  interferometry, 97, 157
  ultrasonics, 1, 7, 27, 33, 73, 79, 85, 97, 111, 239, 323, 329, 341
  ultrasound, 1, 21, 67, 73, 111, 117, 211, 323, 329, 335
    testing of non-metals, 47
Leakage field, 787
Leaky Rayleigh wave, 707
Leaky wave, 775
Line-focus transducer, 401, 799
Liquid level, 39
Liquid-solid interfaces, 371
Lithium glass neutron detector, 535
Loading, 311
Local variations, 757, 763
Lock-in thermography, 757
Long term strength, 523
Longitudinal sound velocity, 763
Low carbon steel, 665
Low cost, 739

Machining tool sensor, 205
Magnetic field measurement, 493
Magnetic flux leakage testing (MFL), 793
Magnetic technique, 511
Magnetization, 469
Magnetostriction, 197, 665, 733
Maincrack formation, 629
Martensite, 223, 701
Matched filtering, 67
Material characterization, 269, 401, 775
Material slab, 297
Mechanical properties, 269
Medical element engineering and simulation, 757, 763
Metallography, 229
Michelson, 341
Michelson interferometer, 825
Microfracture, 615
Microfracture process, 629
Micromagnetics, 701
Microplasticity, 157
Microstructure, 145, 223, 229, 409, 653, 701
Microwaves, 285, 291, 297, 305, 311
Mode selection analysis, 695
Modulus of elasticity, 581
Moisture, 757
Molten materials, 33
Mouth, 763
Mult-channel techniques, 377
Multi-color pyrometry, 263
Multi-energy radioscopy, 455
Multi-layered media, 769

Near-field optics, 53
Near-surface elastic properties, 817
Neural network classification, 347
Neutron, 535
Neutron diffraction, 475, 529, 689
Neutron scattering, 689
Nichel superalloys, 431
Nodular growth, 677
Non-contact, 285
  inspection, 239
  ultrasonics, 111
  ultrasound, 39
Non-contacting evaluation, 139
Nondestructive, 21, 335, 475, 535, 549, 683
Nondestructive control, 169
Nondestructive evaluation (NDE), 85, 239, 317, 487, 511
Nondestructive process control, 229
Nondestructive testing, 125, 133, 455, 555
Non-quadratic yield functions, 751
Non-zero optical penetration, 47
Noninvasive, 393
Nonlinear acoustics, 125
Nonlinear interactions, 133, 555
Nonlinear ultrasonics, 825
Normal incidence, 763
Nucleation and initial growth, 371
Numerical simulations, 567

Open-ended sensors, 311
Optical detection, 13
Optical techniques, 7

Parametric formulae, 587
Phase conjugator, 33
Phase image, 757
Phase modulation, 13
Phase transformation, 323
Photo induced-EMF, 21, 335
Photoacoustic effect, 105
Photodiode array detectors, 461
Photorefractivity, 79
Photothermal radiometry, 757
Physical properties, 757, 763
Pipeline inspection, 575
Plasma, 39
Plastic deformation, 169, 493, 659, 665
Plastic strain, 437
Plasticity, 541, 751
Plate waves, 189
Ply-cords, 601
Poor conductors, 739
Porosity, 757, 763
Porosity content, 713
Portable optical emission spectrometer, 461
Portland cement hydration, 535
Positron annihilation, 475, 701
Powder injection molding, 347
Predictive measurement, 621
Process control, 175, 257
Process monitoring, 347

Pulse compression, 1
Pulse-echo, 239
Pulse-echo diagnostics, 763
PVDF, 647
Pyrometry, 377

Quality assurance, 269
Quantitative nondestructive evaluation (QNDE), 97
Quartz, 27
QUASI-elastic neutron scattering, 535

Radial tires, 601
Raman spectroscopy, 811
Rayleigh wave, 775, 781
Reaction kinetics, 257
Reconstruction, 757, 763
Recovery, 145
Recrystallization, 145
Relaxation transition, 157
Reliability, 507
Research reactor, 689
Residual life, 601
Residual stress, 659, 677
Resolution, 417
Resonance, 145, 671
Resonant frequency, 277
Resonant ultrasound spectroscopy, 475
Retreads, 601
Ring burst test, 641
Rock fracturing, 517
Rolled metallic sheet, 263
Rolled steel, 285

SAFT, 67
Sagnac interferometer, 97
Scan probe microscopy, 53
Semiconductor crystal growth, 371
Semi-rigid connections, 587
Semi-rigid joints, 587
Sensor, 647
17-4PH stainless or (4340 alloy), 277
Sheet texture, 189
Signal characteristics, 235
Signal processing, 719
Silicon nitride, 443
Silicon nitride (Si_3N_4) ceramics, 365
Simplex control algorithms, 347
Simulation of porosity, 713
Single crystal, 217, 423
Single laser, 1
Sintering, 251
Sizing, 787
Snoek-type relaxation, 163
Sonic, 549
Sound speed, 393
Spark plasma activated sintering, 251
Speckle beam, 13
Speckle hunting, 97
Spectra frequency, 595
SQUID, 683
SQUID gradiometer, 493
Steam generator deposits, 811
Steel, 27, 323, 329, 475
Steel cables, 511
Steel strips, 175
Stiffness tensor, 91
Stochastic process, 615
Stress, 481
Stress measurement, 665, 671
Stress-wave-factors, 601
Structure, 449
Super plastic forming/diffusion bonding (SPF/DB), 239
Superconducting quantum interference devices, 487
Support conditions, 581
Surface acoustic wave methods, 817
Surface and near-surface defects, 365
Surface defects, 285
Surface waves in anisotropic solids, 799
Surface waves, 67, 443, 707
Synchrotron, 431
Synchrotron radiation, 449
Synthetic aperture focusing, 67
System architecture, 499

Temperature, 341
Tempered martensite, 825
Testing, 507
Textron, 341
Texture, 359, 653, 751
Texture analysis, 529
Texture coefficients, 229
Thermal, 747
  cycle, 257
  diffusivity, 739, 805
  neutrons, 689
  waves, 757
Thermography, 257, 377
Thickness, 575, 725, 757, 763
Thickness-independence, 725
3-D surface crack, 787
3-dimensional stress mapping, 653
Through-transmission, 239
Timber, 561
Tire casings, 601
Tires, 601
Tissue phantoms, 105
Tokamak, 469
Tooth, 757, 763
Topography, 423, 431
Trapped torsional mode, 475
True color, 455
Tubes, 725
Turbine blades, 423, 431
Two-port measurement, 291

Ultrasonic, 53, 59, 91, 145, 205, 481
  anisotropy, 189
  attenuation, 139, 323, 329
  imaging, 251, 725
  interferometry, 393

Ultrasonic (*cont.*)
measurement, 211
monitoring, 245
signal analysis, 713
sounding, 529
sound speed, 383
techniques, 695
testing, 541
of anisotropic materials, 799
velocity, 329, 775
Ultrasound, 13, 67, 197, 223, 609, 757, 763
US-imaging, 763

Vacuum vessel shell, 469
Velocity, 223, 707, 725
dispersion, 695
measurements, 33
Vibration, 79, 133, 555
Visualization of temperature, 251
Void, 305

Waveform capture, 499
Waveguides, 561
Weld metal, 257
Welding stress, 653
Winding tension, 641
Wire, 733

X-ray, 431, 449, 455
diffraction, 217, 423, 437
technique, 659
tomography, 443
topography, 417

Zipper, 601